TABLES OF THERMODYNAMIC DATA

DIAGRAMS OF THERMODYNAMIC DATA

All the thermodynamic data diagrams that are listed above appear as Adobe PDF files on the CD-ROM accompanying this book, and may be enlarged and printed for easier reading and for use in solving problems.

Chemical, Biochemical, and Engineering Thermodynamics

Fourth Edition

Stanley I. Sandler
University of Delaware

John Wiley & Sons, Inc.

ASSOCIATE PUBLISHER	Dan Sayre
ACQUISITIONS EDITOR	Jennifer Welter
EDITORIAL ASSISTANT	Mary Moran-McGee
SENIOR PRODUCTION EDITOR	Ken Santor
MARKETING DIRECTOR	Frank Lyman
ILLUSTRATION COORDINATOR	Mary Alma
DESIGNER	Hope Miller

This book was set in LaTeX by Publication Services and printed and bound by Courier-Kendallville. The cover was printed by Phoenix Color Corporation.

This book is printed on acid free paper. ∞

To order books or for customer service, please call 1-800-CALL WILEY (225-5945).

Mathcad and Mathsoft are registered trademarks of Mathsoft Engineering and Education, Inc., http://www.mathsoft.com

Library of Congress Cataloging-in-Publication Data

Sandler, Stanley I., 1940-,
    Chemical, biochemical, and engineering thermodynamics / Stanly I. Sandler.—4th ed.
        p. cm.
    Rev. Ed. of: Chemical and engineering thermodynamics. 3rd ed. c1999.
    Includes index
    ISBN-13: 978-0-471-66174-0 (cloth : acid-free paper)

    1. Thermodynamics–Textbooks. 2. Chemical engineering–Textbooks. 3. Biochemical engineering–Textbooks. I. Sandler, Stanley I., 1940- Chemical and engineering thermodynamics. II. Title.
    QD504.S25 2006
    541.3'69–dc22

                                                                    2055054235

Printed in the United States of America

15 16 17 18

About the Author

STANLEY I. SANDLER earned the B.Ch.E. degree in 1962 from the City College of New York, and the Ph.D. in chemical engineering from the University of Minnesota in 1966. He was then a National Science Foundation Postdoctoral Fellow at the Institute for Molecular Physics at the University of Maryland for the 1966–67 academic year. He joined the faculty of the University of Delaware in 1967 as an assistant professor, and was promoted to associate professor in 1970, professor in 1973, and Henry Belin du Pont Professor of Chemical Engineering in 1982. He was department chairman from 1982 to 1986. He currently is also professor of chemistry and biochemistry at the University of Delaware and founding director of its Center for Molecular and Engineering Thermodynamics. He has been a visiting professor at Imperial College (London), the Technical University of Berlin, the University of Queensland (Australia), the University of California–Berkeley, and the University of Melbourne (Australia).

In addition to this book, Professor Sandler is the author of 325 research papers and a monograph, and he is the editor of a book on thermodynamic modeling and five conference proceedings. He is also the editor of the *AIChE Journal*. Among his many awards and honors are a Faculty Scholar Award (1971) from the Camille and Henry Dreyfus Foundation; a Research Fellowship (1980) and a U.S. Senior Scientist Award (1988) from the Alexander von Humboldt Foundation (Germany); the 3M Chemical Engineering Lectureship Award (1988) from the American Society for Engineering Education; the Professional Progress (1984), Warren K. Lewis (1996), and Founders (2004) Awards from the American Institute of Chemical Engineers; the E. V. Murphree Award (1996) from the American Chemical Society, the Rossini Lectureship Award (1997) from the International Union of Pure and Applied Chemistry, and election to the U.S. National Academy of Engineering (1996). He is a Fellow of the American Institute of Chemical Engineers and the Institution of Chemical Engineers (Britian and Australia), and a Chartered Engineer.

Preface

INTENDED AUDIENCE AND OBJECTIVES

This book is intended as the text for a course in thermodynamics for undergraduate students in chemical engineering. It has been used in this manner at the University of Delaware for more than 20 years, originally in a course for third-year students and currently for use by sophomores. I had two objectives in writing the first edition of this book, which have been retained in the succeeding editions. The first was to develop a modern applied thermodynamics text, especially for chemical engineering students, that was relevant to other parts of the curriculum—specifically courses in separations processes, chemical reactor analysis, and process design. The other objective was to organize and present material in sufficient detail and in such a way that the student can obtain a good understanding of the principles of thermodynamics and a proficiency in applying these principles to the solution of a large variety of energy flow and equilibrium problems.

Though this is designed to be an undergraduate textbook, I also use the material in this book when I teach the one-semester graduate thermodynamics course. I am frequently asked what I teach in that course, as the graduate thermodynamics course in different schools is probably the least defined and most heterogeneous course in graduate programs. The material I cover there appears in my book *An Introduction to Applied Statistical Thermodynamics* published by John Wiley & Sons, Inc. in 2011.

Chemical engineering faculties are increasingly examining their curricula to find space to add new subjects, such as biochemical engineering, and at the same time to meet institutional requirements of reduced credit hours. I believe that breadth is more important than redundancy, especially when redundancy has little positive effect. It is very important that we keep the right balance between breadth and depth, and critically examine the need for redundancy in our core program. If we omit introducing our students to some area, we make it much more difficult for them to independently learn about this subject later in their careers, since they will start without any basic knowledge, or without even knowing the nomenclature. Thermodynamics, in some form, is typically presented to students in general chemistry, physical chemistry, and physics courses, before the chemical engineering course. I believe this redundancy has had little positive, and perhaps even a negative, effect. Because of the types of systems considered (open systems in chemical engineering, but only simple closed systems elsewhere), and frequently the use of quite different notation among the different subject areas, I find that chemical engineering students are more confused by this redundancy than aided by it. Therefore, when I teach undergraduates I begin by telling them to forget what they have been taught about thermodynamics elsewhere.

Consequently, I suggest that thermodynamics should be studied only in the general chemistry course and in a chemical engineering thermodynamics course, eliminating the physical chemistry and physics exposures. As this textbook contains sufficient information on colligative properties for chemical engineers, I suggest eliminating the one-semester course of the physical chemistry program (retaining the course on statistical mechanics and quantum mechanics), and perhaps reducing by one credit the physics course most chemical engineering students take by eliminating thermodynamics there as well.

I am also frequently asked why I have not included statistical mechanics in this book or in my undergraduate course. This is an especially poignant question for me, since statistical mechanics is one of my research areas. My answer is as follows. I find that students have sufficient difficulty with macroscopic thermodynamics and its application that moving to the microscopic or statistical mechanics level adds little, except perhaps some confusion, and usually detracts from the flow of the course. Further, we generally do not go to the molecular level in other core undergraduate courses in chemical engineering (i.e., rarely is kinetic theory introduced in fluid mechanics or mass transfer courses). I do believe exposure to statistical and quantum mechanics is important, and that is why the physical chemistry course that deals with these should be retained.

NEW TO THIS EDITION

This fourth edition of *Chemical and Engineering Thermodynamics*, renamed *Chemical, Biochemical, and Engineering Thermodynamics*, is a significant revision to the previous editions. The most evident changes are the increase in the number of chapters from 9 to 15, and the inclusion of material on some of the biochemical applications of thermodynamics; but a number of additional changes have been made to enable this text to continue to evolve as a teaching and learning tool.

- *Instructional objectives and notation.* Two items have been added at the beginning of each chapter. The first is a list of instructional objectives. This provides students with my expectations for what they should learn from the chapter, and can serve as a checklist that they can use to monitor progress in their studies (especially for examinations). The second is a list of the important new notation introduced in that chapter. This is meant to provide an easy reference for students and to indicate the important quantities being introduced in the chapter.
- *Number of chapters increased from 9 to 15.* The increase in the number of the chapters was done largely to reduce the length of some chapters to units more easily digestible by students. Thus, the content of the very long Chapter 8 of the previous editions, dealing with all types of phase equilibria, in this new edition has been separated into Chapter 10, which considers only vapor-liquid equilibrium; Chapter 11, which discusses other types of fluid-phase equilibria (liquid-liquid, vapor-liquid-liquid, supercritical extraction, etc.); and Chapter 12, on phase equilibria involving a solid. Similarly, the two subjects of the previous Chapter 9, chemical reaction equilibria and the energy balances on reacting systems, has been split into Chapters 13 and 14. Also, the content of Chapters 2 and 3 of the previous edition has been reorganized into four chapters. Having the book organized in this way should make it easier for the professor to choose among the material to be covered, if he or she decides to do so.
- *Material related to biochemical applications of thermodynamics.* The only new chapter is Chapter 15, dealing with biochemical processes, though a number of biochemical examples also appear in Chapters 11 and 12, and elsewhere in the book. The new Chapter 15 deals with the application of thermodynamics to some biochemical processes. Since pH is important in many such processes, the first section of this chapter deals with this subject. It should be a review, perhaps in a little more depth, of what the student has learned in courses in general and physical chemistry. Section 15.2 discusses how pH affects the ionization of biochemicals, including how the charge on a protein changes with pH. Section 15.3 deals with the solubilities of weak acids and bases as a function of pH, and the role this can play in the formulation of pharmaceuticals. Ligand binding, a different type of biochemical reaction, is considered in Sec. 15.4, with some detail on the unique case of oxygen binding to the four-site hemoglobin molecule that makes our form of life possible. Other biochemical reactions are considered in Sec. 15.5, including the unfolding (denaturation) of proteins as a function of temperature and pressure. The ultracentrifugation of proteins is considered briefly in Sec. 15.6, and Sec. 15.7 contains a discussion of Gibbs-Donnan osmotic equilibrium and how this is affected by pH. The coupling of chemical reactions, with particular attention to the ATP-ADP reaction, is considered in Sec. 15.8. There the discussion is a very macroscopic one, ignoring the details of the metabolic pathways, which are best considered in a biochemistry course. This chapter and the book conclude with the thermodynamic analysis of a fermenter, including the second law (or availability) constraint that applies to biochemical reactions and the operation of fermenters.
- *Brief introductions (and simple examples) of advanced and emerging applications of thermodynamics.* Thermodynamics is central to the practice of chemical engineering and to the curriculum; for example, phase equilibria is the basis for most separation and purification processes, and energy balances are needed in many areas, including reactor design. However, in my teaching experience, I have found that students do

not appreciate this connection until later in their studies, and many feel that thermodynamics is too abstract while they are studying this subject. I have tried to dispel this notion by providing some brief introductions (and simple examples) of the applications of thermodynamics to subjects chemical engineering students will encounter later in their studies. For example, in Sec. 10.1, on vapor-liquid equilibria of ideal mixtures, there is also a brief introduction to distillation (only several pages in length, and a simple McCabe-Thiele type discussion); this is intended to motivate an understanding of vapor-liquid equilibrium, not meant to replace a course in mass transfer operations. Distillation is revisited briefly in Sec. 10.2, on the vapor-liquid equilibrium of nonideal mixtures, to explain why azeotropes cause difficulties in distillation, and one way to deal with this. Additional "practical considerations" include brief examples on Rayleigh distillation, air stripping (to remove radon from ground water), staged liquid-liquid extraction, and others. Also, in Chapter 12 there is a brief introduction to the thermodynamics of the new field of product engineering. These introductions are all short and designed merely to motivate the student's interest in thermodynamics, and to provide an introduction to courses that will follow in the chemical engineering curriculum. Other practical applications that appeared in the previous editions, such as the liquefaction of gases, computing heat loads on chemical reactors, and the application of thermodynamics to safety and environmental problems, have been retained.

- *Descriptions of laboratory equipment.* Chapter 10 also includes brief descriptions of the types of laboratory equipment used to measure fluid phase equilibria, so that students will have a better understanding of how such data are obtained, and so they gain some familiarity with the equipment they may encounter in typical junior and senior level laboratory courses.

- *Help in choosing the appropriate thermodynamic model in solving problems, and in using process simulators.* I have put more emphasis on helping students choose the appropriate thermodynamic model for their separations and design courses, and on the use of process simulators, such as Aspen, Hysys, SimSci, and Chemcad. This is especially evident in Sec. 9.11.

- *Connections with other courses in the curriculum.* I believe that as educators, we should consider each of our courses to be a tree with roots and branches in other parts of the chemical engineering curriculum, rather than a silo of knowledge disconnected from the rest. It is for this reason in this book that I have made connections to courses in fluid mechanics, separations processes, reactor design, as well as the new chapter related to biochemical engineering.

- *New problems and illustrations.* Almost 200 new or revised problems and new illustrations have been included.

- *Answers to selected problems.* I have also added an appendix with answers to selected problems that students can use to check their work. However, it is only the final answers that are provided, not the complete solution. Students should use these answers with some discretion. Unlike a simple algebra problem, for which there will be an exact solution, here students should remember that if they read a thermodynamic property chart somewhat differently than I have, they will get a slightly different answer. Likewise, if they use one equation of state in the solution and I have used another, or if they used one activity coefficient model and I have used another, the final answers will be somewhat different. Therefore, except for the simplest problems (for example, those involving ideal gases), a student may not get the exact same answers I did.

- *IUPAC notation.* Another, lesser change I have made in this revision, largely suggested by my European colleagues, is to adopt some of the thermodynamic notation recommended by the International Union of Pure and Applied Chemistry (IUPAC). However, I have not been consistent in doing this, as I have found that some of the IUPAC-recommended notation did not meet my needs and might be confusing to the student.

- *Updated computational tools.* As in previous editions, I provide many opportunities for the use of personal computers in problem solving. The third edition of this book included MATHCAD worksheets and DOS-based Basic language (and compiled) programs. To these programs I have added Windows-friendly Visual Basic versions of those earlier programs that are easier to use and have greater capabilities. There is also a new Visual Basic pure-component database that can be used as a stand-alone program or accessed by the equation-of-state and UNIFAC programs to import the needed parameters. Also, the UNIFAC program has been updated to the latest version (as of the beginning of 2005) with the most recently published parameters. Several MATLAB programs have also been added. I have used the CD-ROM symbol in the margin, as here, to indicate when reference is made to using one or more of these programs, and to suggest to the professor and student when these programs can be used in problem solving.

These programs and worksheets are included in the CD-ROM that accompanies this book, and they can also be downloaded from the Web site www.wiley.com/college/sandler. It is my intention to add to and update these programs on this Web site. Appendix B on the CD-ROM and Web site describes the use of these programs and worksheets, and provides installation instructions for their use. While there is no necessity to use these computational aids to learn thermodynamics with this book, their availability greatly facilitates the solution of many interesting and practical problems that would otherwise be possible only after tedious programming.

- *MATHCAD 13.* A 120-day trial version of MATHCAD 13 is included on the CD-ROM that accompanies this text.

ADDITIONAL FEATURES

Margin notes. The margin notes I have added are meant to emphasize the concepts I believe to be important, as well as to make it easier for the student to find those concepts again at a later time. Since I frequently write notes in the margins of books that I use, I wanted to provide a place for students to add notes of their own.

Boxed equations. I have placed boxes around important equations so that the reader can easily identify the equations that are the end results of sometimes quite detailed analysis. It is hoped that in this way the student will easily see the important tree in what may appear to be a forest of equations.

Descriptive titles for illustrations. I have provided a short title or description to indicate what is to be learned from or seen in each illustration.

Realistic problems. Realistic problems are employed to familiarize students with the types of challenges they will encounter in industry and graduate research.

Introduction of environmental and safety applications of thermodynamics. An introduction to these topics provides course material useful for ABET accreditation.

SI units. SI units are used throughout.

STUDENT RESOURCES

The following resources are available from the book Web site at www.wiley.com/college/sandler. Visit the Student section of the Web site.

- *MATHCAD worksheets* to solve pure-fluid and mixture thermodynamics problems
- *Visual Basic programs* that run in the Windows environment to solve pure fluid and mixture thermodynamics problems and to obtain graphs of the results
- *DOS-based programs* (from the third edition) that can easily be modified by students with access to an inexpensive Basic compiler
- *MATLAB programs and library* to solve problems in the thermodynamics of pure fluids and mixtures
- *Adobe PDF versions of important data graphs* that students can enlarge, print out, use in problem solving and then hand in as part of their homework assignments.

These resources are also available on the CD-ROM accompanying the text. Updates to these resources will be included on the Web site.

INSTRUCTOR RESOURCES

All student resources are also available on the Instructor section of the Web site at www.wiley.com/college/sandler. The following supplements are available only to instructors who adopt the text:

- *Solutions Manual:* All solutions available as Word documents, and many solutions also available as MATH-CAD worksheets
- *Image gallery of text figures*
- *Text figures in PowerPoint format*

An important advantage of the MATHCAD worksheets is that a faculty member can create variations to the problems in this book by changing the conditions or input parameters in the worksheet, and immediately obtain the answer to the revised problem. In this way the recycling of problem solutions can be minimized. Also, the ability to change parameters in a MATHCAD worksheet allows "what if" questions to be answered quickly. For example, the question of how the phase behavior changes in a mixture with changes in temperature or pressure is quickly answered by changing a parameter or input variable in a MATHCAD worksheet. These instructor-only resources are password-protected. Visit the Instructor section of the book Web site to register for a password to access these materials.

PROPOSED SYLLABI

As with most textbooks, there is more material here than can be covered in a two-semester or two-quarter course. This allows the instructor to tailor the course to his or her interests, and to the needs of the curriculum. The material not covered in class should still be useful to students as reference material for other courses in the curriculum, and for use later in their professional careers. Some suggested syllabi are given next.

Two-semester undergraduate chemical engineering thermodynamics course
Cover as much of the book as possible. If omissions are necessary, I would *not* cover the following material:

Secs. 2.4 and 3.6	Secs. 12.3, 12.4, and 12.5
Secs. 6.9 and 6.10	Chapter 14
Sec. 9.9	Chapter 15, if there is no interest in biochemical engineering

Two-quarter undergraduate chemical engineering thermodynamics course
I suggest *not* covering the following chapters and sections:

Secs. 2.4 and 3.6	Sec. 9.9
Sec. 5.3	Secs. 12.3, 12.4, and 12.5
Secs. 6.6, 6.9, and 6.10	Chapter 14
Sec. 7.8	Chapter 15, if there is no interest in biochemical engineering

One-semester undergraduate chemical engineering thermodynamics course following a one-semester general or mechanical engineering course
I suggest quickly reviewing the notation in Chapters 2, 3, and 4, and then starting with Chapter 8. With the limited time available, I suggest *not* covering the following chapters and sections:

Sec. 9.9	Chapter 14
Secs. 12.3, 12.4, and 12.5	Chapter 15, if there is no interest in biochemical engineering

One-quarter undergraduate chemical engineering thermodynamics course following a general or mechanical engineering thermodynamics course
Quickly review the notation in Chapters 2, 3, and 4, and then go directly to Chapter 8. With the very limited time

available, I suggest *not* covering the following chapters and sections:

Sec. 9.9	Chapter 14
Secs. 11.3, 11.4, and 11.5	Chapter 15, if there is no interest in biochemical engineering
Chapter 12	

One-semester graduate thermodynamics course
Review much of the material in this book in about 60 percent of the semester, a blistering pace, and then provide an introduction to statistical mechanics in the remainder of the course.

FOR STUDENTS

Applied thermodynamics, as considered in this textbook, is one of the subjects that is the foundation for the practice of chemical engineering. A major part of the equipment and operating costs of processes developed by chemical engineers is based on design methods that use the principles of applied thermodynamics. While this will be demonstrated in courses you will take in mass transfer, reaction engineering, and process design, some brief introductory examples are provided in this book.

At this point in your education, you have probably been exposed to some aspects of thermodynamics in courses in general chemistry, physical chemistry and physics. My recommendation is that you completely forget what you have "learned" about thermodynamics in those courses. The notation in this book is different, and much more like that in other chemical engineering courses. Also, in those non-engineering courses, thermodynamics was usually applied only to a closed system (for example, a fixed mass of a substance), while engineering applications generally involve open systems, that is, those with mass flows into and/or out of the system. Also, you may have been introduced to entropy using a device such as a Carnot cycle. Please immediately expunge from your mind the connection between entropy and such devices. Entropy, like energy, is a very general concept, independent of any such device. However, entropy is different from energy (and momentum) in that it is not a conserved property. Energy is conserved, entropy is not.

As you will see (in Chapter 4), even though it is a non-conserved property, entropy is very important. For example, if two metal blocks, one hot and another cold, are put into contact with each other, the concept of entropy leads us to the conclusion that after a while, the two blocks will be at the same temperature, which is in agreement with our experience. However, the principle of energy conservation tells us only that the total energy of the system finally will equal the total energy initially, not that the blocks need to be at the same temperature. This is an illustration of how we frequently have to use both the concepts of energy conservation and entropy (and in open systems that total mass can not be created or destroyed) to solve problems in thermodynamics.

There are two general ways in which thermodynamics is used. One is the calculation of heat and work (or more generally, energy) flows—for example, in determining the conversion of heat to work in various types of engines, determining the heat flows accompanying chemical reactions, or in changes from one state of system to another. The second important type of thermodynamic calculation is the determination the equilibrium state, for example, in calculating the equilibrium compositions of the vapor and liquid of a complex mixture needed in order to design a method for purifying the components, or the equilibrium composition of a chemically reacting system. You should be able to do all such calculations after completing the material in this textbook, as well as some computations relating to biochemical processing, safety, and the distribution of chemicals in the environment.

Chemical engineering, and chemical engineering thermodynamics in this book, deals with real substances, and therein lie two of the difficulties. The first is that the properties of real substances are not completely known from experiment at all temperatures and pressures (and for mixtures at all compositions) and are approximately described by model equations—for example, a volumetric equation of state that interrelates pressure, volume, and temperature (the ideal gas equation of state applies only to gases at very low pressures), or equations that relate activity coefficients to composition. Any one of several different equations may be used to describe a pure substance or mixture, and each will result in a slightly different answer in solving a problem. However, within the accuracy of the underlying equations, all the solutions are likely to be correct. This may be disconcerting, as from other courses you may be used to solving algebraic equations that have only a single correct answer. The situation here is the one continually faced by practicing engineers of needing to solve a problem even though the description of the properties is imperfect, and a choice of equation of state or activity coefficient model must be made. (Some guidance in making such choices for mixtures is given in Sec. 9.11.)

The second problem is that the equations of state and activity coefficient models used in thermodynamics are not linear algebraic equations, so that computations involving them can be difficult. It is for this reason that I provide a collection of computer aids on the CD-ROM accompanying this book. Included on this CD-ROM are MATHCAD worksheets, Visual Basic programs (as code and stand-alone executable modules), MATLAB programs (as code and essentially stand-alone programs), and older DOS Basic programs (as code and stand-alone executable programs). These computer aids are described in Appendix B. I recommend the use of the MATHCAD worksheets (as the form of the code is the same as that in which the equations are normally written so that it is easily understood, and it is easy to make changes) or the Visual Basic programs (as they are simple to use and have nice graphics). To use the MATHCAD worksheets, the MATHCAD engine is required. A 120-day evaluation version of this software is included on the CD-ROM. The Visual Basic programs do not require any other software; however, the Visual Basic compiler is needed if you wish to change the programs. To run the MATLAB programs you need to install the MCR library, also on the CD-ROM; the complete MATLAB program is needed to make any changes.

I have also provided several instructional aids to help you in your study of thermodynamics using this book. First, every chapter begins with instructional objectives listing the important items to be learned. I suggest reading these objectives before starting a chapter, and then reviewing them while preparing for examinations. Second, important equations in the book are shown in boxes, and the really important equations also are indicated by name or description in the margin. Third, there are many problems at the end of each chapter (or, in the case of Chapters 10 though 12, at the end of each section) for you to hone your problem-solving skills. Finally, in Appendix C there are answers to selected problems. However, only the final answers appear, not the complete solution with the steps to get to that answer. Keep in mind that you may be solving a problem correctly but get a slightly different numerical answer than the one I have provided because you read a graph of thermodynamic properties slightly differently than I did, or used a correct but different equation of state or activity coefficient model than I did. So if your answer and mine differ only slightly, it is likely that both are correct.

Good luck in your study of thermodynamics.

Stanley I. Sandler
Newark, Delaware
June 10, 2005

ACKNOWLEDGMENTS

I want to thank a collection of people who have contributed to this book in many important ways. First, and foremost, my family, who have put up with me closeted in my office, typing away on my computer instead of spending time with them. Next are my faculty colleagues, present and former, who have supported me in this book writing activity in so many, many ways, including providing many useful comments, criticisms, and suggestions. Special thanks go to Michael Paulaitis and Norman Wagner, who have contributed problems to this edition, and to Abraham Lenhoff, who both provided problems and corrected my many initial errors in Chapter 15. I am also pleased to acknowledge Jürgen Gmehling of the University Oldenburg (Germany) and the UNIFAC Consortium for providing the group information and parameters for use in a program that accompanies this book. Finally, I wish to acknowledge the contributions of the students, both at the University of Delaware and (by e-mail) worldwide, who have used previous versions of the book and who have pointed out errors and typos, and asked questions, that have resulted in some of the changes you see in this edition.

Stanley I. Sandler
May 10, 2005

Contents

Chapter 1

Introduction

A major objective of any field of pure or applied science is to summarize a large amount of experimental information with a few basic principles. The hope, then, is that any new experimental measurement or phenomenon can be easily understood in terms of the established principles, and that predictions based on these principles will be accurate. This book demonstrates how a collection of general experimental observations can be used to establish the principles of an area of science called thermodynamics, and then shows how these principles can be used to study a wide variety of physical, chemical, and biochemical phenomena.

Questions the reader of this book might ask include, what is thermodynamics and why should one study it? The word *thermodynamics* consists of two parts: the prefix *thermo,* referring to heat and temperature, and *dynamics,* meaning motion. Initially, thermodynamics had to do with the flow of heat to produce mechanical energy that could be used for industrial processes and locomotion. This was the study of heat engines, devices to operate mechanical equipment, drive trains and cars, and perform many other functions that accelerated progress in the industrial age. These started with steam engines and progressed to internal combustion engines, turbines, heat pumps, air conditioners, and other devices. This part of thermodynamics is largely the realm of mechanical engineers. However, because such equipment is also used in chemical processing plants, it is important for chemical engineers to have an understanding of the fundamentals of such equipment. Therefore, such equipment is considered briefly in Chapters 4 and 5 of this book. These applications of thermodynamics generally require an understanding the properties of pure fluids, such as steam and various refrigerants, and gases such as oxygen and nitrogen.

More central to chemical engineering is the study of mixtures. The production of chemicals, polymers, pharmaceuticals and other biological materials, and oil and gas processing, all involve chemical or biochemical reactions (frequently in a solvent) that produce a mixture of reaction products. These must separated from the mixture and purified to result in products of societal, commercial, or medicinal value. It is in these areas that thermodynamics plays a central role in chemical engineering. Separation processes, of which distillation is the most commonly used in the chemical industry, are designed based on information from thermodynamics. Of particular interest in the design of separation and purification processes is the compositions of two phases that are in equilibrium. For example, when a liquid mixture boils, the vapor coming off can be of a quite different composition than the liquid from which it was obtained. This is the basis for distillation, and the design of a distillation column is based on predictions from thermodynamics. Similarly, when partially miscible components are brought together, two (or more) liquid phases of very different composition will form, and other components added to this two-phase mixture will partition differently between the phases. This phenomenon is the basis for liquid-liquid extraction, another commonly used separation process, especially for chemicals and biochemicals that cannot be distilled because they do not vaporize appreciably or because they break down on heating. The design of such processes is also based on predictions from thermodynamics. Thus, thermodynamics plays a central role in chemical process design. Although this subject is properly considered in other courses in the chemical engineering curriculum, we will provide very brief introductions to distillation, air stripping, liquid-liquid extraction, and other processes so that the student can appreciate why the study of thermodynamics is central to chemical engineering.

Other applications of thermodynamics considered in this book include how chemicals distribute when released to the environment, determining safety by estimating the possible impact (or energy release) of mechanical and chemical explosions, the analysis of biochemical processes, and product design, that is, identifying a chemical or mixture that has the properties needed for a specific application.

A generally important feature of engineering design is making estimates when specific information on a fluid or fluid mixture is not available, which is almost always the case. To understand why this is so, consider the fact that there are several hundred chemicals commonly used in industry, either as final products or intermediates. If this number were, say, 200, there would be about 20,000 possible binary mixtures, 1.3 million possible ternary mixtures, 67 million possible four-component mixtures, and so on. However, in the history of mankind the vapor-liquid equilibria of considerably fewer than 10,000 different mixtures have been measured. Further, even if we were interested in one of the mixtures for which data exist, it is unlikely that the measurements were done at exactly the temperature and pressure in which we are interested. Therefore, many times engineers have to make estimates by extrapolating the limited data available to the conditions (temperature, pressure, and composition) of interest to them, or predict the behavior of multicomponent mixtures based only on sets of two-component mixture data. In other cases predictions may have to be made for mixtures in which the chemical identity of one or more of the components is not known. One example of this is petroleum or crude oil; another is the result of a polymerization reaction or biochemical process. In these cases, many components of different molecular weights are present that will not, and perhaps cannot, be identified by chemical analytic methods, and yet purification methods have to be designed, so approximations are made.

Although the estimation of thermodynamic properties, especially of mixtures, is not part of the theoretical foundation of chemical engineering thermodynamics, it is necessary for its application to real problems. Therefore, various estimation methods are interspersed with the basic theory, especially in Chapters 6, 8, and 11, so that the theory can be applied.

This book can be considered as consisting of two parts. The first is the study of pure fluids, which begins after this introductory chapter. In Chapter 2 is a review of the use of mass balance, largely for pure fluids, but with a digression to reacting mixtures in order to explain the idea of nonconserved variables. Although mass balances should be familiar to a chemical engineering student from a course on stoichiometry or chemical process principles, it is reviewed here to introduce the different forms of the mass balance that will be used, the rate-of-change and difference forms (as well as the microscopic form for the advanced student), and some of the subtleties in applying the mass balance to systems in which flow occurs. The mass balance is the simplest of the balance equations we will use, and it is important to understand its application before proceeding to the use of other balance equations. We then move on to the development of the framework of thermodynamics and its application to power cycles and other processes involving only pure fluids, thereby avoiding the problems of estimating the properties of mixtures.

However, in the second part of this book, which begins in Chapter 8 and continues to the end of the book, the thermodynamic theory of mixtures, the properties of mixtures, and many different types of phase equilibria necessary for process design are considered, as are chemical reaction equilibria. It is this part of the book that is the essential background for chemical engineering courses in equipment and process design. We end the book with a chapter on the application of thermodynamics to biological and biochemical processes, though other such examples have been included in several of the preceding chapters.

Before proceeding, it is worthwhile to introduce a few of the terms used in applying the balance equations; other, more specific thermodynamic terms and definitions appear elsewhere in this book.

Glossary

Adiabatic system: A well-insulated system in which there are no heat flows in or out.

Closed system: A system in which there are no mass flows in or out.

Isolated system: A system that is closed to the flow of mass and energy in the form of work flows and heat flows (i.e., is adiabatic).

Steady-state system: A system in which flows of mass, heat, and work may be present but in such a way that the system properties do not change over time.

Cyclic process: A process that follows a periodic path so that the system has the same properties at any point in the cycle as it did at that point in any preceding or succeeding cycle.

The chapters in this book are all organized in a similar manner. First there is a paragraph or two describing the contents of the chapter and where it fits in to the general subject of thermodynamics. This introduction is followed by some specific instructional objectives or desired educational outcomes that the student is expected to develop from the chapter. Next is a brief list of the new terms or nomenclature introduced within the chapter. After these preliminaries, the real work starts.

INSTRUCTIONAL OBJECTIVES FOR CHAPTER 1

The goals of this chapter are for the student to:

- Know the basic terminology of thermodynamics, such as internal energy, potential energy, and kinetic energy; system, phase, and thermal and mechanical contact; adiabatic and isolated systems; and the difference between a system and a phase
- Be able to use the SI unit system which is used in this book and throughout the world
- Understand the concepts of absolute temperature and pressure
- Understand the difference between heat and work, and between mechanical and thermal energies
- Understand the general concept of equilibrium, which is very important in the application of thermodynamics in chemical engineering
- Understand the difference between intensive and extensive variables
- Understand that total mass and total energy are conserved in any process

IMPORTANT NOTATION INTRODUCED IN THIS CHAPTER

M Mass (g)

N Number of moles (mol)

P Absolute pressure (kPa or bar)

R Gas constant (J/mol K)

T Absolute temperature (K)

U Internal energy (J)

$\hat{U}$ Internal energy per unit mass (J/g)

$\underline{U}$ Internal energy per mole (J/mol)

V Volume (m^3)

$\hat{V}$ Specific volume, volume per unit mass (m^3/g)

$\underline{V}$ Volume per mole (m^3/mol)

1.1   THE CENTRAL PROBLEMS OF THERMODYNAMICS

Thermodynamics is the study of the changes in the state or condition of a substance when changes in its temperature, state of aggregation, or **internal energy** are important. By internal energy we mean the energy of a substance associated with the motions, interactions, and bonding of its constituent molecules, as opposed to the **external energy** associated with the velocity and location of its center of mass, which is of primary interest in mechanics. Thermodynamics is a macroscopic science; it deals with the average changes that occur among large numbers of molecules rather than the detailed changes that occur in a single molecule. Consequently, this book will quantitatively relate the internal energy of a substance not to its molecular motions and interaction, but to other, macroscopic variables such as temperature, which is primarily related to the extent of molecular motions, and density, which is a measure of how closely the molecules are packed and thus largely determines the extent of molecular interactions. The total energy of any substance is the sum of its internal energy and its bulk potential and kinetic energy; that is, it is the sum of the internal and external energies.

Our interest in thermodynamics is mainly in changes that occur in some small part of the universe, for example, within a steam engine, a laboratory beaker, or a chemical or biochemical reactor. The region under study, which may be a specified volume in space or a quantity of matter, is called the **system**; the rest of the universe is its **surroundings**. Throughout this book the term **state** refers to the thermodynamic state of a system as characterized by its density, refractive index, composition, pressure, temperature, or other variables to be introduced later. The state of agglomeration of the system—that is, whether it is a gas, liquid, or solid—is called its **phase**.

A system is said to be in contact with its surroundings if a change in the surroundings can produce a change in the system. Thus, a thermodynamic system is in **mechanical contact** with its surroundings if a change in pressure in the surroundings results in a pressure change in the system. Similarly, a system is in **thermal contact** with its surroundings if a temperature change in the surroundings can produce a change in the system. If a system does not change as a result of changes in its surroundings, the system is said to be **isolated**. Systems may be partially isolated from their surroundings. An **adiabatic** system is one that is thermally isolated from its surroundings; that is, it is a system that is not in thermal contact, but may be in mechanical contact, with its surroundings. If mass can flow into or out of a thermodynamic system, the system is said to be **open**; if not, the system is **closed**. Similarly, if heat can be added to the system or work done on it, we say the system is open to heat or work flows, respectively.[1]

---

[1] Both heat and work will be defined shortly.

An important concept in thermodynamics is the equilibrium state, which will be discussed in detail in the following sections. Here we merely note that if a system is not subjected to a continual forced flow of mass, heat, or work, the system will eventually evolve to a time-invariant state in which there are no internal or external flows of heat or mass and no change in composition as a result of chemical or biochemical reactions. This state of the system is the **equilibrium state**. The precise nature of the equilibrium state depends on both the character of the system and the **constraints** imposed on the system by its immediate surroundings and its container (e.g., a constant-volume container fixes the system volume, and a thermostatic bath fixes the system temperature; see Problem 1.1).

Using these definitions, we can identify the two general classes of problems that are of interest in thermodynamics. In the first class are problems concerned with computing the amount of work or the flow of heat either required or released to accomplish a specified change of state in a system or, alternatively, the prediction of the change in thermodynamic state that occurs for given heat or work flows. We refer to these problems as **energy flow** problems.

The second class of thermodynamic problems are those involving equilibrium. Of particular interest here is the identification or prediction of the equilibrium state of a system that initially is not in equilibrium. The most common problem of this type is the prediction of the new equilibrium state of a system that has undergone a change in the constraints that had been maintaining it in a previous state. For example, we will want to predict whether a single liquid mixture or two partially miscible liquid phases will be the equilibrium state when two pure liquids (the initial equilibrium state) are mixed (the change of constraint; see Chapter 11). Similarly, we will be interested in predicting the final temperatures and pressures in two gas cylinders after opening the connecting valve (change of constraint) between a cylinder that was initially filled and another that was empty (see Chapters 3 and 4).

It is useful to mention another class of problems related to those referred to in the previous paragraphs, but that is not considered here. We do not try to answer the question of how fast a system will respond to a change in constraints; that is, we do not try to study system dynamics. The answers to such problems, depending on the system and its constraints, may involve chemical kinetics, heat or mass transfer, and fluid mechanics, all of which are studied elsewhere. Thus, in the example above, we are interested in the final state of the gas in each cylinder, but not in computing how long a valve of given size must be held open to allow the necessary amount of gas to pass from one cylinder to the other. Similarly, when, in Chapters 10, 11, and 12, we study phase equilibrium and, in Chapter 13, chemical equilibrium, our interest is in the prediction of the equilibrium state, not in how long it will take to achieve this equilibrium state.

Shortly we will start the formal development of the principles of thermodynamics, first qualitatively and then, in the following chapters, in a quantitative manner. First, however, we make a short digression to discuss the system of units used in this text.

1.2 A SYSTEM OF UNITS

The study of thermodynamics involves mechanical variables such as force, pressure, and work, and thermal variables such as temperature and energy. Over the years many definitions and units for each of these variables have been proposed; for example, there are several values of the calorie, British thermal unit, and horsepower. Also, whole

Table 1.2-1 The SI Unit System

Unit	Name	Abbreviation	Basis of Definition
Length	meter	m	The distance light travels in a vacuum in $1/299\,792\,458$ second
Mass	kilogram	kg	Platinum-iridium prototype at the International Bureau of Weights and Measures, Sèvres, France
Time	second	s	Proportional to the period of one cesium-133 radiative transition
Electric current	ampere	A	Current that would produce a specified force between two parallel conductors in a specified geometry
Temperature	kelvin	K	$1/273.16$ of the thermodynamic temperature (to be defined shortly) of water at its triple point (see Chapter 7)
Amount of substance	mole	mol	Amount of a substance that contains as many elementary entities as there are atoms in 0.012 kilogram of carbon-12 (6.022×10^{23}, which is Avogadro's number)
Luminous intensity	candela	cd	Related to the black-body radiation from freezing platinum (2045 K)

systems of units, such as the English and cgs systems, have been used. The problem of standardizing units was studied, and the Système International d'Unités (abbreviated SI units) was agreed on at the Eleventh General Conference on Weights and Measures in 1960. This conference was one of a series convened periodically to obtain international agreement on questions of metrology, so important in international trade. The SI unit system is used throughout this book, with some lapses to the use of common units such as volume in liters and frequently pressure in bar.

In the SI system the seven basic units listed in Table 1.2-1 are identified and their values are assigned. From these seven basic well-defined units, the units of other quantities can be derived. Also, certain quantities appear so frequently that they have been given special names and symbols in the SI system. Those of interest here are listed in Table 1.2-2. Some other derived units acceptable in the SI system are given in Table 1.2-3, and Table 1.2-4 lists the acceptable scaling prefixes. [It should be pointed out

Table 1.2-2 Derived Units with Special Names and Symbols Acceptable in SI Units

Quantity	Name	Symbol	Expression in SI Units	Expression in Derived Units
Force	newton	N	$m\,kg\,s^{-2}$	$J\,m^{-1}$
Energy, work, or quantity of heat	joule	J	$m^2\,kg\,s^{-2}$	$N\,m$
Pressure or stress	pascal	Pa	$m^{-1}\,kg\,s^{-2}$	N/m^2
Power	watt	W	$m^2\,kg\,s^{-3}$	J/s
Frequency	hertz	Hz	s^{-1}	

Table 1.2-3 Other Derived Units in Terms of Acceptable SI Units

Quantity	Expression in SI Units	Symbol
Concentration of substance	mol m^{-3}	mol/m^3
Mass density ($\rho = m/V$)	kg m^{-3}	kg/m^3
Heat capacity or entropy	m^2 kg s^{-1} K^{-1}	J/K
Heat flow rate ($\dot{Q}$)	m^2 kg s^{-3}	W or J/s
Molar energy	m^2 kg s^{-2} mol^{-1}	J/mol
Specific energy	m^2 s^{-2}	J/kg
Specific heat capacity or specific entropy	m^2 s^{-2} K^{-1}	J/(kg K)
Specific volume	m^3 kg^{-1}	m^3/kg
Viscosity (absolute or dynamic)	m^{-1} kg s^{-1}	Pa s
Volume	m^3	m^3
Work, energy (W)	m^2 kg s^{-2}	J or N m

Table 1.2-4 Prefixes for SI Units

Multiplication Factor	Prefix	Symbol
10^{12}	tera	T
10^{9}	giga	G
10^{6}	mega	M
10^{3}	kilo	k (e.g., kilogram)
10^{2}	hecto	h
10	deka	da
10^{-1}	deci	d
10^{-2}	centi	c (e.g., centimeter)
10^{-3}	milli	m
10^{-6}	micro	μ
10^{-9}	nano	n
10^{-12}	pico	p
10^{-15}	femto	f

that, except at the end of a sentence, a period is never used after the symbol for an SI unit, and the degree symbol is not used. Also, capital letters are not used in units that are written out (e.g., pascals, joules, or meters) except at the beginning of a sentence. When the units are expressed in symbols, the first letter is capitalized only when the unit name is that of a person (e.g., Pa and J, but m).]

Appendix A.I presents approximate factors to convert from various common units to acceptable SI units. In the SI unit system, energy is expressed in joules, J, with 1 joule being the energy required to move an object 1 meter when it is opposed by a force of 1 newton. Thus 1 J = 1 N m = 1 kg m^2 s^{-2}. A pulse of the human heart, or lifting this book 0.1 meters, requires approximately 1 joule. Since this is such a small unit of energy, kilojoules (kJ = 1000 J) are frequently used. Similarly, we frequently use bar = 10^5 Pa = 0.987 atm as the unit of pressure.

1.3 THE EQUILIBRIUM STATE

As indicated in Section 1.1, the equilibrium state plays a central role in thermodynamics. The general characteristics of the equilibrium state are that (1) it does not vary with time; (2) the system is uniform (that is, there are no internal temperature, pressure, velocity, or concentration gradients) or is composed of subsystems each of which is uniform; (3) all flows of heat, mass, or work between the system and its surroundings are zero; and (4) the net rate of all chemical reactions is zero.

At first it might appear that the characteristics of the equilibrium state are so restrictive that such states rarely occur. In fact, the opposite is true. The equilibrium state will always occur, given sufficient time, as the terminal state of a system closed to the flow of mass, heat, or work across its boundaries. In addition, systems open to such flows, depending on the nature of the interaction between the system and its surroundings, may also evolve to an equilibrium state. If the surroundings merely impose a value of temperature, pressure, or volume on the system, the system will evolve to an equilibrium state. If, on the other hand, the surroundings impose a mass flow into and out of the system (as a result of a pumping mechanism) or a heat flow (as would occur if one part of the system were exposed to one temperature and another part of the system to a different temperature), the system may evolve to a time-invariant state only if the flows are steady. The time-invariant states of these driven systems are not equilibrium states in that the systems may or may not be uniform (this will become clear when the continuous-flow stirred tank and plug-flow chemical reactors are considered in Chapter 14) and certainly do not satisfy part or all of criterion (3). Such time-invariant states are called **steady states** and occur frequently in continuous chemical and physical processing. Steady-state processes are of only minor interest in this book.

Nondriven systems reach equilibrium because all spontaneous flows that occur in nature tend to dissipate the driving forces that cause them. Thus, the flow of heat that arises in response to a temperature difference occurs in the direction that dissipates the temperature difference, the mass diffusion flux that arises in response to a concentration gradient occurs in such a way that a state of uniform concentration develops, and the flux of momentum that occurs when a velocity gradient is present in a fluid tends to dissipate that gradient. Similarly, chemical reactions occur in a direction that drives the system toward equilibrium (Chapter 13). At various points throughout this book it will be useful to distinguish between the flows that arise naturally and drive the system to equilibrium, which we will call **natural flows**, and flows imposed on the system by its surroundings, which we term **forced flows**.

An important experimental observation in thermodynamics is that any system free from forced flows will, given sufficient time, evolve to an equilibrium state. This empirical fact is used repeatedly in our discussion.

It is useful to distinguish between two types of equilibrium states according to their response to small disturbances. To be specific, suppose a system in equilibrium is subjected to a small disturbance that is then removed (e.g., temperature fluctuation or pressure pulse). If the system returns to its initial equilibrium state, this state of the system is said to be **stable** with respect to small disturbances. If, however, the system does not return to the initial state, that state is said to have been **unstable**.

There is a simple mechanical analogy, shown in Fig. 1.3-1, that can be used to illustrate the concept of stability. Figure 1.3-1a, b, and c represent equilibrium positions of a block on a horizontal surface. The configuration in Fig. 1.3-1c is, however, precarious;

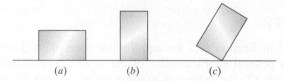

(a) (b) (c)

Figure 1.3-1 Blocks in states (a) and (b) are stable to small mechanical disturbances; the delicately balanced block in state (c) is not.

an infinitesimal movement of the block in any direction (so that its center of gravity is not directly over the pivotal point) would cause the block to revert to the configuration of either Fig. 1.3-1a or b. Thus, Fig. l.3-1c represents an unstable equilibrium position. The configurations of Fig. 1.3-1a and b are not affected by small disturbances, and these states are stable. Intuition suggests that the configuration of Fig. 1.3-1a is the most stable; clearly, it has the lowest center of gravity and hence the lowest potential energy. To go from the configuration of Fig. 1.3-1b to that of Fig. 1.3-1a, the block must pass through the still higher potential energy state indicated in Fig. 1.3-1c. If we use $\Delta\varepsilon$ to represent the potential energy difference between the configurations of Fig. 1.3-1b and c, we can say that the equilibrium state of Fig. 1.3-1b is stable to energy disturbances less than $\Delta\varepsilon$ in magnitude and is unstable to larger disturbances.

Certain equilibrium states of thermodynamic systems are stable to small fluctuations; others are not. For example, the equilibrium state of a simple gas is stable to all fluctuations, as are most of the equilibrium states we will be concerned with. It is possible, however, to carefully prepare a subcooled liquid, that is, a liquid below its normal solidification temperature, that satisfies the equilibrium criteria. This is an unstable equilibrium state because the slightest disturbance, such as tapping on the side of the containing vessel, will cause the liquid to freeze. One sometimes encounters mixtures that, by the chemical reaction equilibrium criterion (see Chapter 13), should react; however, the chemical reaction rate is so small as to be immeasurable at the temperature of interest. Such a mixture can achieve a state of thermal equilibrium that is stable with respect to small fluctuations of temperature and pressure. If, however, there is a sufficiently large, but temporary, increase in temperature (so that the rate of the chemical reaction is appreciable for some period of time) and then the system is quickly cooled, a new thermal equilibrium state with a chemical composition that differs from the initial state will be obtained. The initial equilibrium state, like the mechanical state in Fig. 1.3-1b, is then said to be stable with respect to small disturbances, but not to large disturbances.

Unstable equilibrium states are rarely encountered in nature unless they have been specially prepared (e.g., the subcooled liquid mentioned earlier). The reason for this is that during the approach to equilibrium, temperature gradients, density gradients, or other nonuniformities that exist within a system are of a sufficient magnitude to act as disturbances to unstable states and prevent their natural occurrence.

In fact, the natural occurrence of an unstable thermodynamic equilibrium state is about as likely as the natural occurrence of the unstable mechanical equilibrium state of Fig. 1.3-1c. Consequently, our concern in this book is mainly with stable equilibrium states.

If an equilibrium state is stable with respect to all disturbances, the properties of this state cannot depend on the past history of the system or, to be more specific, on the path followed during the approach to equilibrium. Similarly, if an equilibrium state is stable with respect to small disturbances, its properties do not depend on the path followed in the immediate vicinity of the equilibrium state. We can establish the validity of the latter statement by the following thought experiment (the validity of the first statement follows from a simple generalization of the argument). Suppose a system in a stable equilibrium state is subjected to a small temporary disturbance of a completely arbitrary nature. Since the initial state was one of stable equilibrium, the system will return to precisely that state after the removal of the disturbance. However, since any type of small disturbance is permitted, the return to the equilibrium state may be along a path that is different from the path followed in initially achieving the stable equilibrium state. The fact that the system is in exactly the same state as before means that all the properties of the system that characterize the equilibrium state must have their previous values; the fact that different paths were followed in obtaining this equilibrium state implies that none of these properties can depend on the path followed.

Another important experimental observation for the development of thermodynamics is that a system in a stable equilibrium state will never *spontaneously* evolve to a state of nonequilibrium. For example, any temperature gradients in a thermally conducting material free from a forced flow of heat will eventually dissipate so that a state of uniform temperature is achieved. Once this equilibrium state has been achieved, a measurable temperature gradient will never spontaneously occur in the material.

The two observations that (1) a system free from forced flows will evolve to an equilibrium state and (2) once in equilibrium a system will never spontaneously evolve to a nonequilibrium state, are evidence for a unidirectional character of natural processes. Thus we can take as a general principle that the direction of natural processes is such that systems evolve toward an equilibrium state, not away from it.

1.4   PRESSURE, TEMPERATURE, AND EQUILIBRIUM

Most people have at least a primitive understanding of the notions of temperature, pressure, heat, and work, and we have, perhaps unfairly, relied on this understanding in previous sections. Since these concepts are important for the development of thermodynamics, each will be discussed in somewhat more detail here and in the following sections.

The concept of pressure as the total force exerted on an element of surface divided by the surface area should be familiar from courses in physics and chemistry. Pressure—or, equivalently, force—is important in both mechanics and thermodynamics because it is closely related to the concept of mechanical equilibrium. This is simply illustrated by considering the two piston-and-cylinder devices shown in Fig. 1.4-1a and b. In each case we assume that the piston and cylinder have been carefully machined so that there is no friction between them. From elementary physics we know that for the systems in these figures to be in mechanical equilibrium (as recognized by the absence

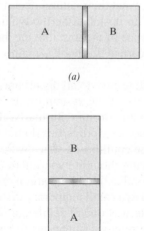

(a)

(b)

Figure 1.4-1 The piston separating gases A and B and the cylinder containing them have been carefully machined so that the piston moves freely in the cylinder.

of movement of the piston), there must be no unbalanced forces; the pressure of gas A must equal that of gas B in the system of Fig. 1.4-1a, and in the system of Fig. 1.4-1b it must be equal to the sum of the pressure of gas B and the force of gravity on the piston divided by its surface area. Thus, the requirement that a state of mechanical equilibrium exist is really a restriction on the pressure of the system.

Since pressure is a force per unit area, the direction of the pressure scale is evident; the greater the force per unit area, the greater the pressure. To measure pressure, one uses a pressure gauge, a device that produces a change in some indicator, such as the position of a pointer, the height of a column of liquid, or the electrical properties of a specially designed circuit, in response to a change in pressure. Pressure gauges are calibrated using devices such as that shown in Fig. 1.4-2. There, known pressures are created by placing weights on a frictionless piston of known weight. The pressure at the gauge P_g due to the metal weight and the piston is

$$P_g = \frac{M_w + M_p}{A} g \tag{1.4-1}$$

Here g is the local acceleration of gravity on an element of mass; the standard value is 9.80665 m/s^2. The position of the indicator at several known pressures is recorded, and the scale of the pressure gauge is completed by interpolation.

There is, however, a complication with this calibration procedure. It arises because the weight of the air of the earth's atmosphere produces an average pressure of 14.696 pounds force per square inch, or 101.3 kilopascals, at sea level. Since atmospheric pressure acts equally in all directions, we usually are not aware of its presence, so that in most nonscientific uses of pressure the zero of the pressure scale is the sea-level atmospheric pressure (i.e., the pressure of the atmosphere is neglected in the pressure gauge calibration). Thus, when the recommended inflation pressure of an automobile tire is 200 kPa, what is really meant is 200 kPa above atmospheric pressure. We refer to pressures on such a scale as gauge pressures. Note that gauge pressures may be negative (in partially or completely evacuated systems), zero, or positive, and errors in pressure measurement result from changes in atmospheric pressure from the

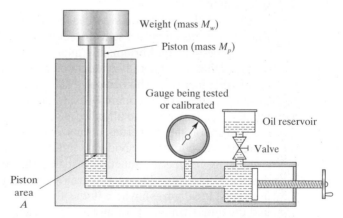

Weight (mass M_w)

Piston (mass M_p)

Gauge being tested or calibrated

Oil reservoir

Valve

Piston area A

Figure 1.4-2 A simple deadweight pressure tester. (The purpose of the oil reservoir and the system volume adjustment is to maintain equal heights of the oil column in the cylinder and gauge sections, so that no corrections for the height of the liquid column need be made in the pressure calibration.)

gauge calibration conditions (e.g., using a gauge calibrated in New York for pressure measurements in Denver).

We define the total pressure P to be equal to the sum of the gauge pressure P_g and the ambient atmospheric pressure P_{atm}. By accounting for the atmospheric pressure in this way we have developed an **absolute pressure** scale, that is, a pressure scale with zero as the lowest pressure attainable (the pressure in a completely evacuated region of space). One advantage of such a scale is its simplicity; the pressure is always a positive quantity, and measurements do not have to be corrected for either fluctuations in atmospheric pressure or its change with height above sea level. We will frequently be concerned with interrelationships between the temperature, pressure, and specific volume of fluids. These interrelationships are simplest if the absolute pressure is used. Consequently, unless otherwise indicated, the term *pressure* in this book refers to absolute pressure.

Although pressure arises naturally from mechanics, the concept of temperature is more abstract. To the nonscientist, temperature is a measure of hotness or coldness and as such is not carefully defined, but rather is a quantity related to such things as physical comfort, cooking conditions, or the level of mercury or colored alcohol in a thermometer. To the scientist, temperature is a precisely defined quantity, deeply rooted in the concept of equilibrium and related to the energy content of a substance.

The origin of the formal definition of temperature lies in the concept of thermal equilibrium. Consider a thermodynamic system composed of two subsystems that are in thermal contact but that do not interchange mass (e.g., the two subsystems may be two solids in contact, or liquids or gases separated by a thin, impenetrable barrier or membrane) and are isolated from their surroundings. When this composite system achieves a state of equilibrium (detected by observing that the properties of each system are time invariant), it is found that the property measured by the height of fluid in a given thermometer is the same in each system, although the other properties of the subsystems, such as their density and chemical composition, may be different. In accord with this observation, **temperature** is defined to be that system property which, if it has the same value for any two systems, indicates that these systems are in thermal equilibrium if they are in contact, or would be in thermal equilibrium if they were placed in thermal contact.

Although this definition provides the link between temperature and thermal equilibrium, it does not suggest a scale for temperature. If temperature is used only as an indicator of thermal equilibrium, any quantification or scale of temperature is satisfactory provided that it is generally understood and reproducible, though the accepted convention is that increasing hotness of a substance should correspond to increasing values of temperature. An important consideration in developing a thermodynamic scale of temperature is that it, like all other aspects of thermodynamics, should be general and not depend on the properties of any one fluid (such as the specific volume of liquid mercury). Experimental evidence indicates that it should be possible to formulate a completely universal temperature scale. The first indication came from the study of gases at densities so low that intermolecular interactions are unimportant (such gases are called **ideal gases**), where it was found that the product of the absolute pressure P and the molar volume $\underline{V}$ of any low-density gas away from its condensation line (see Chapter 7) increases with increasing hotness. This observation has been used as the basis for a temperature scale by defining the temperature $\mathcal{T}$ to be linearly proportional to the product of $P\underline{V}$ for a particular low-density gas, that is, by choosing $\mathcal{T}$ so that

$$P\underline{V} = A + R\mathcal{T} \qquad (1.4\text{-}2)$$

where A and R are constants. In fact, without any loss of generality one can define a new temperature $T = \mathcal{T} + (A/R)$, which differs from the choice of Eq. 1.4-2 only by an additive constant, to obtain

$$P\underline{V} = RT \tag{1.4-3}$$

Since neither the absolute pressure nor the molar volume of a gas can ever be negative, the temperature defined in this way must always be positive, and therefore the ideal gas temperature scale of Eq. 1.4-3 is an absolute scale (i.e., $T \geq 0$).

To complete this low-density gas temperature scale, it remains to specify the constant R, or equivalently the size of a unit of temperature. This can be done in two equivalent ways. The first is to specify the value of T for a given value of $P\underline{V}$ and thus determine the constant R; the second is to choose two reproducible points on a hotness scale and to decide arbitrarily how many units of T correspond to the difference in the $P\underline{V}$ products at these two fixed points. In fact, it is the latter procedure that is used; the ice point temperature of water[2] and the boiling temperature of water at standard atmospheric pressure (101.3 kPa) provide the two reproducible fixed-point temperatures. What is done, then, is to allow a low-density gas to achieve thermal equilibrium with water at its ice point and measure the product $P\underline{V}$, and then repeat the process at the boiling temperature. One then decides how many units of temperature correspond to this measured difference in the product $P\underline{V}$; the choice of 100 units or degrees leads to the Kelvin temperature scale, whereas the use of 180 degrees leads to the Rankine scale. With either of these choices, the constant R can be evaluated for a given low-density gas. *The important fact for the formulation of a universal temperature scale is that the constant R and hence the temperature scales determined in this way are the same for all low-density gases!* Values of the gas constant R in SI units are given in Table 1.4-1.

For the present we assume this low-density or ideal gas Kelvin (denoted by K) temperature scale is equivalent to an absolute universal thermodynamic temperature scale; this is proven in Chapter 6.

More common than the Kelvin temperature scale for nonscientific uses of temperature are the closely related Fahrenheit and Celsius scales. The size of the degree is the same in both the Celsius ($^\circ$C) and Kelvin temperature scales. However, the zero point of the Celsius scale is arbitrarily chosen to be the ice point temperature of water. Consequently, it is found that

$$T(\mathrm{K}) = T(^\circ\mathrm{C}) + 273.15 \tag{1.4-4a}$$

Table 1.4-1 The Gas Constant

$R = 8.314$ J/mol K
$= 8.314$ N m/mol K
$= 8.314 \times 10^{-3}$ kPa m^3/mol K
$= 8.314 \times 10^{-5}$ bar m^3/mol K
$= 8.314 \times 10^{-2}$ bar m^3/kmol K
$= 8.314 \times 10^{-6}$ MPa m^3/mol K

---

[2]The freezing temperature of water saturated with air at 101.3 kPa. On this scale the triple point of water is 0.01°C.

In the Fahrenheit (°F) temperature scale the ice point and boiling point of water (at 101.3 kPa) are 32°F and 212°F, respectively. Thus

$$T(\text{K}) = \frac{T(°\text{F}) + 459.67}{1.8} \qquad \textbf{(1.4-4b)}$$

Since we are assuming, for the present, that only the ideal gas Kelvin temperature scale has a firm thermodynamic basis, we will use it, rather than the Fahrenheit and Celsius scales, in all thermodynamic calculations.[3] (Another justification for the use of an absolute-temperature scale is that the interrelation between pressure, volume, and temperature for fluids is simplest when absolute temperature is used.) Consequently, if the data for a thermodynamic calculation are not given in terms of absolute temperature, it will generally be necessary to convert these data to absolute temperatures using Eqs. 1.4-4.

The product of $P\underline{V}$ for a low-density gas is said to be a **thermometric property** in that to each value of $P\underline{V}$ there corresponds only a single value of temperature. The ideal gas thermometer is not convenient to use, however, because of both its mechanical construction (see Fig. 1.4-3) and the manipulation required to make a measurement. Therefore, common thermometers make use of thermometric properties of other materials—for example, the single-valued relation between temperature and the specific volume of liquid mercury (Problem 1.2) or the electrical resistance of platinum wire. There are two steps in the construction of thermometers based on these other thermometric properties: first, fabrication of the device, such as sealing liquid mercury in an otherwise evacuated tube; and second, the calibration of the thermometric indi-

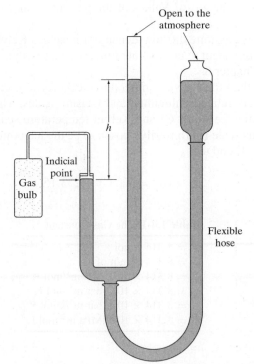

Open to the
atmosphere

Indicial
point

Gas
bulb

h

Flexible
hose

Figure 1.4-3 A simplified diagram of a constant-volume ideal gas thermometer. In this thermometer the product PV for a gas at various temperatures is found by measuring the pressure P at constant volume. For each measurement the mercury reservoir is raised or lowered until the mercury column at the left touches an index mark. The pressure of the gas in the bulb is then equal to the atmospheric pressure plus the pressure due to the height of the mercury column.

[3]Of course, for calculations involving only temperature differences, any convenient temperature scale may be used, since a temperature difference is independent of the zero of the scale.

cator with a known temperature scale. To calibrate a thermometer, its readings (e.g., the height of a mercury column) are determined at a collection of known temperatures, and its scale is completed by interpolating between these fixed points. The calibration procedure for a common mercury thermometer is usually far simpler. The height of the mercury column is determined at only two fixed points (e.g., the temperature of an ice-water bath and the temperature of boiling water at atmospheric pressure), and the distance between these two heights is divided into equal units; the number of units depends on whether the Rankine or Kelvin degree is used, and whether a unit is to represent a fraction of a degree, a degree, or several degrees. Since only two fixed points are used in the calibration, intermediate temperatures recorded on such a thermometer may be different from those that would be obtained using an ideal gas thermometer because (1) the specific volume of liquid mercury has a slightly nonlinear dependence on temperature, and (2) the diameter of the capillary tube may not be completely uniform (so that the volume of mercury will not be simply related to its height in the tube; see Problem 1.2).

1.5 HEAT, WORK, AND THE CONSERVATION OF ENERGY

As we have already indicated, two systems in thermal contact but otherwise isolated from their surroundings will eventually reach an equilibrium state in which the systems have the same temperature. During the approach to this equilibrium state the temperature of the initially low-temperature system increases while the temperature of the initially high-temperature system decreases. We know that the temperature of a substance is directly related to its internal energy, especially the energy of molecular motion. Thus, in the approach to equilibrium, energy has been transferred from the high-temperature system to the one of lower temperature. This transfer of energy as a result of only a temperature difference is called a flow of **heat**.

It is also possible to increase the total energy (internal, potential, and kinetic) of a system by mechanical processes involving motion. In particular, the kinetic or potential energy of a system can change as a result of motion without deformation of the system boundaries, as in the movement of a solid body acted on by an external force, whereas the internal energy and temperature of a system may change when external forces result in the deformation of the system boundaries, as in the compression of a gas. Energy transfer by mechanical motion also occurs as a result of the motion of a drive shaft, push rod, or similar device across the system boundaries. For example, mechanical stirring of a fluid first results in fluid motion (evidence of an increase in fluid kinetic energy) and then, as this motion is damped by the action of the fluid viscosity, in an increase in the temperature (and internal energy) of the fluid. Energy transfer by any mechanism that involves mechanical motion of or across the system boundaries is called **work**.

Finally, it is possible to increase the energy of a system by supplying it with electrical energy in the form of an electrical current driven by a potential difference. This electrical energy can be converted to mechanical energy if the system contains an electric motor, it can increase the temperature of the system if it is dissipated through a resistor (resistive heating), or it can be used to cause an electrochemical change in the system (e.g., recharging a lead storage battery). Throughout this book we consider the flow of electrical energy to be a form of work. The reason for this choice will become clear shortly.

The amount of mechanical work is, from mechanics, equal to the product of the force exerted times the distance moved in the direction of the applied force, or, alternatively, to the product of the applied pressure and the displaced volume. Similarly, the electrical work is equal to the product of the current flow through the system, the potential difference across the system, and the time interval over which the current flow takes place. Therefore, the total amount of work supplied to a system is frequently easy to calculate.

An important experimental observation, initially made by James Prescott Joule between the years 1837 and 1847, is that a specified amount of energy can always be used in such a way as to produce the same temperature rise in a given mass of water, regardless of the precise mechanism or device used to supply the energy, and regardless of whether this energy is in the form of mechanical work, electrical work, or heat. Rather than describe Joule's experiments, consider how this hypothesis could be proved in the laboratory. Suppose a sample of water at temperature T_1 is placed in a well-insulated container (e.g., a Dewar flask) and, by the series of experiments in Table 1.5-1, the amount of energy expended in producing a final equilibrium temperature T_2 is measured. Based on the experiments of Joule and others, we would expect to find that this

Table 1.5-1 Experiments Designed to Prove the Energy Equivalence of Heat and Work

Form in Which Energy Is Transferred to Water	Mechanism Used	Form of Energy Supplied to Mechanism	Method of Measuring Energy Input	Corrections That Must Be Made to Energy Input Data
(1) Mechanical energy	Stirring: Paddlewheel driven by electric motor	Electrical energy	Product of voltage, current, and time	Electrical energy loss in motor and circuit, temperature rise of paddlewheel
(2) Mechanical energy	Stirring: Paddlewheel driven by pulley and falling weight	Mechanical energy	Change in potential energy of weight: product of mass of weight, change in height, and the gravitational constant g	Temperature rise of paddlewheel
(3) Heat flow	Electrical energy converted to heat in a resistor	Electrical energy	Product of voltage, current, and time	Temperature rise of resistor and electrical losses in circuit
(4) Heat flow	Mechanical energy of falling weight is converted to heat through friction of rubbing two surfaces together, as with a brake on the axle of a pulley	Mechanical energy	Change in potential energy of weight: product of mass of weight, change in height, and g	Temperature rise of mechanical brakes, etc.

energy (determined by correcting the measurements of column 4 for the temperature rise of the container and the effects of column 5) is precisely the same in all cases.

By comparing the first two experiments with the third and fourth, we conclude that there is an equivalence between mechanical energy (or work) and heat, in that precisely the same amount of energy was required to produce a given temperature rise, independent of whether this energy was delivered as heat or work. Furthermore, since both mechanical and electrical energy sources have been used (see column 3), there is a similar equivalence between mechanical and electrical energy, and hence among all three energy forms. This conclusion is not specific to the experiments in Table 1.5-1 but is, in fact, a special case of a more general experimental observation; that is, any change of state in a system that occurs solely as a result of the addition of heat can also be produced by adding the same amount of energy as work, electrical energy, or a combination of heat, work, and electrical energy.

Returning to the experiments of Table 1.5-1, we can now ask what has happened to the energy that was supplied to the water. The answer, of course, is that at the end of the experiment the temperature, and hence the molecular energy, of the water has increased. Consequently, the energy added to the water is now present as increased internal energy. It is possible to extract this increased internal energy by processes that return the water to its original temperature. One could, for example, use the warm water to heat a metal bar. The important experimental observation here is that if you measured the temperature rise in the metal, which occurred in returning the water to its initial state, and compared it with the electrical or mechanical energy required to cause the same temperature rise in the metal, you would find that all the energy added to the water in raising its temperature could be recovered as heat by returning the water to its initial state. Thus, total energy has been conserved in the process.

The observation that energy has been conserved in this experiment is only one example of a general energy conservation principle that is based on a much wider range of experiments. The more general principle is that in any change of state, the total energy, which is the sum of the internal, kinetic, and potential energies of the system, heat, and electrical and mechanical work, is conserved. A more succinct statement is that energy is neither created nor destroyed, but may change in form.

Although heat, mechanical work, and electrical work are equivalent in that a given energy input, in any form, can be made to produce the same internal energy increase in a system, there is an equally important difference among the various energy forms. To see this, suppose that the internal energy of some system (perhaps the water in the experiments just considered) has been increased by increasing its temperature from T_1 to a higher temperature T_2, and we now wish to recover the added energy by returning the system to its initial state at temperature T_1. It is clear that we can recover the added energy completely as a heat flow merely by putting the system in contact with another system at a lower temperature. There is, however, no process or device by which it is possible to convert all the added internal energy of the system to mechanical energy and restore both the system and the surroundings to their initial states, even though the increased internal energy may have resulted from adding only mechanical energy to the system. In general, only a portion of the increased internal energy can be recovered as mechanical energy, the remainder appearing as heat. This situation is not specific to the experiments discussed here; it occurs in all similar efforts to convert both heat and internal energy to work or mechanical energy.

We use the term **thermal energy** to designate energy in the form of internal energy and heat, and **mechanical energy** to designate mechanical and electrical work and the external energy of a system. This distinction is based on the general experimental observation that while, in principle, any form of mechanical energy can be completely converted to other forms of mechanical energy or thermal energy, only a fraction of the thermal energy can be converted to mechanical energy in any **cyclic process** (that is, a process that at the end of the cycle restores the system and surroundings to their states at the beginning of the cycle), or any process in which the only change in the universe (system and surroundings) is the conversion of thermal energy to mechanical energy.

The units of mechanical work arise naturally from its definition as the product of a force and a distance. Typical units of mechanical work are foot–pound force, dyne-centimeter, and erg, though we will use the newton-meter (N m), which is equal to one joule; and from the formulation of work as pressure times displaced volume, pascal-meter^3, which is also equal to one joule. The unit of electrical work is the volt-ampere-second or, equivalently, the watt-second (again equal to one joule). Heat, however, not having a mechanical definition, has traditionally been defined experimentally. Thus, the heat unit calorie was defined as the amount of heat required to raise the temperature of 1 gram of water from 14.5°C to 15.5°C, and the British thermal unit (Btu) was defined to be the amount of heat required to raise 1 pound of water from 59°F to 60°F. These experimental definitions of heat units have proved unsatisfactory because the amount of energy in both the calorie and Btu have been subject to continual change as measurement techniques improved. Consequently, there are several different definitions of the Btu and calorie (e.g., the thermochemical calorie, the mean calorie, and the International Table calorie) that differ by less than one and one-half parts in a thousand. Current practice is to recognize the energy equivalence of heat and work and to use a common energy unit for both. We will use only the joule, which is equal to 0.2390 calorie (thermochemical) or 0.9485×10^{-3} Btu (thermochemical).

1.6   SPECIFICATION OF THE EQUILIBRIUM STATE; INTENSIVE AND EXTENSIVE VARIABLES; EQUATIONS OF STATE

Since our main interest throughout this book is with stable equilibrium states, it is important to consider how to characterize the equilibrium state and, especially, what is the minimum number of properties of a system at equilibrium that must be specified to fix the values of all its remaining properties completely.[4] To be specific, suppose we had 1 kilogram of a pure gas, say oxygen, at equilibrium whose temperature is some value T, pressure some value P, volume V, refractive index R, electrical permitivity ε, and so on, and we wanted to adjust *some* of the equilibrium properties of a second sample of oxygen so that *all* the properties of the two samples would be identical. The questions we are asking, then, are what sorts of properties, and how many properties, must correspond if all of the properties of the two systems are to be identical?

The fact that we are interested only in stable equilibrium states is sufficient to decide the types of properties needed to specify the equilibrium state. First, since gradients in velocity, pressure, and temperature cannot be present in the equilibrium state, they do not enter into its characterization. Next, since, as we saw in Sec. 1.3, the properties of a stable equilibrium state do not depend on the history of the system or its approach to equilibrium, the *stable equilibrium state is characterized only by equilibrium properties of the system.*

---

[4]Throughout this book, we implicitly assume that a system contains a large number of molecules (at least several tens of thousands), so that the surface effects present in small systems are unimportant. See, however, Sec. 7.8.

The remaining question, that of how many equilibrium properties are necessary to specify the equilibrium state of the system, can be answered only by experiment. The important experimental observation here is that an equilibrium state of a single-phase, one-component system in the absence of external electric and magnetic fields is completely specified if its mass and two other thermodynamic properties are given. Thus, going back to our example, if the second oxygen sample also weighs 1 kilogram, and if it were made to have the same temperature and pressure as the first sample, it would also be found to have the same volume, refractive index, and so forth. If, however, only the temperature of the second 1-kg sample was set equal to that of the first sample, neither its pressure nor any other physical property would necessarily be the same as that of the first sample. Consequently, the values of the density, refractive index, and, more generally, all thermodynamic properties of an equilibrium single-component, single-phase fluid are completely fixed once the mass of the system and the values of at least two other system parameters are given. (The specification of the equilibrium state of multiphase and multicomponent systems is considered in Chapters 7 and 8.)

The specification of an equilibrium system can be made slightly simpler by recognizing that the variables used in thermodynamic descriptions are of two different types. To see this, consider a gas of mass M that is at a temperature T and pressure P and is confined to a glass bulb of volume V. Suppose that an identical glass bulb is also filled with mass M of the same gas and heated to the same temperature T. Based on the previous discussion, since the values of T, V, and M are the same, the pressure in the second glass bulb is also P. If these two bulbs are now connected to form a new system, the temperature and pressure of this composite system are unchanged from those of the separated systems, although the volume and mass of this new system are clearly twice those of the original single glass bulb. The pressure and temperature, because of their size-independent property, are called **intensive variables**, whereas the mass, volume, and total energy are **extensive variables**, or variables dependent on the size or amount of the system. Extensive variables can be transformed into intensive variables by dividing by the total mass or total number of moles so that a specific volume (volume per unit mass or volume per mole), a specific energy (energy per unit mass or per mole), and so forth are obtained. By definition, the term **state variable** refers to any of the intensive variables of an equilibrium system: temperature, pressure, specific volume, specific internal energy, refractive index, and other variables introduced in the following chapters. Clearly, from the previous discussion, the value of any state variable depends only on the equilibrium state of the system, not on the path by which the equilibrium state was reached.

With the distinction now made between intensive and extensive variables, it is possible to rephrase the requirement for the complete specification of a thermodynamic state in a more coherent manner. The experimental observation is that the specification of two state variables uniquely determines the values of all other state variables of an equilibrium, single-component, single-phase system. [Remember, however, that to determine the size of the system, that is, its mass or total volume, one must also specify the mass of the system, or the value of one other extensive parameter (total volume, total energy, etc.).] The implication of this statement is that for each substance there exist, in principle, equations relating each state variable to two others. For example,

$$P = P(T, \hat{V})$$
$$\hat{U} = \hat{U}(T, \hat{V}) \qquad\qquad \textbf{(1.6-1)}$$

Here P is the pressure, T the temperature, $\hat{U}$ the internal energy per unit mass, and $\hat{V}$ the volume per unit mass.[5] The first equation indicates that the pressure is a function of the temperature and volume per unit mass; the second indicates that the internal energy is a function of temperature and volume. Also, there are relations of the form

$$\hat{U} = \hat{U}(T, P)$$
$$\hat{U} = \hat{U}(P, \hat{V})$$
$$P = P(\hat{U}, \hat{V}) \tag{1.6-2}$$

Similar equations are valid for the additional thermodynamic properties to be introduced later.

The interrelations of the form of Eqs. 1.6-1 and 1.6-2 are always obeyed in nature, though we may not have been sufficiently accurate in our experiments, or clever enough in other ways to have discovered them. In particular, Eq. 1.6-1 indicates that if we prepare a fluid such that it has specified values T and $\hat{V}$, it will always have the same pressure P. What is this value of the pressure P? To know this we have to either have done the experiment sometime in the past or know the exact functional relationship between T, $\hat{V}$, and P for the fluid being considered. What is frequently done for fluids of scientific or engineering interest is to make a large number of measurements of P, $\hat{V}$, and T and then to develop a **volumetric equation of state** for the fluid, that is, a mathematical relationship between the variables P, $\hat{V}$, and T. Similarly, measurements of $\hat{U}$, $\hat{V}$, and T are made to develop a **thermal equation of state** for the fluid. Alternatively, the data that have been obtained may be presented directly in graphical or tabular form. (In fact, as will be shown later in this book, it is more convenient to formulate volumetric equations of state in terms of P, $\underline{V}$, and T than in terms of P, $\hat{V}$, and T, since in this case the same gas constant of Eq. 1.4-3 can be used for all substances. If volume on a per-mass basis $\hat{V}$ was used, the constant in the ideal gas equation of state would be R divided by the molecular weight of the substance.)

There are some complications in the description of thermodynamic states of systems. For certain idealized fluids, such as the ideal gas and the incompressible liquid (both discussed in Sec. 3.3), the specification of any two state variables may not be sufficient to fix the thermodynamic state of the system. To be specific, the internal energy of the ideal gas is a function only of its temperature, and not of its pressure or density. Thus, the specification of the internal energy and temperature of an ideal gas contains no more information than specifying only its temperature and therefore is insufficient to determine its pressure. Similarly, if a liquid is incompressible, its molar volume will depend on temperature but not on the pressure exerted on it. Consequently, specifying the temperature and the specific volume of an incompressible liquid contains no more information than specifying only its temperature. The ideal gas and the incompressible liquid are limiting cases of the behavior of real fluids, so that although the internal energy of a real gas depends on density and temperature, the density dependence may be weak; also the densities of most liquids are only weakly dependent on their pressure.

---

[5]In this book we use letters with carets to indicate properties per unit mass, such as $\hat{U}$ and $\hat{V}$, and letters with underbars, such as $\underline{U}$ and $\underline{V}$, to indicate properties per mole, which are referred to as molar properties. When, in later chapters, we consider mixtures and have to distinguish between species, the notation will become a bit more complicated in that $\underline{U}_i$ and $\underline{V}_i$ will be used to designate the molar internal energy and volume, respectively, of pure species i. Also, when necessary, within parentheses we can indicate the temperature and/or pressure (and in later chapters the composition) of the substance. In these cases, notation such as $\underline{U}_i(T, P)$ and $\underline{V}_i(T, P)$ will be used.

Therefore, although in principle any two state variables may be used to describe the thermodynamic state of a system, one should not use $\hat{U}$ and T as the independent state variables for gases or $\hat{V}$ and T as the independent state variables for liquids and solids.

As was pointed out in the previous paragraphs, two state variables are needed to fix the thermodynamic state of an equilibrium system. The obvious next question is, how does one specify the thermodynamic state of a nonequilibrium system? This is clearly a much more complicated question, and the detailed answer would involve a discussion of the relative time scales for changes imposed on the system and the changes that occur within the system (as a result of chemical reaction, internal energy flows, and fluid motion). Such a discussion is beyond the scope of this book. The important observation is that if we do not consider very fast system changes (as occur within a shock wave), or systems that relax at a very slow but perceptible rate (e.g., molten polymers), the equilibrium relationships between the fluid properties, such as the volumetric and thermal equations of state, are also satisfied in nonequilibrium flows on a point-by-point basis. That is, even though the temperature and pressure may vary in a flowing fluid, as long as the changes are not as sharp as in a shock wave and the fluid internal relaxation times are rapid,[6] the properties at each point in the fluid are interrelated by the same equations of state as for the equilibrium fluid. This situation is referred to as **local equilibrium**. This is an important concept, since it allows us to consider not only equilibrium phenomena in thermodynamics but also many flow problems involving distinctly nonequilibrium processes.

1.7 A SUMMARY OF IMPORTANT EXPERIMENTAL OBSERVATIONS

An objective of this book is to present the subject of thermodynamics in a logical, coherent manner. We do this by demonstrating how the complete structure of thermodynamics can be built from a number of important experimental observations, some of which have been introduced in this chapter, some of which are familiar from mechanics, and some of which are introduced in the following chapters. For convenience, the most important of these observations are listed here.

From classical mechanics and chemistry we have the following two observations.

Experimental observation 1. In any change of state (except one involving a nuclear reaction, which is not considered in this book) total mass is conserved.

Experimental observation 2. In any change of state total momentum is a conserved quantity.

In this chapter the following eight experimental facts have been mentioned.

Experimental observation 3 (Sec. 1.5). In any change of state the total energy—which includes internal, potential, and kinetic energy, heat, and work—is a conserved quantity.

Experimental observation 4 (Sec. 1.5). A flow of heat and a flow of work are equivalent in that supplying a given amount of energy to a system in either of these forms can be made to result in the same increase in its internal energy. Heat and work, or more generally, thermal and mechanical energy, are not equivalent in the sense

---

[6]The situation being considered here is not as restrictive as it appears. In fact, it is by far the most common case in engineering. It is the only case that is considered in this book.

that mechanical energy can be completely converted to thermal energy, but thermal energy can be only partially converted to mechanical energy in a cyclic process.

Experimental observation 5 (Sec. 1.3). A system that is not subject to forced flows of mass or energy from its surroundings will evolve to a time-invariant state that is uniform or composed of uniform subsystems. This is the equilibrium state.

Experimental observation 6 (Sec. 1.3). A system in equilibrium with its surroundings will never spontaneously revert to a nonequilibrium state.

Experimental observation 7 (Sec. 1.3). Equilibrium states that arise naturally are stable to small disturbances.

Experimental observation 8 (Secs. 1.3 and 1.6). The stable equilibrium state of a system is completely characterized by values of only equilibrium properties (and not properties that describe the approach to equilibrium). For a single-component, single-phase system the values of only two intensive, independent state variables are needed to fix the thermodynamic state of the equilibrium system completely; the further specification of one extensive variable of the system fixes its size.

Experimental observation 9 (Sec. 1.6). The interrelationships between the thermodynamic state variables for a fluid in equilibrium also apply locally (i.e., at each point) for a fluid not in equilibrium, provided the internal relaxation processes are rapid with respect to the rate at which changes are imposed on the system. For fluids of interest in this book, this condition is satisfied.

Although we are not, in general, interested in the detailed description of nonequilibrium systems, it is useful to note that the rates at which natural relaxation processes (i.e., heat fluxes, mass fluxes, etc.) occur are directly proportional to the magnitude of the driving forces (i.e., temperature gradients, concentration gradients, etc.) necessary for their occurrence.

Experimental observation 10. The flow of heat $\dot{Q}$ (units of J/s or W) that arises because of a temperature difference ΔT is linearly proportional to the magnitude of the temperature difference:[7]

$$\dot{Q} = -h\,\Delta T \qquad \qquad \textbf{(1.7-1)}$$

Here h is a positive constant, and the minus sign in the equation indicates that the heat flow is in the opposite direction to the temperature difference; that is, the flow of heat is from a region of high temperature to a region of low temperature. Similarly, on a microscopic scale, the heat flux in the x-coordinate direction, denoted by q_x (with units of J/m^2 s), is linearly related to the temperature gradient in that direction:

$$q_x = -k\frac{\partial T}{\partial x} \qquad \qquad \textbf{(1.7-2)}$$

The mass flux of species A in the x direction, $j_A|_x$ (kg/m^2 s), relative to the fluid mass average velocity is linearly related to its concentration gradient,

$$j_A|_x = -\rho D\frac{\partial w_A}{\partial x} \qquad \qquad \textbf{(1.7-3)}$$

---

[7]Throughout this book we use a dot, as on $\dot{Q}$, to indicate a flow term. Thus, $\dot{Q}$ is a flow of heat with units of J/s, and $\dot{M}$ is a flow of mass with units of kg/s. Also, radiative heat transfer is more complicated, and is not considered here.

and for many fluids, the flux of the x-component of momentum in the y-coordinate direction is

$$\tau_{yx} = -\mu \frac{\partial v_x}{\partial y} \qquad \textbf{(1.7-4)}$$

In these equations T is the temperature, ρ the mass density, w_A the mass fraction of species A, and v_x the x-component of the fluid velocity vector. The parameter k is the thermal conductivity, D the diffusion coefficient for species A, and μ the fluid viscosity; from experiment the values of these parameters are all greater than or equal to zero (this is, in fact, a requirement for the system to evolve toward equilibrium). Equation 1.7-2 is known as Fourier's law of heat conduction, Eq. 1.7-3 is called Fick's first law of diffusion, and Eq. 1.7-4 is Newton's law of viscosity.

1.8 A COMMENT ON THE DEVELOPMENT OF THERMODYNAMICS

The formulation of the principles of thermodynamics that is used in this book is a reflection of the author's preference and experience, and is not an indication of the historical development of the subject. This is the case in most textbooks, as a good textbook should present its subject in an orderly, coherent fashion, even though most branches of science have developed in a disordered manner marked by both brilliant, and frequently unfounded, generalizations and, in retrospect, equally amazing blunders. It would serve little purpose to relate here the caloric or other theories of heat that have been proposed in the past, or to describe all the futile efforts that went into the construction of perpetual motion machines. Similarly, the energy equivalence of heat and work seems obvious now, though it was accepted by the scientific community only after 10 years of work by J. P. Joule. Historically, this equivalence was first pointed out by a medical doctor, J. R. Mayer. However, it would be foolish to reproduce in a textbook the stages of his discovery, which started with the observation that the venous blood of sailors being bled in Java was unusually red, made use of a theory of Lavoisier relating the rate of oxidation in animals to their heat losses, and ultimately led to the conclusion that heat and work were energetically equivalent.

The science of thermodynamics as we now know it is basically the work of the experimentalist, in that each of its principles represents the generalization of a large amount of varied experimental data and the life's work of many. We have tried to keep this flavor by basing our development of thermodynamics on a number of key experimental observations. However, the presentation of thermodynamics in this book, and especially in the introduction of entropy in Chapter 4, certainly does not parallel its historical development.

PROBLEMS

1.1 For each of the cases that follow, list as many properties of the equilibrium state as you can, especially the constraints placed on the equilibrium state of the system by its surroundings and/or its container.

a. The system is placed in thermal contact with a thermostatic bath maintained at temperature T.

b. The system is contained in a constant-volume container and thermally and mechanically isolated from its surroundings.

c. The system is contained in a frictionless piston and cylinder exposed to an atmosphere at pressure P and thermally isolated from its surroundings.

d. The system is contained in a frictionless piston and cylinder exposed to an atmosphere at pressure P and is in thermal contact with a thermostatic bath maintained at temperature T.

e. The system consists of two tanks of gas connected by tubing. A valve between the two tanks is fully opened for a short time and then closed.

1.2 The following table lists the volumes of 1 gram of water and 1 gram of mercury as functions of temperature.

a. Discuss why water would not be an appropriate thermometer fluid between 0°C and 10°C.

b. Because of the slightly nonlinear temperature dependence of the specific volume of liquid mercury, there is an inherent error in using a mercury-filled thermometer that has been calibrated against an ideal gas thermometer at only 0°C and 100°C. Using the data in the table, prepare a graph of the error, ΔT, as a function of temperature.

c. Why does a common mercury thermometer consist of a large-volume mercury-filled bulb attached to a capillary tube?

T (°C)	Volume of 1 gram of H_2O (cm^3)	Volume of 1 gram of Hg (cm^3)
0	1.0001329	0.0735560
1	1.0000733	0.0735694
2	1.0000321	0.0735828
3	1.0000078	0.0735961
4	1.0000000	0.0736095
5	1.0000081	0.0736228
6	1.0000318	0.0736362
7	1.0000704	0.0736496
8	1.0001236	0.0736629
9	1.0001909	0.0736763
10	1.0002719	0.0736893
20	1.0015678	0.0738233
30	1.0043408	0.0739572
40	1.0078108	0.0740910
50	1.012074	0.0742250
60	1.017046	0.0743592
70	1.022694	0.0744936
80	1.028987	0.0746282
90	1.035904	0.0747631
100	1.043427	0.0748981

*Based on data in R. H. Perry and D. Green, eds., *Chemical Engineers' Handbook,* 6th ed., McGraw-Hill, New York, 1984, pp. 3-75–3-77.

Chapter **2**

Conservation of Mass

In this chapter we start the quantitative development of thermodynamics using one of the qualitative observations of the previous chapter, that mass is conserved. Here we begin by developing the balance equations for the total mass of a system (a piece of equipment, a defined volume in space, or whatever is convenient for the problem at hand) by considering systems of only a single component. In this case, mass of the single species being considered is also the total mass, which is conserved, and we can write the balance equation either based on mass or by dividing by the molecular weight, on the number of moles. We develop two forms of these mass balance equations—one for computing the rate at which mass in a system changes with time, and the second set, obtained by integrating these rate equations over an interval of time, to compute only the change in mass (or number of moles) in that time interval.

We next consider the mass balances for a mixture. In this case while total mass is conserved, there will be a change of mass of some or all species if one or more chemical reactions occur. For this case, it is more convenient to develop the mass balance for mixtures on a molar basis, as chemical reaction stoichiomentry is much easier to write on a molar basis than on a mass basis. In this chapter we will consider only the case of a single chemical reaction; in later chapters the more general case of several chemical reactions occuring simultaneously will be considered.

The most important goals of this chapter are for the student to understand when to use the rate-of-change form of the mass balance equation and when to use the difference form, and how to use these equations to solve problems. Mastering the use of the mass balance equations here will make it easier to use the more complicated energy and other balance equations that will be introduced in the following two chapters.

INSTRUCTIONAL OBJECTIVES FOR CHAPTER 2

The goals of this chapter are for the student to:

- Be able to use the rate-of-change form of the pure component mass balance in problem solving (Sec. 2.2)
- Be able to use the difference form of the pure component mass balance in problem solving (Sec. 2.2)
- Be able to use the rate-of-change form of the multicomponent mass balance in problem solving (Sec. 2.3)
- Be able to use the difference form of the multicomponent mass balance in problem solving (Sec. 2.3)
- Be able to solve mass balance problems involving a single chemical reaction (Sec. 2.3)

IMPORTANT NOTATION INTRODUCED IN THIS CHAPTER

M_i Mass of species i (g)
$\dot{M}_k$ Mass flow rate at location k (g/s)
$(\dot{M}_i)_k$ Mass flow rate of species i at location k (g/s)
N_i Moles of species i (mol)
$\dot{N}_k$ Molar flow rate at location k (g/s)
$(\dot{N}_i)_k$ Molar flow rate of species i at location k (g/s)
t Time (s)
$\underline{x}$ Set of mole fractions of all species $x_1, x_2, x_3, \ldots$
$\underline{X}$ Molar extent of reaction (mol)
ν_i Stoichiometric coefficient of species i

2.1   A GENERAL BALANCE EQUATION AND CONSERVED QUANTITIES

The balance equations used in thermodynamics are conceptually simple. Each is obtained by choosing a system, either a quantity of mass or a region of space (e.g., the contents of a tank), and equating the change of some property of this system to the amounts of the property that have entered and left the system and that have been produced within it. We are interested both in the change of a system property over a time interval and in its instantaneous rate of change; therefore, we will formulate equations of change for both. Which of the two formulations of the equations of change is used for the description of a particular physical situation will largely depend on the type of information desired or available.

To illustrate the two types of descriptions and the relationship between them, as well as the idea of using balance equations, consider the problem of studying the total mass of water in Lake Mead (the lake behind Hoover Dam on the Colorado River). If you were interested in determining, at any instant, whether the water level in this lake was rising or falling, you would have to ascertain whether the water flows into the lake were greater or less than the flows of water out of the lake. That is, at some instant you would determine the rates at which water was entering the lake (due to the flow of the Colorado River and rainfall) and leaving it (due to flow across the dam, evaporation from the lake surface, and seepage through the canyon walls), and then use the equation

$$\begin{pmatrix} \text{Rate of change of} \\ \text{amount of water} \\ \text{in the lake} \end{pmatrix} = \begin{pmatrix} \text{Rate at which} \\ \text{water flows} \\ \text{into the lake} \end{pmatrix} - \begin{pmatrix} \text{Rate at which} \\ \text{water flows out} \\ \text{of the lake} \end{pmatrix} \qquad \textbf{(2.1-1)}$$

to determine the precise rate of change of the amount of water in the lake.

If, on the other hand, you were interested in determining the change in the amount of water for some period of time, say the month of January, you could use a balance equation in terms of the total amounts of water that entered and left the lake during this time:

$$\begin{pmatrix} \text{Change in amount} \\ \text{of water in the} \\ \text{lake during the} \\ \text{month of January} \end{pmatrix} = \begin{pmatrix} \text{Amount of water that} \\ \text{flowed into the lake} \\ \text{during the month of} \\ \text{January} \end{pmatrix} - \begin{pmatrix} \text{Amount of water that} \\ \text{flowed out of} \\ \text{the lake during the} \\ \text{month of January} \end{pmatrix}$$

$$\text{(2.1-2)}$$

Notice that Eq. 2.1-1 is concerned with an instantaneous rate of change, and it requires data on the rates at which flows occur. Equation 2.1-2, on the other hand, is for computing the total change that has occurred and requires data only on the total flows over the time interval. These two equations, one for the instantaneous rate of change of a system property (here the amount of water) and the other for the change over an interval of time, illustrate the two types of change-of-state problems that are of interest in this book and the forms of the balance equations that are used in their solution.

There is, of course, an interrelationship between the two balance equations. If you had information on each water flow rate at each instant of time for the whole month of January, you could integrate Eq. 2.1-1 over that period of time to obtain the same answer for the total change in the amount of water as would be obtained directly from Eq. 2.1-2 using the much less detailed information on the total flows for the month.

The example used here to illustrate the balance equation concept is artificial in that although water flows into and out of a lake are difficult to measure, the amount of water in the lake can be determined directly from the water level. Thus, Eqs. 2.1-1 and 2.1-2 are not likely to be used. However, the system properties of interest in thermodynamics and, indeed, in most areas of engineering are frequently much more difficult to measure than flow rates of mass and energy, so that the balance equation approach may be the only practical way to proceed.

While our interest here is specifically in the mass balance, to avoid having to repeat the analysis leading to equation 2.1-4, which follows, for other properties, such as energy (see next chapter), we will develop a general balance equation for any extensive property θ. We will then replace θ with the total mass. In the next section θ will be the mass of only one of the species (which may undergo chemical reaction), and in the next chapter θ will be replaced by the total energy. In the remainder of this section, the balance equations for an unspecified extensive property θ of a thermodynamic system are developed.

With the balance equations formulated in a general manner, they will be applicable (by appropriate simplification) to all systems studied in this book. In this way it will not be necessary to rederive the balance equations for each new problem; we will merely simplify the general equations. Specific choices for θ, such as total mass, mass (or number of moles) of a single species, and energy, are considered in Sec. 2.2, 2.3, and 3.2, respectively.

We consider a general system that may be moving or stationary, in which mass and energy may flow across its boundaries at one or more places, and the boundaries of which may distort. Since we are concerned with equating the total change within the system to flows across its boundaries, the details of the internal structure of the system will be left unspecified. This "black-box" system is illustrated in Fig. 2.1-1, and characteristics of this system are

1. Mass may flow into one, several, all, or none of the K entry ports labeled 1, 2, . . . , K (i.e., the system may be either open or closed to the flow of mass). Since we are concerned with pure fluids here, only one molecular species will be involved, although its temperature and pressure may be different at each entry port. The

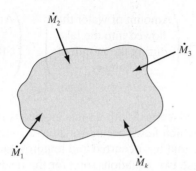

Figure 2.1-1 A single-component system with several mass flows.

mass flow rate *into* the system at the kth entry port will be $\dot{M}_k$, so that $\dot{M}_k > 0$ for flow into the system, and $\dot{M}_k < 0$ for flow out of the system.

2. The boundaries of the black-box system may be stationary or moving. If the system boundaries are moving, it can be either because the system is expanding or contracting, or because the system as a whole is moving, or both.

The following two characteristics are important for the energy balance.

3. Energy in the form of heat may enter or leave the system across the system boundaries.

4. Energy in the form of work (mechanical shaft motion, electrical energy, etc.) may enter or leave the system across the system boundaries.

Throughout this book we will use the convention that a flow into the system, whether it be a mass flow or an energy flow, is positive and a flow out of the system is negative.

The balance equation for the total amount of any extensive quantity θ in this system is obtained by equating the change in the amount of θ in the system between times t and $t + \Delta t$ to the flows of θ into and out of the system, and the generation of θ within the system, in the time interval Δt. Thus,

$$\begin{pmatrix} \text{Amount of } \theta \text{ in the} \\ \text{system at time } t + \Delta t \end{pmatrix} - \begin{pmatrix} \text{Amount of } \theta \text{ in the} \\ \text{system at time } t \end{pmatrix} = \begin{pmatrix} \text{Amount of } \theta \text{ that entered the system across} \\ \text{system boundaries between } t \text{ and } t + \Delta t \end{pmatrix}$$

$$- \begin{pmatrix} \text{Amount of } \theta \text{ that left the system across} \\ \text{system boundaries between } t \text{ and } t + \Delta t \end{pmatrix}$$

$$+ \begin{pmatrix} \text{Amount of } \theta \text{ generated within the} \\ \text{system between } t \text{ and } t + \Delta t \end{pmatrix} \tag{2.1-3}$$

The meaning of the first two terms on the right side of this equation is clear, but the last term deserves some discussion. If the extensive property θ is equal to the total mass, total energy, or total momentum, quantities that are conserved (experimental observations 1 to 3 of Sec. 1.7), then the internal generation of θ is equal to zero. This is easily seen as follows for the special case of a system isolated from its environment (so that the flow terms across the system boundaries vanish); here Eq. 2.1-3 reduces to

$$\begin{pmatrix} \text{Amount of } \theta \text{ in the} \\ \text{system at time} \\ t + \Delta t \end{pmatrix} - \begin{pmatrix} \text{Amount of } \theta \text{ in the} \\ \text{system at time } t \end{pmatrix} = \begin{pmatrix} \text{Amount of } \theta \text{ generated} \\ \text{within the system} \\ \text{between } t \text{ and } t + \Delta t \end{pmatrix} \tag{2.1-3a}$$

Since neither total mass, total momentum, nor total energy can be spontaneously produced, if θ is any of these quantities, the internal generation term must be zero. If, however, θ is some other quantity, the internal generation term may be positive (if θ is produced within the system), negative (if θ is consumed within the system), or zero. For example, suppose the black-box system in Fig. 2.1-1 is a closed (batch) chemical reactor in which cyclohexane is partially dehydrogenated to benzene and hydrogen according to the reaction

$$C_6H_{12} \rightarrow C_6H_6 + 3H_2$$

If θ is set equal to the total mass, then, by the principle of conservation of mass, the internal generation term in Eq. 2.1-3a would be zero. If, however, θ is taken to be the mass of benzene in the system, the internal generation term for benzene would be positive, since benzene is produced by the chemical reaction. Conversely, if θ is taken to be the mass of cyclohexane in the system, the internal generation term would be negative. In either case the magnitude of the internal generation term would depend on the rate of reaction.

The balance equation (Eq. 2.1-3) is useful for computing the change in the extensive property θ over the time interval Δt. We can also obtain an equation for computing the instantaneous rate of change of θ by letting the time interval Δt go to zero. This is done as follows. First we use the symbol $\theta(t)$ to represent the amount of θ in the system at time t, and we recognize that for a very small time interval Δt (over which the flows into and out of the system are constant) we can write

$$\begin{pmatrix} \text{Amount of } \theta \text{ that enters the} \\ \text{system across system boundaries} \\ \text{between } t \text{ and } t + \Delta t \end{pmatrix} \quad \text{as} \quad \begin{pmatrix} \text{Rate at which } \theta \text{ enters} \\ \text{the system across system} \\ \text{boundaries} \end{pmatrix} \Delta t$$

with similar expressions for the outflow and generation terms. Next we rewrite Eq. 2.1-3 as

$$\frac{\theta(t + \Delta t) - \theta(t)}{\Delta t} = \begin{pmatrix} \text{Rate at which } \theta \text{ enters the system} \\ \text{across system boundaries} \end{pmatrix}$$

$$- \begin{pmatrix} \text{Rate at which } \theta \text{ leaves the system} \\ \text{across system boundaries} \end{pmatrix}$$

$$+ \begin{pmatrix} \text{Rate at which } \theta \text{ is generated} \\ \text{within the system} \end{pmatrix}$$

Finally, taking the limit as $\Delta t \rightarrow 0$ and using the definition of the derivative from calculus,

$$\frac{d\theta}{dt} = \lim_{\Delta t \rightarrow 0} \frac{\theta(t + \Delta t) - \theta(t)}{\Delta t}$$

we obtain

$$\frac{d\theta}{dt} = \begin{pmatrix} \text{Rate of change of} \\ \theta \text{ in the system} \end{pmatrix} = \begin{pmatrix} \text{Rate at which } \theta \text{ enters the} \\ \text{system across system boundaries} \end{pmatrix}$$

$$- \begin{pmatrix} \text{Rate at which } \theta \text{ leaves the} \\ \text{system across system boundaries} \end{pmatrix}$$

$$+ \begin{pmatrix} \text{Rate at which } \theta \text{ is generated} \\ \text{within the system} \end{pmatrix} \qquad \textbf{(2.1-4)}$$

Balance equation 2.1-4 is general and applicable to conserved and nonconserved quantities. There is, however, the important advantage in dealing with conserved quantities that the internal generation term is zero. For example, to use the total mass balance to compute the rate of change of mass in the system, we need know only the mass flows into and out of the system. On the other hand, to compute the rate of change of the mass of cyclohexane undergoing a dehydrogenation reaction in a chemical reactor, we also need data on the rate of reaction in the system, which may be a function of concentration, temperature, catalyst activity, and internal characteristics of the system. Thus, additional information may be needed to use the balance equation for the mass of cyclohexane, and, more generally, for any nonconserved quantity. The applications of thermodynamics sometimes require the use of balance equations for nonconserved quantities.

2.2 CONSERVATION OF MASS FOR A PURE FLUID

The first balance equation of interest in thermodynamics is the conservation equation for total mass. If θ is taken to be the total mass in the system, designated by the symbol M, we have, from Eq. 2.1-3

$$M(t + \Delta t) - M(t) = \begin{pmatrix} \text{Amount of mass that} \\ \text{entered the system} \\ \text{across the system} \\ \text{boundaries between} \\ t \text{ and } t + \Delta t \end{pmatrix} - \begin{pmatrix} \text{Amount of mass} \\ \text{that left the system} \\ \text{across the system} \\ \text{boundaries between} \\ t \text{ and } t + \Delta t \end{pmatrix} \quad \textbf{(2.2-1a)}$$

where we have recognized that the total mass is a conserved quantity and that the only mechanism by which mass enters or leaves a system is by a mass flow. Using $\dot{M}_k$ to represent the mass flow rate into the system at the kth entry point, we have, from Eq. 2.1-4, the equation for the instantaneous rate of change of mass in the system:

Rate-of-change mass balance

$$\boxed{\frac{dM}{dt} = \sum_{k=1}^{K} \dot{M}_k} \quad \textbf{(2.2-1b)}$$

Equations 2.2-1 are general and valid regardless of the details of the system and whether the system is stationary or moving.

Since we are interested only in pure fluids here, we can divide Eqs. 2.2-1 by the molecular weight of the fluid and use the fact that N, the number of moles in the system, is equal to M/mw, where mw is the molecular weight, and $\dot{N}_k$, the molar flow rate into the system at the kth entry port, is $\dot{M}_k/\text{mw}$, to obtain instead of Eq. 2.2-1a a similar equation in which the term *moles* replaces the word *mass* and, instead of Eq. 2.2-1b,

Rate-of-change mass balance: molar basis

$$\boxed{\frac{dN}{dt} = \sum_{k=1}^{K} \dot{N}_k} \quad \textbf{(2.2-2)}$$

We introduce this equation here because it is frequently convenient to do calculations on a molar rather than on a mass basis.

In Sec. 2.1 it was indicated that the equation for the change of an extensive state variable of a system in the time interval Δt could be obtained by integration over the time interval of the equation for the rate of change of that variable. Here we demonstrate how this integration is accomplished. For convenience, t_1 represents the beginning of the time interval and t_2 represents the end of the time interval, so that $\Delta t = t_2 - t_1$. Integrating Eq. 2.2-1b between t_1 and t_2 yields

$$\int_{t_1}^{t_2} \frac{dM}{dt}\, dt = \sum_{k=1}^{K} \int_{t_1}^{t_2} \dot{M}_k\, dt \tag{2.2-3}$$

The left side of the equation is treated as follows:

$$\int_{t_1}^{t_2} \frac{dM}{dt}\, dt = \int_{M(t_1)}^{M(t_2)} dM = M(t_2) - M(t_1) = \begin{pmatrix} \text{Change in total mass} \\ \text{of system between} \\ t_1 \text{ and } t_2 \end{pmatrix}$$

where $M(t)$ is the mass in the system at time t. The term on the right side of the equation may be simplified by observing that

$$\int_{t_1}^{t_2} \dot{M}_k\, dt = \begin{pmatrix} \text{Mass that entered the} \\ \text{system at the } k\text{th entry} \\ \text{port between } t_1 \text{ and } t_2 \end{pmatrix} \equiv \Delta M_k$$

Thus

Integral mass balance

$$\boxed{M(t_2) - M(t_1) = \sum_{k=1}^{K} \Delta M_k} \tag{2.2-4}$$

This is the symbolic form of Eq. 2.2-1a.

Equation 2.2-4 may be written in a simpler form when the mass flow rates are steady, that is, independent of time. For this case

$$\int_{t_1}^{t_2} \dot{M}_k\, dt = \dot{M}_k \int_{t_1}^{t_2} dt = \dot{M}_k \Delta t$$

so that

$$M(t_2) - M(t_1) = \sum_{k=1}^{K} \dot{M}_k \Delta t \quad \text{(steady flows)} \tag{2.2-5}$$

The equations of this section that will be used throughout this book are listed in Table 2.2-1.

ILLUSTRATION 2.2-1

Use of the Difference Form of the Mass Balance

A tank of volume 25 m³ contains 1.5×10^4 kg of water. Over a two-day period the inlet to the tank delivers 2.0×10^3 kg, 1.3×10^3 kg leaves the tank through the outlet port, and 50 kg of

Table 2.2-1 The Mass Conservation Equation

	Mass Basis	Molar Basis
Rate-of-change form of the mass balance		
General equation	$\dfrac{dM}{dt} = \displaystyle\sum_{k=1}^{K} \dot{M}_k$	$\dfrac{dN}{dt} = \displaystyle\sum_{k=1}^{K} \dot{N}_k$
Special case:		
Closed system	$\dfrac{dM}{dt} = 0$	$\dfrac{dN}{dt} = 0$
	$M = \text{constant}$	$N = \text{constant}$
*Difference form of the mass balance**		
General equation	$M_2 - M_1 = \displaystyle\sum_{k=1}^{K} \Delta M_k$	$N_2 - N_1 = \displaystyle\sum_{k=1}^{K} \Delta N_k$
Special cases:		
Closed system	$M_2 = M_1$	$N_2 = N_1$
Steady flow	$M_2 - M_1 = \displaystyle\sum_{k=1}^{K} \dot{M}_k \, \Delta t$	$N_2 - N_1 = \displaystyle\sum_{k=1}^{K} \dot{N}_k \, \Delta t$

*Here we have used the abbreviated notation $M_i = M(t_i)$ and $N_i = N(t_i)$.

water leaves the tank by evaporation. How much water is in the tank at the end of the two-day period?

SOLUTION

Since we are interested only in the change in the mass of water in the tank over the two-day period, and not in the rate of change, we will use the difference form of the mass balance over the period from the initial time (which we take to be $t = 0$) until two days later ($t = 2$ days). We use Eq. 2.2-4, recognizing that we have three flow terms: M_1 (flow into the tank) $= +2.0 \times 10^3$ kg, M_2 (flow from the tank) $= -1.3 \times 10^3$ kg, and M_3 (evaporation) $= -50$ kg. (Remember, in our notation the $+$ sign is for flow into the system, the tank, and the $-$ sign is for flow out of the system.)

Therefore,

$$M(t = 2 \text{ days}) - M(t = 0) = M_1 + M_2 + M_3$$
$$M(t = 2 \text{ days}) - 1.5 \times 10^4 \text{ kg} = 2.0 \times 10^3 - 1.3 \times 10^3 - 50$$
$$M(t = 2 \text{ days}) = 1.5 \times 10^4 + 2.0 \times 10^3 - 1.3 \times 10^3 - 50$$
$$= 1.565 \times 10^4 \text{ kg}$$

∎[1]

ILLUSTRATION 2.2-2
Use of the Rate-of-Change Form of the Mass Balance

A storage tank is being used in a chemical plant to dampen fluctuations in the flow to a downstream chemical reactor. The exit flow from this tank will be kept constant at 1.5 kg/s; if the instantaneous flow into the tank exceeds this, the level in the tank will rise, while if the instantaneous flow is less, the level in the tank will drop. If the instantaneous flow into the storage tank is 1.2 kg/s, what is the rate of change of mass in the tank?

---

[1] Throughout this text the symbol ∎ will be used to indicate the end of an illustration.

SOLUTION

Since we are interested in the rate of change of mass, here we use the rate-of-change form of the mass balance (Eq. 2.2-1b):

$$\frac{dM}{dt} = \sum_k \dot{M}_k$$

or, in this case,

$$\frac{dM}{dt} = -1.5 + 1.2 = -0.3\frac{\text{kg}}{\text{s}}$$

Thus, at the instant the measurements were made, the amount of liquid in the tank was decreasing by 0.3 kg/s or 300 g/s.

COMMENT

Remember, if we are interested in the rate of change of mass, as we are here, we use the rate-of-change form of the mass balance, Eq. 2.2-1b. However, if we are interested only in the change of total mass over a period of time, we use Eq. 2.2-4. ■

ILLUSTRATION 2.2-3

Use of the Rate-of-Change Form of the Mass Balance

Gas is being removed from a high-pressure storage tank through a device that removes 1 percent of the current contents of the tank each minute. If the tank initially contains 1000 mols of gas, how much will remain at the end of 20 minutes?

SOLUTION

Since 1 percent of the gas is removed at any time, the rate at which gas leaves the tank will change with time. For example, initially gas is leaving at the rate of 0.01×1000 mol/min $=$ 10 mol/min. However, later when only 900 mol of gas remain in the tank, the exiting flow rate will be 0.01×900 mol/min $=9$ mol/min. In fact, the exiting flow rate is continuously changing with time. Therefore, we have to use the rate-of-change or differential form of the mass (mole) balance. Starting from the rate-of-change form of the mass balance (Eq. 2.2-1b) around the tank that has only a single flow term, we have

$$\frac{dN}{dt} = \dot{N} \quad \text{where} \quad \dot{N} = -0.01 \times N \quad \text{so that} \quad \frac{d\ln N}{dt} = -0.01$$

The solution to this first-order differential equation is

$$\ln\left(\frac{N(t)}{N(t=0)}\right) = -0.01t \quad \text{or} \quad N(t) = N(t=0)e^{-0.01t}$$

Therefore,

$$N(t=20) = N(t=0)e^{-0.01 \times 20} = 1000e^{-0.2} = 818.7 \text{ mol}$$

Note that if we had merely (and incorrectly) used the initial rate of 10 mol/min we would have obtained the incorrect answer of 800 mol remaining in the tank.

COMMENT

To solve any problem in which the mass (or molar) flow rate changes with time, we need to use the differential or rate-of-change form of the mass balance. For problems in which all of the flow terms are constant, we can use the general difference form of the mass balance (which has been obtained from the rate-of-change form by integration over time), or we can use the rate-of-change form and then integrate over time. However, it is important to emphasize that if one (or more) flow rates are changing with time, the rate-of-change form must be used. ∎

ILLUSTRATION 2.2-4

Another Problem Using the Rate-of-Change Form of the Mass Balance

An open cylindrical tank with a base area of 1 m^2 and a height of 10 m contains 5 m^3 of water. As a result of corrosion, the tank develops a leak at its bottom. The rate at which water leaves the tank through the leak is

$$\text{Leak rate} \quad \left(\frac{\text{m}^3}{\text{s}}\right) = 0.5\sqrt{\Delta P}$$

where ΔP is the pressure difference in bar between the fluid at the base of the tank and the atmosphere. (You will learn about the origin of this equation in a course dealing with fluid flow.)

Determine the amount of water in the tank at any time.

SOLUTION

Note that the pressure at the bottom of the tank is equal to the atmosphere pressure plus the hydrostatic pressure due to the water above the leak; that is, $P = 1.013 \text{ bar} + \rho g h$, where ρ is the density of water and h is the height of water above the leak. Therefore, $\Delta P = (1.013 + \rho g h) - 1.013 = \rho g h$ and

$$\Delta P = 10^3 \, \frac{\text{kg}}{\text{m}^3} \times h \, \text{m} \times 9.807 \, \frac{\text{m}}{\text{s}^2} \times 1 \, \frac{\text{Pa}\,\text{m}\,\text{s}}{\text{kg}} \times 10^{-5} \, \frac{\text{bar}}{\text{Pa}} = 0.09807h \text{ bar}$$

Since the height of fluid in the tank is changing with time, the flow rate of the leak will change with time. Therefore, to solve the problem, we must use the rate-of-change form of the mass balance. The mass of water in the tank at any time is

$$M(t) = \rho A h(t) = 10^3 \, \frac{\text{kg}}{\text{m}^3} \cdot 1 \, \text{m}^2 \cdot h(t) \, \text{m} = 10^3 h(t) \text{ kg}$$

The mass balance on the contents of the tank at any time is

$$\frac{dM(t)}{dt} = 10^3 \frac{dh(t)}{dt} = \dot{M} = -0.5\sqrt{0.098\,07h(t)} = -0.1566\sqrt{h(t)}$$

where the negative sign arises because the flow is out of the tank. Integrating this equation between $t = 0$ and any later time t yields

$$2\sqrt{h(t)} - 2\sqrt{h(0)} = 2\sqrt{h(t)} - 2\sqrt{5} = -0.1566 \times 10^{-3}t$$

or

$$\sqrt{h(t)} = \sqrt{5} - \left(\frac{0.1566}{2}\right) \times 10^{-3}t$$

which can be rearranged to

$$h(t) = \left(\sqrt{5} - 0.7829 \times 10^{-4}t\right)^2 \quad \text{and} \quad M(t) = \left(\sqrt{5} - 0.7829 \times 10^{-4}t\right)^2 \times 10^3 \text{ kg}$$

From this equation, one finds that the tank will be completely drained in 28 580 s or 7.938 hr.

COMMENT

Since the rate of flow of water out of the tank depends on the hydrostatic pressure due to the water column above the leak, and since the height of this column changes with time, again we must use the rate-of-change form of the mass balance to solve the problem.

It is important in any problem to be able to recognize whether the flows are steady, in which case the difference form of the mass balance can be used, or the flows vary with time, as is the case here, in which case the rate-of-change form of the mass balance must be used. ∎

2.3 THE MASS BALANCE EQUATIONS FOR A MULTICOMPONENT SYSTEM WITH A CHEMICAL REACTION

When chemical reactions occur, the mass (or mole) balance for each species is somewhat more complicated since the amount of the species can increase or decrease as a result of the reactions. Here we will consider mass balances when there is only a single chemical reaction; in Chapter 8 and later chapters the more general case of several chemical reactions occurring simultaneously is considered. Also, we will write the mass balances using number of moles only, since the stoichiometry of chemical reactions is usually written in terms of the number of moles of each species that undergoes chemical reaction rather than the mass of each species that reacts. Using the notation $\left(\dot{N_i}\right)_k$ for the rate at which moles of species i enter (if positive) or leave (if negative) in flow stream k, we have the differential or rate-of-change form of the mass balance on species i as

Rate-of-change mass balance with chemical reaction on a molar basis

$$\boxed{\frac{dN_i}{dt} = \sum_{k=1}^{K} (\dot{N_i})_k + \left(\frac{dN_i}{dt}\right)_{\text{rxn}}} \tag{2.3-1}$$

where the last term is new and describes the rate at which species i is produced (if positive) or consumed (if negative) within the system by chemical reaction. The difference form of this equation, obtained by integrating over the time period from t_1 to t_2, is

Difference form of the mass balance

$$\boxed{N_i(t_2) - N_i(t_1) = \sum_{k=1}^{K} \int_{t_1}^{t_2} (\dot{N_i})_k \, dt + (\Delta N_i)_{\text{rxn}} = \sum_{k=1}^{K} \Delta N_k + (\Delta N_i)_{\text{rxn}}}$$

$$\tag{2.3-2}$$

where the summation terms after the equal signs are the changes in the number of moles of the species due to the flow streams, and the second terms are the result of the chemical reaction. Note that only if the flow rate of a stream is steady (i.e., $\left(\dot{N_i}\right)_k$ is constant), then

If a flow rate is steady

$$(\Delta N_i)_k = \left(\dot{N_i}\right)_k \Delta t$$

Now consider the mass (mole) balances for a reactor in which the following chemical reaction occurs

$$C_2H_4 + Cl_2 \rightarrow C_2H_4Cl_2$$

but in which neither ethylene nor chlorine is completely consumed. The mass balances for these species, in which stream 1 is pure ethylene and stream 2 is pure chlorine, are

$$C_2H_4: \quad \left(\frac{dN_{C_2H_4}}{dt}\right) = \left(\dot{N}_{C_2H_4}\right)_1 + \left(\dot{N}_{C_2H_4}\right)_3 + \left(\frac{dN_{C_2H_4}}{dt}\right)_{rxn}$$

$$Cl_2: \quad \left(\frac{dN_{Cl_2}}{dt}\right) = \left(\dot{N}_{Cl_2}\right)_2 + \left(\dot{N}_{Cl_2}\right)_3 + \left(\frac{dN_{Cl_2}}{dt}\right)_{rxn} \qquad \text{(2.3-3)}$$

$$C_2H_4Cl_2: \quad \left(\frac{dN_{C_2H_4Cl_2}}{dt}\right) = \left(\dot{N}_{C_2H_4Cl_2}\right)_3 + \left(\frac{dN_{C_2H_4Cl_2}}{dt}\right)_{rxn}$$

where all the exit streams [i.e., $(\ldots)_3$] will be negative in value.

From the stoichiometry of this reaction, all the reaction rate terms are interrelated. In the case here, the rate at which ethylene chloride is created, that is, the number of moles per second, is equal to the rate at which ethylene is consumed, which is also equal to the rate at which chlorine is consumed. That is,

$$\left(\frac{dN_{C_2H_4Cl_2}}{dt}\right)_{rxn} = -\left(\frac{dN_{C_2H_4}}{dt}\right)_{rxn} = -\left(\frac{dN_{Cl_2}}{dt}\right)_{rxn}$$

so we can simplify Eqs. 2.3-3 by replacing the three different reaction rates with a single one.

We can generalize this discussion of the interrelationships between reaction rates by introducing the following convenient notation for the description of chemical reactions. Throughout this book the chemical reaction

$$\alpha A + \beta B + \cdots \rightleftharpoons \rho R + \cdots$$

where $\alpha, \beta, \ldots$, are the molar stoichiometric coefficients, will be written as

$$\rho R + \cdots - \alpha A - \beta B - \cdots = 0$$

or

$$\sum_i \nu_i I = 0 \qquad \text{(2.3-4)}$$

Here ν_i is the **stoichiometric coefficient** of species I, defined so that ν_i is positive for reaction products, negative for reactants, and equal to zero for inert species. In this notation the electrolytic dissociation reaction $H_2O = H_2 + \frac{1}{2}O_2$ is written as $H_2 + \frac{1}{2}O_2 - H_2O = 0$, so that $\nu_{H_2O} = -1$, $\nu_{H_2} = +1$, and $\nu_{O_2} = +\frac{1}{2}$.

We will use N_i to represent the number of moles of species i in the system at any time t and $N_{i,0}$ to be the initial number of moles of species i. For a closed system, N_i and $N_{i,0}$ are related through the reaction variable X, called the **molar extent of reaction**, and the stoichiometric coefficient ν_i by the equation

**Molar extent of
reaction**

$$N_i = N_{i,0} + \nu_i X \qquad\qquad (2.3\text{-}5a)$$

or

$$X = \frac{N_i - N_{i,0}}{\nu_i} \qquad\qquad (2.3\text{-}5b)$$

An important characteristic of the reaction variable X defined in this way is that it has the same value for each molecular species involved in a reaction; this is illustrated in the following example. Thus, given the initial mole numbers of all species and X (or the number of moles of one species from which the molar extent of reaction X can be calculated) at time t, one can easily compute all other mole numbers in the system. In this way the complete progress of a chemical reaction (i.e., the change in mole numbers of all the species involved in the reaction) is given by the value of the single variable X.

ILLUSTRATION 2.3-1
Using the Molar Extent of Reaction Notation

The electrolytic decomposition of water to form hydrogen and oxygen occurs as follows: $H_2O \longrightarrow H_2 + \frac{1}{2}O_2$. Initially, only 3.0 mol of water are present in a closed system. At some later time it is found that 1.2 mol of H_2 and 1.8 mol of H_2O are present.

a. Show that the molar extents of reaction based on H_2 and H_2O are equal.
b. Compute the number of moles of O_2 in the system.

SOLUTION

a. The reaction $H_2O \rightarrow H_2 + \frac{1}{2}O_2$ is rewritten as

$$H_2 + \tfrac{1}{2}O_2 - H_2O = 0$$

so that

$$\nu_{H_2O} = -1 \qquad \nu_{H_2} = +1 \quad \text{and} \quad \nu_{O_2} = +\tfrac{1}{2}$$

From the H_2 data,

$$X = \frac{1.2 - 0.0}{+1} = +1.2 \text{ mol}$$

From the H_2O data,

$$X = \frac{1.8 - 3.0}{-1} = +1.2 \text{ mol}$$

b. Starting from $N_i = N_{i,0} + \nu_i X$, we have

$$N_{O_2} = 0 + (+\tfrac{1}{2})(1.2) = 0.6 \text{ mol}$$

COMMENT

Note that the molar extent of reaction is *not* a fractional conversion variable, and therefore its value is *not* restricted to lie between 0 and 1. As defined here X, which has units of number of moles, is the number of moles of a species that has reacted divided by the stoichiometric coefficient for the species. In fact, X may be negative if the reaction proceeds in the reverse direction to that indicated (e.g., if hydrogen and oxygen react to form water).   ∎

The rate of change of the number of the moles of species i resulting from a chemical reaction is

$$\left(\frac{dN_i}{dt}\right)_{\mathrm{rxn}} = \nu_i \dot{X} \tag{2.3-6}$$

where the subscript rxn indicates that this is the rate of change of the number of moles of species i due to chemical reaction alone, and $\dot{X}$ is the rate of change of the molar extent of reaction. Using this notation, the balance equation for species i is

Rate-of-change form of the mass balance with chemical reaction: molar basis

$$\frac{dN_i}{dt} = \sum_{k=1}^{K} (\dot{N}_i)_k + \nu_i \frac{dX}{dt} \tag{2.3-7}$$

The difference form of this equation, obtained by integrating over the time period from t_1 to t_2, is

Difference form of mass balance with chemical reaction

$$N_i(t_2) - N_i(t_1) = \sum_{k=1}^{K} \int_{t_1}^{t_2} (\dot{N}_i)_k \, dt + (\Delta N_i)_{\mathrm{rxn}} = \sum_{k=1}^{K} (\Delta N_i)_k + \nu_i \Delta X \tag{2.3-8}$$

Using this notation for the description of the production of ethylene dichloride considered earlier, we have

$$C_2H_4: \quad \left(\frac{dN_{C_2H_4}}{dt}\right) = \left(\dot{N}_{C_2H_4}\right)_1 + \left(\dot{N}_{C_2H_4}\right)_3 - \frac{dX}{dt}$$

$$Cl_2: \quad \left(\frac{dN_{Cl_2}}{dt}\right) = \left(\dot{N}_{Cl_2}\right)_2 + \left(\dot{N}_{Cl_2}\right)_3 - \frac{dX}{dt} \tag{2.3-9}$$

$$C_2H_4Cl_2: \quad \left(\frac{dN_{C_2H_4Cl_2}}{dt}\right) = \left(\dot{N}_{C_2H_4Cl_2}\right)_3 + \frac{dX}{dt}$$

ILLUSTRATION 2.3-2

Mass Balance for a Mixture with Chemical Reaction

At high temperatures acetaldehyde (CH_3CHO) dissociates into methane and carbon monoxide by the following reaction

$$CH_3CHO \longrightarrow CH_4 + CO$$

At 520°C the rate at which acetaldehyde dissociates is

$$\frac{dC_{CH_3CHO}}{dt} = -0.48 C_{CH_3CHO}^2 \quad \frac{m^3}{kmol \ s}$$

where C is concentration in kmol/m^3. The reaction occurs in a constant-volume, 1-L vessel, and the initial concentration of acetaldehyde is 10 kmol/m^3.

a. If 5 mols of the acetaldehyde reacts, how much methane and carbon monoxide is produced?

b. Develop expressions for the amounts of acetaldehyde, methane, and carbon monoxide present at any time, and determine how long it would take for 5 mol of acetaldehyde to have reacted.

SOLUTION

First we must determine the initial amount of acetaldehyde present. Since the initial concentration is

$$C_{CH_3CHO} = 10 \frac{mol}{L} = 10 \frac{kmol}{m^3}$$

it follows that initially

$$N_{CH_3CHO} = 10 \frac{mol}{L} \times 1\,L = 10\,mol$$

Next we write the stoichiometry for the reaction in terms of the molar extent of reaction X as follows:

$$N_{CH_3CHO} = 10 - X \qquad N_{CH_4} = X \quad and \quad N_{CO} = X \qquad \text{(a)}$$

a. To determine the amounts of each species after a given amount of acetaldehyde has reacted, we can use the difference form of the mass balance for this system with no flows of species into or out of the reactor:

$$N_i(t) - N_i(t = 0) = (\Delta N_i)_{rxn} = \nu_i \Delta X$$

Therefore, for acetaldehyde

$$N_{CH_3CHO}(t) - N_{CH_3CHO}(t = 0) = 5 - 10\,mol = -5\,mol = -X$$

so that $X = 5$ mol. Then amounts of the other species are

$$N_{CH_4}(t) = X = 5\,mol \quad and \quad N_{CO}(t) = X = 5\,mol$$

b. To determine the amount of each species as a function of time is more difficult and must be done using the rate-of-change form of the mass balance since the rate of reaction and therefore the value of X change with time. However, because the amounts of the species are always related by the stoichiometry of Eq. a, we can use the mass balance for one of the species to determine the time variation of X, and then can use the expression for $X(t)$ to obtain the compositions of all species in the reaction as a function of time. Since the rate expression is written for acetaldehyde, we will use this substance to determine the time dependence of X. Since there are no flows into or out of the reactor, Eq. 2.3-7 is

$$\frac{dN_{CH_3CHO}}{dt} = \left(\frac{dN_{CH_3CHO}}{dt}\right)_{rxn} = \nu_{CH_3CHO}\frac{dX}{dt} = -\frac{dX}{dt}$$

Next the reaction rate expression can be written as

$$\frac{d\left(\frac{N_{CH_3CHO}}{V}\right)}{dt} = \frac{1}{V}\frac{dN_{CH_3CHO}}{dt} = -0.48 C_{CH_3CHO}^2 = -0.48 \frac{N_{CH_3CHO}^2}{V^2} \frac{m^3}{kmol\,s}$$

which, after using $N_{CH_3CHO} = 10 - X$, becomes

$$\frac{dX}{dt} = \frac{0.48(10 - X)^2}{V} \frac{mol}{L\,s} = 0.48(10 - X)^2 \frac{mol}{s}$$

Now integrating from $t = 0$, at which $X = 0$, to some other time t gives

$$\frac{1}{10 - X(t)} - \frac{1}{10} = 0.48t \quad \text{or} \quad X(t) = 10 - \frac{1}{0.48t + 0.1} \text{ mol}$$

Therefore,

$$N_{CH_3CHO}(t) = \frac{1}{0.48t + 0.1} \quad \text{and} \quad N_{CH_4}(t) = N_{CO}(t) = 10 - \frac{1}{0.48t + 0.1}$$

Finally, solving this equation for $X(t) = 5$ mol gives $t = 0.208$ s; that is, half of the acetaldehyde dissociates within approximately two-tenths of a second.

COMMENT

Notice again that solving the rate-of-change form of the mass balance requires more information (here the rate of reaction) and more effort than solving the difference form of the mass balance. However, we also get more information—the amount of each species present as a function of time. In Sec. 2.4, which is optional and more difficult, we consider another, even deeper level of description, where not only is time allowed to vary, but the system is not spatially homogeneous; that is, the composition in the reactor varies from point to point. However, this section is not for the faint-hearted and is best considered after a course in fluid mechanics. ∎

ILLUSTRATION 2.3-3

Mass Balance for a Liquid Mixture with a Reversible Reaction

The ester ethyl acetate is produced by the reversible reaction

$$CH_3COOH + C_2H_5OH \underset{k'}{\overset{k}{\rightleftharpoons}} CH_3COOC_2H_5 + H_2O$$

in the presence of a catalyst such as sulfuric or hydrochloric acid. The rate of ethyl acetate production has been found, from the analysis of chemical kinetics data, to be given by the following equation:

$$\frac{dC_{EA}}{dt} = kC_A C_E - k'C_{EA} C_W$$

where the subscripts EA, A, E, and W denote ethyl acetate, acetic acid, ethanol, and water, respectively, and the concentration of each species in units of kmol/m^3. The values of the reaction rate constants at 100°C and the catalyst concentration of interest are

$$k = 4.76 \times 10^{-4} \text{ m}^3/\text{kmol min}$$

and

$$k' = 1.63 \times 10^{-4} \text{ m}^3/\text{kmol min}$$

Develop expressions for the number of moles of each species as a function of time if the feed to the reactor is 1 m^3 of an aqueous solution that initially contains 250 kg of acetic acid and 500 kg of ethyl alcohol. The density of the solution may be assumed to be constant and equal to 1040 kg/m^3, and the reactor will be operated at a sufficiently high pressure that negligible amounts of reactants or products vaporize. Compute the number of moles of each species present 100 minutes after the reaction has started, and at infinite time when the reaction will have stopped and the system is at equilibrium.

SOLUTION

Since the reaction rate expression is a function of the compositions, which are changing as a function of time, the mass balance for each species must be written in the rate-of-change form. Since the species mole numbers and concentrations are functions of the molar extent of reaction, X, we first determine how X varies with time by solving the mass balance for one species. We will use ethyl acetate since the reaction rate is given for that species. Once the amount of ethyl acetate is known, the other species mole numbers are easily computed as shown below.

The initial concentration of each species is

$$C_A = \frac{250 \text{ kg/m}^3}{60 \text{ g/mol}} = 4.17 \text{ kmol/m}^3$$

$$C_E = \frac{500 \text{ kg/m}^3}{46 \text{ g/mol}} = 10.9 \text{ kmol/m}^3$$

$$C_W = \frac{(1040 - 250 - 500) \text{ kg/m}^3}{18 \text{ g/mol}} = 16.1 \text{ kmol/m}^3$$

Since there is 1 m^3 of solution, the initial amount of each species is

$$N_A = 4.17 \text{ kmol}$$
$$N_E = 10.9 \text{ kmol}$$
$$N_W = 16.1 \text{ kmol}$$
$$N_{EA} = 0 \text{ kmol}$$

and by the reaction stoichiometry, the amount of each species present at any time (in kmol) and its concentration (since 1 m^3 of volume is being considered) is

$$N_A = 4.17 - X \qquad C_A = 4.17 - X$$

$$N_E = 10.9 - X \qquad C_E = 10.9 - X$$

$$N_W = 16.1 + X \qquad C_W = 16.1 + X$$

and

$$N_{EA} = X \qquad C_{EA} = X$$

Because the concentration of a species is equal to the number of moles N divided by the volume V, the chemical reaction rate equation can be written as

$$\frac{d}{dt}\left(\frac{N_{EA}}{V}\right) = k\frac{N_A}{V}\frac{N_E}{V} - k'\frac{N_{EA}}{V}\frac{N_W}{V}$$

Now using $V = 1$ m^3 and the mole numbers, we have

$$\frac{d}{dt}X = k(4.17 - X)(10.9 - X) - k'X(16.1 + X)$$

or

$$\frac{dX}{dt} = 4.76 \times 10^{-4}(4.17 - X)(10.9 - X) - 1.63 \times 10^{-4}(16.1 + X)X$$

$$= 2.163 \times 10^{-2}(1 - 0.4528X - 0.01447X^2) \text{ kmol/m}^3 \text{ min}$$

which can be rearranged to

$$2.163 \times 10^{-2}\, dt = \frac{dX}{1 - 0.4528X + 0.01447X^2}$$

Integrating this equation between $t = 0$ and time t yields[2]

$$2.163 \times 10^{-2} \int_0^t dt = 2.163 \times 10^{-2} \times t = \int_0^X \frac{dX}{1 - 0.4528X + 0.01447X^2}$$

or

$$\ln\left(\frac{0.02894X - 0.8364}{0.02894X - 0.0692}\right) - \ln\left(\frac{0.8364}{0.0692}\right) = 0.8297 \times 10^{-2}t \qquad \text{(c)}$$

and on rearrangement

$$X(t) = 2.3911 \frac{e^{0.008297t} - 1}{e^{0.008297t} - 0.08274}$$

for t in minutes and X in kmol.

Therefore,

$$N_A(t) = C_A(t) = 4.17 - X = 4.17 - 2.3911 \frac{e^{0.008297t} - 1}{e^{0.008297t} - 0.08274}$$

$$N_E(t) = C_E(t) = 10.9 - X = 10.9 - 2.3911 \frac{e^{0.008297t} - 1}{e^{0.008297t} - 0.08274}$$

$$N_W(t) = C_W(t) = 16.1 + X = 16.1 + 2.3911 \frac{e^{0.008297t} - 1}{e^{0.008297t} - 0.08274}$$

and

$$N_{EA}(t) = C_{EA}(t) = X = 2.3911 \frac{e^{0.008297t} - 1}{e^{0.008297t} - 0.08274}$$

At 100 minutes, $X = 1.40$ kmol so that

$$N_A = 4.17 - X = 4.17 - 1.4 = 2.77 \qquad N_E = 9.5, \qquad N_W = 17.5, \qquad N_{EA} = 1.4$$

Also, at infinite time $X = 2.39$ (actually 2.3911) and

$$N_A = 4.17 - X = 4.17 - 2.39 = 1.78, \qquad N_E = 8.51, \qquad N_W = 18.49, \qquad N_{EA} = 2.39$$

■

ILLUSTRATION 2.3-4

Mass Balance Modeling of a Simple Environmental Problem

Water in a lake initially contains a pollutant at a parts-per-million concentration. This pollutant is no longer present in the water entering the lake. The rate of inflow of water to the lake from a creek is constant and equal to the rate of outflow, so the lake volume does not change.

---

[2]Note that

$$\int \frac{dX}{a + bX + cX^2} = \frac{1}{\sqrt{2}} \ln\left[\frac{2cX + b - \sqrt{-q}}{2cX + b + \sqrt{-q}}\right]$$

where $q = 4ac - b^2$.

a. Assuming the water in the lake is well mixed, so its composition is uniform and the pollutant concentration in the exit stream is the same as in the lake, estimate the number of lake volumes of water that must be added to the lake and then leave in order for the concentration of the pollutant in the water to decrease to one-half of its initial concentration.

b. How many lake volumes would it take for the concentration of the pollutant in the lake to decrease to one-tenth of its initial concentration?

c. If the volume of water in the lake is equal to the inflow for a one-year period, assuming the inflow of water is uniform in time, how long would it take for the concentration of the pollutant in the lake to decrease to one-half and one-tenth of its initial concentration?

SOLUTION

a. In writing the overall mass balance for the lake, which we take to be the system, we will use that the flow rates into and out of the lake are constant and equal, and that the concentration of the pollutant is so low that its change has a negligible effect on the total mass of the water in the lake. With these simplifications the mass balance is

$$\frac{dM}{dt} = 0 = (\dot{M})_1 + (\dot{M})_2, \quad \text{so that} \quad (\dot{M})_1 = -(\dot{M})_2 = \dot{M}$$

That is, the rate of mass flow out of the lake is equal in magnitude and opposite in sign to the rate of mass flow into the lake. The mass balance on the pollutant is

$$\frac{dM_p}{dt} = \frac{d\left(C_p M\right)}{dt} = M \frac{d\left(C_p\right)}{dt} = \left(\dot{M}_p\right)_2 = (\dot{M})_2 C_p = -(\dot{M})_1 C_p$$

where we have used that the total amount of pollutant is equal to the product of its concentration per unit mass C_p and the total mass M of water in the lake. Therefore,

$$\frac{dC_p}{C_p} = -\frac{\dot{M}}{M} dt \quad \text{which has the solution} \quad C_p(t) = C_p(t=0) e^{-(\dot{M}/M)t}$$

$$\frac{C_p(t)}{C_p(t=0)} = 0.5 = \exp\left(-\frac{\dot{M}t}{M}\right) \quad \text{or} \quad \dot{M}t = 0.693 M$$

Now $\dot{M}t = \triangle M$, which is the amount of water that entered the lake over the time interval from 0 to t. Therefore, when the amount of fresh water that has entered the lake $\dot{M}t = \triangle M$ equals 69.3 percent of the initial (polluted) water in the lake, the concentration of the pollutant in the lake will have decreased to half its initial value.

b. We proceed as in part (a), except that we now have

$$\frac{C_p(t)}{C_p(t=0)} = 0.1 = \exp\left(-\frac{\dot{M}t}{M}\right) \quad \text{or} \quad \dot{M}t = 2.303 M$$

so that when when an amount of water enters that equals 2.303 times the initial volume of water in the lake, the concentration of pollutant will have decreased to one-tenth its initial value.

c. It will take 0.693 years (253 days) and 2.303 years (840 days) for the concentration of the pollutant to decrease to 50 percent and 10 percent of its inital concentration, respectively. ∎

2.4   THE MICROSCOPIC MASS BALANCE EQUATIONS IN THERMODYNAMICS AND FLUID MECHANICS[3]

This section appears on the CD that accompanies this text.

---

[3]This section is optional—only for graduate and advanced undergraduate students.

PROBLEMS

2.1 As a result of a chemical spill, benzene is evaporating at the rate of 1 gram per minute into a room that is 6 m $\times$ 6 m $\times$ 3 m in size and has a ventilation rate of 10 m^3/min.

   a. Compute the steady-state concentration of benzene in the room.

   b. Assuming the air in the room is initially free of benzene, compute the time necessary for benzene to reach 95 percent of the steady-state concentration.

2.2 The insecticide DDT has a half-life in the human body of approximately 7 years. That is, in 7 years its concentration decreases to half its initial concentration. Although DDT is no longer in general use in the United States, it was estimated that 25 years ago the average farm worker had a body DDT concentration of 22 ppm (parts per million by weight). Estimate what the farm worker's present concentration would be.

2.3 At high temperatures phosphine (PH$_3$) dissociates into phosphorus and hydrogen by the following reaction:

$$4PH_3 \longrightarrow P_4 + 6H_2$$

At 800°C the rate at which phosphine dissociates is

$$\frac{dC_{PH_3}}{dt} = -3.715 \times 10^{-6} C_{PH_3}$$

for t in seconds. The reaction occurs in a constant-volume, 2-L vessel, and the initial concentration of phosphine is 5 kmol/m^3

   a. If 3 mol of the phosphine reacts, how much phosphorus and hydrogen is produced?

   b. Develop expressions for the number of moles of phosphine, phosphorus, and hydrogen present at any time, and determine how long it would take for 3 mol of phosphine to have reacted.

2.4 The following reaction occurs in air:

$$2NO + O_2 \longrightarrow 2NO_2$$

At 20°C the rate of this reaction is

$$\frac{dC_{NO}}{dt} = -1.4 \times 10^{-4} C_{NO}^2 C_{O_2}$$

for t in seconds and concentrations in kmol/m^3. The reaction occurs in a constant-volume, 2-L vessel, and the initial concentration of NO is 1 kmol/m^3 and that of O$_2$ is 3 kmol/m^3

   a. If 0.5 mol of NO reacts, how much NO$_2$ is produced?

   b. Determine how long it would take for 0.5 mol of NO to have reacted.

Chapter 3

Conservation of Energy

In this chapter we continue the quantitative development of thermodynamics by deriving the energy balance, the second of the three balance equations that will be used in the thermodynamic description of physical, chemical, and (later) biochemical processes. The mass and energy balance equations (and the third balance equation, to be developed in the following chapter), together with experimental data and information about the process, will then be used to relate the change in a system's properties to a change in its thermodynamic state. In this context, physics, fluid mechanics, thermodynamics, and other physical sciences are all similar, in that the tools of each are the same: a set of balance equations, a collection of experimental observations (equation-of-state data in thermodynamics, viscosity data in fluid mechanics, etc.), and the initial and boundary conditions for each problem. The real distinction between these different subject areas is the class of problems, and in some cases the portion of a particular problem, that each deals with.

One important difference between thermodynamics and, say, fluid mechanics and chemical reactor analysis is the level of description used. In fluid mechanics one is usually interested in a very detailed microscopic description of flow phenomena and may try to determine, for example, the fluid velocity profile for flow in a pipe. Similarly, in chemical reactor analysis one is interested in determining the concentrations and rates of chemical reaction everywhere in the reactor. In thermodynamics the description is usually more primitive in that we choose either a region of space or an element of mass as the system and merely try to balance the change in the system with what is entering and leaving it. The advantage of such a description is that we can frequently make important predictions about certain types of processes for which a more detailed description might not be possible. The compromise is that the thermodynamic description yields information only about certain overall properties of the system, though with relatively little labor and simple initial information.

In this chapter we are concerned with developing the equations of energy conservation to be used in the thermodynamic analysis of systems of pure substances. (The thermodynamics of mixtures is more complicated and will be considered in later chapters.) To emphasize both the generality of these equations and the lack of detail necessary, we write these energy balance equations for a general black-box system. For contrast, and also because a more detailed description will be useful in Chapter 4, the rudiments of the more detailed microscopic description are provided in the final, optional section of this chapter. This microscopic description is not central to our development of thermodynamic principles, is suitable only for advanced students, and may be omitted.

45

To use the energy balances, we will need to relate the energy to more easily measurable properties, such as temperature and pressure (and in later chapters, when we consider mixtures, to composition as well). The interrelationships between energy, temperature, pressure, and composition can be complicated, and we will develop this in stages. In this chapter and in Chapters 4, 5, and 6 we will consider only pure fluids, so composition is not a variable. Then, in Chapters 8 to 15, mixtures will be considered. Also, here and in Chapters 4 and 5 we will consider only the simple ideal gas and incompressible liquids and solids for which the equations relating the energy, temperature, and pressure are simple, or fluids for which charts and tables interrelating these properties are available. Then, in Chapter 6, we will discuss how such tables and charts are prepared.

INSTRUCTIONAL OBJECTIVES FOR CHAPTER 3

The goals of this chapter are for the student to:

- Be able to use the differential form of the pure component energy balance in problem solving
- Be able to use the difference form of the pure component energy balance in problem solving
- Be able to compute changes in energy with changes in temperature and pressure for the ideal gas
- Be able to compute changes in energy with changes in temperature and pressure of real fluids using tables and charts of thermodynamic properties

NOTATION INTRODUCED IN THIS CHAPTER

C_P Constant-pressure molar heat capacity (J/mol K)

C_P^* Ideal gas constant-pressure molar heat capacity (J/mol K)

C_V Constant-volume molar heat capacity (J/mol K)

C_V^* Ideal gas constant-volume molar heat capacity (J/mol K)

H Enthalpy (J)

$\underline{H}$ Enthalpy per mole, or molar enthalpy (J/mol)

$\hat{H}$ Enthalpy per unit mass, or specific enthalpy (J/g)

$\triangle_{\mathrm{fus}}\underline{H}$ Molar enthalpy of melting or fusion (J/mol)

$\triangle_{\mathrm{sub}}\underline{H}$ Molar enthalpy of sublimation (J/mol)

$\triangle_{\mathrm{vap}}\underline{H}$ Molar enthalpy of vaporization (J/mol)

$\triangle_{\mathrm{vap}}\hat{H}$ Enthalpy of vaporization per unit mass (J/g)

$\dot{Q}$ Rate of flow of heat into the system (J/s)

Q Heat that has flowed into the system (J)

T_R Reference temperature for internal energy of enthalpy (K)

$\underline{U}$ Internal energy per mole, or molar internal energy (J/mol)

$\hat{U}$ Internal energy per unit mass, or specific internal energy (J/g)

$\underline{V}$ Volume per mole, or molar volume (m³/mol)

$\dot{W}$ Rate at which work is being done on the system (J/s)

W Work that has been done on the system (J)

W_s Shaft work that has been done on the system (J)

$\dot{W}_s$ Rate at which shaft work is being done on the system (J/s)

ω^{I} Mass fraction of phase I (quality for steam)
ψ Potential energy per unit of mass (J/g)

3.1 CONSERVATION OF ENERGY

To derive the energy conservation equation for a single-component system, we again use the black-box system of Figure 2.1-1 and start from the general balance equation, Eq. 2.1-4. Taking θ to be the sum of the internal, kinetic, and potential energy of the system,

$$\theta = U + M\left(\frac{v^2}{2} + \psi\right)$$

Here U is the total internal energy, $v^2/2$ is the kinetic energy per unit mass (where v is the center of mass velocity), and ψ is the potential energy per unit mass.[1] If gravity is the only force field present, then $\psi = gh$, where h is the height of the center of mass with respect to some reference, and g is the force of gravity per unit mass. Since energy is a conserved quantity, we can write

$$\frac{d}{dt}\left\{U + M\left(\frac{v^2}{2} + \psi\right)\right\} = \left(\begin{array}{c}\text{Rate at which energy}\\ \text{enters the system}\end{array}\right) - \left(\begin{array}{c}\text{Rate at which energy}\\ \text{leaves the system}\end{array}\right)$$

$$(3.1\text{-}1)$$

To complete the balance it remains only to identify the various mechanisms by which energy can enter and leave the system. These are as follows.

Energy flow accompanying mass flow. As a fluid element enters or leaves the system, it carries its internal, potential, and kinetic energy. This energy flow accompanying the mass flow is simply the product of a mass flow and the energy per unit mass,

$$\sum_{k=1}^{K} \dot{M}_k \left(\hat{U} + \frac{v^2}{2} + \psi\right)_k \qquad (3.1\text{-}2)$$

where $\hat{U}_k$ is the internal energy per unit mass of the kth flow stream, and $\dot{M}_k$ is its mass flow rate.

Heat. We use $\dot{Q}$ to denote the total rate of flow of heat *into* the system, by both conduction and radiation, so that $\dot{Q}$ is positive if energy in the form of heat flows into the system and negative if heat flows from the system to its surroundings. If heat flows occur at several different places, the total rate of heat flow into the system is

$$\dot{Q} = \sum \dot{Q}_j$$

where $\dot{Q}_j$ is the heat flow at the jth heat flow port.

Work. The total energy flow into the system due to work will be divided into several parts. The first part, called shaft work and denoted by the symbol W_s, is the mechanical energy flow that occurs without a deformation of the system boundaries. For example,

---

[1] In writing this form for the energy term, it has been assumed that the system consists of only one phase, that is, a gas, a liquid, or a solid. If the system consists of several distinct parts—for example, gas and a liquid, or a gas and the piston and cylinder containing it—the total energy, which is an extensive property, is the sum of the energies of the constituent parts.

if the system under consideration is a steam turbine or internal combustion engine, the rate of shaft work $\dot{W}_s$ is equal to the rate at which energy is transferred across the stationary system boundaries by the drive shaft or push rod. Following the convention that energy flow into the system is positive, $\dot{W}_s$ is positive if the surroundings do work on the system and negative if the system does work on its surroundings.

For convenience, the flow of electrical energy into or out of the system will be included in the shaft work term. In this case, $\dot{W}_s = \pm EI$, where E is the electrical potential difference across the system and I is the current flow through the system. The positive sign applies if electrical energy is being supplied to the system, and the negative sign applies if the system is the source of electrical energy.

Work also results from the movement of the system boundaries. The rate at which work is done when a force F is moved through a distance in the direction of the applied force dL in the time interval dt is

$$\dot{W} = F \frac{dL}{dt}$$

Here we recognize that pressure is a force per unit area and write

$$\dot{W} = -P \frac{dV}{dt} \tag{3.1-3}$$

where P is the pressure exerted by the system at its boundaries.[2] The negative sign in this equation arises from the convention that work done on a system in compression (for which dV/dt is negative) is positive, and work done by the system on its surroundings in an expansion (for which dV/dt is positive) is negative. The pressure at the boundaries of a nonstationary system will be opposed by (1) the pressure of the environment surrounding the system, (2) inertial forces if the expansion or compression of the system results in an acceleration or deceleration of the surroundings, and (3) other external forces, such as gravity. As we will see in Illustration 3.4-7, the contribution to the energy balance of the first of these forces is a term corresponding to the work done against the atmosphere, the second is a work term corresponding to the change in kinetic energy of the surroundings, and the last is the work done that changes the potential energy of the surroundings.

Work of a flowing fluid against pressure. One additional flow of energy for systems open to the flow of mass must be included in the energy balance equation; it is more subtle than the energy flows just considered. This is the energy flow that arises from the fact that as an element of fluid moves, it does work on the fluid ahead of it, and the fluid behind it does work on it. Clearly, each of these work terms is of the $P\Delta V$ type. To evaluate this energy flow term, which occurs only in systems open to the flow of mass, we will compute the net work done as one fluid element of mass ΔM_1 enters a system, such as the valve in Fig. 3.1-1, and another fluid element of

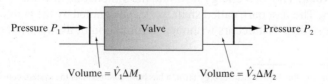

Pressure $P_1 \longrightarrow$ Valve $\longrightarrow$ Pressure P_2

Volume $= \hat{V}_1 \Delta M_1$ Volume $= \hat{V}_2 \Delta M_2$

Figure 3.1-1 A schematic representation of flow through a valve.

[2]In writing this form for the work term, we have assumed the pressure to be uniform at the system boundary. If this is not the case, Eq. 3.1-3 is to be replaced with an integral over the surface of the system.

mass ΔM_2 leaves the system. The pressure of the fluid at the inlet side of the valve is P_1, and the fluid pressure at the outlet side is P_2, so that we have

$$\left(\begin{array}{c} \text{Work done } by \text{ surrounding fluid in} \\ \text{pushing fluid element of mass } (M)_1 \\ \text{into the valve} \end{array}\right) = P_1 \hat{V}_1 \Delta M_1$$

$$\left(\begin{array}{c} \text{Work done } on \text{ surrounding fluid} \\ \text{by movement of fluid element of} \\ \text{mass } (\Delta M)_2 \text{ out of the valve (since} \\ \text{this fluid element is pushing the} \\ \text{surrounding fluid)} \end{array}\right) = -P_2 \hat{V}_2 \Delta M_2$$

$$\left(\begin{array}{c} \text{Net work done on the system due to} \\ \text{movement of fluid} \end{array}\right) = P_1 \hat{V}_1 \Delta M_1 - P_2 \hat{V}_2 \Delta M_2$$

For a more general system, with several mass flow ports, we have

$$\left(\begin{array}{c} \text{Net work done on the system due} \\ \text{to the pressure forces acting on} \\ \text{the fluids moving into and out of} \\ \text{the system} \end{array}\right) = \sum_{k=1}^{K} \Delta M_k P \hat{V}_k$$

Finally, to obtain the net *rate* at which work is done, we replace each mass flow ΔM_k with a mass flow rate $\dot{M}_k$, so that

$$\left(\begin{array}{c} \text{Net } rate \text{ at which work is done on} \\ \text{the system due to pressure forces} \\ \text{acting on fluids moving into and out} \\ \text{of the system} \end{array}\right) = \sum_{k=1}^{K} \dot{M}_k (P\hat{V})_k$$

where the sign of each term of this energy flow is the same as that of $\dot{M}_k$.

One important application of this pressure-induced energy flow accompanying a mass flow is hydroelectric power generation, schematically indicated in Fig. 3.1-2. Here a water turbine is being used to obtain mechanical energy from the flow of water through the base of a dam. Since the water velocity, height, and temperature are approximately the same at both sides of the turbine (even though there are large velocity changes within the turbine), the mechanical (or electrical) energy obtained is a result of only the mass flow across the pressure difference at the turbine.

Collecting all the energy terms discussed gives

$$\boxed{\begin{array}{l} \dfrac{d}{dt}\left\{ U + M\left(\dfrac{v^2}{2} + \psi\right)\right\} = \displaystyle\sum_{k=1}^{K} \dot{M}_k \left(\hat{U} + \dfrac{v^2}{2} + \psi\right)_k + \dot{Q} \\[2em] \qquad\qquad\qquad + \dot{W}_s - P\dfrac{dV}{dt} + \displaystyle\sum_{k=1}^{K} \dot{M}_k (P\hat{V})_k \end{array}} \qquad \textbf{(3.1-4)}$$

This equation can be written in a more compact form by combining the first and last terms on the right side and introducing the notation

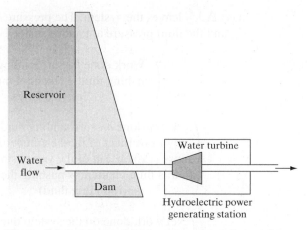

Figure 3.1-2 A hydroelectric power generating station: a device for obtaining work from a fluid flowing across a large pressure drop.

$$H = U + PV$$

where the function H is called the **enthalpy**, and by using the symbol $\dot{W}$ to represent the combination of shaft work $\dot{W}_s$ and expansion work $-P(dV/dt)$, that is, $\dot{W} = \dot{W}_s - P(dV/dt)$. Thus we have

Complete energy balance, frequently referred to as the first law of thermodynamics

$$\frac{d}{dt}\left\{ U + M\left(\frac{v^2}{2} + \psi \right) \right\} = \sum_{k=1}^{K} \dot{M}_k \left(\hat{H} + \frac{v^2}{2} + \psi \right)_k + \dot{Q} + \dot{W} \qquad \textbf{(3.1-4a)}$$

It is also convenient to have the energy balance on a molar rather than a mass basis. This change is easily accomplished by recognizing that $\dot{M}_k \hat{H}_k$ can equally well be written as $\dot{N}_k \underline{H}_k$, where $\underline{H}$ is the enthalpy per mole or molar enthalpy,[3] and $M(v^2/2 + \psi) = Nm(v^2/2 + \psi)$, where m is the molecular weight. Therefore, we can write the energy balance as

$$\frac{d}{dt}\left\{ U + Nm\left(\frac{v^2}{2} + \psi \right) \right\} = \sum_{k=1}^{K} \dot{N}_k \left\{ \underline{H} + m\left(\frac{v^2}{2} + \psi \right) \right\}_k + \dot{Q} + \dot{W} \qquad \textbf{(3.1-4b)}$$

Several special cases of Eqs. 3.1-4 are listed in Table 3.1-1.

The changes in energy associated with either the kinetic energy or potential energy terms, especially for gases, are usually very small compared with those for the thermal (internal) energy terms, unless the fluid velocity is near the velocity of sound, the change in height is very large, or the system temperature is nearly constant. This point will become evident in some of the illustrations and problems (see particularly Illustration 3.4-2). Therefore, it is frequently possible to approximate Eqs. 3.1-4 by

---

[3] $\underline{H} = \underline{U} + P\underline{V}$, where $\underline{U}$ and $\underline{V}$ are the molar internal energy and volume, respectively.

**Commonly used forms
of energy balance**

$$\frac{dU}{dt} = \sum_{k=1}^{K}(\dot{M}\hat{H})_k + \dot{Q} + \dot{W} \quad \text{(mass basis)}$$ **(3.1-5a)**

$$\frac{dU}{dt} = \sum_{k=1}^{K}(\dot{N}\underline{H})_k + \dot{Q} + \dot{W} \quad \text{(molar basis)}$$ **(3.1-5b)**

As with the mass balance, it is useful to have a form of the energy balance applicable to a change from state 1 to state 2. This is easily obtained by integrating Eq. 3.1-4 over the time interval t_1 to t_2, the time required for the system to go from state 1 to state 2.

Table 3.1-1 Differential Form of the Energy Balance

General equation

$$\frac{d}{dt}\left\{U + M\left(\frac{v^2}{2} + \psi\right)\right\} = \sum_{k=1}^{K}\dot{M}_k\left(\hat{H} + \frac{v^2}{2} + \psi\right)_k + \dot{Q} + \dot{W} \tag{a}$$

Special cases:
  (i) Closed system

$$\dot{M}_k = 0, \qquad \frac{dM}{dt} = 0$$

so

$$\frac{dU}{dt} + M\frac{d}{dt}\left(\frac{v^2}{2} + \psi\right) = \dot{Q} + \dot{W} \tag{b}$$

  (ii) Adiabatic process

in Eqs. a, b, and d
$$\dot{Q} = 0 \tag{c}$$

  (iii) Open and steady-state system

$$\frac{dM}{dt} = 0, \qquad \frac{dV}{dt} = 0, \qquad \frac{d}{dt}\left\{U + M\left(\frac{v^2}{2} + \psi\right)\right\} = 0$$

so

$$0 = \sum_{k=1}^{K}\dot{M}_k\left(\hat{H} + \frac{v^2}{2} + \psi\right)_k + \dot{Q} + \dot{W}_s \tag{d}$$

  (iv) Uniform system

In Eqs. a and b
$$U = M\hat{U} \tag{e}$$

Note: To obtain the energy balance on a molar basis, make the following substitutions:

Replace	*with*
$M\left(\dfrac{v^2}{2} + \psi\right)$	$Nm\left(\dfrac{v^2}{2} + \psi\right)$
$\dot{M}_k\left(\hat{H} + \dfrac{v^2}{2} + \psi\right)_k$	$\dot{N}_k\left\{\underline{H} + m\left(\dfrac{v^2}{2} + \psi\right)\right\}_k$
$M\hat{U}$	$N\underline{U}$

The result is

$$\left\{ U + M \left(\frac{v^2}{2} + \psi \right) \right\}_{t_2} - \left\{ U + M \left(\frac{v^2}{2} + \psi \right) \right\}_{t_1}$$

$$= \sum_{k=1}^{K} \int_{t_1}^{t_2} \dot{M}_k \left(\hat{H} + \frac{v^2}{2} + \psi \right)_k \, dt + Q + W \qquad \textbf{(3.1-6)}$$

where

$$Q = \int_{t_1}^{t_2} \dot{Q} \, dt \qquad W_s = \int_{t_1}^{t_2} \dot{W}_s \, dt \qquad \int_{V(t_1)}^{V(t_2)} P \, dV = \int_{t_1}^{t_2} P \frac{dV}{dt} \, dt$$

and

$$W = W_s - \int_{V(t_1)}^{V(t_2)} P \, dV$$

The first term on the right side of Eq. 3.1-6 is usually the most troublesome to evaluate because the mass flow rate and/or the thermodynamic properties of the flowing fluid may change with time. However, if the thermodynamic properties of the fluids entering and leaving the system are independent of time (even though the mass flow rate may depend on time), we have

$$\sum_{k=1}^{K} \int_{t_1}^{t_2} \dot{M}_k \left(\hat{H} + \frac{v^2}{2} + \psi \right)_k \, dt = \sum_{k=1}^{K} \left(\hat{H} + \frac{v^2}{2} + \psi \right)_k \int_{t_1}^{t_2} \dot{M}_k \, dt$$

$$\textbf{(3.1-7)}$$

$$= \sum_{k=1}^{K} \Delta M_k \left(\hat{H} + \frac{v^2}{2} + \psi \right)_k$$

If, on the other hand, the thermodynamic properties of the flow streams change with time in some arbitrary way, the energy balance of Eq. 3.1-6 may not be useful since it may not be possible to evaluate the integral. The usual procedure, then, is to try to choose a new system (or subsystem) for the description of the process in which these time-dependent flows do not occur or are more easily handled (see Illustration 3.4-5).

Table 3.1-2 lists various special cases of Eq. 3.1-6 that will be useful in solving thermodynamic problems.

For the study of thermodynamics it will be useful to have equations that relate the differential change in certain thermodynamic variables of the system to differential changes in other system properties. Such equations can be obtained from the differential form of the mass and energy balances. For processes in which kinetic and potential energy terms are unimportant, there is no shaft work, and there is only a single mass flow stream, these equations reduce to

$$\frac{dM}{dt} = \dot{M}$$

and

$$\frac{dU}{dt} = \dot{M}\hat{H} + \dot{Q} - P\frac{dV}{dt}$$

which can be combined to give

$$\frac{dU}{dt} = \hat{H}\frac{dM}{dt} + \dot{Q} - P\frac{dV}{dt} \qquad \text{(3.1-8)}$$

where $\hat{H}$ is the enthalpy per unit mass entering or leaving the system. (Note that for a system closed to the flow of mass, $dM/dt = \dot{M} = 0$.) Defining $Q = \dot{Q}\,dt$ to be equal to the heat flow into the system in the differential time interval dt, and $dM = \dot{M}\,dt = (dM/dt)\,dt$ to be equal to the mass flow in that time interval, we obtain the following expression for the change of the internal energy in the time interval dt:

$$dU = \hat{H}\,dM + Q - P\,dV \qquad \text{(3.1-9a)}$$

Table 3.1-2 Difference Form of the Energy Balance

General equation

$$\left\{ U + M\left(\frac{v^2}{2} + \psi\right) \right\}_{t_2} - \left\{ U + M\left(\frac{v^2}{2} + \psi\right) \right\}_{t_1} = \sum_{k=1}^{K} \int_{t_1}^{t_2} \dot{M}_k\left(\hat{H} + \frac{v^2}{2} + \psi\right)_k dt + Q + W \quad \text{(a)}$$

Special cases:
  (i) Closed system

$$\left\{ U + M\left(\frac{v^2}{2} + \psi\right) \right\}_{t_2} - \left\{ U + M\left(\frac{v^2}{2} + \psi\right) \right\}_{t_1} = Q + W \qquad \text{(b)}$$

  and

$$M(t_1) = M(t_2)$$

 (ii) Adiabatic process

In Eqs. a and b
$$Q = 0 \qquad \text{(c)}$$

(iii) Open system, flow of fluids of constant thermodynamic properties

$$\sum_{k=1}^{K} \int_{t_1}^{t_2} \dot{M}_k\left(\hat{H} + \frac{v^2}{2} + \psi\right)_k dt = \sum_{k=1}^{K} \Delta M_k\left(\hat{H} + \frac{v^2}{2} + \psi\right)_k \qquad \text{(d)}$$

in Eq. a

(iv) Uniform system

$$\left\{ U + M\left(\frac{v^2}{2} + \psi\right) \right\} = M\left(\hat{U} + \frac{v^2}{2} + \psi\right) \qquad \text{(e)}$$

in Eqs. a and b

Note: To obtain the energy balance on a molar basis, make the following substitutions:

Replace	*with*
$M\left(\dfrac{v^2}{2} + \psi\right)$	$Nm\left(\dfrac{v^2}{2} + \psi\right)$
$\displaystyle\int_{t_1}^{t_2} \dot{M}_k\left(\hat{H} + \frac{v^2}{2} + \psi\right)_k dt$	$\displaystyle\int_{t_1}^{t_2} \dot{N}_k\left\{\underline{H} + m\left(\frac{v^2}{2} + \psi\right)\right\}_k dt$
$M\left(\hat{U} + \dfrac{v^2}{2} + \psi\right)$	$N\left\{\underline{U} + m\left(\dfrac{v^2}{2} + \psi\right)\right\}$

For a closed system this equation reduces to

$$dU = Q - P \, dV \qquad \qquad \textbf{(3.1-9b)}$$

Since the time derivative operator d/dt is mathematically well defined, and the operator d is not, it is important to remember in using Eqs. 3.1-9 that they are abbreviations of Eq. 3.1-8. It is part of the traditional notation of thermodynamics to use $d\theta$ to indicate a differential change in the property θ, rather than the mathematically more correct $d\theta/dt$.

3.2   SEVERAL EXAMPLES OF USING THE ENERGY BALANCE

The energy balance equations developed so far in this chapter can be used for the description of any process. As the first step in using these equations, it is necessary to choose the system for which the mass and energy balances are to be written. The important fact for the student of thermodynamics to recognize is that processes occurring in nature are in no way influenced by our mathematical description of them. Therefore, if our descriptions are correct, they must lead to the same final result for the system and its surroundings regardless of which system choice is made. This is demonstrated in the following example, where the same result is obtained by choosing for the system first a given mass of material and then a specified region in space. Since the first system choice is closed and the second open, this illustration also establishes the way in which the open-system energy flow $PV\dot{M}$ is related to the closed-system work term $P(dV/dt)$.

ILLUSTRATION 3.2-1

Showing That the Final Result Should Not Depend on the Choice of System

A compressor is operating in a continuous, steady-state manner to produce a gas at temperature T_2 and pressure P_2 from one at T_1 and P_1. Show that for the time interval Δt

$$Q + W_s = (\hat{H}_2 - \hat{H}_1) \, \Delta M$$

where ΔM is the mass of gas that has flowed into or out of the system in the time Δt. Establish this result by (a) first writing the balance equations for a closed system consisting of some convenient element of mass, and then (b) by writing the balance equations for the compressor and its contents, which is an open system.

SOLUTION

a. *The closed-system analysis*

Here we take as the system the gas in the compressor and the mass of gas ΔM, that will enter the compressor in the time interval Δt. The system is enclosed by dotted lines in the figure.

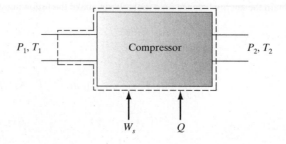

At the later time $t + \Delta t$, the mass of gas we have chosen as the system is as shown below.

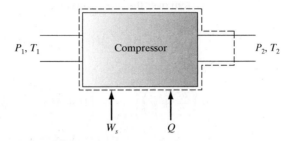

P_1, T_1 Compressor P_2, T_2

W_s Q

We use the subscript c to denote the characteristics of the fluid in the compressor, the subscript 1 for the gas contained in the system that is in the inlet pipe at time t, and the subscript 2 for the gas in the system that is in the exit pipe at time $t + \Delta t$. With this notation the mass balance for the closed system is

$$M_2(t + \Delta t) + M_c(t + \Delta t) = M_1(t) + M_c(t)$$

Since the compressor is in steady-state operation, the amount of gas contained within it and the properties of this gas are constant. Thus, $M_c(t + \Delta t) = M_c(t)$ and

$$M_2(t + \Delta t) = M_1(t) = \Delta M$$

The energy balance for this system, neglecting the potential and kinetic energy terms (which, if retained, would largely cancel), is

$$M_2\hat{U}_2|_{t+\Delta t} + M_c\hat{U}_c|_{t+\Delta t} - M_1\hat{U}_1|_t - M_c\hat{U}_c|_t = W_s + Q + P_1\hat{V}_1 M_1 - P_2\hat{V}_2 M_2 \qquad \textbf{(a)}$$

In writing this equation we have recognized that the flow terms vanish for the closed system and that there are two contributions of the $\int P \, dV$ type, one due to the deformation of the system boundary at the compressor inlet and another at the compressor outlet. Since the inlet and exit pressures are constant at P_1 and P_2, these terms are

$$-\int P \, dV = -P_1 \int dV|_{\text{inlet}} \qquad -P_2 \int dV|_{\text{outlet}}$$
$$= -P_1\{V_1(t + \Delta t) - V_1(t)\} - P_2\{V_2(t + \Delta t) - V_2(t)\}$$

However, $V_1(t + \Delta t) = 0$ and $V_2(t) = 0$, so that

$$-\int P \, dV = +P_1 V_1 - P_2 V_2 = P_1\hat{V}_1 M_1 - P_2\hat{V}_2 M_2$$

Now, using the energy balance and Eq. a above, and recognizing that since the compressor is in steady-state operation,

$$M_c\hat{U}_c|_{t+\Delta t} = M_c\hat{U}_c|_t$$

we obtain

$$\Delta M(\hat{U}_2 - \hat{U}_1) = W_s + Q + P_1\hat{V}_1 \Delta M - P_2\hat{V}_2 \Delta M$$

or

$$\Delta M(\hat{U}_2 + P_2\hat{V}_2 - \hat{U}_1 - P_1\hat{V}_1) = \Delta M(\hat{H}_2 - \hat{H}_1) = W_s + Q$$

b. *The open-system analysis*

Here we take the contents of the compressor at any time to be the system. The mass balance for this system over the time interval Δt is

$$M(t + \Delta t) - M(t) = \int_t^{t+\Delta t} \dot{M}_1 \, dt + \int_t^{t+\Delta t} \dot{M}_2 \, dt = (M)_1 + (M)_2$$

and the energy balance is

$$\left\{ U + M\left(\frac{v^2}{2} + \psi\right) \right\}_{t+\Delta t} - \left\{ U + M\left(\frac{v^2}{2} + \psi\right) \right\}_t$$

$$= \int_t^{t+\Delta t} \dot{M}_1(\hat{H}_1 + v_1^2/2 + \psi_1) \, dt + \int_t^{t+\Delta t} \dot{M}_2(\hat{H}_2 + v_2^2/2 + \psi_2) \, dt + Q + W$$

These equations may be simplified as follows:

1. Since the compressor is operating continuously in a steady-state manner, its contents must, by definition, have the same mass and thermodynamic properties at all times. Therefore,

$$M(t + \Delta t) = M(t)$$

and

$$\left\{ U + M\left(\frac{v^2}{2} + \psi\right) \right\}_{t+\Delta t} = \left\{ U + M\left(\frac{v^2}{2} + \psi\right) \right\}_t$$

2. Since the thermodynamic properties of the fluids entering and leaving the turbine do not change in time, we can write

$$\int_t^{t+\Delta t} \dot{M}_1(\hat{H}_1 + v_1^2/2 + \psi_1) \, dt = (\hat{H}_1 + v_1^2/2 + \psi_1) \int_t^{t+\Delta t} \dot{M}_1 \, dt$$

$$= (\hat{H}_1 + v_1^2/2 + \psi_1) \, \Delta M_1$$

   with a similar expression for the compressor exit stream.

3. Since the volume of the system here, the contents of the compressor, is constant

$$\int_{V_1}^{V_2} P \, dV = 0$$

   so that

$$W = W_s$$

4. Finally, we will neglect the potential and kinetic energy changes of the entering and exiting fluids.

With these simplifications, we have

$$0 = \Delta M_1 + \Delta M_2 \quad \text{or} \quad \Delta M_1 = -\Delta M_2 = \Delta M$$

and

$$0 = \Delta M_1 \hat{H}_1 + \Delta M_2 \hat{H}_2 + Q + W_s$$

Combining these two equations, we obtain

$$Q + W_s = (\hat{H}_2 - \hat{H}_1) \, \Delta M$$

This is the same result as in part (a).

COMMENT

Notice that in the closed-system analysis the surroundings are doing work on the system (the mass element) at the inlet to the compressor, while the system is doing work on its surroundings at the outlet pipe. Each of these terms is a $\int P\,dV$–type work term. For the open system this work term has been included in the energy balance as a $P\hat{V}\,\Delta M$ term, so that it is the enthalpy, rather than the internal energy, of the flow streams that appears in the equation. The explicit $\int P\,dV$ term that does appear in the open-system energy balance represents only the work done if the system boundaries deform; for the choice of the compressor and its contents as the system here this term is zero unless the compressor (the boundary of our system) explodes.   ■

This illustration demonstrates that the sum $Q + W_s$ is the same for a fluid undergoing some change in a continuous process regardless of whether we choose to compute this sum from the closed-system analysis on a mass of gas or from an open-system analysis on a given volume in space. In Illustration 3.2-2 we consider another problem, the compression of a gas by two *different* processes, the first being a closed-system piston-and-cylinder process and the second being a flow compressor process. Here we will find that the sum $Q + W$ is different in the two processes, but the origin of this difference is easily understood.

ILLUSTRATION 3.2-2

Showing That Processes in Closed and Open Systems Are Different

A mass M of gas is to be compressed from a temperature T_1 and a pressure P_1 to T_2 and P_2 in (a) a one-step process in a frictionless piston and cylinder,[4] and (b) a continuous process in which the mass M of gas is part of the feed stream to the compressor of the previous illustration. Compute the sum $Q + W$ for each process.

SOLUTION

a. *The piston-and-cylinder process*

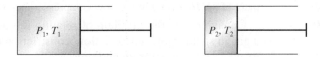

Here we take the gas within the piston and cylinder as the system. The energy balance for this closed system is

$$M(\hat{U}_2 - \hat{U}_1) = Q + W \quad \text{(piston-cylinder process)}$$

It is useful to note that $W_s = 0$ and $W = -\int P\,dV$.

b. *The flow compressor process* (see the figures in Illustration 3.2-1)

If we take the contents of the compressor as the system and follow the analysis of the previous illustration, we obtain

$$M(\hat{H}_2 - \hat{H}_1) = Q + W \quad \text{(flow compressor)}$$

where, since $\int P\,dV = 0$, $W = W_s$.

COMMENT

From these results it is evident that the sum $Q + W$ is different in the two cases, since the two processes are different. The origin of the difference in the flow and nonflow energy changes

---

[4]Since the piston is frictionless, the pressure of the gas is equal to the pressure applied by the piston.

accompanying a change of state is easily identified by considering two different ways of compressing a mass M of gas in a piston and cylinder from (T_1, P_1) to (T_2, P_2). The first way is merely to compress the gas in situ. The sum of heat and work flows needed to accomplish the change of state is, from the preceding computations,

$$Q + W = M(\hat{U}_2 - \hat{U}_1)$$

A second way to accomplish the compression is to open a valve at the side of the cylinder and use the piston movement (at constant pressure P_1) to inject the gas into the compressor inlet stream, use the compressor to compress the gas, and then withdraw the gas from the compressor exit stream by moving the piston against a constant external pressure P_2. The energy required in the compressor stage is the same as that found above:

$$(Q + W)_c = M(\hat{H}_2 - \hat{H}_1)$$

To this we must add the work done in using the piston movement to pump the fluid into the compressor inlet stream,

$$W_1 = \int P \, dV = P_1 V_1 = P_1 \hat{V}_1 M$$

(this is the work done by the system on the gas in the inlet pipe to the compressor), and subtract the work obtained as a result of the piston movement as the cylinder is refilled,

$$W_2 = -\int P \, dV = -P_2 V_2 = -P_2 \hat{V}_2 M$$

(this is the work done on the system by the gas in the compressor exit stream). Thus the total energy change in the process is

$$Q + W = (Q + W)_c + W_1 + W_2$$
$$= M(\hat{H}_2 - \hat{H}_1) + P_1 \hat{V}_1 M - P_2 \hat{V}_2 M = M(\hat{U}_2 - \hat{U}_1)$$

which is what we found in part (a). Here, however, it results from the sum of an energy requirement of $M(\hat{H}_2 - \hat{H}_1)$ in the flow compressor and the two pumping terms. ∎

Consider now the problem of relating the downstream temperature and pressure of a gas in steady flow across a flow constriction (e.g., a valve, orifice, or porous plug) to its upstream temperature and pressure.

ILLUSTRATION 3.2-3

A Joule-Thomson or Isenthalpic Expansion

A gas at pressure P_1 and temperature T_1 is steadily exhausted to the atmosphere at pressure P_2 through a pressure-reducing valve. Find an expression relating the downstream gas temperature T_2 to P_1, P_2, and T_1. Since the gas flows through the valve rapidly, one can assume that there is no heat transfer to the gas. Also, the potential and kinetic energy terms can be neglected.

$T_1, P_1 \longrightarrow \qquad \longrightarrow T_2 = ?, P_2$

SOLUTION

The flow process is schematically shown in the figure. We will consider the region of space that includes the flow obstruction (indicated by the dashed line) to be the system, although, as in Illustration 3.2-1, a fixed mass of gas could have been chosen as well. The pressure of the gas exiting the reducing valve will be P_2, the pressure of the surrounding atmosphere. (It is not completely obvious that these two pressures should be the same. However, in the laboratory we find that the velocity of the flowing fluid will always adjust in such a way that the fluid exit pressure and the pressure of the surroundings are equal.) Now recognizing that our system (the valve and its contents) is of constant volume, that the flow is steady, and that there are no heat or work flows and negligible kinetic and potential energy changes, the mass and energy balances (on a molar basis) yield

$$0 = \dot{N}_1 + \dot{N}_2 \quad \text{or} \quad \dot{N}_2 = -\dot{N}_1$$

and

$$0 = \dot{N}_1 \underline{H}_1 + \dot{N}_2 \underline{H}_2 = \dot{N}_1(\underline{H}_1 - \underline{H}_2)$$

Thus

Isenthalpic or Joule-Thomson expansion

$$\boxed{\underline{H}_1 = \underline{H}_2}$$

or, to be explicit,

$$\underline{H}(T_1, P_1) = \underline{H}(T_2, P_2)$$

so that the initial and final states of the gas have the same enthalpy. Consequently, if we knew how the enthalpy of the gas depended on its temperature and pressure, we could use the known values of T_1, P_1, and P_2 to determine the unknown downstream temperature T_2.

COMMENTS

1. The equality of enthalpies in the upstream and downstream states is the only information we get from the thermodynamic balance equations. To proceed further we need constitutive information, that is, an equation of state or experimental data interrelating $\underline{H}$, T, and P. Equations of state are discussed in the following section and in much of Chapter 6.
2. The experiment discussed in this illustration was devised by William Thomson (later Lord Kelvin) and performed by J. P. Joule to study departures from ideal gas behavior. The **Joule-Thomson expansion**, as it is called, is used in the liquefaction of gases and in refrigeration processes (see Chapter 5). ■

3.3 THE THERMODYNAMIC PROPERTIES OF MATTER

The balance equations of this chapter allow one to relate the mass, work, and heat flows of a system to the change in its thermodynamic state. From the experimental observations discussed in Chapter 1, the change of state for a single-component, single-phase system can be described by specifying the initial and final values of any two independent intensive variables. However, certain intensive variables, especially temperature and pressure, are far easier to measure than others. Consequently, for most problems we will want to specify the state of a system by its temperature and pressure rather than by its specific volume, internal energy, and enthalpy, which appear in the energy balance. What are needed, then, are interrelations between the fluid properties that allow one to eliminate some thermodynamic variables in terms of other, more easily measured ones. Of particular interest is the volumetric equation of state, which is a relation between temperature, pressure, and specific volume, and the thermal equation of state, which is usually either in the form of a relationship between internal energy, temperature, and specific (or molar) volume, or between specific or molar enthalpy, temperature, and pressure. Such information may be available in either of two forms. First, there are analytic equations of state, which provide an algebraic relation between the thermodynamic state variables. Second, experimental data, usually in graphical or tabular form, may be available to provide the needed interrelationships between the fluid properties.

Equations of state for fluids are considered in detail in Chapter 6. To illustrate the use of the mass and energy balance equations in a simple form, we briefly consider here the equation of state for the ideal gas and the graphical and tabular display of the thermodynamic properties of several real fluids.

An **ideal gas** is a gas at such a low pressure that there are no interactions among its molecules. For such gases it is possible to show, either experimentally or by the methods of statistical mechanics, that at all absolute temperatures and pressures the volumetric equation of state is

$$P\underline{V} = RT \tag{3.3-1}$$

(as indicated in Sec. 1.4) and that the enthalpy and internal energy are functions of temperature only (and not pressure or specific volume). We denote this latter fact by $\underline{H} = \underline{H}(T)$ and $\underline{U} = \underline{U}(T)$. This simple behavior is to be compared with the enthalpy for a real fluid, which is a function of temperature and pressure [i.e., $\underline{H} = \underline{H}(T, P)$] and the internal energy, which is usually written as a function of temperature and specific volume [$\underline{U} = \underline{U}(T, \underline{V})$], as will be discussed in Chapter 6.

The temperature dependence of the internal energy and enthalpy of all substances (not merely ideal gases) can be found by measuring the temperature rise that accompanies a heat flow into a closed stationary system. If a sufficiently small quantity of heat is added to such a system, it is observed that the temperature rise produced, ΔT, is linearly related to the heat added and inversely proportional to N, the number of moles in the system:

$$\frac{Q}{N} = C\,\Delta T = C\{T(t_2) - T(t_1)\}$$

where C is a parameter and Q is the heat added to the system between the times t_1 and t_2. The object of the experiment is to accurately measure the parameter C for a very small temperature rise, since C generally is also a function of temperature. If the measurement is made at constant volume and with $W_s = 0$, we have, from the energy balance and the foregoing equation,

$$U(t_2) - U(t_1) = Q = NC_V\{T(t_2) - T(t_1)\}$$

Thus

$$C_V = \frac{U(t_2) - U(t_1)}{N\{T(t_2) - T(t_1)\}} = \frac{\underline{U}(t_2) - \underline{U}(t_1)}{T(t_2) - T(t_1)}$$

where the subscript V has been introduced to remind us that the parameter C was determined in a constant-volume experiment. In the limit of a very small temperature difference, we have

Constant-volume heat capacity definition

$$C_V(T, V) = \lim_{T(t_2)-T(t_1)\to 0} \frac{\underline{U}(t_2) - \underline{U}(t_1)}{T(t_2) - T(t_1)} = \left(\frac{\partial \underline{U}}{\partial T}\right)_V = \left(\frac{\partial \underline{U}(T, \underline{V})}{\partial T}\right)_V \tag{3.3-2}$$

so that the measured parameter C_V is, in fact, equal to the temperature derivative of the internal energy at constant volume. Similarly, if the parameter C is determined in a constant-pressure experiment, we have

$$
\begin{aligned}
Q &= U(t_2) - U(t_1) + P\{V(t_2) - V(t_1)\} = U(t_2) + P(t_2)V(t_2) - U(t_1) - P(t_1)V(t_1) \\
&= H(t_2) - H(t_1) = NC_P\{T(t_2) - T(t_1)\}
\end{aligned}
$$

where we have used the fact that since pressure is constant, $P = P(t_2) = P(t_1)$. Then

Constant-pressure heat capacity definition

$$
\boxed{C_P(T, P) = \left(\frac{\partial \underline{H}}{\partial T}\right)_P = \left(\frac{\partial \underline{H}(T, P)}{\partial T}\right)_P}
\tag{3.3-3}
$$

so that the measured parameter here is equal to the temperature derivative of the enthalpy at constant pressure.

The quantity C_V is called the **constant-volume heat capacity**, and C_P is the **constant-pressure heat capacity**; both appear frequently throughout this book. Partial derivatives have been used in Eqs. 3.3-2 and 3.3-3 to indicate that although the internal energy is a function of temperature and density or specific (or molar) volume, C_V has been measured along a path of constant volume; and although the enthalpy is a function of temperature and pressure, C_P has been evaluated in an experiment in which the pressure was held constant.

For the special case of the ideal gas, the enthalpy and internal energy of the fluid are functions only of temperature. In this case the partial derivatives above become total derivatives, and

$$
C_P^*(T) = \frac{d\underline{H}}{dT} \quad \text{and} \quad C_V^*(T) = \frac{d\underline{U}}{dT}
\tag{3.3-4}
$$

so that the ideal gas heat capacities, which we denote using asterisks as $C_P^*(T)$ and $C_V^*(T)$, are only functions of T as well. The temperature dependence of the ideal gas heat capacity can be measured or, in some cases, computed using the methods of statistical mechanics and detailed information about molecular structure, bond lengths, vibrational frequencies, and so forth. For our purposes $C_P^*(T)$ will either be considered to be a constant or be written as a function of temperature in the form

Ideal gas heat capacity

$$
\boxed{C_P^*(T) = a + bT + cT^2 + dT^3 + \cdots}
\tag{3.3-5}
$$

Since $\underline{H} = \underline{U} + P\underline{V}$, and for the ideal gas $P\underline{V} = RT$, we have $\underline{H} = \underline{U} + RT$ and

$$
\boxed{C_P^*(T) = \frac{d\underline{H}}{dT} = \frac{d(\underline{U} + RT)}{dT} = C_V^*(T) + R}
$$

so that $C_V^*(T) = C_P^*(T) - R = (a - R) + bT + cT^2 + dT^3 + \cdots$. The constants in Eq. 3.3-5 for various gases are given in Appendix A.II.

The enthalpy and internal energy of an ideal gas at a temperature T_2 can be related to their values at T_1 by integration of Eqs. 3.3-4 to obtain

$$
\boxed{\underline{H}^{IG}(T_2) = \underline{H}^{IG}(T_1) + \int_{T_1}^{T_2} C_P^*(T)\, dT}
$$

and

$$\underline{U}^{IG}(T_2) = \underline{U}^{IG}(T_1) + \int_{T_1}^{T_2} C_V^*(T)\,dT \qquad \text{(3.3-6)}$$

where the superscript IG has been introduced to remind us that these equations are valid only for ideal gases.

Our interest in the first part of this book is with energy flow problems in single-component systems. Since the only energy information needed in solving these problems is the change in internal energy and/or enthalpy of a substance between two states, and since determination of the absolute energy of a substance is not possible, what is done is to choose an easily accessible state of a substance to be the **reference state**, for which $\underline{H}^{IG}$ is arbitrarily set equal to zero, and then report the enthalpy and internal energy of all other states relative to this reference state. (That there is a state for each substance for which the enthalpy has been arbitrarily set to zero does lead to difficulties when chemical reactions occur. Consequently, another energy convention is introduced in Chapter 8.)

If, for the ideal gas, the temperature T_R is chosen as the reference temperature (i.e., $\underline{H}^{IG}$ at T_R is set equal to zero), the enthalpy at temperature T is then

$$\underline{H}^{IG}(T) = \int_{T_R}^{T} C_P^*(T)\,dT \qquad \text{(3.3-7)}$$

Similarly, the internal energy at T is

$$\underline{U}^{IG}(T) = \underline{U}^{IG}(T_R) + \int_{T_R}^{T} C_V^*(T)\,dT = \{\underline{H}^{IG}(T_R) - RT_R\} + \int_{T_R}^{T} C_V^*(T)\,dT$$

$$= \int_{T_R}^{T} C_V^*(T)\,dT - RT_R$$

$$\text{(3.3-8)}$$

One possible choice for the reference temperature T_R is absolute zero. In this case,

$$\underline{H}^{IG}(T) = \int_{0}^{T} C_P^*(T)\,dT \quad \text{and} \quad \underline{U}^{IG}(T) = \int_{0}^{T} C_V^*(T)\,dT$$

However, to use these equations, heat capacity data are needed from absolute zero to the temperature of interest. These data are not likely to be available, so a more convenient reference temperature, such as $0°C$, is frequently used.

For the special case in which the constant-pressure and constant-volume heat capacities are independent of temperature, we have, from Eqs. 3.3-7 and 3.3-8,

$$\underline{H}^{IG}(T) = C_P^*(T - T_R) \qquad \text{(3.3-7')}$$

and

$$\underline{U}^{IG}(T) = C_V^*(T - T_R) - RT_R = C_V^*T - C_P^*T_R \qquad \text{(3.3-8')}$$

which, when T_R is taken to be absolute zero, simplify further to

$$\underline{H}^{IG}(T) = C_P^*T \quad \text{and} \quad \underline{U}^{IG}(T) = C_V^*T$$

As we will see in Chapters 6 and 7, very few fluids are ideal gases, and the mathematics of relating the enthalpy and internal energy to the temperature and pressure of a real fluid is much more complicated than indicated here. Therefore, for fluids of industrial and scientific importance, detailed experimental thermodynamic data have been collected. These data can be presented in tabular form (see Appendix A.III for a table of the thermodynamic properties of steam) or in graphical form, as in Figs. 3.3-1 to 3.3-4 for steam, methane, nitrogen, and the environmentally friendly refrigerant HFC-134a. (Can you identify the reference state for the construction of the steam tables in Appendix A.III?) With these detailed data one can, given values of temperature and pressure, easily find the enthalpy, specific volume, and entropy (a thermodynamic quantity that is introduced in the next chapter). More generally, given any two intensive variables of a single-component, single-phase system, the remaining properties can be found.

Notice that different choices have been made for the independent variables in these figures. Although the independent variables may be chosen arbitrarily,[5] some choices are especially convenient for solving certain types of problems. Thus, as we will see, an enthalpy-entropy ($\underline{H}$-$\underline{S}$) or Mollier diagram,[6] such as Fig. 3.3-1a, is useful for problems involving turbines and compressors; enthalpy-pressure ($\underline{H}$-P) diagrams (for example Figs. 3.3-2 to 3.3-4) are useful in solving refrigeration problems; and temperature-entropy (T-$\underline{S}$) diagrams, of which Fig. 3.3-1b is an example, are used in the analysis of engines and power and refrigeration cycles (see Sec. 3.4 and Chapter 5).

An important characteristic of real fluids is that at sufficiently low temperatures they condense to form liquids and solids. Also, many applications of thermodynamics of interest to engineers involve either a range of thermodynamic states for which the fluid of interest undergoes a phase change, or equilibrium multiphase mixtures (e.g., steam and water at 100°C and 101.3 kPa). Since the energy balance equation is expressed in terms of the internal energy and enthalpy per unit mass of the system, this equation is valid regardless of which phase or mixture of phases is present. Consequently, there is no difficulty, in principle, in using the energy balance (or other equations to be introduced later) for multiphase or phase-change problems, provided thermodynamic information is available for each of the phases present. Figures 3.3-1 to 3.3-4 and the steam tables in Appendix A.III provide such information for the vapor and liquid phases and, within the dome-shaped region, for vapor-liquid mixtures. Similar information for many other fluids is also available. Thus, you should not hesitate to apply the equations of thermodynamics to the solution of problems involving gases, liquids, solids, and mixtures thereof.

There is a simple relationship between the thermodynamic properties of a two-phase mixture (e.g., a mixture of water and steam), the properties of the individual phases, and the mass distribution between the phases. If $\hat{\theta}$ is any intensive property, such as internal energy per unit mass or volume per unit mass, its value in an equilibrium two-phase mixture is

Properties of a two-phase mixture: the lever rule

$$\hat{\theta} = \omega^{\mathrm{I}}\hat{\theta}^{\mathrm{I}} + \omega^{\mathrm{II}}\hat{\theta}^{\mathrm{II}} = \omega^{\mathrm{I}}\hat{\theta}^{\mathrm{I}} + (1 - \omega^{\mathrm{I}})\hat{\theta}^{\mathrm{II}}$$

(3.3-9)

Here ω^{I} is the mass fraction of the system that is in phase I, and $\hat{\theta}^{\mathrm{I}}$ is the value of the variable in that phase. Also, by definition of the mass fraction, $\omega^{\mathrm{I}} + \omega^{\mathrm{II}} = 1$.

---

[5]See, however, the comments made in Sec. 1.6 concerning the use of the combinations U and T, and V and T as the independent thermodynamic variables.

[6]Entropy, denoted by the letter S, is a thermodynamic variable to be introduced in Chapter 4.

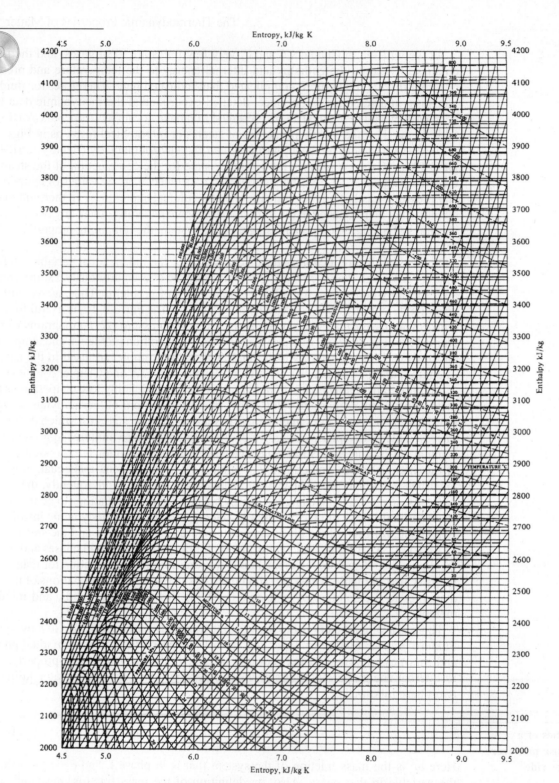

Figure 3.3-1 (*a*) Enthalpy-entropy of Mollier diagram for steam. [*Source:* ASME Steam Tables in SI (Metric) Units for Instructional Use, American Society of Mechanical Engineers, New York, 1967. Used with permission.] (This figure appears as an Adobe PDF file on the CD-ROM accompanying this book, and may be enlarged and printed for easier reading and for use in solving problems.)

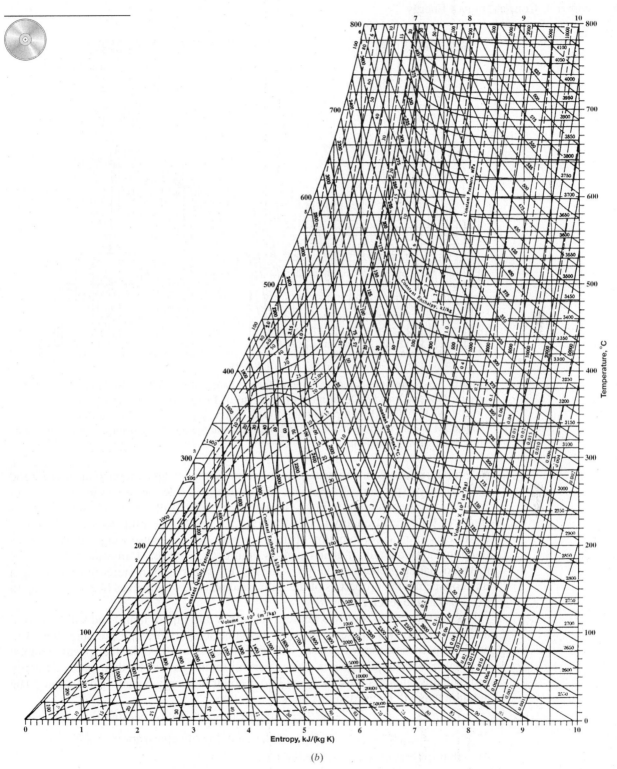

(b)

Figure 3.3-1(b) Temperature-entropy diagram for steam. [*Source:* J. H. Keenan, F. G. Keyes, P. G. Hill, and J. G. Moore, *Steam Tables* (International Edition—Metric Units). Copyright 1969. John Wiley & Sons, Inc., New York. Used with permission.] (This figure appears as an Adobe PDF file on the CD-ROM accompanying this book, and may be enlarged and printed for easier reading and for use in solving problems.)

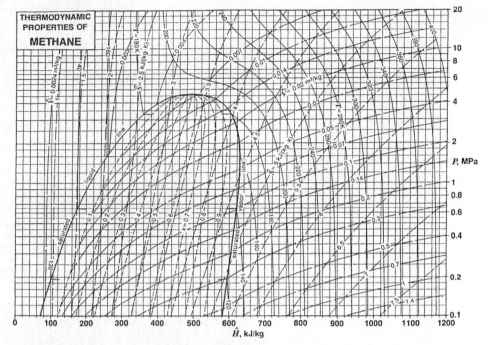

Figure 3.3-2 Pressure-enthalpy diagram for methane. (*Source:* W. C. Reynolds, *Thermodynamic Properties in SI,* Department of Mechanical Engineering, Stanford University, Stanford, CA, 1979. Used with permission.) (This figure appears as an Adobe PDF file on the CD-ROM accompanying this book, and may be enlarged and printed for easier reading and for use in solving problems.)

(For mixtures of steam and water, the mass fraction of steam is termed the *quality* and is frequently expressed as a percent, rather than as a fraction; for example, a steam-water mixture containing 0.02 kg of water for each kg of mixture is referred to as steam of 98 percent quality.) Note that Eq. 3.3-9 gives the property of a two-phase mixture as a linear combination of the properties of each phase weighted by its mass fraction. Consequently, if charts such as Figures 3.3-1 to 3.3-4 are used to obtain the thermodynamic properties, the properties of the two-phase mixture will fall along a line connecting the properties of the individual phases, which gives rise to referring to Eq. 3.3-9 as the **lever rule**.

If a mixture consists of two phases (i.e., vapor and liquid, liquid and solid, or solid and vapor), the two phases will be at the same temperature and pressure; however, other properties of the two phases will be different. For example, the specific volume of the vapor and liquid phases can be very different, as will be their internal energy and enthalpy, and this must be taken into account in energy balance calculations. The notation that will be used in this book is as follows:

$$\triangle_{vap}\hat{H} = \hat{H}^V - \hat{H}^L = \text{enthalpy of vaporization per unit mass}$$

$$\triangle_{vap}\underline{H} = \underline{H}^V - \underline{H}^L = \text{molar enthalpy of vaporization}$$

Also (but for brevity, only on a molar basis),

$$\triangle_{fus}\underline{H} = \underline{H}^L - \underline{H}^S = \text{molar enthalpy of melting or fusion}$$

$$\triangle_{sub}\underline{H} = \underline{H}^V - \underline{H}^S = \text{molar enthalpy of sublimation}$$

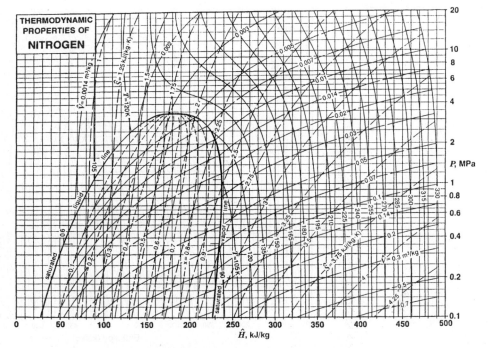

Figure 3.3-3 Pressure-enthalpy diagram for nitrogen. (*Source:* W. C. Reynolds, *Thermodynamic Properties in SI,* Department of Mechanical Engineering, Stanford University, Stanford, CA, 1979. Used with permission.) (This figure appears as an Adobe PDF file on the CD-ROM accompanying this book, and may be enlarged and printed for easier reading and for use in solving problems.)

Similar expressions can be written for the volume changes and internal energy changes on a phase change.

It is also useful to note that several simplifications can be made in computing the thermodynamic properties of solids and liquids. First, because the molar volumes of condensed phases are small, the product $P\underline{V}$ can be neglected unless the pressure is high. Thus, for solids and liquids,

Solids or liquids at low pressure

$$\underline{H} \approx \underline{U} \qquad (3.3\text{-}10)$$

A further simplification commonly made for liquids and solids is to assume that they are also incompressible; that is, their volume is only a function of temperature, so that

Idealized incompressible fluid or solid

$$\left(\frac{\partial \underline{V}}{\partial P}\right)_T = 0 \qquad (3.3\text{-}11)$$

In Chapter 6 we show that for incompressible fluids, the thermodynamic properties $\underline{U}$, C_P, and C_V are functions of temperature only. Since, in fact, solids, and most liquids away from their critical point (see Chapter 7) are relatively incompressible, Eqs. 3.3-10 and 3.3-11, together with the assumption that these properties depend only on temperature, are reasonably accurate and often used in thermodynamic studies involving liquids and solids. Thus, for example, the internal energy of liquid water at a temperature T_1 and pressure P_1 is, to a very good approximation, equal to the internal

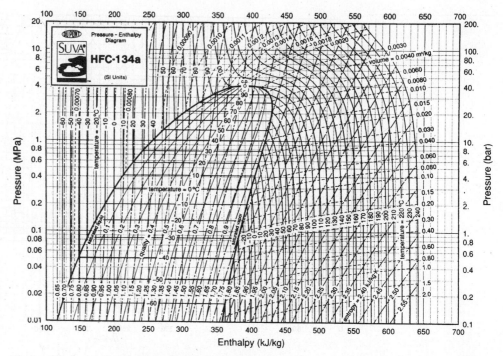

Figure 3.3-4 Pressure–enthalpy diagram for HFC-134a. (Used with permission of DuPont Fluoroproducts.) (This figure appears as an Adobe PDF file on the CD-ROM accompanying this book, and may be enlarged and printed for easier reading and for use in solving problems.)

energy of liquid water at the temperature T_1 and any other pressure. Consequently, the entries for the internal energy of liquid water for a variety of temperatures (at pressures corresponding to the vapor-liquid coexistence or saturation pressures at each temperature) given in the saturation steam tables of Appendix A.III can also be used for the internal energy of liquid water at these same temperatures and higher pressures.

Although we will not try to quantitatively relate the interactions between molecules to their properties—that is the role of statistical mechanics, not thermodynamics—it is useful to make some qualitative observations. The starting point is that the interactions between a pair of simple molecules (for example, argon or methane) depend on the separation distance between their centers of mass, as shown in Fig. 3.3-5. There we see that if the molecules are far apart, the interaction energy is very low, and it vanishes at infinite separation. As the molecules are brought closer together, they attract each other, which decreases the energy of the system. However, if the molecules are brought too close together (so that their electrons overlap), the molecules repel each other, resulting in a positive energy that increases rapidly as one attempts to bring the molecules closer.

We can make the following observations from this simple picture of molecular interactions. First, if the molecules are widely separated, as occurs in a dilute gas, there will be no energy of interaction between the molecules; this is the case of an ideal gas. Next, as the density increases, and the molecules are somewhat closer together, molecular attractions become more important, and the energy of the system decreases. Next, at liquid densities, the average distance between the molecules will be near the deepest (most attractive) part of the interaction energy curve shown in Fig. 3.3-5. (Note that the molecules will not be at the very lowest value of the interaction energy curve as a result of their thermal motion, and because the behavior of a large collection of molecules

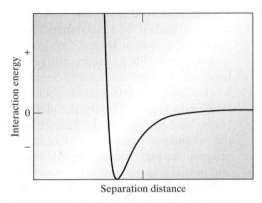

Figure 3.3-5 The interaction energy between two molecules as a function of their separation distance. Since the molecules cannot overlap, there is a strong repulsion (positive interaction energy) at small separation distances. At larger separation distances the interactions between the electrons result in an attraction between the molecules (negative interaction energy), which vanishes at very large separations.

is more complicated than can be inferred by examining the interaction between only a pair of molecules.) Consequently, a liquid has considerably less internal energy than a gas. The energy that must be added to a liquid to cause its molecules to move farther apart and vaporize is the heat of vaporization $\Delta_{vap}H$. In solids, the molecules generally are located very close to the minimum in the interaction energy function in an ordered lattice, so that a solid has even less internal energy than a liquid. The amount of energy required to slightly increase the separation distances between the molecules in a solid and form a liquid is the heat of melting or the heat of fusion $\Delta_{fus}H$.

As mentioned earlier, the constant-pressure heat capacity of solids is a function of temperature; in fact, C_P goes to zero at the absolute zero of temperature and approaches a constant at high temperatures. An approximate estimate for C_P of solids for temperatures of interest to chemical engineers comes from the empirical law (or observation) of DuLong and Petit that

$$C_P = 3NR = 24.942N \ \frac{J}{mol \ K} \tag{3.3-12}$$

where N is the number of atoms in the formula unit. For comparison, the constant-pressure heat capacities of lead, gold, and aluminum at 25°C are 26.8, 25.2, and 24.4 J/(mol K), respectively. Similarly, Eq. 3.3-12 gives a prediction of 49.9 for gallium arsenide (used in the electronics industry), which is close to the measured value of 47.0 J/mol K. For Fe_3C the prediction is 99.8 J/mol K and the measured value is 105.9. So we see that the DuLong-Petit law gives reasonable though not exact values for the heat capacities of solids.

3.4 APPLICATIONS OF THE MASS AND ENERGY BALANCES

In many thermodynamics problems one is given some information about the initial equilibrium state of a substance and asked to find the final state if the heat and work flows are specified, or to find the heat or work flows accompanying the change to a specified final state. Since we use thermodynamic balance equations to get the information needed to solve this sort of problem, the starting point is always the same: the identification of a convenient thermodynamic system. The main restriction on the choice of a system is that the flow terms into and out of the system must be of a simple form—for example, not varying with time, or perhaps even zero. Next, the forms of the mass and energy balance equations appropriate to the system choice are written, and any information about the initial and final states of the system and the flow terms is used. Finally, the thermal equation of state is used to replace the internal energy and enthalpy in the balance equations with temperature, pressure, and volume; the volumetric equation of state may then be used to eliminate volume in terms of temperature and pressure. In this way equations are obtained that contain temperature and pressure as the only state variables.

The volumetric equation of state may also provide another relationship between the temperature, pressure, mass, and volume when the information about the final state of the system is presented in terms of total volume, rather than volume per unit mass or molar volume (see Illustration 3.4-5).

By using the balance equations and the equation-of-state information, we will frequently be left with equations that contain only temperature, pressure, mass, shaft work (W_s), and heat flow (Q). If the number of equations equals the number of unknowns, the problem can be solved. The mass and energy balance equations, together with equation-of-state information, are sufficient to solve many, but not all, energy flow problems. In some situations we are left with more unknowns than equations. In fact, we can readily identify a class of problems of this sort. The mass and energy balance equations together can, at most, yield new information about only one intensive variable of the system (the internal energy or enthalpy per unit mass) or about the sum of the heat and work flows if only the state variables are specified. Therefore, we are not, at present, able to solve problems in which (1) there is no information about any intensive variable of the final state of the system, (2) both the heat flow (Q) and the shaft work (W_s) are unspecified, or (3) one intensive variable of the final state and either Q or W_s are unknown, as in Illustration 3.4-4. To solve these problems, an additional balance equation is needed; such an equation is developed in Chapter 4.

The seemingly most arbitrary step in thermodynamic problem solving is the choice of the system. Since the mass and energy balances were formulated with great generality, they apply to any choice of system, and, as was demonstrated in Illustration 3.2-1, the solution of a problem is independent of the system chosen in obtaining the solution. However, some system choices may result in less effort being required to obtain a solution. This is demonstrated here and again in Chapter 4.

ILLUSTRATION 3.4-1

Joule-Thomson Calculation Using a Mollier Diagram and Steam Tables

Steam at 400 bar and 500°C undergoes a Joule-Thomson expansion to 1 bar. Determine the temperature of the steam after the expansion using

   a. Fig. 3.3-1*a*
   b. Fig. 3.3-1*b*
   c. The steam tables in Appendix A.III

SOLUTION

(Since only one thermodynamic state variable—here the final temperature—is unknown, from the discussion that precedes this illustration we can expect to be able to obtain a solution to this problem.)

We start from Illustration 3.2-3, where it was shown that

$$\hat{H}_1 = \hat{H}(T_1, P_1) = \hat{H}(T_2, P_2) = \hat{H}_2$$

for a Joule-Thomson expansion. Since T_1 and P_1 are known, $\hat{H}_1$ can be found from either Fig. 3.3-1 or the steam tables. Then, since $\hat{H}_2$ ($= \hat{H}_1$, from the foregoing) and P_2 are known, T_2 can be found.

   a. Using Fig. 3.3-1*a*, the Mollier diagram, we first locate the point $P = 400$ bar $= 40\,000$ kPa and $T = 500°$C, which corresponds to $\hat{H}_1 = 2900$ kJ/kg. Following a line of constant enthalpy (a horizontal line on this diagram) to $P = 1$ bar $= 100$ kPa, we find that the final temperature is about 214°C.

   b. Using Fig. 3.3-1*b*, we locate the point $P = 400$ bar and $T = 500°$C (which is somewhat easier to do than it was using Fig. 3.3-1*a*) and follow the curved line of constant enthalpy to a pressure of 1 bar to see that $T_2 = 214°$C.

   c. Using the steam tables of Appendix A.III, we have that at $P = 400$ bar $= 40$ MPa and $T = 500°$C, $\hat{H} = 2903.3$ kJ/kg. At $P = 1$ bar $= 0.1$ MPa, $\hat{H} = 2875.3$ kJ/kg at $T = 200°$C and $\hat{H} = 2974.3$ kJ/kg at $T = 250°$C. Assuming that the enthalpy varies linearly with temperature between 200 and 250°C at $P = 1$ bar, we have by interpolation

$$T = 200 + (250 - 200) \times \frac{2903.3 - 2875.3}{2974.3 - 2875.3} = 214.1°\text{C}$$

COMMENT

For many problems a graphical representation of thermodynamic data, such as Figure 3.3-1, is easiest to use, although the answers obtained are approximate and certain parts of the graphs may be difficult to read accurately. The use of tables of thermodynamic data, such as the steam tables, generally leads to the most accurate answers; however, one or more interpolations may be required. For example, if the initial conditions of the steam had been 475 bar and 530°C instead of 400 bar and 500°C, the method of solution using Fig. 3.3-1 would be unchanged; however, using the steam tables, we would have to interpolate with respect to both temperature and pressure to get the initial enthalpy of the steam.

   One way to do this is first, by interpolation between temperatures, to obtain the enthalpy of steam at 530°C at both 400 bar $= 40$ MPa and 500 bar. Then, by interpolation with respect to pressure between these two values, we obtain the enthalpy at 475 bar. That is, from

$$\hat{H}(40 \text{ MPa}, 500°\text{C}) = 2903.3 \text{ kJ/kg} \qquad \hat{H}(50 \text{ MPa}, 500°\text{C}) = 2720.1 \text{ kJ/kg}$$
$$\hat{H}(40 \text{ MPa}, 550°\text{C}) = 3149.1 \text{ kJ/kg} \qquad \hat{H}(50 \text{ MPa}, 550°\text{C}) = 3019.5 \text{ kJ/kg}$$

and the interpolation formula

$$\Theta(x + \Delta) = \Theta(x) + \Delta \frac{\Theta(y) - \Theta(x)}{y - x}$$

where Θ is any tabulated function, and x and y are two adjacent values at which Θ is available, we have

$$\hat{H}(40 \text{ MPa}, 530°\text{C}) = \hat{H}(40 \text{ MPa}, 500°\text{C}) + 30 \times \frac{\hat{H}(40 \text{ MPa}, 550°\text{C}) - \hat{H}(40 \text{ MPa}, 500°\text{C})}{550 - 500}$$

$$= 2903.3 + 30 \times \frac{3149.1 - 2903.3}{50} = 2903.3 + 147.5$$

$$= 3050.8 \text{ kJ/kg}$$

and

$$\hat{H}(50 \text{ MPa}, 530°C) = 2720.1 + 30 \times \frac{3019.5 - 2720.1}{50} = 2899.7 \text{ kJ/kg}$$

Then

$$\hat{H}(47.5 \text{ MPa}, 530°C) = \hat{H}(40 \text{ MPa}, 530°C) + 7.5 \times \frac{\hat{H}(50 \text{ MPa}, 530°C) - \hat{H}(40 \text{ MPa}, 530°C)}{50 - 40}$$

$$= 3050.8 + \frac{7.5}{10} \times (2881.7 - 3050.8) = 2924.0 \text{ kJ/kg} \qquad ■$$

ILLUSTRATION 3.4-2

Application of the Complete Energy Balance Using the Steam Tables

An adiabatic steady-state turbine is being designed to serve as an energy source for a small electrical generator. The inlet to the turbine is steam at 600°C and 10 bar, with a mass flow rate of 2.5 kg/s through an inlet pipe that is 10 cm in diameter. The conditions at the turbine exit are $T = 400°C$ and $P = 1$ bar. Since the steam expands through the turbine, the outlet pipe is 25 cm in diameter. Estimate the rate at which work can be obtained from this turbine.

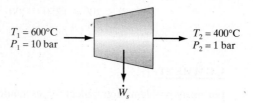

SOLUTION

(This is another problem in which there is only a single thermodynamic unknown, the rate at which work is obtained, so we can expect to be able to solve this problem.)

The first step in solving any energy flow problem is to choose the thermodynamic system; the second step is to write the balance equations for the system. Here we take the turbine and its contents to be the system. The mass and energy balance equations for this adiabatic, steady-state system are

$$\frac{dM}{dt} = 0 = \dot{M}_1 + \dot{M}_2 \qquad \text{(a)}$$

and

$$\frac{d}{dt}\left\{ U + M\left(\frac{v^2}{2} + gh\right)\right\} = 0 = \dot{M}_1\left(\hat{H}_1 + \frac{v_1^2}{2}\right) + \dot{M}_2\left(\hat{H}_2 + \frac{v_2^2}{2}\right) + \dot{W}_s \qquad \text{(b)}$$

In writing these equations we have set the rate of change of mass and energy equal to zero because the turbine is in steady-state operation; Q is equal to zero because the process is adiabatic, and $P(dV/dt)$ is equal to zero because the volume of the system is constant (unless the turbine explodes). Finally, since the schematic diagram indicates that the turbine is positioned horizontally, we have assumed there is no potential energy change in the flowing steam.

There are six unknowns—$\dot{M}_2$, $\hat{H}_1$, $\hat{H}_2$, $\dot{W}_s$, v_1, and v_2—in Eqs. a and b. However, both velocities will be found from the mass flow rates, pipe diameters, and volumetric equation-of-state information (here the steam tables in Appendix A.III). Also, thermal equation-of-state information (again the steam tables in Appendix A.III) relates the enthalpies to temperature and pressure, both of which are known. Thus $\dot{M}_2$ and $\hat{H}_2$ are the only real unknowns, and these may be found from the balance equations above. From the mass balance equation, we have

$$\dot{M}_2 = -\dot{M}_1 = -2.5 \text{ kg/s}$$

From the steam tables or, less accurately from Fig. 3.3-1, we have

$$\hat{H}_1 = 3697.9 \text{ kJ/kg} \qquad \hat{H}_2 = 3278.2 \text{ kJ/kg}$$
$$\hat{V}_1 = 0.4011 \text{ m}^3/\text{kg} \qquad \hat{V}_2 = 3.103 \text{ m}^3/\text{kg}$$

The velocities at the inlet and outlet to the turbine are calculated from

$$\text{Volumetric flow rate} = \dot{M}\hat{V} = \frac{\pi d^2}{4} v$$

where d is the pipe diameter. Therefore,

$$v_1 = \frac{4\dot{M}_1\hat{V}_1}{\pi d_{\text{in}}^2} = \frac{4 \cdot 2.5 \, \frac{\text{kg}}{\text{s}} \cdot 0.4011 \, \frac{\text{m}^3}{\text{kg}}}{3.14159 \cdot (0.1 \text{ m})^2} = 127.7 \, \frac{\text{m}}{\text{s}}$$

$$v_2 = \frac{4\dot{M}_2\hat{V}_2}{\pi d_{\text{out}}^2} = 158.0 \, \frac{\text{m}}{\text{s}}$$

Therefore, the energy balance yields

$$\dot{W}_s = -\dot{M}_2\left(\hat{H}_2 + \frac{v_2^2}{2}\right) - \dot{M}_1\left(\hat{H}_1 + \frac{v_1^2}{2}\right)$$

$$= -2.5 \, \frac{\text{kg}}{\text{s}} \left\{ (\hat{H}_1 - \hat{H}_2) + \frac{1}{2}(v_1^2 - v_2^2) \right\} \frac{\text{kJ}}{\text{kg}}$$

$$= -2.5 \, \frac{\text{kg}}{\text{s}} \left\{ 419.7 \, \frac{\text{kJ}}{\text{kg}} + \frac{1}{2}(127.7^2 - 158.0^2) \, \frac{\text{m}^2}{\text{s}^2} \cdot \frac{1 \, \frac{\text{J}}{\text{kg}}}{\frac{\text{m}^2}{\text{s}^2}} \cdot \frac{1 \text{ kJ}}{1000 \text{ J}} \right\}$$

$$= -2.5 \, \frac{\text{kg}}{\text{s}} \{419.7 - 4.3\} \, \frac{\text{kJ}}{\text{kg}} = -1038.5 \, \frac{\text{kJ}}{\text{s}} \qquad (= -1329 \text{ hp})$$

COMMENT

If we had completely neglected the kinetic energy terms in this calculation, the error in the work term would be 4.3 kJ/kg, or about 1%. Generally, the contribution of kinetic and potential energy terms can be neglected when there is a significant change in the fluid temperature, as was suggested in Sec. 3.2. ∎

ILLUSTRATION 3.4-3
Use of Mass and Energy Balances with an Ideal Gas

A compressed-air tank is to be repressurized to 40 bar by being connected to a high-pressure line containing air at 50 bar and 20°C. The repressurization of the tank occurs so quickly that the process can be assumed to be adiabatic; also, there is no heat transfer from the air to the tank. For this illustration, assume air to be an ideal gas with $C_V^* = 21$ J/(mol K).

   a. If the tank initially contains air at 1 bar and 20°C, what will be the temperature of the air in the tank at the end of the filling process?

b. After a sufficiently long period of time, the gas in the tank is found to be at room temperature (20°C) because of heat exchange with the tank and the atmosphere. What is the new pressure of air in the tank?

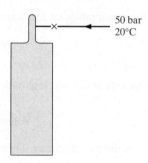

50 bar
20°C

SOLUTION

(Each of these problems contains only a single unknown thermodynamic property, so solutions should be possible.)

a. We will take the contents of the tank to be the system. The difference form of the mass (or rather mole) and energy balances for this open system are

$$N_2 - N_1 = \Delta N \tag{a}$$

$$N_2 \underline{U}_2 - N_1 \underline{U}_1 = (\Delta N) \underline{H}_{\text{in}} \tag{b}$$

In writing the energy balance we have made the following observations:

1. The kinetic and potential energy terms are small and can be neglected.
2. Since the tank is connected to a source of gas at constant temperature and pressure, $\underline{H}_{\text{in}}$ is constant.
3. The initial process is adiabatic, so $Q = 0$, and the system (the contents of the tank) is of constant volume, so $\Delta V = 0$.

Substituting Eq. a in Eq. b, and recognizing that for the ideal gas $\underline{H}(T) = C_P^*(T - T_R)$ and $\underline{U}(T) = C_V^*(T - T_R) - RT_R$, yields

$$N_2\{C_V^*(T_2 - T_R) - RT_R\} - N_1\{C_V^*(T_1 - T_R) - RT_R\} = (N_2 - N_1)C_P^*(T_{\text{in}} - T_R)$$

or

$$N_2 C_V^* T_2 - N_1 C_V^* T_1 = (N_2 - N_1)C_P^* T_{\text{in}}$$

(Note that the reference temperature T_R cancels out of the equation, as it must, since the final result cannot depend on the arbitrarily chosen reference temperature.) Finally, using the ideal gas equation of state to eliminate N_1 and N_2, and recognizing that $V_1 = V_2$, yields

$$\frac{P_2}{T_2} = \frac{P_1}{T_1} + \frac{C_V^*}{C_P^*}\left(\frac{P_2 - P_1}{T_{\text{in}}}\right) \quad \text{or} \quad T_2 = \frac{P_2}{\dfrac{P_1}{T_1} + \dfrac{C_V^*}{C_P^*}\left(\dfrac{P_2 - P_1}{T_{\text{in}}}\right)}$$

The only unknown in this equation is T_2, so, formally, the problem is solved. The answer is $T = 405.2$ K $= 132.05°$C.

Before proceeding to the second part of the problem, it is interesting to consider the case in which the tank is initially evacuated. Here $P_1 = 0$, and so

$$T_2 = \frac{C_P^*}{C_V^*} T_{in}$$

independent of the final pressure. Since C_P is always greater than C_V, the temperature of the gas in the tank at the end of the filling process will be greater than the temperature of gas in the line. Why is this so?

b. To find the pressure in the tank after the heat transfer process, we use the mass balance and the equation of state. Again, choosing the contents of the tank as the system, the mass (mole) balance is $N_2 = N_1$, since there is no transfer of mass into or out of the system during the heat transfer process (unless, of course, the tank is leaking; we do not consider this complication here). Now using the ideal gas equation of state, we have

$$\frac{P_2}{T_2} = \frac{P_1}{T_1} \quad \text{or} \quad P_2 = P_1 \frac{T_2}{T_1}$$

Thus, $P_2 = 28.94$ bar. ∎

ILLUSTRATION 3.4-4

Example of a Thermodynamics Problem That Cannot Be Solved with Only the Mass and Energy Balances[7]

A compressor is a gas pumping device that takes in gas at low pressure and discharges it at a higher pressure. Since this process occurs quickly compared with heat transfer, it is usually assumed to be adiabatic; that is, there is no heat transfer to or from the gas during its compression. Assuming that the inlet to the compressor is air [which we will take to be an ideal gas with $C_P^* = 29.3$ J/(mol K)] at 1 bar and 290 K and that the discharge is at a pressure of 10 bar, estimate the temperature of the exit gas and the rate at which work is done on the gas (i.e., the power requirement) for a gas flow of 2.5 mol/s.

SOLUTION

(Since there are two unknown thermodynamic quantities, the final temperature and the rate at which work is being done, we can anticipate that the mass and energy balances will not be sufficient to solve this problem.)

The system will be taken to be the gas contained in the compressor. The differential form of the molar mass and energy balances for this open system are

$$\frac{dN}{dt} = \dot{N}_1 + \dot{N}_2$$

$$\frac{dU}{dt} = \dot{N}_1 \underline{H}_1 + \dot{N}_2 \underline{H}_2 + \dot{Q} + \dot{W}$$

where we have used the subscript 1 to indicate the flow stream into the compressor and 2 to indicate the flow stream out of the compressor.

Since the compressor operates continuously, the process may be assumed to be in a steady state,

$$\frac{dN}{dt} = 0 \quad \text{or} \quad \dot{N}_1 = -\dot{N}_2$$

$$\frac{dU}{dt} = 0$$

---

[7]We return to this problem in the next chapter after formulating the balance equation for an additional thermodynamic variable, the entropy.

that is, the time variations of the mass of the gas contained in the compressor and of the energy content of this gas are both zero. Also, $\dot{Q} = 0$ since there is no heat transfer to the gas, and $\dot{W} = \dot{W}_s$ since the system boundaries (the compressor) are not changing with time. Thus we have

$$\dot{W}_s = \dot{N}_1 \underline{H}_2 - \dot{N}_1 \underline{H}_1 = \dot{N}_1 C_P^* (T_2 - T_1)$$

or

$$\underline{W}_s = C_P^* (T_2 - T_1)$$

where $\underline{W}_s = \dot{W}_s / \dot{N}_1$ is the work done per mole of gas. Therefore, the power necessary to drive the compressor can be computed once the outlet temperature of the gas is known, or the outlet temperature can be determined if the power input is known.

We are at an impasse; we need more information before a solution can be obtained. It is clear by comparison with the previous examples why we cannot obtain a solution here. In the previous cases, the mass balance and the energy balance, together with the equation of state of the fluid and the problem statement, provided the information necessary to determine the final state of the system. However, here we have a situation where the energy balance contains two unknowns, the final temperature and $\underline{W}_s$. Since neither is specified, we need additional information about the system or process before we can solve the problem. This additional information will be obtained using an additional balance equation developed in the next chapter. ∎

ILLUSTRATION 3.4-5

Use of Mass and Energy Balances to Solve an Ideal Gas Problem[8]

A gas cylinder of 1 m³ volume containing nitrogen initially at a pressure of 40 bar and a temperature of 200 K is connected to another cylinder of 1 m³ volume that is evacuated. A valve between the two cylinders is opened until the pressures in the cylinders equalize. Find the final temperature and pressure in each cylinder if there is no heat flow into or out of the cylinders or between the gas and the cylinder. You may assume that the gas is ideal with a constant-pressure heat capacity of 29.3 J/(mol K).

SOLUTION

This problem is more complicated than the previous ones because we are interested in changes that occur in two separate cylinders. We can try to obtain a solution to this problem in two different ways. First, we could consider each tank to be a separate system, and so obtain two mass balance equations and two energy balance equations, which are coupled by the fact that the mass flow rate and enthalpy of the gas leaving the first cylinder are equal to the like quantities entering the second cylinder.[9] Alternatively, we could obtain an equivalent set of equations by choosing a composite system of the two interconnected gas cylinders to be the first system and the second system to be either one of the cylinders. In this way the first (composite) system is closed and the second system is open. We will use the second system choice here; you are encouraged to explore the first system choice independently and to verify that the same solution is obtained.

The difference form of the mass and energy balance equations (on a molar basis) for the two-cylinder composite system are

$$N_1^i = N_1^f + N_2^f \qquad \text{(a)}$$

_____

[8] An alternative solution to this problem giving the same answer is given in the next chapter.

[9] That the enthalpy of the gas leaving the first cylinder is equal to that entering the second, even though the two cylinders are at different pressures, follows from the fact that the plumbing between the two can be thought of as a flow constriction, as in the Joule-Thomson expansion. Thus the analysis of Illustration 3.2-3 applies to this part of the total process.

and

$$N_1^i \underline{U}_1^i = N_1^f \underline{U}_1^f + N_2^f \underline{U}_2^f \tag{b}$$

Here the subscripts 1 and 2 refer to the cylinders, and the superscripts i and f refer to the initial and final states. In writing the energy balance equation we have recognized that for the system consisting of both cylinders there is no mass flow, heat flow, or change in volume.

Now using, in Eq. a, the ideal gas equation of state written as $N = PV/RT$ and the fact that the volumes of both cylinders are equal yields

$$\frac{P_1^i}{T_1^i} = \frac{P_1^f}{T_1^f} + \frac{P_2^f}{T_2^f} \tag{a'}$$

Using the same observations in Eq. b and further recognizing that for a constant heat capacity gas we have, from Eq. 3.3-8, that

$$\underline{U}(T) = C_V^* T - C_P^* T_R$$

yields

$$\frac{P_1^i}{T_1^i}\{C_V^* T_1^i - C_P^* T_R\} = \frac{P_1^f}{T_1^f}\{C_V^* T_1^f - C_P^* T_R\} + \frac{P_2^f}{T_2^f}\{C_V^* T_2^f - C_P^* T_R\}$$

which, on rearrangement, gives

$$-\left\{\frac{P_1^i}{T_1^i} - \frac{P_1^f}{T_1^f} - \frac{P_2^f}{T_2^f}\right\} C_P^* T_R + C_V^*\{P_1^i - P_1^f - P_2^f\} = 0$$

Since the bracketed quantity in the first term is identically zero (see Eq. a'), we obtain

$$P_1^i = P_1^f + P_2^f \tag{c}$$

(Note that the properties of the reference state have canceled. This is to be expected, since the solution to a change-of-state problem must be independent of the arbitrarily chosen reference state. This is an important point. In nature, the process will result in the same final state independent of our arbitrary choice of reference temperature, T_R. Therefore, if our analysis is correct, T_R must not appear in the final answer.)

Next, we observe that from the problem statement $P_1^f = P_2^f$; thus

$$P_1^f = P_2^f = \tfrac{1}{2}P_1^i = 20 \text{ bar}$$

and from Eq. a'

$$\frac{1}{T_1^f} + \frac{1}{T_2^f} = \frac{2}{T_1^i} \tag{c'}$$

Thus we have one equation for the two unknowns, the two final temperatures. We cannot assume that the final gas temperatures in the two cylinders are the same because nothing in the problem statement indicates that a transfer of heat between the cylinders necessary to equalize the gas temperatures has occurred.

To get the additional information necessary to solve this problem, we write the mass and energy balance equations for the initially filled cylinder. The rate-of-change form of these equations for this system are

$$\frac{dN_1}{dt} = \dot{N} \tag{d}$$

and

$$\frac{d(N_1 \underline{U}_1)}{dt} = \dot{N} \underline{H}_1 \qquad \text{(e)}$$

In writing the energy balance equation, we have made use of the fact that $\dot{Q}$, $\dot{W}_s$, and dV/dt are all zero. Also, we have assumed that while the gas temperature is changing with time, it is spatially uniform within the cylinder, so that at any instant the temperature and pressure of the gas leaving the cylinder are identical with those properties of the gas in the cylinder. Thus, the molar enthalpy of the gas leaving the cylinder is

$$\underline{H} = \underline{H}(T_1, P_1) = \underline{H}_1$$

Since our interest is in the change in temperature of the gas that occurs as its pressure drops from 40 bar to 20 bar due to the escaping gas, you may ask why the balance equations here have been written in the rate-of-change form rather than in terms of the change over a time interval. The answer is that since the properties of the gas within the cylinder (i.e., its temperature and pressure) are changing with time, so is $\underline{H}_1$, the enthalpy of the exiting gas. Thus, if we were to use the form of Eq. e integrated over a time interval (i.e., the difference form of the energy balance equation),

$$N_1^f \underline{U}_1^f - N_1^i \underline{U}_1^i = \int \dot{N} \underline{H}_1 \, dt$$

we would have no way of evaluating the integral on the right side. Consequently, the difference equation provides no useful information for the solution of the problem. However, by starting with Eqs. d and e, it is possible to obtain a solution, as will be evident shortly.

To proceed with the solution, we first combine and rearrange the mass and energy balances to obtain

$$\frac{d(N_1 \underline{U}_1)}{dt} \equiv N_1 \frac{d\underline{U}_1}{dt} + \underline{U}_1 \frac{dN_1}{dt} = \dot{N} \underline{H}_1 = \underline{H}_1 \frac{dN_1}{dt}$$

so that we have

$$N_1 \frac{d\underline{U}_1}{dt} = (\underline{H}_1 - \underline{U}_1)\frac{dN_1}{dt}$$

Now we use the following properties of the ideal gas (see Eqs. 3.3-7 and 3.3-8)

$$N = PV/RT \qquad \underline{H} = C_P^*(T - T_R)$$

and

$$\underline{U} = C_V^*(T - T_R) - RT_R$$

to obtain

$$\frac{P_1 V}{RT_1} C_V^* \frac{dT_1}{dt} = RT_1 \frac{d}{dt}\left(\frac{P_1 V}{RT_1}\right)$$

Simplifying this equation yields

$$\frac{C_V^*}{R} \frac{1}{T_1} \frac{dT_1}{dt} = \frac{T_1}{P_1} \frac{d}{dt}\left(\frac{P_1}{T_1}\right)$$

or

$$\frac{C_V^*}{R} \frac{d \ln T_1}{dt} = \frac{d}{dt} \ln\left(\frac{P_1}{T_1}\right)$$

Now integrating between the initial and final states, we obtain

$$\left(\frac{T_1^f}{T_1^i}\right)^{C_V^*/R} = \left(\frac{P_1^f}{P_1^i}\right)\left(\frac{T_1^i}{T_1^f}\right)$$

or

$$\left(\frac{T_1^f}{T_1^i}\right)^{C_P^*/R} = \left(\frac{P_1^f}{P_1^i}\right) \tag{f}$$

where we have used the fact that for the ideal gas $C_P^* = C_V^* + R$. Equation f provides the means to compute T_1^f, and T_2^f can then be found from Eq. c'. Finally, using the ideal gas equation of state we can compute the final number of moles of gas in each cylinder using the relation

$$N_J^f = \frac{V_{cylJ} P_J^f}{R T_J^f} \tag{g}$$

where the subscript J refers to the cylinder number. The answers are

$$T_1^f = 164.3 \text{ K} \qquad N_1^f = 1.464 \text{ kmol}$$
$$T_2^f = 255.6 \text{ K} \qquad N_2^f = 0.941 \text{ kmol}$$

COMMENTS

The solution of this problem for real fluids is considerably more complicated than for the ideal gas. The starting points are again

$$P_1^f = P_2^f \tag{h}$$

and

$$N_1^f + N_2^f = N_1^i \tag{i}$$

and Eqs. d and e. However, instead of Eq. g we now have

$$N_1^f = \frac{V_{cyl1}}{\underline{V}_1^f} \tag{j}$$

and

$$N_2^f = \frac{V_{cyl2}}{\underline{V}_2^f} \tag{k}$$

where $\underline{V}_1^f$ and $\underline{V}_2^f$ are related to (T_1^f, P_1^f) and (T_2^f, P_2^f), respectively, through the equation of state or tabular PVT data of the form

$$\underline{V}_1^f = \underline{V}_1^f(T_1^f, P_1^f) \tag{l}$$
$$\underline{V}_2^f = \underline{V}_2^f(T_2^f, P_2^f) \tag{m}$$

The energy balance for the two-cylinder composite system is

$$N_1^i \underline{U}_1^i = N_1^f \underline{U}_1^f + N_2^f \underline{U}_2^f \tag{n}$$

Since a thermal equation of state or tabular data of the form $\underline{U} = \underline{U}(T, \underline{V})$ are presumed available, Eq. n introduces no new variables.

Thus we have seven equations among eight unknowns (N_1^f, N_2^f, T_1^f, T_2^f, P_1^f, P_2^f, $\underline{V}_1^f$, and $\underline{V}_2^f$). The final equation needed to solve this problem can, in principle, be obtained by the manipulation and integration of Eq. e, as in the ideal gas case, but now using the real fluid equation of state or tabular data and numerical integration techniques. Since this analysis is difficult, and a simpler method of solution (discussed in Chapter 6) is available, the solution of this problem for the real fluid case is postponed until Sec. 6.5. ■

ILLUSTRATION 3.4-6

Showing That the Change in State Variables between Fixed Initial and Final States Is Independent of the Path Followed

It is possible to go from a given initial equilibrium state of a system to a given final equilibrium state by a number of different paths, involving different intermediate states and different amounts of heat and work. Since the internal energy of a system is a state property, its change between any two states must be independent of the path chosen (see Sec. 1.3). The heat and work flows are, however, path-dependent quantities and can differ on different paths between given initial and final states. This assertion is established here by example. One mole of a gas at a temperature of 25°C and a pressure of 1 bar (the initial state) is to be heated and compressed in a frictionless piston and cylinder to 300°C and 10 bar (the final state). Compute the heat and work required along each of the following paths.

Path A. Isothermal (constant temperature) compression to 10 bar, and then isobaric (constant pressure) heating to 300°C
Path B. Isobaric heating to 300°C followed by isothermal compression to 10 bar
Path C. A compression in which $P\underline{V}^\gamma$ = constant, where $\gamma = C_P^*/C_V^*$, followed by an isobaric cooling or heating, if necessary, to 300°C.

For simplicity, the gas is assumed to be ideal with $C_P^* = 38$ J/(mol K).

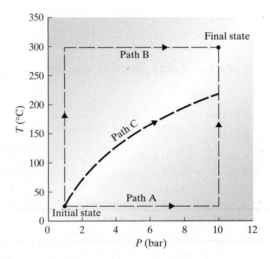

SOLUTION

The 1-mol sample of gas will be taken as the thermodynamic system. The difference form of the mass balance for this closed, deforming volume of gas is

$$N = \text{constant} = 1 \text{ mol}$$

and the difference form of the energy balance is

$$\Delta U = Q - \int P \, dV = Q + W$$

Path A

i. *Isothermal compression*

$$W_{\mathrm{i}} = -\int_{\underline{V}_1}^{\underline{V}_2} P \, dV = -\int_{\underline{V}_1}^{\underline{V}_2} RT \frac{dV}{\underline{V}} = -RT \int_{\underline{V}_1}^{\underline{V}_2} \frac{dV}{\underline{V}} = -RT \ln \frac{\underline{V}_2}{\underline{V}_1} = RT \ln \frac{P_2}{P_1}$$

$$= 8.314 \text{ J/(mol K)} \times 298.15 \text{ K} \times \ln \frac{10}{1} = 5707.7 \text{ J/mol}$$

Since

$$\Delta \underline{U} = \int_{T_1}^{T_2} C_V^* \, dT = C_V^*(T_2 - T_1) \quad \text{and} \quad T_2 = T_1 = 25^\circ\text{C}$$

we have

$$\Delta \underline{U} = 0 \quad \text{and} \quad Q_{\mathrm{i}} = -W_{\mathrm{i}} = -5707.7 \text{ J/mol}$$

ii. *Isobaric heating*

$$W_{\mathrm{ii}} = -\int_{\underline{V}_2}^{\underline{V}_3} P_2 \, dV = -P_2 \int_{\underline{V}_2}^{\underline{V}_3} dV = -P_2(\underline{V}_3 - \underline{V}_2) = -R(T_3 - T_2)$$

$$\Delta \underline{U} = \int_{T_2}^{T_3} C_V^* \, dT = C_V^*(T_3 - T_2)$$

and

$$Q_{\mathrm{ii}} = \Delta \underline{U} - W_{\mathrm{ii}} = C_V^*(T_3 - T_2) + R(T_3 - T_2) = (C_V^* + R)(T_3 - T_2)$$
$$= C_P^*(T_3 - T_2)$$

[This is, in fact, a special case of the general result that at constant pressure for a closed system, $Q = \int C_P^* \, dT$. This is easily proved by starting with

$$\dot{Q} = \frac{d\underline{U}}{dt} + P \frac{d\underline{V}}{dt}$$

and using the fact that P is constant to obtain

$$\dot{Q} = \frac{d\underline{U}}{dt} + \frac{d}{dt}(P\underline{V}) = \frac{d}{dt}(\underline{U} + P\underline{V}) = \frac{d\underline{H}}{dt} = C_P^* \frac{dT}{dt}$$

Now setting $Q = \int \dot{Q} \, dt$ yields $Q = \int C_P^* \, dT$.]
   Therefore,

$$W_{\mathrm{ii}} = -8.314 \text{ J/(mol K)} \times 275 \text{ K} = -2286.3 \text{ J/mol}$$
$$Q_{\mathrm{ii}} = 38 \text{ J/(mol K)} \times 275 \text{ K} = 10\,450 \text{ J/mol}$$
$$Q = Q_{\mathrm{i}} + Q_{\mathrm{ii}} = -5707.7 + 10\,450 = 4742.3 \text{ J/mol}$$
$$W = W_{\mathrm{i}} + W_{\mathrm{ii}} = 5707.7 - 2286.3 = 3421.4 \text{ J/mol}$$

Path B

i. *Isobaric heating*

$$Q_i = C_P^*(T_2 - T_1) = 10\,450 \text{ J/mol}$$
$$W_i = -R(T_2 - T_1) = -2286.3 \text{ J/mol}$$

ii. *Isothermal compression*

$$W_{ii} = RT \ln \frac{P_2}{P_1} = 8.314 \times 573.15 \ln\left(\frac{10}{1}\right) = 10\,972.2 \text{ J/mol}$$

$$Q_{ii} = -W_{ii} = -10\,972.2 \text{ J/mol}$$
$$Q = 10\,450 - 10\,972.2 = -522.2 \text{ J/mol}$$
$$W = -2286.3 + 10\,972.2 = 8685.9 \text{ J/mol}$$

Path C

i. *Compression with $P\underline{V}^\gamma = constant$*

$$W_i = -\int_{\underline{V}_1}^{\underline{V}_2} P \, d\underline{V} = -\int_{\underline{V}_1}^{\underline{V}_2} \frac{constant}{\underline{V}^\gamma} \, d\underline{V} = -\frac{constant}{1-\gamma}(\underline{V}_2^{1-\gamma} - \underline{V}_1^{1-\gamma})$$

$$= -\frac{1}{1-\gamma}(P_2\underline{V}_2 - P_1\underline{V}_1) = \frac{-R(T_2 - T_1)}{1-\gamma} = \frac{-R(T_2 - T_1)}{1 \quad (C_P^*/C_V^*)} = C_V^*(T_2 - T_1)$$

where T_2 can be computed from

$$P_1\underline{V}_1^\gamma = P_1\left(\frac{RT_1}{P_1}\right)^\gamma = P_2\underline{V}_2^\gamma = P_2\left(\frac{RT_2}{P_2}\right)^\gamma$$

or

$$\frac{T_2}{T_1} = \left(\frac{P_2}{P_1}\right)^{(\gamma-1)/\gamma}$$

Now

$$\gamma = \frac{C_P^*}{C_V^*} = \frac{38}{38 - 8.314} = 1.280$$

so that

$$T_2 = 298.15 \text{ K } (10)^{0.280/1.280} = 493.38 \text{ K}$$

and

$$W_i = C_V^*(T_2 - T_1) = (38 - 8.314) \text{ J/(mol K)} \times (493.38 - 298.15) \text{ K}$$
$$= 5795.6 \text{ J/mol}$$
$$\Delta U_i = C_V^*(T_2 - T_1) = 5795.6 \text{ J/mol}$$
$$Q_i = \Delta U_i - W_i = 0$$

ii. *Isobaric heating*

$$Q_{ii} = C_P^*(T_3 - T_2) = 38 \text{ J/(mol K)} \times (573.15 - 493.38) \text{ K} = 3031.3 \text{ J/mol}$$
$$W_{ii} = -R(T_3 - T_2) = -8.314 \text{ J/(mol K)} \times (573.15 - 493.38) \text{ K} = -663.2 \text{ J/mol}$$

and

$$Q = 0 + 3031.3 = 3031.3 \text{ J/mol}$$
$$W = 5795.6 - 663.2 = 5132.4 \text{ J/mol}$$

SOLUTION

Path	Q (J/mol)	W (J/mol)	Q + W = ΔU(J/mol)
A	4742.3	3421.4	8163.7
B	−522.2	8685.9	8163.7
C	3031.3	5132.4	8163.7

COMMENT

Notice that along each of the three paths considered (and, in fact, any other path between the initial and final states), the sum of Q and W, which is equal to ΔU, is 8163.7 J/mol, even though Q and W separately are different along the different paths. This illustrates that whereas the internal energy is a state property and is path independent (i.e., its change in going from state 1 to state 2 depends only on these states and not on the path between them), the heat and work flows depend on the path and are therefore path functions. ■

ILLUSTRATION 3.4-7

Showing That More Work Is Obtained If a Process Occurs without Friction

An initial pressure of 2.043 bar is maintained on 1 mol of air contained in a piston-and-cylinder system by a set of weights $\mathcal{W}$, the weight of the piston, and the surrounding atmosphere. Work is obtained by removing some of the weights and allowing the air to isothermally expand at 25°C, thus lifting the piston and the remaining weights. The process is repeated until all the weights have been removed. The piston has a mass of $\omega = 5$ kg and an area of 0.01 m^2. For simplicity, the air can be considered to be an ideal gas. Assume that as a result of sliding friction between the piston and the cylinder wall, all oscillatory motions of the piston after the removal of a weight will eventually be damped.

Compute the work obtained from the isothermal expansion and the heat required from external sources for each of the following:

    a. The weight $\mathcal{W}$ is taken off in one step.
    b. The weight is taken off in two steps, with $\mathcal{W}/2$ removed each time.
    c. The weight is taken off in four steps, with $\mathcal{W}/4$ removed each time.
    d. The weight is replaced by a pile of sand (of total weight $\mathcal{W}$), and the grains of sand are removed one at a time.

Processes b and d are illustrated in the following figure.[10]

SOLUTION

I. *Analysis of the problem.* Choosing the air in the cylinder to be the system, recognizing that for an ideal gas at constant temperature U is constant so that $\Delta U = 0$, and neglecting the kinetic and potential energy terms for the gas (since the mass of 1 mol of air is only 29 g), we obtain the following energy balance equation:

$$0 = Q - \int P \, dV \tag{a}$$

[10]From H. C. Van Ness, *Understanding Thermodynamics,* McGraw-Hill, New York, 1969. Used with permission of the McGraw-Hill Book Co.

The total work done by the gas in lifting and accelerating the piston and the weights against the frictional forces, and in expanding the system volume against atmospheric pressure is contained in the $-\int P\, dV$ term. To see this we recognize that the laws of classical mechanics apply to the piston and weights, and equate, at each instant, all the forces on the piston and weights to their acceleration,

$$\text{Forces on piston and weights} = \begin{pmatrix} \text{Mass of piston} \\ \text{and weights} \end{pmatrix} \times \text{Acceleration}$$

and obtain

$$[P \times A - P_{\text{atm}} \times A - (\mathcal{W} + \omega)g + F_{\text{fr}}] = (\mathcal{W} + \omega)\frac{dv}{dt} \tag{b}$$

Here we have taken the vertical upward $(+z)$ direction as being positive and used P and P_{atm} to represent the pressure of the gas and atmosphere, respectively; A the piston area; ω its mass; $\mathcal{W}$ the mass of the weights on the piston at any time; v the piston velocity; and F_{fr} the frictional force, which is proportional to the piston velocity. Recognizing that the piston velocity v is equal to the rate of change of the piston height h or the gas volume V, we have

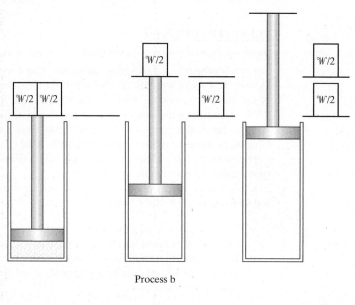

Process b

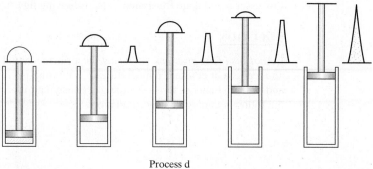

Process d

$$v = \frac{dh}{dt} = \frac{1}{A}\frac{dV}{dt}$$

Also we can solve Eq. b for the gas pressure:

$$P = P_{atm} + \frac{(W+\omega)}{A}g - \frac{F_{fr}}{A} + \frac{(W+\omega)}{A}\frac{dv}{dt} \tag{c}$$

At mechanical and thermodynamic equilibrium (i.e., when $dv/dt = 0$ and $v = 0$), we have

$$P = P_{atm} + \frac{(W+\omega)}{A}g \tag{d}$$

With these results, the total work done by the gas can be computed. In particular,

$$\int P\,dV = \int \left[P_{atm} + \left(\frac{W+\omega}{A}\right)g - \frac{F_{fr}}{A} + \frac{(W+\omega)}{A}\frac{dv}{dt} \right] dV$$

$$= \left[P_{atm} + \frac{(W+\omega)}{A}g \right] \Delta V - \frac{1}{A}\int F_{fr}\,dV + \frac{(W+\omega)}{A}\int \frac{dv}{dt}\,dV$$

This equation can be simplified by rewriting the last integral as follows:

$$\frac{1}{A}\int \frac{dv}{dt}\,dV = \frac{1}{A}\int \frac{dv}{dt}\frac{dV}{dt}\,dt = \int \frac{dv}{dt}v\,dt = \frac{1}{2}\int \frac{dv^2}{dt}\,dt = \Delta\left(\frac{1}{2}v^2\right)$$

where the symbol Δ indicates the change between the initial and final states. Next, we recall from mechanics that the force due to sliding friction, here F_{fr}, is in the direction opposite to the relative velocity of the moving surfaces and can be written as

$$F_{fr} = -k_{fr}v$$

where k_{fr} is the coefficient of sliding friction. Thus, the remaining integral can be written as

$$\frac{1}{A}\int F_{fr}\,dV = -\frac{1}{A}\int k_{fr}v\frac{dV}{dt}\,dt = -k_{fr}\int v^2\,dt$$

The energy balance, Eq. a, then becomes

$$Q = \int P\,dV = P_{atm}\,\Delta V + (W+\omega)g\,\Delta h + k_{fr}\int v^2\,dt + (W+\omega)\,\Delta\left(\tfrac{1}{2}v^2\right) \tag{e}$$

This equation relates the heat flow into the gas (to keep its temperature constant) to the work the gas does against the atmosphere in lifting the piston and weights (hence increasing their potential energy) against friction and in accelerating the piston and weights (thus increasing their kinetic energy).

The work done against frictional forces is dissipated into thermal energy, resulting in a higher temperature at the piston and cylinder wall. This thermal energy is then absorbed by the gas and appears as part of Q. Consequently, the net flow of heat from a temperature bath to the gas is

$$Q^{NET} = Q - k_{fr}\int v^2\,dt \tag{f}$$

Since heat will be transferred to the gas, and the integral is always positive, this equation establishes that less heat will be needed to keep the gas at a constant temperature if the expansion occurs with friction than in a frictionless process.

Also, since the expansion occurs isothermally, the total heat flow to the gas is, from Eq. a,

$$Q = \int_{V_i}^{V_f} P\,dV = \int_{V_i}^{V_f} \frac{NRT}{V}\,dV = NRT\ln\frac{V_f}{V_i} \tag{g}$$

Combining Eqs. e and f, and recognizing that our interest here will be in computing the heat and work flows between states for which the piston has come to rest ($v = 0$), yields

$$Q^{\text{NET}} = Q - k_{\text{fr}} \int v^2 \, dt = P_{\text{atm}} \, \Delta V + (\mathcal{W} + \omega) g \, \Delta h = P \Delta V = -W^{\text{NET}} \qquad \textbf{(h)}$$

where P is the equilibrium final pressure given by Eq. d. Also from Eqs. f, g, and h, we have

$$Q^{\text{NET}} = -W^{\text{NET}} = NRT \ln \left(\frac{V_f}{V_i} \right) - k_{\text{fr}} \int v^2 \, dt \qquad \textbf{(i)}$$

Here W^{NET} represents the net work obtained by the expansion of the gas (i.e., the work obtained in raising the piston and weights and in doing work against the atmosphere).

The foregoing equations can now be used in the solution of the problem. In particular, as a weight is removed, the new equilibrium gas pressure is computed from Eq. d, the resulting volume change from the ideal gas law, W^{NET} and Q^{NET} from Eq. h, and the work against friction from Eq. i. There is, however, one point that should be mentioned before we proceed with this calculation. If there were no mechanism for the dissipation of kinetic energy to thermal energy (that is, sliding friction between the piston and cylinder wall, and possibly also viscous dissipation on expansion and compression of the gas due to its bulk viscosity), then when a weight was removed the piston would be put into a perpetual oscillatory motion. The presence of a dissipative mechanism will damp the oscillatory motion. (As will be seen, the value of the coefficient of sliding friction, k_{fr}, does not affect the amount of kinetic energy ultimately dissipated as heat. Its value does, however, affect the dynamics of the system and thus determine how quickly the oscillatory motion is damped.)

II. *The numerical solution.* First, the mass $\mathcal{W}$ of the weights is computed using Eq. d and the fact that the initial pressure is 2.043 bar. Thus,

$$P = 2.043 \text{ bar} = 1.013 \text{ bar} + \frac{(5 + \mathcal{W}) \text{ kg}}{0.01 \text{ m}^2} \times 9.807 \, \frac{\text{m}}{\text{s}^2} \times \frac{1 \text{ Pa}}{\text{kg}/(\text{m s}^2)} \times \frac{1 \text{ bar}}{10^5 \text{ Pa}}$$

or

$$(5 + \mathcal{W}) \text{ kg} = 105.0 \text{ kg}$$

so that $\mathcal{W} = 100$ kg.

The ideal gas equation of state for 1 mol of air at $25°$C is

$$PV = NRT = 1 \text{ mol} \times 8.314 \times 10^{-5} \, \frac{\text{bar m}^3}{\text{mol K}} \times (25 + 273.15) \text{ K} \qquad \textbf{(j)}$$

$$= 2.479 \times 10^{-2} \text{ bar m}^3 = 2479 \text{ J}$$

and the initial volume of the gas is

$$V = \frac{2.479 \times 10^{-2} \text{ bar m}^3}{2.043 \text{ bar}} = 1.213 \times 10^{-2} \text{ m}^3$$

Process a

The 100-kg weight is removed. The equilibrium pressure of the gas (after the piston has stopped oscillating) is

$$P_1 = 1.013 \text{ bar} + \frac{5 \text{ kg} \times 9.807 \text{ m/s}^2}{0.01 \text{ m}^2} \times \frac{10^{-5} \text{ bar}}{\text{kg}/(\text{m s}^2)} = 1.062 \text{ bar}$$

and the gas volume is

$$V_1 = \frac{2.479 \times 10^{-2} \text{ bar m}^3}{1.062 \text{ bar}} = 2.334 \times 10^{-2} \text{ m}^3$$

Thus

$$\Delta V = (2.334 - 1.213) \times 10^{-2} \text{ m}^3 = 1.121 \times 10^{-2} \text{ m}^3$$
$$- W^{\text{NET}} = 1.062 \text{ bar} \times 1.121 \times 10^{-2} \text{ m}^3 \times 10^5 \frac{\text{J}}{\text{bar m}^3}$$
$$= 1190.5 \text{ J} = Q^{\text{NET}}$$

and

$$Q = NRT \ln \frac{V_1}{V_0} = 2479 \text{ J} \ln \frac{2.334 \times 10^{-2}}{1.213 \times 10^{-2}} = 1622.5 \text{ J}$$

(since $NRT = 2479$ J from Eq. j). Consequently, the work done against frictional forces (and converted to thermal energy), which we denote by W_{fr}, is

$$-W_{\text{fr}} = Q - Q^{\text{NET}} = (1622.5 - 1190.5) \text{ J} = 432 \text{ J}$$

The total useful work obtained, the net heat supplied, and the work against frictional forces are given in Table 1. Also, the net work, $P\,\Delta V$, is shown as the shaded area in the accompanying figure, together with the line representing the isothermal equation of state, Eq. j. Note that in this case the net work is that of raising the piston and pushing back the atmosphere.

Table 1

Process	$-W^{\text{NET}} = Q^{\text{NET}}$ (J)	Q (J)	$-W_{\text{fr}}$ (J)
a	1190.5	1622.5	432.0
b	1378.7	1622.5	243.8
c	1493.0	1622.5	129.5
d	1622.5	1622.5	0

Process b

The situation here is similar to that of process a, except that the weight is removed in two 50-kg increments. The pressure, volume, work, and heat flows for each step of the process are given in Table 2, and $-W_i^{\text{NET}} = P_i(\Delta V)_i$, the net work for each step, is given in the figure.

Process c

Here the weight is removed in four 25-kg increments. The pressure, volume, work, and heat flows for each step are given in Table 2 and summarized in Table 1. Also, the net work for each step is given in the figure.

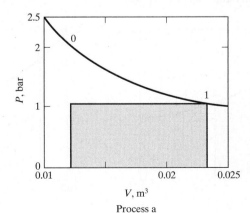

Process a

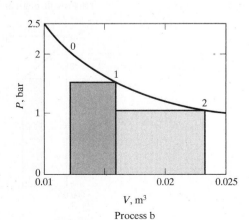

Process b

Table 2

Process b

Stage	P (bar)	V (m³)	$-W_i^{NET} = P_i(\Delta V)_i$ (J)	$Q = NRT \ln \dfrac{V_i}{V_{i-1}}$ (J)	$-W_{fr}$ (J)
0	2.043	1.213×10^{-2}			
1	1.552	1.597×10^{-2}	596.0	681.8	85.8
2	1.062	2.334×10^{-2}	782.7	940.7	158.0
Total			1378.7	1622.5	243.8

Process c

Stage	P (bar)	V (m³)	$-W_i^{NET} = P_i(\Delta V)_i$ (J)	$Q = NRT \ln \dfrac{V_i}{V_{i-1}}$ (J)	$-W_{fr}$ (J)
0	2.043	1.213×10^{-2}			
1	1.798	1.379×10^{-2}	298.5	318.0	19.5
2	1.552	1.597×10^{-2}	338.3	363.8	25.5
3	1.307	1.897×10^{-2}	392.1	426.8	34.7
4	1.062	2.334×10^{-2}	464.1	513.9	49.8
Total			1493.0	1622.5	129.5

Process d

The computation here is somewhat more difficult since the number of stages to the calculation is almost infinite. However, recognizing that in the limit of the mass of a grain of sand going to zero there is only a differential change in the pressure and volume of the gas and negligible velocity or acceleration of the piston, we have

$$-W^{NET} = \sum_i P_i(\Delta V_i) \rightarrow \int P\,dV = NRT \ln \frac{V_f}{V_i} = Q^{NET}$$

and $W_{fr} = 0$, since the piston velocity is essentially zero at all times. Thus,

$$-W^{NET} = Q^{NET} = Q = 1\ \text{mol} \times 8.314\ \frac{\text{J}}{\text{mol K}} \times 298.15\ \text{K} \times \ln \frac{2.334 \times 10^{-2}}{1.213 \times 10^{-2}}$$
$$= 1622.5\ \text{J}$$

This result is given in Table 1 and the figure.

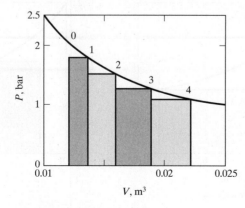

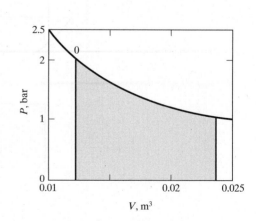

COMMENTS

Several points are worth noting about this illustration. First, although the initial and final states of the gas are the same in all three processes, the useful or net work obtained and the net heat required differ. Of course, by the energy conservation principle, it is true that $-W^{\mathrm{NET}} = Q^{\mathrm{NET}}$ for each process. It is important to note that the most useful work is obtained for a given change of state if the change is carried out in differential steps, so that there is no frictional dissipation of mechanical energy to thermal energy (compare process d with processes a, b, and c). Also, in this case, if we were to reverse the process and compress the gas, it would be found that the minimum work required for the compression is obtained when weights are added to the piston in differential (rather than finite) steps. (See Problem 3.29.)

It should also be pointed out that in each of the four processes considered here the gas did 1622.5 J of work on its surroundings (the piston, the weights, and the atmosphere) and absorbed 1622.5 J of heat (from the thermostatic bath maintaining the system temperature constant and from the piston and cylinder as a result of their increased temperature due to frictional heating). We can see this from Table 1, since $-(W^{\mathrm{NET}} + W_{\mathrm{fr}}) = Q = 1622.5$ J for all four processes. However, the fraction of the total work of the gas obtained as useful work versus work against friction varies among the different processes. ∎

At first glance it might appear that in the process in illustration 3.4.7 the proscription in Chapter 1 that thermal energy (here heat) cannot be completely converted to mechanical energy (here work) has been violated. However, that statement included the requirement that such a complete conversion was not possible in a cyclic process or in a process in which there were no changes in the universe (system plus surroundings). In the illustration here heat has been completely converted to work, but this can only be done once since the system, the gas contained within the cylinder, has not been restored to its initial pressure. Consequently, we cannot continue to add heat and extract the same amount of energy as work.

ILLUSTRATION 3.4-8

Computing the Pressure, and Heat and Work Flows in a Complicated Ideal Gas Problem

Consider the cylinder filled with air and with the piston and spring arrangement shown below. The external atmospheric pressure is 1 bar, the initial temperature of the air is 25°C, the no-load length of the spring is 50 cm, the spring constant is 40 000 N/m, the frictionless piston weighs 500 kg, and the constant-volume heat capacity of air can be taken to be constant at 20.3 J/(mol K). Assume air is an ideal gas. Compute

a. The initial pressure of the gas in the cylinder
b. How much heat must be added to the gas in the cylinder for the spring to compress 2 cm

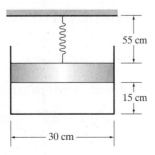

SOLUTION

a. The pressure of the gas inside the cylinder is a result of the atmospheric pressure (1 bar), the force exerted on the gas as a result of the weight of the piston, and the force of the spring. The contribution from the piston is

$$P_{\text{Piston}} = \frac{F}{A} = \frac{M \text{ kg} \times 9.807 \, \frac{\text{m}}{\text{s}^2} \times \frac{1 \text{ Pa}}{\text{kg}/(\text{m s}^2)} \times \frac{1 \text{ bar}}{10^5 \text{ Pa}}}{\pi(d^2/4) \text{ m}^2}$$

$$= \frac{M}{d^2} \times 1.2487 \times 10^{-4} \text{ bar} = \frac{500}{(0.3)^2} \times 1.2487 \times 10^{-4} \text{ bar} = 0.6937 \text{ bar}$$

The contribution due to the spring is

$$P_{\text{Spring}} = \frac{F}{A} = \frac{-k(x - x_0)}{\pi(d^2/4)} =$$

$$= \frac{-40\,000 \, \frac{\text{N}}{\text{m}} \times (55 - 50) \times 10^{-2} \text{ m}}{\pi[(0.3)^2/4] \text{ m}^2} \times 1 \frac{\text{m kg}}{\text{s}^2 \text{ N}} \times 1 \frac{\text{Pa}}{\text{kg}/(\text{m s}^2)} \times \frac{1 \text{ bar}}{10^5 \text{ Pa}}$$

$$= -0.2829 \text{ bar}$$

Therefore, the pressure of the air in the cylinder is

$$P = P_{\text{Atmosphere}} + P_{\text{Piston}} + P_{\text{Spring}}$$
$$= 1 + 0.6937 - 0.2829 \text{ bar} = 1.4108 \text{ bar}$$

For later reference we note that the number of moles of gas in the cylinder is, from the ideal gas law,

$$N = \frac{PV}{RT} = \frac{1.4108 \text{ bar} \times \frac{\pi}{4}(0.3)^2(0.15) \text{ m}^3}{298.15 \text{ K} \times 8.314 \times 10^{-5} \frac{\text{bar m}^3}{\text{mol K}}}$$

$$= 0.6035 \text{ mol} \quad \text{which is equal to } 17.5 \text{ g} = 0.0175 \text{ kg}$$

b. The contribution to the total pressure due to the spring after heating is computed as above, except that now the spring extension is 53 cm (55 cm − 2 cm). Thus

$$P_{\text{Spring}} = \frac{-40\,000 \, \frac{\text{N}}{\text{m}} \times (53 - 50) \times 10^{-2} \text{ m}}{\pi \dfrac{(0.3)^2 \text{ m}^2}{4}} = -0.1697 \text{ bar}$$

So the final pressure of the air in the cylinder is

$$P = 1 + 0.6937 - 0.1697 \text{ bar} = 1.524 \text{ bar}$$

[Note that the pressure at any extension of the spring, x, is

$$P = 1.6937 - 5.658(x - 0.50) \text{ bar} \quad \text{where } x \text{ is the extension of spring (m)}]$$

The volume change on expansion of the gas (to raise the piston by 2 cm) is

$$\Delta V = \frac{\pi}{4}(0.3)^2 \text{ m}^2 \times 0.02 \text{ m} = 1.414 \times 10^{-3} \text{ m}^3$$

Also, the final temperature of the gas, again from the ideal gas law, is

$$T = \frac{PV}{NR} = \frac{1.524 \text{ bar} \times \frac{\pi}{4} 0.3^2 \text{ m}^2 \times 0.17 \text{ m}}{0.6035 \text{ mol} \times 8.314 \times 10^{-5} \frac{\text{bar m}^3}{\text{mol K}}} = 365.0 \text{ K}$$

The work done can be computed in two ways, as shown below. For simplicity, we first compute the individual contributions, and then see how these terms are combined.

Work done by the gas against the atmosphere is

$$W_{\text{Atmosphere}} = -P_{\text{Atmosphere}} \times \Delta V = -1 \text{ bar} \times 1.414 \times 10^{-3} \text{ m}^3 \times 10^5 \frac{\text{J}}{\text{bar m}^3} = -141.4 \text{ J}$$

(The minus sign indicates that the gas did work on the surrounding atmosphere.)

Work done by the gas in compressing the spring is

$$W_{\text{Spring}} = -\frac{k}{2}[(x_2 - x_0)^2 - (x_1 - x_0)^2]$$

$$= -\frac{40\,000 \text{ N}}{2 \text{ m}}[3^2 \times 10^{-4} \text{ m}^2 - 5^2 \times 10^{-4} \text{ m}^2]$$

$$= -2 \times (9 - 25) = 32 \text{ N m} \times 1 \frac{\text{J}}{\text{N m}} = 32 \text{ J}$$

Work done by the gas in raising the 500-kg piston is

$$W_{\text{Piston}} = -500 \text{ kg} \times 0.02 \text{ m} \times 9.807 \frac{\text{m}}{\text{s}^2} \times \frac{1 \text{ J}}{\text{m}^2 \text{ kg/s}^2} = -98.07 \text{ J}$$

Work done by the gas in raising its center of mass by 1 cm (why does the center of mass of the gas increase by only 1 cm if the piston is raised 2 cm?) is

$$W_{\text{gas}} = -17.5 \text{ g} \times 0.01 \text{ m} \times 9.807 \frac{\text{m}}{\text{s}^2} \times \frac{1 \text{ J}}{\text{m}^2 \text{ kg/s}^2} \times \frac{1 \text{ kg}}{1000 \text{ g}} = -0.0017 \text{ J}$$

which is negligible compared with the other work terms. If we consider the gas in the cylinder to be the system, the work done by the gas is

$W =$ Work to raise piston + Work against atmosphere + Work to compress spring
$ = -98.1 \text{ J} - 141.4 \text{ J} + 32.0 \text{ J} = -207.5 \text{ J}$

[Here we have recognized that the gas is doing work by raising the piston and by expansion against the atmosphere, but since the spring is extended beyond its no-load point, it is doing work on the gas as it contracts. (The opposite would be true if the spring were initially compressed to a distance less than its no-load point.)]

An alternative method of computing this work is as follows. The work done by the gas on expanding (considering only the gas in the system) is

$$W = -\int P \, dV = -A \int_{0.15 \text{ m}}^{0.17 \text{ m}} P \, dy = -A \int_{0.15 \text{ m}}^{0.17 \text{ m}} [1.6937 - 5.658(x - 0.5)] \, dy$$

where y is the height of the bottom of the piston at any time. The difficulty with the integral above is that two different coordinate systems are involved, since y is the height of the piston and x is the extension of the spring. Therefore, we need to make a coordinate transformation. To do this we note that when $y = 0.15$, $x = 0.55$, and when $y = 0.17$, $x = 0.53$. Consequently, $x = 0.7 - y$ and

$$W = -A \int_{0.15 \text{ m}}^{0.17 \text{ m}} [1.6937 - 5.658(0.7 - y - 0.5)]\, dy$$

$$= -A \left[1.6937 \times 0.02 - 5.658 \int_{0.15 \text{ m}}^{0.17 \text{ m}} (0.2 - y)\, dy \right]$$

$$= -A \left[1.6937 \times 0.02 - 5.658 \left(0.2 \times 0.02 - \tfrac{1}{2}(0.17^2 - 0.15^2) \right) \right]$$

$$= -7.068 \times 10^{-2}[0.03387 - 0.02263 + 0.01811]$$

$$= -0.2075 \times 10^{-2} \text{ bar m}^3 \times 10^5 \frac{\text{J}}{\text{bar m}^3} = -207.5 \text{ J}$$

This is identical to the result obtained earlier.

Finally, we can compute the heat that must be added to raise the piston. The difference form of the energy balance on the closed system consisting of the gas is

$$U(\text{final state}) - U(\text{initial state}) = Q + W = NC_V(T_f - T_i)$$

so that

$$Q = NC_V(T_f - T_i) - W$$

$$= 0.6035 \text{ mol} \times 20.3 \frac{\text{J}}{\text{mol K}} \times (365 - 298.15) \text{ K} + 207.5 \text{ J}$$

$$= 819.0 \text{ J} + 207.5 \text{ J} = 1026.5 \text{ J}$$

Therefore, to accomplish the desired change, 1026.5 J of heat must be added to the gas. Of this amount, 819 J are used to heat the gas, 98 J to raise the 500-kg piston, and 141.4 J to push back the atmosphere; during the process, the spring supplies 32 J.   ■

3.5   CONSERVATION OF MOMENTUM

Based on the discussion of Sec. 3.4 and Illustration 3.4-4, we can conclude that the equations of mass and energy conservation are not sufficient to obtain the solution to all the problems of thermodynamics in which we might be interested. What is needed is a balance equation for an additional thermodynamic state variable. The one conservation principle that has not yet been used is the conservation of momentum. If, in Eq. 2.1-4, θ is taken to be the momentum of a black-box system, we have

$$\frac{d}{dt}(M\mathbf{v}) = \begin{pmatrix} \text{Rate at which momentum} \\ \text{enters system by all} \\ \text{mechanisms} \end{pmatrix} - \begin{pmatrix} \text{Rate at which momentum} \\ \text{leaves system by all} \\ \text{mechanisms} \end{pmatrix} \quad \textbf{(3.5-1)}$$

where $\mathbf{v}$ is the center-of-mass velocity vector of the system, and M its total mass. We could now continue the derivation by evaluating all the momentum flows; however, it is clear by looking at the left side of this equation that we will get an equation for the rate of change of the center-of-mass velocity of the system, *not* an equation of change for a thermodynamic state variable. Consequently, the conservation-of-momentum equation will not lead to the additional balance equation we need, so this derivation will not be completed. The development of an additional, useful balance equation is not a straightforward task, and will be delayed until Chapter 4.

3.6   THE MICROSCOPIC ENERGY BALANCE (OPTIONAL)

This section appears on the CD that accompanies this text.

PROBLEMS

3.1 a. A bicyclist is traveling at 20 km/hr when he encounters a hill 70 m in height. Neglecting road and wind resistance (a poor assumption), what is the maximum vertical elevation gain the bicyclist could achieve without pedaling?

b. If the hill is a down hill, what speed would the bicyclist achieve without pedaling, again neglecting road and wind resistance?

3.2 Water in the Niagara River approaches the falls at a velocity of 8 km/hr. Niagara Falls is approximately 55 m high. Neglecting air resistance, estimate the velocity of the water at the bottom of the falls.

3.3 a. One kilogram of steam contained in a horizontal frictionless piston and cylinder is heated at a constant pressure of 1.013 bar from 125°C to such a temperature that its volume doubles. Calculate the amount of heat that must be added to accomplish this change, the final temperature of the steam, the work the steam does against its surroundings, and the internal energy and enthalpy changes of the steam for this process.

b. Repeat the calculation of part (a) if the heating occurs at a constant volume to a pressure that is twice the initial pressure.

c. Repeat the calculation of part (a) assuming that steam is an ideal gas with a constant-pressure heat capacity of 34.4 J/mol K.

d. Repeat the calculation of part (b) assuming steam is an ideal gas as in part (c).

3.4 In Joule's experiments, the slow lowering of a weight (through a pulley and cable arrangement) turned a stirrer in an insulated container of water. As a result of viscosity, the kinetic energy transferred from the stirrer to the water eventually dissipated. In this process the potential energy of the weight was first converted to kinetic energy of the stirrer and the water, and then as a result of viscous forces, the kinetic energy of the water was converted to thermal energy apparent as a rise in temperature. Assuming no friction in the pulleys and no heat losses, how large a temperature rise would be found in 1 kg of water as a result of a 1-kg weight being lowered 1 m?

3.5 Steam at 500 bar and 600°C is to undergo a Joule-Thomson expansion to atmospheric pressure. What will be the temperature of the steam after the expansion? What would be the downstream temperature if the steam were replaced by an ideal gas?

3.6 Water in an open metal drum is to be heated from room temperature (25°C) to 80°C by adding steam slowly enough that all the steam condenses. The drum initially contains 100 kg of water, and steam is supplied at 3.0 bar and 300°C. How many kilograms of steam should be added so that the final temperature of the water in the tank is exactly 80°C? Neglect all heat losses from the water in this calculation.

3.7 Consider the following statement: "The adiabatic work necessary to cause a given change of state in a closed system is independent of the path by which that change occurs."

a. Is this statement true or false? Why? Does this statement contradict Illustration 3.4-6, which establishes the path dependence of work?

b. Show that if the statement is false, it would be possible to construct a machine that would generate energy.

3.8 A nonconducting tank of negligible heat capacity and 1 m³ volume is connected to a pipeline containing steam at 5 bar and 370°C, filled with steam to a pressure of 5 bar, and disconnected from the pipeline.

a. If the tank is initially evacuated, how much steam is in the tank at the end of the filling process, and what is its temperature?

b. If the tank initially contains steam at 1 bar and 150°C, how much steam is in the tank at the end of the filling process, and what is its temperature?

3.9 a. A 1-kg iron block is to be accelerated through a process that supplies it with 1 kJ of energy. Assuming all this energy appears as kinetic energy, what is the final velocity of the block?

b. If the heat capacity of iron is 25.10 J/(mol K) and the molecular weight of iron is 55.85, how large a temperature rise would result from 1 kJ of energy supplied as heat?

3.10 The voltage drop across an electrical resistor is 10 volts and the current through it is 1 ampere. The total heat capacity of the resistor is 20 J/K, and heat is dissipated from the resistor to the surrounding air according to the relation

$$\dot{Q} = -h(T - T_{am})$$

where T_{am} is the ambient air temperature, 25°C; T is the temperature of the resistor; and h, the heat transfer coefficient, is equal to 0.2 J/(K s). Compute the steady-state temperature of the resistor, that is, the temperature of the resistor when the energy loss from the resistor is just equal to the electrical energy input.

3.11 The frictionless piston-and-cylinder system shown here is subjected to 1.013 bar external pressure. The piston mass is 200 kg, it has an area of 0.15 m², and the initial volume of the entrapped ideal gas is 0.12 m³. The piston and cylinder do not conduct heat, but heat can be added to the gas by a heating coil. The gas has a constant-volume heat capacity of 30.1 J/(mol K) and an initial temperature of 298 K, and 10.5 kJ of energy are to be supplied to the gas through the heating coil.

a. If stops placed at the initial equilibrium position of the piston prevent it from rising, what will be the final temperature and pressure of the gas?

b. If the piston is allowed to move freely, what will be the final temperature and volume of the gas?

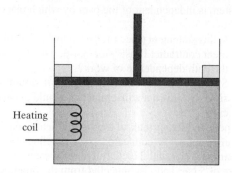

3.12 As an energy conservation measure in a chemical plant, a 40-m³ tank will be used for temporary storage of exhaust process steam. This steam is then used in a later stage of the processing. The storage tank is well insulated and initially contains 0.02 m³ of liquid water at 50°C; the remainder of the tank contains water vapor in equilibrium with this liquid. Process steam at 1.013 bar and 90 percent quality enters the storage tank until the pressure in the tank is 1.013 bar. How many kilograms of wet steam enter the tank during the filling process, and how much liquid water is present at the end of the process? Assume that there is no heat transfer between the steam or water and the tank walls.

3.13 The mixing tank shown here initially contains 50 kg of water at 25°C. Suddenly the two inlet valves and the single outlet valve are opened, so that two water streams, each with a flow rate of 5 kg/min, flow into the tank, and a single exit stream with a flow rate of 10 kg/min leaves the tank. The temperature of one inlet stream is 80°C, and that of the other is 50°C. The tank is well mixed, so that the temperature of the outlet stream is always the same as the temperature of the water in the tank.

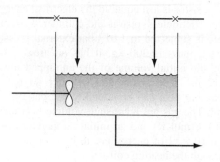

a. Compute the steady-state temperature that will finally be obtained in the tank.

b. Develop an expression for the temperature of the fluid in the tank at any time.

3.14 A well-insulated storage tank of 60 m³ contains 200 L of liquid water at 75°C. The rest of the tank contains steam in equilibrium with the water. Spent process steam at 2 bar and 90 percent quality enters the storage tank until the pressure in the tank reaches 2 bar. Assuming that the heat losses from the system to the tank and the environment are negligible, calculate the total amount of steam that enters the tank during the filling process and the fraction of liquid water present at the end of the process.

3.15 An isolated chamber with rigid walls is divided into two equal compartments, one containing gas and the other evacuated. The partition between the compartments ruptures. After the passage of a sufficiently long period of time, the temperature and pressure are found to be uniform throughout the chamber.

a. If the filled compartment initially contains an ideal gas of constant heat capacity at 1 MPa and 500 K, what are the final temperature and pressure in the chamber?

b. If the filled compartment initially contains steam at 1 MPa and 500 K, what are the final temperature and pressure in the compartment?

c. Repeat part (a) if the second compartment initially contains an ideal gas, but at half the pressure and 100 K higher temperature.

d. Repeat part (b) if the second compartment initially contains steam, but at half the pressure and 100 K higher temperature.

3.16 a. An adiabatic turbine expands steam from 500°C and 3.5 MPa to 200°C and 0.3 MPa. If the turbine generates 750 kW, what is the flow rate of steam through the turbine?

b. If a breakdown of the thermal insulation around the turbine allows a heat loss of 60 kJ per kg of steam, and the exiting steam is at 150°C and 0.3 MPa, what will be the power developed by the turbine if the inlet steam conditions and flow rate are unchanged?

3.17 Intermolecular forces play an important role in determining the thermodynamic properties of fluids. To see this, consider the vaporization (boiling) of a liquid such as water in the frictionless piston-and-cylinder device shown.

a. Compute the work obtained from the piston when 1 kg of water is vaporized to steam at 100°C (the vapor and liquid volumes of steam at the boiling point can be found in the steam tables).

b. Show that the heat required for the vaporization of the steam is considerably greater than the work done. (Note that the enthalpy change for the vaporization is given as 2257 kJ/kg in the steam tables in Appendix A.III.)

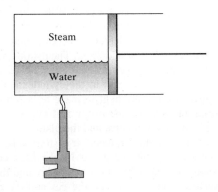

3.18 It is sometimes necessary to produce saturated steam from superheated steam (steam at a temperature higher than the vapor-liquid coexistence temperature at the given pressure). This change can be accomplished in a desuperheater, a device in which just the right amount of water is sprayed into superheated steam to produce dry saturated steam. If superheated steam at 3.0 MPa and 500°C enters the desuperheater at a rate of 500 kg/hr, at what rate should liquid water at 2.5 MPa and 25°C be added to the desuperheater to produce saturated steam at 2.25 MPa?

3.19 Nitrogen gas leaves a compressor at 2.0 MPa and 120°C and is collected in three different cylinders, each of volume 0.3 m³. In each case the cylinder is to be filled to a pressure of 2.0 MPa. Cylinder 1 is initially evacuated, cylinder 2 contains nitrogen gas at 0.1 MPa and 20°C, and cylinder 3 contains nitrogen at 1 MPa and 20°C. Find the final temperature of nitrogen in each of the cylinders, assuming nitrogen to be an ideal gas with $C_P^* = 29.3$ J/mol K. In each case assume the gas does not exchange heat with the cylinder walls.

3.20 A clever chemical engineer has devised the thermally operated elevator shown in the accompanying diagram. The elevator compartment is made to rise by electrically heating the air contained in the piston-and-cylinder drive mechanism, and the elevator is lowered by opening a valve at the side of the cylinder, allowing the air in the cylinder to slowly escape. Once the elevator compartment is back to the lower level, a small pump forces out the air remaining in the cylinder and replaces it with air at 20°C and a pressure just sufficient to support the elevator compartment. The cycle can then be repeated. There is no heat transfer between the piston, cylinder, and gas; the weight of the piston, elevator, and elevator

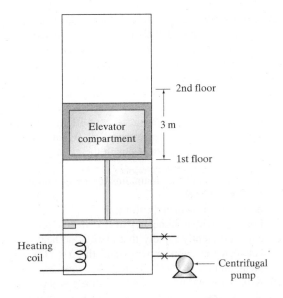

contents is 4000 kg; the piston has a surface area of 2.5 m²; and the volume contained in the cylinder when the elevator is at its lowest level is 25 m³. There is no friction between the piston and the cylinder, and the air in the cylinder is assumed to be an ideal gas with $C_P^* = 30$ J/(mol K).

a. What is the pressure in the cylinder throughout the process?

b. How much heat must be added to the air during the process of raising the elevator 3 m, and what is the final temperature of the gas?

c. What fraction of the heat added is used in doing work, and what fraction is used in raising the temperature of the gas?

d. How many moles of air must be allowed to escape in order for the elevator to return to the lowest level?

3.21 The elevator in the previous problem is to be designed to ascend and descend at the rate of 0.2 m/s, and to rise a total of 3 m.

a. At what rate should heat be added to the cylinder during the ascent?

b. How many kilomoles per second of air should be removed from the cylinder during the descent?

3.22 Nitrogen gas is being withdrawn at the rate of 4.5 g/s from a 0.15-m³ cylinder, initially containing the gas at a pressure of 10 bar and 320 K. The cylinder does not conduct heat, nor does its temperature change during the emptying process. What will be the temperature and pressure of the gas in the cylinder after 5 minutes? What will be the rate of change of the gas temperature at this time? Nitrogen can be considered to be an ideal gas with $C_P^* = 30$ J/(mol K).

3.23 In Illustration 3.4-6 we considered the compression of an ideal gas in which $P\underline{V}^\gamma$ = constant, where $\gamma = C_P^*/C_V^*$. Show that such a pressure-volume relationship is obtained in the adiabatic compression of an ideal gas of constant heat capacity.

3.24 Air in a 0.3-m³ cylinder is initially at a pressure of 10 bar and a temperature of 330 K. The cylinder is to be emptied by opening a valve and letting the pressure drop to that of the atmosphere. What will be the temperature and mass of gas in the cylinder if this is accomplished?

   a. In a manner that maintains the temperature of the gas at 330 K?

   b. In a well-insulated cylinder?

   For simplicity assume, in part (b), that the process occurs sufficiently rapidly that there is no heat transfer between the cylinder walls and the gas. The gas is ideal, and $C_P^* = 29$ (J/mol K).

3.25 A 0.01-m³ cylinder containing nitrogen gas initially at a pressure of 200 bar and 250 K is connected to another cylinder 0.005 m³ in volume, which is initially evacuated. A valve between the two cylinders is opened until the pressures in the cylinders equalize. Find the final temperature and pressure in each cylinder if there is no heat flow into or out of the cylinder. You may assume that there is no heat transfer between the gas and the cylinder walls and that the gas is ideal with a constant-pressure heat capacity of 30 J/(mol K).

3.26 Repeat the calculation of Problem 3.25, but now assume that sufficient heat transfer occurs between the gas in the two cylinders that both final temperatures and both final pressures are the same.

3.27 Repeat the calculation in Problem 3.25, but now assume that the second cylinder, instead of being evacuated, is filled with nitrogen gas at 20 bar and 160 K.

3.28 A 1.5 kW heater is to be used to heat a room with dimensions 3.5 m × 5.0 m × 3.0 m. There are no heat losses from the room, but the room is not airtight, so the pressure in the room is always atmospheric. Consider the air in the room to be an ideal gas with $C_P^* = 29$ J/(mol K), and its initial temperature is 10°C.

   a. Assuming that the rate of heat transfer from the air to the walls is low, what will be the rate of increase of the temperature in the room when the heater is turned on?

   b. What would be the rate of increase in the room temperature if the room were hermetically sealed?

3.29 The piston-and-cylinder device of Illustration 3.4-7 is to be operated in reverse to isothermally compress the 1 mol of air. Assume that the weights in the illustration have been left at the heights they were at when they were removed from the piston (i.e., in process b the first 50-kg weight is at the initial piston height and the second is at $\Delta h = \Delta V/A = 0.384$ m above

the initial piston height). Compute the minimum work that must be done by the surroundings and the net heat that must be withdrawn to return the gas, piston, and weights to their initial states. Also compute the total heat and the total work for each of the four expansion and compression cycles and comment on the results.

3.30 The piston-and-cylinder device shown here contains an ideal gas at 20 bar and 25°C. The piston has a mass of 300 kg and a cross-sectional area of 0.05 m². The initial volume of the gas in the cylinder is 0.03 m³, the piston is initially held in place by a pin, and the external pressure on the piston and cylinder is 1 bar. The pin suddenly breaks, and the piston moves 0.6 m farther up the cylinder, where it is stopped by another pin. Assuming that the gas is ideal with a constant-pressure heat capacity of 30 J/(mol K), and that there is no heat transfer between the gas and the cylinder walls or piston, estimate the piston velocity, and the temperature and pressure of the gas just before the piston hits the second pin. Do this calculation assuming

   a. No friction between the piston and the cylinder

   b. Friction between the piston and the cylinder

   List and defend all assumptions you make in solving this problem.

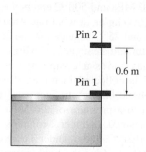

3.31 A 0.6 m diameter gas pipeline is being used for the long-distance transport of natural gas. Just past a pumping station, the gas is found to be at a temperature of 25°C and a pressure of 3.0 MPa. The mass flow rate is 125 kg/s, and the gas flow is adiabatic. Forty miles down the pipeline is another pumping station. At this point the pressure is found to be 2.0 MPa. At the pumping station the gas is first adiabatically compressed to a pressure of 3.0 MPa and then isobarically (i.e., at constant pressure) cooled to 25°C.

   a. Find the temperature and velocity of the gas just before it enters the pumping station.

   b. Find the rate at which the gas compressor in the pumping station does work on the gas, the temperature of the gas leaving the compressor, and the heat load on the gas cooler. You may assume that the compressor exhaust is also a 0.6-m pipe. (Explain why you cannot solve this problem. You will have another chance in Chapter 4.)

Natural gas can be assumed to be pure methane [molecular weight $= 16$, $C_P^* = 36.8$ J/(mol K)], and an ideal gas at the conditions being considered here. Note that the mass flow rate M is $\rho v A$, where ρ is the mass density of the gas, v is the average gas velocity, and A is the area of the pipe.

3.32 Nitrogen can be liquefied using a Joule-Thomson expansion process. This is done by rapidly and adiabatically expanding cold nitrogen gas from high pressure to a low pressure. If nitrogen at 135 K and 20 MPa undergoes a Joule-Thomson expansion to 0.4 MPa,

a. Estimate the fraction of vapor and liquid present after the expansion, and the temperature of this mixture using the pressure-enthalpy diagram for nitrogen.

b. Repeat the calculation assuming nitrogen to be an ideal gas with $C_P^* = 29.3$ J/(mol K).

3.33 A very large mass M of hot porous rock equal to 10^{12} kg is to be utilized to generate electricity by injecting water and using the resulting hot steam to drive a turbine. As a result of heat extraction, the temperature of the rock drops according to $\dot{Q} = -MC_P \, dT/dt$, where C_P is the specific heat of the rock which is assumed to be independent of temperature. If the plant produces 1.36×10^9 kW hr of energy per year, and only 25 percent of the heat extracted from the rock can be converted to work, how long will it take for the temperature of the rock to drop from 600°C to 110°C? Assume that for the rock $C_P = 1$ J/(g K).

3.34 The human body generates heat by the metabolism of carbohydrates and other food materials. Metabolism provides energy for all biological activities (e.g., muscle contractions). The metabolic processes also generate heat, and there are special cells in the body whose main function is heat generation. Now let us assume that our friend Joe BlueHen ingests 1 L of ice, which he allows to melt in his mouth before swallowing.

a. How much energy is required to melt the ice and warm the water to the body temperature?

b. If 1 g of fat when metabolized releases approximately 42 kJ of energy, how much fat will Joe burn by ingesting the water?

c. Suppose that, instead of ice, Joe drank 1 L of water at 0°C. How would the answers to parts (a) and (b) change?

Several years ago there was a story circulating the Internet that a good way to lose weight is to drink a lot of very cold water, since considerable energy would be expended within the body in heating up the cold water. Based on your calculations above; is that a reasonable method of weight loss? (The reason this claim was widely circulated is a result of the sloppy use of units. Some countries report the biologically accessible energy in food as being in units of calories, when in fact the number reported is kilocalories. For example, in the United States a teaspoon of sugar is reported to contain 16 calories, when it is actually 16 kilocalories. *Calories* is incorrectly being used as an abbreviation for *kilocalories*.)

3.35 Water is to be heated from its pipeline temperature of 20°C to 90°C using superheated steam at 450°C and 2.5 MPa in a steady-state process to produce 10 kg/s of heated water. In each of the processes below, assume there is no heat loss.

a. The heating is to be done in a mixing tank by direct injection of the steam, all of which condenses. Determine the two inlet mass flows needed to meet the desired hot water flow rate.

b. Instead of direct mixing, a heat exchanger will be used in which the water to be heated will flow inside copper tubes and the steam will partially condense on the outside of the tubes. In this case heat will flow from the steam to the water, but the two streams are not mixed. Calculate the steam flow rate if the steam leaves the heat exchanger at 50 percent quality at 100°C.

3.36 People partially cool themselves by sweating, which releases water that evaporates. If during exercise a human "burns" 1000 kcal (4184 kJ) in one hour of exercise, how many grams of water must evaporate at a body temperature of 37°C? Assuming only 75 percent of the sweat evaporates (the rest being retained by the exercise clothes), how many grams of sweat must actually be produced?

3.37 It is thought that people develop respiratory infections during air travel because much of the airplane cabin air is recirculated. Airlines claim that using only fresh air in the cabins is too costly since at an altitude of 30 000 feet the outside conditions are −50°C and 0.1 bar, so that the air would have to be compressed and heated before being introduced into the cabin. The airplane cabin has a volume of 100 m³ with air at the in-flight conditions of 25°C and 0.8 bar. What would be the cost of completely refreshing the air every minute if air has a heat capacity of $C_P^* = 30$ J/(mol K) and energy costs $0.2 per kW hr?

Chapter 4

Entropy: An Additional Balance Equation

Several of the illustrations and problems in Chapter 3 show that the equations of mass and energy conservation are not sufficient to solve all the thermodynamic energy flow problems in which we might be interested. To be more specific, these two equations are not always sufficient to determine the final values of two state variables, or the heat and work flows for a system undergoing a change of state. What is needed is a balance equation for an additional state variable. As we have seen, the principle of momentum conservation does not provide this additional equation. Although a large number of additional state variables could be defined and could serve as the basis for a new balance equation, these variables would have the common feature that they are not conserved quantities. Thus the internal generation term for each of these variables would, in general, be nonzero and would have to be evaluated for the balance equation to be of use. Clearly, the most useful variable to introduce as the basis for a new balance equation is one that has an internal generation rate that can be specified and has some physical significance.

Another defect in our present development of thermodynamics has to do with the unidirectional character of natural processes that was considered in Sec. 1.3. There it was pointed out that all spontaneous or natural processes proceed only in the direction that tends to dissipate the gradients in the system and thus lead to a state of equilibrium, and never in the reverse direction. This characteristic of natural processes has not yet been included in our thermodynamic description.

To complete our thermodynamic description of pure component systems, it is therefore necessary that we (1) develop an additional balance equation for a state variable and (2) incorporate into our description the unidirectional character of natural processes. In Sec. 4.1 we show that both these objectives can be accomplished by introducing a single new thermodynamic function, the entropy. The remaining sections of this chapter are concerned with illustrating the properties and utility of this new variable and its balance equation.

INSTRUCTIONAL OBJECTIVES FOR CHAPTER 4

The goals of this chapter are for the student to:

- Be able to use the rate-of-change form of the pure component entropy balance in problem solving

- Be able to use the difference form of the pure component entropy balance in problem solving
- Be able to calculate the entropy change between two states of an ideal gas
- Be able to calculate the entropy change of a real fluid using thermodynamic properties charts and tables

NOTATION INTRODUCED IN THIS CHAPTER

A Helmholtz energy $= U - TS$ (J)
$\underline{A}$ Molar Helmholtz energy (J/mol)
G Gibbs energy $= H - TS$ (J)
$\underline{G}$ Molar Gibbs energy (J/mol)
$\dot{Q}_R$ Radiant heat flux (J/s)
S Entropy (J/K)
$\underline{S}$ Entropy per mole [J/(mol K)]
$\hat{S}$ Entropy per unit mass [J/(kg K)]
$\dot{S}_{gen}$ Rate at which entropy is generated within the system [J/(mol K s)]
S_{gen} Entropy generated within the system [J/(mol K)]
T_R Temperature of body emitting radiation (K)
W^{rev} Work in a reversible process (J)
γ $= C_P/C_V$

4.1 ENTROPY: A NEW CONCEPT

We take as the starting point for the identification of an additional thermodynamic variable the experimental observation that all spontaneous processes that occur in an isolated constant-volume system result in the evolution of the system to a state of equilibrium (this is a special case of experimental observation 5, Sec. 1.7). The problem is to quantify this qualitative observation. We can obtain some insight into how to do this by considering the general balance equation (Eq. 2.1-4) for any extensive variable θ of a closed, isolated, constant-volume system

$$\frac{d\theta}{dt} = \begin{pmatrix} \text{Rate of change of} \\ \theta \text{ in the system} \end{pmatrix} = \begin{pmatrix} \text{Rate at which } \theta \text{ is generated} \\ \text{within the system} \end{pmatrix} \qquad \textbf{(4.1-1)}$$

Alternatively, we can write Eq. 4.1-1 as

$$\frac{d\theta}{dt} = \dot{\theta}_{gen} \qquad \textbf{(4.1-2)}$$

where $\dot{\theta}_{gen}$ is the rate of internal generation of the yet-unspecified state variable θ. Now, if the system under consideration were in a true time-invariant equilibrium state, $d\theta/dt = 0$ (since, by definition of a time-invariant state, no state variable can change with time). Thus

$$\dot{\theta}_{gen} = 0 \quad \text{at equilibrium} \qquad \textbf{(4.1-3)}$$

Equations 4.1-2 and 4.1-3 suggest a way of quantifying the qualitative observation of the unidirectional evolution of an isolated system to an equilibrium state. In particular, suppose we could identify a thermodynamic variable θ whose rate of internal generation $\dot{\theta}_{gen}$ was positive,[1] except at equilibrium, where $\dot{\theta}_{gen} = 0$. For this variable

$$\frac{d\theta}{dt} > 0 \qquad \text{away from equilibrium}$$

$$\left.\begin{array}{l} \dfrac{d\theta}{dt} = 0 \\[2mm] \text{or} \quad \theta = \text{constant} \end{array}\right\} \quad \text{at equilibrium} \qquad \textbf{(4.1-4)}$$

Furthermore, since the function θ is increasing in the approach to equilibrium, θ must be a maximum at equilibrium subject to the constraints of constant mass, energy, and volume for the isolated constant-volume system.[2]

Thus, if we could find a thermodynamic function with the properties given in Eq. 4.1-4, the experimental observation of unidirectional evolution to the equilibrium state would be built into the thermodynamic description through the properties of the function θ. That is, the unidirectional evolution to the equilibrium state would be mathematically described by the continually increasing value of the function θ and the occurrence of equilibrium in an isolated, constant-volume system to the attainment of a maximum value of the function θ.

The problem, then, is to identify a thermodynamic state function θ with a rate of internal generation, $\dot{\theta}_{gen}$, that is always greater than or equal to zero. Before searching for the variable θ, it should be noted that the property we are looking for is $\dot{\theta}_{gen} \geq 0$; this is clearly not as strong a statement as $\dot{\theta}_{gen} = 0$ always, which occurs if θ is a conserved variable such as total mass or total energy, *but it is as strong a general statement as we can expect for a nonconserved variable.*

We could now institute an extensive search of possible thermodynamic functions in the hope of finding a function that is a state variable and also has the property that its rate of internal generation is a positive quantity. Instead, we will just introduce this new thermodynamic property by its definition and then show that the property so defined has the desired characteristics.

Definition

The entropy (denoted by the symbol S) is a state function. In a system in which there are flows of both heat by conduction ($\dot{Q}$) and work [$\dot{W}_s$ and $P(dV/dt)$] across the system boundaries, the conductive heat flow, but not the work flow, causes a change in the entropy of the system; this rate of entropy change is $\dot{Q}/T$, where T is the absolute thermodynamic temperature of the system at the point of the heat flow. If, in addition, there are mass flows across the system boundaries, the total entropy of the system will also change due to this

---

[1] If we were to choose the other possibility, $\dot{\theta}_{gen}$ being less than zero, the discussion here would still be valid except that θ would monotonically decrease to a minimum, rather than increase to a maximum, in the evolution to the equilibrium state. The positive choice is made here in agreement with standard thermodynamic convention.

[2] Since an isolated constant-volume system has fixed mass, internal energy, and volume, you might ask how θ can vary if M, U, and V or alternatively, the two state variables $\underline{U}$ and $\underline{V}$ are fixed. The answer is that the discussion of Secs. 1.3 and 1.6 established that two state variables completely fix the state of a *uniform* one-component, one-phase system. Consequently, θ (or any other state variable) can vary for fixed $\underline{U}$ and $\underline{V}$ in (1) a nonuniform system, (2) a multicomponent system, or (3) a multiphase system. The first case is of importance in the approach to equilibrium in the presence of internal relaxation processes, and the second and third cases for chemical reaction equilibrium and phase equilibrium, which are discussed later in this book.

convected flow. That is, each element of mass entering or leaving the system carries with it its entropy (as well as internal energy, enthalpy, etc.).

Using this definition and Eq. 2.1-4, we have the following as the balance equation for entropy:[3]

$$\frac{dS}{dt} = \sum_{k=1}^{K} \dot{M}_k \hat{S}_k + \frac{\dot{Q}}{T} + \dot{S}_{\text{gen}}$$

(4.1-5a)

where

$$\sum_{k=1}^{K} \dot{M}_k \hat{S}_k = \text{net rate of entropy flow due to the flows of mass into and out of the system } (\hat{S} = \text{entropy per unit mass})$$

$$\frac{\dot{Q}}{T} = \text{rate of entropy flow due to the flow of heat across the system boundary}$$

$$\dot{S}_{\text{gen}} = \text{rate of internal generation of entropy within the system.}$$

For a system closed to the flow of mass (i.e., all $\dot{M}_k = 0$), we have

$$\frac{dS}{dt} = \frac{\dot{Q}}{T} + \dot{S}_{\text{gen}}$$

(4.1-5b)

Based on the discussion above, we then have (from Eqs. 4.1-3 and 4.1-4)

$$\dot{S}_{\text{gen}} \geq 0, \quad \text{and} \quad \frac{dS}{dt} = 0 \text{ at equilibrium}$$

(4.1-5c)

Equations 4.1-5 are usually referred to as the second law of thermodynamics.

Before we consider how the entropy balance, Eq. 4.1-5a, will be used in problem solving, we should establish (1) that the entropy function is a state variable, and (2) that it has a positive rate of internal generation, that is, $\dot{S}_{\text{gen}} \geq 0$. Notice that Eq. 4.1-5 cannot provide general information about the internal generation of entropy since it is an equation for the black-box description of a system, whereas $\dot{S}_{\text{gen}}$ depends on the detailed internal relaxation processes that occur within the system. In certain special cases, however, one can use Eq. 4.1-5 to get some insight into the form of $\dot{S}_{\text{gen}}$. To see this, consider the thermodynamic system of Fig. 4.1-1, which is a composite of two subsystems, A and B. These subsystems are well insulated except at their interface, so that the only heat transfer that occurs is a flow of heat from the high-temperature subsystem A to the low-temperature subsystem B. We assume that the resistance to heat transfer at this interface is large relative to the internal resistances of the subsystems (which would occur if, for example, the subsystems were well-mixed liquids or highly conducting solids), so that the temperature of each subsystem is uniform, but varying with time.

---

[3]For simplicity, we have assumed that there is only a single heat flow into the system. If there are multiple heat flows by conduction, the term $\dot{Q}/T$ is to be replaced by a $\sum \dot{Q}_j/T_j$, where $\dot{Q}_j$ is the heat flow and T_j the temperature at the jth heat flow port into the system.

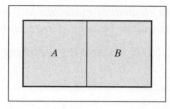

Figure 4.1-1 Systems A and B are free to interchange energy, but the composite system $(A + B)$ is isolated from the environment.

In this situation the heat transfer process occurs in such a way that, at any instant, each subsystem is in a state of internal thermal equilibrium (i.e., if the two subsystems were suddenly separated, they would be of uniform but different temperatures and would not change with time); however, the composite system consisting of both subsystems (at different temperatures) is not in thermal equilibrium, as evidenced by the fact that though the composite system is isolated from the environment, its properties are changing with time (as heat is transferred from A to B and the temperatures of these subsystems are changing). The rate-of-change form of the entropy balances for subsystems A and B, which are passing through a succession of equilibrium states,[4] and therefore have no internal generation of entropy, are

$$\frac{dS_A}{dt} - \frac{\dot{Q}_A}{T_A} = -h\left(\frac{T_A - T_B}{T_A}\right) \tag{4.1-6a}$$

and

$$\frac{dS_B}{dt} = \frac{\dot{Q}_B}{T_B} = +h\left(\frac{T_A - T_B}{T_B}\right) \tag{4.1-6b}$$

In writing these equations we have recognized that the amount of heat that leaves subsystem A enters subsystem B, and that the heat flow from A to B is proportional to the temperature difference between the two systems, that is,

$$\dot{Q}_A = -\dot{Q}_B = -h\,(T_A - T_B)$$

where h is the heat transfer coefficient (experimental observation 10 of Sec. 1.7).

The entropy balance for the isolated ($\dot{Q} = 0$), nonequilibrium composite system composed of subsystems A and B is

$$\frac{dS}{dt} = \dot{S}_{\text{gen}} \tag{4.1-7}$$

Since the total entropy is an extensive property, $S = S_A + S_B$, and Eqs. 4.1-6 and 4.1-7 can be combined to yield

$$\dot{S}_{\text{gen}} = -h\left(\frac{T_A - T_B}{T_A}\right) + h\left(\frac{T_A - T_B}{T_B}\right) = \frac{h(T_A - T_B)^2}{T_A T_B} = \frac{h(\Delta T)^2}{T_A T_B} \tag{4.1-8}$$

Since h, T_A, and T_B are positive, Eq. 4.1-8 establishes that for this simple example the entropy generation term is positive. It is also important to note that $\dot{S}_{\text{gen}}$ is proportional to the second power of the system nonuniformity, here $(\Delta T)^2$. Thus, $\dot{S}_{\text{gen}}$ is positive away from equilibrium (when $T_A \neq T_B$) and equal to zero at equilibrium.

---

[4]A process in which a system goes through a succession of equilibrium states is termed a **quasistatic** process.

This expression for the rate of entropy generation was obtained by partitioning a black-box system into two subsystems, resulting in a limited amount of information about processes internal to the overall system. In Sec. 4.6 a more general derivation of the entropy generation term is given, based on the detailed microscopic description introduced in Sec. 2.4, and it is shown that the entropy is indeed a state function and that $\dot{S}_{gen}$ is positive, except in the equilibrium state, where it is equal to zero. It is also established that $\dot{S}_{gen}$ is proportional to the second power of the gradients of temperature and velocity; thus the rate of generation of entropy is related to the square of the departure from the equilibrium state.

Table 4.1-1 gives several special cases of the entropy balance equation, on both a mass and a molar basis, for situations similar to those considered for the mass and energy balance equations in Tables 2.2-1, 3.1-1, and 3.1-2.

Frequently, one is interested in the change in entropy of a system in going from state 1 to state 2, rather than the rate of change of entropy with time. This entropy change can be determined by integrating Eq. 4.1-5 over the time interval t_1 to t_2, where $(t_2 - t_1)$ is the (perhaps unknown) time required to go between the two states. The result is

Difference form of entropy balance

$$\boxed{S_2 - S_1 = \sum_k \int_{t_1}^{t_2} \dot{M}_k \hat{S}_k \, dt + \int_{t_1}^{t_2} \frac{\dot{Q}}{T} \, dt + S_{gen}} \tag{4.1-9}$$

where

$$S_{gen} = \text{total entropy generated} = \int_{t_1}^{t_2} \dot{S}_{gen} \, dt$$

Since $\dot{S}_{gen}$ is always greater than or equal to zero in any process, it follows that its integral S_{gen} must also be greater than or equal to zero.

There are two important simplifications of Eq. 4.1-9. First, if the entropy per unit mass of each stream entering and leaving the system is constant in time (even though

Table 4.1-1 Rate-of-Change Form of the Entropy Balance

General equation:	$\dfrac{dS}{dt} = \sum_{k=1}^{K} \dot{M}_k \hat{S}_k + \dfrac{\dot{Q}}{T} + \dot{S}_{gen}$	(a)
Special cases:		
(i) Closed system	set $\dot{M}_k = 0$	
so	$\dfrac{dS}{dt} = \dfrac{\dot{Q}}{T} + \dot{S}_{gen}$	(b)
(ii) Adiabatic process	set $\dot{Q} = 0$ in Eqs. a, b, and e	(c)
(iii) Reversible process	set $\dot{S}_{gen} = 0$ in Eqs. a, b, and e	(d)
(iv) Open steady-state system		
	$\dfrac{dS}{dt} = 0$	
so	$0 = \sum_{k=1}^{K} \dot{M}_k \hat{S}_k + \dfrac{\dot{Q}}{T} + \dot{S}_{gen}$	(e)
(v) Uniform system	$S = M\hat{S}$ in Eqs. a and b	(f)

Note: To obtain the entropy balance on a molar basis, replace $\dot{M}_k \hat{S}_k$ by $\dot{N}_k \underline{S}_k$, and $M\hat{S}$ by $N\underline{S}$, where $\underline{S}$ is the entropy per mole of fluid.

the flow rates may vary), we have

$$\sum_k \int_{t_1}^{t_2} \dot{M}_k \hat{S}_k \, dt = \sum_k \hat{S}_k \int_{t_1}^{t_2} \dot{M}_k \, dt = \sum_k \Delta M_k \hat{S}_k$$

where $(\Delta M)_k = \int_{t_1}^{t_2} \dot{M}_k \, dt$ is the total mass that has entered the system from the kth stream. Next, if the temperature is constant at the location where the heat flow occurs, then

$$\int_{t_1}^{t_2} \frac{\dot{Q}}{T} \, dt = \frac{1}{T} \int_{t_1}^{t_2} \dot{Q} \, dt = \frac{Q}{T}$$

where $Q = \int_{t_1}^{t_2} \dot{Q} \, dt$ is the total heat flow into the system between t_1 and t_2. If either of these simplifications is not valid, the respective integrals must be evaluated if Eq. 4.1-9 is to be used. This may be a difficult or impossible task, so that, as with the energy balance, the system for which the entropy balance is to be written must be chosen with care, as illustrated later in this chapter. Table 4.1-2 summarizes various forms of the integrated entropy balance.

It should be pointed out that we have introduced the entropy function in an axiomatic and mathematical fashion. In the history of thermodynamics, entropy has been presented in many different ways, and it is interesting to read about these alternative approaches. One interesting source is *The Second Law* by P. W. Atkins (W. H. Freeman, New York, 1984).

The axiom that is used here, $S_{gen} \geq 0$, is a statement of what has historically been called the **second law of thermodynamics.** The principle of conservation of energy discussed previously is referred to as the **first law.** It is interesting to compare the form of the second law used here with two other forms that have been proposed in the history of thermodynamics, both of which deal with the transformation of heat to work.

Table 4.1-2 Difference Form of the Entropy Balance

General equation	$S_2 - S_1 = \sum\limits_{k=1}^{K} \int_{t_1}^{t_2} \dot{M}_k \hat{S}_k \, dt + \int_{t_1}^{t_2} \frac{\dot{Q}}{T} \, dt + S_{gen}$	(a)
Special cases:		
(i) Closed system	set $\dot{M}_k = 0$ in Eq. a	
so	$S_2 - S_1 = \int_{t_1}^{t_2} \frac{\dot{Q}}{T} \, dt + S_{gen}$	(b)
(ii) Adiabatic process		
	set $\int_{t_1}^{t_2} \frac{\dot{Q}}{T} \, dt = 0$ in Eq. a	(c)
(iii) Reversible process	set $S_{gen} = 0$ in Eqs. a and b	(d)
(iv) Open system: Flow of fluids of constant thermodynamic properties		
	set $\sum\limits_{k=1}^{K} \int_{t_1}^{t_2} \dot{M}_k \hat{S}_k \, dt = \sum\limits_{k=1}^{K} \Delta M_k \hat{S}_k$ in Eq. a	(e)
(v) Uniform system	$S = M\hat{S}$ in Eq. a	(f)

Note: To obtain the entropy balance on a molar basis, replace $M\hat{S}$ by $N\underline{S}$, $\int_{t_1}^{t_2} \dot{M}_k \hat{S}_k \, dt$ by $\int_{t_1}^{t_2} \dot{N}_k \underline{S}_k \, dt$, and $\Delta M_k \hat{S}_k$ with $\Delta N_k \underline{S}_k$.

ILLUSTRATION 4.1-1

Clausius Statement of the Second Law

The second-law statement of Rudolf Clausius (1822–1888) is that it is not possible to construct a device that operates in a cycle and whose sole effect is to transfer heat from a colder body to a hotter body. Show from the axiom $S_{gen} \geq 0$ that the process below is impossible, so that the Clausius statement of the second law is consistent with what has been presented in this book.

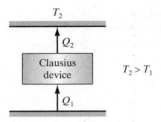

SOLUTION

The energy balance over one complete cycle, so that the device is in the same state at the beginning and at the end of the process, is

$$U_f - U_i = 0 = Q_1 + Q_2 + \cancel{W}^{0}$$

since there is no work produced or absorbed in the device. Therefore, the energy balance yields $Q_2 = -Q_1$. Thus the energy balance tells us that the process may be possible as long as the heat flows in and out balance, that is, if energy is conserved.

The entropy balance over one complete cycle is

$$S_f - S_i = 0 = \frac{Q_1}{T_1} + \frac{Q_2}{T_2} + S_{gen}$$

or

$$S_{gen} = -\frac{Q_2}{T_2} - \frac{Q_1}{T_1} = \frac{Q_1}{T_2} - \frac{Q_1}{T_1} = Q_1 \left(\frac{1}{T_2} - \frac{1}{T_1} \right)$$

The second-law statement we have used is $S_{gen} > 0$. However, since $T_2 > T_1$, if Q_1 were positive (heat flow into device from cold body), $S_{gen} < 0$, which is a violation of our axiom. If Q_1 were negative, so that the heat flow was from high temperature to low temperature (or equivalently $T_1 > T_2$), the process would be possible. Consequently, our statement of the second law, $S_{gen} \geq 0$, is consistent with the Clausius version, but much more general. ∎

ILLUSTRATION 4.1-2

Kelvin-Planck Statement of the Second Law

An alternative statement of the second law, due to Lord Kelvin (William Thomson, 1824–1907) and Max Planck (1858–1947), is that it is not possible to construct a device operating in a cycle that results in no effect other than the production of work by transferring heat from a single body. A schematic diagram of a Kelvin-Planck device is shown below.

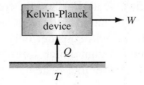

Show from our axiom $S_{\text{gen}} \geq 0$ that this process is indeed impossible.

SOLUTION

The energy balance over one complete cycle is

$$U_f - U_i = 0 = Q + W \quad \text{or} \quad W = -Q$$

As long as the work produced equals the heat absorbed, this device satisfies the principle of energy conservation, and is possible by the first law of thermodynamics. The entropy balance over one complete cycle is

$$S_f - S_i = 0 = \frac{Q}{T} + S_{\text{gen}}$$

or

$$S_{\text{gen}} = -\frac{Q}{T}$$

Since T is positive (remember, we are using absolute temperature), for the entropy generation to be positive, Q must be negative. That is, to be consistent with our statement of the second law, the device cannot absorb heat and convert all of it to work. However, the reverse process, in which the device receives work and converts all that work to heat, is possible. Therefore, our statement of the second law is consistent with that of Kelvin and Planck, but again more general. ∎

An alternative wording of the Kelvin-Planck statement is that heat cannot be completely converted to work in a cyclic process. However, it is possible, as shown above, to do the converse and completely convert work to heat. Since heat cannot be completely converted to work, heat is sometimes considered a less useful form of energy than work. When work or mechanical energy is converted to heat, for example, by friction, it is said to be degraded.

Finally, notice the very different roles of the first law (energy conservation) and second law ($S_{\text{gen}} \geq 0$) in the analyses above. The first law did not provide a definitive result as to whether a process was possible or not, but merely set some constraints (i.e., $Q_2 = -Q_1$ for the Clausius device and $W = -Q$ for the Kelvin-Planck device). However, the second law is definitive in that it establishes that it is impossible to construct devices that would operate in the manners proposed.

Heat flow by radiation

So far we have considered the heat flow $\dot{Q}$ to be due only to conduction across the system boundaries. However, heat can also be transmitted to a system by radiation—for example, as the sun heats the earth. If there is a radiative heat flow, it can be incorporated into the energy balance as an additional energy flow that can be combined with the conductive heat flow. The change in entropy accompanying radiative heat transfer is more complicated than for conductive heat transfer, and must be treated separately. In particular, the entropy flow depends on the distribution of energy in the radiation, which in turn depends on the source and temperature of the radiation. Radiation can be monochromatic, that is, of a single frequency, when produced by a laser, or it can have a broad spectrum of energies, for example, following the Stefan-Boltzmann law for typical radiating bodies, such as the sun, a red-hot metal, or even our bodies (which do radiate heat). Further, the frequency distribution (or energy spectrum) emitted from fluorescent and incandescent lamps is different.

Here we will consider only the entropic contribution from the most common form of radiation, that which follows the Stefan-Boltzmann law. In this case, for a radiant energy transfer to a system of $\dot{Q}_{\text{R}}$ the rate of entropy change is

$$\text{Entropy flow accompanying radiation} = \frac{4\dot{Q}_R}{3T_R}$$

where T_R is the temperature of the body *emitting* the radiation, not the temperature of the receiving body. Therefore, the entropy balance for a system, including heat transfer by radiation (from a body or device emitting radiation with a Stefan-Boltzmann energy distribution), is, on a mass basis

$$\frac{dS}{dt} = \sum_{k=1}^{K} \dot{M}_k \hat{S}_k + \frac{\dot{Q}}{T} + \frac{4\dot{Q}_R}{3T_R} + \dot{S}_{\text{gen}} \qquad \textbf{(4.1-10a)}$$

and on a molar basis

$$\frac{dS}{dt} = \sum_{k=1}^{K} \dot{N}_k \underline{S}_k + \frac{\dot{Q}}{T} + \frac{4\dot{Q}_R}{3T_R} + \dot{S}_{\text{gen}} \qquad \textbf{(4.1-10b)}$$

Since radiative heat transfer is not generally important in chemical engineering applications of thermodynamics, Eqs. 4.1-10 will not be used much in this book.

4.2 THE ENTROPY BALANCE AND REVERSIBILITY

Reversible processes

An important class of processes is that for which the rate of generation of entropy is always zero. Such processes are called **reversible** processes and are of special interest in thermodynamics. Since $\dot{S}_{\text{gen}}$ is proportional to the square of the temperature gradients and velocity gradients in the system, such gradients must vanish in a process in which $\dot{S}_{\text{gen}}$ is zero. Notice, however, that although the rate of entropy generation is second order in the system gradients, the internal relaxation processes that occur in the approach to equilibrium are linearly proportional to these gradients (i.e., the heat flux **q** is proportional to the temperature gradient ∇T, the stress tensor is proportional to the velocity gradient, etc. as shown in Sec. 1.7). Therefore, if there is a very small temperature gradient in the system, the heat flux **q**, which depends on ∇T, will be small, and $\dot{S}_{\text{gen}}$, which depends on $(\nabla T)^2$, may be so small as to be negligible. Similarly, the rate of entropy generation $\dot{S}_{\text{gen}}$ may be negligible for very small velocity gradients. Processes that occur with such small gradients in temperature and velocity that $\dot{S}_{\text{gen}}$ is essentially zero can also be considered to be reversible.

The designation *reversible* arises from the following observation. Consider the change in state of a general system open to the flow of mass, heat, and work, between two equal time intervals, 0 to t_1 and t_1 to t_2, where $t_2 = 2t_1$. The mass, energy, and entropy balances for this system are, from Eqs. 2.2-4, 3.1-6, and 4.1-9,

$$M_2 - M_0 = \sum_k \int_0^{t_1} \dot{M}_k \, dt + \sum_k \int_{t_1}^{t_2} \dot{M}_k \, dt$$

$$U_2 - U_0 = \int_0^{t_1} \left[\sum_k \dot{M}_k \hat{H}_k - P\frac{dV}{dt} + \dot{W}_s + \dot{Q} \right] dt$$

$$+ \int_{t_1}^{t_2} \left[\sum_k \dot{M}_k \hat{H}_k - P\frac{dV}{dt} + \dot{W}_s + \dot{Q} \right] dt$$

and

$$S_2 - S_0 = \int_0^{t_1} \left[\sum_k \dot{M}_k \hat{S}_k + \frac{\dot{Q}}{T} \right] dt + \int_{t_1}^{t_2} \left[\sum_k \dot{M}_k \hat{S}_k + \frac{\dot{Q}}{T} \right] dt$$

$$+ \int_0^{t_1} \dot{S}_{gen} \, dt + \int_{t_1}^{t_2} \dot{S}_{gen} \, dt$$

Now, suppose that all of the mass, heat, and work flows are just reversed between t_1 and t_2 from what they had been between 0 and t_1, so that

$$\int_0^{t_1} \dot{M}_k \, dt = - \int_{t_1}^{t_2} \dot{M}_k \, dt \qquad \int_0^{t_1} \dot{Q} \, dt = - \int_{t_1}^{t_2} \dot{Q} \, dt$$

$$\int_0^{t_1} \dot{M}_k \hat{H}_k \, dt = - \int_{t_1}^{t_2} \dot{M}_k \hat{H}_k \, dt \qquad \int_0^{t_1} \frac{\dot{Q}}{T} \, dt = - \int_{t_1}^{t_2} \frac{\dot{Q}}{T} \, dt$$

$$\int_0^{t_1} \dot{M}_k \hat{S}_k \, dt = - \int_{t_1}^{t_2} \dot{M}_k \hat{S}_k \, dt \qquad \int_0^{t_1} \dot{W}_s \, dt = - \int_{t_1}^{t_2} \dot{W}_s \, dt \qquad \text{(4.2-1)}$$

$$\int_0^{t_1} P \frac{dV}{dt} \, dt = - \int_{t_1}^{t_2} P \frac{dV}{dt} \, dt$$

In this case the equations reduce to

$$M_2 = M_0 \qquad \text{(4.2-2a)}$$
$$U_2 = U_0 \qquad \text{(4.2-2b)}$$
$$S_2 = S_0 + \int_0^{t_1} \dot{S}_{gen} \, dt + \int_{t_1}^{t_2} \dot{S}_{gen} \, dt \qquad \text{(4.2-2c)}$$

In general, $\dot{S}_{gen} \geq 0$, so the two integrals in Eq. 4.2-2c will be positive, and the entropy of the initial and final states will differ. Thus the initial and final states of the system will have different entropies and therefore must be different. If, however, the changes were accomplished in such a manner that the gradients in the system over the whole time interval are infinitesimal, then $\dot{S}_{gen} = 0$ in each time interval, and $S_2 = S_0$. In this case the system has been returned to its initial state from its state at t_1 by a process in which the work and each of the flows were reversed. Such a process is said to be reversible. If $\dot{S}_{gen}$ had not been equal to zero, then $S_2 > S_0$, and the system would not have been returned to its initial state by simply reversing the work and other flows; the process is then said to be **irreversible**.

The main characteristic of a reversible process is that it proceeds with infinitesimal gradients within the system. Since transport processes are linearly related to the gradients in the system, this requirement implies that a reversible change occurs slowly on the time scale of macroscopic relaxation times. Changes of state in real systems can be approximated as being reversible if there is no appreciable internal heat flow or viscous dissipation; they are irreversible if such processes occur. Consequently, expansions and compressions that occur uniformly throughout a fluid, or in well-designed turbines, compressors, and nozzles for which viscous dissipation and internal heat flows are unimportant, can generally be considered to occur reversibly (i.e., $\dot{S}_{gen} = 0$). Flows through pipes, through flow constrictions (e.g., a valve or a porous plug), and through shock waves all involve velocity gradients and viscous dissipation and hence are irreversible. Table 4.2-1 contains some examples of reversible and irreversible processes.

Table 4.2-1 Examples of Reversible and Irreversible Processes

Reversible process

Reversible process: A process that occurs with no (appreciable) internal temperature, pressure, or velocity gradients, and therefore no internal flows or viscous dissipation

Examples:
  Fluid flow in a well-designed turbine, compressor, or nozzle
  Uniform and slow expansion or compression of a fluid
  Many processes in which changes occur sufficiently slowly that gradients do not appear in the system

Irreversible process

Irreversible process: A process that occurs with internal temperature, pressure, and/or velocity gradients so that there are internal flows and/or viscous dissipation

Examples:
  Flow of fluid in a pipe or duct in which viscous forces are present
  Flow of fluid through a constriction such as a partially open valve, a small orifice, or a porous plug (i.e., the Joule-Thomson expansion)
  Flow of a fluid through a sharp gradient such as a shock wave
  Heat conduction process in which a temperature gradient exists
  Any process in which friction is important
  Mixing of fluids of different temperatures, pressures, or compositions

Another characteristic of a reversible process is that if the surroundings are extracting work from the system, the maximum amount of work is obtained for a given change of state if the process is carried out reversibly (i.e., so that $\dot{S}_{gen} = 0$). A corollary to this statement is that if the surroundings are doing work on the system, a minimum amount of work is needed for a given change of state if the change occurs reversibly. The first of these statements is evident from Illustration 3.4-7, where it was found that the maximum work (W^{NET}) was extracted from the expansion of a gas in a piston-and-cylinder device between given initial and final states if the expansion was carried out reversibly (by removing grains of sand one at a time), so that only differential changes were occurring, and there was no frictional dissipation of kinetic energy to thermal energy in the work-producing device. It will be shown, in Illustration 4.5-8, that such a process is also one for which $S_{gen} = 0$.

Although few processes are truly reversible, it is sometimes useful to model them to be so. When this is done, it is clear that any computations made based on Eqs. 4.1-5 or 4.1-9, with $\dot{S}_{gen} = S_{gen} = 0$, will only be approximate. However, these approximate results may be very useful, since the term neglected (the entropy generation) is of known sign, so we will know whether our estimate for the heat, work, or any state variable is an upper or lower bound to the true value. To see this, consider the energy and entropy balances for a closed, isothermal, constant-volume system:

$$U_2 = U_1 + Q + W_s \tag{4.2-3}$$

and

$$S_2 = S_1 + \frac{Q}{T} + S_{gen} \tag{4.2-4}$$

Eliminating Q between these two equations and using $T_1 = T_2 = T$ gives

$$W_s = (U_2 - T_2 S_2) - (U_1 - T_1 S_1) + T S_{gen}$$
$$= A_2 - A_1 + T S_{gen} \qquad \text{(4.2-5)}$$

Here we have defined a new thermodynamic state variable, the **Helmholtz energy**, as

Helmholtz energy

$$\boxed{A = U - TS} \qquad \text{(4.2-6)}$$

(Note that A must be a state variable since it is a combination of state variables.) The work required to bring the system from state 1 to state 2 by a reversible (i.e., $S_{gen} = 0$), isothermal, constant-volume process is

Reversible work at
constant N, V, and T

$$W_s^{rev} = A_2 - A_1 \qquad \text{(4.2-7)}$$

while the work in an irreversible (i.e., $S_{gen} > 0$), isothermal, constant-volume process between the same initial and final states is

$$W_s = A_2 - A_1 + T S_{gen} = W_s^{rev} + T S_{gen}$$

Since $T S_{gen} > 0$, this equation establishes that more work is needed to drive the system from state 1 to state 2 if the process is carried out irreversibly than if it were carried out reversibly. Conversely, if we are interested in the amount of work the system can do on its surroundings at constant temperature and volume in going from state 1 to state 2 (so that W_s is negative), we find, by the same argument, that more work is obtained if the process is carried out reversibly than if it is carried out irreversibly.

One should not conclude from Eq. 4.2-7 that the reversible work for any process is equal to the change in Helmholtz energy, since this result was derived only for an isothermal, constant-volume process. The value of W_s^{rev}, and the thermodynamic functions to which it is related, depends on the constraints placed on the system during the change of state (see Problem 4.3). For example, consider a process occurring in a closed system at fixed temperature and pressure. Here we have

$$U_2 = U_1 + Q + W_s - (P_2 V_2 - P_1 V_1)$$

where $P_2 = P_1$, $T_2 = T_1$, and

$$S_2 = S_1 + \frac{Q}{T} + S_{gen}$$

Thus

$$W_s^{rev} = G_2 - G_1$$

Reversible work at
constant N, P, and T   where

$$\boxed{G \equiv U + PV - TS = H - TS} \qquad \text{(4.2-8)}$$

Gibbs energy

G is called the **Gibbs energy**. Therefore, for the case of a closed system change at constant temperature and pressure, we have

$$W_s = G_2 - G_1 + T S_{gen} = W_s^{rev} + T S_{gen}$$

The quantity $T S_{gen}$ in either process can be interpreted as the amount of mechanical energy that has been converted to thermal energy by viscous dissipation and other system irreversibilities. To see this, consider a reversible process ($S_{gen} = 0$) in a closed system. The energy balance is

$$Q^{rev} = U_2 - U_1 - W^{rev}$$

(here W is the sum of the shaft work and $P \, \Delta V$ work). If, on the other hand, the process was carried out between the same initial and final states so that macroscopic gradients arose in the system (that is, irreversibly, so $S_{gen} > 0$), then

$$Q = U_2 - U_1 - W = U_2 - U_1 - W^{rev} - T S_{gen}$$

or

$$Q = Q^{rev} - T S_{gen} \qquad (4.2\text{-}9a)$$

and

$$W = W^{rev} + T S_{gen} \qquad (4.2\text{-}9b)$$

Since $T S_{gen}$ is greater than zero, less heat and more work are required to accomplish a given change of state in the second (irreversible) process than in the first.[5] This is because the additional mechanical energy supplied in the second case has been converted to thermal energy. It is also generally true that system gradients that lead to heat flows, mass flows, and viscous dissipation result in the conversion of mechanical energy to thermal energy and in a decrease in the amount of work that can be obtained from a work-producing device between fixed initial and final states.

Finally, since the evaluation of the changes in the thermodynamic properties of a system accompanying its change of state are important in thermodynamics, it is useful to have an expression relating the entropy change to changes in other state variables. To obtain this equation we start with the rate-of-change form of the mass, energy, and entropy balances for a system in which the kinetic and potential energy terms are unimportant, there is only one mass flow stream, and the mass and heat flows occur at the common temperature T:

$$\frac{dM}{dt} = \dot{M}$$

$$\frac{dU}{dt} = \dot{M} \hat{H} + \dot{Q} - P \frac{dV}{dt} + \dot{W}_s$$

and

$$\frac{dS}{dt} = \dot{M} \hat{S} + \frac{\dot{Q}}{T} + \dot{S}_{gen}$$

where $\hat{H}$ and $\hat{S}$ are the enthalpy per unit mass and entropy per unit mass of the fluid entering or leaving the system. Eliminating $\dot{M}$ between these equations and multiplying the third equation by T yields

$$\frac{dU}{dt} = \hat{H} \frac{dM}{dt} + \dot{Q} - P \frac{dV}{dt} + \dot{W}_s$$

---

[5]The argument used here assumes that work is required to drive the system from state 1 to state 2. You should verify that the same conclusions would be reached if work were obtained in going from state 1 to state 2.

and

$$T\frac{dS}{dt} = T\hat{S}\frac{dM}{dt} + \dot{Q} + T\dot{S}_{\text{gen}}$$

Of particular interest is the form of these equations applicable to the change of state occurring in the differential time interval dt. As in Sec. 3.1, we write

$$Q = \dot{Q}\,dt = \text{heat flow into the system in the time interval } dt$$
$$W_s = \dot{W}_s\,dt = \text{shaft work into the system in the time interval } dt$$
$$S_{\text{gen}} = \dot{S}_{\text{gen}}\,dt = \text{entropy generated in the system in the time interval } dt$$

and obtain the following balance equations for the time interval dt:

$$dU = \hat{H}\,dM + Q - P\,dV + W_s \qquad\qquad \textbf{(4.2-10a)}$$

and

$$T\,dS = T\hat{S}\,dM + Q + T\,S_{\text{gen}} \qquad\qquad \textbf{(4.2-10b)}$$

Equations 4.2-10 can be used to interrelate the differential changes in internal energy, entropy, and volume that occur between fixed initial and final states that are only slightly different. This is accomplished by first solving Eq. 4.2-10b for Q,

$$Q = T\,dS - T\,S_{\text{gen}} - T\hat{S}\,dM \qquad\qquad \textbf{(4.2-11a)}$$

and using this result to eliminate the heat flow from Eq. 4.2-10a, to obtain[6]

$$dU = T\,dS - T\,S_{\text{gen}} - P\,dV + W_s + (\hat{H} - T\hat{S})\,dM$$
$$= T\,dS - T\,S_{\text{gen}} - P\,dV + W_s + \hat{G}\,dM \qquad\qquad \textbf{(4.2-11b)}$$

For a reversible process ($S_{\text{gen}} = 0$), these equations reduce to

$$Q^{\text{rev}} = T\,dS - T\hat{S}\,dM \qquad\qquad \textbf{(4.2-12a)}$$

and

$$dU = T\,dS + (-P\,dV + W_s)^{\text{rev}} + \hat{G}\,dM \qquad\qquad \textbf{(4.2-12b)}$$

In writing these equations we have recognized that since the initial and final states of the system are fixed, the changes in the path-independent functions are the same for reversible and irreversible processes. The heat and work terms that depend on the path followed are denoted by the superscript rev.

Equating the changes in the (state variables) internal energy and entropy for the reversible process (Eqs. 4.2-12) and the irreversible process (Eqs. 4.2-11) yields

$$Q^{\text{rev}} = Q^{\text{irrev}} + T\,S_{\text{gen}}$$

and

$$(-P\,dV + W_s)^{\text{rev}} = (-P\,dV + W_s)^{\text{irrev}} - T\,S_{\text{gen}}$$

---

[6]Alternatively, this equation and those that follow can be written in terms of the molar Gibbs energy and the mole number change by replacing $\hat{G}\,dM$ by $\underline{G}\,dN$.

Substituting these results into Eqs. 4.2-11 again gives Eqs. 4.2-12, which are now seen to apply for any process (reversible or irreversible) for which kinetic and potential changes are negligible. There is a subtle point here. That is, Eq. 4.2-12b can be used to interrelate the internal energy and entropy changes for any process, reversible or irreversible. However, to relate the heat flow Q and the entropy, Eq. 4.2-11a is always valid, whereas Eq. 4.2-12a can be used only for reversible processes.

Although both Eqs. 4.2-11b and 4-2.12b provide relationships between the changes in internal energy, entropy, mass, and the work flow for any real process, we will, in fact use only Eq. 4.2-12b to interrelate these changes since, in many cases, this simplifies the computation. Finally, we note that for a system with no shaft work[7]

Differential entropy change for an open system

$$dU = T\,dS - P\,dV + \hat{G}\,dM \qquad \text{(4.2-13a)}$$

and, further, if the system is closed to the flow of mass,

$$dU = T\,dS - P\,dV$$

or

Differential entropy change for a closed system

$$d\underline{U} = T\,d\underline{S} - P\,d\underline{V} \qquad \text{(4.2-13b)}$$

Since we are always free to choose the system for a given change of state such that there is no shaft work (only $P\,\Delta V$ work), Eq. 4.2-13a can generally be used to compute entropy changes in open systems. Similarly, Eq. 4.2-13b can generally be used to compute entropy changes in closed systems. This last point needs to be emphasized. In many cases it is necessary to compute the change in thermodynamic properties between two states of a substance. Since, for this computation, we can choose the system in such a way that it is closed and shaft work is excluded, Eq. 4.2-13b can be used in the calculation of the change in thermodynamic properties of the substance between the given initial and final states, regardless of the device or path used to accomplish this change.

4.3 HEAT, WORK, ENGINES, AND ENTROPY

The discussion of Sec. 4.2, and especially Eqs. 4.2-9, reveals an interesting distinction between mechanical energy (and work) and heat or thermal energy. In particular, while both heat and work are forms of energy, the relaxation processes (i.e., heat flow and viscous dissipation) that act naturally to reduce any temperature and velocity gradients in the system result in the conversion of mechanical energy in the form of work or the potential to do work, to heat or thermal energy. Friction in any moving object, which reduces its speed and increases its temperature, is the most common example of this phenomenon.

A problem of great concern to scientists and engineers since the late eighteenth century has been the development of devices (engines) to accomplish the reverse transformation, the conversion of heat (from the combustion of wood, coal, oil, natural gas, and other fuels) to mechanical energy or work. Much effort has been spent on developing engines of high efficiency, that is, engines that convert as large a fraction of the heat supplied to useful work as is possible.

---

[7]Equivalently, on a molar basis, we have $dU = T\,dS - P\,dV + \underline{G}\,dN$.

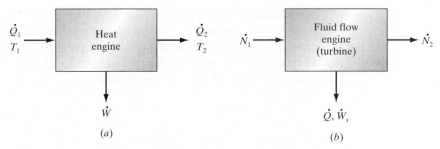

Figure 4.3-1 (*a*) Schematic diagram of a simple heat engine. (*b*) Schematic diagram of a fluid flow engine.

Such engines are schematically represented in Fig. 4.3-1*a*, where $\dot{Q}_1$ is the heat flow rate into the engine from the surroundings at temperature T_1, $\dot{Q}_2$ is the heat flow rate from the engine to its surroundings at temperature T_2, and $\dot{W}$ is the rate at which work is done by the engine. Remember that from the Clausius and Kelvin-Planck statements discussed earlier that $\dot{Q}_2$ cannot be zero; that is, some of the heat entering the engine cannot be converted to work and must be expelled at a lower temperature.

The engine in Fig. 4.3-1a may operate either in a steady-state fashion, in which case $\dot{Q}_1$, $\dot{Q}_2$, and $\dot{W}$ are independent of time, or cyclically. If the energy and entropy balances for the engine are integrated over a time interval Δt, which is the period of one cycle of the cyclic engine, or any convenient time interval for the steady-state engine, one obtains[8]

$$0 = Q_1 + Q_2 + W \tag{4.3-1}$$

$$0 = \frac{Q_1}{T_1} + \frac{Q_2}{T_2} + S_{\text{gen}} \tag{4.3-2}$$

where $Q = \int_{t_1}^{t_2} \dot{Q}\, dt$ and $W = \int_{t_1}^{t_2} (\dot{W}_s - P(dV/dt))\, dt$. The left sides of Eqs. 4.3-1 and 4.3-2 are zero for the cyclic and steady-state engines, as both engines are in the same state at time t_2 as they were at t_1. Eliminating Q_2 between Eqs. 4.3-1 and 4.3-2 yields

$$\text{Work done by the engine} = -W = Q_1 \left(\frac{T_1 - T_2}{T_1} \right) - T_2 S_{\text{gen}} \tag{4.3-3}$$

Based on the discussion of the previous section, to obtain the maximum work from an engine operating between fixed temperatures T_1 and T_2 it is necessary that all processes be carried out reversibly, so that $S_{\text{gen}} = 0$. In this case

$$\text{Maximum work done by the engine} = -W = Q_1 \left(\frac{T_1 - T_2}{T_1} \right) \tag{4.3-4}$$

and

$$\text{Engine efficiency} = \left(\begin{array}{c} \text{Fraction of heat} \\ \text{supplied that is} \\ \text{converted to work} \end{array} \right) = \frac{-W}{Q_1} = \frac{T_1 - T_2}{T_1} \tag{4.3-5}$$

---

[8]Independent of the way the arrows have been drawn in these figures, we still use the sign convention that a flow of heat, work, or mass into the system or device is positive, and a flow out is negative. The arrows in this and the following figures are drawn to remind the reader of the expected directions of the flows. Consequently, $\dot{Q}_2$ in Fig. 4.3-1a will be negative in value.

This is a surprising result since it establishes that, independent of the engine design, there is a maximum engine efficiency that depends only on the temperature levels between which the engine operates. Less than ideal design or irreversible operation of the engine will, of course, result in the efficiency of an engine being less than the maximum efficiency given in Eq. 4.3-5. Industrial heat engines, due to design and operating limitations, heat losses, and friction, typically operate at about half this efficiency.

Another aspect of the conversion of heat to work can be illustrated by solving for Q_2, the heat leaving the engine, in Eqs. 4.3-1 and 4.3-3 to get

$$-Q_2 = Q_1 \frac{T_2}{T_1} + T_2 S_{\text{gen}} \qquad (4.3\text{-}6)$$

This equation establishes that it is impossible to convert all the heat supplied to an engine to work (that is, for Q_2 to equal zero) in a continuous or cyclic process, unless the engine has a lower operating temperature (T_2) of absolute zero. In contrast, the inverse process, that of converting work or mechanical energy completely to heat, can be accomplished completely at any temperature, and unfortunately occurs frequently in natural processes. For example, friction can convert mechanical energy completely to heat, as when stopping an automobile by applying the brake, which converts mechanical energy (here kinetic energy) to heat in the brake drum.

Therefore, it is clear that although heat and work are equivalent in the sense that both are forms of energy (experimental observations 3 and 4 of Sec. 1.7), there is a real distinction between them in that work or mechanical energy can spontaneously (naturally) be converted completely to heat or thermal energy, but thermal energy can, with some effort, be only partially converted to mechanical energy. It is in this sense that mechanical energy is regarded as a higher form of energy than thermal energy.

At this point one might ask if it is possible to construct an engine having the efficiency given by Eq. 4.3-5. Nicolas Léonard Sadi Carnot (1796–1832) described such a cyclic engine in 1824. The **Carnot engine** consists of a fluid enclosed in a frictionless piston-and-cylinder device, schematically shown in Fig. 4.3-2. Work is extracted from this engine by the movement of the piston. In the first step of the four-part cycle, the fluid is isothermally and reversibly expanded from volume V_a to volume V_b at a constant temperature T_1 by adding an amount of heat Q_1 from the first heat source. The mechanical work obtained in this expansion is $\int_{V_a}^{V_b} P\, dV$. The next part of the cycle is a reversible adiabatic expansion of the fluid from the state (P_b, V_b, T_1) to the state (P_c, V_c, T_2). The work obtained in this expansion is $\int_{V_b}^{V_c} P\, dV$ and is gotten by directly converting the internal energy of the fluid to work. The next step in the cycle is to reversibly and isothermally compress the fluid to the state (P_d, V_d, T_2).

The work done on the fluid in this process is $\int_{V_c}^{V_d} P\, dV$, and the heat *removed* is Q_2. The final step in the cycle is a reversible adiabatic compression to the initial state (P_a, V_a, T_1). The work done on the fluid in this part of the process is $\int_{V_d}^{V_a} P\, dV$. The complete work cycle is summarized in the following table.

The energy and entropy balances for one complete cycle are

$$0 = W + Q_1 + Q_2 \qquad (4.3\text{-}7)$$

$$0 = \frac{Q_1}{T_1} + \frac{Q_2}{T_2} \qquad (4.3\text{-}8)$$

where

Carnot cycle work

$$W = -\int_{V_a}^{V_b} P\, dV - \int_{V_b}^{V_c} P\, dV - \int_{V_c}^{V_d} P\, dV - \int_{V_d}^{V_a} P\, dV$$

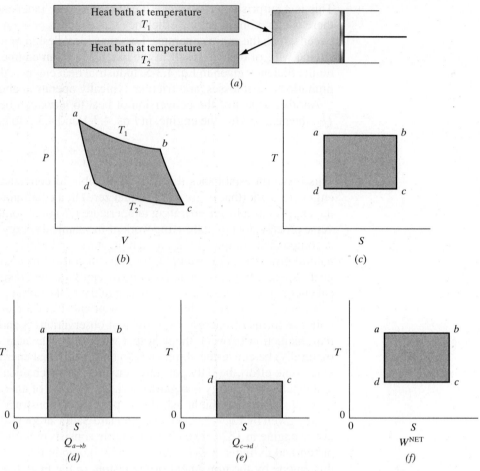

Figure 4.3-2 The Carnot cycle. (*a*) Schematic diagram of a Carnot engine. (*b*) The Carnot cycle on a pressure-volume plot. (*c*) The Carnot cycle on a temperature-entropy plot. (*d*) Heat flow into the cycle going from point *a* to *b*. (*e*) Heat flow into the cycle going from point *c* to point *d*. (*f*) Net work flow.

Path		Work Done on the Fluid	Heat Added to the Fluid
$(P_a, V_a, T_1) \xrightarrow[\text{expansion}]{\substack{\text{reversible} \\ \text{isothermal}}} (P_b, V_b, T_1)$		$-\int_{V_a}^{V_b} P\,dV$	Q_1
$(P_b, V_b, T_1) \xrightarrow[\substack{\text{adiabatic} \\ \text{expansion}}]{\text{reversible}} (P_c, V_c, T_2)$		$-\int_{V_b}^{V_c} P\,dV$	0
$(P_c, V_c, T_2) \xrightarrow[\text{compression}]{\substack{\text{reversible} \\ \text{isothermal}}} (P_d, V_d, T_2)$		$-\int_{V_c}^{V_d} P\,dV$	Q_2
$(P_d, V_d, T_2) \xrightarrow[\substack{\text{adiabatic} \\ \text{compression}}]{\text{reversible}} (P_a, V_a, T_1)$		$-\int_{V_d}^{V_a} P\,dV$	0

Now using Eq. 4.3-8 to eliminate Q_2 from Eq. 4.3-7 yields

$$-W = Q_1 - \left(\frac{T_2}{T_1}Q_1\right) = \frac{T_1 - T_2}{T_1}Q_1 \qquad \textbf{(4.3-9)}$$

and

Carnot cycle efficiency

$$\frac{-W}{Q_1} = \frac{T_1 - T_2}{T_1}$$

which is exactly the result of Eq. 4.3-4. Equations 4.3-5 and 4.3-6 then follow directly. Thus we conclude that the Carnot engine is the most efficient possible in the sense of extracting the most work from a given flow of heat between temperature baths at T_1 and T_2 in a cyclic or continuous manner.

Perhaps the most surprising aspect of the Carnot cycle engine (or Eq. 4.3-4, for that matter) is that the work obtained depends only on T_1, T_2, and the heat flow Q_1, and does not depend on the working fluid, that is, which fluid is used in the piston-and-cylinder device. Consequently, the efficiency, $-W/Q_1$, of the Carnot cycle depends only on the temperatures T_1 and T_2, and is the same for all fluids.

Since the work supplied or obtained in each step of the Carnot cycle is expressible in the form $-\int P\,dV$, the enclosed area on the P-V diagram of Fig. 4.3-2b is equal to the total work *supplied* by the Carnot engine to its surroundings in one complete cycle. (You should verify that if the Carnot engine is driven in reverse, so that the cycle in Fig. 4.3-2b is traversed counterclockwise, the enclosed area is equal to the work *absorbed* by the engine from its surroundings in one cycle.[9])

Since the differential entropy change dS and the heat flow Q for a reversible process in a closed system are related as

$$dS = \frac{Q}{T} \quad \text{or} \quad Q = T\,dS$$

the heat flows (and the work produced) can be related to areas on the T-S diagram for the process as follows. The heat flow into the Carnot cycle in going from point a to point b is equal to

$$Q_{a \to b} = \int T\,dS = T_1 \Delta S_{a \to b} = T_1 \cdot (S_b - S_a) \quad \text{since the temperature is constant}$$

This heat flow is given by the area shown in Fig. 4.3-2d. Similarly, the heat flow from point c to point d is equal in magnitude to the area shown in Fig. 4.3-2e, but negative in value (since S_c is larger than S_d). Finally, since from Eq. 4.3-7 the net work flow (work produced less work used in the isothermal compression step) is equal to the difference between the two heat flows into the Carnot cycle, this work flow is given by the rectangular area in Fig. 4.3-2f. A similar graphical analysis of the heat and work flows can be used for other cycles, as will be shown in Sec. 5.2.

Thus, for reversible cycles, the P-V diagram directly supplies information about the net work flow, and the T-S diagram provides information about the net heat flow. For

---

[9]Note that if the Carnot heat engine is operated as shown in Fig. 4.3-2 it absorbs heat from the high-temperature bath, exhausts heat to the low-temperature bath, and produces work. However, if the engine is operated in reverse, it accepts work, absorbs heat from the low-temperature bath, and exhausts heat to the high-temperature bath. In this mode it is operating as a refrigerator, air conditioner, or heat pump.

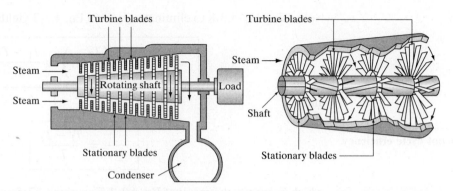

Figure 4.3-3 Illustration of a steam turbine, adapted from The World Book Encyclopedia © 1976 Field Enterprises Educational Corp. By permission of World Book, Inc. All rights reserved. This book may not be reproduced in whole or in part in any form without prior written permission of the publisher. www.worldbookonline.com.

irreversible processes (i.e., processes for which $S_{gen} \neq 0$), the heat flow and entropy change are not simply related as above, and the area on a T-S diagram is not directly related to the heat flow.

In addition to the Carnot heat engine, other cycles and devices may be used for the conversion of thermal energy to mechanical energy or work, although the conversion efficiencies for these other cycles, because of to the paths followed, are less than that of the Carnot engine. Despite their decreased efficiency, these other engines offer certain design and operating advantages over the Carnot cycle, and hence are more widely used. The efficiencies of some other cycles are considered in Sec. 5.2.

Another class of work-producing devices are engines that convert the thermal energy of a *flowing fluid* to mechanical energy. Examples of this type of engine are the nozzle-turbine systems of Fig. 4.3-3. Here a high-pressure, high-temperature fluid, frequently steam, is expanded through a nozzle to obtain a low-pressure, high-velocity gas. This gas then impinges on turbine blades, where the kinetic energy of the gas is transferred to the turbine rotor, and thus is available as shaft work. The resulting low-pressure, low-velocity gas leaves the turbine. Of course, many other devices can be used to accomplish the same energy transformation. All of these devices can be schematically represented as shown in Fig. 4.3-1*b*.

The steady-state mass, energy, and entropy balances on a molar basis for such heat engines are

$$0 = \dot{N}_1 + \dot{N}_2 \quad \text{or} \quad \dot{N}_2 = -\dot{N}_1$$
$$0 = (\underline{H}_1 - \underline{H}_2)\dot{N}_1 + \dot{Q} + \dot{W}_s \tag{4.3-10}$$

and

$$0 = (\underline{S}_1 - \underline{S}_2)\dot{N}_1 + \frac{\dot{Q}}{T} + \dot{S}_{gen} \tag{4.3-11}$$

Here we have assumed that the kinetic and potential energy changes of the fluid entering and leaving the device cancel or are negligible (regardless of what happens internally) and that the heat flow $\dot{Q}$ can be identified as occurring at a single temperature T.[10] Solving these equations for the heat flow $\dot{Q}$ and the work flow $\dot{W}_s$ yields

---

[10]If this is not the case, a sum of $\dot{Q}/T$ terms or an integral of the heat flux divided by the temperature over the surface of the system is needed.

$$\dot{Q} = -T(\underline{S}_1 - \underline{S}_2)\dot{N}_1 - T\dot{S}_{gen} \tag{4.3-12}$$

and

$$\dot{W}_s = -\dot{N}_1[(\underline{H}_1 - T\underline{S}_1) - (\underline{H}_2 - T\underline{S}_2)] + T\dot{S}_{gen} \tag{4.3-13}$$

Note that the quantity $(\underline{H}_1 - T\underline{S}_1)$ is *not* equal to the Gibbs energy unless the temperature T at which heat transfer occurs is equal to the inlet fluid temperature (that is, $\underline{G}_1 = \underline{H}_1 - T_1\underline{S}_1$); a similar comment applies to the term $(\underline{H}_2 - T\underline{S}_2)$.

Several special cases of these equations are important. First, for the isothermal flow engine (i.e., for an engine in which the inlet temperature T_1, the outlet temperature T_2, and the operating temperature T are all equal), Eq. 4.3-13 reduces to

$$\dot{W}_s = -\dot{N}_1(\underline{G}_1 - \underline{G}_2) + T\dot{S}_{gen}$$

The maximum rate at which work can be obtained from such an engine for fixed inlet and exit pressures and fixed temperature occurs when the engine is reversible; in this case

$$\dot{W}_s^{rev} = -\dot{N}_1(\underline{G}_1 - \underline{G}_2)$$

and the heat load for reversible, isothermal operation is

$$\dot{Q}^{rev} = -T\dot{N}_1(\underline{S}_1 - \underline{S}_2)$$

Second, for adiabatic operation ($\dot{Q} = 0$) of a flow engine, we have from Eq. 4.3-10 that

$$\dot{W}_s = -\dot{N}_1(\underline{H}_1 - \underline{H}_2) \tag{4.3-14}$$

and

$$0 = \dot{N}_1(\underline{S}_1 - \underline{S}_2) + \dot{S}_{gen} \tag{4.3-15}$$

so that the work flow is proportional to the difference in the enthalpies of the inlet and exiting fluids. Usually, the inlet temperature and pressure and the exit pressure of the adiabatic flow engine can be specified by the design engineer; the exit temperature cannot be specified, but instead adjusts so that Eq. 4.3-15 is satisfied. Thus, although the entropy generation term does not explicitly appear in the work term of Eq. 4.3-14, it is contained implicitly through the exit temperature and therefore the exit enthalpy $\underline{H}_2$. By example (Problems 4.9 and 4.24), one can establish that a reversible adiabatic engine has the lowest exit temperature and enthalpy for fixed inlet and exit pressures, and thereby achieves the best conversion of fluid thermal energy to work.

Finally, we want to develop an expression in terms of the pressure and volume for the maximum rate at which work is obtained, or the minimum rate at which work must be added, to accomplish a given change of state in continuous-flow systems such as turbines and compressors. Figure 4.3-4 is a generic diagram of a device through which fluid is flowing continuously. The volume element in the figure contained within the dashed lines is a very small region of length ΔL in which the temperature and pressure of the fluid can be taken to be approximately constant (in fact, shortly we will consider the limit in which $\Delta L \rightarrow 0$). The mass, energy, and entropy balances for this steady-state system are

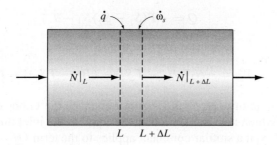

Figure 4.3-4 Device with fluid, heat, and work flows.

$$\frac{dN}{dt} = 0 = \dot{N}|_L - \dot{N}|_{L+\Delta L} \tag{4.3-16a}$$

$$\frac{dU}{dt} = 0 = \dot{N}\underline{H}|_L - \dot{N}\underline{H}|_{L+\Delta L} + \dot{q}\,\Delta L + \dot{\omega}_s\,\Delta L \tag{4.3-16b}$$

$$\frac{dS}{dt} = 0 = \dot{N}\underline{S}|_L - \dot{N}\underline{S}|_{L+\Delta L} + \frac{\dot{q}}{T}\,\Delta L + \dot{\sigma}_{\text{gen}}\,\Delta L \tag{4.3-16c}$$

In these equations, $\dot{q}$, $\dot{\omega}_s$, and $\dot{\sigma}_{\text{gen}}$ are, respectively, the heat and work flows and the rate of entropy generation per unit length of the device.

Dividing by ΔL, taking the limit as $\Delta L \to 0$, and using the definition of the total derivative from calculus gives

$$\lim_{\Delta L \to 0} \frac{\dot{N}|_{L+\Delta L} - \dot{N}|_L}{\Delta L} = \frac{d\dot{N}}{dL} = 0 \quad \text{or} \quad \dot{N} = \text{constant} \tag{4.3-17a}$$

$$\dot{N} \lim_{\Delta L \to 0} \left(\frac{\underline{H}|_{L+\Delta L} - \underline{H}|_L}{\Delta L} \right) = \dot{N}\frac{d\underline{H}}{dL} = \dot{q} + \dot{\omega}_s \tag{4.3-17b}$$

$$\dot{N} \lim_{\Delta L \to 0} \left(\frac{\underline{S}|_{L+\Delta L} - \underline{S}|_L}{\Delta L} \right) = \dot{N}\frac{d\underline{S}}{dL} = \frac{\dot{q}}{T} + \dot{\sigma}_{\text{gen}} \tag{4.3-17c}$$

From the discussion of the previous section, the maximum work that can be obtained, or the minimum work required, in a given change of state occurs in a reversible process. Setting $\dot{\sigma}_{\text{gen}} = 0$ yields

$$\dot{N}\frac{d\underline{S}}{dL} = \frac{\dot{q}}{T}$$

$$\dot{\omega}_s = \dot{N}\left(\frac{d\underline{H}}{dL} - T\frac{d\underline{S}}{dL} \right)$$

Now, from $\underline{H} = \underline{U} + P\underline{V}$,

$$d\underline{H} = d\underline{U} + d(P\underline{V}) = d\underline{U} + P\,d\underline{V} + \underline{V}\,dP$$

and from Eq. 4.2-13b,

$$d\underline{U} = T\,d\underline{S} - P\,d\underline{V}$$

we have

$$d\underline{H} - T\,d\underline{S} = T\,d\underline{S} - P\,d\underline{V} + P\,d\underline{V} + \underline{V}\,dP - T\,d\underline{S} = \underline{V}\,dP$$

or

$$\frac{d\underline{H}}{dL} - T\frac{d\underline{S}}{dL} = \underline{V}\frac{dP}{dL}$$

and

$$\dot{\omega}_s = \dot{N}\underline{V}\frac{dP}{dL}$$

Further, integrating over the length of the device gives

$$\dot{W}_s = \int \dot{\omega}_s \, dL = \dot{N}\int \underline{V}\frac{dP}{dL}\,dL$$

or

$$\underline{W}_s = \frac{\dot{W}_s}{\dot{N}} = \int \underline{V}\,dP \tag{4.3-18}$$

Equation 4.3-18 is the desired result.

It is useful to consider several applications of Eq. 4.3-18. For the ideal gas undergoing an isothermal change, so that $P\underline{V} = RT = $ constant,

$$\underline{W}_s = \int \underline{V}\,dP = \int \frac{RT}{P}\,dP = RT\int \frac{dP}{P} = RT\,\ln\frac{P_2}{P_1} \tag{4.3-19}$$

For an expansion or compression for which

$$P_1\underline{V}_1^{\gamma} = P_2\underline{V}_2^{\gamma} = \text{constant} \tag{4.3-20}$$

$$\begin{aligned}
\underline{W}_s &= \int \underline{V}\,dP = (\text{constant})^{1/\gamma}\int \frac{dP}{P^{1/\gamma}} \\
&= \frac{(\text{constant})^{1/\gamma}\left(P_2^{1-1/\gamma} - P_1^{1-1/\gamma}\right)}{1 - \dfrac{1}{\gamma}} \\
&= \frac{(P_2\underline{V}_2^{\gamma})^{1/\gamma}\,P_2^{1-1/\gamma} - (P_1\underline{V}_1^{\gamma})^{1/\gamma}\,P_1^{1-1/\gamma}}{1 - \dfrac{1}{\gamma}}
\end{aligned}$$

Work in a polytropic process

$$\boxed{\underline{W}_s = \frac{\gamma\,(P_2\underline{V}_2 - P_1\underline{V}_1)}{\gamma - 1}} \tag{4.3-21}$$

A process that obeys Eq. 4.3-20 is referred to as a **polytropic process**. For an ideal gas it is easy to show that

$$\gamma = 0 \quad \text{for an isobaric process}$$
$$\gamma = 1 \quad \text{for an isothermal process}$$
$$\gamma = \infty \quad \text{for a constant-volume (isochoric) process}$$

Also, we will show later that

$$\gamma = C_P^*/C_V^* \quad \text{for a constant-entropy (isentropic) process}$$
$$\text{in an ideal gas of constant heat capacity}$$

Note that since entropy is a state property, once two properties of a one-phase system, such as temperature and pressure, are fixed, the value of the entropy is also fixed. Consequently, the entropy of steam can be found in the steam tables or the Mollier diagram, and that of methane, nitrogen, and HFC-134a in the appropriate figures in Chapter 3. In the next section we consider entropy changes for an ideal gas, and in Chapter 6 we develop the equations to be used to compute entropy changes for nonideal fluids.

In preparation for the discussion of how entropy changes accompanying a change of state can be computed, it is useful to consider the thermodynamic balance equations for a change of state of a closed system. The difference form of the energy balance is

$$U_f - U_i = Q + W$$

Consequently we see that the change in the state variable U, the internal energy, is related to two path variables, the heat and the work. For a change between a given initial and final state, the internal energy change will always be the same, even though, as we have shown in the previous chapter, the heat and work flows along different paths will be different. Similarly, the entropy balance is

$$S_f - S_i = \frac{Q}{T} + S_{\text{gen}}$$

and we have a similar situation as above, where a change in a state function between two states is related to two path functions, here Q/T and the entropy generation.

Use of a reversible path to calculate the change in state variables

This discussion brings up a subtle, sometimes confusing, but very important point. Since properties such as the internal energy, enthalpy, and entropy are state functions, the changes in their values with a change of state depend only on the initial and final states, not on the path used to go between these two states. Therefore, in calculating a change in a state property, such as the internal energy or entropy, between fixed initial and final states, any convenient path can be used; to some degree this was demonstrated in Illustration 3.4-6. Frequently, a reversible path will be the most convenient for computing the change in the state variables between given initial and final states even though the actual system change is not reversible. However, for path variables such as heat flows, work flows, and entropy generation, the path is important, and different values for these quantities will be obtained along different paths. Therefore, to accurately compute the heat flows, the work flows, and the entropy generation, the actual path of the system change must be followed.

To conclude this section, we consider a brief introduction to the thermodynamic limits on the conversion of sunlight to electrical (or mechanical) energy.

ILLUSTRATION 4.3-1

Conversion of Radiant Energy to Mechanical or Electrical Energy

Show that a solar or photovoltaic cell that converts solar energy to mechanical or electrical energy must emit some of the energy of the incident radiation as heat.

SOLUTION

The way we will prove that a solar cell must emit heat is to assume that it does not, and show that this would be a violation of the entropy generation principle (that is, the second law of thermodynamics). The steady-state energy balance on a solar cell absorbing radiation but not releasing any heat is

$$0 = \dot{Q}_R + \dot{W}$$

or

$$-\dot{W} = \dot{Q}_R$$

and the entropy balance is

$$0 = \frac{4}{3} \frac{\dot{Q}_R}{T_R} + \dot{S}_{gen}$$

so that

$$\dot{S}_{gen} = -\frac{4}{3} \frac{\dot{Q}_R}{T_R}$$

The only way that $\dot{S}_{gen}$ can be greater than or equal to zero, which is necessary to satisfy the second law of thermodynamics, is if $\dot{Q}_R \leq 0$; that is, radiant energy must be released rather than absorbed. Therefore, we see that a solar cell also obeys the Kelvin-Planck statement of the second law (see Illustration 4.1-2). However, note that the cell can operate in the reverse manner in that it can receive electrical energy and completely convert it to radiant energy. A light-emitting diode (LED) or a metal wire electrically heated until it is red hot are two examples of complete conversion of electrical energy to radiant energy. ∎

ILLUSTRATION 4.3-2

Maximum Conversion of Solar Energy to Mechanical or Electrical Energy

Based on analysis of the frequency distribution of radiation from the sun, it can be considered to be emitting radiant energy with a Stefan-Boltzmann distribution at a temperature of 6000 K. Estimate the maximum efficiency with which this radiant energy can be converted to electrical (or mechanical) energy using solar cells (commonly called photovoltaic cells). For this analysis, assume that the solar cell is operating in steady state and is receiving radiant energy, that its surface temperature is 300 K, and that it is losing heat by conduction to the environment.

SOLUTION

The energy balance on this solar cell is

$$0 = \dot{Q}_1 + \dot{Q}_2 + \dot{W} = \dot{Q}_R + \dot{Q}_2 + \dot{W}$$

or

$$-\dot{W} = \dot{Q}_R + \dot{Q}_2$$

and the entropy balance is

$$0 = \frac{4}{3} \frac{\dot{Q}_R}{T_R} + \frac{\dot{Q}_2}{T} + \dot{S}_{gen}$$

For maximum conversion efficiency, $\dot{S}_{gen} = 0$. Therefore,

$$\dot{Q}_2 = -\frac{4}{3}\frac{\dot{Q}_R}{T_R}T$$

and

$$-\dot{W} = \dot{Q}_R - \frac{4}{3}\frac{\dot{Q}_R}{T_R}T = \dot{Q}_R\left(1 - \frac{4}{3}\frac{T}{T_R}\right)$$

and the maximum efficiency is

$$\frac{-\dot{W}}{\dot{Q}_R} = 1 - \frac{4}{3}\frac{T}{T_R}$$

With $T_R = 6000$ K and $T = 300$ K, the maximum efficiency for solar energy conversion is 0.9333, or 93.33 percent.

COMMENTS

The efficiency of commercial solar cells is generally 10 percent or less, rather the 93.33 percent from the calculation above. The most important reason for this is that is solar cells (and also biological cells in photosynthesis) can use radiant energy in just a small portion of the radiation frequency spectrum. Consequently, most of the radiation received by a solar cell is merely absorbed as heat, which is why many solar systems combine photovoltaic cells for the production of electricity with panels for heating water. Another loss factor is that some of the radiant energy received is reradiated to the sky (which can be considered a black body at a temperature of 0 K.)

Also, one should note that the efficiency calculated above is somewhat less than the Carnot efficiency,

$$\eta = 1 - \frac{300}{6000} = 0.95$$

operating between the same two temperatures. However, as there is no way to transfer heat at 6000 K from the the sun to the earth by conduction, the Carnot efficiency is not applicable. ∎

4.4 ENTROPY CHANGES OF MATTER

Equation 4.2-13b provides the basis for computing entropy changes for real fluids, and it will be used in that manner in Chapter 6. However, to illustrate the use of the entropy balance here in a simple way, we consider the calculation of the entropy change accompanying a change of state for 1 mol of an ideal gas, and for incompressible liquids and solids.

From the discussion of Sec. 3.3 the internal energy change and pressure of an ideal gas are

$$d\underline{U} = C_V^* \, dT \quad \text{and} \quad P = RT/\underline{V}$$

respectively, so that for 1 mol of an ideal gas we have

$$d\underline{S} = \frac{1}{T}\,d\underline{U} + \frac{P}{T}\,d\underline{V} = \frac{C_V^*}{T}\,dT + \frac{R}{\underline{V}}\,d\underline{V} \tag{4.4-1}$$

If C_V^* is independent of temperature, we can immediately integrate this equation to obtain

Ideal gas entropy change with T and $\underline{V}$ as independent variables if C_V^* is a constant

$$\begin{aligned}\underline{S}(T_2, \underline{V}_2) - \underline{S}(T_1, \underline{V}_1) &= C_V^* \int_{T_1}^{T_2} \frac{dT}{T} + R \int_{\underline{V}_1}^{\underline{V}_2} \frac{d\underline{V}}{\underline{V}} \\ &= C_V^* \ln\left(\frac{T_2}{T_1}\right) + R \ln\left(\frac{\underline{V}_2}{\underline{V}_1}\right)\end{aligned}$$

(4.4-2)

Using the ideal gas law, we can eliminate either the temperature or the volume in this equation and obtain expressions for the change in entropy with changes in temperature and pressure,

$$\begin{aligned}\underline{S}(T_2, P_2) - \underline{S}(T_1, P_1) &= C_V^* \ln\left(\frac{T_2}{T_1}\right) + R \ln \frac{RT_2/P_2}{RT_1/P_1} \\ &= (C_V^* + R) \ln\left(\frac{T_2}{T_1}\right) - R \ln\left(\frac{P_2}{P_1}\right)\end{aligned}$$

Entropy change of an ideal gas with T and P as independent variables if C_P^* is a constant

$$\underline{S}(T_2, P_2) - \underline{S}(T_1, P_1) = C_P^* \ln\left(\frac{T_2}{T_1}\right) - R \ln\left(\frac{P_2}{P_1}\right)$$

(4.4-3)

and pressure and volume:

$$\begin{aligned}\underline{S}(P_2, \underline{V}_2) - \underline{S}(P_1, \underline{V}_1) &= C_V^* \ln \frac{P_2 \underline{V}_2/R}{P_1 \underline{V}_1/R} + R \ln\left(\frac{\underline{V}_2}{\underline{V}_1}\right) \\ &= (C_V^* + R) \ln\left(\frac{\underline{V}_2}{\underline{V}_1}\right) + C_V^* \ln\left(\frac{P_2}{P_1}\right) \\ &= C_P^* \ln\left(\frac{\underline{V}_2}{\underline{V}_1}\right) + C_V^* \ln\left(\frac{P_2}{P_1}\right)\end{aligned}$$

(4.4-4)

The evaluation of the entropy change for an ideal gas in which the heat capacity is a function of temperature (see Eq. 3.3-5) leads to more complicated equations than those given here. It is left to the reader to develop the appropriate expressions (Problem 4.13).

For liquids or solids we can generally write

$$d\underline{S} = \frac{1}{T} d\underline{U} + \frac{P}{T} d\underline{V} \approx \frac{1}{T} d\underline{U}$$

(4.4-5)

since the molar volume is very weakly dependent on either temperature or pressure (i.e., $d\underline{V}$ is generally small). Furthermore, since for a liquid or a solid $C_V \approx C_P$, we have

$$d\underline{U} = C_V\, dT \approx C_P\, dT$$

for these substances, so that

$$d\underline{S} = C_P \frac{dT}{T}$$

and

Entropy change for a solid or liquid

$$\underline{S}(T_2) - \underline{S}(T_1) = \int_{T_1}^{T_2} C_P \frac{dT}{T}$$

(4.4-6)

Finally, if thermodynamic tables and charts that include entropy have been prepared for real fluids, the entropy changes accompanying a change in state can easily be calculated. In this way the entropy changes on a change of state for methane, nitrogen, HFC-134a, and steam can be calculated using the figures of Chapter 3 and the tables in Appendix A.III for steam.

ILLUSTRATION 4.4-1

Calculation of Entropy Generation for a Process

Compute the entropy generated on mixing 1 kg of steam at 1 bar and 200°C (state 1) with 1 kg of steam at 1 bar and 300°C (state 2).

SOLUTION

Considering the 2 kg of steam to be a closed system, the mass balance is

$$M_f - (M_1 + M_2) = 0 \quad \text{or} \quad M_f = M_1 + M_2 = 2 \text{ kg}$$

The energy balance is

$$M_f \hat{U}_f - M_1 \hat{U}_1 - M_2 \hat{U}_2 = -P(M_f \hat{V}_f - M_1 \hat{V}_1 - M_2 \hat{V}_2)$$

Since the pressure is constant, the U and PV terms can be combined to give

$$M_f \hat{H}_f - M_1 \hat{H}_1 - M_2 \hat{H}_2 = 0$$

or

$$2 \text{ kg} \cdot \hat{H}(T_f =?, P = 1 \text{ bar})$$
$$= 1 \text{ kg} \cdot \hat{H}(T = 200°C, P = 1 \text{ bar}) + 1 \text{ kg} \cdot \hat{H}(T = 300°C, P = 1 \text{ bar})$$

Therefore,

$$\hat{H}(T_f =?, P = 1 \text{ bar}) = \tfrac{1}{2}[\hat{H}(T = 200°C, P = 1 \text{ bar}) + \hat{H}(T = 300°C, P = 1 \text{ bar})]$$
$$= \tfrac{1}{2}[2875.3 + 3074.3] \frac{\text{kJ}}{\text{kg}} = 2974.8 \frac{\text{kJ}}{\text{kg}}$$

and from the steam tables we find for this value of the enthalpy at a pressure of 1 bar that $T_f = 250°C$. Now from the entropy balance, we have

$$M_f \hat{S}_f - M_1 \hat{S}_1 - M_2 \hat{S}_2 = S_{\text{gen}}$$

or

$$S_{\text{gen}} = 2 \text{ kg} \cdot \hat{S}(T = 250°C, P = 1 \text{ bar}) - 1 \text{ kg} \cdot \hat{S}(T = 200°C, P = 1 \text{ bar})$$
$$- 1 \text{ kg} \cdot \hat{S}(T = 300°C, P = 1 \text{ bar})$$
$$= \left(2 \text{ kg} \cdot 8.0333 \frac{\text{kJ}}{\text{kg K}} - 1 \text{ kg} \cdot 7.8343 \frac{\text{kJ}}{\text{kg K}} - 1 \text{ kg} \cdot 8.2158 \frac{\text{kJ}}{\text{kg K}}\right)$$
$$= 0.0165 \frac{\text{kJ}}{\text{K}}$$

So we see that mixing two fluids of the same pressure but different temperatures generates entropy and therefore is an irreversible process.   ∎

ILLUSTRATION 4.4-2

Illustration 3.4-1 Continued

Compute the entropy generated by the flow of 1 kg/s of steam at 400 bar and 500°C undergoing a Joule-Thomson expansion to 1 bar.

SOLUTION

In Illustration 3.4-1, from the energy balance, we found that

$$\hat{H}_1 = \hat{H}(T_1 = 500°C, P_1 = 400 \text{ bar}) = \hat{H}_2 = \hat{H}(T_2 = ?, P_2 = 1 \text{ bar})$$

and then by interpolation of the information in the steam tables that $T_2 = 214.1°C$. From the entropy balance on this steady-state system, we have

$$\frac{dS}{dt} = 0 = \dot{M}_1 \hat{S}_1 + \dot{M}_2 \hat{S}_2 + \dot{S}_{gen} = \dot{M}_1(\hat{S}_1 - \hat{S}_2) + \dot{S}_{gen}$$

so that

$$\dot{S}_{gen} = \dot{M}_1(\hat{S}_2 - \hat{S}_1)$$

From the steam tables (using interpolation to obtain the entropy of steam at 214.1°C and 1 bar), we have

$$\hat{S}_1 = \hat{S}(T = 500°C, P = 400 \text{ bar}) = 5.4700 \frac{\text{kJ}}{\text{kg K}}$$

and

$$\hat{S}_2 = \hat{S}(T = 214.1°C, P = 1 \text{ bar}) = 7.8904 \frac{\text{kJ}}{\text{kg K}}$$

Therefore

$$\dot{S}_{gen} = \dot{M}_1(\hat{S}_2 - \hat{S}_1) = 1 \frac{\text{kg}}{\text{s}} \cdot (7.8904 - 5.4700) \frac{\text{kJ}}{\text{kg K}} = 2.4202 \frac{\text{kJ}}{\text{K s}}$$

Since $\dot{S}_{gen} > 0$, the Joule-Thomson expansion is also an irreversible process. ∎

4.5 APPLICATIONS OF THE ENTROPY BALANCE

In this section we show, by example, that the entropy balance provides a useful additional equation for the analysis of thermodynamic problems. In fact, some of the examples considered here are continuations of the illustrations of the previous chapter, to emphasize that the entropy balance can provide the information needed to solve problems that were unsolvable using only the mass and energy balance equations, or, in some cases, to develop a simpler solution method for problems that were solvable (see Illustration 4.5-2).

ILLUSTRATION 4.5-1

Illustration 3.4-4 Continued, Using the Entropy Balance

In Illustration 3.4-4 we tried to estimate the exit temperature and power requirements for a gas compressor. From the steady-state mass balance we found that

$$\dot{N}_1 = -\dot{N}_2 = \dot{N} \tag{a}$$

and from the steady-state energy balance we had

$$\dot{W}_s = \dot{N} C_p^*(T_2 - T_1) \tag{b}$$

which resulted in one equation (Eq. b) with two unknowns, $\dot{W}_s$ and T_2. Now, writing a molar entropy balance for the same system yields

$$0 = (\underline{S}_1 - \underline{S}_2)\dot{N} + \dot{S}_{gen} \tag{c}$$

To obtain an estimate of the exit temperature and the power requirements, we assume that the compressor is well designed and operates reversibly, that is

$$\dot{S}_{gen} = 0 \tag{d}$$

Thus, we have

$$\underline{S}_1 = \underline{S}_2 \tag{e}$$

which is the additional relation for a state variable needed to solve the problem. Now using Eq. 4.4-3,

$$\underline{S}(P_2, T_2) - \underline{S}(P_1, T_1) = C_P^* \ln\left(\frac{T_2}{T_1}\right) - R \ln\left(\frac{P_2}{P_1}\right)$$

and recognizing that $\underline{S}(P_2, T_2) = \underline{S}(P_1, T_1)$ yields

$$\left(\frac{T_2}{T_1}\right) = \left(\frac{P_2}{P_1}\right)^{R/C_P^*} \tag{f}$$

or

$$T_2 = T_1 \left(\frac{P_2}{P_1}\right)^{R/C_P^*} = 290 \text{ K} \left(\frac{10}{1}\right)^{8.314/29.3} = 557.4 \text{ K} \quad \text{or} \quad 284.2°\text{C} \tag{g}$$

Thus T_2 is known, and hence $\underline{W}_s$ can be computed:

$$\underline{W}_s = C_P^*(T_2 - T_1) = 29.3 \times (557.4 - 290) = 7834.8 \text{ J/mol}$$

and

$$\dot{W}_s = \dot{N}\underline{W}_s = 2.5 \frac{\text{mol}}{\text{s}} \times 7834.8 \frac{\text{J}}{\text{mol}} = 19.59 \frac{\text{kJ}}{\text{s}}$$

Before considering the problem to be solved, we should try to assess the validity of the assumption $\dot{S}_{gen} = 0$. However, this can be done only by experiment. One method is to measure the inlet and exit temperatures and pressures for an adiabatic turbine and see if Eq. e is satisfied. Experiments of this type indicate that Eq. e is reasonably accurate, so that reversible operation is a reasonable approximation for a gas compressor. ∎

ILLUSTRATION 4.5-2

An Alternative Way to Solve One Problem

Sometimes it is possible to solve a thermodynamic problem several ways, based on different choices of the system. To see this, we consider Illustration 3.4-5, which was concerned with the partial evacuation of a compressed gas cylinder into an evacuated cylinder of equal volume. Suppose we now choose for the system of interest only that portion of the contents of the first cylinder that remains in the cylinder when the pressures have equalized (see Fig. 4.5-1, where the thermodynamic system of interest is within the dashed lines). Note that with this choice the system is closed, but of changing volume. Furthermore, since the gas on one side of the imaginary boundary has precisely the same temperature as the gas at the other side, we can assume there is no heat transfer across the boundary, so that the system is adiabatic. Also, with the exception of the region near the valve (which is *outside* what we have taken to be the

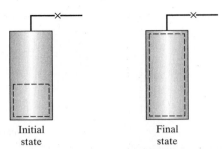

Initial state	Final state	

Figure 4.5-1 The dashed lines enclose a system consisting of gas initially in the first cylinder that remains in that cylinder at the end of the process.

system), the gas in the cylinder is undergoing a uniform expansion so there will be no pressure, velocity, or temperature gradients in the cylinder. Therefore, we can assume that the changes taking place in the system occur reversibly.

The mass, energy, and entropy balances (on a molar basis) for this system are

$$N_1^f = N_1^i \tag{a}$$

$$N_1^f \underline{U}_1^f = N_1^i \underline{U}_1^i - \int_{V_1^i}^{V_1^f} P\, dV \tag{b}$$

and

$$N_1^f \underline{S}_1^f = N_1^i \underline{S}_1^i \tag{c}$$

Now the important observation is that by combining Eqs. a and c, we obtain

$$\underline{S}_1^f = \underline{S}_1^i \tag{d}$$

so the process is **isentropic** (i.e., occurs at constant entropy) for the system we have chosen. Using Eq. 4.4-3,

$$\underline{S}_1(P^f, T^f) - \underline{S}_1(P^i, T^i) = C_P^* \ln\left(\frac{T_1^f}{T_1^i}\right) - R\ln\left(\frac{P_1^f}{P_1^i}\right)$$

and Eq. d yields

$$\left(\frac{T_1^f}{T_1^i}\right)^{C_P^*/R} = \left(\frac{P_1^f}{P_1^i}\right)$$

This is precisely the result obtained in Eq. f of Illustration 3.4-5 using the energy balance on the open system consisting of the total contents of cylinder 1. The remainder of the problem can now be solved in exactly the same manner used in Illustration 3.4-5.

Although the system choice used in this illustration is an unusual one, it is one that leads quickly to a useful result. This demonstrates that sometimes a clever choice for the thermodynamic system can be the key to solving a thermodynamic problem with minimum effort. However, one also has to be careful about the assumptions in unusual system choices. For example, consider the two cylinders connected as in Fig. 4.5-2, where the second cylinder is not initially evacuated. Here we have chosen to treat that part of the initial contents of cylinder 1 that will be in that cylinder at the end of the process as one system and the total initial contents of cylinder 2 as a second system. The change that occurs in the first system, as we already discussed, is adiabatic and reversible, so that

$$\left(\frac{T_1^f}{T_1^i}\right)^{C_P^*/R} = \left(\frac{P_1^f}{P_1^i}\right) \tag{e}$$

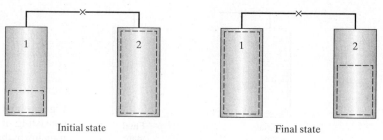

Figure 4.5-2 Incorrect system choice for gas contained in cylinder 2.

One might expect that a similar relation would hold for the system shown in cylinder 2. This is not the case, however, since the gas entering cylinder 2 is not necessarily at the same temperature as the gas already there (hydrodynamics will ensure that the pressures are the same). Therefore, temperature gradients will exist within the cylinder, and our system, the partial contents of cylinder 2, will not be adiabatic. Consequently, Eq. e will not apply.   ∎

ILLUSTRATION 4.5-3

Illustration 3.4-5 Continued, Using the Entropy Balance Instead of the Energy Balance

A gas cylinder of 1 m³ volume containing nitrogen initially at a pressure of 40 bar and a temperature of 200 K is connected to another cylinder of 1 m³ volume, which is evacuated. A valve between the two cylinders is opened only until the pressure in both cylinders equalizes and then is closed. Find the final temperature and pressure in each cylinder if there is no heat flow into or out of the cylinders, or between the gas and the cylinder walls. The properties of nitrogen gas are given in Fig. 3.3-3.

SOLUTION

Except for the fact that nitrogen is now being considered to be a real rather than an ideal gas, the problem here is the same as in Illustration 3.4-5. In fact, Eqs. h–n in the comments to that illustration apply here. The additional equation needed to solve this problem is obtained in the same manner as in Illustration 4.5-2. Thus, for the cylinder initially filled we have

$$\underline{S}_1^i = \underline{S}_1^f \tag{o}$$

For an ideal gas of constant heat capacity (which is not the case here) this reduces to

$$\left(\frac{P_1^f}{P_1^i}\right) = \left(\frac{T_1^f}{T_1^i}\right)^{C_P^*/R}$$

by Eq. 4.4-3. For real nitrogen gas, Eq. o requires that the initial and final states in cylinder 1 be connected by a line of constant entropy in Fig. 3.3-3.

With eight equations (Eqs. h–o) and eight unknowns, this problem can be solved, though the solution is a trial-and-error process. In general, a reasonable first guess for the pressure in the nonideal gas problem is the ideal gas solution, which was

$$P_1^f = P_2^f = 20 \text{ bar}$$

Locating the initial conditions ($P = 40$ bar $= 4$ MPa and $T = 200$ K) in Fig. 3.3-3 yields[11] $\hat{H}_1^i \approx 337$ kJ/kg and $\hat{V}_1^i \approx 0.0136$ m³/kg, so that

---

[11] Since the thermodynamic properties in Fig. 3.3-3 are on a mass basis, all calculations here will be on a mass, rather than a molar, basis.

$$M_1^i = \frac{V_1}{\hat{V}_1^i} = \frac{1 \text{ m}^3}{0.136 \dfrac{\text{m}^3}{\text{kg}}} = 73.5 \text{ kg}$$

Now using Eq. o (in the form $\hat{S}_1^i = \hat{S}_1^f$) and Fig. 3.3-3, we find $T_1^f \approx 165$ K, $H_1^f = 310$ kJ/kg, and $\hat{V}_1^f \approx 0.0224$ m^3/kg, so that

$$M_1^f = \frac{V_1}{\hat{V}_1^f} = \frac{1 \text{ m}^3}{0.0224 \dfrac{\text{m}^3}{\text{kg}}} = 44.6 \text{ kg}$$

and $M_2^f = M_1^i - M_1^f = 28.9$ kg, which implies

$$\hat{V}_2^f = \frac{1 \text{ m}^3}{28.9 \text{ kg}} = 0.0346 \frac{\text{m}^3}{\text{kg}}$$

Locating $P = 20$ bar (2 MPa) and $\hat{V} = 0.0346$ m^3/kg on Fig. 3.3-3 gives a value for T_2^f of about 240 K and $\hat{H}_2^f = 392$ kJ/kg.

Finally, we must check whether the energy balance is satisfied for the conditions computed here based on the final pressure *we have assumed*. To do this we must first compute the internal energies of the initial and final states as follows:

$$\hat{U}_1^i = \hat{H}_1^i - P_1^i \hat{V}_1^i$$
$$= 337 \frac{\text{kJ}}{\text{kg}} - 40 \text{ bar} \times 0.0136 \frac{\text{m}^3}{\text{kg}} \times 10^5 \frac{\text{Pa}}{\text{bar}} \times \frac{1 \text{ J}}{\text{m}^3 \text{ Pa}} \times \frac{1 \text{ kJ}}{1000 \text{ J}} = 282.6 \frac{\text{kJ}}{\text{kg}}$$

Similarly,

$$\hat{U}_1^f = 265.2 \text{ kJ/kg}$$

and

$$\hat{U}_2^f = 322.8 \text{ kJ/kg}$$

The energy balance (Eq. n of Illustration 3.4-5) on a mass basis is

$$M_1^i \hat{U}_1^i = M_1^f \hat{U}_1^f + M_2^f \hat{U}_2^f$$

or

$$73.5 \times 282.6 \text{ kJ} = 44.6 \times 265.2 + 28.9 \times 322.8 \text{ kJ}$$
$$2.0771 \times 10^4 \text{ kJ} \approx 2.1157 \times 10^4 \text{ kJ}$$

Thus, to the accuracy of our calculations, the energy balance can be considered to be satisfied and the problem solved. Had the energy balance not been satisfied, it would have been necessary to make another guess for the final pressures and repeat the calculation.

It is interesting to note that the solution obtained here is essentially the same as that for the ideal gas case. This is not generally true, but occurs here because the initial and final pressures are sufficiently low, and the temperature sufficiently high, that nitrogen behaves as an ideal gas. Had we chosen the initial pressure to be higher, say several hundred bars, the ideal gas and real gas solutions would have been significantly different (see Problem 4.22). ∎

ILLUSTRATION 4.5-4

Illustration 3.4-6 Continued, Showing That Entropy Is a State Function

Show that the entropy S is a state function by computing ΔS for each of the three paths of Illustration 3.4-6.

SOLUTION

Since the piston-and-cylinder device is frictionless (see Illustration 3.4-6), each of the expansion processes will be reversible (see also Illustration 4.5-8). Thus, the entropy balance for the gas within the piston and cylinder reduces to

$$\frac{dS}{dt} = \frac{\dot{Q}}{T}$$

Path A

i. *Isothermal compression.* Since T is constant,

$$\Delta \underline{S}_A = \frac{Q_A}{T} = -\frac{5707.7 \text{ J/mol}}{298.15 \text{ K}} = -19.14 \text{ J/(mol K)}$$

ii. *Isobaric heating*

$$\dot{Q} = C_P^* \frac{dT}{dt}$$

so

$$\frac{d\underline{S}}{dt} = \frac{\dot{Q}}{T} = \frac{C_P^*}{T} \frac{dT}{dt}$$

and

$$\Delta \underline{S}_B = C_P^* \ln \frac{T_2}{T_1} = 38 \frac{\text{J}}{\text{mol K}} \times \ln \frac{573.15}{298.15} = 24.83 \text{ J/(mol K)}$$

$$\Delta \underline{S} = \Delta \underline{S}_A + \Delta \underline{S}_B = -19.14 + 24.83 = 5.69 \text{ J/(mol K)}$$

Path B

i. *Isobaric heating*

$$\Delta \underline{S}_A = C_P^* \ln \frac{T_2}{T_1} = 24.83 \text{ J/(mol K)}$$

ii. *Isothermal compression*

$$\Delta \underline{S}_B = \frac{Q}{T} = -\frac{10\,972.2}{573.15} = -19.14 \text{ J/(mol K)}$$

$$\Delta \underline{S} = 24.83 - 19.14 = 5.69 \text{ J/(mol K)}$$

Path C

i. *Compression with* $PV^\gamma = constant$

$$\Delta \underline{S}_A = \frac{Q}{T} = 0$$

ii. *Isobaric heating*

$$\Delta \underline{S}_B = C_P^* \ln \frac{T_3}{T_2} = 38 \frac{\text{J}}{\text{mol K}} \times \ln \frac{573.15}{493.38} = 5.69 \text{ J/(mol K)}$$

$$\Delta \underline{S} = 5.69 \text{ J/(mol K)}$$

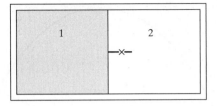

Figure 4.5-3 A well-insulated box divided into two equal compartments.

COMMENT

This example verifies, at least for the paths considered here, that the entropy is a state function. For reversible processes in closed systems, the rate of change of entropy and the ratio $\dot{Q}/T$ are equal. Thus, for reversible changes, $\dot{Q}/T$ is also a state function, even though the total heat flow $\dot{Q}$ is a path function. ∎

ILLUSTRATION 4.5-5
Showing That the Entropy Reaches a Maximum at Equilibrium in a Closed, Isolated System

(In Sec. 4.1 we established that the entropy function will be a maximum at equilibrium in an isolated system. This is illustrated by example for the system shown here.)

Figure 4.5-3 shows a well-insulated box of volume 6 m^3 divided into two equal volumes. The left-hand cell is initially filled with air at 100°C and 2 bar, and the right-hand cell is initially evacuated. The valve connecting the two cells will be opened so that gas will slowly pass from cell 1 to cell 2. The wall connecting the two cells conducts heat sufficiently well that the temperature of the gas in the two cells will always be the same. Plot on the same graph (1) the pressure in the second tank versus the pressure in the first tank, and (2) the change in the total entropy of the system versus the pressure in tank 1. At these temperatures and pressures, air can be considered to be an ideal gas of constant heat capacity.

SOLUTION

For this system

$$\text{Total mass} = N = N_1 + N_2$$
$$\text{Total energy} = U = U_1 + U_2$$
$$\text{Total entropy} = S = S_1 + S_2 = N_1 \underline{S}_1 + N_2 \underline{S}_2$$

From the ideal gas equation of state and the fact that $V = N\underline{V}$, we have

$$N_1^i = \frac{PV}{RT} = \frac{2 \text{ bar} \times 3 \text{ m}^3}{8.314 \times 10^{-5} \, \dfrac{\text{bar m}^3}{\text{mol K}} \times 373.15 \text{ K}} = 193.4 \text{ mol} = 0.1934 \text{ kmol}$$

Now since $U = $ constant, $T_1 = T_2$, and for the ideal gas $\underline{U}$ is a function of temperature only, we conclude that $T_1 = T_2 = 100°C$ at all times. This result greatly simplifies the computation. Suppose that the pressure in cell 1 is decreased from 2 bar to 1.9 bar by transferring some gas from cell 1 to cell 2. Since the temperature in cell 1 is constant, we have, from the ideal gas law,

$$N_1 = 0.95 N_1^i$$

and by mass conservation, $N_2 = 0.05 N_1^i$. Applying the ideal gas relation, we obtain $P_2 = 0.1$ bar.

For any element of gas, we have, from Eq. 4.4-3,

$$\underline{S}^f - \underline{S}^i = C_P^* \ln \frac{T^f}{T^i} - R \ln \frac{P^f}{P^i} = -R \ln \frac{P^f}{P^i}$$

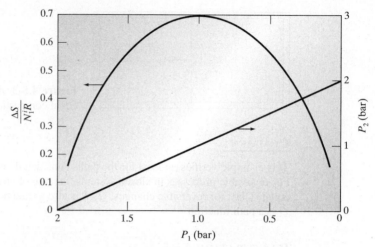

Figure 4.5-4 The system entropy change and the pressure in cell 2 as a function of the pressure in cell 1.

since temperature is constant. Therefore, to compute the change in entropy of the system, we visualize the process of transferring $0.05N_1^i$ moles of gas from cell 1 to cell 2 as having two effects:

1. To decrease the pressure of the $0.95N_1^i$ moles of gas remaining in cell 1 from 2 bar to 1.9 bar
2. To decrease the pressure of the $0.05N_1^i$ moles of gas that have been transferred from cell 1 to cell 2 from 2 bar to 0.1 bar

Thus

$$S^f - S^i = -0.95N_1^i R \ \ln(1.9/2) - 0.05N_1^i R \ \ln(0.1/2)$$

or

$$\frac{\Delta S}{N_1^i R} = -0.95 \ln 0.95 - 0.05 \ln 0.05 \qquad \left\{ \begin{array}{l} P_1 = 1.9 \text{ bar} \\ \\ P_2 = 0.1 \text{ bar} \end{array} \right.$$
$$= 0.199$$

Similarly, if $P_1 = 1.8$ bar,

$$\frac{\Delta S}{N_1^i R} = -0.9 \ln 0.9 - 0.1 \ln 0.1 \qquad \left\{ \begin{array}{l} P_1 = 1.8 \text{ bar} \\ \\ P_2 = 0.2 \text{ bar} \end{array} \right.$$
$$= 0.325$$

and so forth. The results are plotted in Fig. 4.5-4.

From this figure it is clear that ΔS, the change in entropy from the initial state, and therefore the total entropy of the system, reaches a maximum value when $P_1 = P_2 = 1$ bar. Consequently, the equilibrium state of the system under consideration is the state in which the pressure in both cells is the same, as one would expect. (Since the use of the entropy function leads to a solution that agrees with one's intuition, this example should reinforce confidence in the use of the entropy function as a criterion for equilibrium in an isolated constant-volume system.)   ∎

ILLUSTRATION 4.5-6

Showing That the Energy and Entropy Balances Can Be Used to Determine Whether a Process Is Possible

An engineer claims to have invented a steady-flow device that will take air at 4 bar and 20°C and separate it into two streams of equal mass, one at 1 bar and −20°C and the second at 1 bar and 60°C. Furthermore, the inventor states that his device operates adiabatically and does not require (or produce) work. Is such a device possible? [Air can be assumed to be an ideal gas with a constant heat capacity of $C_P^* = 29.3$ J/(mol K)].

SOLUTION

The three principles of thermodynamics—(1) conservation of mass, (2) conservation of energy, and (3) $\dot{S}_{gen} \geq 0$—must be satisfied for this or any other device. These principles can be used to test whether any device can meet the specifications given here.

The steady-state mass balance equation for the open system consisting of the device and its contents is $dN/dt = 0 = \sum_k \dot{N}_k = \dot{N}_1 + \dot{N}_2 + \dot{N}_3$. Since, from the problem statement, $\dot{N}_2 = \dot{N}_3 = -\frac{1}{2}\dot{N}_1$, mass is conserved. The steady-state energy balance for this device is

$$\frac{dU}{dt} = 0 = \sum_k \dot{N}_k \underline{H}_k = \dot{N}_1 \underline{H}_1 - \tfrac{1}{2}\dot{N}_1 \underline{H}_2 - \tfrac{1}{2}\dot{N}_1 \underline{H}_3$$

$$= \dot{N}_1 C_P^* (293.15 \text{ K} - \tfrac{1}{2} \times 253.15 \text{ K} - \tfrac{1}{2} \times 333.15 \text{ K}) = 0$$

so the energy balance is also satisfied. Finally, the steady-state entropy balance is

$$\frac{dS}{dt} = 0 = \sum_K \dot{N}_k \underline{S}_k + \dot{S}_{gen} = \dot{N}_1 \underline{S}_1 - \tfrac{1}{2}\dot{N}_1 \underline{S}_2 - \tfrac{1}{2}\dot{N}_1 \underline{S}_3 + \dot{S}_{gen}$$

$$= \tfrac{1}{2}\dot{N}_1 [(\underline{S}_1 - \underline{S}_2) + (\underline{S}_1 - \underline{S}_3)] + \dot{S}_{gen}$$

Now using Eq. 4.4-3, we have

$$\dot{S}_{gen} = -\frac{1}{2}\dot{N}_1 \left(C_P^* \ln \frac{T_1}{T_2} - R \ln \frac{P_1}{P_2} + C_P^* \ln \frac{T_1}{T_3} - R \ln \frac{P_1}{P_3} \right)$$

$$= -\frac{1}{2}\dot{N}_1 \left(29.3 \ln \frac{293.15 \times 293.15}{253.15 \times 333.15} - 8.314 \ln \frac{4 \times 4}{1 \times 1} \right) = 11.25\dot{N}_1 \frac{\text{J}}{\text{K s}}$$

Therefore, we conclude, on the basis of thermodynamics, that it *is* possible to construct a device with the specifications claimed by the inventor. Thermodynamics, however, gives us no insight into how to design such a device. That is an engineering problem.

Two possible devices are indicated in Fig. 4.5-5. The first device consists of an air-driven turbine that extracts work from the flowing gas. This work is then used to drive a heat pump (an air conditioner or refrigerator) to cool part of the gas and heat the rest. The second device, the Hilsch-Ranque vortex tube, is somewhat more interesting in that it accomplishes the desired change of state with only a valve and no moving parts. In this device the air expands as it enters the tube, thus gaining kinetic energy at the expense of internal energy (i.e., at the end of the expansion process we have high-velocity air of both lower pressure and lower temperature than the incoming air). Some of this cooled air is withdrawn from the center of the vortex tube. The rest of the air swirls down the tube, where, as a result of viscous dissipation, the kinetic energy is dissipated into heat, which increases the internal energy (temperature) of the air. Thus, the air being withdrawn at the valve is warmer than the incoming air.

(a) Turbine-heat pump system

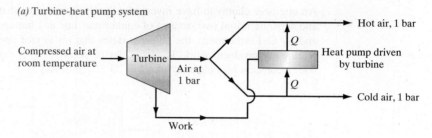

(b) Hilsch-Ranque vortex tube

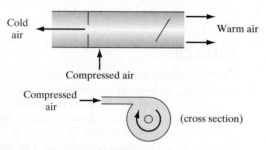

Figure 4.5-5 Two devices to separate compressed air into two low-pressure air streams of different temperature.   ■

ILLUSTRATION 4.5-7

Another Example of Using the Entropy Balance in Problem Solving

A steam turbine operates at the following conditions:

	Inlet	Outlet
Velocity (m/min)	2000	7500
T (K)	800	440
P (MPa)	3.5	0.15
Flow rate (kg/hr)	10 000	
Heat loss (kJ/hr)	125 000	

a. Compute the horsepower developed by the turbine and the entropy change of the steam.

b. Suppose the turbine is replaced with one that is well insulated, so that the heat loss is eliminated, and well designed, so that the expansion is reversible. If the exit pressure and velocity are maintained at the previous values, what are the outlet steam temperature and the horsepower developed by the turbine?

SOLUTION

The steady-state mass and energy balances on the turbine and its contents (the system) yield

$$\frac{dM}{dt} = 0 = \dot{M}_1 + \dot{M}_2 \qquad \dot{M}_2 = -\dot{M}_1 = -10\ 000\ \text{kg/hr}$$

$$\frac{d}{dt}\left[U + M\left(\frac{v^2}{2} + gh\right)\right] = 0 = \dot{M}_1\left(\hat{H}_1 + \frac{v_1^2}{2}\right) + \dot{M}_2\left(\hat{H}_2 + \frac{v_2^2}{2}\right) + \dot{W}_s + \dot{Q}$$

a. From the Mollier diagram of Fig. 3.3-1 (or Appendix A.III),

$$\hat{H}_1 \simeq 3510 \text{ J/g}$$

and

$$\hat{S}_1 \simeq 7.23 \text{ J/(g K)}$$

Also,

$$\frac{v_1^2}{2} = \frac{1}{2}\left(\frac{2000 \text{ m/min}^2}{60 \text{ s/min}}\right)^2 \times \frac{1 \text{ J/kg}}{\text{m}^2/\text{s}^2} \times \frac{1 \text{ kg}}{1000 \text{ g}} = 0.56 \text{ J/g}$$

so that

$$\hat{H}_1 + \frac{v_1^2}{2} = 3510.6 \text{ J/g}$$

Similarly,

$$\hat{H}_2 \simeq 2805 \text{ J/g} \qquad \hat{H}_2 + \frac{v_2^2}{2} = 2805 + 7.8 = 2812.8 \text{ J/g}$$

and

$$\hat{S}_2 \simeq 7.50 \text{ J/(g K)}$$

Therefore,

$$-\dot{W}_s = (3510.6 - 2812.8)\frac{\text{J}}{\text{g}} \times 10\,000\,\frac{\text{kg}}{\text{hr}} \times 1000\,\frac{\text{g}}{\text{kg}} - 12.5 \times 10^4\,\frac{\text{kJ}}{\text{hr}} \times 1000\,\frac{\text{J}}{\text{kJ}}$$

$$= 6.853 \times 10^9\,\frac{\text{J}}{\text{hr}} \times \frac{1 \text{ kJ}}{1000 \text{ J}} \times \frac{1 \text{ hr}}{3600 \text{ s}} = 1903.6 \text{ kJ/s} = 1.9036 \times 10^6 \text{ W}$$

$$= 2553 \text{ hp}$$

Also,

$$\Delta\hat{S} = (\hat{S}_2 - \hat{S}_1) = 0.27 \text{ J/(g K)}$$

b. The steady-state entropy balance for the turbine and its contents is

$$\frac{dS}{dt} = 0 = \dot{M}_1\hat{S}_1 - \dot{M}_1\hat{S}_2 + \dot{S}_{\text{gen}}$$

since $\dot{Q} = 0$, and $\dot{M}_2 = -\dot{M}_1$. Also, the turbine operates reversibly so that $\dot{S}_{\text{gen}} = 0$, and $\hat{S}_1 = \hat{S}_2$; that is, the expansion is isentropic. We now use Fig. 3.3-1, the entropy-enthalpy plot (Mollier diagram) for steam, to solve this problem. In particular, we locate the initial steam conditions ($T = 800$ K, $P = 3.5$ MPa) on the chart and follow a line of constant entropy (a vertical line on the Mollier diagram) to the exit pressure (0.15 MPa), to obtain the enthalpy of the exiting steam ($\hat{H}_2 \approx 2690$ J/g) and its final temperature ($T \approx 373$ K). Since the exit velocity is known, we can immediately compute the horsepower generated by the turbine:

$$-\dot{W}_s = [(3510 + 0.6) - (2690 + 7.8)]\frac{\text{J}}{\text{g}} \times 10\,000\frac{\text{kg}}{\text{hr}} \times \frac{1000 \text{ g/kg}}{1000 \text{ J/kJ}} \times \frac{1 \text{ hr}}{3600 \text{ s}}$$

$$= 2257.8 \text{ kJ/s} = 2.2578 \times 10^6 \text{ W} = 3028 \text{ hp}$$

COMMENTS

1. Here, as before, the kinetic energy term is of negligible importance compared with the internal energy term.
2. Notice from the Mollier diagram that the turbine exit steam is right at the boundary of a two-phase mixture of vapor and liquid. For the solution of this problem, no difficulties arise if the exit steam is a vapor, a liquid, or a two-phase vapor-liquid mixture since our mass, energy, and entropy balances are of general applicability. In particular, the information required to use these balance equations is the internal energy, enthalpy, and entropy per unit mass of each of the flow streams. Provided we have this information, the balance equations can be used independent of whether the flow streams consist of single or multiple phases, or, in fact, single or multiple components. Here the Mollier diagram provides the necessary thermodynamic information, and the solution of this problem is straightforward.
3. Finally, note that more work is obtained from the turbine by operating it in a reversible and adiabatic manner. ∎

ILLUSTRATION 4.5-8

Showing That $S_{gen} = 0$ for a Reversible Process, and $S_{gen} > 0$ for an Irreversible Process

a. By considering only the gas contained within the piston-and-cylinder device of Illustration 3.4-7 to be the system, show that the gas undergoes a reversible expansion in each of the four processes considered in that illustration. That is, show that $S_{gen} = 0$ for each process.
b. By considering the gas, piston, and cylinder to be the system, show that processes a, b, and c of Illustration 3.4-7 are not reversible (i.e., $S_{gen} > 0$), and that process d is reversible.

SOLUTION

a. The entropy balance for the 1 mol of gas contained in the piston and cylinder is

$$\underline{S}_f - \underline{S}_i = \frac{Q}{T} + S_{gen}$$

where T is the constant temperature of this isothermal system and Q is the total heat flow (from both the thermostatic bath and the cylinder walls) to the gas. From Eq. g of Illustration 3.4-7, we have for the 1 mol of gas

$$Q = RT \ln \frac{\underline{V}_f}{\underline{V}_i}$$

and from Eq. 4.4-2, we have

$$\underline{S}_f - \underline{S}_i = R \ln \frac{\underline{V}_f}{\underline{V}_i}$$

since the temperature of the gas is constant. Thus

$$S_{gen} = \underline{S}_f - \underline{S}_i - \frac{Q}{T} = R \ln \frac{\underline{V}_f}{\underline{V}_i} - \frac{1}{T} \left\{ RT \ln \frac{\underline{V}_f}{\underline{V}_i} \right\} = 0$$

so that the gas undergoes a reversible expansion in all four processes.

b. The entropy balance for the isothermal system consisting of 1 mol of gas and the piston and cylinder is

$$S_f - S_i = \frac{Q}{T} + S_{gen}$$

where Q is the heat flow to the piston, cylinder, and gas (Q^{NET} of Illustration 3.4-7) and $S_f - S_i$ is the entropy change for that composite system:

$$S_f - S_i = (\underline{S}_f - \underline{S}_i)_{\text{gas}} + (S_f - S_i)_{\text{piston-cylinder}}$$

Since the system is isothermal,

$$(\underline{S}_f - \underline{S}_i)_{\text{gas}} = R \ln \frac{\underline{V}_f}{\underline{V}_i} \quad \text{(see Eq. 4.4-2)}$$

and

$$(S_f - S_i)_{\text{piston-cylinder}} = 0 \quad \text{(see Eq. 4.4-6)}$$

Consequently,

$$S_{\text{gen}} = R \ln \frac{\underline{V}_f}{\underline{V}_i} - \frac{Q}{T} = \frac{1622.5 - Q^{\text{NET}}}{298.15} \text{ J/K}$$

so we find, using the entries in Table 1 of Illustration 3.4-7, that

$$S_{\text{gen}} = \begin{cases} 1.4473 \text{ J/K} & \text{for process a} \\ 0.8177 \text{ J/K} & \text{for process b} \\ 0.4343 \text{ J/K} & \text{for process c} \\ 0 & \text{for process d} \end{cases}$$

Thus, we conclude that for the piston, cylinder, and gas system, processes a, b, and c are not reversible, whereas process d is reversible.

COMMENT

From the results of part (a) we find that for the gas all expansion processes are reversible (i.e., there are no dissipative mechanisms within the gas). However, from part (b), we see that when the piston, cylinder, and gas are taken to be the system, the expansion process is irreversible unless the expansion occurs in differential steps. The conclusion, then, is that the irreversibility, or the dissipation of mechanical energy to thermal energy, occurs between the piston and the cylinder. This is, of course, obvious from the fact that the only source of dissipation in this problem is the friction between the piston and the cylinder wall.   ∎

4.6   THE MICROSCOPIC ENTROPY BALANCE (OPTIONAL)

This section appears on the CD that accompanies this text.

PROBLEMS

4.1 A 5-kg copper ball at 75°C is dropped into 12 kg of water, initially at 5°C, in a well-insulated container.
  a. Find the common temperature of the water and copper ball after the passage of a long period of time.
  b. What is the entropy change of the water in going from its initial to final state? Of the ball? Of the composite system of water and ball?

Data:
$C_P(\text{copper}) = 0.5$ J/(g K)
$C_P(\text{water}) = 4.2$ J/(g K)

4.2 In a foundry, metal castings are cooled by quenching in an oil bath. Typically, a casting weighing 20 kg and at a temperature of 450°C is cooled by placing it in a 150-kg involatile oil bath initially at 50°C. If C_P of the metal is 0.5 J/(kg K), and C_P of the oil 2.6 J/(kg K), determine the common final temperature of the oil and casting after quenching if there are no heat losses. Also, find the entropy change in this process.

4.3 **a.** Show that the rate at which shaft work is obtained or required for a reversible change of state in a closed system at constant internal energy and volume is equal to the negative of the product of the temperature and the rate of change of the entropy for the system.

b. Show that the rate at which shaft work is obtained or required for a reversible change of state in a closed system at constant entropy and pressure is equal to the rate of change of enthalpy of the system.

4.4 Steam at 700 bar and 600°C is withdrawn from a steam line and adiabatically expanded to 10 bar at a rate of 2 kg/min. What is the temperature of the steam that was expanded, and what is the rate of entropy generation in this process?

4.5 Two metal blocks of equal mass M of the same substance, one at an initial temperature T_1^i and the other at an initial temperature T_2^i, are placed in a well-insulated (adiabatic) box of constant volume. A device that can produce work from a flow of heat across a temperature difference (i.e., a heat engine) is connected between the two blocks. Develop expressions for the maximum amount of work that can be obtained from this process and the common final temperature of the blocks when this amount of work is obtained. You may assume that the heat capacity of the blocks does not vary with temperature.

4.6 The compressor discussed in Illustrations 3.4-4 and 4.5-1 is being used to compress air from 1 bar and 290 K to 10 bar. The compression can be assumed to be adiabatic, and the compressed air is found to have an outlet temperature of 575 K.

a. What is the value of ΔS for this process?

b. How much work, W_s, is needed per mole of air for the compression?

c. The temperature of the air leaving the compressor here is higher than in Illustration 4.5-1. How do you account for this?

In your calculations you may assume air is an ideal gas with $C_P^* = 29.3$ J/(mol K).

4.7 A block of metal of total heat capacity C_P is initially at a temperature of T_1, which is higher than the ambient temperature T_2. Determine the maximum amount of work that can be obtained on cooling this block to ambient temperature.

4.8 The Ocean Thermal Energy Conversion (OTEC) project in Hawaii produces electricity from the temperature difference between water near the surface of the ocean (about 27°C) and the 600 m deep water at 5°C that surrounds the island. Estimate the maximum net work (total work less the work of pumping the water to the surface) that can be obtained from each kilogram of water brought to the surface, and the overall efficiency of the process.

4.9 **a.** A steam turbine in a small electric power plant is designed to accept 4500 kg/hr of steam at 60 bar and 500°C and exhaust the steam at 10 bar. Assuming that the turbine is adiabatic and has been well designed (so that $\dot{S}_{gen} = 0$), compute the exit temperature of the steam and the power generated by the turbine.

b. The efficiency of a turbine is defined to be the ratio of the work actually obtained from the turbine to the work that would be obtained if the turbine operated isentropically between the same inlet and exit pressures. If the turbine in part (a) is adiabatic but only 80 percent efficient, what would be the exit temperature of the steam? At what rate would entropy be generated within the turbine?

c. In off-peak hours the power output of the turbine in part (a) (100 percent efficient) is decreased by adjusting a throttling valve that reduces the turbine inlet steam pressure to 30 bar (see diagram) while keeping the flow rate constant. Compute T_1, the steam temperature to the turbine, T_2, the steam temperature at the turbine exit, and the power output of the turbine.

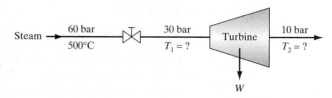

4.10 Complete part (b) of Problem 3.31, assuming the compressor operates reversibly and adiabatically.

4.11 Steam is produced at 70 bar and some unknown temperature. A small amount of steam is bled off just before entering a turbine and goes through an adiabatic throttling valve to atmospheric pressure. The temperature of the steam exiting the throttling valve is 400°C. The unthrottled steam is fed into the turbine, where it is adiabatically expanded to atmospheric pressure.

a. What is the temperature of the steam entering the turbine?

b. What is the maximum work per kilogram of steam that can be obtained using the turbine in its present mode of operation?

c. Tests on the turbine exhaust indicate that the steam leaving is a saturated vapor. What is the efficiency of the turbine and the entropy generated per kilogram of steam?

d. If the ambient temperature is 25°C and the ambient pressure is 1 bar, what is the maximum possible work that could be obtained per kilogram of steam in any continuous process?

4.12 A well-insulated, 0.7-m³ gas cylinder containing natural gas (which can be considered to be pure methane) at 70 bar and 300 K is exhausted until the pressure drops to 3.5 bar. This process occurs fast enough that there is no heat transfer between the cylinder walls and the gas, but not so rapidly as to produce large velocity or temperature gradients in the gas within the cylinder. Compute the number of moles of gas withdrawn and the final temperature of the gas in the cylinder if

a. Methane gas is assumed to be ideal with $C_P^* = 36$ J/(mol K).

b. Methane is considered to be a real gas with the properties given in Fig. 3.3-2.

4.13 If the heat capacity of an ideal gas is given by

$$C_V^* = (a - R) + bT + cT^2 + dT^3 + e/T^2$$

show that

$$S(T_2, \underline{V}_2) - S(T_1, \underline{V}_1) = (a - R) \ln\left(\frac{T_2}{T_1}\right) + b(T_2 - T_1)$$
$$+ \frac{c}{2}(T_2^2 - T_1^2) + \frac{d}{3}(T_2^3 - T_1^3)$$
$$- \frac{e}{2}(T_2^{-2} - T_1^{-2}) + R \ln\left(\frac{\underline{V}_2}{\underline{V}_1}\right)$$

Also develop expressions for this fluid that replace Eqs. 4.4-3 and 4.4-4.

4.14 a. Steam at 35 bar and 600 K enters a throttling valve that reduces the steam pressure to 7 bar. Assuming there is no heat loss from the valve, what is the exit temperature of the steam and its change in entropy?

b. If air [assumed to be an ideal gas with $C_P^* = 29.3$ J/(mol K)] entered the valve at 35 bar and 600 K and left at 7 bar, what would be its exit temperature and entropy change?

4.15 A tank contains 20 percent liquid water and 80 percent steam by volume at 200°C. Steam is withdrawn from the top of the tank until the fluid remaining in the tank is at a temperature of 150°C. Assuming the tank is adiabatic and that only vapor is withdrawn, compute

a. The pressure in the tank finally

b. The fraction of vapor and liquid in the tank finally

c. The fraction of the total water present initially that was withdrawn

4.16 One mole of carbon dioxide is to be compressed adiabatically from 1 bar and 25°C to 10 bar. Because of

irreversibilities and poor design of the compressor, the compressor work required is found to be 25 percent greater than that for a well-designed (reversible) compressor. Compute the outlet temperature of the carbon dioxide and the work that must be supplied to the compressor for both the reversible and irreversible compressors for the two cases below.

a. Carbon dioxide is an ideal gas with a constant-pressure heat capacity of 37.151 J/(mol K).

b. Carbon dioxide is an ideal gas with the constant-pressure heat capacity given in Appendix A.II.

4.17 Hydrogen has an auto-ignition temperature of 853 K; that is, hydrogen will ignite spontaneously at that temperature if exposed to oxygen. Hydrogen is to be adiabatically and reversibly compressed from 1 bar and 300 K to a high pressure. To avoid accidental explosions in case of a leak, the maximum allowed exit temperature from the compressor will be 800 K. Compute the maximum pressure that can be obtained in the compressor. You may consider hydrogen to be an ideal gas with the heat capacity given in Appendix A.II.

4.18 If it is necessary to compress hydrogen to a higher pressure than is possible with the single-compression step above, an alternative is to use two compressors (or a two-stage compressor) with intercooling. In such a process the hydrogen is compressed in the first stage of the compressor, then cooled at constant pressure to a lower temperature, and then compressed further in a second compressor or stage. Although it may not be economical to do so, more than two stages can be used.

a. Compute the maximum pressure that can be obtained in a two-stage compression with intercooling to 300 K between the stages, assuming hydrogen to be an ideal gas with the heat capacity given in Appendix A.II.

b. Repeat the calculation above for a three-stage compression with intercooling to 300 K.

4.19 Joe Unidel claims to have invented a steady-state flow device in which the inlet is steam at 300°C and 5 bar, the outlet is saturated steam at 100°C and 1 bar, the device is adiabatic and produces approximately 388 kJ per kilogram of steam passed through the device. Should we believe his claim?

4.20 Steam at 20 bar and 300°C is to be continuously expanded to 1 bar.

a. Compute the final temperature, the entropy generated, the heat required, and the work obtained per kilogram of steam if this expansion is done by passing the steam through an adiabatic expansion valve. Will the final state be a vapor, a liquid, or a vapor-liquid mixture?

b. Compute the final temperature, the entropy generated, the heat required, and the work obtained per kilogram of steam if this expansion is done by passing the steam through a well-designed, adiabatic turbine. Will the final state be a vapor, a liquid, or a vapor-liquid mixture?

   c. Compute the final temperature, the entropy generated, the heat required, and the work obtained per kilogram of steam if this expansion is done by passing the steam through a well-designed, isothermal turbine. Will the final state be a vapor, a liquid, or a vapor-liquid mixture?

4.21 In a large refrigeration plant it is necessary to compress a fluid, which we will assume to be an ideal gas with constant heat capacity, from a low pressure P_1 to a much higher pressure P_2.

   a. If the compression is done in a single compressor that operates reversibly and adiabatically, obtain an expression for the work needed for the compression in terms of the mass flow rate, P_1, P_2, and the initial temperature, T_1.

   b. If the compression is to be done in two stages, first compressing the gas from P_1 to P^*, then cooling the gas at constant pressure down to the compressor inlet temperature T_1, and then compressing the gas to P_2, develop an expression for the work needed for the compression. What should the value of the intermediate pressure be to accomplish the compression with minimum work?

4.22 Repeat Problem 3.25, now considering nitrogen to be a real gas with the thermodynamic properties given in Fig. 3.3-3.

4.23 An isolated chamber with rigid walls is divided into two equal compartments, one containing steam at 10 bar and 370°C, and the other evacuated. A valve between the compartments is opened to permit steam to pass from one chamber to the other.

   a. After the pressures (but not the temperatures) in the two chambers have equalized, the valve is closed, isolating the two systems. What are the temperature and pressure in each cylinder?

   b. If the valve were left open, an equilibrium state would be obtained in which each chamber has the same temperature and pressure. What are this temperature and pressure?

(*Note:* Steam is not an ideal gas under the conditions here.)

4.24 An adiabatic turbine is operating with an ideal gas working fluid of fixed inlet temperature and pressure, T_1 and P_1, respectively, and a fixed exit pressure, P_2. Show that

   a. The minimum outlet temperature, T_2, occurs when the turbine operates reversibly, that is, when $S_{gen} = 0$.

   b. The maximum work that can be extracted from the turbine is obtained when $S_{gen} = 0$.

4.25 a. Consider the following statement: "Although the

entropy of a given system may increase, decrease, or remain constant, the entropy of the universe cannot decrease." Is this statement true? Why?

   b. Consider any two states, labeled 1 and 2. Show that if state 1 is accessible from state 2 by a real (irreversible) adiabatic process, then state 2 is inaccessible from state 1 by a real adiabatic process.

4.26 A very simple solar engine absorbs heat through a collector. The collector loses some of the heat it absorbs by convection, and the remainder is passed through a heat engine to produce electricity. The heat engine operates with one-half the Carnot efficiency with its low-temperature side at ambient temperature T_{amb} and its high-temperature side at the steady-state temperature of the collector, T_c. The expression for the heat loss from the collector is

$$\text{Rate of heat loss from the collector} = k\,(T_c - T_{amb})$$

where T_1 is the temperature at which heat enters the solar collector, and k is the overall heat transfer coefficient; all temperatures are absolute. What collector temperature produces the maximum rate at which work is produced for a given heat flux to the collector?

4.27 It is necessary to estimate how rapidly a piece of equipment can be evacuated. The equipment, which is 0.7 m^3 in volume, initially contains carbon dioxide at 340 K and 1 bar pressure. The equipment will be evacuated by connecting it to a reciprocating constant-displacement vacuum pump that will pump out 0.14 m^3/min of gas at any conditions. At the conditions here carbon dioxide can be considered to be an ideal gas with $C_P^* = 39$ J/(mol K).

   a. What will be the temperature and pressure of the carbon dioxide inside the tank after 5 minutes of pumping if there is no exchange of heat between the gas and the process equipment?

   b. The gas exiting the pump is always at 1 bar pressure, and the pump operates in a reversible adiabatic manner. Compute the temperature of the gas exiting the pump after 5 minutes of operation.

4.28 A 0.2-m^3 tank containing helium at 15 bar and 22°C will be used to supply 4.5 moles per minute of helium at atmospheric pressure using a controlled adiabatic throttling valve.

   a. If the tank is well insulated, what will be the pressure in the tank and the temperature of the gas stream leaving the throttling valve at any later time t?

   b. If the tank is isothermal, what will be the pressure in the tank as a function of time?

You may assume helium to be an ideal gas with $C_P^* = 22$ J/(mol K), and that there is no heat transfer between the tank and the gas.

4.29 A portable engine of nineteenth-century design used a tank of compressed air and an "evacuated" tank as its power source. The first tank had a capacity of 0.3 m³ and was initially filled with air at 14 bar and a temperature of 700°C. The "evacuated" tank had a capacity of 0.75 m³. Unfortunately, nineteenth-century vacuum techniques were not very efficient, so the "evacuated" tank contained air at 0.35 bar and 25°C. What is the maximum total work that could be obtained from an air-driven engine connected between the two tanks if the process is adiabatic? What would be the temperature and pressure in each tank at the end of the process? You may assume that air is an ideal gas with $C_P^* = 7R/2$.

4.30 A heat exchanger is a device in which heat flows between two fluid streams brought into thermal contact through a barrier, such as a pipe wall. Heat exchangers can be operated in either the cocurrent (both fluid streams flowing in the same direction) or countercurrent (streams flowing in opposite direction) configuration; schematic diagrams are given here.

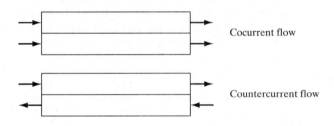

Cocurrent flow

Countercurrent flow

The heat flow rate from fluid 1 to fluid 2 per unit length of the heat exchanger, $\dot{Q}$, is proportional to the temperature difference $(T_1 - T_2)$:

$$\dot{Q} = \begin{pmatrix} \text{Heat flow rate from fluid 1} \\ \text{to fluid 2 per unit length of} \\ \text{heat exchanger} \end{pmatrix} = \kappa(T_1 - T_2)$$

where κ is a constant of proportionality with units of J/(m s K). The fluids in the two streams are the same and their flow rates are equal. The initial and final temperatures of stream 1 will be 35°C and 15°C, respectively, and those for stream 2 will be −15°C and 5°C.

a. Write the balance equations for each fluid stream in a portion of the heat exchanger of length dL and obtain differential equations by letting $dL \to 0$.

b. Integrate the energy balance equations over the length of the exchanger to obtain expressions for the temperature of each stream at any point in the exchanger for each flow configuration. Also compute the length of the exchanger, in units of $L_0 = \dot{M}C_P/2\kappa$ (where $\dot{M}$ is the mass flow rate of either stream), needed to accomplish the desired heat transfer.

c. Write an expression for the change of entropy of stream 1 with distance for any point in the exchanger.

4.31 One measure of the thermodynamic efficiency of a complex process (e.g., an electrical generating station or an automobile engine) is the ratio of the useful work obtained for a specified change of state to the maximum useful work obtainable with the ambient temperature T_0 and pressure P_0. Here by **useful work** we mean the total work done by the system less the work done in expansion of the system boundaries against the ambient pressure.

Clearly, to obtain the maximum useful work all processes should occur reversibly, and all heat transferred to the surroundings should leave the system at T_0, since heat available at any other temperature could be used with a Carnot or similar engine to obtain additional work. For a similar reason, the feed and exit streams in an open process should also be at the ambient conditions.

a. Show that the maximum useful work for a closed-system change of state is

$$W_u^{max} = \mathcal{A}_2 - \mathcal{A}_1$$

where $\mathcal{A} = U + P_0 V - T_0 S$ is the closed-system **availability** function.

b. For a steady-state or cyclic-flow system, show that the maximum useful work is

$$W_u^{max} = M(\hat{B}_2 - \hat{B}_1)$$

where $\hat{B} = \hat{H} - T_0\hat{S}$ is the flow availability function and M is the mass that has entered the system in any convenient time interval in a steady-flow process or in one complete cycle in a cyclic process. [Note that since the inlet and exit temperatures and pressures are equal, $\hat{B}_1$ will equal $\hat{B}_2$ and W_u^{max} will equal zero unless a chemical reaction, (for example, the burning of coal or gasoline, a phase change, or some other change in the composition of the inlet and exit streams) occurs.]

c. Compute the maximum useful work that can be obtained when 1 kg of steam undergoes a closed-system change of state from 30 bar and 600°C to 5 bar and 300°C when the atmospheric temperature and pressure are as given in the data.

d. A coal-fired power generation station generates approximately 2.2 kWh of electricity for each kilogram of coal (composition given below) burned.

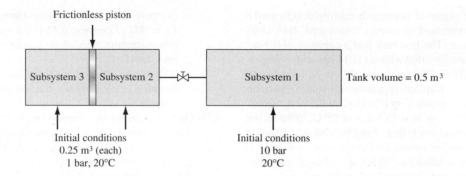

Frictionless piston

Subsystem 3 | Subsystem 2 — ⋈ — | Subsystem 1 | Tank volume = 0.5 m³

Initial conditions
0.25 m³ (each)
1 bar, 20°C

Initial conditions
10 bar
20°C

Using the availability concept, compute the thermodynamic efficiency of the station.

Data:
Ambient conditions: 25°C and 1.013 bar

Chemical composition of coal (simplified):

Carbon	70 weight percent
Water (liquid)	15 weight percent
Inorganic matter	15 weight percent

	Enthalpy at 25°C, 1.013 bar (kJ/mol)	Entropy at 25°C 1.013 bar [kJ/(mol K)]
O_2	0.0	0.0
C	0.0	0.0
CO_2	−393.51	−394.4
H_2O (liq)	−285.84	−237.1
H_2O (vap)	−241.84	−228.6

Assume that the inorganic matter (ash) passes through the power station unchanged, that all the water leaves as vapor in the flue gas, and that all the carbon is converted to carbon dioxide.

4.32 Two tanks are connected as shown here. Tank 1 initially contains an ideal gas at 10 bar and 20°C, and both parts of tank 2 contain the same gas at 1 bar and 20°C. The valve connecting the two tanks is opened long enough to allow the pressures in the tanks to equilibrate and is then shut. There is no transfer of heat from the gas to the tanks or the frictionless piston, and the constant-pressure heat capacity of the gas is $4R$. Compute the temperature and pressure of the gas in each part of the system at the end of the process and the work done on the gas behind the piston (i.e., the gas in subsystem 3).

4.33 An important factor in determining the extent of air pollution in urban areas is whether the atmosphere is stable (poor mixing, accumulation of pollutants) or unstable (good mixing and dispersion of pollutants). Whether the atmosphere is stable or unstable depends on how the temperature profile (the so-called lapse rate) in the atmosphere near ground level compares with the "adiabatic lapse rate." The adiabatic lapse rate is the atmospheric temperature that would result at each elevation if a packet of air (for example, as contained in a balloon) rose from ground level without any heat or mass transfer into or out of the packet. Assuming that air is an ideal gas, with a molecular weight of 29 and a constant-pressure heat capacity of 29.2 J/(mol K), obtain expressions for the pressure profile and the adiabatic lapse rate in the atmosphere.

4.34 A device is being marketed to your company. The device takes in hot, high-pressure water and generates work by converting it to two outlet streams: steam and low-pressure water. You are asked to evaluate the device to see if your firm should buy it. All you are told is that 10 kW of work is obtained under the following operating conditions:

	Inlet	Outlet Stream 1	Outlet Stream 2
Mass flow rate (kg/s)	1.0		0.5
T (°C)	300	100	300
P (MPa)	5.0		
Steam	Liquid	Quality unknown	Sat. liquid

a. What is the quality of the steam in outlet stream 1?
b. Is this device thermodynamically feasible?

4.35 Diesel engines differ from gasoline engines in that the fuel is not ignited by a spark plug. Instead, the air in the cylinder is first compressed to a higher pressure than in a gasoline engine, and the resulting high temperature results in the spontaneous ignition of the fuel when it is injected into the cylinder.

a. Assuming that the compression ratio is 25:1 (that is, the final volume of the air is 1/25 times the initial volume), that air can be considered an ideal gas with $C_V^* = 21$J/(mol K), that the initial conditions of the air are 1 atm and 30°C, and that the compression is adiabatic and reversible, determine the temperature of the air in the cylinder before the fuel is injected, and the minimum amount of work that must be done in the compression.

b. The compression ratio in a supercharged gasoline engine is usually 10:1 (or less). Repeat the calculations of part (a) for this compression ratio.

4.36 A rigid, isolated container 300 L in volume is divided into two parts by a partition. One part is 100 L in volume and contains nitrogen at a pressure of 200 kPa and a temperature of 100°C. The other part, 200 L in volume, contains nitrogen at a pressure of 2 MPa and a temperature of 200°C. If the partition breaks and a sufficiently long time elapses for the temperature and pressure to become uniform within the container, assuming that there is no heat transfer to the container and that nitrogen is an ideal gas with $C_P^* = 30\,\text{J/(mol K)}$

   a. What is the final temperature and pressure of the gas in the container?

   b. How much entropy is generated in this process?

4.37 In normal operation, a paper mill generates excess steam at 20 bar and 400°C. It is planned to use this steam as the feed to a turbine to generate electricity for the mill. There are 5000 kg/hr of steam available, and it is planned that the exit pressure of the steam will be 2 bar. Assuming that the turbine is well insulated, what is the maximum power that can be generated by the turbine?

4.38 Atmospheric air is to be compressed and heated as shown in the figure before being fed into a chemical reactor. The ambient air is at 20°C and 1 bar, and is first compressed to 5 bar and then heated at constant pressure to 400°C. Assuming air can be treated as a single-component ideal gas with $C_V^* = 21\ \text{J/(mol K)}$, and that the compressor operates adiabatically and reversibly,

   a. Determine the amount of work that must be supplied to the compressor and the temperature of the gas leaving the compressor.

   b. Determine the heat load on the heat exchanger.

4.39 One kilogram of saturated liquid methane at 160 K is placed in an adiabatic piston-and-cylinder device, and the piston will be moved slowly and reversibly until 25 percent of the liquid has vaporized. Compute the maximum work that can be obtained.

4.40 A sugar mill in Florida has been disposing of the bagasse (used sugar cane) by open-air burning. You, as a new chemical engineer, determine that by using the dried bagasse as boiler fuel in the mill, you can generate 5000 kg/hr of surplus steam at 20 bar and 600°C. You propose to your supervisor that the mill invest in a steam turbine–electrical generator assembly to generate electricity that can be sold to the local power company.

   a. If you purchase a new, state-of-the-art adiabatic and isentropic steam turbine, and the turbine exit pressure is 1 bar, what is the temperature of the exit steam, how much electrical energy (in kW) can be generated, and what fraction of the steam exiting the turbine is vapor?

   b. Instead, a much cheaper used turbine is purchased, which is adiabatic but not isentropic, that at the same exit pressure of 1 bar produces 90 percent of the work produced by the state-of-the-art turbine. What is the temperature of the exit steam, what fraction of the steam exiting the turbine is vapor, and what is the rate of entropy generation?

4.41 A well-insulated cylinder fitted with a frictionless piston initially contains nitrogen at 15°C and 0.1 MPa. The piston is placed such that the volume to the right of the piston is 0.5 m³, and there is negligible volume to the left of the piston and between the piston and the tank, which is also well insulated. Initially the tank of volume 0.25 m³ contains nitrogen at 400 kPa and 200°C. For the calculations below, assume that nitrogen is an ideal gas with $C_P^* = 29.3\,\text{J/(mol K)}$

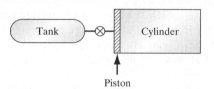

Piston

   a. The piston is a perfect thermal insulator, and the valve between the cylinder and the tank is opened only long enough for the pressure in the tank and cylinder to equalize. What will the equalized pressure be, and what will be the temperatures of the nitrogen in the tank, in the cylinder to the left of the piston, and in the cylinder to the right of the piston?

   b. What will the volumes in the cylinder be to the left and to the right of the piston?

   c. How much entropy has been generated by the process?

4.42 If the piston in the previous problem is replaced by one that is a perfect conductor, so that after the valve is closed the temperatures of the gas on the two sides of the piston equalize (clearly, the piston will move in this process),

   a. What will the equalized pressure in the cylinder be, and what will be the temperature of the nitrogen in the cylinder?

   b. What will the volumes in the cylinder be to the left and to the right of the piston?

   c. How much entropy has been generated in going

from the initial state (before the valve was opened) to the final state (equalized temperature and pressure on the two sides of the piston)?

4.43 A tank of 0.1 m³ volume initially containing nitrogen at 25°C and 1 bar will be filled with compressed nitrogen at a rate of 20 mol/s. The nitrogen coming from the compressor and into the tank is at an absolute pressure of 110 bar and a temperature of 80°C. The filling process occurs sufficiently rapidly that there is negligible heat transfer between the gas and the tank walls, and a valve is closed to stop the filling process when the pressure in the tank reaches 100 bar. Assuming nitrogen is an ideal gas with $C_P^* = 29.3 \text{ J/(mol K)}$,

   a. What is the nitrogen temperature immediately after the filling process ends?

   b. How long did it take for the pressure of the gas in the tank to reach 100 bar?

   c. After a sufficiently long period of time, due to heat transfer with the surroundings, the temperature of the gas drops to 25°C. What is the pressure of the nitrogen in the tank?

4.44 A stream of hot water at 85°C and a rate of 1 kg/s is needed for the pasteurizing unit in a milk-bottling plant. Such a stream is not readily available, and will be produced in a well-insulated mixing tank by directly injecting steam from the boiler plant at 10 bar and 200°C into city water available at 1 bar and 20°C.

   a. Calculate the flow rates of city water and steam needed.

   b. Calculate the rate of entropy production in the mixing tank.

4.45 Low-density polyethylene is manufactured from ethylene at medium to high pressure in a radical chain polymerization process. The reaction is exothermic, and on occasion, because of cooling system failure or operator error, there are runaway reactions that raise the pressure and temperature in the reactor to dangerous levels. For safety, there is a relief valve on the reactor that will open when the total pressure reaches 200 bar and discharge the high-pressure ethylene into a holding tank with a volume that is four times larger than that of the reactor, thereby reducing the risk of explosion and discharge into the environment. Assuming the volume occupied by the polymer is very small, that the gas phase is pure ethylene, that in case of a runaway reaction the temperature in the reactor will rise to 400°C, and that the holding tank is initially evacuated, what will be the temperature and pressure in the reactor and the holding tank if ethylene can be considered to be an ideal gas with $C_P^* = 73.2 \text{ J/(mol K)}$?

4.46 An inventor has proposed a flow device of secret design for increasing the superheat of steam. He claims that a feed steam at 1 bar and 100°C is converted to superheated steam at 1 bar and 250°C and saturated water (that is, water at its boiling point) also at 1 bar, and that no additional heat or work is needed. He also claims that 89.9 percent of the outlet product is superheated steam, and the remainder is saturated water. Will the device work as claimed?

Chapter **5**

Liquefaction, Power Cycles, and Explosions

In this chapter we consider several practical applications of the energy and entropy balances that we have developed. These include the liquefaction of a gas, and the analysis of cycles used to convert heat to work and to provide refrigeration and air conditioning. Finally, as engineers we have a social responsibility to consider safety as a paramount issue in anything we design or operate. Therefore, methods to estimate the energy that could be released in different types of nonchemical explosions are presented.

INSTRUCTIONAL OBJECTIVES FOR CHAPTER 5

The goals of this chapter are for the student to:

- Be able to solve problems involving the liquefaction of gases
- Be able to compute the work that can be obtained from different types of power cycles and using different working fluids
- Be able to compute the work required for the operation of refrigeration cycles
- Be able to compute the the energy release resulting from the uncontrolled expansion of a gas
- Be able to compute the the energy release resulting from an explosion that involves a boiling liquid

NOTATION INTRODUCED IN THIS CHAPTER

C.O.P.	Coefficient of performance of a refrigeration cycle
$\triangle_{\text{vap}}\hat{H}$	Enthalpy change (or heat) of vaporization per unit mass (J/kg)
$\triangle_{\text{vap}}\underline{H}$	Molar enthalpy change on vaporization (J/mol)
T_{b}	Boiling temperature of a liquid at atmospheric pressure (K)
$\triangle_{\text{vap}}\hat{U}$	Internal energy change on vaporization per unit mass (J/kg)
$\triangle_{\text{vap}}\underline{U}$	Molar internal energy change on vaporization (J/mol)

5.1 LIQUEFACTION

An important industrial process is the liquefaction of gases, such as natural gas (to produce LNG), propane, and refrigerant gases, to name a few. One way to liquefy a gas

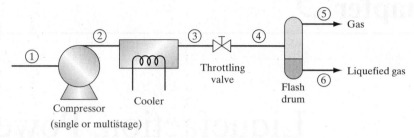

Figure 5.1-1 A simple liquefaction process without recycle.

is to cool it below its boiling-point temperature at the desired pressure. However, this would require refrigeration equipment capable of producing very low temperatures. Therefore, such a direct liquefaction process is not generally used. What is more commonly done is to start with a gas at low pressure, compress it to high pressure (which increases its temperature since work has been done on it), cool this high-temperature gas at the constant high pressure, and then expand it to low pressure and low temperature using a Joule-Thomson expansion, which produces a mixture of liquid and vapor. In this way the cooling is done at a higher temperature (and pressure), so that low-temperature refrigeration is not needed. (Such a process is also used internally in many refrigeration cycles, including your home refrigerator, as discussed in the next section.) The vapor and liquid are then separated in a flash drum (an insulated, constant-pressure container). The process just described is shown schematically in Fig. 5.1-1.

The efficiency of this process, that is, the amount of liquefied gas produced for each unit of work done in the compressor, can be improved upon by better engineering design. For example, instead of merely discarding the low-temperature, low-pressure gas leaving the flash drum (stream 5), the gas can be used to cool the high-pressure gas upstream of the throttle valve and then returned to the compressor, so that none of the gas is wasted or exhausted to the atmosphere. This process, referred to as the **Linde process**, is shown in Fig. 5.1-2. In this way the only stream leaving the liquefaction plant is liquefied gas, and, as shown in Illustration 5.1-1, more liquefied gas is produced per unit of energy expended in the compressor.

ILLUSTRATION 5.1-1

Comparing the Efficiency of the Simple and Linde Liquefaction Processes

It is desired to produce liquefied natural gas (LNG), which we consider to be pure methane, from that gas at 1 bar and 280 K (conditions at point 1 in Figs. 5.1-1 and 5.1-2). Leaving the cooler, methane is at 100 bar and 210 K (point 3). The flash drum is adiabatic and operates

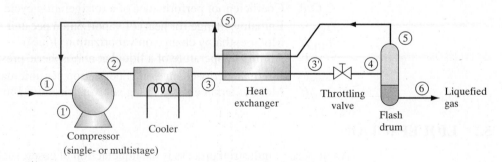

Figure 5.1-2 The more efficient Linde liquefaction process.

at 1 bar, and the compressor can be assumed to operate reversibly and adiabatically. However, because of the large pressure change, a three-stage compressor with intercooling is used. The first stage compresses the gas from 1 bar to 5 bar, the second stage from 5 bar to 25 bar, and the third stage from 25 bar to 100 bar. Between stages the gas is isobarically cooled to 280 K.

a. Calculate the amount of work required for each kilogram of methane that passes through the compressor in the simple liquefaction process.
b. Calculate the fractions of vapor and liquid leaving the flash drum in the simple liquefaction process of Fig. 5.1-1 and the amount of compressor work required for each kilogram of LNG produced.
c. Assuming that the recycled methane leaving the heat exchanger in the Linde process (Fig. 5.1-2) is at 1 bar and 200 K, calculate the amount of compressor work required for each kilogram of LNG produced.

Data: The thermodynamic properties of methane are given in Fig. 3.3-2

SOLUTION

a. For each stage of compression, assuming steady-state and reversible adiabatic operation, we have from the mass, energy, and entropy balances, respectively:

$$0 = \dot{M}_{in} + \dot{M}_{out} \quad \text{or} \quad \dot{M}_{out} = -\dot{M}_{in} = -\dot{M}$$
$$0 = \dot{M}_{in}\hat{H}_{in} + \dot{M}_{out}\hat{H}_{out} + \dot{W} \quad \text{or} \quad \dot{W} = \dot{M}(\hat{H}_{out} - \hat{H}_{in})$$
$$0 = \dot{M}_{in}\hat{S}_{in} + \dot{M}_{out}\hat{S}_{out} \quad \text{or} \quad \hat{S}_{out} = \hat{S}_{in}$$

So through each compressor (but not intercooler) stage, one follows a line of constant entropy in Fig. 3.3-2 (which is redrawn here with all the stages indicated). For the first

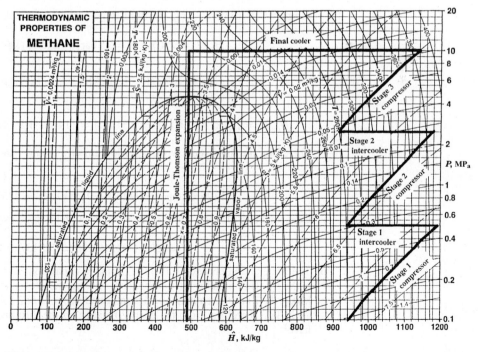

Figure 5.1-3 Pressure-enthalpy diagram for methane. (*Source:* W. C. Reynolds, *Thermodynamic Properties in SI,* Department of Mechanical Engineering, Stanford University, Stanford, CA, 1979. Used with permission.)

compressor stage, we have

$$\hat{H}_{in}(T = 280\,\text{K}, P = 1\,\text{bar}) = 940\,\text{kJ/kg} \quad \text{and}$$

$$\hat{S}_{in}(T = 280\,\text{K}, P = 1\,\text{bar}) = 7.2\,\text{kJ/(kg K)}$$

$$\hat{H}_{out}(\hat{S} = 7.2\,\text{kJ/(kg K)}, P = 5\,\text{bar}) = 1195\,\text{kJ/kg} \quad \text{and} \quad T_{out} = 388\,\text{K}$$

so that the first-stage work per kilogram of methane flowing through the compressor is

$$\dot{W}(\text{first stage}) = (1195 - 940)\,\text{kJ/kg} = 225\,\text{kJ/kg}$$

After cooling, the temperature of the methane stream is 280 K, so that for the second compressor stage, we have

$$\hat{H}_{in}(T = 280\,\text{K}, P = 5\,\text{bar}) = 938\,\text{kJ/kg} \quad \text{and}$$

$$\hat{S}_{in}(T = 280\,\text{K}, P = 5\,\text{bar}) = 6.35\,\text{kJ/(kg K)}$$

$$\hat{H}_{out}(\hat{S} = 6.35\,\text{kJ/(kg K)}, P = 25\,\text{bar}) = 1180\,\text{kJ/kg} \quad \text{and} \quad T_{out} = 386\,\text{K}$$

Therefore, the second-stage work per kilogram of methane flowing through the compressor is

$$\dot{W}(\text{second stage}) = (1180 - 938)\,\text{kJ/kg} = 242\,\text{kJ/kg}$$

Similarly, after intercooling, the third-stage compressor work is found from

$$\hat{H}_{in}(T = 280\,\text{K}, P = 25\,\text{bar}) = 915\,\text{kJ/kg} \quad \text{and}$$

$$\hat{S}_{in}(T = 280\,\text{K}, P = 25\,\text{bar}) = 5.5\,\text{kJ/(kg K)}$$

$$\hat{H}_{out}(\hat{S} = 5.5\,\text{kJ/(kg K)}, P = 100\,\text{bar}) = 1140\,\text{kJ/kg} \quad \text{and} \quad T_{out} = 383\,\text{K}$$

Therefore, the third-stage work per kilogram of methane flowing through the compressor is

$$\dot{W}(\text{third stage}) = (1140 - 915)\,\text{kJ/kg} = 225\,\text{kJ/kg}$$

Consequently, the total compressor work through all three stages is

$$\dot{W} = 255 + 242 + 225 = 722\,\text{kJ/kg}$$

b. The liquefaction process is a Joule-Thomson expansion, and therefore occurs at constant enthalpy. The enthalpy of the methane leaving the cooler at 100 bar and 210 K is 493 kJ/kg. At 1 bar the enthalpy of the saturated vapor is 582 kJ/kg, and that of the liquid is 71 kJ/kg. Therefore, from the energy balance on the throttling valve and flash drum, we have[1]

$$\hat{H}_{in} = \hat{H}_{out} \quad \text{or}$$

---

[1] In this equation *saturated liquid* refers to the fact that the phase is at its equilibrium (boiling) temperature at the specified pressure.

$$\hat{H}(210\,\text{K},\,100\,\text{bar}) = (1-\omega)\hat{H}(\text{saturated liquid, 1 bar}) + \omega\hat{H}(\text{saturated vapor, 1 bar})$$

$$493\,\frac{\text{kJ}}{\text{kg}} = (1-\omega)\cdot 71\,\frac{\text{kJ}}{\text{kg}} + \omega\cdot 582\,\frac{\text{kJ}}{\text{kg}}$$

so that $\omega = 0.826$ is the fraction of vapor leaving the flash drum, and $(1-\omega) = 0.174$ is the fraction of the methane that has been liquefied. Therefore, for each kilogram of methane that enters the simple liquefaction unit, 826 g of methane are lost as vapor, and only 174 g of LNG are produced. Further, since 722 kJ of work are required in the compressor to produce 174 g of LNG, approximately 4149 kJ of compressor work are required for each kilogram of LNG produced.

c. While the Linde process looks more complicated, it can be analyzed in a relatively simple manner. First, we choose the system for writing balance equations to be the subsystem consisting of the heat exchanger, throttle valve, and flash drum (though other choices could be made). The mass and energy balances for this subsystem (since there are no heat losses to the outside or any work flows) are

$$\dot{M}_3 = \dot{M}_{5'} + \dot{M}_6$$

or, taking $\dot{M}_3 = 1$ and letting ω be the mass fraction of vapor,

$$1 = (1-\omega) + \omega$$

for the mass balance, and

$$\dot{M}_3\hat{H}_3 = \dot{M}_{5'}\hat{H}_{5'} + \dot{M}_6\hat{H}_6$$

$$\hat{H}(T = 210\,\text{K},\ P = 100\,\text{bar})$$
$$= (1-\omega)\cdot\hat{H}(\text{saturated liquid},\ P = 1\,\text{bar}) + \omega\cdot\hat{H}(\text{saturated vapor},\ P = 1\,\text{bar})$$

$$493\,\frac{\text{kJ}}{\text{kg}} = \omega\cdot 770\,\frac{\text{kJ}}{\text{kg}} + (1-\omega)\cdot 71\,\frac{\text{kJ}}{\text{kg}}$$

for the energy balance. The solution to this equation is $\omega = 0.604$ as the fraction of vapor that is recycled, and 0.396 as the fraction of liquid.

The balance equations for mixing the streams immediately before the compressor are

$$\dot{M}_{5'} + \dot{M}_1 = \dot{M}_{1'}$$

and basing the calculation on 1 kg of flow into the compressor,

$$\dot{M}_{1'} = 1 \qquad \dot{M}_{5'} = 0.604 \quad\text{and}\quad \dot{M}_1 = 0.396$$

for the mass balance, and

$$\dot{M}_{1'}\hat{H}_{1'} + \dot{M}_{5'}\hat{H}_{5'} = \dot{M}_{1'}\hat{H}_{1'}$$

$$0.396\cdot\hat{H}(T = 280\,\text{K},\ P = 1\,\text{bar})$$
$$+\,0.604\cdot\hat{H}(T = 200\,\text{K},\ P = 1\,\text{bar}) = \hat{H}_{1'}(T = ?,\ P = 1\,\text{bar})$$

$$0.396\cdot 940\,\frac{\text{kJ}}{\text{kg}} + 0.604\cdot 770\,\frac{\text{kJ}}{\text{kg}} = 837.3\,\frac{\text{kJ}}{\text{kg}} = \hat{H}(T = ?,\ P = 1\,\text{bar})$$

for the energy balance. Using the redrawn Fig. 3.3-2 here, we conclude that the temperature of a stream entering the compressor is 233 K. To complete the solution to this problem, we need to compute the compressor work load. Following the solution for part (a), we find for the first stage that

$$\hat{H}_{in}(T = 233\,\text{K},\ P = 1\,\text{bar}) = 837\,\text{kJ/kg} \quad \text{and}$$

$$\hat{S}_{in}(T = 233\,\text{K},\ P = 1\,\text{bar}) = 6.8\,\text{kJ/(kg K)}$$

$$\hat{H}_{out}(\hat{S} = 6.8\,\text{kJ/(kg K)},\ P = 5\,\text{bar}) = 1020\,\text{kJ/kg} \quad \text{and} \quad T_{out} = 388\,\text{K}$$

Therefore, now the first-stage work per kilogram of methane flowing through the compressor is

$$\dot{W}(\text{first stage}) = 1020 - 837\,\text{kJ/kg} = 183\,\text{kJ/kg}$$

Since the methane leaving the intercooler is at 280 K and 5 bar, the work in each of the second and third stages of the compressor is the same as in part (a). Therefore, the total compressor work is

$$\dot{W} = 183 + 242 + 225 = 650\,\text{kJ/kg of methane through the compressor}$$

However, each kilogram of methane through the compressor results in only 0.396 kg of LNG, as the remainder of the methane is recycled. Consequently, the compressor work required per kilogram of LNG produced is (650 kJ/kg)/0.396 kg = 1641 kJ/kg of LNG produced. This is to be compared with 4149 kJ/kg of LNG produced in the simple liquefaction process.

COMMENT

By comparing the Linde process with the simple liquefaction process we see the importance and advantages of clever engineering design. In particular, the Linde process uses only about 40 percent of the energy required in the simple process to produce a kilogram of LNG. Also, unlike the simple process, the Linde process releases no gaseous methane, which is a greenhouse gas and contributes to global warming, into the atmosphere. Finally, it should be pointed out that since a graph of thermodynamic properties, rather than a more detailed table of values, was used, the numerical values of the properties read from the diagram are approximate. ■

5.2 POWER GENERATION AND REFRIGERATION CYCLES

So far we have considered only the Carnot cycle for converting heat (produced by burning coal, oil, or natural gas) to work (usually electricity). While this cycle is the most efficient possible for converting heat to work, in practice it is rarely used because of the large amount of work that must be supplied during the isothermal compression step. One of the simplest and most widely used commercial cycles is the Rankine cycle, shown in Fig. 5.2-1 with water as the operating or so-called working fluid, though other fluids may also be used.

   In this cycle the turbine and the pump are considered to operate isentropically, the condenser operates isobarically, and the fluid in the boiler is heated at constant pressure. The properties and path for this cycle are:

Rankine cycle

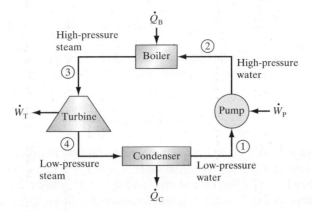

Figure 5.2-1 Rankine cycle.

Point	Path to Next Point	T	P	$\hat{S}$	$\hat{H}$	Energy Flow
1		T_1	P_1	$\hat{S}_1$	$\hat{H}_1$	
	Isentropic		$\downarrow$			$\dot{W}_P = \dot{M}\hat{V}_1(P_2 - P_1)$
2		T_2	P_2	$\hat{S}_2 = \hat{S}_1$	$\hat{H}_2$	
	Isobaric		$\downarrow$			$\dot{Q}_B$
3		T_3	$P_3 = P_2$	$\hat{S}_3$	$\hat{H}_3$	
	Isentropic		$\downarrow$			$\dot{W}_T = \dot{M}(\hat{H}_4 - \hat{H}_3)$
4		T_4	P_4	$\hat{S}_4 = \hat{S}_3$	$\hat{H}_4$	
	Isobaric		$\downarrow$			$\dot{Q}_C$
1		T_1	$P_1 = P_4$	$\hat{S}_1$	$\hat{H}_1$	

The work flows in this process are computed as follows.

Step 1 → 2 (pump)
The mass and energy balances on the open system consisting of the pump and its contents operating at steady state are

$$\frac{dM}{dt} = 0 = \dot{M}_1 + \dot{M}_2 \quad \text{or} \quad \dot{M}_1 = -\dot{M}_2 = \dot{M}$$

and

$$\frac{dU}{dt} = 0 = \dot{M}_1\hat{H}_1 + \dot{M}_2\hat{H}_2 + \dot{W}_P$$

so that

$$\dot{W}_P = -\dot{M}_2\hat{H}_2 - \dot{M}_1\hat{H}_1 = \dot{M}(\hat{H}_2 - \hat{H}_1)$$
$$= \dot{M}_1(\hat{U}_2 + P_2\hat{V}_2 - \hat{U}_1 - P_1\hat{V}_1)$$
$$= \dot{M}\hat{V}_1(P_2 - P_1)$$

since the molar volume and internal energy of the liquid are essentially independent of pressure at constant temperature.

Step 3 → 4 (turbine)

$$\frac{dM}{dt} = 0 = \dot{M}_3 + \dot{M}_4 \quad \text{or} \quad \dot{M}_3 = -\dot{M}_4 = \dot{M}$$

since the mass flow rate is constant throughout the process, and

$$\frac{dU}{dt} = 0 = \dot{M}_3 \hat{H}_3 - \dot{M}_4 \hat{H}_4 + \dot{W}_\text{T} \quad \text{or} \quad \dot{W}_\text{T} = \dot{M}(\hat{H}_4 - \hat{H}_3)$$

The efficiency η of this process (and other cyclic processes as well) is defined to be the ratio of the net work obtained, here $-(\dot{W}_\text{T} + \dot{W}_\text{P})$, where $\dot{W}_\text{T}$ is negative and $\dot{W}_\text{P}$ is positive in value, to the heat input $\dot{Q}_\text{B}$; that is,

$$\eta = \frac{-(\dot{W}_\text{T} + \dot{W}_\text{P})}{\dot{Q}_\text{B}}$$

The plot of this cycle on a temperature–entropy diagram is such that $1 \to 2$ is a line of constant entropy (vertical line), $2 \to 3$ is a line of constant pressure, $3 \to 4$ is another line of constant entropy, and $4 \to 1$ closes the cycle with a line of constant temperature. Several different locations for these lines, depending largely on the operation of the boiler, are shown in Fig. 5.2-2.

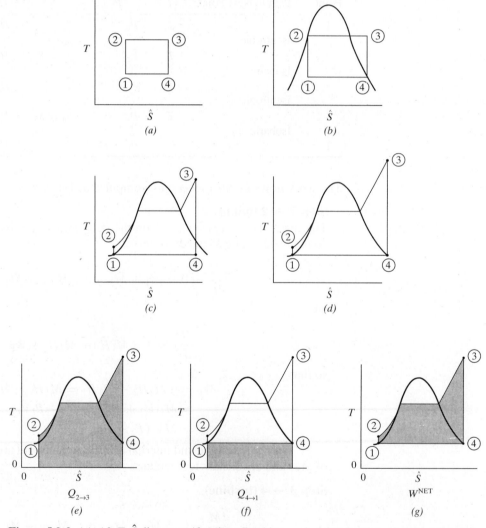

Figure 5.2-2 (*a*)–(*d*) T-$\hat{S}$ diagrams of various Rankine cycles. (*e*) Area equal to heat flow into boiler. (*f*) Area equal to heat removed in condenser. (*g*) Area equal to net work produced in cycle.

Figure 5.2-2*a* looks like the T-$\hat{S}$ diagram for the Carnot cycle (Fig. 4.3-2*c*), and indeed at these conditions the Rankine cycle has the same efficiency as the Carnot cycle. The practical difficulty is that the high-speed turbine blades will severely erode if impacted by liquid droplets. Therefore, step $3 \rightarrow 4$ should be completely in the vapor region, as in Fig. 5.2-2*b* and *d*. Also, to minimize the work required for the pump, the fluid passing through it should be all liquid, as in Fig. 5.2-2*c* and *d*. Therefore, of the choices in Fig. 5.2-2, the conditions in Fig. 5.2-2*d* are used for practical operation of a Rankine power cycle. In this case, the efficiency is less than that of the Carnot cycle. Also, the condenser operating temperature is generally determined by the available cooling water. However, lower temperatures and pressures in the condenser improve the efficiency of the cycle (see Problems 5.2 and 5.3).

As pointed out in Sec. 4.3, the differential entropy change dS and the heat flow Q for a reversible process in a closed system are related as follows:

$$dS = \frac{Q}{T} \quad \text{or} \quad Q = T \, dS$$

so that the heat flows (and the work produced) are related to areas on the T-S diagram for the Rankine power cycle as follows. The heat flow into the cycle on going from point 2 to 3 is equal to

$$Q_{2\to3} = \int_2^3 T \, dS$$

This heat flow is given by the area shown in Fig. 5.2-2*e*. Similarly, the heat flow from point 4 to 1 is equal in magnitude to the area shown in Fig 5.2-2*f*, but negative in value (since S_4 is larger than S_1). Finally, since from the energy balance, the net work flow is equal to the difference of the two heat flows into the cycle, this work flow is given by the area in Fig. 5.2-2*g*.

Similar graphical analyses can be done for the other cycles considered in this section, though we will not do so.

ILLUSTRATION 5.2-1

Calculating the Efficiency of a Steam Power Cycle

A Rankine power generation cycle using steam operates at a temperature of 100°C in the condenser, a pressure of 3.0 MPa in the evaporator, and a maximum temperature of 600°C. Assuming the pump and turbine operate reversibly, plot the cycle on a T-$\hat{S}$ diagram for steam, and compute the efficiency of the cycle.

SOLUTION

We start by plotting the cycle on the temperature-entropy diagram for steam (Fig. 3.3-1*b*, repeated below). Then we construct the table of thermodynamic properties (which follows) for each point in the cycle using the known operating conditions and the path from each point in the cycle to the next point. Based on this information, we can then compute the efficiency of the cycle

$$\eta = \frac{-(\dot{W}_\text{T} + \dot{W}_\text{P})}{\dot{Q}_\text{B}} = \frac{-(-947.6 + 3.03)}{3260.2} = \frac{944.3}{3260.2} = 0.290 \quad \text{or 29.0 percent}$$

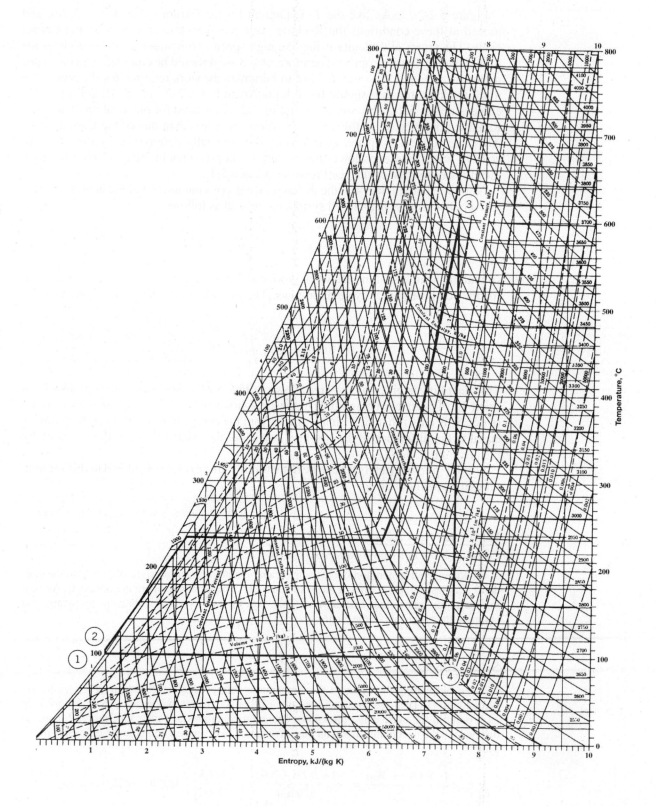

Point	(State) Path to Next Point	T (°C)	P (MPa)	$\hat{S}$ (kJ/kgK)	$\hat{H}$ (kJ/kg)	$\hat{V}$ (m³/kg)	Energy Flow
1	(Saturated liquid)	**100**	*0.10135*	*1.3069* ↓	*419.04*	*0.001044*	
	Isentropic						$\hat{W}_P = \hat{V}_1(P_2 - P_1)$ $= 0.001044 \text{ m}^3/\text{kg}$ $\times (3.0 - 0.10135) \text{ MPa}$ $= 3.03 \text{ kJ/kg}$
2		~ 103	**3.0** ↑	**1.3069**	422.07		
	Isobaric						$\hat{Q}_B = 3682.3 - 422.1$ $= 3260.2 \text{ kJ/kg}$
3		**600**	**3.0**	*7.5085* ↓	*3682.3*		
	Isentropic						$\hat{W}_T = (\hat{H}_4 - \hat{H}_3)$ $= 2734.7 - 3682.3$ $= -947.6 \text{ kJ/kg}$
4		~ 126.2	**0.10135** ↑	**7.5085**	~ 2734.7		
	Isobaric						$\hat{Q}_C = \hat{H}_1 - \hat{H}_4$ $= 419.04 - 2734.7$ $= -2315.7 \text{ kJ/kg}$
1	Saturated liquid	**100**	*0.10135*	*1.3069*	*419.04*		

Note: In preparing this table we have used bold notation to indicate data given in the problem statement (e.g., **100**°C) or from information on an adjacent point in the cycle (also indicated by arrows connecting two points), italics to indicate numbers found in the steam tables (e.g., *0.10135* MPa), and a tilde to indicate numbers obtained from the steam tables by interpolation (e.g., ~2734.7 kJ/kg).

■

The Rankine power cycle above receives heat at a high temperature, emits it at a lower temperature, and produces work. (Does this cycle violate the Kelvin-Planck statement of the second law? The answer is no since there are two heat flows, one into the cycle from a hot body, and the second from the cycle into a cooler body or cooling water. Consequently, not all the heat available at high temperature is converted into work, only a portion of it.) A similar cycle, known as the Rankine refrigeration cycle, operates essentially in reverse by using work to pump heat from a low-temperature region to a high-temperature region. This refrigeration cycle is shown in Fig. 5.2-3a.

In a home refrigerator the condenser is the air-cooled coil usually found at the bottom or back of the refrigerator, and the evaporator is the cooling coil located in the freezer section. Here the compressor and the turbine are considered to operate isentropically, and the condenser and evaporator to operate at constant pressure. Frequently, and this is the case we consider here, the condenser receives a two-phase mixture and emits a saturated liquid at the same pressure. The properties and path for this cycle are given in the following table.

Schematic diagrams of the path of this cycle on both T-$\hat{S}$ and P-$\hat{H}$ diagrams are shown in Fig. 5.2-3b.

As in the previous case, the mass flow is the same in all parts of the cycle, and we designate this by $\dot{M}$. The steady-state energy balance around the evaporator and its contents yields

$$0 = \dot{M}\hat{H}_2 - \dot{M}\hat{H}_3 + \dot{Q}_B \quad \text{or} \quad \dot{Q}_B = \dot{M}(\hat{H}_3 - \hat{H}_2)$$

Refrigeration cycle

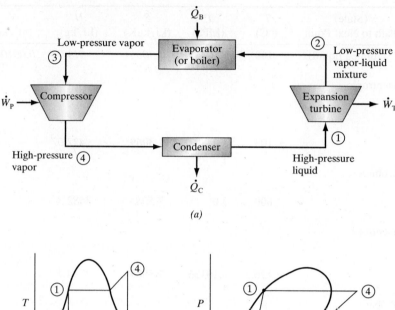

Figure 5.2-3 (*a*) The Rankine refrigeration cycle. (*b*) The Rankine refrigeration cycle on $T\text{-}\hat{S}$ and $P\text{-}\hat{H}$ diagrams.

The energy balance on the compressor is

$$0 = \dot{M}\hat{H}_3 - \dot{M}\hat{H}_4 + \dot{W}_P \quad \text{or} \quad \dot{W}_P = \dot{M}(\hat{H}_4 - \hat{H}_3)$$

The energy balance on the condenser is

$$0 = \dot{M}\hat{H}_4 - \dot{M}\hat{H}_1 + \dot{Q}_C \quad \text{or} \quad \dot{Q}_C = \dot{M}(\hat{H}_1 - \hat{H}_4)$$

Point	(State) Path to Next Point	T	P	$\hat{S}$	$\hat{H}$	Energy Flow
1	(Saturated liquid)	T_1	P_1	$\hat{S}_1$	$\hat{H}_1$	
	Isentropic			$\downarrow$		$\dot{W}_T$
2	(Vapor-liquid mix.)	T_2	P_2	$\hat{S}_2 = \hat{S}_1$	$\hat{H}_2$	
	Isobaric [also isothermal in this case]	$\downarrow$	$\downarrow$			$\dot{Q}_B$
3	(Saturated vapor)	$T_3 = T_2$	$P_3 = P_2$	$\hat{S}_3$	$\hat{H}_3$	
	Isentropic			$\downarrow$		$\dot{W}_P$
4	(Superheated vapor)	T_4	P_4	$\hat{S}_4 = \hat{S}_3$	$\hat{H}_4$	
	Isobaric			$\downarrow$		$\dot{Q}_C$
1		T_1	$P_1 = P_4$	$\hat{S}_1$	$\hat{H}_1$	

and the energy balance on the turbine is

$$0 = \dot{M}\hat{H}_1 - \dot{M}\hat{H}_2 + \dot{W}_T \quad \text{or} \quad \dot{W}_T = \dot{M}(\hat{H}_2 - \hat{H}_1)$$

In refrigeration it is common to use the term **coefficient of performance** (C.O.P.), defined as

$$\text{C.O.P.} = \frac{\text{Heat removed from low-temperature region}}{\text{Net work required for heat removal}} = -\frac{\dot{Q}_B}{\dot{W}_P + \dot{W}_T}$$

rather than efficiency (as was used in the Rankine power generation cycle). For the cycle shown here, we have

$$\text{C.O.P.} = \frac{\dot{Q}_B}{\dot{W}_P + \dot{W}_T} = \frac{\hat{H}_3 - \hat{H}_2}{(\hat{H}_4 - \hat{H}_3) + (\hat{H}_2 - \hat{H}_1)}$$

The amount of work recovered in the expansion turbine of the Rankine refrigeration cycle is small. Consequently, to simplify the design and operation of refrigerators and small air-conditioning units, and to reduce the number of moving parts in such equipment, it is common to replace the expansion turbine with an expansion valve, capillary tube, orifice, or other partial obstruction so that the refrigerant undergoes a Joule-Thomson expansion between points 1 and 2 in the cycle rather than an isentropic expansion. Such a cycle is shown in Fig. 5.2-4a. The properties and path for this cycle are given in the table below. This cycle is referred to as the vapor-compression refrigeration cycle, and is more commonly used than the Rankine refrigeration cycle. Note that the net process in both the Rankine and vapor-compression refrigeration cycles is to use work to remove heat from a low-temperature source and exhaust it to a reservoir at a higher temperature.

The compressor in this cycle is considered to operate isentropically, the expansion valve results in a Joule-Thomson (isenthalpic) expansion, and the condenser and evaporator operate at constant pressure. Here we consider the case in which the condenser receives a two-phase mixture and emits a saturated liquid at the same pressure; consequently, the condenser also operates at constant temperature. The properties and path for this cycle are given in the table.

Point	(State) Path to Next Point	T	P	$\hat{S}$	$\hat{H}$	Energy Flow
1	(Saturated liquid)	T_1	P_1	$\hat{S}_1$	$\hat{H}_1$	
	Isenthalpic				↓	0
2	(Vapor-liquid mixture)	T_2	P_2	$\hat{S}_2$	$\hat{H}_2 = \hat{H}_1$	
	Isobaric heating [also isothermal in this case]	↓	↓			$\dot{Q}_B$
3	(Saturated vapor)	$T_3 = T_2$	$P_3 = P_2$	$\hat{S}_3$	$\hat{H}_3$	
	Isentropic			↓		$\dot{W}_P$
4	(Superheated vapor)	T_4	P_4	$\hat{S}_4 = \hat{S}_3$	$\hat{H}_4$	
	Isobaric		↓			$-\dot{Q}_C$
1		T_1	$P_1 = P_4$	$\hat{S}_1$	$\hat{H}_1$	

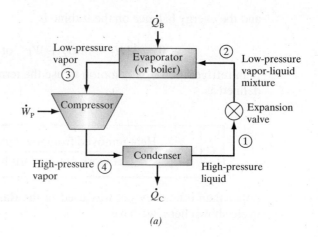

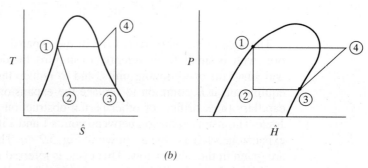

(b)

Figure 5.2-4 (*a*) The vapor-compression refrigeration cycle. (*b*) The vapor-compression refrigeration cycle on T-$\hat{S}$ and P-$\hat{H}$ diagrams.

Schematic T-$\hat{S}$ and P-$\hat{H}$ diagrams of the path of this cycle are shown in Fig. 5.2-4*b*. The mass flow is the same in all parts of the cycle, and we designate this by $\dot{M}$. The steady-state energy balance around the evaporator and its contents yields

$$0 = \dot{M}\hat{H}_2 - \dot{M}\hat{H}_3 + \dot{Q}_B \quad \text{or} \quad \dot{Q}_B = \dot{M}(\hat{H}_3 - \hat{H}_2)$$

The energy balance on the compressor is

$$0 = \dot{M}\hat{H}_3 - \dot{M}\hat{H}_4 + \dot{W}_P \quad \text{or} \quad \dot{W}_P = \dot{M}(\hat{H}_4 - \hat{H}_3)$$

and the energy balance on the condenser is

$$0 = \dot{M}\hat{H}_4 - \dot{M}\hat{H}_1 + \dot{Q}_C \quad \text{or} \quad \dot{Q}_C = \dot{M}(\hat{H}_1 - \hat{H}_4)$$

The coefficient of performance (C.O.P.) for the vapor-compression refrigeration cycle is

$$\text{C.O.P.} = \frac{\text{Heat removed from low-temperature region}}{\text{Net work required for that heat removal}}$$

$$= \frac{\dot{Q}_B}{\dot{W}_P} = \frac{\hat{H}_3 - \hat{H}_2}{\hat{H}_4 - \hat{H}_3}$$

ILLUSTRATION 5.2-2

Calculating the Coefficient of Performance of an Automobile Air Conditioner

An automobile air conditioner uses a vapor-compression refrigeration cycle with the environmentally friendly refrigerant HFC-134a as the working fluid. The following data are available for this cycle.

Point	Fluid State	Temperature
1	Saturated liquid	55°C
2	Vapor-liquid mixture	
3	Saturated vapor	5°C
4	Superheated vapor	

a. Supply the missing temperatures and the pressures in the table.
b. Evaluate the coefficient of performance for the refrigeration cycle described in this problem.

SOLUTION

The keys to being able to solve this problem are (1) to identify the paths between the various locations and (2) to be able to use the thermodynamic properties chart of Fig. 3.3-4 for HFC-134a to obtain the properties at each of the locations. The following figure shows the path followed in the cycle, and the table provides the thermodynamic properties for each stage of the cycle as read from the figure.

With this information, the C.O.P. is

$$\text{C.O.P.} = \frac{\dot{Q}_B}{\dot{W}} = \frac{\hat{H}_3 - \hat{H}_2}{\hat{H}_4 - \hat{H}_3} = \frac{402 - 280}{432 - 402} = 4.07$$

∎

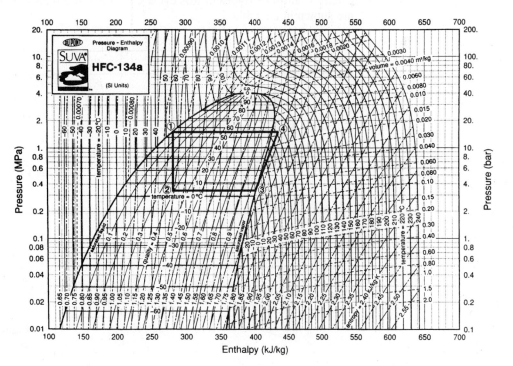

Point	State	Path	T (°C)	P (MPa)	$\hat{H}$ (kJ/kg)	$\hat{S}$ (kJ/kg K)
1	Saturated liquid		**55**	*1.493*	*280*	*1.262*
		Isenthalp to 5°C			$\downarrow$	
2	Vapor-liquid mixture		**5**	**0.350**	**280**	*1.30*
		P constant		$\uparrow$		
3	Saturated vapor		**5**	*0.350*	*402*	*1.725*
		Isentrope to 1493 kPa				$\downarrow$
4	Superheated vapor		*60*	**1.493**	*432*	**1.725**
		P constant		$\uparrow$		
1	Saturated liquid		**55**	**1.493**	*280*	*1.262*

In this table the numbers in boldface are properties known from the problem statement or from an adjacent step in the cycle (i.e., a process that is at constant pressure as in steps 2 → 3 and 4 → 1, at constant enthalpy as in step 1 → 2, or at constant entropy as in step 3 → 4). The numbers in italics are found from Fig. 3.3-4 with the state of the refrigerant fixed from the values of the known variables.

Stirling cycle

A large number of other cycles and variations to the standard cycles considered above have been proposed. We consider only a few additional cycles here. The Stirling cycle, shown in Fig. 5.2-5, operates with a vapor-phase working fluid, rather than a two-phase mixture as considered above. In this process the compressor and turbine, which are on the same shaft, are cooled and heated, respectively, in order to operate isothermally. The heat exchanger operates isochorically (that is, at constant volume). The P-$\hat{V}$ and T-$\hat{S}$ traces of this cycle are shown in Fig. 5.2-6. The properties and path are shown in the table.

Point	Path to Next Point	T	P	$\hat{V}$	$\hat{H}$	Energy Flow
1		T_1	P_1	$\hat{V}_1$	$\hat{H}_1$	
	$\hat{V}$ = constant			$\downarrow$		$\dot{Q}_{12}$
2		T_2	P_2	$\hat{V}_2 = \hat{V}_1$	$\hat{H}_2$	
	T = constant	$\downarrow$				$\dot{Q}_T, \ -\dot{W}_{23}$
3		$T_3 = T_2$	P_3	$\hat{V}_3$	$\hat{H}_3$	
	$\hat{V}$ = constant			$\downarrow$		$-\dot{Q}_{34} = \dot{Q}_{12}$
4		T_4	P_4	$\hat{V}_4 = \hat{V}_3$	$\hat{H}_4$	
	T = constant	$\downarrow$				$-\dot{Q}_C, \ -\dot{W}_{41}$
1		$T_1 = T_4$	P_1	$\hat{V}_1$	$\hat{H}_1$	

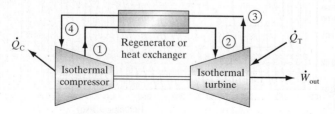

Figure 5.2-5 The Stirling cycle.

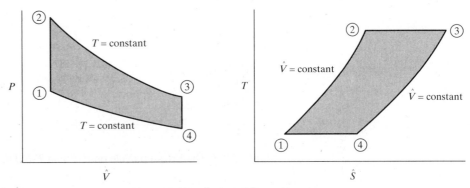

Figure 5.2-6 The Stirling cycle on P-$\hat{V}$ and T-$\hat{S}$ diagrams.

From an energy balance on stream 3 in the regenerator, we have (at steady state)

$$0 = \dot{M}\hat{H}_3 - \dot{M}\hat{H}_4 + \dot{Q}_{34}$$

and on stream 1

$$0 = \dot{M}\hat{H}_1 - \dot{M}\hat{H}_2 + \dot{Q}_{12}$$

Adding these two equations and noting that $\dot{Q}_{12} + \dot{Q}_{34} = 0$, since there is no net heat flow into or out of the heat exchanger, gives

$$(\hat{H}_3 - \hat{H}_4) + (\hat{H}_1 - \hat{H}_2) = 0$$

An energy balance on the compressor ($4 \rightarrow 1$) gives

$$\dot{M}(\hat{H}_4 - \hat{H}_1) + \dot{W}_{41} + \dot{Q}_C = 0$$

and on the turbine ($2 \rightarrow 3$) gives

$$\dot{M}(\hat{H}_2 - \hat{H}_3) + \dot{W}_{23} + \dot{Q}_T = 0$$

Adding these last two equations together gives

$$\dot{M}(\hat{H}_4 - \hat{H}_1) + \dot{M}(\hat{H}_2 - \hat{H}_3) + (\dot{W}_{41} + \dot{W}_{23}) + \dot{Q}_C + \dot{Q}_T = 0$$

From the balance equation on the heat exchanger, the first two terms cancel. Also, $\dot{W}_{41} + \dot{W}_{23}$ is equal to $\dot{W}_{\text{out}}$. Thus we obtain

$$\dot{W}_{\text{out}} + \dot{Q}_C + \dot{Q}_T = 0$$

Therefore, the efficiency of this Stirling cycle, η, is

$$\eta = \frac{-\dot{W}_{\text{out}}}{\dot{Q}_T} = \frac{\dot{W}_{\text{out}}}{\dot{W}_{\text{out}} + \dot{Q}_C} = \frac{-\dot{W}_{23} - \dot{W}_{41}}{\dot{Q}_T}$$

Ericsson cycle

The Ericsson cycle is similar to the Stirling cycle except that each of the streams in the regenerator is at a constant (but different) pressure rather than at a constant volume. The P-$\hat{V}$ and T-$\hat{S}$ traces are shown in Fig. 5.2-7.

The properties and path are given in the following table.

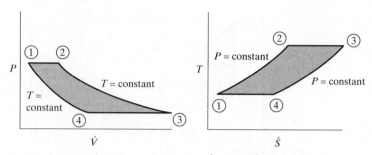

Figure 5.2-7 The Ericsson cycle on P-$\hat{V}$ and T-$\hat{S}$ diagrams.

Point	Path to Next Point	T	P	$\hat{H}$	Energy Flow
1		T_1	P_1	$\hat{H}_1$	
	P = constant		$\downarrow$		$\dot{Q}_{12}$
2		T_2	$P_2 = P_1$	$\hat{H}_2$	
	T = constant	$\downarrow$			$\dot{Q}_T, -\dot{W}_{23}$
3		$T_3 = T_2$	P_3	$\hat{H}_3$	
	P = constant		$\downarrow$		$-\dot{Q}_{34} = \dot{Q}_{12}$
4		T_4	$P_4 = P_3$	$\hat{H}_4$	
	T = constant	$\downarrow$			$-\dot{Q}_C, -\dot{W}_{41}$
1		$T_1 = T_4$	P_1	$\hat{H}_1$	

The balance equations are exactly as for the Stirling cycle, so that the efficiency is

$$\eta = \frac{-\dot{W}_{\text{out}}}{\dot{Q}_T} = \frac{-\dot{W}_{23} - \dot{W}_{41}}{\dot{Q}_T} = \frac{-W_{\text{out}}}{\dot{Q}_C + W_{\text{out}}}$$

Indeed, the only computational difference between the Stirling and Ericsson cycles is that the thermodynamic properties at the various points of the cycle are slightly different because of the difference between the constant-pressure and constant-volume paths.

ILLUSTRATION 5.2-3

Calculating the Efficiency of an Ericsson Cycle

An Ericsson cycle with air as the working fluid is operating between a low temperature of 70°C and a high temperature of 450°C, a low pressure of 2 bar, and a pressure compression ratio of 8 (so that the high pressure is 16 bar). Assume that at these conditions air can be considered to be an ideal gas with C_P^* constant. Compute the properties at each point in each of these cycles and the cycle efficiencies.

SOLUTION

We first compute the energy flows in each step of the process. In this analysis the process is at steady state so that all the time derivatives are equal to zero and the mass flow rate $\dot{M}$ (actually we will use the molar flow rate $\dot{N}$) is constant throughout the process.

$1 \rightarrow 2$ Through the regenerative heat exchanger
Energy balance:

$$0 = \dot{N}\underline{H}_1 - \dot{N}\underline{H}_2 + \dot{Q}_{12} \quad \text{or} \quad \frac{\dot{Q}_{12}}{N} = (\underline{H}_2 - \underline{H}_1) = C_P^*(T_2 - T_1)$$

2 → 3 Through the isothermal turbine
Energy balance:

$$0 = \dot{N}\underline{H}_2 - \dot{N}\underline{H}_3 + \dot{Q}_T + \dot{W}_{23}$$

Entropy balance:

$$0 = \dot{N}\underline{S}_2 - \dot{N}\underline{S}_3 + \frac{\dot{Q}_T}{T}$$

or

$$\frac{\dot{Q}_T}{\dot{N}} = T_2(\underline{S}_3 - \underline{S}_2) = T_2\left[C_P^* \ln\frac{T_3}{T_2} - R\ln\frac{P_3}{P_2}\right] = -RT_2\ln\frac{P_3}{P_2} \equiv -RT_2 \ln K_T$$

where $K_T = P_3/P_2$ is the compression ratio of the turbine. Consequently,

$$\frac{\dot{W}_{23}}{\dot{N}} = \underline{H}_3 - \underline{H}_2 - \frac{\dot{Q}_T}{\dot{N}} = C_P^*(T_3 - T_2) + RT_2\ln K_T = RT_2\ln K_T$$

3 → 4 Through the regenerative heat exchanger
Energy balance:

$$0 = \dot{N}\underline{H}_3 - \dot{N}\underline{H}_4 + \dot{Q}_{34} \qquad \text{or} \qquad \frac{\dot{Q}_{34}}{\dot{N}} = \underline{H}_4 - \underline{H}_3 = C_P^*(T_4 - T_3)$$

However,

$$\dot{Q}_{12} = -\dot{Q}_{34}$$

which implies (since C_P^* is constant) that $T_2 - T_1 = T_3 - T_4$. Also, since both the compressor and turbine are isothermal (so that $T_2 = T_3$ and $T_4 = T_1$), the temperature decrease of stream 1 exactly equals the temperature increase of stream 3 for this case of an ideal gas of constant heat capacity.

4 → 1 Isothermal compressor
As for the isothermal turbine, we find

$$\frac{\dot{Q}_C}{\dot{N}} = RT_1\ln\frac{P_1}{P_4} \qquad \text{and} \qquad \frac{\dot{W}_{41}}{\dot{N}} = RT_1\ln\frac{P_1}{P_4}$$

The efficiency of this cycle is

$$\eta = \frac{-\dot{W}_{out}}{\dot{Q}_T} = \frac{-(\dot{W}_{23} + \dot{W}_{41})}{\dot{Q}_T} = \frac{\left[RT_2\ln\dfrac{P_3}{P_2} + RT_1\ln\dfrac{P_1}{P_4}\right]}{RT_2\ln\dfrac{P_3}{P_2}}$$

Since each of the streams in the heat exchanger operates at constant pressure, it follows that $P_3 = P_4$ and $P_2 = P_1$. Therefore,

$$\eta = \frac{\left[RT_2\ln\dfrac{P_3}{P_2} + RT_1\ln\dfrac{P_1}{P_3}\right]}{RT_2\ln\dfrac{P_3}{P_2}} = \frac{T_2 - T_1}{T_2}$$

As T_2 is the high temperature of the cycle and T_1 is the low temperature, this efficiency is exactly equal to the Carnot efficiency. (Indeed, it can be shown that for any cycle, if the two heat transfers to the surroundings occur at different but constant temperatures, as is the case here with an isothermal compressor and an isothermal turbine, the efficiency of the cycle will be that of a Carnot cycle.)

Therefore, the efficiency of the Ericsson cycle considered here is

$$\eta = \frac{(450 + 273.15) - (70 + 273.15)}{(450 + 273.15)} = 0.525$$

Note that the efficiency is dependent on the temperatures of the various parts of the cycle, but not on the pressures or the compression ratio. However, the pressure ratio does affect the heat and work flows in each part of the cycle. ∎

ILLUSTRATION 5.2-4

Calculating the Efficiency of a Stirling Cycle

Compute the efficiency of a Stirling cycle operating under the same conditions as the Ericsson cycle above.

SOLUTION

The terms for each of the steps in the Stirling cycle are the same as for the Ericsson cycle; however, the properties are slightly different, since in this case the heat exchanger (regenerator) operates such that each of its streams is at constant volume, not constant pressure. Therefore, rather than $P_2 = P_1$ as in the Ericsson cycle, here we have (by the ideal gas law) $P_2 = P_1 T_2 / T_1$. Similarly, $P_4 = P_3 T_4 / T_3$. Therefore,

$$\eta = \frac{-\dot{W}_{out}}{\dot{Q}_T} = \frac{RT_2 \ln \dfrac{P_3}{P_2} + RT_1 \ln \dfrac{P_1}{P_4}}{RT_2 \ln \dfrac{P_3}{P_2}} = \frac{RT_2 \ln \dfrac{P_3}{P_2} + RT_1 \ln \dfrac{P_2 T_1 / T_2}{P_3 T_4 / T_3}}{RT_2 \ln \dfrac{P_3}{P_2}}$$

$$= \frac{RT_2 \ln \dfrac{P_3}{P_2} + RT_1 \ln \dfrac{P_2}{P_3}}{RT_2 \ln \dfrac{P_3}{P_2}} = \frac{T_2 - T_1}{T_2}$$

Since $T_2 = T_3$ and $T_1 = T_4$, both the compressor and turbine operate isothermally. Therefore, the efficiency of the Stirling cycle is also equal to that of the Carnot cycle. For the operating conditions here, $\eta = 0.525$. ∎

A common method of producing mechanical work (usually for electrical power generation) is to use a gas turbine and the Brayton or gas turbine power cycles. This open cycle consists of a compressor (on the same shaft as the turbine), a combustor in which fuel is added and ignited to heat the gas, and a turbine that extracts work from the high-temperature, high-pressure gas, which is then exhausted to the atmosphere. The open-cycle gas turbine is used, for example, in airplane jet engines and in some trucks. This cycle is shown in Fig. 5.2-8.

In this cycle the compressor and turbine are assumed to operate isentropically, and the gas flow through the combustor is at constant pressure. Note that the temperatures of streams 1 and 4 may not be the same, though the pressures are both atmospheric.

The closed gas turbine cycle is shown in Fig. 5.2-9. Closed cycles are typically used in nuclear power plants, where the heating fluid in the high-temperature heat exchanger

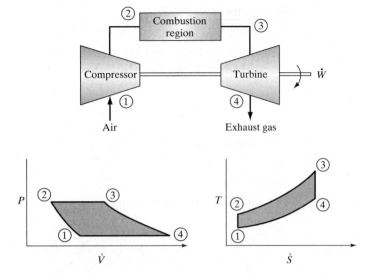

Figure 5.2-8 The open Brayton power cycle.

is the reactor cooling fluid, which for safety and environmental reasons must be contained within the plant, and the low-temperature coolant is river water. The properties and path for this cycle are given below.

Point	Path to Next Point	T	P	$\underline{S}$	$\underline{H}$	Energy Flow
1		T_1	P_1	$\underline{S}_1$	$\underline{H}_1$	
	Isentropic			$\downarrow$		$\dot{W}_{12}$
2		T_2	P_2	$\underline{S}_2 = \underline{S}_1$	$\underline{H}_2$	
	Isobaric			$\downarrow$		$\dot{Q}_{23}$
3		T_3	$P_3 = P_2$	$\underline{S}_3$	$\underline{H}_3$	
	Isentropic			$\downarrow$		$\dot{W}_{34}$
4		T_4	P_4	$\underline{S}_4 = \underline{S}_3$	$\underline{H}_4$	
	Isobaric			$\downarrow$		$\dot{Q}_{41}$
1		T_1	$P_1 = P_4$	$\underline{S}_1$	$\underline{H}_1$	

Closed Brayton cycle

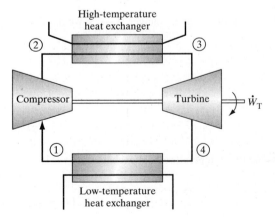

Figure 5.2-9 The closed Brayton power generation cycle.

ILLUSTRATION 5.2-5

Calculating the Efficiency of a Brayton Cycle

Assuming the working fluid in a Brayton power cycle is an ideal gas of constant heat capacity, show that the efficiency of the cycle is

$$\eta = \frac{-(\dot{W}_{34} - \dot{W}_{12})}{\dot{Q}_{23}}$$

and obtain an explicit expression for the efficiency in terms of the cycle temperatures.

SOLUTION

For path $1 \to 2$, through the compressor, we have from the energy balance that

$$\frac{\dot{W}_{12}}{\dot{N}} = \underline{H}_2 - \underline{H}_1 = C_P^*(T_2 - T_1)$$

Also from $\underline{S}_2 = \underline{S}_1$, we have

$$0 = C_P^* \ln \frac{T_2}{T_1} - R \ln \frac{P_2}{P_1} \quad \text{or} \quad \frac{T_2}{T_1} = \left(\frac{P_2}{P_1}\right)^{R/C_P^*}$$

Similarly, across the gas turbine, $3 \to 4$, we have

$$\frac{\dot{W}_{34}}{\dot{N}} = \underline{H}_4 - \underline{H}_3 = C_P^*(T_4 - T_3)$$

and

$$\frac{T_4}{T_3} = \left(\frac{P_4}{P_3}\right)^{R/C_P^*}$$

But $P_3 = P_2$ and $P_4 = P_1$, so that

$$\frac{T_4}{T_3} = \left(\frac{P_4}{P_3}\right)^{R/C_P^*} = \left(\frac{P_1}{P_2}\right)^{R/C_P^*} = \frac{T_1}{T_2}$$

Across the high-temperature heat exchanger, $2 \to 3$, we have

$$0 = \dot{N}\underline{H}_2 - \dot{N}\underline{H}_3 + \dot{Q}_{23}$$

or

$$\frac{\dot{Q}_{23}}{\dot{N}} = \underline{H}_3 - \underline{H}_2 = C_P^*(T_3 - T_2)$$

Similarly, across the low-temperature heat exchanger, $4 \to 1$,

$$\frac{\dot{Q}_{41}}{\dot{N}} = \underline{H}_1 - \underline{H}_4 = C_P^*(T_1 - T_4)$$

Therefore,

$$\eta = -\frac{(\dot{W}_{34} + \dot{W}_{12})}{\dot{Q}_{23}} = \frac{-C_P^*(T_4 - T_3) - C_P^*(T_2 - T_1)}{C_P^*(T_3 - T_2)}$$

$$= \frac{(T_3 - T_4) + (T_1 - T_2)}{T_3 - T_2} = \frac{(T_3 - T_2) - (T_4 - T_1)}{(T_3 - T_2)} = 1 - \frac{(T_4 - T_1)}{(T_3 - T_2)}$$

$$= 1 - \frac{T_1}{T_2} \frac{\left(\dfrac{T_4}{T_1} - 1\right)}{\left(\dfrac{T_3}{T_2} - 1\right)}$$

However, from $T_4/T_3 = T_1/T_2$, we have that $T_4/T_1 = T_3/T_2$, so that

$$\eta = 1 - \frac{T_1}{T_2} = 1 - \left(\frac{P_1}{P_2}\right)^{R/C_P^*} = 1 - K_T^{-R/C_P^*}$$

where $K_T = P_2/P_1$ is the compression ratio. ∎

The final cyclic device we consider is the heat pump, that is, a device used to pump heat from a low-temperature source to a high-temperature sink by expending work. Refrigerators and air conditioners are examples of heat pumps. Heat pumps (with appropriate valving) are now being installed in residential housing so that they can be used for both winter heating (by pumping heat to the house from its surroundings) and summer cooling (by pumping heat from the house to the surroundings). The surroundings may be either the atmosphere, the water in a lake, or, with use of underground coils, the earth. The term *pumping* is used since in both the winter and summer modes, heat is being taken from a region of low temperature and exhausted to a region of higher temperature. In principle, any power generation or refrigeration cycle that can be made to operate in reverse can serve as a heat pump.

A heat pump that uses the vapor-compression refrigeration cycle and two connected (so that they operate together) three-way valves is schematically shown in Fig. 5.2-10. In this way the indoor coil is the condenser during the winter months and the evaporator or boiler during the summer. Similarly, the outdoor coil is the evaporator (boiler) during

Heat pump

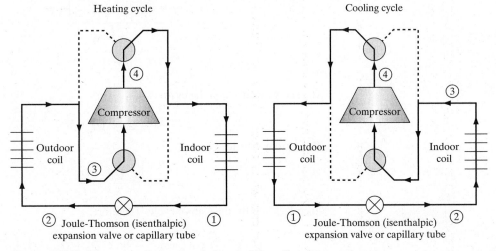

Figure 5.2-10 Heat pump in heating (winter) and cooling (summer) cycles.

the winter and the condenser during the summer. The vapor compression refrigeration cycle was discussed earlier in this section, and the paths on T-$\hat{S}$ and P-$\hat{H}$ plots are given in Fig. 5.2-4b.

ILLUSTRATION 5.2-6

Calculating the Coefficients of Performance of a Heat Pump

Consider a residential heat pump that uses lake water as a heat source in the winter and as a heat sink in the summer. The house is to be maintained at a winter temperature of 18°C and a summer temperature of 25°C. To do this efficiently, it is found that the indoor coil temperature should be at 50°C in the winter and 5°C in the summer. The outdoor coil temperature can be assumed to be 5°C during the winter months and 35°C during the summer.

a. Compute the coefficient of performance of a reverse Carnot cycle (Carnot cycle heat pump) in the winter and summer if it is operating between the temperatures listed above.

b. Instead of the reverse Carnot cycle, a vapor-compression cycle will be used for the heat pump with HFC-134a as the working fluid. Compute the winter and summer coefficients of performance for this heat pump. Assume that the only pressure changes in the cycle occur across the compressor and the expansion valve, and that the only heat transfer to and from the refrigerant occurs in the indoor and outdoor heat transfer coils.

Point	Heating Temperature	Cooling Temperature	Fluid State
1	50°C	35°C	Saturated liquid
2			Vapor-liquid mixture
3	5°C	5°C	Saturated vapor
4			Superheated vapor

SOLUTION

a. *Carnot cycle coefficients of performance:*

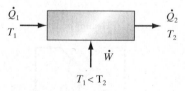

The energy balance on the cycle is

$$0 = \dot{Q}_1 + \dot{Q}_2 + \dot{W}$$

The entropy balance is

$$0 = \frac{\dot{Q}_1}{T_1} + \frac{\dot{Q}_2}{T_2}$$

since $S_{\text{gen}} = 0$ for the Carnot cycle. Therefore,

$$\dot{Q}_2 = \frac{-T_2}{T_1}\dot{Q}_1 \quad \text{and} \quad -\dot{W} = \dot{Q}_2 + \dot{Q}_1 = \dot{Q}_1\left(\frac{T_1 - T_2}{T_1}\right)$$

Winter Operation (Heating Mode)

$$\text{C.O.P. (winter)} = \frac{-\dot{Q}_1}{\dot{W}} = \frac{T_1}{T_2 - T_1} = \frac{(5 + 273.15)}{45} = 6.18$$

Summer Operation (Cooling Mode):

$$\text{C.O.P. (summer)} = \frac{\dot{Q}_1}{\dot{W}} = \frac{T_1}{T_2 - T_1} = \frac{(5 + 273.15)}{30} = 9.27$$

b. *Vapor compression cycle coefficients of performance:*

The following procedure is used to locate the cycles on the pressure-enthalpy diagram of Fig. 3.3-4, a portion of which is repeated here. First, point 1 is identified as corresponding to the saturated liquid at the condenser temperature. (In this illustration the condenser is the indoor coil at 50°C in the heating mode, and the outdoor coil at 35°C in the cooling mode.) Since the path from point 1 to 2 is an isenthalpic (Joule-Thomson) expansion, a downward vertical line is drawn to the evaporator temperature (corresponding to the outdoor coil temperature of 5°C in the heating mode, or in the cooling mode to the indoor coil temperature, which also happens to be 5°C in this illustration). This temperature and the enthalpy of point 1 fix the location of point 2. Next, point 3 is found by drawing a horizontal line to the saturated vapor at the evaporator pressure. The path from point 2 to 3 describes the vaporization of the vapor-liquid mixture that resulted from the Joule-Thomson expansion. The evaporator is the outdoor coil in the heating mode and the indoor coil in the cooling mode.

The path from point 3 to 4 is through the compressor, which is assumed to be isentropic, and so corresponds to a line of constant entropy on the pressure-enthalpy diagram. Point 4 is located on this line at its intersection with the horizontal line corresponding to the pressure of the evaporator. The fluid at point 4 is a superheated vapor. The path from 4 to 1 is a horizontal line of constant pressure terminating at the saturated liquid at the temperature and pressure of the condenser.

The tables below give the values of those properties (read from the pressure-enthalpy diagram) at each point in the cycle needed for the calculation of the coefficient of performance in the heating and cooling modes of the heat pump. In these tables the properties known at each point from the problem statement or from an adjacent point appear in boldface, and the properties found from the chart at each point appear in italics.

Heating (winter operation)

Point	T (°C)	P (MPa)	Condition	$\hat{H}\left(\dfrac{\text{kJ}}{\text{kg}}\right)$	$\hat{S}\left(\dfrac{\text{kJ}}{\text{kg K}}\right)$
1	**50**	*1.319*	Saturated liquid	*271.9*	
				↓	
2	**5**	*0.3499*	Vapor-liquid mixture	**271.9**	
		↓			
3	**5**	**0.3499**	Saturated vapor	*401.7*	*1.7252*
					↓
4	*58*	**1.319**	Superheated vapor	*432*	**1.7252**

Therefore,

$$\frac{-\dot{Q}_{23}}{\dot{M}} = \hat{H}_3 - \hat{H}_2 = (407.1 - 271.9)\,\frac{\text{kJ}}{\text{kg}} = 129.8\,\frac{\text{kJ}}{\text{kg}}$$

$$\frac{\dot{W}_{34}}{\dot{M}} = \hat{H}_4 - \hat{H}_3 = (432 - 401.7)\,\frac{\text{kJ}}{\text{kg}} = 30.3\,\frac{\text{kJ}}{\text{kg}}$$

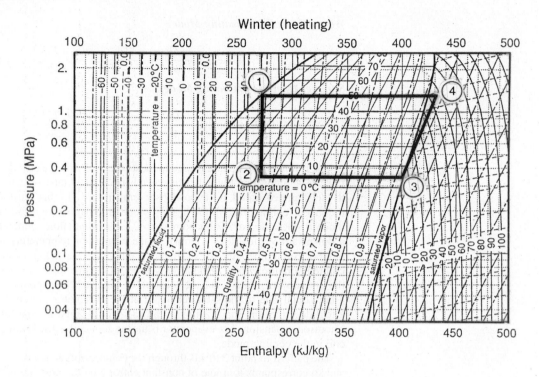

Winter (heating)

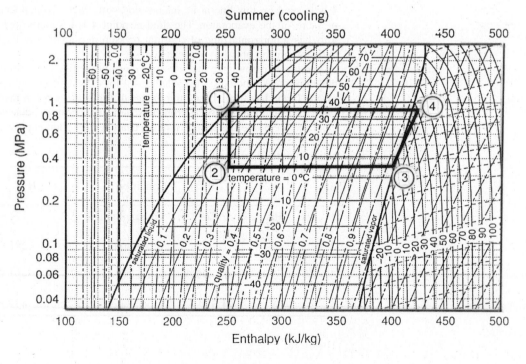

Summer (cooling)

and

$$\text{C.O.P.} = \frac{129.8}{30.3} = 4.28$$

Cooling (summer operation)

Point	T (°C)	P (MPa)	Condition	$\hat{H}\left(\dfrac{\text{kJ}}{\text{kg}}\right)$	$\hat{S}\left(\dfrac{\text{kJ}}{\text{kg K}}\right)$
1	**35**	*0.8879*	Saturated liquid	*249.2*	
				↓	
2	5	*0.3499*	Vapor-liquid mixture	**249.2**	
		↓			
3	5	**0.3499**	Saturated vapor	*401.7*	*1.7252*
					↓
4	*45*	**0.8879**	Superheated vapor	*436*	**1.7252**

Consequently, here

$$\frac{\dot{Q}_{23}}{\dot{M}} = \hat{H}_3 - \hat{H}_2 = (401.7 - 249.2)\,\frac{\text{kJ}}{\text{kg}} = 152.5\,\frac{\text{kJ}}{\text{kg}}$$

$$\frac{\dot{W}_{34}}{\dot{M}} = \hat{H}_4 - \hat{H}_3 = (436 - 401.7)\,\frac{\text{kJ}}{\text{kg}} = 34.3\,\frac{\text{kJ}}{\text{kg}}$$

and

$$\text{C.O.P.} = \frac{152.5}{34.3} = 4.45$$

Note that in both the heating and cooling modes, the heat pump in this illustration has a lower coefficient of performance (and therefore lower efficiency) than a reverse Carnot cycle operating between the same temperatures. Finally, it should be mentioned that the thermodynamic properties listed above were obtained from a detailed thermodynamic properties table for HFC-134a, akin to the steam tables for water in Appendix A.III; values cannot be obtained from Fig. 3.3-4 to this level of accuracy. ■

5.3 THE THERMODYNAMICS OF MECHANICAL EXPLOSIONS

In this section we are interested in predicting the uncontrolled energy release of an explosion that occurs without a chemical reaction. That is, we are interested in energy released from an explosion that results from the bursting of an overpressurized tank, or the rapid depressurization of a hot liquid that leads to its partially boiling (flashing) to a vapor-liquid mixture. We do not consider chemical explosions, for example, the detonation of TNT or a natural gas explosion, both of which involve multiple chemical reactions and require the estimate of properties of mixtures that we have not yet considered. Chemical explosions are considered in Chapter 13.

First we consider the impact of an explosion involving a high-pressure gas, and then we consider the explosion of a flashing liquid due to rapid depressurization. A primary characteristic of an explosion is that it is rapid, indeed so rapid that there is insufficient time for the transfer of heat or mass to or from the exploding material. Instead, the

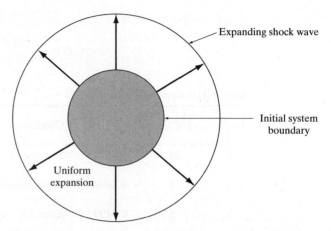

Figure 5.3-1 Explosion—a rapid, uniform expansion. If the shock wave moves at less than the speed of sound, the event is classified as a deflagration. If the shock wave travels at the speed of sound, the event is an explosion.

exploding matter undergoes a rapid expansion, pushing away the surrounding air. This is schematically illustrated in Fig. 5.3-1.

During a detonation, a shock wave is produced; outside this shock wave the pressure is ambient, but inside the shock wave the pressure is above ambient. This shock wave continues to travel outward until, as a result of the expansion of the gas behind the shock front, the pressure inside the wave falls to ambient pressure and the temperature is below the initial temperature of the gas. If the initial pressure is sufficiently high, the shock wave will travel at, but not exceed, the speed of sound. This is the strict definition of an explosion. If the shock wave travels at a slower speed, the event is referred to as a deflagration. It is the overpressurization or pressure difference resulting from the passage of a shock wave that causes most of the damage in an explosion.

We compute the energy released from an explosion, since the energy release is related to the extent of overpressurization and the amount of damage. We do this by writing the thermodynamic balance equations for a closed system consisting of the initial contents of an exploding tank. This is a system with an expanding boundary (as the explosion occurs) that is closed to the transfer of mass and is adiabatic (i.e., $Q = 0$) since the expansion occurs too rapidly for heat transfer to occur. We also assume that the expansion occurs uniformly, so that there are no temperature or pressure gradients, except at the boundary of the shock wave.

The difference form of the mass balance for this closed system inside the shock wave is

$$M_f - M_i = 0 \quad \text{or} \quad M_f = M_i = M \tag{5.3-1}$$

Here the subscript i denotes the initial state, before the explosion has occurred, and f denotes the final state, when the system volume has expanded to such an extent that the pressure inside the shock wave is the ambient pressure. The energy balance for this system is

$$M_f \hat{U}(T_f, P_f) - M_i \hat{U}(T_i, P_i) = W \tag{5.3-2}$$

Using the mass balance and the fact that the final pressure is ambient gives

$$M[\hat{U}(T_f, P = \text{ambient}) - \hat{U}(T_i, P_i)] = W \qquad \text{(5.3-3)}$$

where $-W$ is the energy that the exploding system imparts to its surroundings.

This equation contains two unknowns, W and $\hat{U}(T_f, P = \text{ambient})$, that we cannot evaluate since T_f is unknown. Therefore, we need another equation, which we get from the entropy balance. The entropy balance for this system is

$$M[\hat{S}(T_f, P = \text{ambient}) - \hat{S}(T_i, P_i)] = S_{\text{gen}} \qquad \text{(5.3-4)}$$

Here S_{gen} is the entropy generated during the explosion. Since the expansion is a uniform expansion (that is, there are no spatial gradients except at the shock wave boundary), the only generation of entropy occurs across the shock wave. If we neglect this entropy generation, the work we compute will be somewhat too high. However, in safety problems we prefer to be conservative and err on the side of overpredicting an energy release resulting from an explosion, since we are usually interested in estimating the maximum energy release and the maximum damage that could result. Further, we really do not have a good way of computing the amount of entropy generated during an explosion. Consequently, we will set $S_{\text{gen}} = 0$, and from Eqs. 5.3-2 and 4 obtain

$$\hat{S}(T_f, P = \text{ambient}) = \hat{S}(T_i, P_i) \qquad \text{(5.3-5)}$$

which provides the additional equation necessary to find T_f and $\hat{U}_f$.

Equations 5.3-3 and 5.3-5 will be used to compute the work or explosive energy imparted to the surroundings as a result of a gas phase explosion. This work is transferred to the surroundings by the damage-producing overpressurization of the shock wave. The procedure to be used to compute the energy transfer to the surroundings is as follows:

1. Knowing the initial temperature and pressure of the system, compute the initial specific volume $\hat{V}_i$, specific internal energy $\hat{U}_i$, and specific entropy $\hat{S}_i$.
2. If the mass of the system is known, proceed to step 3. If the mass of the system is not known, but the total initial volume of the system V_S is known, compute the mass from

$$M = \frac{V_S}{\hat{V}_i(T_i, P_i)}$$

3. Since the initial and final specific entropies of the system are the same (see Eq. 5.3-5), compute the final temperature of the system from the known final specific entropy and the ambient pressure.
4. From the ambient pressure and computed final temperature, calculate the final specific internal energy $\hat{U}_f$.
5. Finally, use Eq. 5.3-3 to determine how much work the exploding system does on its surroundings.

To use the procedure described above, we need to be able to compute the thermodynamic properties of the fluid. We may have access to tables of thermodynamic properties for the fluid. For example, if the fluid is water vapor, we could use the steam tables. Other possibilities include using graphs for the properties of the gas under consideration, the ideal gas law, or if available an accurate equation of state for the fluid (to be discussed in the next chapter).

ILLUSTRATION 5.3-1

Energy Released on the Explosion of a Steam Tank

A tank of volume 0.1 m³ that contains steam at 600°C and 1 MPa bursts. Estimate the energy of the blast. For comparison, the blast energy of trinitrotoluene (TNT) is 4600 kJ/kg.

SOLUTION

From the steam tables we have at the initial conditions that $\hat{V} = 0.4011$ m³/kg, $\hat{U} = 3296.8$ kJ/kg, and $\hat{S} = 8.0290$ kJ/(kg K). At $P = 0.1$ MPa $= 1$ bar we have that $\hat{S} = 8.0290$ kJ/(kg K) at $T = 248°C$, where $\hat{U} \approx 2731$ kJ/kg. Consequently,

$$M = \frac{0.1 \text{ m}^3}{0.4011 \dfrac{\text{m}^3}{\text{kg}}} = 0.2493 \text{ kg}$$

and

$$-W = 0.2493 \text{ kg} \times (3296.8 - 2731) \frac{\text{kJ}}{\text{kg}} = 141.1 \text{ kJ}$$

This is equivalent to the energy released by 30.7 g of TNT. ∎

The simplest and least accurate procedure for calculating the blast energy is to assume that the fluid is an ideal gas with a constant heat capacity. In this case we have (using a molar basis rather than per unit mass)

$$N = \text{number of moles} = \frac{P_i V_S}{R T_i} \tag{5.3-6}$$

where V_S is the initial system volume (i.e., the volume of the tank). Also,

$$\underline{S}(T_f, P = \text{ambient}) - \underline{S}(T_i, P_i) = 0 = C_P^* \ln \frac{T_f}{T_i} - R \ln \frac{(P = \text{ambient})}{P_i}$$

or

$$T_f = T_i \left(\frac{P = \text{ambient}}{P_i} \right)^{R/C_P^*} \tag{5.3-7}$$

Then

$$W = N[\underline{U}(T_f, P = \text{ambient}) - \underline{U}(T_i, P_i)] = N C_V^*(T_f - T_i)$$

$$= \frac{P_i V_S}{R T_i} C_V^* T_i \left[\left(\frac{P = \text{ambient}}{P_i} \right)^{R/C_P^*} - 1 \right] = \frac{P_i V_S}{R} C_V^* \left[\left(\frac{P = \text{ambient}}{P_i} \right)^{R/C_P^*} - 1 \right]$$

$$\tag{5.3-8}$$

Finally, using $\gamma = C_P^* / C_V^*$, we have

$$-W = \frac{P_i V_S}{\gamma - 1}\left[1 - \left(\frac{P = \text{ambient}}{P_i}\right)^{(\gamma - 1)/\gamma}\right] \tag{5.3-9}$$

Here P_i is the initial pressure, that is, the pressure at which the tank bursts, V_S is the initial total volume of the system or tank, P is the ambient pressure, and $-W$ (which is a positive number) is the work or energy the explosion imparts to its surroundings, which is what does the damage.

ILLUSTRATION 5.3-2

Energy Released on the Explosion of a Compressed Air Tank

A compressed air tank has a volume of 0.167 m^3 and contains air at 25°C and 650 bar when it explodes. Estimate the amount of work done on the surroundings in the explosion. Compute the TNT equivalent of the compressed air tank blast.

Data: For air $C_P^* = 29.3$ J/(mol K) and $\gamma = C_P^*/C_V^* = 1.396$.

SOLUTION

$$-W = \frac{650 \text{ bar} \times 0.167 \text{ m}^3}{(1.396 - 1)}\left[1 - \left(\frac{1 \text{ bar}}{650 \text{ bar}}\right)^{(1.396-1)/1.396}\right]$$

$$= 230.2 \text{ bar m}^3 \times 10^5 \frac{\text{Pa}}{\text{bar}} \times \frac{1 \frac{\text{J}}{\text{m}^3}}{\text{Pa}} \times \frac{1 \text{ kJ}}{10^3 \text{ J}} = 23\,020 \text{ kJ}$$

Therefore, the blast energy is equivalent to

$$\frac{23\,020 \text{ kJ}}{4600 \frac{\text{kJ}}{\text{kg TNT}}} = 5.0 \text{ kg of TNT}$$

Clearly, this is a sizable explosion. In fact, a blast of 5 kg of TNT will cause the total destruction of structures not reinforced to withstand blasts within a circle of radius 7 meters from the blast site, substantial damage out to a radius of 14 meters, minor structural damage out to 55 meters, and broken windows out to 130 meters. Also, eardrum ruptures can be expected to 10 meters from the site of the explosion. ∎

So far we have considered the blast energy when the fluid is a gas. If the fluid contained in the vessel is a liquid, the same equations apply:

$$\hat{S}(T_f, P = \text{ambient}) = \hat{S}(T_i, P_i) \tag{5.3-5}$$

and

$$W = M[\hat{U}(T_f, P = \text{ambient}) - \hat{U}(T_i, P_i)] \tag{5.3-3}$$

However, for a liquid the entropy and internal energy are, to an excellent approximation, independent of pressure. This implies that there is no entropy or internal energy change from the failure of a vessel containing a liquid (as long as the liquid does not vaporize). Consequently, there is no blast energy resulting from the change in thermodynamic properties of the fluid. However, there will be some release of energy that had been stored as elastic or strain energy in the walls of the container. Since the evaluation of this contribution is based on solid mechanics and not thermodynamics, it is not discussed here. The important observation is that compressed gases can result in devastating explosions; however, compressed liquids, as long as they do not vaporize, contain little energy to release on an explosion.[2] It is for this reason that liquids, not gases, are used to test high-pressure vessels before they are put into use.

A vaporizing liquid is considered next. If a tank initially under high pressure ruptures, it is possible that as a result of the sudden depressurization the liquid or liquid-vapor mixture initially in the tank may partially or completely vaporize. Such explosions are known as boiling liquid–evaporating vapor explosions (BLEVEs). Here we estimate the energy released on a BLEVE.

For generality we assume that both the initial and final states include two phases, and we use ω_i and ω_f to represent the fraction of vapor in the initial and final states, respectively. The mass balance for this system is

$$M_i[\omega_i + (1 - \omega_i)] = M_f[\omega_f + (1 - \omega_f)] \qquad \textbf{(5.3-10a)}$$

or

$$M_i = M_f = M \qquad \textbf{(5.3-10b)}$$

The energy balance is

$$\boxed{\begin{aligned} &M[\omega_f \hat{U}^{\mathrm{V}}(T_f, P = \text{ambient}) + (1 - \omega_f)\hat{U}^{\mathrm{L}}(T_f, P = \text{ambient})] \\ &- M[\omega_i \hat{U}^{\mathrm{V}}(T_i, P_i) + (1 - \omega_i)\hat{U}^{\mathrm{L}}(T_i, P_i)] = W \end{aligned}} \qquad \textbf{(5.3-11)}$$

and the entropy balance is

Balance equations for a vapor-liquid explosion

$$\boxed{\begin{aligned} &\omega_f \hat{S}^{\mathrm{V}}(T_f, P = \text{ambient}) + (1 - \omega_f)\hat{S}^{\mathrm{L}}(T_f, P = \text{ambient}) \\ &= \omega_i \hat{S}^{\mathrm{V}}(T_i, P_i) + (1 - \omega_i)\hat{S}^{\mathrm{L}}(T_i, P_i) \end{aligned}} \qquad \textbf{(5.3-12)}$$

Here the superscripts V and L indicate the vapor and liquid phases, respectively. Also, it should be remembered that if two phases are present, the temperature must be the saturation temperature of the fluid at the specified pressure. For example, if the fluid is water and the final condition is ambient pressure (1 atm or 1.013 bar), then T_f must be 100°C.

ILLUSTRATION 5.3-3

Energy Released from a Steam-Water Explosion

Joe Udel decides to install his own 52-gallon (0.197 m^3) hot-water heater. However, being cheap, he ignores safety codes and neglects to add safety devices such as a thermostat and a rupture

---

[2]It should be noted that we have considered only the internal energy change of an explosion. For air or steam this is all we need to consider. However, if the tank contents are combustible, the result can be more devastating. There are numerous examples of a tank of combustible material rupturing and producing a flammable vapor cloud. At some point distant from the initial explosion, this vapor cloud comes in contact with sufficient oxygen and an ignition source, which results in a second, chemical explosion. In chemical plants such secondary explosions are usually more devastating than the initial explosion.

disk or pressure relief valve. In operation, the hot-water heater is almost completely filled with liquid water and the pressure in the heater tank is the waterline pressure of 1.8 bar or the water saturation pressure at the heater temperature, whichever is greater. The water heater tank will rupture at a pressure of 20 bar. Several hours after Joe completes the installation of the water heater, it explodes.

    a. What was the temperature of the water when the tank exploded?
    b. Estimate the energy released in the blast.

SOLUTION

    a. From Appendix A.III the thermodynamic properties of saturated liquid water at 20 bar (2 MPa) are

$$T_i = 212.42°\text{C} \qquad \hat{V}_i^{\text{L}} = 0.001177 \ \frac{\text{m}^3}{\text{kg}} \qquad \hat{U}_i^{\text{L}} = 906.44 \ \frac{\text{kJ}}{\text{kg}}$$

and

$$\hat{S}_i^{\text{L}} = 2.4474 \ \frac{\text{kJ}}{\text{kg K}}$$

Therefore, the water temperature when the tank explodes is 212.42°C. Also, as indicated in the problem statement, the tank contains only liquid, so $\omega_i = 0$ and

$$M = \frac{V_{\text{T}}}{\hat{V}_i^{\text{L}}} = \frac{0.197 \ \text{m}^3}{0.001177 \ \dfrac{\text{m}^3}{\text{kg}}} = 167.24 \ \text{kg}$$

    b. After the explosion we expect to have a vapor-liquid mixture. Since the pressure is 1 atm (actually, we use 1 bar), the temperature is 100°C and the other thermodynamic properties are

$$\hat{U}^{\text{L}} = 417.36 \ \frac{\text{kJ}}{\text{kg}} \qquad \hat{U}^{\text{V}} = 2506.1 \ \frac{\text{kJ}}{\text{kg}}$$

$$\hat{S}^{\text{L}} = 1.3026 \ \frac{\text{kJ}}{\text{kg K}} \qquad \hat{S}^{\text{V}} = 7.3594 \ \frac{\text{kJ}}{\text{kg K}}$$

We first use the entropy balance to determine the fractions of vapor and liquid present

$$\hat{S}_i^{\text{L}} = \omega_f \hat{S}_f^{\text{V}} + (1 - \omega_f)\hat{S}_f^{\text{L}}$$

or

$$2.4474 = \omega_f 7.3594 + (1 - \omega_f)1.3026$$

which gives

$$\omega_f = 0.1890$$

Next, to calculate the blast energy we use the energy balance, Eq. 5.3-11:

$$\begin{aligned} -W &= M[\hat{U}_i^{\text{L}} - \hat{U}_f] = M[\hat{U}_i^{\text{L}} - \{\omega_f \hat{U}_f^{\text{V}} + (1 - \omega_f)\hat{U}_f^{\text{L}}\}] \\ &= 167.24 \ \text{kg} \ [906.44 - \{0.1890 \times 2506.1 + 0.8110 \times 417.36\}] \ \text{kJ/kg} \\ &= 15 \ 772 \ \text{kJ} \end{aligned}$$

This is equivalent to 3.43 kg of TNT.

COMMENT

In January 1982 a large hot-water tank exploded in an Oklahoma school, killing 7 people and injuring 33 others. The tank was found 40 meters from its original location, and part of the school cafeteria was destroyed. It was estimated that the tank failed at a pressure of only 7 bar.■

If thermodynamic tables for the fluid in the explosion are not available, it may still be possible to make an estimate of the fraction of vapor and liquid present after the explosion and the energy released knowing just the heat of vaporization of the fluid and its liquid heat capacity. To do this, we first write the unsteady-state mass and energy balances on the open system consisting of the liquid, shown in Fig. 5.3-2. In this figure $\dot{M}$ is the molar flow rate of the liquid being vaporized. The mass and energy balances for this open, constant-volume system are as follows:

Mass balance

$$\frac{dM}{dt} = -\dot{M} \tag{5.3-13}$$

Energy balance

$$\frac{d}{dt}(M\hat{U}^{L}) = -\dot{M}\hat{H}^{V} \tag{5.3-14}$$

Here we have used the superscripts L and V to indicate the vapor and liquid phases. Also, we will use that for liquids at moderate pressure $\hat{U}^{L} \approx \hat{H}^{L}$, as discussed earlier. Therefore, using this result and combining the two above equations, we have

$$M\frac{d\hat{H}^{L}}{dt} + \hat{H}^{L}\frac{dM}{dt} = \frac{dM}{dt}\hat{H}^{V}$$

or

$$MC_{P}^{L}\frac{dT}{dt} = \frac{dM}{dt}\Delta_{vap}\hat{H} \tag{5.3-15}$$

Here we have used $d\hat{H}^{L} = C_{P}^{L}\,dT$. Also, we have defined the heat of vaporization of a liquid, $\Delta_{vap}\hat{H}$, to be the difference between the enthalpies of the vapor and liquid when both are at the boiling temperature (i.e., $\Delta_{vap}\hat{H} = \hat{H}^{V} - \hat{H}^{L}$).

Integrating this equation between the initial and final states, we have

$$\int_{M_{i}}^{M_{f}} \frac{1}{M}\,dM = \int_{T_{i}}^{T_{f}} \frac{C_{P}^{L}}{\Delta_{vap}\hat{H}}\,dT \cong \frac{C_{P}^{L}}{\Delta_{vap}\hat{H}} \int_{T_{i}}^{T_{f}} dT \tag{5.3-16}$$

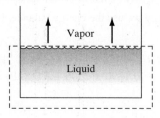

Figure 5.3-2 Boiling liquid–evaporating vapor event. The dashed line indicates the system for which the mass and energy balances are being written.

where, in writing the last of these equations, we have assumed that both the liquid heat capacity and the heat of vaporization are independent of temperature. We then obtain

$$\ln \frac{M_f}{M_i} = \frac{C_P^L}{\Delta_{vap}\hat{H}}(T_f - T_i) \tag{5.3-17}$$

Note that the final temperature, T_f, is the normal boiling temperature of the fluid, T_b. Also, the initial and final masses in this equation are those of the liquid. Further, if the pre-explosion state is only liquid ($\omega_i = 0$), then M_f/M_i is the fraction of liquid initially present that is *not* vaporized in the BLEVE. For this case we have

$$1 - \omega_f = \exp\left[\frac{C_P^L}{\Delta_{vap}\hat{H}}(T_b - T_i)\right]$$

or

$$\omega_f = 1 - \exp\left[\frac{C_P^L}{\Delta_{vap}\hat{H}}(T_b - T_i)\right] \tag{5.3-18}$$

To proceed further, we need to compute the change in internal energy between the initial and final states. For simplicity in this calculation, we choose the reference state to be one in which the internal energy is zero for the liquid at its normal boiling point, and compute the internal energies of the other states relative to this reference state since our interest is only with changes in internal energy. (Note that this choice of reference state is just a matter of convenience and not a necessity. Can you show that any state can be chosen as the reference state for the calculation of the energy differences without affecting the final result?) Consequently, we have

$$\hat{U}^L(T_b, P = 1 \text{ bar}) = 0 \tag{5.3-19a}$$

$$\hat{U}^V(T_b, P = 1 \text{ bar}) = \hat{U}^L(T_b, P = 1 \text{ bar}) + \Delta_{vap}\hat{U} \tag{5.3-19b}$$

$$= 0 + \left(\Delta_{vap}\hat{H} - \frac{RT_b}{MW}\right) = \Delta_{vap}\hat{H} - \frac{RT_b}{MW}$$

and

$$\hat{U}^L(T_i, P_i) = \int_{T_b}^{T_i} C_P^L\, dT \cong C_P^L(T_i - T_b) \tag{5.3-19c}$$

Therefore, the energy released in an explosion is

| Simplified equation for estimating the energy released on a two-phase explosion | $$-W = M[\hat{U}^L(T_i, P_i) - \omega_f \hat{U}^V(T_b, P = \text{ambient}) - (1 - \omega_f)\hat{U}^L(T_b, P = \text{ambient})]$$ $$= M\left[C_P^L(T_i - T_b) - \omega_f\left(\Delta_{vap}\hat{H} - \frac{RT_b}{MW}\right)\right]$$ |

$$\tag{5.3-20}$$

where ω_f is given by Eq. 5.3-18.

ILLUSTRATION 5.3-4

Use of the Simplified Equation to Calculate the Energy Released in a Two-Phase Explosion (BLEVE)

Rework the previous illustration using only the fact that the density of water at 212.4°C is about 0.85 g/cm³, and that over the temperature range

$$C_P^L = 4.184 \ \frac{J}{g\,K} \quad \text{and} \quad \Delta_{vap}\hat{H} = 2250 \ \frac{J}{g} = 2250 \ \frac{kJ}{kg}$$

SOLUTION

Given the density of water above, we have

$$M = 0.197 \text{ m}^3 \times 0.85 \ \frac{g}{cc} \times 10^6 \ \frac{cc}{m^3} \times 10^{-3} \ \frac{kg}{g} = 167.5 \text{ kg}$$

Next, we use

$$\omega_f = 1 - \exp\left[\frac{C_P^L}{\Delta_{vap}\hat{H}}(T_b - T_i)\right]$$

$$= 1 - \exp\left[\frac{4.184}{2250}(100 - 212.42)\right] = 0.189$$

which is equal to the value found in the previous illustration using accurate thermodynamic tables.

Next we have

$$-W = M\left[4.184(212.42 - 100) - 0.189\left(2250 - \frac{8.314 \times 373.15}{18}\right)\right]$$

$$= 167.5 \text{ kg}\,[470.37 - 392.68] \ \frac{kJ}{kg} = 13\,013 \text{ kJ}$$

compared with 15 772 kJ from the more accurate calculation of the previous example. Since the total mass and the fraction vaporized have been correctly calculated, all the error is a result of the approximate calculation of the internal energies. Nonetheless, such approximate calculations are useful when detailed thermodynamic data are not available.   ■

PROBLEMS

5.1 a. An automobile air conditioner uses the vapor-compression refrigeration cycle with HFC-134a as the refrigerant. The operational temperature of the evaporator is 7°C and that of the condenser is 45°C. Determine the coefficient of performance of this air-conditioning system and the amount of work needed for each kilojoule of cooling provided by the air conditioner.

b. For service in high-temperature areas, the condenser temperature may go up to 65°C. How would the answers to part (a) change in this case?

5.2 It is desired to improve the thermal efficiency of the Rankine power generation cycle. Two possibilities have been suggested. One is to increase the evaporator temperature, and the second is to decrease the condenser temperature (and consequently the pressure) of the low-pressure part of the cycle.

a. Draw a T-$\hat{S}$ diagram for the Rankine cycle similar to that in Fig. 5.2-2d, but with a higher evaporator temperature. Show from a comparison of those two diagrams that the efficiency of the Rankine power generation cycle increases with increasing evaporator temperature.

b. Repeat part (a) for a lower condenser temperature.

5.3 Using a T-$\hat{S}$ diagram, discuss the effect of subcooling in the condenser and superheating in the evaporator on the efficiency of a Rankine (or other) power generation cycle.

5.4 A power plant using a Rankine power generation cycle and steam operates at a temperature of 80°C in the condenser, a pressure of 2.5 MPa in the evaporator, and a maximum evaporator temperature of 700°C. Draw the two cycles described below on a temperature-entropy diagram for steam, and answer the following questions.

a. What is the efficiency of this power plant, assuming the pump and turbine operate adiabatically and reversibly? What is the temperature of the steam leaving the turbine?

b. If the turbine is found to be only 85 percent efficient, what is the overall efficiency of the cycle? What is the temperature of the steam leaving the turbine in this case?

5.5 Forest cabins, remote mobile homes, Amish farms, and residential structures in locations where electricity is not available are often equipped with absorption refrigerators that rely on changes from absorption at low temperatures to desorption at high temperatures to produce pressure changes in a refrigeration system instead of a compressor. The energy source for such refrigeration systems is a flame, typically produced by propane or another fuel. The most common absorption refrigeration working fluid is the ammonia-water system. The only patent ever awarded to Albert Einstein was for an absortion refrigeration design.

The simplest representation of an absorption refrigeration is given in the figure below. The energy flows in such a device are a high-temperature (T_H) heat flow ($\dot{Q}_H$) that supplies the energy for the refrigerator, a low-temperature (T_L) heat flow ($\dot{Q}_L$) into the condenser of the refrigeration cycle extracted from the cold box of the refrigerator, and a moderate-temperature (T_M) heat flow ($\dot{Q}_M$) out of the refrigerator.

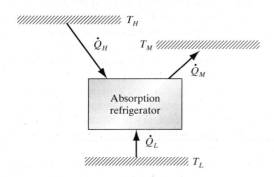

a. Compute the coefficient of performance for an absorption refrigerator defined as

$$\text{C.O.P.} = \frac{\dot{Q}_L}{\dot{Q}_H}$$

assuming that the absorption refrigeration cycle is reversible.

b. Calculate the maximum coefficient of performance that can be achieved if heat transfer from the flame occurs at 750°C, the ambient temperature near the refrigerator is 27°C, and the temperature inside the refrigerator is −3°C.

5.6 a. Nitrogen can be liquefied using a simple Joule-Thomson expansion process. This is done by rapidly and adiabatically expanding cold nitrogen from a high-pressure gas to a low-temperature, low-pressure vapor-liquid mixture. To produce the high pressure, nitrogen initially available at 0.1 MPa and 135 K is reversibly and adiabatically compressed to 2 MPa, isobarically cooled to 135 K, recompressed to 20 MPa, and again isobarically cooled before undergoing the Joule-Thomson expansion to 0.1 MPa. What is the temperature of the liquid nitrogen, and how much compressor work is required per kilogram of liquid nitrogen produced?

b. If, to improve efficiency, the Linde process is used with the same two-stage compressor as in part (a) and with nitrogen vapor leaving the heat exchanger at 0.1 MPa and 125 K, how much compressor work is required per kilogram of liquid nitrogen produced?

5.7 A Rankine steam cycle has been proposed to generate work from burning fuel. The temperature of the burning fuel is 1100°C, and cooling water is available at 15°C. The steam leaving the boiler is at 20 bar and 700°C, and the condenser produces a saturated liquid at 0.2 bar. The steam lines are well insulated, the turbine and pump operate reversibly and adiabatically, and some of the mechanical work generated by the turbine is used to drive the pump.

a. What is the net work obtained in the cycle per kilogram of steam generated in the boiler?

b. How much heat is discarded in the condenser per kilogram of steam generated in the boiler?

c. What fraction of the work generated by the turbine is used to operate the pump?

d. How much heat is absorbed in the boiler per kilogram of steam generated?

e. Calculate the engine efficiency and compare it with the efficiency of a Carnot cycle receiving heat at 1100°C and discharging heat at 15°C.

5.8 As in Illustration 5.1-1 it is desired to produce liquefied methane; however, the conditions are now changed so that the gas is initially available at 1 bar and 200 K, and methane leaving the cooler will be at 100 bar and 200 K. The flash drum is adiabatic and operates at 1 bar, and each compressor stage can be assumed to operate reversibly and adiabatically. A three-stage compressor will be used, with the first stage compressing the gas from 1 bar to 5 bar, the second stage from 5 bar to 25 bar, and the third stage from 25 bar to 100 bar. Between stages the gas will be isobarically cooled to 200 K.

a. Calculate the amount of work required for each kilogram of methane that passes through the compressor in the simple liquefaction process.

b. Calculate the fractions of vapor and liquid leaving the flash drum in the simple liquefaction process of Fig. 5.1-1 and the amount of compressor work required for each kilogram of LNG produced.

c. Assuming that the recycled methane leaving the heat exchanger in the Linde process (Fig. 5.1-2) is at 1 bar and 200 K, calculate the amount of compressor work required per kilogram of LNG produced.

5.9 High-pressure helium is available from gas producers in 0.045-m³ cylinders at 400 bar and 298 K. Calculate the explosion equivalent of a tank of compressed helium in terms of kilograms of TNT. Assume helium is an ideal gas.

5.10 The "Quick Fill" bicycle tire filling system consists of a small (2 cm diameter, 6.5 cm long) cylinder filled with nitrogen to a pressure of 140 bar. Estimate the explosion equivalent of the gas contained in the cylinder in grams of TNT. Assume nitrogen is an ideal gas.

5.11 A tank containing liquid water in equilibrium with a small amount of vapor at 25 bar suddenly ruptures. Estimate the fraction of liquid water in the tank that flash vaporizes, and the explosive energy released per kilogram of water initially in the tank.

5.12 Electrical power is to be produced from a steam turbine connected to a nuclear reactor. Steam is obtained from the reactor at 540 K and 36 bar, the turbine exit pressure is 1.0 bar, and the turbine is adiabatic.

a. Compute the maximum work per kilogram of steam that can be obtained from the turbine.

A clever chemical engineer has suggested that the single-stage turbine considered here be replaced by a two-stage adiabatic turbine, and that the steam exiting from the first stage be returned to the reactor and re-

heated, at constant pressure, to 540 K, and then fed to the second stage of the turbine. (See the figure.)

b. Compute the maximum work obtained per kilogram of steam if the two-stage turbine is used and the exhaust pressure of the first stage is $P^* = \frac{1}{2}(36 + 1.0) = 18.5$ bar.

c. Compute the maximum work obtained per kilogram of steam if the two-stage turbine is used and the exhaust pressure of the first stage is $P^* = \sqrt{36 \times 1} = 6.0$ bar.

d. Compute the heat absorbed per kilogram of steam in the reheating steps in parts (b) and (c).

5.13 A coal-fired power plant had been operating using a standard Rankine cycle to produce power. The operating conditions are as given in Illustration 5.2-1. However, the boiler is aging and will need to be replaced. While waiting for the replacement, it has been suggested that for safety the operating temperature be reduced from 600°C to 400°C. The plant operates with steam, with a condenser temperature of 100°C, and in this emergency mode the boiler would operate at 3.0 MPa and 400°C.

a. Can the plant function in this mode? Why or why not?

b. A clever operator suggests that a Joule-Thompson valve could be placed after the boiler and before the turbine. This valve is to be designed such that the exhaust from the turbine is at the same conditions as in Illustration 5.2-1, 0.10135 MPa and approximately 126°C. Assuming that the pump and turbine operate adiabatically and reversibly, fill in the missing thermodynamic properties in the table below. [Values that are unchanged from Illustration 5.2-1 are given in bold.]

c. Determine the heat and work flows per kg of steam in the pump, boiler, turbine and condenser.

d. What is the efficiency of this new cycle? How does it compare with the efficiency of the cycle in Illustration 5.2-1?

5.14 During methane liquefaction, about 1000 kg of methane are stored at a pressure of 10 MPa and 180 K. The plant manager is worried about the possibility of explosion. Determine the energy released by a sudden rupture of this storage tank and the temperature and physical state of the methane immediately after the rupture.

5.15 A Rankine power generation cycle is operated with water as the working fluid. It is found that 100 MW of power is produced in the turbine by 89 kg/s of steam that enters the turbine at 700°C and 5 MPa and leaves at 0.10135 MPa. Saturated liquid water exits the condenser and is pumped back to 5 MPa and fed to the boiler, which operates isobarically.

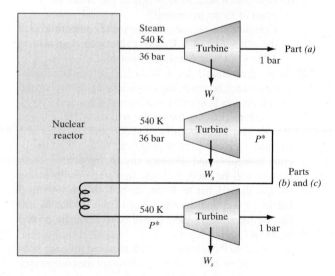

Point	T (°C)	P (MPa)	$\hat{S}$ [kJ/(kg K)]	$\hat{H}$ (kJ/kg)
1 Exit from condenser	100	0.10135	1.3069	419.04
entry to pump	100	0.10135	1.3069	419.04
2 Exit from pump				
entry to boiler	103	3.0	1.3069	422.07
3 Exit from boiler				
entry to J-T valve		3.0		
4 Exit from J-T valve				
entry to turbine				
5 Exit from turbine				
entry to condenser	126.2	0.10135	7.5085	2734.7

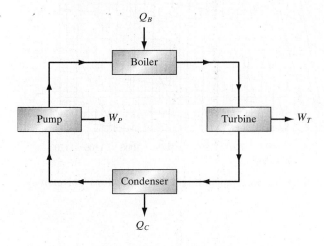

a. The turbine operates adiabatically, but not reversibly. Find the temperature of the steam exiting the turbine.

b. Determine the rate of entropy generation in the turbine and compute the efficiency of the turbine.

c. How much work must be supplied to the pump?

5.16 The United States produces about 2700 megawatts (MW) of electricity from geothermal energy, which is comparable to burning 60 million barrels of oil each year. Worldwide about 7000 MW of geothermal electricity are produced. The process is that naturally occurring steam or hot water that is not far below the earth's surface (especially in places such as Yellowstone National Park and other volcanic and geothermal areas) is brought to the surface and used to heat a working fluid in a binary fluid power generation cycle, such as that shown in Fig 5.2-9. (Geothermal steam and water are not directly injected into a turbine, as

the dissolved salts and minerals would precipitate and quickly damage the equipment.) For geothermal water at temperatures less than 200°C, isobutane is used as the working fluid. Isobutane is vaporized and superheated to 480 K and 10 MPa in the heat exchanger by the geothermal water, and is then passed through a turbine (which we will assume to be adiabatic and isentropic) connected to an electrical generator. The isobutane next passes through an isobaric condenser that produces a saturated liquid at 320 K. A pressure-enthalpy diagram for isobutane follows.

a. At what pressure does the condenser operate?

b. What are the temperature and pressure of the isobutane leaving the turbine?

c. Determine the work produced by the turbine per kilogram of isobutane circulating, and the flow rate of isobutane necessary to produce 3 MW of electricity.

d. Draw the cycle of the process on the isobutane pressure-enthalpy diagram.

e. Obtain the heat or work requirements for the four units of the cycle in the table below.

Unit	Heat flow (kJ/kg)	Work flow (kJ/kg)
Pump	0	?
Boiler	?	0
Turbine	0	?
Condenser	?	0

f. What is the efficiency of the proposed cycle?

©Department of Mechanical Engineering, Stanford University

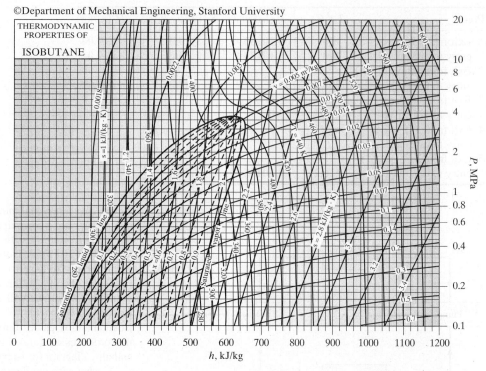

(Used with permission of the Department of Mechanical Engineering, Stanford University.)

Chapter **6**

The Thermodynamic Properties of Real Substances

In Chapters 2, 3, and 4 we derived a general set of balance equations for mass, energy, and entropy that can be used to compute energy changes, and heat or work requirements, for the changes of state of any substance. However, these balance equations are in terms of the internal energy, enthalpy, and entropy rather than the pressure and temperature, the variables most easily measured and thus most often used to specify the thermodynamic state of the system. To illustrate the use of the balance equations in the simplest manner, examples were given using either ideal gases or fluids whose thermodynamic properties were available in graphical and tabular form. Unfortunately, no gas is ideal over the whole range of pressure and temperature, and thermodynamic properties tables are not always available, so a necessary ingredient of many thermodynamic computations is the calculation of the thermodynamic properties of real substances in any state. The main topic of this chapter is establishing how to solve thermodynamic problems for real substances given heat capacity data and the relationship between pressure, volume, and temperature. The problem of constructing a thermodynamic properties chart from such data is also considered. The discussion of the relationship between the ideal gas and absolute temperature scales, which began in Chapter 1, is completed here. Finally, the principle of corresponding states and generalized equations of state are considered, as is their application.

INSTRUCTIONAL OBJECTIVES FOR CHAPTER 6

The goals of this chapter are for the student to:

- Be able to evaluate the partial derivative of a thermodynamic variable with respect to one variable (e.g., temperature) while holding a second variable constant (e.g., pressure) (Sec. 6.2)
- Be able to interrelate the partial derivatives that arise in thermodynamics (Sec. 6.2)
- Be able to obtain volumetric equation of state parameters from critical properties (Sec. 6.4)

187

- Be able to solve problems for real fluids using volumetric equations of state (e.g., van der Waals or Peng-Robinson) (Secs. 6.4 and 6.7)
- Be able to construct tables and charts of thermodynamic properties (Sec. 6.4).

NOTATION INTRODUCED IN THIS CHAPTER

$a(T)$	Equation-of-state parameter (dimensions depend on equation)
A	Dimensionless form of equation-of-state parameter a
$b(T)$	Equation-of-state parameter (m^3/mol)
B	Dimensionless form of equation-of-state parameter b
$B(T)$	Second virial coefficient (m^3/mol)
$C(T)$	Third virial coefficient (m^3/mol)^2
P_c	Pressure at the critical point (kPa)
P_r	Reduced pressure $= P/P_c$
T_c	Temperature at the critical point (K)
T_r	reduced temperature $= T/T_c$
$\underline{V}_c$	Molar volume at the critical point (m^3/mol)
V_r	Reduced volume $= \underline{V}/\underline{V}_c$
Z	Compressibility factor $= \dfrac{PV}{RT}$
Z_c	Compressibility factor at the critical point
α	coefficient of thermal expansion $= \dfrac{1}{\underline{V}}\left(\dfrac{\partial \underline{V}}{\partial T}\right)_P$ (K^{-1})
$\alpha(T)$	Temperature-dependent term in equation of state
κ_T	isothermal compressibility $= -\dfrac{1}{\underline{V}}\left(\dfrac{\partial \underline{V}}{\partial P}\right)_T$ (kPa^{-1})
κ	Parameter in temperature dependence of $a(T)$ in equation of state
μ	Joule-Thomson coefficient $= \left(\dfrac{\partial T}{\partial P}\right)_H$ (K kPa^{-1})
ω	Acentric factor

6.1　SOME MATHEMATICAL PRELIMINARIES

In the previous chapters eight thermodynamic state variables (P, T, $\underline{V}$, $\underline{S}$, $\underline{U}$, $\underline{H}$, $\underline{A}$, and $\underline{G}$), which frequently appear in thermodynamic calculations, were introduced. If values of any two of these variables are given, the thermodynamic state of a pure, single-phase system is fixed, as are the values of the remaining six variables (experimental observation 8 of Sec. 1.7). Mathematically we describe this situation by saying that any two variables may be chosen as the independent variables for the single-component, one-phase system, and the remaining six variables are dependent variables. If, for example, T and $\underline{V}$ are taken as the independent variables, then any other variable, such as the internal energy $\underline{U}$, is a dependent variable; this is denoted by $\underline{U} = \underline{U}(T, \underline{V})$ to indicate that the internal energy is a function of temperature and specific volume. The change in internal energy $d\underline{U}$, which results from differential changes in T and $\underline{V}$, can be computed using the chain rule of differentiation:

$$d\underline{U} = \left(\frac{\partial \underline{U}}{\partial T}\right)_V dT + \left(\frac{\partial \underline{U}}{\partial \underline{V}}\right)_T d\underline{V}$$

where the subscript on each derivative indicates the variable being held constant; that is, $(\partial \underline{U}/\partial T)_V$ denotes the differential change in molar internal energy accompanying a differential change in temperature in a process in which the molar volume is constant. Note that both $(\partial \underline{U}/\partial T)_{\underline{V}}$ and $(\partial \underline{U}/\partial \underline{V})_T$ are partial derivatives of the type

$$\left(\frac{\partial \underline{X}}{\partial \underline{Y}}\right)_{\underline{Z}} \qquad\qquad\text{(6.1-1)}$$

where $\underline{X}$, $\underline{Y}$, and $\underline{Z}$ are used here to denote the state variables P, T, $\underline{V}$, $\underline{S}$, $\underline{U}$, $\underline{H}$, $\underline{A}$, and $\underline{G}$. Since derivatives of this type occur whenever one tries to relate a change in one thermodynamic function (here $\underline{U}$) to changes in two others (here T and $\underline{V}$), one of the goals of this chapter is to develop methods for computing numerical values for these derivatives, as these will be needed in many thermodynamic calculations.

In open systems extensive properties, such as the total internal energy U, the total enthalpy H, and the total entropy S, are functions of three variables, usually taken to be two thermodynamic variables (either intensive or extensive) and the total mass (M) or mole number (N). Thus, for example, the total internal energy can be considered to be a function of temperature, total volume, and number of moles, so that for a differential change in T, V, and N

$$dU = \left(\frac{\partial U}{\partial T}\right)_{V,N} dT + \left(\frac{\partial U}{\partial V}\right)_{T,N} dV + \left(\frac{\partial U}{\partial N}\right)_{T,V} dN$$

where two subscripts are now needed to indicate the variables being held constant for each differential change.

The derivatives of extensive properties at constant mole number or mass are simply related to the analogous derivatives among the state variables, that is, to derivatives of the form of Eq. 6.1-1. To see this, we note that since any extensive property X can be written as $N\underline{X}$, where $\underline{X}$ is an intensive (or molar) property, the derivative of an extensive property with respect to an intensive property (e.g., temperature, pressure, or specific volume) at constant mole number is

$$\left(\frac{\partial X}{\partial \underline{Y}}\right)_{\underline{Z},N} = \left(\frac{\partial (N\underline{X})}{\partial \underline{Y}}\right)_{\underline{Z},N} = N\left(\frac{\partial \underline{X}}{\partial \underline{Y}}\right)_{\underline{Z}}$$

Thus,

$$\left(\frac{\partial U}{\partial T}\right)_{V,N} = N\left(\frac{\partial \underline{U}}{\partial T}\right)_{\underline{V}} = NC_V$$

where C_V is the constant-volume heat capacity defined in Chapter 3. Similarly, the derivative of an extensive property with respect to an extensive property at a constant mole number is found from the observation that

$$\left(\frac{\partial X}{\partial Y}\right)_{\underline{Z},N} = \left(\frac{\partial (N\underline{X})}{\partial (N\underline{Y})}\right)_{\underline{Z},N} = \frac{N}{N}\left(\frac{\partial \underline{X}}{\partial \underline{Y}}\right)_{\underline{Z}} = \left(\frac{\partial \underline{X}}{\partial \underline{Y}}\right)_{\underline{Z}}$$

so that, for example,

$$\left(\frac{\partial U}{\partial V}\right)_{T,N} = \left(\frac{\partial \underline{U}}{\partial \underline{V}}\right)_T$$

Consequently, once a method is developed to obtain numerical values for derivatives in the form of Eq. 6.1-1,

$$\left(\frac{\partial \underline{X}}{\partial \underline{Y}}\right)_Z$$

it can also be used to evaluate derivatives of the form

$$\left(\frac{\partial X}{\partial Y}\right)_{Z,N} \quad \text{and} \quad \left(\frac{\partial X}{\partial \underline{Y}}\right)_{Z,N}$$

The derivative of an extensive property with respect to mole number [e.g., the derivative $(\partial U/\partial N)_{T,V}$ in this discussion] is not of the form of Eq. 6.1-1. Such derivatives are considered in the following sections and in Sec. 6.10.

We start the analysis of partial derivatives of the type indicated in Eq. 6.1-1 by listing several of their important properties. First, their numerical value depends on the path followed, that is, on which variable is being held constant. Thus,

$$\left(\frac{\partial \underline{U}}{\partial T}\right)_V \neq \left(\frac{\partial \underline{U}}{\partial T}\right)_P$$

as will be verified shortly (Illustration 6.2-1). If two intensive variables are held constant in a one-component, single-phase system, all derivatives, such as

$$\left(\frac{\partial \underline{U}}{\partial T}\right)_{V,P}$$

must equal zero, since by fixing the values of two intensive variables, one also fixes the values of all the remaining variables.

We will assume that all the thermodynamic variables in which we are interested exist and are well behaved in some mathematical sense that we will leave unspecified. In this case, the derivative of Eq. 6.1-1 has the properties that

$$\left(\frac{\partial \underline{X}}{\partial \underline{Y}}\right)_Z = \frac{1}{(\partial \underline{Y}/\partial \underline{X})_Z} \tag{6.1-2}$$

and

$$\frac{\partial}{\partial \underline{Z}}\bigg|_Y \left(\frac{\partial \underline{X}}{\partial \underline{Y}}\right)_Z = \frac{\partial}{\partial \underline{Y}}\bigg|_Z \left(\frac{\partial \underline{X}}{\partial \underline{Z}}\right)_Y \tag{6.1-3}$$

The last equation, the commutative property, states that in a mixed second derivative the order of taking derivatives is unimportant. Also,

$$\boxed{\left(\frac{\partial \underline{X}}{\partial \underline{Y}}\right)_X = 0} \tag{6.1-4a}$$

and

$$\boxed{\left(\frac{\partial \underline{X}}{\partial \underline{X}}\right)_Z = 1} \tag{6.1-4b}$$

since the first derivative is the change in $\underline{X}$ along a path of constant $\underline{X}$, and the second derivative is the change in $\underline{X}$ in response to an imposed change in $\underline{X}$.

If $\underline{X}$ is any dependent variable, and $\underline{Y}$ and $\underline{Z}$ the two independent variables, by the chain rule of differentiation we can write $\underline{X} = \underline{X}(\underline{Y}, \underline{Z})$, and

$$d\underline{X} = \left(\frac{\partial \underline{X}}{\partial \underline{Y}}\right)_{\underline{Z}} d\underline{Y} + \left(\frac{\partial \underline{X}}{\partial \underline{Z}}\right)_{\underline{Y}} d\underline{Z} \tag{6.1-5}$$

Now from Eqs. 6.1-4a and 6.1-5, we have

$$\left(\frac{\partial \underline{X}}{\partial \underline{Y}}\right)_{\underline{X}} = 0 = \left(\frac{\partial \underline{X}}{\partial \underline{Y}}\right)_{\underline{Z}} \left(\frac{\partial \underline{Y}}{\partial \underline{Y}}\right)_{\underline{X}} + \left(\frac{\partial \underline{X}}{\partial \underline{Z}}\right)_{\underline{Y}} \left(\frac{\partial \underline{Z}}{\partial \underline{Y}}\right)_{\underline{X}}$$

and using Eq. 6.1-4b we obtain

$$\boxed{\left(\frac{\partial \underline{X}}{\partial \underline{Y}}\right)_{\underline{Z}} \left(\frac{\partial \underline{Z}}{\partial \underline{X}}\right)_{\underline{Y}} \left(\frac{\partial \underline{Y}}{\partial \underline{Z}}\right)_{\underline{X}} = -1} \tag{6.1-6a}$$

or

Triple-product rule

$$\boxed{\left(\frac{\partial \underline{X}}{\partial \underline{Y}}\right)_{\underline{Z}} \left(\frac{\partial \underline{Z}}{\partial \underline{X}}\right)_{\underline{Y}} = -\left(\frac{\partial \underline{Z}}{\partial \underline{Y}}\right)_{\underline{X}}} \tag{6.1-6b}$$

Equations 6.1-6 are known as the triple product rule and will be used frequently in this book; this rule is easily remembered by noting the symmetric form of Eq. 6.1-6a in that each variable appears in each derivative position once and only once.

There are two other important results to be derived from Eq. 6.1-5. The first is the expansion rule, obtained by introducing any two additional thermodynamic properties $\underline{K}$ and $\underline{L}$,

$$\left(\frac{\partial \underline{X}}{\partial \underline{K}}\right)_{\underline{L}} = \left(\frac{\partial \underline{X}}{\partial \underline{Y}}\right)_{\underline{Z}} \left(\frac{\partial \underline{Y}}{\partial \underline{K}}\right)_{\underline{L}} + \left(\frac{\partial \underline{X}}{\partial \underline{Z}}\right)_{\underline{Y}} \left(\frac{\partial \underline{Z}}{\partial \underline{K}}\right)_{\underline{L}} \tag{6.1-7}$$

and the second is a special case of the first in which $\underline{L} = \underline{Z}$, so that, by Eq. 6.1-4a,

Chain rule

$$\boxed{\left(\frac{\partial \underline{X}}{\partial \underline{K}}\right)_{\underline{Z}} = \left(\frac{\partial \underline{X}}{\partial \underline{Y}}\right)_{\underline{Z}} \left(\frac{\partial \underline{Y}}{\partial \underline{K}}\right)_{\underline{Z}}} \tag{6.1-8}$$

This equation is known as the chain rule. (You should compare Eq. 6.1-6b with Eq. 6.1-8 and note the difference between them.) Thus, if for some reason it is convenient, we can use this equation interpose a new variable in evaluating a partial derivative, as is the case with the variable $\underline{Y}$ in Eq. 6.1-8.

Finally, we note that for an open system it is usually convenient to use the mass M or mole number N, and two variables from among T, P, and the extensive variables U, V, S, G, H, and A, as the independent variables. Letting X, Y, and Z represent variables from among the set (U, V, S, G, H, A, T, and P), we have $X = X(Y, Z, N)$ and

$$dX = \left(\frac{\partial X}{\partial Y}\right)_{Z,N} dY + \left(\frac{\partial X}{\partial Z}\right)_{Y,N} dZ + \left(\frac{\partial X}{\partial N}\right)_{Y,Z} dN \tag{6.1-9}$$

6.2   THE EVALUATION OF THERMODYNAMIC PARTIAL DERIVATIVES

Whenever the change in a thermodynamic property is related to changes in two others, derivatives of the type of Eq. 6.1-1 occur. Four of these partial derivatives occur so frequently in both experiment and calculation that they have been given special designations:

$$\left(\frac{\partial \underline{U}}{\partial T}\right)_V = C_V = \text{constant-volume heat capacity (see Sec. 3.3)} \tag{6.2-1}$$

$$\left(\frac{\partial \underline{H}}{\partial T}\right)_P = C_P = \text{constant-pressure heat capacity (see Sec. 3.3)} \tag{6.2-2}$$

Definitions

$$\frac{1}{\underline{V}}\left(\frac{\partial \underline{V}}{\partial T}\right)_P = \alpha = \text{coefficient of thermal expansion} \tag{6.2-3}$$

$$-\frac{1}{\underline{V}}\left(\frac{\partial \underline{V}}{\partial P}\right)_T = \kappa_T = \text{isothermal compressibility} \tag{6.2-4}$$

The starting point of the analysis of other thermodynamic derivatives is Eq. 4.2-13a for open systems,

$$dU = T\,dS - P\,dV + \underline{G}\,dN \tag{6.2-5a}$$

and Eq. 4.2-13b for closed systems:

$$d\underline{U} = T\,d\underline{S} - P\,d\underline{V} = \left(\frac{\partial \underline{U}}{\partial \underline{S}}\right)_{\underline{V}} d\underline{S} + \left(\frac{\partial \underline{U}}{\partial \underline{V}}\right)_{\underline{S}} d\underline{V} \tag{6.2-5b}$$

which leads to $\left(\frac{\partial \underline{U}}{\partial \underline{S}}\right)_{\underline{V}} = T$ and $\left(\frac{\partial \underline{U}}{\partial \underline{V}}\right)_{\underline{S}} = -P$. Alternatively, these equations can be rearranged to

$$dS = \frac{1}{T}\,dU + \frac{P}{T}\,dV - \frac{G}{T}\,dN \tag{6.2-5c}$$

and

$$d\underline{S} = \frac{1}{T}\,d\underline{U} + \frac{P}{T}\,d\underline{V} \tag{6.2-5d}$$

By definition, $H = U + PV$, so that

$$\begin{aligned} dH &= dU + V\,dP + P\,dV = T\,dS - P\,dV + \underline{G}\,dN + V\,dP + P\,dV \\ &= T\,dS + V\,dP + \underline{G}\,dN \end{aligned} \tag{6.2-6a}$$

and for the closed system

$$d\underline{H} = T\,d\underline{S} + \underline{V}\,dP \tag{6.2-6b}$$

Similarly, from Eq. 4.2-6 we have $A = U - TS$, so that $dA = dU - S\,dT - T\,dS$. Using Eq. 6.2-5 yields

$$dA = -P\,dV - S\,dT + \underline{G}\,dN \tag{6.2-7a}$$

which for the closed system becomes

$$dA = -P\,dV - S\,dT \tag{6.2-7b}$$

Finally, from Eq. 4.2-8, we have $G = H - TS$, so that

$$dG = V\,dP - S\,dT + G\,dN \tag{6.2-8a}$$

and

$$dG = V\,dP - S\,dT \tag{6.2-8b}$$

Next, we note the analogy between Eqs. 6.1-5 and 6.1-9 and Eqs. 6.2-5 through 6.2-8 and obtain the following relations:

$$\left(\frac{\partial U}{\partial S}\right)_{V,N} = \left(\frac{\partial U}{\partial S}\right)_{V} = T \tag{6.2-9a}$$

$$\left(\frac{\partial U}{\partial V}\right)_{S,N} = \left(\frac{\partial U}{\partial V}\right)_{S} = -P \tag{6.2-9b}$$

$$\left(\frac{\partial U}{\partial N}\right)_{S,V} = G \tag{6.2-9c}$$

$$\left(\frac{\partial S}{\partial N}\right)_{U,V} = -\frac{G}{T} \tag{6.2-9d}$$

$$\left(\frac{\partial H}{\partial S}\right)_{P,N} = \left(\frac{\partial H}{\partial S}\right)_{P} = T \tag{6.2-10a}$$

$$\frac{1}{N}\left(\frac{\partial H}{\partial P}\right)_{S,N} = \left(\frac{\partial H}{\partial P}\right)_{S} = V \tag{6.2-10b}$$

$$\left(\frac{\partial H}{\partial N}\right)_{P,S} = G \tag{6.2-10c}$$

$$\left(\frac{\partial A}{\partial V}\right)_{T,N} = \left(\frac{\partial A}{\partial V}\right)_{T} = -P \tag{6.2-11a}$$

$$\frac{1}{N}\left(\frac{\partial A}{\partial T}\right)_{V,N} = \left(\frac{\partial A}{\partial T}\right)_{V} = -S \tag{6.2-11b}$$

$$\left(\frac{\partial A}{\partial N}\right)_{T,V} = G \tag{6.2-11c}$$

$$\frac{1}{N}\left(\frac{\partial G}{\partial P}\right)_{T,N} = \left(\frac{\partial G}{\partial P}\right)_{T} = V \tag{6.2-12a}$$

$$\frac{1}{N}\left(\frac{\partial G}{\partial T}\right)_{P,N} = \left(\frac{\partial G}{\partial T}\right)_{P} = -S \tag{6.2-12b}$$

and

$$\left(\frac{\partial G}{\partial N}\right)_{T,P} = G \tag{6.2-12c[1]}$$

---

[1] Note that comparing Eqs. 6.2-9c and 6.2-9d, 6.2-10c, 6.2-11c, and 6.2-12c we have

$$\left(\frac{\partial U}{\partial N}\right)_{S,V} = \left(\frac{\partial H}{\partial N}\right)_{P,S} = \left(\frac{\partial A}{\partial N}\right)_{T,V} = \left(\frac{\partial G}{\partial N}\right)_{T,P} = -T\left(\frac{\partial S}{\partial N}\right)_{U,V} = G$$

The multicomponent analogues of these equations are given in Sec. 8.2.

To get expressions for several additional thermodynamic derivatives we next use Eq. 6.1-3, the commutative property of mixed second derivatives. In this way, starting with Eq. 6.2-9 we obtain

$$\frac{\partial}{\partial \underline{V}}\bigg|_{\underline{S}} \left(\frac{\partial \underline{U}}{\partial \underline{S}}\right)_{\underline{V}} = \left(\frac{\partial T}{\partial \underline{V}}\right)_{\underline{S}}$$

$$\frac{\partial}{\partial \underline{S}}\bigg|_{\underline{V}} \left(\frac{\partial \underline{U}}{\partial \underline{V}}\right)_{\underline{S}} = -\left(\frac{\partial P}{\partial \underline{S}}\right)_{\underline{V}}$$

so that from Eq. 6.1-3 we have

Maxwell relations

$$\left(\frac{\partial T}{\partial \underline{V}}\right)_{\underline{S}} = -\left(\frac{\partial P}{\partial \underline{S}}\right)_{\underline{V}} \qquad (6.2\text{-}13)$$

Similarly, from Eqs. 6.2-10 through 6.2-12 we obtain

$$\left(\frac{\partial T}{\partial P}\right)_{\underline{S}} = \left(\frac{\partial \underline{V}}{\partial \underline{S}}\right)_{P} \qquad (6.2\text{-}14)$$

$$\left(\frac{\partial P}{\partial T}\right)_{\underline{V}} = \left(\frac{\partial \underline{S}}{\partial \underline{V}}\right)_{T} \qquad (6.2\text{-}15)$$

and

$$\left(\frac{\partial \underline{V}}{\partial T}\right)_{P} = -\left(\frac{\partial \underline{S}}{\partial P}\right)_{T} \qquad (6.2\text{-}16)$$

Equations 6.2-13 through 6.2-16 are known as the **Maxwell relations.** (It is left to you to derive the Maxwell relations for open systems; see Problem 6.27.)

Equation 6.2-5d relates the change in entropy to changes in internal energy and volume. Since temperature and pressure, or temperature and volume, are, because of the ease with which they can be measured, more common choices for the independent variables than $\underline{U}$ and $\underline{V}$, it would be useful to have expressions relating $d\underline{S}$ to dT and $d\underline{V}$, or to dT and dP. We can derive such expressions by first writing $\underline{S}$ as a function of T and $\underline{V}$,

$$\underline{S} = \underline{S}(T, \underline{V})$$

and then using the chain rule of partial differentiation to obtain

$$d\underline{S} = \left(\frac{\partial \underline{S}}{\partial T}\right)_{\underline{V}} dT + \left(\frac{\partial \underline{S}}{\partial \underline{V}}\right)_{T} d\underline{V} \qquad (6.2\text{-}17)$$

From the application of first Eq. 6.1-8, then Eqs. 6.1-2, and finally Eqs. 6.2-1 and 6.2-9a, we obtain

$$\left(\frac{\partial \underline{S}}{\partial T}\right)_{\underline{V}} = \left(\frac{\partial \underline{S}}{\partial \underline{U}}\right)_{\underline{V}} \left(\frac{\partial \underline{U}}{\partial T}\right)_{\underline{V}} = \left(\frac{\partial \underline{U}}{\partial T}\right)_{\underline{V}} \left[\left(\frac{\partial \underline{U}}{\partial \underline{S}}\right)_{\underline{V}}\right]^{-1} = C_{\mathrm{V}}/T \qquad (6.2\text{-}18)$$

and from Eqs. 6.2-15, we have

$$\left(\frac{\partial \underline{S}}{\partial \underline{V}}\right)_{T} = \left(\frac{\partial P}{\partial T}\right)_{\underline{V}}$$

Thus

$$dS = \frac{C_V}{T} dT + \left(\frac{\partial P}{\partial T} \right)_V dV \qquad \textbf{(6.2-19)}$$

Consequently, given heat capacity data as a function of T and P or T and V, volumetric equation-of-state information, that is, a relationship between P, V, and T, and a value of the entropy at some value of T and V, it is possible to compute the entropy at any other value of T and V by integration of Eq. 6.2-19. Similarly, starting from $S = S(T, P)$, one can easily show that

$$\left(\frac{\partial S}{\partial T} \right)_P = \frac{C_P}{T}, \qquad \left(\frac{\partial S}{\partial P} \right)_T = -\left(\frac{\partial V}{\partial T} \right)_P$$

and

$$dS = \frac{C_P}{T} dT - \left(\frac{\partial V}{\partial T} \right)_P dP \qquad \textbf{(6.2-20)}$$

Equations 6.2-19 and 6.2-20 can now be used in Eqs. 6.2-5 and 6.2-6 to get

$$
\begin{aligned}
dU &= T\, dS - P\, dV \\
&= T \left[\frac{C_V}{T} dT + \left(\frac{\partial P}{\partial T} \right)_V dV \right] - P\, dV \\
&= C_V\, dT + \left[T \left(\frac{\partial P}{\partial T} \right)_V - P \right] dV
\end{aligned}
\qquad \textbf{(6.2-21)}
$$

and

$$dH = C_P\, dT + \left[V - T \left(\frac{\partial V}{\partial T} \right)_P \right] dP \qquad \textbf{(6.2-22)}$$

From these last two equations we obtain

$$\left(\frac{\partial U}{\partial V} \right)_T = T \left(\frac{\partial P}{\partial T} \right)_V - P \qquad \textbf{(6.2-23)}$$

and

$$\left(\frac{\partial H}{\partial P} \right)_T = V - T \left(\frac{\partial V}{\partial T} \right)_P \qquad \textbf{(6.2-24)}$$

Table 6.2-1 summarizes the definitions used and some of the thermodynamic identities developed so far in this chapter. The equations in this table can be useful in obtaining information about some thermodynamic derivatives, as indicated in Illustration 6.2-1.

ILLUSTRATION 6.2-1

Showing That Similar Partial Derivatives with Different State Variables Held Constant Are Not Equal

Obtain expressions for the two derivatives $(\partial U/\partial T)_V$ and $(\partial U/\partial T)_P$, and show that they are not equal.

Important table

Table 6.2-1 Some Useful Definitions and Thermodynamic Identities

Definitions

Constant-volume heat capacity $= C_V = \left(\dfrac{\partial \underline{U}}{\partial T}\right)_{\underline{V}} = T\left(\dfrac{\partial \underline{S}}{\partial T}\right)_{\underline{V}}$

Constant-pressure heat capacity $= C_P = \left(\dfrac{\partial \underline{H}}{\partial T}\right)_P = T\left(\dfrac{\partial \underline{S}}{\partial T}\right)_P$

Isothermal compressibility $= \kappa_T = -\dfrac{1}{\underline{V}}\left(\dfrac{\partial \underline{V}}{\partial P}\right)_T$

Coefficient of thermal expansion $= \alpha = \dfrac{1}{\underline{V}}\left(\dfrac{\partial \underline{V}}{\partial T}\right)_P$

Joule-Thomson coefficient $= \mu = -\dfrac{\left[\underline{V} - T\left(\dfrac{\partial \underline{V}}{\partial T}\right)_P\right]}{C_P}$

Maxwell relations

$$\left(\frac{\partial T}{\partial \underline{V}}\right)_{\underline{S}} = -\left(\frac{\partial P}{\partial \underline{S}}\right)_{\underline{V}} \qquad \left(\frac{\partial T}{\partial P}\right)_{\underline{S}} = \left(\frac{\partial \underline{V}}{\partial \underline{S}}\right)_P$$

$$\left(\frac{\partial P}{\partial T}\right)_{\underline{V}} = \left(\frac{\partial \underline{S}}{\partial \underline{V}}\right)_T \qquad \left(\frac{\partial \underline{V}}{\partial T}\right)_P = -\left(\frac{\partial \underline{S}}{\partial P}\right)_T$$

Thermodynamic identities

$$\left(\frac{\partial \underline{H}}{\partial \underline{S}}\right)_P = \left(\frac{\partial \underline{U}}{\partial \underline{S}}\right)_{\underline{V}} = T \qquad \left(\frac{\partial \underline{G}}{\partial P}\right)_T = \left(\frac{\partial \underline{H}}{\partial P}\right)_{\underline{S}} = \underline{V}$$

$$\left(\frac{\partial \underline{U}}{\partial \underline{V}}\right)_{\underline{S}} = \left(\frac{\partial \underline{A}}{\partial \underline{V}}\right)_T = -P \qquad \left(\frac{\partial \underline{A}}{\partial T}\right)_{\underline{V}} = \left(\frac{\partial \underline{G}}{\partial T}\right)_P = -\underline{S}$$

Thermodynamic functions

$$d\underline{U} = T\,d\underline{S} - P\,d\underline{V} = C_V\,dT + \left[T\left(\frac{\partial P}{\partial T}\right)_{\underline{V}} - P\right]d\underline{V}$$

$$d\underline{H} = T\,d\underline{S} + \underline{V}\,dP = C_P\,dT + \left[\underline{V} - T\left(\frac{\partial \underline{V}}{\partial T}\right)_P\right]dP$$

$$d\underline{A} = -P\,d\underline{V} - \underline{S}\,dT$$

$$d\underline{G} = \underline{V}\,dP - \underline{S}\,dT$$

Miscellaneous

$$\left(\frac{\partial \underline{U}}{\partial \underline{V}}\right)_T = T\left(\frac{\partial P}{\partial T}\right)_{\underline{V}} - P = \frac{T\alpha}{\kappa_T} - P \qquad \frac{\partial}{\partial T}\bigg|_P\left(\frac{\underline{G}}{T}\right) = -\frac{\underline{H}}{T^2} \qquad \frac{\partial}{\partial\frac{1}{T}}\bigg|_P\left(\frac{\underline{G}}{T}\right) = \underline{H}$$

$$\left(\frac{\partial \underline{H}}{\partial P}\right)_T = \underline{V} - T\left(\frac{\partial \underline{V}}{\partial T}\right)_P = \underline{V}(1 - T\alpha) \qquad \frac{\partial}{\partial T}\bigg|_{\underline{V}}\left(\frac{\underline{A}}{T}\right) = -\frac{\underline{U}}{T^2} \qquad \frac{\partial}{\partial\frac{1}{T}}\bigg|_{\underline{V}}\left(\frac{\underline{A}}{T}\right) = \underline{U}$$

$$\underline{G} = \left(\frac{\partial G}{\partial N}\right)_{T,P} = \left(\frac{\partial A}{\partial N}\right)_{T,V} = \left(\frac{\partial H}{\partial N}\right)_{P,S} = \left(\frac{\partial U}{\partial N}\right)_{S,V} = -T\left(\frac{\partial S}{\partial N}\right)_{U,V}$$

SOLUTION

Starting from Eq. 6.2-21,

$$dU = C_V \, dT + \left[T \left(\frac{\partial P}{\partial T} \right)_V - P \right] dV$$

Using Eq. 6.1-7 yields

$$\left(\frac{\partial U}{\partial T} \right)_V = C_V \left(\frac{\partial T}{\partial T} \right)_V + \left[T \left(\frac{\partial P}{\partial T} \right)_V - P \right] \left(\frac{\partial V}{\partial T} \right)_V$$

which, with Eq. 6.1-4, reduces to

$$\left(\frac{\partial U}{\partial T} \right)_V = C_V$$

Again starting with Eq. 6.2-21, we obtain

$$\left(\frac{\partial U}{\partial T} \right)_P = C_V + \left[T \left(\frac{\partial P}{\partial T} \right)_V - P \right] \left(\frac{\partial V}{\partial T} \right)_P \qquad \textbf{(6.2-25)}$$

Clearly, then,

$$\left(\frac{\partial U}{\partial T} \right)_V \neq \left(\frac{\partial U}{\partial T} \right)_P$$

(see Problem 6.3). ■

The form of Eqs. 6.2-19 through 6.2-22 is nice for two reasons. First the equations relate the change in entropy, internal energy, and enthalpy to changes in only P, V, and T. Next, the right sides of these equations contain only C_P, C_V, P, V, and T, and partial derivatives involving P, V, and T. Thus, given heat capacity data and the volumetric equation of state for the fluid, the changes in S, U, and H accompanying a change in system temperature, pressure, or volume can be computed.

Ideally, we would like to develop equations similar to Eqs. 6.2-19 through 6.2-22 for all the thermodynamic variables of interest and, more generally, to be able to relate numerically the change in any thermodynamic property to the changes in any two others. To do this we must be able to obtain a numerical value for any derivative of the form $(\partial X / \partial Y)_Z$. Since engineers generally use two variables from among pressure, temperature, and volume as the independent variables, and they also have most information about the interrelationship between these variables, the discussion that follows centers on reducing all partial derivatives to functions of P, V, and T; their mutual derivatives; and the heat capacity, as in the equations already derived. Unfortunately, it is not possible to reduce all partial derivatives to functions of only these variables because certain partial derivatives introduce the entropy (see Eqs. 6.2-11 and 6.2-12). Since, using Eqs. 6.2-19 and 6.2-20, entropy can be evaluated from heat capacity and volumetric equation-of-state data, its inclusion introduces no real difficulty. Thus, we will be satisfied if we can reduce any partial derivative to a form containing P, V, T, S, and C_P or C_V, and derivatives containing only P, V, and T. In fact, as we shall see later in this section (Eq. 6.2-30), there are only three independent partial derivatives from among the four in Eqs. 6.2-1 through 6.2-4, for example, C_P, α, and κ_T, so that is possible to reduce the partial derivatives encountered in this chapter to functions of only P, V, T, S, α, κ_T, and C_P or C_V.

Since eight different variables (T, P, $\underline{V}$, $\underline{U}$, $\underline{H}$, $\underline{S}$, $\underline{A}$, and $\underline{G}$) may be used in the thermodynamic description of a one-component system, there are $8 \times 7 \times 6 = 336$ possible nonzero derivatives of the form of Eq. 6.1-1 to be considered. (In a binary mixture there is a ninth variable, composition, and three independent variables—temperature, specific volume, and composition—so that the thermodynamic partial derivatives of possible interest number in the thousands.) Therefore, it is necessary that a systematic procedure be developed for reducing any such derivative. Although not needed for the discussion that follows, such a procedure is introduced in Sec. 6.10.

As mentioned earlier, the Joule-Thomson expansion is a process that occurs at constant enthalpy. The Joule-Thomson coefficient μ is defined to be the change in temperature accompanying a differential change in pressure at constant enthalpy; that is,

$$\mu = \left(\frac{\partial T}{\partial P}\right)_{\underline{H}} \tag{6.2-26}$$

From Eq. 6.2-22 we have that at constant enthalpy

$$0 = C_P \, dT|_{\underline{H}} + \left[\underline{V} - T\left(\frac{\partial \underline{V}}{\partial T}\right)_P\right] dP|_{\underline{H}}$$

or

$$\mu = \left(\frac{\partial T}{\partial P}\right)_{\underline{H}} = -\frac{\left[\underline{V} - T\left(\frac{\partial \underline{V}}{\partial T}\right)_P\right]}{C_P} \tag{6.2-27}$$

(*Note:* The procedure used above can also be used with Eqs. 6.2-5 through 6.2-8, 6.2-19 through 6.2-22, and elsewhere in the evaluation of thermodynamic partial derivatives.)

For later reference, we also note

$$\frac{\partial}{\partial T}\bigg|_P \frac{\underline{G}}{T} = \frac{1}{T}\left(\frac{\partial \underline{G}}{\partial T}\right) - \frac{\underline{G}}{T^2} = -\frac{\underline{S}}{T} - \frac{(\underline{H} - T\underline{S})}{T^2} = -\frac{\underline{H}}{T^2} \tag{6.2-28}$$

and

$$\frac{\partial}{\partial T}\bigg|_{\underline{V}} \frac{\underline{A}}{T} = \frac{1}{T}\left(\frac{\partial \underline{A}}{\partial T}\right)_{\underline{V}} - \frac{\underline{A}}{T^2} = -\frac{\underline{S}}{T} - \frac{(\underline{U} - T\underline{S})}{T^2} = -\frac{\underline{U}}{T^2} \tag{6.2-29}$$

Also, since

$$\frac{\partial}{\partial\left(\frac{1}{T}\right)} = -\frac{\partial}{(1/T^2)\partial T} = -T^2 \frac{\partial}{\partial T}$$

we have

$$\frac{\partial(\underline{G}/T)}{\partial(1/T)}\bigg|_P = \underline{H} \quad \text{and} \quad \frac{\partial(\underline{A}/T)}{\partial(1/T)}\bigg|_{\underline{V}} = \underline{U} \tag{6.2-30}$$

ILLUSTRATION 6.2-2

Showing Again That Similar Partial Derivatives with Different State Variables Held Constant Have Different Values

For the discussion of the difference between the constant-pressure heat capacity C_P and the constant-volume heat capacity C_V, it is useful to have an expression for the derivative $(\partial \underline{S}/\partial T)_P$ in which T and $\underline{V}$ are the independent variables. Derive such an expression.

SOLUTION

Starting from Eq. 6.2-19,

$$dS = \frac{C_V}{T} dT + \left(\frac{\partial P}{\partial T}\right)_{\underline{V}} d\underline{V}$$

we have

$$\left(\frac{d\underline{S}}{\partial T}\right)_P = \frac{C_V}{T} \left(\frac{\partial T}{\partial T}\right)_P + \left(\frac{\partial P}{\partial T}\right)_{\underline{V}} \left(\frac{\partial \underline{V}}{\partial T}\right)_P$$

$$= \frac{C_V}{T} + \left(\frac{\partial P}{\partial T}\right)_{\underline{V}} \left(\frac{\partial \underline{V}}{\partial T}\right)_P \qquad \text{(6.2-31)}$$

Now using the triple-product rule,

$$\left(\frac{\partial P}{\partial T}\right)_{\underline{V}} \left(\frac{\partial \underline{V}}{\partial P}\right)_T \left(\frac{\partial T}{\partial \underline{V}}\right)_P = -1 \qquad \text{(6.2-32)}$$

we get

$$\left(\frac{\partial \underline{S}}{\partial T}\right)_P = \frac{C_V}{T} - \left(\frac{\partial P}{\partial \underline{V}}\right)_T \left(\frac{\partial \underline{V}}{\partial T}\right)_P^2 \qquad \text{(6.2-33a)}$$

$$= \frac{C_V}{T} - \left(\frac{\partial \underline{V}}{\partial P}\right)_T \left(\frac{\partial P}{\partial T}\right)_{\underline{V}}^2 \qquad \text{(6.2-33b)}$$

and finally

$$\left(\frac{\partial \underline{S}}{\partial T}\right)_P = \frac{C_V}{T} + \frac{\underline{V}\alpha^2}{\kappa_T} \qquad \text{(6.2-34)}$$

Compare this with

$$\left(\frac{\partial \underline{S}}{\partial T}\right)_{\underline{V}} = \frac{C_V}{T} \qquad \blacksquare$$

In Eqs. 6.2-1 through 6.2-4, four partial derivatives that frequently occur were introduced. As has already been indicated, only three of these derivatives are independent in that given the values of three derivatives, we can easily compute the value of the fourth. We establish this here by deriving an equation that relates C_P to C_V, α, and κ_T. The starting point is the relation $C_P = T(\partial \underline{S}/\partial T)_P$ and the results developed in Illustration 6.2-2, which yield

$$C_P = T \left(\frac{\partial \underline{S}}{\partial T}\right)_P = C_V + T \left(\frac{\partial P}{\partial T}\right)_{\underline{V}} \left(\frac{\partial \underline{V}}{\partial T}\right)_P$$

$$= C_V - T \left(\frac{\partial P}{\partial \underline{V}}\right)_T \left(\frac{\partial \underline{V}}{\partial T}\right)_P^2$$

$$= C_V - T \left(\frac{\partial \underline{V}}{\partial P}\right)_T \left(\frac{\partial P}{\partial T}\right)_{\underline{V}}^2 \qquad \text{(6.2-35)}$$

$$= C_V + T \underline{V} \alpha^2 / \kappa_T$$

establishing that C_P, C_V, α, and κ_T are all interrelated.

For the discussion of the following section, we need to know the dependence of the constant-volume heat capacity on specific volume (or density) at constant temperature. To obtain $(\partial C_V/\partial \underline{V})_T$, we start with

$$dU = C_V\, dT + \left[T\left(\frac{\partial P}{\partial T} \right)_{\underline{V}} - P \right] d\underline{V}$$

and note that

$$\left(\frac{\partial \underline{U}}{\partial T} \right)_{\underline{V}} = C_V \quad \text{and} \quad \left(\frac{\partial \underline{U}}{\partial \underline{V}} \right)_T = T\left(\frac{\partial P}{\partial T} \right)_{\underline{V}} - P$$

Use of the commutative property, Eq. 6.1-3, yields the desired result:

$$\left. \frac{\partial}{\partial \underline{V}} \right|_T \left(\frac{\partial \underline{U}}{\partial T} \right)_{\underline{V}} = \left(\frac{\partial C_V}{\partial \underline{V}} \right)_T = T\left(\frac{\partial^2 P}{\partial T^2} \right)_{\underline{V}} = \left. \frac{\partial}{\partial T} \right|_{\underline{V}} \left(\frac{\partial \underline{U}}{\partial \underline{V}} \right)_T \qquad \textbf{(6.2-36)}$$

In a similar fashion, starting with Eq. 6.2-22, one obtains

$$\left(\frac{\partial C_P}{\partial P} \right)_T = -T\left(\frac{\partial^2 \underline{V}}{\partial T^2} \right)_P \qquad \textbf{(6.2-37)}$$

ILLUSTRATION 6.2-3
Use of Partial Derivative Interrelations to Obtain Useful Results

Develop expressions for the coefficient of thermal expansion α, the isothermal compressibility κ_T, the Joule-Thomson coefficient μ, and the difference $C_P - C_V$ for (a) the ideal gas and (b) the gas that obeys the volumetric equation of state

$$\left(P + \frac{a}{\underline{V}^2} \right) (\underline{V} - b) = RT \qquad \textbf{(6.2-38a)}$$

where a and b are constants. (This equation of state was developed by J. D. van der Waals in 1873, and fluids that obey this equation of state are called van der Waals fluids.)

SOLUTION

a. For the ideal gas $P\underline{V} = RT$; thus

$$\left(\frac{\partial \underline{V}}{\partial T} \right)_P = \frac{R}{P} = \frac{\underline{V}}{T} \quad \text{so that} \quad \alpha = \frac{1}{\underline{V}} \left(\frac{\partial \underline{V}}{\partial T} \right)_P = \frac{1}{T}$$

and

$$\left(\frac{\partial \underline{V}}{\partial P} \right)_T = -\frac{\underline{V}}{P} = -\frac{RT}{P^2} \quad \text{so that} \quad \kappa_T = -\frac{1}{\underline{V}} \left(\frac{\partial \underline{V}}{\partial P} \right)_T = \frac{1}{P}$$

From Eq. 6.2-27 we have

$$\left(\frac{\partial T}{\partial P} \right)_{\underline{H}} = \mu = -\frac{\left[\underline{V} - T\left(\frac{\partial \underline{V}}{\partial T} \right)_P \right]}{C_P} = -\frac{\underline{V}}{C_P} [1 - T\alpha]$$

For the ideal gas,

$$\mu = \left(\frac{\partial T}{\partial P}\right)_{\underline{H}} = -\frac{\underline{V}}{C_P^*}\left[1 - T\cdot\frac{1}{T}\right] = 0$$

and

$$C_P^* = C_V^* + \frac{T\underline{V}\alpha^2}{\kappa_T} = C_V^* + T\underline{V}\frac{1}{T^2}P = C_V^* + \frac{P\underline{V}}{T} = C_V^* + R$$

b. For the van der Waals fluid, we first rewrite the equation of state as

$$T = \frac{P\underline{V}}{R} - \frac{Pb}{R} + \frac{a}{\underline{V}R} - \frac{ab}{R\underline{V}^2}$$

so that

$$\left(\frac{\partial T}{\partial \underline{V}}\right)_P = \frac{P}{R} - \frac{a}{R\underline{V}^2} + \frac{2ab}{R\underline{V}^3}$$

and

$$(\alpha)^{-1} = \underline{V}\left(\frac{\partial T}{\partial \underline{V}}\right)_P = \frac{P\underline{V}}{R} - \frac{a}{R\underline{V}} + \frac{2ab}{R\underline{V}^2}$$

Now rewriting the van der Waals equation as

$$P = \frac{RT}{\underline{V} - b} - \frac{a}{\underline{V}^2} \tag{6.2-38b}$$

allows us to eliminate P from the expression for α to obtain

$$(\alpha)^{-1} = \frac{T\underline{V}}{(\underline{V} - b)} - \frac{2a}{R\underline{V}} + \frac{2ab}{R\underline{V}^2} = \frac{T\underline{V}}{(\underline{V} - b)} - \frac{2a}{R\underline{V}^2}(\underline{V} - b)$$

An expression for κ_T is obtained as follows:

$$\left(\frac{\partial P}{\partial \underline{V}}\right)_T = -\frac{RT}{(\underline{V} - b)^2} + \frac{2a}{\underline{V}^3}$$

or

$$(\kappa_T)^{-1} = \frac{RT\underline{V}}{(\underline{V} - b)^2} - \frac{2a}{\underline{V}^2} = \frac{R}{(\underline{V} - b)}\left[\frac{T\underline{V}}{(\underline{V} - b)} - \frac{2a}{R\underline{V}^2}(\underline{V} - b)\right]$$
$$= \frac{R}{(\underline{V} - b)}\alpha^{-1}$$

Consequently,

$$\mu = \left(\frac{\partial T}{\partial P}\right)_{\underline{H}} = -\frac{\underline{V}}{C_P}(1 - T\alpha) = -\frac{\underline{V}}{C_P}\left[1 - \frac{1}{\dfrac{\underline{V}}{\underline{V} - b} - \dfrac{2a}{RT}\left(\dfrac{\underline{V} - b}{\underline{V}^2}\right)}\right]$$

and

$$C_P = C_V + \frac{T\underline{V}\alpha^2}{\kappa_T} = C_V + T\underline{V}\alpha^2\left(\frac{R}{(\underline{V} - b)\alpha}\right)$$
$$= C_V + \frac{R}{1 - \dfrac{2a}{RT}\dfrac{(\underline{V} - b)^2}{\underline{V}^3}}$$

∎

Finally, it is useful to note that although the molar internal energy can be considered to be a function of any two state variables, an equation of state that gives the internal energy as a function of entropy and volume is, in principle, more useful than an equation of state for the internal energy in terms of temperature and volume or any other pair of state variables. To see that this is so, suppose we had thermal equations of state of the form $\underline{U} = \underline{U}(\underline{S}, \underline{V})$ and $\underline{U} = \underline{U}(T, \underline{V})$. From Eq. 6.2-5b it is evident that we could differentiate the first equation of state to get other thermodynamic functions directly, for example,

$$T = \left(\frac{\partial \underline{U}}{\partial \underline{S}}\right)_{\underline{V}} \quad \text{and} \quad P = -\left(\frac{\partial \underline{U}}{\partial \underline{V}}\right)_{\underline{S}}$$

However, using the second equation of state $\underline{U}(T, \underline{V})$ and Eq. 6.2-21, we obtain

$$\left(\frac{\partial \underline{U}}{\partial T}\right)_{\underline{V}} = C_{\mathrm{V}} \quad \text{and} \quad \left(\frac{\partial \underline{U}}{\partial \underline{V}}\right)_{T} = T\left(\frac{\partial P}{\partial T}\right)_{\underline{V}} - P$$

In this case, on differentiation, we do not obtain thermodynamic state functions directly, but rather derivatives of state functions or combinations of state functions and their derivatives.

In a similar fashion, it is possible to show that an equation of state that relates $\underline{H}$, $\underline{S}$, and P, or $\underline{A}$, $\underline{V}$, and T, or $\underline{G}$, P, and T is more useful than other equations of state. Equations of state relating $(\underline{S}, \underline{U}, \text{and } \underline{V})$, $(\underline{H}, \underline{S}, \text{and } P)$, $(\underline{A}, \underline{V}, \text{and } T)$, or $(\underline{G}, T, \text{and } P)$ are called **fundamental equations of state**, a term first used by the American physicist Josiah Willard Gibbs in 1873. Unfortunately, in general we do not have the information to obtain or construct a fundamental equation of state. More commonly, we have only a volumetric equation of state, that is, an equation relating P, $\underline{V}$, and T.

ILLUSTRATION 6.2-4

Showing That a Fundamental Equation of State Contains All the Information about a Fluid

Show that from an equation of state relating the Gibbs energy, temperature, and pressure, equations of state for all other state functions (and their derivatives as well) can be obtained by appropriate differentiation.

SOLUTION

Suppose we had an equation of state of the form $\underline{G} = \underline{G}(T, P)$. The entropy and volume, as a function of temperature and pressure, are then immediately obtained using Eqs. 6.2-12:

$$\underline{S}(T, P) = -\left(\frac{\partial \underline{G}}{\partial T}\right)_{P} \quad \text{and} \quad \underline{V}(T, P) = \left(\frac{\partial \underline{G}}{\partial P}\right)_{T}$$

Since we have $\underline{G}$ as a function of T and P, these derivatives can be evaluated. Next, the enthalpy and internal energy can be found as follows:

$$\underline{H}(T, P) = \underline{G}(T, P) + T\underline{S}(T, P) = \underline{G}(T, P) - T\left(\frac{\partial \underline{G}}{\partial T}\right)_{P}$$

and

$$\underline{U}(T, P) = \underline{G}(T, P) + T\underline{S}(T, P) - P\underline{V}(T, P) = \underline{G}(T, P) - T\left(\frac{\partial \underline{G}}{\partial T}\right)_{P} - P\left(\frac{\partial \underline{G}}{\partial P}\right)_{T}$$

The Hemholtz energy is obtained from

$$\underline{A}(T, P) = \underline{G}(T, P) - P\underline{V} = \underline{G}(T, P) - P\left(\frac{\partial \underline{G}}{\partial P}\right)_T$$

and the constant-pressure and constant-volume heat capacities can then be found as follows:

$$C_P(T, P) = \left(\frac{\partial \underline{H}}{\partial T}\right)_P = \frac{\partial}{\partial T}\Big|_P\left[\underline{G}(T, P) - T\left(\frac{\partial \underline{G}}{\partial T}\right)_P\right] = -T\left(\frac{\partial^2 \underline{G}}{\partial T^2}\right)_P$$

and

$$C_V(T, P) = C_P + \frac{T(\partial \underline{V}/\partial T)_P^2}{(\partial \underline{V}/\partial P)_T} = -T\left(\frac{\partial^2 \underline{G}}{\partial T^2}\right)_P - T\left(\frac{\partial^2 \underline{G}}{\partial T \partial P}\right)^2\left(\frac{\partial^2 \underline{G}}{\partial P^2}\right)_T^{-1}$$

Finally, the isothermal compressibility κ_T and coefficient of thermal expansion α are found as

$$\kappa_T = -\frac{1}{\underline{V}}\left(\frac{\partial \underline{V}}{\partial P}\right)_T = -\frac{(\partial^2 \underline{G}/\partial P^2)_T}{(\partial \underline{G}/\partial P)_T}$$

and

$$\alpha = \frac{1}{\underline{V}}\left(\frac{\partial \underline{V}}{\partial T}\right)_P = \frac{(\partial^2 \underline{G}/\partial P \partial T)}{(\partial \underline{G}/\partial P)_T}$$

Thus, if the fundamental equation of state for a substance in the form $\underline{G} = \underline{G}(T, P)$ were available, we could, using these relations, obtain explicit equations relating all other state variables for this substance to temperature and pressure by taking the appropriate derivatives of the fundamental equation.

QUESTIONS

1. Why do we need two equations, a volumetric equation of state $P = P(T, \underline{V})$ and a thermal equation of state $\underline{U} = \underline{U}(T, \underline{V})$, to define an ideal gas?
2. Can you develop a single equation of state that would completely specify all the properties of an ideal gas? (See Problem 7.7.) ■

Although fundamental equations of state are, in principle, the most useful thermodynamic descriptions of any substance, it is unlikely that such equations will be available for all fluids of interest to engineers. In fact, frequently only heat capacity and $P\underline{V}T$ data are available; in Sec. 6.4 we consider how these more limited data are used in thermodynamic problem solving.

ILLUSTRATION 6.2-5

Calculation of the Joule-Thomson Coefficient of Steam from Data in the Steam Tables

Use the steam tables in Appendix A.III to evaluate the Joule-Thomson coefficient of steam at 600°C and 0.8 MPa.

SOLUTION

From Eq. 6.2-27,

$$\mu = \left(\frac{\partial T}{\partial P}\right)_{\underline{H}} = -\frac{\left[\underline{V} - T\left(\frac{\partial \underline{V}}{\partial T}\right)_P\right]}{C_P} = -\frac{\left[\hat{V} - T\left(\frac{\partial \hat{V}}{\partial T}\right)_P\right]}{\hat{C}_P}$$

and

$$\hat{C}_{\mathrm{P}} = \left(\frac{\partial \hat{H}}{\partial T} \right)_P$$

From the superheated vapor section of the steam tables we have the following entries at $P = 0.8$ MPa:

T (°C)	$\hat{V}$ (m³/kg)	$\hat{H}$ (kJ/kg)
500	0.4433	3480.6
600	0.5018	3699.4
700	0.5601	3924.2

So

$$\hat{C}_{\mathrm{P}} = \left(\frac{\partial \hat{H}}{\partial T} \right)_P \approx \frac{\hat{H}(700°\mathrm{C}) - \hat{H}(500°\mathrm{C})}{(700-500)°\mathrm{C}} = \frac{(3924.2 - 3480.6)\, \frac{\mathrm{kJ}}{\mathrm{kg}}}{200\ \mathrm{K}} = 2.218\, \frac{\mathrm{kJ}}{\mathrm{kg\ K}}$$

and

$$\left(\frac{\partial \hat{V}}{\partial T} \right)_P \approx \frac{\hat{V}(700°\mathrm{C}) - \hat{V}(500°\mathrm{C})}{(700-500)°\mathrm{C}} = \frac{(0.5601 - 0.4433)\, \frac{\mathrm{m}^3}{\mathrm{kg}}}{200\ \mathrm{K}} = 5.84 \times 10^{-4}\, \frac{\mathrm{m}^3}{\mathrm{kg\ K}}$$

Therefore,

$$\mu = -\frac{[0.5018 - (600 + 273.15) \times 5.84 \times 10^{-4}]\, \frac{\mathrm{m}^3}{\mathrm{kg}}}{2.218\, \frac{\mathrm{kJ}}{\mathrm{kg\ K}}} = 3.66 \times 10^{-3}\, \frac{\mathrm{m}^3\ \mathrm{K}}{\mathrm{kJ}}$$

Since 1 kJ $= 10^{-3}$ MPa m³, and 1 MPa $= 10$ bar,

$$\mu = 3.66\, \frac{\mathrm{K}}{\mathrm{MPa}} = 0.366\, \frac{\mathrm{K}}{\mathrm{bar}}$$

Since μ is positive, the temperature of steam at these conditions will decrease if the pressure is reduced, that is, if the steam is isenthalpically expanded.

Note: An alternative method of doing this calculation is to use the definition of the Joule-Thomson coefficient, $(\partial T / \partial P)_H$, directly. For example, at 0.6 MPa we have

T (°C)	$\hat{H}$ (kJ/kg)
500	3482.8
600	3700.9

By interpolation we can find the temperature at 0.6 MPa at which: $\hat{H}(T, 0.6\ \mathrm{MPa}) = 3699.4\ \mathrm{kJ/kg} = \hat{H}(600°\mathrm{C}, 0.8\ \mathrm{MPa})$.

$$\frac{\hat{H}(600°\mathrm{C}, 0.6\ \mathrm{MPa}) - \hat{H}(T, 0.6\ \mathrm{MPa})}{\hat{H}(600°\mathrm{C}, 0.6\ \mathrm{MPa}) - \hat{H}(500°\mathrm{C}, 0.6\ \mathrm{MPa})} = \frac{3700.9 - 3699.4}{3700.9 - 3482.8} = \frac{600 - T}{600 - 500}$$

which gives $T = 599.3°\mathrm{C}$ or $\Delta T = -0.6977$ K. Therefore,

$$\mu = \left(\frac{\partial T}{\partial P} \right)_H = \frac{-0.6977\ \mathrm{K}}{-0.2\ \mathrm{MPa}} = 3.49\, \frac{\mathrm{K}}{\mathrm{MPa}} = 0.349\, \frac{\mathrm{K}}{\mathrm{bar}}$$

which is in reasonably good agreement with the value obtained above. ∎

ILLUSTRATION 6.2-6

Calculation of C_P for Water

The constant-pressure heat capacity of liquid water at 25°C at 1 bar is 4.18 J/(g K). Estimate the heat capacity of liquid water at 25°C at 100 bar.

Data:

$$\alpha = \frac{1}{\underline{V}}\left(\frac{\partial \underline{V}}{\partial T}\right)_P = \frac{1}{\hat{V}}\left(\frac{\partial \hat{V}}{\partial T}\right)_P = 2.56 \times 10^{-4} \text{ K}^{-1}, \quad \left(\frac{\partial \alpha}{\partial T}\right)_P = 9.6 \times 10^{-6} \text{ K}^{-2}$$

and $\hat{V} = 1.003 \text{ cm}^3/\text{g}$.

SOLUTION

From Eq. 6.2-37, we have

$$\left(\frac{\partial C_P}{\partial P}\right)_T = -T\left(\frac{\partial^2 \underline{V}}{\partial T^2}\right)_P \quad \text{or equivalently} \quad \left(\frac{\partial \hat{C}_P}{\partial P}\right)_T = -T\left(\frac{\partial^2 \hat{V}}{\partial T^2}\right)_P$$

Now

$$\left(\frac{\partial \hat{V}}{\partial T}\right)_P = \hat{V}\alpha \quad \text{so} \quad \left(\frac{\partial^2 \hat{V}}{\partial T^2}\right)_P = \left(\frac{\partial \hat{V}}{\partial T}\right)_P \alpha + \hat{V}\left(\frac{\partial \alpha}{\partial T}\right)_P$$

and

$$\left(\frac{\partial \hat{C}_P}{\partial P}\right)_T = -T\left[\alpha\left(\frac{\partial \hat{V}}{\partial T}\right)_P + \hat{V}\left(\frac{\partial \alpha}{\partial T}\right)_P\right] = -T\left[\alpha^2 \hat{V} + \hat{V}\left(\frac{\partial \alpha}{\partial T}\right)_P\right]$$

$$= -298.15 \text{ K} \times 1.003 \frac{\text{cm}^3}{\text{g}}[6.5536 \times 10^{-8} + 9.6 \times 10^{-6}] \text{ K}^{-2}$$

$$= -2.890 \times 10^{-3} \frac{\text{cm}^3}{\text{g K}} \times \frac{1 \text{ J}}{10 \text{ bar cm}^3} = -2.890 \times 10^{-4} \frac{\text{J}}{\text{g K bar}}$$

Assuming that $(\partial \hat{C}_P/\partial P)_T$ is reasonably constant with pressure, we have

$$\hat{C}_P(T, P + \Delta P) - \hat{C}_P(T, P) = \left(\frac{\partial \hat{C}_P}{\partial P}\right)_T \Delta P$$

and

$$\hat{C}_P(25°C, 100 \text{ bar}) - \hat{C}_P(25°C, 1 \text{ bar}) = -2.890 \times 10^{-4} \frac{\text{J}}{\text{g K bar}} \times 99 \text{ bar}$$

$$= -0.0286 \frac{\text{J}}{\text{g K}}$$

Therefore,

$$\hat{C}_P(25°C, 100 \text{ bar}) = 4.18 - 0.0286 = 4.1514 \frac{\text{J}}{\text{g K}}$$

and we see that the heat capacity of water at these conditions (and indeed that of most liquids and solids) is only very weakly dependent on pressure. ∎

Finally, there are two other equations based on Eq. 6.2-12b that will be useful in analyzing phase behavior. The first of these equations is

$$\left.\frac{\partial}{\partial T}\right|_{P,N}\left(\frac{G}{T}\right) = -\frac{H}{T^2} \tag{6.2-39}$$

The second equation, which is related to the one above, is

$$\left[\frac{\partial\left(\dfrac{G}{T}\right)}{\partial\left(\dfrac{1}{T}\right)}\right]_{P,N} = H \tag{6.2-40}$$

Various forms of Eqs. 6.2-28 and 6.2-30 are used later in this book.

6.3 THE IDEAL GAS AND ABSOLUTE TEMPERATURE SCALES

In Chapters 1 and 3, we introduced the concept of the ideal gas and suggested, without proof, that if the ideal gas were used to establish a scale of temperature, an absolute and universal, or thermodynamic scale would be obtained. We now have developed sufficient thermodynamic theory to prove this to be the case.

We take as the starting point for this discussion the facts that the product $P\underline{V}$ and the internal energy $\underline{U}$ of an ideal gas are both unspecified but increasing functions of the absolute temperature T and are independent of density, pressure, or specific volume (see Sec. 3.3). To be perfectly general at this point we denote these characteristics by

$$P\underline{V} = RT^{\text{IG}} = \Theta_1(T) \tag{6.3-1}$$

and

$$\underline{U} = \Theta_2(T) \tag{6.3-2}$$

where T^{IG} is the temperature on the ideal gas temperature scale. To prove the equality of the ideal gas and thermodynamic temperature scales it is necessary to establish that $\Theta_1(T)$ is a linear function of T as was suggested in Eqs. 1.4-2 and 3.3-1.

It is important to note that in the balance equations the thermodynamic temperature first appears in the introduction of the entropy function—in particular, in the $\dot{Q}/T$ term in Eq. 4.1-5. Thus it is the thermodynamic temperature T that appears in all equations derived from Eq. 4.1-5, and therefore in the equations of Sec. 6.2. From Eq. 6.2-21 we have, in general,

$$d\underline{U} = C_V\,dT + \left[T\left(\frac{\partial P}{\partial T}\right)_{\underline{V}} - P\right]d\underline{V}$$

whereas from Eq. 6.3-2 we have, for the ideal gas,

$$d\underline{U} = \frac{d\Theta_2(T)}{dT}\,dT \tag{6.3-3}$$

The only way to reconcile these two equations is for the coefficient of the $d\underline{V}$ term in Eq. 6.2-21 to be zero for the ideal gas, that is,

$$P = T \left(\frac{\partial P}{\partial T}\right)_{\underline{V}}$$

This implies that for changes at constant volume,

$$\frac{dP}{P}\bigg|_{\underline{V}} = \frac{dT}{T}\bigg|_{\underline{V}}$$

or

$$\frac{P_2}{P_1} = \frac{T_2}{T_1} \tag{6.3-4}$$

for any two states 1 and 2 with the same specific volume. However, from Eq. 6.3-1 we have, that under these circumstances,

$$\frac{P_2}{P_1} = \frac{\Theta_1(T_2)}{\Theta_1(T_1)} \tag{6.3-5}$$

To satisfy Eqs. 6.3-4 and 6.3-5, $\Theta_1(T)$ must be a linear function of temperature, that is,

$$\Theta_1(T) = RT \tag{6.3-6}$$

where R is a constant related to the unit of a degree (see Sec. 1.4). Using Eq. 6.3-6 in Eq. 6.3-1 yields

$$P\underline{V} = RT$$

which establishes that the ideal gas temperature scale is also a thermodynamic temperature scale.

6.4 THE EVALUATION OF CHANGES IN THE THERMODYNAMIC PROPERTIES OF REAL SUBSTANCES ACCOMPANYING A CHANGE OF STATE

The Necessary Data

In order to use the energy and entropy balances for any real substance for which thermodynamic tables are not available, we must be able to compute the changes in its internal energy, enthalpy, and entropy for any change of state. The equations of Sec. 6.2 provide the basis for such computations. However, before we discuss these calculations it is worthwhile to consider the minimum amount of information needed and the form in which this information is likely to be available.

Volumetric Equation-of-State Information

Clearly, to use Eqs. 6.2-19 through 6.2-22 we need volumetric equation-of-state data, that is, information on the interrelationship between P, $\underline{V}$, and T. This information may be available as tables of experimental data, or, more frequently, as approximate equations with parameters that have been fitted to experimental data. Hundreds of analytic equations of state have been suggested for the correlation of $P\underline{V}T$ data.

The equation

$$\left(P + \frac{a}{\underline{V}^2}\right)(\underline{V} - b) = RT \tag{6.2-38a}$$

or equivalently,

The van der Waals
equation of state

$$P = \frac{RT}{\underline{V} - b} - \frac{a}{\underline{V}^2} \tag{6.2-38b}$$

with a and b constants, was proposed by J. D. van der Waals in 1873 to describe the volumetric or $P\underline{V}T$ behavior of both vapors and liquids, work for which he was awarded the Nobel Prize in Physics in 1910. The constants a and b can be determined either by fitting this equation to experimental data or, more commonly, from critical-point data as will be described in Sec. 6.6. Values for the parameters for several gases appear in Table 6.4-1. These parameters were computed from critical-point data as described in Sec. 6.6.

The van der Waals equation of state is not very accurate and is mainly of historic interest in that it was the first equation capable of predicting the transition between vapor and liquid; this will be discussed in Sec. 7.3. It also is the prototype for modern, more accurate equations of state, such as those of Redlich-Kwong (1949);[2]

$$P = \frac{RT}{\underline{V} - b} - \frac{a}{T^{1/2}\underline{V}(\underline{V} + b)} \tag{6.4-1}$$

of Soave (1972),[3] in which the $a/T^{1/2}$ term in Eq. 6.4-1 is replaced with $a(T)$, a function of temperature; and of Peng and Robinson (1976),[4]

The Peng-Robinson
equation of state

$$P = \frac{RT}{\underline{V} - b} - \frac{a(T)}{\underline{V}(\underline{V} + b) + b(\underline{V} - b)} \tag{6.4-2}$$

Table 6.4-1 Parameters for the van der Waals Equation of State

Gas	a (Pa m^6/mol^2)	b [(m^3/mol) $\times 10^5$]
O_2	0.1381	3.184
N_2	0.1368	3.864
H_2O	0.5542	3.051
CH_4	0.2303	4.306
CO	0.1473	3.951
CO_2	0.3658	4.286
NH_3	0.4253	3.737
H_2	0.0248	2.660
He	0.00346	2.376

[2]O. Redlich and J. N. S. Kwong, *Chem. Rev.* **44**, 233 (1949).

[3]G. Soave, *Chem. Eng. Sci.* **27**, 1197 (1972).

[4]D.-Y. Peng and D. B. Robinson, *IEC Fundam.* **15**, 59 (1976).

Equations 6.2-38b, 6.4-1, and 6.4-2 are special cases of the general class of equations of state

$$P = \frac{RT}{\underline{V} - b} - \frac{(\underline{V} - \eta)\theta}{(\underline{V} - b)(\underline{V}^2 + \delta\underline{V} + \varepsilon)} \tag{6.4-3}$$

where each of the five parameters b, θ, δ, ε, and η can depend on temperature. In practice, however, generally only θ is taken to be a function of temperature, and it is adjusted to give the correct boiling temperature as a function of pressure; this will be discussed in Sec. 7.5. Table 6.4-2 gives the parameters of Eq. 6.4-3 for some common equations of state from among the hundreds of this class that have been published. Clearly, many other choices are possible.

Numeric values for equation-of-state parameters are commonly obtained in one of two ways. First, parameters can be obtained by fitting the equation to $P\underline{V}T$ and other data for the fluid of interest; this leads to the most accurate values, but is very tedious. Second, as will be discussed in Sec. 6.7, general relations can be obtained between the equation-of-state parameters and critical-point properties. From these equations, somewhat less accurate parameter values are easily obtained from only critical-point properties.

Each of the equations of state discussed here can be written in the form

Cubic form of the above equations

$$\boxed{Z^3 + \alpha Z^2 + \beta Z + \gamma = 0} \tag{6.4-4}$$

Table 6.4-2 Parameters for Cubic Equations of State $P = \dfrac{RT}{\underline{V} - b} - \Delta$

Author	Year	θ	η	δ	ε	Δ
van der Waals	1873	a	b	0	0	$\dfrac{a}{\underline{V}^2}$
Clausius	1880	a/T	b	$2c$	c^2	$\dfrac{a/T}{(\underline{V} + c)^2}$
Berthelot	1899	a/T	b	0	0	$\dfrac{a/T}{\underline{V}^2}$
Redlich-Kwong	1949	$a/\sqrt{T}$	b	b	0	$\dfrac{a/\sqrt{T}}{\underline{V}(\underline{V} + b)}$
Soave	1972	$\theta_S(T)$	b	b	0	$\dfrac{\theta_S(T)}{\underline{V}(\underline{V} + b)}$
Lee-Erbar-Edmister	1973	$\theta_L(T)$	$\eta(T)$	b	0	$\dfrac{\theta_L(T)[\underline{V} - \eta(T)]}{(\underline{V} - b)(\underline{V} + b)}$
Peng-Robinson	1976	$\theta_{PR}(T)$	b	$2b$	$-b^2$	$\dfrac{\theta_{PR}(T)}{\underline{V}(\underline{V} + b) + b(\underline{V} - b)}$
Patel-Teja	1981	$\theta_{PT}(T)$	b	$b + c$	$-cb$	$\dfrac{\theta_{PT}(T)}{\underline{V}(\underline{V} + b) + c(\underline{V} - b)}$

Note: If $\eta = b$, Eq. 6.4-3 reduces to

$$P = \frac{RT}{\underline{V} - b} - \frac{\theta}{\underline{V}^2 + \delta\underline{V} + \varepsilon}$$

Table 6.4-3 Parameters in Eq. 6.4-4 for the Three Equations of State

	van der Waals	Redlich-Kwong and Soave	Peng-Robinson
α	$-1-B$	-1	$-1+B$
β	A	$A - B - B^2$	$A - 3B^2 - 2B$
γ	$-AB$	$-AB$	$-AB + B^2 + B^3$

where
$$Z = P\underline{V}/RT$$
$$B = bP/RT$$

and

$$A = \begin{cases} aP/(RT)^2 & \text{in the van der Waals, Soave, and} \\ & \text{Peng-Robinson equations of state} \\ aP/R^2T^{2.5} & \text{in the Redlich-Kwong equation of} \\ & \text{state} \end{cases}$$

where $Z = P\underline{V}/RT$ is the **compressibility factor**, and the parameters α, β, and γ for some representative equations of state are given in Table 6.4-3. Consequently, these equations are said to be cubic equations of state. Many such equations have been suggested in the scientific literature. One should remember that all cubic equations of state are approximate; generally, they provide a reasonable description of the $P\underline{V}T$ behavior in both the vapor and liquid regions for hydrocarbons, and of the vapor region only for many other pure fluids. The Soave–Redlich-Kwong and the Peng-Robinson equations are, at present, the most commonly used cubic equations of state.

A different type of equation of state is the virial equation

Virial equation of state

$$\frac{P\underline{V}}{RT} = 1 + \frac{B(T)}{\underline{V}} + \frac{C(T)}{\underline{V}^2} + \cdots \tag{6.4-5}$$

where $B(T)$ and $C(T)$ are the temperature-dependent second and third virial coefficients. Although higher-order terms can be defined in a similar fashion, data generally are available only for the second virial coefficient.[5]

The virial equation was first used by H. Kamerlingh Onnes in 1901 and is of theoretical interest since it can be derived from statistical mechanics, with explicit expressions obtained for the virial coefficients in terms of the interaction energy between molecules. The virial equation of state is a power series expansion in density (that is, reciprocal volume) about the ideal gas result ($P\underline{V}/RT = 1$). With a sufficient number of coefficients, the virial equation can give excellent vapor-phase predictions, but it is not applicable to the liquid phase. When truncated at the $B(T)$ term, as is usually the case because of a lack of higher virial coefficient data, the virial equation of state can be used only at low densities; as a general rule, it should not be used at pressures above 10 bar for most fluids.

There are other, more complicated equations of state that accurately predict the $P\underline{V}T$ behavior in most of the vapor and liquid regions; such equations contain the reciprocal

---

[5]A recent tabulation, *The Virial Coefficients of Pure Gases and Mixtures: A Critical Evaluation*, by J. H. Dymond and E. B. Smith, Clarendon Press, Oxford, 1980, provides second virial coefficient data for more than 250 compounds.

volume in both integral powers (like the virial equation) and exponential functions. One example is the equation of Benedict, Webb, and Rubin (1940)[6],

$$\frac{P\underline{V}}{RT} = 1 + \left(B - \frac{A}{RT} - \frac{C}{RT^3}\right)\frac{1}{\underline{V}} + \left(b - \frac{a}{RT}\right)\frac{1}{\underline{V}^2}$$
$$+ \frac{a\alpha}{RT\underline{V}^5} + \frac{\beta}{RT^3\underline{V}}\left(1 + \frac{\gamma}{\underline{V}^2}\right)\exp(-\gamma/\underline{V}^2)$$

(6.4-6a)

where the eight constants a, b, A, B, C, α, β, and γ are specific to each fluid and are obtained by fitting the equation of state to a variety of experimental data. The exponential term in this equation (and others in this class of equations) is meant to compensate for the truncation of the virial series since, if expanded in a Taylor series around $\underline{V} = \infty$, the exponential function generates terms of higher order in reciprocal volume. The 20-constant Bender equation (1970)[7],

$$P = \frac{T}{\underline{V}}\left[R + \frac{B}{\underline{V}} + \frac{C}{\underline{V}^2} + \frac{D}{\underline{V}^3} + \frac{E}{\underline{V}^4} + \frac{F}{\underline{V}^5} + \left(G + \frac{H}{\underline{V}^2}\right)\frac{1}{\underline{V}^2}\exp(-a_{20}/\underline{V}^2)\right]$$

(6.4-6b)

with

$$B = a_1 - a_2/T - a_3/T^2 - a_4/T^3 - a_5/T^4$$
$$C = a_6 + a_7/T + a_8/T^2$$
$$D = a_9 + a_{10}/T$$
$$E = a_{11} + a_{12}/T$$
$$F = a_{13}/T$$
$$G = a_{14}/T^3 + a_{15}/T^4 + a_{16}/T^5$$
$$H = a_{17}/T^3 + a_{18}/T^4 + a_{19}/T^5$$

is another example of an equation of this type. Although such equations provide more accurate descriptions of fluid behavior, including the vapor-liquid phase transition, than simple equations of state, they are useful only with digital computers. Furthermore, because of the amount of data needed the coefficients that appear in these equations are known only for light hydrocarbons and a few other substances.

Some recent equations of state are expressed in the form of the Helmholtz energy as a function of temperature and specific volume. The advantage of such an interrelationship is that it is a fundamental equation of state, as discussed in Sec. 4.2. An example of this is the 1995 International Association for the Properties of Water and Steam (IAPWS) formulation[8] which contains 56 parameters to very accurately describe the properties of water and steam. Such high accuracy in the thermodynamic properties is needed in a number of applications, including evaluating the efficiency of newly installed steam turbines, which is frequently a contract specification between the power company and the turbine manufacturer. Few other fluids are of sufficient industrial interest to justify the expense of measuring the very large amount of data needed to develop such high-accuracy equations.

For an evaluation of volumetric equations of state important in engineering, refer to J. M. Prausnitz, B. E. Poling, and J. P. O'Connell, *The Properties of Gases and*

---

[6]M. Benedict, G. B. Webb, and L. C. Rubin, *J. Chem. Phys.* **8.** 334 (1940), and later papers by the same authors.
[7]E. Bender, *5th Symposium on Thermophysical Properties*, ASME, New York (1970), p. 227, and later papers by the same author.
[8]W. Wagner and A. Pruss, *J. Phys. Chem. Ref. Data*, **31**, 387 (2002)

Liquids, 5th ed. (McGraw-Hill, New York, 2001). In the discussion that follows, we will generally assume that a volumetric equation of state in analytic form is available.

Heat Capacity Data

It is also evident from Eqs. 6.2-19 through 6.2-22 that data for C_P and C_V are needed to compute changes in thermodynamic properties. At first glance it might appear that we need data for C_V as a function of T and $\underline{V}$, and C_P as a function of T and P for each fluid over the complete range of conditions of interest. However, from Eqs. 6.2-36 and 6.2-37 it is clear that our need for heat capacity data is much more modest once we have volumetric equation-of-state information. To see this, consider the situation in which we have data for C_P as a function of temperature at a pressure P_1, and want C_P as a function of temperature at another pressure, P_2. At each temperature we can integrate Eq. 6.2-37 to obtain the desired result:

$$\int_{P_1,T}^{P_2,T} dC_P = C_P(P_2, T) - C_P(P_1, T) = -T \int_{P_1,T}^{P_2,T} \left(\frac{\partial^2 \underline{V}}{\partial T^2}\right)_P dP \tag{6.4-7}$$

or

$$C_P(P_2, T) = C_P(P_1, T) - T \int_{P_1,T}^{P_2,T} \left(\frac{\partial^2 \underline{V}}{\partial T^2}\right)_P dP \tag{6.4-8}$$

Here we have included T in the limits of integration to stress that the integration is carried out over pressure at a fixed value of temperature. Similarly, for the constant-volume heat capacity, one obtains (from Eq. 6.2-36)

$$C_V(\underline{V}_2, T) = C_V(\underline{V}_1, T) + T \int_{\underline{V}_1,T}^{\underline{V}_2,T} \left(\frac{\partial^2 P}{\partial T^2}\right)_{\underline{V}} d\underline{V} \tag{6.4-9}$$

Therefore, given the volumetric equation of state (or, equivalently, a numerical tabulation of the volumetric data for a fluid) and heat capacity data as a function of temperature at a single pressure or volume, the value of the heat capacity in any other state can be computed.

In practice, heat capacity data are tabulated only for states of very low pressure or, equivalently, large specific volume, where all fluids are ideal gases.[9] Therefore, if P_1 and $\underline{V}_1$ are taken as 0 and ∞, respectively, in Eqs. 6.4-8 and 6.4-9, we obtain

$$\boxed{C_P(P, T) = C_P^*(T) - T \int_{P=0,T}^{P,T} \left(\frac{\partial^2 \underline{V}}{\partial T^2}\right)_P dP} \tag{6.4-10}$$

and

$$C_V(\underline{V}, T) = C_V^*(T) + T \int_{\underline{V}=\infty,T}^{\underline{V},T} \left(\frac{\partial^2 P}{\partial T^2}\right)_{\underline{V}} d\underline{V} \tag{6.4-11}$$

where we have used the notation

$$C_P^*(T) = C_P(P = 0, T)$$

---

[9]That all fluids become ideal gases at large specific volumes is easily verified by observing that all volumetric equations of state (e.g., Eqs. 6.2-33, 6.4-1, and 6.4-3) reduce to $P\underline{V} = RT$ in the limit of $\underline{V} \to \infty$.

and

$$C_V^*(T) = C_V(\underline{V} = \infty, T)$$

where the asterisk denotes the ideal gas heat capacity, as in Chapter 3.

Data for C_P^* and C_V^* are available in most data reference books, such as the *Chemical Engineers' Handbook*[10] or *The Handbook of Chemistry and Physics*.[11] This information is frequently presented in the form

Ideal gas heat capacity

$$C_P^*(T) = a + bT + cT^2 + dT^3 + \cdots$$

See Appendix A.II for C_P^* data for some compounds.

The Evaluation of $\Delta\underline{H}$, $\Delta\underline{U}$, and $\Delta\underline{S}$

To compute the change in enthalpy in going from the state (T_1, P_1) to the state (T_2, P_2), we start from

$$\Delta\underline{H} = \underline{H}(T_2, P_2) - \underline{H}(T_1, P_1) = \int_{T_1,P_1}^{T_2,P_2} d\underline{H} \tag{6.4-12}$$

and note that since enthalpy is a state function, we can compute its change between two states by evaluating the integral along any convenient path. In particular, if the path indicated by the solid line in Fig. 6.4-1 is used (isothermal expansion followed by isobaric heating and isothermal compression), we have, from Eq. 6.2-22,

Enthalpy change between two states

$$\Delta\underline{H} = \int_{P_1,T_1}^{P=0,T_1} \left[\underline{V} - T\left(\frac{\partial\underline{V}}{\partial T}\right)_P\right] dP + \int_{T_1,P=0}^{T_2,P=0} C_P^* \, dT$$
$$+ \int_{P=0,T_2}^{P_2,T_2} \left[\underline{V} - T\left(\frac{\partial\underline{V}}{\partial T}\right)_P\right] dP \tag{6.4-13}$$

Alternatively, we could compute the enthalpy change using the path indicated by the dashed line in Fig. 6.4-1—isobaric heating followed by isothermal compression. For

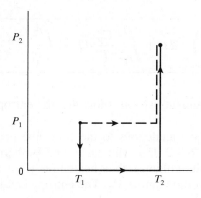

Figure 6.4-1 Two paths for the integration of Eq. 6.4-12.

---

[10]R. H. Perry and D. Green, eds., *Chemical Engineers' Handbook*, 6th ed., McGraw-Hill, New York (1984).

[11]R. C. Weast, ed., *The Handbook of Chemistry and Physics*, Cleveland Chemical Rubber Co. This handbook is updated annually.

this path

$$\Delta \underline{H} = \int_{P_1,T_1}^{P_1,T_2} C_P\, dT + \int_{P_1,T_2}^{P_2,T_2} \left[\underline{V} - T \left(\frac{\partial \underline{V}}{\partial T} \right)_P \right] dP$$

$$= \int_{T_1}^{T_2} C_P^*\, dT - \int_{T_1}^{T_2} T \left\{ \int_0^{P_1} \left(\frac{\partial^2 \underline{V}}{\partial T^2} \right)_P dP \right\} dT + \int_{P_1,T_2}^{P_2,T_2} \left[\underline{V} - T \left(\frac{\partial \underline{V}}{\partial T} \right)_P \right] dP$$

(6.4-14a)

where in going from the first to the second of these equations we have used Eq. 6.4-10.

The equality of Eqs. 6.4-13 and 6.4-14 is easily established as follows. First we note that

$$\int_{T_1}^{T_2} T \left\{ \int_0^{P_1} \left(\frac{\partial^2 \underline{V}}{\partial T^2} \right)_P dP \right\} dT$$

$$= \int_0^{P_1} \left\{ \int_{T_1}^{T_2} T \left(\frac{\partial^2 \underline{V}}{\partial T^2} \right)_P dT \right\} dP$$

$$= \int_{P=0}^{P_1} \left\{ \int_{T_1}^{T_2} \left. \frac{\partial}{\partial T} \right|_P \left[T \left(\frac{\partial \underline{V}}{\partial T} \right)_P - \underline{V} \right] dT \right\} dP$$

$$= \int_{P=0,T_2}^{P_1,T_2} \left[T \left(\frac{\partial \underline{V}}{\partial T} \right)_P - \underline{V} \right] dP - \int_{P=0,T_1}^{P_1,T_1} \left[T \left(\frac{\partial \underline{V}}{\partial T} \right)_P - \underline{V} \right] dP$$

(6.4-15)

where we have used the fact that the order of integration with respect to T and P can be interchanged, and then recognized that $T(\partial^2 \underline{V}/\partial T^2)$ has an exact differential. Next, substituting Eq. 6.4-15 into Eq. 6.4-14 yields Eq. 6.4-13, verifying that the enthalpy change between given initial and final states is independent of the path used in its computation.

Using the solid-line path in Fig. 6.4-1 and Eq. 6.2-20, we obtain

Entropy change between two states

$$\boxed{\Delta \underline{S} = - \int_{P_1,T_1}^{P=0,T_1} \left(\frac{\partial \underline{V}}{\partial T} \right)_P dP + \int_{T_1}^{T_2} \frac{C_P^*}{T}\, dT - \int_{P=0,T_2}^{P_2,T_2} \left(\frac{\partial \underline{V}}{\partial T} \right)_P dP}$$ **(6.4-16a)**

By following the same argument, we can show that the entropy function is also path independent.

The path in the $\underline{V}$-T plane analogous to the solid-line path in the P-T plane of Fig. 6.4-1 is shown in Fig. 6.4-2. Here the gas is first isothermally expanded to zero pressure (and, hence, infinite volume), then heated (at $\underline{V} = \infty$) from T_1 to T_2, and finally compressed to a specific volume $\underline{V}_2$. The entropy and internal energy changes from Eqs. 6.2-19 and 6.2-21 are

$$\Delta \underline{S} = \int_{\underline{V}_1,T_1}^{\underline{V}=\infty,T_1} \left(\frac{\partial P}{\partial T} \right)_{\underline{V}} d\underline{V} + \int_{T_1}^{T_2} \frac{C_V^*}{T}\, dT + \int_{\underline{V}=\infty,T_2}^{\underline{V}_2,T_2} \left(\frac{\partial P}{\partial T} \right)_{\underline{V}} d\underline{V}$$ **(6.4-17a)**

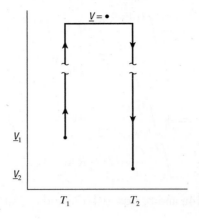

Figure 6.4-2 Integration path in the $\underline{V}$-T plane.

and

$$\Delta \underline{U} = \int_{\underline{V}_1, T_1}^{\underline{V}=\infty, T_1} \left[T \left(\frac{\partial P}{\partial T} \right)_{\underline{V}} - P \right] d\underline{V} + \int_{T_1}^{T_2} C_{\mathrm{V}}^* dT + \int_{\underline{V}=\infty, T_2}^{\underline{V}_2, T_2} \left[T \left(\frac{\partial P}{\partial T} \right)_{\underline{V}} - P \right] d\underline{V}$$

(6.4-18)

It can be easily shown that alternative paths lead to identical results for $\Delta \underline{S}$ and $\Delta \underline{U}$.

Note that only C_{P}^*, C_{V}^*, and terms related to the volumetric equation of state appear in Eqs. 6.4-13, 6.4-14a, 6.4-16a, 6.4-17a, and 6.4-18. Thus, as has already been pointed out, we do not need heat capacity data at all densities, but merely in the low-density (ideal gas) limit and volumetric equation-of-state information.

Given C_{P}^* or C_{V}^* data and volumetric equation-of-state data (in either analytic or tabular form), it is possible to compute $\Delta \underline{H}$, $\Delta \underline{U}$, and $\Delta \underline{S}$ for any two states of a fluid; given the value of $\underline{S}$ in any one state, it is also possible to compute $\Delta \underline{G}$ and $\Delta \underline{A}$. Thus, we now have the equations necessary to construct complete tables or charts interrelating $\underline{H}$, $\underline{U}$, $\underline{S}$, T, P, and $\underline{V}$, such as those in Chapter 3. Although the process of constructing thermodynamic properties tables and charts is tedious (see Illustration 6.4-1), their availability, as we saw in Chapters 3, 4 and 5, makes it possible to use the balance equations to solve thermodynamic problems for real fluids quickly and with good accuracy.

It is interesting to compare the equations just derived with the analogous results for an ideal gas. Since

Ideal gas results

$$\left(\frac{\partial \underline{V}}{\partial T} \right)_P^{\mathrm{IG}} = \left(\frac{\partial \underline{V}}{\partial T} \right)_P^{\mathrm{IG}} = \frac{\underline{V}}{T} = \frac{R}{P} \quad \text{so} \quad \underline{V}^{\mathrm{IG}} - T \left(\frac{\partial \underline{V}}{\partial T} \right)_P^{\mathrm{IG}} = 0$$

and

$$\left(\frac{\partial P}{\partial T} \right)_{\underline{V}}^{\mathrm{IG}} = \frac{R}{\underline{V}} = \frac{P}{T}$$

we have

$$\underline{H}^{\mathrm{IG}}(T_2, P_2) - \underline{H}^{\mathrm{IG}}(T_1, P_1) = \int_{T_1, P=0}^{T_2, P=0} C_{\mathrm{P}}^* dT \qquad \textbf{(6.4-14b)}$$

$$\underline{S}^{IG}(T_2, P_2) - \underline{S}^{IG}(T_1, P_1) = -\int_{T_1, P_1}^{T_1, P=0} \frac{R}{P} dP + \int_{T_1}^{T_2} \frac{C_P^*}{T} dT - \int_{T_2, P=0}^{T_2, P_2} \frac{R}{P} dP$$

$$= \int_{T_1}^{T_2} \frac{C_P^*}{T} dT - R \ln \frac{P_2}{P_1} \qquad \text{(6.4-16b)}$$

$$\underline{S}^{IG}(T_2, \underline{V}_2) - \underline{S}^{IG}(T_1, \underline{V}_1) = +\int_{T_1, \underline{V}_1}^{T_1, \underline{V}=\infty} \frac{R}{\underline{V}} d\underline{V} + \int_{T_1}^{T_2} \frac{C_V^*}{T} dT + \int_{T_2, \underline{V}=\infty}^{T_2, \underline{V}_2} \frac{R}{\underline{V}} d\underline{V}$$

$$= \int_{T_1}^{T_2} \frac{C_V^*}{T} dT + R \ln \frac{\underline{V}_2}{\underline{V}_1} \qquad \text{(6.4-17b)}$$

Thus, comparing Eqs. 6.4-14a and b, Eqs. 6.4-17a and b, and Eqs. 6.4-16a and b, we have for the real fluid

Enthalpy and entropy changes in terms of departure functions

$$\boxed{\begin{aligned}
\underline{H}&(T_2, P_2) - \underline{H}(T_1, P_1) \\
&= \underline{H}^{IG}(T_2, P_2) - \underline{H}^{IG}(T_1, P_1) \\
&\quad + \int_{T_1, P_1}^{T_1, P=0} \left[\underline{V} - T\left(\frac{\partial \underline{V}}{\partial T}\right)_P\right] dP + \int_{T_2, P=0}^{T_2, P_2} \left[\underline{V} - T\left(\frac{\partial \underline{V}}{\partial T}\right)_P\right] dP \\
&= \underline{H}^{IG}(T_2, P_2) - \underline{H}^{IG}(T_1, P_1) + (\underline{H} - \underline{H}^{IG})_{T_2, P_2} - (\underline{H} - \underline{H}^{IG})_{T_1, P_1}
\end{aligned}}$$

$$\text{(6.4-19)}$$

$$\boxed{\begin{aligned}
\underline{S}&(T_2, P_2) - \underline{S}(T_1, P_1) \\
&= \underline{S}^{IG}(T_2, P_2) - \underline{S}^{IG}(T_1, P_1) \\
&\quad - \int_{T_1, P_1}^{T_1, P=0} \left[\left(\frac{\partial \underline{V}}{\partial T}\right)_P - \frac{R}{P}\right] dP - \int_{T_2, P=0}^{T_2, P_2} \left[\left(\frac{\partial \underline{V}}{\partial T}\right)_P - \frac{R}{P}\right] dP \\
&= \underline{S}^{IG}(T_2, P_2) - \underline{S}^{IG}(T_1, P_1) + (\underline{S} - \underline{S}^{IG})_{T_2, P_2} - (\underline{S} - \underline{S}^{IG})_{T_1, P_1}
\end{aligned}}$$

$$\text{(6.4-20)}$$

and

$$\begin{aligned}
\underline{S}&(T_2, \underline{V}_2) - \underline{S}(T_1, \underline{V}_1) \\
&= \underline{S}^{IG}(T_2, \underline{V}_2) - \underline{S}^{IG}(T_1, \underline{V}_1) \\
&\quad + \int_{T_1, \underline{V}_1}^{T_1, \underline{V}_1=\infty} \left[\left(\frac{\partial P}{\partial T}\right)_{\underline{V}} - \frac{R}{\underline{V}}\right] d\underline{V} + \int_{T_2, \underline{V}=\infty}^{T_2, \underline{V}_2} \left[\left(\frac{\partial P}{\partial T}\right)_{\underline{V}} - \frac{R}{\underline{V}}\right] d\underline{V} \\
&= \underline{S}^{IG}(T_2, \underline{V}_2) - \underline{S}^{IG}(T_1, \underline{V}_1) + (\underline{S} - \underline{S}^{IG})_{T_2, \underline{V}_2} - (\underline{S} - \underline{S}^{IG})_{T_1, \underline{V}_1}
\end{aligned}$$

$$\text{(6.4-21)}$$

where

Departure functions

$$(\underline{H} - \underline{H}^{IG})_{T, P} = \int_{T, P=0}^{T, P} \left[\underline{V} - T\left(\frac{\partial \underline{V}}{\partial T}\right)_P\right] dP \qquad \text{(6.4-22)}$$

$$(\underline{S} - \underline{S}^{IG})_{T,P} = -\int_{T,P=0}^{T,P} \left[\left(\frac{\partial \underline{V}}{\partial T} \right)_P - \frac{R}{P} \right] dP \tag{6.4-23}$$

and

$$(\underline{S} - \underline{S}^{IG})_{T,\underline{V}} = S(T, \underline{V}) - \underline{S}^{IG}(T, \underline{V})$$

$$= \int_{T,\underline{V}=\infty}^{T,\underline{V}} \left[\left(\frac{\partial P}{\partial T} \right)_{\underline{V}} - \frac{R}{\underline{V}} \right] d\underline{V} \tag{6.4-24}$$

The interpretation of Eqs. 6.4-19, 6.4-20, and 6.4-21 is clear: The changes in enthalpy and entropy of a real fluid are equal to those for an ideal gas undergoing the same change of state plus the departure of the fluid from ideal gas behavior at the end state less the departure from ideal gas behavior of the initial state. These **departure functions**, given by Eqs. 6.4-22, 6.4-23, and 6.4-24, can be computed once the fluid equation of state is known.

Before leaving this subject, we note that although Eqs. 6.4-22, 6.4-23, and 6.4-24 are useful for calculating the enthalpy and entropy departures from ideal gas behavior for some equations of state, their form is less helpful for the van der Waals, Peng-Robinson and other equations of state considered in this section in which $\underline{V}$ and T are the convenient independent variables.[12]

In such cases it is useful to have alternative expressions for the departure functions at fixed temperature and pressure. To obtain such expressions, we start with Eqs. 6.4-22 and 6.4-23 and use

$$dP = \frac{1}{\underline{V}} d(P\underline{V}) - \frac{P}{\underline{V}} d\underline{V} \tag{6.4-25}$$

and the triple product rule (Eq. 6.1-6a) in the form

$$\left(\frac{\partial \underline{V}}{\partial T} \right)_P \left(\frac{\partial P}{\partial \underline{V}} \right)_T \left(\frac{\partial T}{\partial P} \right)_{\underline{V}} = -1 \quad \text{or} \quad \left(\frac{\partial \underline{V}}{\partial T} \right)_P dP \bigg|_T = - \left(\frac{\partial P}{\partial T} \right)_{\underline{V}} d\underline{V} \bigg|_T \tag{6.4-26}$$

to obtain

$$\int_{P=0}^{P} \left[\underline{V} - T \left(\frac{\partial \underline{V}}{\partial T} \right)_P \right] dP = \int_{P\underline{V}=RT}^{P\underline{V}(T,P)} d(P\underline{V}) + \int_{\underline{V}=\infty}^{\underline{V}=\underline{V}(T,P)} \left[T \left(\frac{\partial P}{\partial T} \right)_{\underline{V}} - P \right] d\underline{V}$$

$$= (P\underline{V} - RT) + \int_{\underline{V}=\infty}^{\underline{V}=\underline{V}(T,P)} \left[T \left(\frac{\partial P}{\partial T} \right)_{\underline{V}} - P \right] d\underline{V}$$

therefore

Useful form of enthalpy departure equation

$$\underline{H}(T, P) - \underline{H}^{IG}(T, P) = RT(Z-1) + \int_{\underline{V}=\infty}^{\underline{V}=\underline{V}(T,P)} \left[T \left(\frac{\partial P}{\partial T} \right)_{\underline{V}} - P \right] d\underline{V} \tag{6.4-27}$$

---

[12]That is, with such equations of state, it is easier to solve for P given $\underline{V}$ and T than for $\underline{V}$ given P and T. Consequently, derivatives of P with respect to $\underline{V}$ or T are more easily found in terms of $\underline{V}$ and T than are derivatives of $\underline{V}$ with respect to P and T (try it!).

where $Z = P\underline{V}/RT$. Similarly, from

$$\frac{R}{P}\,dP = R\frac{d(P\underline{V})}{P\underline{V}} - R\frac{d\underline{V}}{\underline{V}} = Rd\ln(P\underline{V}) - \frac{R}{\underline{V}}\,d\underline{V}$$

and Eq. 6.4-23 we obtain

$$\int_{P=0}^{P}\left[\frac{R}{P} - \left(\frac{\partial\underline{V}}{\partial T}\right)_P\right]dP = R\int_{P\underline{V}=RT}^{P\underline{V}(T,P)}d\ln(P\underline{V}) + \int_{\underline{V}=\infty}^{\underline{V}=\underline{V}(T,P)}\left[\left(\frac{\partial P}{\partial T}\right)_{\underline{V}} - \frac{R}{\underline{V}}\right]d\underline{V}$$

$$= R\ln\left(\frac{P\underline{V}}{RT}\right) + \int_{\underline{V}=\infty}^{\underline{V}=\underline{V}(T,P)}\left[\left(\frac{\partial P}{\partial T}\right)_{\underline{V}} - \frac{R}{\underline{V}}\right]d\underline{V}$$

and

$$\underline{S}(T,P) - \underline{S}^{\mathrm{IG}}(T,P) = R\ln Z + \int_{\underline{V}=\infty}^{\underline{V}=\underline{V}(T,P)}\left[\left(\frac{\partial P}{\partial T}\right)_{\underline{V}} - \frac{R}{\underline{V}}\right]d\underline{V} \qquad \textbf{(6.4-28)}$$

The desired equations, Eqs. 6.4-27 and 6.4-28, are general and can be used with any equation of state. Using, for example, the Peng-Robinson equation of state, one obtains (see Problem 6.2)

Enthalpy and entropy changes for the Peng-Robinson equation of state

$$\boxed{\underline{H}(T,P) - \underline{H}^{\mathrm{IG}}(T,P) = RT(Z-1) + \frac{T\left(\dfrac{da}{dT}\right) - a}{2\sqrt{2}b}\ln\left[\frac{Z + (1+\sqrt{2})B}{Z + (1-\sqrt{2})B}\right]}$$

$$\textbf{(6.4-29)}$$

and

$$\boxed{\underline{S}(T,P) - \underline{S}^{\mathrm{IG}}(T,P) = R\ln(Z-B) + \frac{\dfrac{da}{dT}}{2\sqrt{2}b}\ln\left[\frac{Z + (1+\sqrt{2})B}{Z + (1-\sqrt{2})B}\right]} \qquad \textbf{(6.4-30)}$$

where

$$Z = P\underline{V}/RT \quad\text{and}\quad B = Pb/RT$$

Demonstrating how an equation of state is used in the construction of a thermodynamics properties chart

ILLUSTRATION 6.4-1

Construction of a Thermodynamic Properties Chart

As an introduction to the problem of constructing a chart or table of the thermodynamic properties of a real fluid, develop a thermodynamic properties chart for oxygen over the temperature range of $-100°C$ to $+150°C$ and a pressure range of 1 to 100 bar. (A larger temperature range, including the vapor-liquid two-phase region, will be considered in Chapter 7.) In particular,

calculate the compressibility factor, specific volume, molar enthalpy, and molar entropy as a function of temperature and pressure. Also, prepare a pressure-volume plot, a pressure-enthalpy plot (see Fig. 3.4-2), and a temperature-entropy plot (see Fig. 3.4-3) for oxygen.
Data: For simplicity we will assume oxygen obeys the Peng-Robinson equation of state and has an ideal gas heat capacity given by

$$C_P^* \left(\frac{J}{mol\ K} \right) = 25.46 + 1.519 \times 10^{-2} T - 0.7151 \times 10^{-5} T^2 + 1.311 \times 10^{-9} T^3$$

We will choose the reference state of oxygen to be the ideal gas state at 25°C and 1 bar:

$$\underline{H}^{IG}(T = 25°C, P = 1\ bar) = 0$$

and

$$\underline{S}^{IG}(T = 25°C, P = 1\ bar) = 0$$

As will be explained shortly, the Peng-Robinson parameters for oxygen are

$$a(T) = 0.45724 \frac{R^2 T_c^2}{P_c} \alpha(T)$$

$$b(T) = 0.07780 \frac{R T_c}{P_c}$$

$$[\alpha(T)]^{1/2} = 1 + \kappa \left(1 - \sqrt{\frac{T}{T_c}} \right)$$

where $\kappa = 0.4069$, $T_c = 154.6$ K is the critical temperature of oxygen, and $P_c = 5.046$ MPa is its critical pressure.

SOLUTION

I. Volume
The volume is the easiest of the properties to calculate. For each value of temperature we compute values for $a(T)$ and b, and then at each pressure solve the equation

$$P = \frac{RT}{\underline{V} - b} - \frac{a(T)}{\underline{V}(\underline{V} + b) + b(\underline{V} - b)}$$

for the volume, or equivalently (and preferably), solve the equation

$$Z^3 + (-1 + B)Z^2 + (A - 3B^2 - 2B)Z + (-AB + B^2 + B^3) = 0$$

with $B = bP/RT$ and $A = aP/(RT)^2$ (see Eq. 6.4-4 and Table 6.4-3) for the compressibility factor $Z = P\underline{V}/RT$, from which the volume is easily calculated as $\underline{V} = ZRT/P$. Repeating the calculation a number of times leads to the entries in Table 6.4-4. These calculations can be done using the Visual Basic computer program described in Appendix B.I-2, the DOS-based program PR1 described in Appendix B.II-1, the MATHCAD worksheet described in Appendix B.III, or the MATLAB program described in Appendix B.IV and included on the CD-ROM accompanying this book.

Since the pressure and specific volume vary by several orders of magnitude over the range of interest, these results have been plotted in Fig. 6.4-3 as ln P versus ln $\underline{V}$. Remember that for the ideal gas $P\underline{V} = RT$, so that an ideal gas isotherm on a log-log plot is a straight line. (Note that Fig. 6.4-3 also contains isotherms for temperatures below −100°C. The calculation of these will be discussed in Sec. 7.5.)

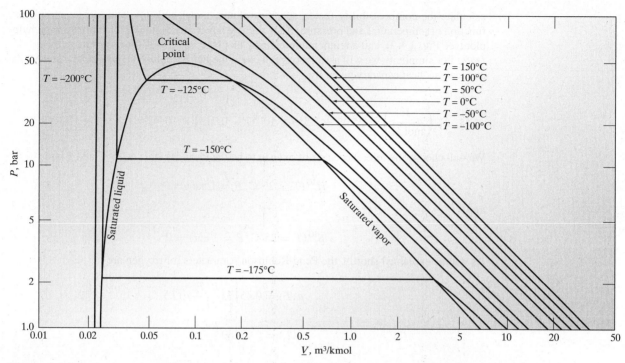

Figure 6.4-3 Pressure-volume diagram for oxygen calculated using the Peng-Robinson equation of state.

II. Enthalpy

To compute the difference in the enthalpy of oxygen between a state of temperature T and pressure P and the ideal gas reference state at 25°C and 1 bar, we use

$$\underline{H}(T, P) - \underline{H}^{IG}(T = 25°C, P = 1 \text{ bar})$$
$$= \underline{H}(T, P) - \underline{H}^{IG}(T, P) + \underline{H}^{IG}(T, P) - \underline{H}^{IG}(T = 25°C, P = 1 \text{ bar})$$
$$= (\underline{H} - \underline{H}^{IG})_{T,P} + \int_{T=298.15K}^{T} C_P^* \, dT$$
$$= (\underline{H} - \underline{H}^{IG})_{T,P} + 25.46(T - 298.15) + \frac{1.519 \times 10^{-2}}{2}(T^2 - 298.15^2)$$
$$- \frac{0.7151 \times 10^{-5}}{3}(T^3 - 298.15^3) + \frac{1.311 \times 10^{-9}}{4}(T^4 - 298.15^4)$$

The quantity $(\underline{H} - \underline{H}^{IG})_{T,P}$ is computed using Eq. 6.4-29 and the value of the compressibility factor Z found earlier at each set of (T, P) values. The values of enthalpy computed in this manner appear in Table 6.4-4 and have been plotted in Fig. 6.4-4 as a pressure and enthalpy diagram. The Peng-Robinson program (Appendices B.I-3 and B.II-1) was used for this calculation as well. [Note that at $T = 25°C$ and $P = 1$ bar, oxygen is not quite an ideal gas; $Z = 0.9991$, not unity. Consequently, $(\underline{H} - \underline{H}^{IG})_{T=25°C, P=1 \text{ bar}} = -9.44$ J/mol, so that the enthalpy of oxygen at $T = 25°C$, $P = 1$ bar is -9.44 J/mol, whereas if it were an ideal gas at these conditions its enthalpy would be zero.] Once again, lower-temperature results, including the two-phase region, appear in Fig 6.4-6. The basis for those calculations will be discussed in Sec. 7.5.

Table 6.4-4 Thermodynamic Properties of Oxygen Calculated Using the Peng-Robinson Equation of State

$T(°C)$	−100	−75	−50	−25	0	25	50	75	100	125	150
$P = 1$ bar											
Z	0.9945	0.9963	0.9974	0.9982	0.9987	0.9991	0.9994	0.9996	0.9997	0.9998	0.9999
$\underline{V}$	14.3200	16.4200	18.5100	20.5900	22.6800	24.7700	26.8500	28.9300	31.0200	33.1000	35.1800
$\underline{H}$	−3603.75	−2898.69	−2186.80	−1480.83	−742.14	−9.44	730.04	1476.18	2228.83	2987.86	3753.10
$\underline{S}$	−15.59	−11.81	−8.43	−5.38	−2.59	−0.02	2.36	4.58	6.67	8.64	10.50
$P = 2$ bar											
Z	0.9889	0.9925	0.9948	0.9963	0.9974	0.9982	0.9987	0.9991	0.9994	0.9996	0.9998
$\underline{V}$	7.1190	8.1760	9.2290	10.2800	11.3300	12.3700	13.4200	14.4600	15.5000	16.5500	17.5900
$\underline{H}$	−3626.18	−2917.04	−2202.08	−1480.03	−753.13	−18.88	721.89	1469.10	2222.68	2982.49	3748.42
$\underline{S}$	−21.38	−17.63	−14.23	−11.17	−8.38	−5.81	−3.43	−1.19	0.87	2.87	4.73
$P = 5$ bar											
Z	0.9722	0.9813	0.9871	0.9909	0.9936	0.9954	0.9968	0.9978	0.9986	0.9991	0.9996
$\underline{V}$	2.7990	3.2330	3.6630	4.0890	4.5130	4.9350	5.3560	5.7760	6.1960	6.6150	7.0330
$\underline{H}$	−3694.40	−2972.47	−2248.04	−1519.52	−786.05	−47.12	697.51	1447.97	2204.29	2966.37	3734.44
$\underline{S}$	−29.33	−25.44	−22.00	−18.90	−16.09	−13.50	−11.10	−8.86	−6.77	−4.79	−2.92
$P = 10$ bar											
Z	0.9439	0.9626	0.9743	0.9820	0.9872	0.9910	0.9937	0.9957	0.9972	0.9983	0.9992
$\underline{V}$	1.3590	1.5860	1.8080	2.0260	2.2420	2.4570	2.6700	2.8820	3.0940	3.3050	3.5150
$\underline{H}$	−3811.36	−3066.16	−2325.11	−1584.04	−840.75	−93.94	657.16	1413.02	2173.92	2940.01	3711.36
$\underline{S}$	−35.55	−31.53	−28.00	−24.85	−21.99	−19.38	−16.97	−14.71	−12.60	−10.61	−8.73
$P = 20$ bar											
Z	0.8856	0.9252	0.9491	0.9646	0.9751	0.9825	0.9878	0.9918	0.9947	0.9969	0.9987
$\underline{V}$	0.6375	0.7621	0.8804	0.9951	1.1070	1.2180	1.3270	1.4350	1.5430	1.6500	1.7570
$\underline{H}$	−4059.21	−3258.71	−2480.87	−1713.17	−949.56	−186.68	577.48	1344.15	2114.16	2888.07	3666.08
$\underline{S}$	−42.27	−37.95	−34.25	−30.99	−28.06	−25.39	−22.92	−20.64	−18.50	−16.50	−14.60
$P = 30$ bar											
Z	0.8245	0.8878	0.9246	0.9479	0.9636	0.9745	0.9824	0.9882	0.9925	0.9958	0.9983
$\underline{V}$	0.3957	0.4876	0.5718	0.6519	0.7295	0.8053	0.8798	0.9535	1.0260	1.0990	1.1710
$\underline{H}$	−4329.71	−3458.44	−2638.59	−1842.18	−1057.37	−278.07	499.24	1276.72	2055.76	2837.29	3621.95
$\underline{S}$	−46.72	−42.01	−38.11	−34.73	−31.72	−28.89	−26.48	−24.17	−22.00	−20.00	−18.07

$P = 40$ bar											
$\underline{Z}$	0.9981	0.9949	0.9906	0.9850	0.9774	0.9671	0.9528	0.9320	0.9008	0.8508	0.7602
$\underline{V}$	0.8779	0.8233	0.7683	0.7128	0.6565	0.5994	0.5409	0.4807	0.4178	0.3504	0.2736
$\underline{H}$	3578.97	2787.83	1998.73	1210.75	422.53	−367.98	−1163.97	−1970.75	−2797.93	−3665.63	−4629.51
$\underline{S}$	−20.56	−22.50	−24.54	−26.72	−29.06	−31.61	−34.40	−37.50	−41.01	−45.14	−50.35
$P = 50$ bar											
$\underline{Z}$	0.9981	0.9942	0.9890	0.9821	0.9729	0.9603	0.9426	0.9170	0.8781	0.8143	0.6922
$\underline{V}$	0.7023	0.6582	0.6137	0.5686	0.5228	0.4761	0.4282	0.3784	0.3258	0.2683	0.1993
$\underline{H}$	3537.15	2739.64	1943.10	1146.30	347.92	−456.28	−1269.13	−2098.53	−2958.41	−3880.25	−4968.26
$\underline{S}$	−22.51	−24.45	−26.52	−28.73	−31.11	−33.70	−36.54	−39.72	−43.39	−47.77	−53.66
$P = 60$ bar											
$\underline{Z}$	0.9983	0.9937	0.9877	0.9796	0.9698	0.9541	0.9332	0.9029	0.8564	0.7788	0.6202
$\underline{V}$	0.5854	0.5483	0.5107	0.4726	0.4338	0.3942	0.3532	0.3105	0.2648	0.2139	0.1448
$\underline{H}$	3496.47	2692.73	1888.88	1083.39	273.98	−542.81	−1372.64	−2225.13	−3119.39	−4101.78	−5358.88
$\underline{S}$	−24.12	−26.07	−28.16	−30.39	−32.80	−35.44	−38.34	−41.62	−45.42	−50.09	−56.90
$P = 70$ bar											
$\underline{Z}$	0.9987	0.9935	0.9866	0.9775	0.9652	0.9484	0.9246	0.8898	0.8361	0.7450	0.5471
$\underline{V}$	0.5020	0.4698	0.4373	0.4042	0.3705	0.3358	0.3000	0.2623	0.2216	0.1753	0.1125
$\underline{H}$	3456.93	2647.08	1836.08	1022.07	202.27	−627.54	−1472.25	−2350.13	−3280.08	−4328.89	−5878.46
$\underline{S}$	−25.49	−27.46	−29.57	−31.82	−34.26	−36.94	−39.90	−43.27	−47.23	−52.21	−60.27
$P = 80$ bar											
$\underline{Z}$	0.9993	0.9935	0.9859	0.9757	0.9621	0.9433	0.9167	0.8778	0.8173	0.7135	0.4833
$\underline{V}$	0.4395	0.4111	0.3823	0.3530	0.3231	0.2923	0.2602	0.2264	0.1896	0.1469	0.0870
$\underline{H}$	3418.53	2602.71	1784.71	962.34	132.35	−710.26	−1573.76	−2473.08	−3439.53	−4559.08	−6296.72
$\underline{S}$	−26.69	−28.68	−30.80	−33.08	−35.55	−38.27	−41.29	−44.75	−48.86	−54.19	−63.65
$P = 90$ bar											
$\underline{Z}$	1.0001	0.9937	0.9854	0.9743	0.9593	0.9387	0.9096	0.8669	0.8003	0.6852	0.4444
$\underline{V}$	0.3909	0.3655	0.3397	0.3134	0.2864	0.2586	0.2295	0.1987	0.1650	0.1254	0.0711
$\underline{H}$	3381.24	2559.61	1734.77	904.24	64.27	−790.91	−1670.95	−2593.54	−3596.54	−4788.49	−6719.52
$\underline{S}$	−27.76	−29.76	−31.90	−34.20	−36.70	−39.46	−42.54	−46.09	−50.35	−56.03	−66.53
$P = 100$ bar											
$\underline{Z}$	1.0010	0.9941	0.9851	0.9732	0.9570	0.9348	0.9034	0.8571	0.7851	0.6611	0.4297
$\underline{V}$	0.3522	0.3291	0.3056	0.2817	0.2571	0.2317	0.2052	0.1768	0.1457	0.1089	0.0619
$\underline{H}$	3345.07	2517.77	1686.27	847.77	−1.92	−869.40	−1765.61	−2711.04	−3750.23	−5012.01	−7026.48
$\underline{S}$	−28.72	−30.74	−32.89	−35.22	−37.75	−40.55	−43.69	−47.32	−51.74	−57.75	−68.69

$\underline{V}$ [=] m³/kmol; $\underline{H}$ [=] J/mol = kJ/kmol; $\underline{S}$ [=] J/(mol K) = kJ/(kmol K).

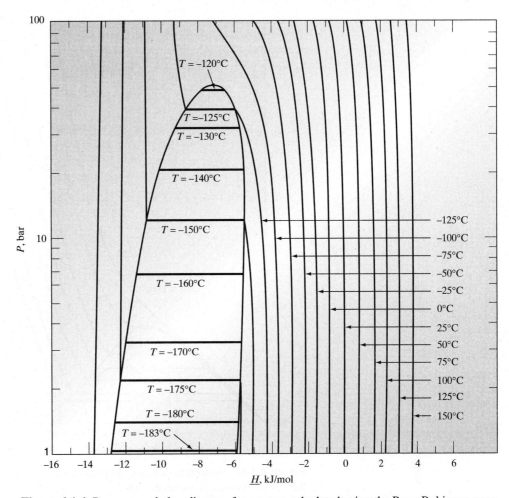

Figure 6.4-4 Pressure-enthalpy diagram for oxygen calculated using the Peng-Robinson equation of state.

III. Entropy

To compute the difference in the entropy of oxygen between the state (T, P) and the ideal gas reference state at 25°C and 1 bar, we use

$$\underline{S}(T, P) - \underline{S}^{IG}(T = 25°C, P = 1 \text{ bar})$$

$$= \underline{S}(T, P) - \underline{S}^{IG}(T, P) + \underline{S}^{IG}(T, P) - \underline{S}^{IG}(T = 25°C, P = 1 \text{ bar})$$

$$= (\underline{S} - \underline{S}^{IG})_{T,P} + \int_{T=298.15K}^{T} \frac{C_P^*}{T} \, dT - R \, \ln\left(\frac{P}{1 \text{ bar}}\right)$$

$$= (\underline{S} - \underline{S}^{IG})_{T,P} + 25.46 \ln \frac{T}{298.15} + 1.519 \times 10^{-2}(T - 298.15)$$

$$- \frac{0.7151 \times 10^{-5}}{2}(T^2 - 298.15^2) + \frac{1.311 \times 10^{-9}}{3}(T^3 - 298.15^3)$$

$$- R \ln\left(\frac{P}{1 \text{ bar}}\right)$$

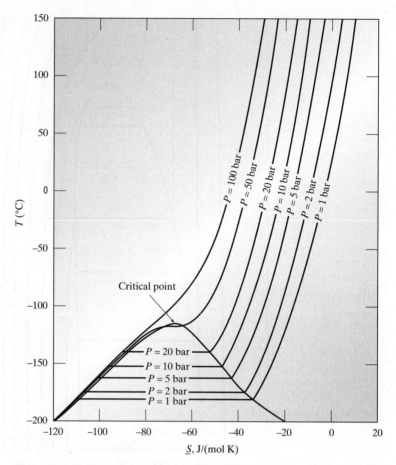

Figure 6.4-5 Temperature-entropy diagram for oxygen calculated using the Peng-Robinson equation of state.

The quantity $(\underline{S} - \underline{S}^{IG})_{T,P}$ is computed using Eq. 6.4-30, and the value of the compressibility factor Z is found earlier at each T and P. The values of the entropy computed in this manner with the programs in either Appendix B.I or B.II appear in Table 6.4-4 and as a temperature-entropy diagram in Fig. 6.4-5.

COMMENTS

For illustrative purposes, the Peng-Robinson equation of state was used here. Generally, if one were going to take the time and effort to construct tables or plots such as those developed here, much more complicated equations of state, for example, the Bender equation (Eq. 6.4-6b), would be used to obtain more accurate results.

Note that in this illustration we chose the hypothetical ideal gas at 25°C and 1 bar as the reference state for the calculation, rather than the real gas at those conditions. Since the actual state of oxygen from the Peng-Robinson equation is not quite ideal, we have

$$Z(1 \text{ bar}, 25°C) = 0.9991$$
$$\underline{H}(1 \text{ bar}, 25°C) = -9.44 \text{ kJ/mol}$$
$$\underline{S}(1 \text{ bar}, 25°C) = -0.02 \text{ kJ/(mol K)}$$

At first glance it might appear that to avoid such differences one could choose zero pressure as the reference state, since all gases are then ideal. The reason this is not done is that the entropy of a gas is infinite at zero pressure (see, for example, Eq. 4.4-3). ∎

Table 6.4-5 Molar Volume, Coefficient of Thermal Expansion, and Coefficient of Isothermal Compressiblity at 20°C for Some Condensed Phases

	$\underline{V}$ (cc/mol)	$\alpha \times 10^4$ (K^{-1})	$\kappa_T \times 10^6$ (bar^{-1})
Solids			
Copper	7.12	0.492	0.77
Graphite	5.31	0.24	2.96
Platinum	9.10	0.265	0.375
Silver	10.27	0.583	1.0
Sodium chloride	27.02	1.21	4.15
Liquids			
Benzene	88.87	12.4	92.8
Carbon tetrachloride	96.44	12.4	102
Ethanol	58.40	11.2	109
Methanol	40.47	12.0	118
Water	18.02	2.07	44.7
Mercury	14.81	1.81	3.80

In Chapter 2 we made the assumption that some thermodynamic properties of condensed phases, that is, liquids and solids, were independent of pressure. We now have the equations that allow us to examine and improve upon this assumption. We will do this using the data in Table 6.4-5, which includes the molar volume $\underline{V}$, the coefficient of thermal expansion $\alpha = (1/\underline{V})(\partial \underline{V}/\partial T)_P$, and the isothermal compressibility $\kappa_T = -(1/\underline{V})(\partial \underline{V}/\partial P)_T$ for a number of substances. Here we are interested in the change in thermodynamic properties with pressure when α and κ_T are the only information available, as is typically the case for a liquid or a solid.

To compute the change in volume of a condensed phase from the pressure P to the pressure $P + \Delta P$, assuming that the isothermal compressibility is independent of pressure, we use

$$\underline{V}(T, P + \Delta P) = \underline{V}(T, P) + \left(\frac{\partial \underline{V}}{\partial P}\right)_T \Delta P$$

$$= \underline{V}(T, P) - \kappa_T \underline{V}(T, P)\Delta P = \underline{V}(T, P)[1 - \kappa_T \Delta P]$$

(6.4-31)

Note that for an incompressible fluid $\kappa_T = 0$, so that $\underline{V}(T, P + \Delta P) = \underline{V}(T, P)$.

ILLUSTRATION 6.4-2

Effect of Pressure on the Volume of Liquids and Solids

Compute the change in the molar volume of copper, sodium chloride, ethanol, and mercury for the very large pressure change of going from 1 bar to 1000 bar at 20°C.

SOLUTION

Using the equation, we have

$$\underline{V}_{Cu}(20°C, 1000 \text{ bar}) = 7.12 \, \frac{cc}{mol} \times [1 - 0.77 \times 10^{-6} \text{ bar}^{-1} \times 999 \text{ bar}]$$

$$= 7.12 \, \frac{cc}{mol} \times [1 - 0.000769] = 7.115 \, \frac{cc}{mol}$$

$$\underline{V}_{\text{NaCl}}(20°\text{C}, 1000 \text{ bar}) = 27.02 \frac{\text{cc}}{\text{mol}} \times [1 - 4.15 \times 10^{-6} \text{ bar}^{-1} \times 999 \text{ bar}] = 26.91 \frac{\text{cc}}{\text{mol}}$$

$$\underline{V}_{\text{EtOH}}(20°\text{C}, 1000 \text{ bar}) = 58.40 \frac{\text{cc}}{\text{mol}} [1 - 109 \times 10^{-6} \text{ bar}^{-1} \times 999 \text{ bar}] = 52.04 \frac{\text{cc}}{\text{mol}}$$

$$\underline{V}_{\text{Hg}}(20°\text{C}, 1000 \text{ bar}) = 14.81 \frac{\text{cc}}{\text{mol}} [1 - 3.80 \times 10^{-6} \text{ bar}^{-1} \times 999 \text{ bar}] = 14.75 \frac{\text{cc}}{\text{mol}}$$

COMMENT

We see that the change in volume with a factor of 1000 change in pressure is very small for the two solids and the liquid metal considered, but that there is an 8.9 percent change in volume for the more compressible liquid ethanol. ∎

From Eq. 6.2-22 we have that the change in enthalpy with pressure at constant temperature is

$$d\underline{H} = \left[\underline{V} - T \left(\frac{\partial \underline{V}}{\partial T} \right)_P \right] dP = [\underline{V} - T\underline{V}\alpha]dP \qquad (6.4\text{-}32)$$

Therefore, assuming that $\underline{V}$ and α are essentially constant with pressure gives

$$\boxed{\underline{H}(T, P + \Delta P) - \underline{H}(T, P) = \underline{V}[1 - T\alpha]\Delta P} \qquad (6.4\text{-}33)$$

ILLUSTRATION 6.4-3
Effect of Pressure on the Enthalpy of Liquids and Solids

Compute the change in enthalpy of the four substances considered in the previous example on going from 1 bar to 1000 bar at 20°C.

SOLUTION

$$\underline{H}_{\text{Cu}}(20°\text{C}, 1000 \text{ bar}) - \underline{H}_{\text{Cu}}(20°\text{C}, 1 \text{ bar})$$

$$= 7.12 \frac{\text{cc}}{\text{mol}} [1 - 293.15 \text{ K} \times 0.492 \times 10^{-4} \text{ K}^{-1}] \times 999 \text{ bar}$$

$$= 7.12 \frac{\text{cc}}{\text{mol}} [1 - 0.0144] \times 999 \text{ bar} = 7010.3 \frac{\text{cc bar}}{\text{mol}} = 701 \frac{\text{J}}{\text{mol}}$$

$$\underline{H}_{\text{NaCl}}(20°\text{C}, 1000 \text{ bar}) - \underline{H}_{\text{NaCl}}(20°\text{C}, 1 \text{ bar})$$

$$= 27.02 \frac{\text{cc}}{\text{mol}} \times [1 - 293.15 \times 1.21 \times 10^{-4}] \times 999 \text{ bar}$$

$$= 2604 \frac{\text{J}}{\text{mol}}$$

$\underline{H}_{\text{EtOH}}(20°\text{C}, 1000 \text{ bar}) - \underline{H}_{\text{EtOH}}(20°\text{C}, 1 \text{ bar})$

$$= 58.40 \, \frac{\text{cc}}{\text{mol}} \times [1 - 293.15 \times 11.2 \times 10^{-4}] \times 999 \text{ bar}$$

$$= 3919 \, \frac{\text{J}}{\text{mol}}$$

and

$\underline{H}_{\text{Hg}}(20°\text{C}, 1000 \text{ bar}) - \underline{H}_{\text{Hg}}(20°\text{C}, 1 \text{ bar})$

$$= 14.81 \, \frac{\text{cc}}{\text{mol}} \times [1 - 293.15 \times 1.81 \times 10^{-4}] \times 999 \text{ bar}$$

$$= 1401 \, \frac{\text{J}}{\text{mol}}$$

COMMENT

Note that in each case the enthalpy change due to the pressure change is large and cannot be neglected. In all cases the contribution from the $\underline{V}\Delta P$ dominates, with the contribution from the isothermal compressibility being only 1.4 percent of the total for copper, 3.6 percent for sodium chloride, and 5.3 percent for liquid mercury. However, this contribution is 32.8 percent for liquid ethanol. ∎

The easiest way to compute the change in internal energy with pressure at constant temperature for a condensed phase is to note that $\underline{U} = \underline{H} - P\underline{V}$. Therefore, if we can assume the volume of a condensed phase does not change much with pressure, we have

$$\underline{U}(T, P + \Delta P) - \underline{U}(T, P) = \underline{H}(T, P + \Delta P) - (P + \Delta P)\underline{V}(T, P)$$

$$- [\underline{H}(T, P) - P\underline{V}(T, P)]$$

$$= \underline{H}(T, P + \Delta P) - \underline{H}(T, P) - \underline{V}(T, P)\Delta P \tag{6.4-34}$$

$$= \underline{V}\Delta P - \underline{V}T\alpha\Delta P - \underline{V}\Delta P = -\underline{V}T\alpha\Delta P$$

ILLUSTRATION 6.4-4

Effect of Pressure on the Internal Energy of Liquids and Solids

Compute the change in internal energy of the four substances considered in the previous illustrations on going from 1 bar to 1000 bar at 20°C.

SOLUTION

$\underline{U}_{\text{Cu}}(20°\text{C}, 1000 \text{ bar}) - \underline{U}_{\text{Cu}}(20°\text{C}, 1 \text{ bar})$

$$= -\underline{V}T\alpha\Delta P$$

$$= -7.12 \, \frac{\text{cc}}{\text{mol}} \times 293.15 \text{ K} \times 0.492 \times 10^{-4} \text{ K}^{-1} \times 999 \text{ bar} \times \frac{1 \text{ J}}{10 \text{ cc bar}}$$

$$= -10.3 \, \frac{\text{J}}{\text{mol}}$$

$$\underline{U}_{\text{NaCl}}(20°\text{C, 1000 bar}) - \underline{U}_{\text{NaCl}}(20°\text{C, 1 bar}) = -95.7 \, \frac{\text{J}}{\text{mol}}$$

$$\underline{U}_{\text{EtOH}}(20°\text{C, 1000 bar}) - \underline{U}_{\text{EtOH}}(20°\text{C, 1 bar}) = -1915.5 \, \frac{\text{J}}{\text{mol}}$$

$$\underline{U}_{\text{Hg}}(20°\text{C, 1000 bar}) - \underline{U}_{\text{Hg}}(20°\text{C, 1 bar}) = -78.5 \, \frac{\text{J}}{\text{mol}}$$

COMMENT

With the exception of ethanol, the changes in internal energy are small, much smaller than the changes in enthalpy for these substances. Indeed, for moderate changes in pressure, rather than the large pressure change considered here, the change in internal energy with pressure can be neglected. ∎

The change in entropy with pressure at constant temperature is, from Eq. 6.2-20,

$$d\underline{S} = -\left(\frac{\partial \underline{V}}{\partial T}\right)_P dP \tag{6.4-35}$$

Assuming the coefficient of thermal expansion is independent of pressure, we obtain

$$\underline{S}(T, P + \Delta P) - \underline{S}(T, P) = -\alpha \underline{V} \Delta P \tag{6.4-36}$$

ILLUSTRATION 6.4-5
Effect of Pressure on the Entropy of Liquids and Solids

Compute the change in entropy of the four substances considered in the previous example on going from 1 bar to 1000 bar.

SOLUTION

$$\underline{S}_{\text{Cu}}(20°\text{C, 1000 bar}) - \underline{S}_{\text{Cu}}(20°\text{C, 1 bar})$$

$$= 7.12 \, \frac{\text{cc}}{\text{mol}} \times 0.492 \times 10^{-4} \, \text{K}^{-1} \times 999 \, \text{bar} \times \frac{1 \, \text{J}}{10 \, \text{cc bar}}$$

$$= -0.035 \, \frac{\text{J}}{\text{mol K}}$$

$$\underline{S}_{\text{NaCl}}(20°\text{C, 1000 bar}) - \underline{S}_{\text{NaCl}}(20°\text{C, 1 bar}) = -0.327 \, \frac{\text{J}}{\text{mol K}}$$

$$\underline{S}_{\text{EtOH}}(20°\text{C, 1000 bar}) - \underline{S}_{\text{EtOH}}(20°\text{C, 1 bar}) = -6.534 \, \frac{\text{J}}{\text{mol K}}$$

$$\underline{S}_{\text{Hg}}(20°\text{C, 1000 bar}) - \underline{S}_{\text{Hg}}(20°\text{C, 1 bar}) = -0.268 \, \frac{\text{J}}{\text{mol K}}$$

COMMENT

Again, with the exception of ethanol, the changes in entropy are quite small. For moderate changes in pressure the entropy change of a condensed phase with pressure is negligible. ∎

The illustrations above show that unless the pressure change is large, and the condensed phase is quite compressible (as is the case for ethanol here), the changes in volume, internal energy, and entropy with changing pressure are quite small. However, the change in enthalpy can be significant and must be accounted for. Indeed, it is this enthalpy difference that is important in computing the work required to pump a liquid from a lower pressure to a higher pressure.

ILLUSTRATION 6.4-6

Energy Required to Pressurize a Liquid

Compute the minimum work required and the heat that must be removed to isothermally pump liquid ethanol from 20°C and 1 bar to 1000 bar in a continuous process.

SOLUTION

Consider the pump and its contents to be the system around which balance equations are to be written.

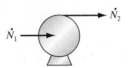

The steady-state mass balance (on a molar basis) is

$$\frac{dN}{dt} = 0 = \dot{N}_1 + \dot{N}_2 \quad \text{or} \quad \dot{N}_2 = -\dot{N}_1 = -\dot{N}$$

The steady-state energy balance is

$$\frac{dU}{dt} = 0 = \dot{N}_1 \underline{H}_1 + \dot{N}_2 \underline{H}_2 + \dot{Q} + \dot{W}$$

or

$$\frac{\dot{W}}{\dot{N}} = -\frac{\dot{Q}}{\dot{N}} + (\underline{H}_2 - \underline{H}_1)$$

The steady-state entropy balance (with $\dot{S}_{\text{gen}} = 0$ for minimum work) is

$$\frac{dS}{dt} = 0 = \dot{N}_1 \underline{S}_1 + \dot{N}_2 \underline{S}_2 + \frac{\dot{Q}}{T}$$

or

$$\frac{\dot{Q}}{\dot{N}} = T(\underline{S}_2 - \underline{S}_1)$$

From the previous illustrations, we have

$$\underline{S}_2 - \underline{S}_1 = \underline{S}(20°C, 1000 \text{ bar}) - \underline{S}(20°C, 1 \text{ bar}) = -6.534 \text{ J/(mol K)}$$

and

$$\underline{H}_2 - \underline{H}_1 = \underline{H}(20°C, 1000 \text{ bar}) - \underline{H}(20°C, 1 \text{ bar}) = 3919 \text{ J/mol}$$

Therefore,

$$\frac{\dot{Q}}{\dot{N}} = 293.15 \text{ K} \times (-6.534)\frac{\text{J}}{\text{mol K}} = -1915\frac{\text{J}}{\text{mol}}$$

and

$$\frac{\dot{W}}{\dot{N}} = 1915 + 3919 = 5834\frac{\text{J}}{\text{mol}}$$

COMMENT

If we neglected the effect of thermal expansion, that is, set $\alpha = 0$, then from Eq. 6.4-33,

$$\underline{H}(T, P + \Delta P) - \underline{H}(T, P) = \underline{V}[1 - T\alpha]\Delta P = \underline{V}\Delta P$$

and from Eq. 6.4-36,

$$\underline{S}(T, P + \Delta P) - \underline{S}(T, P) = -\alpha\underline{V}\Delta P = 0$$

Then, for the problem here,

$$\underline{H}(20°\text{C}, 1000 \text{ bar}) - \underline{H}(20°\text{C}, 1 \text{ bar}) = 58.40\frac{\text{cc}}{\text{mol}} \times 999 \text{ bar} \times \frac{1\text{J}}{10 \text{ cc bar}} = 5834\frac{\text{J}}{\text{mol}}$$

and

$$\underline{S}(20°\text{C}, 1000 \text{ bar}) - \underline{S}(20°\text{C}, 1 \text{ bar}) = 0$$

so that

$$\frac{\dot{Q}}{\dot{N}} = 0 \quad \text{and} \quad \frac{\dot{W}}{\dot{N}} = 5834\frac{\text{J}}{\text{mol}}$$

Note that the total work is the same as before for this isothermal pressurization (why?), but in this case no heat would have to be removed to keep the liquid at constant temperature.

To pressurize liquid mercury, which is closer to being incompressible than ethanol, over the same conditions one obtains (without making the assumption that $\alpha = 0$)

$$\frac{\dot{Q}}{\dot{N}} = 293.15 \text{ K} \times \left(-0.286\frac{\text{J}}{\text{mol K}}\right) = -78.6\frac{\text{J}}{\text{mol}}$$

and

$$\frac{\dot{W}}{\dot{N}} = 78.6 + 1401\frac{\text{J}}{\text{mol}} = 1478.6\frac{\text{J}}{\text{mol}}$$

Much less work is needed for this less compressible fluid since very little work is being done to change the volume of the liquid; the work is largely going to change the fluid pressure. ∎

ILLUSTRATION 6.4-7

Computing the Difference between C_P and C_V

Compute the difference between the constant-pressure and constant-volume heat capacities at 20°C for the four condensed phases considered in the previous illustrations.

SOLUTION

From Eq. 6.2-35 we have

$$C_P - C_V = T\underline{V}\frac{\alpha^2}{\kappa_T}$$

$$(C_P - C_V)_{Cu} = 293.15 \text{ K} \times 7.12\,\frac{\text{cc}}{\text{mol}} \times \frac{(0.492 \times 10^{-4})^2}{\text{K}^2}\,\frac{\text{bar}}{0.77 \times 10^{-6}} \times \frac{1\text{ J}}{10\text{ bar cc}}$$

$$= 0.656\,\frac{\text{J}}{\text{mol K}}$$

$$(C_P - C_V)_{NaCl} = 2.795\,\frac{\text{J}}{\text{mol K}}$$

$$(C_P - C_V)_{EtOH} = 19.70\,\frac{\text{J}}{\text{mol K}}$$

$$(C_P - C_V)_{Hg} = 3.742\,\frac{\text{J}}{\text{mol K}}$$

COMMENT

Note that the difference between the constant-pressure and constant-volume heat capacities is significant, and in fact can be quite large if the fluid has a large coefficient of thermal expansion, as is the case for ethanol. For comparison, the difference between C_P and C_V for an ideal gas is $R = 8.314$ J/(mol K). ∎

ILLUSTRATION 6.4-8

Effect of Pressure on the Enthalpy of Liquid Water

The coefficient of thermal expansion of liquid water at 25°C was given in Illustration 6.2-6 to be

$$\alpha = \frac{1}{\hat{V}}\left(\frac{\partial \hat{V}}{\partial T}\right)_P = 2.56 \times 10^{-4}\text{ K}$$

Use this information and the fact that $\hat{V}(25°\text{C, saturated liquid at 3.169 kPa}) = 0.001003$ m³/kg and $\hat{H}(25°\text{C, saturated liquid at 3.169 kPa}) = 104.89$ kJ/kg to determine the enthalpy of liquid water at (a) 25°C and 1 bar and (b) 25°C and 10 bar.

SOLUTION

Using Eq. 6.4-33, on a mass basis we have

$$\hat{H}(25°\text{C, 1 bar}) = \hat{H}(25°\text{C, 3.169 kPa}) + \hat{V}[1 - T\alpha](100 - 3.169)\text{ kPa}$$

$$= 104.89\,\frac{\text{kJ}}{\text{kg}} + 1.003 \times 10^{-3}\,\frac{\text{m}^3}{\text{kg}}[1 - 298.15 \times 2.56 \times 10^{-4}] \times 96.83\text{ kPa}$$

$$= 104.89\,\frac{\text{kJ}}{\text{kg}} + 0.0897\,\frac{\text{Pa m}^3}{\text{kg}} \times 1\,\frac{\text{kJ}}{\text{kPa m}^3} = 104.98\,\frac{\text{kJ}}{\text{kg}}$$

(a)

and

$$\hat{H}(25°C, 10 \text{ bar}) = 104.89 + 1.003 \times 10^{-3}[1 - 298.15 \times 2.56 \times 10^{-4}](1000 - 3.169)$$
$$= 104.89 + 0.92 = 105.81 \frac{\text{kJ}}{\text{kg}} \tag{b}$$

COMMENT

We see from this that for modest pressure changes, the liquid enthalpy changes little with pressure. Therefore, little error is incurred if, at fixed temperature, one uses the saturated liquid enthalpy in the steam tables at higher pressures. However, this will lead to errors near the critical point, where the density of the liquid changes rapidly with small changes in both temperature and pressure. ∎

6.5 AN EXAMPLE INVOLVING THE CHANGE OF STATE OF A REAL GAS

Given low-density heat capacity data and volumetric equation-of-state information for a fluid, we can use the procedures developed in this chapter to calculate the change in the thermodynamic properties of the fluid accompanying any change in its state. Thus, at least in principle, we can use the mass, energy, and entropy balances to solve energy flow problems for all fluids. The starting point for solving such problems is exactly the same here as that used in Chapters 2, 3, and 4, in that the balance equations are written for a convenient choice of the thermodynamic system. For real fluids, tedious calculations may be necessary to eliminate the internal energy, enthalpy, and entropy, which appear in the balance equations in terms of the pressure, temperature, and specific volume. If the fluid under study is likely to be of continual interest, it is probably worthwhile to construct a complete thermodynamic properties chart, as in Figs. 3.3-1 through 3.3-3 and in Illustration 6.4-1, so that all problems can be solved rapidly, as was demonstrated in Chapters 2, 3, and 4. If, on the other hand, the fluid is of limited interest, it is more logical to use the heat capacity and volumetric equation-of-state data only to solve the problem at hand. This type of calculation is illustrated in the following example which uses the van der Waals equation of state merely as a prototype for the equations of state of real fluids. This example is also considered, and enlarged upon, in the following two sections. Of course, the use of the Benedict-Webb-Rubin and other, more complicated equations of state would lead to predictions of greater accuracy, but with a great deal more computational effort.

ILLUSTRATION 6.5-1

Comparing Solutions to a Problem Assuming the Gas Is Ideal, Using a Thermodynamic Properties Chart, and Assuming That the Gas Can Be Described by the van der Waals Equation of State

Nitrogen gas is being withdrawn from a 0.15-m^3 cylinder at the rate of 10 mol/min. The cylinder initially contains the gas at a pressure of 100 bar and 170 K. The cylinder is well insulated, and there is a negligible heat transfer between the cylinder walls and the gas. How many moles of gas will be in the cylinder at any time? What will the temperature and pressure of the gas in the cylinder be after 50 minutes?

   a. Assume that nitrogen is an ideal gas.
   b. Use the nitrogen properties chart of Fig. 3.3-3.
   c. Assume that nitrogen is a van der Waals fluid.

Data: For parts (a) and (c) use

$$C_p^* \ [\text{J/(mol K)}] = 27.2 + 4.2 \times 10^{-3} T \ (\text{K})$$

SOLUTION

The mass and energy balances for the contents of the cylinder are

$$\frac{dN}{dt} = \dot{N} \quad \text{or} \quad N(t) = N(t=0) + \dot{N}t \tag{a}$$

and

$$\frac{d(N\underline{U})}{dt} = \dot{N}\underline{H} \tag{b}$$

where $\dot{N} = -10$ mol/min.

Following Illustration 4.5-2, the result of writing an entropy balance for that portion of the gas that always remains in the cylinder is

$$\frac{dS}{dt} = 0 \quad \text{or} \quad \underline{S}(t=0) = \underline{S}(t) \tag{c}$$

Also, from $V = N\underline{V}$, we have

$$N(t=0) = \frac{V_{cyl}}{\underline{V}(t=0)} = \frac{0.15 \text{ m}^3}{\underline{V}(t=0)} \tag{d}$$

Equations a, b, c, and d apply to both ideal and nonideal gases.
Computation of $N(t)$

a. *Using the ideal gas equation of state*

$$\underline{V}(t=0) = \frac{RT(t=0)}{P(t=0)} = \frac{8.314 \times 10^{-5} \text{ bar m}^3/\text{mol K} \times 170 \text{ K}}{100 \text{ bar}}$$

$$= 1.4134 \times 10^{-4} \text{ m}^3/\text{mol}$$

so that

$$N(t=0) = 1061.3 \text{ mol}$$

and

$$N(t) = 1061.3 - 10t \text{ mol}$$

b. *Using Figure 3.3-3*

$$\hat{V}(T = 170 \text{ K}, P = 100 \text{ bar}) = \hat{V}(T = 170 \text{ K}, P = 10 \text{ MPa}) \approx 0.0035 \text{ m}^3/\text{kg}$$

$$\underline{V}(T = 170 \text{ K}, P = 10 \text{ MPa}) = 28.014 \frac{\text{g}}{\text{mol}} \times 0.0035 \frac{\text{m}^3}{\text{kg}} \times \frac{1 \text{ kg}}{1000 \text{ g}}$$

$$= 9.80 \times 10^{-5} \text{ m}^3/\text{mol}$$

so that

$$N(t=0) = \frac{0.15 \text{ m}^3}{9.80 \times 10^{-5} \text{ m}^3/\text{mol}} = 1529.8 \text{ mol}$$

and

$$N(t) = 1529.8 - 10t \text{ mol}$$

c. *Using the van der Waals equation of state*
The van der Waals equation is

$$P = \frac{RT}{\underline{V} - b} - \frac{a}{\underline{V}^2}$$

With the constants given in Table 6.4-1, we have

$$100 \text{ bar} = 1 \times 10^7 \text{ Pa} = \frac{8.314 \times 170}{\underline{V}(t=0) - 3.864 \times 10^{-5}} - \frac{0.1368}{[\underline{V}(t=0)]^2}$$

Solving this cubic equation, we obtain

$$\underline{V}(t=0) = 9.435 \times 10^{-5} \text{ m}^3/\text{mol}$$

so that

$$N(t=0) = 1589.9 \text{ mol}$$

and

$$N(t) = 1589.9 - 10t \text{ mol}$$

Note also that since the volume of the cylinder is constant (0.15 m³), and as $N(t)$ is known, we can compute the molar volume of the gas at any later time from

$$\underline{V}(t) = \frac{0.15 \text{ m}^3}{N(t)} \tag{e}$$

In particular, we can use this equation to compute $\underline{V}(t = 50 \text{ min})$. These results, together with other information gathered so far, and some results from the following sections, are listed in Table 6.5-1.

Since we know the specific volume after 50 minutes, we need to determine only one further state property to have the final state of the system completely specified. In principle, either the energy or entropy balance could be used to find this property. The entropy balance is more convenient to use, especially for the nonideal gas calculations. Thus, all the calculations here are based on the fact (see Eq. c) that

$$\underline{S}(t=0) = \underline{S}(t = 50 \text{ min})$$

Computation of $T(t = 50\text{min})$ *and* $P(t = 50 \text{ min})$

a. *Using the ideal gas equation of state*
From Eq. 4.4-1,

$$d\underline{S} = C_V \frac{dT}{T} + R \frac{d\underline{V}}{\underline{V}}$$

Now for the ideal gas,

Table 6.5-1

Equation of State	$N(t=0)$	$N(t)$	$\underline{V}(t = 50 \text{ min})$ (m³/mol)
Ideal gas	1061.3	$1061.3 - 10t$	2.672×10^{-4}
Fig 3.3-3	1529.8	$1529.8 - 10t$	1.457×10^{-4}
Van der Waals	1589.9	$1589.9 - 10t$	1.376×10^{-4}
Corresponding states (Illustration 6.6-2)	1432.7	$1432.7 - 10t$	1.608×10^{-4}
Peng-Robinson (Illustration 6.7-1)	1567.9	$1567.9 - 10t$	1.405×10^{-4}

$$C_V = C_V^* = C_P^* - R = (27.2 - 8.314) + 4.2 \times 10^{-3} T$$
$$= 18.9 + 4.2 \times 10^{-3} T \; \text{J/(mol K)}$$

Also, since $\underline{S} = $ constant, or $\Delta \underline{S} = 0$, we have

$$\int_{T=170}^{T} \frac{(18.9 + 4.2 \times 10^{-3} T)}{T} dT + 8.314 \int_{\underline{V}=1.413 \times 10^{-4}}^{\underline{V}=2.672 \times 10^{-4}} \frac{d\underline{V}}{\underline{V}} = 0$$

or

$$18.9 \ln \frac{T}{170} + 4.2 \times 10^{-3} (T - 170) + 8.314 \ln \frac{2.672 \times 10^{-4}}{1.413 \times 10^{-4}} = 0$$

The solution to this equation is 130.3 K, so that

$$P(t = 50 \text{ min}) = \frac{RT(t = 50 \text{ min})}{\underline{V}(t = 50 \text{ min})} = 40.5 \text{ bar}$$

b. *Using Fig. 3.3-3*

$$\underline{V}(t = 50 \text{ min}) = 1.457 \times 10^{-4} \frac{\text{m}^3}{\text{mol}} \times \frac{1 \text{ mol}}{28.014 \text{ g}} \times 1000 \frac{\text{g}}{\text{kg}}$$

$$= 5.20 \times 10^{-3} \frac{\text{m}^3}{\text{kg}}$$

To find the final temperature and pressure using Fig. 3.3-3, we locate the initial point ($T = 170$ K, $P = 10$ MPa) and follow a line of constant entropy (dashed line) through this point to the intersection with a line of constant specific volume $\hat{V} = 5.2 \times 10^{-3} \text{m}^3/\text{kg}$. This intersection gives the pressure and temperature of the $t = 50$ min state. We find

$$T(t = 50 \text{ min}) \approx 133 \text{ K}$$
$$P(t = 50 \text{ min}) \approx 3.9 \text{ MPa} = 39 \text{ bar}$$

This construction is shown in the following diagram, which is a portion of Fig. 3.3-3.

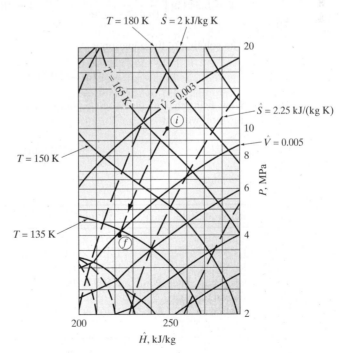

c. *Using the van der Waals equation of state*
Here we start with Eq. 6.2-19,

$$d\underline{S} = \frac{C_V}{T}dT + \left(\frac{\partial P}{\partial T}\right)_V d\underline{V}$$

and note that for the van der Waals gas

$$\left(\frac{\partial P}{\partial T}\right)_V = \frac{R}{\underline{V} - b}$$

The integration path to be followed is

i. $T(t=0), \underline{V}(t=0) \rightarrow T(t=0), \underline{V} = \infty$
ii. $T(t=0), \underline{V} = \infty \rightarrow T(t=50\text{ min}), \underline{V} = \infty$ ($C_V = C_V^*$ for this step)
iii. $T(t=50\text{ min}), \underline{V} = \infty \rightarrow T(t=50\text{ min}), \underline{V}(t=50\text{ min})$

so that

$$\underline{S}(t=50\text{ min}) - \underline{S}(t=0) = 0 = R\int_{\underline{V}(t=0)}^{\underline{V}=\infty} \frac{d\underline{V}}{\underline{V}-b} + \int_{T=170\text{K}}^{T(t=50)} \frac{C_V^*}{T}dT + R\int_{\underline{V}=\infty}^{\underline{V}(t=50)} \frac{d\underline{V}}{\underline{V}-b}$$

or

$$0 = R\int_{\underline{V}(t=0)}^{\underline{V}(t=50)} \frac{d\underline{V}}{\underline{V}-b} + \int_{T=170\text{K}}^{T(t=50)} \frac{18.9 + 4.2 \times 10^{-3}T}{T}dT$$

Thus, using $b = 3.864 \times 10^{-5}$ m³/mol, we have

$$0 = 8.314\ln\left[\frac{(13.77 - 3.864) \times 10^{-5}}{(9.437 - 3.864) \times 10^{-5}}\right] + 18.9\ln\frac{T}{170} + 4.2 \times 10^{-3}(T - 170)$$

The solution to the equation is

$$T(t=50\text{ min}) = 133.1\text{ K}$$

Now, using the van der Waals equation of state with $T = 133.1$ K and $\underline{V} = 13.76 \times 10^{-5}$ m³/mol, we find

$$P(t=50\text{ min}) = 39.5\text{ bar}$$

Summary

Equation of State	$\underline{V}(t=50\text{ min})$ (m³/mol)	$T(t=50\text{ min})$ (K)	$P(t=50\text{ min})$ (bar)
Ideal gas	2.672×10^{-4}	129.6	40.3
Fig. 3.3-3	1.457×10^{-4}	133	39
Van der Waals	1.376×10^{-4}	133.1	39.5
Corresponding states (Illustration 6.6-2)	1.608×10^{-4}	136	41
Peng-Robinson (Illustration 6.7-1)	1.405×10^{-4}	134.7	40.6

COMMENTS

It is clear from the results of this problem that one cannot assume that a high-pressure gas is ideal and expect to make useful predictions. Of the techniques that have been considered for the thermodynamic calculations involving real fluids, the use of a previously prepared thermodynamic properties chart for the fluid is the most rapid way to proceed. The alternatives, the use of a simple volumetric equation of state for the fluid or of corresponding-states correlations (discussed in the following section), are always tedious, and their accuracy is hard to assess. However, if you do not have a thermodynamic properties chart available for the working fluid, you have no choice but to use the more tedious methods.   ◼

6.6   THE PRINCIPLE OF CORRESPONDING STATES

The analysis presented in Sec. 6.4 makes it possible to solve thermodynamic problems for real substances or to construct tables and charts of their thermodynamic properties given only the low-density (ideal gas) heat capacity and analytic or tabular information on the volumetric equation of state. Unfortunately, these can be tedious tasks, and the necessary volumetric equation-of-state information is not available for all fluids. Thus, we consider here the principle of corresponding states, which allows one to predict some thermodynamic properties of fluids from generalized property correlations based on available experimental data for similar fluids.

Before we introduce the concept of generalized fluid properties correlations, it is useful to consider which properties we can hope to get from this sort of correlation scheme, and which we cannot. The two types of information needed in thermodynamic calculations are the volumetric equation of state and the ideal gas heat capacity. The ideal gas heat capacity is determined solely by the intramolecular structure (e.g., bond lengths, vibration frequencies, configuration of constituent atoms) of only a single molecule, as there is no intermolecular interaction energy in an ideal gas. As the structure of each molecular species is sufficiently different from that of others, and since experimental low-density heat capacity data are frequently available, we will not attempt to develop correlations for heat capacity data or for the ideal gas part of the enthalpy, internal energy, or entropy, which can be computed directly from C_P^* and C_V^*.

However, the volumetric equation of a state of a fluid is determined solely by the interactions of each molecule with its neighbors. An interesting fact that has emerged from the study of molecular behavior (i.e., statistical mechanics) is that as far as molecular interactions are concerned, molecules can be grouped into classes, such as spherical molecules, nonspherical molecules, molecules with permanent dipole moments, and so forth, and that within any one class, molecular interactions are similar. It has also been found that the volumetric equations of state obeyed by all members of a class are similar in the sense that if a given equation of state (e.g., Peng-Robinson or Benedict-Webb-Rubin) fits the volumetric data for one member of a class, the same equation of state, with different parameter values, is likely to fit the data for other molecular species in the same class.

The fact that a number of different molecular species may be represented by a volumetric equation of state of the same form suggests that it might be possible to construct generalized correlations, that is, correlations applicable to many different molecular species, for both the volumetric equation of state and the density- (or pressure-) dependent contribution to the enthalpy, entropy, or other thermodynamic properties. Historically, the first generalized correlation arose from the study of the van der Waals equation of state. Although *present correlations are largely based on experimental data,* it is useful to review the van der Waals generalized correlation scheme since, al-

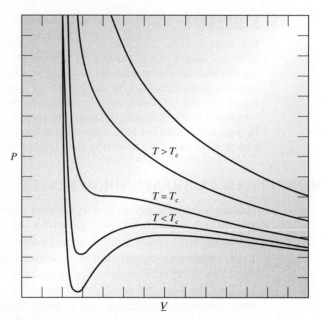

Figure 6.6-1 The pressure-volume behavior of the van der Waals equation of state.

though not very accurate, it does indicate both the structure and correlative parameters used in modern fluid property correlations.

A plot of P versus $\underline{V}$ for various temperatures for a van der Waals fluid is given in Fig. 6.6-1. An interesting characteristic of this figure is that isotherms below the one labeled T_c exhibit a local minimum followed by a local maximum with increasing $\underline{V}$ over part of the specific volume-pressure range. As we will see in Chapter 7, such behavior is associated with a vapor-liquid transition. For $T > T_c$ this structure in the P-$\underline{V}$ plot has vanished, whereas at $T = T_c$ this minimum and maximum have converged to a single point. At this point the first and second derivatives of P with respect to $\underline{V}$ at constant T are zero. For the present discussion, this point is taken to be the **critical point** of the fluid, that is, the point of highest temperature at which a liquid can exist (justification for this identification is given in Chapter 7). A brief list of measured critical temperatures, critical pressures P_c, and critical volumes $\underline{V}_c$ (i.e., the pressure and volume of the highest-temperature liquid) for various fluids is given in Table 6.6-1.

To identify the critical point for the van der Waals fluid analytically, we make use of the following mathematical requirements for the occurrence of an inflection point on an isotherm in the P-$\underline{V}$ plane:

Critical-point conditions

$$\left(\frac{\partial P}{\partial \underline{V}}\right)_{T_c} = 0 \quad \text{and} \quad \left(\frac{\partial^2 P}{\partial \underline{V}^2}\right)_{T_c} = 0 \quad \text{at } P_c \text{ and } V_c \qquad \textbf{(6.6-1)}$$

(Also, the first nonzero derivative should be odd and negative in value, but this is not used here.) Using Eqs. 6.6-1 together with Eq. 6.2-38, we find that at the critical point

$$P_c = \frac{RT_c}{\underline{V}_c - b} - \frac{a}{\underline{V}_c^2} \qquad \textbf{(6.6-2a)}$$

$$\left(\frac{\partial P}{\partial \underline{V}}\right)_T \Bigg|_{\substack{\text{evaluated at} \\ T_c, P_c, \underline{V}_c}} = 0 = -\frac{RT_c}{(\underline{V}_c - b)^2} + \frac{2a}{\underline{V}_c^3} \tag{6.6-2b}$$

and

$$\left(\frac{\partial^2 P}{\partial \underline{V}^2}\right)_T \Bigg|_{\substack{\text{evaluated at} \\ T_c, P_c, \underline{V}_c}} = 0 = \frac{2RT_c}{(\underline{V}_c - b)^3} - \frac{6a}{\underline{V}_c^4} \tag{6.6-2c}$$

Thus, three equations interrelate the two unknowns a and b. Using Eqs. 6.6-2b and c to ensure that the critical-point conditions of Eqs. 6.6-1 are satisfied, we obtain (Problem 6.5)

$$a = \frac{9\underline{V}_c RT_c}{8} \tag{6.6-3a}$$

and

$$b = \frac{\underline{V}_c}{3}$$

Using these results in Eq. 6.6-2a yields

$$P_c = \frac{a}{27b^2} \tag{6.6-3b}$$

By direct substitution we find that the compressibility at the critical point is

For the van der Waals equation only

$$\boxed{Z_c = \frac{P_c \underline{V}_c}{RT_c} = \frac{3}{8} = 0.375} \tag{6.6-3c}$$

This last equation can be used to obtain expressions for the van der Waals parameters in terms of (1) the critical temperature and pressure,

For the van der Waals equation only

$$a = \frac{27R^2 T_c^2}{64 P_c} \quad \text{and} \quad b = \frac{RT_c}{8P_c} \tag{6.6-4a}$$

or (2) the critical pressure and volume:

$$a = 3P_c \underline{V}_c^2 \quad \text{and} \quad b = \frac{\underline{V}_c}{3} \tag{6.6-4b}$$

Substituting either of these sets of parameters into Eq. 6.2-38 yields

$$\left[\frac{P}{P_c} + 3\left(\frac{\underline{V}_c}{\underline{V}}\right)^2\right]\left[3\left(\frac{\underline{V}}{\underline{V}_c}\right) - 1\right] = 8\frac{T}{T_c}$$

Now defining a dimensionless or reduced temperature T_r, reduced pressure P_r, and reduced volume V_r by

$$T_r = \frac{T}{T_c} \qquad P_r = \frac{P}{P_c} \quad \text{and} \quad V_r = \frac{\underline{V}}{\underline{V}_c} \tag{6.6-5}$$

we obtain

$$\left[P_r + \frac{3}{V_r^2}\right][3V_r - 1] = 8T_r \tag{6.6-6}$$

The form of Eq. 6.6-6 is very interesting since it suggests that for all fluids that obey the van der Waals equation, V_r is the same function of T_r and P_r. That is, at given values of the **reduced temperature** ($T_r = T/T_c$) and **reduced pressure** ($P_r = P/P_c$), all van der Waals fluids will have the same numerical value of **reduced volume** ($V_r = \underline{V}/\underline{V}_c$). This does not mean that at the same value of T and P all van der Waals fluids have the same value of $\underline{V}$; this is *certainly not* the case, as can be seen from the illustration that follows. Two fluids that have the same values of reduced temperature and pressure and therefore the same reduced volume, are said to be in **corresponding states**.

ILLUSTRATION 6.6-1
Simple Example of the Use of Corresponding States

Assume that oxygen ($T_c = 154.6$ K, $P_c = 5.046 \times 10^6$ Pa, and $\underline{V}_c = 7.32 \times 10^{-5}$ m^3/mol) and water ($T_c = 647.3$ K, $P_c = 2.205 \times 10^7$ Pa, and $\underline{V}_c = 5.6 \times 10^{-5}$ m^3/mol) can be considered van der Waals fluids.

   a. Find the value of the reduced volume both fluids would have at $T_r = 3/2$ and $P_r = 3$.
   b. Find the temperature, pressure, and volume of each gas at $T_r = 3/2$ and $P_r = 3$.
   c. If O_2 and H_2O are both at a temperature of 200°C and a pressure of 2.5×10^6 Pa, find their specific volumes.

SOLUTION

   a. Using $T_r = 3/2$ and $P_r = 3$ in Eq. 6.6-6, one finds that $V_r = 1$ for both fluids.
   b. Since $V_r = 1$ at these conditions [see part (a)], the specific volume of each fluid is equal to the critical volume of that fluid. So

$$\underline{V}_{O_2} = 7.32 \times 10^{-5} \text{ m}^3/\text{mol} \qquad \underline{V}_{H_2O} = 5.6 \times 10^{-5} \text{ m}^3/\text{mol}$$
at
$$P_{O_2} = 3 \times P_{c,O_2} = 1.514 \times 10^7 \text{ Pa} \qquad P_{H_2O} = 3 \times P_{c,H_2O} = 6.615 \times 10^7 \text{ Pa}$$
and
$$T_{O_2} = 1.5 \times T_{c,O_2} = 232.2 \text{ K} \qquad T_{H_2O} = 1.5 \times T_{c,H_2O} = 971 \text{ K}$$

   c. For oxygen we have

$$P_r = \frac{2.5 \times 10^6}{5.046 \times 10^6} = 0.495$$

and $T_r = 473.2/154.6 = 3.061$. Therefore, $V_r = 16.5$ and $\underline{V} = 1.208 \times 10^{-3}$ m^3/mol. Similarly for water,

$$P_r = \frac{2.5 \times 10^6}{2.205 \times 10^7} = 0.113$$

$T_r = 473.2/647.3 = 0.731$, so that $V_r = 15.95$ and $\underline{V} = 8.932 \times 10^{-4}$ m^3/mol. ∎

As has been already indicated, the accuracy of the van der Waals equation is not very good. This may be verified by comparing the results of the previous illustration with experimental data, or by comparing the compressibility factor $Z = P\underline{V}/RT$ for the van der Waals equation of state with experimental data for a variety of gases. In particular, at the critical point (see Eq. 6.6-3c)

$$Z_c\big|_{\text{van der Waals}} = P_c\underline{V}_c/RT_c = 0.375$$

while for most fluids the critical compressibility Z_c is in the range 0.23 to 0.31 (Table 6.6-1), so that the van der Waals equation fails to predict accurate critical-point behav-

Table 6.6-1 The Critical and Other Constants for Selected Fluids

Substance	Symbol	Molecular Weight (g mol^{-1})	T_c(K)	P_c(MPa)	V_c(m^3/kmol)	Z_c	ω	T_{boil}(K)
Acetylene	C_2H_2	26.038	308.3	6.140	0.113	0.271	0.184	189.2
Ammonia	NH_3	17.031	405.6	11.28	0.0724	0.242	0.250	239.7
Argon	Ar	39.948	150.8	4.874	0.0749	0.291	−0.004	87.3
Benzene	C_6H_6	78.114	562.1	4.894	0.259	0.271	0.212	353.3
n-Butane	C_4H_{10}	58.124	425.2	3.800	0.255	0.274	0.193	272.7
Isobutane	C_4H_{10}	58.124	408.1	3.648	0.263	0.283	0.176	261.3
1-Butene	C_4H_8	56.108	419.6	4.023	0.240	0.277	0.187	266.9
Carbon dioxide	CO_2	44.010	304.2	7.376	0.0940	0.274	0.225	194.7
Carbon monoxide	CO	28.010	132.9	3.496	0.0931	0.295	0.049	81.7
Carbon tetrachloride	CCl_4	153.823	556.4	4.560	0.276	0.272	0.194	349.7
n-Decane	$C_{10}H_{22}$	142.286	617.6	2.108	0.603	0.247	0.490	447.3
n-Dodecane	$C_{12}H_{26}$	170.340	658.3	1.824	0.713	0.24	0.562	489.5
Ethane	C_2H_6	30.070	305.4	4.884	0.148	0.285	0.098	184.5
Ethyl ether	$C_4H_{10}O$	74.123	466.7	3.638	0.280	0.262	0.281	307.7
Ethylene	C_2H_4	28.054	282.4	5.036	0.129	0.276	0.085	169.4
Helium	He	4.003	5.19	0.227	0.0573	0.301	−0.387	4.21
n-Heptane	C_7H_{16}	100.205	540.2	2.736	0.304	0.263	0.351	371.6
n-Hexane	C_6H_{14}	86.178	507.4	2.969	0.370	0.260	0.296	341.9
Hydrogen	H_2	2.016	33.2	1.297	0.065	0.305	−0.22	20.4
Hydrogen fluoride	HF	20.006	461.0	6.488	0.069	0.12	0.372	292.7
Hydrogen sulfide	H_2S	34.080	373.2	8.942	0.0985	0.284	0.100	212.8
Methane	CH_4	16.043	190.6	4.600	0.099	0.288	0.008	111.7
Naphthalene	$C_{10}H_8$	128.174	748.4	4.05	0.410	0.267	0.302	491.1
Neon	Ne	20.183	44.4	2.756	0.0417	0.311	0	27.0
Nitric oxide	NO	30.006	180.0	6.485	0.058	0.250	0.607	121.4
Nitrogen	N_2	28.013	126.2	3.394	0.0895	0.290	0.040	77.4
n-Octane	C_8H_{18}	114.232	568.8	2.482	0.492	0.259	0.394	398.8
Oxygen	O_2	31.999	154.6	5.046	0.0732	0.288	0.021	90.2
n-Pentane	C_5H_{12}	72.151	469.6	3.374	0.304	0.262	0.251	309.2
Isopentane	C_5H_{12}	72.151	460.4	3.384	0.306	0.271	0.227	301.0
Propane	C_3H_8	44.097	369.8	4.246	0.203	0.281	0.152	231.1
Propylene	C_3H_6	42.081	365.0	4.620	0.181	0.275	0.148	225.4
Refrigerant R12	CCl_2F_2	120.914	385.0	4.124	0.217	0.280	0.176	243.4
Refrigerant HFC-134a	CH_2FCF_3	102.03	374.23	4.060	0.198	0.258	0.332	247.1
Sulfur dioxide	SO_2	64.063	430.8	7.883	0.122	0.268	0.251	263
Toluene	C_7H_8	92.141	591.7	4.113	0.316	0.264	0.257	383.8
Water	H_2O	18.015	647.3	22.048	0.056	0.229	0.344	373.2
Xenon	Xe	131.300	289.7	5.836	0.118	0.286	0.002	165.0

Source: Adapted from R. C. Reid, J. M. Prausnitz, and B. E. Poling, *The Properties of Gases and Liquids,* 4th ed., McGraw-Hill, New York, 1986, Appendix A and other sources.

ior. (It is, however, a great improvement over the ideal gas equation of state, which predicts that $Z = 1$ for all conditions.)

The fact that the critical compressibility of the van der Waals fluid is not equal to that for most real fluids also means that different values for the van der Waals parameters are obtained for any one fluid, depending on whether Eqs. 6.6-3a, Eqs. 6.6-4a, or Eqs. 6.6-4b are used to relate these parameters to the critical properties. In practice, the critical volume of a fluid is known with less experimental accuracy than either the critical temperature or critical pressure, so that Eqs. 6.6-4a and critical-point data are most frequently used to obtain the van der Waals parameters. Indeed, the entries in Tables 6.4-1 and 6.6-1 are related in this way. Thus, if the parameters in Table 6.4-1 are used in the van der Waals equation, the critical temperature and pressure will be correctly predicted, but the critical volume will be too high by the factor

$$\frac{Z_c|_{\text{van der Waals}}}{Z_c} = \frac{3}{8Z_c}$$

where Z_c is the real fluid critical compressibility.

Although the van der Waals equation is not accurate, the idea of a correspondence of states to which it historically led is both appealing and, as we will see, useful. Attempts at using the corresponding-states concept over the last 40 years have been directed toward representing the compressibility factor Z as a function of the reduced pressure and temperature, that is,

$$\boxed{Z = P\underline{V}/RT = Z(P_r, T_r)} \tag{6.6-7}$$

where the functional relationship between T_r, P_r, and Z is *determined from experimental data*, or from a very accurate equation of state. That such a procedure has some merit is evident from Fig. 6.6-2, where the compressibility data for different fluids have been made to almost superimpose by plotting each as a function of its reduced temperature and pressure.

A close study of Fig. 6.6-2 indicates that there are systematic deviations from the simple corresponding-states relation of Eq. 6.6-7. In particular, the compressibility factors for the inorganic fluids are almost always below those for the hydrocarbons. Furthermore, if Eq. 6.6-7 were universally valid, all fluids would have the same value of the critical compressibility $Z_c = Z(P_r = 1, T_r = 1)$; however, from Table 6.6-1, it is clear that Z_c for most real fluids ranges from 0.23 to 0.31. These failings of Eq. 6.6-7 have led to the development of more complicated corresponding-states principles. The simplest generalization is the suggestion that there should not be a single $Z = Z(P_r, T_r)$ relationship for all fluids, but rather a family of relationships for different values of Z_c. Thus, to find the value of the compressibility factor for given values T_r and P_r, it would be necessary to use a chart of $Z = Z(P_r, T_r)$ prepared from experimental data for fluids with approximately the same value of Z_c. This is equivalent to saying that Eq. 6.6-7 is to be replaced by

$$Z = Z(T_r, P_r, Z_c)$$

Alternatively, fluid characteristics other than Z_c can be used as the additional parameter in the generalization of the simple corresponding-states principle. In fact, since for many substances the critical density, and hence Z_c, is known with limited accuracy, if at all, there is some advantage in avoiding the use of Z_c.

Pitzer[13] has suggested that for nonspherical molecules the **acentric factor** ω be used as the third correlative parameter, where ω is defined to be

Acentric factor

$$\boxed{\omega = -1.0 - \log_{10}[P^{\text{vap}}(T_r = 0.7)/P_c]}$$

---

[13]See Appendix 1 of K. S. Pitzer, *Thermodynamics,* 3rd ed., McGraw-Hill, New York, 1995.

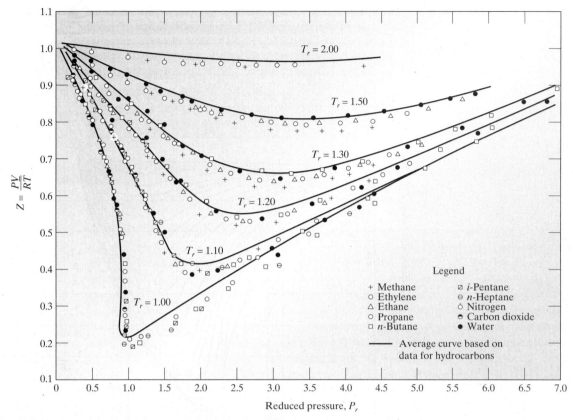

Figure 6.6-2 Compressibility factors for different fluids as a function of the reduced temperature and pressure. [Reprinted with permission from G.-J. Su, *Ind. Eng. Chem.* **38.** 803 (1946). Copyright American Chemical Society.]

Here $P^{vap}(T_r = 0.7)$ is the vapor pressure of the fluid at $T_r = 0.7$, a temperature near the normal boiling point. In this case the corresponding-states relation would be of the form

$$Z = Z(T_r, P_r, \omega)$$

Three-parameter corresponding states

Even these extensions of the corresponding-states concept, which are meant to account for molecular structure, cannot be expected to be applicable to fluids with permanent dipoles and quadrupoles. Since molecules with strong permanent dipoles interact differently than molecules without dipoles, or than molecules with weak dipoles, one would expect the volumetric equation of state for polar fluids to be a function of the dipole moment. In principle, the corresponding-states concept could be further generalized to include this new parameter, but we will not do so here. Instead, we refer you to the book by Reid, Prausnitz, and Poling for a detailed discussion of the corresponding-states correlations commonly used by engineers.[14]

The last several paragraphs have emphasized the shortcomings of a single corresponding-states principle when dealing with fluids of different molecular classes. However, it is useful to point out that a corresponding-states correlation can be an accurate rep-

[14]J. M. Prausnitz, B. E. Poling, and J. P. O'Connell, *Properties of Gases and Liquids,* 5th ed., McGraw-Hill, New York, 2001.

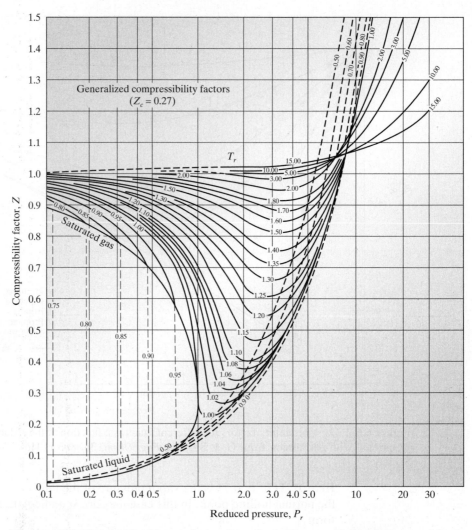

Figure 6.6-3 (Reprinted with permission from O. A. Hougen, K. M. Watson, and R. A. Ragatz, *Chemical Process Principles Charts*, 2nd ed., John Wiley & Sons, New York, 1960. This figure appears as an Adobe PDF file on the CD-ROM accompanying this book, and may be enlarged and printed for easier reading and for use in solving problems.)

resentation of the equation-of-state behavior within any one class of similar molecules. Indeed, the volumetric equation-of-state behavior of many simple fluids and most hydrocarbons is approximately represented by the plot in Fig. 6.6-3, which was developed from experimental data for molecules with $Z_c = 0.27$.

The existence of an accurate corresponding-states relationship of the type $Z = Z(T_r, P_r)$ (or perhaps a whole family of such relationships for different values of Z_c or ω) allows one to also develop corresponding-states correlations for that contribution to the thermodynamic properties of the fluid that results from molecular interactions, or nonideal behavior, that is, the departure functions of Sec. 6.4. For example, starting with Eq. 6.4-22, we have

$$\underline{H}(T, P) - \underline{H}^{\text{IG}}(T, P) = \int_{P=0, T}^{P, T} \left[\underline{V} - T \left(\frac{\partial \underline{V}}{\partial T} \right)_P \right] dP$$

and using the corresponding-states relation, Eq. 6.6-7, yields

$$\underline{V} = \frac{RT}{P} Z(T_r, P_r)$$

and

$$\underline{V} - T \left(\frac{\partial \underline{V}}{\partial T} \right)_P = -\frac{RT^2}{P} \left(\frac{\partial Z}{\partial T} \right)_P$$

so that

$$\underline{H}(T, P) - \underline{H}^{\text{IG}}(T, P)$$

$$= -\int_{P=0,T}^{P,T} \frac{RT^2}{P} \left(\frac{\partial Z}{\partial T} \right)_P dP = -RT_c \int_{P_r=0,T_r}^{P_r,T_r} \frac{T_r^2}{P_r} \left(\frac{\partial Z}{\partial T_r} \right)_{P_r} dP_r$$

or

$$\frac{\underline{H}(T, P) - \underline{H}^{\text{IG}}(T, P)}{T_c} = -RT_r^2 \int_{P_r=0,T_r}^{P_r,T_r} \frac{1}{P_r} \left(\frac{\partial Z}{\partial T_r} \right)_{P_r} dP_r \qquad \textbf{(6.6-8)}$$

The important thing to notice about this equation is that nothing in the integral depends on the properties of a specific fluid, so that when the integral is evaluated using the corresponding-states equation of state, the result will be applicable to *all* corresponding-states fluids.

Figure 6.6-4 contains in detailed graphical form the corresponding-states prediction for the enthalpy departure from ideal gas behavior computed from Fig. 6.6-3 (for fluids with $Z_c = 0.27$) and Eq. 6.6-8.[15]

The enthalpy change of a real fluid in going from $(T_0, P = 0)$ to (T, P) can then be computed using Fig. 6.6-4 as indicated here:

$$H(T, P) - \underline{H}(T_0, P = 0) = \underline{H}^{\text{IG}}(T, P) - \underline{H}(T_0, P = 0) + \{\underline{H}(T, P) - \underline{H}^{\text{IG}}(T, P)\}$$

$$= \int_{T_0}^{T} C_{\text{P}}^* dT + T_c \left[\frac{\underline{H}(T, P) - \underline{H}^{\text{IG}}(T, P)}{T_c} \right]_{\text{from Fig.6.6-4}}$$

$$\textbf{(6.6-9)}$$

Similarly, the enthalpy change in going from any state (T_1, P_1) to state (T_2, P_2) can be computed from the repeated application of Eq. 6.6-9, which yields

Enthalpy change from corresponding states

$$\underline{H}(T_2, P_2) - \underline{H}(T_1, P_1) = \int_{T_1}^{T_2} C_{\text{P}}^* dT + T_c \left\{ \left[\frac{\underline{H}(T, P) - H^{\text{IG}}(T, P)}{T_c} \right]_{T_{r_2}, P_{r_2}} \right.$$

$$\left. - \left[\frac{\underline{H}(T, P) - \underline{H}^{\text{IG}}(T, P)}{T_c} \right]_{T_{r_1}, P_{r_1}} \right\}_{\text{from Fig. 6.6-4}}$$

$$\textbf{(6.6-10)}$$

[15] Note that Eqs. 6.6-8, 6.6-9, and 6.6-10 contain the term $(\underline{H} - \underline{H}^{\text{IG}})/T_c$, whereas Fig. 6.6-4 gives $(\underline{H}^{\text{IG}} - \underline{H})/T_c$.

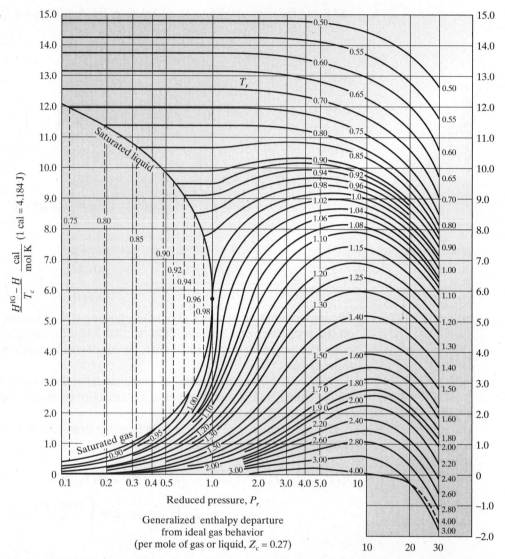

Figure 6.6-4 (Reprinted with permission from O. A. Hougen, K. M. Watson, and R. A. Ragatz, *Chemical Process Principles Charts,* 2nd ed., John Wiley & Sons, New York, 1960. This figure appears as an Adobe PDF file on the CD-ROM accompanying this book, and may be enlarged and printed for easier reading and for use in solving problems.)

The form of Eqs. 6.6-9 and 6.6-10 makes good physical sense in that each consists of two terms with well-defined meanings. The first term depends only on the ideal gas heat capacity, which is a function of the molecular structure and is specific to the molecular species involved. The second term, on the other hand, represents the nonideal behavior of the fluid due to intermolecular interactions that do not exist in the ideal gas, but whose contribution can be estimated from the generalized correlation.

In a manner equivalent to that just used, it is also possible to show (see Problem 6.6) that

$$\underline{S}(T, P) - \underline{S}^{IG}(T, P) = -R \int_{P_r=0,T_r}^{P_r,T_r} \left[\frac{Z - 1}{P_r} + \frac{T_r}{P_r} \left(\frac{\partial Z}{\partial T_r} \right)_{P_r} \right] dP_r \qquad \textbf{(6.6-11)}$$

This equation and Fig. 6.6-3 are the bases for the entropy departure plot given in Fig. 6.6-5.[16] The change in entropy between any two states (T_1, P_1) and (T_2, P_2) can then be computed from

Entropy change from corresponding states

$$\underline{S}(T_2, P_2) - \underline{S}(T_1, P_1) = \int_{T_1}^{T_2} \frac{C_P^*}{T}\, dT - R \int_{P_1}^{P_2} \frac{dP}{P}$$

$$+ \left\{ [\underline{S} - \underline{S}^{IG}]_{T_{r_2}, P_{r_2}} - [\underline{S} - \underline{S}^{IG}]_{T_{r_1}, P_{r_1}} \right\}_{\text{from Fig. 6.6-5}}$$

(6.6-12)

Similarly, corresponding-states plots could be developed for the other thermodynamic properties $\underline{U}$, $\underline{A}$, and $\underline{G}$, though these properties are usually computed from the relations

$$\underline{U} = \underline{H} - P\underline{V}$$
$$\underline{A} = \underline{U} - T\underline{S}$$
$$\underline{G} = \underline{H} - T\underline{S}$$

(6.6-13)

and the corresponding-states figures already given.

ILLUSTRATION 6.6-2

Using Corresponding States to Solve a Real Gas Problem

Rework Illustration 6.5-1, assuming that nitrogen obeys the generalized correlations of Figs. 6.6-3, 6.6-4, and 6.6-5.

SOLUTION

From Table 6.6-1 we have for nitrogen $T_c = 126.2$ K and $P_c = 33.94$ bar, and from the initial conditions of the problem,

$$T_r = \frac{170}{126.2} = 1.347 \quad \text{and} \quad P_r = \frac{100}{33.94} = 2.946$$

From Fig. 6.6-3, $Z = 0.741$, so

$$\underline{V}(t = 0) = Z \frac{RT(t = 0)}{P(t = 0)} = Z\underline{V}^{IG}(t = 0) = 1.047 \times 10^{-4} \text{ m}^3/\text{mol}$$

Therefore, following Illustration 6.5-1,

$$N(t = 0) = 1432.7 \text{ mol}$$
$$N(t) = 1432.7 - 10t \text{ mol}$$

and

$$\underline{V}(t = 50 \text{ min}) = \frac{0.15 \text{ m}^3}{(1432.7 - 500) \text{ mol}} = 1.6082 \times 10^{-4} \text{ m}^3/\text{mol}$$

To compute the temperature and pressure at the end of the 50 minutes, we use

$$\underline{S}(t = 0) = \underline{S}(t = 50 \text{ min})$$

---

[16]Note that Eqs. 6.6-11 and 6.6-12 contain the term $\underline{S} - \underline{S}^{IG}$, whereas Fig. 6.6-5 gives $\underline{S}^{IG} - \underline{S}$.

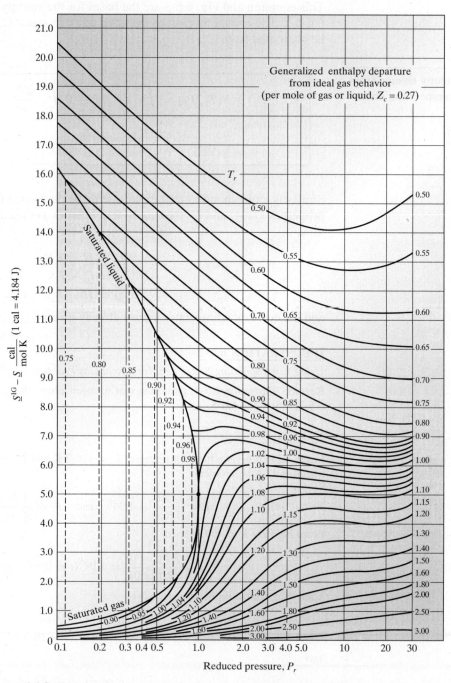

Figure 6.6-5 (Reprinted with permission from O. A. Hougen, K. M. Watson, and R. A. Ragatz, *Chemical Process Principles Charts,* 2nd ed., John Wiley & Sons, New York, 1960. This figure appears as an Adobe PDF file on the CD-ROM accompanying this book, and may be enlarged and printed for easier reading and for use in solving problems.)

and recognize that

$$\underline{S}(t = 0) = \underline{S}^{IG}(T, \underline{V})_{t=0} + (\underline{S} - \underline{S}^{IG})_{t=0}$$

and

$$\underline{S}(t = 50 \text{ min}) = \underline{S}^{IG}(T, \underline{V})_{t=50 \text{ min}} + (\underline{S} - \underline{S}^{IG})_{t=50 \text{ min}}$$

so that

$$\underline{S}(t = 50 \text{ min}) - \underline{S}(t = 0) = \underline{S}^{IG}(T, P)_{t=50 \text{ min}} - \underline{S}^{IG}(T, P)_{t=0}$$
$$+ (\underline{S} - \underline{S}^{IG})_{t=50 \text{ min}} - (\underline{S} - \underline{S}^{IG})_{t=0}$$

where

$$\underline{S}^{IG}(T, P)_{t=50 \text{ min}} - \underline{S}^{IG}(T, P)_{t=0}$$
$$= 27.2 \ln \frac{T(t = 50 \text{ min})}{170}$$
$$+ 4.2 \times 10^{-3}[T(t = 50 \text{ min}) - 170] - 8.314 \ln \frac{P(t = 50 \text{ min})}{100}$$

Both of the $(\underline{S} - \underline{S}^{IG})$ terms are obtained from the corresponding-states charts. $(\underline{S} - \underline{S}^{IG})_{t=0}$ is easily evaluated, since the initial state is known; that is, $T_r = 1.347$ and $P_r = 2.946$, so that, from Fig. 6.6-5 $(\underline{S} - \underline{S}^{IG})_{t=0} = -2.08$ cal/(mol K) $= -8.70$ J/(mol K). To compute $(\underline{S} - \underline{S}^{IG})$ at $t = 50$ minutes is more difficult because neither T_r nor P_r is known. The procedure to be followed is

1. $\underline{V}(t = 50 \text{ min})$ is known, so guess a value of $T(t = 50 \text{ min})$. (A reasonable first guess is the ideal gas solution obtained earlier.)
2. Use $\underline{V}(t = 50 \text{ min})$ and $T(t = 50 \text{ min})$ to compute, by trial and error, $P(t = 50 \text{ min})$ from

$$P = \frac{RT}{\underline{V}} Z\left(\frac{T}{T_c}, \frac{P}{P_c}\right) = \frac{RT}{\underline{V}} Z(T_r, P_r)$$

3. Use the values of P and T from steps 1 and 2 to compute $(\underline{S} - \underline{S}^{IG})_{t=50 \text{ min}}$.
4. Determine whether $\underline{S}(t = 50 \text{ min}) = \underline{S}(t = 0)$ is satisfied with the trial values of T and P. If not, guess another value of $T(t = 50)$ and go back to step 2.

Our solution after a number of trials is

$$T(t = 50 \text{ min}) = 136 \text{ K}$$
$$P(t = 50 \text{ min}) = 41 \text{ bar}$$

COMMENT

Because of the inaccuracy in reading numerical values from the corresponding-states graphs, this solution cannot be considered to be of high accuracy. ∎

It should be pointed out that although the principle of corresponding states and Eqs. 6.6-7, 6.6-8, and 6.6-11 appear simple, the application of these equations can become tedious, as is evident from this illustration. Also, the use of generalized correlations will lead to results that are not as accurate as those obtained using tabulations of the thermodynamic properties for the fluid of interest. Therefore, the corresponding-states principle is used in calculations only when reliable thermodynamic data are not available.

6.7 GENERALIZED EQUATIONS OF STATE

Although the discussion of the previous section focused on the van der Waals equation and corresponding-states charts for both the compressibility factor Z and the thermodynamic departure functions, the modern application of the corresponding states idea is to use **generalized equations of state.** The concept is most easily demonstrated by again using the van der Waals equation of state. From Eqs. 6.2-38,

$$P = \frac{RT}{\underline{V} - b} - \frac{a}{\underline{V}^2} \qquad (6.2\text{-}38b)$$

and the result of the inflection point analysis of Sec. 6.6, the constants a and b can be obtained from the fluid critical properties using

$$a = \frac{27 R^2 T_c^2}{64 P_c} \quad \text{and} \quad b = \frac{RT_c}{8 P_c} \qquad (6.6\text{-}4a)$$

The combination of Eqs. 6.2-38b and 6.6-4a is an example of a generalized equation of state, since we now have an equation of state that is presumed to be valid for a class of fluids with parameters (a and b) that have not been fitted to a whole collection of experimental data, but rather are obtained only from the fluid critical properties. The important content of these equations is that they permit the calculation of the $P\underline{V}T$ behavior of a fluid knowing only its critical properties, as was the case in corresponding-states theory.

It must be emphasized that the van der Waals equation of state is *not very accurate* and has been used here merely for demonstration because of its simplicity. It is never used for engineering design predictions, though other cubic equations of state are used. To illustrate the use of generalized equations of state, we will consider only the Peng-Robinson equation, which is commonly used to represent hydrocarbons and inorganic gases such as nitrogen, oxygen, and hydrogen sulfide. The generalized form of the Peng-Robinson equation of state is

Complete generalized Peng-Robinson equation of state

$$P = \frac{RT}{\underline{V} - b} - \frac{a(T)}{\underline{V}(\underline{V} + b) + b(\underline{V} - b)} \qquad (6.4\text{-}2)$$

with

$$a(T) = 0.45724 \frac{R^2 T_c^2}{P_c} \alpha(T) \qquad (6.7\text{-}1)$$

$$b = 0.07780 \frac{RT_c}{P_c} \qquad (6.7\text{-}2)$$

$$\sqrt{\alpha} = 1 + \kappa \left(1 - \sqrt{\frac{T}{T_c}} \right) \qquad (6.7\text{-}3)$$

and

$$\kappa = 0.37464 + 1.54226\omega - 0.26992\omega^2 \qquad (6.7\text{-}4)$$

where ω is the acentric factor defined earlier and given in Table 6.6-1.

Equations 6.7-1 through 6.7-4 were obtained in the following manner. First, the critical-point restrictions of Eqs. 6.6-1 were used, which leads to (see Problem 6.11)

$$a(T_c) = 0.45724 \frac{R^2 T_c^2}{P_c} \quad \text{and} \quad b = 0.07780 \frac{RT_c}{P_c}$$

ve the predictions of the boiling pressure as a function of temperature,
or pressure (which will be discussed in Sec. 7.5), Peng and Robinson
nal temperature-dependent term to their equation by setting

$$a(T) = a(T_c)\alpha(T)$$

fy the critical-point restrictions, $\alpha(T = T_c)$ must equal unity, as does
7-3. The specific form of α given by Eqs. 6.7-3 and 6.7-4 was chosen
essure data for many fluids.

oints to be noted in comparing the generalized van der Waals and
uations of state. First, although the parameter a is a constant in the
ation, in the Peng-Robinson equation it is a function of tempera-
ed temperature, $T_r = T/T_c$) through the temperature dependence
eneralized parameters of the Peng-Robinson equation of state are
cal temperature, the critical pressure, *and* the acentric factor ω of
tly, the Peng-Robinson equation of state, as generalized here, is
rameter (T_c, P_c, ω) equation of state, whereas the van der Waals
y two parameters, T_c and P_c.

orm of the Peng-Robinson equation of state (or other equations
used to compute not only the compressibility, but also the departure
functions for the other thermodynamic properties. This is done using Eqs. 6.4-29 and
6.4-30. In particular, to obtain numerical values for the enthalpy or entropy departure
for a fluid that obeys the Peng-Robinson equation of state, one uses the following
procedure:

1. Use the critical properties and acentric factor of the fluid to calculate b, κ, and
   the temperature-independent part of a using Eqs. 6.7-1, 6.7-2, and 6.7-4.
2. At the temperature of interest, compute numerical values for α and a using Eqs.
   6.7-1 and 6.7-3.
3. Solve the equation of state, Eq. 6.4-2, for $\underline{V}$ and compute $Z = P\underline{V}/RT$. Alternatively, solve for Z directly from the equivalent equation

$$Z^3 - (1 - B)Z^2 + (A - 3B^2 - 2B)Z - (AB - B^2 - B^3) = 0 \qquad \textbf{(6.7-5)}$$

   where $B = Pb/RT$ and $A = aP/R^2T^2$.
4. Use the computed value of Z and

**For Peng-Robinson
equation of state**

$$\boxed{\frac{da}{dT} = -0.45724 \frac{R^2 T_c^2}{P_c} \kappa \sqrt{\frac{\alpha}{T T_c}}}$$

to compute $[\underline{H}(T, P) - \underline{H}^{\text{IG}}(T, P)]$ and/or $[\underline{S}(T, P) - \underline{S}^{\text{IG}}(T, P)]$ as desired,
using Eqs. 6.4-29 and 6.4-30.

The enthalpy and entropy departures from ideal gas behavior calculated in this way can
be used to solve thermodynamic problems in the same manner as the similar functions
obtained from the corresponding-states graphs were used in the previous section.

It is clear that the calculation outlined here using the Peng-Robinson equation of state is, when doing computations by hand, more tedious than merely calculating the reduced temperature and pressure and using the graphs in Sec. 6.6. However, the equations here have some important advantages with the use of digital computers. First, this analytic computation avoids putting the three corresponding-states graphs in a computer memory in numerical form. Second, the values of the compressibility factor and departures from ideal gas properties obtained in the present three-parameter calculation should be more accurate than those obtained from the simple two-parameter (T_c, P_c) corresponding-states method of the previous section because of the additional fluid parameter (acentric factor) involved. Also, there is an absence of interpolation errors. Finally, if, at some time in the future, it is decided to use a different equation of state, only a few lines of computer code need be changed, as opposed to one's having to draw a complete new series of corresponding-states graphs.

Before there was easy access to electronic calculators and computers, it was common practice to apply the corresponding-states principle by using tables and graphs as illustrated in the previous section. Now, however, the usual industrial practice is to directly incorporate the corresponding-states idea that different fluids obey the same form of the equation of state, by using digital computer programs and generalized equations of state such as the one discussed here. This is demonstrated in the following illustration.

ILLUSTRATION 6.7-1

Using the Peng-Robinson Equation of State to Solve a Real Gas Problem

Rework Illustration 6.5-1 assuming that nitrogen can be described using the Peng-Robinson equation of state.

SOLUTION

The equations for solving this problem are the same as in Illustration 6.5-1. Specifically, the final temperature and pressure in the tank should be such that the molar entropy of gas finally in the tank is equal the initial molar entropy (Eq. c of Illustration 6.5-1), and the final molar volume should be such that the correct number of moles of gas remains in the tank (Eq. e of Illustration 6.5-1). The main difference here is that the Peng-Robinson equation of state is to be used in the solution. The general procedure used to calculate thermodynamic properties from the Peng-Robinson equation of state, and specifically for this problem, is as follows:

1. Choose values of T and P (these are known for the initial state and here will have to be found by trial and error for the final state).
2. Calculate a and b using Eqs. 6.7-1 through 6.7-4 and then $A = aP/R^2T^2$ and $B = Pb/RT$.
3. Find the compressibility Z or the molar volume (for the vapor phase in this problem) by solving the cubic equation, Eq. 6.7-5 (here for the largest root).
4. Using the value of Z found above and Eq. 6.4-30 to calculate the entropy departure from ideal gas behavior, $\underline{S} - \underline{S}^{IG}$. (Note that though they are not needed in this problem, the enthalpy departure and other properties can also be computed once the compressibility is known.)

The equations to be solved are first the Peng-Robinson equation of Eqs. 6.7-1 to 6.7-4 for the initial molar volume or compressibility, and then the initial number of moles in the tank using Eq. d of Illustration 6.5-2. The results, using the Visual Basic computer program described in Appendix B.I-2, the DOS-based program PR1 described in Appendix B.II-1, the MATHCAD worksheet described in Appendix B.III, or the MATLAB program described in B.IV included on the CD-ROM accompanying this book, are

$$Z = 0.6769 \qquad \underline{V}(t = 0) = 0.9567 \times 10^{-4} \text{ m}^3/\text{mol}$$

$$N(t = 0) = 1567.9 \text{ mol} \quad \text{and} \quad (\underline{S} - \underline{S}^{IG})_{t=0} = -9.18 \text{ J/(mol K)}$$

Consequently,

$$N(t = 50) = 1567.9 - 10 \times 50 = 1067.9 \text{ mol} \quad \text{and}$$

$$\underline{V}(t = 50) = 0.15 \text{ m}^3/1067.9 \text{ mol} = 1.4046 \times 10^{-4} \text{ m}^3/\text{mol}$$ **(a)**

Also

$$\underline{S}(t = 50) - \underline{S}(t = 0) = 0$$

$$= \underline{S}^{IG}(t = 50) - \underline{S}^{IG}(t = 0) + (\underline{S} - \underline{S}^{IG})_{t=50} - (\underline{S} - \underline{S}^{IG})_{t=0}$$

$$= \int_{T(t=0)}^{T(t=50)} \frac{C_P^*(T)}{T} dT - R \ln\left(\frac{P(t = 50)}{P(t = 0)}\right) + (\underline{S} - \underline{S}^{IG})_{t=50} - (\underline{S} - \underline{S}^{IG})_{t=0}$$

$$= 27.2 \ln\left(\frac{T(t = 50)}{170}\right) + 0.0042 \times (T(t = 50) - 170)$$

$$- R \ln\left(\frac{P(t = 50)}{P(t = 0)}\right) + (\underline{S} - \underline{S}^{IG})_{t=50} - (\underline{S} - \underline{S}^{IG})_{t=0}$$

(b)

Equations a and b are to be solved together with the Peng-Robinson equation of state and Eq. 6.4-30 for the entropy departure. This can be done in several ways. The simplest is to use an equation-solving program; this is illustrated using the MATHCAD worksheet described in Appendix B III. A somewhat more tedious method is to use one of the other Peng-Robinson equation-of-state programs described in Appendix B in an iterative fashion. That is, one could use the following procedure:

1. Guess the final temperature of the expansion process (the ideal gas result is generally a good initial guess).
2. Using the Peng-Robinson equation of state, iterate on the pressure until the correct value of $\underline{V}(t = 50)$ is obtained for the guessed value of $T(t = 50)$.
3. With the values of $T(t = 50)$ and $P(t = 50)$ so obtained, check whether Eq. b above is satisfied. If it is, the correct solution has been obtained. If not, adjust the guessed value of $T(t = 50)$ and repeat the calculation.

The result, directly from the MATHCAD worksheet or after several iterations with the other programs following the procedure above, is

$$T(t = 50 \text{ min}) = 134.66 \text{ K} \quad \text{and} \quad P(t = 50 \text{ min}) = 40.56 \text{ bar}$$

[Note that at these conditions $(\underline{S} - \underline{S}^{IG})_{t=50} = -10.19$ J/mol.] ∎

Though we will usually use the generalized Peng-Robinson equation of state for calculations and illustrations in this text, it is of interest to also list the generalized version of the Soave-Redlich-Kwong equation of state since it is also widely used in industry:

Soave-Redlich-Kwong equation of state

$$P = \frac{RT}{\underline{V} - b} - \frac{a(T)}{\underline{V}(\underline{V} + b)}$$ **(6.4-1b)**

with

$$a(T) = 0.42748 \frac{R^2 T_c^2}{P_c} \alpha(T) \qquad (6.7\text{-}6)$$

$$b = 0.08664 \frac{R T_c}{P_c} \qquad (6.7\text{-}7)$$

$$\sqrt{\alpha(T)} = 1 + \kappa \left(1 - \sqrt{\frac{T}{T_c}} \right) \qquad (6.7\text{-}8)$$

and

$$\kappa = 0.480 + 1.574\omega - 0.176\omega^2 \qquad (6.7\text{-}9)$$

6.8 THE THIRD LAW OF THERMODYNAMICS

In most treatises on thermodynamics, it is usual to refer to the laws of thermodynamics. The conservation of energy is referred to as the **First Law of Thermodynamics**, and this principle was discussed in detail in Chapter 3. The positive-definite nature of entropy generation used in Chapter 4, or any of the other statements such as those of Clausius or Kelvin and Planck, are referred to as the **Second Law of Thermodynamics**. The principle of conservation of mass precedes the development of thermodynamics, and therefore is not considered to be a law of thermodynamics.

There is a **Third Law of Thermodynamics**, though it is less generally useful than the first two. One version of the third law is

The entropy of all substances in the perfect crystalline state (for solids) or the perfect liquid state (for example, for helium) is zero at the absolute zero of temperature (0 K).

Third law of thermodynamics

Before we can use this statement, the perfect state must be defined. Here by "perfect" we mean without any disturbance in the arrangement of the atoms. That is, the substance must be without any vacancies, dislocations, or defects in the structure of the solid (or liquid) and not contain any impurities. The statement of the third law here is somewhat too constraining. A more correct statement is that all substances in the perfect state mentioned above should have the same value of entropy at 0 K, not necessarily a value of zero. It is mostly for convenience in the preparation of thermodynamic tables that a value of entropy of zero at 0 K is chosen.

There are several implications of the above statement. The first obvious one is that there will be no entropy change on a chemical reaction at 0 K if each of the reacting substances is in a perfect state, to produce one or more products in perfect states. In fact, it was this observation that led to the formulation of the third law. A second implication, which is less obvious and is sometimes used as an alternative statement of the third law, is

It is impossible to obtain a temperature of absolute zero.

This statement is proved as follows. From Eq. 6.2-20, we have

$$d\underline{S} = \frac{C_P}{T} dT - \left(\frac{\partial \underline{V}}{\partial T} \right)_P dP = \frac{C_P}{T} dT \qquad (6.2\text{-}20)$$

for a change at constant pressure. To continue we divide by dT and take the limits as $T \to 0$ to obtain

$$\lim_{T \to 0} \left(\frac{\partial S}{\partial T} \right)_P = \lim_{T \to 0} \frac{C_P}{T} \qquad \textbf{(6.8-1)}$$

There is experimental evidence showing that the constant-pressure heat capacity is finite and positive in value at all temperatures, and zero at absolute zero. Since C_P is positive and T is zero, we have

$$\lim_{T \to 0} \left(\frac{\partial S}{\partial T} \right)_P = \infty \qquad \textbf{(6.8-2)}$$

There are two possible conclusions from this equation. One is that as T decreases to absolute zero (so dT is negative), the entropy of any substance will become negative infinity. However, the experimental evidence is that the Gibbs energy of a substance, $G = H - TS$, converges to its enthalpy as absolute zero is approached, which means that the entropy must be finite. Therefore, the second, alternative conclusion is that it is not possible to reach a temperature of 0 K. This interpretation is the correct one, and in fact 0 K has not been attained in the laboratory. (However, with considerable effort temperatures of the order of 20 to 100 nK have been obtained.)

6.9 ESTIMATION METHODS FOR CRITICAL AND OTHER PROPERTIES

To use either the generalized equations of state (such as the Peng-Robinson and Soave–Redlich-Kwong equations) or the method of corresponding states, one needs information on the critical and other properties of the fluids of interest. The discussion so far has been concerned with molecules for which such data are available. However, an issue that arises is what to do when one does not have such data, either because the data are not available, (e.g., the compound one wishes to study has not yet been made in the laboratory) or perhaps because one is interested in a preliminary identification of which compounds or which class of compounds might have certain desired properties. This is especially the case in "product engineering," where one is interested in creating a compound or mixture of compounds with certain desired properties. This might be done by using a fast computational method to narrow the search of compounds with the desired properties, and then going to the library, searching the Web, or doing measurements in the laboratory to determine if the compounds so identified actually do have the desired properties. This task of identifying compounds with specific properties to make a new product is different from the usual job of a chemical engineer, which is "process engineering," that is, designing a process to make a desired product. Also, the thermodynamic properties of most pharmaceuticals and naturally occurring biologically-produced chemicals are unknown, and this is another case where being able to make some estimates, even very approximate ones, can be useful in developing purification methods.

The most common way to make properties estimates in the absence of experimental data is to use various group contribution methods. The basis of the method is that a molecule is thought of as a collection of functional groups, each of which makes an additive, though not necessarily linear, contribution to the properties of the molecule. Then as a result of summing up the contributions of each of the functional groups, the properties of the molecule are obtained. The underlying idea is that all molecules can be assembled from a limited number of functional groups (much in the same way that all of English literature can be created from only the 26 letters or functional groups in the alphabet). We will consider only one simple group contribution method for estimating pure component properties here, that of Joback[17], though a number of other methods exist.

However, before we proceed, a word of caution. Any group contribution method is inherently approximate and has some shortcomings. For example, it is assumed that a functional group makes the same contribution to the properties of a molecule independent of the molecule, and also which other functional groups it is bound to. This is a serious assumption, and one that is not generally true. For example, a methylene group, $-CH_2-$, makes a different contribution to the properties of a molecule if its binding partners are other methylene groups, halogens, or alcohols. Also, since simple group contribution methods are based on merely counting the number of each type of functional group, and not on their location within the molecule, they do not distinguish between isomers. In principle, group contribution methods can be improved by accounting for the first nearest neighbors, or the first and second nearest neighbors of each functional group. However, this makes the method much more difficult to apply, and would require extensive high-accuracy data and complicated data regression methods to obtain the contributions of each group. Instead, generally only simple group contributions methods are used, with the understanding that the results will be of uncertain accuracy and only of use for preliminary analysis, not for engineering design.

The Joback group contribution method uses the following equations:

$$T_c \, (\text{K}) = \frac{T_b \, (\text{K})}{0.584 + 0.965 \cdot \sum_i \nu_i \triangle T_{c,i} - \left(\sum_i \nu_i \triangle T_{c,i} \right)^2}$$

$$P_c \, (\text{bar}) = \frac{1}{\left(0.113 + 0.0032 \cdot n - \sum_i \nu_i \cdot \triangle P_{c,i} \right)^2}$$

$$\underline{V}_c \left(\frac{\text{cm}^3}{\text{mol}} \right) = 17.5 + \sum_i \nu_i \cdot \triangle V_{c,i}$$

$$T_b \, (\text{K}) = 198 + \sum_i \nu_i \cdot \triangle T_{b,i}$$

$$T_f \, (\text{K}) = 122 + \sum_i \nu_i \cdot \triangle T_{f,i}$$

In these equations, the subscripts c, b, and f indicate the critical point, boiling point, and freezing point, respectively; the $\triangle$ terms are the contributions of the group to each of the specified properties given in the following table; and ν_i is the number of functional groups of type i in the molecule.

To use the generalized form of, for example, the Peng-Robinson or Soave–Redlich-Kwong equations of state, one also needs the acentric factor. If the vapor pressure of the substance is known as a function of temperature, and the critical properties are known, the acentric factor can be computed from its definition,

$$\omega = -\log \frac{P^{\text{vap}} \, (T = 0.7 T_c)}{P_c} - 1$$

---

[17] See, for example, Chapter 2 of B. E. Poling, J. M. Prausnitz, and J. P. O'Connell, *The Properties of Gases and Liquids*, 5th ed., McGraw-Hill, New York, 2001

Table 6.9-1 Joback Group Contributions to Pure Component Properties

Group	$\triangle T_c$	$\triangle P_c$	$\triangle V_c$	$\triangle T_b$	$\triangle T_f$
$-CH_3$ nonring	0.0141	-0.0012	65	23.58	-5.1
CH_2 nonring	0.0189	0.0000	56	22.88	11.27
$-CH_2-$ ring	0.0100	0.0025	48	27.15	7.75
$CH-$ nonring	0.0164	0.0020	41	21.74	12.64
$CH-$ ring	0.0122	0.0004	38	21.78	19.88
C nonring	0.0067	0.0043	27	18.25	46.43
C ring	0.0042	0.0061	27	21.32	60.15
$=CH_2$ nonring	0.0113	-0.0028	56	18.18	-4.32
$=CH-$ nonring	0.0129	-0.0006	46	24.96	8.73
$=CH-$ ring	0.0082	0.0011	41	26.73	8.13
$=C$ nonring	0.0117	0.0011	38	24.14	11.14
$=C$ ring	0.0143	0.0008	32	31.01	37.02
$=C=$ nonring	0.0026	0.0028	36	26.15	17.78
$\equiv CH$ nonring	0.0027	-0.0008	46	9.2	-11.18
$\equiv C-$ nonring	0.0020	0.0016	37	27.38	64.32
$-F$ all	0.0111	-0.0057	27	-0.03	-15.78
$-Cl$ all	0.0105	-0.0049	58	38.13	13.55
$-Br$ all	0.0133	0.0057	71	66.86	43.43
$-I$ all	0.0068	-0.0034	97	93.84	41.69
$-OH$ alcohol	0.0741	0.0112	28	92.88	44.45
$-OH$ phenol	0.0240	0.0184	-25	76.34	82.83
$-O-$ nonring	0.0168	0.0015	18	22.42	22.23
$-O-$ ring	0.0098	0.0048	13	31.22	23.05
$C=O$ nonring	0.0380	0.0031	62	76.75	61.2
$C=O$ ring	0.0284	0.0028	55	94.97	75.97
$O=CH-$ aldehyde	0.0379	0.003	82	72.24	36.9
$-COOH$ acid	0.0791	0.0077	89	169.09	155.5
$-COO-$ nonring	0.0481	0.0005	82	81.1	53.6
$=O$ other	0.0143	0.0101	36	-10.5	2.08
$-NH_2$ all	0.0243	0.0109	38	-10.5	2.08
NH nonring	0.0295	0.0077	35	50.17	52.66
NH ring	0.0130	0.0114	29	52.82	101.51
$N-$ nonring	0.0169	0.0074	9	11.74	48.84
$-N=$ nonring	0.0255	-0.0099	0	74.6	0
$-N=$ ring	0.0085	0.0076	34	57.55	68.4
$-CN$ all	0.0496	-0.0101	91	125.66	59.89
$-NO_2$ all	0.0437	0.0064	91	152.54	127.24
$-SH$ all	0.0031	0.0084	63	63.56	20.09
$-S-$ nonring	0.0119	0.0049	54	68.78	34.4
$-S-$ ring	0.0019	0.0051	38	52.1	79.93

When such information is not available, the following approximate equation can be used:

$$\omega = \frac{3}{7}\frac{T_{br}}{1 - T_{br}}\log P_c - 1$$

where $T_{br} = T_b/T_c$. A very crude approximation, which was not suggested by Joback, is to use an estimated rather than a measured value for the normal boiling-point temperature in this equation should experimental data not be available.

ILLUSTRATION 6.9-1

Group Contribution Estimate of the Properties of a Pure Fluid

Use the methods described above to estimate the properties of *n*-octane that has a boiling point of 398.8 K and ethylene glycol (1,2-ethanediol) that has a boiling point of 470.5 K. Also compare the estimates of using and not using the measured boiling points.

SOLUTION

The results for *n*-octane are as follows:

	Experiment	Using T_b	Not Using T_b
T_b (K)	398.8	398.8	382.4
T_f (K)	216.4	179.4	179.4
T_c (K)	568.8	569.2	545.9
P_c (bar)	24.9	25.35	25.35
V_c (cc/mol)	492	483.5	483.5
ω	0.392	0.402	0.402

The results for ethylene glycol are

	Experiment	Using T_b	Not Using T_b
T_b (K)	470.5	470.5	429.5
T_f (K)	260.2	233.4	233.4
T_c (K)	645.0	645.5	589.3
P_c (bar)	77.0	66.5	66.5
V_c (cc/mol)	?	185.5	185.5
ω	?	1.094	1.094

We see that while none of the results are perfect, the estimates for *n*-octane are reasonably good, while those for ethylene glycol are less accurate. However, all the predictions are good enough to provide at least a qualitative estimate of the properties of these fluids. That is especially important when experimental data are not available, as is the case here for the critical volume and acentric factor of ethylene glycol. Note also that if a measured value of the normal boiling temperature is available and used, it results in a considerably more accurate value of the critical temperature than is the case if the information is not available.

COMMENT

Another situation in which approximate group contribution methods are especially useful is in product design where an engineer can ask (on a computer) what changes in properties would result if one or more functional groups were added, removed or replaced in a molecule. An example of a product design will be given later in this book

However, it is important to remember that all group contribution methods are very approximate. Therefore, while the methods discussed here can be used to obtain a preliminary estimate of the properties of a molecule, they should be verified against measurements before being used in an engineering design. ■

6.10   MORE ABOUT THERMODYNAMIC PARTIAL DERIVATIVES (OPTIONAL)

This section appears on the CD that accompanies this text.

PROBLEMS

Note that when the CDR icon appears next to a problem, it will be helpful to use the computer programs and/or MATHCAD worksheets on the CD-ROM accompanying this book and described in Appendix B in solving part of all of the problem.

6.1 For steam at 500°C and 10 MPa, using the Mollier diagram,
  a. Compute the Joule-Thomson coefficient $\mu = (\partial T/\partial P)_H$.
  b. Compute the coefficient $\kappa_S = (\partial T/\partial P)_S$.
  c. Relate the ratio $(\partial \underline{H}/\partial \underline{S})_T/(\partial \underline{H}/\partial \underline{S})_P$ to μ and κ_S, and compute its value for steam at the same conditions.

6.2 Derive Eqs. 6.4-29 and 6.4-30.

6.3 Evaluate the difference

$$\left(\frac{\partial \underline{U}}{\partial T}\right)_P - \left(\frac{\partial \underline{U}}{\partial T}\right)_{\underline{V}}$$

for the ideal and van der Waals gases, and for a gas that obeys the virial equation of state.

6.4 One of the beauties of thermodynamics is that it provides interrelationships between various state variables and their derivatives so that information from one set of experiments can be used to predict the results of a completely different experiment. This is illustrated here.
  a. Show that

$$C_P = \frac{T^2}{\mu}\left(\frac{\partial (\underline{V}/T)}{\partial T}\right)_P$$

Thus, if the Joule-Thomson coefficient μ and the volumetric equation of state (in analytic or tabular form) are known for a fluid, C_P can be computed. Alternatively, if C_P and μ are known, $(\partial(\underline{V}/T)/\partial T)_P$ can be calculated, or if C_P and $(\partial(\underline{V}/T)/\partial T)_P$ are known, μ can be calculated.

  b. Show that

$$\underline{V}(P, T_2) = \frac{T_2}{T_1}\underline{V}(P, T_1) + T_2\int_{P,T_1}^{P,T_2} \frac{\mu C_P}{T^2}\, dT$$

so that if μ and C_P are known functions of temperature at pressure P, and $\underline{V}$ is known at P and T_1, the specific volume at P and T_2 can be computed.

6.5 Derive Eqs. 6.6-2 and 6.6-3, and show that $Z_c|_{\text{van der Waals}} = 3/8$.

6.6 Derive Eq. 6.6-11.

6.7 One hundred cubic meters of carbon dioxide initially at 150°C and 50 bar is to be isothermally compressed in a frictionless piston-and-cylinder device to a final pressure of 300 bar. Calculate
  i. The volume of the compressed gas
  ii. The work done to compress the gas
  iii. The heat flow on compression
  assuming carbon dioxide
  a. Is an ideal gas
  b. Obeys the principle of corresponding states of Sec. 6.6
  c. Obeys the Peng-Robinson equation of state

6.8 By measuring the temperature change and the specific volume change accompanying a small pressure change in a reversible adiabatic process, one can evaluate the derivative

$$\left(\frac{\partial T}{\partial P}\right)_{\underline{S}}$$

and the adiabatic compressibility

$$\kappa_S = -\frac{1}{\underline{V}}\left(\frac{\partial \underline{V}}{\partial P}\right)_{\underline{S}}$$

Develop an expression for $(\partial T/\partial P)_S$ in terms of T, $\underline{V}$, C_P, α, and κ_T, and show that

$$\frac{\kappa_S}{\kappa_T} = \frac{C_V}{C_P}$$

6.9 Prove that the following statements are true.

   a. $(\partial \underline{H}/\partial \underline{V})_T$ is equal to zero if $(\partial \underline{H}/\partial P)_T$ is equal to zero.

   b. The derivative $(\partial \underline{S}/\partial \underline{V})_P$ for a fluid has the same sign as its coefficient of thermal expansion α and is inversely proportional to it.

6.10 By measuring the temperature change accompanying a differential volume change in a free expansion across a valve and separately in a reversible adiabatic expansion, the two derivatives $(\partial T/\partial \underline{V})_H$ and $(\partial T/\partial \underline{V})_S$ can be experimentally evaluated.

   a. Develop expressions for these derivatives in terms of the more fundamental quantities.

   b. Evaluate these derivatives for a van der Waals fluid.

6.11 a. Show for the Peng-Robinson equation of state (Eq. 6.4-2) that

$$a(T_c) = 0.45724 R^2 T_c^2/P_c$$

and

$$b = 0.07780 RT_c/P_c$$

   b. Determine the critical compressibility of the Peng-Robinson equation of state.

6.12 Ethylene at 30 bar and 100°C passes through a heater-expander and emerges at 20 bar and 150°C. There is no flow of work into or out of the heater-expander, but heat is supplied. Assuming that ethylene obeys the Peng-Robinson equation of state, compute the flow of heat into the heater-expander per mole of ethylene.

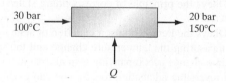

6.13 A natural gas stream (essentially pure methane) is available at 310 K and 14 bar. The gas is to be compressed to 345 bar before transmission by underground pipeline. If the compression is carried out adiabatically and reversibly, determine the compressor outlet temperature and the work of compression per mole of methane. You may assume that methane obeys the Peng-Robinson equation of state. See Appendix A.II for heat capacity data.

6.14 Values of the virial coefficients B and C at a fixed

temperature can be obtained from experimental $P\underline{V}T$ data by noting that

$$\lim_{\substack{P\to 0 \\ (\underline{V}\to\infty)}} \frac{P\underline{V}}{RT} = 1$$

$$\lim_{\substack{P\to 0 \\ (\underline{V}\to\infty)}} \underline{V}\left(\frac{P\underline{V}}{RT} - 1\right) = B$$

$$\lim_{\substack{P\to 0 \\ (\underline{V}\to\infty)}} \underline{V}^2\left(\frac{P\underline{V}}{RT} - 1 - \frac{B}{\underline{V}}\right) = C$$

   a. Using these formulas, show that the van der Waals equation leads to the following expressions for the virial coefficients.

$$B = b - \frac{a}{RT}$$

$$C = b^2$$

   b. The temperature at which

$$\lim_{\substack{P\to 0 \\ (\text{or } \underline{V}\to\infty)}} \underline{V}\left(\frac{P\underline{V}}{RT} - 1\right) = B = 0$$

   is called the Boyle temperature. Show that for the van der Waals fluid

$$T_{\text{Boyle}} = 3.375 T_c$$

   where T_c is the critical temperature of the van der Waals fluid given by Eqs. 6.6-3. (For many real gases T_{Boyle} is approximately $2.5T_c$!)

6.15 From experimental data it is known that at moderate pressures the volumetric equation of state may be written as

$$P\underline{V} = RT + BP$$

   where the virial coefficient B is a function of temperature only. Data for nitrogen are given in the table.

   a. Identify the Boyle temperature (the temperature at which $B = 0$) and the inversion temperature [the temperature at which $\mu = (\partial T/\partial P)_H = 0$] for gaseous nitrogen.

   b. Show, from the data in the table, that at temperatures above the inversion temperature the gas temperature increases in a Joule-Thomson expansion, whereas it decreases if the initial temperature is below the inversion temperature.

   c. Describe how you would find the inversion temper-

T (K)	B (cm^3/mol)
75	−274
100	−160
125	−104
150	−71.5
200	−35.2
250	−16.2
300	−4.2
400	+9.0
500	+16.9
600	+21.3
700	+24.0

Source: J. H. Dymond and E. B. Smith, *The Virial Coefficients of Gases* (1969). Figure from p. 188. By permission of Oxford University Press.

ature as a function of pressure for nitrogen using Fig. 3.3-3 and for methane using Fig. 3.3-2.

6.16 Eighteen kilograms of the refrigerant HFC-134a at 150°C is contained in a 0.03-m^3 tank. Compare the prediction you can make for the pressure in the tank with that obtained using Fig. 3.3-4. For data, see Table 6.6-1.

6.17 Calculate the molar volume, enthalpy, and entropy of carbon tetrachloride at 300°C and 35 bar using the Peng-Robinson equation of state and the principle of corresponding states of Sec. 6.6. The following data are available:

$$\underline{H}(T = 16°C, \text{ ideal gas at } P = 0.1 \text{ bar}) = 0$$
$$\underline{S}(T = 16°C, \text{ ideal gas at } P = 0.1 \text{ bar}) = 0$$
$$T_c = 283.2°C$$
$$P_c = 45.6 \text{ bar}$$
$$Z_c = 0.272 \qquad \omega = 0.194$$
$$C_P^*: \text{ see Appendix A.II}$$

6.18 The Clausius equation of state is

$$P(\underline{V} - b) = RT$$

a. Show that for this volumetric equation of state

$$C_P(P, T) = C_V(P, T) + R$$
$$C_P(P, T) = C_P^*(T)$$
$$\text{and}$$
$$C_V(\underline{V}, T) = C_V^*(T)$$

b. For a certain process the pressure of a gas must be reduced from an initial pressure P_1 to the final pressure P_2. The gas obeys the Clausius equation

of state, and the pressure reduction is to be accomplished either by passing the gas through a flow constriction, such as a pressure-reducing valve, or by passing it through a small gas turbine (which we can assume to be both reversible and adiabatic). Obtain expressions for the final gas temperature in each of these cases in terms of the initial state and the properties of the gas.

6.19 A tank is divided into two equal chambers by an internal diaphragm. One chamber contains methane at a pressure of 500 bar and a temperature of 20°C, and the other chamber is evacuated. Suddenly, the diaphragm bursts. Compute the final temperature and pressure of the gas in the tank after sufficient time has passed for equilibrium to be attained. Assume that there is no heat transfer between the tank and the gas and that methane:

a. is an ideal gas;

b. obeys the principle of corresponding states of Sec. 6.6;

c. obeys the van der Waals equation of state;

d. obeys the Peng-Robinson equation of state;

Data: For simplicity you may assume $C_P^* = 35.56$ J/(mol K).

6.20 The divided tank of the preceding problem is replaced with two interconnected tanks of equal volume; one tank is initially evacuated, and the other contains methane at 500 bar and 20°C. A valve connecting the two tanks is opened only long enough to allow the pressures in the tanks to equilibrate. If there is no heat transfer between the gas and the tanks, what are the temperature and pressure of the gas in each tank after the valve has been shut? Assume that methane

a. Is an ideal gas

b. Obeys the principle of corresponding states of Sec. 6.6

c. Obeys the Peng-Robinson equation of state

6.21 Ammonia is to be isothermally compressed in a specially designed flow turbine from 1 bar and 100°C to 50 bar. If the compression is done reversibly, compute the heat and work flows needed per mole of ammonia if

a. Ammonia obeys the principle of corresponding states of Sec. 6.6.

b. Ammonia satisfies the Clausius equation of state $P(\underline{V} - b) = RT$ with $b = 3.730 \times 10^{-2}$ m^3/kmol.

c. Ammonia obeys the Peng-Robinson equation of state.

6.22 A tank containing carbon dioxide at 400 K and 50 bar is vented until the temperature in the tank falls to 300 K. Assuming there is no heat transfer between the gas and the tank, find the pressure in the tank at the end of the venting process and the fraction of the initial mass of gas remaining in the tank for each of the following cases.

a. The equation of state of carbon dioxide is

$$P(\underline{V} - b) = RT \quad \text{with} \quad b = 0.0441 \text{ m}^3/\text{kmol}$$

b. Carbon dioxide obeys the law of corresponding states of Sec. 6.6.
c. Carbon dioxide obeys the Peng-Robinson equation of state.
  The low-pressure (ideal gas) heat capacity of CO_2 is given in Appendix A.II.

6.23 Derive the equations necessary to expand Illustration 6.4-1 to include the thermodynamic state variables internal energy, Gibbs energy, and Helmholtz energy.

6.24 Draw lines of constant Gibbs and Helmholtz energies on the diagrams of Illustration 6.4-1.

6.25 The speed of propagation of a small pressure pulse or sound wave in a fluid, v_S, can be shown to be equal to

$$v_S = \sqrt{\left(\frac{\partial P}{\partial \rho}\right)_S}$$

where ρ is the molar density.

a. Show that an alternative expression for the sonic velocity is

$$v_S = \sqrt{\gamma \underline{V}^2 \left(\frac{\partial T}{\partial \underline{V}}\right)_P \left(\frac{\partial P}{\partial T}\right)_{\underline{V}}}$$

where $\gamma = C_P/C_V$.

b. Show that $\gamma = 1 + R/C_V$ for both the ideal gas and a gas that obeys the Clausius equation of state

$$P(\underline{V} - b) = RT$$

and that γ is independent of specific volume for both gases.

c. Develop expressions for v_S for both the ideal and the Clausius gases that do not contain any derivatives other than C_V and C_P.

6.26 The force required to maintain a polymeric fiber at a length L when its unstretched length is L_0 has been observed to be related to its temperature by

$$F = \gamma T (L - L_0)$$

where γ is a positive constant. The heat capacity of the fiber measured at the constant length L_0 is given by

$$C_L = \alpha + \beta T$$

where α and β are parameters that depend on the fiber length.

a. Develop an equation that relates the change in entropy of the fiber to changes in its temperature and length, and evaluate the derivatives $(\partial S/\partial L)_T$ and $(\partial S/\partial T)_L$ that appear in this equation.
b. Develop an equation that relates the change in internal energy of the fiber to changes in its temperature and length.
c. Develop an equation that relates the entropy of the fiber at a temperature T_0 and an extension L_0 to its entropy at any other temperature T and extension L.
d. If the fiber at $T = T_i$ and $L = L_i$ is stretched slowly and adiabatically until it attains a length L_f, what is the fiber temperature at T_f?
e. In polymer science it is common to attribute the force necessary to stretch a fiber to energetic and entropic effects. The energetic force (i.e., that part of the force that, on an isothermal extension of the fiber, increases its internal energy) is $F_U = (\partial U/\partial L)_T$, and the entropic force is $F_S = -T(\partial S/\partial L)_T$. Evaluate F_U and F_S for the fiber being considered here.

6.27 Derive the following Maxwell relations for open systems.

a. Starting from Eq. 6.2-5a,

$$\left(\frac{\partial T}{\partial V}\right)_{S,N} = -\left(\frac{\partial P}{\partial S}\right)_{V,N}$$

$$\left(\frac{\partial T}{\partial N}\right)_{S,V} = \left(\frac{\partial G}{\partial S}\right)_{V,N}$$

$$\left(\frac{\partial P}{\partial N}\right)_{S,V} = -\left(\frac{\partial G}{\partial V}\right)_{S,N}$$

b. Starting from Eq. 6.2-6a,

$$\left(\frac{\partial T}{\partial P}\right)_{S,N} = \left(\frac{\partial V}{\partial S}\right)_{P,N}$$

$$\left(\frac{\partial T}{\partial N}\right)_{S,P} = \left(\frac{\partial G}{\partial S}\right)_{P,N}$$

$$\left(\frac{\partial V}{\partial N}\right)_{S,P} = \left(\frac{\partial G}{\partial P}\right)_{S,N}$$

c. Starting from Eq. 6.2-7a,

$$\left(\frac{\partial S}{\partial V}\right)_{T,N} = \left(\frac{\partial P}{\partial T}\right)_{V,N}$$

$$\left(\frac{\partial S}{\partial N}\right)_{T,V} = -\left(\frac{\partial G}{\partial T}\right)_{V,N}$$

$$\left(\frac{\partial P}{\partial N}\right)_{T,V} = -\left(\frac{\partial G}{\partial V}\right)_{T,N}$$

d. Starting from Eq. 6.2-8a,

$$\left(\frac{\partial S}{\partial P}\right)_{T,N} = -\left(\frac{\partial V}{\partial T}\right)_{P,N}$$

$$\left(\frac{\partial S}{\partial N}\right)_{T,P} = -\left(\frac{\partial G}{\partial T}\right)_{P,N}$$

$$\left(\frac{\partial V}{\partial N}\right)_{T,P} = \left(\frac{\partial G}{\partial P}\right)_{T,N}$$

6.28 For real gases the Joule-Thomson coefficient is greater than zero at low temperatures and less than zero at high temperatures. The temperature at which μ is equal to zero at a given pressure is called the inversion temperature.

a. Show that the van der Waals equation of state exhibits this behavior, and develop an equation for the inversion temperature of this fluid as a function of its specific volume.

b. Show that the van der Waals prediction for the inversion temperature can be written in the corresponding-states form

$$T_r^{\text{inv}} = \frac{3(3V_r - 1)^2}{4V_r^2}$$

c. The following graph shows the inversion temperature of nitrogen as a function of pressure.[18] Plot on this graph the van der Waals prediction for the inversion curve for nitrogen.

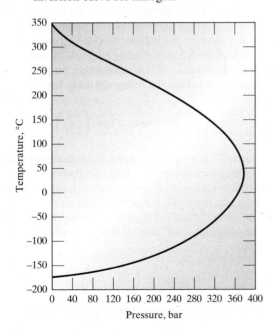

Temperature, °C (vertical axis); Pressure, bar (horizontal axis)

6.29 Nitrogen is to be isothermally compressed at 0°C from 1 bar to 100 bar. Compute the work required for this compression; the change in internal energy, enthalpy; Helmholtz and Gibbs energies of the gas; and the heat that must be removed to keep the gas at constant temperature if

a. The gas is an ideal gas.

b. The gas obeys the virial equation of state

$$\frac{P\underline{V}}{RT} = 1 + \frac{B}{\underline{V}} + \frac{C}{\underline{V}^2}$$

with $B = -10.3 \times 10^{-6}$ m³/mol and $C = 1517 \times 10^{-12}$ m⁶/mol².

c. The gas is described by the van der Waals equation

$$P = \frac{RT}{\underline{V} - b} - \frac{a}{\underline{V}^2}$$

or equivalently

$$\frac{P\underline{V}}{RT} = \frac{\underline{V}}{\underline{V} - b} - \frac{a}{\underline{V}RT}$$

with $a = 0.1368$ Pa m⁶/mol² and $b = 3.864 \times 10^{-5}$ m³/mol.

d. The gas is described by the Peng-Robinson equation of state.

6.30 For an isothermal process involving a fluid described by the Redlich-Kwong equation of state, develop expressions for the changes in

a. Internal energy

b. Enthalpy

c. Entropy

in terms of the initial temperature and the initial and final volumes.

6.31 Steam is continuously expanded from a pressure of 25 bar and 300°C to 1 bar through a Joule-Thomson expansion valve. Calculate the final temperature and the entropy generated per kilogram of steam using

a. The ideal gas law

b. The van der Waals equation of state

c. The Peng-Robinson equation of state

d. The steam tables

6.32 Repeat the calculations of Problem 6.13 if the mechanical efficiency of the adiabatic turbine is only 85 percent.

6.33 In statistical mechanics one tries to find an equation for the partition function of a substance. The canonical partition function, $Q(N, V, T)$, is used for a closed system at constant temperature, volume, and number of particles N. This partition function can be written

[18]From B. F. Dodge, *Chemical Engineering Thermodynamics,* McGraw-Hill. Used with permission of McGraw-Hill Book Company.

as a product of terms as follows:

$$Q(N, V, T) = \frac{f(T)^N \cdot V^N \cdot Z(N/V, T)}{N!}$$

where $f(T)$ is the part that depends only on the properties of a single molecule, and $Z(N/V, T)$ is a normalized configuration integral that is unity for an ideal gas and depends on the interaction energies among the molecules for a real gas. Also, the Helmholtz energy is related to the canonical partition function as follows:

$$A(N, V, T) = -kT \ln Q(N/V, T)$$

where k is Boltzmann's constant (the gas constant divided by Avogadro's number). Write expressions for all the thermodynamic properties of a fluid in terms of its canonical partition function and its derivatives.

6.34 Any residual property $\underline{\theta}$ is defined to be

$$\underline{\theta}^{res}(T, P) = \underline{\theta}(T, P) - \underline{\theta}^{IG}(T, P)$$

where IG denotes the same property in the ideal gas state. Such a quantity is also referred to as a departure function.

a. Develop general expressions for $d\underline{H}^{res}$, $d\underline{U}^{res}$, $d\underline{S}^{res}$, and $d\underline{G}^{res}$ with temperature and pressure as the independent variables.

b. Recognizing that as $P \to 0$ all fluids become ideal gases, so that their residual properties are zero, develop explicit expressions for the residual properties $\underline{H}^{res}$, $\underline{U}^{res}$, $\underline{S}^{res}$, and $\underline{G}^{res}$ as functions of temperature and pressure for the van der Waals equation of state.

c. Repeat part (a) for the Peng-Robinson equation of state.

d. Repeat part (a) for the Redlich-Kwong equation of state.

Problems Involving the Redlich-Kwong Equation of State (6.35–6.48).

6.35 a. Show for the Soave–Redlich-Kwong equation of state (Eq. 6.4-1) that

$$a(T) = 0.427\,48 \frac{R^2 T_c^2}{P_c} \alpha(T)$$

$$b = 0.086\,64 \frac{R T_c}{P_c}$$

b. Show that the critical compressibility of the Soave–Redlich-Kwong equation of state is 1/3.

6.36 Derive the expressions for the enthalpy and entropy departures from ideal gas behavior (that is, the analogues of Eqs. 6.4-29 and 6.4-30) for the Soave–Redlich-Kwong equation of state.

6.37 Repeat Illustration 6.4-1 using the Soave–Redlich-Kwong equation of state.

6.38 Repeat Illustration 6.7-1 using the Soave–Redlich-Kwong equation of state.

6.39 Redo Problem 6.7 with the Soave–Redlich-Kwong equation of state.

6.40 Redo Problem 6.12 with the Soave–Redlich-Kwong equation of state.

6.41 Redo Problem 6.22 with the Soave–Redlich-Kwong equation of state.

6.42 Using the Redlich-Kwong equation of state, compute the following quantities for nitrogen at 298.15 K.

a. The difference $C_P - C_V$ as a function of pressure from low pressures to very high pressures

b. C_P as a function of pressure from low pressures to very high pressures. [*Hint:* It is easier to first compute both $C_P - C_V$ and C_V for chosen values of volume, then compute the pressure that corresponds to those volumes, and finally calculate C_P as the sum of $(C_P - C_V) + C_V$.]

6.43 The Boyle temperature is defined as the temperature at which the second virial coefficient B is equal to zero.

a. Recognizing that any equation of state can be expanded in virial form, find the Boyle temperature for the Redlich-Kwong equation of state in terms of the parameters in that equation.

b. The Redlich-Kwong parameters for carbon dioxide are $a = 6.466 \times 10^{-4}$ m^6 MPa K$^{0.5}$ mol^{-2} and $b = 2.971 \times 10^{-3}$ m^3 mol^{-1}. Estimate the Boyle temperature for carbon dioxide, assuming that it obeys the Redlich-Kwong equation of state.

6.44 In the calculation of thermodynamic properties, it is convenient to have the following partial derivatives:

$$\lim_{P \to 0} \left(\frac{\partial Z}{\partial P} \right)_T \quad \text{and} \quad \lim_{P \to \infty} \left(\frac{\partial Z}{\partial P} \right)_T$$

where $Z = (P\underline{V}/RT)$ is the compressibility factor. Develop expressions for these two derivatives for the Redlich-Kwong equation of state in terms of temperature and the Redlich-Kwong parameters.

6.45 The second virial coefficient B can be obtained from experimental PVT data or from an equation of state from

$$B = \lim_{P \to 0} \underline{V} \left(\frac{P\underline{V}}{RT} - 1 \right)$$

a. Show that for the Redlich-Kwong equation

$$P = \frac{RT}{\underline{V} - b} - \frac{a}{\sqrt{T}\underline{V}(\underline{V} + b)}$$

the second virial coefficient is

$$B = b - \frac{a}{RT^{3/2}}$$

b. Compute the second virial coefficient of n-pentane as a function of temperature from the Redlich-Kwong equation of state.

6.46 The Joule-Thomson coefficient, μ, given by

$$\mu = \left(\frac{\partial T}{\partial P}\right)_H = -\frac{V}{C_P}[1 - T\alpha]$$

is a function of temperature. The temperature at which $\mu = 0$ is known as the inversion temperature.

a. Use the van der Waals equation of state to determine the inversion temperature of H_2, O_2, N_2, CO, and CH_4. The van der Waals parameters for these gases can be found in Table 6.4-1.

b. Repeat part (a) using the Redlich-Kwong equation of state.

6.47 Using the Redlich-Kwong equation of state, compute and plot (on separate graphs) the pressure of nitrogen as a function of specific volume at the two temperatures:

a. 110 K

b. 150 K

6.48 Use the information in Illustration 6.4-1 and the Soave–Redlich-Kwong equation of state to compute the thermodynamic properties of oxygen along the following two isotherms:

a. 155 K

b. 200 K

6.49 Repeat Problem 5.9 assuming that helium is described by the Peng-Robinson equation of state.

6.50 Repeat Problem 5.10 assuming that nitrogen is described by the Peng-Robinson equation of state.

6.51 In this chapter, from the third law of thermodynamics, it has been shown that the entropy of all substances approaches a common value at 0 K (which for convenience we have taken to be zero). This implies that the value of the entropy at 0 K is not a function of volume or pressure. Use this information to show the following:

$$C_V = 0$$

at $T = 0$ K and that the coefficient of thermal expansion

$$\alpha = \frac{1}{V}\left(\frac{\partial V}{\partial T}\right)_P$$

is zero at 0 K.

6.52 A fluid is described by the Clausius equation of state

$$P = \frac{RT}{V - b}$$

where b is a constant. Also, the ideal gas heat capacity is given by

$$C_P^* = \alpha + \beta T + \gamma T^2$$

For this fluid, obtain explicit expressions for

a. A line of constant enthalpy as a function of pressure and temperature

b. A line of constant entropy as a function of temperature and pressure

c. Does this fluid have a Joule-Thomson inversion temperature?

6.53 A piston-and-cylinder device contains 10 kmol of n-pentane at $-35.5°$C and 100 bar. Slowly the piston is moved until the vapor pressure of n-pentane is reached, and then further moved until 5 kmol of the n-pentane is evaporated. This complete process takes place at the constant temperature of $-35.5°$C. Assume n-pentane can be described by the Peng-Robinson equation of state.

a. What is the volume change for the process?

b. How much heat must be supplied for the process to be isothermal?

6.54 Develop an expression for how the constant-volume heat capacity varies with temperature and specific volume for the Peng-Robinson fluid.

6.55 A manuscript recently submitted to a major journal for publication gave the following volumetric and thermal equations of state for a solid:

$$\underline{V}(T, P) = a + bT - cP \quad \text{and} \quad \underline{U}(T, P) = dT + eP$$

where a, b, c, d, and e are constants. Are these two equations consistent with each other?

6.56 The following equation of state describes the behavior of a certain fluid:

$$P\left(\underline{V} - b\right) = RT + \frac{aP^2}{T}$$

where the constants are $a = 10^{-3}$ m^3 K/(bar mol) = 10^2 (J K)/(bar^2 mol) and b = 8×10^{-5} m^3/mol. Also, for this fluid the mean ideal gas constant-pressure heat capacity, C_P, over the temperature range of 0 to 300°C at 1 bar is 33.5 J/(mol K).

a. Estimate the mean value of C_P over the temperature range at 12 bar.

b. Calculate the enthalpy change of the fluid for a change from $P = 4$ bar, $T = 300$ K to $P = 12$ bar and $T = 400$ K.

c. Calculate the entropy change of the fluid for the same change of conditions as in part (b).

6.57 The van der Waals equation of state with $a = 0.1368$ Pa m^3/mol^2 and $b = 3.864 \times 10^{-5}$ m^3/mol can be used to describe nitrogen.

a. Using the van der Waals equation of state, prepare a graph of pressure (P) of nitrogen as a function of log $\underline{V}$, where $\underline{V}$ is molar volume at temperatures of 100 K, 125 K, 150 K, and 175 K. The range of the plot should be $\underline{V} = 1 \times 10^{-4}$ to 25 m^3/mol.

b. One mole of nitrogen is to be isothermally compressed from 100 kPa to 10 MPa at 300 K. Determine (1) the molar volume at the initial and final conditions and (2) the amount of work necessary to carry out the isothermal compression.

c. Repeat the calculation of part (b) for an ideal gas.

6.58 A bicycle pump can be treated as a piston-and-cylinder system that is connected to the tire at the "closed" end of the cylinder. The connection is through a valve that is initially closed, while the cylinder is filled with air at atmospheric pressure, following which the pumping in a cycle occurs as the plunger (piston) is pushed further into the cylinder. When the air pressure in the cylinder increases until it reaches the same pressure as that in the tire, the valve opens and further movement of the piston forces air from the pump into the tire. The whole process occurs quickly enough that there is no significant heat flow to or from the surroundings to the pump or tire during the pumping process. Thus, a single pumping cycle can be considered a three-step process. First, air is drawn into the pump cylinder at constant pressure. Second, as the pumping begins, the valve is closed so that the pressure increases without a flow of air from the pump. Third, the valve opens and the pumping action forces air into the tire. Here we are interested in calculations only for the second step if the tire pressure is initially 60 psig (5.15 bar absolute). For reference, the cylinder of a bicycle hand pump is about 50 cm long and about 3 cm in diameter.

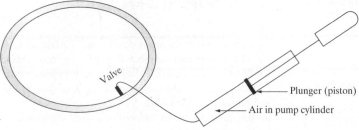

a. If the ambient temperature is 25°C and air is treated as an ideal gas with $C_P^* = 29.3$ J/(mol K),

how far down will the piston have moved before the valve opens? What is the temperature of the gas?

b. Repeat the calculation of part (a) if the Peng-Robinson equation is used to describe air, assuming that the critical constants and acentric factor of air are the same as for nitrogen.

6.59 The Euken coefficient ξ is defined as $\xi = (\partial T/\partial \underline{V})_U$.

a. Show that the Euken factor is also equal to

$$\xi = -\left(\frac{T\,(\partial P/\partial T)_{\underline{V}} - P}{C_V}\right)$$

b. What is the value of the Euken coefficient for an ideal gas?

c. Develop an explicit expression for the Euken coefficient for a gas that is described by the truncated virial equation

$$\frac{P\underline{V}}{RT} = Z = 1 + \frac{B(T)}{\underline{V}}$$

d. Develop an expression for the Euken coefficient for a gas that is described by the Peng-Robinson equation of state.

6.60 Redo Problem 4.45 if ethylene is described by the truncated virial equation with $B(T) = 5.86 \cdot 10^{-5} - 0.056/T$ m^3/mol and T in K.

6.61 Redo Problem 4.45 if ethylene is described by the Peng-Robinson equation of state.

6.62 a. Show that

$$\frac{(\partial \underline{V}/\partial T)_S}{(\partial \underline{V}/\partial T)_P} = -\frac{C_V}{\kappa_T \underline{V} T}\left(\frac{\partial T}{\partial P}\right)_{\underline{V}}^2$$

b. Use the result of part (a) to show that for a stable system at equilibrium $(\partial \underline{V}/\partial T)_S$ and $(\partial \underline{V}/\partial T)_P$ must have opposite signs.

c. Two separate measurements are to be performed on a gas enclosed in a piston-and-cylinder device. In the first measurement the device is well insulated so there is no flow of heat to or from the gas, and the piston is slowly moved inward, compressing the gas, and its temperature is found to increase. In the second measurement the piston is free to move and the external pressure is constant. A small amount of heat is added to the gas in the cylinder, resulting in the expansion of the gas. Will the temperature of the gas increase or decrease?

6.63 Gasoline vapor is to be recovered at a filling station rather than being released into the atmosphere. In the scheme we will analyze, the vapor is first condensed to a liquid and then pumped back to a storage tank. Determine the work necessary to adiabatically pump the liquid gasoline from 0.1 MPa and 25°C to 25 MPa, and the final temperature of the liquid gasoline. Consider gasoline to be *n*-octane and $C_P^* = 122.2$ J/(mol K) independent of temperature.

6.64 A 3-m³ tank is in the basement of your house to store propane that will be used for home and water heating and cooking. Your basement, and therefore the contents of the tank, remain at 20°C all year. Initially, except for a small vapor space the tank is completely full of liquid propane, but by January 60 percent of the propane has been used. Compute

a. The fraction of the remaining propane that is present as vapor
b. The total flow of heat into the tank over the period in which the propane has been used

6.65 Pipelines are used to transport natural gas over thousands of miles, with compressors (pumping stations) at regular intervals along the pipeline to compensate for the pressure drop due to flow. Because of the long distances involved and the good heat transfer, the gas remains at an ambient temperature of 20°C. At each pumping station the gas in compressed with a single-stage compressor to 4 MPa, and then isobarically cooled back to 20°C. Because of safety and equipment constraints, the gas temperature during compression should not exceed 120°C. Assuming natural gas is methane that is described by the Peng-Robinson equation of state, how low a pressure is allowed before the gas should be recompressed?

6.66 In a continuous manufacturing process, chlorodifluoromethane ($CHClF_2$), initially at 10 bar and 420 K, passes through an adiabatic pressure-reducing valve so that its pressure drops to 0.1 bar (this last pressure low enough that $CHClF_2$ can be considered to be an ideal gas). At these operating conditions, the gas can be represented by a one-term virial equation of state:

$$\frac{PV}{RT} = 1 + \frac{B(T)}{V}$$

The following data are available for this gas:

$$B(T) = 297.6 - \frac{256100}{T} \frac{cm^3}{mol}$$

and

$$C_P^* = 27.93 + 0.093T \frac{J}{mol\,K}$$

What is the temperature of the chlorodifluoromethane exiting the valve?

6.67 How much entropy is generated per mole of chlorodifluoromethane that passes through the pressure-reducing valve of the previous problem?

6.68 One kilogram of saturated liquid methane at 160 K is placed in an adiabatic piston-and-cylinder device, and the piston will be moved slowly and reversibly until 25 percent of the liquid has vaporized. Compute the maximum work that can be obtained, assuming that methane is described by the Peng-Robinson equation of state. Compare your results with the solution to Problem 4.39.

6.71 The thermoelastic effect is the temperature change that results from stretching an elastic material or fiber. The work done on the material is given by

$$\dot{W} = Force \times Rate\ of\ change\ of\ distance = A \cdot L \cdot \sigma \cdot \frac{d\varepsilon}{dt}$$

where

$$\sigma = Stress = \frac{Force}{Cross\text{-}sectional\ area} = \frac{Force}{A}$$

and

$$\frac{d\varepsilon}{dt} = Rate\ of\ strain = \frac{1}{L}\frac{dL}{dt}$$

In the situation being considered here, the thermodynamic properties such as the heat capacity depend on the stress (just as for a gas the heat capacity depends on the pressure).

a. Develop the energy and entropy balances for this one-dimensional material, assuming that there is no additional work term other than the stretching work, and that the volume of the material does not change on stretching (that is, the length change is compensated for by a change in cross-sectional area).
b. An elastic material has the following properties at 733 K and a stress of 103 MPa.

$$\alpha_\sigma = \frac{1}{L}\left(\frac{\partial L}{\partial T}\right)_\sigma = 16.8 \times 10^{-6}\ K^{-1}$$

and

$$C_\sigma = T\left(\frac{\partial S}{\partial T}\right)_\sigma = 38.1 \frac{J}{(mol\,K)}$$

Estimate the change in temperature per unit of strain at these conditions.

Chapter 7

Equilibrium and Stability in One-Component Systems

Now that the basic principles of thermodynamics have been developed and some computational details considered, we can study one of the fundamental problems of thermodynamics: the prediction of the equilibrium state of a system. In this chapter we examine the conditions for the existence of a stable equilibrium state in a single-component system, with particular reference to the problem of phase equilibrium. Equilibrium in multicomponent systems will be considered in following chapters.

INSTRUCTIONAL OBJECTIVES FOR CHAPTER 7

The goals of this chapter are for the student to:

- Identify the criterion for equilibrium in systems subject to different constraints (Sec. 7.1)
- Use the concept of stability to identify when phase splitting into vapor and liquid phases will occur (Sec. 7.2)
- Identify the conditions of phase equilibrium (Sec. 7.3)
- Be able to identify the critical point of a fluid (Sec. 7.3)
- Calculate the fugacity of a pure substance that is a gas or a liquid when a volumetric equation of state is available (Sec. 7.4)
- Calculate the fugacity of a pure liquid or solid when a volumetric equation of state is not available (Sec. 7.4)
- Use fugacity in the calculation of vapor-liquid equilibrium (Sec. 7.5)
- Be able to determine the number degrees of freedom for a pure fluid (Sec. 7.6)

NOTATION INTRODUCED IN THIS CHAPTER

f Fugacity (kPa)
$\mathcal{F}$ Number of degrees of freedom
ϕ Fugacity coefficient $= f/P$
$f_{\text{sat}}^{\text{I}}(T)$ Fugacity of phase I at a phase change at temperature T (kPa)
$\underline{H}^{\text{L}}$ Molar enthalpy of a liquid (kJ/mol)
$\underline{H}^{\text{S}}$ Molar enthalpy of a solid (kJ/mol)
$\underline{H}^{\text{V}}$ Molar enthalpy of a vapor (kJ/mol)

$\triangle_{\text{vap}}\underline{H}$ Enthalpy change (or heat) of vaporization $= \underline{H}^V - \underline{H}^L$ (kJ/mol)

$\triangle_{\text{fus}}\underline{H}$ Enthalpy change (or heat) of melting $= \underline{H}^L - \underline{H}^S$ (kJ/mol)

$\triangle_{\text{sub}}\underline{H}$ Enthalpy change (or heat) of sublimation $= \underline{H}^V - \underline{H}^S$ (kJ/mol)

P_t Triple-point pressure (kPa)

$P^{\text{sub}}(T)$ Sublimation pressure at temperature T (kPa)

$P^{\text{vap}}(T)$ Vapor pressure at temperature T (kPa)

T_t Triple-point temperature (K)

7.1 THE CRITERIA FOR EQUILIBRIUM

In Chapter 4 we established that the entropy function provides a means of mathematically identifying the state of equilibrium in a closed, isolated system (i.e., a system in which M, U, and V are constant). The aim in this section is to develop a means of identifying the equilibrium state for closed thermodynamic systems subject to other constraints, especially those of constant temperature and volume, and constant temperature and pressure. This will be done by first reconsidering the equilibrium analysis for the closed, isolated system used in Sec. 4.1, and then extending this analysis to the study of more general systems.

The starting points for the analysis are the energy and entropy balances for a closed system:

Balance equations for a closed system

$$\boxed{\frac{dU}{dt} = \dot{Q} - P\frac{dV}{dt}} \tag{7.1-1}$$

and

$$\boxed{\frac{dS}{dt} = \frac{\dot{Q}}{T} + \dot{S}_{\text{gen}}} \tag{7.1-2}$$

with

Second law of thermodynamics

$$\boxed{\dot{S}_{\text{gen}} \geq 0 \quad \text{(the equality holding at equilibrium or for reversible processes)}}$$

$$\tag{7.1-3}$$

Here we have chosen the system in such a way that the only work term is that of the deformation of the system boundary. For a constant-volume system exchanging no heat with its surroundings, Eqs. 7.1-1 and 7.1-2 reduce to

$$\frac{dU}{dt} = 0$$

so that

$$U = \text{constant}$$

(or, equivalently, $\underline{U} = $ constant, since the total number of moles, or mass, is fixed in a closed, one-component system) and

$$\boxed{\frac{dS}{dt} = \dot{S}_{\text{gen}} \geq 0}$$

(7.1-4)

Since the entropy function can only increase in value during the approach to equilibrium (because of the sign of $\dot{S}_{\text{gen}}$), the entropy must be a maximum at equilibrium. Thus, the equilibrium criterion for a closed, isolated system is

$$\left.\begin{array}{l} S = \text{maximum} \\ \text{or} \\ \underline{S} = \text{maximum} \end{array}\right\} \quad \text{at equilibrium in a closed system at constant } U \text{ and } V$$

(7.1-5)

At this point you might ask how $\underline{S}$ can achieve a maximum value if $\underline{U}$ and $\underline{V}$ are fixed, since the specification of any two intensive variables completely fixes the values of all others. The answer to this question was given in Chapter 4, where it was pointed out that the specification of two intensive variables fixes the values of all other state variables in the uniform equilibrium state of a single-component, single-phase system. Thus, the equilibrium criterion of Eq. 7.1-5 can be used for identifying the final equilibrium state in a closed, isolated system that is initially nonuniform, or in which several phases or components are present.

To illustrate the use of this equilibrium criterion, consider the very simple, initially nonuniform system shown in Fig. 7.1-1. There a single-phase, single-component fluid in an adiabatic, constant-volume container has been divided into two subsystems by an imaginary boundary. Each of these subsystems is assumed to contain the same chemical species of uniform thermodynamic properties. However, these subsystems are open to the flow of heat and mass across the imaginary internal boundary, and their temperature and pressure need not be the same. For the composite system consisting of the two subsystems, the total mass (though, in fact, we will use number of moles), internal energy, volume, and entropy, all of which are extensive variables, are the sums of these respective quantities for the two subsystems, that is,

$$\begin{array}{l} N = N^{\text{I}} + N^{\text{II}} \\ U = U^{\text{I}} + U^{\text{II}} \\ V = V^{\text{I}} + V^{\text{II}} \end{array}$$

(7.1-6)

and

$$S = S^{\text{I}} + S^{\text{II}}$$

Now considering the entropy to be a function of internal energy, volume, and mole number, we can compute the change in the entropy of system I due to changes in N^{I}, U^{I}, and V^{I} from Eq. 6.2-5c.

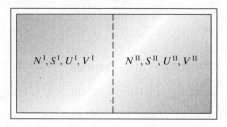

Figure 7.1-1 An isolated nonequilibrium system.

$$dS^{\mathrm{I}} = \left(\frac{\partial S^{\mathrm{I}}}{\partial U^{\mathrm{I}}}\right)_{V^{\mathrm{I}},N^{\mathrm{I}}} dU^{\mathrm{I}} + \left(\frac{\partial S^{\mathrm{I}}}{\partial V^{\mathrm{I}}}\right)_{U^{\mathrm{I}},N^{\mathrm{I}}} dV^{\mathrm{I}} + \left(\frac{\partial S^{\mathrm{I}}}{\partial N^{\mathrm{I}}}\right)_{U^{\mathrm{I}},V^{\mathrm{I}}} dN^{\mathrm{I}}$$

$$= \frac{1}{T^{\mathrm{I}}} dU^{\mathrm{I}} + \frac{P^{\mathrm{I}}}{T^{\mathrm{I}}} dV^{\mathrm{I}} - \frac{\underline{G}^{\mathrm{I}}}{T^{\mathrm{I}}} dN^{\mathrm{I}}$$

In a similar fashion dS^{II} can be computed. The entropy change of the composite system is

$$dS = dS^{\mathrm{I}} + dS^{\mathrm{II}} = \frac{1}{T^{\mathrm{I}}} dU^{\mathrm{I}} + \frac{1}{T^{\mathrm{II}}} dU^{\mathrm{II}} + \frac{P^{\mathrm{I}}}{T^{\mathrm{I}}} dV^{\mathrm{I}} + \frac{P^{\mathrm{II}}}{T^{\mathrm{II}}} dV^{\mathrm{II}} - \frac{\underline{G}^{\mathrm{I}}}{T^{\mathrm{I}}} dN^{\mathrm{I}} - \frac{\underline{G}^{\mathrm{II}}}{T^{\mathrm{II}}} dN^{\mathrm{II}}$$

However, the total number of moles, total internal energy, and total volume are constant by the constraints on the system, so that

$$dN = dN^{\mathrm{I}} + dN^{\mathrm{II}} = 0 \quad \text{or} \quad dN^{\mathrm{I}} = -dN^{\mathrm{II}}$$
$$dU = dU^{\mathrm{I}} + dU^{\mathrm{II}} = 0 \quad \text{or} \quad dU^{\mathrm{I}} = -dU^{\mathrm{II}} \qquad \text{(7.1-7)}$$
$$dV = dV^{\mathrm{I}} + dV^{\mathrm{II}} = 0 \quad \text{or} \quad dV^{\mathrm{I}} = -dV^{\mathrm{II}}$$

Consequently,

$$dS = \left(\frac{1}{T^{\mathrm{I}}} - \frac{1}{T^{\mathrm{II}}}\right) dU^{\mathrm{I}} + \left(\frac{P^{\mathrm{I}}}{T^{\mathrm{I}}} - \frac{P^{\mathrm{II}}}{T^{\mathrm{II}}}\right) dV^{\mathrm{I}} - \left(\frac{\underline{G}^{\mathrm{I}}}{T^{\mathrm{I}}} - \frac{\underline{G}^{\mathrm{II}}}{T^{\mathrm{II}}}\right) dN^{\mathrm{I}} \qquad \text{(7.1-8)}$$

Now since $S = \text{maximum}$ or $dS = 0$ for all system variations at constant N, U, and V (here all variations of the independent variables dU^{I}, dV^{I}, and dN^{I} at constant total number of moles, total internal energy, and total volume), we conclude that

$$\left(\frac{\partial S}{\partial U^{\mathrm{I}}}\right)_{V^{\mathrm{I}},N^{\mathrm{I}}} = 0 \quad \text{so that} \quad \frac{1}{T^{\mathrm{I}}} = \frac{1}{T^{\mathrm{II}}} \quad \text{or} \quad T^{\mathrm{I}} = T^{\mathrm{II}} \qquad \text{(7.1-9a)}$$

$$\left(\frac{\partial S}{\partial V^{\mathrm{I}}}\right)_{U^{\mathrm{I}},N^{\mathrm{I}}} = 0 \quad \text{so that} \quad \frac{P^{\mathrm{I}}}{T^{\mathrm{I}}} = \frac{P^{\mathrm{II}}}{T^{\mathrm{II}}} \quad \text{or} \quad P^{\mathrm{I}} = P^{\mathrm{II}} \qquad \text{(7.1-9b)}$$

(since it has already been shown that $T^{\mathrm{I}} = T^{\mathrm{II}}$)

$$\left(\frac{\partial S}{\partial N^{\mathrm{I}}}\right)_{U^{\mathrm{I}},V^{\mathrm{I}}} = 0 \quad \text{so that} \quad \frac{\underline{G}^{\mathrm{I}}}{T^{\mathrm{I}}} = \frac{\underline{G}^{\mathrm{II}}}{T^{\mathrm{II}}} \quad \text{or} \quad \underline{G}^{\mathrm{I}} = \underline{G}^{\mathrm{II}} \qquad \text{(7.1-9c)}$$

Therefore, the equilibrium condition for the system illustrated in Fig. 7.1-1 is satisfied if both subsystems have the same temperature, the same pressure, and the same molar Gibbs energy. For a single-component, single-phase system, this implies that the composite system should be uniform. This is an obvious result, and it is reassuring that it arises so naturally from our development.

The foregoing discussion illustrates how the condition $dS = 0$ may be used to identify a possible equilibrium state of the closed, isolated system, that is, a state for which S is a maximum. From calculus we know that $dS = 0$ is a necessary but not sufficient condition for S to achieve a maximum value. In particular, $dS = 0$ at a minimum value or an inflection point of S, as well as when S is a maximum. The condition $d^2S < 0$, when $dS = 0$, assures us that a maximum value of the entropy, and hence a true equilibrium state, has been identified rather than a metastable state (an inflection point) or an unstable state (minimum value of S). Thus, the sign of $d^2 S$ determines the stability of the state found from the condition that $dS = 0$. The implications of this stability condition are considered in the next section.

It is also possible to develop the equilibrium and stability conditions for systems subject to other constraints. For a closed system at constant temperature and volume, the energy and entropy balances are

$$\frac{dU}{dt} = \dot{Q}$$

and

$$\frac{dS}{dt} = \frac{\dot{Q}}{T} + \dot{S}_{\text{gen}}$$

Eliminating $\dot{Q}$ between these two equations, and using the fact that $T \, dS = d(TS)$, since T is constant, gives

$$\frac{d(U - TS)}{dt} = \frac{dA}{dt} = -T\dot{S}_{\text{gen}} \leq 0$$

Here we have also used the facts that $T \geq 0$ and $\dot{S}_{\text{gen}} \geq 0$, so that $(-T\dot{S}_{\text{gen}}) \leq 0$. Using the same argument that led from Eq. 7.1-4 to Eq. 7.1-5 here yields

$$\left.\begin{array}{l} A = \text{minimum} \\ \text{or} \\ \underline{A} = \text{minimum} \end{array}\right\} \quad \text{for equilibrium in a closed system at constant } T \text{ and } V$$

$$(7.1\text{-}10)$$

It should be noted that this result is also a consequence of $\dot{S}_{\text{gen}} \geq 0$.

If we were to use Eq. 7.1-10 to identify the equilibrium state of an initially nonuniform, single-component system, such as that in Fig. 7.1-1, but now maintained at a constant temperature and volume, we would find, following the previous analysis, that at equilibrium the pressures of the two subsystems are equal, as are the molar Gibbs energies (cf. Eqs. 7.1-9 and Problem 7.4); since temperature is being maintained constant, it would, of course, be the same in the two subsystems.

For a closed system maintained at constant temperature and pressure, we have

$$\frac{dU}{dt} = \dot{Q} - P\frac{dV}{dt} = \dot{Q} - \frac{d}{dt}(PV)$$

and

$$\frac{dS}{dt} = \frac{\dot{Q}}{T} + \dot{S}_{\text{gen}}$$

Again eliminating $\dot{Q}$ between these two equations gives

$$\frac{dU}{dt} + \frac{d}{dt}(PV) - \frac{d}{dt}(TS) = \frac{d}{dt}(U + PV - TS) = \frac{dG}{dt} = -T\dot{S}_{\text{gen}} \leq 0 \quad (7.1\text{-}11)$$

so that the equilibrium criterion here is

Most important of the
equilibrium criteria

$$\left.\begin{array}{r} G = \text{minimum} \\[2mm] \text{or} \\[2mm] \underline{G} = \text{minimum} \end{array}\right\} \quad \text{for equilibrium in a closed system at constant } T \text{ and } P$$

$$\textbf{(7.1-12)}$$

This equation also leads to the equality of the molar Gibbs energies (Eq. 7.1-9c) as a condition for equilibrium. Of course, since both temperature and pressure are held constant, the system is also at uniform temperature and pressure at equilibrium.

Finally, the equilibrium criterion for a system consisting of an element of fluid moving with the velocity of the fluid around it is also of interest, as such a choice of system arises in the study of continuous processing equipment used in the chemical industry. The tubular chemical reactor discussed in Chapter 14 is perhaps the most common example. Since each fluid element is moving with the velocity of the fluid surrounding it, there is no convected flow of mass into or out of this system. Therefore, each such element of mass in a pure fluid is a system closed to the flow of mass, and consequently is subject to precisely the same equilibrium criteria as the closed systems discussed above (i.e., Eq. 7.1-5, 7.1-10, or 7.1-12, depending on the constraints on the system).

The equilibrium and stability criteria for systems and constraints that are of interest in this book are collected in Table 7.1-1. As we will see in Chapter 8, these equilibrium and stability criteria are also valid for multicomponent systems. Thus the entries in this table form the basis for the analysis of phase and chemical equilibrium problems to be considered during much of the remainder of this book.

Using the method of analysis indicated here, it is also possible to derive the equilibrium criteria for systems subject to other constraints. However, this task is left to you (Problem 7.2).

As a simple example of the use of the equilibrium conditions, we consider a problem to which we intuitively know the solution, but want to show that the answer arises naturally from the analysis of this section.

ILLUSTRATION 7.1-1

Using the Equilibrium Condition to Solve a Simple Problem

Two identical metal blocks of mass M with initial temperatures $T_{1,i}$ and $T_{2,i}$, respectively, are in contact with each other in a well-insulated (adiabatic), constant-volume enclosure. Find the

Table 7.1-1 Equilibrium and Stability Criteria

System	Constraint	Equilibrium Criterion	Stability Criterion
Isolated, adiabatic fixed-boundary system	$U = \text{constant}$ $V = \text{constant}$	$S = \text{maximum}$ $dS = 0$	$d^2 S < 0$
Isothermal closed system with fixed boundaries	$T = \text{constant}$ $V = \text{constant}$	$A = \text{minimum}$ $dA = 0$	$d^2 A > 0$
Isothermal, isobaric closed system	$T = \text{constant}$ $P = \text{constant}$	$G = \text{minimum}$ $dG = 0$	$d^2 G > 0$
Isothermal, isobaric open system moving with the fluid velocity	$T = \text{constant}$ $P = \text{constant}$ $M = \text{constant}$	$G = \text{minimum}$ $dG = 0$	$d^2 G > 0$

final equilibrium temperatures of the metal blocks. (Of course, intuituively, we know the solution, $T_{1,f} = T_{2,f}$. However, we want to show that this arises naturally, though with some work, from the equilibrium condition, as we will encounter many problems, especially when we deal with mixtures, where our intuition will not help us identify the equilibrium state.)

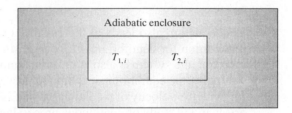

SOLUTION

We will choose the two metal blocks as the system for writing the balance equations. There is no exchange of mass between the blocks, so the mass balance does not provide any useful information. The energy balance for the system is

$$M\hat{U}_{1,f} + M\hat{U}_{2,f} - M\hat{U}_{1,i} - M\hat{U}_{2,i} = 0$$

or

$$M\hat{C}_V\left(T_{1,f} - T_{1,\text{ref}}\right) + M\hat{C}_V\left(T_{2,f} - T_{2,\text{ref}}\right) - M\hat{C}_V\left(T_{1,i} - T_{1,\text{ref}}\right) - M\hat{C}_V\left(T_{2,i} - T_{2,\text{ref}}\right) = 0$$

which reduces to

$$T_{1,f} - T_{1,i} + T_{2,f} - T_{2,i} = 0 \tag{a}$$

The entropy balance cannot be used directly to provide useful information since the entropy generation due to heat transfer cannot be evaluated at this stage in the calculation. However, we can use the fact that for a system in which M, U, and V are constant, at equilibrium the entropy is a maximum with respect to the independent variations within the system. The final entropy of the system is

$$S = M\hat{S}_{1,f} + M\hat{S}_{2,f} = M\hat{C}_V \ln\left(\frac{T_{1,f}}{T_{1,\text{ref}}}\right) + M\hat{C}_V \ln\left(\frac{T_{2,f}}{T_{2,\text{ref}}}\right)$$

The only possible variations are in the two final temperatures, $T_{1,f}$ and $T_{2,f}$. However, as the energy balance connects these two temperatures, only one of them can be considered to be independent, say $T_{1,f}$. Therefore, to identify the equilibrium state we set the differential of the entropy with respect to $T_{1,f}$ equal to zero:

$$\frac{dS}{dT_{1,f}} = 0 = \frac{d}{dT_{1,f}}\left[M\hat{C}_V \ln\left(\frac{T_{1,f}}{T_{1,\text{ref}}}\right) + M\hat{C}_V \ln\left(\frac{T_{2,f}}{T_{2,\text{ref}}}\right)\right]$$

$$= \frac{M\hat{C}_V}{T_{1,f}} + \frac{M\hat{C}_V}{T_{1,f}}\frac{dT_{2,f}}{dT_{1,f}}$$

or

$$\frac{dT_{2,f}}{dT_{1,f}} = -\frac{T_{2,f}}{T_{1,f}} \tag{b}$$

Now taking the derivative of the energy balance with respect to $T_{1,f}$, we obtain

$$\frac{d}{dT_{1,f}} \left[M\hat{C}_V (T_{1,f} - T_{1,i}) + M\hat{C}_V (T_{2,f} - T_{2,i}) \right] = 0 = M\hat{C}_V + M\hat{C}_V \frac{dT_{2,f}}{dT_{1,f}}$$

or

$$\frac{dT_{2,f}}{dT_{1,f}} = -1 \qquad \text{(c)}$$

Comparing Eqs. b and c, we have

$$\frac{dT_{2,f}}{dT_{1,f}} = -1 = -\frac{T_{1,f}}{T_{2,f}}$$

Clearly, the only way this combined equation can be satisfied is if $T_{1,f} = T_{2,f} = T_f$; that is, the final temperature of the two blocks must be equal, which is the intuitively obvious solution.

To calculate this final equilibrium temperature, we use the energy balance, Eq. a,

$$T_f - T_{1,i} + T_f - T_{2,i} = 2T_f - T_{1,i} - T_{2,i} = 0$$

which has the solution

$$T_f = \frac{T_{1,i} + T_{2,i}}{2}$$

Knowing the final state of the system, it is of interest to calculate the amount of entropy generated in the process of achieving equilibrium. We do this using the difference form of the entropy balance,

$$S_f - S_i = S_{\text{gen}}$$

$$
\begin{aligned}
S_{\text{gen}} = S_f - S_i \\
&= M\hat{C}_V \left[\ln \left(\frac{T_f}{T_{1,\text{ref}}} \right) + \ln \left(\frac{T_f}{T_{2,\text{ref}}} \right) \right] - M\hat{C}_V \left[\ln \left(\frac{T_{1,i}}{T_{1,\text{ref}}} \right) + \ln \left(\frac{T_{2,i}}{T_{2,\text{ref}}} \right) \right] \\
&= M\hat{C}_V \ln \left[\frac{T_{1,i} + T_{2,i}}{2T_{1,i}} \right] + M\hat{C}_V \ln \left[\frac{T_{1,i} + T_{2,i}}{2T_{2,i}} \right] = M\hat{C}_V \ln \left[\frac{(T_{1,i} + T_{2,i})^2}{4T_{1,i}T_{2,i}} \right]
\end{aligned}
$$

■

ILLUSTRATION 7.1-2

Proving the Equality of Gibbs Energies for Vapor-Liquid Equilibrium

Use the information in the steam tables of Appendix A.III to show that Eq. 7.1-9c is satisfied at 100°C and 0.101 35 MPa.

SOLUTION

From the saturated steam temperature table, we have at 100°C and 0.101 35 MPa

$$\hat{H}^L = 419.04 \text{ kJ/kg} \qquad \hat{H}^V = 2676.1 \text{ kJ/kg}$$
$$\hat{S}^L = 1.3069 \text{ kJ/(kg K)} \qquad \hat{S}^V = 7.3549 \text{ kJ/(kg K)}$$

Since $\hat{G} = \hat{H} - T\hat{S}$, we have further that

$$\hat{G}^L = 419.04 - 373.15 \times 1.3069 = -68.6 \text{ kJ/kg}$$

and

$$\hat{G}^V = 2676.1 - 373.15 \times 7.3549 = -68.4 \text{ kJ/kg}$$

which, to the accuracy of the calculations here, confirms that $\hat{G}^L = \hat{G}^V$ (or, equivalently, that $\underline{G}^L = \underline{G}^V$) at this vapor-liquid phase equilibrium state.   ∎

7.2   STABILITY OF THERMODYNAMIC SYSTEMS

In the previous section we used the result $dS = 0$ to identify the equilibrium state of an initially nonuniform system constrained to remain at constant mass, internal energy, and volume. In this section we explore the information content of the stability criterion

$$d^2S < 0 \quad \text{at constant } M, \ U, \text{ and } V$$

[The stability analysis for closed systems subject to other constraints (i.e., constant T and V or constant T and P) is similar to, and, in fact, somewhat simpler than the analysis here, and so it is left to you (Problem 7.3).]

By studying the sign of the second differential of the entropy, we are really considering the following question: Suppose that a small fluctuation in a fluid property, say temperature or pressure, occurs in some region of a fluid that was initially at equilibrium; is the character of the equilibrium state such that $d^2S < 0$, and the fluctuation will dissipate, or such that $d^2S > 0$, in which case the fluctuation grows until the system evolves to a new equilibrium state of higher entropy?

In fact, since we know that fluids exist in thermodynamically stable states (experimental observation 7 of Sec. 1.7), we will take as an empirical fact that $d^2S < 0$ for all real fluids at equilibrium, and instead establish the restrictions placed on the equations of state of fluids by this stability condition. We first study the problem of the intrinsic stability of the equilibrium state in a pure single-phase fluid, and then the mutual stability of two interacting systems or phases.

We begin the discussion of intrinsic stability by considering further the example of Figure 7.1-1 of the last section, equilibrium in a pure fluid at constant mass (actually, we will use number of moles), internal energy, and volume. Using the (imaginary) subdivision of the system into two subsystems, and writing the extensive properties N, U, V, and S as sums of these properties for each subsystem, we were able to show in Sec. 7.1 that the condition

$$dS = dS^I + dS^{II} = 0 \tag{7.2-1}$$

for all system variations consistent with the constraints (i.e., all variations in dN^I, dV^I, and dU^I) led to the requirements that at equilibrium

$$T^I = T^{II}$$
$$P^I = P^{II}$$

and

$$\underline{G}^I = \underline{G}^{II}$$

Continuing, we write an expression for the stability requirement $d^2S < 0$ for this system and obtain

$$d^2S = S^I_{UU}(dU^I)^2 + 2S^I_{UV}(dU^I)(dV^I) + S^I_{VV}(dV^I)^2$$
$$+ 2S^I_{UN}(dU^I)(dN^I) + 2S^I_{VN}(dV^I)(dN^I) + S^I_{NN}(dN^I)^2$$
$$+ S^{II}_{UU}(dU^{II})^2 + 2S^{II}_{UV}(dU^{II})(dV^{II}) + S^{II}_{VV}(dV^{II})^2$$
$$+ 2S^{II}_{UN}(dU^{II})(dN^{II}) + 2S^{II}_{VN}(dV^{II})(dN^{II}) + S^{II}_{NN}(dN^{II})^2 < 0 \quad \textbf{(7.2-2a)}$$

In this equation we have used the abbreviated notation

$$S^I_{UU} = \left(\frac{\partial^2 S}{\partial U^2}\right)^I_{V,N} \qquad S^I_{UV} = \frac{\partial}{\partial U}\bigg|_{V,N}\left(\frac{\partial S}{\partial V}\right)^I_{U,N} \qquad \text{etc.}$$

Since the total number of moles, total internal energy, and total volume of the composite system are fixed, we have, as in Eqs. 7.1-7,

$$dN^I = -dN^{II} \qquad dU^I = -dU^{II} \qquad dV^I = -dV^{II}$$

and

$$d^2S = (S^I_{UU} + S^{II}_{UU})(dU^I)^2 + 2(S^I_{UV} + S^{II}_{UV})(dU^I)(dV^I)$$
$$+ (S^I_{VV} + S^{II}_{VV})(dV^I)^2 + 2(S^I_{UN} + S^{II}_{UN})(dU^I)(dN^I)$$
$$+ 2(S^I_{VN} + S^{II}_{VN})(dV^I)(dN^I) + (S^I_{NN} + S^{II}_{NN})(dN^I)^2 \quad \textbf{(7.2-2b)}$$

Furthermore, since the same fluid in the same state of aggregation is present in regions I and II, and since we have already established that the temperature, pressure, and molar Gibbs energy each have the same value in the two regions, the value of any state property must be the same in the two subsystems. It follows that any thermodynamic derivative that can be reduced to combinations of intensive variables must have the same value in the two regions of the fluid. The second derivatives (Eq. 7.2-2b), as we will see shortly, are combinations of intensive and extensive variables. However, the quantities NS_{xy}, where x and y denote U, V, or N, are intensive variables. Therefore, it follows that

$$N^I S^I_{xy} = N^{II} S^{II}_{xy} \quad \textbf{(7.2-3)}$$

Using Eqs. 7.1-7 and 7.2-3 in Eq. 7.2-2 yields

$$d^2S = \left(\frac{N^I + N^{II}}{N^I N^{II}}\right)[N^I S^I_{UU}(dU^I)^2 + 2N^I S^I_{UV}(dU^I)(dV^I)$$
$$+ N^I S^I_{VV}(dV^I)^2 + 2N^I S^I_{UN}(dU^I)(dN^I)$$
$$+ 2N^I S^I_{VN}(dV^I)(dN^I) + N^I S^I_{NN}(dN^I)^2] < 0 \quad \textbf{(7.2-4a)}$$

The term $(N^I + N^{II})/N^I N^{II}$ must be greater than zero since mole numbers can only be positive. Also, we can eliminate the superscripts from the products NS_{xy}, as they are equal in the two regions. Therefore, the inequality Eq. 7.2-4a can be rewritten as

$$NS_{UU}(dU^I)^2 + 2NS_{UV}(dU^I)(dV^I) + NS_{VV}(dV^I)^2 + 2NS_{UN}(dU^I)(dN^I)$$
$$+ 2NS_{VN}(dV^I)(dN^I) + NS_{NN}(dN^I)^2 < 0 \quad \textbf{(7.2-4b)}$$

Equations 7.2-3 and 4 must be satisfied for all variations in N^{I}, U^{I}, and V^{I} if the fluid is to be stable. In particular, since Eq. 7.2-4b must be satisfied for all variations in U^{I} ($dU^{\mathrm{I}} \neq 0$) at fixed values of N^{I} and V^{I} (i.e., $dN^{\mathrm{I}} = 0$ and $dV^{\mathrm{I}} = 0$), stable fluids must have the property that

$$NS_{\mathrm{UU}} < 0 \tag{7.2-5a}$$

Similarly, by considering variations in volume at fixed internal energy and mole number, and variations in mole number at fixed volume and internal energy, we obtain

$$NS_{\mathrm{VV}} < 0 \tag{7.2-5b}$$

and

$$NS_{\mathrm{NN}} < 0 \tag{7.2-5c}$$

as additional conditions for fluid stability.

More severe restrictions on the equation of state result from demanding that Eq. 7.2-4b be satisfied for all possible and simultaneous variations in internal energy, volume, and mole number and not merely for variations in one property with the others held fixed. Unfortunately, the present form of Eq. 7.2-4b is not well suited for studying this more general situation since the cross-terms (i.e., $dU^{\mathrm{I}}dV^{\mathrm{I}}$, $dU^{\mathrm{I}}dN^{\mathrm{I}}$, and $dV^{\mathrm{I}}dN^{\mathrm{I}}$) may be positive or negative depending on the sign of the variations dU^{I}, dV^{I}, and dN^{I}, so that little can be said about the coefficients of these terms.

By much algebraic manipulation (Problem 7.32), Eq. 7.2-4b can be written as

$$\theta_1(dX_1)^2 + \theta_2(dX_2)^2 + \theta_3(dX_3)^2 < 0 \tag{7.2-6}$$

where

$$\theta_1 = NS_{\mathrm{UU}}$$
$$\theta_2 = (NS_{\mathrm{UU}}NS_{\mathrm{VV}} - N^2 S_{\mathrm{UV}}^2)/NS_{\mathrm{UU}}$$
$$\theta_3 = \frac{(NS_{\mathrm{UU}}NS_{\mathrm{NN}} - N^2 S_{\mathrm{UN}}^2)}{NS_{\mathrm{UU}}} - \frac{(NS_{\mathrm{UU}}NS_{\mathrm{VN}} - NS_{\mathrm{UV}}NS_{\mathrm{UN}})^2}{NS_{\mathrm{UU}}(NS_{\mathrm{UU}}NS_{\mathrm{VV}} - N^2 S_{\mathrm{UV}}^2)}$$
$$dX_1 = dU^{\mathrm{I}} + \frac{S_{\mathrm{UV}}}{S_{\mathrm{UU}}}dV^{\mathrm{I}} + \frac{S_{\mathrm{UN}}}{S_{\mathrm{UU}}}dN^{\mathrm{I}}$$
$$dX_2 = dV^{\mathrm{I}} + \frac{(S_{\mathrm{UU}}S_{\mathrm{VN}} - S_{\mathrm{UV}}S_{\mathrm{UN}})}{(S_{\mathrm{UU}}S_{\mathrm{VV}} - S_{\mathrm{UV}}^2)}dN^{\mathrm{I}}$$

and

$$dX_3 = dN^{\mathrm{I}}$$

The important feature of Eq. 7.2-6 is that it contains only square terms in the system variations. Thus, $(dX_1)^2$, $(dX_2)^2$, and $(dX_3)^2$ are greater than or equal to zero regardless of whether dU^{I}, dV^{I}, and dN^{I} individually are positive or negative. Consequently, if

$$\theta_1 = NS_{\mathrm{UU}} \leq 0 \tag{7.2-7a}$$
$$\theta_2 = \frac{NS_{\mathrm{UU}}NS_{\mathrm{VV}} - N^2 S_{\mathrm{UV}}^2}{NS_{\mathrm{UU}}} \leq 0 \tag{7.2-7b}$$
$$\theta_3 \leq 0 \tag{7.2-7c}$$

Eq. 7.2-6, and hence Eq. 7.2-4b, will be satisfied for all possible system variations. Equations 7.2-7 provide more restrictive conditions for fluid stability than Eqs. 7.2-5.

It now remains to evaluate the various entropy derivatives, so that the stability restrictions of Eqs. 7.2-7 can be put into a more usable form. Starting from

$$NS_{UU} = N\frac{\partial}{\partial U}\bigg|_{N,V}\left(\frac{\partial S}{\partial U}\right)_{N,V} = N\left(\frac{\partial(1/T)}{\partial U}\right)_{N,V} = -\frac{N}{T^2}\left(\frac{\partial T}{\partial U}\right)_{N,V}$$

and using that for the open system that

$$dU = NC_V\, dT + \left[T\left(\frac{\partial P}{\partial T}\right)_V - P\right]dV + \underline{G}\, dN \tag{7.2-8}$$

leads to

$$\left(\frac{\partial U}{\partial T}\right)_{V,N} = NC_V$$

$$NS_{UU} = -\frac{1}{T^2 C_V} < 0$$

Since T is absolute temperature and positive, one condition for the existence of a stable equilibrium state of a fluid is that

First or thermal stability criterion

$$\boxed{C_V > 0} \tag{7.2-9}$$

That is, the constant-volume heat capacity must be positive, so that internal energy increases as the fluid temperature increases.

Next, we note that

$$NS_{UV} = N\frac{\partial}{\partial U}\bigg|_{V,N}\left(\frac{\partial S}{\partial V}\right)_{N,U} = N\frac{\partial}{\partial V}\bigg|_{U,N}\left(\frac{\partial S}{\partial U}\right)_{V,N}$$

$$= N\frac{\partial}{\partial V}\bigg|_{U,N}\left(\frac{1}{T}\right) = -\frac{N}{T^2}\left(\frac{\partial T}{\partial V}\right)_{U,N}$$

and from Eq. 7.2-8,

$$\left(\frac{\partial T}{\partial V}\right)_{U,N} = -\frac{\left[T\left(\frac{\partial P}{\partial T}\right)_V - P\right]}{NC_V}$$

to obtain

$$NS_{UV} = \frac{\left[T\left(\frac{\partial P}{\partial T}\right)_V - P\right]}{C_V T^2} \tag{7.2-10a}[1]$$

---

[1] Note, from these equations, that NS_{UU}, NS_{UV}, and NS_{VV} are intensive variables, as was suggested earlier, whereas S_{UU}, S_{UV}, and S_{VV} are proportional to N^{-1}.

Similarly, but with a great deal more algebra, we can show that

$$NS_{VV} = \frac{1}{T}\left(\frac{\partial P}{\partial \underline{V}}\right)_T - \frac{1}{C_V T^2}\left[T\left(\frac{\partial P}{\partial T}\right)_{\underline{V}} - P\right]^2 \tag{7.2-10b}[1]$$

and

$$\theta_2 = \frac{NS_{UU}NS_{VV} - N^2 S_{UV}^2}{NS_{UU}} = \frac{1}{T}\left(\frac{\partial P}{\partial \underline{V}}\right)_T$$

Thus, a further stability restriction on the equation of state of a substance, since T is positive, is that

**Second or mechanical
stability criterion**

$$\boxed{\left(\frac{\partial P}{\partial \underline{V}}\right)_T < 0}$$

or

$$\kappa_T = -\frac{1}{\underline{V}}\left(\frac{\partial \underline{V}}{\partial P}\right)_T > 0 \tag{7.2-11}$$

where κ_T is the isothermal compressibility of the fluid, introduced in Sec. 6.2. This result indicates that if a fluid is to be stable, its volumetric equation of state must be such that the fluid volume decreases as the pressure increases at constant temperature. As we will see shortly, this restriction has important implications in the interpretation of phase behavior from the equation of state of a fluid.

Finally, and with a *great* deal more algebra (Problem 7.32), one can show that θ_3 is identically equal to zero and thus does not provide any further restrictions on the equation of state.

The main conclusion from this exercise is that if a fluid is to exist in a stable equilibrium state, that is, an equilibrium state in which all small internal fluctuations will dissipate rather than grow, the fluid must be such that

$$C_V > 0 \tag{7.2-12}$$

and

$$\left(\frac{\partial P}{\partial \underline{V}}\right)_T < 0 \quad \text{or} \quad \kappa_T > 0 \tag{7.2-13}$$

Alternatively, since all real fluids exist in thermodynamically stable states, Eqs. 7.2-12 and 7.2-13 must be satisfied for real fluids. In fact, no real fluid state for which either $(\partial P/\partial \underline{V})_T > 0$ or $C_V < 0$ has yet been found.

Equations 7.2-12 and 7.2-13 may be thought of as part of the philosophical content of thermodynamics. In particular, thermodynamics alone does not give information on the heat capacity or the equation of state of any substance; such information can be gotten only from statistical mechanics or experiment. However, thermodynamics does provide restrictions or consistency relations that must be satisfied by such data; Eqs. 7.2-12 and 7.2-13 are examples of this. (Consistency relations for mixtures are discussed in later chapters.)

ILLUSTRATION 7.2-1

Using the Steam Tables to Show That the Stability Conditions Are Satisfied for Steam

Show that Eqs. 7.2-12 and 7.2-13 are satisfied by superheated steam.

SOLUTION

It is easiest to use Eq. 7.2-13 in the form $(\partial P/\partial \underline{V})_T < 0$, which requires that the volume decrease as the pressure increases at constant temperature. This is seen to be true by using the superheated steam table and observing that $\hat{V}$ decreases as P increases at fixed temperature. For example, at 1000°C

P (MPa)	0.50	0.8	1.0	1.4	1.8	2.5
$\hat{V}$ (m³/kg)	1.1747	0.7340	0.5871	0.4192	0.3260	0.2346

Proving that $C_V > 0$ or $\hat{C}_V > 0$ is a bit more difficult since

$$\hat{C}_V = \left(\frac{\partial \hat{U}}{\partial T} \right)_V$$

and the internal energy is not given at constant volume. Therefore, interpolation methods must be used. As an example of how the calculation is done, we start with the following data from the superheated vapor portion of the steam tables.

T (°C)	$P = 1.80$ MPa		$P = 2.00$ MPa	
	$\hat{V}$ (m³/kg)	$\hat{U}$ (kJ/kg)	$\hat{V}$ (m³/kg)	$\hat{U}$ (kJ/kg)
800	0.2742	3657.6	0.2467	3657.0
900	0.3001	3849.9	0.2700	3849.3
1000	0.3260	4048.5	0.2933	4048.0

To proceed, we need values of the internal energy at two different temperatures and the same specific volume. We will use $P = 2.00$ MPa and $T = 1000$°C as one point; at these conditions $\hat{V} = 0.2933$ m³/kg and $\hat{U} = 4048.0$ kJ/kg. We now need to find the temperature at which $\hat{V} = 0.2933$ m³/kg at $P = 1.80$ MPa. We use linear interpolation for this:

$$\frac{T - 800}{900 - 800} = \frac{\hat{V}(T, 1.80 \text{ MPa}) - \hat{V}(800°C, 1.80 \text{ MPa})}{\hat{V}(900°C, 1.80 \text{ MPa}) - \hat{V}(800°C, 1.80 \text{ MPa})} = \frac{0.2933 - 0.2742}{0.3001 - 0.2742}$$

so that $T = 873.75$°C. Next we need the internal energy $\hat{U}$ at $T = 873.75$°C and $P = 1.80$ MPa (since at these conditions $\hat{V} = 0.2933$ m³/kg). Again using linear interpolation,

$$\frac{873.75 - 800}{900 - 800} = \frac{\hat{U}(873.75°C, 1.80 \text{ MPa}) - \hat{U}(800°C, 1.80 \text{ MPa})}{\hat{U}(900°C, 1.80 \text{ MPa}) - \hat{U}(800°C, 1.80 \text{ MPa})}$$

$$= \frac{\hat{U}(873.75°C, 1.80 \text{ MPa}) - 3657.6}{3849.9 - 3657.6}$$

we find that

$$\hat{U}(873.75°C, 1.80 \text{ MPa}) = \hat{U}(873.75°C, 0.2933 \text{ m}^3/\text{kg}) = 3799.4 \text{ kJ/kg}$$

Finally, replacing the derivative by a finite difference, and for the average temperature [i.e., $T = (1000 + 873.75)/2 = 936.9°C$], we have

$$\hat{C}_V(T = 936.9°C, \hat{V} = 0.2933 \text{ m}^3/\text{kg})$$

$$\approx \frac{\hat{U}(1000°C, 0.2933 \text{ m}^3/\text{kg}) - \hat{U}(873.75°C, 0.2933 \text{ m}^3/\text{kg})}{1000°C - 873.75°C}$$

$$= \frac{4048.0 - 3799.4}{1000 - 873.75} = 1.969 \frac{\text{kJ}}{\text{kg K}} > 0$$

Similarly, we would find that $\hat{C}_V > 0$ at all other conditions.   ■

Next we consider the problems of identifying the equilibrium state for two interacting phases of the same molecular species but in different states of aggregation, and of determining the requirements for the stability of this state. An example of this is a vapor and a liquid in equilibrium. To be general, we again consider a composite system isolated from its environment, except that here the boundary between the two subsystems is the real interface between the phases. For this system, we have

$$\begin{aligned}
S &= S^I + S^{II} \\
N &= N^I + N^{II} = \text{constant} \\
V &= V^I + V^{II} = \text{constant} \\
U &= U^I + U^{II} = \text{constant}
\end{aligned} \qquad \textbf{(7.2-14)}$$

Since N, U, and V are fixed, the equilibrium condition is that the entropy should attain a maximum value. Now, however, we allow for the fact that the states of aggregation in regions I and II are different, so that the fluids in these regions may follow different equations of state (or the different vapor and liquid branches of the same equation of state).

Using the analysis of Eqs. 7.1-6, 7.1-7, and 7.1-8, we find that at equilibrium (i.e., when $dS = 0$),

Important criteria for phase equilibria under all constraints

$$\begin{aligned}
T^I &= T^{II} & \textbf{(7.2-15a)} \\
P^I &= P^{II} & \textbf{(7.2-15b)} \\
\underline{G}^I &= \underline{G}^{II} & \textbf{(7.2-15c)}
\end{aligned}$$

Here Eqs. 7.2-15a and b provide the obvious conditions for equilibrium, and since two different phases are present, Eq. 7.2-15c provides a less obvious condition for equilibrium.

Next, from the stability condition $d^2S < 0$, we obtain (following the analysis that led to Eq. 7.2-2b)

$$\begin{aligned}
d^2S = &\{S_{VV}^I + S_{VV}^{II}\}(dV^I)^2 + 2\{S_{UV}^I + S_{UV}^{II}\}(dU^I)(dV^I) \\
&+ \{S_{UU}^I + S_{UU}^{II}\}(dU^I)^2 + 2\{S_{VN}^I + S_{VN}^{II}\}(dV^I)(dN^I) \\
&+ \{S_{NN}^I + S_{NN}^{II}\}(dN^I)^2 + 2\{S_{UN}^I + S_{UN}^{II}\}(dU^I)(dN^I) \qquad \textbf{(7.2-16)}
\end{aligned}$$

Here, however, the two partial derivatives in each of the bracketed terms need not be equal, since the two phases are in different states of aggregation and thus obey different equations of state, or different roots of the same equation of state. It is clear from a comparison with Eq. 7.2-2b that a sufficient condition for Eq. 7.2-16 to be satisfied is that each phase to be intrinsically stable; that is, Eq. 7.2-16 is satisfied if, for *each* of the coexisting phases, the equations

$$C_V > 0 \quad \text{and} \quad \left(\frac{\partial P}{\partial \underline{V}}\right)_T < 0$$

are satisfied. Therefore, a condition for the mutual stability of two interacting subsystems is that each subsystem be intrinsically stable.

7.3   PHASE EQUILIBRIA: APPLICATION OF THE EQUILIBRIUM AND STABILITY CRITERIA TO THE EQUATION OF STATE

Figure 7.3-1 shows the shapes of different isotherms computed using a typical equation of state (for illustration we have used the van der Waals equation of state). In this figure the isotherms are labeled so that $T_5 > T_4 > T_3 > T_2 > T_1$. The isotherm T_3 has a single point, C, for which $(\partial P/\partial \underline{V})_T = 0$; at all other points on this isotherm $(\partial P/\partial \underline{V})_T < 0$. On the isotherms T_4 and T_5, $(\partial P/\partial \underline{V})_T < 0$ everywhere, whereas on the isotherms T_1 and T_2, $(\partial P/\partial \underline{V})_T < 0$ in some regions and $(\partial P/\partial \underline{V})_T > 0$ in other regions (i.e., between the points A and B on isotherm T_1, and the points A' and B' on T_2). The criterion for fluid stability requires that $(\partial P/\partial \underline{V})_T < 0$, which is satisfied for the isotherms T_4 and T_5, but not in the aforementioned regions of the T_1 and T_2 isotherms. Thus we conclude that the regions A to B and A' to B' of the isotherms T_1 and T_2, respectively, are not physically realizable; that is, they will not be observed in any experiment since a fluid in these states is not stable, and instead would go to an appropriate stable state.

This observation raises some question about the interpretation to be given to the T_1 and T_2 isotherms. We cannot simply attribute these oddities to a peculiarity of the

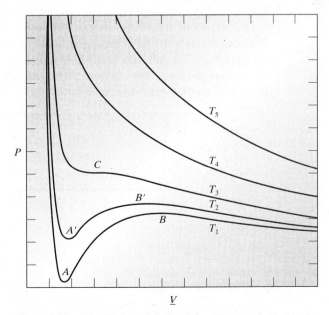

Figure 7.3-1 Isotherms of the van der Waals equation in the pressure-volume plane.

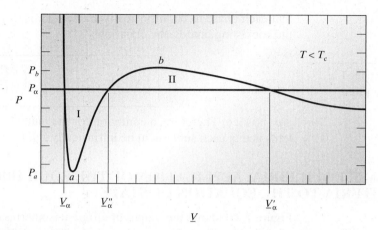

Figure 7.3-2 A low-temperature isotherm of the van der Waals equation.

van der Waals equation because many other, more accurate equations of state show essentially the same behavior. Some insight into the physical meaning of isotherms such as T_1 can be gained from Fig. 7.3-2, which shows this isotherm separately. If we look at any isobar (constant-pressure line) between P_A and P_B in this figure, such as P_α, we see that it intersects the equation of state three times, corresponding to the fluid volumes $\underline{V}_\alpha$, $\underline{V}'_\alpha$, and $\underline{V}''_\alpha$. One of these, $\underline{V}''_\alpha$, is on the part of the isotherm that is unattainable by the stability criterion. However, the other two intersections, at $\underline{V}_\alpha$ and $\underline{V}'_\alpha$, are physically possible. This suggests that at a given pressure and temperature the system can have two different volumes, a conclusion that apparently contradicts the experimental observation of Chapter 1 that two state variables completely determine the state of a single-component, single-phase system. However, this can occur if equilibrium can exist between two phases of the same species that are in different states of aggregation (and hence density). The equilibrium between liquid water and steam at 100°C and 101.325 kPa (1 atm) is one such example.

One experimental observation in phase equilibrium is that the two coexisting equilibrium phases must have the same temperature and pressure. Clearly, the arguments given in Secs. 7.1 and 7.2 establish this. Another experimental observation is that as the pressure is lowered along an isotherm on which a liquid-vapor phase transition occurs, the actual volume-pressure behavior is as shown in Fig. 7.3-3, and not as in Fig. 7.3-2.

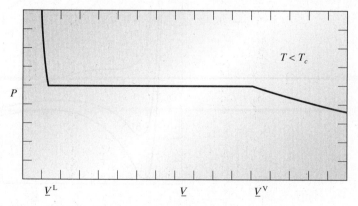

Figure 7.3-3 A low-temperature isotherm of a real fluid.

That is, there is a portion of the isotherm where the specific volume varies continuously at fixed temperature and pressure; this is the two-phase coexistence region, here the vapor-liquid coexistence region.

The variation of the overall (or two-phase) specific volume in this region arises from the fact that although the specific volumes of the vapor and liquid phases are fixed (since the temperature and pressure are fixed in each of the one-phase equilibrium subsystems), the fraction of the mixture that is vapor, ω^V, can vary continuously from 0 to 1. In the two-phase region the specific volume of the two-phase mixture is given by

$$\underline{V} = \omega^V \underline{V}^V + \omega^L \underline{V}^L = \omega^V \underline{V}^V + (1 - \omega^V)\underline{V}^L \tag{7.3-1a}$$

where ω^V and ω^L are fractions of vapor and liquid, respectively, on a molar basis. (Equation 7.3-1a could also be written using volumes per unit mass and mass fractions.) These fractions can vary between 0 and 1 subject to the condition that $\omega^V + \omega^L = 1$. Solving for ω^V yields

Maxwell or lever rule

$$\omega^V = \frac{\underline{V} - \underline{V}^L}{\underline{V}^V - \underline{V}^L} \tag{7.3-1b}$$

and

$$\frac{\omega^V}{1 - \omega^V} = \frac{\underline{V} - \underline{V}^L}{\underline{V}^V - \underline{V}} \tag{7.3-1c}$$

Equations analogous to those here also hold for the $\underline{H}$, $\underline{U}$, $\underline{G}$, $\underline{S}$, and $\underline{A}$. Equations of the form of Eq. 7.3-1c are the **Maxwell's rules** or **lever rules** first discussed in Sec. 3.3.

ILLUSTRATION 7.3-1

Computing the Properties of a Two-Phase Mixture

Compute the total volume, total enthalpy, and total entropy of 1 kg of water at 100°C, half by weight of which is steam and the remainder liquid water.

SOLUTION

From the saturated steam temperature table at 100°C, the equilibrium pressure is 0.101 35 MPa and

$$\hat{V}^L = 0.001\ 004\ \text{m}^3/\text{kg} \qquad \hat{V}^V = 1.6729\ \text{m}^3/\text{kg}$$
$$\hat{H}^L = 419.04\ \text{kJ/kg} \qquad \hat{H}^V = 2676.1\ \text{kJ/kg}$$
$$\hat{S}^L = 1.3069\ \text{kJ/(kg K)} \qquad \hat{S}^V = 7.3549\ \text{kJ/(kg K)}$$

Using Eq. 7.3-1a on a mass basis gives

$$\hat{V} = 0.5 \times 0.001\ 004 + 0.5 \times 1.6729 = 0.836\ 45\ \text{m}^3/\text{kg}$$

The analogous equation for enthalpy is

$$\hat{H} = \omega^L \hat{H}^L + \omega^V \hat{H}^V = 0.5 \times 419.04 + 0.5 \times 2676.1 = 1547.6\ \frac{\text{kJ}}{\text{kg}}$$

and that for entropy is

$$\hat{S} = \omega^L \hat{S}^L + \omega^V \hat{S}^V = 0.5 \times 1.3069 + 0.5 \times 7.3549 = 4.3309 \ \frac{kJ}{kg \ K}$$

∎

Consequently, the continuous variation of specific volume of the vapor-liquid mixture at fixed temperature and pressure is a result of the continuous change in the fraction of the mixture that is vapor. The conclusion, then, is that an isotherm such as that shown in Fig. 7.3-2 is an approximate representation of the real phase behavior (shown in Fig. 7.3-3) by a relatively simple analytic equation of state. In fact, it is impossible to represent the discontinuities in the derivative $(\partial P / \partial \underline{V})_T$ that occur at $\underline{V}^L$ and $\underline{V}^V$ with *any* analytic equation of state. By its sigmoidal behavior in the two-phase region, the van der Waals equation of state is somewhat qualitatively and crudely exhibiting the essential features of vapor-liquid phase equilibrium; historically, it was the first equation of state to do so.

We can improve the representation of the two-phase region when using the van der Waals or other analytic equations of state by recognizing that all **van der Waals loops**, such as those in Fig. 7.3-2, should be replaced by horizontal lines (isobars), as shown in Fig. 7.3-3. This construction ensures that the equilibrium phases will have the same temperature and pressure (see Eqs. 7.1-9a and b). The question that remains is at which pressure should the isobar be drawn, since any pressure such that

$$P_a < P < P_b$$

will yield an isotherm like that in Fig. 7.3-3. The answer is that the pressure chosen must satisfy the last condition for equilibrium, $\underline{G}^I = \underline{G}^{II}$.

To identify the equilibrium pressure, we start from Eq. 4.2-8b,

$$d\underline{G} = \underline{V} \, dP - \underline{S} \, dT$$

and recognize that for the integration between any two points along an isotherm of the equation of state, we have

$$\Delta \underline{G} = \int_{P_1}^{P_2} \underline{V} \, dP$$

Thus, for a given equation of state we can identify the equilibrium pressure for each temperature by arbitrarily choosing pressures P_α along the van der Waals loop, until we find one for which

Vapor-liquid coexistence pressure or vapor pressure

$$\underline{G}^V - \underline{G}^L = 0 = \int_{P_a}^{P_a} \underline{V} \, dP + \int_{P_a}^{P_b} \underline{V} \, dP + \int_{P_b}^{P_a} \underline{V} \, dP \qquad (7.3\text{-}2)$$

Here the specific volume in each of the integrations is to be computed from the equation of state for the appropriate part of the van der Waals loop. Alternatively, we can find the equilibrium pressure graphically by noting that Eq. 7.3-2 requires that areas I and II between the van der Waals loop and the constant pressure line in Fig. 7.3-2 be equal at the pressure at which the vapor and liquid exist in equilibrium. This vapor-liquid coexistence pressure, which is a function of temperature, is called the **vapor pressure** of the liquid and will be denoted by $P^{vap}(T)$.

We can continue in the manner described here to determine the phase behavior of the fluid for all temperatures and pressures. For the van der Waals fluid, this result is

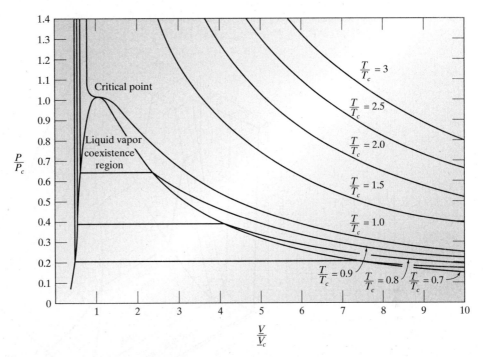

Figure 7.3-4 The van der Waals fluid with the vapor-liquid coexistence region identified.

shown in Fig. 7.3-4. An important feature of this figure is the dome-shaped, two-phase coexistence region. The inflection point C of Fig. 7.3-1 is the peak of this dome, and therefore is the highest temperature at which the condensed phase (the liquid) can exist; this point is called the **critical point** of the liquid.

It is worthwhile retracing the steps followed in identifying the existence and location of the two-phase region in the P-V plane:

1. The stability condition $(\partial P/\partial \underline{V})_T < 0$ was used to identify the unstable region of an isotherm and thereby establish the existence of a two-phase region.
2. The conditions $T^{\mathrm{I}} = T^{\mathrm{II}}$ and $P^{\mathrm{I}} = P^{\mathrm{II}}$ were then used to establish the shape (but not the location) of the horizontal coexistence line in the P-V plane.
3. Finally, the equilibrium condition $\underline{G}^{\mathrm{I}} = \underline{G}^{\mathrm{II}}$ was used to locate the position of the coexistence line.

A more detailed representation of phase equilibrium in a pure fluid, including the presence of a single solid phase,[2] is given in the three-dimensional PVT phase diagram of Fig. 7.3-5. Such complete phase diagrams are rarely available, although data may be available in the form of Fig. 7.3-4, which is a projection of the more complete diagram onto the P-V plane, and Fig. 7.3-6, which is the projection onto the P-T plane.

The concepts of phase equilibrium and the critical point can also be considered from a somewhat different point of view. Presume it were possible to compute the Gibbs energy as a function of temperature and pressure for any phase, either from an equation of state, experimental data, or statistical mechanics. Then, at fixed pressure, one could

---

[2]If several solid phases occur corresponding to different crystal structures, as is frequently the case, the solid region is partitioned into several regions.

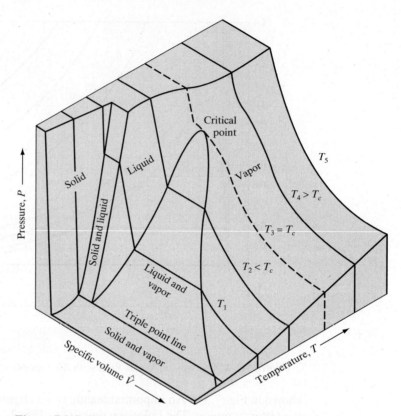

Figure 7.3-5 The PVT phase diagram for a substance with a single solid phase. [Adapted from J. Kestin, *A Course in Thermodynamics,* vol. 1. © 1966 by Blaisdell Publishing Co. (John Wiley & Sons, Inc.) Used with permission of John Wiley & Sons, Inc.]

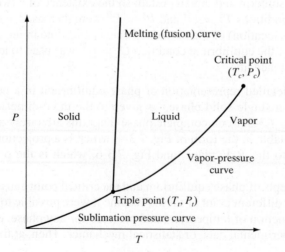

Figure 7.3-6 Phase diagram in the *P-T* plane.

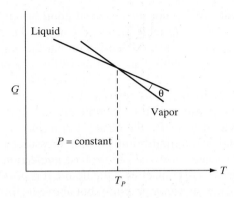

Figure 7.3-7 The molar Gibbs energy as a function of temperature for the vapor and liquid phases of the same substance.

plot $\underline{G}$ as a function of T for each phase, as shown in Fig. 7.3-7 for the vapor and liquid phases. From the equilibrium condition that $\underline{G}$ be a minimum, one can conclude that the liquid is the equilibrium phase at temperatures below T_P, that the vapor is the equilibrium phase above T_P, and that both phases are present at the phase transition temperature T_P.

If such calculations are repeated for a wide range of temperatures and pressures, it is observed that the angle of intersection θ between the liquid and vapor Gibbs energy curves decreases as the pressure (and temperature) at which the intersection occurs increases (provided $P \le P_c$). At the critical pressure, the two Gibbs energy curves intersect, with $\theta = 0$; that is, the two curves are collinear for some range of T around the critical temperature T_c. Thus, at the critical point,

$$\left(\frac{\partial \underline{G}^L}{\partial T}\right)_P = \left(\frac{\partial \underline{G}^V}{\partial T}\right)_P$$

Further, since $(\partial \underline{G}/\partial T)_P = -\underline{S}$, we have that at the critical point

$$\underline{S}^L(T_c, P_c) = \underline{S}^V(T_c, P_c)$$

Also, for the coexisting phases at equilibrium, we have

$$\underline{G}^L(T_c, P_c) = \underline{G}^V(T_c, P_c)$$

by Eq. 7.2-15c. Since the molar Gibbs energy, molar entropy, temperature, and pressure each have the same value in the vapor and the liquid phases, the values of all other state variables must be identical in the two equilibrium phases at the critical point. Consequently, the vapor and liquid phases become indistinguishable at the critical point. This is exactly what is experimentally observed. At all temperatures higher than the critical temperature, regardless of the pressure, only the vapor phase exists. This is the reason for the abrupt terminus of the vapor-liquid coexistence line in the pressure-temperature plane (Fig. 7.3-6). [Thus, we have two ways of recognizing the fluid critical point: first, as the peak in the vapor-liquid coexistence curve in the P-V plane (Fig. 7.3-4), and second, as the terminus of the vapor-liquid coexistence curve in the P-T plane.]

Also interesting is the fluid **triple point,** which is the intersection of the solid-liquid, liquid-vapor, and solid-vapor coexistence curves. It is the only point on the phase diagram where the solid, liquid, and vapor coexist at equilibrium. Since the solid-liquid coexistence curve generally has a steep slope (see Fig. 7.3-6), the triple-point temperature for most fluids is close to the normal melting temperature, that is, the melting temperature at atmospheric pressure (see Problem 7.10).

Although, in general, we are not interested in equilibrium states that are unstable to large perturbations (the metastable states of Chapter 1), superheated liquids and subcooled vapors do occur and are sufficiently familiar that we will briefly relate these states to the equilibrium and stability discussions of this chapter. For convenience, the van der Waals equation of state and Fig. 7.3-2 are the basis for this discussion, though the concepts involved are by no means restricted to this equation. We start by noticing that although the liquid phase is thermodynamically stable along the isotherm shown in Fig. 7.3-2 down to a pressure P_a, the phase equilibrium analysis indicates that the vapor, and not the liquid, is the equilibrium phase at pressures below the vapor pressure $P_a = P^{\text{vap}}(T_1)$. If care is taken to avoid vapor-phase nucleation sites, such as by having only clean, smooth surfaces in contact with the liquid, it is possible to maintain a liquid at fixed temperature below its vapor pressure (but above its limit of stability, P_a), or at fixed pressure at a temperature higher than its boiling temperature, without having the liquid boil. Such a liquid is said to be **superheated**. The metastability of this state is illustrated by the fact that a superheated liquid, if slightly perturbed, may vaporize with explosive violence. (To prevent this occurrence, "boiling stones" are used in chemistry laboratory experiments.) It is also possible, if no condensation nucleation sites, such as dust particles, are present, to prepare a vapor at a pressure higher than the liquid-vapor coexistence pressure or vapor pressure at the given temperature, but below its limit of stability [i.e., between $P_a = P^{\text{vap}}(T_1)$ and P_b in Fig. 7.3-2] or, at a temperature lower than the liquid boiling temperature. Such a vapor is termed **subcooled** and is also metastable. (See Problem 7.8.) Why superheating and subcooling occur is discussed in Sec. 7.8.

At sufficiently low temperatures, the van der Waals equation predicts that the limit of stability of the liquid phase occurs at negative values of pressure, that is, that a liquid could support a tensile force. In fact, such behavior has been observed with water in capillary tubes and is thought to be important in the vascular system of plants.[3]

7.4 THE MOLAR GIBBS ENERGY AND FUGACITY OF A PURE COMPONENT

In this section we consider how one uses an equation of state to identify the states of vapor-liquid equilibrium in a pure fluid. The starting point is the equality of molar Gibbs energies in the coexisting phases,

$$\underline{G}^{\text{L}}(T, P) = \underline{G}^{\text{V}}(T, P) \tag{7.1-9c}$$

To proceed, we note that from Eq. 6.2-8b,

$$d\underline{G} = -\underline{S}\,dT + \underline{V}\,dP$$

so that

$$\left(\frac{\partial \underline{G}}{\partial T}\right)_P = -\underline{S} \tag{7.4-1}$$

and

$$\left(\frac{\partial \underline{G}}{\partial P}\right)_T = \underline{V} \tag{7.4-2}$$

---

[3]For a review of water under tension, especially in biological systems, see P. F. Scholander, *Am. Sci.* **60**, 584 (1972).

Since our presumption here is that we have an equation of state from which we can compute $\underline{V}$ as a function of T and P, only Eq. 7.4-2 will be considered further.

Integration of Eq. 7.4-2 between any two pressures P_1 and P_2 (at constant temperature) yields

$$\underline{G}(T_1, P_2) - \underline{G}(T_1, P_1) = \int_{P_1}^{P_2} \underline{V}\, dP \tag{7.4-3}$$

If the fluid under consideration were an ideal gas, then $\underline{V}^{\mathrm{IG}} = RT/P$, so that

$$\underline{G}^{\mathrm{IG}}(T_1, P_2) - \underline{G}^{\mathrm{IG}}(T_1, P_1) = \int_{P_1}^{P_2} \frac{RT}{P}\, dP \tag{7.4-4}$$

Subtracting Eq. 7.4-4 from Eq. 7.4-3 gives

$$[\underline{G}(T_1, P_2) - \underline{G}^{\mathrm{IG}}(T_1, P_2)] - [\underline{G}(T_1, P_1) - \underline{G}^{\mathrm{IG}}(T_1, P_1)] = \int_{P_1}^{P_2} \left(\underline{V} - \frac{RT}{P} \right) dP \tag{7.4-5a}$$

Further, (1) setting P_1 equal to zero, (2) recognizing that at $P = 0$ all fluids are ideal gases so that $\underline{G}(T_1, P = 0) = \underline{G}^{\mathrm{IG}}(T_1, P = 0)$, and (3) omitting all subscripts yields

$$\underline{G}(T, P) - \underline{G}^{\mathrm{IG}}(T, P) = \int_0^P \left(\underline{V} - \frac{RT}{P} \right) dP \tag{7.4-5b}$$

For convenience, we define a new thermodynamic function, the **fugacity**, denoted by the symbol f, as

Definition of fugacity

$$\boxed{ f = P \exp \left\{ \frac{\underline{G}(T, P) - \underline{G}^{\mathrm{IG}}(T, P)}{RT} \right\} = P \exp \left\{ \frac{1}{RT} \int_0^P \left(\underline{V} - \frac{RT}{P} \right) dP \right\} } \tag{7.4-6a}$$

and the related **fugacity coefficient** ϕ by

Definition of the fugacity coefficient

$$\boxed{ \phi = \frac{f}{P} = \exp \left\{ \frac{\underline{G}(T, P) - \underline{G}^{\mathrm{IG}}(T, P)}{RT} \right\} = \exp \left\{ \frac{1}{RT} \int_0^P \left(\underline{V} - \frac{RT}{P} \right) dP \right\} } \tag{7.4-6b}$$

From this definition it is clear that the fugacity has units of pressure, and that $f \to P$ as $P \to 0$; that is, the fugacity becomes equal to the pressure at pressures low enough that the fluid approaches the ideal gas state.[4] Similarly, the fugacity coefficient $\phi = f/P \to 1$ as $P \to 0$. Both the fugacity and the fugacity coefficient will be used extensively throughout this book.

---

[4]It is tempting to view the fugacity as a sort of corrected pressure; it is, however, a well-defined function related to the exponential of the difference between the real and ideal gas Gibbs energies.

The fugacity function has been introduced because its relation to the Gibbs energy makes it useful in phase equilibrium calculations. The present criterion for equilibrium between two phases (Eq. 7.1-9c) is $\underline{G}^{I} = \underline{G}^{II}$, with the restriction that the temperature and pressure be constant and equal in the two phases. Using this equality and the definition of the fugacity (Eq. 7.4-6) gives

$$\underline{G}^{IG}(T, P) + RT \ln \frac{f^{I}(T, P)}{P} = \underline{G}^{IG}(T, P) + RT \ln \frac{f^{II}(T, P)}{P}$$

Now recognizing that the ideal gas molar Gibbs energy at fixed temperature and pressure has the same value regardless of which phase is considered yields as the condition for phase equilibrium

$$\ln \frac{f^{I}(T, P)}{P} = \ln \frac{f^{II}(T, P)}{P}$$

or in terms of the fugacity,

Important forms of equilibrium criterion

$$\boxed{f^{I}(T, P) = f^{II}(T, P)} \tag{7.4-7a}$$

and in terms of the fugacity coefficient,

$$\boxed{\phi^{I}(T, P) = \phi^{II}(T, P)} \tag{7.4-7b}$$

Since these equations follow directly from the equality of the molar Gibbs energy in each phase at phase equilibrium, Eqs. 7.4-7 can be used as criteria for equilibrium. They will be used for this purpose in this book.

Since the fugacity is, by Eq. 7.4-6, related to the equation of state, the equality of fugacities provides a direct way of doing phase equilibrium calculations using the equations of state. In practice, however, Eq. 7.4-6 is somewhat difficult to use because although the molar volume $\underline{V}$ is needed as a function of T and P, it is difficult to solve the equations of state considered in Sec. 6.4 explicitly for volume. In fact, all these equations of state are in a form in which pressure is an explicit function of volume and temperature. Therefore, it is useful to have an equation relating the fugacity to an integral over volume (rather than pressure). We obtain such an equation by starting with Eq. 7.4-6b and using Eq. 6.4-25 at constant temperature in the form

$$dP = \frac{1}{\underline{V}} d(P\underline{V}) - \frac{P}{\underline{V}} d\underline{V} = \frac{P}{Z} dZ - \frac{P}{\underline{V}} d\underline{V}$$

to change the variable of integration to obtain (Problem 7.14)

$$\ln \frac{f(T, P)}{P} = \ln \phi = \frac{1}{RT} \int_{\underline{V}=\infty}^{\underline{V}} \left[\frac{RT}{\underline{V}} - P \right] d\underline{V} - \ln Z + (Z - 1) \tag{7.4-8}$$

where $Z = P\underline{V}/RT$ is the compressibility factor defined earlier. [Alternatively, Eq. 7.4-8 can be obtained from Eq. 7.4-6 using

$$\underline{G}(T, P) - \underline{G}^{IG}(T, P) = [\underline{H}(T, P) - \underline{H}^{IG}(T, P)] - T[\underline{S}(T, P) - \underline{S}^{IG}(T, P)]$$

and Eqs. 6.4-27 and 6.4-28.]

For later reference we note that from Eqs. 7.4-2 and 7.4-6 we have

$$RT \left(\frac{\partial \ln f}{\partial P} \right)_T = \underline{V} = \left(\frac{\partial \underline{G}}{\partial P} \right)_T \qquad \textbf{(7.4-9a)}$$

Also, the temperature dependence of the fugacity is usually given as the temperature dependence of the logarithm of the fugacity coefficient (f/P), which is computed as

$$\frac{\partial}{\partial T} \left(\ln \frac{f}{P} \right)_P = \frac{\partial}{\partial T} \bigg|_P \left\{ \frac{\underline{G}(T, P) - \underline{G}^{\mathrm{IG}}(T, P)}{RT} \right\}$$

$$= \frac{1}{RT} \frac{\partial}{\partial T} \{ \underline{G}(T, P) - \underline{G}^{\mathrm{IG}}(T, P) \} - \left\{ \frac{\underline{G}(T, P) - \underline{G}^{\mathrm{IG}}(T, P)}{RT^2} \right\}$$

$$= -\frac{1}{RT} \{ \underline{S}(T, P) - \underline{S}^{\mathrm{IG}}(T, P) \} - \left\{ \frac{\underline{G}(T, P) - \underline{G}^{\mathrm{IG}}(T, P)}{RT^2} \right\}$$

$$= -\frac{1}{RT^2} \{ [\underline{G}(T, P) - T\underline{S}(T, P)] - [\underline{G}^{\mathrm{IG}}(T, P) - T\underline{S}^{\mathrm{IG}}(T, P)] \}$$

$$= - \left\{ \frac{\underline{H}(T, P) - \underline{H}^{\mathrm{IG}}(T, P)}{RT^2} \right\}$$

$$\textbf{(7.4-9b)}$$

In deriving this result we have used the relations $(\partial \underline{G}/\partial T)_P = -\underline{S}$ and $\underline{G} = \underline{H} - T\underline{S}$.

Since the fugacity function is of central importance in phase equilibrium calculations, we consider here the computation of the fugacity for pure gases, liquids, and solids.

a. Fugacity of a Pure Gaseous Species

To compute the fugacity of a pure gaseous species we will always use a volumetric equation of state and Eq. 7.4-6b or

General fugacity coefficient equation

$$\ln \frac{f^{\mathrm{V}}(T, P)}{P} = \frac{1}{RT} \int_{\underline{V}=\infty}^{\underline{V}=Z^{\mathrm{V}} RT/P} \left(\frac{RT}{\underline{V}} - P \right) d\underline{V} - \ln Z^{\mathrm{V}} + (Z^{\mathrm{V}} - 1) \qquad \textbf{(7.4-8)}$$

where the superscript V is used to designate the fugacity and compressibility of the vapor phase. Thus, given a volumetric equation of state of a gas applicable up to the pressure of interest, the fugacity of a pure gas can be computed by integration of Eq. 7.4-8. At very low pressures, where a gas can be described by the ideal gas equation of state

$$P\underline{V} = RT \qquad \text{or} \qquad Z^{\mathrm{V}} = 1$$

we have

$$\ln \frac{f^{\mathrm{V}}(T, P)}{P} = 0 \qquad \text{or} \qquad f^{\mathrm{V}}(T, P) = P \qquad \textbf{(7.4-10)}$$

Thus, for a low-pressure gas, the fugacity of a species is just equal to the total pressure.

The calculation of the fugacity when detailed thermodynamic data, such as the steam tables, are available is demonstrated in the following illustrations.

ILLUSTRATION 7.4-1

Computing Fugacity from Volumetric Data

Use the volumetric information in the steam tables of Appendix A.III to compute the fugacity of superheated steam at 300°C and 8 MPa.

SOLUTION

With tabulated volumetric data, as in the steam tables, it is most convenient to use Eq. 7.4-6a:

$$f = P \exp\left[\frac{1}{RT} \int_0^P \left(\underline{V} - \frac{RT}{P}\right) dP\right]$$

From the superheated vapor steam tables at 300°C, we have

P (MPa)	$\hat{V}$ (m³/kg)	$\underline{V}$ (m³/mol)	$[\underline{V} - RT/P]$ (m³/mol) × 10⁴
0.01	26.445	0.476 41	−1.02
0.05	5.284	0.095 191	−1.121
0.10	2.639	0.047 542	−1.101
0.20	1.316 2	0.023 711	−1.145
0.30	0.875 3	0.015 769	−1.154
0.40	0.654 8	0.011 796	−1.167
0.50	0.522 6	0.009 414 6	−1.157
0.60	0.433 4	0.007 807 7	−1.342
0.80	0.324 1	0.005 838 7	−1.178
1.0	0.257 9	0.004 646 1	−1.191
1.2	0.213 8	0.003 851 6	−1.194
1.4	0.182 28	0.003 283 8	−1.199
1.6	0.158 62	0.002 857 5	−1.207
1.8	0.140 21	0.002 525 9	−1.214
2.0	0.125 47	0.002 260 3	−1.222
2.5	0.098 90	0.001 781 7	−1.244
3.0	0.081 14	0.001 461 7	−1.267
3.5	0.068 42	0.001 232 6	−1.289
4.0	0.058 84	0.001 060 0	−1.313
4.5	0.051 35	0.000 925 07	−1.339
5.0	0.045 32	0.000 816 44	−1.366
6.0	0.036 16	0.000 651 42	−1.428
7.0	0.029 47	0.000 530 90	−1.498
8.0	0.024 26	0.000 437 04	−1.586

Numerically evaluating the integral using the data above, we find

$$\int_0^{8 \text{ MPa}} \left(\underline{V} - \frac{RT}{P}\right) dP \approx -1.093 \times 10^{-3} \frac{\text{m}^3 \text{ MPa}}{\text{mol}}$$

and

$$f = 8 \text{ MPa} \exp\left[\frac{-1.093 \times 10^{-3} \dfrac{\text{m}^3 \text{ MPa}}{\text{mol}}}{573.15 \text{ K} \times 8.314 \times 10^{-6} \dfrac{\text{MPa m}^3}{\text{mol K}}}\right] = 8 \exp(-0.2367) \text{ MPa}$$

$$= 8 \times 0.7996 \text{ MPa} = 6.397 \text{ MPa}$$

Also, the fugacity coefficient, ϕ, in this case is

$$\phi = \frac{f}{P} = 0.7996$$

COMMENT

Had the same calculation been done at a much higher temperature, the steam would be closer to an ideal vapor, and the fugacity coefficient would be closer to unity in value. For example, the result of a similar calculation at 1000°C and 10 MPa yields $f = 9.926$ MPa and $\phi = f/P = 0.9926$. ∎

ILLUSTRATION 7.4-2
Alternative Fugacity Calculation

Use other data in the superheated vapor steam tables to calculate the fugacity of steam at 300°C and 8 MPa, and check the answer obtained in the previous illustration.

SOLUTION

At 300°C and 0.01 MPa we have from the steam tables $\hat{H} = 3076.5$ kJ/kg and $\hat{S} = 9.2813$ kJ/(kg K). Therefore,

$$\hat{G}(300°\text{C}, 0.01\ \text{MPa}) = \hat{H} - T\hat{S}$$
$$= 3076.5 - 573.15 \times 9.2813 = -2243.1\ \text{kJ/kg}$$

and $\underline{G}(300°\text{C}, 0.01\ \text{MPa}) = -2243.1$ J/g $\times$ 18.015 g/mol $= -40\,409$ J/mol. Since the pressure is so low (0.01 MPa) and well away from the saturation pressure of steam at 300°C (which is 8.581 MPa), we assume steam is an ideal gas at these conditions. Then using Eq. 7.4-3 for an ideal gas, we have

$$\underline{G}^{\text{IG}}(300°\text{C}, 8\ \text{MPa}) = \underline{G}^{\text{IG}}(300°\text{C}, 0.01\ \text{MPa}) + \int_{0.01\text{MPa}}^{8\text{MPa}} \underline{V}^{\text{IG}}\,dP$$

$$= -40\,409 + \int_{0.01\text{MPa}}^{8\text{MPa}} \frac{RT}{P}\,dP$$

$$= -40\,409 + RT \ln \frac{8}{0.01}$$

$$= -40\,409 + 8.134 \times 573.15 \ln 800$$

$$= -8555.7\ \frac{\text{J}}{\text{mol}}$$

For real steam at 300°C and 8 MPa, we have, from the steam tables, $\hat{H} = 2785.0$ kJ/kg and $\hat{S} = 5.7906$ kJ/(kg K), so that

$$\hat{G}(300°\text{C}, 8\ \text{MPa}) = 2785.0 - 573.15 \times 5.7906 = -533.88\ \frac{\text{kJ}}{\text{kg}}$$

and

$$\underline{G}(300°\text{C}, 8\ \text{MPa}) = -9617.9\ \frac{\text{J}}{\text{mol}}$$

Now using Eq. 7.4-6a in the form

$$f(T, P) = P \exp\left[\frac{\underline{G}(T, P) - \underline{G}^{\text{IG}}(T, P)}{RT}\right]$$

results in

$$f(300°C, 8 \text{ MPa}) = 8 \text{ MPa} \exp\left[\frac{-9617.9 - (-8555.7)}{8.314 \times 573.15}\right]$$

$$= 8 \text{ MPa} \exp[-0.2229]$$

$$= 6.402 \text{ MPa}$$

which is in excellent agreement with the results obtained in the previous illustration. ∎

ILLUSTRATION 7.4-3

Calculation of the Fugacity of Saturated Steam

Compute the fugacity of saturated steam at 300°C.

SOLUTION

The saturation pressure of steam at 300°C is 8.581 MPa, and from Illustration 7.3-1 we have $\hat{G}^V = -520.5$ kJ/kg and $\underline{G}^V = -9376.8$ J/mol. Following the previous illustration, we have

$$\underline{G}^{IG}(300°C, 8.581 \text{ MPa}) = \underline{G}^{IG}(300°C, 0.01\text{MPa}) + \int_{0.01 \text{ MPa}}^{8.581 \text{ MPa}} \frac{RT}{P} dP$$

$$= -40\,409 + 8.314 \times 573.15 \ln 858.1$$

$$= -8221.6 \frac{J}{mol}$$

Therefore,

$$f^V(300°C, 8.581 \text{ MPa}) = 8.581 \text{ MPa} \exp\left[\frac{-9376.8 - (-8221.6)}{8.314 \times 573.15}\right]$$

$$= 8.581 \text{ MPa} \times 0.7847$$

$$= 6.7337 \text{ MPa}$$

COMMENT

Note that since, at equilibrium, $f^V = f^L$, it is also true that

$$f^L(300°C, 8.581 \text{ MPa}) = 6.7337 \text{ MPa}$$

∎

At low to moderate pressures, the virial equation of state truncated after the second virial coefficient,

$$\frac{P\underline{V}}{RT} = Z = 1 + \frac{B(T)}{\underline{V}} \tag{7.4-11}$$

may be used, if data for $B(T)$ are available. Using Eq. 7.4-11 in Eq. 7.4-8, we obtain (Problem 7.14)

Fugacity coefficient: virial equation of state

$$\ln \frac{f^V(T, P)}{P} = \frac{2B(T)}{\underline{V}} - \ln Z = \frac{2PB(T)}{ZRT} - \ln Z \tag{7.4-12}$$

where

$$Z = 1 + \frac{B(T)}{\underline{V}} = \frac{1}{2}\left[1 + \sqrt{1 + \frac{4B(T)P}{RT}}\right]$$

ILLUSTRATION 7.4-4

Calculation of the Fugacity of a Gas Using the Virial Equation

Compute the fugacities of pure ethane and pure butane at 373.15 K and 1, 10, and 15 bar, assuming the virial equation of state can describe these gases at these conditions.

Data:

$$B_{ET}(373.15 \text{ K}) = -1.15 \times 10^{-4} \text{ m}^3/\text{mol}$$
$$B_{BU}(373.15 \text{ K}) = -4.22 \times 10^{-4} \text{ m}^3/\text{mol}$$

[*Source:* E. M. Dontzler, C. M. Knobler, and M. L. Windsor, *J. Phys. Chem.* **72,** 676 (1968).]

SOLUTION

Using Eqs. 7.4-12, we find

	Ethane		Butane	
P (bar)	Z	f (bar)	Z	f (bar)
1	0.996	0.996	0.986	0.986
10	0.961	9.629	0.838	8.628
15	0.941	14.16	0.714	11.86

Since the pressures are not very high, these results should be reasonably accurate. However, the virial equation with only the second virial coefficient will be less accurate as the pressure increases. In fact, at slightly above 15 bar and 373.15 K, *n*-butane condenses to form a liquid. In this case the virial equation description is inappropriate, as it does not show a phase change or describe liquids. ∎

At higher pressure, a more complicated equation of state (or higher terms in the virial expansion) must be used. By using the (not very accurate) van der Waals equation, one obtains

Fugacity coefficient: van der Waals equation of state

$$\ln \frac{f^V}{P} = \ln \frac{\underline{V}}{\underline{V} - b} - \frac{a}{RT\underline{V}} + \left(\frac{P\underline{V}}{RT} - 1\right) - \ln\left(\frac{P\underline{V}}{RT}\right)$$

or

$$= (Z^V - 1) - \ln(Z^V - B) - \frac{A}{Z^V} \tag{7.4-13}$$

For hydrocarbons and simple gases, the Peng-Robinson equation (Eq. 6.4-2) provides a more accurate description. In this case we have

Fugacity coefficient: Peng-Robinson equation of state

$$\ln \frac{f^V}{P} = (Z^V - 1) - \ln\left(Z^V - \frac{bP}{RT}\right) - \frac{a}{2\sqrt{2}bRT} \ln\left[\frac{Z^V + (1+\sqrt{2})bP/RT}{Z^V + (1-\sqrt{2})bP/RT}\right]$$

$$= (Z^V - 1) - \ln(Z^V - B) - \frac{A}{2\sqrt{2}B} \ln\left[\frac{Z^V + (1+\sqrt{2})B}{Z^V + (1-\sqrt{2})B}\right]$$

$$\tag{7.4-14a}$$

where in Eqs. 7.4-13 and 7.4-14a, $A = aP/(RT)^2$ and $B = Pb/RT$. Of course, other equations of state could be used for the fugacity calculations starting from Eq. 7.4-8, though we do not consider such calculations here.

To use either the virial, van der Waals, Peng-Robinson, or other equation of state to calculate the fugacity of a gaseous species, the following procedure is used: (1) For a given value of T and P, use the chosen equation of state to calculate the molar volume $\underline{V}$ or, equivalently, the compressibility factor Z. When using cubic or more complicated equations of state, it is the low-density (large $\underline{V}$ or Z) solution that is used. (2) This value of $\underline{V}$ or Z is then used in Eq. 7.4-12, 7.4-13, or 7.4-14, as appropriate, to calculate the species fugacity coefficient, f/P, and thus the fugacity. The Peng-Robinson equation-of-state programs or MATHCAD worksheets described in Appendix B can be used for this calculation.

ILLUSTRATION 7.4-5
Calculation of the Fugacity of a Gas Using the Peng-Robinson Equation of State

Compute the fugacities of pure ethane and pure butane at 373.15 K and 1, 10, and 15 bar, assuming the Peng-Robinson equation describes these gases at these conditions.

SOLUTION

Using the Visual Basic computer program described in Appendix B.I-2, the DOS-based program PR1 described in Appendix B.II-1, the MATHCAD worksheet described in Appendix B.III, or the MATLAB program described in Appendix B.IV and the data in Table 4.6-1, we obtain the following:

	Ethane		Butane	
P (bar)	Z	f (bar)	Z	f (bar)
1	0.996	1.00	0.985	0.99
10	0.957	9.58	0.837	8.57
15	0.935	14.06	0.733	11.81

COMMENT

We see that these results are only slightly different from those computed with the virial equation of state. The differences would become larger as the pressure increases or the temperature decreases. ∎

For hand calculations it is simpler, but less accurate, to compute the fugacity of a species using a specially prepared corresponding-states fugacity chart. To do this, we note that since, for simple gases and hydrocarbons, the compressibility factor Z obeys a corresponding-states relation (see Sec. 6.6), the fugacity coefficient f/P given by Eq. 7.4-6 can also be written in corresponding-states form as follows:

$$\frac{f^{\mathrm{V}}}{P} = \exp\left\{ \frac{1}{RT} \int_0^P (\underline{V} - \underline{V}^{\mathrm{IG}})\, dP \right\} = \exp\left\{ \int_0^P \left(\frac{P\underline{V}}{RT} - 1 \right) \frac{dP}{P} \right\}$$

$$= \exp\left\{ \int_0^{P_r} [Z(T_r, P_r, \omega) - 1]\, d\ln P_r \right\} \tag{7.4-15a}$$

or

Fugacity coefficient corresponding states

$$\ln \frac{f^{\mathrm{V}}}{P} = \left\{ \int_0^{P_r} [Z(T_r, P_r, \omega) - 1]\, d\ln P_r \right\} \tag{7.4-15b}$$

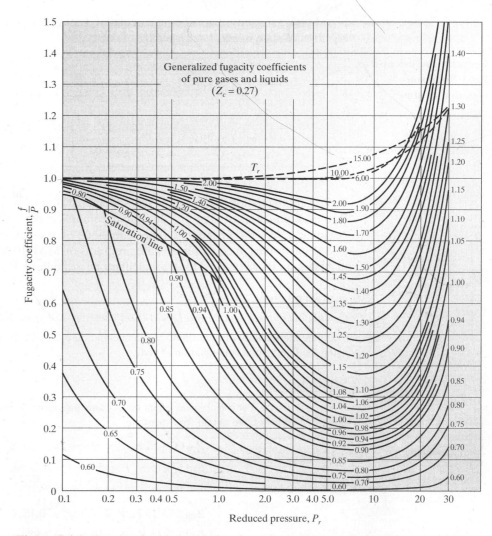

Figure 7.4-1 (Reprinted with permission from O. A. Hougen, K. M. Watson, and R. A. Ragatz, *Chemical Process Principles Charts*, 2nd ed., John Wiley & Sons, New York, 1960.)

Consequently, the fugacity coefficient can be tabulated in the corresponding-states manner. The corresponding-states correlation for the fugacity coefficient of nonpolar gases and liquids given in Fig. 7.4-1 was obtained using Eq. 7.4-15b and the compressibility correlation (Fig. 6.6-3).

b. The Fugacity of a Pure Liquid

The fugacity of a pure liquid species can be computed in a number of ways, depending on the data available. If the equation of state for the liquid is known, we can again start from Eq. 7.4-8, now written as

$$\ln \frac{f^{L}(T, P)}{P} = \frac{1}{RT} \int_{\underline{V}=\infty}^{\underline{V}=Z^{L} RT/P} \left(\frac{RT}{\underline{V}} - P \right) d\underline{V} - \ln Z^{L} + (Z^{L} - 1) \quad \textbf{(7.4-16)}$$

where the superscript L is used to indicate that the liquid-phase compressibility (high density, small $\underline{V}$ and Z) solution of the equation of state is to be used, and it is the liquid-phase fugacity that is being calculated. Using, for example, the Peng-Robinson equation of state in Eq. 7.4-8 yields

Fugacity coefficient Peng-Robinson equation of state

$$
\ln \frac{f^{\mathrm{L}}}{P} = (Z^{\mathrm{L}} - 1) - \ln \left(Z^{\mathrm{L}} - \frac{bP}{RT} \right) - \frac{a}{2\sqrt{2}bRT} \ln \left[\frac{Z^{\mathrm{L}} + (1 + \sqrt{2})bP/RT}{Z^{\mathrm{L}} + (1 - \sqrt{2})bP/RT} \right]
$$

$$
= (Z^{\mathrm{L}} - 1) - \ln(Z^{\mathrm{L}} - B) - \frac{A}{2\sqrt{2}B} \ln \left[\frac{Z^{\mathrm{L}} + (1 + \sqrt{2})B}{Z^{\mathrm{L}} + (1 - \sqrt{2})B} \right]
$$

$$(7.4\text{-}14b)$$

To use this equation, we first, at the specified value of T and P, solve the Peng-Robinson equation of state for the liquid compressibility, Z^{L}, and use this value to compute $f^{\mathrm{L}}(T, P)$. Of course, other equations of state could be used in Eq. 7.4-8. The Peng-Robinson equation-of-state programs or MATHCAD worksheets described in Appendix B can be used for this calculation for the Peng-Robinson equation of state.

ILLUSTRATION 7.4-6

Calculation of the Fugacity of a Liquid Using the Peng-Robinson Equation of State

Compute the fugacity of pure liquid n-pentane and pure liquid benzene at 373.15 K and 50 bar using the Peng-Robinson equation of state.

SOLUTION

Using one of the Peng-Robinson equation-of-state programs in Appendix B with the liquid (high-density) root, we obtain for n-pentane

$$
Z_{\mathrm{PE}}(373.15 \text{ K}, P = 50 \text{ bar}) = 0.2058
$$

$$
f_{\mathrm{PE}}(373.15 \text{ K}, P = 50 \text{ bar}) = 6.22 \text{ bar}
$$

and for benzene

$$
Z_{\mathrm{BZ}}(373.15 \text{ K}, P = 50 \text{ bar}) = 0.1522
$$

$$
f_{\mathrm{BZ}}(373.15 \text{ K}, P = 50 \text{ bar}) = 1.98 \text{ bar} \qquad \blacksquare
$$

If one has some liquid volume data, but not an equation of state, it is more convenient to start with Eq. 7.4-6, which can be rearranged to

$$
RT \ln \frac{f^{\mathrm{L}\prime}}{P} = \int_0^P \left[\underline{V} - \frac{RT}{P} \right] dP \qquad (7.4\text{-}16a)
$$

and perform the integration. However, one has to recognize that a phase change from a vapor to a liquid occurs within the integration range as the pressure is increased from zero pressure to the vapor pressure, and that the molar volume of a fluid is discontinuous at this phase transition. Thus, the result of the integration is

$$
RT \ln \left(\frac{f^{\mathrm{L}}}{P} \right) = \underline{G}(T, P) - \underline{G}^{\mathrm{IG}}(T, P)
$$

$$
= \int_{P=0}^{P^{\mathrm{vap}}(T)} \left(\underline{V} - \frac{RT}{P} \right) dP + RT \Delta \left(\ln \frac{f}{P} \right)_{\substack{\text{phase} \\ \text{change}}} + \int_{P^{\mathrm{vap}}(T)}^{P} \left(\underline{V} - \frac{RT}{P} \right) dP
$$

$$(7.4\text{-}16b)$$

The first term on the right side of this equation is the difference between the real and ideal gas Gibbs energy changes of compressing the vapor from zero pressure to the vapor pressure of the substance at temperature T. The second term allows for the Gibbs energy change at the phase transition. The last term is the difference between the liquid and ideal gas Gibbs energy changes on compression of the liquid from its vapor pressure to the pressure of interest.

From Eq. 7.4-7 and the fact that the pressure is continuous and the fugacities are equal at a phase change, we have

$$\Delta \left(\ln \frac{f}{P} \right)_{\substack{\text{phase} \\ \text{change}}} = 0$$

and from Eq. 7.4-6, we further have that

$$\int_0^{P^{\text{vap}}(T)} \left(\underline{V} - \frac{RT}{P} \right) dP = RT \ln \left(\frac{f}{P} \right)_{\text{sat},T} \tag{7.4-17}$$

where $(f/P)_{\text{sat},T}$ is the fugacity coefficient of the saturated fluid (either vapor or liquid at the phase transition pressure, since these fugacities are equal) at temperature T. Finally, the last term in Eq. 7.4-16 can be partially integrated as follows:

$$\int_{P^{\text{vap}}(T)}^{P} \left(\underline{V} - \frac{RT}{P} \right) dP = \int_{P^{\text{vap}}(T)}^{P} \underline{V}\, dP - RT \int_{P^{\text{vap}}(T)}^{P} \frac{1}{P}\, dP$$

$$= \int_{P^{\text{vap}}(T)}^{P} \underline{V}\, dP - RT \ln \frac{P}{P^{\text{vap}}(T)}$$

Combining these terms yields the following expressions for the fugacity of a pure liquid:

Poynting correction

$$f^{\text{L}}(T, P) = P^{\text{vap}}(T) \left(\frac{f}{P} \right)_{\text{sat},T} \exp \left[\frac{1}{RT} \int_{P^{\text{vap}}(T)}^{P} \underline{V}\, dP \right]$$

$$= f_{\text{sat}}(T) \exp \left[\frac{1}{RT} \int_{P^{\text{vap}}(T)}^{P} \underline{V}\, dP \right] \tag{7.4-18}$$

The exponential term in this equation, known as the **Poynting pressure correction**, accounts for the increase in fugacity due to the fact that the system pressure is higher than the vapor pressure of the liquid. Since the molar volume of a liquid is generally much less than that of a gas (so that $P\underline{V}^{\text{L}}/RT \ll 1$), the Poynting term is only important at high pressures. (An exception to this is for cryogenic systems, where T is very low.)

Equation 7.4-18 lends itself to several levels of approximation. The simplest approximation is to neglect the Poynting and $(f/P)_{\text{sat}}$ terms and set the fugacity of the liquid equal to its vapor pressure at the same temperature, that is,

Simplest approximation for f^{L}

$$f^{\text{L}}(T, P) = P^{\text{vap}}(T) \tag{7.4-19}$$

The result is valid only when the vapor pressure is low and the liquid is at a low total pressure. Although this equation applies to most fluids at low pressure, it is not correct for fluids that associate, that is, form dimers in the vapor phase, such as acetic acid. A more accurate approximation is

Better approximation for f^L

$$f^L(T, P) = f_{sat}^L(T) = f^V(T, P^{vap}) = P^{vap}(T)\left(\frac{f}{P}\right)_{sat,T} \qquad \text{(7.4-20)}$$

which states that the fugacity of a liquid at a given temperature and any pressure is equal to its fugacity at its vapor pressure (as a vapor or liquid). This approximation is valid provided the system pressure is not greatly different from the species vapor pressure at the temperature of interest (so that the Poynting term is negligible). To evaluate $f^V(T, P^{vap})$, any of the methods considered in Sec. 7.4a may be used. For example, if second virial coefficient data are available, Eq. 7.4-13 can be used, or if an equation of state is available for the vapor, but not for the liquid (as may be the case for nonhydrocarbon fluids), Eq. 7.4-8 (or its integrated form for the equations of state considered in Sec. 7.4a) can be used to estimate $f^V(T, P^{vap})$. Alternatively, but less accurately, $(f/P)_{sat,T}$ can be gotten from the corresponding-states diagram of Fig. 7.4-1 using the saturation line and the reduced temperature of interest.

Another approximation that can be made in Eq. 7.4-18 is to take into account the Poynting pressure correction, but to assume that the liquid is incompressible (Fig. 7.3-4 for the van der Waals equation of state, for example, does indicate that at constant temperature the liquid volume is only slightly pressure dependent). In this case

Best approximation for f^L

$$f^L(T, P) = P^{vap}(T)\left(\frac{f}{P}\right)_{sat,T} \exp\left[\frac{\underline{V}(P - P^{vap})}{RT}\right] \qquad \text{(7.4-21)}$$

where the term on the right, before the exponential, is the same as in Eq. 7.4-20. Alternatively, one can use

$$f^L(T, P) = P\left(\frac{f}{P}\right)_{T,P} \qquad \text{(7.4-22)}$$

where the fugacity coefficient is evaluated using the principle of corresponding states and Fig. 7.4-1. However, if an equation of state for the liquid is available, one should use Eq. 7.4-8.

ILLUSTRATION 7.4-7
Calculation of the Fugacity of Liquid Water from Density Data

Compute the fugacity of liquid water at 300°C and 25 MPa.

SOLUTION

From Eq. 7.4-18, we have

$$f^L(T, P) = f^{sat}(T)\exp\left[\frac{1}{RT}\int_{P^{vap}(T)}^{P} \underline{V}^L\, dP\right]$$

Since liquids are not very compressible, we can assume (away from the critical point of steam) that $\underline{V}^L \sim \underline{V}_{sat}^L$, so that

$$f^L(T, P) = f^{sat}(T)\exp\left[\frac{\underline{V}_{sat}^L(P - P^{vap}(T))}{RT}\right]$$

From Illustration 7.4-3 we have $f^{sat}(300°C) = 6.7337$ MPa and from the steam tables, at $300°C$, $P^{vap} = 8.581$ MPa and $\hat{V}^L = 0.001\ 404$ m^3/kg, so that

$$\underline{V}_{sat} = 1.404 \times 10^{-3}\ \frac{m^3}{kg} \times \frac{1\ kg}{10^3\ g} \times 18.015\ \frac{g}{mol}$$

$$= 2.5293 \times 10^{-5}\ \frac{m^3}{mol}$$

Consequently,

$$f^L(300°C,\ 25\ MPa) = 6.7337\ MPa\ exp\left[\frac{2.5293 \times 10^{-5}\ \dfrac{m^3}{mol} \times (25 - 8.581)\ MPa}{8.314 \times 10^{-6}\ \dfrac{MPa\ m^3}{mol\ K} \times 573.15\ K}\right]$$

$$= 6.7337\ exp[0.08715]\ MPa$$

$$= 6.7337 \times 1.0911\ MPa = 7.347\ MPa$$

∎

c. Fugacity of a Pure Solid Phase

The extension of the previous discussion to solids is relatively simple. In fact, if we recognize that a solid phase may undergo several phase transitions, and let $\underline{V}^J$ be the molar volume of the Jth phase and P^J be the pressure above which this phase is stable at the temperature T, we have

$$f^S(T, P) = P^{sat}(T)\left(\frac{f}{P}\right)_{sat,T} exp\left[\frac{1}{RT}\sum_{J=1}\int_{P^J}^{P^{J+1}} \underline{V}^J\ dP\right] \qquad \textbf{(7.4-23)}$$

Here P^{sat} is generally equal to the sublimation pressure of the solid.[5] Since the sublimation (or vapor) pressure of a solid is generally small, so that the fugacity coefficient can be taken equal to unity, it is usually satisfactory to approximate Eq. 7.4-23 at low total pressures by

$$f^S(T, P) = P^{sat}(T) \qquad \textbf{(7.4-24a)}$$

or, for a solid subject to moderate or high total pressures, by

Fugacity of a solid

$$\boxed{f^S(T, P) = P^{sat}(T)\ exp\left[\frac{\underline{V}^S(P - P^{sat}(T))}{RT}\right]} \qquad \textbf{(7.4-24b)}$$

ILLUSTRATION 7.4-8

Calculation of the Fugacity of Ice from Density Data

Compute the fugacity of ice at $-5°C$ and pressures of 0.1 MPa, 10 MPa, and 100 MPa.

---

[5]Below the triple-point temperature, as the pressure is reduced at constant temperature, a solid will sublimate directly to a vapor. However, near the triple-point temperature some solids first melt to form liquids and then vaporize as the ambient pressure is reduced at constant temperature (see Fig. 7.3-6). In such cases $P^{sat}(T)$ in Eq. 7.4-23 is taken to be the vapor pressure of the liquid.

Data:

At $-5°C$, the sublimation pressure of ice is 0.4 kPa and its specific gravity is 0.915.

SOLUTION

We assume that ice is incompressible over the range from 0.4 kPa to 100 MPa, so that its molar volume over this pressure range is

$$\underline{V} = \frac{18.015 \text{ g/mol}}{0.915 \text{ g/cc}} = 19.69 \text{ cc/mol} = 1.969 \times 10^{-5} \text{ m}^3/\text{mol}$$

From Eq. 7.4-24b, we then have

$$f_{\text{ice}}(-5°C, 0.1 \text{ MPa}) = 0.4 \text{ kPa} \exp\left[\frac{1.969 \times 10^{-5} \dfrac{\text{m}^3}{\text{mol}} \times (0.1 - 0.0004) \text{ MPa}}{268.15 \text{ K} \times 8.314 \times 10^{-6} \dfrac{\text{MPa m}^3}{\text{mol K}}}\right]$$

$$= 0.4 \exp[8.797 \times 10^{-4}] \text{ kPa}$$
$$= 0.4 \times 1.000 \, 88 \text{ kPa}$$
$$= 0.4004 \text{ kPa}$$

Similarly,

$$f_{\text{ice}}(-5°C, 10 \text{ MPa}) = 0.4 \text{ kPa} \exp\left[\frac{1.969 \times 10^{-5} \dfrac{\text{m}^3}{\text{mol}} \times (10 - 0.0004) \text{ MPa}}{268.15 \text{ K} \times 8.314 \times 10^{-6} \dfrac{\text{MPa m}^3}{\text{mol K}}}\right]$$

$$= 0.4369 \text{ kPa}$$

and

$$f_{\text{ice}}(-5°C, 100 \text{ MPa}) = 0.4 \text{ kPa} \exp\left[\frac{1.969 \times 10^{-5} \dfrac{\text{m}^3}{\text{mol}} \times (100 - 0.0004) \text{ MPa}}{268.15 \text{ K} \times 8.314 \times 10^{-6} \dfrac{\text{MPa m}^3}{\text{mol K}}}\right]$$

$$= 0.9674 \text{ kPa}$$

COMMENT

Generally, the change in fugacity of a condensed species (liquid or solid) with small pressure changes is small. Here we find that the fugacity of ice increases by 10 percent for an increase in pressure from the sublimation pressure to 100 bar, and by a factor of 2.5 as the pressure on ice increases to 1000 bar. ∎

ILLUSTRATION 7.4-9

Calculation of the Fugacity of Naphthalene from Density Data

The saturation pressure of solid naphthalene can be represented by

$$\log_{10} P^{\text{sat}}(\text{bar}) = 8.722 - \frac{3783}{T} \qquad (T \text{ in K})$$

The density of the solid is 1.025 g/cm^3, and the molecular weight of naphthalene is 128.19. Estimate the fugacity of solid naphthalene at 35°C at its vapor pressure, and also at 1, 20, 30, 40, 50, and 60 bar. (This information is used in Sec. 12.1 in a study of supercritical phase behavior.)

SOLUTION

We start with Eq. 7.4-23 for a single solid phase with the assumption that the solid is incompressible:

$$f^S(T, P) = P^{\text{sat}}(T) \left(\frac{f}{P}\right)_{\text{sat},T} \exp\left[\frac{\underline{V}(P - P^{\text{sat}}(T))}{RT}\right]$$

Using the data in the problem statement,

$$P^{\text{sat}}(35°C) = 10^{[8.722-(3783/(273.15+35))]} = 2.789 \times 10^{-4} \text{ bar}$$

Since the sublimation pressure is so low, we can use (see Fig. 7.4-1)

$$\left(\frac{f}{P}\right)_{\text{sat}} = 1$$

Therefore,

$$f^S(T = 35°C, P^{\text{sat}} = 2.789 \times 10^{-4} \text{ bar}$$

At any higher pressure, we have

$$f^S(T = 35°C, P) = 2.789 \times 10^{-4} \text{ (bar)}$$

$$\times \exp\left[\frac{\dfrac{128.19}{1.025} \dfrac{\text{cm}^3}{\text{mol}} \times (P - 2.879 \times 10^{-4}) \text{ bar} \times 10^{-6} \dfrac{\text{m}^3}{\text{cm}^3}}{8.314 \times 10^{-5} \dfrac{\text{bar m}^3}{\text{mol K}} \times 308.15 \text{ K}}\right]$$

The result is

P (bar)	$f^S(35°C, P)$ (bar)
2.789×10^{-4}	2.789×10^{-4}
1	2.803×10^{-4}
10	2.933×10^{-4}
20	3.083×10^{-4}
30	3.241×10^{-4}
40	3.408×10^{-4}
50	3.582×10^{-4}
60	3.766×10^{-4}

7.5 THE CALCULATION OF PURE FLUID-PHASE EQUILIBRIUM: THE COMPUTATION OF VAPOR PRESSURE FROM AN EQUATION OF STATE

Now that the fugacity (or equivalently, the molar Gibbs energy) of a pure fluid can be calculated, it is instructive to consider how one can compute the vapor-liquid equilibrium pressure of a pure fluid, that is, the vapor pressure, as a function of temperature, using a volumetric equation of state. Such calculations are straightforward in principle, but, because of the iterative nature of the computation involved, are best done on a computer. One starts by choosing a temperature between the melting point and critical point of the fluid of interest and guessing the vapor pressure. Next, for these values of T and P, the equation of state is solved to find the liquid (smaller $\underline{V}$ and Z) and the vapor (larger $\underline{V}$ and Z) roots, which we denote by Z^L and Z^V, respectively. These values are then used in the fugacity coefficient expression appropriate to the equation of state to obtain f^L and f^V. Finally, these values are compared. If f^L is, within a specified tolerance, equal to f^V, the guessed pressure is the correct vapor pressure for the temperature of interest. If, however, the liquid-phase fugacity is greater than the vapor-phase fugacity (i.e., $f^L > f^V$), the guessed pressure is too low; it is too high if $f^V > f^L$. In either case, a new guess must be made for the pressure, and the calculation repeated.

Figure 7.5-1 is a flow diagram for the calculation of the vapor pressure using the Peng-Robinson equation-of-state (the Peng-Robinson equation-of-state programs in Appendix B can be used for this calculation); clearly, other equations of state could have been used. Also, the algorithm could be modified slightly so that pressure is chosen, and the boiling temperature at this pressure found (in this case remember that, from Eq. 6.7-1, the a parameter in the equation of state is temperature dependent).

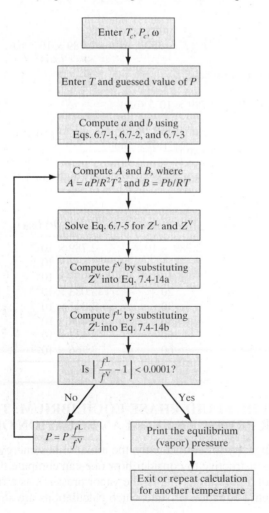

Figure 7.5-1 Flow sheet of a computer program for the calculation of the vapor pressure of a fluid using the Peng-Robinson equation of state.

Figure 7.5-2 contains experimental vapor pressure versus temperature data for *n*-butane, together with vapor pressure predictions from (1) the van der Waals equation; (2) the Peng-Robinson equation, but with $a = 0.457\,24R^2T_c^2/P_c$ rather than the correct expression of Eq. 6.7-1; and (3) the complete Peng-Robinson equation of state, that is, with the a parameter being the function of temperature given in Eq. 6.7-1. From this figure we see that the vapor pressure predictions of the van der Waals equation are not very good, nor are the predictions of the simplified Peng-Robinson equation with the a parameter independent of temperature. However, the predictions with the complete Peng-Robinson equation are excellent. Indeed, the specific form of the temperature dependence of the $\alpha(T)$ term in the a parameter was chosen so that good vapor pressure predictions would be obtained at all temperatures, and so that $\alpha(T_c) = 1$ to ensure that the critical-point conditions are met.

Finally, it should be pointed out that in the calculation scheme suggested here, the initial guess for the vapor pressure at the chosen temperature (or temperature at fixed pressure) must be made with some care. In particular, the pressure (or temperature) must be within the range of the van der Waals loop of the equation of state, so that separate solutions for the vapor and liquid densities (or compressibilities) are obtained. If the guessed pressure is either too high so that only the high-density root exists, or too low so that only the low-density root exists, the presumed vapor and liquid phases will be identical, and the algorithm of Fig. 7.5-1 will accept the guessed pressure as the vapor pressure, even though it is an incorrect one-phase solution, not the correct two-phase solution to the problem. The incorrect "solution" so obtained is referred to as the **trivial solution** (in which the two phases are identical) rather than the actual solution

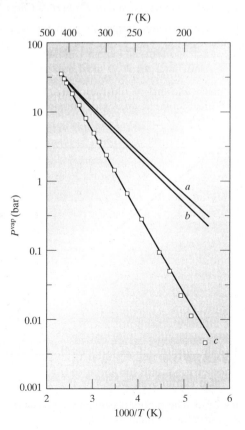

Figure 7.5-2 The vapor pressure of *n*-butane as a function of temperature. The points □ are the experimental data. Line *a* is the prediction of the van der Waals equation, line *b* is the prediction of the Peng-Robinson equation with $\alpha = 1$, and line *c* is the prediction of the complete Peng-Robinson equation [i.e., $\alpha = \alpha(T)$]. The reason for plotting $\ln P^{\text{vap}}$ versus $1/T$ rather than P^{vap} versus T is discussed in Sec. 7.7.

to the problem, in which vapor and liquid phases exist. The best method of avoiding this difficulty is to make a very good initial guess, though this becomes increasingly harder to do as one approaches the critical point of the fluid, and the van der Waals loop becomes very small (remember, it vanishes at the critical point).

Of course, using an equation of state, not only can the vapor pressure of a fluid be calculated, but so can other thermodynamic properties along the vapor-liquid phase boundary. This is demonstrated in the following illustration, which is a continuation of Illustration 6.4-1, dealing with the thermodynamic properties of oxygen.

ILLUSTRATION 7.5-1 (Illustration 6.4-1 continued)

Using an Equation of State to Calculate the Vapor Pressure of a Compound

Using the data in Illustration 6.4-1, and the same reference state, compute the vapor pressure of oxygen over the temperature range of $-200°C$ to the critical temperature, and also compute the specific volume, enthalpy, and entropy along the vapor-liquid equilibrium phase envelope. Add these results to Figs. 6.4-3, 6.4-4, and 6.4-5.

SOLUTION

Using the Peng-Robinson equation-of-state programs or MATHCAD worksheets described in Appendix B, we obtain the results in Table 7.5-1. The vapor pressure as a function of temperature is plotted in Fig. 7.5-3. The specific volumes and molar enthalpies and entropies of the coexisting phases have been added as the two-phase envelopes in Figs. 6.4-3, 6.4-4, and 6.4-5.■

ILLUSTRATION 7.5-2 (Illustration 6.4-1 concluded)

Completing the Construction of a Thermodynamic Properties Chart Using an Equation of State

Complete the calculated thermodynamic properties chart for oxygen by considering temperatures between $-100°C$ and $-200°C$.

SOLUTION

The calculation here is much like that of Illustration 6.4-1 *except* that for some pressures at temperatures below the critical temperature, three solutions for the compressibility or specific

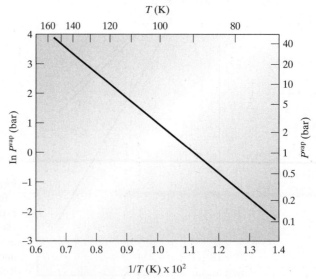

Figure 7.5-3 The vapor pressure of oxygen calculated using the Peng-Robinson equation of state.

Table 7.5-1 Thermodynamic Properties of Oxygen Along the Vapor-Liquid Phase Boundary Calculated Using the Peng-Robinson Equation of State

T (°C)		Vapor	Liquid	T (°C)		Vapor	Liquid
	$P = 0.11$ bar				$P = 12.30$ bar		
	Z	0.9945	0.0004		Z	0.8062	0.0371
−200	$\underline{V}$	53.3256	0.0232	−150	$\underline{V}$	0.6713	0.0309
	$\underline{H}$	−6311.60	−13 501.04		$\underline{H}$	−5475.94	−10 829.37
	$\underline{S}$	−20.87	−119.14		$\underline{S}$	−48.49	−91.95
	$P = 0.47$ bar				$P = 20.54$ bar		
	Z	0.9832	0.0016		Z	0.7170	0.0641
−190	$\underline{V}$	14.5751	0.0241	−140	$\underline{V}$	0.3864	0.0346
	$\underline{H}$	−6064.97	−13 013.02		$\underline{H}$	−5534.19	−10 131.95
	$\underline{S}$	−29.34	−112.90		$\underline{S}$	−52.20	−86.72
	$P = 1.39$ bar				$P = 32.15$ bar		
	Z	0.9611	0.0045		Z	0.5983	0.1107
−180	$\underline{V}$	5.3617	0.0252	−130	$\underline{V}$	0.2215	0.0410
	$\underline{H}$	−5842.71	−12 517.98		$\underline{H}$	−5780.50	−9273.89
	$\underline{S}$	−35.65	−107.30		$\underline{S}$	−56.43	−80.84
	$P = 2.19$ bar				$P = 39.44$ bar		
	Z	0.9452	0.0070		Z	0.5191	0.1501
−175	$\underline{V}$	3.5151	0.0259	−125	$\underline{V}$	0.1621	0.0469
	$\underline{H}$	−5744.89	−12 261.51		$\underline{H}$	−6043.66	−8716.50
	$\underline{S}$	−38.25	−104.64		$\underline{S}$	−59.19	−77.23
	$P = 3.31$ bar				$P = 47.85$ bar		
	Z	0.9256	0.0103		Z	0.4000	0.2253
−170	$\underline{V}$	2.3964	0.0266	−120	$\underline{V}$	0.1065	0.0600
	$\underline{H}$	−5658.53	−11 997.44		$\underline{H}$	−6599.85	−7885.19
	$\underline{S}$	−40.60	−102.05		$\underline{S}$	−63.62	−72.01
	$P = 6.76$ bar						
	Z	0.8746	0.0204				
−160	$\underline{V}$	1.2175	0.0284				
	$\underline{H}$	−5529.07	−11 440.52				
	$\underline{S}$	−44.75	−96.99				

$\underline{V}$ [=] m^3/kmol; $\underline{H}$ [=] J/mol = kJ/kmol; $\underline{S}$ [=] J/mol K = kJ/(kmol K).

volume are obtained. To choose the correct solution in such cases, the vapor pressure calculated in the previous illustration is needed. If three solutions are obtained and the system pressure is above the vapor pressure, the liquid is the stable phase and the smallest compressibility (or specific volume) is the correct solution, and it should be used in the calculation of all other thermodynamic properties. Conversely, if the system pressure is below the calculated vapor pressure, the vapor is the stable phase and the largest compressibility or specific volume root is to be used.

Using this calculational procedure, which is incorporated into the programs on the CD-ROM and discussed in Appendix B, the entries in Table 7.5-2 were obtained. These values also appear in Figs. 6.4-3, 6.4-4, and 6.4-5.

Table 7.5-2 The Thermodynamic Properties of Oxygen in the Low-Temperature Range Calculated Using the Peng-Robinson Equation of State

| | \multicolumn{4}{c}{T (°C)} | | | |
	-125	-150	-175	-200
$P = 1$ bar				
Z	0.9915	0.9863	0.9756	0.0038
$\underline{V}$	12.2138	10.0988	7.9621	0.0232
$\underline{H}$	-4302.26	-4995.07	-5684.10	$-13\,491.36$
$\underline{S}$	-19.97	-25.09	-31.35	-119.04
$P = 2$ bar				
Z	0.9830	0.9723	0.9502	0.0076
$\underline{V}$	6.0543	4.9777	3.8775	0.0232
$\underline{H}$	-4330.39	-5031.67	-5734.76	$-13\,489.39$
$\underline{S}$	-25.85	-31.04	-37.42	-119.04
$P = 5$ bar				
Z	0.9569	0.9285	0.0158	0.0191
$\underline{V}$	2.3574	1.9014	0.0258	0.0232
$\underline{H}$	-4416.92	-5146.94	$-12\,257.99$	$-13\,484.22$
$\underline{S}$	-33.84	-39.24	-104.68	-119.07
$P = 10$ bar				
Z	0.9116	0.8478	0.0316	0.0382
$\underline{V}$	1.1229	0.8681	0.0258	0.0232
$\underline{H}$	-4569.30	-5362.75	$-12\,251.78$	$-13\,475.52$
$\underline{S}$	-40.27	-46.14	-104.75	-119.11
$P = 20$ bar				
Z	0.8119	0.0598	0.0630	0.0762
$\underline{V}$	0.5000	0.03063	0.0257	0.0232
$\underline{H}$	-4915.26	$-10\,834.90$	$-12\,239.17$	$-13\,458.10$
$\underline{S}$	-47.61	-92.19	-104.88	-119.19
$P = 30$ bar				
Z	0.6912	0.0889	0.0943	0.1142
$\underline{V}$	0.2838	0.0304	0.0256	0.0232
$\underline{H}$	-5356.44	$-10\,840.08$	$-12\,226.32$	$-13\,440.62$
$\underline{S}$	-53.14	-92.48	-105.01	-119.26
$P = 40$ bar				
Z	0.1512	0.1176	0.1254	0.1521
$\underline{V}$	0.0466	0.0301	0.0256	0.0231
$\underline{H}$	-8734.00	$-10\,843.32$	$-12\,213.23$	$-13\,423.11$
$\underline{S}$	-77.37	-92.75	-105.14	-119.34
$P = 50$ bar				
Z	0.1740	0.1458	0.1563	0.1899
$\underline{V}$	0.0429	0.0299	0.0255	0.0231
$\underline{H}$	-8937.76	$-10\,844.89$	$-12\,199.92$	$-13\,405.54$
$\underline{S}$	-79.04	-93.01	-105.27	-119.42

(continued)

Table 7.5-2 (*Continued*)

	T (°C)			
	-125	-150	-175	-200
$P = 60$ bar				
Z	0.1987	0.1737	0.1871	0.2277
$\underline{V}$	0.0408	0.0296	0.0256	0.0231
$\underline{H}$	-9054.31	$-10\,845.02$	$-12\,186.40$	$-13\,387.93$
$\underline{S}$	-80.11	-93.25	-105.39	-119.49
$P = 70$ bar				
Z	0.2235	0.2013	0.2178	0.2654
$\underline{V}$	0.0393	0.0294	0.0254	0.0231
$\underline{H}$	-9134.97	$-10\,843.88$	$-12\,172.67$	$-13\,370.28$
$\underline{S}$	-80.92	-93.48	-105.51	-119.57
$P = 80$ bar				
Z	0.2481	0.2285	0.2483	0.3030
$\underline{V}$	0.0382	0.0293	0.0253	0.0230
$\underline{H}$	-9195.27	$-10\,841.61$	$-12\,158.77$	$-13\,352.59$
$\underline{S}$	-81.59	-93.70	-105.62	-119.64
$P = 90$ bar				
Z	0.2724	0.2555	0.2787	0.3405
$\underline{V}$	0.0373	0.0291	0.0253	0.0230
$\underline{H}$	-9242.27	$-10\,838.35$	$-12\,144.68$	$-13\,334.86$
$\underline{S}$	-82.16	-93.91	-105.74	-119.71
$P = 100$ bar				
Z	0.2964	0.2823	0.3089	0.3780
$\underline{V}$	0.0365	0.0289	0.0252	0.0230
$\underline{H}$	-9279.82	$-10\,834.20$	$-12\,130.43$	$-13\,317.09$
$\underline{S}$	-82.67	-94.11	-105.85	-119.78

$\underline{V}$ [=] m³/mol; $\underline{H}$ [=] J/mol = kJ/kmol; $\underline{S}$ [=] J/mol K = kJ/(kmol K).

NOTE

This completes the calculation of the thermodynamic properties of oxygen. You should remember, however, that these results were obtained using only a simple, generalized three-parameter (T_c, P_c, ω) cubic equation of state. Therefore, although they are good estimates, the results are not of as high an accuracy as would be obtained using a considerably more complicated equation of state. For example, an equation with 59 constants specific to water was used to compute the very accurate properties of steam in the steam tables (Appendix A.III). Clearly, a very large amount of carefully obtained experimental data, and sophisticated numerical analysis, must be used to obtain the 59 equation-of-state constants. This was done for water because in various industrial applications, especially in the design and evaluation of steam boilers and turbines, the properties of steam are needed to high accuracy. ∎

It must be emphasized that the generalization of the Peng-Robinson equation-of-state parameters given by Eqs. 6.7-2, 6.7-3, and 6.7-4 is useful only for hydrocarbons and inorganic gases (O_2, N_2, CO_2, etc.). For polar fluids (water, organic acids, alcohols, etc.), this simple generalization is not accurate, especially at low temperatures and pressures. A number of alternative procedures have been suggested in this case. One is to add additional species properties in the generalization for the equation-of-state parameters, such as the dipole moment, polarizability, and/or other properties. A simpler and generally more accurate procedure is to make the equation-of-state parameters specific for each fluid. One such procedure, and the only one that we shall introduce here, is due to Stryjek and Vera,[6] which is to replace Eq. 6.7-4 with

$$\kappa = \kappa_0 + \kappa_1(1 + T_r^{0.5})(0.7 - T_r) \tag{7.5-1}$$

where

$$\kappa_0 = 0.378893 + 1.4897153\omega - 0.17131848\omega^2 + 0.0196554\omega^3 \tag{7.5-2}$$

Here κ_1 is a parameter specific to each pure compound that is optimized to accurately fit low-temperature vapor pressure information. We refer to this modification of the Peng-Robinson equation as the PRSV equation. We also note (for later reference) that in this case

$$\frac{da(T)}{dT}$$

$$= 0.457\,24 \frac{R^2 T_c}{P_c} \left(\frac{1 - \sqrt{T_r}}{T_r} \right) \left[\left\{ \left(\frac{0.7 - T_r}{2} \right) - (1 + \sqrt{T_r})\sqrt{T_r} \right\} \kappa_1(1 - \sqrt{T_r}) - \kappa \right] \tag{7.5-3}$$

ILLUSTRATION 7.5-3
Vapor Pressure Calculations for Water with the Peng-Robinson Equation of State

  a. Compare the predictions for the vapor pressure of water from the Peng-Robinson equation of state with generalized coefficients with data in the saturated steam tables.
  b. Use the PRSV equation of state with Eqs. 7.5-1 and 7.5-2 with $\kappa_1 = -0.0665$ to calculate the vapor pressure of water, and compare the results with data in the steam tables.

SOLUTION

In the table below are the vapor pressure data from the steam tables and as calculated from the Peng-Robinson (PR) and PRSV equations of state.

COMMENT

Clearly, the PRSV equation of state, with the fluid-specific parameter κ_1, leads to more accurate water-vapor pressure predictions than the Peng-Robinson equation of state with generalized parameters. This is not surprising since the PRSV equation has an extra parameter that can be fit to data for each fluid. In general, the difference in accuracy between the Peng-Robinson and PRSV equations will be larger the more different the fluid is from a simple hydrocarbon, with the PRSV equation being more accurate. For comparison, values of the vapor pressure calculated from the Peng-Robinson equation of state with α equal to a constant value of unity are also given. These values are in poor agreement with the experimental data, and demonstrate why α has been made a function of temperature. Finally, since the critical properties were used in determining the parameters in the Peng-Robinson (with α a constant or a function of temperature)

---

[6]R. Stryjek and J. H. Vera, *Can. J. Chem. Eng. Sci.* **64**, 334, 820 (1986).

	Vapor Pressure (kPa)			
T (K)	Steam Tables	PR ($\alpha(T)$)	PR ($\alpha = 1$)	PRSV ($\kappa_1 = -0.0665$)
273.16	0.6113	0.4827	121.7	0.6094
283.15	1.2276	0.9969	164.8	1.2233
293.15	2.339	1.947	218.6	2.330
303.15	4.246	3.617	284.8	4.229
323.15	12.35	10.94	460.7	12.30
343.15	31.19	28.50	705.9	31.08
373.15	101.4	95.98	1 233	101.0
393.15	198.5	191.3	1 712	198.0
423.15	475.8	467.4	2 626	474.4
448.15	892.0	885.9	3 668	889.5
474.15	1 554	1 555	4 976	1 550
523.15	3 973	4 012	8 228	3 970
573.15	8 581	8 690	12 750	8 602
623.15	16 513	16 646	18 654	16 579
643.15	21 030	21 030	21 440	21 030

and PRSV equations, both equations must give the same critical point, so that the difference between these equations is greatest far away from the critical point, at low temperatures and pressures.   ■

In Chapter 10 we consider vapor-liquid equilibria in mixtures. For such calculations it is important to have the correct pure component vapor pressures if the mixture behavior is to be predicted correctly. Therefore, for equation-of-state calculations involving polar fluids, the PRSV equation will be used.

7.6 SPECIFICATION OF THE EQUILIBRIUM THERMODYNAMIC STATE OF A SYSTEM OF SEVERAL PHASES: THE GIBBS PHASE RULE FOR A ONE-COMPONENT SYSTEM

As we have already indicated, to completely fix the equilibrium thermodynamic state of a one-component, single-phase system, we must specify the values of two state variables. For example, to fix the thermodynamic state in either the vapor, liquid, or solid region of Fig. 7.3-6, both the temperature and pressure are needed. Thus, we say that a one-component, single-phase system has two **degrees of freedom**. In addition, to fix the total size or extent of the system we must also specify its mass or one of its extensive properties, such as total volume or total energy, from which the mass can be calculated.

In this section we are interested in determining the amount of information, and its type, that must be specified to completely fix the thermodynamic state of an equilibrium single-component, multiphase system. That is, we are interested in obtaining answers to the following questions:

1. How many state variables must be specified to completely fix the thermodynamic state of each phase when several phases are in equilibrium (i.e., how many degrees of freedom are there in a single-component, multiphase system)?
2. How many additional variables need be specified, and what type of variable should they be, to fix the distribution of mass (or number of moles) between the phases, and thereby fix the overall molar properties of the composite, multiphase system?

3. What additional information is needed to fix the total size of the multiphase system?

To specify the thermodynamic state of any one phase of a single-component, multiphase system, two thermodynamic state variables of that phase must be specified; that is, each phase has two degrees of freedom. Thus, it might appear that if $\mathcal{P}$ phases are present, the system would have $2\mathcal{P}$ degrees of freedom. The actual number of degrees of freedom is considerably fewer, however, since the requirement that the phases be in equilibrium puts constraints on the values of certain state variables in each phase. For example, from the analysis of Secs. 7.1 and 7.2 it is clear that at equilibrium the temperature in each phase must be the same. Thus, there are $\mathcal{P} - 1$ relations of the form

$$T^{\mathrm{I}} = T^{\mathrm{II}}$$
$$T^{\mathrm{I}} = T^{\mathrm{III}}$$
$$\vdots$$

that must be satisfied. Similarly, at equilibrium the pressure in each phase must be the same, so that there are an additional $\mathcal{P} - 1$ restrictions on the state variables of the form

$$P^{\mathrm{I}} = P^{\mathrm{II}}$$
$$P^{\mathrm{I}} = P^{\mathrm{III}}$$
$$\vdots$$

Finally, at equilibrium, the molar Gibbs energies must be the same in each phase, so that

$$\underline{G}^{\mathrm{I}}(T, P) = \underline{G}^{\mathrm{II}}(T, P)$$
$$\underline{G}^{\mathrm{I}}(T, P) = \underline{G}^{\mathrm{III}}(T, P)$$
$$\vdots$$

which provides an additional $\mathcal{P} - 1$ restrictions on the phase variables.

Since there are a total of $3(\mathcal{P} - 1)$ restrictions on the $2\mathcal{P}$ state variables needed to fix the thermodynamic state of each of the $\mathcal{P}$ phases, the number of degrees of freedom for the single-component, multiphase system is

Gibbs phase rule for a single-component system

$$
\mathcal{F} = \begin{array}{c}\text{Number of degrees}\\\text{of freedom}\end{array} = \left(\begin{array}{c}\text{Number of state}\\\text{variables needed to}\\\text{fix the state of each}\\\text{of the } \mathcal{P} \text{ phases}\end{array}\right) - \left(\begin{array}{c}\text{Restrictions on}\\\text{these state variables}\\\text{as a result of each of the}\\\text{phases being in}\\\text{equilibrium}\end{array}\right)
$$

$$= 2\mathcal{P} - 3(\mathcal{P} - 1)$$

$$= 3 - \mathcal{P}$$

(7.6-1)

Thus, the specification of $3 - \mathcal{P}$ state variables of the individual phases is all that is needed, in principle, to completely fix the thermodynamic state of each of the phases in a one-component, multiphase system. Of course, to fix the thermodynamic states of the phases, we would, in fact, need appropriate equation-of-state information.

It is easy to demonstrate that Eq. 7.6-1 is in agreement with our experience and with the phase diagram of Fig. 7.3-6. For example, we have already indicated that a single-phase ($\mathcal{P} = 1$) system has two degrees of freedom; this follows immediately from Eq. 7.6-1. To specify the thermodynamic state of each phase in a two-phase system (i.e., vapor-liquid, vapor-solid, or solid-liquid coexistence regions), it is clear from Fig. 7.3-6 that we need specify only the temperature *or* the pressure of the system; the value of the other variable can then be obtained from the appropriate coexistence curve. Setting $\mathcal{P}$ equal to 2 in Eq. 7.6-1 confirms that a two-phase system has only a single degree of freedom. Finally, since the solid, liquid, and vapor phases coexist at only a single point, the triple point, a single-component, three-phase system has no degrees of freedom. This also follows from Eq. 7.6-1 with $\mathcal{P}$ equal to 3.

The character of the variable to be specified as a degree of freedom is not completely arbitrary. To see this, consider Eq. 7.3-1a, which gives the molar volume of a two-phase mixture as a function of ω^{V} and the two single-phase molar volumes. Clearly, a specification of either $\underline{V}^{\mathrm{V}}$ or $\underline{V}^{\mathrm{L}}$ is sufficient to fix the thermodynamic state of both phases because the two-phase system has only one degree of freedom. However, the specification of the two-phase molar volume $\underline{V}$ can be satisfied by a range of choices of temperatures along the coexistence curve by suitably adjusting ω^{V}, so that $\underline{V}$ or any other molar property of the two phases combined is not suitable for a degree-of-freedom specification. Consequently, to fix the thermodynamic state of each of the $\mathcal{P}$ phases in equilibrium, we must specify $3 - \mathcal{P}$ properties of the *individual phases*.

Next we want to consider how many variables, and of what type, must be specified to also fix the distribution of mass or number of moles between the phases, so that the molar thermodynamic properties of the system consisting of several phases in equilibrium can be determined. If there are $\mathcal{P}$ phases, there are $\mathcal{P}$ values of ω^{i}, the mass fraction in phase i, which must be determined. Since the mass fractions must sum to unity, we have

$$\omega^{\mathrm{I}} + \omega^{\mathrm{II}} + \cdots = \sum_{\mathrm{i}=1}^{\mathcal{P}} \omega^{\mathrm{i}} = 1 \tag{7.6-2}$$

as one of the relations between the mass fractions. Thus, $\mathcal{P} - 1$ additional equations of the form

$$\sum_{\mathrm{i}=1}^{\mathcal{P}} \omega^{\mathrm{i}} \hat{V}^{\mathrm{i}} = \hat{V}$$

$$\sum_{\mathrm{i}=1}^{\mathcal{P}} \omega^{\mathrm{i}} \hat{H}^{\mathrm{i}} = \hat{H} \tag{7.6-3a}$$

or generally,

$$\sum_{\mathrm{i}=1}^{\mathcal{P}} \omega^{\mathrm{i}} \hat{\theta}^{\mathrm{i}} = \hat{\theta} \tag{7.6-3b}$$

are needed. In these equations, $\hat{\theta}^{\mathrm{i}}$ is the property per unit mass in phase i, and $\hat{\theta}$ is the property per unit mass of the multiphase mixture.

From these equations, it is evident that to determine the mass distribution between the phases, we need to specify a sufficient number of variables of the individual phases to fix the thermodynamic state of each phase (i.e., the degrees of freedom $\mathcal{F}$) and $\mathcal{P} - 1$ thermodynamic properties of the multiphase system in the form of Eq. 7.6-3. For example, if we know that steam and water are in equilibrium at some temperature T (which fixes the single-degree freedom of this two-phase system), the equation of state or the steam tables can be used to obtain the equilibrium pressure, specific enthalpy, entropy, and volume of each of the phases, but not the mass distribution between the phases. If, in addition, the volume (or enthalpy or entropy, etc.) per unit mass of the two-phase mixture were known, this would be sufficient to determine the distribution of mass between the two phases, and then all the other overall thermodynamic properties.

Once the thermodynamic properties of all the phases are fixed (by specification of the $\mathcal{F} = 3 - \mathcal{P}$ degrees of freedom) and the distribution of mass determined (by the specification of an additional $\mathcal{P} - 1$ specific properties of the multiphase system), the value of any one extensive variable (total volume, total enthalpy, etc.) of the multiphase system is sufficient to determine the total mass and all other extensive variables of the multiphase system.

Thus, to determine the thermodynamic properties per unit mass of a single-component, two-phase mixture, we need to specify the equivalent of one single-phase state variable (the one degree of freedom) and one variable that provides information on the mass distribution. The additional specification of one extensive property is needed to determine the total mass or size of the system. Similarly, to fix the thermodynamic properties of a single-component, three-phase mixture, we need not specify any single state variable (since the triple point is unique), but two variables that provide information on the distribution of mass between the vapor, liquid, and solid phases and one extensive variable to determine the total mass of the three-phase system.

ILLUSTRATION 7.6-1

Use of the Gibbs Phase Rule

    a. Show, using the steam tables, that fixing the equilibrium pressure of a steam-water mixture at 1.0135 bar is sufficient to completely fix the thermodynamic states of each phase. (This is an experimental verification of the fact that a one-component, two-phase system has only one degree of freedom.)

    b. Show that additionally fixing the specific volume of the two-phase system $\hat{V}$ at 1 m^3/kg is sufficient to determine the distribution of mass between the two phases.

    c. What is the total enthalpy of 3.2 m^3 of this steam-water mixture?

SOLUTION

    a. Using the saturation steam tables of Appendix A.III, we see that fixing the pressure at 1.0135 bar is sufficient to determine the temperature and the thermodynamic properties of each phase:

$$
\begin{aligned}
T &= 100°C \\
\hat{V}^{\mathrm{L}} &= 0.001\ 044 \text{ m}^3/\text{kg} & \hat{V}^{\mathrm{V}} &= 1.6729 \text{ m}^3/\text{kg} \\
\hat{H}^{\mathrm{L}} &= 419.04 \text{ kJ/kg} & \hat{H}^{\mathrm{V}} &= 2676.1 \text{ kJ/kg} \\
\hat{S}^{\mathrm{L}} &= 1.3069 \text{ kJ/(kg K)} & \hat{S}^{\mathrm{V}} &= 7.3549 \text{ kJ/(kg K)}
\end{aligned}
$$

Alternatively, specifying only the temperature, the specific volume, the specific enthalpy, or in fact any one other intensive variable of one of the phases would be sufficient to fix the thermodynamic properties of both phases. However, a specification of only the system pressure, temperature, or any one-phase state variable is not sufficient to determine the relative amounts of the vapor and liquid phases.

b. To determine the relative amounts of each of the phases, we need information from which ω^V and ω^L can be determined. Here the specific volume of the two-phase mixture is given, so we can use Eq. 7.3-1,

$$\hat{V} = \omega^V \hat{V}^V + (1 - \omega^V)\hat{V}^L$$

and the relation

$$\omega^V + \omega^L = 1$$

to find the distribution of mass between the two phases. For the situation here,

$$1 \text{ m}^3/\text{kg} = \omega^V(1.6729 \text{ m}^3/\text{kg}) + (1 - \omega^V)(0.001\,044 \text{ m}^3/\text{kg})$$

so that

$$\omega^V = 0.5975$$

and

$$\omega^L = 0.4025$$

c. Using the data in the problem statement, we know that the total mass of the steam-water mixture is

$$M = \frac{V}{\hat{V}} = \frac{3.2 \text{ m}^3}{1.0 \text{ m}^3/\text{kg}} = 3.2 \text{ kg}$$

From the results in parts (a) and (b) we can compute the enthalpy per unit mass of the two-phase mixture:

$$\hat{H} = \omega^L \hat{H}^L + \omega^V \hat{H}^V$$
$$= 0.4025(419.04) + 0.5975(2676.1) = 1767.7 \text{ kJ/kg}$$

Therefore,

$$H = M\hat{H} = 3.2 \times 1767.7 = 5303 \text{ kJ} \qquad \blacksquare$$

7.7 THERMODYNAMIC PROPERTIES OF PHASE TRANSITIONS

In this section several general properties of phase transitions are considered, as well as a phase transition classification system. The discussion and results of this section are applicable to all phase transitions (liquid-solid, solid-solid, vapor-solid, vapor-liquid, etc.), although special attention is given to vapor-liquid equilibrium.

One phase transition property important to chemists and engineers is the slope of the coexistence curves in the P-T plane; the slope of the vapor-liquid equilibrium line gives the rate of change of the vapor pressure of the liquid with temperature, the slope of the vapor-solid coexistence line is equal to the change with temperature of the vapor pressure of the solid (called the **sublimation pressure**), and the inverse of the slope of the liquid-solid coexistence line gives the change of the melting temperature of the solid with pressure. The slope of any of these phase equilibrium coexistence curves can be found by starting with Eq. 7.2-15c,

$$\underline{G}^I(T, P) = \underline{G}^{II}(T, P) \qquad \text{(7.7-1)}$$

where the superscripts label the phases. For a small change in the equilibrium temperature dT, the corresponding change of the coexistence pressure dP (i.e., the change in pressure following the coexistence curve) can be computed from the requirement that since Eq. 7.7-1 must be satisfied all along the coexistence curve, the changes in Gibbs energies in the two phases corresponding to the temperature and pressure changes must be equal, that is,

$$d\underline{G}^{\mathrm{I}} = d\underline{G}^{\mathrm{II}} \tag{7.7-2}$$

Using Eq. 6.2-8b in Eq. 7.7-2 gives

$$\underline{V}^{\mathrm{I}} dP - \underline{S}^{\mathrm{I}} dT = \underline{V}^{\mathrm{II}} dP - \underline{S}^{\mathrm{II}} dT$$

or, since P and T are the same in both phases for this equilibrium system (see Sec. 7.2)

$$\left(\frac{\partial P}{\partial T}\right)_{\underline{G}^{\mathrm{I}}=\underline{G}^{\mathrm{II}}} = \left(\frac{\underline{S}^{\mathrm{I}} - \underline{S}^{\mathrm{II}}}{\underline{V}^{\mathrm{I}} - \underline{V}^{\mathrm{II}}}\right) = \frac{\Delta \underline{S}}{\Delta \underline{V}} \tag{7.7-3}$$

From Eq. 7.7-1, we also have

$$\underline{G}^{\mathrm{I}} = \underline{H}^{\mathrm{I}} - T\underline{S}^{\mathrm{I}} = \underline{G}^{\mathrm{II}} = \underline{H}^{\mathrm{II}} - T\underline{S}^{\mathrm{II}}$$

or

$$\underline{S}^{\mathrm{I}} - \underline{S}^{\mathrm{II}} = \frac{\underline{H}^{\mathrm{I}} - \underline{H}^{\mathrm{II}}}{T}$$

so that Eq. 7.7-3 can be rewritten as

Clapeyron equation

$$\boxed{\left(\frac{\partial P^{\mathrm{sat}}}{\partial T}\right)_{\underline{G}^{\mathrm{I}}=\underline{G}^{\mathrm{II}}} = \frac{\Delta \underline{S}}{\Delta \underline{V}} = \frac{\Delta \underline{H}}{T \Delta \underline{V}}} \tag{7.7-4}$$

where $\Delta\underline{\theta} = \underline{\theta}^{\mathrm{I}} - \underline{\theta}^{\mathrm{II}}$, and we have used P^{sat} to denote the equilibrium coexistence pressure.[7] Equation 7.7-4 is the **Clapeyron equation**; it relates the slope of the coexistence curve to the enthalpy and volume changes at a phase transition.

Figure 7.3-6 is, in many ways, a typical phase diagram. From this figure and Eqs. 7.7-3 and 7.7-4 one can make several observations about property changes at phase transitions. First, since none of the coexistence lines has zero slope, neither the entropy change nor the enthalpy change is equal to zero for solid-liquid, liquid-vapor, or solid-vapor phase transitions. Also, since the coexistence lines do not have infinite slope, $\Delta \underline{V}$ is not generally equal to zero. (In general, both the heat of fusion, $\Delta_{\mathrm{fus}}\underline{H} = \underline{H}^{\mathrm{L}} - \underline{H}^{\mathrm{S}}$, and the volume change on melting, $\Delta_{\mathrm{fus}}\underline{V} = \underline{V}^{\mathrm{L}} - \underline{V}^{\mathrm{S}}$, are greater than zero for the liquid-solid transition, so that the liquid-solid coexistence line is as shown in the figure. One exception to this is water, for which $\Delta_{\mathrm{fus}}\underline{H} > 0$ but $\Delta_{\mathrm{fus}}\underline{V} < 0$, so the ice-water coexistence line has a negative slope.) From Sec. 7.3 we know that at the fluid critical point the coexisting phases are indistinguishable. Therefore, we can conclude that $\Delta\underline{H}$, $\Delta\underline{V}$, and $\Delta\underline{S}$ are all nonzero away from the fluid critical point and approach zero as the critical point is approached.

---

[7] P^{vap} is used to denote the vapor pressure of the liquid, P^{sub} to denote the vapor pressure of the solid, and P^{sat} to designate a general equilibrium coexistence pressure; equations containing P^{sat} are applicable to both the vapor pressures and sublimation pressures.

Of particular interest is the application of Eq. 7.7-4 to the vapor-liquid coexistence curve because this gives the change in vapor pressure with temperature. At temperatures for which the vapor pressure is not very high, it is found that $\underline{V}^V \gg \underline{V}^L$ and $\Delta \underline{V} \approx \underline{V}^V$. If, in addition, the vapor phase is ideal, we have $\Delta_{vap}\underline{V} = \underline{V}^V = RT/P$, so that

Clausius-Clapeyron equation

$$\frac{dP^{vap}}{dT} = \frac{P^{vap}\,\Delta_{vap}\underline{H}}{RT^2} \qquad \text{or} \qquad \boxed{\frac{d\ln P^{vap}}{dT} = \frac{\Delta_{vap}\underline{H}}{RT^2}} \qquad (7.7\text{-}5a)$$

and

$$\ln\frac{P^{vap}(T_2)}{P^{vap}(T_1)} = \int_{T_1}^{T_2}\frac{\Delta_{vap}\underline{H}}{RT^2}\,dT \qquad (7.7\text{-}5b)$$

which relates the fluid vapor pressures at two different temperatures to the heat of vaporization, $\Delta_{vap}\underline{H} = \underline{H}^V - \underline{H}^L$. Equation 7.7-5a is referred to as the **Clausius-Clapeyron equation.** The heat of vaporization is a function of temperature; however, if it is assumed to be independent of temperature, Eq. 7.7-5 can be integrated to give

Approximate integrated Clausius-Clapeyron equation

$$\boxed{\ln\frac{P^{vap}(T_2)}{P^{vap}(T_1)} = -\frac{\Delta_{vap}\underline{H}}{R}\left(\frac{1}{T_2} - \frac{1}{T_1}\right)} \qquad (7.7\text{-}6)$$

a result that is also valid over small temperature ranges even when $\Delta_{vap}\underline{H}$ is temperature dependent. Equation 7.7-6 has been found to be fairly accurate for correlating the temperature dependence of the vapor pressure of liquids over limited temperature ranges. (Note that Eq. 7.7-6 indicates that $\ln P^{vap}$ should be a linear function of $1/T$, where T is the absolute temperature. It is for this reason that Figs. 7.5-2 and 7.5-3 were plotted as $\ln P^{vap}$ versus $1/T$.)

ILLUSTRATION 7.7-1

Use of the Clausius-Clapeyron Equation

The vapor pressure of liquid 2,2,4-trimethyl pentane at various temperatures is given below. Estimate the heat of vaporization of this compound at 25°C.

Vapor pressure (kPa)	0.667	1.333	2.666	5.333	8.000	13.33	26.66	53.33	101.32
Temperature (°C)	−15.0	−4.3	7.5	20.7	29.1	40.7	58.1	78.0	99.2

SOLUTION

Over a relatively small range of temperature (say from 20.7 to 29.1°C), $\Delta_{vap}\underline{H}$ may be taken to be constant. Using Eq. 7.7-6, we obtain

$$\frac{\Delta_{vap}\underline{H}}{R} = \frac{-\ln[P^{vap}(T_2)/P^{vap}(T_1)]}{\dfrac{1}{T_2} - \dfrac{1}{T_1}} = \frac{-\ln(8.000/5.333)}{\dfrac{1}{302.25} - \dfrac{1}{293.85}} = 4287.8 \text{ K}$$

so that

$$\Delta_{vap}\underline{H} = 35.649 \text{ kJ/mol}$$

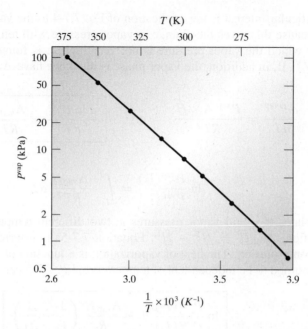

Figure 7.7-1 The vapor pressure of 2,2,4-trimethyl pentane as a function of temperature.

One can obtain an estimate of the temperature variation of the heat of vaporization by noting that the integration of Eq. 7.7-5 can be carried out as an indefinite rather than definite integral. In this case we obtain

$$\ln P^{\text{vap}}(T) = -\frac{\Delta_{\text{vap}}\underline{H}}{RT} + C$$

where C is a constant. Therefore, if we were to plot $\ln P^{\text{vap}}$ versus $1/T$, we should get a straight line with a slope equal to $-\Delta_{\text{vap}}\underline{H}/R$ if the heat of vaporization is independent of temperature, and a curve if $\Delta_{\text{vap}}\underline{H}$ varies with temperature. Figure 7.7-1 is a vapor pressure-temperature plot for the 2,2,4-trimethyl pentane system. As is evident from the linearity of the plot, $\Delta_{\text{vap}}\underline{H}$ is virtually constant over the whole temperature range. ∎

This illustration is a nice example of the utility of thermodynamics in providing interrelationships between properties. In this case we see how data on the temperature dependence of the vapor pressure of a fluid can be used to determine its heat of vaporization.

The equation developed in the illustration can be rewritten as

$$\ln P^{\text{vap}}(T) = A - \frac{B}{T} \tag{7.7-7}$$

with $B = \Delta_{\text{vap}}\underline{H}/R$, and it is reasonably accurate for estimating the temperature dependence of the vapor pressure over small temperature ranges. More commonly, the **Antoine equation**

Antoine equation

$$\boxed{\ln P^{\text{vap}}(T) = A - \frac{B}{T + C}} \tag{7.7-8}$$

is used to correlate vapor pressures accurately over the range from 1 to 200 kPa. Antoine constants for many substances are given by Poling, Prausnitz, and O'Connell.[8] Other commonly used vapor pressure correlations include the **Riedel equation**

$$\ln P^{\text{vap}}(T) = A + \frac{B}{T} + C \ln T + DT^6 \tag{7.7-9}$$

---

[8]B. E. Poling, J. M. Prausnitz, and J. P. O'Connell, *The Properties of Gases and Liquids,* 5th ed., McGraw-Hill, New York (2001).

and the **Harlecher-Braun equation,**

$$\ln P^{\mathrm{vap}}(T) = A + \frac{B}{T} + C\ln(T) + \frac{DP^{\mathrm{vap}}(T)}{T^2} \qquad \textbf{(7.7-10)}$$

which must be solved iteratively for the vapor pressure, but is reasonably accurate from low vapor pressure up to the critical pressure.

ILLUSTRATION 7.7-2

Interrelating the Thermodynamic Properties of Phase Changes

The following vapor pressure data are available

	T (°C)	P^{vap} (mm Hg)
Ice	−4	3.280
	−2	3.880
Water	+2	5.294
	+4	6.101

Estimate each of the following:

a. Heat of sublimation of ice
b. Heat of vaporization of water
c. Heat of fusion of ice
d. The triple point of water

SOLUTION

a. Here we use Eq. 7.7-6 in the form

$$\frac{\Delta_{\mathrm{sub}}\underline{H}}{R} = -\frac{\ln[P^{\mathrm{sub}}(T_2)/P^{\mathrm{sub}}(T_1)]}{\dfrac{1}{T_2} - \dfrac{1}{T_1}} = -\frac{\ln\left(\dfrac{3.880}{3.280}\right)}{\dfrac{1}{271.15\ \mathrm{K}} - \dfrac{1}{269.15\ \mathrm{K}}} = 6130\ \mathrm{K}$$

so that

$$\Delta_{\mathrm{sub}}\underline{H} = 6130\ \mathrm{K} \times 8.314\ \frac{\mathrm{J}}{\mathrm{mol\ K}} = 50\,965\ \frac{\mathrm{J}}{\mathrm{mol}} = 50.97\ \frac{\mathrm{kJ}}{\mathrm{mol}}$$

b. Similarly, here we have

$$\frac{\Delta_{\mathrm{vap}}\underline{H}}{R} = -\frac{\ln[P^{\mathrm{vap}}(T_2)/P^{\mathrm{vap}}(T_1)]}{\dfrac{1}{T_2} - \dfrac{1}{T_1}} = -\frac{\ln\left(\dfrac{6.101}{5.294}\right)}{\dfrac{1}{277.15\ \mathrm{K}} - \dfrac{1}{275.15\ \mathrm{K}}} = 5410\ \mathrm{K}$$

or

$$\Delta_{\mathrm{vap}}\underline{H} = 5410\ \mathrm{K} \times 8.314\ \frac{\mathrm{J}}{\mathrm{mol\ K}} = 44\,979\ \frac{\mathrm{J}}{\mathrm{mol}} = 44.98\ \frac{\mathrm{kJ}}{\mathrm{mol}}$$

c. Since

$$\Delta_{\mathrm{sub}}\underline{H} = \underline{H}(\text{vapor}) - \underline{H}(\text{solid})$$

and

$$\Delta_{\mathrm{vap}}\underline{H} = \underline{H}(\text{vapor}) - \underline{H}(\text{liquid})$$

it then follows that

$$\Delta_{\mathrm{fus}}\underline{H} = \underline{H}(\text{liquid}) - \underline{H}(\text{solid}) = \Delta_{\mathrm{sub}}\underline{H} - \Delta_{\mathrm{vap}}\underline{H} = 50.97 - 44.98 = 5.99 \ \frac{\text{kJ}}{\text{mol}}$$

d. At the triple-point temperature T_t the sublimation pressure of the solid and the vapor pressure of the liquid are equal; we denote this triple-point pressure as P_t. Using Eq. 7.7-6 for both the solid and liquid phases gives

$$\frac{\Delta_{\mathrm{sub}}\underline{H}}{R} = 6130 = -\frac{\ln\left(\dfrac{P_t}{P^{\mathrm{sub}}(T)}\right)}{\dfrac{1}{T_t} - \dfrac{1}{T}} = -\frac{\ln\left(\dfrac{P_t}{3.880}\right)}{\dfrac{1}{T_t} - \dfrac{1}{271.15}}$$

and

$$\frac{\Delta_{\mathrm{vap}}\underline{H}}{R} = 5410 = -\frac{\ln\left(\dfrac{P_t}{P^{\mathrm{vap}}(T)}\right)}{\dfrac{1}{T_t} - \dfrac{1}{T}} = -\frac{\ln\left(\dfrac{P_t}{5.294}\right)}{\dfrac{1}{T_t} - \dfrac{1}{275.15}}$$

The solution to this pair of equations is $T_t = 273.279$ K, and $P_t = 4.627$ mm Hg. The reported triple point is 273.16 K and 4.579 mm Hg, so our estimate is quite good.

COMMENT

This example illustrates the value of thermodynamics in interrelating properties in that from two sublimation pressure and two vapor pressure data points, we were able to estimate the heat of sublimation, the heat of vaporization, the heat of fusion, and the triple point. Further, we can now use the information we have obtained and write the equations

$$\ln\left(\frac{P^{\mathrm{sub}}}{3.880}\right) = -6130\left(\frac{1}{T} - \frac{1}{271.15}\right)$$

and

$$\ln\left(\frac{P^{\mathrm{vap}}}{5.294}\right) = -5410\left(\frac{1}{T} - \frac{1}{275.15}\right)$$

Consequently, we can also calculate the sublimation pressure and vapor pressure of ice and water, respectively, at other temperatures. Using these equations, we find

$$P^{\mathrm{sub}}(-10^\circ\text{C}) = 1.951 \text{ mm Hg}$$

which compares favorably with the measured value of 1.950 mm Hg. Also,

$$P^{\mathrm{vap}}(+10^\circ\text{C}) = 9.227 \text{ mm Hg}$$

compared with the measured value of 9.209 mm Hg. ∎

An important characteristic of the class of phase transitions we have been considering so far is that, except at the critical point, certain thermodynamic properties are discontinuous across the coexistence line; that is, these properties have different values in the two coexisting phases. For example, for the phase transitions indicated in Fig. 7.3-6 there is an enthalpy change, an entropy change, and a volume change on crossing the coexistence line. Also, the constant-pressure heat capacity $C_P = (\partial \underline{H}/\partial T)_P$ becomes infinite at a phase transition for a pure component because temperature is constant and the enthalpy increases across the coexistence line. In contrast, the Gibbs energy is, by Eq. 7.2-15c, continuous at a phase transition.

These observations may be summarized by noting that for the phase transitions considered here, the Gibbs energy is continuous across the coexistence curve, but its first derivatives

$$\left(\frac{\partial \underline{G}}{\partial T}\right)_P = -\underline{S} \quad \text{and} \quad \left(\frac{\partial \underline{G}}{\partial P}\right)_T = \underline{V}$$

are discontinuous, as are all higher derivatives. This class of phase transition is called a **first-order phase transition**. The concept of higher-order phase transitions follows naturally. A second-order phase transition (at constant T and P) is one in which $\underline{G}$ and its first derivatives are continuous, but derivatives of second order and higher are discontinuous. Third-order phase transition is defined in a similar manner.

One example of a second-order phase change is the structural rearrangement of quartz, where $\underline{G}$, $\underline{S}$, and $\underline{V}$ are continuous across the coexistence line, but the constant-pressure heat capacity, which is related to the second temperature derivative of the Gibbs energy,

$$\left(\frac{\partial^2 \underline{G}}{\partial T^2}\right)_P = -\left(\frac{\partial \underline{S}}{\partial T}\right)_P = -\frac{C_P}{T}$$

is not only discontinuous but has a singularity at the phase change (see Fig. 7.7-2). Another example of a second-order phase transition is the change from the ferromagnetic to paramagnetic states in some materials. No phase transitions higher than second order have been observed in the laboratory.

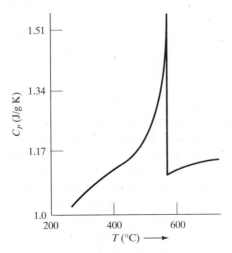

Figure 7.7-2 The specific heat of quartz near a second-order phase transition. (Reprinted with permission from H. B. Callen, *Thermodynamics*, John Wiley & Sons, New York, 1960.)

7.8 THERMODYNAMIC PROPERTIES OF SMALL SYSTEMS, OR WHY SUBCOOLING AND SUPERHEATING OCCUR

A drop of liquid in a vapor consists of molecules in the bulk of the liquid interacting with many other molecules, and molecules at the surface that on the liquid side of the interface interact with many molecules and on the vapor side interact with only a few molecules. Consequently, the energy of interaction (a part of the internal energy) of the molecules at the interface between the two phases is different from that of molecules in the bulk. Other cases of molecules at the surface having different energies than those in the bulk occur at liquid-liquid, liquid-solid, and solid-vapor interfaces. Except for very small drops that have large surface areas relative to their volume, the number of molecules at the surface is very much smaller than the number of molecules in the bulk, so that the effect of the difference in energies between molecules in the bulk and surface molecules can be neglected. However, for very small drops, this surface effect is important.

The contribution to the energy from the surface is usually written as σA, where σ is the surface tension for the liquid-vapor interface (or the interfacial tension for a liquid-liquid interface) and A is the surface area. The σA contribution is the two-dimensional analogue of the PV term for bulk fluids. Including the effect of changing surface area in the energy balance, just as we have included the effect of changing volume, gives for a closed system without shaft work

$$\frac{dU}{dt} = \dot{Q} - P\frac{dV}{dt} - \sigma\frac{dA}{dt} \qquad (7.8\text{-}1)$$

where P is the external pressure on the system. The negative sign on the $\sigma(dA/dt)$ term is due to the fact that the system must do work to increase its surface area, just as the system must do work against the surroundings to expand (increase its volume).

Our interest here will be in spherical drops, as droplets that occur in nature are generally spherical or almost so. In this case $V = \frac{4}{3}\pi r^3$ and $A = 4\pi r^2$, where r is the drop radius. Therefore, the energy balance is

$$\frac{dU}{dt} = \dot{Q} - 4\pi r^2 P\frac{dr}{dt} - 8\pi r\sigma\frac{dr}{dt} = \dot{Q} - 4\pi r^2\left(P + \frac{2\sigma}{r}\right)\frac{dr}{dt} = \dot{Q} - \left(P + \frac{2\sigma}{r}\right)\frac{dV}{dt}$$

or simply

$$\frac{dU}{dt} = \dot{Q} - P_{\text{int}}\frac{dV}{dt} \qquad (7.8\text{-}2)$$

where $P_{\text{int}} = (P + 2\sigma/r)$ is the internal pressure in the droplet as a result of the external pressure and the surface tension acting on the surface of the droplet. (The same relation between the internal pressure and the external pressure can be obtained from a force balance.) If the drop is very large (i.e., $r \to \infty$), the external pressure and the internal pressure are equal. However, for a very small drop (that is, as $r \to 0$), the internal pressure is very much larger than the external pressure. (The expression for the internal pressure indicates that it becomes infinite in the limit of a drop of the size of a single molecule. However, when only one or a few molecules are involved, one must use a statistical mechanical description, not the macroscopic thermodynamic description in this book. Therefore, we will not consider the case of r being of molecular dimensions.)

Table 7.8-1 Surface Tension at a Liquid-Vapor Interface

Liquid	Temperature (°C)	σ (dyne/cm)*
Water	20	72.9
	25	72.1
Methanol	20	22.7
Ethanol	20	22.1
1-Octanol	20	27.6
Benzene	20	28.9
Aniline	20	43.4
Glycerol	20	64.0
Perfluorohexane	20	11.9
n-Heptane	20	20.4
n-Octane	20	21.6
Propionic acid	20	26.7
Mercury	20	487
Sodium	139	198
Sodium chloride	1073	115

*Divide by 1000 for J/m^2.

Table 7.8-2 Interfacial Tension at a Liquid-Liquid Interface

Liquid	Temperature (°C)	σ (dyne/cm)*
Water/n-butyl alcohol	20	1.8
Water/mercury	20	415
Water/benzaldehyde	20	15.5
Water/diethylene glycol	25	57
Mercury/n-hexane	20	378

*Divide by 1000 for J/m^2.

Some values of the surface tension (for vapor-liquid interfaces) and interfacial tensions (for liquid-liquid interfaces) are given in Table 7.8-1.

We now consider two cases. The first is a bulk liquid at a temperature T subject to an external pressure of P. The fugacity of this liquid, $f^L(T, P)$, can be computed by any of the methods described earlier in this chapter. The second case is that of a small droplet of the same liquid at the same temperature and external pressure. The fugacity of this droplet is

$$f_{\text{drop}}^L(T, P) = f^L(T, P_{\text{int}}) = f^L\left(T, P + \frac{2\sigma}{r}\right) = f^L(T, P)\exp\left(\frac{2\sigma}{rRT}\underline{V}^L\right)$$

$$(7.8\text{-}3)$$

where $\underline{V}^L$ is the molar volume of the liquid, and we have introduced the Poynting correction (so that the fugacity of the liquid drop can be expressed as the product of the fugacity of the bulk liquid at the same temperature and same external pressure) and a correction factor for the surface effect. Clearly, the contribution from the surface term is always greater than unity, so that the fugacity of the drop is always greater than the fugacity of the bulk liquid at the same temperature and external pressure. Further, this difference will be very large if the drop is very small, as shown in Fig. 7.8-1.

There are several implications of this result. The first is that since at a given temperature the fugacity of a drop is higher than that of a bulk liquid, it will attain a value of

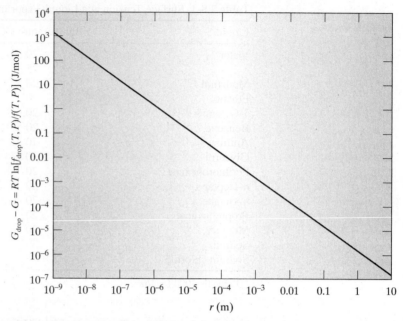

Figure 7.8-1 Difference in Gibbs energy between a bubble of radius r and the bulk fluid for water at 20°C.

1.013 bar at a lower temperature than that of the bulk liquid; that is, the boiling temperature of small drops at atmospheric pressure will be lower than that of the bulk liquid (Problem 7.72). Or, more generally, small drops will vaporize more easily than the bulk fluid because of their higher fugacity (a thermodynamic effect) and their greater surface area per unit mass (promoting mass transfer). Also, as vaporization of the drop occurs and the drop radius becomes smaller, both the fugacity of the liquid in the drop and the surface-to-volume ratio increase, so that the vaporization process accelerates. It is for this reason that to vaporize a liquid quickly in consumer products—for example, in an air freshener—an atomization process is used that produces small droplets.

From the relationship between fugacity and molar Gibbs energy, we know that the molar Gibbs energy of a droplet is higher than that of the bulk fluid at the same temperature and external pressure. This is shown in Fig. 7.8-1. In particular, as the droplet radius goes to zero, this difference in molar Gibbs energy becomes infinite.

Instead of vaporization, which is the disappearance of a phase, consider the opposite process of condensation, in which a liquid phase is created. The liquid will form when the fugacity (or, equivalently, molar Gibbs energy) of the liquid is less than that of the vapor. However, we see from Fig. 7.8-1 that the first liquid droplet that forms, of infinitesimal size, will have a very large Gibbs energy. Therefore, a vapor can be cooled below its normal condensation temperature (as long as there is no dust or other nucleation sites present) but may not condense, because while the Gibbs energy of the bulk liquid will be less than that of the vapor, the Gibbs energy of the very small droplets that will form first is greater than that of the vapor. This is the phenomenon of subcooling. As the vapor continues to be cooled, the difference between the vapor and bulk liquid Gibbs energies will become large enough to overcome the Gibbs energy penalty required for the formation of the first droplet of liquid. Once this first droplet is formed, since the Gibbs energy of a droplet decreases by the droplet growing in size, droplet growth (condensation) then occurs rapidly.

Similarly, to initiate boiling in a liquid, an infinitesimal bubble of vapor must first be created. However, here too there is a very large Gibbs energy barrier to the formation of the first, very small bubble, and unless boiling chips or other devices are used to induce the nucleation of bubbles, on heating the normal boiling temperature will be exceeded before boiling starts. This is superheating. Note that the cause of both subcooling and superheating is the cost in Gibbs energy of producing a new phase (the liquid droplet in subcooling and a vapor bubble in superheating).

Similar phenomena occur in other phase transitions, such as solidification or crystallization. That is, on cooling a liquid, initially a temperature below its normal freezing point will be reached without freezing (liquid-solid subcooling), and then suddenly a significant amount of freezing or crystallization will occur.

There is a related process that occurs after nucleation of a new phase and is especially evident in crystallization. Initially, crystals of many different sizes form when crystallization (or freezing) first occurs. If the system is adiabatic (so that no further heating or cooling is supplied the small crystals will decrease in size and eventually disappear, while the large crystals will grow, until there may be only a few (or even just one) large crystals. This phenomenon is referred to as **Ostwald ripening**. We can easily understand why this occurs by examining Fig. 7.8-1, where we see that crystals of large size (radius) have a lower Gibbs energy than crystals of small size. Therefore, in the evolution toward equilibrium, the state of lowest Gibbs energy, the Gibbs energy of the system is lowered by replacing many small crystals with a few larger ones.

PROBLEMS

7.1 By doing some simple calculations and plotting several graphs, one can verify some of the statements made in this chapter concerning phase equilibrium and phase transitions. All the calculations should be done using the steam tables.

a. Establish, by direct calculation, that

$$\underline{G}^{L} = \underline{G}^{V}$$

for steam at 2.5 MPa and $T = 224°C$.

b. Calculate $\underline{G}^{V}$ at $P = 2.5$ MPa for a collection of temperatures between 225 and $400°C$ and extrapolate this curve below $224°C$.

c. Find $\underline{G}^{L}$ at 160, 170, 180, 190, 200, and $210°C$. Plot this result on the same graph as used for part (b) and extrapolate above $224°C$. (*Hint:* For a liquid $\underline{H}$ and $\underline{S}$ can be taken to be independent of pressure. Therefore, the values of $\underline{H}$ and $\underline{S}$ for the liquid at any pressure can be gotten from the data for the saturated liquid at the same temperature.) How does this graph compare with Fig. 7.3-7?

d. Plot $\underline{V}$ versus T at $P = 2.5$ MPa for the temperature range of 150 to $400°C$, and show that $\underline{V}$ is discontinuous.

e. Plot C_P versus T at $P = 2.5$ MPa over the temperature range of 150 to $400°C$, and thereby establish that C_P is discontinuous.

f. Using the data in the steam tables, show that the Gibbs energies per unit mass of steam and liquid water in equilibrium at $300°C$ are equal.

7.2 a. Show that the condition for equilibrium in a closed system at constant entropy and volume is that the internal energy U achieve a minimum value subject to the constraints.

b. Show that the condition for equilibrium in a closed system at constant entropy and pressure is that the enthalpy H achieve a minimum value subject to the constraints.

7.3 a. Show that the intrinsic stability analysis for fluid equilibrium at constant temperature and volume leads to the single condition that

$$\left(\frac{\partial P}{\partial \underline{V}}\right)_T < 0$$

b. Show that intrinsic stability analysis for fluid equilibrium at constant temperature and pressure does not lead to any restrictions on the equation of state.

7.4 a. Show that the conditions for vapor-liquid equilibrium at constant N, T, and V are $\underline{G}^{V} = \underline{G}^{L}$ and $P^{V} = P^{L}$.

b. Show that the condition for vapor-liquid equilibrium at constant N, T, and P is $\underline{G}^{V} = \underline{G}^{L}$.

7.5 Prove that $C_P \geq C_V$ for any fluid, and identify those conditions for which $C_P = C_V$.

7.6 Show that if the polymer fiber of Problem 6.26 is to be thermodynamically stable at all temperatures, the parameters α, β, and γ must be positive.

7.7 The entropy of a certain fluid has been found to be related to its internal energy and volume in the following way:

$$\underline{S} = \underline{S}^\circ + \alpha \ln \frac{\underline{U}}{\underline{U}^\circ} + \beta \ln \frac{\underline{V}}{\underline{V}^\circ}$$

where $\underline{S}^\circ$, $\underline{U}^\circ$, and $\underline{V}^\circ$ are, respectively, the molar entropy, internal energy, and volume of the fluid for some appropriately chosen reference state, and α and β are positive constants.

a. Develop an interrelationship between internal energy, temperature, and specific volume (the thermal equation of state) for this fluid.

b. Develop an interrelationship between pressure, temperature, and volume (the volumetric equation of state) for this fluid.

c. Show that this fluid does not have a first-order phase transition by establishing that the fluid is stable in all thermodynamic states.

7.8 Figure 7.3-4 is the phase diagram for a van der Waals fluid. Within the vapor-liquid coexistence envelope one can draw another envelope representing the limits of supercooling of the vapor and superheating of the liquid that can be observed in the laboratory; along each isotherm these are the points for which

$$\left(\frac{\partial P}{\partial \underline{V}} \right)_T = 0$$

Obtain this envelope for the van der Waals fluid. This is the spinodal curve. The region between the coexistence curve and the curve just obtained is the metastable region of the fluid. Notice also that the critical point of the fluid is metastable.

7.9 Derive the following two independent equations for a second-order phase transition:

$$\left. \frac{\partial P}{\partial T} \right|_{\substack{\text{along phase} \\ \text{transition} \\ \text{curve}}} = \frac{C_P^{\mathrm{I}} - C_P^{\mathrm{II}}}{T \left\{ \left(\frac{\partial \underline{V}}{\partial T} \right)_P^{\mathrm{I}} - \left(\frac{\partial \underline{V}}{\partial T} \right)_P^{\mathrm{II}} \right\}}$$

and

$$\left. \frac{\partial P}{\partial T} \right|_{\substack{\text{along phase} \\ \text{transition} \\ \text{curve}}} = \frac{\left\{ \left(\frac{\partial \underline{V}}{\partial T} \right)_P^{\mathrm{I}} - \left(\frac{\partial \underline{V}}{\partial T} \right)_P^{\mathrm{II}} \right\}}{\left\{ \left(\frac{\partial \underline{V}}{\partial P} \right)_T^{\mathrm{I}} - \left(\frac{\partial \underline{V}}{\partial P} \right)_T^{\mathrm{II}} \right\}} = \frac{\alpha^{\mathrm{I}} - \alpha^{\mathrm{II}}}{\kappa_T^{\mathrm{I}} - \kappa_T^{\mathrm{II}}}$$

These equations, which are analogues of the Clapeyron equation, are sometimes referred to as the Ehrenfest equations.

Also, show that these two equations can be derived by applying L'Hopital's rule to the Clapeyron equation for a first-order phase transition.

7.10 a. The heat of fusion $\Delta_{\mathrm{fus}} \hat{H}$ for the ice-water phase transition is 335 kJ/kg at 0°C and 1 bar. The density of water is 1000 kg/m³ at these conditions, and that of ice is 915 kg/m³. Develop an expression for the change of the melting temperature of ice with pressure.

b. The heat of vaporization for the steam-water phase transition is 2255 kJ/kg at 100°C and 1 bar. Develop an expression for the change in the boiling temperature of water with pressure.

c. Compute the freezing and boiling points of water in Denver, Colorado, where the mean atmospheric pressure is 84.6 kPa.

7.11 The triple point of iodine, I_2, occurs at 112.9°C and 11.57 kPa. The heat of fusion at the triple point is 15.27 kJ/mol, and the following vapor pressure data are available for solid iodine:

Vapor pressure (kPa)	2.67	5.33	8.00
Temperature (°C)	84.7	97.5	105.4

Estimate the normal boiling temperature of molecular iodine.

7.12 The following data are available for water:

$$\ln P^{\mathrm{sub}}(\mathrm{ice}) = 28.8926 - 6140.1/T \quad P \text{ in Pa}$$
$$\ln P^{\mathrm{vap}}(\mathrm{water}) = 26.3026 - 5432.8/T \quad T \text{ in K}$$

a. Compute the triple-point temperature and pressure of water.

b. Compute the heat of vaporization, the heat of sublimation, and the heat of fusion of water at its triple point.

7.13 a. The following data have been reported for the vapor pressure of ethanol as a function of temperature.[9]

P^{vap} (kPa)	0.6667	1.333	2.667	5.333	8.00	13.33
T (°C)	−12.0	−2.3	8.0	19.0	26.0	34.9

Use these data to calculate the heat of vaporization of ethanol at 17.33°C.

b. Ackermann and Rauh have measured the vapor pressure of liquid plutonium using a clever mass

---

[9]*Reference:* R. H. Perry, D. W. Green, and J. O. Maloney, eds., *The Chemical Engineers' Handbook,* 6th ed., McGraw-Hill, New York (1984) pp. 3–55.

effusion technique.[10] Some of their results are given here:

T (K)	1343	1379	1424	1449
P^{vap} (bar)	7.336 $\times 10^{-9}$	1.601 $\times 10^{-8}$	3.840 $\times 10^{-8}$	5.654 $\times 10^{-8}$

Estimate the heat of vaporization of liquid plutonium at 1400 K.

7.14 a. Derive Eq. 7.4-8.
   b. Derive Eq. 7.4-12.
   c. Obtain an expression for the fugacity of a pure species that obeys the van der Waals equation of state in terms of Z, $B = Pb/RT$, and $A = aP/(RT)^2$ (i.e., derive Eq. 7.4-13).
   d. Repeat the derivation with the Peng-Robinson equation of state (i.e., derive Eq. 7.4-14a).

7.15 a. Calculate the fugacity of liquid hydrogen sulfide in contact with its saturated vapor at 25.5°C and 20 bar.
   b. The vapor pressure of pure water at 310.6 K is 6.455 kPa. Compute the fugacity of pure liquid water at 310.6 K when it is under a pressure of 100 bar, 500 bar, and 1000 bar. [Volume 7 of the *International Critical Tables* (McGraw-Hill, New York, 1929) gives values of 6.925, 9.175, and 12.966 kPa, respectively, for these conditions.]

7.16 a. Using only the steam tables, compute the fugacity of steam at 400°C and 2 MPa, and at 400°C and 50 MPa.
   b. Compute the fugacity of steam at 400°C and 2 MPa using the principle of corresponding states. Repeat the calculation at 400°C and 50 MPa.
   c. Repeat the calculations using the Peng-Robinson equation of state.
   Comment on the causes of the differences among these predictions.

7.17 a. Show that at moderately low pressures and densities the virial equation of state can be written as

$$\frac{PV}{RT} = 1 + B\left(\frac{P}{RT}\right) + (C - B^2)\left(\frac{P}{RT}\right)^2 + \cdots$$

   b. Prove that the fugacity coefficient for this form of the virial equation of state is

$$\frac{f}{P} = \exp\left\{B\left(\frac{P}{RT}\right) + \frac{(C - B^2)}{2}\left(\frac{P}{RT}\right)^2 + \cdots\right\}$$

   c. The first two virial coefficients for methyl fluoride at 50°C are $B = -0.1663$ m³/kmol and $C =$

0.012 92 (m³/kmol)². Plot the ratio f/P as a function of pressure at 50°C for pressures up to 150 bar. Compare the results with the corresponding-states plot of f/P versus P/P_c.

7.18 The following data are available for carbon tetrachloride:

$T_c = 283.3$°C	$P_c = 4.56$ MPa	$Z_c = 0.272$

Vapor pressure (MPa)	0.5065	1.013	2.026
T (°C)	141.7	178.0	222.0

   a. Compute the heat of vaporization of carbon tetrachloride at 200°C using only these data.
   b. Derive the following expression, which can be used to compute the heat of vaporization from the principle of corresponding states:

$$\Delta_{vap}\underline{H} = T_c\left[\left(\frac{H - H^{IG}}{T_c}\right)_{\substack{sat \\ vap}} - \left(\frac{H - H^{IG}}{T_c}\right)_{\substack{sat \\ liq}}\right]$$

   c. Compute the heat of vaporization of carbon tetrachloride at 200°C using the principle of corresponding states.
   d. Comment on the reasons for the difference between the heats of vaporization computed in parts (a) and (c) and suggest a way to correct the results to improve the agreement.

7.19 An article in *Chemical and Engineering News* (Sept. 28, 1987) describes a hydrothermal autoclave. This device is of constant volume, is evacuated, and then water is added so that a fraction x of the total volume is filled with liquid water and the remainder is filled with water vapor. The autoclave is then heated so that the temperature and pressure in the sealed vessel increase. It is observed that if x is greater than a "critical fill" value, x_c, the liquid volume fraction increases as the temperature increases, and the vessel becomes completely filled with liquid at temperatures below the critical temperature. On the other hand, if $x < x_c$, the liquid evaporates as temperature is increased, and the autoclave becomes completely filled with vapor below the water critical temperature. If, however, $x = x_c$, the volume fraction of liquid in the autoclave remains constant as the temperature increases, and the temperature-pressure trajectory passes through the water critical point. Assuming the hydrothermal autoclave is to be loaded at 25°C, calculate the critical fill x_c
   a. Using the steam tables
   b. Assuming the water obeys the Peng-Robinson equation of state

---

[10]R. J. Ackermann and E. G. Rauh, *J. Chem. Thermodyn.* **7**, 211 (1975).

7.20 a. Quantitatively explain why ice skates slide along the surface of ice. Can it get too cold to ice skate?

b. Is it possible to ice skate on other materials, such as frozen carbon dioxide?

c. What is the approximate lowest temperature at which a good snowball can be made? Why can't a snowball be made if the temperature is too low?

7.21 The effect of pressure on the melting temperature of solids depends on the heat of fusion and the volume change on melting. The heat of fusion is always positive (that is, heat must be added to melt the solid), while the volume change on melting will be positive if the solid is a close-packed structure, and negative for solids that crystallize into an open structure. Metallic elements generally crystallize into close-packed structures, so that $\Delta_{fus}V$ is positive for most metals. Using the data below, calculate the change in the melting temperature T_{mp} for each substance for a pressure increase from 1 bar to 1001 bar, and compare the results with the experimentally

Substance	T_{mp} (K)	$\Delta_{fus}H$ (J/g)	$\Delta_{fus}V$ (cc/g)	ΔT (K)
Water	273.2	333.8	−0.0906	−7.4
Acetic acid	289.8	187.0	+0.01595	+2.44
Tin	505.0	58.6	+0.00389	+3.3
Bismuth	544	52.7	−0.00342	−3.55

7.22 The vapor pressure of some materials can be represented by the equation

$$\ln P \text{ (mm Hg)} = \frac{a}{T} + b \ln T + cT + d$$

Values of the constants in this equation are given below.

Compute the heats of sublimation of the solids and the heats of vaporization of the liquids above over the temperature range specified.

7.23 The metal tin undergoes a transition from a gray phase to a white phase at 286 K and ambient pressure. Given that the enthalpy change of this transition is 2090 J/mol and that the volume change of this transition is

−4.35 cm^3/mol, compute the temperature at which this transition occurs at 100 bar.

7.24 Estimate the triple-point temperature and pressure of benzene. The following data are available:

		Vapor Pressure		
T (°C)	−36.7	−19.6	−11.5	−2.6
P^{vap} (Pa)	1.333	6.667	13.33	26.67

Melting point at atmospheric pressure = 5.49°C
Heat of fusion at 5.49°C = 127 J/g
Liquid volume at 5.49°C = 0.901×10^{-3} m^3/kg
Volume change on melting = 0.1317×10^{-3} m^3/kg

7.25 The following data are available for the thermodynamic properties of graphite and diamond:

Property	Graphite	Diamond
$\underline{G}(T = 298$ K$)$	$0 \frac{\text{J}}{\text{mol}}$	$2900 \frac{\text{J}}{\text{mol}}$
$\underline{S}(T = 298$ K$)$	$5.740 \frac{\text{J}}{\text{mol K}}$	$2.377 \frac{\text{J}}{\text{mol K}}$
Density	$2220 \frac{\text{kg}}{\text{m}^3}$	$3510 \frac{\text{kg}}{\text{m}^3}$

Assuming that the entropies and densities are approximately independent of temperature and pressure, determine the range of conditions for which diamonds can be produced from graphite. (The procedure proposed would be to hold the graphite at high temperatures and pressures until diamonds formed, followed by rapid cooling and depressurizing so that the diamonds can not revert to graphite.)

7.26 A thermally insulated (adiabatic) constant-volume bomb has been very carefully prepared so that half its volume is filled with water vapor and half with subcooled liquid water, both at −10°C and 0.2876 kPa (the saturation pressure of the subcooled liquid). Find the temperature, pressure, and fraction of water in each phase after equilibrium has been established in

Substance	a	b	$c \times 10^3$	d	Temperature Range (K)
Ag(s)	−14 710	−0.328		11.66	298–1234
Ag(l)	−14 260	−0.458		12.23	1234–2485
BeO(s)	−34 230	−0.869		18.50	298–2800
Ge(s)	−20 150	−0.395		13.28	298–1210
Mg(s)	−7 780	−0.371		11.41	298–924
Mg(l)	−7 750	−0.612		12.79	924–1380
NaCl(s)	−12 440	−0.391	−0.46	14.31	298–1074
Si(s)	−18 000	−0.444		12.83	1200–1683

the bomb. What is the entropy change for the process?

For simplicity, neglect the heat capacity of the bomb, assume the vapor phase is ideal, and for the limited temperature range of interest here, assume that each of the quantities below is independent of temperature.

Data:

$$\Delta \hat{H}(\text{solid} \rightarrow \text{liquid}) = 335 \text{ J/g}$$
$$\Delta \hat{H}(\text{liquid} \rightarrow \text{vapor}, \ T \sim 0°\text{C}) = 2530 \text{ J/g}$$

$C_P(\text{liquid}) = 4.22 \text{ J/(g °C)}$ $\hat{V}(\text{liquid}) = 1 \times 10^{-3} \text{ m}^3/\text{kg}$
$C_P(\text{solid}) = 2.1 \text{ J/(g °C)}$ $\hat{V}(\text{solid}) = 1.11 \times 10^{-3} \text{ m}^3/\text{kg}$
$C_P(\text{vapor}) = 2.03 \text{ J/(g °C)}$

7.27 One question that arises in phase equilibrium calculations and experiments is how many phases can be in equilibrium simultaneously, since this determines how many phases one should search for. Consider a one-component system.

   a. If only solid phases in different crystal forms are present, what is the maximum number of equilibrium solid phases that can coexist?

   b. If a liquid is present, what is the maximum number of solid phases that can also exist at equilibrium?

   c. If a vapor is present, what is the maximum number of solid phases that can also exist at equilibrium?

   d. If a vapor and a liquid are present, what is the maximum number of solid phases that can also exist at equilibrium?

7.28 Many thermodynamic and statistical mechanical theories of fluids lead to predictions of the Helmholtz energy $\underline{A}$ with T and $\underline{V}$ as the independent variables; that is, the result of the theory is an expression of the form $\underline{A} = \underline{A}(T, \underline{V})$. The following figure is a plot of $\underline{A}$ for one molecular species as a function of specific volume at constant temperature. The curve on the left has been calculated assuming the species is present as a liquid, and the curve on the right assuming the species is a gas.

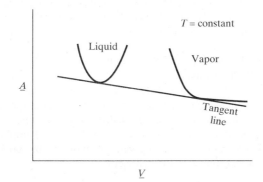

Prove that for the situation indicated in the figure, the vapor and liquid can coexist at equilibrium, that the specific volumes of the two coexisting phases are given by the points of tangency of the Helmholtz energy curves with the line that is tangent to both curves, and that the slope of this tangent line is equal to the negative of the equilibrium (vapor) pressure.

7.29 The principle of corresponding states has far greater applicability in thermodynamics than was indicated in the discussion of Sec. 6.6. For example, it is possible to construct corresponding-states relations for both the vapor and liquid densities along the coexistence curve, and for the vapor pressure and enthalpy change on vaporization ($\Delta_{\text{vap}}\underline{H}$) as a function of reduced temperature. This will be demonstrated here using the van der Waals equation as a model equation of state.

   a. Show that at vapor-liquid equilibrium

$$\int_{\underline{V}^{\text{L}}}^{\underline{V}^{\text{V}}} \underline{V}\left(\frac{\partial P}{\partial \underline{V}}\right)_T d\underline{V} = 0$$

   b. Show that the solution to this equation for the van der Waals fluid is

$$\ln\left[\frac{3V_r^{\text{L}} - 1}{3V_r^{\text{V}} - 1}\right] + (3V_r^{\text{V}} - 1)^{-1} + (3V_r^{\text{L}} - 1)^{-1}$$
$$+ \frac{9}{4T_r V_r^{\text{L}}}\left(1 - \frac{V_r^{\text{L}}}{V_r^{\text{V}}}\right) = 0$$

     where $V_r^{\text{L}} = \underline{V}^{\text{L}}/\underline{V}_{\text{c}}$ and $V_r^{\text{V}} = \underline{V}^{\text{V}}/\underline{V}_{\text{c}}$.

   c. Construct a corresponding-states curve for the reduced vapor pressure P_r^{vap} of the van der Waals fluid as a function of its reduced temperature.

   d. Describe how a corresponding-states curve for the enthalpy change on vaporization as a function of the reduced temperature can be obtained for the van der Waals fluid.

7.30 a. Show that it is not possible for a pure fluid to have a quartenary point where the vapor, liquid, and two solid phases are all in equilibrium.

   b. Show that the specification of the overall molar volume of a one-component, two-phase mixture is not sufficient to identify the thermodynamic state of either phase, but that the specification of both the molar volume and the molar enthalpy (or any two other molar properties) is sufficient to determine the thermodynamic states of each of the phases and the distribution of mass between the phases.

7.31 From Eqs. 6.2-18 and 6.2-20, we have the following as definitions of the heat capacities at constant volume and constant pressure:

$$C_V = T \left(\frac{\partial S}{\partial T} \right)_V$$

and

$$C_P = T \left(\frac{\partial S}{\partial T} \right)_P$$

More generally, we can define a heat capacity subject to some other constraint X by

$$C_X = T \left(\frac{\partial S}{\partial T} \right)_X$$

One such heat capacity of special interest is C_{LV}, the heat capacity along the vapor-liquid equilibrium line.

a. Show that

$$C_{LV}^i = C_P^i - \alpha^i \underline{V}^i \frac{\Delta_{vap} \underline{H}}{\Delta_{vap} \underline{V}}$$

where α^i is the coefficient of thermal expansion for phase i (liquid or vapor), C_P^i is its constant-pressure heat capacity, $\underline{V}^i$ is its molar volume, and $\Delta_{vap} \underline{H}$ and $\Delta_{vap} \underline{V}$ are the molar enthalpy and volume changes on vaporization.

b. Show that

(i) $C_{LV}^L \approx C_P^L$

(ii) C_{LV}^V may be negative by considering saturated steam at its normal boiling point and at 370°C.

7.32 a. By multiplying out the various terms, show that Eqs. 7.2-4 and 7.2-6 are equivalent.

Also show that

b. $S_{UV} = \frac{\partial}{\partial U} \Big|_{N,V} \left(\frac{\partial S}{\partial N} \right)_{U,V} = -\frac{1}{T^2} \left(\frac{\partial T}{\partial N} \right)_{U,V}$

$$= \frac{-\underline{V}}{NC_V T} \left(\frac{\partial P}{\partial T} \right)_V + \frac{H}{NT^2 C_V}$$

c. $S_{VN} = \frac{1}{T} \left(\frac{\partial P}{\partial N} \right)_{U,V} - \frac{P}{T^2} \left(\frac{\partial T}{\partial N} \right)_{U,V}$

$$= -\frac{1}{NC_V T^2} \left\{ H - \underline{V} T \left(\frac{\partial P}{\partial T} \right)_V \right\}$$

$$\left\{ T \left(\frac{\partial P}{\partial T} \right)_V - P \right\} - \frac{\underline{V}}{NT} \left(\frac{\partial P}{\partial \underline{V}} \right)_T$$

d. $S_{NN} = \frac{2 \underline{H} \underline{V}}{NC_V T} \left(\frac{\partial P}{\partial T} \right)_V - \frac{H^2}{NC_V T^2}$

$$- \frac{\underline{V}^2}{NC_V} \left(\frac{\partial P}{\partial T} \right)_V^2 + \frac{\underline{V}^2}{NT} \left(\frac{\partial P}{\partial \underline{V}} \right)_T$$

e. $\theta_3 = 0$

7.33 Assuming that nitrogen obeys the Peng-Robinson equation of state, develop tables and charts of its thermodynamic properties such as those in Illustrations 6.4-1, 7.5-1, and 7.5-2. Compare your results with those in Figure 3.3-3 and comment on the differences.

7.34 Assuming that water obeys the Peng-Robinson equation of state, develop tables and charts of its thermodynamic properties such as those in Illustrations 6.4-1, 7.5-1, and 7.5-2. Compare your results with those in Figures 3.3-1 and Appendix A.III.

7.35 A well-insulated gas cylinder, containing ethylene at 85 bar and 25°C, is exhausted until the pressure drops to 10 bar. This process occurs fast enough that no heat transfer occurs between the gas and the cylinder walls, but not so rapidly as to produce large velocity or temperature gradients within the cylinder.

a. Calculate the final temperature of the ethylene in the cylinder.

b. What fraction of the ethylene remaining in the cylinder is present as a liquid?

7.36 The following vapor-liquid equilibrium data are available for methyl ethyl ketone:

Heat of vaporization at 75°C: 31 602 J/mol
Molar volume of saturated liquid at 75°C: 9.65×10^{-2} m³/kmol

$$\ln P^{vap} = 43.552 - \frac{5622.7}{T} - 4.705\,04 \ln T$$

where P^{vap} is the vapor pressure in bar and T is in K. Assuming the saturated vapor obeys the volume-explicit form of the virial equation,

$$\underline{V} = \frac{RT}{P} + B$$

Calculate the second virial coefficient, B, for methyl ethyl ketone at 75°C.

7.37 The freezing point of n-hexadecane is approximately 18.5°C, where its vapor pressure is very low. By how much would the freezing point of n-hexadecane be depressed if n-hexadecane were under a 200-bar pressure? For simplicity in this calculation you may assume that the change in heat capacity on fusion of n-hexadecane is zero, and that its heat of fusion is 48.702 kJ/mol.

7.38 Ten grams of liquid water at 95°C are contained in the insulated container shown. The pin holding the frictionless piston in place breaks, and the volume available to the water increases to 1×10^{-3} m³. During the expansion some of the water evaporates, but no heat is transferred to the cylinder.

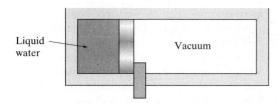

Liquid water

Vacuum

Find the temperature, pressure, and amounts of vapor, liquid, and solid water present after the expansion.

7.39 At a subcritical temperature, the branch of the Peng-Robinson equation of state for $\underline{V} > b$ exhibits a van der Waals loop. However, there is also interesting behavior of the equation in the ranges $\underline{V} < b$ and $\underline{V} < 0$, though these regions do not have any physical meaning. At supercritical temperatures the van der Waals loop disappears, but much of the structure in the $\underline{V} < 0$ region remains. Establish this by drawing a P-$\underline{V}$ plot for n-butane at temperatures of $0.6\ T_c$, $0.8\ T_c$, $1.0\ T_c$, $1.5\ T_c$, and $5\ T_c$ for all values of $\underline{V}$.

7.40 The van't Hoff corollary to the third law of thermodynamics is that whenever two solid forms of a substance are known, the one with the greater specific heat will be the more stable one at higher temperatures. Explain why this is so.

7.41 Consider the truncated virial equation of state

$$\frac{P\underline{V}}{RT} = 1 + \frac{B(T)}{\underline{V}}$$

where $B(T)$ is the second virial coefficient. Obtain the constraint on $B(T)$ if the fluid is to be thermodynamically stable.

7.42 Show that the requirement for stability of a closed system at constant entropy and pressure leads to the condition that $C_P > 0$ for all stable fluids. (*Hint:* You do not need to make a change of variable to show this.)

7.43 A fluid obeys the Clausius equation of state

$$P = \frac{RT}{\underline{V} - b(T)}$$

where $b(T) = b_0 + b_1 T$. The fluid undergoes a Joule-Thomson expansion from $T_1 = 120.0°C$ and $P_1 = 5.0$ MPa to a final pressure $P_2 = 1.0$ MPa. Given that $C_P = 20.97\,\mathrm{J/(mol\ K)}$, $b_0 = 4.28 \times 10^{-5}$ m^3/mol, and $b_1 = 1.35 \times 10^{-7}$ m^3/(mol K), determine the final temperature of the fluid.

7.44 Does a fluid obeying the Clausius equation of state have a vapor-liquid transition?

7.45 Can a fluid obeying the virial equation of state have a vapor-liquid transition?

7.46 Derive the expression for the fugacity coefficient of a species described by the Soave–Redlich-Kwong equation of state (the analogue of Eq. 7.4-14b).

7.47 Using the Redlich-Kwong equation of state, compute and plot (on separate graphs) the pressure and fugacity of nitrogen as a function of specific volume at the two temperatures
  a. 110 K
  b. 150 K

7.48 The vapor pressure of liquid ethanol at 126°C is 505 kPa, and its second virial coefficient at this temperature is -523 cm^3/mol.
  a. Calculate the fugacity of ethanol at saturation at 126°C assuming ethanol is an ideal gas.
  b. Calculate the fugacity of ethanol at saturation at 126°C assuming ethanol is described by the virial equation truncated after the second virial coefficient.
(*Note:* Since this calculation is being done at saturation conditions, the value computed is for both the vapor and liquid, even though the calculation has been done using only the vapor pressure and vapor-phase properties.)

7.49 Liquid ethanol, for which $\kappa_T = 1.09 \times 10^{-6}$ kPa^{-1}, has a relatively high isothermal compressibility, which can have a significant effect on its fugacity at high pressures. Using the data in the previous problem, calculate the fugacity of liquid ethanol at 25 MPa and 126°C
  a. Assuming liquid ethanol is incompressible
  b. Using the value for the isothermal compressibility of liquid ethanol given above

7.50 Redo Illustration 7.4-6 using the Soave–Redlich-Kwong equation of state.

7.51 Redo Illustration 7.5-1 using the Soave–Redlich-Kwong equation of state.

7.52 Redo Illustration 7.5-2 using the Soave–Redlich-Kwong equation of state.

7.53 Adapt the program PR1, or one of the other Peng-Robinson programs, or develop a program of your own to use the Soave–Redlich-Kwong equation of state rather than the Peng-Robinson equation to calculate the thermodynamic properties of a fluid.

7.54 Adapt the program PR1, or one of the other Peng-Robinson programs, or develop a program of your own using the Peng-Robinson equation of state to do the calculations for a Joule-Thomson expansion of a liquid under pressure to produce a vapor-liquid mixture at ambient pressure. The output results should include the outlet temperature and the fractions of the outlet stream that are liquid and vapor, respectively.

7.55 Adapt the program PR1, or one of the other Peng-Robinson programs, or develop a program of your own using the Peng-Robinson equation of state to do the calculations for an isentropic expansion of a liquid under pressure to produce a vapor-liquid mixture at ambient pressure. The output results should include the outlet temperature and the fractions of the outlet stream that are liquid and vapor, respectively.

7.56 Redo Problem 7.54 with the Soave–Redlich-Kwong equation of state.

7.57 Redo Problem 7.55 with the Soave–Redlich-Kwong equation of state.

7.58 Redo Illustration 7.5-3 with the Soave–Redlich-Kwong equation of state. Compare the results obtained with those for the Peng-Robinson equation in the illustration. Also, repeat the calculations for the Soave–Redlich-Kwong equation but with the simplifying assumption that the α function of Eqs. 6.7-6 and 6.7-8 is a constant and equal to unity, rather than a function of temperature.

7.59 Redo Illustration 6.7-1 using the Soave–Redlich-Kwong equation of state.

7.60 A fluid obeys the equation of state

$$\frac{P\underline{V}}{RT} = 1 + \frac{B}{\underline{V}} + \frac{C}{\underline{V}^2}$$

where B and C are constants. Also,

$$C_V^* = a + bT$$

a. For what values of the constants B and C will this fluid undergo a vapor-liquid phase transition?

b. What is the molar internal energy change if this fluid is heated at a constant pressure P from T_1 to T_2, and how does this compare with the molar internal energy change for an ideal gas with the same ideal gas heat capacity undergoing the same change?

c. Develop an expression relating the differential change in temperature accompanying a differential change in volume for a reversible adiabatic expansion of this fluid. What would be the analogous expression if the fluid were an ideal gas?

7.61 It has been suggested that a simple cubic equation of state can also be used to describe a solid. However, since in a solid certain molecular contacts are preferred, compared with a fluid (vapor or liquid) in which molecules interact in random orientations, it has also been suggested that different equation-of-state parameters be used for the fluid and solid phases. Assume both the fluid and solid phases obey the equation of state

$$P = \frac{RT}{\underline{V}} - \frac{a}{\underline{V}^2}$$

with the parameters a_f and a_s, respectively. Develop expressions for the fugacity coefficients of the gas, liquid, and solid. Discuss how you would use such

an equation to calculate the vapor-liquid, liquid-solid, vapor-solid, and vapor-liquid-solid equilibria.

7.62 Proteins can exist in one of two states, the active, folded state and the inactive, unfolded state. Protein folding is sometimes thought of as a first-order phase transition from folded to unfolded (denaturation) with increasing temperature. (We will revisit this description in Chapter 15.) Denaturation is accompanied by an increase in molecular volume and a significant positive latent heat (i.e., $\triangle_{F \rightarrow D}\underline{V} > 0$ and $\triangle_{F \rightarrow D}\underline{H} > 0$).

a. Sketch the Gibbs energy function for both D (denatured) and F (folded) proteins, and identify T^*, the phase transition temperature. Which state has the higher entropy?

b. Given the observed volume and enthalpy changes upon denaturation, how would increasing the pressure affect T^*? Should biopharmaceuticals (i.e., folded proteins) be stored under high pressure?

7.63 Calculate the vapor pressure of methane as a function of temperature using the Peng-Robinson equation of state. Compare your results with (a) literature values and (b) predictions using the Peng-Robinson equation of state in which the temperature-dependent parameter $\alpha(T)$ is set equal to unity at all temperatures.

7.64 Calculate the vapor pressure of n-butane as a function of temperature using the Peng-Robinson equation of state. Compare your results with (a) literature values and (b) predictions using the Peng-Robinson equation of state in which the temperature-dependent parameter $\alpha(T)$ is set equal to unity at all temperatures.

7.65 Calculate the vapor pressure of n-decane as a function of temperature using the Peng-Robinson equation of state. Compare your results with (a) literature values and (b) predictions using the Peng-Robinson equation of state in which the temperature-dependent parameter $\alpha(T)$ is set equal to unity at all temperatures.

7.66 The critical temperature of benzene is $289°C$ and its critical pressure is 4.89 MPa. At $220°C$ its vapor pressure is 1.91 MPa.

a. Calculate the fugacity of liquid benzene in equilibrium with its pure vapor at $220°C$.

b. Repeat the calculation for the case in which liquid benzene is under an atmosphere of hydrogen such that the total pressure is 13.61 MPa. The average density of liquid benzene at these conditions is 0.63 g/cc, and you can assume that hydrogen is insoluble in benzene at this temperature. The molecular weight of benzene is 78.02.

7.67 The sublimation pressure of carbon dioxide as a function of temperature is

T (K)	130	155	185	194.5	205
P (kPa)	0.032	1.674	44.02	101.3	227

and the molar volume of CO_2 is 2.8×10^{-5} m³/mol.
a. Determine the heat of sublimation of CO_2 at 190 K.
b. Estimate the fugacity of solid CO_2 at 190 K and 200 bar.

7.68 The figure below shows a pressure-enthalpy diagram submitted by Joe Udel as part of a homework assignment. On the diagram isotherms ($T_1 < T_2 < T_3$), isochores ($\underline{V}_1 < \underline{V}_2 < \underline{V}_3$), and isentropes ($\underline{S}_1 < \underline{S}_2 < \underline{S}_3$) are shown. Is this figure consistent with requirements for a stable equilibrium system?

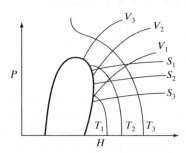

7.69 A relatively simple cubic equation of state is

$$P = \frac{RT}{\underline{V}} - b - \frac{a}{\sqrt{T}\underline{V}^2}$$

where a is a constant.
a. Determine the relationship between the a and b parameters in this equation of state and the critical temperature and pressure of the fluid.
b. Obtain an expression for the constant-volume heat capacity as a function of temperature, volume, the critical temperature and pressure, and the ideal heat capacity.
c. The constant-volume heat capacities of real fluids diverge as the critical point is approached. Does the constant-volume heat capacity obtained from this equation of state diverge as the critical point is approached?

7.70 a. Show that

$$\frac{\left(\frac{\partial \underline{V}}{\partial T}\right)_{\underline{S}}}{\left(\frac{\partial \underline{V}}{\partial T}\right)_P} = -\frac{C_V}{T\underline{V}\kappa_T}\left(\frac{\partial T}{\partial P}\right)_{\underline{V}}^2$$

b. Use this result to show that, for a stable system at equilibrium, $(\partial \underline{V}/\partial T)_{\underline{S}}$ and $(\partial \underline{V}/\partial T)_P$ must have opposite signs.
c. Two separate measurements are to be performed on a gas enclosed in a piston-and-cylinder device. In the first measurement the device is well insulated so there is no flow of heat to or from the gas, and the piston is slowly moved inward, compressing the gas, and its temperature is found to increase. In the second measurement the piston is free to move and the external pressure is constant. A small amount of heat is added to the gas in the cylinder, resulting in the expansion of the gas. Will the temperature of the gas increase or decrease?

7.71 Using a Knudsen effusion cell, Svec and Clyde[11] measured the sublimation pressures of several amino acids as a function of temperature.
a. The results they reported for glycine are

T (K)	453	457	466	471
P (torr)	5.87 $\times 10^{-5}$	4.57 $\times 10^{-5}$	1.59 $\times 10^{-4}$	2.43 $\times 10^{-4}$

Use these data to estimate the heat of sublimation of glycine over this temperature range.
b. The results they reported for *l*-alanine are

T (K)	453	460	465	469
P (torr)	5.59 $\times 10^{-5}$	1.22 $\times 10^{-4}$	2.03 $\times 10^{-4}$	2.58 $\times 10^{-4}$

Use these data to estimate the heat of sublimation of *l*-alanine over this temperature range. [1 torr = 133.32 Pa]

7.72 Determine the boiling temperature of a water droplet at 1.013 bar as a function of the droplet radius.

[11]H. J. Svec and D. C. Clyde, *J. Chem. Eng. Data* **10**, 151 (1965).

Chapter 8

The Thermodynamics of Multicomponent Mixtures

In this chapter the study of thermodynamics, which was restricted to pure fluids in the first part of this book, is extended to mixtures. First, the problem of relating the thermodynamic properties of a mixture, such as the volume, enthalpy, Gibbs energy, and entropy, to its temperature, pressure, and composition is discussed; this is the problem of the thermodynamic description of mixtures. Next, the equations of change for mixtures are developed. Using the equations of change and the positive definite nature of the entropy generation term, the general equilibrium criteria for mixtures are considered and shown to be formally identical to the pure fluid equilibrium criteria. These criteria are then used to establish the conditions of phase and chemical equilibrium in mixtures, results that are of central importance in the remainder of this book.

INSTRUCTIONAL OBJECTIVES FOR CHAPTER 8

The goals of this chapter are for the student to:

- Understand the difference between partial molar properties and pure component properties (Sec. 8.1)
- Be able to use the Gibbs-Duhem equation to simplify equations (Sec. 8.2)
- Be able to identify a set of independent reactions
- Be able to use the mass, energy, and entropy balance equations for mixtures (Secs. 8.4 and 8.5)
- Be able to compute partial molar properties from experimental data (Sec. 8.6)
- Be able to derive the criteria for phase and chemical equilibria in multicomponent systems (Secs. 8.7 and 8.8)

NOTATION INTRODUCED IN THIS CHAPTER

$\bar{A}_i$ Partial molar Helmholtz energy of species i (J/mol)
$\mathcal{C}$ Number of components
$\bar{C}_{P,i}$ Partial molar heat capacity of species i (J/mol K)
$\bar{f}_i$ Fugacity of species i in a mixture (kPa)
$\bar{G}_i$ Partial molar Gibbs energy of species i (J/mol)

$\Delta_f \underline{G}_i^\circ$	Standard state molar Gibbs energy of formation of species i (kJ)
$\Delta_{mix} G$	Gibbs energy change on mixing (kJ)
$\Delta_{rxn} G$	Gibbs energy change of reaction (kJ)
$\Delta_{rxn} G^\circ$	Standard state Gibbs energy change on reaction (kJ)
$\bar{H}_i$	Partial molar enthalpy of species i (J/mol)
$\Delta_{mix} H$	Enthalpy energy change on mixing (kJ)
$\Delta_f \underline{H}_i^\circ$	Standard State molar enthalpy of formation of species i (kJ)
$\Delta_{rxn} H$	Enthalpy change (or heat) of reaction (kJ)
$\Delta_{rxn} H^\circ$	Standard state enthalpy change on reaction (kJ)
$\mathcal{M}$	Number of independent chemical reactions
$\mathcal{P}$	Number of phases
$\bar{S}_i$	Partial molar entropy of species i [J/(mol K)]
$\bar{V}_i$	Partial molar volume of species i (m³/mol)
$\Delta_{mix} V$	Volume change on mixing (m³)
X_j	Molar extent of reaction of the jth independent chemical reaction (mol)
w_i	Mass fraction of species i in a phase
w^I	Fraction of total system mass in phase I
$\bar{\phi}_i$	Fugacity coefficient of species i in a mixture ($= \bar{f}_i / P$)
$\nu_{i,j}$	Stoichiometric coefficient of species i in independent chemical reaction j

8.1 THE THERMODYNAMIC DESCRIPTION OF MIXTURES

In Chapter 1 we pointed out that the thermodynamic state of a pure, single-phase system is completely specified by fixing the values of two of its intensive state variables (e.g., T and P or T and $\underline{V}$), and that its size and thermodynamic state is fixed by a specification of its mass or number of moles and two of its state variables (e.g., N, T, and P or N, T, and $\underline{V}$). For a single-phase mixture of $\mathcal{C}$ components, one intuitively expects that the thermodynamic state will be fixed by specifying the values of two of the intensive variables of the system *and* all but one of the species mole fractions (e.g., T, P, x_1, x_2, ..., $x_{\mathcal{C}-1}$), the remaining mole fraction being calculable from the fact that the mole fractions must sum to 1. (Though we shall largely use mole fractions in the discussion, other composition scales, such as mass fraction, volume fraction, and molality, could be used. Indeed, there are applications in which each is used.) Similarly, one anticipates that the size and state of the system will also be fixed by specifying the values of two state variables and the mole numbers of all species present. That is, for a $\mathcal{C}$-component mixture we expect equations of state of the form

$$U = U(T, P, N_1, N_2, \ldots, N_{\mathcal{C}}) \quad \text{or} \quad \underline{U} = \underline{U}(T, P, x_1, x_2, \ldots, x_{\mathcal{C}-1})$$
$$V = V(T, P, N_1, N_2, \ldots, N_{\mathcal{C}}) \quad \text{or} \quad \underline{V} = \underline{V}(T, P, x_1, x_2, \ldots, x_{\mathcal{C}-1})$$

and so on. Here x_i is the mole fraction of species i—that is, $x_i = N_i / N$, where N_i is the number of moles of species i and N is the total number of moles—and $\underline{U}$ and $\underline{V}$ are the internal energy and volume per mole of the mixture.[1]

Our main interest here will be in determining how the thermodynamic properties of the mixture depend on species concentrations. That is, we would like to know the concentration dependence of the mixture equations of state.

---

[1]In writing these equations we have chosen T and P as the independent state variables; other choices could have been made. However, since temperature and pressure are easily measured and controlled, they are the most practical choice of independent variables for processes of interest to chemists and engineers.

A naive expectation is that each thermodynamic mixture property is a sum of the analogous property for the pure components at the same temperature and pressure weighted with their fractional compositions. That is, if $\underline{U}_i$ is the internal energy per mole of pure species i at temperature T and pressure P, then it would be convenient if, for a $\mathcal{C}$-component mixture,

$$\underline{U}(T, P, x_1, \ldots, x_{\mathcal{C}-1}) = \sum_{i=1}^{\mathcal{C}} x_i \underline{U}_i(T, P) \tag{8.1-1}$$

Alternatively, if $\hat{U}_i$ is the internal energy per unit mass of pure species i, then the development of a mixture equation of state would be simple if

$$\hat{U}(T, P, w_1, \ldots, w_{\mathcal{C}-1}) = \sum_{i=1}^{\mathcal{C}} w_i \hat{U}_i(T, P) \tag{8.1-2}$$

where w_i is the mass fraction of species i.

Unfortunately, relations as simple as Eqs. 8.1-1 and 8.1-2 are generally not valid. This is evident from Fig. 8.1-1, which shows the enthalpy of a sulfuric acid–water mixture at various temperatures. If equations of the form of Eq. 8.1-2 were valid, then

$$\hat{H} = w_1 \hat{H}_1 + w_2 \hat{H}_2 = w_1 \hat{H}_1 + (1 - w_1)\hat{H}_2$$

and the enthalpy of the mixture would be a linear combination of the two pure-component enthalpies; this is indicated by the dashed line connecting the pure-component enthalpies at 0°C. The actual thermodynamic behavior of the mixture, given by the solid lines in the figure, is quite different. In Fig. 8.1-2 are plotted the **volume change on mixing**, defined to be[2]

Definition of the volume change on mixing

$$\boxed{\Delta_{\text{mix}}\underline{V} = \underline{V}(T, P, \underline{x}) - x_1\underline{V}_1(T, P) - x_2\underline{V}_2(T, P)}$$

and the **enthalpy change on mixing**,

Definition of the enthalpy change on mixing

$$\boxed{\Delta_{\text{mix}}\underline{H} = \underline{H}(T, P, \underline{x}) - x_1\underline{H}_1(T, P) - x_2\underline{H}_2(T, P)}$$

for several mixtures. (The origin of such data is discussed in Sec. 8.6.) If equations like Eq. 8.1-1 were valid for these mixtures, then both $\Delta_{\text{mix}}\underline{V}$ and $\Delta_{\text{mix}}\underline{H}$ would be equal to zero at all compositions; the data in the figures clearly show that this is not the case.

There is a simple reason why such elementary mixture equations as Eqs. 8.1-1 and 8.1-2 fail for real mixtures. We have already pointed out that one contribution to the internal energy (and other thermodynamic properties) of a fluid is the interactions between its molecules. For a pure fluid this contribution to the internal energy is the result of molecules interacting only with other molecules of the same species. However, in a binary fluid mixture of species 1 and 2, the total interaction energy is a result of 1-1, 2-2, and 1-2 interactions. Although the 1-1 and 2-2 molecular interactions that occur in the mixture are approximately taken into account in Eqs. 8.1-1 and 8.1-2 through the pure-component properties, the effects of the 1-2 interactions are not. Therefore, it is

---

[2]In these equations, and most of those that follow, the abbreviated notation $\theta(T, P, x_1, x_2, \ldots, x_{\mathcal{C}-1}) = \theta(T, P, \underline{x})$ will be used.

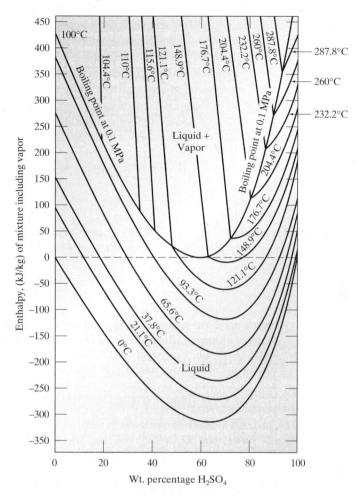

Figure 8.1-1 Enthalpy-concentration diagram for aqueous sulfuric acid at 0.1 MPa. The sulfuric acid percentage is by weight (which, in the two-phase region, includes the vapor). *Reference states:* The enthalpies of the pure liquids at 0°C and their vapor pressures are zero. (Based on Fig. 81, p. 325 in O. A. Hougen, K. M. Watson, and R. A. Ragatz, *Chemical Process Principles: I. Material and Energy Balances,* 2nd ed. Reprinted with permission from John Wiley & Sons, New York, 1954.) This figure appears as an Adobe PDF file on the CD-ROM accompanying this book, and may be enlarged and printed for easier reading and for use in solving problems.

not surprising that these equations are not good approximations for the properties of most real fluid mixtures.

What we must do, instead of making simple guesses such as Eqs. 8.1-1 and 8.1-2, is develop a general framework for relating the thermodynamic properties of a mixture to the composition variables. Since the interaction energy of a mixture is largely determined by the number of molecular interactions of each type, the mole fractions, or, equivalently, the mole numbers, are the most logical composition variables for the analysis. We will use the symbol $\underline{\theta}$ to represent any molar property (molar volume, molar enthalpy, etc.) of a mixture consisting of N_1 moles of species 1, N_2 moles of species 2, and so forth, $N = \sum_i^C N_i$ being the total number of moles in the system. For the present our main interest is in the composition variation of $\underline{\theta}$ at fixed T and P, so

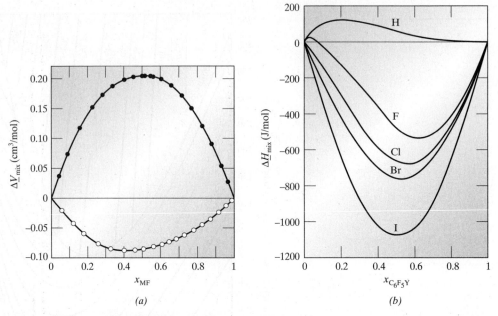

Figure 8.1-2 (*a*) Volume change on mixing at 298.15 K: ○ methyl formate + methanol, ● methyl formate + ethanol. [Reprinted from *The Journal of Chemical Thermodynamics*, **4**, 469, J. Polak and B. C.-Y Lu, Copyright (1972), with permission from Elsevier.] (*b*) Enthalpy change on mixing at 298.15 K for mixtures of benzene (C_6H_6) and aromatic fluorocarbons (C_6F_5Y), with Y = H, F, Cl, Br, and I. [Reprinted from *The Journal of Chemical Thermodynamics*, **5**, 227, D. V. Fenby and S. Ruenkrairergsa, Copyright (1973), with permission from Elsevier.]

we write

$$N\theta = \phi(N_1, N_2, \ldots, N_C) \quad \text{at constant } T \text{ and } P \tag{8.1-3}$$

to indicate that $N\theta$ is a function of all the mole numbers.[3]

If the number of moles of each species in the mixture under consideration were doubled, so that $N_1 \rightarrow 2N_1$, $N_2 \rightarrow 2N_2$, ..., and $N \rightarrow 2N$ (thereby keeping the relative amounts of each species, and the relative numbers of each type of molecular interaction in the mixture, unchanged), the value of θ *per mole of mixture*, θ, would remain unchanged; that is, $\theta(T, P, N_1, N_2, \ldots) = \theta(T, P, 2N_1, 2N_2, \ldots)$. Rewriting Eq. 8.1-3 for this situation gives

$$2N\theta = \phi(2N_1, 2N_2, \ldots, 2N_C) \tag{8.1-4}$$

while multiplying Eq. 8.1-3 by the factor 2 yields

$$2N\theta = 2\phi(N_1, N_2, \ldots, N_C) \tag{8.1-5}$$

so that

$$\phi(2N_1, 2N_2, \ldots, 2N_C) = 2\phi(N_1, N_2, \ldots, N_C)$$

---

[3]The total property $\theta = N\theta$ is a function of T, P, and the mole numbers; that is,

$$N\theta = N\theta(T, P, x_1, x_2, \ldots) = \theta(T, P, N_1, N_2, \ldots)$$

This relation indicates that as long as the relative amounts of each species are kept constant, ϕ is a linear function of the total number of moles. This suggests ϕ should really be considered to be a function of the mole ratios. Thus, a more general way of indicating the concentration dependence of θ is as follows:

$$N\underline{\theta} = N_1\phi_1 \left(\frac{N_2}{N_1}, \frac{N_3}{N_1}, \ldots, \frac{N_C}{N_1} \right) \tag{8.1-6}$$

where we have introduced a new function, ϕ_1, of only $C - 1$ mole ratios in a C-component mixture (at constant T and P). In this equation the multiplicative factor N_1 establishes the extent or total mass of the mixture, and the function ϕ_1 is dependent only on the relative amounts of the various species. (Of course, any species other than species 1 could have been singled out in writing Eq. 8.1-6.)

To determine how the property $N\underline{\theta}$ varies with the number of moles of each species present, we look at the derivatives of $N\underline{\theta}$ with respect to mole number. The change in the value of $N\underline{\theta}$ as the number of moles of species 1 changes (all other mole numbers, T, and P being held constant) is

$$\frac{\partial}{\partial N_1}(N\underline{\theta})\bigg|_{T,P,N_2,N_3,\ldots} = \phi_1\left(\frac{N_2}{N_1},\ldots\right) + N_1 \sum_{i=2}^{C} \frac{\partial\phi_1}{\partial\left(\frac{N_i}{N_1}\right)} \frac{\partial\left(\frac{N_i}{N_1}\right)}{\partial N_1} \tag{8.1-7}$$

where the last term arises from the chain rule of partial differentiation and the fact that ϕ_1 is a function of the mole ratios. Now multiplying by N_1, and using the fact that

$$\frac{\partial\left(\frac{N_i}{N_1}\right)}{\partial N_1}\bigg|_{i\neq 1} = -\frac{N_i}{N_1^2}$$

gives

$$N_1\frac{\partial}{\partial N_1}(N\underline{\theta})\bigg|_{T,P,N_{j\neq 1}} = N_1 \left\{ \phi_1\left(\frac{N_2}{N_1},\ldots\right) - \frac{1}{N_1}\sum_{i=2}^{C} N_i \frac{\partial\phi_1}{\partial\left(\frac{N_i}{N_1}\right)} \right\} \tag{8.1-8a}$$

$$= N\underline{\theta} - \sum_{i=2}^{C} N_i \frac{\partial\phi_1}{\partial\left(\frac{N_i}{N_1}\right)}$$

or

$$N\underline{\theta} = N_1\frac{\partial}{\partial N_1}(N\underline{\theta})\bigg|_{T,P,N_{j\neq 1}} + \sum_{i=2}^{C} N_i \frac{\partial\phi_1}{\partial\left(\frac{N_i}{N_1}\right)} \tag{8.1-8b}$$

Here we have introduced the notation $N_{j\neq 1}$ in the partial derivative to indicate that all mole numbers except N_1 are being held constant. Similarly, for the derivative with respect to N_i, $i \neq 1$, we have

$$\left. \frac{\partial (N\underline{\theta})}{\partial N_i} \right|_{T,P,N_{j\neq i}} = N_1 \sum_{j=2}^{c} \frac{\partial \phi_1}{\partial \left(\frac{N_j}{N_1} \right)} \frac{\partial \left(\frac{N_j}{N_1} \right)}{\partial N_i} = \frac{\partial \phi_1}{\partial \left(\frac{N_i}{N_1} \right)} \tag{8.1-9}$$

since $(\partial N_j / \partial N_i)$ is equal to 1 for $i = j$, and 0 for $i \neq j$. Using Eq. 8.1-9 in Eq. 8.1-8b yields

$$N\underline{\theta} = N_1 \frac{\partial}{\partial N_1} (N\underline{\theta}) \Big|_{T,P,N_{j\neq 1}} + \sum_{i=2}^{c} N_i \frac{\partial (N\underline{\theta})}{\partial N_i} \Big|_{T,P,N_{j\neq i}}$$

or

$$N\underline{\theta} = \sum_{i=1}^{c} N_i \frac{\partial (N\underline{\theta})}{\partial N_i} \Big|_{T,P,N_{j\neq i}} \tag{8.1-10}$$

Dividing by $N = \sum N_i$ gives

$$\underline{\theta} = \sum_{i=1}^{c} x_i \frac{\partial (N\underline{\theta})}{\partial N_i} \Big|_{T,P,N_{j\neq i}} \tag{8.1-11}$$

Common thermodynamic notation is to define a **partial molar thermodynamic property** $\bar{\theta}_i$ as[4]

Definition of a partial molar property

$$\boxed{\bar{\theta}_i = \bar{\theta}_i(T, P, \underline{x}) = \frac{\partial (N\underline{\theta})}{\partial N_i} \Big|_{T,P,N_{j\neq i}}} \tag{8.1-12}$$

so that Eq. 8.1-11 can be written as

Mixture property in terms of partial molar properties

$$\boxed{\underline{\theta} = \sum_{i=1}^{c} x_i \bar{\theta}_i(T, P, \underline{x})} \tag{8.1-13}$$

Although this equation is similar in form to the naive assumption of Eq. 8.1-1, there is the very important difference that $\bar{\theta}_i$, which appears here, is a *true mixture property* to be evaluated experimentally for each mixture (see Sec. 8.6), whereas $\underline{\theta}_i$, which appears in Eq. 8.1-1, is a pure component property. It should be emphasized that generally $\bar{\theta}_i \neq \underline{\theta}_i$; that is, *the partial molar and pure component thermodynamic properties are not equal!* Some common partial molar thermodynamic properties are listed in Table 8.1-1. An important problem in the thermodynamics of mixtures is the measurement or estimation of these partial molar properties; this will be the focus of part of this chapter and all of Chapter 9.

The fact that the partial molar properties of each species in a mixture are different from the pure-component molar properties gives rise to changes in thermodynamic

---

[4]Note that the temperature and pressure (as well as certain mole numbers) are being held constant in the partial molar derivative. Later in this chapter we will be concerned with similar derivatives in which T and/or P are not held constant; such derivatives are not partial molar quantities!

Table 8.1-1

Partial Molar Property			Molar Propert of the Mixture
$\bar{U}_i =$	Partial molar internal energy	$= \left(\dfrac{\partial(N\underline{U})}{\partial N_i} \right)_{T,P,N_{j\neq i}}$	$\underline{U} = \sum x_i \bar{U}_i$
$\bar{V}_i =$	Partial molar volume	$= \left(\dfrac{\partial(N\underline{V})}{\partial N_i} \right)_{T,P,N_{j\neq i}}$	$\underline{V} = \sum x_i \bar{V}_i$
$\bar{H}_i =$	Partial molar enthalpy	$= \left(\dfrac{\partial(N\underline{H})}{\partial N_i} \right)_{T,P,N_{j\neq i}}$	$\underline{H} = \sum x_i \bar{H}_i$
$\bar{S}_i =$	Partial molar entropy	$= \left(\dfrac{\partial(N\underline{S})}{\partial N_i} \right)_{T,P,N_{j\neq i}}$	$\underline{S} = \sum x_i \bar{S}_i$
$\bar{G}_i =$	Partial molar Gibbs energy	$= \left(\dfrac{\partial(N\underline{G})}{\partial N_i} \right)_{T,P,N_{j\neq i}}$	$\underline{G} = \sum x_i \bar{G}_i$
$\bar{A}_i =$	Partial molar Helmholtz energy	$= \left(\dfrac{\partial(N\underline{A})}{\partial N_i} \right)_{T,P,N_{j\neq i}}$	$\underline{A} = \sum x_i \bar{A}_i$

properties during the process of forming a mixture from the pure components. For example, consider the formation of a mixture from N_1 moles of species 1, N_2 moles of species 2, and so on. The total volume and enthalpy of the *unmixed* pure components are

$$V = \sum_i^C N_i \underline{V}_i(T, P)$$

and

$$H = \sum_i^C N_i \underline{H}_i(T, P)$$

whereas the volume and enthalpy of the mixture at the same temperature and pressure are, from Eq. 8.1-13,

$$V(T, P, N_1, N_2, \ldots) = N\underline{V}(T, P, \underline{x}) = \sum_i^C N_i \bar{V}_i(T, P, \underline{x})$$

and

$$H(T, P, N_1, N_2, \ldots) = N\underline{H}(T, P, \underline{x}) = \sum_i^C N_i \bar{H}_i(T, P, \underline{x})$$

Therefore, the isothermal volume change on mixing, $\Delta_{\text{mix}} V$, and the isothermal enthalpy change on mixing or the **heat of mixing**, $\Delta_{\text{mix}} H$, are

Volume change on mixing in terms of partial molar volumes

$$\Delta_{\text{mix}} V(T, P, N_1, N_2, \ldots) = V(T, P, N_1, N_2, \ldots) - \sum_{i=1}^{c} N_i \underline{V}_i(T, P)$$

$$= \sum_{i=1}^{c} N_i[\overline{V}_i(T, P, \underline{x}) - \underline{V}_i(T, P)]$$

(8.1-14)

and

Heat of mixing in terms of partial molar enthalpies

$$\Delta_{\text{mix}} H(T, P, N_1, N_2, \ldots) = H(T, P, N_1, N_2, \ldots) - \sum_{i=1}^{c} N_i \underline{H}_i(T, P)$$

$$= \sum_{i=1}^{c} N_i[\overline{H}_i(T, P, \underline{x}) - \underline{H}_i(T, P)]$$

(8.1-15)

Other thermodynamic property changes on mixing can be defined similarly.

In Chapter 4 the Helmholtz and Gibbs energies of pure components were introduced by the relations

$$\underline{A} = \underline{U} - T\underline{S} \tag{8.1-16a}$$
$$\underline{G} = \underline{H} - T\underline{S} \tag{8.1-16b}$$

These definitions are also valid for mixtures, provided the values of $\underline{U}$, $\underline{H}$, and $\underline{S}$ used are those for the mixture. That is, the Helmholtz and Gibbs energies of a mixture bear the same relation to the mixture internal energy, enthalpy, and entropy as the pure component Gibbs energies do to the pure component internal energy, enthalpy, and entropy.

Equations like Eqs. 8.1-16a and b are also satisfied by the partial molar properties. To see this, we multiply Eq. 8.1-16a by the total number of moles N and take the derivative with respect to N_i at constant T, P, and $N_{j \neq i}$ to obtain

$$\frac{\partial}{\partial N_i}(N\underline{A})\Big|_{T,P,N_{j\neq i}} = \frac{\partial}{\partial N_i}(N\underline{U})\Big|_{T,P,N_{j\neq i}} - T\frac{\partial}{\partial N_i}(N\underline{S})\Big|_{T,P,N_{j\neq i}}$$

or

$$\overline{A}_i = \overline{U}_i - T\overline{S}_i$$

Similarly, starting with Eq. 8.1-16b, one can easily show that $\overline{G}_i = \overline{H}_i - T\overline{S}_i$.

The constant-pressure heat capacity of a mixture is given by

$$C_P = \left(\frac{\partial \underline{H}}{\partial T}\right)_{P,N_j}$$

where $\underline{H}$ is the mixture enthalpy, and the subscript N_j indicates that the derivative is to be taken at a constant number of moles of all species present; the partial molar heat capacity for species i in a mixture is defined to be

$$\overline{C}_{P,i} = \frac{\partial}{\partial N_i}(NC_P)\Big|_{T,P,N_{j\neq i}}$$

Now

$$NC_P = \frac{\partial(N\underline{H})}{\partial T}\bigg|_{P,N_j}$$

and

$$\bar{C}_{P,i} = \frac{\partial(NC_P)}{\partial N_i}\bigg|_{T,P,N_{j\neq i}} = \frac{\partial}{\partial N_i}\bigg|_{T,P,N_{j\neq i}} \frac{\partial}{\partial T}(N\underline{H})\bigg|_{P,N_j}$$

$$= \frac{\partial}{\partial T}\bigg|_{P,N_i} \frac{\partial(N\underline{H})}{\partial N_i}\bigg|_{T,P,N_{j\neq i}} = \left(\frac{\partial \bar{H}_i}{\partial T}\right)_{P,N_j}$$

so that $\bar{C}_{P,i} = (\partial \bar{H}_i/\partial T)_{P,N_j}$ for species i in a mixture, just as $C_{P,i} = (\partial \underline{H}_i/\partial T)_{P,N}$ for the pure species i.

In a similar fashion a large collection of relations among the partial molar quantities can be developed. For example, since $(\partial \underline{G}_i/\partial T)_{P,N} = -\underline{S}_i$ for a pure fluid, one can easily show that $(\partial \bar{G}_i/\partial T)_{P,N_i} = -\bar{S}_i$ for a mixture. In fact, by extending this argument to other mixture properties, one finds that for each relationship among the thermodynamic variables in a pure fluid, there exists an identical relationship for the partial molar thermodynamic properties in a mixture!

ILLUSTRATION 8.1-1

Calculating the Energy Release of an Exothermic Mixing Process

Three moles of water and one mole of sulfuric acid are mixed isothermally at 0°C. How much heat must be absorbed or released to keep the mixture at 0°C?

SOLUTION

Water has a molecular weight of 18.015, and that of sulfuric acid is 98.078. Therefore, the mixture will contain $3 \times 18.015 + 1 \times 98.078 = 152.12$ g, and will have a composition of

$$\frac{98.078 \text{ g}}{152.12 \text{ g}} \times 100\% = 64.5 \text{ wt \% sulfuric acid}$$

From Fig. 8.1-1 the enthalpy of the mixture is about -315 kJ/kg. Therefore, when 3 mol water and 1 mol sulfuric acid are mixed isothermally,

$$\Delta_{\text{mix}}\hat{H} = \hat{H}_{\text{mix}} - w_1\hat{H}_1 - w_2\hat{H}_2 = -315 \frac{\text{kJ}}{\text{kg}}$$

since $\hat{H}_1(T = 0°C) = 0$ and $\hat{H}_2(T = 0°C) = 0$, so that a total of -315 kJ/kg $\times$ 0.152 kg $= -47.9$ kJ of energy must be removed to keep the mixture at a constant temperature of 0°C.

COMMENT

Sulfuric acid and water are said to mix exothermically since energy must be released to the environment to mix these two components at constant temperature. The temperature rise that occurs when these two components are mixed adiabatically is considered in Illustration 8.4-1. Note also that to solve this problem we have, in effect, used an energy balance without explicitly writing a detailed balance equation. We will consider the balance equations (mass, energy, and entropy) for mixtures in Sec. 8.4. ■

The meaning of a partial molar property may be understood as follows. For a property $\theta(T, P, N_1, N_2)$ in a binary mixture, we have

$$\theta(T, P, N_1, N_2) = N_1\bar{\theta}_1(T, P, \underline{x}) + N_2\bar{\theta}_2(T, P, \underline{x}) \tag{8.1-17a}$$

Now suppose we add a small amount of species 1, ΔN_1, which is so small compared with the total number of moles of species 1 and 2 that x_1 and x_2 are essentially unchanged. In this case we have

$$\theta(T, P, N_1 + \Delta N_1, N_2) = (N_1 + \Delta N_1)\bar{\theta}_1(T, P, \underline{x}) + N_2\bar{\theta}_2(T, P, \underline{x}) \tag{8.1-17b}$$

Subtracting Eq. 8.1-17a from Eq. 8.1-17b, we find that the change in the property θ is

$$\Delta\theta = \theta(T, P, N_1 + \Delta N_1, N_2) - \theta(T, P, N_1, N_2) = \Delta N_1\bar{\theta}_1(T, P, \underline{x})$$

Therefore, the amount by which a small addition of a species to a mixture changes the mixture property is equal to the product of the amount added and its partial molar property, that is, how the species behaves in a mixture, and not its pure component property.

8.2 THE PARTIAL MOLAR GIBBS ENERGY AND THE GENERALIZED GIBBS-DUHEM EQUATION

Since the Gibbs energy of a multicomponent mixture is a function of temperature, pressure, and each species mole number, the total differential of the Gibbs energy function can be written as

$$dG = \left(\frac{\partial G}{\partial T}\right)_{P,N_i} dT + \left(\frac{\partial G}{\partial P}\right)_{T,N_i} dP + \sum_{i=1}^{C}\left(\frac{\partial G}{\partial N_i}\right)_{T,P,N_{j\neq i}} dN_i$$

$$= -S\,dT + V\,dP + \sum_{i=1}^{C}\bar{G}_i\,dN_i \tag{8.2-1}$$

Chemical potential

Here the first two derivatives follow from Eqs. 6.2-12 for the pure fluid, and the last from the definition of the partial molar Gibbs energy. Historically, the partial molar Gibbs energy has been called the **chemical potential** and designated by the symbol μ_i.

Since the enthalpy can be written as a function of entropy and pressure (see Eq. 6.2-6), we have

$$dH = \left(\frac{\partial H}{\partial P}\right)_{S,N_i} dP + \left(\frac{\partial H}{\partial S}\right)_{P,N_i} dS + \sum_{i=1}^{C}\left(\frac{\partial H}{\partial N_i}\right)_{P,S,N_{j\neq i}} dN_i$$

$$= V\,dP + T\,dS + \sum_{i=1}^{C}\left(\frac{\partial H}{\partial N_i}\right)_{P,S,N_{j\neq i}} dN_i \tag{8.2-2}$$

In this equation it should be noted that $(\partial H/\partial N_i)_{P,S,N_{j\neq i}}$ is not equal to the partial molar enthalpy, which is $\bar{H}_i = (\partial H/\partial N_i)_{T,P,N_{j\neq i}}$ (see Problem 8.1).

However, from $H = G + TS$ and Eq. 8.2-1, we have[5]

$$dH = dG + T\,dS + S\,dT = -S\,dT + V\,dP + \sum_i \bar{G}_i\,dN_i + T\,dS + S\,dT$$

$$= V\,dP + T\,dS + \sum_i \bar{G}_i\,dN_i$$

$$(8.2\text{-}3)$$

Comparing Eqs. 8.2-2 and 8.2-3 establishes that

$$\left(\frac{\partial H}{\partial N_i}\right)_{P,S,N_{j\neq i}} = \bar{G}_i = \left(\frac{\partial G}{\partial N_i}\right)_{T,P,N_{j\neq i}}$$

so that the derivative $(\partial H/\partial N_i)_{P,S,N_{j\neq i}}$ is equal to the partial molar Gibbs energy. Using the procedure established here, it is also easily shown (Problem 8.1) that

$$dU = T\,dS - P\,dV + \sum_i^c \bar{G}_i\,dN_i \qquad (8.2\text{-}4)$$

$$dA = -P\,dV - S\,dT + \sum_i^c \bar{G}_i\,dN_i \qquad (8.2\text{-}5)$$

with

$$\bar{G}_i = \left(\frac{\partial U}{\partial N_i}\right)_{S,V,N_{j\neq i}} = \left(\frac{\partial A}{\partial N_i}\right)_{T,V,N_{j\neq i}}$$

From these equations we see that the partial molar Gibbs energy assumes special importance in mixtures, as the molar Gibbs energy does in pure fluids.

Based on the discussion of partial molar properties in the previous section, any thermodynamic function θ can be written as

$$N\underline{\theta} = \sum N_i \bar{\theta}_i$$

so that

$$d(N\underline{\theta}) = \sum N_i\,d\bar{\theta}_i + \sum \bar{\theta}_i\,dN_i \qquad (8.2\text{-}6)$$

by the product rule of differentiation. However, $N\underline{\theta}$ can also be considered to be a function of T, P, and each of the mole numbers, in which case we have, following Eq. 8.2-1,

$$d(N\underline{\theta}) = \left.\frac{\partial(N\underline{\theta})}{\partial T}\right|_{P,N_i} dT + \left.\frac{\partial(N\underline{\theta})}{\partial P}\right|_{T,N_i} dP + \sum_{i=1}^c \left.\frac{\partial(N\underline{\theta})}{\partial N_i}\right|_{T,P,N_{j\neq i}} dN_i$$

$$= N\left(\frac{\partial\underline{\theta}}{\partial T}\right)_{P,N_i} dT + N\left(\frac{\partial\underline{\theta}}{\partial P}\right)_{T,N_i} dP + \sum_{i=1}^c \bar{\theta}_i\,dN_i$$

$$(8.2\text{-}7)$$

---

[5]You should compare Eqs. 8.2-3, 8.2-4, and 8.2-5 with Eqs. 6.2-6a, 6.2-5a, and 6.2-7a, respectively.

Subtracting Eq. 8.2-7 from Eq. 8.2-6 gives the relation

$$-N\left(\frac{\partial \underline{\theta}}{\partial T}\right)_{P,N_i} dT - N\left(\frac{\partial \underline{\theta}}{\partial P}\right)_{T,N_i} dP + \sum_{i=1}^{C} N_i \, d\bar{\theta}_i = 0 \qquad \text{(8.2-8a)}$$

and dividing by the total number of moles N yields

Generalized Gibbs-Duhem equation

$$\boxed{-\left(\frac{\partial \underline{\theta}}{\partial T}\right)_{P,N_i} dT - \left(\frac{\partial \underline{\theta}}{\partial P}\right)_{T,N_i} dP + \sum_{i=1}^{C} x_i \, d\bar{\theta}_i = 0} \qquad \text{(8.2-8b)}$$

These results are forms of the **generalized Gibbs-Duhem equation**.

For changes at constant temperature and pressure, the Gibbs-Duhem equation reduces to

Gibbs-Duhem equation at constant T and P

$$\boxed{\sum_{i=1}^{C} N_i d\bar{\theta}_i \Big|_{T,P} = 0} \qquad \text{(8.2-9a)}$$

and

$$\boxed{\sum_{i=1}^{C} x_i d\bar{\theta}_i |_{T,P} = 0} \qquad \text{(8.2-9b)}$$

Finally, for a change in any property Y (except mole fraction[6]) at constant temperature and pressure, Eq. 8.2-9a can be rewritten as

$$\sum_{i}^{C} N_i \left(\frac{\partial \bar{\theta}_i}{\partial Y}\right)_{T,P} = 0 \qquad \text{(8.2-10)}$$

whereas for a change in the number of moles of species j at constant temperature, pressure, and all other mole numbers, we have

$$\sum_{i}^{C} N_i \left(\frac{\partial \bar{\theta}_i}{\partial N_j}\right)_{T,P,N_{k\neq j}} = 0 \qquad \text{(8.2-11)}$$

Since in much of the remainder of this book we are concerned with equilibrium at constant temperature and pressure, the Gibbs energy will be of central interest. The Gibbs-Duhem equations for the Gibbs energy, obtained by setting $\theta = G$ in Eqs. 8.2-8, are

$$0 = S \, dT - V \, dP + \sum_{i=1}^{C} N_i \, d\bar{G}_i \qquad \text{(8.2-12a)}$$

$$0 = \underline{S} \, dT - \underline{V} \, dP + \sum_{i=1}^{C} x_i \, d\bar{G}_i \qquad \text{(8.2-12b)}$$

---

[6]Since mole fractions can be varied in several ways—for example, by varying the number of moles of only one species while holding all other mole numbers fixed, by varying the mole numbers of only two species, and so on, a more careful derivation than that given here must be used to obtain the mole fraction analogue of Eq. 8.2-10. The result of such an analysis is Eq. 8.2-18.

and the relations analogous to Eqs. 8.2-9, 8.2-10, and 8.2-11 are

$$\sum_{i=1}^{C} N_i \, d\bar{G}_i \Big|_{T,P} = 0 \tag{8.2-13a}$$

$$\sum_{i=1}^{C} x_i \, d\bar{G}_i \Big|_{T,P} = 0 \tag{8.2-13b}$$

$$\sum_{i=1}^{C} N_i \left(\frac{\partial \bar{G}_i}{\partial Y} \right)_{T,P} = 0 \tag{8.2-14}$$

and

$$\sum_{i=1}^{C} N_i \left(\frac{\partial \bar{G}_i}{\partial N_j} \right)_{T,P,N_{k \neq j}} = 0 \tag{8.2-15}$$

Each of these equations will be used later.

The Gibbs-Duhem equation is a thermodynamic consistency relation that expresses the fact that among the set of $C + 2$ state variables, T, P, and C partial molar Gibbs energies in a C-component system, only $C+1$ of these variables are independent. Thus, for example, although temperature, pressure, $\bar{G}_1$, $\bar{G}_2$, ..., and $\bar{G}_{C-1}$ can be independently varied, $\bar{G}_C$ cannot also be changed at will; instead, its change $d\bar{G}_C$ is related to the changes dT, dP, $d\bar{G}_1$, ..., $d\bar{G}_{C-1}$ by Eq. 8.2-12a, so that

$$d\bar{G}_C = \frac{1}{N_C} \left\{ -S \, dT + V \, dP - \sum_{i=1}^{C-1} N_i \, d\bar{G}_i \right\}$$

Thus, the interrelationships provided by Eqs. 8.2-8 through 8.2-15 are really restrictions on the mixture equation of state. As such, these equations are important in minimizing the amount of experimental data necessary in evaluating the thermodynamic properties of mixtures, in simplifying the description of multicomponent systems, and in testing the consistency of certain types of experimental data (see Chapter 10). Later in this chapter we show how the equations of change for mixtures and the Gibbs-Duhem equations provide a basis for the experimental determination of partial molar properties.

Although Eqs. 8.2-8 through 8.2-11 are well suited for calculations in which temperature, pressure, and the partial molar properties are the independent variables, it is usually more convenient to have T, P, and the mole fractions x_i as the independent variables. A change of variables is accomplished by realizing that for a C-component mixture there are only $C - 1$ independent mole fractions (since $\sum_{i=1}^{C} x_i = 1$). Thus we write

$$d\bar{\theta}_i = \left(\frac{\partial \bar{\theta}_i}{\partial T} \right)_{P,\underline{x}} dT + \left(\frac{\partial \bar{\theta}_i}{\partial P} \right)_{T,\underline{x}} dP + \sum_{j=1}^{C-1} \left(\frac{\partial \bar{\theta}_i}{\partial x_j} \right)_{T,P} dx_j \tag{8.2-16}$$

where T, P, x_1, ..., x_{C-1} have been chosen as the independent variables. Substituting this expansion in Eq. 8.2-8b gives

$$0 = -\left(\frac{\partial \underline{\theta}}{\partial T}\right)_{P,N_i} dT - \left(\frac{\partial \underline{\theta}}{\partial P}\right)_{T,N_i} dP$$

$$+ \sum_{i=1}^{c} x_i \left[\left(\frac{\partial \bar{\theta}_i}{\partial T}\right)_{P,\underline{x}} dT + \left(\frac{\partial \bar{\theta}_i}{\partial P}\right)_{T,\underline{x}} dP + \sum_{j=1}^{c-1} \left(\frac{\partial \bar{\theta}_i}{\partial x_j}\right)_{T,P} dx_j\right]$$

(8.2-17)

Since

$$\sum_{i=1}^{c} x_i \left(\frac{\partial \bar{\theta}_i}{\partial T}\right)_{P,\underline{x}} dT = \frac{\partial}{\partial T}\bigg|_{P,\underline{x}} \left(\sum_{i=1}^{c} x_i \bar{\theta}_i\right) dT = \left(\frac{\partial \underline{\theta}}{\partial T}\right)_{P,\underline{x}} dT$$

and since holding all the mole numbers constant is equivalent to keeping all the mole fractions fixed, the first and third terms in Eq. 8.2-17 cancel. Similarly, the second and fourth terms cancel. Thus we are left with

$$\sum_{i=1}^{c} x_i \sum_{j=1}^{c-1} \left(\frac{\partial \bar{\theta}_i}{\partial x_j}\right)_{T,P} dx_j = 0$$

(8.2-18)

which for a binary mixture reduces to

$$\sum_{i=1}^{2} x_i \left(\frac{\partial \bar{\theta}_i}{\partial x_1}\right)_{T,P} dx_1 = 0$$

(8.2-19a)

or, equivalently

$$\boxed{x_1 \left(\frac{\partial \bar{\theta}_1}{\partial x_1}\right)_{T,P} + x_2 \left(\frac{\partial \bar{\theta}_2}{\partial x_1}\right)_{T,P} = 0}$$

(8.2-19b)

Gibbs-Duhem equations for binary mixture at constant T and P

For the special case in which θ is equal to the Gibbs energy, we have

$$\boxed{x_1 \left(\frac{\partial \bar{G}_1}{\partial x_1}\right)_{T,P} + x_2 \left(\frac{\partial \bar{G}_2}{\partial x_1}\right)_{T,P} = 0}$$

(8.2-20)

Finally, note that several different forms of Eqs. 8.2-19 and 8.2-20 can be obtained by using $x_2 = 1 - x_1$ and $dx_2 = -dx_1$.

8.3 A NOTATION FOR CHEMICAL REACTIONS

Since our interest in this chapter is with the equations of change and the equilibrium criteria for general thermodynamic systems, we need a convenient notation for the description of chemical reactions. Here we will generalize the notation for a chemical reaction introduced in Sec. 2.3. There we wrote the reaction

$$\alpha A + \beta B + \cdots \rightleftharpoons \rho R + \cdots$$

where $\alpha, \beta, \ldots$ are the molar stoichiometric coefficients, as

$$\rho R + \cdots - \alpha A - \beta B - \cdots = 0$$

or

$$\sum_i \nu_i I = 0 \tag{8.3-1}$$

In this notation, from Chapter 2, ν_i is the **stoichiometric coefficient** of species I, so defined that ν_i is positive for reaction products, negative for reactants, and equal to zero for inert species. (You should remember that in this notation the reaction $H_2 + \frac{1}{2}O_2 = H_2O$ is written as $H_2O - H_2 - \frac{1}{2}O_2 = 0$, so that $\nu_{H_2O} = +1$, $\nu_{H_2} = -1$, and $\nu_{O_2} = -\frac{1}{2}$.)

Using N_i to represent the number of moles of species i in a closed system at any time t, and $N_{i,0}$ for the initial number of moles of species i, then N_i and $N_{i,0}$ are related through the reaction variable X, the molar extent of reaction, and the stoichiometric coefficient ν_i by

$$N_i = N_{i,0} + \nu_i X \tag{8.3-2a}$$

and

$$X = \frac{N_i - N_{i,0}}{\nu_i} \tag{8.3-2b}$$

The reason for introducing the reaction variable X is that it has the same value for each molecular species involved in a reaction, as was shown in Illustration 2.3-1. Thus, given the initial mole numbers of all species and X (or the number of moles of one species from which X can be calculated) at time t, one can easily compute all other mole numbers in the system. In this way the complete progress of a chemical reaction (i.e., the change in mole numbers of all the species involved in the reaction) is given by the value of the single variable X. Also, the rate of change of the number of moles of species i resulting from the chemical reaction is

$$\left(\frac{dN_i}{dt}\right)_{rxn} = \nu_i \dot{X} \tag{8.3-3}$$

where the subscript rxn indicates that this is the rate of change of species i attributable to chemical reaction alone, and $\dot{X}$ is the rate of change of the molar extent of reaction.[7] The total number of moles in a closed system at any time (i.e., for any extent of reaction) is

$$N = \sum_{i=1}^{c} N_i = \sum_{i=1}^{c} (N_{i,0} + \nu_i X) = \sum_{i=1}^{c} N_{i,0} + X \sum_{i=1}^{c} \nu_i \tag{8.3-4}$$

However, our interest can also extend to situations in which there are multiple chemical reactions, and the notation introduced in Sec. 2.3 and reviewed here needs to be extended to such cases. Before we do this, it is necessary to introduce the concept of **independent chemical reactions.** The term *independent reactions* is used here to designate the smallest collection of reactions that, on forming various linear combinations,

[7]For an open system the number of moles of species i may also change due to mass flows into and out of the system.

includes all possible chemical reactions among the species present. Since we have used the adjective *smallest* in defining a set of independent reactions, it follows that no reaction in the set can itself be a linear combination of the others. For example, one can write the following three reactions between carbon and oxygen:

$$C + O_2 = CO_2$$
$$2C + O_2 = 2CO$$
$$2CO + O_2 = 2CO_2$$

If we add the second and third of these equations, we get twice the first, so these three reactions are not independent. In this case, any two of the three reactions form an independent set.

As we will see in Chapter 13, to describe a chemically reacting system it is not necessary to consider all the chemical reactions that can occur between the reactant species, only the independent reactions. Furthermore, the molar extent of reaction for any chemical reaction among the species can be computed from an appropriate linear combination of the known extents of reaction for the set of independent chemical reactions (see Sec. 13.3). Thus, in the carbon-oxygen reaction system just considered, only two reaction variables (and the initial mole numbers) need be specified to completely fix the mole numbers of each species; the specification of a third reaction variable would be redundant and may result in confusion.

To avoid unnecessary complications in the analysis of multiple reactions, we restrict the following discussion to sets of independent chemical reactions. Consequently, we need a method of identifying a set of independent reactions from a larger collection of reactions. When only a few reactions are involved, as in the foregoing example, this can be done by inspection. If many reactions occur, the methods of matrix algebra[8] can be used to determine a set of independent reactions, though we will employ a simpler procedure developed by Denbigh.[9]

In the Denbigh procedure one first writes the stoichiometric equations for the formation, from its constituent atoms, of each of the molecular species present in the chemical reaction system. One of the equations that contains an atomic species not actually present in the atomic state at the reaction conditions is then used, by addition and/or subtraction, to eliminate that atomic species from the remaining equations. In this way the number of stoichiometric equations is reduced by one. The procedure is repeated until all atomic species not present have been eliminated. The equations that remain form a set of independent chemical reactions; the molar extents of reaction for these reactions are the variables to be used for the description of the multiple reaction system and to follow the composition changes in the mixture.[10]

As an example of this method, consider again the oxidation of carbon. We start by writing the following equations for the formation of each of the compounds:

$$2O = O_2$$
$$C + O = CO$$
$$C + 2O = CO_2$$

[8]N. R. Amundson, *Mathematical Methods in Chemical Engineering,* Prentice Hall, Englewood Cliffs, N.J. (1966), p. 50.

[9]K. Denbigh, *Principles of Chemical Equilibrium,* 4th ed., Cambridge University Press, Cambridge (1981), pp. 169–172.

[10]When considering chemical reactions involving isomers—for example, ortho-, meta-, and para-xylene—one proceeds as described here, treating each isomer as a separate chemical species.

Since a free oxygen atom does not occur, the first equation is used to eliminate O from the other two equations to obtain

$$2C + O_2 = 2CO$$
$$C + O_2 = CO_2$$

All the remaining atomic species in these equations (here only carbon) are present in the reaction system, so no further reduction of the equations is possible, and these two equations form a set of independent reactions. Note that two other, different sets of independent reactions are obtained by using instead the second or third equation to eliminate the free oxygen atom. Though different sets of independent reactions are obtained, each would contain only two reactions, and either set could be used as the set of independent reactions.

For the multiple-reaction case, X_j is defined to be the molar extent of reaction (or simply the reaction variable) for the jth independent reaction, and ν_{ij} the stoichiometric coefficient for species i in the jth reaction. The number of moles of species i present at any time (in a closed system) is

$$N_i = N_{i,0} + \sum_{j=1}^{\mathcal{M}} \nu_{ij} X_j \qquad (8.3\text{-}5)$$

where the summation is over the set of $\mathcal{M}$ independent chemical reactions. The instantaneous rate of change of the number of moles of species i due only to chemical reaction is

$$\left(\frac{dN_i}{dt}\right)_{rxn} = \sum_{j=1}^{\mathcal{M}} \nu_{ij} \dot{X}_j \qquad (8.3\text{-}6)$$

Finally, with $\hat{X}_j = X_j/V$ defined to be the molar extent of reaction per unit volume, Eq. 8.3-6 can be written as

$$\left(\frac{dN_i}{dt}\right)_{rxn} = \sum_{j=1}^{\mathcal{M}} \nu_{ij} \frac{d}{dt}(V\hat{X}_j) = \frac{dV}{dt} \sum_{j=1}^{\mathcal{M}} \nu_{ij} \hat{X}_j + V \sum_{j=1}^{\mathcal{M}} \nu_{ij} r_j \qquad (8.3\text{-}7)$$

where $r_j = d\hat{X}_j/dt$, the rate of reaction per unit volume, is the reaction variable most frequently used by chemists and chemical engineers in chemical reactor analysis.

8.4 THE EQUATIONS OF CHANGE FOR A MULTICOMPONENT SYSTEM

The next step in the development of the thermodynamics of multicomponent systems is the formulation of the equations of change. These equations can, in completely general form, be considerably more complicated than the analogous pure component equations since (1) the mass or number of moles of each species may not be conserved due to chemical reactions, and (2) the diffusion of one species relative to the others may occur if concentration gradients are present. Furthermore, there is the computational difficulty that each thermodynamic property depends, in a complicated fashion, on the temperature, pressure, and composition of the mixture.

To simplify the development of the equations, we will neglect all diffusional processes, since diffusion has very little effect on the thermodynamic state of the system. (This assumption is equivalent to setting the average velocity of each species equal to the mass average velocity of the fluid.) In Chapter 2 the kinetic and potential energy terms were found to make a small contribution to the pure component energy balance;

the relative importance of these terms to the energy balance for a reacting mixture is even less, due to the large energy changes that accompany chemical transformations. Therefore, we will neglect the potential and kinetic energy terms in the formulation of the mixture energy balance. This omission will cause serious errors only in the analysis of rocket engines and similar high-speed devices involving chemical reaction.

With these assumptions, the formulation of the equations of change for a multicomponent reacting mixture is not nearly so formidable a task as it might first appear. In fact, merely by making the proper identifications, the equations of change for a mixture can be written as simple generalizations of the equations of change for a single-component system. The starting point is Eq. 2.1-4, which is rewritten as

$$\frac{d\theta}{dt} = \begin{pmatrix} \text{Rate of change of} \\ \theta \text{ in the system} \end{pmatrix} = \begin{pmatrix} \text{Rate at which } \theta \text{ enters} \\ \text{the system across} \\ \text{system boundaries} \end{pmatrix}$$
$$- \begin{pmatrix} \text{Rate at which } \theta \text{ leaves} \\ \text{the system across} \\ \text{system boundaries} \end{pmatrix} \qquad \textbf{(2.1-4)}$$
$$+ \begin{pmatrix} \text{Rate at which } \theta \text{ is} \\ \text{generated within} \\ \text{the system} \end{pmatrix}$$

The balance equation for species i is gotten by setting θ equal to N_i, the number of moles of species i; letting $(\dot{N}_i)_k$ equal the molar flow rate of species i into the system at the kth port; and recognizing that species i may be generated within the system by chemical reaction;

Species mass balance for a reacting system

$$\frac{dN_i}{dt} = \sum_{k=1}^{K} (\dot{N}_i)_k + \begin{pmatrix} \text{Rate at which species i is being} \\ \text{produced by chemical reaction} \end{pmatrix}$$
$$= \sum_{k=1}^{K} (\dot{N}_i)_k + \left(\frac{dN_i}{dt}\right)_{\text{rxn}} \qquad \textbf{(8.4-1a)}$$
$$= \sum_{k=1}^{K} (\dot{N}_i)_k + \sum_{j=1}^{\mathcal{M}} \nu_{ij} \dot{X}_j$$

We can obtain a balance equation on the total number of moles in the system by summing Eq. 8.4-1a over all species i, recognizing that $\sum_{i=1}^{C} N_i = N$ is the total number of moles, and that

$$\sum_{i=1}^{C} \sum_{k=1}^{K} (\dot{N}_i)_k = \sum_{k=1}^{K} \sum_{i=1}^{C} (\dot{N}_i)_k = \sum_{k=1}^{K} \dot{N}_k$$

where $\dot{N}_k = \sum_{i=1}^{C} (\dot{N}_i)_k$ is the total molar flow rate at the kth entry port, so that

Total mass balance for a reacting system

$$\boxed{\frac{dN}{dt} = \sum_{k=1}^{K} \dot{N}_k + \sum_{i=1}^{C} \sum_{j=1}^{\mathcal{M}} \nu_{ij} \dot{X}_j} \qquad \textbf{(8.4-1b)}$$

Since neither the number of moles of species i nor the total number of moles is a conserved quantity, we need information on the rates at which all chemical reactions occur to use Eqs. 8.4-1. That is, we need detailed information on the reaction processes internal to the black-box system (unless, of course, no reactions occur, in which case $\dot{X}_j = 0$ for all j).

Although we will be interested in the equations of change mainly on a molar basis, for completeness and for several illustrations that follow, some of the equations of change will also be given on a mass basis. To obtain a balance equation for the mass of species i, we need only multiply Eq. 8.4-1a by the molecular weight of species i, m_i, and use the notation $M_i = m_i N_i$ and $(\dot{M}_i)_k = m_i(\dot{N}_i)_k$ to get

$$\frac{dM_i}{dt} = \sum_{k=1}^{K}(\dot{M}_i)_k + \sum_{j=1}^{\mathcal{M}} m_i \nu_{ij} \dot{X}_j \qquad \textbf{(8.4-2a)}$$

Also, summing the equation over species i, we get the overall mass balance equation

$$\frac{dM}{dt} = \sum_{k=1}^{K} \dot{M}_k \qquad \textbf{(8.4-2b)}$$

where $\dot{M}_k = \sum_{i=1}^{C}(\dot{M}_i)_k$ is the total mass flow at the kth entry port. In Eq. 8.4-2b the chemical reaction term vanishes since total mass is a conserved quantity (Problem 8.5).

If θ in Eq. 2.1-4 is now taken to be the total energy of the system (really only the internal energy, since we are neglecting the kinetic and potential energy terms), and the same energy transfer mechanisms identified in Sec. 3.1 are used here, we obtain

$$\boxed{\frac{dU}{dt} = \sum_{k=1}^{K}(\dot{N}\underline{H})_k + \dot{Q} + \dot{W}_s - P\frac{dV}{dt}} \quad \text{(molar basis)} \qquad \textbf{(8.4-3)}$$

Energy balance for a reacting system

and

$$\frac{dU}{dt} = \sum_{k=1}^{K}(\dot{M}\hat{H})_k + \dot{Q} + \dot{W}_s - P\frac{dV}{dt} \quad \text{(mass basis)}$$

Here $(\dot{N}\underline{H})_k$ is the product of the molar flow rate and the molar enthalpy of the fluid, and $(\dot{M}\hat{H})_k$ the product of mass flow rate and enthalpy per unit mass, both at the kth entry port. Since some or all of the flow streams may be mixtures, to evaluate a term such as $(\dot{N}\underline{H})_k$ the relation

$$(\dot{N}\underline{H})_k = \sum_{i=1}^{C}(\dot{N}_i \overline{H}_i)_k \qquad \textbf{(8.4-4)}$$

must be used, where $(\dot{N}_i)_k$ is the molar flow rate of species i in the kth flow stream, and $(\overline{H}_i)_k$ is its partial molar enthalpy. Similarly, the internal energy of the system must be obtained from

$$U = \sum_{i=1}^{C} N_i \overline{U}_i$$

where N_i is the number of moles of species i and $\overline{U}_i$ is its partial molar internal energy.

At this point you might be concerned that the heat of reaction (i.e., the energy released or absorbed on reaction, since the total energy of the chemical bonds in the reactant and product molecules are not equal) does not explicitly appear in the energy balance equation. Since the only assumptions made in deriving this equation were to neglect the potential and kinetic energy terms and diffusion, neither of which involves chemical reaction, the heat of reaction is in fact contained in the energy balance, though it appears implicitly rather than explicitly. Although energy balances on chemical reactors are studied in detail in Chapter 14, to demonstrate that the heat of reaction is contained in Eq. 8.4-3, we will consider its application to the well-stirred, steady-state chemical reactor in Fig. 8.4-1. Here by steady state we mean that the rate at which each species i leaves the reactor just balances the rate at which species i enters the reactor and is produced within it, so that N_i does not vary with time, that is,

$$\frac{dN_i}{dt} = 0 = (\dot{N}_i)_{\text{in}} + (\dot{N}_i)_{\text{out}} + \nu_i \dot{X} \tag{8.4-5a}$$

and, further, that the rates of flow of energy into and out of the reactor just balance, so that the internal energy of the contents of the reactor does not change with time

Energy balance for a continuous-flow stirred-tank reactor

$$\frac{dU}{dt} = 0 = \sum_{i=1}^{c} (\dot{N}_i \overline{H}_i)_{\text{in}} + \sum_{i=1}^{c} (\dot{N}_i \overline{H}_i)_{\text{out}} + \dot{Q} \tag{8.4-5b}$$

For the very simple isothermal case in which the inlet and outlet streams and the reactor contents are all at temperature T, and with the assumption that the partial molar enthalpy of each species is just equal to its pure-component enthalpy, we obtain

$$\dot{Q} = -\sum_{i=1}^{c} [(\dot{N}_i)_{\text{out}} + (\dot{N}_i)_{\text{in}}] \underline{H}_i$$

Now if there were no chemical reaction $(\dot{N}_i)_{\text{out}} = -(\dot{N}_i)_{\text{in}}$ and the heat flow rate $\dot{Q}$ should be equal to zero to maintain the constant temperature T. However, when a chemical reaction occurs, $(\dot{N}_i)_{\text{out}}$ and $(\dot{N}_i)_{\text{in}}$ are not equal in magnitude, and the steady heat flow $\dot{Q}$ required to keep the reactor at constant temperature is

$$\dot{Q} = \sum_{i=1}^{c} [(\dot{N}_i)_{\text{out}} + (\dot{N}_i)_{\text{in}}] \underline{H}_i = \sum_{i=1}^{c} \dot{X} \nu_i \underline{H}_i$$

$$= \dot{X} \sum_{i=1}^{c} \nu_i \underline{H}_i = \Delta_{\text{rxn}} H \dot{X}$$

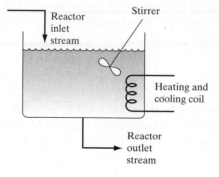

Figure 8.4-1 A simple stirred-tank reactor.

so that the heat flow rate is equal to the product of $\Delta_{rxn}H = \sum \nu_i \underline{H}_i$, the **isothermal heat of reaction**, and $\dot{X}$, the reaction rate. Thus, we see that the heat of reaction is implicitly contained in the energy balance equation through the difference in species mole numbers in the inlet and outlet flow streams.

For nonideal mixtures (i.e., mixtures for which $\overline{H}_i \neq \underline{H}_i$) and nonisothermal reactors, the expression for the heat of reaction is buried somewhat deeper in the energy balance equation; this is discussed in Chapter 14.

To derive the macroscopic entropy balance equation for mixtures we now set $\theta = S$ in Eq. 2.1-4 and make the same identification for the entropy flow terms as was made in Sec. 4.1, to obtain

$$\frac{dS}{dt} = \sum_{k=1}^{K} (\dot{N}\underline{S})_k + \frac{\dot{Q}}{T} + \dot{S}_{gen} \tag{8.4-6}$$

where

$$(\dot{N}\underline{S})_k = \sum_{i=1}^{C} (\dot{N}_i \bar{S}_i)_k$$

Thus the entropy balance, like the energy balance equation, is of the same form for the pure fluid and for mixtures.

As the final step in the development of the entropy balance, an expression for the entropy generation term should be obtained here, as was done for the pure fluid in Sec. 4.6. The derivation of such an expression is tedious and would require us to develop detailed microscopic equations of change for a mixture. Rather than doing this, we will merely write down the final result, referring the reader to the book by de Groot and Mazur for the details.[11] Their result, with the slight modification of writing the chemical reaction term on a molar rather than a mass basis, is

$$\dot{S}_{gen} = \int \dot{\sigma}_s \, dV$$

where

$$\dot{\sigma}_s = \frac{\lambda}{T^2}(\nabla T)^2 + \frac{\mu}{T}\phi^2 - \frac{1}{T}\sum_{j=1}^{M}\sum_{i=1}^{C} \nu_{ij}\overline{G}_i \dot{X}_j \tag{8.4-7}$$

(see Secs. 3.6 and 4.6 for the definitions of λ, ϕ, and the gradient operator ∇). The last term, which is new, represents the contribution of chemical reactions to the entropy generation rate. One can establish that this term makes a positive contribution to the entropy generation term. It is of interest to note that had diffusional processes also been considered, there would be an additional contribution to $\dot{S}_{gen}$ due to diffusion. That term would be in the form of a diffusional flux times a driving force for diffusion and would also be greater than or equal to zero.

In Chapter 4 we defined a reversible process to be a process for which $\dot{S}_{gen} = 0$. This led to the conclusion that in a reversible process in a one-component system only infinitesimal temperature and velocity gradients could be present. Clearly, on the basis of Eq. 8.4-7, a reversible process in a multicomponent system is one in which only

---

[11]S. R. de Groot and P. Mazur, *Non-equilibrium Thermodynamics,* North-Holland, Amsterdam (1962), Chapter 3.

infinitesimal gradients in temperature, velocity, and concentration may be present, and in which all chemical reactions proceed at only infinitesimal rates.

The differential form of the multicomponent mass, energy, and entropy balances can be integrated over time to get difference forms of the balance equations, as was done in Chapters 2, 3, and 4 for the pure-component equations. The differential and difference forms of the balance equations are listed in Tables 8.4-1 and 8.4-2, respectively. It is left to the reader to work out the various simplifications of these equations that arise for special cases of closed systems, adiabatic processes, and so forth.

With the equations of change for mixtures, and given mixture thermodynamic data, such as the enthalpy data for sulfuric acid–water mixtures in Fig. 8.1-1, it is possible to solve many thermodynamic energy flow problems for mixtures. One example is given in the following illustration.

Table 8.4-1 The Differential Form of the Equations of Change for a Multicomponent System

Species balance

$$\frac{dN_i}{dt} = \sum_{k=1}^{K}(\dot{N}_i)_k + \sum_{j=1}^{M} \nu_{ij}\dot{X}_j \qquad \text{(molar basis)}$$

$$\frac{dM_i}{dt} = \sum_{k=1}^{K}(\dot{M}_i)_k + \sum_{j=1}^{M} m_i\nu_{ij}\dot{X}_j \qquad \text{(mass basis)}$$

Overall mass balance

$$\frac{dN}{dt} = \sum_{k=1}^{K}\dot{N}_k + \sum_{i=1}^{c}\sum_{j=1}^{M} \nu_{ij}\dot{X}_j \qquad \text{(molar basis)}$$

$$\frac{dM}{dt} = \sum_{k=1}^{K}\dot{M}_k \qquad \text{(mass basis)}$$

Energy balance

$$\frac{dU}{dt} = \sum_{k=1}^{K}(\dot{N}\underline{H})_k + \dot{Q} + \dot{W}_s - P\frac{dV}{dt} \qquad \text{(molar basis)}$$

$$\frac{dU}{dt} = \sum_{k=1}^{K}(\dot{M}\hat{H})_k + \dot{Q} + \dot{W}_s - P\frac{dV}{dt} \qquad \text{(mass basis)}$$

Entropy balance

$$\frac{dS}{dt} = \sum_{k=1}^{K}(\dot{N}\underline{S})_k + \frac{\dot{Q}}{T} + \dot{S}_{gen} \qquad \text{(molar basis)}$$

$$\frac{dS}{dt} = \sum_{k=1}^{K}(\dot{M}\hat{S})_k + \frac{\dot{Q}}{T} + \dot{S}_{gen} \qquad \text{(mass basis)}$$

where

$$\dot{N}_k = \sum_{i=1}^{c}(\dot{N}_i)_k \qquad (\dot{N}\underline{H})_k = \sum_{i=1}^{c}(\dot{N}_i\bar{H}_i)_k \qquad (\dot{N}\underline{S})_k = \sum_{i=1}^{c}(\dot{N}_i\bar{S}_i)_k$$

and

$$\dot{M}_k = \sum_{i=1}^{c}(\dot{M}_i)_k$$

Table 8.4-2 The Difference Form of the Equations of Change for a Multicomponent Mixture

Species balance

$$N_i(t_2) - N_i(t_1) = \sum_{k=1}^{K} \int_{t_1}^{t_2} (\dot{N}_i)_k \, dt + \sum_{j=1}^{\mathcal{M}} \nu_{ij} X_j$$

$$M_i(t_2) - M_i(t_1) = \sum_{k=1}^{K} \int_{t_1}^{t_2} (\dot{M}_i)_k \, dt + \sum_{j=1}^{\mathcal{M}} \nu_{ij} m_i X_j$$

Overall mass balance

$$N(t_2) - N(t_1) = \sum_{k=1}^{K} \int_{t_1}^{t_2} \dot{N}_k \, dt + \sum_{i=1}^{c} \sum_{j=1}^{\mathcal{M}} \nu_{ij} X_j$$

$$M(t_2) - M(t_1) = \sum_{k=1}^{K} \int_{t_1}^{t_2} \dot{M}_k \, dt$$

Energy balance

$$U(t_2) - U(t_1) = \sum_{k=1}^{K} \int_{t_1}^{t_2} (\dot{N}\underline{H})_k \, dt + Q + W_s - \int P \, dV$$

$$U(t_2) - U(t_1) = \sum_{k=1}^{K} \int_{t_1}^{t_2} (\dot{M}\hat{H})_k \, dt + Q + W_s - \int P \, dV$$

Entropy balance

$$S(t_2) - S(t_1) = \sum_{k=1}^{K} \int_{t_1}^{t_2} (\dot{N}\underline{S})_k \, dt + \int_{t_1}^{t_2} \frac{\dot{Q}}{T} \, dt + S_{\text{gen}}$$

$$S(t_2) - S(t_1) = \sum_{k=1}^{K} \int_{t_1}^{t_2} (\dot{M}\hat{S})_k \, dt + \int_{t_1}^{t_2} \frac{\dot{Q}}{T} \, dt + S_{\text{gen}}$$

where $(\dot{N})_k$, $(\dot{N}\underline{H})_k$, $(\dot{N}\underline{S})_k$, and $(\dot{M})_k$ are defined in Table 8.4-1

ILLUSTRATION 8.4-1

Temperature Change on Adiabatic Mixing of an Acid and Water

Three moles of water and one mole of sulfuric acid, each at 0°C, are mixed adiabatically. Use the data in Fig. 8.1-1 and the information in Illustration 8.1-1 to estimate the final temperature of the mixture.

SOLUTION

From the closed-system mass balance, we have

$$M^f = M_{H_2O} + M_{H_2SO_4} = 3 \times 18.015 + 1 \times 98.078 = 152.12 \text{ g}$$

and from the energy balance, we have

$$U^f = H^f = H^i = M^f \hat{H}_{\text{mix}} = M_{H_2O} \hat{H}_{H_2O} + M_{H_2SO_4} \hat{H}_{H_2SO_4} = 3 \times 0 + 1 \times 0 = 0 \text{ kJ}$$

Thus finally we have a mixture of 64.5 wt % sulfuric acid that has an enthalpy of 0 kJ/kg. (Note that we have used the fact that for liquids and solids at low pressure, the internal energy and enthalpy are essentially equal.) From Fig. 8.1-1 we see that a mixture containing 64.5 wt % sulfuric acid has an enthalpy of 0 kJ/kg at about 150°C. Therefore, if water and sulfuric acid are adiabatically mixed in the ratio of 3:1, the mixture will achieve a temperature of 150°C, which is just below the boiling point of the mixture. This large temperature rise, and the fact that the mixture is just below its boiling point, makes mixing sulfuric acid and water an operation that must be done with extreme care.

COMMENT

If instead of starting with refrigerated sulfuric acid and water (at 0°C), one started with these components at 21.2°C and mixed them adiabatically, the resulting 3:1 mixture would be in the liquid + vapor region; that is, the mixture would boil (and splatter). Also note that because of the shape of the curves on the enthalpy-concentration diagram, adding sulfuric acid to water adiabatically (i.e., moving to the right from the pure water edge of the diagram) results in a more gradual temperature rise than adding water to sulfuric acid (i.e., moving to the left from the pure–sulfuric acid edge). Therefore, whenever possible, sulfuric acid should be added to water, and not vice versa.   ∎

ILLUSTRATION 8.4-2

Mass and Energy Balances on a Nonreacting System

A continuous-flow steam-heated mixing kettle will be used to produce a 20 wt % sulfuric acid solution at 65.6°C from a solution of 90 wt % sulfuric acid at 0°C and pure water at 21.1°C. Estimate

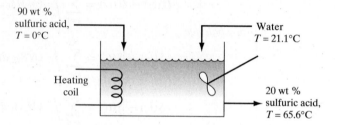

a. The kilograms of pure water needed per kilogram of initial sulfuric acid solution to produce a mixture of the desired concentration
b. The amount of heat needed per kilogram of initial sulfuric acid solution to heat the mixture to the desired temperature
c. The temperature of the kettle effluent if the mixing process is carried out adiabatically

SOLUTION

We choose the contents of the mixing kettle as the system. The difference form of the equations of change will be used for a time interval in which 1 kg of concentrated sulfuric acid enters the kettle.

a. Since there is no chemical reaction, and the mixing tank operates continuously, the total and sulfuric acid mass balances reduce to

$$0 = \sum_{k=1}^{3} \dot{M}_k \quad \text{and} \quad \sum_{k=1}^{3} (M_{H_2SO_4})_k = 0$$

Denoting the 90 wt % acid stream by the subscript 1 and its mass flow by $\dot{M}_1$, the water stream by the subscript 2, and the dilute acid stream by subscript 3, we have, from the total mass balance,

$$0 = \dot{M}_1 + \dot{M}_2 + \dot{M}_3 = \dot{M}_1 + Z\dot{M}_1 + \dot{M}_3$$

or

$$\dot{M}_3 = -(1 + Z)\dot{M}_1$$

where Z is equal to the number of kilograms of water used per kilogram of the 90 wt % acid. Also, from the mass balance on sulfuric acid, we have

$$0 = 0.90\dot{M}_1 + 0\dot{M}_2 + 0.20\dot{M}_3 = 0.90\dot{M}_1 - 0.20(1 + Z)\dot{M}_1$$

Therefore,

$$1 + Z = \frac{0.90}{0.20} = 4.5 \quad \text{and} \quad Z = 3.5$$

so that 3.5 kg of water must be added to each 1 kg of 90 wt % acid solution to produce a 20 wt % solution.

b. The steady-state energy balance is

$$0 = \sum_k (\dot{M}\hat{H})_k + \dot{Q}$$

since $W_s = 0$ and $\int P\,dV = 0$. From the mass balance of part (a),

$$\dot{M}_2 = 3.5\dot{M}_1$$
$$\dot{M}_3 = -4.5\dot{M}_1$$

From the enthalpy-concentration chart, Fig. 8.1-1, we have

$$\hat{H}_1 = \hat{H}(90 \text{ wt } \% \text{ H}_2\text{SO}_4, T = 0°\text{C}) = -183 \text{ kJ/kg}$$
$$\hat{H}_2 = \hat{H}(\text{pure H}_2\text{O}, T = 21.1°\text{C}) = 91 \text{ kJ/kg}$$
$$\hat{H}_3 = \hat{H}(20 \text{ wt } \% \text{ H}_2\text{SO}_4, T = 65.56°\text{C}) = 87 \text{ kJ/kg}$$

so that

$$Q = (4.5 \times 87 - 3.5 \times 91 - 1 \times (-183))$$
$$= (391.5 - 318.5 + 183) = 256 \text{ kJ/kg of initial acid solution}$$

c. For adiabatic operation, the energy balance is

$$0 = \sum_k (\dot{M}\hat{H})_k$$

or

$$0 = 4.5\hat{H}_3 - 3.5(91) - 1(-183)$$
$$4.5\hat{H}_3 = 135.5 \quad \text{and} \quad \hat{H}_3 = 30.1 \text{ kJ/kg}$$

Referring to the enthalpy-concentration diagram, we find that $T \sim 50°\text{C}$. ■

8.5 THE HEAT OF REACTION AND A CONVENTION FOR THE THERMODYNAMIC PROPERTIES OF REACTING MIXTURES

In the first part of this book we were interested in the change in the internal energy, enthalpy, and entropy accompanying a change in the thermodynamic state of a pure substance. For convenience in such calculations, one state of each substance was chosen as the reference state, with zero values for both enthalpy and entropy, and the values of the thermodynamic properties of the substance in other states were given relative to this reference state. Thus in the steam tables in Appendix A.III the internal energy and entropy are zero for liquid water at the triple point, whereas the zero values of these thermodynamic properties for methane (Fig. 2.4-2) and nitrogen (Fig. 2.4-3) correspond to different conditions of temperature and pressure. The reference states for sulfuric acid and water in the enthalpy-concentration diagram of Fig. 8.1-1 were also chosen on the basis of convenience. Such arbitrary choices for the reference state were satisfactory because our interest was only with the changes in thermodynamic properties in going from one thermodynamic state to another in a nonreacting system. However, when chemical reactions occur, the reference state for the thermodynamic properties of each species must be chosen with great care. In particular, the thermodynamic properties for each species must be such that the differences between the reactant and product species in any chemical transformation will be equal to those that would be measured in the appropriate experiment. For example, in the ideal gas-phase reaction

$$H_2 + \tfrac{1}{2}O_2 \rightarrow H_2O$$

it is observed that 241.82 kJ are liberated for each mole of water vapor produced when this reaction is run in an isothermal, constant-pressure calorimeter at 25°C and 1 bar with all species in the vapor phase. Clearly, then, the enthalpies of the reacting molecules must be related as follows:

$$
\begin{aligned}
\Delta_{\mathrm{rxn}} H(T = 25°C, P = 1 \text{ bar}) = {} & \underline{H}_{H_2O}(\text{vapor}, T = 25°C, P = 1 \text{ bar}) \\
& - \underline{H}_{H_2}(\text{gas}, T = 25°C, P = 1 \text{ bar}) \\
& - \tfrac{1}{2}\underline{H}_{O_2}(\text{gas}, T = 25°C, P = 1 \text{ bar}) \\
& = -241.82 \, \frac{\text{kJ}}{\text{mol } H_2O \text{ produced}}
\end{aligned}
$$

so that we are not free to choose the values of the enthalpy of hydrogen, oxygen, and water vapor all arbitrarily. A similar argument applies to the other thermodynamic properties such as the entropy and Gibbs free energy.

By considering a large collection of chemical reactions, one finds that, to be consistent with the heat-of-reaction data for all possible reactions, the zero value of internal energy or enthalpy can be set *only once for each element.* Thus, one could set the enthalpy of each element to be zero at 25°C and 1 bar (or some other reference state) and then determine the enthalpy of every compound by measuring the heat liberated or absorbed during its production, by isothermal chemical reaction, from its elements.

Although such a procedure is a conceptually pleasing method of devising a thermodynamic enthalpy scale, it is experimentally difficult. For example, to determine the enthalpy of water vapor, one would have to measure the heat liberated on its formation from hydrogen and oxygen atoms. However, since hydrogen and oxygen molecules, and not their atoms, are the thermodynamically stable species at 25°C, one would first have to dissociate the oxygen and hydrogen molecules into their constituent atoms and then accurately measure the heat evolved when the atoms combine to form a water molecule—a very difficult task. Instead, the thermodynamic energy scale that is most frequently used is based on choosing as the zero state of both enthalpy and Gibbs energy[12] for each atomic species its simplest thermodynamically stable state at 25°C and 1 bar. Thus, the reference state for argon and helium is the atomic gases; for oxygen, nitrogen, and hydrogen, it is the molecular gases; for mercury and bromine, it is the atomic and molecular liquids, respectively; whereas for iron and carbon, it is as the α-crystalline and graphite solids, respectively, all at 25°C and 1 bar.

Starting from this basis, and using the measured data from a large number of heats of reaction, heats of mixing, and chemical equilibrium measurements, the enthalpy and Gibbs energy content of all other molecular species relative to their constituent atoms in their reference states can be determined; these quantities are called the **enthalpy of formation**, $\Delta_f \underline{H}^\circ$ and the **Gibbs energy of formation,** $\Delta_f \underline{G}^\circ$. Thus, by definition,

$$\Delta_f \underline{H}^\circ_{H_2O}(\text{vapor, }25°C, 1\text{ bar}) = \underline{H}_{H_2O}(\text{vapor, }25°C, 1\text{ bar})$$

$$- \underline{H}_{H_2}(\text{gas, }25°C, 1\text{ bar})$$

$$- \tfrac{1}{2}\underline{H}_{O_2}(\text{gas, }25°C, 1\text{ bar})$$

Appendix A.IV contains a listing of $\Delta_f \underline{H}^\circ$ and $\Delta_f \underline{G}^\circ$ for a large collection of substances in their normal states of aggregation at 25°C and 1 bar.

Isothermal heats (enthalpies) and Gibbs energies of formation of species may be summed to compute the enthalpy change and Gibbs free energy change that would occur if the molecular species at 25°C, 1 bar, and the state of aggregation listed in Appendix A.IV reacted to form products at 25°C, 1 bar, and their listed state of aggregation. We will denote these changes by $\Delta_{rxn}H^\circ(25°C, 1$ bar$)$ and $\Delta_{rxn}G^\circ(25°C,$ 1 bar$)$, respectively. For example, for the gas-phase reaction

$$3NO_2 + H_2O = 2HNO_3 + NO$$

we have

$$\Delta_{rxn}H^\circ(25°C, 1\text{ bar}) = 2\underline{H}_{HNO_3}(25°C, 1\text{ bar}) + \underline{H}_{NO}(25°C, 1\text{ bar})$$

$$- 3\underline{H}_{NO_2}(25°C, 1\text{ bar}) - \underline{H}_{H_2O}(25°C, 1\text{bar})$$

$$= 2[\Delta_f \underline{H}^\circ_{HNO_3} + \tfrac{3}{2}\underline{H}_{O_2} + \tfrac{1}{2}\underline{H}_{H_2} + \tfrac{1}{2}\underline{H}_{N_2}]_{25°C, 1\text{bar}}$$

$$+ [\Delta_f \underline{H}^\circ_{NO} + \tfrac{1}{2}\underline{H}_{O_2} + \tfrac{1}{2}\underline{H}_{N_2}]_{25°C, 1\text{bar}}$$

$$- 3[\Delta_f \underline{H}^\circ_{NO_2} + \underline{H}_{O_2} + \tfrac{1}{2}\underline{H}_{N_2}]_{25°C, 1\text{bar}}$$

$$- [\Delta_f \underline{H}^\circ_{H_2O} + \underline{H}_{H_2} + \tfrac{1}{2}\underline{H}_{O_2}]_{25°C, 1\text{bar}}$$

$$= [2\Delta_f \underline{H}^\circ_{HNO_3} + \Delta_f \underline{H}^\circ_{NO} - 3\Delta_f \underline{H}^\circ_{NO_2} - \Delta_f \underline{H}^\circ_{H_2O}]_{25°C, 1\text{bar}}$$

$$= \sum \nu_i \Delta_f \underline{H}^\circ_i (25°C, 1\text{ bar})$$

Note that the enthalpies of the reference state atomic species cancel; by the conservation of atomic species, this will always occur. Thus, we have, as general results,

Standard state heat of reaction

$$\boxed{\Delta_{rxn}H^\circ(25°C, 1\text{ bar}) = \sum_i \nu_i \Delta_f \underline{H}^\circ_i(25°C, 1\text{ bar})} \qquad (8.5\text{-}1)$$

---

[12]For chemical equilibrium the Gibbs energy, rather than the entropy, is of central importance. Therefore, generally the enthalpy and the Gibbs energy are set equal to zero in the reference state.

and

$$\Delta_{rxn}G^\circ(25^\circ C, 1 \text{ bar}) = \sum_i \nu_i \Delta_f \underline{G}_i^\circ(25^\circ C, 1 \text{ bar})$$

(8.5-2)

We use the term **standard state** of a substance to indicate the state of aggregation given in Appendix A.IV (or, in certain cases, the reference states discussed in Chapter 13), at the temperature T and pressure of 1 bar, with the following qualifications. For a gas here (as in Chapter 6) we use the ideal gas at 1 bar, not the real gas. As was mentioned in Chapter 6, the $P = 0$ state cannot be used since the entropy would then be positive infinity and the Gibbs energy would be negative infinity. For substances that are typically solutes—sodium chloride, for example—properties in two or more standard states are given. One standard state is as a solid (in the crystalline state) and the other is in solution. The latter case is a bit complicated in that the enthalpy of formation may be given at a specified concentration whereas the Gibbs energy formation is that in a (hypothetical) ideal 1-molal solution (see Sec. 9.8). The pure liquid cannot be used as the standard state since the solute does not exist as a liquid at these conditions, and the very high (infinite) dilution state cannot be used since the Gibbs energy of the solute goes to negative infinity at infinite dilution. What is used, instead, is a hypothetical ideal state of finite concentration, here 1 molal, with properties obtained from the extrapolation of data obtained for the solute highly diluted in aqueous solution, as will be discussed in Chapter 9.

The **standard heat of reaction** at any temperature T, $\Delta_{rxn}H^\circ(T)$, is defined as the change in enthalpy that results when a stoichiometric amount of the reactants, each in their standard states at the temperature T, chemically react to form the reaction products in their standard states at the temperature T. Thus, in analogy with Eqs. 8.5-1 and 8.5-2,

$$\Delta_{rxn}H^\circ(T, P = 1 \text{ bar}) = \sum_i \nu_i \Delta_f \underline{H}_i^\circ(T, P = 1 \text{ bar})$$

(8.5-3)

and

$$\Delta_{rxn}G^\circ(T, P = 1 \text{ bar}) = \sum_i \nu_i \Delta_f \underline{G}_i^\circ(T, P = 1 \text{ bar})$$

(8.5-4)

where the subscript i indicates the species and the superscript $^\circ$ denotes the standard state. From

$$\underline{H}_i^\circ(T, P = 1 \text{ bar}) = \underline{H}_i^\circ(T_0, P = 1 \text{ bar}) + \int_{T_0}^{T} C_{P,i}^\circ(T, P = 1 \text{ bar}) \, dT$$

we have

$$\Delta_{rxn}H^\circ(T, 1 \text{ bar}) = \sum_i \nu_i \Delta_f \underline{H}_i^\circ(25^\circ C, 1 \text{ bar}) + \sum_i \nu_i \int_{T=25^\circ C}^{T} C_{P,i}^\circ \, dT$$

$$= \Delta_{rxn}H^\circ(25^\circ C, 1 \text{ bar}) + \sum_i \nu_i \int_{T=25^\circ C}^{T} C_{P,i}^\circ \, dT$$

(8.5-5)

where $C_{P,i}^\circ$ is the heat capacity of species i in its standard state.

In certain instances, such as in the study of large organic compounds that require a complicated synthesis procedure or of biochemical molecules, it is not possible to measure the heat of formation directly. A substitute procedure in these cases is to measure the energy change for some other reaction, usually the heat of combustion, that is, the energy liberated when the compound is completely oxidized (all the carbon is oxidized to carbon dioxide, all the hydrogen to water, etc.). The **standard heat of combustion** $\Delta_c \underline{H}^\circ(T)$ is defined to be the heat of combustion with both the reactants and the products in their standard states, at the temperature T. The standard heat of combustion at 25°C and 1 bar is listed in Appendix A.V for a number of compounds. (Note that there are two entries in this table, one corresponding to the liquid phase and the other to the vapor phase, being the standard state for water.) Given the standard heat of combustion, the standard heat of reaction is computed as indicated in Illustration 8.5-1.

ILLUSTRATION 8.5-1

Calculation of the Standard Heat of Reaction at 25°C

Compute the standard heat of reaction for the hydrogenation of benzene to cyclohexane,

$$C_6H_6 + 3H_2 \rightarrow C_6H_{12}$$

from the standard-heat-of-combustion data.

SOLUTION

The standard heat of reaction can, in principle, be computed from Eq. 8.5-3; however, for illustration, we will use the heat of combustion for cyclohexane. From the standard-heat-of-combustion data in Appendix A.V, we have $\Delta_c \underline{H}^\circ = -3\,919\,906$ J/mol of cyclohexane for the following reaction:

$$C_6H_{12}(l) + 9O_2 \rightarrow 6CO_2 + 6H_2O(l)$$

Thus

$$\Delta_c \underline{H}^\circ_{C_6H_{12}} = 6\underline{H}_{CO_2} + 6\underline{H}_{H_2O} - \underline{H}_{C_6H_{12}} - 9\underline{H}_{O_2} = -3\,919\,906 \text{ J/mol}$$

or

$$\underline{H}_{C_6H_{12}} = -\Delta_c \underline{H}^\circ_{C_6H_{12}} - 9\underline{H}_{O_2} + 6\underline{H}_{CO_2} + 6\underline{H}_{H_2O}$$

Similarly,

$$\underline{H}_{C_6H_6} = -\Delta_c \underline{H}^\circ_{C_6H_6} - 7\tfrac{1}{2}\underline{H}_{O_2} + 6\underline{H}_{CO_2} + 3\underline{H}_{H_2O}$$

and

$$3\underline{H}_{H_2} = -3\Delta_c \underline{H}^\circ_{H_2} - 1\tfrac{1}{2}\underline{H}_{O_2} + 3\underline{H}_{H_2O}$$

Therefore,

$$\begin{aligned}
\Delta_{rxn} \underline{H}^\circ &= -\Delta_c \underline{H}^\circ_{C_6H_{12}} - 9\underline{H}_{O_2} + 6\underline{H}_{CO_2} + 6\underline{H}_{H_2O} + \Delta_c \underline{H}^\circ_{C_6H_6} \\
&\quad + 7\tfrac{1}{2}\underline{H}_{O_2} - 6\underline{H}_{CO_2} - 3\underline{H}_{H_2O} + 3\Delta_c \underline{H}^\circ_{H_2} + 1\tfrac{1}{2}\underline{H}_{O_2} - 3\underline{H}_{H_2O} \\
&= -\Delta_c \underline{H}^\circ_{C_6H_{12}} + \Delta_c \underline{H}^\circ_{C_6H_6} + 3\Delta_c \underline{H}^\circ_{H_2} = -\sum_i \nu_i \Delta_c \underline{H}^\circ_i \\
&= 3\,919\,906 - 3\,267\,620 - 3 \times 285\,840 = -205\,234 \; \frac{\text{J}}{\text{mol benzene}} \\
&= -205.23 \; \frac{\text{kJ}}{\text{mol benzene}}
\end{aligned}$$

COMMENT

Note that in the final equation, $\Delta_{rxn}H^\circ = -\sum \nu_i \Delta_c \underline{H}_i^\circ$, the enthalpies of the reference-state atomic species cancel, as they must due to conservation of atomic species on chemical reaction.

∎

The equation developed in this illustration,

$$\Delta_{rxn}H^\circ(T = 25°C, 1\ \text{bar}) = -\sum_i \nu_i \Delta_c \underline{H}_i^\circ(T = 25°C, P = 1\ \text{bar}) \qquad \textbf{(8.5-6)}$$

is always valid and provides a way of computing the standard heat of reaction from standard-heat-of-combustion data.

ILLUSTRATION 8.5-2

Calculation of the Standard Heat of Reaction as a Function of Temperature

Compute the standard-state heat of reaction for the gas-phase reaction $N_2O_4 = 2NO_2$ over the temperature range of 200 to 600 K.[13]

Data: See Appendices A.II and A.IV.

SOLUTION

The heat of reaction at a temperature T can be computed from

$$\Delta_{rxn}H^\circ(T) = \sum_i \nu_i \Delta_f \underline{H}_i^\circ(T)$$

At $T = 25°C$ we find, from the data in Appendix A.IV, that for each mole of N_2O_4 reacted,

$$\Delta_{rxn}H^\circ(T = 25°C) = [2 \times 33.18 - 9.16]\ \text{kJ/mol}$$
$$= 57.20\ \text{kJ/mol}$$

To compute the heat of reaction at any temperature T, we start from Eq. 8.5-5 and note that since the standard state for each species is a low-pressure gas, $C_{P,i}^\circ = C_{P,i}^*$. Therefore,

$$\Delta_{rxn}H^\circ(T) = \Delta_{rxn}H^\circ(T = 25°C) + \int_{T=298.2K}^{T} \sum_i \nu_i C_{P,i}^* \, dT$$

For the case here we have, from Appendix A.II,

$$\sum_i \nu_i C_{P,i}^* = 2C_{P,NO_2}^* - C_{P,N_2O_4}^*$$
$$= 12.804 - 7.239 \times 10^{-2}T + 4.301 \times 10^{-5}T^2 + 1.5732 \times 10^{-8}T^3 \ \frac{\text{J}}{\text{mol K}}$$

---

[13]Since, as we will see, $\overline{H}_i = \underline{H}_i$ for an ideal gas mixture, the standard state heat of reaction and the actual heat of reaction are identical in this case.

Thus

$$
\begin{aligned}
\Delta_{\mathrm{rxn}}\underline{H}^{\circ}(T) &= 57\,200 + \int_{298.2}^{T} (12.804 - 7.239 \times 10^{-2}T \\
&\quad + 4.301 \times 10^{-5}T^2 + 1.5732 \times 10^{-8}T^3)\,dT \\
&= 57\,200 + 12.804(T - 298.15) - \frac{7.239}{2} \times 10^{-2}(T^2 - 298.15^2) \\
&\quad + \frac{4.301}{3} \times 10^{-5}(T^3 - 298.15^3) + \frac{1.5732}{4} \times 10^{-8}(T^4 - 298.15^4) \\
&= 56\,189 + 12.804T - 3.619 \times 10^{-2}T^2 + 1.4337 \times 10^{-5}T^3 \\
&\quad + 3.933 \times 10^{-9}T^4 \ \mathrm{J/mol}\ N_2O_4
\end{aligned}
$$

Values of $\Delta_{\mathrm{rxn}}\underline{H}^{\circ}$ for various values of T are given in the following table.

T (K)	200	300	400	500	600
$\Delta_{\mathrm{rxn}}\underline{H}^{\circ}$ (kJ/mol N_2O_4)	57.423	57.192	56.538	55.580	54.448

COMMENT

The heat of reaction can be, and in fact usually is, a much stronger function of temperature than is the case here. ∎

Third law reference state

In some databases—for example, in the very extensive NIST Chemistry Web-Book[14]—the data reported for each substance are the the standard state heat of formation $\Delta_{\mathrm{f}}H^{\circ}$ and the absolute entropy $\underline{S}^{\circ}$, both at 25°C. Here by absolute entropy is meant entropy based on the third law of thermodynamics as defined in Sec. 6.8. The reason for reporting these two quantities is that they are determined directly by thermal or calorimetric measurements, unlike the Gibbs energy of formation, which is obtained by measuring chemical equilibrium constants.

If the standard state heat of formation and the absolute entropy of each substance are known, the Gibbs energy of reaction can be computed as follows. First, the heat of reaction is computed using

$$
\Delta_{\mathrm{rxn}}H^{\circ} = \sum_{i=1}^{\mathcal{C}} \nu_i \Delta_{\mathrm{f}}\underline{H}_i^{\circ}
$$

and then the entropy change for the reaction is computed from

$$
\Delta_{\mathrm{rxn}}S^{\circ} = \sum_{i=1}^{\mathcal{C}} \nu_i \underline{S}^{\circ}
$$

The standard-state Gibbs energy change on reaction at $T = 25°C$ can then be computed from

$$
\Delta_{\mathrm{rxn}}G^{\circ} = \Delta_{\mathrm{rxn}}H^{\circ} - 298.15 \cdot \Delta_{\mathrm{rxn}}S^{\circ} = \sum_{i=1}^{\mathcal{C}} \nu_i \left[\Delta_{\mathrm{f}}\underline{H}_i^{\circ} - 298.15 \cdot \underline{S}_i^{\circ} \right]
$$

[14]http://webbook.nist.gov/chemistry/

Then using the heat capacities reported in the NIST Chemistry WebBook and Eq. 8.5-5, the Gibbs free energy of reaction at any other temperature can be obtained.

As an example of the use of data in this form, we return to the gas-phase reaction of hydrogen and oxygen to form water considered at the beginning of this section. Using the NIST Chemistry WebBook, we obtain the following data.

Species	$\Delta_f \underline{H}^\circ \; \dfrac{kJ}{mol}$	$\underline{S}^\circ \; \dfrac{J}{mol\,K}$
H_2	0	130.680
O_2	0	205.150
H_2O	-241.826	188.835

This results in $\Delta_{rxn}H = -241.826$ kJ/mol, $\Delta_{rxn}S = 44.43$ J/(molK), and $\Delta_{rxn}G = -228.579$ kJ/mol. This last value agrees with the Gibbs energy of formation for water as a vapor in Appendix A.IV.

8.6   THE EXPERIMENTAL DETERMINATION OF THE PARTIAL MOLAR VOLUME AND ENTHALPY

Experimental values for some of the partial molar quantities can be obtained from laboratory measurements on mixtures. In particular, mixture density measurements can be used to obtain partial molar volumes, and heat-of-mixing data yield information on partial molar enthalpies. Both of these measurements are considered here. In Chapter 10 phase equilibrium measurements that provide information on the partial molar Gibbs energy of a component in a mixture are discussed. Once the partial molar enthalpy and partial molar Gibbs energy are known at the same temperature, the partial molar entropy can be computed from the relation $\bar{S}_i = (\bar{G}_i - \bar{H}_i)/T$.

Table 8.6-1 contains data on the mixture density, at constant temperature and pressure, for the water(1)–methanol(2) system. Column 4 of this table contains calculated values for the molar volume change on mixing (i.e., $\Delta_{mix}\underline{V} = \underline{V} - x_1\underline{V}_1 - x_2\underline{V}_2$) at various compositions; these data have also been plotted in Fig. 8.6-1. The slope of the

Table 8.6-1 Density Data for the Water(1)–Methanol(2) System at $T = 298.15$ K

x_1	ρ (kg/m^3)	$\underline{V}$ (m^3/mol) $\times 10^6$ *	$\Delta_{mix}\underline{V}$ (m^3/mol) $\times 10^6$
0.0	786.846	40.7221	0
0.1162	806.655	37.7015	-0.3883
0.2221	825.959	35.0219	-0.6688
0.2841	837.504	33.5007	-0.7855
0.3729	855.031	31.3572	-0.9174
0.4186	864.245	30.2812	-0.9581
0.5266	887.222	27.7895	-1.0032
0.6119	905.376	25.9108	-0.9496
0.7220	929.537	23.5759	-0.7904
0.8509	957.522	20.9986	-0.4476
0.9489	981.906	19.0772	-0.1490
1.0	997.047	18.0686	0

*Note that the notation $\underline{V}$ (m^3/mol) $\times 10^6$ means that the entries in the table have been multiplied by the factor 10^6. Therefore, for example, the volume of pure methanol $\underline{V}(x_1 = 0)$ is 40.7221×10^{-6} m^3/mol = 40.7221 cm^3/mol.

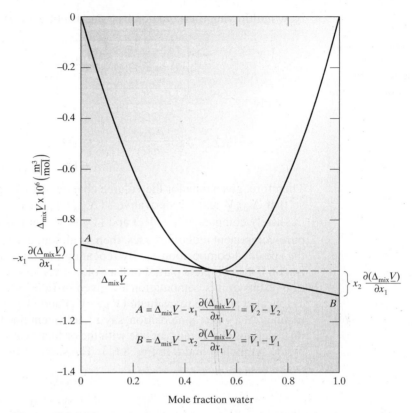

Figure 8.6-1 The isothermal volume change on mixing for the water(1)–methanol(2) system.

$\Delta_{mix}\underline{V}$ versus mole fraction curve at any mole fraction is related to the partial molar volumes. To obtain this relationship we start from

$$\Delta_{mix}\underline{V} = (x_1\overline{V}_1 + x_2\overline{V}_2) - x_1\underline{V}_1 - x_2\underline{V}_2 = x_1(\overline{V}_1 - \underline{V}_1) + x_2(\overline{V}_2 - \underline{V}_2) \quad \textbf{(8.6-1)}$$

so that

$$\frac{\partial(\Delta_{mix}\underline{V})}{\partial x_1}\bigg|_{T,P} = (\overline{V}_1 - \underline{V}_1) + x_1\left(\frac{\partial \overline{V}_1}{\partial x_1}\right)_{T,P} - (\overline{V}_2 - \underline{V}_2) + x_2\left(\frac{\partial \overline{V}_2}{\partial x_1}\right)_{T,P} \quad \textbf{(8.6-2)}$$

Here we have used the facts that the pure-component molar volumes are independent of mixture composition (i.e., $(\partial \underline{V}_i/\partial x_1)_{T,P} = 0$), and that for a binary mixture $x_2 = 1 - x_1$, so $(\partial x_2/\partial x_1) = -1$.

From the Gibbs-Duhem equation, Eq. 8.2-19b with $\bar{\theta}_i = \overline{V}_i$, we have

$$x_1\left(\frac{\partial \overline{V}_1}{\partial x_1}\right)_{T,P} + x_2\left(\frac{\partial \overline{V}_2}{\partial x_1}\right)_{T,P} = 0$$

so that Eq. 8.6-2 becomes

$$\frac{\partial(\Delta_{mix}\underline{V})}{\partial x_1}\bigg|_{T,P} = (\overline{V}_1 - \underline{V}_1) - (\overline{V}_2 - \underline{V}_2) \quad \textbf{(8.6-3)}$$

Now multiplying this equation by x_1 and subtracting the result from Eq. 8.6-1 gives

Equation for the calculation of the partial molar volume

$$\boxed{\Delta_{\mathrm{mix}}\underline{V} - x_1 \frac{\partial(\Delta_{\mathrm{mix}}\underline{V})}{\partial x_1}\bigg|_{T,P} = (\bar{V}_2 - \underline{V}_2)} \tag{8.6-4a}$$

Similarly,

$$\Delta_{\mathrm{mix}}\underline{V} - x_2 \frac{\partial(\Delta_{\mathrm{mix}}\underline{V})}{\partial x_2}\bigg|_{T,P} = \Delta\underline{V}_{\mathrm{mix}} + x_2 \frac{\partial(\Delta_{\mathrm{mix}}\underline{V})}{\partial x_1}\bigg|_{T,P} \tag{8.6-4b}$$
$$= (\bar{V}_1 - \underline{V}_1)$$

Therefore, given data for the volume change on mixing as a function of concentration, so that $\Delta_{\mathrm{mix}}\underline{V}$ and the derivative $\partial(\Delta_{\mathrm{mix}}\underline{V})/\partial x_1$ can be evaluated at x_1, we can immediately compute $(\bar{V}_1 - \underline{V}_1)$ and $(\bar{V}_2 - \underline{V}_2)$ at this composition. Knowledge of the pure-component molar volumes, then, is all that is necessary to compute $\bar{V}_1$ and $\bar{V}_2$ at the specified composition x_1. By repeating the calculation at other values of the mole fraction, the complete partial molar volume versus composition curve can be obtained. The results of this computation are given in Table 8.6-2.

It is also possible to evaluate $(\bar{V}_1 - \underline{V}_1)$ and $(\bar{V}_2 - \underline{V}_2)$ in a more direct, graphical manner. At a given composition, say x_1^*, a tangent line to the $\Delta_{\mathrm{mix}}\underline{V}$ curve is drawn; the intersections of this tangent line with the ordinates at $x_1 = 0$ and $x_1 = 1$ are designated by the symbols A and B in Fig. 8.6-1. The slope of the this tangent line is

$$\frac{\partial(\Delta_{\mathrm{mix}}\underline{V})}{\partial x_1}\bigg|_{x_1^*}$$

so that

$$-x_1^* \frac{\partial(\Delta_{\mathrm{mix}}\underline{V})}{\partial x_1}\bigg|_{x_1^*}$$

is the distance indicated in Fig. 8.6-1. Referring to the figure, it is evident that the numerical value on the ordinate at point A is equal to the left side of Eq. 8.6-4a and,

Table 8.6-2 The Partial Molar Volumes of the Water(1)–Methanol(2) System at $T = 298.15$ K

x_1	$\bar{V}_1 - \underline{V}_1$ $(\mathrm{m}^3/\mathrm{mol}) \times 10^6$	$\bar{V}_1$ $(\mathrm{m}^3/\mathrm{mol}) \times 10^6$	$\bar{V}_2 - \underline{V}_2$ $(\mathrm{m}^3/\mathrm{mol}) \times 10^6$	$\bar{V}_2$ $(\mathrm{m}^3/\mathrm{mol}) \times 10^6$
0	−3.8893	14.180*	0	40.722†
0.1162	−2.9741	15.095	−0.0530	40.669
0.2221	−2.3833	15.686	−0.1727	40.549
0.2841	−2.0751	15.994	−0.2773	40.445
0.3729	−1.6452	16.424	−0.4884	40.234
0.4186	−1.4260	16.643	−0.6321	40.090
0.5266	−0.9260	17.143	−1.0822	39.640
0.6119	−0.5752	17.494	−1.5464	39.176
0.7220	−0.2294	17.840	−2.2363	38.486
0.8509	−0.0254	18.044	−2.9631	37.759
0.9489	−0.0026	18.072	−3.1689	37.553
1.0	0	18.069†	−3.0348	37.687*

*Value of partial molar volume at infinite dilution.
†Value of pure-component molar volume.

therefore, equal to the value of $(\bar{V}_2 - \underline{V}_2)$ at x_1^*. Similarly, the intersection of the tangent line with the ordinate at $x_1 = 1$ (point B) gives the value of $(\bar{V}_1 - \underline{V}_1)$ at x_1^*. Thus, both $(\bar{V}_1 - \underline{V}_1)$ and $(\bar{V}_2 - \underline{V}_2)$ are obtained by a simple graphical construction.

For more accurate calculations of the partial molar volume (or any other partial molar property), an analytical, rather than graphical, procedure is used. First, one fits the volume change on mixing, $\Delta_{mix}\underline{V}$, with a polynomial in mole fraction, and then the necessary derivative is found analytically. Since $\Delta_{mix}\underline{V}$ must equal zero at $x_1 = 0$ and $x_1 = 1$ ($x_2 = 0$), it is usually fit with a polynomial of the **Redlich-Kister** form:

Redlich-Kister expansion (can be used for the change on mixing of any molar property)

$$\Delta_{mix}\underline{V} = x_1 x_2 \sum_{i=0}^{n} a_i (x_1 - x_2)^i \qquad \text{(8.6-5a)}$$

(Similar expansions are also used for $\Delta_{mix}\underline{H}$, $\Delta_{mix}\underline{U}$, and the other excess properties to be defined in Chapter 9.) Then, rewriting Eq. 8.6-5a, we have

$$\Delta_{mix}\underline{V} = x_1(1 - x_1) \sum_{i=0}^{n} a_i (2x_1 - 1)^i$$

$$\left.\frac{\partial \Delta_{mix}\underline{V}}{\partial x_1}\right|_{T,P} = -\sum_{i=0}^{n} a_i (2x_1 - 1)^{i+1} + 2x_1(1 - x_1) \sum_{i=0}^{n} a_i i (2x_1 - 1)^{i-1}$$

$$\text{(8.6-5b)}$$

and

$$\Delta_{mix}\underline{V} - x_1 \left.\frac{\partial(\Delta_{mix}\underline{V})}{\partial x_1}\right|_{T,P} = \bar{V}_2 - \underline{V}_2$$

$$= x_1^2 \sum_i a_i[(x_1 - x_2)^i - 2ix_2(x_1 - x_2)^{i-1}] \qquad \text{(8.6-6a)}$$

Also

$$\bar{V}_1 - \underline{V}_1 = \Delta_{mix}\underline{V} + x_2 \left.\frac{\partial(\Delta_{mix}\underline{V})}{\partial x_1}\right|_{T,P}$$

$$= x_2^2 \sum_i a_i[(x_1 - x_2)^i + 2ix_1(x_1 - x_2)^{i-1}] \qquad \text{(8.6-6b)}$$

An accurate representation of the water-methanol data has been obtained using Eq. 8.6-5 with (in units of m^3/mol)

$$a_0 = -4.0034 \times 10^{-6}$$
$$a_1 = -0.177\,56 \times 10^{-6}$$
$$a_2 = 0.541\,39 \times 10^{-6}$$
$$a_3 = 0.604\,81 \times 10^{-6}$$

and the partial molar volumes in Table 8.6-2 have been computed using these constants and Eqs. 8.6-6.

Finally, we note that for the water-methanol system the volume change on mixing was negative, as were $(\bar{V}_1 - \underline{V}_1)$ and $(\bar{V}_2 - \underline{V}_2)$. This is not a general characteristic in

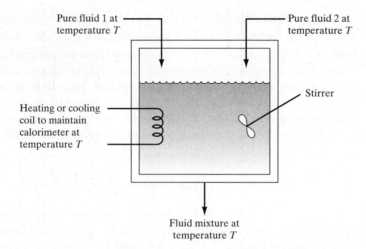

Pure fluid 1 at temperature T

Pure fluid 2 at temperature T

Stirrer

Heating or cooling coil to maintain calorimeter at temperature T

Fluid mixture at temperature T

Figure 8.6-2 An isothermal flow calorimeter.

that, depending on the system, these three quantities can be positive, negative, or even positive over part of the composition range and negative over the rest.

The partial molar enthalpy of a species in a binary mixture can be obtained by a similar analysis, but using enthalpy change on mixing (or heat of mixing) data. Such measurements are frequently made using the steady-state flow calorimeter schematically indicated in Fig. 8.6-2. Two streams, one of pure fluid 1 and the second of pure fluid 2, both at a temperature T and a pressure P, enter this steady-state mixing device, and a single mixed stream, also at T and P, leaves. Heat is added or removed to maintain the temperature of the outlet stream. Taking the contents of the calorimeter to be the system, the mass and energy balances (Eqs. 8.4-1 and 8.4-3) are

$$0 = \dot{N}_1 + \dot{N}_2 + \dot{N}_3 \qquad (8.6\text{-}7)$$

and

$$0 = \dot{N}_1 \underline{H}_1 + \dot{N}_2 \underline{H}_2 + \dot{N}_3 \underline{H}_{\text{mix}} + \dot{Q} \qquad (8.6\text{-}8)$$

Thus

$$\dot{Q} = [\dot{N}_1 + \dot{N}_2]\underline{H}_{\text{mix}} - \dot{N}_1 \underline{H}_1 - \dot{N}_2 \underline{H}_2 = [\dot{N}_1 + \dot{N}_2]\Delta_{\text{mix}}\underline{H}$$

and

$$\Delta_{\text{mix}}\underline{H} = \dot{Q}/[\dot{N}_1 + \dot{N}_2]$$

where $\Delta_{\text{mix}}\underline{H} = \underline{H}_{\text{mix}} - x_1\underline{H}_1 - x_2\underline{H}_2$. Therefore, by monitoring $\dot{N}_1$ and $\dot{N}_2$ as well as the heat flow rate $\dot{Q}$ necessary to maintain constant temperature, the heat of mixing $\Delta_{\text{mix}}\underline{H} = \dot{Q}/[\dot{N}_1 + \dot{N}_2]$ can be determined at the composition $x_1 = \dot{N}_1/[\dot{N}_1 + \dot{N}_2]$. Measurements at a collection of values of the ratio $\dot{N}_1/\dot{N}_2$ give the complete heat of mixing versus composition curve at fixed T and P.

Once the composition dependence of the heat of mixing is known, $\overline{H}_1$ and $\overline{H}_2$ may be computed in a manner completely analogous to the procedure used for the partial molar volumes. In particular, it is easily established that

$$\Delta_{\text{mix}}\underline{H} - x_1 \left.\frac{\partial(\Delta_{\text{mix}}\underline{H})}{\partial x_1}\right|_{T,P} = (\overline{H}_2 - \underline{H}_2) \qquad (8.6\text{-}9\text{a})$$

Table 8.6-3 Heat of Mixing Data for the Water(1)–Methanol(2)
System at $T = 19.69°C$

x_1	Q^+ (kJ/mol MeOH)	Q (kJ/mol) $= \Delta_{mix}\underline{H}$
0.05	−0.134	−0.127
0.10	−0.272	−0.245
0.15	−0.419	−0.356
0.20	−0.569	−0.455
0.25	−0.716	−0.537
0.30	−0.862	−0.603
0.35	−1.017	−0.661
0.40	−1.197	−0.718
0.45	−1.398	−0.769
0.50	−1.632	−0.816
0.55	−1.896	−0.853
0.60	−2.218	−0.887
0.65	−2.591	−0.907
0.70	−3.055	−0.917
0.75	−3.666	−0.917
0.80	−4.357	−0.871
0.85	−5.114	−0.767
0.90	−5.989	−0.599
0.95	−6.838	−0.342

Source: International Critical Tables, Vol. 5, Elsevier, 1929, p. 159.
$Q = (1 - x_1)Q^+$.

and

$$\Delta_{mix}\underline{H} - x_2 \frac{\partial(\Delta_{mix}\underline{H})}{\partial x_2}\bigg|_{T,P} = \Delta_{mix}\underline{H} + x_2 \frac{\partial(\Delta_{mix}\underline{H})}{\partial x_1}\bigg|_{T,P} \qquad \textbf{(8.6-9b)}$$
$$= (\overline{H}_1 - \underline{H}_1)$$

so that either the computational or graphical technique may also be used to calculate the partial molar enthalpy.

Table 8.6-3 and Fig. 8.6-3 contain the heat of mixing data for the water–methanol system. These data have been used to compute the partial molar enthalpies given in

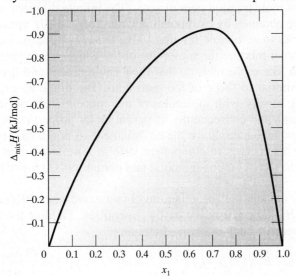

Figure 8.6-3 Heat of mixing data for the water(1)–methanol(2) system at $T = 19.69°C$

Table 8.6-4 The Difference between the Partial Molar and
Pure Component Enthalpies for the
Water(1)–Methanol(2) System at $T = 19.69°$ C

x_1	$\overline{H}_1 - \underline{H}_1$ (kJ/mol)	$\overline{H}_2 - \underline{H}_2$ (kJ/mol)
0	−2.703*	0
0.05	−2.482	−0.006
0.10	−2.251	−0.025
0.15	−2.032	−0.056
0.20	−1.838	−0.097
0.25	−1.678	−0.143
0.30	−1.551	−0.191
0.35	−1.456	−0.237
0.40	−1.383	−0.280
0.45	−1.325	−0.323
0.50	−1.270	−0.373
0.55	−1.209	−0.441
0.60	−1.131	−0.548
0.65	−1.028	−0.719
0.70	−0.898	−0.992
0.75	−0.740	−1.412
0.80	−0.560	−2.036
0.85	−0.371	−2.935
0.90	−0.193	−4.192
0.95	−0.056	−5.905
1.00	0	−8.188*

*Indicates value at infinite dilution.

Table 8.6-4. Note that one feature of the heat of mixing data of Fig. 8.6-3 is that it is skewed, with the largest absolute value at $x_1 = 0.73$ (and not $x_1 = 0.5$).

The entries in Tables 8.6-2 and 8.6-4 are interesting in that they show that the partial molar volume and partial molar enthalpy of a species in a mixture are very similar to the pure component molar quantities when the mole fraction of that species is near unity and are most different from the pure component values at infinite dilution, that is, as the species mole fraction goes to zero. (The infinite dilution values in Tables 8.6-2 and 8.6-4 were obtained by extrapolating both the $\overline{V}$ and $\overline{H}$ versus mole fraction data for each species to zero mole fraction.) This behavior is reasonable because in a strongly nonequimolar mixture the molecules of the concentrated species are interacting most often with like molecules, so that their environment and thus their molar properties are very similar to those of the pure fluid. The dilute species, on the other hand, is interacting mostly with molecules of the concentrated species, so that its molecular environment, and consequently its partial molar properties, will be unlike those of its pure component state. Since the environment around a molecule in a mixture is most dissimilar from its pure component state at infinite dilution, the greatest difference between the pure component molar and partial molar properties usually occurs in this limit.

Finally, the analyses used here to obtain expressions relating $\overline{V}_i$ and $\overline{H}_i$ to $\Delta_{mix}V$ and $\Delta_{mix}H$, respectively, are easily generalized, yielding the following for the partial molar property of any extensive function θ:

General equation relating the partial molar property to the pure component property and the property change on mixing

$$\bar{\theta}_1(T, P, \underline{x}) - \underline{\theta}_1(T, P) = \Delta_{mix}\underline{\theta}(T, P, \underline{x}) - x_2 \frac{\partial(\Delta_{mix}\underline{\theta})}{\partial x_2}\bigg|_{T,P} \quad \textbf{(8.6-10a)}$$

and

$$\bar{\theta}_2(T, P, \underline{x}) - \underline{\theta}_2(T, P) = \Delta_{\text{mix}}\underline{\theta}(T, P, \underline{x}) - x_1 \left.\frac{\partial(\Delta_{\text{mix}}\underline{\theta})}{\partial x_1}\right|_{T,P} \tag{8.6-10b}$$

One can also show that if $\underline{\theta}_{\text{mix}}$ is any molar property of the mixture (not the change on mixing, which is $\Delta_{\text{mix}}\underline{\theta}$), we have

$$\boxed{\bar{\theta}_1(T, P, \underline{x}) = \underline{\theta}_{\text{mix}}(T, P, \underline{x}) - x_2 \left.\frac{\partial(\underline{\theta}_{\text{mix}})}{\partial x_2}\right|_{T,P}} \tag{8.6-11a}$$

and

$$\boxed{\bar{\theta}_2(T, P, \underline{x}) = \underline{\theta}_{\text{mix}}(T, P, \underline{x}) - x_1 \left.\frac{\partial(\underline{\theta}_{\text{mix}})}{\partial x_1}\right|_{T,P}} \tag{8.6-11b}$$

where, for a binary mixture, $\underline{x}$ is used to represent one pair of mole fractions x_1 and x_2.

ILLUSTRATION 8.6-1

Calculations of Partial Molar Enthalpies from Experimental Data

Using the data in Fig. 8.1-1, determine the partial molar enthalpy of sulfuric acid and water at 50 mol % sulfuric acid and 65.6°C.

SOLUTION

First we must obtain values of enthalpy versus concentration at 65.6°C. The values read from this figure and converted to a molar basis are given below.

wt % H_2SO_4	$\hat{H} \left(\dfrac{\text{kJ}}{\text{kg}}\right)$	$\Delta_{\text{mix}}\hat{H} \left(\dfrac{\text{kJ}}{\text{kg}}\right)$	$x_{H_2SO_4}$	$\dfrac{\text{mol}}{\text{kg}}$	$\Delta_{\text{mix}}\underline{H} \left(\dfrac{\text{kJ}}{\text{mol}}\right)$
0	+278	0	0		0
20	+85	−155.8	0.0439	46.45	−3.354
40	−78	−281.6	0.1091	37.38	−7.533
60	−175	−341.4	0.2160	28.32	−12.055
80	−153	−282.2	0.4235	19.26	−14.652
90	−60	−170.6	0.6231	14.73	−11.582
100	92	0	1.000		0

These data are fit reasonably well with the simple expression

$$\begin{aligned}\Delta_{\text{mix}}\underline{H} \left(\frac{\text{kJ}}{\text{mol}}\right) &= x_{H_2SO_4}x_{H_2O}(-82.795 + 56.683x_{H_2SO_4}) \\ &= x_{H_2SO_4}(1 - x_{H_2SO_4})(-82.795 + 56.683x_{H_2SO_4}) \\ &= -82.795x_{H_2SO_4} + 139.478x_{H_2SO_4}^2 - 56.683x_{H_2SO_4}^3\end{aligned}$$

and

$$\frac{d\Delta_{\text{mix}}\underline{H}}{dx_{H_2SO_4}} = -82.795 + 278.965x_{H_2SO_4} - 170.049x_{H_2SO_4}^2$$

Therefore,

$$\left. \frac{d\,\Delta_{\text{mix}}\underline{H}}{dx_{\text{H}_2\text{SO}_4}} \right|_{x_{\text{H}_2\text{SO}_4}=0.5} = 14.17\ \frac{\text{kJ}}{\text{mol}}$$

and

$$\left. \frac{d\,\Delta_{\text{mix}}\underline{H}}{dx_{\text{H}_2\text{O}}} \right|_{x_{\text{H}_2\text{SO}_4}=0.5} = -\left. \frac{d\,\Delta_{\text{mix}}\underline{H}}{dx_{\text{H}_2\text{SO}_4}} \right|_{x_{\text{H}_2\text{SO}_4}=0.5} = -14.17\ \frac{\text{kJ}}{\text{mol}}$$

Also,

$$\Delta_{\text{mix}}\underline{H}\,(x_{\text{H}_2\text{SO}_4}=0.5) = -13.61\ \frac{\text{kJ}}{\text{mol}}$$

$$\underline{H}_{\text{H}_2\text{SO}_4} = 92\ \frac{\text{kJ}}{\text{kg}} \times \frac{1\ \text{kg}}{1000\ \text{g}} \times \frac{98.708\ \text{g}}{\text{mol}} = 9.02\ \frac{\text{kJ}}{\text{mol}}$$

and

$$\underline{H}_{\text{H}_2\text{O}} = 278\ \frac{\text{kJ}}{\text{kg}} \times \frac{1\ \text{kg}}{1000\ \text{g}} \times \frac{18.015\ \text{g}}{\text{mol}} = 5.01\ \frac{\text{kJ}}{\text{mol}}$$

Finally, from Eq. 8.6-9b, we have

$$\overline{H}_{\text{H}_2\text{SO}_4}(x_{\text{H}_2\text{SO}_4} = 0.5,\, T = 65.6°\text{C})$$

$$= \underline{H}_{\text{H}_2\text{SO}_4}(T = 65.6°\text{C}) + \Delta_{\text{mix}}\underline{H}\,(x_{\text{H}_2\text{SO}_4} = 0.5) + x_{\text{H}_2\text{O}} \left. \frac{\partial\,\Delta_{\text{mix}}\underline{H}}{\partial x_{\text{H}_2\text{SO}_4}} \right|_{x_{\text{H}_2\text{SO}_4}=0.5}$$

$$= 9.02 - 13.61 + 0.5(-14.17) = -11.68\ \frac{\text{kJ}}{\text{mol}}$$

and

$$\overline{H}_{\text{H}_2\text{O}}(x_{\text{H}_2\text{SO}_4} = 0.5,\, T = 65.6°\text{C})$$

$$= \underline{H}_{\text{H}_2\text{O}}(T = 65.6°\text{C}) + \Delta_{\text{mix}}\underline{H}\,(x_{\text{H}_2\text{SO}_4} = 0.5) + x_{\text{H}_2\text{SO}_4} \left. \frac{\partial\,\Delta_{\text{mix}}\underline{H}}{\partial x_{\text{H}_2\text{O}}} \right|_{x_{\text{H}_2\text{SO}_4}=0.5}$$

$$= 5.01 - 13.61 + 0.5(14.17) = -1.52\ \frac{\text{kJ}}{\text{mol}}$$

Note that in this case the pure component and partial molar enthalpies differ considerably. Consequently, we say that this solution is quite nonideal, where, as we shall see in Chapter 9, an ideal solution is one in which some partial molar properties (in particular the enthalpy, internal energy, and volume) are equal to the pure component values. Further, here the solution is so nonideal that at the temperature chosen the pure component and partial molar enthalpies are even of different signs for both water and sulfuric acid. For later reference we note that, at $x_{\text{H}_2\text{SO}_4} = 0.5$, we have

$$\overline{H}_{\text{H}_2\text{SO}_4} - \underline{H}_{\text{H}_2\text{SO}_4} = -20.7\ \frac{\text{kJ}}{\text{mol}} \quad \text{and} \quad \overline{H}_{\text{H}_2\text{O}} - \underline{H}_{\text{H}_2\text{O}} = -6.5\ \frac{\text{kJ}}{\text{mol}} \qquad \blacksquare$$

Generally, any partial molar property differs most from the pure component property in the limit of the component being in high dilution, or at infinite dilution. Therefore, except for components that associate to form dimers, for example, the largest differences between the partial molar and pure component molar properties are

$$\bar{\theta}_1(T, P, x_1 \to 0) - \underline{\theta}_1(T, P) = -\left.\frac{\partial(\Delta_{\text{mix}}\underline{\theta})}{\partial x_2}\right|_{x_2=1} = +\left.\frac{\partial(\Delta_{\text{mix}}\underline{\theta})}{\partial x_1}\right|_{x_1=0}$$

(8.6-12a)

and

$$\bar{\theta}_2(T, P, x_2 \to 0) - \underline{\theta}_2(T, P) = -\left.\frac{\partial(\Delta_{\text{mix}}\underline{\theta})}{\partial x_1}\right|_{x_1=1} = +\left.\frac{\partial(\Delta_{\text{mix}}\underline{\theta})}{\partial x_2}\right|_{x_2=0}$$

(8.6-12b)

ILLUSTRATION 8.6-2
Calculation of Infinite Dilution Partial Molar Enthalpies from Experimental Data

Compute the difference between the infinite dilution partial molar enthalpy and the pure component molar enthalpy for sulfuric acid and water at 65.6°C using the information in the previous illustration.

SOLUTION

From the previous illustration

$$\left(\frac{\partial \Delta_{\text{mix}}\underline{H}}{\partial x_{\text{H}_2\text{SO}_4}}\right)_{T,P} = -82.795 + 278.965 x_{\text{H}_2\text{SO}_4} - 170.049 x_{\text{H}_2\text{SO}_4}^2$$

Therefore,

$$\left(\frac{\partial \Delta_{\text{mix}}\underline{H}}{\partial x_{\text{H}_2\text{SO}_4}}\right)_{x_{\text{H}_2\text{SO}_4}=1} = +26.11 \frac{\text{kJ}}{\text{mol}} \quad \text{and} \quad \left(\frac{\partial \Delta_{\text{mix}}\underline{H}}{\partial x_{\text{H}_2\text{SO}_4}}\right)_{x_{\text{H}_2\text{SO}_4}=0} = -82.80 \frac{\text{kJ}}{\text{mol}}$$

so that

$$\bar{H}_{\text{H}_2\text{SO}_4}(T = 65.6°\text{C}, x_{\text{H}_2\text{SO}_4} = 0) - \underline{H}_{\text{H}_2\text{SO}_4}(T = 65.6°\text{C}) = -82.80 \frac{\text{kJ}}{\text{mol}}$$

in which case

$$\bar{H}_{\text{H}_2\text{SO}_4}(T = 65.6°\text{C}, x_{\text{H}_2\text{SO}_4} = 0) = 9.02 - 82.80 = -73.80 \frac{\text{kJ}}{\text{mol}}$$

$$\bar{H}_{\text{H}_2\text{O}}(T = 65.6°\text{C}, x_{\text{H}_2\text{O}} = 0) - \underline{H}_{\text{H}_2\text{O}}(T = 65.6°\text{C}) = -26.11 \frac{\text{kJ}}{\text{mol}}$$

and

$$\bar{H}_{\text{H}_2\text{O}}(T = 65.6°\text{C}, x_{\text{H}_2\text{O}} = 0) = 5.01 - 26.11 = -21.1 \frac{\text{kJ}}{\text{mol}}$$

Note that for the sulfuric acid + water system at $T = 65.6°$C the differences between the pure component and partial molar properties at infinite dilution are considerably greater than at the mole fraction of 0.5 in the previous illustration. ∎

There is a simple physical interpretation for partial molar properties at infinite dilution. In general, we have from Eq. 8.1-13 for any total property $\theta(T, P, \underline{x})$ in a binary mixture that

$$\theta(T, P, \underline{x}) = N_1 \bar{\theta}_1(T, P, \underline{x}) + N_2 \bar{\theta}_2(T, P, \underline{x}) \qquad \textbf{(8.6-13)}$$

Now consider the case when $N_2 = 1$ and $N_1 \gg N_2$ so that $x_1 \sim 1$, in which case $\bar{\theta}_1(T, P, x_1 \sim 1) \cong \underline{\theta}_1(T, P)$, since species 1 is essentially at the pure component limit. Also, $\bar{\theta}_2(T, P, x_1 \sim 1)$ is the partial molar property of species 2 at infinite dilution, so that at in this limit

$$\theta(T, P, x_1 \sim 1) = N_1 \underline{\theta}(T, P) + \bar{\theta}_2(T, P, x_2 \sim 0) \qquad \textbf{(8.6-14)}$$

From this equation we see that the infinite dilution partial molar property $\bar{\theta}_2(T, P, x_2 \sim 0)$ is the amount by which the total property θ changes as a result of the addition of one mole of species 2 to an infinitely large amount of species 1 (so that x_2 remains about zero). Note that if the solution were ideal, the total property θ would change by an amount equal to the pure component molar property $\underline{\theta}_2$; however, since most solutions are nonideal, the change is instead equal to $\bar{\theta}_2$.

ILLUSTRATION 8.6-3

Calculation of the Isothermal Enthalpy Change of Mixing

One mole of sulfuric acid at 65.6°C is added to 1000 moles of water at the same temperature. If the mixing is done isothermally, estimate the change in enthalpy of the mixture.

SOLUTION

From Eq. 8.6-14,

$$H(T = 65.6°C, N_{H_2O} = 1000, N_{H_2SO_4} = 1) - H(T = 65.6°C, N_{H_2O} = 1000, N_{H_2SO_4} = 0)$$

$$= \bar{H}_{H_2SO_4}(T = 65.6°C, x_{H_2SO_4} = 0.001 \sim 0) = -73.80 \frac{kJ}{mol}$$

$$= \Delta_{mix}H(1000 \text{ mol } H_2O + 1 \text{ mol } H_2SO_4)$$

[The numerical value for $\bar{H}_{H_2SO_4}(T = 65.6°C, x_{H_2SO_4} \sim 0)$ was obtained from the previous illustration.] ∎

8.7 CRITERIA FOR PHASE EQUILIBRIUM IN MULTICOMPONENT SYSTEMS

An important observation in this chapter is that the equations of change for a multicomponent mixture are identical, in form, to those for a pure fluid. The difference between the two is that the pure fluid equations contain thermodynamic properties ($\underline{U}$, $\underline{H}$, $\underline{S}$, etc.) that can be computed from pure fluid equations of state and heat capacity data, whereas in the multicomponent case these thermodynamic properties can be computed only if the appropriate mixture equation of state and heat capacity data or enthalpy-concentration and entropy-concentration data are given, or if we otherwise have enough information to evaluate the necessary concentration-dependent partial molar quantities at all temperatures, pressures, and compositions of interest. Although this represents an important computational difference between the pure fluid and mixture equations, it has no effect on their fundamental structure. Consequently, for a closed system, we have for both the pure component and multicomponent cases

$$\frac{dU}{dt} = \dot{Q} + \dot{W}_S - P\frac{dV}{dt} \qquad (8.7\text{-}1)$$

and

$$\frac{dS}{dt} = \frac{\dot{Q}}{T} + \dot{S}_{gen} \qquad (8.7\text{-}2)$$

where

$$\begin{aligned} U &= N\underline{U}(T, P) \\ S &= N\underline{S}(T, P) \end{aligned} \quad \left\{ \begin{array}{l} \text{for the pure component} \\ \text{system (molar basis)} \end{array} \right.$$

and

$$\begin{aligned} U &= \sum_{i=1}^{c} N_i \overline{U}_i(T, P, \underline{x}) \\ S &= \sum_{i=1}^{c} N_i \overline{S}_i(T, P, \underline{x}) \end{aligned} \quad \left\{ \begin{array}{l} \text{for a multicomponent} \\ \text{system (molar basis)} \end{array} \right. \qquad (8.7\text{-}3)$$

Since the form of the balance equations is unchanged, we can use, without modification, the analysis of the last chapter to establish that the equilibrium criteria for a closed multicomponent mixture are (Problem 8.23)

$$\begin{aligned} S &= \text{maximum for equilibrium at constant } M,\ U,\ \text{and } V \\ A &= \text{minimum for equilibrium at constant } M,\ T,\ \text{and } V \qquad (8.7\text{-}4) \\ G &= \text{minimum for equilibrium at constant } M,\ T,\ \text{and } P \end{aligned}$$

Thus, although it may be computationally more difficult to identify the equilibrium state in a multicomponent mixture than is the case for a pure fluid, the basic criteria used in this identification are the same.

As the first application of these criteria, consider the problem of identifying the state of equilibrium in a closed, nonreacting multicomponent system at constant internal energy and volume. To be specific, suppose N_1 moles of species 1, N_2 moles of species 2, and so on are put into an adiabatic container that will be maintained at constant volume, and that these species are only partially soluble in one another, but do not chemically react. What we would like to be able to do is to predict the composition of each of the phases present at equilibrium. (A more difficult but solvable problem is to also predict the number of phases that will be present. This problem is briefly considered in Chapter 11.) In the analysis that follows, we develop the equation that will be used in Chapters 10, 11, and 12 to compute the equilibrium compositions.

The starting point for solving this problem are the general equilibrium criteria of Eqs. 8.7-4. In particular, the equilibrium criterion for a closed, adiabatic, constant-volume system is

$$S = \text{maximum}$$

subject to the constraints of constant U, V, and total number of moles of each species N_i. For the two-phase system, each extensive property (e.g., N_i, S, U, V) is the sum of the properties for the individual phases, for example,

$$N_i = N_i^I + N_i^{II}$$

where the superscripts I and II refer to the phase. In general, the problem of finding the extreme value of a function subject to constraints is not a straightforward task, as will become evident later. However, here this can be done easily. We start by setting the differential of the entropy for the two-phase system equal to zero,

$$dS = dS^I + dS^{II} = 0 \tag{8.7-5}$$

and then use Eq. 8.2-4, rearranged as

$$dS = \frac{1}{T} dU + \frac{P}{T} dV - \frac{1}{T} \sum_{i=1}^{c} \overline{G}_i \, dN_i$$

for each phase. Now recognizing that since the total internal energy, total volume, and the number of moles of each species are fixed, we have

$$dU^{II} = -dU^I$$
$$dV^{II} = -dV^I$$
$$dN_i^{II} = -dN_i^I$$

which can be used in Eq. 8.7-5 to obtain

$$dS = \left(\frac{1}{T^I} - \frac{1}{T^{II}}\right) dU^I + \left(\frac{P^I}{T^I} - \frac{P^{II}}{T^{II}}\right) dV^I - \sum_{i=1}^{c} \left(\frac{\overline{G}_i^I}{T^I} - \frac{\overline{G}_i^{II}}{T^{II}}\right) dN_i^I = 0 \tag{8.7-6}$$

The condition for equilibrium is that the differential of the entropy be zero with respect to all variations of the independent and unconstrained variables, here dU^I, dV^I, and each of the dN_i^I. In order for Eq. 8.7-6 to be satisfied, we must have (1) that

$$\left(\frac{\partial S}{\partial U^I}\right)_{V^I, N_i^I} = 0$$

which implies

First criterion for phase equilibrium

(2) that

$$\frac{1}{T^I} = \frac{1}{T^{II}} \quad \text{or simply} \quad \boxed{T^I = T^{II}} \tag{8.7-7a}$$

$$\left(\frac{\partial S}{\partial V^I}\right)_{U^I, N_i^I} = 0$$

which implies

$$\frac{P^I}{T^I} = \frac{P^{II}}{T^{II}}$$

or, in view of Eq. 8.7-7a,

Second criterion for phase equilibrium

$$\boxed{P^I = P^{II}} \tag{8.7-7b}$$

and (3) that $(\partial S / \partial N_i^I)_{U^I, V^I, N_{j \neq i}^I} = 0$ for each species i, which implies

$$\boxed{\bar{G}_i^I = \bar{G}_i^{II} \quad \text{or} \quad \mu_i^I = \mu_i^{II} \quad \text{for each species i}} \qquad (8.7\text{-}7c)$$

since $T^I = T^{II}$. In Eq. 8.7-7c, $\mu_i = \bar{G}_i$ is the chemical potential of species i.

Thus, for phase equilibrium to exist in a closed, nonreacting multicomponent system at constant energy and volume, the pressure must be the same in both phases (so that mechanical equilibrium exists), the temperature must be the same in both phases (so that thermal equilibrium exists), and the partial molar Gibbs energy of each species must be the same in each phase (so that equilibrium with respect to species diffusion exists).[15]

Note that with the replacement of the partial molar Gibbs free energy by the pure component Gibbs energy, Eqs. 8.7-7 become identical to the conditions for phase equilibrium in a one-component system derived in Sec. 7.1 (see Eqs. 7.1-9). In principle, we could now continue to follow the development of Chapter 7 and derive the conditions for stability of the equilibrium state. However, this task is algebraically complicated and will not be considered here.[16]

To derive the conditions for phase equilibrium in a closed system at constant (and, of course, uniform) temperature and pressure, we start from the equilibrium criterion that G be a minimum and set the differential of G for the two-phase system equal to zero, that is,

$$dG = dG^I + dG^{II} = 0 \qquad (8.7\text{-}8)$$

Now recognizing that at constant T and P (from Eq. 8.2-1)

$$dG|_{T,P} = \sum_{i=1}^{c} \bar{G}_i \, dN_i$$

and that the total number of moles of each species is fixed, so that $N_i = N_i^I + N_i^{II}$ or $dN_i^{II} = -dN_i^I$, we obtain

$$dG = \sum_{i=1}^{c} \bar{G}_i^I \, dN_i^I + \sum_{i=1}^{c} \bar{G}_i^{II} \, dN_i^{II} = \sum_{i=1}^{c} (\bar{G}_i^I - \bar{G}_i^{II}) \, dN_i^I = 0$$

Setting the derivative of the Gibbs energy with respect to each of its independent variables (here the mole numbers N_i^I) equal to zero yields

$$\left(\frac{\partial G}{\partial N_i^I} \right)_{N_{j \neq i}^I} = 0 = \bar{G}_i^I - \bar{G}_i^{II} \quad \text{or} \quad \bar{G}_i^I = \bar{G}_i^{II}$$

so that here, as in Chapter 7, we find that the equality of Gibbs free energies is a necessary condition for the existence of phase equilibrium for systems subject to a variety of constraints (see Problem 8.3).

---

[15]Clearly, from these results, it is a species partial molar Gibbs energy difference between phases, rather than a concentration difference, that is the driving force for interphase mass transfer in the approach to equilibrium.

[16]See, for example, of J. W. Tester and M. Modell, *Thermodynamics and Its Applications*, 3rd ed., Prentice Hall, Englewood Cliffs, N.J. (1997) Chapter 7.

Although we will not do so here, it is easy to show that these analyses for two-phase equilibrium are easily generalized to multiphase equilibrium and yield

$$\bar{G}_i^I = \bar{G}_i^{II} = \bar{G}_i^{III} = \cdots$$

(8.7-9)

8.8 CRITERIA FOR CHEMICAL EQUILIBRIUM, AND COMBINED CHEMICAL AND PHASE EQUILIBRIUM

Equations 8.7-4 also provide a means of identifying the equilibrium state when chemical reactions occur. To see this, consider first the case of a single chemical reaction occurring in a single phase (both of these restrictions will be removed shortly) in a closed system at constant temperature and pressure.[17] The total Gibbs energy for this system, using the reaction variable notation introduced in Sec. 8.3, is

$$G = \sum_{i=1}^{C} N_i \bar{G}_i = \sum_{i=1}^{C} (N_{i,0} + \nu_i X) \bar{G}_i$$

Since the only variation possible in a one-phase, closed system at constant temperature and pressure is in the extent of reaction X, the equilibrium criterion is

$$\left(\frac{\partial G}{\partial X} \right)_{T,P} = 0$$

which yields

Criterion for chemical equilibrium of a single reaction

$$0 = \sum_{i=1}^{C} \nu_i \bar{G}_i$$

(8.8-1)

(Note that $\sum_i^C N_i (\partial \bar{G}_i / \partial X)_{T,P}$ is equal to zero by the Gibbs-Duhem equation, Eq. 8.2-14 with $Y = X$.)

It is possible to show that the criterion for chemical equilibrium developed here is also applicable to systems subject to constraints other than constant temperature and pressure (Problem 8.4). In fact, Eq. 8.8-1, like the phase equilibrium criterion of Eq. 8.7-9, is of general applicability. Of course, the difficulty that arises in using either of these equations is translating their simple form into a useful prescription for equilibrium calculations by relating the partial molar Gibbs energies to quantities of more direct interest, such as temperature, pressure, and mole fractions. This problem will be the focus of much of the rest of this book.

The first generalization of the analysis given here is to the case of multiple chemical reactions in a closed, single-phase, constant-temperature and constant-pressure system. Using the notation of Sec. 8.3, the number of moles of species i present at any time is

$$N_i = N_{i,0} + \sum_{j=1}^{M} \nu_{ij} X_j$$

(8.35)

---

[17]Since chemists and chemical engineers are usually interested in chemical and phase equilibria at constant temperature and pressure, the discussions that follow largely concern equilibrium under these constraints.

where the summation is over the *independent* reactions. The total Gibbs energy of the system is

$$\underline{G} = \sum_{i=1}^{\mathcal{C}} N_i \bar{G}_i = \sum_{i=1}^{\mathcal{C}} \left(N_{i,0} + \sum_{j=1}^{\mathcal{M}} \nu_{ij} X_j \right) \bar{G}_i$$

$$= \sum_{i=1}^{\mathcal{C}} N_{i,0} \bar{G}_i + \sum_{i=1}^{\mathcal{C}} \sum_{j=1}^{\mathcal{M}} \nu_{ij} X_j \bar{G}_i$$

(8.8-2)

The condition for chemical equilibrium in this multireaction system is $G = $ minimum or $dG = 0$ for all variations consistent with the stoichiometry at constant temperature, pressure, and total mass. For the present case this implies

$$\left(\frac{\partial G}{\partial X_j} \right)_{T,P,X_{i \neq j}} = 0 \quad j = 1, 2, \ldots, \mathcal{M}$$

(8.8-3)

so that

$$\left(\frac{\partial G}{\partial X_j} \right)_{T,P,X_{i \neq j}} = 0 = \sum_{i=1}^{\mathcal{C}} \nu_{ij} \bar{G}_i + \sum_{i=1}^{\mathcal{C}} N_i \left(\frac{\partial \bar{G}_i}{\partial X_j} \right)_{T,P,X_{k \neq j}} \qquad \begin{array}{l} \text{for all independent} \\ \text{reactions} \\ j = 1, 2, \ldots, \mathcal{M} \end{array}$$

Since the sum $\sum_{i=1}^{\mathcal{C}} N_i (\partial \bar{G}_i / \partial X_j)_{T,P,X_{k \neq j}}$ vanishes by the Gibbs-Duhem equation, the equilibrium criterion is

Chemical equilibrium criteria for multiple reactions

$$\boxed{\sum_{i=1}^{\mathcal{C}} \nu_{ij} \bar{G}_i = 0 \quad j = 1, 2, \ldots, \mathcal{M}}$$

(8.8-4)

This equation is analogous to Eq. 8.8-1 for the single-reaction case. The interpretation of Eq. 8.8-4 is clear: In a system in which several chemical reactions occur, chemical equilibrium is achieved only when each reaction is itself in equilibrium.

The final case to be considered is that of combined phase and chemical equilibrium in a closed system at constant temperature and pressure. At this point you can probably guess the final result: If both phase changes and chemical transformations are possible, equilibrium occurs only when each possible transformation is itself in equilibrium. Thus, Eqs. 8.7-7 and 8.8-4 must be simultaneously satisfied for all species and all reactions in all phases.

To prove this assertion, it is first useful to consider the mathematical technique of Lagrange multipliers, a method used to find the extreme (maximum or minimum) value of a function subject to constraints. Rather than develop the method in complete generality, we merely introduce it by application to the problem just considered: equilibrium in a single-phase, multiple–chemical reaction system.

We identified the equilibrium state for several chemical reactions occurring in a single-phase system at constant temperature and pressure by finding the state for which $G = \sum_{i=1}^{\mathcal{C}} N_i \bar{G}_i$ was equal to a minimum subject to the stoichiometric constraints

$$N_i = N_{i,0} + \sum_{j=1}^{\mathcal{M}} \nu_{ij} X_j \quad \text{for all species i}$$

The procedure used was to incorporate the constraints directly into the Gibbs energy function and then find the minimum value of the resulting unconstrained equation (Eq. 8.8-2) by setting each of the $\mathcal{M}$ derivatives $(\partial G/\partial X_j)_{T,P,X_{i\neq j}}$ equal to zero. However, this direct substitution technique can be very cumbersome when the constraints are complicated, as is the case in the problem of combined chemical and phase equilibrium.

An alternative method of obtaining a solution to the multiple-reaction, single-phase equilibrium problem is to use the method of Lagrange multipliers.[18] Here one first rewrites the constraints as

$$N_i - N_{i,0} - \sum_{j=1}^{\mathcal{M}} \nu_{ij} X_j = 0 \quad i = 1, 2, \ldots, \mathcal{C} \tag{8.8-5}$$

and then creates a new function $\mathcal{G}$ by adding the constraints, each with a multiplying parameter α_i, a Lagrange multiplier, to the original Gibbs function:

$$\mathcal{G} = \sum_{i=1}^{\mathcal{C}} N_i \bar{G}_i + \sum_{i=1}^{\mathcal{C}} \alpha_i \left[N_i - N_{i,0} - \sum_{j=1}^{\mathcal{M}} \nu_{ij} X_j \right] \tag{8.8-6}$$

The independent variables of this new function are $N_1, N_2, \ldots, N_{\mathcal{C}}$; $X_1, X_2, \ldots, X_{\mathcal{M}}$; and $\alpha_1, \alpha_2, \ldots, \alpha_{\mathcal{C}}$. To determine the state for which the Gibbs energy is a minimum subject to the stoichiometric constraints of Eq. 8.8-5, the partial derivatives of this new unconstrained function $\mathcal{G}$ with respect to each of its independent variables are set equal to zero. From this procedure we obtain the following sequence of simultaneous equations to be solved:

$$\left(\frac{\partial \mathcal{G}}{\partial N_k} \right)_{T,P,X_j,N_{j\neq k}} = 0 = \bar{G}_k + \alpha_k + \sum_{i=1}^{\mathcal{C}} N_i \left(\frac{\partial \bar{G}_i}{\partial N_k} \right)_{T,P,X_j,N_{j\neq k}}$$

where the last term is zero by the Gibbs-Duhem equation, so that

$$\alpha_k = -\bar{G}_k \quad k = 1, 2, \ldots, \mathcal{C} \tag{8.8-7}$$

Also,

$$\left(\frac{\partial \mathcal{G}}{\partial X_k} \right)_{T,P,N_i,X_{j\neq k}} = 0 = - \sum_{i=1}^{\mathcal{C}} \nu_{ik} \alpha_i \quad k = 1, 2, \ldots, \mathcal{M} \tag{8.8-8}$$

or, using Eq. 8.8-7,

$$\sum_{i=1}^{\mathcal{C}} \nu_{ik} \bar{G}_i = 0 \quad k = 1, 2, \ldots, \mathcal{M} \tag{8.8-9}$$

and finally,

$$\left(\frac{\partial \mathcal{G}}{\partial \alpha_i} \right)_{T,P,X_j,N_k} = 0 = N_i - N_{i,0} - \sum_{j=1}^{\mathcal{M}} \nu_{ij} X_j \quad i = 1, 2, \ldots \tag{8.8-10}$$

---

[18]A more complete discussion of Lagrange multipliers may be found in M. H. Protter and C. B. Morrey, *College Calculus with Analytic Geometry,* Addison-Wesley, Reading, Mass. (1964), pp. 708–715; and V. G. Jenson and G. V. Jeffreys, *Mathematical Methods in Chemical Engineering,* Academic Press, New York (1963), pp. 482–483.

Clearly, Eq. 8.8-9 gives the same equilibrium requirement as before (see Eq. 8.8-4), whereas Eq. 8.8-10 ensures that the stoichiometric constraints are satisfied in solving the problem. Thus the Lagrange multiplier method yields the same results as the direct substitution or brute-force approach. Although the Lagrange multiplier method appears awkward when applied to the very simple problem here, its real utility is for complicated problems in which the number of constraints is large or the constraints are nonlinear in the independent variables, so that direct substitution is very difficult or impossible.

To derive the criteria for combined chemical and phase equilibrium, the following notation will be used: N_i^k and $\bar{G}_i^k$ are, respectively, the number of moles and partial molar Gibbs energy of species i in the kth phase; $N_{i,0}$ is the initial number of moles of species i in the closed system; and X_j is the overall molar extent of reaction (reaction variable) for the jth independent reaction, regardless of which phase or in how many different phases the reaction occurs. Thus

$$\begin{pmatrix} \text{Total number of} \\ \text{moles of species} \\ \text{i in all } \mathcal{P} \text{ phases} \end{pmatrix} = \sum_{k=1}^{\mathcal{P}} N_i^k = N_{i,0} + \sum_{j=1}^{\mathcal{M}} \nu_{ij} X_j \qquad \textbf{(8.8-11)}$$

and

$$\begin{pmatrix} \text{Total Gibbs free} \\ \text{energy of system} \end{pmatrix} = G = \sum_{k=1}^{\mathcal{P}} \sum_{i=1}^{\mathcal{C}} N_i^k \bar{G}_i^k \qquad \textbf{(8.8-12)}$$

where $\mathcal{P}$ is the number of phases, $\mathcal{C}$ is the number of components, and $\mathcal{M}$ is the number of independent reactions.

The equilibrium state for a closed system at constant temperature and pressure is that state for which the Gibbs energy G achieves a minimum value from among all the states consistent with the reaction stoichiometry. The identification of the equilibrium state is then a problem of minimizing the Gibbs energy subject to the stoichiometric constraints of Eq. 8.8-11. Since the easiest way of solving this problem is to use the method of Lagrange multipliers, we define a set of Lagrange multipliers, $\alpha_1, \alpha_2, \ldots, \alpha_{\mathcal{C}}$ and construct the augmented function

$$\mathcal{G} = \sum_{k=1}^{\mathcal{P}} \sum_{i=1}^{\mathcal{C}} N_i^k \bar{G}_i^k + \sum_{i=1}^{\mathcal{C}} \alpha_i \left\{ \sum_{k=1}^{\mathcal{P}} N_i^k - N_{i,0} - \sum_{j=1}^{\mathcal{M}} \nu_{ij} X_j \right\} \qquad \textbf{(8.8-13)}$$

whose minimum we wish to find for all variations of the independent variables $X_1, X_2, \ldots, X_{\mathcal{M}}, N_1^{\text{I}}, N_1^{\text{II}}, \ldots, N_{\mathcal{C}}^{\mathcal{P}}$, and $\alpha_1, \alpha_2, \ldots, \alpha_{\mathcal{C}}$.

Setting each of the partial derivatives of $\mathcal{G}$ with respect to $N_1^{\text{I}}, N_1^{\text{II}}, \ldots$ equal to zero, remembering that the N's, X's, and α's are now to be treated as independent variables, yields[19]

---

[19]Though we have not listed the variables being held constant, you should recognize that all variables in the set T, P, N_i^k (i = 1, $\ldots$, $\mathcal{C}$; k = 1, $\ldots$, $\mathcal{P}$), X_j (j = 1, $\ldots$, $\mathcal{M}$), and α_i (i = 1, $\ldots$, $\mathcal{C}$), except the one being varied in the derivative, have been held constant.

$$\left(\frac{\partial \mathcal{G}}{\partial N_1^{\mathrm{I}}}\right) = \bar{G}_1^{\mathrm{I}} + \sum_{k=1}^{\mathcal{P}} \sum_{i=1}^{\mathcal{C}} N_i^k \left(\frac{\partial \bar{G}_i^k}{\partial N_1^{\mathrm{I}}}\right) + \alpha_1 = \bar{G}_1^{\mathrm{I}} + \alpha_1 = 0$$

$$\left(\frac{\partial \mathcal{G}}{\partial N_1^{\mathrm{II}}}\right) = \bar{G}_1^{\mathrm{II}} + \sum_{k=1}^{\mathcal{P}} \sum_{i=1}^{\mathcal{C}} N_i^k \left(\frac{\partial \bar{G}_i^k}{\partial N_1^{\mathrm{II}}}\right) + \alpha_1 = \bar{G}_1^{\mathrm{II}} + \alpha_1 = 0 \qquad \textbf{(8.8-14)}$$

$$\vdots$$

In each case the double-summation term vanishes by application of the Gibbs-Duhem equation (Eq. 8.2-15) to each phase. The net information content of Eqs. 8.8-14 is

$$\bar{G}_1^{\mathrm{I}} = \bar{G}_1^{\mathrm{II}} = \cdots = \bar{G}_1^{\mathcal{P}} = -\alpha_1$$

and by generalization,

$$\bar{G}_i^{\mathrm{I}} = \bar{G}_i^{\mathrm{II}} = \cdots = \bar{G}_i^{\mathcal{P}} = -\alpha_i \quad i = 1, 2, \ldots, \mathcal{C} \qquad \textbf{(8.8-15)}$$

These equations establish that one of the equilibrium conditions in a multiple-reaction, multiphase system is that phase equilibrium be established for each of the species among the phases in which the species is present.

Another set of equilibrium criteria is obtained by minimizing $\mathcal{G}$ with respect to each of the reaction variables X_j ($j = 1, \ldots, \mathcal{M}$). Thus[20]

$$\left(\frac{\partial \mathcal{G}}{\partial X_1}\right) = 0 = \sum_{k=1}^{\mathcal{P}} \sum_{i=1}^{\mathcal{C}} N_i^k \left(\frac{\partial \bar{G}_i^k}{\partial X_1}\right) - \sum_{i=1}^{\mathcal{C}} \alpha_i \nu_{i1} = 0 \qquad \textbf{(8.8-16)}$$

The first term on the right side of this equation vanishes by the Gibbs-Duhem equation, and from Eq. 8.8-15, we can set $\bar{G}_i^{\mathrm{I}} = \bar{G}_i^{\mathrm{II}} = \cdots = \bar{G}_i^k = \cdots = -\alpha_i$ and so obtain

$$\left(\frac{\partial \mathcal{G}}{\partial X_1}\right) = 0 = \sum_{i=1}^{\mathcal{C}} \nu_{i1} \bar{G}_i^k \qquad \textbf{(8.8-17)}$$

Similarly, from $\partial \mathcal{G}/\partial X_2 = 0$, $\partial \mathcal{G}/\partial X_3 = 0, \ldots$, we obtain

$$\sum_{i=1}^{\mathcal{C}} \nu_{ij} \bar{G}_i^k = 0 \qquad \begin{array}{l} \text{for all phases k} = \text{I, II}, \ldots, \mathcal{P} \\ \text{and all reactions j} = 1, 2, \ldots, \mathcal{M} \end{array} \qquad \textbf{(8.8-18)}$$

which establishes that a further condition for equilibrium in a multiphase, multireaction system is that each reaction be in equilibrium in every phase. In fact, since at equilibrium the partial molar Gibbs energy of each species is the same in every phase (see Eq. 8.8-15), if Eq. 8.8-18 is satisfied in any phase, it is satisfied in all phases.

Finally, setting the partial derivatives of $\mathcal{G}$ with respect to each of the Lagrange multipliers α_i equal to zero yields the stoichiometric constraints of Eq. 8.8-11. The equilibrium state is that state for which Eqs. 8.8-11, 8.8-15, and 8.8-18 are simultaneously satisfied.

Thus, we have proved the assertion that in the case of combined chemical and phase equilibria the conditions of phase equilibrium must be satisfied for all species in each of the phases and, furthermore, that chemical equilibrium must exist for each reaction

---

[20]See footnote 18.

in each phase. (The fact that each reaction must be in chemical equilibrium in each phase does not imply that each mole fraction will be the same in each phase. This point is demonstrated in Chapter 13.)

8.9 SPECIFICATION OF THE EQUILIBRIUM THERMODYNAMIC STATE OF A MULTICOMPONENT, MULTIPHASE SYSTEM; THE GIBBS PHASE RULE

As has been mentioned several times, the equilibrium state of a single-phase, one-component system is completely fixed by the specification of two independent, intensive variables. From this observation we were able, in Sec. 7.6, to establish a simple relation for determining the number of degrees of freedom for a single-component, multiphase system. Here an analogous equation is developed for determining the number of degrees of freedom in a reacting multicomponent, multiphase system; this relationship is called the **Gibbs phase rule**.

The starting point for the present analysis is the observation in Sec. 8.1 that the equilibrium thermodynamic state of a single-phase C-component system can be fixed by specifying the values of two intensive variables and $C - 1$ mole fractions. Alternatively, the specification of any $C + 1$ independent state variables could be used to fix the state of this system.[21] Thus, we can say that a C-component, single-phase system has $C + 1$ degrees of freedom, that is, we are free to adjust $C + 1$ independent intensive thermodynamic properties of this system; however, once this is done, all the other intensive thermodynamic properties are fixed. This is equivalent to saying that if T, P, $x_1, x_2, \ldots, x_{C-1}$ are taken as the independent variables, there exist equations of state in nature of the form

$$\underline{V} = \underline{V}(T, P, x_1, \ldots, x_{C-1})$$
$$\underline{S} = \underline{S}(T, P, x_1, \ldots, x_{C-1})$$
$$\underline{G} = \underline{G}(T, P, x_1, \ldots, x_{C-1})$$

although we may not have been clever enough in our experiments to have determined the functional relationship between the variables.

Our interest here is in determining the number of degrees of freedom in a general multicomponent, multiphase chemically reacting system consisting of C components distributed among $\mathcal{P}$ phases and in which $\mathcal{M}$ independent chemical reactions occur. Since $C + 1$ variables are required to completely specify the state of each phase, and there are $\mathcal{P}$ phases present, it would appear that a total of $\mathcal{P}(C + 1)$ variables must be specified to fix the state of each of the phases. Actually, the number of variables that must be specified is considerably fewer than this, since the requirement that equilibrium exists provides a number of interrelationships between the state variables in each of the phases. In particular, the fact that the temperature must be the same in all phases,

$$T^{\mathrm{I}} = T^{\mathrm{II}} = T^{\mathrm{III}} = \cdots = T^{\mathcal{P}} \tag{8.9-1}$$

results in $(\mathcal{P} - 1)$ restrictions on the values of the state variables of the phases. Similarly, the requirement that the pressure be the same in all phases,

---

[21]By $C + 1$ independent state variables we mean $C + 1$ nonredundant pieces of information about the thermodynamic state of the system. For example, temperature, pressure, and $C - 1$ mole fractions form a set of $C + 1$ independent variables; temperature and C mole fractions are not independent, however, since $\sum_i^C x_i = 1$, so that only $C - 1$ mole fractions are independent. Similarly, for a gas mixture composed of ideal gases, the enthalpy or internal energy, temperature, and $C - 1$ mole fractions do not form an independent set of variables, since $\underline{H}$ and $\underline{U}$ are calculable from the mole fractions and the temperature. However, $\underline{H}$, P, and $C - 1$ mole fractions are independent.

$$P^{\mathrm{I}} = P^{\mathrm{II}} = P^{\mathrm{III}} = \cdots = P^{\mathcal{P}} \tag{8.9-2}$$

provides an additional $(\mathcal{P} - 1)$ restrictions. Since each of the partial molar Gibbs energies is in principle calculable from equations of state of the form

$$\bar{G}_i^j = \bar{G}_i^j(T, P, x_1^j, x_2^j, \ldots, x_{\mathcal{C}-1}^j) \tag{8.9-3}$$

the condition for phase equilibrium,

$$\bar{G}_i^{\mathrm{I}} = \bar{G}_i^{\mathrm{II}} = \cdots = \bar{G}_i^{\mathcal{P}} \quad i = 1, 2, \ldots, \mathcal{C} \tag{8.9-4}$$

provides $\mathcal{C}(\mathcal{P} - 1)$ additional relationships among the variables without introducing any new unknowns.

Finally, if $\mathcal{M}$ *independent* chemical reactions occur, there are $\mathcal{M}$ additional relations of the form

$$\sum_{i=1}^{\mathcal{C}} \nu_{ij} \bar{G}_i = 0 \quad j = 1, 2, \ldots, \mathcal{M} \tag{8.9-5}$$

(where we have omitted the superscript indicating the phase since, by Eq. 8.9-4, the partial molar Gibbs energy for each species is the same in all phases).

Now designating the number of degrees of freedom by the symbol $\mathcal{F}$, we have

Gibbs phase rule

$$
\boxed{
\begin{aligned}
\mathcal{F} &= \begin{pmatrix} \text{Number of unknown} \\ \text{thermodynamic} \\ \text{parameters} \end{pmatrix} - \begin{pmatrix} \text{Number of independent relations} \\ \text{among the unknown parameters} \end{pmatrix} \\
&= \mathcal{P}(\mathcal{C} + 1) \qquad\qquad - [2(\mathcal{P} - 1) + \mathcal{C}(\mathcal{P} - 1) + \mathcal{M}] \\
&= \mathcal{C} - \mathcal{M} - \mathcal{P} + 2
\end{aligned}
}
\tag{8.9-6}
$$

Therefore, in a $\mathcal{C}$-component, $\mathcal{P}$-phase system in which $\mathcal{M}$ independent chemical reactions occur, the specification of $\mathcal{C} - \mathcal{M} - \mathcal{P} + 2$ state variables of the individual phases completely fixes the thermodynamic state of each of the phases. This result is known as the Gibbs phase rule.

In practice, temperature, pressure, and phase composition are most commonly used to fix the thermodynamic state of multicomponent, multiphase systems, though any other information about the thermodynamic state of the individual phases could be used as well. However, thermodynamic information about the composite multiphase system is not useful in fixing the state of the system. That is, we could use the specific volume of any one of the phases as one of the $\mathcal{C} - \mathcal{M} - \mathcal{P} + 2$ degrees of freedom, but not the molar volume of the multiphase system. Finally, we note that for a pure fluid $\mathcal{C} = 1$ and $\mathcal{M} = 0$, so that Eq. 8.9-6 reduces to

$$\mathcal{F} = 3 - \mathcal{P} \tag{8.9-7}$$

the result found in Sec. 7.6.

ILLUSTRATION 8.9-1

Application of the Gibbs Phase Rule

In Chapter 13 we will consider the reaction equilibrium when styrene is hydrogenated to form ethylbenzene. Depending on the temperature and pressure of the system, this reaction may take place in the vapor phase or in a vapor-liquid mixture. Show that the system has three degrees of freedom if a single phase exists, but only two degrees of freedom if the reactants and products form a two-phase mixture.

SOLUTION

The styrene-hydrogen-ethylbenzene system is a three-component ($C = 3$), single-reaction ($M = 1$) system. Thus

$$\mathcal{F} = C - M - P + 2 = 4 - P$$

Clearly, if only the vapor phase exists ($P = 1$), there are three degrees of freedom; if, however, both the vapor and liquid are present ($P = 2$), the system has only two degrees of freedom.   ∎

ILLUSTRATION 8.9-2

Another Application of the Gibbs Phase Rule

Determine the number of degrees of freedom for each of the following mixtures.

- **a.** A one-component vapor-liquid mixture
- **b.** A nonreacting two-component vapor-liquid mixture
- **c.** A vapor-liquid mixture of ortho-, meta-, and para-xylenes and ethylbenzene at temperatures high enough that the xylenes can undergo isomerization

SOLUTION

- **a.** This system has one component ($C = 1$) and two phases ($P = 2$), and there are no chemical reactions ($M = 0$). Therefore,

$$\mathcal{F} = C - M - P + 2 = 1 - 0 - 2 + 2 = 1$$

Consequently, this system has one degree of freedom. If the temperature is set, the pressure is fixed; or if the pressure is set, the temperature is fixed. We see this when boiling water in a pot open to the atmosphere. At 101.3 kPa (1 atm), the temperature of the boiling water will be 100°C and will remain at this temperature no matter how much of the water boils away, provided the water is pure. In order to change the temperature of this vapor-liquid mixture, the pressure must change.

- **b.** This system has two components ($C = 2$) and two phases ($P = 2$), and there are no chemical reactions occurring ($M = 0$). The number of degrees of freedom in this system is

$$\mathcal{F} = C - M - P + 2 = 2 - 0 - 2 + 2 = 2$$

Therefore, the values of two state parameters—for example, temperature and pressure, temperature and the mole fraction of one of the species, or pressure and the mole fraction of one of the species—must be set to fix the state of this mixture. This suggests that at a fixed pressure the boiling temperature of the mixture will be a function of its composition. To see the implication of this, consider the experiment of preparing a mixture of two species (composition initially known) at one atmosphere and heating this mixture to its boiling point, and removing the vapor as the boiling continues. For most mixtures (an azeotropic mixture, to be discussed in Sec. 10.2, is the exception) as boiling occurs the composition of the vapor will be different from that of the prepared mixture, so that (by a mass balance) the composition of the remaining liquid will change during the boiling process (unless the vapor is continually condensed and replaced). As a result, at fixed pressure, the boiling temperature of this mixture will continually change as the process of boiling continues. This behavior is different from the boiling of a one-component mixture considered above, in which the temperature remains constant as the boiling process continues at fixed pressure. Also, by changing the composition of this mixture, a range of equilibrium temperatures can be obtained at the same pressure, or a range of equilibrium pressures can occur at a fixed temperature.

c. There are three independent reactions for this system. One set of such independent reactions is

$$m\text{-xylene} \leftrightarrow o\text{-xylene}$$
$$m\text{-xylene} \leftrightarrow p\text{-xylene}$$
$$m\text{-xylene} \leftrightarrow \text{ethylbenzene}$$

This system has four components ($\mathcal{C} = 4$) and two phases ($\mathcal{P} = 2$), and there are three independent chemical reactions occurring ($\mathcal{M} = 3$). The number of degrees of freedom in this system is

$$\mathcal{F} = \mathcal{C} - \mathcal{M} - \mathcal{P} + 2 = 4 - 3 - 2 + 2 = 1$$

Therefore, specification of the value of only one state variable—temperature, pressure, or the mole fraction of one of the species in one of the phases—completely fixes the two-phase state of this mixture. For example, consider the experiment of preparing a mixture of these species, and heating the mixture under pressure to a temperature that is high enough that vaporization and chemical reaction occur. In this mixture, once such a temperature is fixed, the pressure, liquid composition, and vapor composition are all fixed. If an additional amount of one of the components is added to this mixture, the total number of moles of vapor and liquid will change, but the pressure, the vapor mole fractions, and the liquid mole fractions will not change. Also for this mixture, as with a pure component, the vapor-liquid equilibrium temperature will change when the pressure is changed, but will remain constant as boiling occurs and the vapor is removed.   ■

It is also of interest to determine the amount and type of additional information needed to fix the relative amounts of each of the phases in equilibrium, once their thermodynamic states are known. We can obtain this from an analysis that equates the number of variables to the number of restrictions on these variables. It is convenient for this discussion to write the specific thermodynamic properties of the multiphase system in terms of the distribution of mass between the phases. The argument could be based on a distribution of numbers of moles; however, it is somewhat more straightforward on a mass basis because total mass, and not total moles, is a conserved quantity. Thus, we will use w^{i} to represent the mass fraction of the ith phase.[22] Clearly the w^{i} must satisfy the equation

$$1 = w^{\mathrm{I}} + w^{\mathrm{II}} + w^{\mathrm{III}} + \cdots + w^{\mathcal{P}} \tag{8.9-8}$$

The total volume per unit mass, $\hat{V}$, the total entropy per unit mass, $\hat{S}$, and so on, are related to the analogous quantities in each of the phases by the equations

$$\hat{V} = w^{\mathrm{I}}\hat{V}^{\mathrm{I}} + w^{\mathrm{II}}\hat{V}^{\mathrm{II}} + \cdots + w^{\mathcal{P}}\hat{V}^{\mathcal{P}}$$
$$\hat{S} = w^{\mathrm{I}}\hat{S}^{\mathrm{I}} + w^{\mathrm{II}}\hat{S}^{\mathrm{II}} + \cdots + w^{\mathcal{P}}\hat{S}^{\mathcal{P}} \tag{8.9-9}$$

In writing these equations we are presuming that the specific volumes, entropies, and so forth for each phase (denoted by the superscript) are known from the equations of

---

[22]I hope the notation is not too confusing. Here w^{I} is the fraction of the total mass of the system in phase I, while earlier in this chapter w_{i} was used to represent the mass fraction of species i in a phase.

state or experimental data for the individual phases and a previous specification of the $C - M - P + 2$ degrees of freedom.

Since there are P unknown mass distribution variables, each of the w^i, it is evident that we need P equations to determine the relative amounts of each of the phases. Therefore, Eq. 8.9-8, together with the specification of $P - 1$ intensive thermodynamic variables for the multiphase system (which can be written in the form of Eq. 8.9-9), are needed. This is in addition to the $C - M - P + 2$ intensive variables of the individual phases that must be specified to completely fix the thermodynamic state of all of the phases. (You should convince yourself that this conclusion is in agreement with Illustration 7.6-1.)

Thus far we have not considered the fact that the initial composition of a chemical or phase equilibrium system may be known. Such information can be used in the formulation of species mass balances and the energy balance, which lead to additional equations relating the phase variables. Depending on the extent of initial information available and the number of phases present, the initial state information may or may not reduce the number of degrees of freedom of the system. This point is most easily demonstrated by reference to specific examples, so the effect of initial state information will be considered in the illustrations of the following chapters, not here.

We should point out that the Gibbs phase rule is of use in deciding whether or not an equilibrium problem is "well posed," that is, whether enough information has been given for the problem to be solvable, but it is not of use in actually solving for the equilibrium state. This too is demonstrated by examples later in this book. The Gibbs phase rule, being general in its scope and application, is regarded as another part of the philosophical content of thermodynamics.

8.10 A CONCLUDING REMARK

The discussion in this chapter essentially concludes our development of thermodynamic theory. The remainder of this book is largely concerned with how this theory is used to solve problems of interest to the chemical process industry. Since the partial molar Gibbs energy has emerged as the central function in equilibrium computations, Chapter 9 is concerned with the techniques used for estimating this quantity in gaseous, liquid, and solid mixtures. Chapters 10 to 15 are devoted to the use of thermodynamics in explaining and predicting the great diversity of physical, chemical, and biochemical equilibria that occur in mixtures.

PROBLEMS

8.1 Prove that

a. $\left(\dfrac{\partial H}{\partial N_i} \right)_{P,S,N_{j \neq i}} = \bar{H}_i - T\bar{S}_i = \bar{G}_i$

b. $\bar{G}_i = \left(\dfrac{\partial U}{\partial N_i} \right)_{S,V,N_{j \neq i}} = \left(\dfrac{\partial A}{\partial N_i} \right)_{V,T,N_{j \neq i}}$

8.2 Derive the analogues of the Gibbs-Duhem equations (Eqs. 8.2-8 and 8.2-9) for the constraints of
a. Constant temperature and volume
b. Constant internal energy and volume
c. Constant entropy and volume

8.3 In Sec. 8.7 we established that the condition for equilibrium between two phases is

$\bar{G}_i^I = \bar{G}_i^{II}$ (for all species present in both phases)

for closed systems either at constant temperature and pressure or at constant internal energy and volume. Show that this equilibrium condition must also be satisfied for closed systems at
a. Constant temperature and volume
b. Constant entropy and volume

8.4 Show that the criterion for chemical equilibrium developed in the text,

$$\sum_{i}^{c} \nu_i \bar{G}_i = 0$$

for a closed system at constant temperature and pressure, is also the equilibrium condition to be satisfied for closed systems subject to the following constraints:
a. Constant temperature and volume
b. Constant internal energy and volume

8.5 Prove that since total mass is conserved during a chemical reaction,

$$\sum_{i=1}^{c} \nu_i m_i = 0 \quad \text{for a single-reaction system}$$

and

$$\sum_{i=1}^{c} \nu_{ij} m_i = 0 \quad j = 1, 2, \ldots, \mathcal{M} \quad \begin{array}{l}\text{for a multiple-reaction}\\ \text{system}\end{array}$$

where m_i is equal to the molecular weight of species i. Also show, by direct substitution, that the first of these equations is satisfied for the reaction

$$H_2O = H_2 + \tfrac{1}{2}O_2$$

8.6 Show that the partial molar volumes computed from Eqs. 8.6-4 and the partial molar enthalpies computed from Eqs. 8.6-9 must satisfy the Gibbs-Duhem equation.

8.7 Compute the partial molar volumes of methyl formate in methanol–methyl formate and ethanol–methyl formate mixtures at 298.15 K for various compositions using the experimental data in Fig. 8.1-2a and the following pure-component data:

$$\underline{V}_{MF} = 0.062\,78 \text{ m}^3/\text{kmol}$$
$$\underline{V}_M = 0.040\,73$$
$$\underline{V}_E = 0.058\,68$$

8.8 Compute the difference between the pure-component and partial molar enthalpies for both components at 298.15 K and various compositions in each of the following mixtures using the data in Fig. 8.1-2b.
a. benzene–C_6F_5H
b. benzene–C_6F_6
c. benzene–C_6F_5Cl
d. benzene–C_6F_5Br
e. benzene–C_6F_5I

8.9 a. In vapor-liquid equilibrium in a binary mixture, both components are generally present in both phases. How many degrees of freedom are there for such a system?
b. The reaction between nitrogen and hydrogen to form ammonia occurs in the gas phase. How many degrees of freedom are there for this system?
c. Steam and coal react at high temperatures to form hydrogen, carbon monoxide, carbon dioxide, and methane. The following reactions have been suggested as being involved in the chemical transformation:

$$C + 2H_2O = CO_2 + 2H_2$$
$$C + H_2O = CO + H_2$$
$$C + CO_2 = 2CO$$
$$C + 2H_2 = CH_4$$
$$CO + H_2O = CO_2 + H_2$$
$$CO + 3H_2 = CH_4 + H_2O$$

How many degrees of freedom are there for this system? [*Hint:* (1) How many independent chemical reactions are there in this sequence? (2) How many phase equilibrium equations are there?]

8.10 a. In vapor-liquid equilibrium, mixtures sometimes occur in which the compositions of the coexisting vapor and liquid phases are the same. Such mixtures are called azeotropes. Show that a binary azeotropic mixture has only one degree of freedom.
b. In osmotic equilibrium, two mixtures at different pressures and separated by a rigid membrane permeable to only one of the species present attain a state of equilibrium in which the two phases have different compositions. How many degrees of freedom are there for osmotic equilibrium in a binary mixture?
c. The phase equilibrium behavior of furfural $(C_5H_4O_2)$–water mixtures is complicated because furfural and water are only partially soluble in the liquid phase.
  (i) How many degrees of freedom are there for the vapor-liquid mixture if only a single liquid phase is present?
  (ii) How many degrees of freedom are there for the vapor-liquid mixture if two liquid phases are present?

8.11 a. What is the maximum number of phases that can coexist for a mixture of two nonreacting components?
b. How would the answer in part (a) change if the two components could react to form a third component?

8.12 Consider a reaction that occurs in a vessel containing a semipermeable membrane that allows only one of the components to pass through it (for example, a small molecule such as hydrogen) but will not allow the passage of large molecules. With such a membrane, the chemical potential of the permeable component can be kept constant in the reaction vessel.

   a. Derive the equilibrium criterion for a one-phase reaction system described above in which the temperature, pressure, and chemical potential of one component are held constant.

   b. Derive the equilibrium criterion for a one-phase reaction system described above in which the temperature, volume, and chemical potential of one component are held constant.

8.13 The following set of reactions is thought to occur between nitrogen and oxygen at high temperatures

$$N_2 + \tfrac{1}{2}O_2 = N_2O$$
$$N_2 + O_2 = 2NO$$
$$N_2 + 2O_2 = N_2O_4$$
$$N_2O_4 = 2NO_2$$
$$2NO + \tfrac{3}{2}O_2 = N_2O_5$$
$$N_2O + NO_2 = 3NO$$

   a. Find an independent set of reactions for the nitrogen-oxygen system.

   b. How many degrees of freedom are there for this system?

   c. If the starting oxygen-to-nitrogen ratio is fixed (as in air), how many degrees of freedom are there?

8.14 The temperature achieved when two fluid streams of differing temperature and/or composition are adiabatically mixed is termed the adiabatic mixing temperature. Compute the adiabatic mixing temperature for the following two cases:

   a. Equal weights of aqueous solutions containing 10 wt % sulfuric acid at 20°C and 90 wt % sulfuric acid at 70°C are mixed.

   b. Equal weights of aqueous solutions containing 10 wt % sulfuric acid at 20°C and 60 wt % sulfuric acid at 0°C are mixed.

   Explain why the adiabatic mixing temperature is greater than that of either of the initial solutions in one of these cases, and intermediate between those of the initial solutions in the other case.

8.15 The molar integral heat of solution $\Delta_s \underline{H}$ is defined as the change in enthalpy that results when 1 mole of solute (component 1) is isothermally mixed with N_2 moles of solvent (component 2) and is given by

$$\Delta_s \underline{H} = (1 + N_2)\underline{H}_{\text{mix}} - \underline{H}_1 - N_2\underline{H}_2$$
$$= \overline{H}_1 + N_2\overline{H}_2 - \underline{H}_1 - N_2\underline{H}_2$$

$\Delta_s \underline{H}$ is easily measured in an isothermal calorimeter by monitoring the heat evolved or absorbed on successive additions of solvent to a given amount of solute. The table below gives the integral heat-of-solution data for 1 mol of sulfuric acid in water at 25°C (the negative sign indicates that heat is evolved in the dilution process).

N_2 (moles of water)	0.25	1.0	1.5	2.33
$-\Delta_s\underline{H}$ (J)	8242	28 200	34 980	44 690

N_2 (moles of water)	4.0	5.44	9.0
$-\Delta_s\underline{H}$ (J)	54 440	58 370	62 800

N_2 (moles of water)	10.1	19.0	20.0
$-\Delta_s\underline{H}$ (J)	64 850	70 710	71 970

   a. Calculate the heat evolved when 100 g of pure sulfuric acid is added isothermally to 100 g of water.

   b. Calculate the heat evolved when the solution prepared in part (a) is diluted with an additional 100 g of water.

   c. Calculate the heat evolved when 100 g of a 60 wt % solution of sulfuric acid is mixed with 75 g of a 25 wt % sulfuric acid solution.

   d. Relate $(\overline{H}_1 - \underline{H}_1)$ and $(\overline{H}_2 - \underline{H}_2)$ to only N_1, N_2, $\Delta\underline{H}_s$, and the derivatives of $\Delta\underline{H}_s$ with respect to the ratio N_2/N_1.

   e. Compute the numerical values of $(\overline{H}_1 - \underline{H}_1)$ and $(\overline{H}_2 - \underline{H}_2)$ in a 50 wt % sulfuric acid solution.

8.16 The following data have been reported for the constant-pressure heat capacity of a benzene–carbon tetrachloride mixture at 20°C.[23]

wt % CCl_4	C_P (J/g °C)	wt % CCl_4	C_P (J/g °C)
0	1.7655	60	1.004
10	1.630	70	0.927
20	1.493	80	0.858
30	1.358	90	0.816
40	1.222	100	0.807
50	1.100		

On a single graph plot the constant-pressure partial molar heat capacity for both benzene and carbon tetrachloride as a function of composition.

---

[23]*International Critical Tables,* Vol. 5, McGraw-Hill, New York (1929).

8.17 A 20 wt % solution of sulfuric acid in water is to be enriched to a 60 wt % sulfuric acid solution by adding pure sulfuric acid.

  a. How much pure sulfuric acid should be added?

  b. If the 20 wt % solution is available at 5°C, and the pure sulfuric acid at 50°C, how much heat will have to be removed to produce the 60 wt % solution at 70°C? How much heat will have to be added or removed to produce the 60 wt % solution at its boiling point?

8.18 Develop a procedure for determining the partial molar properties for each constituent in a three-component (ternary) mixture. In particular, what data would you want, and what would you do with the data? Based on your analysis, do you suppose there is much partial molar property data available for ternary and quaternary mixtures?

8.19 The partial molar enthalpies of species in simple binary mixtures can sometimes be approximated by the following expressions:

$$\overline{H}_1 = a_1 + b_1 x_2^2$$

and

$$\overline{H}_2 = a_2 + b_2 x_1^2$$

  a. For these expressions show that b_1 must equal b_2.

  b. Making use of the fact that

$$\lim_{x_i \to 1} \overline{\theta}_i = \underline{\theta}_i$$

  for any thermodynamic property θ, show that

$$a_1 = \underline{H}_1 \qquad a_2 = \underline{H}_2 \qquad \text{and} \qquad \Delta \underline{H}_{\text{mix}} = b_1 x_1 x_2$$

8.20 A partial molar property of a component in a mixture may be either greater than or less than the corresponding pure-component molar property. Furthermore, the partial molar property may vary with composition in a complicated way. Show this to be the case by computing (a) the partial molar volumes and (b) the partial molar enthalpies of ethanol and water in an ethanol-water mixture. (The data that follow are from Volumes 3 and 5 of the *International Critical Tables*, McGraw-Hill, New York, 1929.)

Alcohol wt %	Density at 20°C (kg m^{-3}) × 10^{-3}	mol % Water	Heat Evolved on Mixing at 17.33°C (kJ/mol of Ethanol)
0	0.9982		
5	0.9894	5	0.042
10	0.9819	10	0.092
15	0.9751	15	0.167
20	0.9686	20	0.251
25	0.9617	25	0.335
30	0.9538	30	0.423
35	0.9449	35	0.519
40	0.9352	40	0.636
45	0.9247	45	0.757
50	0.9138	50	0.946
55	0.9026	55	1.201
60	0.8911	60	1.507
65	0.8795	65	1.925
70	0.8677	70	2.478
75	0.8556	75	3.218
80	0.8434	80	4.269
85	0.8310	85	5.821
90	0.8180	90	7.801
95	0.8042	95	9.818
100	0.7893		

8.21 Using the information in Problems 7.13 and 8.20, estimate the heat of vaporization for the first bit of ethanol from ethanol-water solutions containing 25, 50, and 75 mol % ethanol and from a solution infinitely dilute in ethanol. How do these heats of vaporization compare with that for pure ethanol computed in Problem 7.13? Why is there a difference between the various heats of vaporization?

8.22 The volume of a binary mixture has been reported in the following polynomial form:

$$\underline{V}(T, P, x_1, x_2) = x_1 b_1 + x_2 b_2 + x_1 x_2 \sum_{i=0}^{n} a_i (x_1 - x_2)^i$$

  a. What values should be used for b_1 and b_2?

  b. Derive, from the equation here, expressions for $\overline{V}_1$, $\overline{V}_2$, $\overline{V}_1^{\text{ex}} = \overline{V}_1 - \underline{V}_1$ and $\overline{V}_2^{\text{ex}} = \overline{V}_2 - \underline{V}_2$.

  c. Derive, from the equation here, expressions for the partial molar excess volumes of each species at infinite dilution, that is, $\overline{V}_1^{\text{ex}}(T, P, x_1 \to 0)$ and $\overline{V}_2^{\text{ex}}(T, P, x_2 \to 0)$.

8.23 Prove the validity of Eqs. 8.7-4.

8.24 The definition of a partial molar property is

$$\overline{M}_i = \left(\frac{\partial (N \underline{M})}{\partial N_i} \right)_{T, P, N_{j \neq i}}$$

It is tempting, but incorrect, to assume that this equation can be written as

$$\overline{M}_i = \left(\frac{\partial \underline{M}}{\partial x_i}\right)_{T,P}$$

Prove that the correct result is

$$\overline{M}_i = \underline{M} + \left(\frac{\partial \underline{M}}{\partial x_i}\right)_{T,P,x_{k\neq i}} - \sum x_j \left(\frac{\partial \underline{M}}{\partial x_j}\right)_{T,P,x_{k\neq j}}$$

8.25 In some cases if pure liquid A and pure liquid B are mixed at constant temperature and pressure, two liquid phases are formed at equilibrium, one rich in species A and the other in species B. We have proved that the equilibrium state at constant T and P is a state of minimum Gibbs energy, and the Gibbs energy of a two-phase mixture is the sum of the number of moles times the molar Gibbs energy for each phase. What would the molar Gibbs free energy versus mole fraction curve look like for this system if we could prevent phase separation from occurring? Identify the equilibrium compositions of the two phases on this diagram. The limit of stability of a single phase at constant temperature and pressure can be found from $d^2G = 0$ or

$$\left(\frac{\partial^2 \underline{G}}{\partial x_1^2}\right)_{T,P} = 0$$

Identify the limits of single-phase stability on the Gibbs energy versus mole fraction curve.

8.26 Mattingley and Fenby [*J. Chem. Thermodyn.* **7,** 307 (1975)] have reported that the enthalpies of triethylamine-benzene solutions at 298.15 K are given by

$$\underline{H}_{mix} - [x_B\underline{H}_B + (1 - x_B)\underline{H}_{EA}]$$
$$= x_B(1 - x_B)\{1418 - 482.4(1 - 2x_B) + 187.4(1 - 2x_B)^3\}$$

where x_B is the mole fraction of benzene and $\underline{H}_{mix}$, $\underline{H}_B$, and $\underline{H}_{EA}$ are the molar enthalpies of the mixture, pure benzene, and pure triethylamine, respectively, with units of J/mol.

a. Develop expressions for $(\overline{H}_B - \underline{H}_B)$ and $(\overline{H}_{EA} - \underline{H}_{EA})$.

b. Compute values for $(\overline{H}_B - \underline{H}_B)$ and $(\overline{H}_{EA} - \underline{H}_{EA})$ at $x_B = 0.5$.

c. One mole of a 25 mol % benzene mixture is to be mixed with one mole of a 75 mol % benzene mixture at 298.15 K. How much heat must be added or removed for the process to be isothermal?

[*Note:* $\underline{H}_{mix} - x_B\underline{H}_B - (1 - x_B)\underline{H}_{EA}$ is the enthalpy change on mixing defined in Sec. 8.1.]

8.27 When water and n-propanol are isothermally mixed, heat may be either absorbed ($Q > 0$) or evolved ($Q < 0$), depending on the final composition of the mixture. Volume 5 of the *International Critical Tables* (McGraw-Hill, New York, 1929) gives the following data:

mol % Water	Q, kJ/mol of n-Propanol
5	+0.042
10	+0.084
15	+0.121
20	+0.159
25	+0.197
30	+0.230
35	+0.243
40	+0.243
45	+0.209
50	+0.167
55	+0.084
60	−0.038
65	−0.201
70	−0.431
75	−0.778
80	−1.335
85	−2.264
90	−4.110
95	−7.985

Plot $(\overline{H}_W - \underline{H}_W)$ and $(\overline{H}_{NP} - \underline{H}_{NP})$ over the whole composition range.

8.28 The heat-of-mixing data of Featherstone and Dickinson [*J. Chem. Thermodyn.*, **9,** 75 (1977)] for the n-octanol + n-decane liquid mixture at atmospheric pressure is approximately fit by

$$\Delta_{mix}\underline{H} = x_1x_2(A + B(x_1 - x_2)) \qquad \text{J/mol}$$

where

$$A = -12\,974 + 51.505T$$

and

$$B = +8782.8 - 34.129T$$

with T in K and x_1 being the n-octanol mole fraction.

a. Compute the difference between the partial molar and pure-component enthalpies of n-octanol and n-decane at $x_1 = 0.5$ and $T = 300$ K.

b. Compute the difference between the partial molar and pure-component heat capacities of n-octanol and n-decane at $x_1 = 0.5$ and $T = 300$ K.

c. An $x_1 = 0.2$ solution and an $x_1 = 0.9$ solution are to flow continuously into an isothermal mixer in the mole ratio 2:1 at 300 K. Will heat have to be added or removed to keep the temperature of the solution leaving the mixer at 300 K? What will be the heat flow per mole of solution leaving the mixer?

8.29 Two streams containing pyridine and acetic acid at 25°C are mixed and fed into a heat exchanger. Due to the heat-of-mixing effect, it is desired to reduce the temperature after mixing to 25°C using a stream of chilled ethylene glycol as indicated in the diagram. Calculate the mass flow rate of ethylene glycol needed. The heat capacity of ethylene glycol at these conditions is approximately 2.8 kJ/(kg K), and the enthalpy change of mixing ($\Delta_{mix}\underline{H}$) is given below.

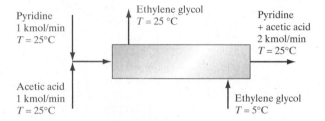

Data: Heat of mixing for pyridine (C_5H_5N) and acetic acid at 25°C [H. Kehlen, F. Herold and H.-J. Rademacher, *Z. Phys. Chem. (Leipzig)*, **261,** 809 (1980)].

Pyridine Mole Fraction	$\Delta_{mix}\underline{H}$ (J/mol)	Pyridine Mole Fraction	$\Delta_{mix}\underline{H}$ (J/mol)
0.0371	−1006	0.4076	−4880
0.0716	−1851	0.4235	−4857
0.1032	−2516	0.4500	−4855
0.1340	−3035	0.4786	−4833
0.1625	−3427	0.5029	−4765
0.1896	−3765	0.5307	−4669
0.2190	−4043	0.5621	−4496
0.2494	−4271	0.5968	−4253
0.2760	−4440	0.6372	−3950
0.3006	−4571	0.6747	−3547
0.3234	−4676	0.7138	−3160
0.3461	−4760	0.7578	−2702
0.3671	−4819	0.8083	−2152
0.3874	−4863	0.8654	−1524
0.3991	−4882	0.9297	−806

8.30 Use the data in problem 8.29 to compute the partial molar enthalpies of pyridine and acetic acid in their mixtures at 25°C over the whole composition range.

8.31 For the study of the oxidation of methane, an engineer devises the following set of possible reactions:

$$CH_4 + 2O_2 \rightarrow CO_2 + 2H_2O \qquad CO + \tfrac{1}{2}O_2 \rightarrow CO_2$$
$$CH_4 + \tfrac{3}{2}O_2 \rightarrow CO + 2H_2O \qquad CH_4 \rightarrow C + 2H_2$$
$$CO + H_2O \rightarrow CO_2 + H_2 \qquad C + \tfrac{1}{2}O_2 \rightarrow CO$$
$$C + O_2 \rightarrow CO_2 \qquad H_2 + \tfrac{1}{2}O_2 \rightarrow H_2O$$

How many independent chemical reactions are there in this system?

8.32 Calculate the standard heats and Gibbs energies of reaction at 25°C for the following reactions:
a. $N_2(g) + 3H_2(g) = 2NH_3(g)$
b. $C_3H_8(g) = C_2H_4(g) + CH_4(g)$
c. $CaCO_3(s) = CaO(s) + CO_2(g)$
d. $4CO(g) + 8H_2(g) = 3CH_4(g) + CO_2(g) + 2H_2O(g)$

8.33 Steele et al. [*J. Phys. Chem. Soc.,* **96,** 4731 (1992)] used bomb calorimetry to compute the standard enthalpy of combustion of solid buckminsterfullerene (C_{60}) at 298.15 K to be 26 033 kJ/mol. Calculate the standard state $\Delta\underline{H}$ of transition from graphite to buckminsterfullerene and its standard enthalpy of formation.

8.34 The following data are available for the isothermal heat of mixing of trichloromethane (1) and ethanol (2) at 30°C [*reference:* Reprinted with permission from J.P. Shatas, M.M. Abbott, and H. C. Van Ness, *J. Chem. Eng. Data,* **20,** 406 (1975). Copyright 1975 American Chemical Society.].

x_1, Mole Fraction	$\Delta_{mix}\underline{H}$ (J/mol)	x_1, Mole Fraction	$\Delta_{mix}\underline{H}$ (J/mol)
0.0120	−68.8	0.5137	−66.1
0.0183	−101.3	0.5391	−1.9
0.0340	−179.1	0.5858	117.1
0.0482	−244.4	0.6172	186.5
0.0736	−344.6	0.6547	266.9
0.1075	−451.1	0.7041	360.3
0.1709	−565.3	0.7519	436.6
0.1919	−581.0	0.7772	470.5
0.2301	−585.0	0.7995	495.9
0.2636	−566.1	0.8239	510.0
0.2681	−561.9	0.8520	515.8
0.2721	−557.8	0.8784	505.3
0.3073	−519.6	0.8963	486.0
0.3221	−508.0	0.9279	420.5
0.3486	−468.5	0.9532	329.2
0.3720	−424.4	0.9778	184.7
0.3983	−369.1	0.9860	123.3
0.4604	−197.1	0.9971	25.1
0.4854	−135.4		

Compute the partial molar enthalpies of trichloromethane and ethanol in their mixtures at 30°C over the whole composition range.

8.35 An equimolar mixture of nitrogen and acetylene enters a steady-flow reactor at 25°C and 1 bar of pressure. The only reaction occurring is

$$N_2(g) + C_2H_2(g) = 2\,HCN(g)$$

The product leaves the reactor at 600°C and contains 24.2 percent mole fraction of HCN. How much heat is supplied to the reactor per mole of HCN?

8.36 Using the data below, calculate the partial molar enthalpies of 1-propanol and water as a function of composition at both 25°C and 50°C.

Mole Fraction Propanol	25°C H^{ex} (J/mol)	Mole Fraction Propanol	50°C H^{ex} (J/mol)
0.027	−223.16	0.031	−76.20
0.034	−290.15	0.043	−121.84
0.054	−329.50	0.082	−97.55
0.094	−384.35	0.098	−52.75
0.153	−275.07	0.206	125.60
0.262	−103.41	0.369	370.53
0.295	−81.22	0.466	435.43
0.349	−11.35	0.587	473.11
0.533	133.98	0.707	460.55
0.602	168.31	0.872	238.23
0.739	177.94		

data: V. P. Belousov, *Vent. Leningrad Univ. Fiz., Khim,* **16**(1), 144 (1961).

8.37 Following are the slightly smoothed heat-of-mixing data of R. P. Rastogi, J. Nath, and J. Misra [Reprinted from *The Journal of Chemical Thermodynamics,* **3,** 307, R. P. Rastogi, J. Nath and R. R. Misra, Copyright (1971), with permission from Elsevier.] for the system trichloromethane (component 1) and 1,2,4-trimethyl benzene at 35°C.

x_1	$\Delta_{mix}H$ (kJ/mol)	x_1	$\Delta_{mix}H$ (kJ/mol)
0.2108	−738	0.5562	−1096
0.2834	−900	0.6001	−1061
0.3023	−933	0.6739	−976
0.4285	−1083	0.7725	−780
0.4498	−1097	0.8309	−622
0.5504	−1095		

a. From the information in this table, calculate the quantity $(\Delta_{mix}\underline{H})/(x_1 x_2)$ at each of the reported compositions.

b. Compute the difference between each of the partial molar and pure-component molar enthalpies for this system at each composition.

8.38 We want to make a simplified estimate of the maximum amount of work that can be obtained from gasoline, which we will assume to be adequately represented by *n*-octane (C_8H_{18}). The processes that occur in the cylinder of an automobile engine are that first the gasoline reacts to form a high-temperature, high-pressure combustion gas consisting of carbon dioxide, water, and the nitrogen initially present in the air (as well as other by-products that we will neglect), and then work is extracted from this combustion gas as its pressure and temperature are reduced.

a. Assuming that *n*-octane (vaporized in the fuel injector) and a stoichiometric amount of air (21 vol. % oxygen, 79 vol. % nitrogen) initially at 1 bar and 25°C react to completion at constant volume, calculate the final temperature and pressure of the combustion gas, assuming further that there is no heat loss to the pistons and cylinders of the automobile engine.

b. Calculate the final temperature of the combustion gas and the work that can be obtained from the gas if it is adiabatically expanded from the pressure found in part (a) to 1 bar.

c. Calculate the maximum amount of additional work that can be obtained from the combustion gas as its temperature is isobarically lowered from the temperature found in part (b) to an exhaust temperature of 150°C.

Data: Consider the gas to be ideal, in which case the partial molar enthalpies of the components are equal to their pure component molar enthalpies at the same temperature and pressure.

	$\Delta_f\underline{H}°$ (kJ/mol)	C_P J/(mol K)
C_8H_{18}	−208.4	
CO_2	−393.5	51.25
H_2O	−241.8	39.75
N_2	0.0	32.43
O_2	0.0	33.45

Note that for the calculations here the molar heat capacity of the mixture is just the mole fraction–weighted sum of the heat capacities of the individual components,

$$C_{P,mixture} = \sum_i x_i C_{P,i}$$

Below is a diagram of the three steps in the process.

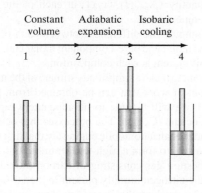

Constant volume	Adiabatic expansion	Isobaric cooling

1   2   3   4

8.39 "Duhem's theorem" states that for any number of components, phases, and chemical reactions in a closed system, if the initial amounts of all the species are specified, the further specification of two independent state variables completely fixes the state of the system. Prove that this theorem is either valid or invalid.

Chapter 9

Estimation of the Gibbs Energy and Fugacity of a Component in a Mixture

The most important ingredient in the thermodynamic analysis of mixtures is information about the partial molar properties of each species in the mixture. The partial molar Gibbs energy is of special interest since it is needed in the study of phase and chemical equilibria, which are considered in great detail in the following chapters. For many mixtures the partial molar property information needed for equilibrium calculations is not available. Consequently, in this chapter we consider methods for estimating the partial molar Gibbs energy and its equivalent, the fugacity. Before proceeding with this detailed study, we will consider two very simple cases, a mixture of ideal gases (Sec. 9.1) and the ideal mixture (Sec. 9.3), for which the partial molar properties are simply related to the pure component properties.

INSTRUCTIONAL OBJECTIVES FOR CHAPTER 9

The goals of this chapter are for the student to:

- Be able to distinguish between ideal mixtures and nonideal mixtures (Sec. 9.3)
- Understand the concepts of excess properties and activity coefficients
- Be able to calculate the fugacity of a component in a vapor mixture and in a liquid mixture if an equation of state is available (Sec. 9.4)
- Be able to calculate the fugacity of a component in a vapor mixture and in a liquid mixture if an equation of state is not available (Sec. 9.4)
- Be able to use correlative activity coefficient models with experimental data (Sec. 9.5)
- Be able to use predictive activity coefficient models when there are no experimental data (Sec. 9.6)
- Be able to compute the fugacity of a species in a mixture when, as a pure component it would be a supercritical gas or a solid (Sec. 9.7)
- Be able to use different standard states in thermodynamic calculations (Secs. 9.7–9.9)
- Be able to do calculations involving electrolyte solutions (Sec. 9.10)

NOTATION INTRODUCED IN THIS CHAPTER

θ^{IGM}	Property θ of an ideal gas mixture
P_i	Partial pressure of species i in a mixture $= y_i P$ (kPa)
$\Delta_{\mathrm{mix}}\underline{\theta}^{\mathrm{IGM}}$	Change in molar property θ on forming an ideal gas mixture
$\underline{\theta}^{\mathrm{IGM}}$	Molar property θ of an ideal gas mixture
θ^{IG}	Pure-component ideal gas property θ
$\underline{\theta}^{\mathrm{IG}}$	Pure-component molar property θ in an ideal gas
$\bar{\theta}_i^{\mathrm{IGM}}$	Partial molar property θ of species i in an ideal gas mixture
θ^{IM}	Property θ of an ideal mixture
$\underline{\theta}^{\mathrm{IM}}$	Molar property θ of an ideal mixture
$\bar{\theta}_i^{\mathrm{IM}}$	Partial molar property θ of species i in an ideal mixture
$\Delta_{\mathrm{mix}}\underline{\theta}^{\mathrm{IM}}$	Change in molar property θ on forming an ideal mixture
$\underline{\theta}_{\mathrm{mix}}$	Molar property θ of a mixture
$\underline{\theta}^{\mathrm{ex}}$	Excess molar property $= \underline{\theta}_{\mathrm{mix}}(T, P, \underline{x}) - \underline{\theta}_{\mathrm{mix}}^{\mathrm{IM}}(T, P, \underline{x})$
$\gamma_i(T, P, \underline{x})$	Activity coefficient of species i at T, P and $\underline{x}$
$\bar{f}_i$	Fugacity of species i in a mixture (kPa)
$\bar{\phi}_i$	Fugacity coefficient of species i in a mixture
$\bar{\theta}_i$	Partial molar property of species i
δ_i	Solubility parameter of species i
ϕ_i	Volume fraction of species i in a mixture
ψ_i	Surface area fraction of species i in a mixture
X_m	Mole fraction of functional group m in a mixture
Φ_m	Volume fraction of functional group m in a mixture
Ψ_m	Surface area fraction of functional group m in a mixture
H_i	Henry's law constant of species i in a mixture (kPa)
M_i	Molality of species i (moles of species i per 1000 g of solvent)
$\mathcal{H}_i$	Molality-based Henry's law constant of species i (kPa/M)
I	Ionic strength (M)
γ_i^*	Activity coefficient of species i in Henry's law based on mole fraction
$\gamma_i^{\square}$	Activity coefficient of species i in Henry's law based on molality
$\gamma_{\pm}$	Mean ionic activity coefficient in an electrolyte solution
$M_{\pm}$	Mean ionic activity molality in an electrolyte solution
z_+, z_-	Charge on cation and anion, respectively, in an electrolyte solution
ν_+, ν_-	Numbers of cations and anions, respectively, from the ionization of an electrolyte

9.1 THE IDEAL GAS MIXTURE

As defined in Chapter 3, the ideal gas is a gas whose volumetric equation of state at all temperatures, pressures, and densities is

$$PV = NRT \quad \text{or} \quad P\underline{V} = RT \tag{9.1-1}$$

and whose internal energy is a function of temperature only. By the methods of statistical mechanics, one can show that such behavior occurs when a gas is sufficiently dilute that interactions between the molecules make a negligible contribution to the total energy of the system. That is, a gas is ideal when each molecule in the gas is (energetically) unaware of the presence of other molecules.

An **ideal gas mixture** is a gas mixture with a density so low that its molecules do not appreciably interact. In this case the volumetric equation of state of the gas mixture will also be of the form of Eq. 9.1-1, and its internal energy will merely be the sum of the internal energies of each of the constituent ideal gases, and thus a function of temperature and mole number only. That is,

Hypothetical ideal gas mixture

$$PV^{\mathrm{IGM}} = (N_1 + N_2 + \cdots)RT = \left(\sum_{j=1}^{c} N_j\right)RT = NRT \qquad \textbf{(9.1-2)}$$

and

$$U^{\mathrm{IGM}}(T, N) = \sum_{j=1}^{c} N_j \underline{U}_j^{\mathrm{IG}}(T) \qquad \textbf{(9.1-3)}$$

Here we have used the superscripts IG and IGM to indicate properties of the ideal gas and the ideal gas mixture, respectively, and taken pressure and temperature to be the independent variables. From Eq. 8.1-12 it then follows that for the ideal gas mixture

Partial molar properties for the ideal gas mixtures

$$\overline{U}_i^{\mathrm{IGM}}(T, \underline{x}) = \left.\frac{\partial U^{\mathrm{IGM}}(T, N)}{\partial N_i}\right|_{T,P,N_{j\neq i}} = \left.\frac{\partial}{\partial N_i}\right|_{T,P,N_{j\neq i}} \sum_{j=1}^{c} N_j \underline{U}_j^{\mathrm{IG}}(T) = \underline{U}_i^{\mathrm{IG}}(T)$$

$$\textbf{(9.1-4)}$$

and

$$\overline{V}_i^{\mathrm{IGM}}(T, P, \underline{x}) = \left.\frac{\partial V^{\mathrm{IGM}}(T, P, N)}{\partial N_i}\right|_{T,P,N_{j\neq i}} = \left.\frac{\partial}{\partial N_i}\right|_{T,P,N_{j\neq i}} \sum_{j} N_j \frac{RT}{P}$$

$$= \frac{RT}{P} = \underline{V}_i^{\mathrm{IG}}(T, P) \qquad \textbf{(9.1-5)}$$

Here and throughout the rest of this book, we use the notation $\underline{x}$ to represent all the mole fractions. That is, any property written as $\theta(T, P, \underline{x})$ is meant to indicate that θ is a function of temperature, pressure, and all the mole fractions $(x_1, x_2, x_3, \ldots,$ etc.).

Equation 9.1-4 indicates that the partial molar internal energy of species i in an ideal gas mixture at a given temperature is equal to the pure component molar internal energy of that component as an ideal gas at the same temperature. Similarly, Eq. 9.1-5 establishes that the partial molar volume of species i in an ideal gas mixture at a given temperature and pressure is identical to the molar volume of the pure component as an ideal gas at that temperature and pressure.

Consider now the process of forming an ideal gas mixture at temperature T and pressure P from a collection of pure ideal gases, all at that temperature and pressure. From the discussion here it is clear that for each species $\overline{V}_i(T, P, \underline{x}) = \underline{V}_i(T, P)$ and $\overline{U}_i(T, \underline{x}) = \underline{U}_i(T)$. It then follows immediately from equations such as Eqs. 8.1-14 and 8.1-15 that $\Delta_{\text{mix}}V = 0$ and $\Delta_{\text{mix}}U = 0$ for this process. Also, $\Delta_{\text{mix}}H \equiv \Delta_{\text{mix}}U + P\,\Delta_{\text{mix}}V = 0$.

The **partial pressure** of species i in a gas mixture, denoted by P_i, is defined for both ideal and nonideal gas mixtures to be the product of the mole fraction of species i and total pressure P, that is,

$$P_i = x_i P \tag{9.1-6}$$

For the ideal gas mixture,

$$P_i^{\text{IGM}}(N, V, T, \underline{x}) = \frac{N_i}{\sum\limits_{j=i}^{c} N_j} P = \frac{N_i}{\sum\limits_{j=i}^{c} N_j} \left\{ \sum_{j=1}^{c} N_j \frac{RT}{V} \right\} = \frac{N_i RT}{V} = P^{\text{IG}}(N_i, V, T)$$

Thus, for the ideal gas mixture, the partial pressure of species i is equal to the pressure that would be exerted if the same number of moles of that species, N_i, alone were contained in the same volume V and maintained at the same temperature T as the mixture.

Since there is no energy of interaction in an ideal gas mixture, the effect on each species of forming an ideal gas mixture at constant temperature and total pressure is equivalent to reducing the pressure from P to its partial pressure in the mixture P_i. Alternatively, the effect is equivalent to expanding each gas from its initial volume $V_i = N_i RT/P$ to the volume of the mixture $V = \sum_i N_i RT/P$. Thus, from Eqs. 6.4-2 and 6.4-3, we have

$$\bar{S}_i^{\text{IGM}}(T, P, \underline{x}) - \underline{S}_i^{\text{IG}}(T, P) = -R \ln \frac{P_i}{P} = -R \ln x_i \tag{9.1-7}$$

or

$$\bar{S}_i^{\text{IGM}}(T, V, \underline{x}) - \underline{S}_i^{\text{IG}}(T, V_i) = R \ln \frac{V}{V_i} = R \ln \frac{\sum N_j RT/P}{N_i RT/P} = -R \ln x_i$$

Consequently,

$$\Delta_{\text{mix}} S^{\text{IGM}} = \sum_{i=1}^{c} N_i [\bar{S}_i^{\text{IGM}}(T, P, \underline{x}) - \underline{S}_i^{\text{IG}}(T, P)] = -R \sum_{i=1}^{c} N_i \ln x_i$$

and

$$\Delta_{\text{mix}}\underline{S}^{\text{IGM}} = \frac{\Delta_{\text{mix}} S^{\text{IGM}}}{N} = -R \sum_{i=1}^{c} x_i \ln x_i \tag{9.1-8}$$

The statistical mechanical interpretation of Eq. 9.1-8 is that an ideal gas mixture is a completely mixed or random mixture. This is discussed in Appendix A9.1.

Using the energy, volume, and entropy changes on mixing given here, one can easily compute the other thermodynamic properties of an ideal gas mixture (Problem 9.1). The results are given in Table 9.1-1. Of particular interest are the expressions for

Table 9.1-1 Properties of an Ideal Gas Mixture (Mixing at Constant T and P)

Internal energy	$\overline{U}_i^{IGM}(T, \underline{x}) = \underline{U}_i^{IG}(T)$	$\Delta_{mix}\underline{U}^{IGM} = 0$
Enthalpy	$\overline{H}_i^{IGM}(T, \underline{x}) = \underline{H}_i^{IG}(T)$	$\Delta_{mix}\underline{H}^{IGM} = 0$
Volume	$\overline{V}_i^{IGM}(T, \underline{x}) = \underline{V}_i^{IG}(T)$	$\Delta_{mix}\underline{V}^{IGM} = 0$
Entropy	$\overline{S}_i^{IGM}(T, P, \underline{x}) = \underline{S}_i^{IG}(T, P) - R \ln x_i$	$\Delta_{mix}\underline{S}^{IGM} = -R \displaystyle\sum_{i=1}^{c} x_i \ln x_i$
Gibbs energy	$\overline{G}_i^{IGM}(T, P, \underline{x}) = \underline{G}_i^{IG}(T, P) + RT \ln x_i$	$\Delta_{mix}\underline{G}^{IGM} = RT \displaystyle\sum_{i=1}^{c} x_i \ln x_i$
Helmholtz energy	$\overline{A}_i^{IGM}(T, P, \underline{x}) = \underline{A}_i^{IG}(T, P) + RT \ln x_i$	$\Delta_{mix}\underline{A}^{IGM} = RT \displaystyle\sum_{i=1}^{c} x_i \ln x_i$

Internal energy	$\underline{U}^{IGM}(T, \underline{x}) = \displaystyle\sum x_i \underline{U}_i^{IG}(T)$
Enthalpy	$\underline{H}^{IGM}(T, P, \underline{x}) = \displaystyle\sum x_i \underline{H}_i^{IG}(T, P)$
Volume	$\underline{V}^{IGM}(T, P, \underline{x}) = \displaystyle\sum x_i \underline{V}_i^{IG}(T, P)$
Entropy	$\underline{S}^{IGM}(T, P, \underline{x}) = \displaystyle\sum x_i \underline{S}_i^{IG}(T, P) - R \sum x_i \ln x_i$
Gibbs energy	$\underline{G}^{IGM}(T, P, \underline{x}) = \displaystyle\sum x_i \underline{G}_i^{IG}(T, P) + RT \sum x_i \ln x_i$
Helmholtz energy	$\underline{A}^{IGM}(T, P, \underline{x}) = \displaystyle\sum x_i \underline{A}_i^{IG}(T, P) + RT \sum x_i \ln x_i$

$\overline{G}_i^{IGM}(T, P, \underline{x})$ and $\Delta_{mix}\underline{G}^{IGM}$:

$$
\begin{aligned}
\overline{G}_i^{IGM}(T, P, \underline{x}) &= \overline{H}_i^{IGM}(T, P, \underline{x}) - T\overline{S}_i^{IGM}(T, P, \underline{x}) \\
&= \underline{H}_i^{IG}(T, P) - T(\underline{S}_i^{IG}(T, P) - R \ln x_i) \\
&= \underline{G}_i^{IG}(T, P) + RT \ln x_i
\end{aligned}
\tag{9.1-9}
$$

$$
\begin{aligned}
\Delta_{mix}\underline{G}^{IGM} &= \sum_{i=1}^{c} x_i\{\overline{G}_i^{IGM}(T, P, \underline{x}) - \underline{G}_i^{IG}(T, P)\} \\
&= RT \sum x_i \ln x_i
\end{aligned}
\tag{9.1-10}
$$

From Eqs. 9.1-1 to 9.1-10, we then have for an ideal gas mixture

$$
\underline{U}^{IGM}(T, \underline{x}) = \sum x_i \underline{U}_i^{IG}(T)
\tag{9.1-11}
$$

$$
\underline{V}^{IGM}(T, P, \underline{x}) = \sum x_i \underline{V}_i^{IG}(T, P)
\tag{9.1-12}
$$

$$
\underline{H}^{IGM}(T, P, \underline{x}) = \sum x_i \underline{H}_i^{IG}(T, P)
\tag{9.1-13}
$$

$$
\underline{S}^{IGM}(T, P, \underline{x}) = \sum x_i \underline{S}_i^{IG}(T, P) - R \sum x_i \ln x_i
\tag{9.1-14}
$$

$$
\underline{G}^{IGM}(T, P, \underline{x}) = \sum x_i \underline{G}_i^{IG}(T, P) + RT \sum x_i \ln x_i
\tag{9.1-15}
$$

$$
\underline{A}^{IGM}(T, P, \underline{x}) = \sum x_i \underline{A}_i^{IG}(T, P) + RT \sum x_i \ln x_i
\tag{9.1-16}
$$

ILLUSTRATION 9.1-1

Examination of Whether a Gas Mixture Is an Ideal Gas Mixture

The following data are available for the compressibility factor $Z = P\underline{V}/RT$ of nitrogen-butane mixtures at 444.3 K.

Mole Fraction Butane	$P = 13.79$ MPa Z_{mix}	$P = 68.95$ MPa Z_{mix}
0	1.06	1.47
0.10	1.04	1.51
0.30	0.96	1.61
0.50	0.81	1.72
0.70	0.64	1.83
0.90	0.52	1.95
1.0	0.51	1.98

Is the nitrogen-butane mixture an ideal gas mixture at these conditions?

SOLUTION

From Eq. 9.1-2 we have, for a gas mixture to be an ideal gas mixture, that

$$\frac{P\underline{V}_{mix}^{IGM}}{NRT} = \frac{P\underline{V}_{mix}^{IGM}}{RT} = Z_{mix} = 1$$

at all temperatures, pressures, and compositions. Clearly, the nitrogen-butane mixture is not an ideal gas mixture at the temperature and two pressures in this illustration. ■

9.2   THE PARTIAL MOLAR GIBBS ENERGY AND FUGACITY

Unfortunately, very few mixtures are ideal gas mixtures, so general methods must be developed for estimating the thermodynamic properties of real mixtures. In the discussion of phase equilibrium in a pure fluid of Sec. 7.4, the fugacity function was especially useful; the same is true for mixtures. Therefore, in an analogous fashion to the derivation in Sec. 7.4, we start from

$$dG = -S\,dT + V\,dP + \sum_{i=1}^{c} \bar{G}_i\,dN_i \qquad (8.2\text{-}1)$$

and, using the commutative property of second derivatives of the thermodynamic functions (cf. Eq. 8.1-3),

$$\left.\frac{\partial}{\partial N_i}\right|_{T,P,N_{j\neq i}}\left(\frac{\partial G}{\partial T}\right)_{P,N_j} = \left.\frac{\partial}{\partial T}\right|_{P,N_j}\left(\frac{\partial G}{\partial N_i}\right)_{T,P,N_{j\neq i}}$$

and

$$\left.\frac{\partial}{\partial N_i}\right|_{T,P,N_{j\neq i}}\left(\frac{\partial G}{\partial P}\right)_{T,N_j} = \left.\frac{\partial}{\partial P}\right|_{T,N_j}\left(\frac{\partial G}{\partial N_i}\right)_{T,P,N_{j\neq i}}$$

to obtain the two equations

$$\bar{S}_i = -\left(\frac{\partial \bar{G}_i}{\partial T}\right)_{P,N_j}$$ (9.2-1)

and

$$\bar{V}_i = \left(\frac{\partial \bar{G}_i}{\partial P}\right)_{T,N_j}$$ (9.2-2)

As in the pure component case, the second of these equations is more useful than the first, and leads to the relation

$$\bar{G}_i(T_1, P_2, \underline{x}) - \bar{G}_i(T_1, P_1, \underline{x}) = \int_{P_1}^{P_2} \bar{V}_i \, dP$$

In analogy with Eq. 7.4-6, the fugacity of species i in a mixture, denoted by $\bar{f}_i$, is defined with reference to the ideal gas mixture as follows:

Fugacity of a species in a mixture

$$
\begin{aligned}
\bar{f}_i(T, P, \underline{x}) &= x_i P \exp\left\{\frac{\bar{G}_i(T, P, \underline{x}) - \bar{G}_i^{IGM}(T, P, \underline{x})}{RT}\right\} \\
&= x_i P \exp\left\{\frac{1}{RT}\int_0^P (\bar{V}_i - \bar{V}_i^{IG}) \, dP\right\} \\
&= P \exp\left\{\frac{\bar{G}_i(T, P, \underline{x}) - \underline{G}_i^{IG}(T, P)}{RT}\right\}
\end{aligned}
$$ (9.2-3a)

so that $\bar{f}_i \to x_i P \equiv P_i$ as $P \to 0$. Here P_i is the partial pressure of species i, and the superscript IGM indicates an ideal gas mixture property. The fact that as the pressure goes to zero all mixtures become ideal gas mixtures (just as all pure fluids become ideal gases) is embedded in this definition. Also, the fugacity coefficient for a component in a mixture, $\bar{\phi}_i$, is defined as

Fugacity coefficient of a species in a mixture

$$
\begin{aligned}
\bar{\phi}_i = \frac{\bar{f}_i}{x_i P} &= \exp\left\{\frac{\bar{G}_i(T, P, \underline{x}) - \bar{G}_i^{IGM}(T, P, \underline{x})}{RT}\right\} \\
&= \exp\left\{\frac{1}{RT}\int_0^P (\bar{V}_i - \bar{V}_i^{IGM}) \, dP\right\}
\end{aligned}
$$ (9.2-3b)

The multicomponent analogue of Eq. 7.4-9a, obtained by differentiating $\ln \bar{f}_i$ with respect to pressure at a constant temperature and composition, is

$$RT\left(\frac{\partial \ln \bar{f}_i}{\partial P}\right)_{T,\underline{x}} = \left(\frac{\partial \bar{G}_i}{\partial P}\right)_{T,\underline{x}} = \bar{V}_i$$ (9.2-4)

To relate the fugacity of pure component i to the fugacity of component i in a mixture, we first subtract Eq. 7.4-9a from Eq. 9.2-4 and then integrate between $P = 0$ and the pressure of interest P, to obtain

$$RT \ln \left\{ \frac{\bar{f}_i(T, P, \underline{x})}{\bar{f}_i(T, P \to 0, \underline{x})} \right\} - RT \ln \left\{ \frac{f_i(T, P)}{f_i(T, P \to 0)} \right\} = \int_{P \to 0}^{P} (\bar{V}_i - \underline{V}_i)\, dP \quad \text{(9.2-5)}$$

We now use the fact that as $P \to 0$, $\bar{f}_i \to x_i P$ and $f_i \to P$, to obtain

$$RT \ln \left\{ \frac{\bar{f}_i(T, P, \underline{x})}{x_i f_i(T, P)} \right\} = \int_{0}^{P} (\bar{V}_i - \underline{V}_i)\, dP \quad \text{(9.2-6)}$$

Therefore, for a mixture in which the pure component and partial molar volumes are identical [i.e., $\bar{V}_i(T, P, \underline{x}) = x_i V_i(T, P)$ *at all conditions*], the fugacity of each species in the mixture is equal to its mole fraction times its pure-component fugacity evaluated at the same temperature and pressure as the mixture $\bar{f}_i(T, P, \underline{x}) = x_i f_i(T, P)$. However, if, as is generally the case, $\bar{V}_i \neq \underline{V}_i$, then $\bar{f}_i$ and f_i are related through the integral over all pressures of the difference between the species partial molar and pure-component molar volumes.

The temperature dependence of the fugacity $\bar{f}_i$ (actually, the fugacity coefficient $\phi_i = \bar{f}_i / x_i P$) can be gotten by differentiating Eq. 9.2-3 with respect to temperature at constant pressure and composition,

$$\left[\frac{\partial \ln(\bar{f}_i / x_i P)}{\partial T} \right]_{P, \underline{x}} = -\frac{(\bar{G}_i - \bar{G}_i^{IG})}{RT^2} + \frac{1}{RT} \left[\frac{\partial (\bar{G}_i - \bar{G}_i^{IGM})}{\partial T} \right]_{P, \underline{x}} \quad \text{(9.2-7)}$$

and then using $d\underline{G} = \underline{V}\, dP - \underline{S}\, dT$ and $\underline{G} = \underline{H} - T\underline{S}$, to obtain

$$\left[\frac{\partial \ln(\bar{f}_i / x_i P)}{\partial T} \right]_{P, \underline{x}} = \left[\frac{\partial \ln \bar{\phi}_i}{\partial T} \right]_{P, \underline{x}} = -\frac{(\bar{H}_i - \bar{H}_i^{IGM})}{RT^2} \quad \text{(9.2-8)}$$

It is useful to have an expression for the change in partial molar Gibbs energy of a species between two states of the same temperature and pressure, but of differing composition. To derive such an equation, we start by writing

$$\Delta \bar{G}_i = \bar{G}_i(T, P, \underline{x}^{II}) - \bar{G}_i(T, P, \underline{x}^{I})$$

where the superscripts I and II denote phases of different composition. Now substituting the logarithm of Eq. 9.2-3,

$$\bar{G}_i(T, P, \underline{x}) = \bar{G}_i^{IGM}(T, P, \underline{x}) + RT \ln \left(\frac{\bar{f}_i(T, P, \underline{x})}{x_i P} \right)$$

$$= \bar{G}_i^{IGM}(T, P, \underline{x}) + RT \ln \bar{\phi}_i(T, P, \underline{x}) \quad \text{(9.2-9)}$$

and using Eq. 9.1-9,

$$\bar{G}_i^{IGM}(T, P, \underline{x}) = \underline{G}_i^{IG}(T, P) + RT \ln x_i$$

yields

$$\bar{G}_i(T, P, \underline{x}^{II}) - \bar{G}_i(T, P, \underline{x}^{I}) = \bar{G}_i^{IGM}(T, P, \underline{x}^{II}) + RT \ln \left\{ \frac{\bar{f}_i(T, P, \underline{x}^{II})}{x_i^{II} P} \right\}$$

$$- \bar{G}_i^{IGM}(T, P, \underline{x}^{I}) - RT \ln \left\{ \frac{\bar{f}_i(T, P, \underline{x}^{I})}{x_i^{I} P} \right\}$$

$$= \underline{G}_i^{IG}(T, P) + RT \ln x_i^{II} + RT \ln \left\{ \frac{\bar{f}_i(T, P, \underline{x}^{II})}{x_i^{II} P} \right\}$$

$$- \underline{G}_i^{IG}(T, P) - RT \ln x_i^{I} - RT \ln \left\{ \frac{\bar{f}_i(T, P, \underline{x}^{I})}{x_i^{I} P} \right\}$$

or, finally,

$$\bar{G}_i(T, P, \underline{x}^{II}) - \bar{G}_i(T, P, \underline{x}^{I}) = RT \ln \left\{ \frac{\bar{f}_i(T, P, \underline{x}^{II})}{\bar{f}_i(T, P, \underline{x}^{I})} \right\} = RT \ln \left\{ \frac{x_i^{II} \bar{\phi}_i(T, P, \underline{x}^{II})}{x_i^{I} \bar{\phi}_i(T, P, \underline{x}^{I})} \right\}$$

$$(9.2\text{-}10a)$$

A special case of this equation is when state I is a pure component:

$$\boxed{\bar{G}_i(T, P, \underline{x}) - \underline{G}_i(T, P) = RT \ln \left\{ \frac{\bar{f}_i(T, P, \underline{x})}{f_i(T, P)} \right\} = RT \ln \left\{ \frac{x_i \bar{\phi}_i(T, P, \underline{x})}{\phi_i(T, P)} \right\}}$$

$$(9.2\text{-}10b)$$

The fugacity function has been introduced because its relation to the Gibbs energy makes it useful in phase equilibrium calculations. The present criterion for equilibrium between two phases is that $\bar{G}_i^{I} = \bar{G}_i^{II}$ for all species i, with the restriction that the temperature and pressure be constant and equal in both phases. Using Eqs. 9.2-10 and the equality of partial molar Gibbs free energies yields

$$\bar{f}_i^{I} = \bar{f}_i(T, P, \underline{x}^{I}) = \bar{f}_i(T, P, \underline{x}^{II}) = \bar{f}_i^{II} \qquad (9.2\text{-}11)$$

Criterion for phase equilibrium

Therefore, at equilibrium, the fugacity of each species must be the same in the two phases. Since this result follows directly from Eq. 8.7-10, it may be substituted for it. Furthermore, since we can make estimates for the fugacity of a species in a mixture in a more direct fashion than for partial molar Gibbs energies, it is more convenient to use Eq. 9.2-11 as the basis for phase equilibrium calculations.

In analogy with Eqs. 7.4-6 and 9.2-3, the overall fugacity of a mixture $\bar{f}$ is defined by the relation

$$\bar{f} = P \exp \left\{ \frac{\underline{G}(T, P, \underline{x}) - \underline{G}^{IGM}(T, P, \underline{x})}{RT} \right\} \qquad (9.2\text{-}12)$$

Note that the fugacity of species i in a mixture, $\bar{f}_i$, as defined by Eq. 9.2-3 is not a partial molar fugacity, that is,

$$\bar{f}_i \neq \left(\frac{\partial(N\bar{f})}{\partial N_i} \right)_{T,P,N_{j\neq i}}$$

Finally, note that Eq. 7.4-6, for computing the fugacity of a pure component, and Eq. 9.2-3, for computing the fugacity of a species in a mixture, both require that the volumetric equation of state be solved explicitly for the volume in terms of the temperature and pressure. However, as was discussed in Sec. 7.4, equations of state are usually pressure explicit (that is, easily solved for pressure as a function of temperature and volume, and not vice versa; see Eqs. 6.4-1 though 6.4-3), so that calculations based on Eqs. 7.4-6 and 9.2-3 can be difficult. Starting from Eq. 9.2-3, using Eq. 6.4-25 in the form

$$dP = \frac{1}{\underline{V}} d(P\underline{V}) - \frac{P}{\underline{V}} d\underline{V} = \frac{P}{Z} dZ - \frac{P}{\underline{V}} d\underline{V}$$

and the triple-product rule (Eq. 6.1-6a),

$$\left(\frac{\partial V}{\partial N_i} \right)_{T,P,N_{j\neq i}} \left(\frac{\partial P}{\partial V} \right)_{T,N_j} \left(\frac{\partial N_i}{\partial P} \right)_{T,V,N_{j\neq i}} = -1$$

in the form

$$\left(\frac{\partial V}{\partial N_i} \right)_{T,P,N_{j\neq i}} dP = - \left(\frac{\partial P}{\partial N_i} \right)_{T,V,N_{j\neq i}} dV$$

we obtain for the fugacity (actually the fugacity coefficient) of a species in a mixture

Fugacity coefficient for a species in a mixture from an equation of state

$$\ln \bar{\phi}_i = \ln \frac{\bar{f}_i(T,P,\underline{x})}{x_i P} = \frac{1}{RT} \int_{\underline{V}=\infty}^{\underline{V}=ZRT/P} \left[\frac{RT}{\underline{V}} - N \left(\frac{\partial P}{\partial N_i} \right)_{T,V,N_{j\neq i}} \right] d\underline{V} - \ln Z$$

$$\text{(9.2-13)}$$

(see Problem 9.24), which is to be compared with the expression we previously obtained for a pure fluid,

$$\ln \phi = \ln \frac{f(T,P)}{P} = \frac{1}{RT} \int_{\underline{V}=\infty}^{\underline{V}=ZRT/P} \left(\frac{RT}{\underline{V}} - P \right) d\underline{V} - \ln Z + (Z-1) \quad \text{(7.4-8)}$$

for the fugacity of a pure component. Equation 9.2-13 is especially useful for computing the fugacity of a species in a mixture from a pressure-explicit equation of state, as we will see in Sec. 9.4.

9.3   IDEAL MIXTURE AND EXCESS MIXTURE PROPERTIES

The estimation of the thermodynamic properties of a real fluid or fluid mixture in the absence of direct experimental data is a very complicated problem involving detailed spectroscopic, structural, and interaction potential data and the use of statistical mechanics. Such a calculation is beyond the scope of this book and would not be of interest to most engineers. Instead, we use procedures for estimating mixture thermodynamic properties that are far simpler than starting from "first principles." In particular, we will either use equations of state or, as in this section, choose a state for each system in which the thermodynamic properties are reasonably well known and then try to estimate how the departure of the real system from the chosen reference state affects the system properties. This last procedure is philosophically similar to the method used in Chapter 6, where the properties of real fluids were computed as the sum of an ideal gas contribution plus the departure from ideal gas behavior.

Clearly, the accuracy of a property estimation technique based on such a procedure increases as the difference between the reference state and actual state of the system diminishes. Therefore, choosing the reference state to be an ideal gas mixture at the same temperature and composition as the mixture under consideration is not very satisfactory because the reference state and actual state, particularly for liquids, may be too dissimilar. It is in this context that we introduce the concept of an ideal mixture.

An **ideal mixture**, which may be either a gaseous or liquid mixture, is defined to be a mixture in which

Ideal mixture

$$\overline{H}_i^{IM}(T, P, \underline{x}) = \underline{H}_i(T, P) \tag{9.3-1a}$$

and

$$\overline{V}_i^{IM}(T, P, \underline{x}) = \underline{V}_i(T, P) \tag{9.3-1b}$$

for *all* temperatures, pressures, and compositions (the superscript IM indicates an ideal mixture property). Note that unlike the ideal gas mixture, here neither $\underline{H}_i$ nor $\underline{V}_i$ is an ideal gas property. From Eqs. 8.1-14 and 8.1-15 it is evident that for such a mixture

$$\Delta_{mix} V^{IM}(T, P, \underline{x}) = \sum_{i=1}^{c} N_i[\overline{V}_i^{IM}(T, P, \underline{x}) - \underline{V}_i(T, P)] = 0 \tag{9.3-2a}$$

and

$$\Delta_{mix} H^{IM}(T, P, \underline{x}) = \sum_{i=1}^{c} N_i[\overline{H}_i^{IM}(T, P, \underline{x}) - \underline{H}_i(T, P)] = 0 \tag{9.3-2b}$$

so that there are no volume or enthalpy changes on the formation of an ideal mixture from its pure components at the same temperature and pressure. Also, since $\overline{V}_i^{IM} = \underline{V}_i$ at all temperatures, pressures, and compositions, we have, from Eq. 9.2-6,

$$\overline{f}_i^{IM}(T, P, \underline{x}) = x_i f_i(T, P) \tag{9.3-3}$$

Next, using Eqs. 9.3-1, it is easily established that

$$\overline{U}_i^{IM}(T, P, \underline{x}) = \underline{U}_i(T, P) \tag{9.3-4a}$$

and from Eq. 9.2-10b,

$$\overline{G}_i^{IM}(T, P, \underline{x}) - \underline{G}_i(T, P) = RT \ln \left\{ \frac{\overline{f}_i^{IM}(T, P, \underline{x})}{f_i(T, P)} \right\}$$

$$= RT \ln \left\{ \frac{x_i f_i(T, P)}{f_i(T, P)} \right\} = RT \ln x_i$$

Consequently, we obtain

$$\overline{G}_i^{IM}(T, P, \underline{x}) = \underline{G}_i(T, P) + RT \ln x_i$$
$$\overline{A}_i^{IM}(T, P, \underline{x}) = \underline{A}_i(T, P) + RT \ln x_i \qquad \textbf{(9.3-4b)}$$
$$\overline{S}_i^{IM}(T, P, \underline{x}) = \underline{S}_i(T, P) - R \ln x_i$$

It then follows from Eqs. 9.3-1 to 9.3-4 that for an ideal mixture

$$\underline{U}^{IM}(T, P, \underline{x}) = \sum x_i \underline{U}_i(T, P) \qquad \textbf{(9.3-5a)}$$

$$\underline{V}^{IM}(T, P, \underline{x}) = \sum x_i \underline{V}_i(T, P) \qquad \textbf{(9.3-5b)}$$

$$\underline{H}^{IM}(T, P, \underline{x}) = \sum x_i \underline{H}_i(T, P) \qquad \textbf{(9.3-5c)}$$

$$\underline{S}^{IM}(T, P, \underline{x}) = \sum x_i \underline{S}_i(T, P) - R \sum x_i \ln x_i \qquad \textbf{(9.3-5d)}$$

$$\underline{G}^{IM}(T, P, \underline{x}) = \sum x_i \underline{G}_i(T, P) + RT \sum x_i \ln x_i \qquad \textbf{(9.3-5e)}$$

$$\underline{A}^{IM}(T, P, \underline{x}) = \sum x_i \underline{A}_i(T, P) + RT \sum x_i \ln x_i \qquad \textbf{(9.3-5f)}$$

ILLUSTRATION 9.3-1

Determining Whether a Mixture Is an Ideal Mixture Even Though It Is Not an Ideal Gas Mixture

Determine whether the nitrogen-butane mixture of Illustration 9.1-1 is an ideal mixture.

SOLUTION

For the nitrogen-butane mixture to be an ideal mixture, from Eq. 9.3-5b at all temperatures, pressures, and compositions,

$$\underline{V}^{IM}(T, P, \underline{x}) = \sum x_i \underline{V}_i(T, P)$$

or, alternatively, after multiplying by P/RT and recognizing that $Z = P\underline{V}/RT$,

$$Z_{mix}^{IM}(T, P, \underline{x}) = \sum x_i Z_i(T, P) \qquad \text{(*)}$$

Below we tabulate Z_{mix} and Z_{mix}^{IM} at the conditions in Illustration 9.1-1.

Mole Fraction Butane	$P = 13.79$ MPa		$P = 68.95$ MPa	
	Z_{mix}	Z_{mix}^{IM}	Z_{mix}	Z_{mix}^{IM}
0	1.06	1.06	1.47	1.47
0.1	1.04	1.01	1.51	1.52
0.3	0.96	0.90	1.61	1.62
0.5	0.81	0.79	1.72	1.73
0.7	0.64	0.68	1.83	1.83
0.9	0.52	0.57	1.95	1.93
1.0	0.51	0.51	1.98	1.98

Clearly, Eq. * is much closer to being satisfied with the higher-pressure data than with the lower-pressure data. However, for an ideal mixture, this equation must be satisfied at all pressures, temperatures, and compositions. Therefore, we conclude that the nitrogen-butane mixture is not quite an ideal mixture at the conditions considered here, and it certainly is not an ideal gas mixture based on Illustration 9.1-1. ∎

Equations 9.3-3 to 9.3-5 resemble those obtained in Sec. 9.1 for the ideal gas mixture. There is an important difference, however. In the present case we are considering an ideal mixture of fluids that are not ideal gases, so each of the pure-component properties here will not be an ideal gas property, but rather a real fluid property that must either be measured or computed using the techniques described in Chapter 6. Thus, the molar volume $\underline{V}_i$ is not equal to RT/P, and the fugacity of each species is not equal to the pressure.

Also, from Eq. 9.3-3,

$$\bar{\phi}_i^{\mathrm{IM}}(T, P, \underline{x}) = \phi_i(T, P)$$

and since all derivatives at constant composition must be equal,

$$\left[\frac{\partial \ln(\bar{f}_i^{\mathrm{IM}}/x_i P)}{\partial T} \right]_{P,\underline{x}} = \left[\frac{\partial \ln(f_i/P)}{\partial T} \right]_P \tag{9.3-6}$$

Note that an ideal mixture identically satisfies the Gibbs-Duhem equation (Eq. 8.2-12b):

$$
\begin{aligned}
\underline{S}^{\mathrm{IM}} \, dT - \underline{V}^{\mathrm{IM}} \, dP &+ \sum x_i \, d\bar{G}_i^{\mathrm{IM}} \\
&= \sum x_i \bar{S}_i^{\mathrm{IM}} \, dT - \sum x_i \bar{V}_i^{\mathrm{IM}} \, dP + \sum x_i \, d(\underline{G}_i + RT \ln x_i) \\
&= \sum x_i \underline{S}_i \, dT - R \sum x_i \ln x_i \, dT - \sum x_i \underline{V}_i \, dP + \sum x_i \, d\underline{G}_i \\
&\quad + R \sum x_i \ln x_i \, dT + RT \sum x_i \, d \ln x_i \\
&= \sum x_i (\underline{S}_i \, dT - \underline{V}_i \, dP + d\underline{G}_i) + RT \sum dx_i \\
&\equiv 0
\end{aligned}
\tag{9.3-7}
$$

since $d\underline{G} = \underline{V} \, dP - \underline{S} \, dT$ for a pure component, and $\sum x_i = 1$, so that $d \sum x_i = \sum dx_i = d(1) = 0$.

Any property change on mixing $\Delta_{\mathrm{mix}} \underline{\theta}$ of a real mixture can be written in terms of the analogous property of an ideal mixture plus an additional term as follows:

$$
\begin{aligned}
\Delta_{\mathrm{mix}} \underline{\theta}(T, P, \underline{x}) &= \Delta_{\mathrm{mix}} \underline{\theta}^{\mathrm{IM}}(T, P, \underline{x}) + [\Delta_{\mathrm{mix}} \underline{\theta}(T, P, \underline{x}) - \Delta_{\mathrm{mix}} \underline{\theta}^{\mathrm{IM}}(T, P, \underline{x})] \\
&= \Delta_{\mathrm{mix}} \underline{\theta}^{\mathrm{IM}} + \underline{\theta}^{\mathrm{ex}}
\end{aligned}
$$

where

Excess mixing property

$$\underline{\theta}^{\mathrm{ex}} = \Delta_{\mathrm{mix}} \underline{\theta}(T, P, \underline{x}) - \Delta_{\mathrm{mix}} \underline{\theta}^{\mathrm{IM}}(T, P, \underline{x}) = \sum_i x_i \bar{\theta}_i - \sum_i x_i \bar{\theta}_i^{\mathrm{IM}} = \sum_i x_i (\bar{\theta}_i - \bar{\theta}_i^{\mathrm{IM}}) \tag{9.3-8}$$

is the **excess mixing property**, that is, the change in $\underline{\theta}$ that occurs on mixing at constant temperature and pressure in addition to that which would occur if an ideal mixture were formed. Excess properties $\underline{\theta}^{ex}$ are generally complicated, nonlinear functions of the composition, temperature, and pressure, and usually must be obtained from experiment. The hope is, however, that the excess properties will be small (particularly when θ is the Gibbs energy or entropy) compared with $\Delta_{mix}\theta^{IM}$ so that even an approximate theory for $\underline{\theta}^{ex}$ may be sufficient to compute $\Delta_{mix}\underline{\theta}$ with reasonable accuracy. Table 9.3-1 also contains a list of excess thermodynamic properties that will be of interest in this book.

Finally, in analogy with Eq. 6.1-12, we define a partial molar excess quantity by the relation

$$\bar{\theta}_i^{ex} = \left. \frac{\partial(N\underline{\theta}^{ex})}{\partial N_i} \right|_{T,P,N_{j\neq i}} = \left. \frac{\partial}{\partial N_i} \right|_{T,P,N_{j\neq i}} \sum_k N_k(\bar{\theta}_k - \bar{\theta}_k^{IM})$$

which reduces to

$$\bar{\theta}_i^{ex} = \bar{\theta}_i - \bar{\theta}_i^{IM} \tag{9.3-9}$$

since

$$\sum_k N_k \frac{\partial}{\partial N_i}(\bar{\theta}_k - \bar{\theta}_k^{IM})\Big|_{T,P,N_{j\neq i}} = 0$$

by the Gibbs-Duhem equation (Eq. 8.2-11). For the case in which $\bar{\theta}_i$ is equal to the partial molar Gibbs energy, we have

$$
\begin{aligned}
\bar{G}_i^{ex} &= (\bar{G}_i - \bar{G}_i^{IM}) = (\bar{G}_i - \bar{G}_i^{IG}) + (\bar{G}_i^{IG} - \bar{G}_i^{IM}) \\
&= (\bar{G}_i - \bar{G}_i^{IG}) + (\underline{G}_i^{IG} + RT \ln x_i - \underline{G}_i - RT \ln x_i) \\
&= (\bar{G}_i - \bar{G}_i^{IG}) + (\underline{G}_i^{IG} - \underline{G}_i) \\
&= RT \ln \frac{\bar{f}_i}{x_i P} - RT \ln \frac{f_i}{P} \\
&= RT \ln\left(\frac{\bar{f}_i}{x_i f_i}\right) = RT \ln\left(\frac{\bar{\phi}_i}{\phi_i}\right) = \int_0^P [\bar{V}_i - \underline{V}_i]\, dP
\end{aligned}
\tag{9.3-10}
$$

If equations of state (or specific volume data) are available for both the pure fluid and the mixture, the integral can be evaluated. However, for liquid mixtures not describable by an equation of state, common thermodynamic notation is to define an **activity coefficient** $\gamma_i(T, P, \underline{x})$, which is a function of temperature, pressure, and composition, by the equation

Definition of the activity coefficient

$$\boxed{\bar{f}_i^L(T, P, \underline{x}) = x_i \gamma_i(T, P, \underline{x}) f_i^L(T, P)} \tag{9.3-11}$$

where the superscript L has been used to denote the liquid phase. Alternatively,

$$\boxed{RT \ln \gamma_i(T, P, \underline{x}) = \bar{G}_i^{ex} = \left(\frac{\partial N\underline{G}^{ex}}{\partial N_i}\right)_{T,P,N_{j\neq i}}} \tag{9.3-12}$$

Table 9.3-1 Mixture Thermodynamic Properties at Constant T and P

Property	Ideal Mixtures	Real Mixtures
Volume	$\overline{V}_i^{IM} = \underline{V}_i$; $\Delta_{mix}\underline{V}^{IM} = 0$; $\underline{V}^{IM} = \sum_{i=1}^{c} x_i \underline{V}_i$	$\overline{V}_i^{ex} = \overline{V}_i - \underline{V}_i$; $\underline{V}^{ex} = \Delta_{mix}\underline{V}$; $\underline{V} = \sum_{i=1}^{c} x_i \underline{V}_i + \Delta_{mix}\underline{V}$
Internal energy	$\overline{U}_i^{IM} = \underline{U}_i$; $\Delta_{mix}\underline{U}^{IM} = 0$; $\underline{U}^{IM} = \sum_{i=1}^{c} x_i \underline{U}_i$	$\overline{U}_i^{ex} = \overline{U}_i - \underline{U}_i$; $\underline{U}^{ex} = \Delta_{mix}\underline{U}$; $\underline{U} = \sum_{i=1}^{c} x_i \underline{U}_i + \Delta_{mix}\underline{U}$
Enthalpy	$\overline{H}_i^{IM} = \underline{H}_i$; $\Delta_{mix}\underline{H}^{IM} = 0$; $\underline{H}^{IM} = \sum_{i=1}^{c} x_i \underline{H}_i$	$\overline{H}_i^{ex} = \overline{H}_i - \underline{H}_i$; $\underline{H}^{ex} = \Delta_{mix}\underline{H}$; $\underline{H} = \sum_{i=1}^{c} x_i \underline{H}_i + \Delta_{mix}\underline{H}$
Entropy	$\overline{S}_i^{IM} = \underline{S}_i - R\ln x_i$; $\Delta_{mix}\underline{S}^{IM} = -R\sum_{i=1}^{c} x_i \ln x_i$; $\underline{S}^{IM} = \sum_{i=1}^{c} x_i \underline{S}_i - R\sum_{i=1}^{c} x_i \ln x_i$	$\overline{S}_i^{ex} = \overline{S}_i - \underline{S}_i + R\ln x_i$; $\underline{S}^{ex} = \Delta_{mix}\underline{S} + R\sum_{i=1}^{c} x_i \ln x_i$; $\underline{S} = \sum_{i=1}^{c} x_i \underline{S}_i - R\sum_{i=1}^{c} x_i \ln x_i + \underline{S}^{ex}$
Gibbs energy	$\overline{G}_i^{IM} = \underline{G}_i + RT\ln x_i$; $\Delta_{mix}\underline{G}^{IM} = RT\sum_{i=1}^{c} x_i \ln x_i$; $\underline{G}^{IM} = \sum_{i=1}^{c} x_i \underline{G}_i + RT\sum_{i=1}^{c} x_i \ln x_i$	$\overline{G}_i^{ex} = \overline{G}_i - \underline{G}_i - RT\ln x_i$; $\underline{G}^{ex} = \Delta_{mix}\underline{G} - RT\sum_{i=1}^{c} x_i \ln x_i$; $\underline{G} = \sum_{i=1}^{c} x_i \underline{G}_i + RT\sum_{i=1}^{c} x_i \ln x_i + \underline{G}^{ex}$
Helmholtz energy	$\overline{A}_i^{IM} = \underline{A}_i + RT\ln x_i$; $\Delta_{mix}\underline{A}^{IM} = RT\sum_{i=1}^{c} x_i \ln x_i$; $\underline{A}^{IM} = \sum_{i=1}^{c} x_i \underline{A}_i + RT\sum_{i=1}^{c} x_i \ln x_i$	$\overline{A}_i^{ex} = \overline{A}_i - \underline{A}_i - RT\ln x_i$; $\underline{A}^{ex} = \Delta_{mix}\underline{A} - RT\sum_{i=1}^{c} x_i \ln x_i$; $\underline{A} = \sum_{i=1}^{c} x_i \underline{A}_i + RT\sum_{i=1}^{c} x_i \ln x_i + \underline{A}^{ex}$

[Note that the fugacity of the pure liquid, $f_i^L(T, P)$, in Eq. 9.3-11 can be found from the methods of Sec. 7.4b.] As will be seen in Chapters 10 to 12, the calculation of the activity coefficient for each species in a mixture is an important step in many phase equilibrium calculations. Therefore, much of this chapter deals with models (equations) for $\underline{G}^{\mathrm{ex}}$ and activity coefficients.

ILLUSTRATION 9.3-2

Analyzing the Gibbs Energy of a Mixture to Determine Whether It Is an Ideal Mixture

Experimentally (as will be described in Sec. 10.2) it has been found that the Gibbs energy for a certain binary mixture has the form

$$\underline{G}_{\mathrm{mix}}(T, P, \underline{x}) = \sum_{i=1}^{2} x_i \underline{G}_i(T, P) + RT \sum_{i=1}^{2} x_i \ln x_i + a x_1 x_2 \qquad (*)$$

where a is a constant. Determine whether this mixture is an ideal mixture.

SOLUTION

For an ideal mixture Eqs. 9.3-1a and b must be satisfied at all conditions. To see if this is so, we start with the relation between the Gibbs energy and volume and use the equation above to obtain

$$\underline{V}_{\mathrm{mix}} = \left(\frac{\partial \underline{G}_{\mathrm{mix}}}{\partial P}\right)_{T,N} = \sum_{i=1}^{2} x_i \left(\frac{\partial \underline{G}_i}{\partial P}\right)_{T,N} = \sum x_i \underline{V}_i(T, P)$$

Therefore,

$$\overline{V}_1(T, P) = \left.\frac{\partial(N\underline{V}_{\mathrm{mix}})}{\partial N_1}\right|_{T,N_2} = \frac{\partial}{\partial N_1} \left(N_1 \underline{V}_1(T, P) + N_2 \underline{V}_2(T, P)\right)_{T,N_2} = \underline{V}_1(T, P)$$

Similarly, $\overline{V}_2(T, P) = \underline{V}_2(T, P)$, so Eq. 9.3-1b is satisfied. We now move on to Eq. 9.3-1a by starting with

$$\left.\frac{\partial}{\partial T}\right|_P \left(\frac{G}{T}\right) = \frac{1}{T}\left(\frac{\partial \underline{G}}{\partial T}\right)_P - \frac{G}{T^2} = \frac{1}{T}(-S) - \frac{(H - TS)}{T^2} = -\frac{H}{T^2}$$

or

$$\underline{H} = -T^2 \left.\frac{\partial}{\partial T}\right|_P \left(\frac{G}{T}\right)$$

Therefore,

$$\underline{H}_{\mathrm{mix}} = -T^2 \left.\frac{\partial}{\partial T}\right|_P \left[\sum x_i \frac{\underline{G}_i}{T} + R \sum x_i \ln x_i + \frac{a}{T} x_1 x_2\right]$$

$$= -T^2 \left[\sum x_i \left(-\frac{\underline{H}_i}{T^2}\right) - \frac{a}{T^2} x_1 x_2\right] = \sum x_i \underline{H}_i + a x_1 x_2$$

Then

$$\bar{H}_1(T, P, \underline{x}) = \left.\frac{\partial}{\partial N_1}\right|_{T,P,N_2} N\underline{H}_{\text{mix}} = \left.\frac{\partial}{\partial N_1}\right|_{T,P,N_2} \left(N_1\underline{H}_1 + N_2\underline{H}_2 + \frac{aN_1N_2}{N_1 + N_2}\right)$$

$$= \underline{H}_1(T, P) + a\left[\frac{N_2}{N_1 + N_2} - \frac{N_1N_2}{(N_1 + N_2)^2}\right]$$

$$= \underline{H}_1(T, P) + a(x_2 - x_1x_2) = \underline{H}_1(T, P) + ax_2(1 - x_1)$$

$$= \underline{H}_1(T, P) + ax_2^2$$

Similarly, $\bar{H}_2(T, P, \underline{x}) = \underline{H}_2(T, P) + ax_1^2$.

Therefore, Eq. 9.3-1a is not satisfied, and the mixture described by Eq. * is not an ideal mixture.

COMMENT

Clearly, for this mixture $\underline{G}^{\text{ex}} = ax_1x_2$ and $\underline{A}^{\text{ex}} = ax_1x_2$. Since the excess Gibbs energies of mixing are not zero, this also shows that the mixture cannot be ideal. ∎

ILLUSTRATION 9.3-3

Developing Expressions for Activity Coefficients and Species Fugacities from the Gibbs Energy

Develop an expression for the activity coefficients and species fugacities using the Gibbs energy function of the previous illustration.

SOLUTION

We start from

$$\underline{G}^{\text{ex}} = \underline{G}_{\text{mix}} - \underline{G}_{\text{mix}}^{\text{IM}}$$

$$= \left(\sum_{i=1}^{2} x_i\underline{G}_i + RT\sum_{i=1}^{2} x_i \ln x_i + ax_1x_2\right) - \left(\sum_{i=1}^{2} x_i\underline{G}_i + RT\sum_{i=1}^{2} x_i \ln x_i\right) = ax_1x_2$$

Then, by definition,

$$\bar{G}_1^{\text{ex}} = \left(\frac{\partial N\underline{G}^{\text{ex}}}{\partial N_1}\right)_{T,P,N_2} = \left.\frac{\partial}{\partial N_1}\right|_{T,P,N_2} Nax_1x_2 = \left.\frac{\partial}{\partial N_1}\right|_{T,P,N_2} \frac{aN_1N_2}{(N_1 + N_2)}$$

$$= a\left[\frac{N_2}{N_1 + N_2} - \frac{N_1N_2}{(N_1 + N_2)^2}\right] = a(x_2 - x_1x_2)$$

$$= ax_2(1 - x_1) = ax_2^2 = RT \ln \gamma_1(x_1)$$

or

$$\gamma_1(x_1) = \exp\left[\frac{ax_2^2}{RT}\right] = \exp\left[\frac{a(1 - x_1)^2}{RT}\right]$$

Similarly,

$$\bar{G}_2^{\text{ex}} = ax_1^2 = a(1 - x_2)^2 \quad \text{and} \quad \gamma_2(x_2) = \exp\left[\frac{ax_1^2}{RT}\right] = \exp\left[\frac{a(1 - x_2)^2}{RT}\right]$$

Finally,

$$\bar{f}_1(T, P, x_1) = x_1 \exp\left[\frac{ax_2^2}{RT}\right] f_1(T, P) = x_1 \exp\left[\frac{a(1 - x_1)^2}{RT}\right] f_1(T, P)$$

and

$$\bar{f}_2(T, P, x_2) = x_2 \exp\left[\frac{a(1 - x_2)^2}{RT}\right] f_2(T, P)$$

COMMENT

The results above for the activity coefficients and fugacities are not general, but are specific to the very simple excess Gibbs energy function we have used,

$$\underline{G}^{ex}(T, P, \underline{x}) = ax_1 x_2$$

As we shall see, in order to have excess Gibbs energy functions that more accurately fit experimental data, more complicated activity coefficient and fugacity expressions are needed. However, the expressions here do show that the activity coefficients are strong functions of composition or mole fraction. Indeed, in this example the activity coefficients are exponential functions of the square of the mole fraction.

There is a very important point in this illustration that is easily overlooked. To obtain the activity coefficient from the expression for the excess Gibbs energy, we used the definition

Correct!

$$\bar{G}_1^{ex} = RT \ln \gamma_1(x_1) = \left(\frac{\partial N \underline{G}^{ex}}{\partial N_1}\right)_{T,P,N_2}$$

We did not use

Incorrect!

$$\bar{G}_1^{ex} = \left(\frac{d\underline{G}^{ex}}{dx_1}\right)_{T,P}$$

which is an incorrect equation and gives the wrong result (try it and see).   ∎

From Eq. 9.3-3, the activity coefficient is equal to unity for species in ideal mixtures, and for nonideal mixtures

$$\gamma_i(T, P, \underline{x}) = \exp\left(\frac{\bar{G}_i^{ex}}{RT}\right) = \exp\left(\frac{1}{RT}\int_0^P \left[\bar{V}_i(T, P, \underline{x}) - \underline{V}_i(T, P)\right] dP\right)$$

$$(9.3\text{-}13)$$

Since both real and ideal mixtures satisfy the Gibbs-Duhem equation of Sec. 8.2,

$$0 = \underline{S}\, dT - \underline{V}\, dP + \sum_{i=1}^{C} x_i \, d\bar{G}_i \qquad (8.2\text{-}12b)$$

and

$$0 = \underline{S}^{IM}\, dT - \underline{V}^{IM}\, dP + \sum_{i=1}^{C} x_i \, d\bar{G}_i^{IM} \qquad (9.3\text{-}7)$$

we can subtract these two equations to obtain a form of the Gibbs-Duhem equation applicable to excess thermodynamic properties:

Gibbs-Duhem equation for excess properties

$$0 = \underline{S}^{ex}\, dT - \underline{V}^{ex}\, dP + \sum_{i=1}^{c} x_i\, d\bar{G}_i^{ex} \tag{9.3-14}$$

Now using Eq. 9.3-12, in the form of $\bar{G}_i^{ex} = RT \ln \gamma_i$, and

$$\underline{G}^{ex} = \sum x_i \bar{G}_i^{ex} = \underline{H}^{ex} - T\underline{S}^{ex}$$

we obtain

$$0 = \frac{\underline{H}^{ex}}{T}\, dT - \underline{V}^{ex}\, dP + RT \sum_{i=1}^{c} x_i\, d \ln \gamma_i \tag{9.3-15}$$

Thus, for changes in composition at constant temperature and pressure, we have

$$0 = \sum_{i=1}^{c} x_i\, d \ln \gamma_i \Big|_{T,P} \tag{9.3-16}$$

which is a special case of Eqs. 8.2-9b and 8.2-13b. For a binary mixture, we have, from Eq. 8.2-20,

Gibbs-Duhem equation for activity coefficients

$$x_1 \left(\frac{\partial \ln \gamma_1}{\partial x_1} \right)_{T,P} + x_2 \left(\frac{\partial \ln \gamma_2}{\partial x_1} \right)_{T,P} = 0 \tag{9.3-17}$$

ILLUSTRATION 9.3-4
Testing Whether an Activity Coefficient Model Satisfies the Gibbs-Duhem Equation

Determine whether the activity coefficient expressions derived in the previous illustration satisfy the Gibbs-Duhem equation, Eq. 9.3-17.

SOLUTION
From the previous illustration, we have

$$\ln \gamma_1(x_1) = \frac{a}{RT} x_2^2 = \frac{a}{RT}(1 - x_1)^2$$

and

$$\ln \gamma_2(x_2) = \frac{a}{RT} x_1^2 = \frac{a}{RT}(1 - x_2)^2$$

Consequently,

$$\left(\frac{\partial \ln \gamma_1}{\partial x_1} \right)_{T,P} = \frac{-2a}{RT}(1 - x_1) = \frac{-2ax_2}{RT} \quad \text{and} \quad \left(\frac{\partial \ln \gamma_2}{\partial x_1} \right)_{T,P} = \frac{2ax_1}{RT}$$

so that

$$x_1 \left(\frac{\partial \ln \gamma_1}{\partial x_1} \right)_{T,P} + x_2 \left(\frac{\partial \ln \gamma_2}{\partial x_1} \right)_{T,P} = -\frac{2ax_1 x_2}{RT} + \frac{2ax_1 x_2}{RT} = 0$$

as required by Eq. 9.3-17.

<u>**COMMENT**</u>

All activity coefficients derived from an excess Gibbs energy expression that satisfies the boundary conditions of being zero at $x_1 = 0$ and 1 will satisfy the Gibbs-Duhem equation. Can you prove this? ∎

To determine the dependence of the activity coefficient γ_i on temperature and pressure, we rewrite Eq. 9.3-13 as

$$\ln \gamma_i(T, P, \underline{x}) = \frac{\overline{G}_i^{ex}(T, P, \underline{x})}{RT} = \frac{1}{RT} \int_0^P \left[\overline{V}_i(T, P, \underline{x}) - \underline{V}_i(T, P) \right] dP \quad \textbf{(9.3-18)}$$

Taking the derivative with respect to pressure at constant temperature and composition, we obtain

$$\left(\frac{\partial \ln \gamma_i(T, P, \underline{x})}{\partial P} \right)_{T,\underline{x}} = \frac{\partial}{\partial P} \left(\frac{1}{RT} \int_0^P \left[\overline{V}_i(T, P, \underline{x}) - \underline{V}_i(T, P) \right] dP \right)_{T,\underline{x}}$$

$$= \frac{\overline{V}_i(T, P, \underline{x}) - \underline{V}_i(T, P)}{RT} = \frac{\overline{V}_i^{ex}(T, P, \underline{x})}{RT}$$

$$\textbf{(9.3-19)}$$

Therefore,

$$\gamma_i(T, P_2, \underline{x}) = \gamma_i(T, P_1, \underline{x}) \exp \left[\int_{P_1}^{P_2} \frac{\overline{V}_i^{ex}(T, P, \underline{x})}{RT} dP \right]$$

$$\simeq \gamma_i(T, P_1, \underline{x}) \exp \left[\frac{\overline{V}_i^{ex}(T, \underline{x})(P_2 - P_1)}{RT} \right]$$

$$\textbf{(9.3-20)}$$

In obtaining the last expression in Eq. 9.3-20 we have assumed that the excess partial molar volume is independent of pressure. (Note that although Eqs. 9.3-18 and 9.3-19 are correct, they are difficult to use in practice since the activity coefficient description is applied to fluid mixtures not well described by an equation of state.) Next, taking the temperature derivative of Eq. 9.3-18 at constant pressure and composition, we obtain

$$\left(\frac{\partial \ln \gamma_i(T, P, \underline{x})}{\partial T} \right)_{P,\underline{x}} = \frac{\partial}{\partial T} \left(\frac{\overline{G}_i^{ex}}{RT} \right)_{P,\underline{x}} = -\frac{\overline{H}_i^{ex}(T, P, \underline{x})}{RT^2} \quad \textbf{(9.3-21)}$$

so that

Temperature variation of activity coefficients

$$\gamma_i(T_2, P, \underline{x}) = \gamma_i(T_1, P, \underline{x}) \exp \left[-\int_{T_1}^{T_2} \frac{\overline{H}_i^{ex}(T, P, \underline{x})}{RT^2} dT \right] \quad \textbf{(9.3-22)}$$

For small temperature changes, or if the excess partial molar enthalpy is independent of temperature, we have

$$\gamma_i(T_2, P, \underline{x}) = \gamma_i(T_1, P, \underline{x}) \exp \left[\frac{\overline{H}_i^{ex}(\underline{x})}{R} \left(\frac{1}{T_2} - \frac{1}{T_1} \right) \right] \quad \textbf{(9.3-23)}$$

These equations, especially Eqs. 9.3-20, 9.3-22, and 9.3-23, are useful when one wants to correct the numerical values of activity coefficients obtained at one temperature and pressure for use at other temperatures and pressures. For example, Eq. 9.3-22 or 9.3-23 would be needed for the prediction of low-temperature liquid-liquid phase equilibrium (Sec. 11.1) if the only activity coefficient data available were those obtained from higher-temperature vapor-liquid equilibrium measurements (Sec. 10.2). Alternatively, if activity coefficients have been measured at several pressures, Eq. 9.3-20 can be used to calculate partial molar excess volumes, or if activity coefficients have been measured at several temperatures, Eq. 9.2-23 can be used to calculate partial molar excess enthalpies (and heats of mixing).

9.4 FUGACITY OF SPECIES IN GASEOUS, LIQUID, AND SOLID MIXTURES

The fugacity function is central to the calculation of phase equilibrium. This should be apparent from the earlier discussion of this chapter and from the calculations of Sec. 7.5, which established that once we had the pure fluid fugacity, phase behavior in a pure fluid could be predicted. Consequently, for the remainder of this chapter we will be concerned with estimating the fugacity of species in gaseous, liquid, and solid mixtures.

9.4-1 Gaseous Mixtures

Several methods are commonly used for estimating the fugacity of a species in a gaseous mixture.[1] The most approximate method is based on the observation that some gaseous mixtures follow Amagat's law, that is,

$$V_{\mathrm{mix}}(T, P, N_1, N_2, \ldots) = \sum_{i=1}^{\mathcal{C}} N_i \underline{V}_i(T, P) \qquad (9.4\text{-}1)$$

which, on multiplying by P/NRT, can be written as

$$Z_{\mathrm{mix}}(T, P, \underline{y}) = \sum_{i=1}^{\mathcal{C}} y_i Z_i(T, P) \qquad (9.4\text{-}2)$$

Here $Z = PV/NRT$ is the compressibility of the mixture, Z_i is the compressibility of the pure component i, and $y_i = N_i/N$ is the mole fraction of species i. (*Hereafter y_i will be used to indicate a gas-phase mole fraction and x_i to denote a mole fraction in the liquid phase or in equations that are applicable to both gases and liquids.*)

The data in Fig. 9.4-1 for the nitrogen-butane system show that the mixture compressibility is nearly a linear function of mole fraction at both low and high pressures, so that Eq. 9.4-2 is approximately satisfied at these conditions, but fails in the intermediate pressure range. If we nonetheless accept Eq. 9.4-2 as being a reasonable approximation over the whole pressure range, we then have, from Eq. 9.4-1 and the definition of the partial molar volume, $\bar{V}_i(T, P, \underline{y}) = \underline{V}_i(T, P)$ and, from Eq. 9.2-6,

Lewis-Randall rule

$$\boxed{\bar{f}_i^{\mathrm{V}}(T, P, \underline{y}) = y_i f_i^{\mathrm{V}}(T, P)} \qquad (9.4\text{-}3)$$

---

[1] See also Secs. 9.7 and 9.9.

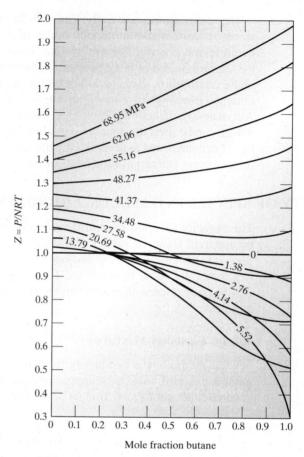

Figure 9.4-1 Compressibility factors for nitrogen-butane mixtures at 444.3 K. Data of R. B. Evans and C. M. Watson, *Chem. Eng. Data Ser.,* **1,** 67 (1956). Reprinted with permission from American Chemical Society, Copyright 1956. PRAUSNITZ, MOLECULAR THERMO-DYNAMICS OF FLUID-PHASE EQUILIBRIA., 1st Edition, © 1969. Reprinted by permission of Pearson Education, Inc., Upper Saddle River, NJ.

This result, which relates the fugacity of a species in a gaseous mixture only to its mole fraction and the fugacity of the pure *gaseous* component at the same temperature and pressure, is known as the **Lewis-Randall rule.**

ILLUSTRATION 9.4-1

Approximate Species Fugacity Calculation Using the Lewis-Randall Rule

Compute the fugacities of ethane and *n*-butane in an equimolar mixture at 323.15 K at 1, 10, and 15 bar total pressure assuming that the Lewis-Randall rule is correct.

SOLUTION

From Illustration 7.4-4, we have

$$f_{ET}(373.15 \text{ K, 1 bar}) = 0.996 \text{ bar} \qquad f_{BU}(373.15 \text{ K, 1 bar}) = 0.986 \text{ bar}$$
$$f_{ET}(373.15 \text{ K, 10 bar}) = 9.629 \text{ bar} \qquad f_{BU}(373.15 \text{ K, 10 bar}) = 8.628 \text{ bar}$$
$$f_{ET}(373.15 \text{ K, 15 bar}) = 14.16 \text{ bar} \qquad f_{BU}(373.15 \text{ K, 15 bar}) = 11.86 \text{ bar}$$

Then, by the Lewis-Randall rule, $\bar{f}_i(T, P, \underline{y}) = y_i f_i(T, P)$, we have

$$\bar{f}_{ET}(373.15 \text{ K, 1 bar, } y_{ET} = 0.5) = 0.5 \times f_{ET}(373.15, 1 \text{ bar})$$
$$= 0.5 \times 0.996 \text{ bar} = 0.498 \text{ bar}$$

$$\bar{f}_{ET}(373.15 \text{ K, 10 bar, } y_{ET} = 0.5) = 4.815 \text{ bar}$$
$$\bar{f}_{ET}(373.15 \text{ K, 15 bar, } y_{ET} = 0.5) = 7.08 \text{ bar}$$

and

$$\bar{f}_{BU}(373.15 \text{ K, 1 bar, } y_{BU} = 0.5) = 0.493 \text{ bar}$$
$$\bar{f}_{BU}(373.15 \text{ K, 10 bar, } y_{BU} = 0.5) = 4.314 \text{ bar}$$
$$\bar{f}_{BU}(373.15 \text{ K, 15 bar, } y_{BU} = 0.5) = 5.930 \text{ bar}$$ ∎

A more accurate way to estimate the fugacity of a species in a gaseous mixture is to start with Eq. 9.2-13,

$$
\ln \frac{\bar{f}_i^V(T, P, \underline{y})}{y_i P} = \ln \bar{\phi}_i^V(T, P, \underline{y})
$$
$$
= \frac{1}{RT} \int_{\underline{V}=\infty}^{\underline{V}=Z^V RT/P} \left[\frac{RT}{\underline{V}} - N \left(\frac{\partial P}{\partial N_i} \right)_{T,V,N_{j\neq i}} \right] d\underline{V} - \ln Z^V
$$

$$(9.2\text{-}13)$$

and use an appropriate equation of state. At low pressures, the truncated virial equation of state

$$
\frac{P\underline{V}}{RT} = 1 + \frac{B_{mix}(T, \underline{y})}{\underline{V}} = Z_{mix}
$$

$$(9.4\text{-}4)$$

can be used if data for the mixture virial coefficient as a function of composition are available.[2] From statistical mechanics it is known that

$$
B_{mix}(T, \underline{y}) = \sum_i \sum_j y_i y_j B_{ij}(T)
$$

$$(9.4\text{-}5)$$

where each $B_{ij}(T)$ is a function only of temperature and $B_{ij} = B_{ji}$. Using Eqs. 9.4-4 and 9.4-5 in Eq. 9.2-13 yields (see Problem 9.6)

Fugacity coefficient from virial equation of state

$$
\boxed{
\begin{aligned}
\ln \frac{\bar{f}_i^V(T, P, \underline{y})}{y_i P} &= \frac{2}{\underline{V}} \sum_j y_j B_{ij}(T) - \ln Z_{mix} \\
&= \frac{2P}{Z_{mix} RT} \sum_j y_j B_{ij}(T) - \ln Z_{mix}
\end{aligned}
}
$$

$$(9.4\text{-}6)$$

[2] The book *The Virial Coefficients of Pure Gases and Mixtures: A Critical Compilation,* by J. H. Dymond and E. B. Smith, Clarendon Press, Oxford, 1980, contains second-virial-coefficient data for more than one thousand mixtures.

where

$$Z_{mix} = \frac{1}{2}\left(1 + \sqrt{1 + \frac{4B_{mix}P}{RT}}\right)$$ (9.4-7)

ILLUSTRATION 9.4-2

Species Fugacity Calculation Using the Virial Equation of State

Compute the fugacities of ethane and *n*-butane in an equimolar mixture at 373.15 K and 1, 10, and 15 bar using the virial equation of state.

Data:

$$B_{ET-ET} = -1.15 \times 10^{-4} \text{ m}^3/\text{mol} \qquad B_{BU-BU} = -4.22 \times 10^{-4} \text{ m}^3/\text{mol}$$

$$B_{ET-BU} = -2.15 \times 10^{-4} \text{ m}^3/\text{mol}$$

SOLUTION

We start by noting from Eq. 9.4-5 that

$$B_{mix} = \sum\sum y_i y_j B_{ij} = y_{ET}^2 B_{ET-ET} + 2y_{ET}y_{BU}B_{ET-BU} + y_{BU}^2 B_{BU-BU}$$

$$= -2.417 \times 10^{-4} \text{ m}^3/\text{mol}$$

Then by using Eqs. 9.4-6 and 9.4-7 we find that

P	Z	$\bar{f}_{ET}(373.15, P, y_{ET} = 0.5)$	$\bar{f}_{BU}(373.15, P, y_{BU} = 0.5)$
1 bar	0.992	0.499 bar	0.494 bar
10 bar	0.915	4.866 bar	4.367 bar
15 bar	0.865	7.211 bar	6.074 bar

COMMENT

These results are slightly higher than those computed from the Lewis-Randall rule, and are more accurate. ∎

At higher pressures, Eq. 9.2-13 can still be used to compute the fugacity of a species in a gaseous mixture, but more accurate equations of state must be used. One can, for example, use the Peng-Robinson equation of state

$$P = \frac{RT}{\underline{V} - b} - \frac{a(T)}{\underline{V}(\underline{V} + b) + b(\underline{V} - b)}$$ (6.4-2)

where the parameters a and b are now those for the mixture, a_{mix} and b_{mix}. To obtain these mixture parameters one starts with the a and b parameters for the pure components obtained from either fitting pure component data or the generalized correlations of Sec. 6.7 and then uses the **mixing rules**

Equation-of-state mixing rule

$$a_{\text{mix}} = \sum_{i=1}^{c} \sum_{j=1}^{c} y_i y_j a_{ij}$$

$$b_{\text{mix}} = \sum_{i=1}^{c} y_i b_i$$

(9.4-8)

where a_{ii} and b_i are the parameters for pure component i, and the **combining rule**

Equation-of-state combining rule

$$a_{ij} = \sqrt{a_{ii} a_{jj}}(1 - k_{ij}) = a_{ji}$$

(9.4-9)

Here a new parameter k_{ij}, known as the **binary interaction parameter**, has been introduced to result in more accurate mixture equation-of-state calculations. This parameter is found by fitting the equation of state to mixture data (usually vapor-liquid equilibrium data, as discussed in Chapter 10). Values of the binary interaction parameter k_{ij} that have been reported for a number of binary mixtures appear in Table 9.4-1. Equations 9.4-8 and 9.4-9 are referred to as the van der Waals one-fluid mixing rules. The term *one-fluid* derives from the fact that the mixture is being described by the same equation of state as the pure fluids, but with concentration-dependent parameters.

As a result of the mole fraction dependence of the equation-of-state parameters, the pressure is a function of mole fraction or, alternatively, of the number of moles of each species present. Evaluating the derivative $(\partial P/\partial N_i)_{T,V,N_{j\neq i}}$, which appears in Eq. 9.2-13, using the Peng-Robinson equation of state yields

Fugacity coefficient from Peng-Robinson equation of state

$$\ln \frac{\bar{f}_i^{\text{V}}(T, P, \underline{y})}{y_i P}$$

$$= \ln \bar{\phi}_i^{\text{V}}(T, P, \underline{y})$$

$$= \frac{b_i}{b_{\text{mix}}}(Z_{\text{mix}}^{\text{V}} - 1) - \ln\left(Z_{\text{mix}}^{\text{V}} - \frac{b_{\text{mix}} P}{RT}\right)$$

$$- \frac{a_{\text{mix}}}{2\sqrt{2}b_{\text{mix}}RT}\left[\frac{2\sum_j y_j a_{ij}}{a_{\text{mix}}} - \frac{b_i}{b_{\text{mix}}}\right]\ln\left[\frac{Z_{\text{mix}}^{\text{V}} + \left(1 + \sqrt{2}\right)\dfrac{b_{\text{mix}} P}{RT}}{Z^{\text{V}} + (1 - \sqrt{2})\dfrac{b_{\text{mix}} P}{RT}}\right]$$

$$= \frac{B_i}{B_{\text{mix}}}(Z_{\text{mix}}^{\text{V}} - 1) - \ln(Z_{\text{mix}}^{\text{V}} - B_{\text{mix}})$$

$$- \frac{A_{\text{mix}}}{2\sqrt{2}B_{\text{mix}}}\left[\frac{2\sum_j y_j A_{ij}}{A_{\text{mix}}} - \frac{B_i}{B_{\text{mix}}}\right]\ln\left[\frac{Z_{\text{mix}}^{\text{V}} + (1 + \sqrt{2})B_{\text{mix}}}{Z_{\text{mix}}^{\text{V}} + (1 - \sqrt{2})B_{\text{mix}}}\right]$$

(9.4-10)

Table 9.4-1 Binary Interaction Parameters k_{12} for the Peng-Robinson Equation of State*

	C_2H_4	C_2H_6	C_3H_6	C_3H_8	$i\text{-}C_4H_{10}$	$n\text{-}C_4H_{10}$	$i\text{-}C_5H_{12}$	$n\text{-}C_5H_{12}$	$n\text{-}C_6H_{14}$	C_6H_6	$c\text{-}C_6H_{12}$	$n\text{-}C_7H_{16}$	$n\text{-}C_8H_{18}$	$n\text{-}C_{10}H_{22}$	N_2	CO	CO_2	SO_2	H_2S
CH_4	0.022	−0.003	0.033	0.016	0.026	0.019	−0.006	0.026	0.040	0.055	0.039	0.035	0.050	0.049	0.030	0.030	0.09	0.136	0.08
C_2H_4		0.010		0.001	−0.007	0.092				0.031		0.014		0.025	0.086	−0.022	0.056		0.086
C_2H_6			0.089	0.007	−0.014	0.010		0.008	−0.04	0.042	0.018	0.007	0.019	0.014	0.044	0.026	0.130		0.08
C_3H_6					−0.007										0.09	0.026	0.093		0.08
C_3H_8						0.003	0.011	0.027	0.001	0.023		0.006	0.0	0.0	0.078	0.03	0.12		0.047
$i\text{-}C_4H_{10}$						0.0									0.10	0.04	0.13		0.07
$n\text{-}C_4H_{10}$								0.017	−0.006					0.008	0.087	0.04	0.135		0.06
$i\text{-}C_5H_{12}$								0.06							0.092	0.04	0.121		0.063
$n\text{-}C_5H_{12}$										0.018	0.004	0.003	0.007		0.10	0.04	0.125		0.06
$n\text{-}C_6H_{14}$										0.010	−0.004	0.007	0.0		0.15	0.04	0.11		
C_6H_6											0.013	−0.008	0.003		0.164	0.11	0.077		
$c\text{-}C_6H_{12}$												0.001		0.1	0.14	0.10	0.105	0.015	
$n\text{-}C_7H_{16}$													0.0		0.1	0.04	0.10		
$n\text{-}C_8H_{18}$															0.1	0.04	0.12		
$n\text{-}C_{10}H_{22}$															0.11	0.04	0.114		0.033
N_2																0.012	−0.02	0.08	0.17
CO																	0.03		
CO_2																		0.136	0.054
SO_2																			0.097
H_2S																			

*Obtained from data in "Vapor-Liquid Equilibria for Mixtures of Low-Boiling Substances," by H. Knapp, R. Döring, L. Oellrich, U. Plöcker, and J. M. Prausnitz, *DECHEMA Chemistry Data Series*, Vol. VI, Frankfurt/Main, 1982, and other sources. Blanks indicate no data are available from which the k_{12} could be evaluated. In such case use estimates from mixtures of similar compounds.

where, again, $A = aP/(RT)^2$, $B = bP/RT$, and the superscript V is used to remind us that the vapor-phase compressibility factor is to be used. Also, a_{mix} and b_{mix} (and therefore A_{mix} and B_{mix}) are computed using the vapor-phase mole fractions.

To calculate the fugacity of each species in a gaseous mixture using Eq. 9.4-10 at specified values of T, P, and mole fractions of all components $y_1, y_2, \ldots, y_C$, the following procedure is used:

1. Obtain the parameters a_{ii} and b_i for each pure component of the mixture either from the generalized correlations given in Sec. 6.7 or, if necessary, from pure component data.
2. Compute the a and b parameters for the mixture using the mixing rules of Eq. 9.4-8. (If the value for any binary interaction parameter, k_{ij}, is not available, either use a value for similar mixtures or assume it is zero.)
3. Solve the cubic equation of state for the vapor compressibility Z^V.
4. Use this value of Z^V to compute the vapor fugacity for each species using Eq. 9.4-10 repeatedly for $i = 1, 2, \ldots, C$, where C is the number of components.

Computer programs and a MATHCAD worksheet to do this calculation are discussed in Appendix B.

ILLUSTRATION 9.4-3

Species Fugacity Calculation Using the Peng-Robinson Equation of State

Compute the fugacities of ethane and *n*-butane in an equimolar mixture at 373.15 K and 1, 10, and 15 bar using the Peng-Robinson equation of state.

SOLUTION

Here we use the DOS-based program VLMU or the Visual Basic Peng-Robinson program for mixtures. Note that from Table 9.4-1, $k_{ET-BU} = 0.010$. The results are given below.

P	Z	$\bar{f}_{ET}(373.15, P, y_{ET} = 0.5)$	$\bar{f}_{BU}(373.15, P, y_{BU} = 0.5)$
1 bar	0.991	0.498 bar	0.493 bar
10 bar	0.910	4.836 bar	4.333 bar
15 bar	0.861	7.143 bar	6.024 bar

COMMENT

These results are similar to those obtained using either the Lewis-Randall rule or the virial equation of state. However, greater differences between the methods occur for gases as the pressure is increased. At higher pressures the results from the Peng-Robinson equation of state are expected to be the most accurate. ∎

ILLUSTRATION 9.4-4

Calculation of the Fugacity of a Species in a Mixture by Two Methods

Compute the fugacity of both carbon dioxide and methane in an equimolar mixture at 500 K and 500 bar using (a) the Lewis-Randall rule and (b) the Peng-Robinson equation of state.

SOLUTION

a. The Lewis-Randall rule is

$$\bar{f}_i(T, P, \underline{y}) = y_i f_i(T, P) = y_i \left(\frac{f}{P}\right)_i P$$

From Fig. 7.4-1, we have

	T_r	P_r	f/P
CO_2	1.644	6.78	~ 0.77
CH_4	2.623	10.87	$\sim 1.01^*$

*Since it is very difficult to read the fugacity coefficient curve in the range of the reduced temperature and pressure for methane in this problem, this entry was obtained from the tables in O. A. Hougen, K. M. Watson, and R. A. Ragatz, *Chemical Process Principles,* Part II, *Thermodynamics,* 2nd ed., John Wiley & Sons, New York (1959).

Thus

$$\bar{f}_{CO_2} = 0.5 \times 0.77 \times 500 \text{ bar} = 192.5 \text{ bar}$$

and

$$\bar{f}_{CH_4} = 0.5 \times 1.01 \times 500 \text{ bar} = 252.5 \text{ bar}$$

b. Using the Peng-Robinson equation of state (Eqs. 6.4-2 and 9.4-8 through 9.4-10) and one of the computer programs discussed in Appendix B, we find, for $k_{12} = 0.0$,

$$\bar{f}_{CO_2} = 208.71 \text{ bar} \quad \text{and} \quad \bar{f}_{CH_4} = 264.72 \text{ bar}$$

and using $k_{12} = 0.09$ (from Table 9.4-1),

$$\bar{f}_{CO_2} = 212.81 \text{ bar} \quad \text{and} \quad \bar{f}_{CH_4} = 269.35 \text{ bar}$$

These last values should be the most accurate.   ∎

9.4-2   Liquid Mixtures

The estimation of species fugacities in liquid mixtures is done in two different ways, depending on the data available and the components in the mixture. For liquid mixtures involving only hydrocarbons and dissolved gases, such as nitrogen, hydrogen sulfide, and carbon dioxide, simple equations of state may be used to describe liquid-state behavior. For example, if the approximate Peng-Robinson equation of state and the van der Waals mixing rules are used, the fugacity of each species in the mixture is, following the same development as for gaseous mixtures, given by

$$\ln \frac{\bar{f}_i^L(T, P, \underline{x})}{x_i P}$$

$$= \ln \bar{\phi}_i^L(T, P, \underline{x})$$

$$= \frac{b_i}{b_{mix}}(Z_{mix}^L - 1) - \ln \left(Z_{mix}^L - \frac{b_{mix} P}{RT} \right)$$

$$- \frac{a_{mix}}{2\sqrt{2} b_{mix} RT} \left[\frac{2 \sum_j x_j a_{ij}}{a_{mix}} - \frac{b_i}{b_{mix}} \right] \ln \left[\frac{Z_{mix}^L + \left(1 + \sqrt{2}\right) \dfrac{b_{mix} P}{RT}}{Z_{mix}^L + (1 - \sqrt{2}) \dfrac{b_{mix} P}{RT}} \right]$$

$$= \frac{B_i}{B_{mix}}(Z_{mix}^L - 1) - \ln(Z_{mix}^L - B_{mix})$$

$$- \frac{A_{mix}}{2\sqrt{2} B_{mix}} \left[\frac{2 \sum_j x_j A_{ij}}{A_{mix}} - \frac{B_i}{B_{mix}} \right] \ln \left[\frac{Z_{mix}^L + (1 + \sqrt{2}) B_{mix}}{Z_{mix}^L + (1 - \sqrt{2}) B_{mix}} \right]$$

$$(9.4\text{-}11)$$

where, again, $A = aP/(RT)^2$, $B = bP/RT$, and Z^L is the liquid (high density or small Z) root for the compressibility factor. Here, however, a_{mix} and b_{mix} (and therefore A_{mix} and B_{mix}) are computed using the liquid-phase mole fractions. Consequently, the calculational scheme for species fugacities in liquid mixtures that can be described by equations of state is similar to that for gaseous mixtures, except that it is the liquid-phase, rather than the vapor-phase, compositions and compressibility factor that are used in the calculations. The Appendix B computer programs or MATHCAD worksheet mentioned previously can be used for this calculation as well.

ILLUSTRATION 9.4-5

Species Fugacity Calculation for a Hydrocarbon Mixture

Compute the fugacities of *n*-pentane and benzene in their equimolar mixture at 373.15 K and 50 bar.

SOLUTION

The simplest estimate is obtained using the Lewis-Randall rule and the results of Illustration 7.4-6. In this case we obtain

$$\bar{f}_{PE}(373.15 \text{ K}, 50 \text{ bar}, x_{PE} = 0.5) = 0.5 \times f_{PE}(373.15 \text{ K}, 50 \text{ bar})$$
$$= 0.5 \times 6.22 \text{ bar} = 3.11 \text{ bar}$$

and

$$\bar{f}_{BZ}(373.15 \text{ K}, 50 \text{ bar}, x_{BZ} = 0.5) = 0.5 \times 1.98 \text{ bar} = 0.99 \text{ bar}$$

A better estimate is obtained using the Peng-Robinson equation of state with a value $k_{PE-BZ} = 0.018$ from Table 9.4-1 and the program VLMU. The results are

$$\bar{f}_{PE}(373.15 \text{ K}, 50 \text{ bar}, x_{PE} = 0.5) = 3.447 \text{ bar}$$
$$\bar{f}_{BZ}(373.15 \text{ K}, 50 \text{ bar}, x_{BZ} = 0.5) = 1.129 \text{ bar}$$

■

For liquid mixtures in which one or more of the components cannot be described by an equation of state in the liquid phase—for example, mixtures containing alcohols, organic or inorganic acids and bases, and electrolytes—another procedure for estimating species fugacities must be used. The most common starting point is Eq. 9.3-11,

$$\bar{f}_i(T, P, \underline{x}) = x_i \gamma_i(T, P, \underline{x}) f_i(T, P) \tag{9.3-11}$$

where γ_i is the activity coefficient of species i, and $f_i(T, P)$ is the fugacity of pure species i as a *liquid* at the temperature and pressure of the mixture.[3] However, since volumetric equation-of-state data are not available, the activity coefficient is obtained not from integration of Eq. 9.3-13, but rather from

$$RT \ln \gamma_i(T, P, \underline{x}) = \bar{G}_i^{ex} = \left(\frac{\partial N \underline{G}^{ex}}{\partial N_i} \right)_{T, P, N_{j \neq i}} \tag{9.4-12}$$

and approximate models for the excess Gibbs energy $\underline{G}^{ex}$ of the liquid mixture. Several such liquid solution models are considered in Sec. 9.5.

9.4-3 Solid Mixtures

The atoms in a crystalline solid are arranged in an ordered lattice structure consisting of repetitions of a unit cell. To form a solid solution, it is necessary to replace some molecules in the lattice with molecules of another component. If the two species are different in size, shape, or in the nature of their intermolecular forces, a solid solution can be obtained only by greatly distorting the crystalline structure. Such a distortion requires a great deal of energy, and, therefore, is energetically unfavorable. That is, the total energy (and in this case the Gibbs energy) of the distorted crystalline solid solution is greater than the energy of the two pure solids (each in its undistorted crystalline state). Consequently, with the exception of metals and metal alloys, the energetically preferred state for many solid mixtures is as heterogeneous mixtures consisting of regions of pure single species in their usual crystalline state. If these regions are large enough that macroscopic thermodynamics can be applied without making corrections for the thermodynamic state of the molecules at the interface between two solid regions, the solid mixture can be treated as an agglomeration of pure species, each having its own *pure component* solid fugacity. Thus

Fugacity of a species in many mixed solids

$$\boxed{\bar{f}_i^S(T, P, \underline{x}) = f_i^S(T, P)} \tag{9.4-13}$$

The point that the discrete pure component regions in a solid mixture should not be too small is worth dwelling on. A collection of molecules can be thought of as being composed of molecules at the surface of the region and molecules in the interior, as was discussed in Sec. 7.8. The interior molecules interact only with similar molecules and therefore are in the same energy state as molecules in a macroscopically large mass of the pure substance. The surface or interfacial molecules, however, interact with both like and unlike molecules, and hence their properties are representative of molecules in a mixture. The assumption being made in Eq. 9.4-13 is that the number of interfacial molecules is small compared with the number of interior molecules, so that even if the

---

[3]If the pure species does not exist as a liquid at the temperature and pressure of interest, as, for example, in the case of a gas or solid dissolved in a liquid, other, more appropriate starting points are used, as discussed in Sec. 9.7.

properties of the interfacial molecules are poorly represented, the effect on the total thermodynamic properties of the system is small.

While Eq. 9.4-13 will generally be used to calculate the fugacity of a species in a solid mixture, especially involving organic compounds, there are cases of true solid mixtures, especially involving metals and inorganic compounds. While we defer considerations of such mixtures until Sec. 12.4, it is useful to note that the activity coefficient for a species in a true solid mixture, following Eq. 9.3-11, can be written as

Fugacity of a species in a true solid mixture

$$\bar{f}_i^S(T, P, \underline{x}^S) = x_i^S \gamma_i^S(T, P, \underline{x}^S) f_i^S(T, P) \tag{9.4-14}$$

9.5 SEVERAL CORRELATIVE LIQUID MIXTURE ACTIVITY COEFFICIENT MODELS

In this section we are interested in mixtures for which equation-of-state models are inapplicable. In such cases attempts are made to estimate $\underline{G}^{ex}$ or $\Delta_{mix}\underline{G}$ directly, either empirically or semitheoretically. In making such estimates it is useful to distinguish between simple and nonsimple liquid mixtures. We define a **simple liquid mixture** to be one that is formed by mixing pure fluids, each of which is a liquid at the temperature and pressure of the mixture. A **nonsimple liquid mixture**, on the other hand, is formed by mixing pure species, at least one of which is not a liquid at the temperature and pressure of the mixture. Examples of nonsimple liquid mixtures are solutions of dissolved solids in liquids (for example, sugar in water) and dissolved gases in liquids (for example, carbon dioxide in soda water). In this section we are concerned with mixture models that are applicable to simple liquid mixtures. Nonsimple liquid mixtures are considered in Sec. 9.7 and elsewhere in this book. Also, for simplicity, most of the excess Gibbs energy models we will consider here are for two-component (binary) mixtures. Multicomponent-mixture versions of these models are listed in Appendix A9.2.

From Eq. 9.3-8, we have that the Gibbs energy change on forming a simple liquid mixture is

$$\Delta_{mix}\underline{G} = \Delta_{mix}\underline{G}^{IM} + \underline{G}^{ex} = RT \sum_{i=1}^{c} x_i \ln x_i + \underline{G}^{ex}(\underline{x})$$

The excess Gibbs energy of mixing for the benzene–2,2,4-trimethyl pentane system (a simple liquid mixture) is shown in Fig. 9.5-1 (the origin of these data is discussed in Sec. 10.2); data for the excess Gibbs energy for several other mixtures are shown in Fig. 9.5-2. Each of the curves in these figures is of simple form. Thus, one approach to estimating the excess Gibbs energy of mixing has been merely to try to fit results, such as those given in Figs. 9.5-1 and 9.5-2, with polynomials in the composition. The hope is that given a limited amount of experimental data, one can determine the parameters in the appropriate polynomial expansion, and then predict the excess Gibbs energy and liquid-phase activity coefficients over the whole composition range. Of course, any expression chosen for the excess Gibbs energy must satisfy the Gibbs-Duhem equation (i.e., Eqs. 9.3-14 through 9.3-17) and, like the data in the figures, go to zero as $x_1 \to 0$ and $x_1 \to 1$.

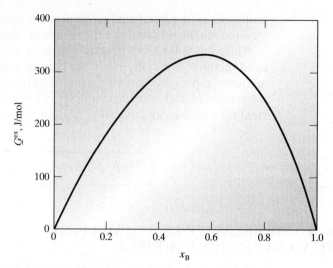

Figure 9.5-1 The excess Gibbs energy for the benzene–2,2,4-trimethyl pentane system at 55°C.

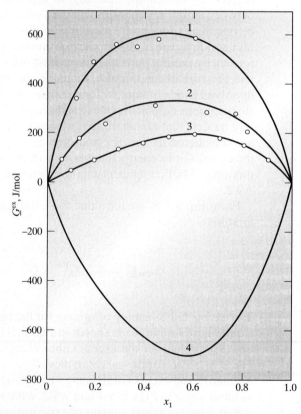

Figure 9.5-2 The excess Gibbs energy for several mixtures. Curve 1: Trimethyl methane (1) and benzene at 0°C. [Data from V. Mathot and A. Desmyter, *J. Chem. Phys.* **21,** 782 (1953).] Curve 2: Trimethyl methane (1) and carbon tetrachloride at 0°C. [Data from ibid.] Curve 3: Methane (1) and propane at 100 K. [Data from A. J. B. Cutler and J. A. Morrison, *Trans. Farad. Soc.,* **61,** 429 (1965).] Curve 4: Water (1) and hydrogen peroxide at 75°C. [From the smoothed data of G. Scatchard, G. M. Kavanagh, and L. B. Ticknor, *J. Am. Chem. Soc.,* **74,** 3715 (1952).]

The simplest polynomial representation of $\underline{G}^{\text{ex}}$ satisfying these criteria is

**One-constant
Margules $\underline{G}^{\text{ex}}$**

$$\boxed{\underline{G}^{\text{ex}} = Ax_1x_2} \tag{9.5-1}$$

for which

$$
\begin{aligned}
\bar{G}_1^{\text{ex}} &= \left.\frac{\partial(N\underline{G}^{\text{ex}})}{\partial N_1}\right|_{T,P,N_2} = \frac{\partial}{\partial N_1}\left(\frac{AN_1N_2}{N_1 + N_2}\right) \\
&= A\left\{\frac{N_2}{N_1 + N_2} - \frac{N_1N_2}{(N_1 + N_2)^2}\right\} = Ax_2^2
\end{aligned} \tag{9.5-2}
$$

so that

$$\gamma_1(\underline{x}) = \exp\left\{\frac{\bar{G}_1^{\text{ex}}}{RT}\right\} = \exp\left\{\frac{Ax_2^2}{RT}\right\} = \exp\left\{\frac{A(1 - x_1)^2}{RT}\right\}$$

and similarly one can show that

$$\gamma_2(\underline{x}) = \exp\left\{\frac{Ax_1^2}{RT}\right\} = \exp\left\{\frac{A(1 - x_2)^2}{RT}\right\} \tag{9.5-3}$$

Consequently, for this solution model we have

$$\bar{f}_i^{\text{L}}(T, P, \underline{x}) = x_i\gamma_i f_i^{\text{L}}(T, P) \tag{9.5-4}$$

where

**One-constant
Margules activity
coefficients**

$$\boxed{\begin{aligned} RT\ln\gamma_1 &= Ax_2^2 \\ RT\ln\gamma_2 &= Ax_1^2 \end{aligned}} \tag{9.5-5}$$

These relations, known as the **one-constant Margules equations**, are plotted in Fig. 9.5-3. Two interesting features of the one-constant Margules equations are apparent from this figure. First, the two species activity coefficients are mirror images of each other as a function of the composition. This is not a general result, but follows from the choice of a symmetric function in the compositions for $\underline{G}^{\text{ex}}$. Second, $\gamma_i \to 1$ as $x_i \to 1$,

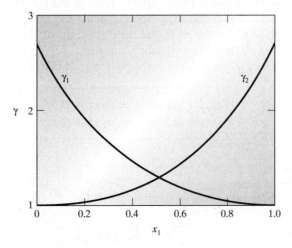

Figure 9.5-3 The activity coefficients for the one-constant Margules equation with $(A/RT) = 1.0$.

so that $\bar{f}_i \rightarrow f_i$ in this limit. This makes good physical sense since one expects the fugacity of a component in a mixture to tend toward that of the pure liquid as the mixture becomes concentrated in that component. Conversely, γ_i exhibits a greater departure from unity (and greater departure from ideal solution behavior) the greater the dilution of species i, as would be expected from the discussion in Chapter 8. Note that the parameter A can be either positive or negative, so that Eq. 9.5-5 can lead to activity coefficients that are greater than 1 ($A > 0$) or less than 1 ($A < 0$).

It is interesting to note that in the ideal solution model (i.e., $\bar{f}_i = x_i f_i$), the only role of components other than the one of interest is as diluents, so that they affect $\bar{f}_i$ only through x_i. No account is taken of the fact that the energy of a species 1–species 2 interaction could be different from that of species 1–species 1 or species 2–species 2 interactions. For the nonideal solution of Eq. 9.5-1 the parameter A is dependent on both species, or more precisely, on the differences in species interaction energies involved. This results in the behavior of species 1 being influenced by both the nature and composition of species 2, and vice versa. The value of the parameter A depends on the macroscopic and molecular properties of both species in the mixture and is difficult to estimate a priori; its value may be either positive or negative and generally is a function of temperature. Over small temperature ranges A may be assumed to be a constant, so that its value found from experiments at one temperature can be used at neighboring temperatures.

The one-constant Margules equation provides a satisfactory representation for activity coefficient behavior only for liquid mixtures containing constituents of similar size, shape, and chemical nature. For more complicated systems, particularly mixtures of dissimilar molecules, simple relations such as Eq. 9.5-1 or 9.5-5 are not valid. In particular, the excess Gibbs energy of a general mixture will not be a symmetric function of the mole fraction, and the activity coefficients of the two species in a mixture should not be expected to be mirror images. One possible generalization of Eq. 9.5-1 to such cases is to set

Redlich-Kister expansion

$$\underline{G}^{ex} = x_1 x_2 \{A + B(x_1 - x_2) + C(x_1 - x_2)^2 + \cdots \} \qquad \text{(9.5-6)}$$

where $A, B, C, \ldots$ are temperature-dependent parameters. This expression for $\underline{G}^{ex}$ is another example of the Redlich-Kister expansion used for the representation of excess thermodynamic properties, which was discussed in Sec. 8.6. The number of terms retained in this expansion depends on the shape of the $\underline{G}^{ex}$ curve as a function of composition, the accuracy of the experimental data, and the goodness of the fit desired. When $A = B = C = \cdots = 0$, the ideal solution model is recovered; for $A \neq 0$, $B = C = \cdots = 0$, the one-constant Margules equation is obtained. For the case in which $A \neq 0$, $B \neq 0$, but $C = D = \cdots = 0$, one obtains (see Problem 9.8)

Two-constant Margules expansion

$$\begin{aligned} RT \ln \gamma_1 &= \alpha_1 x_2^2 + \beta_1 x_2^3 \\ RT \ln \gamma_2 &= \alpha_2 x_1^2 + \beta_2 x_1^3 \end{aligned} \qquad \text{(9.5-7)}$$

where

$$\alpha_i = A + 3(-1)^{i+1} B$$

and

$$\beta_i = 4(-1)^i B$$

In this equation i denotes the species and has values of 1 and 2. These results are known as the **two-constant Margules equations**. In this case the excess Gibbs energy is not symmetric in the mole fractions and the two activity coefficients are not mirror images of each other as a function of concentration.

The expansion of Eq. 9.5-6 is certainly not unique; other types of expansions for the excess Gibbs energy could also be used. Another expansion is that of Wohl,[4]

$$\frac{\underline{G}^{\text{ex}}}{RT(x_1q_1 + x_2q_2)} = 2a_{12}z_1z_2 + 3a_{112}z_1^2z_2 + 3a_{122}z_1z_2^2$$
$$+ 4a_{1112}z_1^3z_2 + 4a_{1222}z_1z_2^3 + 6a_{1122}z_1^2z_2^2$$
$$+ \cdots$$

(9.5-8)

where q_i is some measure of the volume of molecule i (e.g., its liquid molar volume or van der Waals b parameter) and the a's are parameters resulting from the unlike molecule (i.e., species 1–species 2) interactions. The z_i in Eq. 9.5-8 are, essentially, volume fractions defined by

$$z_i = \frac{x_iq_i}{x_1q_1 + x_2q_2}$$

Equation 9.5-8 is modeled after the virial expansion for gaseous mixtures, and, in fact, the constants in the expansion (2, 3, 3, 4, 4, 6, etc.) are those that arise in that equation.

The liquid-phase activity coefficients for the Wohl expansion can be obtained from Eq. 9.5-8 by taking the appropriate derivatives:

$$\ln \gamma_i = \frac{\overline{G}_i^{\text{ex}}}{RT} = \frac{\partial}{\partial N_i}\left(\frac{N\underline{G}^{\text{ex}}}{RT}\right)_{T,P,N_{j\neq i}}$$

In particular, for the case in which we assume $a_{12} \neq 0$ and $a_{112} = a_{122} = \cdots = 0$, we have

$$\frac{\underline{G}^{\text{ex}}}{RT} = (x_1q_1 + x_2q_2)2a_{12}z_1z_2 = \frac{2a_{12}x_1q_1x_2q_2}{x_1q_1 + x_2q_2}$$

and (see Problem 9.8)

van Laar equations

$$\boxed{\ln \gamma_1 = \frac{\alpha}{\left[1 + \dfrac{\alpha}{\beta}\dfrac{x_1}{x_2}\right]^2} \quad \text{and} \quad \ln \gamma_2 = \frac{\beta}{\left[1 + \dfrac{\beta}{\alpha}\dfrac{x_2}{x_1}\right]^2}}$$

(9.5-9)

where $\alpha = 2q_1a_{12}$ and $\beta = 2q_2a_{12}$. Equations 9.5-9 are known as the **van Laar equations**;[5] they are frequently used to correlate activity coefficient data. Other, more complicated, activity coefficient equations can be derived from Eq. 9.5-8 by retaining additional terms in the expansion, though this is not done here.

The values of the parameters in the activity coefficient equations are usually found by fitting these equations to experimental activity coefficient data (see Problem 9.22) over the whole composition range. Alternatively, if only limited data are available, the van Laar equations, which can be written as

---

[4]K. Wohl, *Trans. AIChE,* **42**, 215 (1946).
[5]J. J. van Laar, *Z. Physik. Chem.,* **72**, 723 (1910); **83**, 599 (1913).

$$\alpha = \left(1 + \frac{x_2 \ln \gamma_2}{x_1 \ln \gamma_1}\right)^2 \ln \gamma_1$$

$$\beta = \left(1 + \frac{x_1 \ln \gamma_1}{x_2 \ln \gamma_2}\right)^2 \ln \gamma_2$$

(9.5-10)

so that data for γ_1 and γ_2 at only a single mole fraction can be used to evaluate the two van Laar constants, and hence to compute the activity coefficients at all other compositions. This procedure is used in Chapter 10. However, if activity coefficient data are available at several compositions, a regression procedure can be used to obtain the "best" fit values for α and β. Table 9.5-1 contains values that have been reported for these parameters for a number of binary mixtures.

Table 9.5-1 The van Laar Constants for Some Binary Mixtures

Component 1–Component 2	Temperature Range (°C)	α	β
Acetaldehyde–water	19.8–100	1.59	1.80
Acetone–benzene	56.1–80.1	0.405	0.405
Acetone–methanol	56.1–64.6	0.58	0.56
Acetone–water	25	1.89	1.66
	56.1–100	2.05	1.50
Benzene–isopropanol	71.9–82.3	1.36	1.95
Carbon disulfide–acetone	39.5–56.1	1.28	1.79
Carbon disulfide–Carbon tetrachloride	46.3–76.7	0.23	0.16
Carbon tetrachloride–benzene	76.4–80.2	0.12	0.11
Ethanol–benzene	67.0–80.1	1.946	1.610
Ethanol–cyclohexane	66.3–80.8	2.102	1.729
Ethanol–toluene	76.4–110.7	1.757	1.757
Ethanol–water	25	1.54	0.97
Ethyl acetate–benzene	71.1–80.2	1.15	0.92
Ethyl acetate–ethanol	71.7–78.3	0.896	0.896
Ethyl acetate–toluene	77.2–110.7	0.09	0.58
Ethyl ether–acetone	34.6–56.1	0.741	0.741
Ethyl ether–ethanol	34.6–78.3	0.97	1.27
n-Hexane–ethanol	59.3–78.3	1.57	2.58
Isobutane–furfural	37.8	2.62	3.02
	51.7	2.51	2.83
Isopropanol–water	82.3–100	2.40	1.13
Methanol–benzene	55.5–64.6	0.56	0.56
Methanol–ethyl acetate	62.1–77.1	1.16	1.16
Methanol–water	25	0.58	0.46
	64.6–100	0.83	0.51
Methyl acetate–methanol	53.7–64.6	1.06	1.06
Methyl acetate–water	57.0–100	2.99	1.89
n-Propanol–water	88.0–100	2.53	1.13
Water–phenol	100–181	0.83	3.22

Source: This table is an adaptation of one given in J. H. Perry, ed., *Chemical Engineers' Handbook,* 4th ed., McGraw-Hill, New York (1963), p. 13-7.

Note: When $\alpha = \beta$, the van Laar equation $\ln \gamma_1 = \dfrac{\alpha}{[1 + (\alpha x_1/\beta x_2)]^2}$ reduces to the Margules form $\ln \gamma_1 = \alpha x_2^2$.

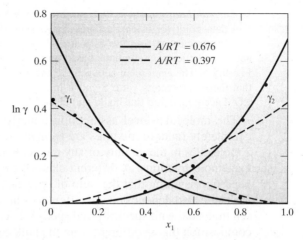

Figure 9.5-4 Experimental activity coefficient data for the benzene–2,2,4-trimethyl pentane mixture and the correlation of these data obtained using the one-constant Margules equation.

ILLUSTRATION 9.5-1

Use of Activity Coefficient Models to Correlate Data

The points in Figs. 9.5-4 and 9.5-5 represent smoothed values of the activity coefficients for both species in a benzene–2,2,4-trimethyl pentane mixture at 55°C taken from the vapor-liquid equilibrium measurements of Weissman and Wood (see Illustration 10.2-4). Test the accuracy of the one-constant Margules equation and the van Laar equations in correlating these data.

SOLUTION

a. The one-constant Margules equation. From the data presented in Fig. 9.5-4 it is clear that the activity coefficient for benzene is not the mirror image of that for trimethyl pentane. Therefore, the one-constant Margules equation cannot be made to fit both sets of activity coefficients simultaneously. (It is interesting to note that the Margules form, $RT \ln \gamma_i = A_i x_j^2$, will fit these data well if A_1 and A_2 are separately chosen. However, this suggestion does not satisfy the Gibbs-Duhem equation! Can you prove this?)

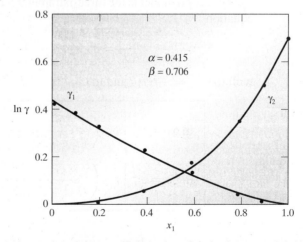

Figure 9.5-5 Experimental activity coefficient data for the benzene–2,2,4-trimethyl pentane mixture and the correlation of these data obtained using the van Laar equation.

b. The van Laar equation. One can use Eqs. 9.5-10 and a single activity coefficient–composition data point (or a least-squares analysis of all the data points) to find values for the van Laar parameters. Using the data at $x_1 = 0.6$, we find that $\alpha = 0.415$ and $\beta = 0.706$. The activity coefficient predictions based on these values of the van Laar parameters are shown in Fig. 9.5-5. The agreement between the correlation and the experimental data is excellent. Note that these parameters were found using the MATHCAD worksheet ACTCOEFF on the CD-ROM accompanying this book, and discussed in Appendix B.III. ∎

The molecular-level assumption underlying the Redlich-Kister expansion is that completely random mixtures are formed, that is, that the ratio of species 1 to species 2 molecules in the vicinity of any molecule is, on the average, the same as the ratio of their mole fractions. A different class of excess Gibbs energy models can be formulated by assuming that the ratio of species 1 to species 2 molecules surrounding any molecule also depends on the differences in size and energies of interaction of the chosen molecule with species 1 and species 2. Thus, around each molecule there is a local composition that is different from the bulk composition. From this picture, the several binary mixture models have been developed.

The first model we consider of this type is the two-parameter $(\Lambda_{12}, \Lambda_{21})$ Wilson equation[6]

$$\frac{G^{\text{ex}}}{RT} = -x_1 \ln(x_1 + x_2\Lambda_{12}) - x_2 \ln(x_2 + x_1\Lambda_{21}) \qquad \textbf{(9.5-11)}$$

for which

Wilson equation

$$\boxed{\begin{aligned} \ln\gamma_1 &= -\ln(x_1 + x_2\Lambda_{12}) + x_2\left[\frac{\Lambda_{12}}{x_1 + x_2\Lambda_{12}} - \frac{\Lambda_{21}}{x_1\Lambda_{21} + x_2}\right] \\ \ln\gamma_2 &= -\ln(x_2 + x_1\Lambda_{21}) - x_1\left[\frac{\Lambda_{12}}{x_1 + x_2\Lambda_{12}} - \frac{\Lambda_{21}}{x_1\Lambda_{21} + x_2}\right] \end{aligned}} \qquad \textbf{(9.5-12a)}$$

The two infinite-dilution activity coefficients in this model are

$$\ln\gamma_1^\infty = -\ln\Lambda_{12} + 1 - \Lambda_{21} \quad \text{and} \quad \ln\gamma_2^\infty = -\ln\Lambda_{21} + 1 - \Lambda_{12} \qquad \textbf{(9.5-12b)}$$

The second model is the three-parameter $(\alpha, \tau_{12}, \tau_{21})$ nonrandom two-liquid (NRTL) equation[7]

$$\frac{G^{\text{ex}}}{RT} = x_1 x_2 \left(\frac{\tau_{21}G_{21}}{x_1 + x_2G_{21}} + \frac{\tau_{12}G_{12}}{x_2 + x_1G_{12}}\right) \qquad \textbf{(9.5-13)}$$

with $\ln G_{12} = -\alpha\tau_{12}$ and $\ln G_{21} = -\alpha\tau_{21}$, for which

NRTL model

$$\boxed{\begin{aligned} \ln\gamma_1 &= x_2^2\left[\tau_{21}\left(\frac{G_{21}}{x_1 + x_2G_{21}}\right)^2 + \frac{\tau_{12}G_{12}}{(x_2 + x_1G_{12})^2}\right] \\ \ln\gamma_2 &= x_1^2\left[\tau_{12}\left(\frac{G_{12}}{x_2 + x_1G_{12}}\right)^2 + \frac{\tau_{21}G_{21}}{(x_1 + x_2G_{21})^2}\right] \end{aligned}} \qquad \textbf{(9.5-14a)}$$

[6]G. M. Wilson, *J. Am. Chem. Soc.,* **86,** 127 (1964).
[7]H. Renon and J. M. Prausnitz, *AIChE J.,* **14,** 135 (1968).

and

$$\ln \gamma_1^\infty = \tau_{21} + \tau_{12}G_{12} = \tau_{21} + \tau_{12}\exp\left(-\alpha\tau_{12}\right)$$
$$\ln \gamma_2^\infty = \tau_{12} + \tau_{21}G_{21} = \tau_{12} + \tau_{21}\exp\left(-\alpha\tau_{21}\right)$$

$$\text{(9.5-14b)}$$

Note that in these models there are different weightings of the mole fractions of the species due to the parameters (Λ_{ij} and τ_{ij}), the values of which depend on differences in size and interaction energies of the molecules in the mixture. (The multicomponent forms of the Wilson and NRTL models are given in Appendix A9.2.)

Another model, the Flory and Huggins model, is meant to apply to mixtures of molecules of very different size, including solutions of polymers. This solution model contains two parts. The first is an expression for the entropy of mixing per mole:

$$\Delta_{\text{mix}}\underline{S} = -R(x_1 \ln \phi_1 + x_2 \ln \phi_2) \qquad \text{(9.5-15a)}$$

or

$$\underline{S}^{\text{ex}} = \Delta_{\text{mix}}\underline{S} - \Delta_{\text{mix}}\underline{S}^{\text{IM}}$$
$$= -R\left(x_1 \ln \frac{\phi_1}{x_1} + x_2 \ln \frac{\phi_2}{x_2}\right) \qquad \text{(9.5-15b)}$$

Here

$$\phi_1 = \frac{x_1 v_1}{x_1 v_1 + x_2 v_2} = \frac{x_1}{x_1 + mx_2} \quad \text{and} \quad \phi_2 = \frac{mx_2}{x_1 + mx_2}$$

are the volume fractions, with v_i being some measure of the volume of species i molecules, and $m = v_2/v_1$. The assumption in Eqs. 9.5-15 is that for molecules of different size it is the volume fractions, rather than the mole fractions, that determine the entropy of mixing. The second part of the model is that the enthalpy of mixing, or excess enthalpy, can be expressed by the simple one-constant term in volume fractions (rather than mole fractions, as in the case of the one-constant Margules equation)

$$\Delta_{\text{mix}}\underline{H} = \underline{H}^{\text{ex}} = \chi RT(x_1 + mx_2)\phi_1\phi_2 \qquad \text{(9.5-16)}$$

where χ is an adjustable parameter referred to as the Flory interaction parameter or simply the Flory parameter (and sometimes as the chi parameter).

Combining Eqs. 9.5-15 and 9.5-16 gives

$$\frac{\underline{G}^{\text{ex}}}{RT} = \frac{\underline{H}^{\text{ex}} - T\underline{S}^{\text{ex}}}{RT}$$
$$= \left[x_1 \ln \frac{\phi_1}{x_1} + x_2 \ln \frac{\phi_2}{x_2}\right] + \chi(x_1 + mx_2)\phi_1\phi_2 \qquad \text{(9.5-17)}$$

which is the **Flory-Huggins** model. The first term on the right side of this equation is the entropic contribution to the excess Gibbs energy, and the second term is the enthalpic contribution. These two terms are also referred to as the combinatorial and residual terms, respectively. The activity coefficient expressions for this model (Problem 9.29) are

Flory-Huggins equations

$$\boxed{\begin{aligned} \ln \gamma_1 &= \ln \frac{\phi_1}{x_1} + \left(1 - \frac{1}{m}\right)\phi_2 + \chi \phi_2^2 \\[2mm] \ln \gamma_2 &= \ln \frac{\phi_2}{x_2} - (m-1)\phi_1 + m\chi \phi_1^2 \end{aligned}}$$

$$\text{(9.5-18)}$$

We will consider only one additional activity coefficient equation here, the UNI-QUAC (universal quasichemical) model of Abrams and Prausnitz.[8] This model, based on statistical mechanical theory, allows local compositions to result from both the size and energy differences between the molecules in the mixture. The result is the expression

$$\frac{G^{\text{ex}}}{RT} = \frac{G^{\text{ex}}(\text{combinatorial})}{RT} + \frac{G^{\text{ex}}(\text{residual})}{RT} \tag{9.5-19}$$

where the first term accounts for molecular size and shape differences, and the second term accounts largely for energy differences. These terms, in multicomponent form, are given by

UNIQUAC equation

$$\frac{G^{\text{ex}}(\text{combinatorial})}{RT} = \sum_i x_i \ln \frac{\phi_i}{x_i} + \frac{z}{2} \sum_i x_i q_i \ln \frac{\theta_i}{\phi_i} \tag{9.5-20}$$

and

$$\frac{G^{\text{ex}}(\text{residual})}{RT} = -\sum_i q_i x_i \ln \left(\sum_j \theta_j \tau_{ji} \right) \tag{9.5-21}$$

where

$$r_i = \text{volume parameter for species i}$$
$$q_i = \text{surface area parameter for species i}$$
$$\theta_i = \text{area fraction of species i} = x_i q_i \Big/ \sum_j x_j q_j$$
$$\phi_i = \text{segment or volume fraction of species i} = x_i r_i \Big/ \sum_j x_j r_j$$
$$\ln \tau_{ij} = -\frac{(u_{ij} - u_{jj})}{RT}$$

with u_{ij} being the average interaction energy for a species i–species j interaction and z being the average coordination number, that is, the number of molecules around a central molecule, usually taken to be 10. Combining Eqs. 9.5-19, 9.5-20, and 9.5-21 gives

UNIQUAC expression
for activity coefficients

$$\ln \gamma_i = \ln \gamma_i(\text{combinatorial}) + \ln \gamma_i(\text{residual}) \tag{9.5-22}$$

$$\ln \gamma_i(\text{combinatorial}) = \ln \frac{\phi_i}{x_i} - \frac{z}{2} q_i \ln \frac{\phi_i}{\theta_i} + l_i - \frac{\phi_i}{x_i} \sum_j x_j l_j \tag{9.5-23a}$$

$$\ln \gamma_i(\text{residual}) = q_i \left[1 - \ln \left(\sum_j \theta_j \tau_{ji} \right) - \sum_j \frac{\theta_j \tau_{ij}}{\sum_k \theta_k \tau_{kj}} \right] \tag{9.5-23b}$$

where $l_i = (r_i - q_i)z/2 - (r_i - 1)$.

Since the size and surface area parameters r_i and q_i can be evaluated from molecular structure information, as will be discussed next, the UNIQUAC equation contains only

---

[8]D. S. Abrams and J. M. Prausnitz, *AIChE J.*, **21**, 116 (1975).

two adjustable parameters, τ_{12} and τ_{21} (or, equivalently, $u_{12}-u_{22}$ and $u_{21}-u_{11}$) for each binary pair. Thus, like the van Laar or Wilson equations, it is a two-parameter activity coefficient model. It does have a better theoretical basis than these models, though it is somewhat more complicated.

Instead of listing the volume (r) and surface area (q) parameters for each molecular species for use in the UNIQUAC model, these parameters are evaluated by a **group contribution** method. The underlying idea is that a molecule can be considered to be a collection of **functional groups**, and that volume R_i and surface area Q_i of functional group i will be approximately the same in any molecule in which that group occurs. For example, we expect the contribution to the total volume and surface area of a molecule from a methyl (CH_3-) group to be the same independent of whether the methyl group is at the end of an ethane, propane, or dodecane molecule. Thus, the volume and surface area parameters r and q of a molecule are obtained from a sum over its functional groups of the R and Q parameters. The advantage of this group contribution approach is that from a relatively small number of functional groups, the properties of the millions upon millions of different molecules can be obtained.

Table 9.5-2 contains the R and Q parameters for 106 functional groups referred to as subgroups in the table (the main group vs. subgroup terminology will be explained in the following section). All the values that appear in the table have been normalized to the properties of a methylene group in polymethylene, and therefore are unitless. There are several things to note about the entries in this table. First, several molecules, such as water and furfural, have such unique properties that they have been treated as functional groups. Such molecule functional groups appear in bold in the table. Second, similar groups, such as $-CH_3$ and $-C-$, may have different surface area or Q parameters. This is because a $-CH_3$ group, being at the end of a molecule, increases its surface area, whereas the $-C-$ group, which is at the interior of a molecule, makes no contribution to its surface area. Finally, as the functional group method continues to evolve, new groups are added and group parameters are subject to change.

ILLUSTRATION 9.5-2

Computation of Volume and Surface Area Fractions for Use in the UNIQUAC Model

One mole each of benzene and 2,2,4-trimethyl pentane are mixed together. Using the data in Table 9.5-2, compute the volume fraction and surface area fractions of benzene and 2,2,4-trimethyl pentane in this mixture.

SOLUTION

Benzene consists of six aromatic CH (ACH) groups. Therefore,

$$r_B = 6 \times 0.3763 = 2.258$$
$$q_B = 6 \times 0.4321 = 2.592$$

The structure of 2,2,4-trimethyl pentane is

This molecule consists of five CH_3 groups, one CH_2 group, one CH group, and one C group. Thus

$$r_{\text{TMP}} = 5 \times 0.6325 + 0.6325 + 0.6325 + 0.6325 = 5.06$$

Table 9.5-2 The Group Volume and Surface Area Parameters, R and Q, for Use with the UNIQUAC and UNIFAC Models*

Main Group	Subgroup	R	Q	Example Assignments
CH_2	CH_3	0.6325	1.0608	n-Hexane: 4 CH_2, 2 CH_3
	CH_2	0.6325	0.7081	n-Heptane: 5 CH_2, 2 CH_3
	CH	0.6325	0.3554	2-Methylpropane: 1 CH, 3 CH_3
	C	0.6325	0.0000	Neopentane: 1 C, 4 CH_3
$C{=}C$	$CH_2{=}CH$	1.2832	1.6016	1-Hexene: 1 $CH_2{=}CH$, 3 CH_2, 1 CH_3
	$CH{=}CH$	1.2832	1.2489	2-Hexene: 1 $CH{=}CH$, 2 CH_3, 2 CH_2
	$CH_2{=}C$	1.2832	1.2489	2-Methyl-1-butene: 1 $CH_2{=}C$, 1 CH_2, 2 CH_3
	$CH{=}C$	1.2832	0.8962	2-Methyl-2-butene: 1 $CH{=}C$, 2 CH_3
	$C{=}C$	1.2832	0.4582	12,3-Dimethyl-2-butene: 1 $C{=}C$, 4 CH_3
ACH	ACH	0.3763	0.4321	Benzene: 6 ACH
	AC	0.3763	0.2113	Styrene: 1 $CH_2{=}C$, 5 ACH_2, 1 ACH
$ACCH_2$	$ACCH_3$	0.9100	0.9490	Toluene: 5 ACH, 1 $ACCH_3$
	$ACCH_2$	0.9100	0.7962	Ethylbenzene: 5 ACH, 1 $ACCH_2$, 1 CH_3
	$ACCH$	0.9100	0.3769	Isopropylbenzene: 5 ACH, 1 $ACCH$, 2 CH_3
OH	$OH(p)$	1.2302	0.8927	1-Propanol: 1 $OH(p)$, 1 CH_3, 2 CH_2
	$OH(s)$	1.0630	0.8663	2-Propanol: 1 $OH(s)$, 2 CH_3, 1 CH
	$OH(t)$	0.6895	0.8345	*tert*-Butanol: 1 $OH(t)$, 3 CH_3, 1 C
CH_3OH	CH_3OH	0.8585	0.9938	**Methanol**
Water	H_2O	1.7334	2.4561	**Water**
$ACOH$	$ACOH$	1.0800	0.9750	Phenol: 1 $ACOH$, 5 ACH
CH_2CO	CH_3CO	1.7048	1.6700	2-Butanone: 1 CH_3CO, 1 CH_3, 1 CH_2
	CH_2CO	1.7048	1.5542	2-Pentanone: 1 CH_2CO, 2 CH_3, 1 CH_2
CHO	CHO	0.7173	0.7710	Propionic aldehyde: 1 CHO, 1 CH_3, 1 CH_2
$CCOO$	CH_3COO	1.2700	1.6286	Butyl acetate: 1 CH_3COO, 1 CH_3, 3 CH_2
	CH_2COO	1.2700	1.4228	Methyl propionate: 1 CH_2COO, 2 CH_3
$HCOO$	$HCOO$	1.9000	1.8000	Ethyl formate: 1 $HCOO$, 1 CH_3, 1 CH_2
CH_2O	CH_3O	1.1434	1.6022	Dimethyl ether: 1 CH_3CO, 1 CH_3
	CH_2O	1.1434	1.2495	Diethyl ether: 1 CH_2O, 2 CH_3, 1 CH_2
	CHO	1.1434	0.8968	Diisopropyl ether: 1 CHO, 4 CH_3, 1 CH
CNH_2	CH_3NH_2	1.6607	1.6904	Methylamine: CH_3NH_2
	CH_2NH_2	1.6607	1.3377	Ethylamine: 1 CH_2NH_2, 1 CH_3
	$CHNH_2$	1.6607	0.9850	Isopropylamine: 1 $CHNH_2$, 2 CH_3
	CNH_2	1.6607	0.9850	*tert*-Butylamine: 1 CNH_2, 3 CH_3
CNH	CH_3NH	1.3680	1.4332	Dimethylamine: CH_3NH, 1 CH_3
	CH_2NH	1.3680	1.0805	Diethylamine: 1 CH_2NH, 2 CH_3, 1 CH_2
	$CHNH$	1.3680	0.7278	Diisopropylamine: 1 $CHNH$, 4 CH_3, 1 CH
$(C)_3N$	CH_3N	1.0746	1.1760	Trimethylamine: 1 CH_3N, 2 CH_3
	CH_2N	1.0746	0.8240	Triethylamine: 1 CH_2N, 2 CH_2, 3 CH_3
$ACNH_2$	$ACNH_2$	1.1849	0.8067	Aniline: 1 $ACNH_2$, 5 ACH
Pyridines	AC_2H_2N	1.4578	0.9022	Pyridine: 1 AC_2H_2N, 3ACH
	AC_2HN	1.2393	0.6330	2-Methylpyridine: 1 AC_2HN, 3 ACH, 1 CH_3
	AC_2N	1.0731	0.3539	2,5-Methylpyridine: 1 AC_2N, 3 ACH, 2 CH_3
CCN	CH_3CN	1.5575	1.5193	**Acetonitrile**
	CH_2CN	1.5575	1.1666	Propionitrile: 1 CH_2CN, 1 CH_3
$COOH$	$COOH$	0.8000	0.9215	Acetic acid: 1 $COOH$, 1 CH_3
$HCOOH$	$HCOOH$	0.8000	1.2742	**Formic acid**

*The parameters for the UNIQUAC and UNIFAC models have been supplied by Prof. J. Gmehling of the University of Oldenburg, Germany. Copyright Wiley-VCH Verlag GmbH & Co. KGaA. Reproduced with permission.

Table 9.5-2 (*Continued*)

Main Group	Subgroup	R	Q	Example Assignments
CCl	CH$_2$Cl	0.9919	1.3654	1-Chlorobutane: 1 CH$_2$Cl, 1 CH$_3$, 2 CH$_2$
	CHCl	0.9919	1.0127	2-Chloropropane: 1 CHCl, 2 CH$_3$
	CCl	0.9919	0.6600	*tert*-Butyl chloride: 1 CHCl, 3 CH$_3$
CCl$_2$	CH$_2$Cl$_2$	1.8000	2.5000	Dichloromethane: 1 CH$_2$Cl$_2$
	CHCl$_2$	1.8000	2.1473	1,1-Dichloroethane: 1 CHCl$_2$, 1 CH$_3$
	CCl$_2$	1.8000	1.7946	2,2-Dichloropropane: 1 CCl$_2$, 2 CH$_3$
CCl$_3$	CHCl$_3$	2.4500	2.8912	**Chloroform**
	CCl$_3$	2.6500	2.3778	1,1,1-Trichloroethane: 1 CCl$_3$, 1 CH$_3$
CCl$_4$	CCl$_4$	2.6180	3.1836	**Tetrachloromethane**
ACCl	ACCl	0.5365	0.3177	Chlorobenzene: 1 ACCl, 5 ACH
CNO$_2$	CH$_3$NO$_2$	2.6440	2.5000	**Nitromethane**
	CH$_2$NO$_2$	2.5000	2.3040	1-Nitropropane: 1 CH$_2$NO$_2$, 1 CH$_3$, 1 CH$_2$
	CHNO$_2$	2.8870	2.2410	2-Nitropropane: 1 CHNO$_2$, 2 CH$_3$
ACNO$_2$	ACNO$_2$	0.4656	0.3589	Nitrobenzene: 1 ACNO$_2$, 5 ACH
CS$_2$	CS$_2$	1.2400	1.0680	**Carbon disulfide**
CH$_3$SH	CH$_3$SH	1.2890	1.7620	**Methanethiol**
	CH$_2$SH	1.5350	1.3160	Ethanethiol: 1 CH$_2$SH, 1 CH$_3$
Furfural	furfural	1.2990	1.2890	**Furfural**
Diol	(CH$_2$OH)$_2$	2.0880	2.4000	**1,2-Ethanediol (ethylene glycol)**
I	I	1.0760	0.9169	Ethyl iodide: 1 I, 1 CH$_3$, 1 CH$_2$
Br	Br	1.2090	1.4000	Ethyl bromide: 1 Br, 1 CH$_3$, 1 CH$_2$
C≡C	CH≡C	0.9214	1.3000	1-Hexyne: 1 CH≡C, 1 CH$_3$, 3 CH$_2$
	C≡C	1.3030	1.1320	2-Hexyne: 1 C≡C, 2 CH$_3$, 2 CH$_2$
DMSO	DMSO	3.6000	2.6920	**Dimethyl sulfoxide**
ACRY	ACRY	1.0000	0.9200	**Acrylonitrile**
Cl(C=C)	Cl(C=C)	0.5229	0.7391	Trichloroethylene: 3 Cl(C=C), 1 CH=C
ACF	ACF	0.8814	0.7269	Hexafluorobenzene: 6 ACF
DMF	DMF	2.0000	2.0930	*N*, *N*-**Dimethylformamide**
	HCON(CH$_2$)$_2$	2.3810	1.5220	*N*, *N*-Diethylformamide: 1 HCON(CH$_2$)$_2$, 2 CH$_3$
CF$_2$	CF$_3$	1.2840	1.2660	1,1,1-Trifluoroethane: 1 CF$_3$, 1 CH$_3$
	CF$_2$	1.2840	1.0980	Perfluorohexane: 4 CF$_2$, 2 CF$_3$
	CF	0.8215	0.5135	Perfluoromethylcyclohexane: 1 CF, 5 CF$_2$, 1 CF$_3$
COO	COO	1.6000	0.9000	Methyl acrylate: 1 COO, 1 CH$_3$, 1 CH$_2$=CH
cy-CH$_2$	cy-CH$_2$	0.7136	0.8635	Cyclohexane: 6 cy-CH$_2$
	cy-CH	0.3479	0.1071	Methylcyclohexane: 1 cy-CH, 5 cy-CH$_2$, 1 CH$_3$
	cy-C	0.3470	0.0000	1,1-Dimethylcyclohexane: 1 cy-C, 5 cy-CH$_2$, 2 CH$_3$
cy-CH$_2$O	cy-CH$_2$OCH$_2$	1.7023	1.8784	Tetrahydrofuran: 1 cy-CH$_2$OCH$_2$, 2 cy-CH$_2$
	cy-CH$_2$O(CH$_2$)$_{1/2}$	1.4046	1.4000	1,3-Dioxane: 2 cy-CH$_2$O(CH$_2$)$_{1/2}$, 1 cy-CH$_2$
	cy-(CH$_2$)$_{1/2}$O(CH$_2$)$_{1/2}$	1.0413	1.0116	1,3,5-Trioxane: 3 cy-(CH$_2$)$_{1/2}$O(CH$_2$)$_{1/2}$
cy-CON-C	cy-CON-CH$_3$	3.9819	3.2000	*N*-Methylpyrrolidone: 1 cy-CON-CH$_3$, 3 cy-CH$_2$
	cy-CON-CH$_2$	3.7543	2.8920	*N*-Ethylpyrrolidone: 1 cy-CON-CH$_2$, 3 cy-CH$_2$, 1 CH$_3$
	cy-CON-CH	3.5268	2.5800	*N*-Isopropylpyrrolidone: 1 cy-CON-CH, 3 cy-CH$_2$, 2 CH$_3$
	cy-CON-C	3.2994	2.3520	*N*-*tert*-Butylpyrrolidone: 1 cy-CON-C, 3 cy-CH$_2$, 3 CH$_3$
ACS	AC$_2$H$_2$S	1.7943	1.3400	Thiophene: 1 AC$_2$H$_2$S, 2 ACH
	AC$_2$HS	1.6282	1.0600	2-Methylthiophene: 1 AC$_2$HS, 2 ACH, 1 CH$_3$
	AC$_2$S	1.4621	0.7800	2,5-Dimethylthiophene: 1 AC$_2$S, 2 ACH, 2 CH$_3$

Note: A (as in ACH) denotes a group in an aromatic ring, cy- denotes a group in a cyclic structure, and functional groups in bold without any example assignments, such as water, formic acid, etc. are specific to that molecule.

and

$$q_{\mathrm{TMP}} = 5 \times 1.0608 + 0.7081 + 0.3554 + 0.0 = 6.368$$

Consequently,

$$\theta_B = \frac{0.5 \times 2.592}{0.5 \times 2.592 + 0.5 \times 6.368} = 0.289 \qquad \theta_{\mathrm{TMP}} = 0.771$$

$$\phi_B = \frac{0.5 \times 2.258}{0.5 \times 2.258 + 0.5 \times 5.060} = 0.309 \qquad \phi_{\mathrm{TMP}} = 0.691 \qquad\blacksquare$$

Finally, it is useful to note that the Gibbs-Duhem equation can be used to get information about the activity coefficient of one component in a binary mixture if the concentration dependence of the activity coefficient of the other species is known. This is demonstrated in the next illustration.

ILLUSTRATION 9.5-3

Interrelating the Two Activity Coefficients in a Binary Mixture

The activity coefficient for species 1 in a binary mixture can be represented by

$$\ln \gamma_1 = ax_2^2 + bx_2^3 + cx_2^4$$

where a, b, and c are concentration-independent parameters. What is the expression for $\ln \gamma_2$ in terms of these same parameters?

SOLUTION

For a binary mixture at constant temperature and pressure we have, from Eq. 8.2-20 or 9.3-17,

$$x_1 \frac{\partial \ln \gamma_1}{\partial x_2} + x_2 \frac{\partial \ln \gamma_2}{\partial x_2} = 0$$

Since $x_1 = 1 - x_2$ and $\ln \gamma_1 = ax_2^2 + bx_2^3 + cx_2^4$, we have

$$\frac{\partial \ln \gamma_2}{\partial x_2} = -\frac{x_1}{x_2} \frac{\partial \ln \gamma_1}{\partial x_2}$$

$$= -\frac{(1 - x_2)}{x_2} (2ax_2 + 3bx_2^2 + 4cx_2^3)$$

$$= -2a + (2a - 3b)x_2 + (3b - 4c)x_2^2 + 4cx_2^3$$

Now, by definition, $\gamma_2(x_2 = 1) = 1$, and $\ln \gamma_2(x_2 = 1) = 0$. Therefore,

$$\int_{x_2=1}^{x_2} \frac{\partial \ln \gamma_2}{\partial x_2} \, dx_2 = \ln \gamma_2(x_2) - \ln \gamma_2(x_2 = 1) = \ln \gamma_2(x_2)$$

$$= \int_{x_2=1}^{x_2} [-2a + (2a - 3b)x_2 + (3b - 4c)x_2^2 + 4cx_2^3] \, dx_2$$

$$= -2a(x_2 - 1) + \frac{(2a - 3b)}{2}(x_2^2 - 1) + \frac{(3b - 4c)}{3}(x_2^3 - 1)$$

$$+ \frac{4c}{4}(x_2^4 - 1)$$

Again using $x_1 + x_2 = 1$, or $x_2 = 1 - x_1$, yields

$$\ln \gamma_2 = -2a(1 - x_1 - 1)$$
$$+ \tfrac{1}{2}(2a - 3b)(1 - 2x_1 + x_1^2 - 1)$$
$$+ \tfrac{1}{3}(3b - 4c)(1 - 3x_1 + 3x_1^2 - x_1^3 - 1)$$
$$+ c(1 - 4x_1 + 6x_1^2 - 4x_1^3 + x_1^4 - 1)$$

$$\ln \gamma_2 = \left(a + \frac{3b}{2} + 2c\right)x_1^2 - (b + \tfrac{8}{3}c)x_1^3 + cx_1^4$$

■

Finally, we should point out that any of the liquid-state activity coefficient models discussed here can also be used for mixtures in the solid state. However, in practice, simpler models, such as the one introduced in Sec. 12.4, are generally used.

9.6 TWO PREDICTIVE ACTIVITY COEFFICIENT MODELS

Because the variety of organic compounds of interest in chemical processing is very large, and the number of possible binary, ternary, and other mixtures is essentially uncountable, situations frequently arise in which engineers need to make activity coefficient predictions for systems or at conditions for which experimental data are not available. Although the models discussed in the previous section are useful for correlating experimental data, or for making predictions given a limited amount of experimental information, they are of little value in making predictions when no experimental data are available. This is because in these models there is no theory that relates the values of the parameters to molecular properties.

Most of the recent theories of liquid solution behavior have been based on well-defined thermodynamic or statistical mechanical assumptions, so that the parameters that appear can be related to the molecular properties of the species in the mixture, and the resulting models have some predictive ability. Although a detailed study of the more fundamental approaches to liquid solution theory is beyond the scope of this book, we consider two examples here: the theory of van Laar,[9] which leads to regular solution theory; and the UNIFAC group contribution model, which is based on the UNIQUAC model introduced in the previous section. Both regular solution theory and the UNIFAC model are useful for estimating solution behavior in the absence of experimental data. However, neither one is considered sufficiently accurate for the design of a chemical process.

The theory of van Laar (i.e., the argument that originally led to Eqs. 9.5-9) is based on the assumptions that (1) the binary mixture is composed of two species of similar size and energies of interaction, and (2) the van der Waals equation of state applies to both the pure fluids and the binary mixture.[10]

The implication of assumption 1 is that the molecules of each species will be uniformly distributed throughout the mixture (see Appendix A9.1) and the intermolecular spacing will be similar to that in the pure fluids. Consequently, it is not unreasonable to expect in this case that at a given temperature and pressure

$$\Delta_{\text{mix}}\underline{V} = 0 \quad \text{or} \quad \underline{V}^{\text{ex}} = 0$$

---

[9] See Footnote 5.
[10] van Laar was a student of van der Waals.

and

$$\Delta_{mix}\underline{S} = -R \sum_{i=1}^{2} x_i \ln x_i \quad \text{or} \quad \underline{S}^{ex} = 0 \qquad \text{(9.6-1)}$$

so that

$$\underline{G}^{ex} = \underline{U}^{ex} + P\underline{V}^{ex} - T\underline{S}^{ex} = \underline{U}^{ex}$$

for such liquid mixtures. Thus to obtain the excess Gibbs energy change, and thereby the activity coefficients for this liquid mixture, we need only compute the excess internal energy change on mixing.

Since the internal energy, and therefore the excess internal energy, is a state function, $\underline{U}^{ex}$ may be computed along any convenient path leading from the pure components to the mixture. The following path is used.

I Start with x_1 moles of pure liquid 1 and x_2 moles of pure liquid 2 (where $x_1 + x_2 = 1$) at the temperature and pressure of the mixture, and, at constant temperature, lower the pressure so that each of the pure liquids vaporizes to an ideal gas.

II At constant temperature and (very low) pressure, mix the ideal gases to form an ideal gas mixture.

III Now compress the gas mixture, at constant temperature, to a liquid mixture at the final pressure P.

The total internal energy change for this process (which is just the excess internal energy change since $\Delta_{mix}\underline{U}^{IM} = 0$) is the sum of the internal energy changes for each of the steps:

$$\underline{G}^{ex} = \underline{U}^{ex} = \Delta\underline{U}^{I} + \Delta\underline{U}^{II} + \Delta\underline{U}^{III} \qquad \text{(9.6-2)}$$

Noting that

$$\left(\frac{\partial \underline{U}}{\partial \underline{V}}\right)_T = T\left(\frac{\partial P}{\partial T}\right)_V - P$$

and using the facts that $\underline{V} \to \infty$ as $P \to 0$ and that $\Delta\underline{U}^{II} = \Delta_{mix}\underline{U}^{IGM} = 0$, one obtains

$$\begin{aligned}
\underline{G}^{ex} = \Delta\underline{U} = x_1 &\left[\int_{\underline{V}_1}^{\infty}\left\{T\left(\frac{\partial P}{\partial T}\right)_V - P\right\}d\underline{V}\right]_{\text{pure fluid 1}} \\
+ x_2 &\left[\int_{\underline{V}_2}^{\infty}\left\{T\left(\frac{\partial P}{\partial T}\right)_V - P\right\}d\underline{V}\right]_{\text{pure fluid 2}} \\
+ &\left[\int_{\infty}^{\underline{V}_{mix}}\left\{T\left(\frac{\partial P}{\partial T}\right)_V - P\right\}d\underline{V}\right]_{\text{mixture}} \qquad \text{(9.6-3)}
\end{aligned}$$

Each of the bracketed terms represents the internal energy change on going from a liquid to an ideal gas and is equal to the internal energy change on vaporization.

Next we use assumption 2, that the van der Waals equation of state is applicable to both pure fluids and the mixture. From

$$P = \frac{RT}{(\underline{V} - b)} - \frac{a}{\underline{V}^2} \tag{9.6-4}$$

it is easily established that

$$\left(\frac{\partial \underline{U}}{\partial \underline{V}} \right)_T = T \left(\frac{\partial P}{\partial T} \right)_{\underline{V}} - P = \frac{a}{\underline{V}^2}$$

so that Eq. 9.6-3 becomes

$$\begin{aligned} \underline{G}^{ex} = \Delta \underline{U} &= x_1 \int_{\underline{V}_1}^{\infty} \frac{a_1}{\underline{V}^2} \, d\underline{V} + x_2 \int_{\underline{V}_2}^{\infty} \frac{a_2}{\underline{V}^2} \, d\underline{V} - \int_{\underline{V}_{mix}}^{\infty} \frac{a_{mix}}{\underline{V}^2} \, d\underline{V} \\ &= x_1 \frac{a_1}{\underline{V}_1} + x_2 \frac{a_2}{\underline{V}_2} - \frac{a_{mix}}{\underline{V}_{mix}} \end{aligned} \tag{9.6-5}$$

where the molar volumes appearing in this equation are those of the liquid. Since the molecules of a liquid are closely packed, liquids are relatively incompressible; that is, enormous pressures are required to produce relatively small changes in the molar volume. Such behavior is predicted by Eq. 9.6-4 if $\underline{V} \approx b$. Making this substitution in Eq. 9.6-5 gives

$$\underline{G}^{ex} = \Delta \underline{U} = x_1 \frac{a_1}{b_1} + x_2 \frac{a_2}{b_2} - \frac{a_{mix}}{b_{mix}}$$

Now using the mixing rules of Eq. 9.4-8 (with $k_{ij} = 0$), we obtain the following expression for the excess internal energy of a van Laar mixture:

$$\underline{U}^{ex} = \underline{G}^{ex} = \Delta \underline{U} = \frac{x_1 x_2 b_1 b_2}{x_1 b_1 + x_2 b_2} \left(\frac{\sqrt{a_1}}{b_1} - \frac{\sqrt{a_2}}{b_2} \right)^2$$

Finally, differentiating this expression with respect to composition yields $\bar{G}_i^{ex}$ and, from Eq. 9.3-12, one obtains the van Laar equations for the activity coefficients,

$$\ln \gamma_1 = \frac{\alpha}{\left[1 + \dfrac{\alpha x_1}{\beta x_2} \right]^2} \quad \text{and} \quad \ln \gamma_2 = \frac{\beta}{\left[1 + \dfrac{\beta x_2}{\alpha x_1} \right]^2}$$

with

$$\alpha = \frac{b_1}{RT} \left[\frac{\sqrt{a_1}}{b_1} - \frac{\sqrt{a_2}}{b_2} \right]^2 \quad \text{and} \quad \beta = \frac{b_2}{RT} \left[\frac{\sqrt{a_1}}{b_1} - \frac{\sqrt{a_2}}{b_2} \right]^2 \tag{9.6-6}$$

This development provides both a justification for the van Laar equations and a method of estimating the van Laar parameters for liquid-phase activity coefficients from the parameters in the van der Waals equation of state. Since we know the van der Waals equation is not very accurate, it is not surprising that if α and β are treated as adjustable parameters, the correlative value of the van Laar equations is greater than when α and β are determined from Eqs. 9.6-6 (Problem 9.9).

Scatchard,[11] based on the observations of Hildebrand,[12] concluded that although $\underline{V}^{ex}$ and $\underline{S}^{ex}$ were approximately equal to zero for some solutions (such mixtures are called **regular solutions**), few obeyed the van der Waals equation of state. Therefore, he suggested that instead of using an equation of state to predict the internal energy change on vaporization, as in the van Laar theory, the experimental internal energy change on vaporization (usually at 25°C) be used. It was also suggested that the internal energy change of vaporization for the mixture be estimated from the approximate mixing rule

$$(\Delta_{vap}\underline{U})_{mix} = \left(x_1 \sqrt{\frac{\underline{V}_1 \Delta_{vap}\underline{U}_1}{\underline{V}_{mix}}} + x_2 \sqrt{\frac{\underline{V}_2 \Delta_{vap}\underline{U}_2}{\underline{V}_{mix}}} \right)^2 \tag{9.6-7}$$

since experimental data on the heat of vaporization of mixtures are rarely available. In this equation all molar volumes are those of the liquids, and $\underline{V}_{mix} = x_1\underline{V}_1 + x_2\underline{V}_2$, since $\Delta_{mix}\underline{V} = 0$ by the first van Laar assumption. Defining the volume fraction Φ_i and **solubility parameter** δ_i of species i by

$$\Phi_i \equiv \frac{x_i \underline{V}_i}{\underline{V}_{mix}} \tag{9.6-8}$$

and

$$\delta_i \equiv \left(\frac{\Delta_{vap}\underline{U}_i}{\underline{V}_i} \right)^{1/2} \tag{9.6-9}$$

one obtains (see Eq. 9.6-3)

$$\underline{G}^{ex} = \underline{U}^{ex} = x_1 \Delta_{vap}\underline{U}_1 + x_2 \Delta\underline{U}_2^{vap} - (\Delta_{vap}\underline{U})_{mix}$$
$$= (x_1\underline{V}_1 + x_2\underline{V}_2)\Phi_1\Phi_2[\delta_1 - \delta_2]^2$$

which, on differentiation, yields

Regular solution model activity coefficients

$$\boxed{\begin{aligned} RT\,\ln\gamma_1 &= \underline{V}_1\Phi_2^2[\delta_1 - \delta_2]^2 \\ RT\,\ln\gamma_2 &= \underline{V}_2\Phi_1^2[\delta_1 - \delta_2]^2 \end{aligned}} \tag{9.6-10}$$

Thus, we have a recipe for estimating the activity coefficients of each species in a binary liquid mixture from a knowledge of the pure-component molar volumes, the mole (or volume) fractions, and the solubility parameters (or internal energy changes on vaporization) of each species. Table 9.6-1 gives a list of the solubility parameters and molar volumes for a number of nonpolar liquids. Note, however, that the assumptions contained in regular solution theory (i.e., $\underline{V}^{ex} = \underline{S}^{ex} = 0$ and Eq. 9.6-7) are not generally applicable to polar substances. Therefore, regular solution theory should be used only for the components listed in Table 9.6-1 and similar compounds.

Regular solution theory is functionally equivalent to the van Laar theory since, with the substitutions

$$\alpha = \frac{\underline{V}_1}{RT}(\delta_1 - \delta_2)^2 \quad \text{and} \quad \beta = \frac{\underline{V}_2}{RT}(\delta_1 - \delta_2)^2$$

[11]G. Scatchard, *Chem Rev.,* **8**, 321 (1931).

[12]J. H. Hildebrand, *J. Am. Chem. Soc.,* **41**, 1067 (1919). This is discussed in J. H. Hildebrand, J. M. Prausnitz, and R. L. Scott, *Regular and Related Solutions,* Van Nostrand-Reinhold, Princeton, N.J. (1970).

Table 9.6-1 Molar Liquid Volumes and Solubility Parameters of Some Nonpolar Liquids

	$\underline{V}^L$ (cc/mol)	δ (cal/cc)$^{1/2}$
Liquefied gases at 90 K		
Nitrogen	38.1	5.3
Carbon monoxide	37.1	5.7
Argon	29.0	6.8
Oxygen	28.0	7.2
Methane	35.3	7.4
Carbon tetrafluoride	46.0	8.3
Ethane	45.7	9.5
Liquid solvents at 25°C		
Perfluoro-*n*-heptane	226	6.0
Neopentane	122	6.2
Isopentane	117	6.8
n-Pentane	116	7.1
n-Hexane	132	7.3
1-Hexene	126	7.3
n-Octane	164	7.5
n-Hexadecane	294	8.0
Cyclohexane	109	8.2
Carbon tetrachloride	97	8.6
Ethyl benzene	123	8.8
Toluene	107	8.9
Benzene	89	9.2
Styrene	116	9.3
Tetrachloroethylene	103	9.3
Carbon disulfide	61	10.0
Bromine	51	11.5

Source: PRAUSNITZ, MOLECULAR THERMODYNAMICS OF FLUID-PHASE EQUILIBRIA., 1st Edition, © 1969. Reprinted by permission of Pearson Education, Inc., Upper Saddle River, NJ.
Note: In regular solution theory the solubility parameter has traditionally been given in the units shown. For this reason the traditional units, rather than SI units, appear in this table.

Eqs. 9.5-9 and 9.6-10 are identical. The important advantage of regular solution theory, however, is that we can calculate its parameters without resorting to activity coefficient measurements. Unfortunately, the parameters derived in this way are not as accurate as those fitted to experimental data.

ILLUSTRATION 9.6-1

Test of the Regular Solution Model

Compare the regular solution theory predictions for the activity coefficients of the benzene–2,2,4-trimethyl pentane mixture with the experimental data given in Illustration 9.5-1.

SOLUTION

From Table 9.6-1 we have $\underline{V}^L = 89$ cc/mol and $\delta = 9.2$ (cal/cc)$^{1/2}$ for benzene. The parameters for 2,2,4-trimethyl pentane are not given. However, the molar volume of this compound is approximately 165 cc/mol, and the solubility parameter can be estimated from

$$\delta = \left(\frac{\Delta_{vap}\underline{U}}{\underline{V}^{\mathrm{L}}} \right)^{1/2} = \left(\frac{\Delta_{vap}\underline{H} - RT}{\underline{V}^{\mathrm{L}}} \right)^{1/2}$$

where in the numerator we have neglected the liquid molar volume with respect to that of the vapor, and further assumed ideal vapor-phase behavior. Using the value of $\Delta_{vap}\underline{H}$ found in Illustration 7.7-1, we obtain $\delta = 6.93 \; (\mathrm{cal/cc})^{1/2}$.

Our predictions for the activity coefficients, together with the experimental data, are plotted in Fig. 9.6-1. It is evident that although the regular solution theory prediction leads to activity coefficient behavior that is qualitatively correct, the quantitative agreement in this case is, in fact, poor. This example should serve as a warning that theoretical predictions may not always be accurate and that experimental data are to be preferred. ∎

The extension of regular solution theory to multicomponent mixtures is an algebraically messy task; the final result is

$$RT \ln \gamma_{\mathrm{i}} = \underline{V}_{\mathrm{i}}(\delta_{\mathrm{i}} - \bar{\delta})^2 \qquad \text{(9.6-11)}$$

where

$$\bar{\delta} = \begin{pmatrix} \text{Volume fraction} \\ \text{averaged solubility} \\ \text{parameter} \end{pmatrix} = \sum_{\mathrm{j}} \Phi_{\mathrm{j}}\delta_{\mathrm{j}}$$

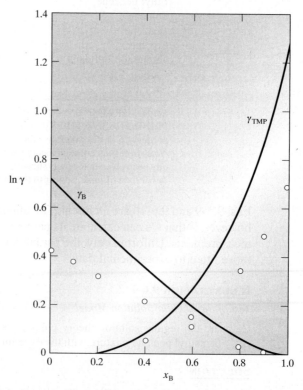

Figure 9.6-1 The regular solution theory predictions for the activity coefficients of the benzene–2,2,4-trimethyl pentane mixture. The points indicated by ○ are the experimental data.

and

$$\Phi_j = \frac{x_j \underline{V}_j}{\sum_k x_k \underline{V}_k}$$

where each of the sums is over all the species in the mixture.

Several other characteristics of regular solution theory are worth noting. The first is that the theory leads only to positive deviations from ideal solution behavior in the sense that $\gamma_i \geq 1$. This result can be traced back to the assumption of Eq. 9.6-7, which requires that $(\Delta_{vap}\underline{U})_{mix}$ always be intermediate to $\Delta_{vap}\underline{U}$ of the two pure components. Also, the solubility parameters are clearly functions of temperature since $\delta \to 0$ as $T \to T_c$ for each species; however, $(\delta_1 - \delta_2)$ is often nearly temperature independent, at least over limited temperature ranges. Therefore, the solubility parameters listed in Table 9.6-1 may be used at temperatures other than the one at which they were obtained (see Eq. 9.3-22, however). Also, referring to Eq. 9.6-10, it is evident that liquids with very different solubility parameters, such as neopentane and carbon disulfide, can be expected to exhibit highly nonideal solution behavior (i.e., $\gamma_i > 1$), whereas adjacent liquids in Table 9.6-1 will form nearly ideal solutions. This is useful information.

The most successful recent activity coefficient prediction methods are based on the idea of group contributions, in which each molecule is considered to be a collection of basic building blocks, the functional groups discussed in the last section. Thus, a mixture of molecules is considered to be a mixture of functional groups, and its properties result from functional group interactions. Because the number of different types of functional groups is very much smaller than the number of different molecular species, it is possible, by the regression of experimental data, to obtain a fairly compact table of parameters for the interaction of each group with all others. From such a table, the activity coefficients in a mixture for which no experimental data are available can be estimated from a knowledge of the functional groups present.

The two most developed group contribution methods are the ASOG (Analytical Solution Of Groups)[13] and UNIFAC (UNIquac Functional-group Activity Coefficient)[14] models, both of which are the subjects of books. We will consider only the UNIFAC model here. UNIFAC is based on the UNIQUAC model of Sec. 9.5. This model, you will remember, has a combinatorial term that depends on the volume and surface area of each molecule, and a residual term that is a result of the energies of interaction between the molecules. In UNIQUAC the combinatorial term was evaluated using group contributions to compute the size parameters, whereas the residual term had two adjustable parameters for each binary system that were to be fit to experimental data. In the UNIFAC model, both the combinatorial and residual terms are obtained using group contribution methods.

When using the UNIFAC model, one first identifies the functional subgroups present in each molecule using the list in Table 9.5-2. Next the activity coefficient for each species is written as

$$\ln \gamma_i = \ln \gamma_i(\text{combinatorial}) + \ln \gamma_i(\text{residual})$$

[13] K. Kojima and T. Tochigi, *Prediction of Vapor-Liquid Equilibrium by the ASOG Method,* Elsevier, Amsterdam (1979).

[14] A. Fredenslund, J. Gmehling, and P. Rasmussen, *Vapor-Liquid Equilibrium Using UNIFAC,* Elsevier, Amsterdam (1977).

The combinatorial term is evaluated from a modified form of the comparable term in the UNIQUAC equation (Eq. 9.5-23a):

$$\ln \gamma_i(\text{combinatorial}) = \ln \frac{\phi_i{}'}{x_i} + 1 - \frac{\phi_i{}'}{x_i} - \frac{z}{2}q_i\left(1 + \ln\frac{\phi_i}{\theta_i} - \frac{\phi_i}{\theta_i}\right) \quad \textbf{(9.6-12a)}$$

where $\phi_i' = x_i r_i^{3/4}\Big/\sum_j x_j r_j^{3/4}$, with the r and q parameters evaluated using the group contribution discussed in the previous section. Here, however, the residual term is also evaluated by a group contribution method, so that the mixture is envisioned as being a mixture of functional groups rather than of molecules. The residual contribution to the logarithm of the activity coefficient of group k in the mixture, $\ln \Gamma_k$, is computed from the group contribution analogue of Eq. 9.5-23b, which is written as

$$\ln \Gamma_k = Q_k\left[1 - \ln\left(\sum_m \Theta_m \Psi_{mk}\right) - \sum_m \frac{\Theta_m \Psi_{km}}{\sum_n \Theta_n \Psi_{nm}}\right] \quad \textbf{(9.6-12b)}$$

where

$$\Theta_m = \begin{pmatrix}\text{Surface area} \\ \text{fraction of} \\ \text{group m}\end{pmatrix} = \frac{X_m Q_m}{\sum_n X_n Q_n}$$

$$X_m = \text{mole fraction of group m in mixture} \quad \textbf{(9.6-13a)}$$

and

$$\Psi_{mn} = \exp\left[\frac{-(u_{mn} - u_{nn})}{kT}\right] = \exp\left[\frac{-a_{mn}}{T}\right] \quad \textbf{(9.6-13b)}$$

where u_{mn} is a measure of the interaction energy between groups m and n, and the sums are over all groups in the mixture.

Finally, the residual part of the activity coefficient of species i is computed from

The UNIFAC model

$$\boxed{\ln \gamma_i(\text{residual}) = \sum_k v_k^{(i)}[\ln \Gamma_k - \ln \Gamma_k^{(i)}]} \quad \textbf{(9.6-14)}$$

Here $v_k^{(i)}$ is the number of k groups present in species i, and $\Gamma_k^{(i)}$ is the residual contribution to the activity coefficient of group k in a pure fluid of species i molecules. This last term is included to ensure that in the limit of pure species i, which is still a mixture of groups (unless species i molecules consist of only a single functional group), $\ln \gamma_i(\text{residual})$ is zero or $\gamma_i(\text{residual}) = 1$.

The combination of Eqs. 9.6-12 through 9.6-14 is the UNIFAC model. Since the volume (R_i) and surface (Q_i) parameters are known (Table 9.5-2), the only unknowns are binary parameters, a_{nm} and a_{mn}, for each pair of functional groups. Continuing with the group contribution idea, it is next assumed that any pair of functional groups m and n will interact in the same manner, that is, have the same value of a_{mn} and a_{nm}, independent of the mixtures in which these two groups occur. Thus, for example, it is assumed that the interaction between an alcohol (—OH) group and a methyl (—CH$_3$) group will be the same, regardless of whether these groups occur in ethanol–n-pentane, isopropanol-decane, or 2-octanol–2,2-4-trimethyl pentane mixtures.

Consequently, by a regression analysis of very large quantities of activity coefficient (or, as we will see in Sec. 10.2, actually vapor-liquid equilibrium) data, the binary parameters a_{nm} and a_{mn} for many group-group interactions can be determined. These parameters can then be used to predict the activity coefficients in mixtures (binary or multicomponent) for which no experimental data are available.

In the course of such an analysis it was found that (1) experimental data existed to determine many of, but not all, the binary group parameters a_{nm} and a_{mn} and (2) that some very similar groups, such as the —CH$_3$, —CH$_2$, —CH, and —C groups, interact with other groups in approximately the same way and, therefore, have the same interaction parameters with other groups. Such very similar subgroups are considered to belong to the same main group. Consequently, in Table 9.5-2 there are 46 main groups among the 106 subgroups. All the subgroups within a main group (i.e., the CH$_3$, CH$_2$, CH, and C subgroups within the CH$_2$ main group) have the same values of the binary parameters for interactions with other main groups and zero values for the interactions with other subgroups in their own main group. Appendix B describes several programs that use the UNIFAC model for activity coefficient predictions that are the on the CD-ROM accompanying this book. The Visual Basic and MATLAB programs use the recent UNIFAC model described in this chapter, whereas the DOS Basic version uses an earlier model (the storage limitations in DOS Basic did not allow updating this program to accommodate the many new parameters that had been added). Therefore, the Visual Basic or MATLAB versions of the program should be used; the DOS-based version has been retained should the reader wish to examine the difference in the two methods.

ILLUSTRATION 9.6-2

Test of the UNIFAC Model

Compare the UNIFAC predictions for the activity coefficients of the benzene–2,2,4-trimethyl pentane mixture with the experimental data given in Illustration 9.5-1.

SOLUTION

Benzene consists of six aromatic CH groups (i.e., subgroup 10 of Table 9.5-2), whereas 2,2,4-trimethyl pentane contains five CH$_3$ groups (subgroup 1), one CH$_2$ group (subgroup 2), one CH group (subgroup 3), and one C group (subgroup 4). Using the UNIFAC program of Appendix B.I, we obtain the results plotted in Fig. 9.6-2.

It is clear from this figure that, for this simple system, the UNIFAC predictions are good—much better than the regular solution predictions of Fig. 9.6-1. Although the UNIFAC predictions for all systems are not always as good as for the benzene–2,2,4-trimethyl pentane system, UNIFAC with its recent improvements is the best activity coefficient prediction method currently available.   ∎

9.7 FUGACITY OF SPECIES IN NONSIMPLE MIXTURES

To estimate the fugacity of a species in a gaseous mixture using the Lewis-Randall rule,

$$\overline{f}_i^{\,V}(T, P, \underline{y}) = y_i f_i^{\,V}(T, P) \tag{9.7-1}$$

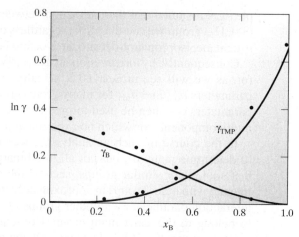

Figure 9.6-2 UNIFAC predictions for the activity coefficients of the benzene–2,2,4-trimethyl pentane system. The points are activity coefficients derived from experimental data.

we need the fugacity of pure species i as a *vapor* at the temperature and pressure of the mixture. In a nonsimple gaseous mixture at least one of the pure components does not exist as a vapor at the mixture temperature and pressure, so $f_i^V(T, P)$ for that species cannot be computed without some sort of approximation or assumption. To estimate the fugacity of a species in a liquid mixture from an activity coefficient model, one uses

$$\bar{f}_i^L(T, P, \underline{x}) = x_i \gamma_i(T, P, \underline{x}) f_i^L(T, P) \qquad \textbf{(9.7-2)}$$

and one would need the fugacity of pure species i as a *liquid* at the temperature and pressure of the mixture. In a nonsimple liquid mixture, at least one of the components, if pure, would be a solid or a vapor at the temperature and pressure of the mixture, so it is not evident how to compute $f_i^L(T, P)$ in this case.

The study of vapor-liquid equilibria (Sec 10.1) of the solubility of gases in liquids (Sec. 11.1), and of the solubility of solids in liquids (Sec. 12.1), all involve nonsimple mixtures. To see why this occurs, consider the criterion for vapor-liquid equilibrium:

$$\bar{f}_i^L(T, P, \underline{x}) = \bar{f}_i^V(T, P, \underline{y})$$

Using Eqs. 9.7-1 and 9.7-2, we have

$$x_i \gamma_i(T, P, \underline{x}) f_i^L(T, P) = y_i f_i^V(T, P)$$

To use this equation we must estimate the pure component fugacity of each species as both a liquid and a vapor at the temperature and pressure of the mixture. However, at this temperature and pressure either the liquid or the vapor will be the stable phase for each species, generally not both.[15]

Consequently, at least one of the phases will be a nonsimple mixture. In most vapor-liquid problems both phases will be nonsimple mixtures, in that species with pure component vapor pressures greater than the system pressure appear in the liquid phase, and those with vapor pressures less than the system pressure appear in the vapor phase.

For those situations one can proceed in several ways. The simplest, and most accurate when it is applicable, is to use equations of state to compute species fugacities in

---

[15]Unless, fortuitously, the system pressure is equal to the vapor pressure of that species.

mixtures, thereby avoiding the use of Eqs. 9.7-1 and 9.7-2 and the necessity of computing a pure component fugacity in a thermodynamic state that does not occur. Thus, when an equation of state (virial, cubic, etc.) can be used for the vapor mixture, a nonsimple gaseous mixture can be treated using the methods of Sec. 9.4. Similarly, if a nonsimple liquid mixture can be described by an equation of state, which is likely to be the case only for hydrocarbons and perhaps hydrocarbons with dissolved gases, it can also be treated by the methods of Sec 9.4.

If an equation of state cannot be used, one can, in principle, proceed in either of two ways. The first procedure is to write the Gibbs energy of mixing as

$$\Delta_{\mathrm{mix}}\underline{G} = \Delta\underline{G}' + \Delta_{\mathrm{mix}}\underline{G}^{\mathrm{IM}} + \underline{G}^{\mathrm{ex}}$$

where $\Delta_{\mathrm{mix}}\underline{G}^{\mathrm{IM}}$ and $\underline{G}^{\mathrm{ex}}$ have their usual meanings, and $\Delta\underline{G}'$ is the Gibbs energy change of converting to the same phase as the mixture those species that exist in other phases as pure substances. For example, $\Delta\underline{G}'$ might be the Gibbs energy change of producing pure liquids from either gases or solids before forming a liquid mixture. The partial molar Gibbs energy and fugacity of each species in the mixture would then be computed directly from $\Delta_{\mathrm{mix}}\underline{G}$. In practice, however, the calculation of $\Delta\underline{G}'$ can be difficult, since it may involve the estimation of the Gibbs energy of substances in hypothetical states (i.e., a liquid above its critical point).

A second, more straightforward procedure is to use Eqs. 9.7-1 and 9.7-2 and the models for γ_i considered in Secs. 9.5 and 9.6, but with estimates for the pure component fugacities of the hypothetical gases and liquids obtained by simple extrapolation procedures. Of course, such extrapolation schemes have no real physical or chemical basis; they are, rather, calculational methods that have been found empirically to lead to satisfactory predictions, provided the extrapolation is not too great. The first extrapolation scheme to be considered is for the liquid-phase fugacity of a species that would be a vapor as a pure component at the temperature and pressure of the mixture. The starting point here is the observation that the fugacity of a pure liquid is equal to its vapor pressure if the vapor pressure is not too high (or the product of P^{vap} and the fugacity coefficient at higher vapor pressures). Consequently, one takes the fugacity of the hypothetical liquid at the temperature T and the pressure P to be equal to the vapor pressure of the real liquid at the same temperature [even though $P^{\mathrm{vap}}(T) > P$]. Thus, in Illustration 10.1-2, where we consider vapor-liquid equilibrium for an n-pentane, n-hexane, and n-heptane mixture at 69°C and 1 bar, the fugacity of "liquid" n-pentane, for use in Eq. 9.7-2, will be taken to be equal to its vapor pressure at 69°C, 2.721 bar.

To calculate the fugacity of a pure vapor from corresponding states that, at the conditions of the mixture, exists only as a liquid, we will use Eq. 7.4-22 with $f_i^{\mathrm{V}}(T, P)$ equal to the total pressure, if the pressure is low enough, or

$$f_i^{\mathrm{V}}(T, P) = P\left(\frac{f}{P}\right)_i$$

at higher pressures. Here, however, the fugacity coefficient is obtained not from Fig. 7.4-1, but rather from Fig. 9.7-1, which is a corresponding states correlation in which the fugacity coefficient for gases has been extrapolated into the liquid region. (You should compare Figs. 7.4-1 and 9.7-1.)

Extrapolation schemes may also be used in some circumstances where the desired phase does not exist at any pressure for the temperature of interest—for example, to estimate the fugacity of a "liquid" not too far above its critical temperature, or not much

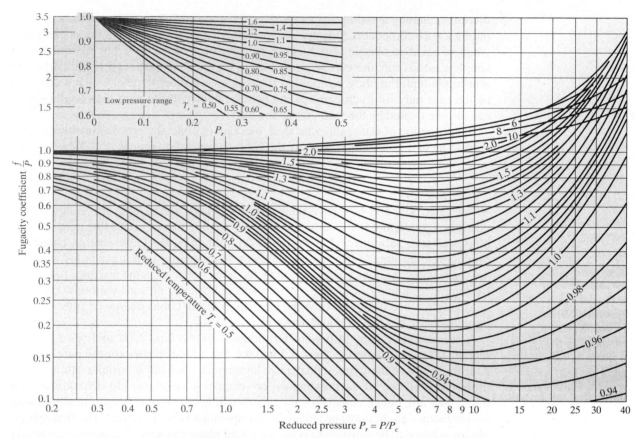

Figure 9.7-1 Fugacity coefficients of gases and vapors. (Reprinted with permission from O. A. Hougen and K. M. Watson, *Chemical Process Principles Charts*, John Wiley & Sons, New York, 1946.) In this figure $Z_c = 0.27$.

below its triple-point temperature. As an example of the methods used, we consider the estimation of the liquid-phase fugacity at temperatures below its triple point (so that the solid is the stable phase) and also at temperatures above its critical temperature (where the gas is the stable phase) for the substance whose pure component phase diagram is given in Fig. 9.7-2a. In either case the first step in the procedure is to extend the vapor pressure curve, either analytically (using the Clausius-Clapeyron equation) or graphically as indicated in Fig. 9.7-2b, to obtain the vapor pressure of the hypothetical liquid.[16]

In the case of the subcooled liquid, which involves an extrapolation into the solid region, the vapor pressure is usually so low that the fugacity coefficient is close to unity, and the fugacity of this hypothetical liquid is equal to the extrapolated vapor pressure. For the supercritical liquid, however, the extrapolation is above the critical temperature of the liquid and yields very high vapor pressures, so that the fugacity of this hypothetical liquid is equal to the product of the extrapolated vapor pressure and the fugacity coefficient (which is taken from the corresponding-states plot of Fig. 9.7-1).

Another way to estimate the "subcooled" liquid fugacity f_1^L below the melting point is to use heat (enthalpy) of fusion data and, if available, the heat capacity data for both

---

[16]For accurate extrapolations $\ln P^{\mathrm{vap}}$ should be plotted versus $1/T$ as in Sec. 7.5.

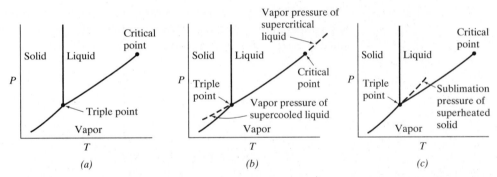

Figure 9.7-2 (*a*) *P-T* phase diagram for a typical substance. (*b*) *P-T* phase diagram with dashed lines indicating extrapolation of the liquid-phase vapor pressure into the solid and supercritical regions. (*c*) *P-T* phase diagram with a dashed line indicating extrapolation of the sublimation pressure of the solid into the liquid region.

the solid and liquid to compute the Gibbs energy of fusion $\Delta_{\text{fus}}\underline{G}(T)$, which is related to the fugacity ratio as follows:

$$\frac{\Delta_{\text{fus}}\underline{G}(T, P)}{RT} = \frac{\underline{G}_1^{\text{L}}(T, P) - \underline{G}_1^{\text{S}}(T, P)}{RT} = \ln \frac{f_1^{\text{L}}(T, P)}{f_1^{\text{S}}(T, P)} \tag{9.7-3}$$

The value of $\Delta_{\text{fus}}\underline{G}(T)$ is computed by separately calculating $\Delta_{\text{fus}}\underline{H}(T)$ and $\Delta_{\text{fus}}\underline{S}(T)$, and then using the relation $\Delta_{\text{fus}}\underline{G}(T) = \Delta_{\text{fus}}\underline{H}(T) - T\Delta_{\text{fus}}\underline{S}(T)$. To compute the enthalpy and entropy changes of fusion, we suppose that the melting of a solid (below its normal melting point) to form a liquid is carried out in the following three-step constant-pressure process:

1. The solid is heated at fixed pressure from the temperature T to its normal melting temperature T_m.
2. The solid is then melted to form a liquid.
3. The liquid is cooled *without* solidification from T_m back to the temperature of the mixture.

The enthalpy and entropy changes for this process are

$$\Delta_{\text{fus}}\underline{H}(T) = \int_T^{T_m} C_{\text{P}}^{\text{S}}\, dT + \Delta_{\text{fus}}\underline{H}(T_m) + \int_{T_m}^T C_{\text{P}}^{\text{L}}\, dT$$

$$= \Delta_{\text{fus}}\underline{H}(T_m) + \int_{T_m}^T \Delta C_{\text{P}}\, dT \tag{9.7-4}$$

$$\Delta_{\text{fus}}\underline{S}(T) = \int_T^{T_m} \frac{C_{\text{P}}^{\text{S}}}{T}\, dT + \Delta_{\text{fus}}\underline{S}(T_m) + \int_{T_m}^T \frac{C_{\text{P}}^{\text{L}}}{T}\, dT$$

$$= \Delta_{\text{fus}}\underline{S}(T_m) + \int_{T_m}^T \frac{\Delta C_{\text{P}}}{T}\, dT \tag{9.7-5}$$

where $\Delta C_{\text{P}} = C_{\text{P}}^{\text{L}} - C_{\text{P}}^{\text{S}}$.

Note that Eqs. 9.7-4 and 9.7-5 relate the enthalpy and entropy changes of fusion at any temperature T to those changes at the melting point at the same pressure. Now since $\underline{G} = \underline{H} - T\underline{S}$, and $\Delta_{\text{fus}}\underline{G}(T = T_m) = 0$, Eq. 9.7-5 can be rewritten as

$$\Delta_{\text{fus}}\underline{S}(T) = \frac{\Delta_{\text{fus}}\underline{H}(T_m)}{T_m} + \int_{T_m}^{T} \frac{\Delta C_P}{T}\, dT \tag{9.7-6}$$

and, therefore,

$$
\begin{aligned}
\Delta_{\text{fus}}\underline{G}(T) &= \Delta_{\text{fus}}\underline{H}(T) - T\,\Delta_{\text{fus}}\underline{S}(T) \\
&= \Delta_{\text{fus}}\underline{H}(T_m)\left[1 - \frac{T}{T_m}\right] + \int_{T_m}^{T} \Delta C_P\, dT - T\int_{T_m}^{T} \frac{\Delta C_P}{T}\, dT \\
&\equiv RT \ln\left[\frac{f_1^{\text{L}}(T, P)}{f_1^{\text{S}}(T, P)}\right]
\end{aligned}
\tag{9.7-7}
$$

Therefore,

$$
\begin{aligned}
&f_1^{\text{L}}(T, P) \\
&= f_1^{\text{S}}(T, P)\exp\left[\frac{1}{RT}\left[\Delta_{\text{fus}}\underline{H}(T)\left(1 - \frac{T}{T_m}\right) + \int_{T_m}^{T} \Delta C_P\, dT - T\int_{T_m}^{T} \frac{\Delta C_P}{T}\, dT\right]\right]
\end{aligned}
\tag{9.7-8a}
$$

As heat capacity data may not be available, this equation is usually simplified to

$$\boxed{f_1^{\text{L}}(T, P) = f_1^{\text{S}}(T, P)\exp\left[\frac{\Delta_{\text{fus}}\underline{H}(T)}{RT}\left(1 - \frac{T}{T_m}\right)\right]} \tag{9.7-8b}$$

Therefore, if the sublimation pressure of the solid, which is equal to the solid fugacity f_1^{S} and the heat of fusion at the melting point, are known, the fugacity of the hypothetical liquid, or liquid below its melting temperature, can be computed.

The fugacity of a hypothetical superheated solid can be estimated by extrapolating the sublimation pressure line into the liquid region of the phase diagram. This is indicated in Fig. 9.7-2c. The fugacity coefficient is usually equal to unity in this case.

For species whose thermodynamic properties are needed in hypothetical states far removed from their stable states, such as a liquid well above its critical point, the extrapolation procedures discussed here are usually inaccurate. In some cases special correlations or prescriptions are used; one such correlation is discussed in Chapter 11. In other cases different procedures, such as those discussed next, are used.

The fugacity of a very dilute species in a liquid mixture (e.g., a dissolved gas or solid of limited solubility) is experimentally found to be linearly proportional to its mole fraction at low mole fractions, that is,

$$\bar{f}_i^{\text{L}}(T, P, \underline{x}) = x_i H_i(T, P) \quad \text{as } x_i \to 0 \tag{9.7-9}$$

The value of the "constant of proportionality," called the **Henry's law constant**, is dependent on the solute-solvent pair, temperature, and pressure. At higher concentrations the linear relationship between $\bar{f}_i^{\text{L}}(T, P, \underline{x})$ and mole fraction fails; a form of Eq.

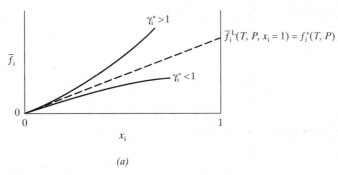

(a)

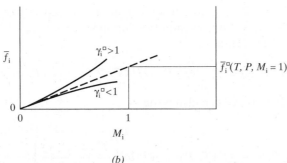

M_i

(b)

Figure 9.7-3 Solute fugacity in real and ideal Henry's law solutions. (*a*) Solute fugacity versus mole fraction. (*b*) Solute fugacity versus molality.

9.7-9 can be used at these higher concentrations by introducing a new activity coefficient $\gamma_i^*(T, P, x_i)$ so that

Henry's law based on mole fraction

$$\bar{f}_i^{\rm L}(T, P, \underline{x}) = x_i\gamma_i^*(T, P, \underline{x})H_i(T, P) \qquad \text{(9.7-10)}$$

The dashed line in Fig. 9.7-3*a* is the fugacity of an ideal Henry's law component, that is, a species that obeys Eq. 9.7-9 over the whole concentration range, and the solid lines represent two real solutions for which $\gamma_i^*(T, P, \underline{x})$ is not equal to unity at all concentrations.

Note that the Henry's law activity coefficient γ_i^* is different from the activity coefficient γ_i defined earlier. In particular, in solutions considered here $\gamma_i^* \to 1$ as $x_i \to 0$, whereas for the usual activity coefficient $\gamma_i \to 1$ as $x_i \to 1$. We can relate these two activity coefficients by comparing Eqs. 9.7-2, 9.7-9, and 9.7-10. First, equating Eqs. 9.7-2 and 9.7-5 gives

$$\bar{f}_i^{\rm L}(T, P, \underline{x}) = x_i\gamma_i^*(T, P, \underline{x})H_i(T, P) = x_i\gamma_i(T, P, \underline{x})f_i^{\rm L}(T, P)$$

or

$$\gamma_i^*(T, P, \underline{x})H_i(T, P) = \gamma_i(T, P, \underline{x})f_i^{\rm L}(T, P) \qquad \text{(9.7-11)}$$

Taking the limit as $x_i \to 0$ [remembering that $\gamma_i^*(T, P, x_i = 0) = 1$] yields

$$H_i(T, P) = \gamma_i(T, P, x_i = 0)f_i^{\rm L}(T, P) \qquad \text{(9.7-12)}$$

Using this relation in Eq. 9.7-11 gives

$$\gamma_i^*(T, P, \underline{x})\gamma_i(T, P, x_i = 0)f_i^{\rm L}(T, P) = \gamma_i(T, P, \underline{x})f_i^{\rm L}(T, P)$$

or simply

$$\gamma_i^*(T, P, \underline{x}) = \frac{\gamma_i(T, P, \underline{x})}{\gamma_i(T, P, x_i = 0)} \tag{9.7-13}$$

so that the activity coefficient γ_i^* is equal to the ratio of the activity coefficient γ_i to its value at infinite dilution.

If a solute-solvent pair were ideal in the Henry's law sense, Eq. 9.7-4 would be satisfied at all mole fractions; in particular, at $x_i = 1$,

$$\bar{f}_i^L(T, P, x_i = 1) = H_i(T, P)$$

Thus, the Henry's law constant is the hypothetical fugacity of a solute species as a pure liquid extrapolated from its infinite-dilution behavior; we will denote this by $f_i^*(T, P)$ (see Fig. 9.7-3a). Thus

$$\bar{f}_i^L(T, P, \underline{x}) = x_i \gamma_i^*(T, P, \underline{x}) H_i(T, P) = x_i \gamma_i^*(T, P, \underline{x}) f_i^*(T, P) \tag{9.7-14}$$

Using Eq. 9.2-10, we can also write

$$\bar{G}_i(T, P, \underline{x}) - \underline{G}_i^*(T, P) = RT \ \ln\left\{ \frac{\bar{f}_i^L(T, P, \underline{x})}{f_i^*(T, P)} \right\} = RT \ \ln\{x_i \gamma_i^*(T, P, \underline{x})\} \tag{9.7-15}$$

where $\underline{G}_i^*(T, P)$ is the (hypothetical) molar Gibbs free energy of the solute species as a pure liquid obtained from extrapolation of its dilute solution behavior.

The fugacity of a very dilute species can also be written as

$$\bar{f}_i^L(T, P, M_i) = M_i \mathcal{H}_i(T, P) \quad \text{as } M_i \to 0 \tag{9.7-16}$$

Here M_i is the **molality** of species i, that is, the number of moles of this species per 1000 g of solvent,[17] and $\mathcal{H}_i$ is the Henry's law constant based on molality; its value depends on the solute-solvent pair, temperature, and pressure. For real solutions the activity coefficient $\gamma_i^\square(T, P, M_i)$ is introduced, so that

Henry's law based on molality

$$\boxed{\bar{f}_i^L(T, P, M_i) = M_i \gamma_i^\square(T, P, M_i) \mathcal{H}_i(T, P)} \tag{9.7-17}$$

Clearly, $\gamma_i^\square(T, P, M_i) \to 1$ as $M_i \to 0$. The behaviors of real and ideal solutions are indicated by dashed and solid lines, respectively, in Fig 9.7-3b.

---

[17]The molality of a solution consisting of n_i moles of solute in n_s moles of a solvent of molecular weight m_s is

$$M_i = \frac{n_i \ 1000}{m_s n_s}$$

whereas the mole fraction of solute i is

$$x_i = \frac{n_i}{n_s + \sum n_j}$$

where the summation is over all solutes. At low solute concentration $n_s \gg \sum_j n_j$, and these equations reduce to $x_i \simeq n_i/n_s$ and $M_i \simeq x_i 1000/m_s$, so that M_i and x_i are linearly related. Therefore, it is not surprising that both Eqs. 9.7-9 and 9.7-16 are satisfied. Furthermore, $H_i = 1000\mathcal{H}_i/m_s$.

From Eq. 9.7-17 we have that the molal Henry's law constant is equal to the fugacity of the solute species at unit molality in an *ideal Henry's law* solution; that is,

$$\mathcal{H}_i(T, P) = \bar{f}_i^\square(T, P, M_i = 1) \tag{9.7-18}$$

where the ideal-solution unit molality fugacity $\bar{f}_i^\square(T, P, M_i = 1)$ is obtained by extrapolation of dilute solution behavior to 1 molal, as indicated in Fig 9.7-3*b*. Using the analysis that led to Eq. 9.7-13, one can show that

$$\gamma_i^\square(M_i) = \frac{x_i\, 1000}{m_s M_i} \frac{\gamma_i(\underline{x})}{\gamma_i(x_i = 0)} = \frac{x_i\, 1000}{m_s M_i} \gamma_i^*(x_i) \tag{9.7-19}$$

and that

$$\bar{G}_i(T, P, M_i) = \bar{G}_i^\square(T, P, M_i = 1) + RT\, \ln\left(\frac{M_i \gamma_i^\square(T, P, M_i)}{M_i = 1}\right) \tag{9.7-20}$$

The value of the partial molar Gibbs energy of species i in the (hypothetical) ideal solution $\bar{G}_i^\square(T, P, M_i = 1)$ is obtained by assuming ideal solution behavior and extrapolating the behavior of $\bar{G}_i(T, P, M_i)$ in very low-molality solutions to one molal. The value of $\bar{G}_i^\square$ (and $\underline{G}_i^*$, as well) obtained in this way depends on temperature, pressure, and the solute-solvent pair.

It is useful to identify the physical significance of the quantities used here and to relate them to the analogous quantities for simple mixtures. In a simple liquid mixture, the properties of the pure components dominate the partial molar properties, and we have

$$\bar{f}_i^L(T, P, \underline{x}) = x_i \gamma_i(T, P, \underline{x}) f_i^L(T, P)$$

where $f_i^L(T, P)$ is the pure component fugacity (i.e., the fugacity of species i when it interacts only with other molecules of the same species), and the explicit mole fraction accounts for its dilution. The activity coefficient γ_i arises because the nature of the interaction between the solute species i and the solvent is different from that between solute molecules, so that γ_i accounts for the effect of replacing solute-solute interactions with solute-solvent interactions. By using a Henry's law description for a nonsimple mixture, we recognize that, for the solute species, the liquid mixture and pure component states are very different. The implication of using

$$\bar{f}_i^L(T, P, \underline{x}) = x_i \gamma_i^*(T, P, \underline{x}) H_i(T, P)$$

or

$$\bar{f}_i^L(T, P, M_i) = M_i \gamma_i^\square(T, P, M_i) \mathcal{H}_i(T, P)$$

is that the properties of the solute species in solution are largely determined by solute molecules interacting only with solvent molecules, which are taken into account by the Henry's law constants H_i and $\mathcal{H}_i$. In this case, the activity coefficients γ_i^* and $\gamma_i^\square$ account for the effect of replacing solute-solvent interactions with solute-solute interactions. Therefore, the values of the Henry's law constants depend on both the solvent and the solute. That is, the Henry's law constant for a solute in different solvents will have different values.

We will have occasion to use the Henry's law descriptions (on both a mole fraction and a molality basis) and the associated activity coefficients several times in this book. The immediate disadvantage of these choices is that $\bar{f}_i^*(T, P, x_i = 1)$ and $\bar{f}_i^{\square}(T, P, M_i = 1)$ can be obtained only by extrapolation of experimental information for very dilute solutions. However, this information may be easier to obtain and more accurate than that obtained by estimating the pure liquid fugacity of a species whose equilibrium state is a supercritical gas or a solid below its triple-point temperature.

It is left as an exercise for you to relate the regular solution and UNIFAC model predictions for γ_i^* and $\gamma_i^{\square}$ to those already obtained for γ_i (Problem 9.10).

Activity coefficients of proteins

Another example of a nonsimple mixture is a protein in aqueous solution. It is a nonsimple mixture because the protein, as a pure species, does not exist as a liquid. A quantity of interest when dealing with proteins is the biological activity, that is, how efficient the proteins are in catalyzing biological processes. A property that is of less importance but may influence the biological activity is the thermodynamic activity of proteins in aqueous solution, and as with any mixture, solution nonidealities arise. However, one is generally interested in the thermodynamic activity of only the protein, not the solvent (usually water). Since the molecular weight of the protein might not be known precisely (for example, because the protein may aggregate, polymerize, or depolymerize), it is common to describe the protein concentrations in mass per unit volume (usually c_i in grams/liter) and the solution nonideality using a virial-type expansion:

$$\ln \gamma_i^{\bullet} = B_i \cdot c_i + C_i \cdot c_i^2 + D_i \cdot c_i^3 + \cdots \qquad (9.7\text{-}21)$$

where the activity coefficient $\gamma_i^{\bullet}$ is defined on a Henry's law basis, so that

$$a_i = c_i \gamma_i^{\bullet} \quad \text{with} \quad \gamma_i^{\bullet} \to 1 \text{ as } c_i \to 0 \qquad (9.7\text{-}22)$$

where a_i is the thermodynamic activity, with units of concentration.

ILLUSTRATION 9.7-1

The Activity Coefficient of Hemoglobin

The following data are available on the activity coefficient of hemoglobin in aqueous solution.[18]

c (g/L)	a (g/L)	c (g/L)	a (g/L)	c (g/L)	a (g/L)
20	23.1	120	330	220	2 060
40	53.5	140	473	240	3 040
60	94.6	160	679	260	4 580
80	150	180	973	280	7 040
100	226	200	1 410	300	11 050

Compute the hemoglobin activity as a function of concentration, and fit those results using Eq. 9.7-21.

SOLUTION

It is the activity coefficients, rather than the activities, that should be fit. So the first step is to calculate the activity coefficient at each concentration from $\gamma(c) = a(c)/c$. The result is

---

[18]A. P. Minton, *J. Molec. Biol.* **110**, 89 (1977).

c (g/l)	$\gamma^{\bullet}$	c (g/l)	$\gamma^{\bullet}$	c (g/l)	$\gamma^{\bullet}$
20	1.155	120	2.750	220	9.364
40	1.337	140	3.379	240	12.67
60	1.577	160	4.244	260	17.62
80	1.875	180	5.406	280	25.14
100	2.260	200	7.050	300	36.83

We find that the equation

$$\ln \gamma_i^{\bullet} = 7.139 \times 10^{-3} \cdot c_i + 6.940 \times 10^{-6} \cdot c_i^2 + 3.116 \times 10^{-8} \cdot c_i^3$$

provides a good fit to the activity coefficients, as shown in the figure below.

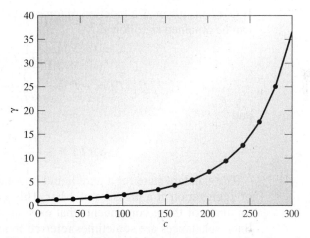

The activity coefficients of hemoglobin as a function of concentration. The line is the fit of Eq. 9.5-24, and the points ● are the reported data.   ■

9.8 SOME COMMENTS ON REFERENCE AND STANDARD STATES

We have referred to a number of reference states and standard states in this book. Therefore, it is useful to review these concepts and build upon them here. The need for a **reference state** arises largely in the development of tables and charts of thermodynamic properties. In this case the reference state is a single state of fixed temperature and pressure at which the values of certain properties (typically either internal energy or enthalpy and either entropy or Gibbs energy) are set equal to zero. One common choice for the reference state is some convenient temperature and pressure (now 1 bar, previously 1 atmosphere) and the fluid as an ideal gas. Another common choice is a pure liquid at 1 bar and some convenient temperature. The values of all other properties are then computed as changes from those in the fixed reference state. Zero pressure cannot be used as the reference state because the entropy and Gibbs and Helmholtz energies diverge in this limit.

The need for a **standard state** arises in the definition of activity coefficients and in chemical transformations. In this case the temperature of the standard state is not fixed as in the definition of a reference state, but is the temperature of the system. The pressure is fixed, generally at 1 bar or 1 atmosphere. However, different states are used depending on the state of aggregation. For the gas phase, the pure component as an ideal gas is used as the standard state. Since pressure and composition are fixed, the standard-state properties are a function only of temperature. If $\underline{G}^\circ(T)$ is the standard-state Gibbs energy in the ideal gas state, and $\mu^\circ(T)$ is the standard-state chemical potential in the ideal gas state, then $\underline{G}^\circ(T) = \mu^\circ(T)$ and

$$\bar{G}_i(T, P, \underline{x}) = \underline{G}_i^\circ(T) + RT \ln \frac{\bar{f}_i(T, P, \underline{x})}{1 \text{ bar}} \tag{9.8-1}$$

and

$$\mu_i(T, P, \underline{x}) = \mu_i^\circ(T) + RT \ln \frac{\bar{f}_i(T, P, \underline{x})}{1 \text{ bar}} \tag{9.8-2}$$

If $\underline{G}_i^\circ(T)$ is given as a function of temperature, the standard-state enthalpy and entropy can be obtained as follows:

$$\underline{H}_i^\circ(T) = -T^2 \frac{d\left(\dfrac{\underline{G}_i^\circ(T)}{T}\right)}{dT} = -T^2 \frac{d\left(\dfrac{\mu_i^\circ(T)}{T}\right)}{dT} \tag{9.8-3}$$

and

$$\underline{S}_i^\circ(T) = -\frac{d\underline{G}_i^\circ(T)}{dT} = -\frac{d\mu_i^\circ(T)}{dT} \tag{9.8-4}$$

The standard state for a pure liquid or solid is taken to be the substance in that state of aggregation at a pressure of 1 bar. This same standard state is also used for liquid mixtures of those components that exist as a liquid at the conditions of the mixture. Such substances are sometimes referred to as "liquids that may act as a solvent." For substances that exist only as a solid or a gas in the pure component state at the temperature of the mixture, sometimes referred to as "substances that can act only as a solute," the situation is more complicated, and standard states based on Henry's law may be used. In this case the pressure is again fixed at 1 bar, and thermal properties such as the standard-state enthalpy and heat capacity are based on the properties of the substance in the solvent at infinite dilution, but the standard-state Gibbs energy and entropy are based on a hypothetical state of unit concentration (either unit molality or unit mole fraction, depending on the form of Henry's law used), with the standard-state fugacity at these conditions extrapolated from infinite-dilution behavior in the solvent, as shown in Fig. 9.7-3a and b. Therefore, just as for a gas, where the ideal gas state at 1 bar is a hypothetical state, the standard state of a substance that can only behave as a solute is a hypothetical state. However, one important characteristic of the solute standard state is that the properties depend strongly upon the solvent used. Therefore, the standard-state properties are a function of the temperature, the solute, and the solvent. This can lead to difficulties when a mixed solvent is used.

9.9 COMBINED EQUATION-OF-STATE AND EXCESS GIBBS ENERGY MODEL

As has already been mentioned, simple cubic equations of state with the van der Waals one-fluid mixing rules of Eqs. 9.4-8 and 9.4-9 are applicable at all densities and temperatures, but only to mixtures of hydrocarbons or hydrocarbons with inorganic gases. That is, this model is applicable to relatively simple mixtures. On the other hand, excess

Gibbs energy or activity coefficient models were developed to describe mixtures of any degree of solution nonideality, including mixtures involving polar organic chemicals, but only in the liquid state. Further, the parameters in activity coefficient models are very temperature dependent, these models are not applicable to expanded liquids (as occurs at high temperatures) or to the vapor phase, and there is also the problem of defining a hypothetical standard state and the standard-state properties for a component that exists as a gas in the pure component state at the temperature and pressure of the mixture, especially if the component is above its critical temperature (see Sec. 9.7). The latter problem results in difficulties in, for example, describing the solubility of hydrogen or nitrogen in liquids. The absence of an accurate gas-phase model for polar organic compounds has resulted in difficulties in describing the vapor-liquid equilibrium of polar mixtures at high temperatures and pressures, and for describing supercritical extraction processes.

Recently methods have been developed that combine an equation of state with an excess Gibbs energy (or, equivalently, activity coefficient) model and that allow simple equations of state to accurately describe all mixtures, including highly nonideal mixtures over large ranges of temperature and pressure without having to deal with hypothetical standard states and the other shortcomings of the direct use of activity coefficient models. The underlying idea of these models is to recognize that cubic equations of state, such as the van der Waals and Peng-Robinson equations, have two constants, a and b, and that this provides an opportunity to satisfy two boundary conditions.

One useful boundary condition is that at low density the composition dependence of the second virial coefficient obtained from an equation of state should agree with the theoretically correct result of Eq. 9.4-5,

$$B_{\mathrm{mix}}(T, \underline{x}) = \sum_i \sum_j x_i x_j B_{ij}(T) \tag{9.9-1}$$

derived from statistical mechanics. Since it has already been shown (Problem 6.14) that the second virial coefficient from a cubic equation of state is

$$B(T) = b - \frac{a(T)}{RT} \tag{9.9-2}$$

the first boundary condition is

$$b - \frac{a}{RT} = \sum_i \sum_j x_i x_j \left(b_{ij} - \frac{a_{ij}}{RT} \right) \tag{9.9-3}$$

Equation 9.9-3 does not give values for the mixture parameters a and b separately, but only for their sum. A second equation comes from requiring that the excess Gibbs energy predicted from an equation of state at liquidlike densities be equivalent to that from excess Gibbs energy or activity coefficient models discussed in Secs. 9.5 and 9.6. Since, from an equation of state, as $P \to \infty$, $\underline{V} \to b$ and $\underline{V}_{\mathrm{mix}} \to b_{\mathrm{mix}}$, so that liquid densities are obtained, the second equation that is used is

$$\underline{A}_{\mathrm{EOS}}^{\mathrm{ex}}(T, P \to \infty, \underline{x}) = \underline{A}_{\gamma}^{\mathrm{ex}}(T, P \to \infty, \underline{x}) \tag{9.9-4}$$

where the subscripts EOS and γ indicate $\underline{A}^{\mathrm{ex}}$ as computed from an equation of state and an activity coefficient model, respectively. The use of the excess Helmholtz energy rather than the excess Gibbs energy deserves some explanation. From the relationship between the Helmholtz and Gibbs energies, we have

$$\underline{G}^{\text{ex}} = \underline{A}^{\text{ex}} + P\underline{V}^{\text{ex}} \tag{9.9-5}$$

Empirically it is found that, at liquid densities, $\underline{A}^{\text{ex}}$ is rather insensitive to pressure, while $\underline{G}^{\text{ex}}$ diverges as $P \to \infty$ because of the $P\underline{V}^{\text{ex}}$ term. Therefore, in the $P \to \infty$ limit it is $\underline{A}^{\text{ex}}$ rather than $\underline{G}^{\text{ex}}$ that should be used.

Further, since $\underline{A}^{\text{ex}}$ is only a weak function of pressure at liquid densities, it is reasonable to assume that the excess Helmholtz energy at infinite pressure can be replaced as follows:

$$\underline{A}^{\text{ex}}(T, P \to \infty, \underline{x}) = \underline{A}^{\text{ex}}(T, P = 1 \text{ bar}, \underline{x}) \tag{9.9-6}$$

or if not 1 bar, at some other pressure at which experimental data are available (obtained as described in Chapter 10). Finally, since the $P\underline{V}^{\text{ex}}$ term makes only a negligible contribution at low pressures in Eq. 9.9-5, we can combine Eqs. 9.9-4, 9.9-5, and 9.9-6 to obtain

$$\underline{A}^{\text{ex}}_{\text{EOS}}(T, P \to \infty, \underline{x}) = \underline{A}^{\text{ex}}_{\gamma}(T, P \to \infty, \underline{x})$$
$$= \underline{A}^{\text{ex}}_{\gamma}(T, P = 1 \text{ bar}, \underline{x}) = \underline{G}^{\text{ex}}_{\gamma}(T, P = 1 \text{ bar}, \underline{x}) \tag{9.9-7}$$

It is the relationship between the first and last terms in this equality that we will use.

To proceed further, we note that as $P \to \infty$ (and $\underline{V}_i \to b_i$, $\underline{V}_{\text{mix}} \to b$) we obtain

$$\underline{A}^{\text{ex}}_{\text{EOS}} = C^*\left[\frac{a}{b} - \sum x_i \frac{a_i}{b_i}\right] \tag{9.9-8}$$

(Problem 9.31), where C^* is a constant whose value depends upon the equation of state used; its value is -1 for the van der Waals equation, and $[\ln(\sqrt{2}-1)]/\sqrt{2} = -0.623\,23$ for the Peng-Robinson equation. Combining all of the equations above, we get the following mixing rules:

$$\frac{a}{RT} = Q\frac{D}{1-D} \qquad b = \frac{Q}{1-D} \tag{9.9-9a}$$

$$Q = \sum_i \sum_j x_i x_j \left(b_{ij} - \frac{a_{ij}}{RT}\right) \tag{9.9-9b}$$

$$D = \sum_i x_i \frac{a_i}{b_i RT} + \frac{G^{\text{ex}}_{\gamma}(T, P, \underline{x})}{C^* RT} \tag{9.9-9c}$$

These equations are usually referred to as the Wong-Sandler[19] mixing rules (Problem 9.32).

The equations above still leave the cross term $(b_{ij} - a_{ij}/RT)$ unspecified. This quantity is usually derived from either of the following combining rules

$$b_{ij} - \frac{a_{ij}}{RT} = \sqrt{\left(b_{ii} - \frac{a_{ii}}{RT}\right)\left(b_{jj} - \frac{a_{jj}}{RT}\right)}(1 - k_{ij}) \tag{9.9-10a}$$

or

$$b_{ij} - \frac{a_{ij}}{RT} = \tfrac{1}{2}(b_{ii} + b_{jj}) - \frac{\sqrt{a_{ii}a_{jj}}}{RT}(1 - k_{ij}) \tag{9.9-10b}$$

---

[19]D. S. H. Wong and S. I. Sandler, *AIChE J.,* **38**, 671–680 (1992).

The second of these equations has the advantage of being more similar to the van der Waals one-fluid mixing rules of Eqs. 9.4-8 and 9.4-9 and, as with those equations, both of the combining rules above introduce a single adjustable parameter, k_{ij}, for each pair of components.

The mixing and combining rule combination introduced here is very useful for describing and even predicting the phase behavior of both moderately nonideal and highly nonideal mixtures over large ranges of temperature and pressure. This will be demonstrated in the next chapter.

The fugacity coefficient for this mixing rule is (Problem 9.33)

$$
\ln \bar{\phi}_i(T, P, \underline{x}) = \ln \frac{\bar{f}_i(T, P, \underline{x})}{x_i P}
$$

$$
= \frac{1}{b}\left(\frac{\partial Nb}{\partial N_i}\right)_{T, N_{j\neq i}} (Z - 1) - \ln\left(Z - \frac{bP}{RT}\right)
$$

$$
- \frac{a}{2\sqrt{2}bRT}\left[\frac{1}{aN}\left(\frac{\partial N^2 a}{\partial N_i}\right)_{T, N_{j\neq i}} - \frac{1}{b}\left(\frac{\partial Nb}{\partial N_i}\right)_{T, N_{j\neq i}}\right] \ln\left[\frac{Z + \left(1 + \sqrt{2}\right)\frac{bP}{RT}}{Z + \left(1 - \sqrt{2}\right)\frac{bP}{RT}}\right]
$$

$$(9.9\text{-}11)$$

where

$$
\left(\frac{\partial Nb}{\partial N_i}\right)_{T, N_{j\neq i}} = \frac{1}{1-D}\frac{1}{N}\left(\frac{\partial N^2 Q}{\partial N_i}\right)_{T, N_{j\neq i}} - \frac{Q}{(1-D)^2}\left[1 - \left(\frac{\partial ND}{\partial N_i}\right)_{T, N_{j\neq i}}\right]
$$

$$
\frac{1}{N}\left(\frac{\partial N^2 a}{\partial N_i}\right)_{T, N_{j\neq i}} = RTD\left(\frac{\partial Nb}{\partial N_i}\right)_{T, N_{j\neq i}} + RTb\left(\frac{\partial ND}{\partial N_i}\right)_{T, N_{j\neq i}}
$$

$$
\frac{1}{N}\left(\frac{\partial N^2 Q}{\partial N_i}\right)_{T, N_{j\neq i}} = 2\sum_j x_j\left(b - \frac{a}{RT}\right)_{ij} \qquad (9.9\text{-}12)
$$

$$
\left(\frac{\partial ND}{\partial N_i}\right)_{T, N_{j\neq i}} = \frac{a_i}{b_i RT} + \frac{1}{C^* RT}\left(\frac{\partial N\underline{G}^{ex}_{\gamma}(T, \underline{x})}{\partial N_i}\right)_{T, N_{j\neq i}} = \frac{a_i}{b_i RT} + \frac{\ln \gamma_i}{C^*}
$$

$$(9.9\text{-}13)$$

Although these equations look complicated, they are in fact easily programmed.

9.10 ELECTROLYTE SOLUTIONS

So far in this chapter we have considered mixtures of electrically neutral molecules. However, liquid solutions containing ionic species, especially aqueous solutions of acids, bases, and salts, occur frequently in chemical and biological processes. Charged particles interact with coulombic forces at small separations and, because of the formation of ion clouds around each ion, with damped coulombic forces at larger separation distances. These forces are stronger and much longer-range than those involved in the interactions of neutral molecules, so that solute ions in solution interact at very low concentrations. Consequently, electrolyte solutions are very nonideal in the sense that the electrolyte Henry's law activity coefficient $\gamma_i^{\square}$ of Eq. 9.7-17 is significantly different from unity at very low electrolyte concentrations; also, the greater the charge on the ions, the stronger their interaction and the more nonideal the solution. Since the solution models discussed in the preceding sections do not allow for the formation of ion clouds, they do not apply to electrolyte solutions.

In this section we discuss certain characteristics of electrolyte solutions and present equations for the prediction or correlation of electrolyte activity coefficients in solution. Since the derivations of these equations are complicated and beyond the scope of this book, they are not given.

An important characteristic of electrolyte solutions is that they are electrically conductive. A useful measure is the equivalent conductance, Ω, the conductance per mole of charge. A strong electrolyte is one that is completely dissociated into ions. In this case the equivalent conductance is high, and decreases only slowly with increasing concentration. A weak electrolyte is only partially dissociated into its constituent ions, and its equivalent conductance is less than that of a strong electrolyte at any concentration but increases rapidly as the concentration decreases. This is because there is more complete dissociation and therefore more ions per mole of electrolyte in solution as the concentration of a weak electrolyte decreases. Sodium chloride, which completely dissociates into sodium and chloride ions,

$$NaCl \rightarrow Na^+ + Cl^-$$

is an example of a strong electrolyte, while acetic acid, which is only partially dissociated into sodium and acetate ions,

$$CH_3COOH \rightleftharpoons H^+ + CH_3OO^-$$

is an example of a weak electrolyte. This partial dissociation is described by a chemical equilibrium constant, as discussed in Chapter 13.

Our interest is with an electrically neutral electrolyte, designated by $A_{\nu_+}B_{\nu_-}$, which, in solution, dissociates as follows:

$$A_{\nu_+}B_{\nu_-} = \nu_+ A^{z+} + \nu_- B^{z-} \qquad \textbf{(9.10-1)}$$

Here ν_+ and ν_- are the numbers of positive ions (cations) and negative ions (anions) obtained from the dissociation of one electrolyte molecule, and z_+ and z_- are the charges of the ions in units of charge of a proton (i.e., z_+ and z_- are the valences of the ions). For an electrically neutral salt, ν_+, ν_-, z_+, and z_- are related by the charge conservation (or electrical neutrality) condition that

$$\nu_+ z_+ + \nu_- z_- = 0 \qquad \textbf{(9.10-2)}$$

An important consideration in the study of electrolytes is that the concentration of any one ionic species is not independently variable because the electrical neutrality of the solution must be maintained. Thus, if N_A and N_B are the numbers of moles of the A^{z+} and B^{z-} ions, respectively, that result from the dissolution and dissociation of $A_{\nu_+}B_{\nu_-}$, N_A and N_B are related by

Electrical neutrality

$$\boxed{z_+ N_A + z_- N_B = 0} \qquad \textbf{(9.10-3)}$$

This restriction has an important implication with regard to the description of electrolyte solutions, as will be evident shortly.

A solution of a single electrolyte in a solvent contains four identifiable species: the solvent, undissociated electrolyte, anions, and cations. Therefore, it might seem appropriate, following Eqs. 8.1-12 and 8.1-13, to write the Gibbs energy of the solution as

$$G = N_S \bar{G}_S + N_{AB} \bar{G}_{AB} + N_A \bar{G}_A + N_B \bar{G}_B \qquad \textbf{(9.10-4)}$$

where N_S and N_{AB} are the mole numbers of solvent and undissociated electrolyte, and $\bar{G}_i$ is the partial molar Gibbs energy of species i, that is,

$$\bar{G}_i = \left(\frac{\partial G}{\partial N_i} \right)_{T,P,N_{j \neq i}} \qquad \textbf{(9.10-5)}$$

Since solutions with low electrolyte concentrations are of most interest, the solute activity coefficients in electrolyte solutions could, in principle, be defined, following Eq. 9.7-20, by

$$\bar{G}_i(T, P, M_i) = \bar{G}_i^\square + RT \ln(\gamma_i^\square M_i/(M_i = 1)) \qquad \textbf{(9.10-6)}$$

where M_i is the molality of species i, $\bar{G}_i^\square$ is its Gibbs energy in an ideal solution of unit molality, and $\gamma_i^\square$ is the activity coefficient defined such that $\gamma_i^\square$ approaches unity as M_i approaches zero. Thus, we have for the undissociated electrolyte, and, in principle, for each of the ions, that

$$\bar{G}_{AB}(T, P, M_{AB}) = \bar{G}_{AB}^\square + RT \ln(\gamma_{AB}^\square M_{AB}/(M_{AB} = 1))$$

$$\bar{G}_A(T, P, M_A) = \bar{G}_A^\square + RT \ln(\gamma_A^\square M_A/(M_A = 1)) \qquad \textbf{(9.10-7)}$$

$$\bar{G}_B(T, P, M_B) = \bar{G}_B^\square + RT \ln(\gamma_B^\square M_B/(M_B = 1))$$

The difficulty with this description is that $\bar{G}_A$ and $\bar{G}_B$ are not separately measurable, because, as a result of Eq. 9.10-3, it is not possible to vary the number of moles of cations holding the number of moles of anions fixed, or vice versa. (Even in mixed electrolyte solutions, that is, solutions of several electrolytes, the condition of overall electrical neutrality makes it impossible to vary the number of only one ionic species.) To maintain the present thermodynamic description of mixtures and, in particular, the concept of the partial molar Gibbs energy, we instead consider a single electrolyte solution to be a three-component system: solvent, undissociated electrolyte, and dissociated electrolyte. Letting $N_{AB,D}$ be the moles of dissociated electrolyte, we then have

$$G = N_S \bar{G}_S + N_{AB} \bar{G}_{AB} + N_{AB,D} \bar{G}_{AB,D} \qquad \textbf{(9.10-8)}$$

where $\bar{G}_{AB,D}$, the partial molar Gibbs energy of the dissociated electrolyte, $\bar{G}_S$, and $\bar{G}_{AB}$ are all measurable.

Comparing Eqs. 9.10-4 and 9.10-8 yields

$$N_{AB,D} \bar{G}_{AB,D} = N_A \bar{G}_A + N_B \bar{G}_B$$

or

$$\bar{G}_{AB,D} = \nu_+ \bar{G}_A + \nu_- \bar{G}_B \qquad \textbf{(9.10-9)}$$

so that

$$\bar{G}_{AB,D} = \nu_+[\bar{G}_A^{\square} + RT\ln(\gamma_A^{\square}M_A/M_A = 1)] + \nu_-[\bar{G}_B^{\square} + RT\ln(\gamma_B^{\square}M_B/M_B = 1)]$$

$$= [\nu_+\bar{G}_A^{\square} + \nu_-\bar{G}_B^{\square}] + RT\ln\left\{\frac{(\gamma_A^{\square}M_A)^{\nu_+}(\gamma_B^{\square}M_B)^{\nu_-}}{(M_A = 1)^{\nu_+}(M_B = 1)^{\nu_-}}\right\}$$

$$(9.10\text{-}10)$$

Finally, we define a **mean ionic activity coefficient**, $\gamma_\pm$, by

Mean ionic activity coefficient

$$\boxed{\gamma_\pm^{\nu} = (\gamma_A^{\square})^{\nu_+}(\gamma_B^{\square})^{\nu_-}} \qquad (9.10\text{-}11)$$

a **mean ionic molality**, $M_\pm$, by

Mean ionic molality

$$\boxed{M_\pm^{\nu} = M_A^{\nu_+} M_B^{\nu_-}} \qquad (9.10\text{-}12)$$

and

$$\bar{G}_{AB,D}^{\square} = \nu_+\bar{G}_A^{\square} + \nu_-\bar{G}_B^{\square} \qquad (9.10\text{-}13)$$

to obtain

$$\bar{G}_{AB,D} = \bar{G}_{AB,D}^{\square} + RT\ln[(M_\pm\gamma_\pm)^{\nu}/(M = 1)^{\nu}] = \bar{G}_{AB,D}^{\square} + \nu RT\ln[(M_\pm\gamma_\pm)/(M = 1)]$$
$$(9.10\text{-}14)$$

where $\nu = \nu_+ + \nu_-$.

As it is the mean activity coefficient, and not the activity coefficients of the individual ions, that is measurable, in the remainder of this section our interest is in formulas for $\gamma_\pm$. Also, since we will be concerned mostly with low electrolyte concentrations in aqueous solutions, in the application of these formulas the distinction between molality (moles per kilogram of solvent) and concentration in molarity (moles per liter of solution) will sometimes be ignored.

P. Debye and E. Hückel,[20] using a statistical mechanical model to obtain the average ion distribution around ions in solution, derived the following expression for the dependence of $\gamma_\pm$ on electrolyte concentration

Debye-Hückel limiting law

$$\boxed{\ln\gamma_\pm = -\alpha|z_+z_-|\sqrt{I}} \qquad (9.10\text{-}15)$$

The bracketed term in this equation is the absolute value of the product of ion valences, α is a parameter that depends on the solvent and the temperature (see Table 9.10-1 for the values of water), and I is the **ionic strength**, defined as

Ionic strength

$$\boxed{I = \frac{1}{2}\sum_{i=\text{ions}} z_i^2 M_i} \qquad (9.10\text{-}16)$$

where the summation is over all ions in solution.

Equation 9.10-15 is exact at very low ionic strengths and is usually referred to as the **Debye-Hückel limiting law.** Unfortunately, significant deviations from this limiting

---

[20]P. Debye and E. Hückel, *Phys. Z.,* **24**, 185 (1923).

Table 9.10-1 Values of the Parameters in the Equations for $\gamma_\pm$ for Aqueous Solutions [21]

T (°C)	α (mol/kg)$^{-1/2}$	β [(mol/kg)$^{1/2}$ Å]$^{-1}$
0	1.129	0.3245
5	1.137	0.3253
10	1.146	0.3261
15	1.155	0.3269
20	1.164	0.3276
25	1.175	0.3284
30	1.184	0.3292
40	1.206	0.3309
50	1.230	0.3326
60	1.255	0.3343
70	1.283	0.3361
80	1.313	0.3380
90	1.345	0.3400
100	1.379	0.3420

law expression are observed at ionic strengths as low as 0.01 molal (see Fig. 9.10-1). For higher electrolyte concentrations, the following empirical and semitheoretical modifications of Eq. 9.10-15 have been proposed:

Correction to the Debye-Hückel limiting law

$$\ln \gamma_\pm = -\frac{\alpha |z_+ z_-| \sqrt{I}}{1 + \beta a \sqrt{I}} \tag{9.10-17}$$

$$\ln \gamma_\pm = -\frac{\alpha |z_+ z_-| \sqrt{I}}{1 + \beta a \sqrt{I}} + \delta I \tag{9.10-18}$$

In these equations β is the parameter given in Table 9.10-1 and a is a constant related to the average hydrated radius of ions, usually about 4 Å. In practice, however, the product βa is sometimes set equal to unity or treated as an adjustable parameter. Similarly, δ is sometimes set to $0.1 |z_+ z_-|$ and sometimes taken to be an adjustable parameter.

ILLUSTRATION 9.10-1

Use of Electrolyte Solution Models

The data below are for the activity coefficients of HCl in aqueous hydrochloric acid solutions as a function of HCl molality at 25°C. Compare the predictions of the Debye-Hückel model (Eqs. 9.10-15 and 9.10-16), and the extended Debye-Hückel models (Eqs. 9.10-17 and 9.10-18) with these data.

M_{HCl}	$\gamma_\pm$	M_{HCl}	$\gamma_\pm$	M_{HCl}	$\gamma_\pm$
0.0005	0.975	0.1	0.796	8	5.90
0.001	0.965	0.5	0.757	10	10.44
0.005	0.928	1.0	0.809	12	17.25
0.01	0.904	3	1.316	14	27.3
0.05	0.830	5	2.38	16	42.4

[21] 1 A° = 1 Angstrom unit = 10^{-10} m. L = liter = $10^{-3} m^3$

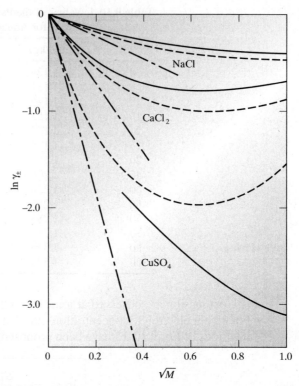

Figure 9.10-1 The activity coefficients of various salts in aqueous solution at 25°C as a function of the salt molarity M. The solid line is the experimental data (from R. A. Robinson and R. H. Stokes, *Electrolyte Solutions,* 2nd ed., Butterworth, London, 1959). The line — · — is the result of the Debye-Hückel limiting law, Eq. 9.10-15, and the dashed line is the prediction of Eq. 9.10-18, with $\beta\alpha = 1$ and $\delta = 0.1|z_+ z_-|$. Note that for NaCl $I = M$; for CaCl$_2$ $I = 3M$; and for CuSO$_4$ $I = 4M$.

SOLUTION

Hydrogen chloride is a strong electrolyte, and is fully ionized. Therefore,

$$I = \tfrac{1}{2}[(+1)^2 M_{H+} + (-1)^2 M_{Cl-}] = \tfrac{1}{2}[M_{H+} + M_{Cl-}] = M_{HCl}$$

In Fig. 9.10-2 the experimental data are plotted as points together with the curves that result from the use of the Debye-Hückel limiting law (Eq. 9.10-15), of the extended version of Eq. 9.10-17 with $\beta a = 1$, of Eq. 9.10-18 with $\beta a = 1$ and $\delta = 0.1$, and finally, of Eq. 9.10-18 with $\beta a = 1$ and $\delta = 0.3$. As can be seen, the last equation provides a good correlation with the experimental mean molal activity coefficient data. ■

ILLUSTRATION 9.10-2
Another Electrolyte Solution Example

The following data are available for the mean ionic activity coefficient of sodium chloride in water at 25°C.

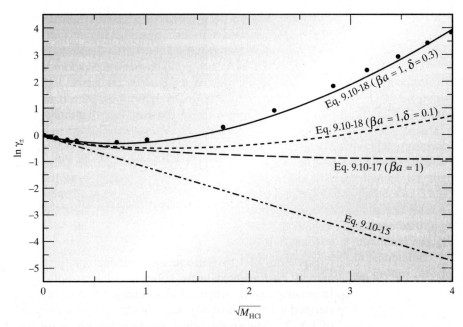

Figure 9.10-2 Mean molar activity coefficient for hydrogen chloride in water at 25°C as a function of the square root of the ionic strength $I = M_{HCl}$. Points are the experimental data, and the lines are the results of various models.

Molality	$\gamma_\pm$	Molality	$\gamma_\pm$
0.0	1.00	2.0	0.669
0.1	0.778	3.0	0.714
0.25	0.720	4.0	0.782
0.5	0.681	5.0	0.873
0.75	0.665	6.0	0.987
1.0	0.657		

a. Compare the predictions of the Debye-Hückel limiting law with these data:

$$\ln \gamma_\pm = -1.178\sqrt{I} \qquad (1)$$

where I is the ionic strength.

b. Compare the predictions of the following version of the extended Debye-Hückel limiting law with these data.

$$\ln \gamma_\pm = -\frac{1.178\sqrt{I}}{1 + \sqrt{I}} \qquad (2)$$

c. Correlate the data with the following expression, in which δ is an adjustable parameter.

$$\ln \gamma_\pm = -\frac{1.178\sqrt{I}}{1 + \sqrt{I}} + \delta I \qquad (3)$$

SOLUTION

The results are given in the figure below. Clearly, only Eq. 3, with a fitted value of $\delta = 0.137$, gives an acceptable description of the experimental data over the whole concentration range.

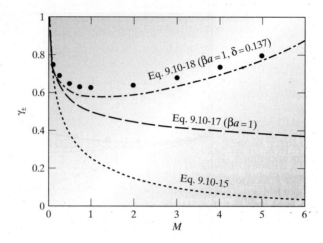

Equation 9.10-18 is frequently satisfactory for the correlation of the mean molal activity coefficient data for solutions of low molality (see Figs. 9.10-1 and 9.10-2). In general, the accuracy of this equation is best for a $z_+ = 1$, $z_- = -1$ electrolyte (termed a 1:1 electrolyte) and becomes progressively less satisfactory for electrolytes of 1:2, 2:2, and so on, which are increasingly more nonideal. This should be kept in mind in using the equations here.

The equations in this section are valid for solutions of several electrolytes, or mixed electrolyte solutions, as well as single electrolytes. In the former case there are several mean activity coefficients, one for each electrolyte, and the equations here are used to compute the value of each. In this calculation the ionic strength I is that computed by summing over all ions in solution and, consequently, is the same for each electrolyte in the solution.

There are few solution models valid for moderate and high concentrations of electrolytes. Perhaps the most successful is the model of Pitzer:[22]

$$\frac{G^{\mathrm{ex}}}{n_w RT} = f(I) + \sum_i \sum_j \lambda_{ij}(I) M_i M_j + \sum_i \sum_j \sum_k \delta_{ijk} M_i M_j M_k \qquad \textbf{(9.10-19)}$$

where G^{ex}/n_w is the excess Gibbs energy per kilogram of solvent, M_i is the molality of each ion or neutral solute present, and $f(I)$ is the Debye-Hückel term. Finally, λ_{ij} and δ_{ijk} are the second and third virial coefficients among the species present, with λ_{ij} also being a function of ionic strength I.

Although Eq. 9.10-19 has a large number of parameters, especially for mixed electrolyte solutions, it has been useful in representing the thermodynamic behavior of electrolyte solutions all the way from dilute solutions to molten salts.

Molality and molarity

Before leaving this section, one comment about the concentration units used should be made. Molality, which we have used here and indicated by the symbol M, is concentration expressed as moles of solute per kilogram of solvent. Molarity, defined as the number of moles of solute per liter of solution, is also a commonly used concentration unit. However, because the volume of a solution varies with composition and with temperature, molarity can be more difficult to deal with than molality, which is

---

[22]See, for example, K. S. Pitzer, "Thermodynamics of Aqueous Systems with Industrial Applications," *ACS Symp. Ser.,* **133**, 451 (1980).

based on a fixed amount of one kilogram of solvent. It is for this reason that molality has been used here, and again in Chapter 15. Note that if the solvent is water, and the solution is dilute in solute (so that one liter of solution contains one kilogram of water), molality and molarity are equivalent.

9.11 CHOOSING THE APPROPRIATE THERMODYNAMIC MODEL

The objective of this chapter has been to develop methods of estimating species fugacities in mixtures. These methods are very important in phase equilibrium calculations, as will be seen in the following chapters. Because of the variety of methods discussed, there may be some confusion as to which fugacity estimation technique applies in a given situation. The comments that follow may be helpful in choosing among the three main methods discussed in this chapter:

1. Equations of state
2. Activity coefficient (or excess Gibbs energy) models
3. Activity coefficient (or excess Gibbs energy) models based on the Henry's law standard state

Electrolyte solutions are special and can only be treated by the methods considered in Sec. 9.10. Therefore, electrolyte solutions are not be considered in this discussion. Also, the Henry's law standard state is used only for a component that does not exist as a pure component at the temperature of interest—for example, a dissolved solid below its melting point, or a dissolved gas much above its critical point. (However, if the liquid mixture can be described by an equation of state—for example, mixtures of hydrocarbons and nitrogen or carbon dioxide—there is no need to use the Henry's law standard state.)

In low- to moderate-density vapors, mixture nonidealities are not very large, and therefore equations of state of the type discussed in this text can generally be used for the prediction of vapor-phase fugacities of all species. [However, mixtures containing species that associate (i.e., form dimers, trimers, etc.) in the vapor phase, such as acetic acid, are generally described using the virial equation of state with experimentally determined virial coefficients.] The Lewis-Randall rule should be used only for approximate calculations; it is best to use an equation of state to calculate the vapor-phase fugacity of vapor mixtures.

At liquid densities, solution nonidealities can be large. In this case, equation-of-state predictions with the van der Waals one-fluid mixing rules will be reasonably accurate only for mixtures composed of relatively simple molecules that are similar. Thus, equations of state with the van der Waals one-fluid mixing rules of Eq. 9.4-8 will be satisfactory only for mixtures involving light hydrocarbons, including hydrocarbons with some dissolved inorganic gases (CO, CO_2, H_2S, N_2, etc.). This equation-of-state method will not be accurate if complicated molecular phenomena are involved—for example, hydrogen bonding, as occurs in mixtures containing water, alcohols, and organic acids—or if chemically dissimilar molecular species are involved. In these cases, activity coefficient models (also referred to as Gibbs energy models) must be used for estimating the liquid-phase species fugacities, even though an equation-of-state model may be used for the vapor phase. Alternatively, the same equation of state can be used for both the liquid and the vapor if the mixing rules of Sec. 9.9 that combine activity coefficient and equation-of-state models are used.

Traditionally, in the petroleum and natural gas industries, where mostly low- and moderate-molecular-weight hydrocarbons (with only small concentrations of aromatic and inorganic compounds) are involved, equation-of-state methods with the van der Waals one-fluid mixing rules are generally used for the prediction of both liquid- and vapor-phase properties. In the chemicals industry, on the other hand, where oxygen, nitrogen, sulfur, or halogen-containing organic compounds and inorganic (electrolyte) compounds are involved, activity coefficient or Gibbs energy models are used for the prediction of liquid-phase properties, with equation-of-state models used for the vapor phase if the pressure is above ambient. However, in recent years the equation-of-state methods of Sec. 9.9 have also been used to model both the vapor and liquid phases of such mixtures.

The critical point of a mixture, to be discussed in the next chapter, is similar to the critical point of a pure component in that it is the temperature and pressure at which the vapor and liquid phases of a mixture of given composition become identical. Although we have not yet discussed phase equilibrium in multicomponent mixtures (see Chapters 10 to 12), we can guess that if we are to predict the point at which two phases become identical (the critical point) with some accuracy, the models we use for the vapor and liquid phases must give identical results for all properties at this point. This will occur only if we use the same equation-of-state model to describe both the vapor and liquid phases, but clearly cannot, in general, be expected to occur if we use different models, such as an activity coefficient model for the liquid phase and an equation of state for the vapor phase. This ability to predict critical phenomena is an important advantage of using the same equation-of-state model for both phases. Another advantage of the equation-of-state description is that the concept of standard states, and especially hypothetical or extrapolated standard states, never arises.

Finally, since a number of different activity coefficient (or excess Gibbs energy) models have been discussed, it is useful to consider their range of application. The most important observation is that none of the completely predictive methods, such as regular solution theory, UNIFAC, or ASOG, can be regarded as highly accurate. Therefore, these methods should be used only when no experimental data are available for the system of interest. Of the predictive methods, UNIFAC is the best developed and regular solution theory is the least accurate.

As we have considered a number of activity coefficient models in this section, it is useful to discuss which might be best for different types of mixtures. To start this discussion, it is useful to classify seven different types of chemicals that one might encounter.

Nonpolar compounds: Chemical species without significant dipole moments, such as most hydrocarbons, and some very symmetric compounds such as tetrachloromethane and hexafluorobenzene.
Weakly polar compounds: Aldehydes, ethers, and ketones
Very polar compounds: Alcohols, amines
Water
Carboxylic acids
Polymers
Electrolyte systems

As a general rule, a mixture containing components from more than one of the classes above exhibits greater nonideality than a mixture containing only components from the same class. For example, a mixture containing an aldehyde and a ketone probably has a lower excess Gibbs energy of mixing than an aldehyde-hydrocarbon mixture. However, it is also true that even for a mixture containing compounds within one class, the mixture may be slightly nonideal if the components are similar in size and chemical functionality, or more nonideal (larger Gibbs energy of mixing) if there is greater dissimilarity in the components. Table 9.11-1 suggests the most appropriate activity coefficient (excess Gibbs energy) model or models to use for different classes of mixtures, and Table 9.11-2 lists the most appropriate equation-of-state model. These tables should be used as references in choosing thermodynamic models—for example, when using a process simulator.

Table 9.11-1 Recommended Activity Coefficient Models

Nonpolar + nonpolar compounds: All of the models (Margules, van Laar, Wilson, UNIQUAC, and NRTL) will give good correlations of data for these mixtures.

Nonpolar + weakly polar compounds: All of the models (Margules, van Laar, Wilson, UNIQUAC, and NRTL) can be used, although the UNIQUAC model is better for mixtures that are more nonideal.

Nonpolar + strongly polar compounds: Although all of the models (Margules, van Laar, Wilson, UNIQUAC, and NRTL) can be used for mixtures that are not too nonideal, the UNIQUAC model appears to give the best correlation for somewhat nonideal systems, wheras the Wilson model may be better for mixtures that are more nonideal (but not so nonideal as to result in liquid-liquid immiscibility; see Sec. 11.2).

Weakly polar + weakly polar compounds: All of the models (Margules, van Laar, Wilson, UNIQUAC, and NRTL) can be used, although it appears that the UNIQUAC model is better for mixtures that are more nonideal.

Weakly polar + strongly polar compounds: All of the models (Margules, van Laar, Wilson, UNIQUAC, and NRTL) can be used, although it appears that the UNIQUAC model is better for mixtures that are more nonideal.

Strongly polar + strongly polar compounds: The UNIQUAC model appears to best correlate data, though all models give reasonable results.

Water + nonpolar compounds: These mixtures generally have limited mutual solubility and are discussed in Sec. 11.2.

Water + weakly polar compounds: These mixtures generally have limited mutual solubility and are discussed in Sec. 11.2.

Water + strongly polar compounds: The UNIQUAC model appears to best correlate data for aqueous mixtures.

Solutions containing carboxylic acids: The Wilson model appears to best correlate data for mixtures containing carboxylic acids if the components are mutually soluble. (The Wilson model does not predict liquid-liquid phase splitting, as discussed in Sec. 11.2). Otherwise, the UNIQUAC, van Laar, or NRTL model should be used.

Solutions containing polymers: Use the Flory-Huggins model.

Solutions containing ionizable salts, strong acids, or bases: Use the extended Debye-Hückel model.

Multicomponent mixtures: Choose the model that, based on the suggestions above, best describes the dominant components in the mixture.

Table 9.11-2 Recommended Equation-of-State Models

Vapor Mixtures

1. **Low pressures, no components that associate (such as carboxylic acids or HF):** Assume the mixture is an ideal gas mixture. $\bar{f}_i^V\left(T, P, \underline{y}\right) = y_i P$

2. **Low pressure for a mixture that contains an associating component:** Use the virial equation of state, retaining only the second virial coefficient, and search for experimental pure component and cross virial coefficient data for all components in the mixture.

3. **Slightly elevated pressures:** Use choice 1 or 2 above. Alternatively, use 4 below if there are no associating components, or 5 if associating components are present.

4. **Elevated pressures for a vapor mixture that contains hydrocarbons, nitrogen, oxygen, carbon dioxide, and/or other inorganic gases (but not HF):** Use an equation of state, such as the Peng-Robinson or Soave–Redlich-Kwong equation with van der Waals one-fluid mixing rules (see Sec. 9.4).

5. **Elevated pressures for a vapor mixture that contains one or more polar and/or associating compounds:** Use an equation of state, such as the Peng-Robinson or Soave–Redlich-Kwong equation with the excess Gibbs energy–based mixing rules (see Sec. 9.9) and the appropriate activity coefficient model (see Table 9.11-1).

Liquid Mixtures

Liquid mixtures at low pressure are generally described using the activity coefficient models as described in Table 9.11-1, and the behavior of a liquid mixture is generally not much affected by pressure, unless the pressure is very high. However, as we will see in Sec. 10.3, for phase equilibrium calculations at high pressures, especially as the critical point of a mixture is approached, there are important advantages to using the same thermodynamic model for both phases. In such cases the same equation-of-state model should be used for the vapor and liquid phases.

1. **Liquid mixture at elevated pressure that contains only hydrocarbons, nitrogen, oxygen, carbon dioxide, and/or other inorganic gases (but not HF):** Use equation of state, such as Peng-Robinson or Soave–Redlich-Kwong with van der Waals one-fluid mixing rules (see Sec. 9.4).

2. **Liquid mixture at elevated pressure that contains that contains one or more polar and/or associating compounds.** Use equation of state, such as Peng-Robinson or Soave–Redlich-Kwong with the excess Gibbs energy–based mixing rules (see Sec. 9.9) and the appropriate activity coefficient model (see Table 9.11-1).

Appendix A9.1   A Statistical Mechanical Interpretation of the Entropy of Mixing in an Ideal Mixture

This section appears on the CD that accompanies this text.

Appendix A9.2   Multicomponent Excess Gibbs Energy (Activity Coefficient) Models

The excess Gibbs energy models for binary mixtures discussed in Sec. 9.5 can be extended to multicomponent mixtures. For example, the Wohl expansion of Eq. 9.5-8 can be extended to ternary mixtures:

$$\frac{G^{\text{ex}}}{RT(x_1q_1 + x_2q_2 + x_3q_3)} = 2a_{12}z_1z_2 + 2a_{13}z_1z_3 + 2a_{23}z_2z_3$$
$$+ 3a_{112}z_1^2z_2 + 3a_{122}z_1z_2^2 + 3a_{113}z_1^2z_3$$
$$+ 3a_{133}z_1z_3^2 + 3a_{223}z_2^2z_3 + 3a_{233}z_2z_3^2$$
$$+ 6a_{123}z_1z_2z_3 + \cdots \tag{A9.2-1}$$

which, neglecting all terms of third and higher order in the volume fractions, yields for species 1

$$\ln \gamma_1 = \frac{\left\{ x_2^2\alpha_{12}\left(\frac{\beta_{12}}{\alpha_{12}}\right)^2 + x_3^2\alpha_{13}\left(\frac{\beta_{13}}{\alpha_{13}}\right)^2 + x_2x_3\frac{\beta_{12}}{\alpha_{12}}\frac{\beta_{13}}{\alpha_{13}}\left(\alpha_{12} + \alpha_{13} - \alpha_{23}\frac{\alpha_{12}}{\beta_{12}}\right) \right\}}{\left[x_1 + x_2\left(\frac{\beta_{12}}{\alpha_{12}}\right) + x_3\left(\frac{\beta_{13}}{\alpha_{13}}\right) \right]^2}$$

$$\tag{A9.2-2}$$

where $\alpha_{ij} = 2q_ia_{ij}$, $\beta_{ij} = 2q_ja_{ij}$, and $a_{ij} = a_{ji}$. (Thus, $\alpha_{ji} = \beta_{ij}$ and $\beta_{ji} = \alpha_{ij}$.) The expression for $\ln \gamma_2$ is obtained by interchanging the subscripts 1 and 2 in Eq. A9.2-2 and that for $\ln \gamma_3$ by interchanging the subscripts 1 and 3. The collection of equations obtained in this way are the van Laar equations for a ternary mixture. Note that there are two parameters, α_{ij} and β_{ij}, for each pair of components in the mixture. This model can be extended to multicomponent mixtures in a similar manner.

The multicomponent form of the van Laar model is obtained by starting from

$$\frac{G^{\text{ex}}}{RT\sum\limits_{i=1}^{c} x_iq_i} = \sum_{i=1}^{c}\sum_{j=1}^{c}\alpha_{ij}z_iz_j + \sum_{i=1}^{c}\sum_{j=1}^{c}\sum_{k=1}^{c}\alpha_{ijk}z_iz_jz_k + \cdots \tag{A9.2-3}$$

and neglecting third- and higher-order terms in the volume fraction. In this case, $\alpha_{ij} = \alpha_{ji} \neq 0$, $\alpha_{ii} = 0$, and $\alpha_{ijk} = \alpha_{ikj} = \alpha_{kij} = \alpha_{iii} = 0$.

The multicomponent form of the Wilson equation is

$$\frac{G^{\text{ex}}}{RT} = -\sum_{i=1}^{c} x_i \ln\left(\sum_{j=1}^{c} x_j\Lambda_{ij}\right) \tag{A9.2-4}$$

for which

$$\ln \gamma_i = 1 - \ln\left(\sum_{j=1}^{c} x_j\Lambda_{ij}\right) - \sum_{j=1}^{c}\frac{x_j\Lambda_{ji}}{\sum\limits_{k=1}^{c} x_k\Lambda_{jk}} \tag{A9.2-5}$$

Since $\Lambda_{ii} = 1$, there are also two parameters, Λ_{ij} and Λ_{ji}, for each binary pair of components in this multicomponent mixture model.

The multicomponent NRTL equation is

$$\frac{G^{\text{ex}}}{RT} = \sum_{i=1}^{c} x_i \frac{\sum_{j=1}^{c} \tau_{ji} G_{ji} x_j}{\sum_{j=1}^{c} G_{ji} x_j} \tag{A9.2-6}$$

with $\ln G_{ij} = -\alpha_{ij}\tau_{ij}$, $\alpha_{ij} = \alpha_{ji}$, and $\tau_{ii} = 0$, for which

$$\ln \gamma_i = \frac{\sum_{j=1}^{c} \tau_{ji} G_{ji} x_j}{\sum_{j=1}^{c} G_{ji} x_j} + \sum_{j=1}^{c} \frac{x_j G_{ij}}{\sum_{k=1}^{c} x_k G_{kj}} \left(\tau_{ij} - \frac{\sum_{k=1}^{c} x_k \tau_{kj} G_{kj}}{\sum_{k=1}^{c} x_k G_{kj}} \right) \tag{A9.2-7}$$

This equation has three parameters, τ_{ij}, τ_{ji}, and α_{ij}, for each pair of components in the multicomponent mixture.

The important feature of each of the equations discussed here is that all the parameters that appear can be determined from activity coefficient data for binary mixtures. That is, by correlating activity coefficient data for the species 1–species 2 mixture using any of the models, the 1–2 parameters can be determined. Similarly, from data for species 2–species 3 and species 1–species 3 binary mixtures, the 2–3 and 1–3 parameters can be found. These coefficients can then be used to estimate the activity coefficients for the ternary 1–2–3 mixture without any experimental data for the three-component system.

One should keep in mind that this ability to predict multicomponent behavior from data on binary mixtures is not an exact result, but rather arises from the assumptions made or the models used. This is most clearly seen in going from Eq. A9.2-1 to Eq. A9.2-2. Had the term $\alpha_{123}z_1z_2z_3$ been retained in the Wohl expansion, Eq. A9.2-2 would contain this α_{123} term, which could be obtained only from experimental data for the ternary mixture. Thus, if this more complete expansion were used, binary data *and* some ternary data would be needed to determine the activity coefficient model parameters for the ternary mixture.

Appendix A9.3   The Activity Coefficient of a Solvent in an Electrolyte Solution

The starting point for the derviation of the activity coefficient of solvent S in a solution with electrolyte AB is the Gibbs-Duhem equation,

$$N_S \, d\bar{G}_S + N_{AB} \, d\bar{G}_{AB} = 0$$

where N_S is the moles of solvent,

$$\bar{G}_S = \underline{G}_S + RT \ln (x_S \gamma_S)$$

and

$$\bar{G}_{AB} = \bar{G}_{AB,D} + RT \ln \left[\frac{M_{\pm}\gamma_{\pm}}{M = 1} \right]^{\nu}$$

with

$$M_{\pm} = M_A^{\nu_+} \cdot M_B^{\nu_-} = \left[\nu_+^{\nu_+} \cdot \nu_-^{\nu_-} \right] M_{AB}$$

We are interested in the case where the number of moles of electrolyte AB is infinitesimally increased at fixed number of moles of solvent N_S, which leads to

$$d\bar{G}_{AB} = \nu RT \cdot d\ln\left[M_\pm\right] + \nu RT \cdot d\ln\left[\gamma_\pm\right]$$

Here we will consider the simplest case, the Debye-Hückel limiting law, in detail and then give expressions for the other cases considered in this book.

$$\ln\gamma_\pm = -\alpha\left|z_+z_-\right|\sqrt{I}$$

where

$$I = \frac{1}{2}\sum_{\text{ions i}} z_i^2 M_i = \frac{1}{2}M_{AB}\sum_{\text{ions i}}\nu_i z_i^2$$

so that

$$d\bar{G}_{AB} = \nu RT \cdot \frac{dM_{AB}}{M_{AB}} + \nu RT\left\{-\alpha\left|z_+z_-\right|\right\}\frac{1}{2}I^{-1/2}\,dI$$

$$= \nu RT \cdot \frac{dM_{AB}}{M_{AB}} + \frac{\nu RT\left\{-\alpha\left|z_+z_-\right|\right\}^{1/2}}{\left\{\dfrac{1}{2}\sum\limits_{\text{ions i}}\nu_i z_i^2 M_{AB}\right\}^{1/2}}\left\{\frac{1}{2}\sum_{\text{ions i}}\nu_i z_i^2\,dM_{AB}\right\}$$

$$= \nu RT \cdot \frac{dM_{AB}}{M_{AB}} - \frac{\nu RT}{2}\alpha\left|z_+z_-\right|\left\{\frac{1}{2}\sum_{\text{ions i}}\nu_i z_i^2\right\}^{1/2}\frac{dM_{AB}}{M_{AB}^{1/2}}$$

and

$$d\bar{G}_{AB} = \nu RT \cdot \frac{dM_{AB}}{M_{AB}} - \nu RT\alpha\left|z_+z_-\right|\left\{\frac{1}{2}\sum_{\text{ions i}}\nu_i z_i^2\right\}^{1/2}dM_{AB}^{1/2}$$

Now using N_{AB} to represent the number of moles of electrolyte, and

$$M_{AB} = \frac{\text{Number of moles of electrolyte}}{\text{kg of solvent}}$$

so that $N_{AB} = M_{AB} \times m_S$, where m_S is the number of kilograms of solvent per mole of solvent ($= 0.018$ for water), then

$$N_{AB}\,d\bar{G}_{AB} = M_{AB} \cdot m_S\,d\bar{G}_{AB}$$

$$= M_{AB} \cdot m_S \cdot \nu RT \cdot \left[\frac{dM_{AB}}{M_{AB}} - \alpha\left|z_+z_-\right|\left\{\frac{1}{2}\sum_{\text{ions i}}\nu_i z_i^2\right\}^{1/2}dM_{AB}^{1/2}\right]$$

$$= m_S \cdot \nu RT \cdot dM_{AB} - m_S \cdot \nu RT \cdot \alpha\left|z_+z_-\right|\left\{\frac{1}{2}\sum_{\text{ions i}}\nu_i z_i^2\right\}^{1/2}M_{AB}dM_{AB}^{1/2}$$

$$= -N_S\,d\bar{G}_S \quad \text{(by the Gibbs-Duhem equation)}.$$

and

$$N_S \int_{M_{AB}=0}^{M_{AB}} d\bar{G}_S = N_S \left(\bar{G}_S - \underline{G}_S\right) = RT \ln \left(x_S \gamma_S\right)$$

$$= -m_S \cdot \nu RT \int_{M_{AB}=0}^{M_{AB}} dM_{AB}$$

$$+ m_S \cdot \nu RT \cdot \alpha \left| z_+ z_- \right| \left\{ \frac{1}{2} \sum_{\text{ions } i} \nu_i z_i^2 \right\}^{1/2} \int_{M_{AB}=0}^{M_{AB}} M_{AB} dM_{AB}^{1/2}$$

Since

$$\int_{M_{AB}=0}^{M_{AB}} dM_{AB} = M_{AB} \quad \text{and} \quad \int_{M_{AB}=0}^{M_{AB}} M_{AB} \, dM_{AB}^{1/2} = \int_{M_{AB}=0}^{\sqrt{M_{AB}}} x^2 \, dx$$

$$= \frac{\left(\sqrt{M_{AB}}\right)}{3} = \frac{M_{AB}^{2/3}}{3}$$

$$RT \ln \left(x_S \gamma_S\right) = -m_S \cdot \nu RT \cdot M_{AB} + m_S \cdot \nu RT \cdot \alpha \left| z_+ z_- \right| \left\{ \frac{1}{2} \sum_{\text{ions } i} \nu_i z_i^2 \right\}^{1/2} \frac{M_{AB}^{3/2}}{3}$$

$$= -m_S \cdot \nu RT \cdot M_{AB} \left[1 - \frac{\alpha}{3} \left| z_+ z_- \right| \sqrt{I} \right]$$

and

$$\ln x_S \gamma_S = -m_S \cdot \nu \cdot M_{AB} \left[1 - \frac{\alpha}{3} \left| z_+ z_- \right| \sqrt{I} \right]$$

where $\nu = \nu_+ + \nu_-$, M_{AB} is the molality of the electrolyte, and m_S is the molecular weight of the solvent in kg/mol.

When using the equation derived here or the ones that follow, it is important to correctly compute the mole fraction of the solvent. To do this one must keep in mind that in an electrolyte solution there are solvent molecules, anions, cations, and also the possibility of undissociated electrolyte. Thus, for example, in an aqueous sodium chloride solution, since NaCl is a strong electrolyte that is fully ionized,

$$x_S = \frac{N_S}{N_S + N_{Na^+} + N_{Cl^-}} = \frac{N_S}{N_S + 2 \cdot N_{NaCl}}$$

and since M = moles of solute per kg of water = moles of solute per 55.51 moles water,

$$x_S = \frac{N_S}{N_S + 2 \cdot N_{NaCl}} \quad \text{which for water is} \quad x_W = \frac{55.51}{55.51 + 2 M_{NaCl}}$$

If instead of the Debye-Hückel limiting law we use the simple extension

$$\ln \gamma_\pm = -\frac{\alpha \left| z_+ z_- \right| \sqrt{I}}{1 + \beta a \sqrt{I}}$$

the result is

$$\ln x_S \gamma_S = -m_S \cdot \nu \cdot M_{AB} \left[1 - \frac{\alpha}{3} \left| z_+ z_- \right| \sqrt{I} \cdot \sigma \left(\beta a \sqrt{I} \right) \right]$$

where

$$\sigma(y) = \frac{1}{y^3} \left[1 + y - 2 \cdot \ln(1 + y) - \frac{1}{1 + y} \right]$$

[*Note:* By repeated use of L'Hopital's rule, one can show that as $\beta a \sqrt{I}$ (or y) $\rightarrow$ 0, $\sigma(y) \rightarrow \frac{1}{3}$, so that in this limit the result agrees with that obtained from the Debye-Hückel limiting law.]

Finally, using

$$\ln \gamma_\pm = -\frac{\alpha \left| z_+ z_- \right| \sqrt{I}}{1 + \beta a \sqrt{I}} + \delta I$$

leads to

$$\ln x_S \gamma_S = -m_S \cdot \nu \cdot M_{AB} \left[1 - \frac{\alpha}{3} \left| z_+ z_- \right| \sqrt{I} \cdot \sigma \left(\beta a \sqrt{I} \right) + \frac{\delta I}{2} \right]$$

ILLUSTRATION A9.3-1

The Activity Coefficient of Water in an Aqueous Sodium Chloride Solution

Over the range of 0 to 6 M, we have shown that the mean ionic activity coefficient of sodium chloride in aqueous solutions at 25°C is well correlated with

$$\ln \gamma_\pm = -\frac{\alpha \left| z_+ z_- \right| \sqrt{I}}{1 + \sqrt{I}} + 0.137 I$$

Determine the activity coefficient of water in aqueous sodium chloride solutions over this range of concentrations using the equation derived in this appendix.

SOLUTION

For sodium chloride solutions $I = \frac{1}{2}(1 + 1) M = M$. Therefore,

$$\ln x_W \gamma_W = \ln \frac{55.51}{55.51 + 2 \cdot M_{NaCl}} + \ln \gamma_W$$

$$= -0.018 \times 2 \times M_{NaCl} \left[1 - \frac{1.178}{3} \sqrt{M_{NaCl}} \cdot \sigma(M_{NaCl}) + \frac{0.137}{2} M_{NaCl} \right]$$

The results are given in the table below.

M_{NaCl}	x_{H_2O}	γ_{H_2O}	M_{NaCl}	x_{H_2O}	γ_{H_2O}
0.0	1.000	1.000	3.0	0.902	0.989
0.1	0.996	1.000	4.0	0.874	0.978
0.25	0.991	1.000	5.0	0.847	0.964
0.5	0.982	1.000	6.0	0.822	0.946
0.75	0.974	1.000	8.0	0.776	0.904
1.0	0.965	1.000	10.0	0.735	0.852
2.0	0.933	0.996			

PROBLEMS

9.1 Show that for mixing of ideal gases at constant temperature and pressure to form an ideal gas mixture,

$$\Delta_{mix}\underline{U} = \Delta_{mix}\underline{H} = \Delta_{mix}\underline{V} = 0$$

and

$$\Delta_{mix}\underline{G} = \Delta_{mix}\underline{A} = RT \sum_i x_i \ln x_i$$

9.2 In Sec. 9.1 we considered the changes in thermodynamic properties on forming an ideal gas mixture from a collection of ideal gases at the same temperature and pressure. A second, less common way of forming an ideal gas mixture is to start with a collection of pure ideal gases, each at the temperature T and volume V, and mix and compress the mixture to produce an ideal gas mixture at temperature T and volume V.

   a. Show that the mixing process described here is mixing at constant partial pressure of each component.

   b. Derive each of the entries in the following table.

9.3 Repeat the derivations of the previous problem for a mixing process in which both pure fluids, initially at a temperature T and pressure P, are mixed at constant temperature and the pressure then adjusted so that the final volume of the mixture is equal to the sum of the initial volumes of the pure components (i.e., there is no volume change on mixing).

9.4 We have the following properties for a certain mixture for mixing at constant temperature and pressure:

$$U(T, P, \underline{x}) = \sum_{i=1}^{C} N_i \underline{U}_i(T, P)$$

$$V(T, P, \underline{x}) = \sum_{i=1}^{C} N_i \underline{V}_i(T, P)$$

$$S(T, P, \underline{x}) = \sum_{i=1}^{C} N_i \underline{S}_i(T, P) - R \sum_{i=1}^{C} N_i \ln x_i$$

where $\underline{S}_i$, the pure-component molar entropy of component i, is given by

$$\underline{S}_i = \underline{S}_i^\circ + C_{V,i} \ln \frac{\underline{U}_i}{\underline{U}_i^\circ} + R \ln \frac{\underline{V}_i}{\underline{V}_i^\circ}$$

Here $\underline{S}_i^\circ$, $\underline{U}_i^\circ$, and $\underline{V}_i^\circ$ are the molar entropy, internal energy, and volume of pure component i in some reference state, and $C_{V,i}$ is its constant-volume heat capacity.

   a. Obtain expressions for the partial molar volume, partial molar internal energy, partial molar entropy, and partial molar Gibbs energy of each component in this mixture in terms of $\underline{S}_i^\circ$, $\underline{U}_i^\circ$, $\underline{V}_i^\circ$, $C_{V,i}$, R, and T.

   b. Obtain expressions for the volumetric and thermal equations of state for this mixture.

   c. Obtain expressions for the enthalpy and the Helmholtz and Gibbs energies of this mixture.

9.5 Assuming that two pure fluids and their mixture can be described by the van der Waals equation of state,

$$P = \frac{RT}{\underline{V} - b} - \frac{a}{\underline{V}^2}$$

Ideal Gas Mixing Properties* at Constant Temperature and Partial Pressure of Each Species

Internal energy	$\overline{U}_i^{IG}(T, x_i) = \underline{U}_i^{IG}(T)$	$\Delta_{mix}\underline{U}^{IG} = 0$
Volume†	$\overline{V}_i^{IGM}(T, P, x_i) = x_i \underline{V}_i^{IG}(T, P_i)$	$\Delta_{mix}\underline{V}^{IGM} = (1 - C)V / \sum_{i=1}^{C} N_i$
Enthalpy	$\overline{H}_i^{IGM}(T, x_i) = \underline{H}_i^{IG}(T)$	$\Delta_{mix}^{IGM}\underline{H} = 0$
Entropy	$\overline{S}_i^{IGM}(T, P, x_i) = \underline{S}_i^{IG}(T, P_i)$	$\Delta_{mix}\underline{S}^{IGM} = 0$
Helmholtz energy	$\overline{A}_i^{IGM}(T, P, x_i) = \underline{A}_i^{IG}(T, P_i)$	$\Delta_{mix}\underline{A}^{IGM} = 0$
Gibbs energy	$\overline{G}_i^{IGM}(T, P, x_i) = \underline{G}_i^{IG}(T, P_i)$	$\Delta_{mix}\underline{G}^{IGM} = 0$

*For mixing at constant temperature and partial pressure of each species, we have the following

$$\Delta_{mix}\underline{\theta}^{IGM} = \sum_i x_i [\overline{\theta}_i(T, P, x_i) - \underline{\theta}_i(T, P_i)]$$

†C = number of components.

and that for the mixture the van der Waals one-fluid mixing rules apply

$$a = \sum_i \sum_j x_i x_j a_{ij} \quad \text{and} \quad b = \sum_i x_i b_i$$

a. Show that the fugacity coefficient for species i in the mixture is

$$\ln \phi_i = \ln \frac{\bar{f}_i}{x_i P} = \frac{B_i}{Z - B} - \ln(Z - B) - \frac{2 \sum_j x_j a_{ij}}{RT \underline{V}}$$

where $B = Pb/RT$.

b. Derive an expression for the activity coefficient of each species.

9.6 Assuming that the van der Waals equation of state,

$$P = \frac{RT}{\underline{V} - b} - \frac{a}{\underline{V}^2}$$

is satisfied by two pure fluids and by their mixture, and that the van der Waals one-fluid rules

$$a = \sum_i \sum_j x_i x_j a_{ij}$$

and

$$b = \sum_i \sum_j x_i x_j b_{ij}$$

with $b_{ij} = b_{ji}$ apply to the mixture, derive expressions for

a. The excess volume change on mixing at constant T and P

b. The excess enthalpy and internal energy changes on mixing at constant T and P

c. The excess entropy change on mixing at constant T and P

d. The excess Helmholtz and Gibbs energy changes on mixing at constant T and P

9.7 The virial equation for a binary mixture is

$$\frac{P\underline{V}}{RT} = 1 + \frac{B_{mix}}{\underline{V}} + \cdots$$

with

$$B_{mix} = y_1^2 B_{11} + y_2^2 B_{22} + 2 y_1 y_2 B_{12}$$

Here B_{11} and B_{22} are the second virial coefficients for pure species 1 and pure species 2, respectively, and B_{12} is the cross second virial coefficient. For a binary mixture

a. Obtain an expression for the fugacity coefficient of a species (Eq. 9.4-6).

b. Show that the activity coefficient for species 1 is

$$\ln \gamma_1 = \delta_{12} y_2^2 P/RT$$

where $\delta_{12} = 2B_{12} - B_{11} - B_{22}$.

c. Generalize the results in part (b) to a multicomponent mixture.

9.8 a. Derive the two-constant Margules equations for the activity coefficients of a binary mixture (Eqs. 9.5-7).

b. Derive Eqs. 9.5-9.

c. Use the results of part (b) to derive van Laar expressions for the activity coefficients of a ternary mixture (Eq. A9.2-2).

9.9 Using the van Laar theory, estimate the activity coefficients for the benzene–2,2,4-trimethyl pentane system at 55°C. Compare the predictions with the results in Illustrations 9.5-1, 9.6-1, and 9.6-2.

Data:

	Benzene	2,2,4-Trimethyl Pentane
Critical temperature	562.1 K	554 K
Critical pressure	4.894 MPa	2.482 MPa
Critical density	301 kg/m^3	235 kg/m^3

9.10 Develop expressions for γ_1^* and $\gamma_1^\square$ using each of the following: the one-constant and two-constant Margules equations, the van Laar equation, regular solution theory, and the UNIFAC model.

9.11 Use the lattice model discussed in Appendix A9.1 to show that the state of maximum entropy for an ideal gas at constant temperature (and therefore energy) and contained in a volume V is the state of uniform density.

9.12 Calculate the fugacity for each species in the following gases at 290 K and 800 bar:

a. Pure oxygen

b. Pure nitrogen

c. Oxygen and nitrogen in a 30 mol % O_2, 70 mol % N_2 mixture using the Lewis-Randall rule

d. Oxygen and nitrogen in the mixture in part (c) using the Peng-Robinson equation of state

9.13 Chemically similar compounds (e.g., ethanol and water or benzene and toluene) generally form mixtures that are close to ideal, as evidenced by activity coefficients that are near unity and by small excess Gibbs energies of mixing. On the other hand, chemically dissimilar species (e.g., benzene and water or toluene and ethanol) form strongly nonideal mixtures. Show, by considering the binary mixtures that can be formed from ethanol, water, benzene, and toluene, that the UNIFAC model predicts such behavior.

9.14 Repeat the calculations of the previous problem with the regular solution model. Compare the two results.

9.15 Develop an expression for the activity coefficient of a species in a mixture from the Peng-Robinson equation of state with the van der Waals one-fluid mixing rules.

9.16 a. Show that the minimum amount of work, W_s^{min}, necessary to separate 1 mole of a binary mixture into its pure components at constant temperature and pressure is

$$W_s^{min} = x_1 RT \ln \frac{f_1(T, P)}{\bar{f}_1(T, P, x_1)}$$
$$+ x_2 RT \ln \frac{f_2(T, P)}{\bar{f}_2(T, P, x_2)}$$

b. Show that this expression reduces to

$$W_s^{min} = -x_1 RT \ln x_1 - x_2 RT \ln x_2$$

for (i) an ideal liquid mixture and (ii) a gaseous mixture for which the Lewis-Randall rule is obeyed.

c. Calculate the minimum amount of work needed to separate a 50/50 mixture of two isomers at 300 K and a pressure of 1 bar into its pure components at the same temperature and pressure. Explicitly state and justify all assumptions.

9.17 There are several possible expressions that can be used for the Gibbs excess energy. One is the Redlich-Kister expansion

$$\underline{G}^{ex} = x_1 x_2 \{A + B(x_1 - x_2) + C(x_1 - x_2)^2\}$$

where $B = 0$, but A and C are nonzero. Find expressions for the activity coefficients for this excess Gibbs energy model in which γ_1 is given solely in terms of x_2 and the parameters A and C, and γ_2 only in terms of x_1, A, and C.

9.18 Experimentally it is observed that

$$\lim_{x_i \to 1} \left(\frac{\partial \ln \gamma_i}{\partial x_i} \right)_{T, P} = 0 \quad \text{for any species i}$$

This equation implies that the activity coefficient γ_i (or its logarithm) is weakly dependent on mole fraction near the pure component limit. Since we also know that $\gamma_i \to 1$ as $x_i \to 1$, this equation further

implies that the activity coefficient is near unity for an almost pure substance.

a. Show that this equation is satisfied by all the liquid solution models discussed in Sec. 9.5.

b. Show that since this equation is satisfied for any substance, we also have

$$\lim_{x_i \to 0} x_i \left(\frac{\partial \ln \gamma_i}{\partial x_i} \right)_{T, P} = 0$$

9.19 The following data are available for mean activity coefficients of single electrolytes in water at 25°C.[23]

Molality	$\gamma_\pm$		
	KCl	CrCl$_3$	Cr$_2$(SO$_4$)$_3$
0.1	0.770	0.331	0.0458
0.2	0.718	0.298	0.0300
0.3	0.688	0.294	0.0238
0.5	0.649	0.314	0.0190
0.6	0.637	0.335	0.0182
0.8	0.618	0.397	0.0185
1.0	0.604	0.481	0.0208

Compare these data with the predictions of the Debye-Hückel limiting law, Eq. 9.10-15, and Eq. 9.10-18 with $\beta a = 1$ and $\delta = 0.1|z_+ z_-|$.

9.20 The data below are for the activity coefficients of lithium bromide in aqueous solutions as a function of molality at 25°C.

M_{LiBr}	$\gamma_\pm$	M_{LiBr}	$\gamma_\pm$	M_{LiBr}	$\gamma_\pm$
0.001	0.967	0.5	0.739	10	19.92
0.005	0.934	1.0	0.774	12	46.3
0.01	0.891	3	1.156	14	104.7
0.05	0.847	5	2.74	16	198.0
0.1	0.790	8	8.61	20	485.0

Compare the predictions of the Debye-Hückel model (Eqs. 9.10-15 and 9.10-16), and the extended Debye-Hückel models (Eqs. 9.10-17 and 9.10-18) with these data.

9.21 Develop a Gibbs-Duhem equation for strong electrolyte-water systems, and use this equation and the data in Illustration 9.10-1 to compute the activity coefficient of water in aqueous hydrochloric acid solutions at 25°C.

---

[23]*Reference:* R. A. Robinson and R. H. Stokes, *Electrolyte Solutions,* 2nd ed., Butterworths, London (1959), Appendix 8.10.

9.22 a. Given experimental data either for the excess Gibbs energy, $\underline{G}^{ex}$, or for species activity coefficients from which $\underline{G}^{ex}$ can be computed, it is sometimes difficult to decide whether to fit the data to the two constant Margules or van Laar expressions for $\underline{G}^{ex}$ and γ_i. One method of making this decision is to plot $\underline{G}^{ex}/x_1 x_2$ versus x_1 and $x_1 x_2/\underline{G}^{ex}$ versus x_1 and determine which of the two plots is most nearly linear. If it is the first, the data are best fit with the two-constant Margules expression; if the second, the van Laar expression should be used. Justify this procedure, and suggest how these plots can be used to obtain the parameters in the activity coefficient equations.

b. The following data have been obtained for the benzene–2,2,4-trimethyl pentane mixture (Illustration 10.1-4). Using the procedure in part (a), decide which of the two solution methods is likely to best fit the data.

x_B	0.0819	0.2192	0.3584	0.3831
$\underline{G}^{ex}$ (J/mol)	83.7	203.8	294.1	302.5
x_B	0.5256	0.8478	0.9872	
$\underline{G}^{ex}$ (J/mol)	351.9	223.8	23.8	

9.23 The excess Gibbs energies for liquid argon–methane mixtures have been measured at several temperatures.[24] The results are

$$\frac{\underline{G}^{ex}}{RT} = x_{Ar}(1 - x_{Ar})\{A - B(1 - 2x_{Ar})\}$$

where numerical values for the parameters are given as

T (K)	A	B
109.0	0.3024	−0.014 53
112.0	0.2929	−0.011 69
115.75	0.2792	+0.051 15

Compute the following:
a. The activity coefficients of argon and methane at 112.0 K and $x_{Ar} = 0.5$
b. The molar isothermal enthalpy change on producing an $x_{Ar} = 0.5$ mixture from its pure components at 112.0 K

c. The molar isothermal entropy change on producing an $x_{Ar} = 0.5$ mixture from its pure components at 112.0 K

9.24 Derive Eq. 9.2-13.

9.25 Wilson[25] has proposed that the excess Gibbs energy of a multicomponent system is given by

$$\underline{G}^{ex} = -RT \sum_{i=1}^{c} x_i \ln\left[\sum_{j=1}^{c} x_j \Lambda_{ij}\right]$$

where

$$\Lambda_{ij} = \frac{\underline{V}_j^L}{\underline{V}_i^L} \exp\left[-\frac{(\lambda_{ij} - \lambda_{ii})}{RT}\right]$$

Note that this equation contains only the interaction parameters Λ_{ij} for binary mixtures. Also, the parameters $(\lambda_{ij} - \lambda_{ii})$ appear to be insensitive to temperature.

Holmes and van Winkle[26] have tested this equation and found it to be accurate for the prediction of binary and ternary vapor–liquid equilibria. They also report values of the parameters $\underline{V}_i^L$ and $(\lambda_{ij} - \lambda_{ii})$ for many binary mixtures. Use the Wilson equation to

a. Derive Eqs. 9.5-12 and 9.5-13 for the activity coefficients of a species in a binary mixture.

b. Obtain the following expression for the activity coefficient of species 1 in a multicomponent mixture

$$\ln \gamma_1 = 1 - \ln\left[\sum_{j=1}^{c} x_j \Lambda_{1j}\right] - \sum_{i=1}^{c} \frac{x_i \Lambda_{i1}}{\displaystyle\sum_{j=1}^{c} x_j \Lambda_{ij}}$$

9.26 The fugacity of a species in a mixture can have a peculiar dependence on composition at fixed temperature and pressure, especially if there is a change of phase with composition. Show this by developing plots of the fugacity of isobutane and of carbon dioxide in their binary mixture as a function of isobutane composition using the Peng-Robinson equation of state for each of the following conditions.
a. $T = 377.6$ K and P in the range from 20 to 80 bar
b. $T = 300$ K and P in the range from 7 to 35 bar

9.27 One expression that has been suggested for the excess Gibbs energy of a binary mixture that is asymmetric in composition is

$$\underline{G}^{ex} = A x_1 x_2 (x_1 - x_2)$$

[24] *Reference:* A. G. Duncan and M. J. Hiza, *I.E.C. Fundam.* **11**, 38 (1972).
[25] G. M. Wilson, *J. Am. Chem. Soc.*, **86**, 127 (1964).
[26] M. J. Holmes and M. van Winkle, *Ind. Eng. Chem.*, **62**, 21 (1970).

   a. Find expressions for the activity coefficients in which γ_1 is specified in terms of x_2 and γ_2 in terms of x_1.

   b. Does this excess Gibbs energy model satisfy the Gibbs-Duhem equation?

9.28 It has been suggested that since the one-parameter Margules expansion is not flexible enough to fit most activity coefficient data, it should be expanded by adding additional constants. In particular, the following have been suggested:

Two-parameter models: $\ln \gamma_1 = Ax_2^2$ $\ln \gamma_2 = Bx_1^2$

                          $\ln \gamma_1 = Ax_2^n$ $\ln \gamma_2 = Ax_1^n$

Three-parameter model: $\ln \gamma_1 = Ax_2^n$ $\ln \gamma_2 = Bx_1^n$

In each case the reference states are the pure components at the temperature and pressure of the mixture.

   a. Which of these models are reasonable?

   b. What are the allowable values for the parameters A, B, and n in each of the models?

9.29 Derive Eqs. 9.5-18.

9.30 At $T = 60°C$ the vapor pressure of methyl acetate is 1.126 bar, and the vapor pressure of methanol is 0.847 bar. Their mixtures can be described by the one-constant Margules equation

$$\underline{G}^{\mathrm{ex}} = Ax_1x_2 \quad \text{with} \quad A = 1.06RT$$

where R is the gas constant and T is temperature in K.

   a. Plot the fugacity of methyl acetate and methanol in their mixtures as a function of composition at this temperature.

   b. The Henry's law coefficient H_i is given by the equation

$$H_i = \lim_{x_i \to 0} \frac{P_i}{x_i}$$

Develop an expression for the Henry's law constant as a function of the A parameter in the Margules expression, the vapor pressure, and composition. Compare the hypothetical pure component fugacity based on the Henry's law standard state with that for the usual pure component standard state.

9.31 Derive Eq. 9.9-8.

9.32 Derive Eqs. 9.9-9.

9.33 Derive Eqs. 9.9-11 to 9.9-13.

9.34 Derive the expression for the fugacity coefficient of the Soave–Redlich-Kwong equation of state (Eq. 4.4-1b) with the van der Waals one-fluid mixing and combining rules of Eqs. 9.4-8 and 9.4-9.

9.35 **a.** A starting point for modeling the thermodynamics of polymers in solution is to use the Flory-Huggins model with the Flory χ parameter assumed to be a constant. For mixtures of polystyrene in toluene, in which $\underline{V}_{\mathrm{PS}} = 1000\underline{V}_{\mathrm{T}}$, $\chi = 0.6$ is a reasonable estimate of the value for that parameter. Plot the activity coefficients of polystyrene and toluene as a function of the mole fraction of toluene at 298 K, assuming that the molecular weight of toluene is 92 and that of the polystyrene in this solution is 90 000.

   b. Plot these activity coefficients versus the toluene volume fraction, and also plot the activity coefficients as a function of toluene mass fraction.

   c. Usually, to accurately fit experimental data, the χ parameter cannot be taken to be a constant, but must be a function of both temperature and composition. One proposed model is

$$\chi = \frac{Ax_2}{T}$$

where T is the absolute temperature in kelvins, x_2 is the mole fraction of polymer, and A is a constant. Using this expression, derive the equations for the activity coefficients and for the excess enthalpy of mixing $\underline{H}^{\mathrm{ex}}$ of polystyrene-toluene solutions as a function of toluene mole fraction.

   d. If the value of the A parameter in the above equation is 1500 K, plot the activity coefficients of polystyrene and toluene as a function of the mole fraction, mass fraction, and volume fraction of toluene at 298 K.

   e. Plot the heat of mixing for polystyrene-toluene mixtures as a function of the mole fraction, mass fraction, and volume fraction of toluene at 298 K.

9.36 **a.** Derive an expression for the minimum amount of work needed to continuously and adiabatically separate two isomers into their pure components at constant temperature and pressure. Explicitly state all assumptions and justify them.

   b. Calculate the minimum work necessary to separate a 50/50 mixture of isomers at 300 K and a constant pressure of 1 bar.

9.37 The activity of a substance, which is a function of temperature, pressure and composition, is defined as follows:

$$a_i \left(T, P, \underline{x} \right) = \frac{\bar{f}_i(T, P, \underline{x})}{\bar{f}_i^{\circ}(T, P^{\circ}, \underline{x}^{\circ})}$$

where $\bar{f}_i^{\circ} \left(T, P^{\circ}, \underline{x}^{\circ} \right)$ is the standard-state fugacity of species i at the standard-state pressure P° and standard-state composition $\underline{x}^\circ$ (which could be the pure component state or one of the various Henry's law standard states).

a. Using this definition of the activity, prove for a binary mixture at constant temperature and pressure that

$$d \ln \left(\frac{a_1}{x_1} \right) = -\frac{x_2}{x_1} d \ln \left(\frac{a_2}{x_2} \right)$$

b. For fixed standard-state temperature, derive an expression for how the species activity changes with temperature at constant pressure and composition.

9.38 Air contains approximately 21 mol % oxygen and 79 mol % nitrogen. An engineer claims to have developed a continuous process in which air is first compressed to 2 bar and 25°C, and then isothermally expanded to atmospheric pressure through a secret device that has no moving parts and results in two gas streams. The first stream is said to contain 99 mol % oxygen, and the second stream contains only 5 mol % oxygen. Prove whether such a device is thermodynamically possible.

9.39 A gas stream at 310 K and 14 bar is to be compressed to 345 bar before transmission by underground pipeline. If the compression is carried out adiabatically and reversibly, determine the compressor outlet temperature and the work of compression for the gas stream, which consists of
a. Pure methane
b. 95 mol % methane and 5 mol % ethane
c. 5 mol % methane and 95 mol % ethane
d. Compare your results with the results given by the ideal gas equation.

9.40 The following data are available for the mean ionic activity coefficients of these salts in water at 25°C.

M	HCl	CaCl$_2$	ZnSO$_4$
0.001	0.966	0.888	0.734
0.005	0.928	0.789	0.477
0.01	0.905	0.732	0.387
0.05	0.830	0.584	0.202
0.1	0.796	0.531	0.148
0.5	0.757	0.457	0.063
1.0	0.809	0.509	0.044
2.0	1.009	0.807	0.035
3.0	1.316	1.055	0.041

a. Fit these data as best you can using the equations in this chapter for the mean ionic activity coefficient.
b. Determine the activity coefficient of water in each of these solutions.

9.41 A thermodynamic property of a mixture is given by

$$\theta(x_1, x_2, x_3, T, P) = \sum_{i=1}^{3} x_i \theta_i(T, P)$$
$$+ \sum_{j=1}^{3} \sum_{i=1}^{3} a_{ij} x_i x_j + a_{123} x_1 x_2 x_3$$

with $a_{ii} = 0$.

a. Develop expressions for the partial molar properties $\bar{\theta}_1$, $\bar{\theta}_2$, and $\bar{\theta}_3$ as a function of the pure component molar properties, the mole fractions, and the parameters a_{12}, a_{13}, a_{12}, and a_{123}.
b. Obtain expressions for the infinite-dilution value of $\bar{\theta}_1$ in solutions of varying concentrations of species 2 and 3. Repeat for $\bar{\theta}_2$ in mixed solvent solutions of 1 and 3, and for $\bar{\theta}_3$ in mixed solvent solutions of 1 and 2.

9.42 The infinite-dilution heat of solution for solid urea (CH$_4$N$_2$O) in water at 25°C is reported in *The Chemical Engineer's Handbook* to be -3609 cal/g. In the same book the heat of formation is reported to be -77.55 kcal/mol for liquid urea and -79.634 kcal/mol for crystalline urea. Compare the heat of melting of urea with its heat of solution.

9.43 Derive the expression for the partial molar volume of a species in a mixture that obeys the Peng-Robinson equation of state and the van der Waals one-fluid mixing rules.

9.44 Derive the expression for the partial molar volume of a species in a mixture that obeys the Peng-Robinson equation of state and the Wong-Sandler mixing rules.

9.45 Derive the expression for the activity coefficient of a species in a mixture that obeys the Peng-Robinson equation of state and the van der Waals one-fluid mixing rules.

9.46 Derive the expression for the activity coefficient of a species in a mixture that obeys the Peng-Robinson equation of state and the Wong-Sandler mixing rules.

9.47 The following data are available for the infinite-dilution activity coefficients in acetone in ethanol:

$T(°C)$	$\gamma_{acetone}^{\infty}$
49.3	2.17
62.6	2.03
75.1	1.92

a. Compute the excess partial molar enthalpy of acetone in ethanol at 62.6°C.

b. Make a thermodynamically based estimate of the value of the infinite-dilution activity coefficient of acetone in ethanol at 100°C. (A simple linear extrapolation is not correct.)

9.48 Use the regular solution model to predict the activity coefficients of benzene and 2,2,4-trimethyl pentane in their mixtures at 55°C. What are the predicted values of the infinite-dilution activity coefficients?

9.49 Use the UNIFAC model to predict the activity coefficients of benzene and 2,2,4-trimethyl pentane in their mixtures at 55°C. What are the predicted values of the infinite-dilution activity coefficients?

9.50 Use the UNIFAC model to predict the activity coefficients of acetone + water in their mixtures at 298 K. What are the predicted values of the two infinite-dilution activity coefficients?

9.51 Using $\underline{G}^{\mathrm{ex}} = ax_1x_2$, show that

$$\left.\frac{\partial \left(N\underline{G}^{\mathrm{ex}}\right)}{\partial N_1}\right|_{T,P} \qquad \left.\frac{\partial \left(\underline{G}^{\mathrm{ex}}\right)}{\partial x_1}\right|_{T,P} \quad \text{and} \quad \left.\frac{\partial \left(\underline{G}^{\mathrm{ex}}\right)}{\partial x_1}\right|_{T,P,x_2}$$

all lead to different results. Note that from definition of a partial molar property, only the first of these derivatives is the partial molar excess Gibbs energy.

9.52 The following simple expression has been suggested for modeling the activity coefficient of species 1 in a binary mixture:

$$RT \ln \gamma_1 = ax_2$$

as it has the proper behavior that $\gamma_1 \to 1$ as $x_1 \to 1$ (so that $x_2 \to 0$). Obtain the expression for γ_2, and

determine whether or not this model is a reasonable one.

9.53 In a binary mixture, the activity coefficient of component 1 has been found to be

$$RT \ln \gamma_1 = Ax_2^2 \quad \text{with} \quad A = a + \frac{b}{T}$$

and

$$\bar{V}_1^{\mathrm{ex}}\left(T, P, \underline{x}\right) = \theta x_2^2$$

where a, b, and θ are constants independent of temperature, pressure, and composition. Find expressions for $\underline{G}$, $\underline{S}$, $\underline{H}$, and $\underline{V}$ in terms of the gas constant R; the temperature T; the parameters a, b, and θ; the pure-component properties; and the mixture composition.

9.54 A 50 mol % mixture of two gases A and B at 300 K and 1 bar is to be isothermally and isobarically separated into its pure components. If the gases form an ideal mixture and $C_{\mathrm{P,A}}^* = 10$ J/(mol K) and $C_{\mathrm{P,B}}^* = 15$ J/(mol K), how much Gibbs energy is required to separate the mixture?

9.55 An oxygen enrichment device is needed for people with impaired respiratory systems. To design such a device, it is necessary to compute the work needed to produce a stream that contains 50 mol % of oxygen from air (21 mol % oxygen) at 300 K and 1 bar. If the exit streams are at the same temperature and pressure as the inlet air, and half of the oxygen in the air is recovered in the enriched oxygen stream, what is the minimum amount of work required to operate whatever device is developed for this process?

Chapter 10

Vapor-Liquid Equilibrium in Mixtures

The objective of this chapter and the two that follow is to illustrate how the principles introduced in Chapter 8 for the thermodynamic description of mixtures together with the calculational procedures of Chapter 9 can be used to study many different types of phase equilibria important in chemical engineering practice. In particular, the following are considered:

1. Vapor-liquid equilibria (this chapter)
2. The solubility of gas in a liquid (Sec. 11.1)
3. The solubility of a liquid in a liquid (Sec. 11.2)
4. Vapor-liquid-liquid equilibria (Sec. 11.3)
5. The distribution of a solute among two liquid phases (Sec. 11.4)
6. Osmotic equilibrium and osmotic pressure (Sec. 11.5)
7. The solubility of a solid in a liquid, gas, or supercritical fluid (Sec. 12.1)
8. The distribution of a solid solute among two liquid phases (Sec. 12.2)
9. The freezing-point depression of a solvent due to the presence of a solute (Sec. 12.3)
10. The phase behavior of solid mixtures (Sec. 12.4)
11. The distribution of chemicals in the environment (Sec. 12.5)

Our interest in phase equilibria is twofold: to make predictions about the equilibrium state for the types of phase equilibria listed above using activity coefficient models and/or equations of state, and to use experimental phase equilibrium data to obtain activity coefficient and other partial molar property information. Also, there are brief introductions to how such information is used in the design of several different types of purification processes, including distillation (this chapter) and liquid-liquid extraction (Chapter 11).

INSTRUCTIONAL OBJECTIVES FOR CHAPTER 10

The goals of this chapter are for the student to:

- Be able to compute the vapor-liquid equilibrium compositions when the liquid is an ideal mixture and the vapor is an ideal gas mixture (that is, to be be able to compute the conditions of vapor-liquid equilibrium and develop *x-y*, *T-x-y*, and *P-x-y* diagrams for ideal mixtures) (Sec. 10.1)

- Be able to correlate the low-pressure vapor-liquid equilibrium data for a nonideal liquid mixture, that is, to be able to compute the conditions of vapor-liquid equilibrium and develop *x-y*, *T-x-y*, and *P-x-y* diagrams for nonideal mixtures using activity coefficient models (the γ-ϕ method) (Sec. 10.2)
- Be able to predict low-pressure vapor-liquid equilibria when no experimental data are available (Sec. 10.2)
- Be able to correlate high-pressure vapor-liquid equilibrium data using an equation of state (that is, to be able to compute the conditions of vapor-liquid equilibrium and develop *x-y*, *T-x-y*, and *P-x-y* diagrams using an equation of state (the ϕ-ϕ method) (Sec. 10.3)
- Be able to predict high-pressure vapor-liquid equilibrium compositions using an equation of state when no experimental data are available (Sec. 10.3)
- Be able to do bubble point, dew point, and partial vaporization calculations for both ideal and nonideal systems (Secs. 10.1, 10.2, and 10.3)
- Have an understanding of the importance of vapor-liquid equilibrium for separations by distillation (Secs. 10.1 and 10.2)

NOTATION INTRODUCED IN THIS CHAPTER

a_i Activity of species i, $a_i = x_i \gamma_i$

B Flow rate of bottoms product from a distillation column

D Flow rate of distillate (overheads product) in a distillation column

F Flow rate of feed to a distillation column or flash unit

L Number of moles of a mixture that are liquid, or the liquid stream in a distillation column or flash unit

P_i Partial pressure of species i $= y_i P$ (kPa)

K_i Separation or K factor for species i, $K_i = y_i / x_i$

V Number of moles of a mixture that are vapor, or the vapor stream in a distillation column or flash unit

x_i Liquid phase mole fraction of species i

$\underline{x}$ Set of liquid phase mole fractions $x_1, x_2, \ldots$

y_i Vapor phase mole fraction of species i

$\underline{y}$ Set of vapor phase mole fractions $y_1, y_2, \ldots$

$z_{i,F}$ Feed composition of species i

γ_i^∞ Activity coefficient of species i when it is at infinite dilution

θ^L Property θ in the liquid phase

θ^V Property θ in the vapor phase

10.0 INTRODUCTION TO VAPOR-LIQUID EQUILIBRIUM

For the analysis of distillation and other vapor-liquid separation processes one must estimate the compositions of the vapor and liquid in equilibrium. This topic is considered in detail in this chapter with particular reference to the preparation of mixture vapor-liquid equilibrium (VLE) phase diagrams, partial vaporization and condensation calculations, and the use of vapor-liquid equilibrium measurements to obtain information on the partial molar properties of mixtures. Also, there is a brief introduction to how such information is used to design purification processes involving distillation.

The starting point for all vapor-liquid calculations is the equilibrium criterion

Starting point for all phase equilibrium calculations

$$\boxed{\bar{f}_i^L(T, P, \underline{x}) = \bar{f}_i^V(T, P, \underline{y})} \tag{10.0-1}$$

where the superscripts L and V refer to the liquid and vapor phases, respectively. From the entries in Table 9.6-2, to compute the fugacity of a species in a vapor, $\bar{f}_i^V(T, P, \underline{y})$, we use either an equation of state or, less accurately, a simplifying assumption such as the Lewis-Randall rule or the ideal gas mixture model. For the fugacity of a species in a liquid, $\bar{f}_i^L(T, P, \underline{x})$, we have two different ways of proceeding—one based on activity coefficient (excess Gibbs energy) models, and the other based on the equation-of-state description of the liquid phase.

If an equation of state is used to describe both phases, the basic equilibrium relation becomes

The equation-of-state or ϕ-ϕ method

$$\bar{f}_i^L(T, P, \underline{x}) = x_i P \bar{\phi}_i^L(T, P, \underline{x}) = \bar{f}_i^V(T, P, \underline{y}) = y_i P \bar{\phi}_i^V(T, P, \underline{y}) \tag{10.0-2}$$

where

$$\bar{\phi}_i^L(T, P, \underline{x}) = \frac{\bar{f}_i^L(T, P, \underline{x})}{x_i P} \quad \text{and} \quad \bar{\phi}_i^V(T, P, \underline{y}) = \frac{\bar{f}_i^V(T, P, \underline{y})}{y_i P} \tag{10.0-3}$$

are the fugacity coefficients for the liquid and vapor phases, respectively, which are computed from an equation of state using Eq. 9.2-13. As a result of this form of relation, the description of vapor-liquid equilibrium using an equation of state for both phases is frequently referred to as the ϕ-ϕ method and will be considered in detail in Sec. 10.3. The other alternative is to use an activity coefficient model for the liquid phase and an equation of state for the vapor phase. At moderate pressures, omitting the Poynting correction, we have

The activity coefficient or γ-ϕ method

$$\bar{f}_i^L(T, P, \underline{x}) = x_i \gamma_i(T, P, \underline{x}) P_i^{sat}(T) \phi_i^{L,sat}(T, P) = \bar{f}_i^V(T, P, \underline{y}) = y_i P \bar{\phi}_i^V(T, P, \underline{y})$$

$$\tag{10.0-4}$$

The description of vapor-liquid equilibrium using an activity coefficient for the liquid phase and an equation of state for the vapor phase is usually referred to as the γ-ϕ method, which is considered in Sec. 10.2. These γ-ϕ and ϕ-ϕ descriptions are two different methods of analysis of the equilibrium problem, and hence we consider them separately.

However, before we proceed with the discussion, it is useful to consider the range of validity of the activity coefficient (γ-ϕ) and equation of state (ϕ-ϕ) models. The activity coefficient (or excess Gibbs energy) models can be used for liquid mixtures of all species. This description generally does not include density, and therefore will not give a good description of an expanded liquid, which occurs near the vapor-liquid critical point of a mixture. Also, when two different models are used—for example, an activity coefficient model for the liquid phase and an equation-of-state model for the vapor phase—the properties of the two phases cannot become identical, so the vapor-liquid critical region behavior is predicted incorrectly. In contrast, equation-of-state (ϕ-ϕ) models can be used at all temperatures, pressures, and densities, including the critical region, but if the van der Waals one-fluid mixing rules are used, only for hydrocarbons and inorganic gases. However, with excess free energy–based mixing rules such as those in Sec. 9.9, the equation-of-state method can be used with all components at all conditions. Nevertheless, the activity coefficient method is simpler to use at low pressures.

10.1 VAPOR-LIQUID EQUILIBRIUM IN IDEAL MIXTURES

At low pressures using the activity coefficient description and the Lewis-Randall rule in Eq. 10.0-4, we obtain

$$x_i \gamma_i(T, P, \underline{x}) P_i^{\text{vap}}(T) \left(\frac{f}{P}\right)_{\text{sat},i} = y_i P \left(\frac{f}{P}\right)_i \tag{10.1-1a}$$

This equation provides a relation between the compositions of the coexisting vapor and liquid equilibrium phases. Summing Eq. 10.1-1a over all species yields

$$\sum_{i=1}^{\mathcal{C}} x_i \gamma_i(T, P, \underline{x}) P_i^{\text{vap}}(T) \left(\frac{f}{P}\right)_{\text{sat},i} = P \sum_{i=1}^{\mathcal{C}} y_i \left(\frac{f}{P}\right)_i \tag{10.1-2a}$$

There are several simplifications that can be made to these equations. First, if the total pressure and the vapor pressure of the species are sufficiently low that all fugacity coefficient corrections are negligible, we have

Low-pressure vapor-liquid equilibrium equation

$$\boxed{x_i \gamma_i(T, P, \underline{x}) P_i^{\text{vap}}(T) = y_i P} \tag{10.1-1b}$$

and on summation, since $\sum y_i = 1$,

$$\sum x_i \gamma_i(T, P, \underline{x}) P_i^{\text{vap}}(T) = P \tag{10.1-2b}$$

Further, if the liquid phase forms an ideal mixture (e.g., $\gamma_i = 1$ for all species), these equations further reduce to

Raoult's law (for ideal liquid mixture + ideal vapor phase)

$$\boxed{x_i P_i^{\text{vap}}(T) = y_i P = P_i} \tag{10.1-3}$$

and

$$\boxed{\sum x_i P_i^{\text{vap}}(T) = \sum P_i = P} \tag{10.1-4}$$

where $P_i = y_i P$ is the **partial pressure** of species i in the vapor phase. Equation 10.1-3, which is known as **Raoult's law**, indicates that the partial pressure of a component in an ideal solution is equal to the product of the species mole fraction and its pure component vapor pressure. Also, from Eq. 10.1-4 the equilibrium pressure of an ideal mixture is equal to the mole fraction–weighted sum of the pure component vapor pressures, and therefore is a linear function of the mole fraction.

Equations 10.1-1 and 10.1-2 or their simplifications here, together with the restrictions that

$$\sum_i x_i = 1 \tag{10.1-5}$$

and

$$\sum_i y_i = 1 \tag{10.1-6}$$

and the mass and energy balance equations, are the basic relations for all vapor-liquid equilibrium calculations we consider in this section.

As the first illustration of the use of these equations, consider vapor-liquid equilibrium in the hexane-triethylamine system at 60°C. These species form an essentially ideal mixture. The vapor pressure of hexane at this temperature is 0.7583 bar and that of triethylamine is 0.3843 bar; these are so low that the fugacity coefficients at saturation and for the vapor phase can be neglected. Consequently, Eqs. 10.1-3 and 10.1-4 should be applicable to this system. The three solid lines in Fig. 10.1-1 represent the two species partial pressures and the total pressure, which were calculated using these equations and all are linear functions of the of liquid-phase mole fraction; the points are the experimental results. The close agreement between the computations and the laboratory data indicates that the hexane-triethylamine mixture is ideal at these conditions. Note that this linear dependence of the partial and total pressures on mole fractions predicted by Eqs. 10.1-2 and 10.1-3 is true only for ideal mixtures; it is not true for nonideal mixtures, as we shall see in Sec. 10.2.

Once the equilibrium total pressure has been computed for a given liquid composition using Eqs. 10.1-2 or 10.1-4, the equilibrium composition of the vapor can be calculated using Eqs. 10.1-1 or 10.1-3, as appropriate. Indeed, we can prepare a complete vapor-liquid equilibrium composition diagram, or x-y diagram, at constant temperature by choosing a collection of values for the composition of one of the phases, say the liquid-phase composition x_i, and then using the vapor pressure data to compute the

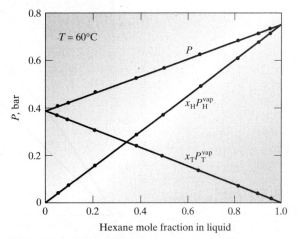

Figure 10.1-1 Equilibrium total pressure and species liquid-phase fugacities $(x_i P_i^{\text{vap}})$ versus mole fraction for the essentially ideal hexane-triethylamine system at 60°C. [Based on data of J. L. Humphrey and M. Van Winkle. *J. Chem. Eng. Data,* **12,** 526 (1967).]

total pressure and value of y_i corresponding to each x_i. For the hexane-triethylamine system, to calculate the composition of the vapor in equilibrium with a 50 mol % hexane mixture at $T = 60°C$, Eq. 10.1-4 is first used to compute the equilibrium pressure,

Construction of an *x-y* diagram

$$P = \sum x_i P_i^{vap} = x_H P_H^{vap} + x_T P_T^{vap}$$
$$= 0.5 \times 0.7583 + 0.5 \times 0.3843 = 0.5713 \text{ bar}$$

and then Eq. 10.1-3 is used to calculate the vapor-phase mole fraction of hexane:

$$y_H = \frac{x_H P_H^{vap}}{P} = \frac{0.5 \times 0.7583}{0.5713} = 0.6637$$

By choosing other liquid compositions and repeating the calculation at a fixed temperature, the complete constant-temperature vapor-liquid equilibrium composition diagram, or *x-y* diagram, can be constructed. The results are shown in Fig. 10.1-2 along with points representing the experimental data. The second line in this figure is the line $x = y$; the greater the difference between the *x-y* curve of the mixture and the $x = y$ line, the greater the difference in composition between the liquid and vapor phases and the easier it is to separate the two components by distillation, as will be discussed later in this section. (Since *x-y* diagrams are most often used in the study of distillation, it is common practice to include the $x = y$ or 45° line.)

An alternative way of presenting vapor-liquid equilibrium data is to plot, on a single figure, the equilibrium pressure and the compositions for both phases at fixed temperature. This has been done for the hexane-triethylamine system in Fig. 10.1-3. In this figure the equilibrium compositions of the vapor and liquid as a function of pressure are given by the curves labeled "vapor" and "liquid," respectively; the compositions of the two coexisting phases at each pressure are given by the intersection of a horizontal line (i.e., a line of constant pressure) with the vapor and liquid curves. The term **tie line** is used here, and generally in this chapter, to indicate a line connecting the equilibrium compositions in two coexisting phases. The tie line drawn in Fig. 10.1-3 shows that at

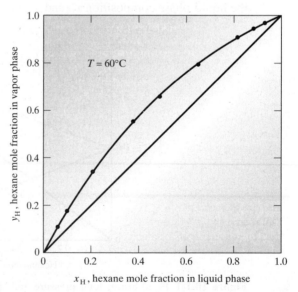

Figure 10.1-2 The *x-y* diagram for the hexane-triethylamine system at $T = 60°C$. [Based on data of J. L. Humphrey and M. Van Winkle. *J. Chem Eng. Data,* **12,** 526 (1967).]

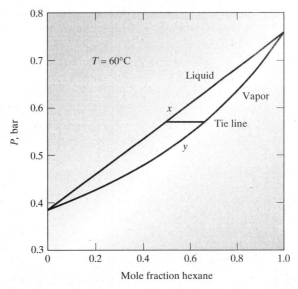

Figure 10.1-3 Pressure-composition diagram for the hexane-triethylamine system at fixed temperature.

$60°C$, a liquid containing 50 mol % hexane is in equilibrium with a vapor containing 66.37 mol % hexane at 0.5713 bar.

So far the discussion has been specific to systems at constant temperature; equivalently, pressure could be fixed and temperature and liquid phase composition taken as the variables. Although much experimental vapor-liquid equilibrium data are obtained in constant-temperature experiments, distillation columns and other vapor-liquid separations equipment in the chemical process industry are operated more nearly at constant pressure. Therefore, it is important that chemical engineers be familiar with both types of calculations.

The vapor-liquid equilibrium temperature for specified pressure and liquid composition is found as the solution to Eqs. 10.1-2 or, if the system is ideal, as the solution to Eq. 10.1-4. However, since the temperature appears only implicitly in these equations through the species vapor pressures,[1] and since there is a nonlinear relationship between the vapor pressure and temperature (cf. the Clausius-Clapeyron equation, Eq. 7.7-5 a), these equations are usually solved by iteration. That is, one guesses a value of the equilibrium temperature, computes the value of the vapor pressure of each species at this temperature, and then tests whether the pressure computed from Eqs. 10.1-2 (or Eq. 10.1-4 if the system is ideal) equals the fixed pressure. If the two are equal, the guessed equilibrium temperature is correct, and the vapor-phase mole fractions can be computed from Eq. 10.1-1 (or, if the system is ideal, from Eq. 10.1-3).[2] If the two pressures do not agree, a new trial temperature is chosen and the calculation repeated. Figure 10.1-4 is a plot, on a single graph, of the equilibrium temperature and mole fractions for the hexane-triethylamine system at 0.7 bar calculated in this way, and

---

[1]In fact, the species activity coefficients also depend on temperature; see Eq. 9.3-22. However, since this temperature dependence is usually small compared with the temperature variation of the vapor pressure, it is neglected here.

[2]If the vapor-phase mole fractions calculated in this way do not sum to 1, only a single phase, vapor or liquid, is present at equilibrium.

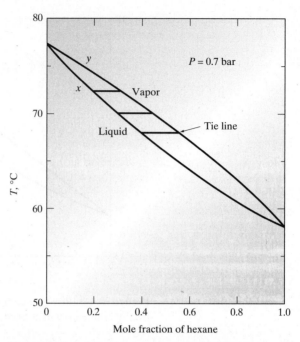

Figure 10.1-4 Temperature-composition diagram for the hexane-triethylamine system at fixed pressure.

Fig. 10.1-5 is the x-y diagram for this system. Note that tie lines drawn on Fig. 10.1-4 are again horizontal lines, though here they are lines of constant temperature.

Bubble point pressure and dew point pressure

The liquid line in vapor-liquid equilibrium diagrams is also referred to as the liquidus, the bubble point curve, or simply the bubble curve. The last two names arise as follows. Consider an equimolar mixture of hexane and triethylamine at 60°C and a pressure of 0.8 bar. Based on Fig. 10.1-3, this mixture is a liquid at these conditions

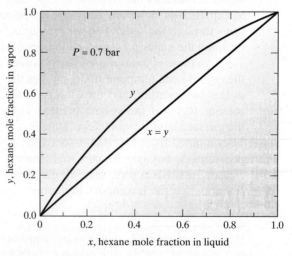

Figure 10.1-5 The x-y diagram for the hexane-triethylamine system at a fixed pressure of 0.7 bar.

(since the pressure is greater than the equilibrium pressure for this composition and temperature). As the pressure is lowered at constant temperature, the mixture remains a liquid until the vapor-liquid equilibrium pressure of 0.5713 bar is reached (the intersection of the 0.5 mole fraction line and the liquid curve). At this pressure the first bubble of vapor forms (containing 0.6637 mole fraction hexane), and this pressure is called the bubble point pressure of this mixture at this temperature. The bubble point pressure versus composition line is referred to as the **bubble point pressure** curve.

In a similar manner if, for example, an equimolar vapor mixture of hexane and triethylamine at low pressure and 60°C is isothermally compressed, at 0.5100 bar the first drop of liquid or dew forms (containing 0.3363 mole fraction hexane). This pressure is called the **dew point pressure**, and the line of dew point pressure versus composition is the dew point pressure curve. The dew point pressure at, for example, a mole fraction of 0.5 is given by the intersection of the vertical 0.5 mole fraction line with the vapor or dew point curve in Fig. 10.1-3.

Note that even for this ideal mixture, the compositions of the two phases in equilibrium at each pressure (or at each temperature) are different. This is because the two pure component vapor pressures are different (see Eq. 10.1-3). Also, at the constant temperature of 60°C, an equimolar hexane-triethylamine mixture begins to vaporize at one pressure (0.5713 bar), while a vapor of that composition starts to condense at a different pressure (0.5100 bar). If we start at low pressure and isothermally compress this mixture, the first drop of liquid forms at 0.5100 bar and then, as the pressure increases, more liquid will form, producing a vapor richer than the initial mixture in hexane and a liquid richer in triethylamine. This process will continue until the pressure of 0.5713 bar is reached at which all the vapor will have condensed to a liquid of the original composition of the vapor. (Can you follow this process in Fig. 10.1-3? Also, can you describe the analogous process if pressure is fixed and temperature varies, as in Fig. 10.1-4?) The behavior described above is unlike that of a pure fluid that undergoes a complete vapor-liquid phase change at a single pressure if the temperature is fixed.

Figures 10.1-3 and 10.1-4 are two-dimensional sections of the three-dimensional phase diagram of Fig. 10.1-6. The intersections of this three-dimensional equilibrium surface with planes of constant temperature (the vertical, unshaded planes) produce two-dimensional figures such as Fig. 10.1-3, whereas the intersection of a plane of constant pressure (horizontal, shaded plane) results in a diagram such as Fig. 10.1-4.

Bubble point temperature and dew point temperature

Next consider the vaporization of a 50 mol % hexane-triethylamine mixture at fixed pressure. As this liquid is heated, a temperature is reached at which the first bubble of vapor is formed; this temperature is termed the **bubble point temperature** of the liquid mixture at the given pressure. Since the composition of the liquid is essentially unchanged by its partial vaporization to form only one small bubble, we can use Fig. 10.1-4 and the initial liquid composition to determine that the composition of this first bubble of vapor is 66 mol % hexane and the bubble point temperature is 66.04°C. As the vapor formed is richer in hexane than the liquid mixture, the liquid will be depleted in hexane as the boiling proceeds. Thus, as more and more liquid vaporizes, the liquid will become increasingly more dilute in hexane and its boiling temperature will increase.

Conversely, we can consider the condensation of a 50 mol % hexane-triethylamine vapor mixture. As the vapor temperature decreases at fixed pressure, the **dew point temperature** is reached, at which the first drop of liquid forms. Since the condensation to form a single small drop of liquid leaves the vapor composition virtually unchanged, we can use Fig. 10.1-4 to determine that, at 0.7 bar, the first drop of condensate will

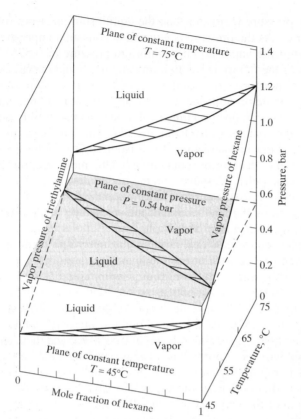

Figure 10.1-6 Vapor-liquid equilibria of hexane-triethylamine mixtures.

appear at about 69.26°C, and its composition will be about 34.2 mol % hexane. Clearly, as the condensation process continues, the vapor will become increasingly richer in hexane, and the equilibrium condensation temperature will decrease.

Thus, at fixed pressure, boiling and condensation phenomena, which occur at a single temperature in a pure fluid, take place over the range of temperatures between the dew point temperature and bubble point temperature in a mixture. Generally, the dew point and bubble point temperatures differ by many degrees (see Illustration 10.1-2); the two are close for the hexane-triethylamine system because the species vapor pressures are close and the components form an ideal mixture.

ILLUSTRATION 10.1-1

Development of Vapor-Liquid Equilibrium Diagrams for a Mixture That Obeys Raoult's Law

Assuming a mixture of *n*-pentane and *n*-heptane is ideal, prepare vapor-liquid equilibrium diagrams for this mixture at

   a. A constant temperature of 50°C
   b. A constant pressure of 1.013 bar

Data:

$$\ln P_5^{\text{vap}} = 10.422 - \frac{26\,799}{RT} \quad \text{and} \quad \ln P_7^{\text{vap}} = 11.431 - \frac{35\,200}{RT}$$

for *P* in bar, *T* in K, and *R* = 8.314 J/(mol K). The subscripts 5 and 7 designate pentane and heptane, respectively.

SOLUTION

Before we attempt to solve this problem, it is useful to check whether this problem is, in fact, solvable. We can get this information from the Gibbs phase rule. At the dew point or bubble point there are two components, two phases, and no chemical reactions, so there are

$$\mathcal{F} = 2 - 0 - 2 + 2 = 2$$

degrees of freedom. In this calculation we will fix either temperature [part (a)] or pressure [part (b)] as one degree of freedom, and then for each liquid-phase composition (the second degree of freedom) calculate the equilibrium conditions. Therefore, the problem is well posed and, in principle, solvable.

a. Using the Antoine equations given above and $T = 50°C = 323.15$ K, we have $P_5^{vap} = 1.564$ bar and $P_7^{vap} = 0.188$ bar. To calculate the equilibrium pressure at each liquid pentane composition x_5, we use

$$P(x_5) = x_5 \cdot P_5^{vap} + x_7 \cdot P_7^{vap} = x_5 \cdot P_5^{vap} + (1 - x_5) \cdot P_7^{vap}$$

and then calculate the vapor phase composition from

$$y_5 = \frac{x_5 \cdot P_5^{vap}}{P(x_5)}$$

Figure a is a plot of the vapor composition versus the liquid composition (that is, an x-y plot) at constant temperature, and Fig. b shows pressure as a function of both the vapor and liquid compositions on a single plot.

b. This calculation is a slightly more complicated, since the equilibrium temperature at each composition is not known, and the vapor pressures are nonlinear functions of temperature. Therefore, at each choice of liquid composition, the following equation must be solved for temperature:

$$x_5 \cdot P_5^{vap}(T) + x_7 \cdot P_7^{vap}(T) = x_5 \cdot P_5^{vap}(T) + (1 - x_5) \cdot P_7^{vap}(T) = 1.013 \text{ bar}$$

The procedure is that at each liquid composition x_5 a guess is made for the equilibrium temperature T, and the the equilibrium pressure is then calculated. If the calculated pressure equals 1.013 bar, the guessed temperature is correct and the vapor composition is computed from

$$y_5 = \frac{x_5 \cdot P_5^{vap}}{1.013}$$

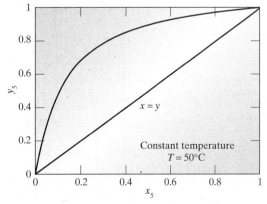

Figure a The x-y diagram for the n-pentane + n-heptane mixture at $T = 50°C$.

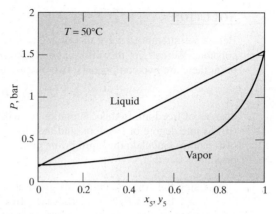

Figure b The P-x-y diagram for the n-pentane + n-heptane mixture at $T = 50°\text{C}$.

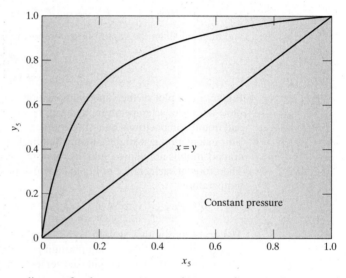

Figure c The x-y diagram for the n-pentane + n-heptane mixture at $P = 1.013$ bar.

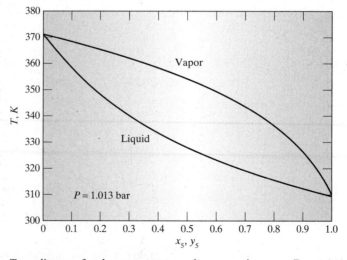

Figure d The T-x-y diagram for the n-pentane + n-heptane mixture at $P = 1.013$ bar.

However, if the calculated pressure is greater than 1.013 bar, a lower temperature is guessed and the calculation repeated, whereas if the calculated pressure is too low, a higher temperature is tried. Figure c is a plot of the vapor composition versus the liquid composition at constant pressure (another x-y plot), calculated in this way, and Fig. d gives the equilibrium temperature as a function of both the vapor and liquid compositions on a single plot.

COMMENT

It would be useful at this point for the reader to compare figures a and c, and figures b and d, and to understand the difference between them—in particular, to understand why the liquid region is at the top part of figure b and at the bottom part of figure d.

Also note that in the x-y diagrams of figures a and c, the equilibrium line relating the composition of the liquid to the composition of the vapor is the curved line. The straight $x = y$ lines in these figures, that is, the 45° lines, are usually added to x-y diagrams to indicate the difference between the vapor and liquid compositions. The difference between the equilibrium and $x = y$ lines is an indication of how easy or difficult it will be to separate the components by distillation. This will be discussed briefly later in this section.

∎

Few liquid mixtures are ideal, so vapor-liquid equilibrium calculations can be more complicated than is the case for the hexane-triethylamine system, and the system phase diagrams can be more structured than Fig. 10.1-6. These complications arise from the (nonlinear) composition dependence of the species activity coefficients. For example, as a result of the composition dependence of γ_i, the equilibrium pressure in a fixed-temperature experiment will no longer be a linear function of mole fraction. Thus nonideal solutions exhibit deviations from Raoult's law. We will discuss this in detail in the following sections of this chapter. However, first, to illustrate the concepts and some of the types of calculations that arise in vapor-liquid equilibrium in the simplest way, we will assume ideal vapor and liquid solutions (Raoult's law) here, and then in Sec. 10.2 consider the calculations for the more difficult case of nonideal solutions.

For a binary (that is, two-component) mixture, if constant-pressure vapor-liquid equilibrium diagrams, such as Fig. 10.1-4 or that of Illustration 10.1-1, have been previously prepared, dewpoint and bubblepoint temperatures can easily be read from these diagrams. For the cases in which such information is not available, or if a multicomponent mixture is of interest, the trial-and-error procedure of Illustration 10.1-2 is used to estimate these temperatures.

ILLUSTRATION 10.1-2

Estimation of Dew Point and Bubble Point Temperatures

Estimate the bubble point and dew point temperatures of a 25 mol % n-pentane, 45 mol % n-hexane, and 30 mol % n-heptane mixture at 1.013 bar.

Data:

$$\ln P_5^{\mathrm{vap}} = 10.422 - \frac{26\,799}{RT} \qquad \delta_5 = 7.02\,(\mathrm{cal/cc})^{1/2}$$

$$\ln P_6^{\mathrm{vap}} = 10.456 - \frac{29\,676}{RT} \qquad \delta_6 = 7.27\,(\mathrm{cal/cc})^{1/2}$$

$$\ln P_7^{\mathrm{vap}} = 11.431 - \frac{35\,200}{RT} \qquad \delta_7 = 7.43\,(\mathrm{cal/cc})^{1/2}$$

for P in bar, T in K, and $R = 8.314$ J/(mol K). The subscripts 5, 6, and 7 designate pentane, hexane, and heptane, respectively.

SOLUTION

Before solving this problem it is again useful to check whether this problem is, in fact, solvable. We can get this information from the Gibbs phase rule. At the dew point or bubble point there are three components, two phases, and no chemical reactions, so there are

$$\mathcal{F} = 3 - 0 - 2 + 2 = 3$$

degrees of freedom. Since the pressure and two independent mole fractions of one phase have been fixed, the problem is well posed and, in principle, solvable.

Since the solubility parameters for these hydrocarbons are sufficiently close, we will assume that this liquid mixture is ideal; that is, $\bar{f}_i^L = x_i P_i^{vap}$. (This assumption should be reasonably accurate here, and simplifies the calculations.) Finally, since the pressure is so low, we will assume the vapor phase is ideal, so that $\bar{f}_i^V = y_i P$. Therefore, the equilibrium relation for each species i is $x_i P_i^{vap} = y_i P$.

a. At the bubble point of the liquid mixture, Eqs. 10.1-3, 10.1-4, and 10.1-6 must be satisfied. Therefore, the procedure is to
  i. Pick a trial value of the bubble point temperature.
  ii. Compute the values of the y_i from

$$y_i = \frac{x_i P_i^{vap}}{P}$$

  iii. If $\sum y_i = 1$, the trial value of T is the bubble point temperature. If $\sum y_i > 1$, T is too high, and if $\sum y_i < 1$, T is too low; in either case, adjust the value of T and go back to step ii.
  Following this calculational procedure, we find

$$T(\text{bubble point}) = 334.6 \text{ K}$$
$$y_5 = 0.554$$
$$y_6 = 0.359$$
$$y_7 = 0.087$$

b. To find the dew point of the vapor mixture, we
  i. Pick a trial value for the dew point temperature.
  ii. Compute the values of the liquid-phase composition from

$$x_i = \frac{y_i P}{P_i^{vap}}$$

  iii. If $\sum x_i = 1$, the trial value of T is the dew-point temperature. If $\sum x_i > 1$, T is too low, and if $\sum x_i < 1$, T is too high; in either case, adjust the value of T and go back to step ii.
  In this case we find

$$T \text{ (dew point)} = 350.5 \text{ } K$$
$$x_5 = 0.073$$
$$x_6 = 0.347$$
$$x_7 = 0.580$$

COMMENT

Note that the dew point and bubble point temperatures differ by 16 K for this mixture. ∎

The next illustration demonstrates that the calculation of the bubble point and dew point pressures, since temperature and therefore pure-component vapor pressures are fixed, is somewhat easier than the computation of bubble point and dew point temperatures as was done above.

ILLUSTRATION 10.1-3

Estimation of Dew Point and Bubble Point Pressures

Estimate the bubble point and dew point pressures for the mixture of Illustration 10.1-2 at 73°C.

SOLUTION

a. At the bubble point pressure of the liquid mixture Eqs. 10.1-3, 10.1-4, and 10.1-6 must be satisfied. However, since this mixture satisfies Raoult's law, the calculation of the bubble point pressure is

$$P \text{ (bubble point)} = \sum_i x_i P_i^{\text{vap}} = x_5 P_5^{\text{vap}} + x_6 P_6^{\text{vap}} + x_7 P_7^{\text{vap}}$$

$$P \text{ (bubble point)} = 0.25 \times 3.034 + 0.45 \times 1.155 + 0.30 \times 0.449 = 1.413 \text{ bar}$$

Then the vapor-phase mole fractions y_i are computed from

$$y_i = \frac{P_i^{\text{vap}}}{P} x_i$$

giving $y_5 = 0.537$, $y_6 = 0.368$, and $y_7 = 0.095$.

b. To find the dew point pressure of the vapor mixture is a bit more complicated, so we
   i. Pick a trial value for the dew point pressure.
   ii. Compute the values of the liquid-phase composition from

$$x_i = y_i \frac{P}{P_i^{\text{vap}}}$$

   iii. If $\sum x_i = 1$, the trial value of P is the dew point pressure of the vapor mixture. If $\sum x_i > 1$, P is too high, and if $\sum x_i < 1$, P is too low; in either case, adjust the value of P and go back to step ii.
   In this case we find

$$P \text{ (dew point)} = 0.877 \text{ bar}$$
$$x_5 = 0.072$$
$$x_6 = 0.342$$
$$x_7 = 0.586$$

COMMENT

Note that at fixed temperature a pure fluid boils and will completely evaporate at a single pressure; for example, water at 1.013 bar boils at 100°C. Therefore, its dew point pressure and bubble point pressure are identical, and we do not use these terms for a pure fluid. However, for the mixture and temperature considered here, the pressures of initial boiling and condensation differ by 0.536 bar. ■

Flash calculations: mass balances for VLE

Another type of vapor-liquid equilibrium problem, and one that is more important for designing separation equipment, is computing the two-phase equilibrium state when either a liquid of known composition is partially vaporized or a vapor is partially condensed as a result of a change in temperature and/or pressure. Such a problem is generically referred to as a **flash calculation**. The term *flash* arises from the fact that if the pressure is suddenly lowered (or the temperature raised) on a mixture that is at its boiling temperature, it will flash-vaporize; that is, there will be a sudden partial vaporization of the liquid. The partial vaporization or partial condensation problem is somewhat more difficult to solve than bubble point and dew point calculations for the following reason: In a bubble point calculation an infinitesimal amount of vapor is produced, so the liquid composition is that of the original mixture; and in a dew point calculation, an infinitesimal amount of liquid forms, so the vapor composition is that of the original mixture. In either case, we know the equilibrium composition of one of the phases. However, in a flash calculation, the final compositions of both phases are unknown. If the flash process occurs at constant temperature (for example, in a heat exchanger), one uses the equilibrium criterion, Eq. 10.1-1, the restrictions of Eqs. 10.1-5 and 10.1-6, and the species mass balance equations. If the flash process does not take place at constant temperature (for example, if there were a sudden rupture of a tank or for flow through an orifice the flash vaporization may occur adiabatically), one would also have to include the energy balance in the calculation, which makes the solution much more tedious. We consider only the constant-temperature (isothermal) case here.

For a process in which 1 mole of a mixture with species mole fractions $z_{1,\mathrm{F}}$, $z_{2,\mathrm{F}}$, ..., $z_{\mathcal{C},\mathrm{F}}$ is, by partial vaporization or condensation, separated into L moles of liquid of composition $x_1, x_2, \ldots, x_{\mathcal{C}}$ and V moles of vapor of composition $y_1, y_2, \ldots, y_{\mathcal{C}}$, the species mass balance yields

$$\boxed{x_i L + y_i V = z_{i,\mathrm{F}} \quad i = 1, 2, \ldots, \mathcal{C}} \tag{10.1-7}$$

(since no chemical reactions occur). From the total mass balance, we also have

$$\boxed{L + V = 1} \tag{10.1-8}$$

though this is not an *independent* equation, since it can be obtained by summing Eq. 10.1-7 over all species, and using Eqs. 10.1-5 and 10.1-6.

Equations 10.1-7 and 10.1-8, together with the equilibrium relations, can be used to solve problems involving partial vaporization and condensation processes at constant temperature. For partial vaporization and condensation processes that occur adiabatically, the final temperature of the vapor-liquid mixture is also unknown and must be found as part of the solution. This is done by including the energy balance among the equations to be solved. Since the isothermal partial vaporization or isothermal flash calculation is already tedious (see Illustration 10.1-4), the adiabatic partial vaporization (or adiabatic flash) problem will not be considered here.[3]

ILLUSTRATION 10.1-4

Partial Vaporization Calculation

A liquid mixture of 25 mol % *n*-pentane, 45 mol % *n*-hexane, and 30 mol % *n*-heptane, initially at 69°C and a high pressure, is partially vaporized by isothermally lowering the pressure to 1.013 bar (1 atm). Find the relative amounts of vapor and liquid in equilibrium and their compositions.

---

[3]Flash vaporization processes are usually considered in courses and books on mass transfer processes and stage-wise operations. See, for example, R. E. Treybal, *Mass Transfer Operations,* 3rd ed., McGraw-Hill, New York (1980), p. 363 et seq.; and C. J. King, *Separation Processes,* 2nd ed., McGraw-Hill, New York (1980), pp. 68–90.

SOLUTION

From the Antoine equation data in Illustration 10.1-2, we have

$$P_5^{\text{vap}} = 2.721 \text{ bar} \quad P_6^{\text{vap}} = 1.024 \text{ bar} \quad \text{and} \quad P_7^{\text{vap}} = 0.389 \text{ bar}$$

Also, using the simplifications for this system introduced in the previous illustrations, the equilibrium relation $\bar{f}_i^L = \bar{f}_i^V$ reduces to $x_i P_i^{\text{vap}} = y_i P$, or

$$\frac{y_i}{x_i} = \frac{P_i^{\text{vap}}}{P}$$

For convenience, we will use the **K factor**, defined by the relation $y_i = K_i x_i$, in the calculations; for the ideal mixtures considered here $K_i = P_i^{\text{vap}}(T)/P$. Thus we obtain the following three equations:

$$y_5 = x_5 K_5 \quad \text{where} \quad K_5 = 2.7406 \tag{1}$$

$$y_6 = x_6 K_6 \qquad\qquad K_6 = 1.0109 \tag{2}$$

$$y_7 = x_7 K_7 \qquad\qquad K_7 = 0.3844 \tag{3}$$

We also have, from Eqs. 10.1-5–10.1-8,

$$x_5 + x_6 + x_7 = 1 \tag{4}$$

$$y_5 + y_6 + y_7 = 1 \tag{5}$$

and

$$x_5 L + y_5 V = 0.25 \tag{6}$$

$$x_6 L + y_6 V = 0.45 \tag{7}$$

$$x_7 L + y_7 V = 0.30 \tag{8}$$

$$L + V = 1.0 \tag{9}$$

Thus we have eight independent equations for eight unknowns (x_5, x_6, x_7, y_5, y_6, y_7, L, and V), and any numerical procedure for solving algebraic equations may be used to solve this set of equations. One method is to use Eqs. 1, 2, and 3 to eliminate the vapor-phase mole fractions and the overall mass balance, Eq. 9, to eliminate the amount of vapor. In this way the eight algebraic equations are reduced to five linear algebraic equations:

$$x_5 + x_6 + x_7 = 1 \tag{4}$$

$$K_5 x_5 + K_6 x_6 + K_7 x_7 = 1 \tag{5'}$$

$$x_5[L + K_5(1 - L)] = x_5[L(1 - K_5) + K_5] = 0.25 \tag{6'}$$

$$x_6[L(1 - K_6) + K_6] = 0.45 \tag{7'}$$

$$x_7[L(1 - K_7) + K_7] = 0.30 \tag{8'}$$

These equations are most easily solved by trial and error. In particular, a value of L is guessed and then used in Eqs. 6'–8' to compute x_5, x_6, and x_7. These trial values of the liquid-phase mole fractions are then tested in Eqs. 4 and 5'. If those equations are satisfied, the guessed value of L and the computed values of the x_i's are correct, and the vapor-phase mole fractions can be computed from Eqs. 1–3. If Eqs. 4 and 5' are not satisfied, a new guess for L is made, and the procedure repeated. Alternatively, one can use an equation-solving computer program such as MATHCAD or MATLAB.

The solution is

$$L = 0.564 \qquad V = 0.436$$
$$x_5 = 0.142 \qquad y_5 = 0.390$$
$$x_6 = 0.448 \qquad y_6 = 0.453$$
$$x_7 = 0.410 \qquad y_7 = 0.158$$

COMMENTS

1. The K-factor formulation introduced in this calculation is frequently useful in solving vapor-liquid equilibrium problems. The procedure is easily generalized to nonideal liquid and vapor phases as follows:

$$\frac{y_i}{x_i} \equiv K_i = \frac{\gamma_i P_i^{\mathrm{vap}}}{\bar{\phi}_i P}$$

In this case K_i is a nonlinear function of the liquid-phase mole fraction through the activity coefficient γ_i, and also a function of the vapor-phase composition through the fugacity coefficient $\bar{\phi}_i$. This makes solving the equations much more difficult.

2. It was not necessary to assume ideal solution behavior to solve this problem. One could, for example, assume that the solution is regular, in which case γ_i (and K_i) would be a nonlinear function of the mole fractions. The calculation of the vapor- and liquid-phase mole fractions is then more complicated than was the case here; however, the basis of the calculation, the equality of the fugacity of each species in both phases, remains unchanged.

3. In some cases it may not be possible to find an *acceptable* solution to the algebraic equations. Here by *acceptable* we mean a solution such that each mole fraction, L, and V are each greater than 0 and less than 1. This difficulty occurs when the vapor and liquid phases cannot coexist in the equilibrium state at the specified conditions. For example, if the flash vaporization temperature were sufficiently high or the total pressure so low that all the K_i's were greater than 1, there would be no set of mole fractions that satisfies both Eqs. 4 and 5'. In this case only the vapor is present. Similarly, if all the K_i's are less than 1 (which occurs at low temperatures or high pressures), only the liquid is present, and again there is no acceptable solution to the equations. It is also possible that there will not be an acceptable solution even with a distribution of values of the K factors if some, but not all, their values are much greater than or much less than unity.

4. For this system $\mathcal{C} = 3$, $\mathcal{P} = 2$, and $\mathcal{M} = 0$, so that, from the Gibbs phase rule, the number of degrees of freedom is

$$\mathcal{F} = 3 - 0 - 2 + 2 = 3$$

Since the equilibrium temperature and pressure were specified, one degree of freedom remains. If no further information about the system were given, that is, if one were asked to determine the equilibrium compositions of vapor and liquid for a pentane–hexane–heptane mixture at 69°C and 1.013 bar with no other restrictions, many different vapor and liquid compositions would be solutions to the problem. A problem that does not have a unique solution is said to be ill posed. With the initial liquid composition given, however, the species mass balances (Eqs. 6, 7, and 8) provide the additional equations that must be satisfied to ensure that there will be no more than one solution to the problem. ∎

ILLUSTRATION 10.1-5

Partial Equilibrium Vaporization Calculation and its Relation to Separation Processes

A liquid mixture of 50 mol % n-pentane and 50 mol % n-heptane, initially at a low temperature, is partially vaporized by heating at a constant pressure of 1.013 bar (1 atm). Find the equilibrium vapor and liquid compositions and the equilibrium temperature as a function of the fraction that is vaporized.

SOLUTION

Here we have for the equilibrium conditions

$$y_i = x_i \frac{P_i^{vap}(T)}{P(x_5, x_7, T)}$$

where $P_i^{vap}(T)$ is obtained from the Antoine equation data in Illustration 10.1-1. From Raoult's law, the total pressure is

$$P(x_5, x_7, T) = x_5 P_5^{vap}(T) + x_7 P_7^{vap}(T)$$

We also have, from Eqs. 10.1-5–10.1-8,

$$x_5 + x_7 = 1$$
$$y_5 + y_7 = 1$$

and

$$x_5 L + y_5 V = 0.50$$
$$x_7 L + y_7 V = 0.50$$
$$L + V = 1.0$$

With pressure fixed, for each value of the fraction of liquid L, the equilibrium temperature and the compositions in vapor and liquid phases can be computed by iteration. The results are given below and shown in Fig. 10.1-7. In this diagram each of the horizontal tie lines shown connecting the vapor and liquid compositions is labeled with its equilibrium temperature. Note that the first bubble of vapor occurs at 327.8 K and has an n-pentane mole fraction of 0.888. As the temperature increases, more of the liquid phase evaporates, and each of the phases becomes increasing more concentrated in n-heptane and less concentrated in n-pentane. Of course, when all the liquid has evaporated ($L = 0$), the vapor will be of the same composition as the initial liquid. Also, the end points of this figure at $L = 0$ and $L = 1$ in fact can be computed somewhat more easily from bubble point and dew point calculations, respectively.

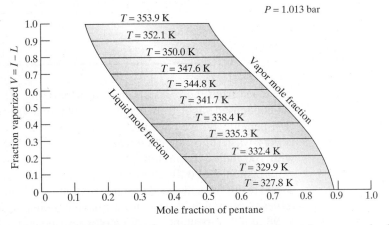

Figure 10.1-7 The fraction of an equimolar n-pentane–n-heptane mixture vaporized at fixed pressure, and the composition of the coexisting phases as a function of temperature.

	$L = 1.0$	$L = 0.9$	$L = 0.8$	$L = 0.7$
x_5	0.500	0.459	0.414	0.367
y_5	0.888	0.869	0.843	0.811
T (K)	327.8	329.9	332.4	335.3
	$L = 0.6$	$L = 0.5$	$L = 0.4$	$L = 0.3$
x_5	0.319	0.274	0.235	0.202
y_5	0.772	0.726	0.677	0.628
T (K)	338.4	341.7	344.8	347.6
	$L = 0.2$	$L = 0.1$	$L = 0.0$	
x_5	0.175	0.154	0.136	
y_5	0.581	0.538	0.500	
T (K)	350.0	352.1	353.9	

■

These results, while specific to this system, have some interesting implications for the purification of mixtures. For example, suppose that by starting with a liquid mixture of 50 mol % n-pentane and 50 mol % n-heptane, we wanted to produce a liquid mixture that contained 98 mol % n-pentane. One way to do this would be to vaporize some of the initial mixture, collect the vapor, and then condense it. However, we see from the results above that the highest concentration of n-pentane we could get in this way is 88.8 mole %, but only if we got an infinitesimal amount of vapor (i.e., $L = 1$). Vaporizing any greater fraction of liquid produces a vapor of lower concentration, as seen above.

Another possibility is to use a two-stage process in which we vaporize some of the liquid to get a vapor enriched in n-pentane, condense this liquid, and then partially vaporize it to produce a vapor that has even a higher concentration of n-pentane. For example, if we vaporized just 10 mol % of the original liquid ($L = 0.9$), we would obtain a vapor containing 86.9 mol % n-pentane. Now condensing this stream to a liquid and using it as the feed to a second partial vaporization process, repeating the calculation above with this new feed, we obtain

	$L = 1.0$	$L = 0.9$	$L = 0.8$	$L = 0.7$
x_5	0.869	0.856	0.841	0.823
y_5	0.982	0.982	0.980	0.977
T (K)	313.4	313.8	314.3	314.9

We see from these results that by vaporizing 20 mol % ($L = 0.8$ for stage 2) of the condensed vapor from stage 1, we obtain a vapor that contains the desired 98 mol % n-pentane. So we have met the concentration specifications. However, this process is very wasteful of the chemicals and not energy efficient. In particular, if we started with 100 moles of the initial feed, only 10 moles would remain after the first partial vaporization, and only 2 moles of the product stream containing 98 mol % n-pentane would result from the second stage. So that by starting with 50 moles of n-pentane (100 moles × 50 mol % n-pentane) we have obtained a product that contains only 2 moles × 98 mol % n-pentane = 1.96 moles of n-pentane. That is, 48.04 moles of n-pentane have not been recovered in the process.

This suggests that a more efficient process of purification is needed than simple partial vaporization and condensation. Chemical engineers have devised a much more efficient method, multistage distillation. This subject is discussed in detail in courses on stagewise operations, unit operations, and/or mass transfer. We will give a very brief discussion of simple, multistage distillation here just to point out the importance of thermodynamics in chemical engineering design.

A tray-type multistage distillation column contains several essential elements. There are a number of trays in a vertical cylindrical column, a boiler at the bottom, and a condenser at the top. Each tray holds up some liquid, and vapor produced in the boiler (called the reboiler) passes up the column and through the liquid on each of the trays. In this process, the vapor reaches equilibrium with the liquid on each tray, and the vapor composition changes. When the vapor reaches the top of the column, it is condensed; some of the condensed vapor is withdrawn as the distillate or overhead product, and the remainder is returned to the distillation column as liquid. This liquid flows down the column from tray to tray and reaches equilibrium with the vapor passing through each tray. When the liquid reaches the reboiler at the bottom of the column, some of it is vaporized and provides the vapor for the distillation column, while the rest is removed as the bottoms product. The overall operation of a distillation column, as shown in Fig. 10.1-8, is that a feed enters the column somewhere between the reboiler and the condenser, and an overhead product containing predominantly the more volatile component and a bottoms product containing predominantly the less volatile component are withdrawn from the column. The important characteristic of a distillation column is that internally there is countercurrent flow between a downward flowing liquid, which is becoming increasingly rich in the less volatile component, and an upward flowing

Relevance of vapor-liquid equilibrium to distillation

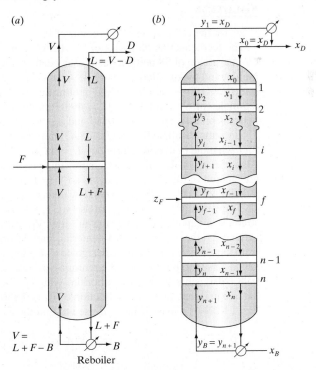

Figure 10.1-8 (*a*) Schematic diagram of a distillation column, showing (*b*) the tray-to-tray flows and compositions.

vapor that is becoming increasingly rich in the more volatile component. (A question that is sometimes asked is why some of the condensed vapor product must be returned to the column. The answer is that if this were not done, trays in the column would be dry, and vapor-liquid equilibrium would not be obtained.)

While the design of distillation columns can be quite complicated, we will consider only the simplest case here. The simplifications we will use are that vapor-liquid equilibrium will be assumed to exist on each tray (or equilibrium stage) and in the reboiler, that the column operates at constant pressure, that the feed is liquid and will enter the distillation column on a tray that has liquid of approximately the same composition as the feed, that the molar flow rate of vapor V is the same throughout the column, and that the liquid flow rate L is constant on all trays above the feed tray, and is constant and equal to $L + F$ below the feed tray, where F is the molar flow rate of the feed to the column, here assumed to be a liquid. The analysis of this simplified distillation column involves only the equilibrium relations and mass balances. This is demonstrated in the illustration below.

ILLUSTRATION 10.1-6

A Very Simple Design of a Distillation Column

Start with the the feed of the previous illustration, consisting of 50 mol % n-pentane and 50 mol % n-heptane, and recover 95 percent of the n-pentane in the feed in a stream that contains 98 mol % n-pentane. At the top of the column, 1 mole of product will be withdrawn for every 9 moles that are returned to the column, and the distillation column will operate at 1.013 bar pressure.

SOLUTION

Basis for the calculations: 100 mol/hr of feed.

The feed contains 50 moles of C_5. To meet the recovery target, $50 \times 0.95 = 47.5$ moles of C_5 in a stream of 98 percent purity is needed, so the distillate flow rate is $D = 47.5/0.98 = 48.47$ mol/hr, and the distillate contains $48.47 - 47.5 = 0.97$ moles of C_7. Therefore, the bottoms flow rate $B = 100 - 48.47 = 51.53$ mol/hr and by a mass balance on C_5 contains the remaining 2.5 moles of C_5 and $50 - 0.97 = 49.03$ mol/hr of C_7. Consequently, the mole fractions of the bottoms product are $x_{C5} = 0.0485$ and $x_{C7} = 0.9515$.

Since the distillate flow rate is 48.47 mol/hr, and 9 moles of liquid are returned to the column for each mole of overhead product, $L = 48.47 \times 9 = 436.23$ mol/hr and $L + F = 536.23$ mol/hr. Also, the vapor flow in the column V must equal the liquid flow at the top of the column plus the amount of distillate withdrawn, so $V = L + D = 436.23 + 48.47 = 484.70$ mol/hr.

With the overall column flows now determined, we next consider what happens on each tray. The top tray, which we refer to as tray 1, is shown schematically in Fig. 10.1-9. It has four streams: a vapor stream leaving the tray V_1, which is in equilibrium with the liquid stream leaving L_1; the vapor stream V_2 entering tray 1 from tray 2 below, and an entering liquid stream from the condenser, which we designate as L_0. On tray 1 we know that since all the vapor leaving is condensed to product and the returning or reflux liquid, the C_5 mole fraction of L_0 and V_1 are 0.98. Also, since L_1 is in equilibrium with V_1, from an equilibrium calculation at $P = 1.013$ bar, we find that the C_5 mole fraction of L_1 is 0.856 and that the equilibrium tray temperature is 313.8 K. Next a mass balance is used to find the C_5 mole fraction of V_2 as follows:

$$y_{C_{5,2}} V + x_{C_{5,0}} L = y_{C_{5,1}} V + x_{C_{5,1}} L$$

$$y_{C_{5,2}} \times 484.70 + 0.98 \times 436.23 = 0.98 \times 484.70 + 0.856 \times 436.23$$

So $y_{C_{5,2}} = 0.868$.

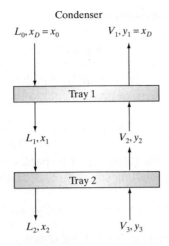

Condenser

$L_0, x_D = x_0$ $\qquad$ $V_1, y_1 = x_D$

Figure 10.1-9 Schematic diagram of top two trays in a distillation column. Note that the vapor and liquid *leaving* the same tray are assumed to be in equilibrium (that is, x_1 is in equilbrium with y_1, x_2 with y_2, etc.). Also, since there is a total condenser on this distillation column, the composition of the vapor leaving the the top tray, y_1, and the liquid returned to the column, x_0, both have the same composition as the distillate x_D.

Now since the liquid leaving tray 2 is in equilibrium with the vapor of composition, $y_{C_5,2} = 0.868$, a dew point calculation can be done to find the tray temperature and liquid composition. The results are $x_{C_5,2} = 0.458$ and the tray temperature is 329.9 K.

Next a mass balance is used to find the vapor composition entering tray 2 from tray 3 below. However, since the liquid composition on this tray is close to the feed composition, the 100 mol/hr should be added to this tray. Therefore, the mass balance is

$$y_{C_5,3} V + x_{C_5,1} L + x_{C_5,F} F = y_{C_5,2} V + x_{C_5,2} (L + F)$$

$$y_{C_5,3} \times 484.70 + 0.856 \times 436.23 + 0.500 \times 100 = 0.868 \times 484.70 + 0.458 \times 536.23$$

Therefore, $y_{C_5,3} = 0.501$.

As the liquid leaving tray 3 is in equilibrium with the vapor of composition, $y_{C_5,3} = 0.501$, a dew point calculation is used find the tray temperature and liquid composition. The results are $x_{C_5,3} = 0.137$ and the tray temperature is 353.8 K.

Proceeding in this manner, a mass balance is then used to find the vapor composition entering tray 3 from tray 4 below.

$$y_{C_5,4} V + x_{C_5,2} (L + F) = y_{C_5,3} V + x_{C_5,3} (L + F)$$

$$y_{C_5,3} \times 484.70 + 0.458 \times 536.23 = 0.501 \times 484.70 + 0.137 \times 536.23$$

Therefore, $y_{C_5,4} = 0.146$ and from an equilibrium calculation the liquid leaving tray 4 is $x_{C_5,4} = 0.029$ and the tray temperature is 366.7 K. As this composition is lower than the specification for the bottoms product of $x_{C5} = 0.0485$, no further stages are needed in the distillation column. Also, it should be pointed out that while the condenser is used to condense all the vapor, the reboiler vaporizes only part of the liquid. Therefore, it is an equilibrium stage, just as are each of the trays. Consequently, the distillation column we have just "designed" requires a total condenser and 4 "trays" or equilibrium stages, which actually consists of three trays and a reboiler. Also, the feed is to be added to the second tray from the top of the distillation column.∎

Alternative distillation calculation

There is another method of analyzing distillation colums. This is done by starting with the overall mass (molar) balance on the distillation column,

$$F = D + B \qquad\qquad \textbf{(10.1-9)}$$

and on either one of the species in a binary mixture:

$$z_F F = x_D D + x_B B \tag{10.1-10}$$

Using these equations, we can relate the amount of distillate D and bottoms B to the amount of feed F and the feed and desired product compositions as follows:

$$D = \frac{z_F - x_B}{x_D - x_B} F \quad \text{and} \quad B = \frac{x_D - z_F}{x_D - x_B} F \tag{10.1-11}$$

Next we do both an overall molar balance and a species molar balance on just a section of the distillation column including stages 1, 2, ..., i above the feed tray:

$$V = L + D \quad \text{and} \quad y_{i+1} V = x_i L + x_D D \tag{10.1-12}$$

Using the definition of the reflux ratio $q = L/D$, we obtain

$$y_{i+1} = \frac{x_D D}{L + D} + \frac{x_i L}{L + D} = \frac{x_D}{1 + q} + \frac{x_i q}{1 + q} \tag{10.1-13}$$

Note that this is a linear equation between the vapor composition entering stage i, y_{i+1}, and the liquid composition leaving that stage, x_i. Also, since the condenser being considered here condenses all of the vapor leaving stage 1, $y_1 = x_D$. Therefore, Eq. 10.1-13 results in a line that starts at x_D and has a slope of $q/(q + 1)$. This line is referred to as the upper operating line, or rectification section operating line, and can be plotted on an x-y diagram, as will be shown in the illustration that follows.

Finally, we do an overall molar balance and a species molar balance on a section of the distillation column including stages below the feed tray (assuming the feed is all liquid) up to stage I:

$$L + F = V + B \quad \text{and} \quad x_{I+1}(L + F) = y_I V + x_B B \tag{10.1-14}$$

where we have used the subscript I to indicate the equilibrium stages below the feed tray. Using the definition of the reflux ratio and the overall mass balance, we obtain

$$y_I = x_{I+1}\left(\frac{q + \frac{F}{D}}{q + 1}\right) - x_B\left(\frac{\frac{F}{D} - 1}{q + 1}\right) \tag{10.1-15}$$

This is also a linear equation between the vapor composition leaving stage I, y_I, and the liquid composition entering that stage from the stage above, x_{I+1}. This equation is referred to as the lower operating line, or stripping section operating line. To plot this equation, we start at the point $x_I = x_B$ and continue it with a slope of $(q + F/D)/(q + 1)$. Since only a portion of the feed exits as distillate (so that $F > D$), this line has a slope greater than unity, as $q + F/D > q + 1$. The following illustration shows how the stripping section and rectification section operating lines can be used in a simple graphical construction to determine the number of equilibrium stages needed to accomplish a separation by distillation of a binary mixture.

(Note that in the analysis here we have assumed that the feed is at its boiling point. If it were below its boiling point, it would condense some of the vapor, so that the vapor flow rate above the feed tray would be less than that below the feed tray. Similarly, if the feed were a two-phase mixture, or a vapor, the vapor and liquid flows would be different from those given here. However, we leave these complications to a course in mass transfer or stagewise operations, which usually follows this course in thermodynamics.)

ILLUSTRATION 10.1-7

Graphical Design of a Distillation Column

Start with the the feed of the previous illustration, consisting of 50 mol % *n*-pentane and 50 mol % *n*-heptane, and determine how many equilibrium stages are needed to recover 95 percent of the *n*-pentane in the feed in a stream that contains 98 mol % *n*-pentane. At the top of the column, 1 mole of product will be withdrawn for every 9 moles that are returned to the column, and the distillation column will operate at 1.013 bar pressure.

SOLUTION

Figure 10.1-10 shows the constant pressure x-y plot for this system. For the example considered here, $q = L/D = 9/1 = 9$. Consequently, the rectification section operating line starts at $x = 0.98$ and has a slope of $q/(q + 1) = 9/(9 + 1) = 0.9$. The stripping section operating line starts at $x = 0.0485$ and has a slope of $(q + F/D)/(q + 1) = (9 + 100/48.47)/(9 + 1) = (9 + 2.063)/10 = 1.106$. Both of these operating lines are drawn on the x-y diagram. Note that these operating lines cross at the feed composition.

The rectification operating line relates the composition of the vapor *entering* an equilibrium stage (from the stage below) to the liquid *leaving* that stage, while the equilibrium line relates the composition of the equilibrium liquid leaving the stage to the equilibrium vapor also leaving that stage and entering the one above it (or the condenser, if it is the top stage). Thus, starting with the vapor composition at the top stage, which is equal to the liquid distillate concentration (that is, the point $x_D = y_1$), a horizontal line is drawn until it intersects the equilibrium curve. This intersection gives the liquid of composition x_1 in equilibrium with y_1. Now drawing a vertical line from this point until it intersects the stripping section operating line gives the vapor composition y_2 entering stage 1. Another horizontal line then gives the liquid composition x_2 in equilibrium with this vapor, and so on. This graphical stage-to-stage construction is repeated until a liquid composition equal to or less than the feed composition is reached. The optimal stage at which to inject feed is the one that has a liquid composition closest to that of the feed. Therefore, once a liquid stage composition is less than that of the feed, we identify that as the

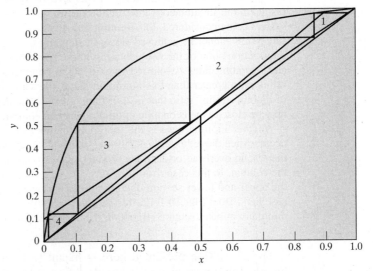

Figure 10.1-10 The graphical stage-to-stage calculation for a distillation column.

feed stage and continue the graphical construction using the stripping (lower) operating line until a composition equal to or less than the desired bottoms composition is reached.

This graphical construction is shown in Fig. 10.1-10. From this figure, we see as before that the desired separation can be achieved in a total of four equilibrium stages, one of which is the reboiler, and that the feed should be on the second stage from the top of the column. The temperature on each tray, the condenser, and the reboiler can now be determined from the T-x-y diagram for this mixture using the liquid composition in each of these locations.

COMMENTS

Note the difference between this method of calculation and the one used in the previous illustration. There we did vapor-liquid equilibrium calculations only for the conditions needed, and then solved the mass balance equations analytically. In this illustration we first had to do vapor-liquid equilibrium calculations for all compositions (to construct the x-y diagram), and then for this binary mixture we were able to do all further calculations graphically. As shown in the following discussion, this makes it easier to consider other reflux ratios than the one used in this illustration.

The choice of reflux ratio $q = L/D$, that is, the ratio of the liquid returned from the condenser to distillate, was chosen arbitrarily to be 9 in this example. Suppose instead a value of 0.5 had been chosen. Then the rectification section operating line would still start at $x = 0.98$ but now have a slope of $q/(q+1) = 0.5/(0.5+1) = 0.333$. The stripping section operating line would again start at $x = 0.0485$ and have a slope of $(q + F/D)/(q+1) = (0.5+100/48.47)/(0.5+1) = 2.563/1.5 = 1.709$. The graphical stage-to-stage calculation for this case is shown below. In this case, we need a total of six equilibrium stages, with the feed on the third stage. Thus by decreasing the reflux ratio, we need two additional stages in the distillation column, which increases the capital cost. However, by decreasing the reflux ratio, for a fixed amount of product, less vapor and liquid are circulating through the column so a column of smaller diameter can be used, which decreases the capital cost. More important, less liquid is being vaporized in the reboiler and less vapor is being condensed in the condenser, resulting in a very significant decrease in utility costs. So by decreasing the reflux ratio, the cost of separation can be greatly reduced.

However, the reflux ratio cannot be reduced indefinitely. There is a value of the reflux ratio below which it is no longer possible to achieve the desired separation. This is the value of q that results in the simultaneous intersection of two operating lines and the equilibrium line. For the system being considered, this limiting reflux ratio is $q = 0.237$. At this reflux ratio the desired separation cannot be achieved since, in stepping of stages, it is not possible to get beyond the intersection point of the two operating lines. This point of intersection between the equilibrium and operating lines is referred to as a pinch point and is shown in Figure 10.1-11. For any reflux ratio greater than this minimum value, the separation is possible. As a (very) rough rule of thumb, a reflux ratio that is 20 percent greater than the minimum required for the separation may be close to the economic optimum between equipment and operating costs.

Another limiting case is the minimum number of stages required for a separation, which occurs when the reflux ratio is infinity, so that all the condensed vapor is returned to the column, there is no overhead or distillate product and therefore no bottoms product and therefore no feed to the column. In this case, from Eq. 10.1-12, $y_{i+1} = x_i$ and from Eq. 10.1-14, $x_{I+1} = y_I$, so that the upper and lower section operating lines are coincident with the $x = y$ or 45° line. For the case here, from Fig. 10.1-12, this results in slightly more than three equilibrium stages as the minimum number required to obtain the desired separation. ∎

The discussion presented here is meant to be a simple introduction merely to illustrate the importance of vapor-liquid equilibrium calculations in the design of distillation columns, and to provide motivation for the study of this subject. The actual

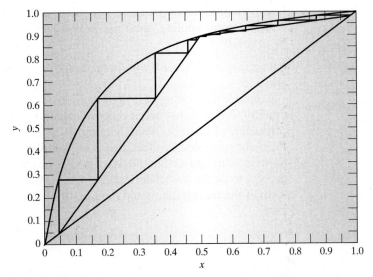

Figure 10.1-11 Stage-to-stage calculation for a distillation column operating at the minimum reflux ratio.

design of distillation columns is much more complicated than can be considered here. Complications include the following:

a. The mixtures involved are not ideal.

b. The assumption of constant molar flow rates of liquid and vapor in the rectifying and stripping sections is not valid, and the mass and energy balances are needed to determine the actual vapor and liquid flow rates on each tray.

c. Equilibrium may not be achieved on each tray (which can be approximately accounted for by an empirical tray efficiency correction factor).

d. The feed may be a liquid at saturation, a subcooled liquid, a vapor-liquid mixture, a vapor, or a subcooled vapor.

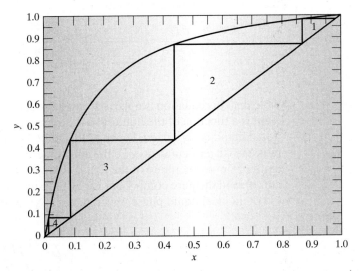

Figure 10.1-12 Stage-to-stage calculation for a distillation column operating at the infinite reflux ratio.

e. A partial rather than total condenser may be used.

f. Intertray cooling, multiple feeds, and other distillation configurations may be used.

All such issues are dealt with in textbooks on distillation, mass transfer, and unit operations.

Differential distillation

Consider the boiling of a liquid mixture in which the vapor that is produced is continually removed. In such a process, which is called a Rayleigh distillation, the concentration of the more volatile component in the liquid will continue to decrease as the boiling proceeds. In fact, this is the simplest type of batch distillation and has been used since antiquity. (Should we consider the earliest distillers of alcohol who used this process to be the first chemical engineers?) For this process the overall rate-of-change form of the mass balance and the rate-of-change form of the mass balance for species i are

$$\frac{dN}{dt} = \dot{N} \quad \text{and} \quad \frac{dN_i}{dt} = \dot{N}_i$$

Combining these equations, we have

$$\frac{dN_i}{dt} = \frac{d(Nx_i)}{dt} = x_i\frac{dN}{dt} + N\frac{dx_i}{dt} = \dot{N}_i = y_i\dot{N} = y_i\frac{dN}{dt}$$

which can be rewritten as

$$\frac{1}{N}\frac{dN}{dt} = \frac{1}{y_i - x_i}\frac{dx_i}{dt}$$

or more simply as

Rayleigh equation

$$\frac{dN}{N} = \frac{dx_i}{y_i - x_i} \tag{10.1-16}$$

which is known as the Rayleigh equation.

For generality, using the activity coefficient description of vapor-liquid equilibrium, the equation can be rewritten as

$$\frac{dN}{N} = \frac{dx_i}{y_i - x_i} = \frac{dx_i}{\dfrac{x_i \cdot \gamma_i\left(\underline{x}\right) P_i^{\mathrm{vap}}\left(T\right)}{P} - x_i} = \frac{dx_i}{x_i\left(\dfrac{\gamma_i\left(\underline{x}\right) P_i^{\mathrm{vap}}\left(T\right)}{P} - 1\right)}$$

Note that this equation is not easily integrated, for two reasons. First, the activity coefficient is a function of the liquid-phase composition, which continually changes as additional liquid is vaporized. Second, differential distillations are usually done at constant pressure (in particular, open to the atmosphere), so that as the composition changes, the equilibrium temperature of the liquid changes (following the bubble point temperature curve), and the pure component vapor pressures are a function of temperature.

For an ideal liquid phase, this reduces slightly to

$$\frac{dN}{N} = \frac{dx_i}{x_i\left(\dfrac{P_i^{\mathrm{vap}}\left(T\right)}{P} - 1\right)}$$

although the temperature dependence of the vapor pressure remains.

An alternative description of differential distillation for a binary mixture is to start from the mass balance on any one species i,

$$\frac{dN_1}{dt} = \dot{N}_1 = y_1\dot{N} = x_1\gamma_1(\underline{x})\,P_1^{\text{vap}}(T)\,\dot{N} = \frac{N_1}{N_1+N_2}\gamma_1(\underline{x})\,P_1^{\text{vap}}(T)\,\dot{N}$$

so that the ratio of the rates of change in the mole numbers of each species in a binary mixture is

$$\frac{dN_1}{dN_2} = \frac{\frac{N_1}{N_1+N_2}\gamma_1(\underline{x})\,P_1^{\text{vap}}(T)}{\frac{N_2}{N_1+N_2}\gamma_2(\underline{x})\,P_2^{\text{vap}}(T)} = \frac{N_1\gamma_1(\underline{x})\,P_1^{\text{vap}}(T)}{N_2\gamma_2(\underline{x})\,P_2^{\text{vap}}(T)} = \alpha_{12}(T,\underline{x})\frac{N_1}{N_2} \quad \textbf{(10.1-17)}$$

where $\alpha_{12}(T,\underline{x}) = [\gamma_1(\underline{x})P_1^{\text{vap}}(T)]/[\gamma_2(\underline{x})P_2^{\text{vap}}(T)]$ is the so-called **relative volatility** of the two components. For the special case where the relative volatility α is constant (which will occur only if the liquid is an ideal mixture in which all the activity coefficients are unity and then only if the ratio of the vapor pressures is a constant) this equation can be integrated to obtain

$$\alpha_{12}\ln\left[\frac{N_2(t)}{N_2(t=0)}\right] = \ln\left[\frac{N_1(t)}{N_1(t=0)}\right]$$

or

$$\frac{N_1(t)}{N_1(t=0)} = \left[\frac{N_2(t)}{N_2(t=0)}\right]^{\alpha_{12}} \quad (10.1\text{-}18)$$

ILLUSTRATION 10.1-8

Rayleigh Distillation

A differential distillation starting at 350 K is to be run on an equimolar mixture of n-heptane and n-octane until 50 percent of the original mixture remains as liquid. Assuming that the relative volatility of n-heptane to n-octane is approximately constant at a value of 2.35 over the temperature range of the distillation, what is the composition of the liquid mixture after 50 percent of the mixture has been vaporized?

SOLUTION

Basis of the calculation: 1 mole of the initial equimolar mixture.

The equations to be solved are

$$N_{C7}(t) + N_{C8}(t) = 0.5$$

and

$$\left(\frac{N_{C8}(t)}{0.5}\right)^{2.35} = \frac{N_{C7}(t)}{0.5}$$

The solution to these equations is $N_{C7} = 0.178$ mol, and $N_{C8} = 0.322$ mol. Therefore, the liquid mole fractions after half of the initial mixture has been vaporized are $x_{C7} = 0.356$ and $x_{C8} = 0.644$. ∎

PROBLEMS FOR SECTION 10.1

10.1-1 The following mixture of hydrocarbons is obtained as one stream in a petroleum refinery.

Component	Mol %	A	B
Ethane	5	817.08	4.402 229
Propane	10	1051.38	4.517 190
n-Butane	40	1267.56	4.617 679
2-Methyl propane	45	1183.44	4.474 013

The parameters A and B in this table are the constants in the equation

$$\log_{10} P^{\mathrm{vap}} = -\frac{A}{T} + B$$

where P^{vap} is the vapor pressure in bar and T is the temperature in K. These paraffinic hydrocarbons form an ideal mixture.

a. Compute the bubble point of the mixture at 5 bar.
b. Compute the dew point of the mixture at 5 bar.
c. Find the amounts and compositions of the vapor and liquid phases that would result if this mixture were isothermally flash-vaporized at 30°C from a high pressure to 5 bar.
d. Set up the equations to be used, and the information needed, to compute the amounts, compositions, and temperature of the vapor and liquid phases that would result if this mixture were *adiabatically* flash-vaporized from a high pressure and 50°C to 5 bar.

10.1-2 The binary mixture of benzene and ethylene chloride forms an ideal solution (i.e., one that obeys Raoult's law) at 49.99°C, as shown by the data of J. von Zawidzki [*Z. Phys. Chem.,* **35,** 129 (1900)]. At this temperature pure benzene boils at 0.357 bar, and pure ethylene chloride boils at 0.315 bar. Develop the analogues of Figs. 10.1-1 to 10.1-3 for this system.

10.1-3 **a.** Calculate the dew point pressure and liquid composition at $T = 69$°C for the mixture of Illustration 10.1-2.
b. Calculate the bubble point pressure and vapor composition at $T = 69$°C for the mixture of Illustration 10.1-2.

10.1-4 Compute the complete vapor-liquid phase behavior for the mixture of Illustration 10.1-2 at $T = 69$°C.

10.1-5 Compute the complete vapor-liquid phase behavior for the mixture of Illustration 10.1-2 at $P = 1.013$ bar.

10.1-6 Relative humidity is the ratio of the partial pressure of water in air to the partial pressure of water in air saturated with water at the same temperature, expressed as a percentage:

$$\text{Relative humidity} = \frac{\text{Partial pressure of water in air}}{\begin{array}{c}\text{Partial pressure of water in air}\\\text{saturated with water at same}\\\text{temperature}\end{array}} \times 100$$

Among aviators it is more common to express the moisture content of the air by giving the air temperature and its dew point, that is, the temperature to which the air must be cooled for the first drop of water to condense. The following atmospheric conditions have been reported:

$$\text{Atmospheric pressure} = 1.011 \text{ bar}$$
$$\text{Air temperature} = 25.6°\text{C}$$
$$\text{Dew point of air} = 20.6°\text{C}$$

What is the relative humidity of the air? (See Problem 7.12 for the necessary data.)

10.1-7 Into an evacuated 30-liter vessel is placed 0.1 mol of n-butane dissolved in 0.9 mol of n-hexane. If the vessel and its contents are kept at 298 K, what is the pressure and the vapor and liquid compositions when equilibrium is achieved? You may assume ideal liquid and vapor behavior.
Data at 298 K: The density of liquid n-butane is 0.575 g/cc and its vapor pressure is 2.428 bar. The density of liquid n-hexane is 0.655 g/cc and its vapor pressure is 0.200 bar.

10.1-8 A stream contains 55 mol % n-pentane, 25 mol % n-hexane, and 20 mol % n-heptane and is to be processed at 69°C. The following data are available.
Data: At 69°C $P_{C5}^{\mathrm{vap}} = 2.755$ bar, $P_{C6}^{\mathrm{vap}} = 1.021$ bar, and $P_{C7}^{\mathrm{vap}} = 0.390$ bar
a. What is the bubble point pressure of this mixture and the vapor composition that results?
b. What are the dew point pressure of this mixture and the liquid composition that results?

10.1-9 The mixture of the previous problem at 69°C is to be isothermally flashed.
a. What pressure will produce a liquid stream that contains one-tenth the number of moles of the feed, and what will be the equilibrium vapor and liquid compositions?
b. What pressure will produce a liquid stream that contains nine-tenths the number of moles of the feed, and what will be the equilibrium vapor and liquid compositions?

10.1-10 The mixture of Problem 10.1-8 at 69°C is to be isothermally flashed.

a. What pressure will produce a liquid stream that contains exactly one-half the number of moles of the feed, and what will be the equilibrium vapor and liquid compositions?

b. What pressure will produce a liquid stream with an n-pentane mole fraction of 0.30, what will be the fraction of the initial feed that is liquid at these conditions, and what will be the equilibrium compositions of the vapor and liquid streams?

10.1-11 Joe Udel wanted to do an isothermal flash on the mixture of Problem 10.1-8 at $69°C$ to produce a vapor that has an n-pentane mole fraction of 0.85, and a liquid-phase n-heptane mole fraction of 0.80. However, he could not find a pressure or vapor-liquid split that would result in the desired separation.

a. By doing a degrees-of-freedom analysis, explain the reason for Joe's failure.

b. What should Joe do to obtain the desired separation?

10.1-12 The lower flammability limit of a combustible material is defined as the partial pressure in an equilibrium vapor in air above the pure liquid that will produce a flammable mixture at a total pressure of 101.325 kPa. (If the partial pressure is below the lower flammability limit, there is insufficient combustible material to sustain combustion. There is also an upper flammability limit at which the vapor contains an amount of flammable material such that there is insufficient oxygen to maintain a flame.) If the lower flammability limit of refrigerant FC152a is 4.35 kPa, at what temperature of the pure liquid will this refrigerant produce a vapor that is flammable? The vapor pressure of FC152a (in MPa) is

$$\ln P^{\text{vap}} = 8.347 - \frac{2644}{T} \quad \text{for } T \text{ in K}$$

10.1-13 A mixture of benzene and 2,2,4-trimethyl pentane is to be purified using distillation. The liquid feed to the column is 10^5 mol/hr with a benzene mole fraction of 0.4 at $55°C$. The distillate is to contain benzene at a mole fraction of 0.99 with 98 percent of the benzene in the feed in this stream. The distillation column operates at 1.013 bar, and the distillate and bottoms products are saturated liquids. What are the amounts and compositions of the distillate and bottoms products?

10.2 LOW-PRESSURE VAPOR-LIQUID EQUILIBRIUM IN NONIDEAL MIXTURES

Since few liquid mixtures are ideal, vapor-liquid equilibrium calculations are somewhat more complicated than for the cases in the previous section, and the phase diagrams for nonideal systems can be more structured than Figs. 10.1-1 to 10.1-6. These complications arise from the (nonlinear) composition dependence of the species activity coefficients. For example, as a result of the composition dependence of γ_i, the vapor-liquid equilibrium pressure in a fixed-temperature experiment will no longer be a linear function of mole fraction, so that nonideal solutions exhibit deviations from Raoult's law. However, all the calculational methods discussed in the previous section for ideal mixtures, including distillation column design, can be used for nonideal mixtures, as long as the composition dependence of the activity coefficients is taken into account.

Several examples of experimental data for nonideal solutions are given in Fig. 10.2-1. It is easy to establish that if

$$P > \sum x_i P_i^{\text{vap}}$$

(see Fig. 10.2-1a and b), then $\gamma_i > 1$ for at least one of the species in the mixture, that is, $P_i > x_i P_i^{\text{vap}}$ (so that positive deviations from Raoult's law occur). Similarly, in real solutions

$$P < \sum x_i P_i^{\text{vap}}$$

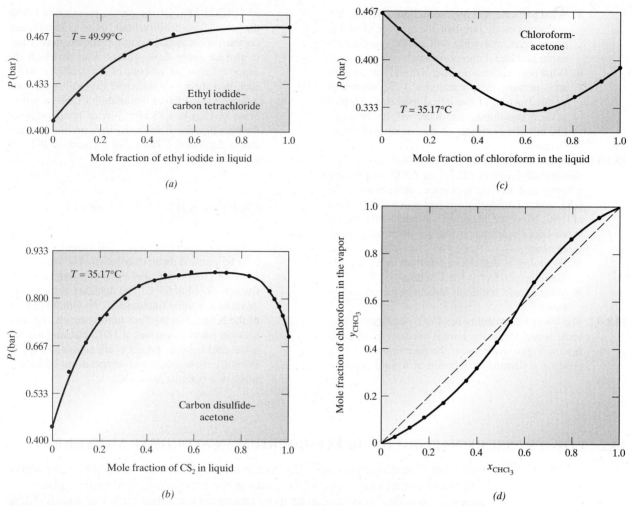

Figure 10.2-1 (*a*) Equilibrium total pressure versus composition for the ethyl iodine–carbon tetrachloride system at $T = 49.99°C$. (*b*) Equilibrium total pressure versus composition for the carbon disulfide–acetone system at $T = 35.17°C$. (*c*) Equilibrium total pressure versus composition for the chloroform-acetone system at $T = 35.17°C$. (*d*) The x-y diagram for the chloroform-acetone system at $T = 35.17°C$.

(see Fig. 10.2-1*c*) occurs when the activity coefficients for one or more of the species in the mixture are less than 1, that is, when there are negative deviations from Raoult's law.

Of special interest are binary mixtures in which the total pressure versus liquid composition curve exhibits an extremum (either a maximum or a minimum), as illustrated in Fig. 10.2-1*b* and *c*. To locate this extremum we set $(\partial P / \partial x_1)_T = 0$, keeping in mind that $x_2 = 1 - x_1$, to get

$$\gamma_1 P_1^{\mathrm{vap}} \left(1 + x_1 \frac{\partial \ln \gamma_1}{\partial x_1} \right) - \gamma_2 P_2^{\mathrm{vap}} \left(1 + x_2 \frac{\partial \ln \gamma_2}{\partial x_2} \right) = 0 \qquad \textbf{(10.2-1)}$$

To proceed further we note that the Gibbs-Duhem equation given in Eq. 9.3-15

$$0 = \frac{H^{\mathrm{ex}}}{T} \, dT - \underline{V}^{\mathrm{ex}} \, dP + RT \sum_{i=1}^{c} x_i \, d \ln \gamma_i$$

can be rewritten as

$$0 = \frac{\underline{H}^{\mathrm{ex}}}{T}\left(\frac{\partial T}{\partial x_1}\right)_T - \underline{V}^{\mathrm{ex}}\left(\frac{\partial P}{\partial x_1}\right)_T + RT\, x_1\left(\frac{\partial \ln \gamma_1}{\partial x_1}\right)_T + RT\, x_2\left(\frac{\partial \ln \gamma_2}{\partial x_1}\right)_T \quad \textbf{(10.2-2)}$$

Next, using $dx_1 = -dx_2$, and observing that $(\partial T/\partial x_1)_T = 0$ since temperature is fixed, and that $(\partial P/\partial x_1)_T = 0$ at an extreme value of the pressure, yields

$$x_1\left(\frac{\partial \ln \gamma_1}{\partial x_1}\right)_T + x_2\left(\frac{\partial \ln \gamma_2}{\partial x_1}\right)_T = 0 \quad \textbf{(10.2-3)}$$

Finally, using this expression in Eq. 10.2-1 and recognizing that $(1 + x_1\,\partial \ln \gamma_1/\partial x_1)_T \neq 0$ (the results of Problem 9.18 provide a verification of this) establishes that when $(\partial P/\partial x_1) = 0$,

$$\gamma_1 P_1^{\mathrm{vap}} - \gamma_2 P_2^{\mathrm{vap}} = 0 \quad \textbf{(10.2-4)}$$

or, from Eq. 10.1-1b,

$$\frac{y_1}{x_1} = \frac{\gamma_1 P_1^{\mathrm{vap}}}{P} = \frac{\gamma_2 P_2^{\mathrm{vap}}}{P} = \frac{y_2}{x_2} \quad \textbf{(10.2-5)}$$

This equation, together with the restrictions

$$x_1 + x_2 = 1 \quad \text{and} \quad y_1 + y_2 = 1$$

has the solution

$$\begin{aligned} x_1 &= y_1 \\ x_2 &= y_2 \end{aligned} \quad \textbf{(10.2-6)}$$

Azeotrope

Thus, at fixed temperature the occurrence of either a maximum or a minimum in the equilibrium pressure versus mole fraction curve at a given composition indicates that both the vapor and liquid are of this composition (this is indicated in Fig. 10.2-1d by the intersection of the x-y and $x = y$ curves). Such a vapor-liquid mixture is called an **azeotrope** or an **azeotropic mixture** and is of special interest (and annoyance) in distillation processes, as will be discussed later.

The occurrence of an azeotrope is also indicated by an interior extreme value of the equilibrium temperature on an equilibrium temperature versus composition curve at fixed pressure (see Problem 10.2-3). If the extreme value of temperature is a maximum (i.e., the vapor-liquid equilibrium temperature of the mixture is greater than the boiling point of either of the pure components at the chosen pressure), then the mixture is said to be a **maximum boiling** azeotrope; a maximum boiling azeotrope occurs when there are negative deviations from Raoult's law, that is, when the activity coefficient of one or more species is less than 1. Similarly, if the azeotropic temperature is below the boiling points of both pure components, the mixture is a **minimum boiling** azeotrope; the activity coefficient of at least one of the species in the mixture is greater than unity, and there are positive deviations from Raoult's law. This is the more common occurrence.

At a low-pressure azeotropic point $\underline{x}^{\mathrm{AZ}}$ the liquid-phase activity coefficients can be calculated from the relation

$$\gamma_i(\underline{x}^{\mathrm{AZ}}) = \frac{P}{P_i^{\mathrm{vap}}} \quad \textbf{(10.2-7)}$$

which results from Eqs. 10.2-5 and 10.2-6. Because of the simplicity of this result, azeotropic data are frequently used to evaluate liquid-phase activity coefficients. In particular, given the azeotropic composition, temperature, and pressure, one can calculate the liquid-phase activity coefficients at the azeotropic composition. This information can be used to obtain the values of the parameters in a liquid solution model (e.g., the van Laar model), which can then be used to calculate the complete pressure-composition and *x-y* diagrams for the system. This procedure is illustrated next.

ILLUSTRATION 10.2-1

Using Azeotropic Data to Predict VLE of a Binary Mixture

Benzene and cyclohexane form an azeotrope at 0.525 mole fraction benzene at a temperature of 77.6°C and a total pressure of 1.013 bar. At this temperature the vapor pressure of pure benzene is 0.993 bar, and that of pure cyclohexane is 0.980 bar.

a. Using the van Laar model, estimate the activity coefficients of benzene and cyclohexane over the whole composition range. Use this activity coefficient information to compute the equilibrium pressure versus liquid composition diagram and the equilibrium vapor composition versus liquid composition diagram at 77.6°C.

b. Make predictions for the activity coefficients of benzene and cyclohexane using regular solution theory, and compare these with the results obtained in part (a).

SOLUTION

a. *The van Laar model.* The starting point is the equilibrium relation

$$\bar{f}_i^L(T, P, \underline{x}) = \bar{f}_i^V(T, P, \underline{y})$$

which, at the pressures here, reduces to

$$x_i \gamma_i P_i^{vap} = y_i P$$

Since $x_i = y_i$ at an azeotropic point, we have $\gamma_i = P/P_i^{vap}$, so that at $x_B = 0.525$

$$\gamma_B(x_B = 0.525) = \frac{1.013}{0.993} = 1.020$$

$$\gamma_C(x_B = 0.525) = \frac{1.013}{0.980} = 1.034$$

Using Eqs. 9.5-10, we obtain $\alpha = 0.125$ and $\beta = 0.0919$. Therefore,

$$\ln \gamma_B = \frac{0.125}{\left[1 + 1.360\left(\dfrac{x_B}{x_C}\right)\right]^2} \quad \text{and} \quad \ln \gamma_C = \frac{0.0919}{\left[1 + 0.736\left(\dfrac{x_C}{x_B}\right)\right]^2} \qquad \textbf{(i)}$$

The values of the activity coefficients for benzene (B) and cyclohexane (C) calculated from these equations are given in the following table and Fig. 1.

To compute the composition of the vapor in equilibrium with the liquid we use Eqs. 10.1-1b

$$x_B \gamma_B P_B^{vap} = y_B P \quad \text{and} \quad x_C \gamma_C P_C^{vap} = y_C P \qquad \textbf{(ii)}$$

and Eq. 10.1-2b

$$x_B \gamma_B P_B^{vap} + x_C \gamma_C P_C^{vap} = P \qquad \textbf{(iii)}$$

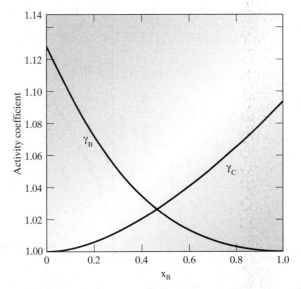

Figure 1 Activity coefficients of benzene and cyclohexane in their mixture at 77.6°C computed from azeotropic data using the van Laar model.

	van Laar				Regular Solution	
x_B	γ_B	γ_C	y_B	P (bar)	γ_B	γ_C
0	1.13	1.00	0	0.980	1.14	1.00
0.1	1.10	1.00	0.110	0.992	1.11	1.00
0.2	1.07	1.01	0.212	1.001	1.09	1.00
0.3	1.05	1.01	0.311	1.006	1.07	1.01
0.4	1.03	1.02	0.406	1.012	1.06	1.02
0.5	1.02	1.03	0.501	1.013	1.04	1.03
0.525	1.02	1.03	0.525	1.013	1.04	1.04
0.6	1.01	1.04	0.596	1.010	1.03	1.05
0.7	1.01	1.05	0.693	1.006	1.02	1.07
0.8	1.00	1.07	0.792	1.005	1.01	1.10
0.9	1.00	1.08	0.894	1.000	1.00	1.13
1.0	1.00	1.10	1.0	0.993	1.00	1.17

In these equations the vapor compositions, y_B and y_C, and the equilibrium pressure P are unknown (the equilibrium pressure is 1.013 bar only at $x_B = 0.525$). The solution is obtained by choosing a value of x_B, using $x_C = 1 - x_B$, and computing γ_B and γ_C from Eqs. i, and the total pressure from Eq. iii. The vapor-phase mole fractions are then computed from Eqs. ii. The results of this calculation are given in the table and Fig. 2.

b. *Regular solution model.* Since benzene and cyclohexane are nonpolar, and their solubility parameters are given in Table 9.6-1, the activity coefficients can be predicted using Eqs. 9.6-10. The results of this calculation are given in the table. The agreement between the correlation of the data using the van Laar model and the predictions (without reference to the experimental data) using the regular solution is good in this case.

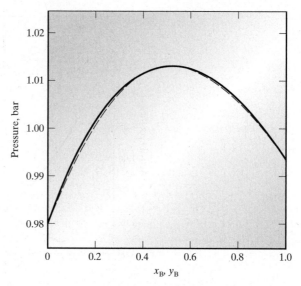

Figure 2 Total pressure versus liquid composition (solid line) and vapor composition (dashed line) of the benzene and cyclohexane mixture at 77.6°C computed from azeotropic data using the van Laar model.

NOTE

From Fig. 2 we see that for this mixture the compositions of the vapor and liquid in equilibrium are very close. An important implication of this is that it will be very difficult to separate benzene and cyclohexane from their mixture using distillation. ∎

ILLUSTRATION 10.2-2

Construction of Vapor-Liquid Equilibrium Diagrams for a Nonideal System

    a. Develop the constant-temperature x-y and P-x-y diagrams for ethyl acetate–benzene mixtures at 75°C.

    b. Develop the constant-pressure x-y and T-x-y diagrams for ethyl acetate–benzene mixtures at 1.013 bar

Data:

$$\ln P_{\mathrm{EA}}^{\mathrm{vap}} \ (\mathrm{bar}) = 9.6830 - \frac{2842.2}{T(\mathrm{K}) - 56.3209}$$

and

$$\ln P_{\mathrm{BZ}}^{\mathrm{vap}} \ (\mathrm{bar}) = 9.3171 - \frac{2810.5}{T(\mathrm{K}) - 51.2586}$$

This mixture can be described by the van Laar equation with the parameters $\alpha = 1.15$ and $\beta = 0.92$ as given in Table 9.5-1.

SOLUTION

    a. *Constant temperature.* At $T = 75°$ C, $P_{\mathrm{EA}}^{\mathrm{vap}} = 0.946$ bar and $P_{\mathrm{BZ}}^{\mathrm{vap}} = 0.862$ bar.

        For any liquid mole fraction of ethyl acetate x_{EA}, the mole fraction of benzene is $x_{\mathrm{BZ}} = 1 - x_{\mathrm{EA}}$. The equilibrium pressure $P(x_{\mathrm{EA}})$ is

$$P\left(x_{\mathrm{EA}}\right) = x_{\mathrm{EA}} \cdot \gamma_{\mathrm{EA}}\left(x_{\mathrm{EA}}\right) \cdot P_{\mathrm{EA}}^{\mathrm{vap}} + x_{\mathrm{BZ}} \cdot \gamma_{\mathrm{BZ}}\left(x_{\mathrm{BZ}}\right) \cdot P_{\mathrm{BZ}}^{\mathrm{vap}}$$

$$P\left(x_{\mathrm{EA}}\right) = x_{\mathrm{EA}} \cdot \exp\left[\frac{\alpha}{\left[1 + \frac{\alpha \cdot x_{\mathrm{EA}}}{\beta \cdot x_{\mathrm{BZ}}}\right]^2}\right] \cdot P_{\mathrm{EA}}^{\mathrm{vap}} + x_{\mathrm{BZ}} \cdot \exp\left[\frac{\beta}{\left[1 + \frac{\beta \cdot x_{\mathrm{BZ}}}{\alpha \cdot x_{\mathrm{EA}}}\right]^2}\right] \cdot P_{\mathrm{BZ}}^{\mathrm{vap}}$$

and

$$y_{\mathrm{EA}} = \frac{x_{\mathrm{EA}} \cdot \exp\left[\dfrac{\alpha}{\left[1 + \frac{\alpha \cdot x_{\mathrm{EA}}}{\beta \cdot x_{\mathrm{BZ}}}\right]^2}\right] \cdot P_{\mathrm{EA}}^{\mathrm{vap}}}{P\left(x_{\mathrm{EA}}\right)}$$

The results of the calculations are shown in Fig. 1. Note that the maximum in the *P-x-y* diagram is indicative of the formation of an azeotrope, which is confirmed by the equilibrium line crossing the $x = y$ line in the *x-y* diagram. From these diagrams we see that at a temperature of 75°C an azeotrope occurs at an ethyl acetate mole fraction of 0.581 and a pressure of 1.165 bar.

b. *Constant pressure.* In this case the pressure is fixed at 1.013 bar and the temperature is unknown, and therefore the vapor pressures are unknown. For any liquid mole fraction of ethyl acetate x_{EA}, the equilibrium pressure $P(x_{\mathrm{EA}}, T)$ is

$$P(x_{\mathrm{EA}}, T) = x_{\mathrm{EA}} \cdot \exp\left[\frac{\alpha}{\left[1 + \frac{\alpha \cdot x_{\mathrm{EA}}}{\beta \cdot x_{\mathrm{BZ}}}\right]^2}\right] \cdot P_{\mathrm{EA}}^{\mathrm{vap}}(T) + x_{\mathrm{BZ}} \cdot \exp\left[\frac{\beta}{\left[1 + \frac{\beta \cdot x_{\mathrm{BZ}}}{\alpha \cdot x_{\mathrm{EA}}}\right]^2}\right] \cdot P_{\mathrm{BZ}}^{\mathrm{vap}}(T)$$

$$= 1.013 \text{ bar}$$

Since there is a nonlinear dependence of the vapor pressures on temperature, an iterative solution (or the use of computer software such as MATHCAD) is required. The vapor phase mole fraction is then calculated from

$$y_{\mathrm{EA}} = \frac{x_{\mathrm{EA}} \cdot \exp\left[\dfrac{\alpha}{\left[1 + \frac{\alpha \cdot x_{\mathrm{EA}}}{\beta \cdot x_{\mathrm{BZ}}}\right]^2}\right] \cdot P_{\mathrm{EA}}^{\mathrm{vap}}(T)}{1.013 \text{ bar}}$$

The results of the calculations are shown in Fig. 2. Note that the mimumum in the *T-x-y* diagram is indicative of the formation of a minimum boiling azeotrope, which is confirmed by the equilibrium line crossing the $x = y$ line in the *x-y* diagram. From these diagrams we see that an azeotrope occurs at an ethyl acetate mole fraction of 0.515 and a temperature of 343.92 K = 70.77°C.

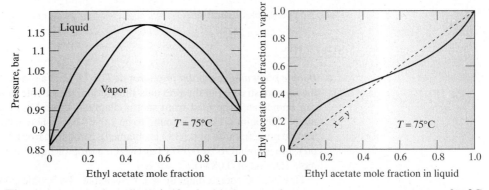

Figure 1 *P-x-y* and *x-y* diagrams for the ethyl acetate–benzene system at a temperature of 75°C.

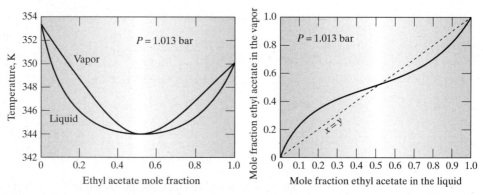

Figure 2 *T-x-y* and *x-y* diagrams for the ethyl acetate–benzene system at a pressure of 1.013 bar

COMMENT

Note that the azeotropic conditions found in parts (a) and (b) are slightly different. This is a result of the difference in temperature between the two cases. Later in this section we will show how this temperature dependence of the azeotropic composition can be used to advantage in distillation.

Not all nonideal liquid mixtures form azeotropes, as will be seen in the some of the other systems studied in this section. However, nonideal liquid mixtures result in a nonlinear dependence of the equilibrium pressure on composition, and do not satisfy Raoult's law.

Finally, notice that the *T-x-y* diagram looks very much like the *P-x-y* diagram turned upside down. There is a simple explanation for this, which is that a mixture that has a higher equilibrium pressure at a fixed temperature will have a lower boiling point at a fixed (for example, atmospheric) pressure. Therefore, a maximum in pressure in a *T-x-y* diagram will result in a minimum in temperature at approximately the same composition on a *P-x-y* diagram. Also note that the upper part of the *P-x-y* diagram corresponds to high pressure and a liquid, while the upper part of a *T-x-y* diagram represents the high-temperature region and a vapor. ∎

ILLUSTRATION 10.2-3
Determination of Dew Point and Bubble Point Pressures

a. Estimate the bubble point and dew point pressures of a 20 mol % ethyl acetate, 80 mol % benzene mixture at 75°C.

b. Estimate the bubble point and dew point temperatures of a 20 mol % ethyl acetate, 80 mol % benzene mixture at 1.013 bar.

Data: See previous illustration.

SOLUTION

a. *Bubble point and dew point pressures at $T = 75°C$.* If a *P-x-y* diagram is available, such as the one constructed in the previous illustration, the bubble point and dew point pressures can immediately be read from such a diagram. From Fig. 1 of the previous illustration, the bubble point pressure is found to be 1.0954 bar (where a bubble of vapor with an ethyl acetate mole fraction of 0.3367 is obtained). Similarly, from the same figure, the dew point pressure is found to be 0.9971 bar (where a drop of liquid is formed with an ethyl acetate mole fraction of 0.0834).

However, if such a diagram is not available, the bubble point and dew point pressures must be found by direct calculation. As the liquid composition is known ($x_{EA} = 0.2$), so that the activity coefficients can be computed, the bubble point is easily found, by solving the equation

$$P_{\text{bubble}} (\text{bar}) = 0.2 \cdot \exp\left[\frac{\alpha}{\left[1 + \frac{\alpha \cdot 0.2}{\beta \cdot 0.8}\right]^2}\right] \cdot 0.946 + 0.8 \cdot \exp\left[\frac{\beta}{\left[1 + \frac{\beta \cdot 0.8}{\alpha \cdot 0.2}\right]^2}\right] \cdot 0.862$$

and then the vapor mole fraction is obtained from

$$y_{\text{EA}} = \frac{0.2 \cdot \exp\left[\frac{\alpha}{\left[1 + \frac{\alpha \cdot 0.2}{\beta \cdot 0.8}\right]^2}\right] \cdot 0.946}{P_{\text{bubble}} (\text{bar})}$$

Note that this is a direct (that is, noniterative) calculation and gives the same bubble point results as were obtained above from the graphical solution.

Finding the dew point pressure is more complicated, as the liquid composition (which also appears in the expressions for the activity coefficients) is unknown. The equation to be solved is

$$y_{\text{EA}} = 0.2$$

$$= \frac{x_{\text{EA}} \cdot \exp\left[\frac{\alpha}{\left[1 + \frac{\alpha \cdot x_{\text{EA}}}{\beta \cdot (1 - x_{\text{EA}})}\right]^2}\right] \cdot 0.946}{P_{\text{dew}} (\text{bar})}$$

$$= \frac{x_{\text{EA}} \cdot \exp\left[\frac{\alpha}{\left[1 + \frac{\alpha \cdot x_{\text{EA}}}{\beta \cdot (1 - x_{\text{EA}})}\right]^2}\right] \cdot 0.946}{x_{\text{EA}} \cdot \exp\left[\frac{\alpha}{\left[1 + \frac{\alpha \cdot x_{\text{EA}}}{\beta \cdot (1 - x_{\text{EA}})}\right]^2}\right] \cdot 0.946 + (1 - x_{\text{EA}}) \cdot \exp\left[\frac{\beta}{\left[1 + \frac{\beta \cdot (1 - x_{\text{EA}})}{\alpha \cdot x_{\text{EA}}}\right]^2}\right] \cdot 0.862}$$

which is solved iteratively (or using a program such as MATHCAD) for the dew point mole fraction of ethyl acetate, x_{EA}. The mole fraction found is then used in the equation

$$P_{\text{dew}} (\text{bar}) = x_{\text{EA}} \cdot \exp\left[\frac{\alpha}{\left[1 + \frac{\alpha \cdot x_{\text{EA}}}{\beta \cdot (1 - x_{\text{EA}})}\right]^2}\right] \cdot 0.946 + (1 - x_{\text{EA}}) \cdot \exp\left[\frac{\beta}{\left[1 + \frac{\beta \cdot (1 - x_{\text{EA}})}{\alpha \cdot x_{\text{EA}}}\right]^2}\right] \cdot 0.862$$

to calculate the bubble point pressure. This leads to the same solution as was found graphically (after the complete *P-x-y* diagram had been constructed.)

b. *Bubble point and dew point temperatures at P= 1.013 bar.* If a *T-x-y* diagram is available, such as the one constructed in the previous illustration, the bubble point and dew point temperatures can immediately be read from such a diagram. Using Fig. 2 of the previous illustration, the bubble point temperature is found to be 345.78 K (when a bubble of vapor with an ethyl acetate mole fraction of 0.3360 is formed). Similarly, from the same figure, the dew point temperature is found to be 348.67 K (where a drop of liquid is formed with an ethyl acetate mole fraction of 0.0833).

However, if such a diagram is not available, the bubble point and dew point temperatures must be found by direct calculation. The bubble point temperature is more easily found, since the liquid composition is known, by solving the following equation

$$P = 1.013 = 0.2 \cdot \exp\left[\cfrac{\alpha}{\left[1 + \cfrac{\alpha \cdot 0.2}{\beta \cdot 0.8}\right]^2}\right] \cdot P_{\text{EA}}^{\text{vap}}(T) + 0.8 \cdot \exp\left[\cfrac{\beta}{\left[1 + \cfrac{\beta \cdot 0.8}{\alpha \cdot 0.2}\right]^2}\right] \cdot P_{\text{BZ}}^{\text{vap}}(T)$$

and using the solution in the equation below to obtain the vapor composition:

$$y_{\text{EA}} = \cfrac{0.2 \cdot \exp\left[\cfrac{\alpha}{\left[1 + \cfrac{\alpha \cdot 0.2}{\beta \cdot 0.8}\right]^2}\right] \cdot P_{\text{EA}}^{\text{vap}}(T)}{P = 1.013 \text{ bar}}$$

Finding the dew point temperature is more complicated, as both the liquid composition and the temperature are unknown. The two equations below must be solved simultaneously for the bubble point temperature and the liquid composition (which is best done with a computer program):

$$P = 1.013 = x_{\text{EA}} \cdot \exp\left[\cfrac{\alpha}{\left[1 + \cfrac{\alpha \cdot x_{\text{EA}}}{\beta \cdot (1 - x_{\text{EA}})}\right]^2}\right] \cdot P_{\text{EA}}^{\text{vap}}(T) + (1 - x_{\text{EA}}) \cdot \exp\left[\cfrac{\beta}{\left[1 + \cfrac{\beta \cdot (1 - x_{\text{EA}})}{\alpha \cdot x_{\text{EA}}}\right]^2}\right] \cdot P_{\text{BZ}}^{\text{vap}}(T)$$

and

$$y_{\text{EA}} = 0.2 = \cfrac{x_{\text{EA}} \cdot \exp\left[\cfrac{\alpha}{\left[1 + \cfrac{\alpha \cdot x_{\text{EA}}}{\beta \cdot (1 - x_{\text{EA}})}\right]^2}\right] \cdot P_{\text{EA}}^{\text{vap}}(T)}{P = 1.013 \text{ bar}}$$

This leads to the same solution as was found graphically (after the complete T-x-y diagram had been constructed.) ■

Azeotropes and distillation

The occurence of azeotropes, such as the one in the ethyl acetate–benzene system above, results in difficulties in separations by distillation. For example, suppose it is desired to produce a distillate containing 98 mol % ethyl acetate and a bottoms product containing 98 mol % benzene from a feed that contains 40 mol % ethyl acetate and 60 mol % benzene. Rather than consider detailed distillation calculations at various reflux ratios, we will consider only the case of total reflux, $q = \infty$, which you should remember results in the minimum number of stages to accomplish a given separation. Also, if the desired separation cannot be made at total reflux, it will not be possible to accomplish the separation at any lower reflux ratio.

Using the graphical method introduced in the previous section, and the constant-pressure x-y diagram in Fig. 10.2-2, we see that by stepping off stages we can obtain a bottoms product of the desired composition, but the maximum ethyl acetate distillate composition we can obtain is that of the azeotrope, $x_{\text{EA}} = 0.518$, since a pinch point occurs where the equilibrium line crosses the operating line, which at the infinite reflux ratio is the same as the $x = y$ or 45° line. Also, note that this situation cannot be resolved by changing the reflux ratio. Any lower value of the reflux ratio will result in more equilibrium stages being required to obtain the bottoms product, while still resulting in a distillate of the azeotropic composition.

(The feed considered above was below the azeotropic composition, and while the desired bottoms product could be obtained by distillation, it is not possible to obtain an overhead product purer than the azeotropic composition. Convince yourself that if

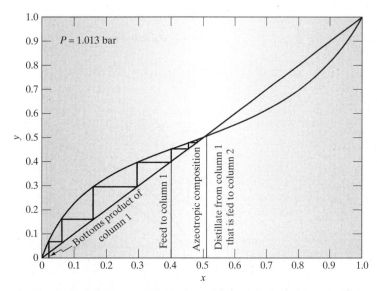

Figure 10.2-2 Graphical tray-to-tray calculation for the ethyl acetate–benzene mixture at 1.013 bar and total reflux, showing that as the feed is of lower composition than the azeotropic point, the highest purity possible of the distillate is the azeotropic composition.

the feed composition was greater than the azeotropic composition, the desired distillate composition could be obtained, but a bottoms product of only the azeotropic composition would be obtained.)

Consequently, when a mixture has an azeotrope, it is not possible to obtain both a distillate and a bottoms product of any desired purity, as is the case with mixtures that do not have azeotropes. When an azeotrope occurs, the maximum purity of either the overhead or bottoms product that can be obtained with a single distillation column (depending on the system and whether the feed is above or below the azeotropic composition) is the azeotropic composition. This is why the occurence of an azeotrope is a problem in distillation.

There are several ways to separate an azeotropic mixture into two components of the desired purities, and these are discussed in other chemical engineering courses. However, one method will be mentioned here, and it is based on the fact that in general the two components will not have the same heat of vaporization, so that by the Clausius-Clapeyron equation the temperature dependence of their vapor pressures will be different. Since the dominant temperature dependence in vapor-liquid equilibrium is that of the pure component vapor pressures, the azeotropic composition will also change with temperature (and pressure). Therefore, what can be done is to use two distillation columns operating at different pressures.

For example, for the case being considered here, the first column could be operated at atmospheric pressure with the feed given above that was below the azeotropic composition, and the column could be designed (i.e., the number of stages and reflux ratio chosen) to produce the desired bottoms product ($x_{BZ} = 0.95$) and a distillate of the azeotropic composition ($x_{EA} = 0.518$). This distillate could then be fed into a second column operated, say, at 0.1013 bar. At this pressure the azeotropic composition is $x_{EA} = 0.458$, which is lower than the feed composition. Therefore, as shown in Fig. 10.2-3, this column could produce the desired distillate of composition ($x_{EA} = 0.95$) and a bottoms product with the azeotropic composition of $x_{EA} = 0.458$ at the operating pressure of this column. This bottoms stream could then be returned to the first

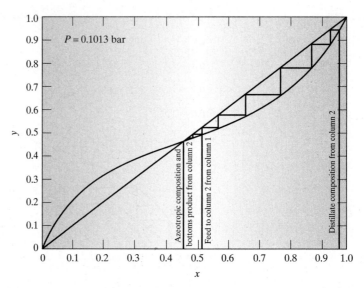

Figure 10.2-3 Graphical tray-to-tray calculation for the ethyl acetate–benzene mixture at 0.1013 bar and total reflux with a feed from column 1 that is of azeotropic composition at the higher pressure of 1.013 bar. Since this composition is higher than the azeotropic composition at 0.1013 bar, a high-purity distillate can be obtained but the bottoms product cannot be less than the azeotropic composition at this lower pressure.

column, operating at 1.013 bar, though since its ethyl acetate composition is higher than that of the feed, it might be fed to the column on a higher tray. This two-column design is schematically shown in Fig. 10.2-4. Note that the only streams leaving this process are the bottoms product of the first column and the distillate product of the of the second column, both at their desired compositions.

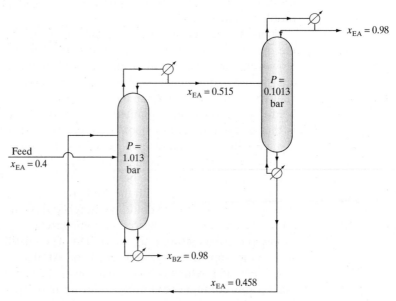

Figure 10.2-4 Proposed two-column design for separation of the ethyl acetate–benzene azeotropic mixture.

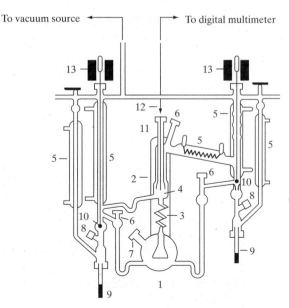

Figure 10.2-5 Schematic diagram of a dynamic still. In this figure, 1 is the boiling flask, 2 is a vacuum jacket so there is no heat loss from the equilibrium chamber, 3 is a device that forces the boiling vapor-liquid mixture into the equilibrium cell 4, items 5 are condensers to insure that no vapor is lost, items 6 are injection ports so that composition changes can be made, 7 is a thermometer port for the bath, items 8 and 10 are valves used to divert liquid to the sampling vials (9) for analysis and actuated by electromagnets (13), 11 is the thermometer well for the equilibrium chamber, and 12 is a very accurate platinum resistance thermometer to measure the temperature in the equilibrium cell.

The point to note here is the important role of thermodynamics, in that the schematic design of the process to separate the feed mixture into two streams of specified purity was based completely on vapor-liquid equilibrium. The fact that an azeotrope occurred dictated that the desired separation could not be obtained with a single distillation column. Then how the azeotropic composition changed with temperature was the basis for whether a second column of higher or lower pressure should be used. This example is just one illustration of the very central role of thermodynamics in general, and vapor-liquid equilibrium in particular, in the design of separations processes in chemical engineering.

The measurement of vapor-liquid equilibrium data

While we have shown various examples of vapor-liquid equilibrium data, we have not discussed the methods by which such data are obtained. There are two general methods, referred to as the dynamic method and the static method. In the dynamic method, the vapor-liquid mixture is boiling, and samples of both the vapor and liquid can be withdrawn and their compositions determined by gas chromatographic or other methods. As the pressure and temperature are also measured, a data point consists consists of a P-T-$\underline{x}$-$\underline{y}$ point. One apparatus for such measurements is shown in Fig. 10.2-5.

Correlation of vapor-liquid equilibrium data

It is by careful use of such equipment that data such as those used in this section are obtained. Though it may not be evident, obtaining accurate data is tedious and requires careful chemical analysis of the samples of liquid and condensed vapor. Once accurate vapor-liquid equilibrium data have been obtained, they can be used to compute liquid-phase activity coefficients and excess Gibbs energies. This is shown in the following illustration.

ILLUSTRATION 10.2-4

Correlation of Vapor-Liquid Equilibrium Data

Weissman and Wood[4] have made very accurate measurements of vapor-liquid equilibria in benzene–2,2,4-trimethyl pentane mixtures over a range of temperatures. Their data for the vapor and liquid compositions and equilibrium total pressures at 55°C are given in the following table:

x_B	y_B	P (bar)
0.0819	0.1869	0.268 92
0.2192	0.4065	0.315 73
0.3584	0.5509	0.354 63
0.3831	0.5748	0.360 88
0.5256	0.6786	0.391 05
0.8478	0.8741	0.432 77
0.9872	0.9863	0.436 41

The vapor pressure of pure benzene at 55°C is 0.435 96 bar, and that of 2,2,4-trimethyl pentane is 0.237 38 bar.

a. Calculate the activity coefficients of benzene and 2,2,4-trimethyl pentane and $\underline{G}^{ex}$ at each of the experimental points.

b. Obtain smoothed values for the excess Gibbs energy and activity coefficients for this system as a function of composition using the following procedure:

  i. Assume an expansion like Eq. 9.5-6 or 9.5-8 for $\underline{G}^{ex}$.

  ii. Use the data from part (a) to compute the coefficients in this expression.

  iii. Use these coefficients to compute smoothed values of $\underline{G}^{ex}$.

  iv. Derive an expression for the activity coefficients for the assumed form of $\underline{G}^{ex}$ and compute smoothed values of the activity coefficients.

SOLUTION

a. The condition for vapor-liquid equilibrium is

$$\bar{f}_i^L(T, P, \underline{x}) = \bar{f}_i^V(T, P, \underline{y})$$

For the conditions here we can assume an ideal gas-phase mixture and neglect all fugacity and Poynting corrections (although Weissman and Wood included these in their analysis of the data), so that

$$x_i \gamma_i P_i^{vap} = y_i P$$

or

$$\gamma_i = y_i P / x_i P_i^{vap}$$

The activity coefficients calculated in this manner are given in the following table. Values of $\underline{G}^{ex}$ computed from

$$\underline{G}^{ex} = x_1 \bar{G}_1^{ex} + x_2 \bar{G}_2^{ex} = RT(x_1 \ln \gamma_1 + x_2 \ln \gamma_2)$$

are also given.

---

[4]S. Weissman and S. E. Wood, *J. Chem. Phys.*, **32**, 1153 (1960).

b. Weissman and Wood used the Redlich-Kister–type expansion for the excess Gibbs energy in the form

$$\underline{G}^{ex} = x_1 x_2 [a + b(x_1 - x_2) + c(x_1 - x_2)^2] \tag{1}$$

Since this equation is linear in the unknown parameters a, b, and c, it is easily fitted, by least-squares analysis, to the experimental data. The results, in J/mol, are

$$a = 1389.0 \qquad b = 419.45 \qquad \text{and} \qquad c = 109.83$$

Now using $x_i = N_i/(N_1 + N_2)$, we obtain

$$G^{ex} = N\underline{G}^{ex} = \frac{N_1 N_2}{(N_1 + N_2)} \left[a + b \left(\frac{N_1 - N_2}{N_1 + N_2} \right) + c \left(\frac{N_1 - N_2}{N_1 + N_2} \right)^2 \right]$$

and from Eqs. 9.3-9 and 9.3-12,

$$\begin{aligned}
\bar{G}_1^{ex} = RT \ln \gamma_1 &= \left(\frac{\partial G^{ex}}{\partial N_1} \right)_{T,P,N_2} \\
&= x_2^2 [a + b(x_1 - x_2) + c(x_1 - x_2)^2] \\
&\quad + 2 x_1 x_2^2 [b + 2c(x_1 - x_2)]
\end{aligned} \tag{2}$$

In a similar fashion, one finds

$$\begin{aligned}
RT \ln \gamma_2 &= x_1^2 [a + b(x_1 - x_2) + c(x_1 - x_2)^2] \\
&\quad - 2 x_1^2 x_2 [b + 2c(x_1 - x_2)]
\end{aligned} \tag{3}$$

Equations 2 and 3, with the parameter values given above, were used to compute the smoothed activity coefficient values and the calculated values for $\underline{G}^{ex}$ given in the following table and plotted in Fig. 1. In Fig. 2 the partial pressures of each species (i.e., $P_B = y_B P$ and $P_{TMP} = y_{TMP} P$) are plotted as a function of the liquid-phase composition. The dashed lines indicate the behavior to be expected if the solution were ideal (i.e., if Raoult's law were obeyed).

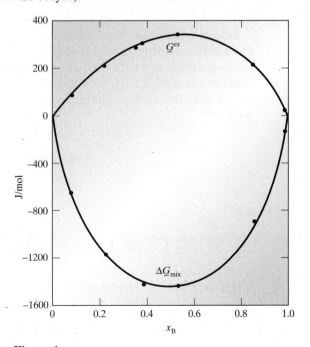

Figure 1

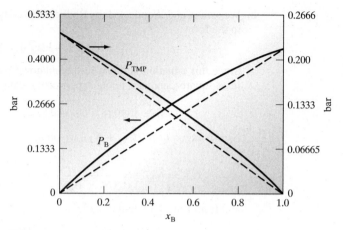

Figure 2

	Experimental Data			Smoothed Data			
x_B	γ_B	γ_{TMP}	$\underline{G}^{ex}$	γ_B	γ_{TMP}	$\underline{G}^{ex}$	$\Delta \underline{G}_{mix}$
0.0819	1.408	1.003	84.68	1.428	1.002	83.85	−689.40
0.2192	1.343	1.011	199.74	1.342	1.013	203.34	−1231.56
0.3584	1.250	1.046	296.73	1.261	1.039	294.09	−1486.24
0.3831	1.242	1.048	305.22	1.246	1.046	306.52	−1509.50
0.5256	1.158	1.116	352.63	1.166	1.107	351.75	−1535.95
0.8478	1.023	1.508	224.30	1.023	1.508	223.72	−940.02
0.9872	1.000	1.968	23.97	1.000	1.961	24.02	−162.88

Note: All Gibbs energy data in units of J/mol.

■

Several further aspects of vapor-liquid equilibria need to be considered. The first is the additional information that can be obtained if vapor-liquid equilibrium measurements are made at a collection of temperatures. Weissman and Wood carried out their experiments at 35, 45, 55, 65, and 75°C and obtained data like that given in Illustration 10.2-4 at all these temperatures. From $\underline{G} = \underline{H} - T\underline{S}$ and the discussion of Chapter 8, we have

$$\Delta_{mix}\underline{G} = \Delta_{mix}\underline{H} - T\,\Delta_{mix}\underline{S}$$

$$\underline{G}^{ex} = \underline{H}^{ex} - T\,\underline{S}^{ex} \tag{10.2-8}$$

and

$$\frac{\partial}{\partial T}\left(\frac{\Delta_{mix}\underline{G}}{T}\right)_{P,\underline{x}} = \frac{1}{T}\left(\frac{\partial \Delta_{mix}\underline{G}}{\partial T}\right)_{P,\underline{x}} - \frac{1}{T^2}(\Delta_{mix}\underline{G}) \tag{10.2-9}$$

Using the Maxwell relation,

$$\left(\frac{\partial \underline{G}}{\partial T}\right)_P = -\underline{S}$$

which for mixtures becomes

$$\left(\frac{\partial \Delta_{\text{mix}}\underline{G}}{\partial T}\right)_{P,x_i} = -\Delta_{\text{mix}}\underline{S} \qquad \left(\frac{\partial \underline{G}^{\text{ex}}}{\partial T}\right)_{P,x_i} = -\underline{S}^{\text{ex}} \qquad \textbf{(10.2-10)}$$

Using Eqs. 10.2-8 and 10.2-10 in Eq. 10.2-9 gives

$$\frac{\partial}{\partial T}\left(\frac{\Delta_{\text{mix}}\underline{G}}{T}\right)_{P,x_i} = -\frac{1}{T}\Delta_{\text{mix}}\underline{S} - \frac{1}{T^2}(\Delta_{\text{mix}}\underline{H} - T\Delta_{\text{mix}}\underline{S})$$

$$= -\frac{\Delta_{\text{mix}}\underline{H}}{T^2} = -\frac{\underline{H}^{\text{ex}}}{T^2} \qquad \textbf{(10.2-11)}$$

Therefore, given $\Delta_{\text{mix}}\underline{G}$ or $\underline{G}^{\text{ex}}$ data at a collection of temperatures, one can obtain information about $\Delta_{\text{mix}}\underline{H}$ (or, equivalently, $\underline{H}^{\text{ex}}$, since $\Delta_{\text{mix}}\underline{H} = \underline{H}^{\text{ex}}$). This was done by Weissman and Wood, and the results of their calculations for 40°C are plotted in Fig. 10.2-3. One would expect the accuracy of these results to be less than that of $\Delta_{\text{mix}}\underline{G}$, since these calculations involve a differentiation of the experimental data. Also $\underline{S}^{\text{ex}}$, calculated from Eq. 10.2-8 and the values of $\underline{G}^{\text{ex}}$ and $\underline{H}^{\text{ex}}$, is plotted in Fig. 10.2-6.

Next, one frequently would like to be able to make some assessment of the accuracy of a set of experimental vapor-liquid or activity coefficient measurements. Basic thermodynamic theory (as opposed to the solution modeling of Chapter 9) provides no means of predicting the values of liquid-phase activity coefficients to which the experimental results could be compared. Also, since the liquid solution models discussed in Chapter 9 only approximate real solution behavior, any discrepancy between these models and experiment is undoubtedly more a reflection of the inadequacy of the model than a test of the experimental results.

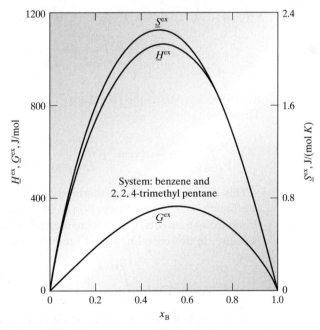

Figure 10.2-6 Excess Gibbs energy, excess enthalpy, and excess entropy as a function of mole fraction for the benzene–2,2,4-trimethyl pentane system.

In Chapter 7 we found that although thermodynamics could not be used to predict the equation of state of a real fluid, it did provide certain consistency tests (i.e., Eqs. 7.2-12 and 7.2-13) that had to be satisfied by any equation of state. The situation is much the same here, in that starting from

$$\underline{G}^{\text{ex}} = \sum_{i=1}^{\mathcal{C}} x_i \bar{G}_i^{\text{ex}} = RT \sum_{i=1}^{\mathcal{C}} x_i \ln \gamma_i \tag{10.2-12}$$

and using the Gibbs-Duhem equation, we can develop a consistency test that must be satisfied by activity coefficient data and thus can be used to accept or reject experimental data. In particular, for a binary mixture, we have, from Eq. 10.2-12,

$$d\left(\frac{\underline{G}^{\text{ex}}}{RT}\right) = x_1 \, d \ln \gamma_1 + \ln \gamma_1 \, dx_1 + x_2 \, d \ln \gamma_2 - \ln \gamma_2 \, dx_1$$

since $dx_2 = -dx_1$. Also, from the Gibbs-Duhem equation, Eq. 9.3-15, we have

$$0 = \frac{\underline{H}^{\text{ex}}}{RT^2} \, dT - \frac{\underline{V}^{\text{ex}}}{RT} \, dP + x_1 \, d \ln \gamma_1 + x_2 \, d \ln \gamma_2$$

Subtracting the second of these equations from the first gives

$$d\left(\frac{\underline{G}^{\text{ex}}}{RT}\right) = \ln \frac{\gamma_1}{\gamma_2} \, dx_1 - \frac{\underline{H}^{\text{ex}}}{RT^2} \, dT + \frac{\underline{V}^{\text{ex}}}{RT} \, dP$$

Now integrating this equation between $x_1 = 0$ and $x_1 = 1$ yields

$$\int_{x_1=0}^{x_1=1} d\left(\frac{\underline{G}^{\text{ex}}}{T}\right) = \frac{\underline{G}^{\text{ex}}}{T}\bigg|_{x_1=1} - \frac{\underline{G}^{\text{ex}}}{T}\bigg|_{x_1=0}$$

$$= +R \int_{x_1=0}^{x_1=1} \ln \frac{\gamma_1}{\gamma_2} \, dx_1 + \int_{P(x_1=0)}^{P(x_1=1)} \frac{\underline{V}^{\text{ex}}}{T} \, dP - \int_{T(x_1=0)}^{T(x_1=1)} \frac{\underline{H}^{\text{ex}}}{T^2} \, dT$$

$$= 0$$

where we have used the fact that $\underline{G}^{\text{ex}}(x_1 = 1) = \underline{G}^{\text{ex}}(x_1 = 0) = 0$ (cf. Section 9.5). Therefore,

$$\int_{x_1=0}^{x_1=1} \ln \frac{\gamma_2}{\gamma_1} \, dx_1 = + \int_{P(x_1=0)}^{P(x_1=1)} \frac{\underline{V}^{\text{ex}}}{RT} \, dP - \int_{T(x_1=0)}^{T(x_1=1)} \frac{\underline{H}^{\text{ex}}}{RT^2} \, dT \tag{10.2-13}$$

This equation provides a thermodynamic consistency test for experimental activity coefficient data. As an illustration of its use, consider its application to the Weissman-Wood measurements, which were carried out at constant temperature but varying total pressure. In this case Eq. 10.2-13 reduces to

$$\int_{x_1=0}^{x_1=1} \ln \frac{\gamma_2}{\gamma_1} \, dx_1 = + \int_{P(x_1=0)}^{P(x_1=1)} \frac{\underline{V}^{\text{ex}}}{RT} \, dP$$

Now since the total pressure variation in the experiments was small, and $\underline{V}^{\text{ex}}$ is usually very small for liquid mixtures, the integral on the right side of this equation can be neglected. Thus to test the thermodynamic consistency of the Weissman-Wood activity coefficient data, we use

$$\int_{x_1=0}^{x_1=1} \ln \frac{\gamma_2}{\gamma_1} \, dx_1 = 0$$

(10.2-14)

Figure 10.2-7 is a plot of $\ln(\gamma_2/\gamma_1)$ versus mole fraction using the activity coefficient data of Illustration 10.2-4. The two areas, I and II, between the curve and the $\ln(\gamma_2/\gamma_1) = 0$ line are virtually equal in size but opposite in sign, so that Eq. 10.2-14 can be considered satisfied. Of course, as a result of experimental error, this equation is never satisfied exactly. One usually considers vapor-liquid (or activity coefficient) data to be thermodynamically consistent if the two areas are such that

$$-0.02 \leq \frac{|\text{area I}| - |\text{area II}|}{|\text{area I}| + |\text{area II}|} \leq 0.02$$

where the vertical bars indicate that absolute values of the areas are to be used.

For activity coefficient data obtained from measurements at constant total pressure, but varying temperature, the appropriate consistency relation is

$$\int_{x_1=0}^{x_1=1} \ln \frac{\gamma_2}{\gamma_1} \, dx_1 = -\int_{T(x_1=0)}^{T(x_1=1)} \frac{\underline{H}^{\text{ex}}}{RT^2} \, dT$$

Depending on the system, and especially the magnitude of the excess enthalpy and the temperature range of the experiments, the integral on the right side of this equation may or may not be negligible.

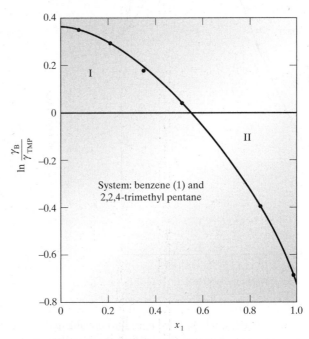

Figure 10.2-7 Thermodynamic consistency test for the activity coefficients of the benzene–2,2,4-trimethyl pentane system.

It is possible, as a result of cancellation, to satisfy the integral test of Eq. 10.2-13 while violating the differential form of the Gibbs-Duhem equation, Eq. 9.3-15, on which Eq. 10.2-13 is based, at some or all data points. In this case the experimental data should be rejected as thermodynamically inconsistent. Thus, the integral consistency test is a necessary, but not sufficient, condition for accepting experimental data.

As another example of low-pressure vapor-liquid equilibrium, we consider the n-pentane–propionaldehyde mixture at 40.0°C. Eng and Sandler[5] took data on this system using the dynamic still of Fig. 10.2-5. The x-y-P-T data in Table 10.2-1 and Fig. 10.2-8a and b were obtained by them. (Such data can be tested for thermodynamic consistency; see Problem 10.2-12.) As is evident, this system is nonideal and has an azeotrope at about 0.656 mole fraction pentane and 1.3640 bar. We will use these data to test the UNIFAC prediction method.

First, we use the UNIFAC program, discussed in Appendix B.I, to compute the activity coefficients in the n-pentane–propionaldehyde mixture over the complete liquid concentration range, and then, using Eqs. 10.2-1b and 10.2-2b, we compute the vapor compositions and equilibrium pressures. The results also appear in Fig. 10.2-8a as a P-x-y diagram and in Fig. 10.2-8b as an x-y diagram. The UNIFAC predictions are in very good agreement with the experimental data, including a reasonably accurate prediction of the azeotropic point. Clearly, an engineer needing information on the

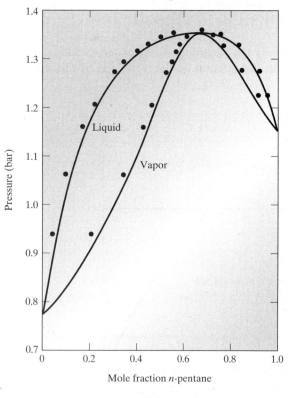

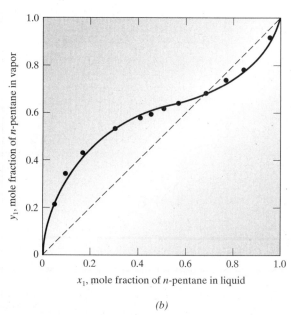

(a)

(b)

Figure 10.2-8 (a) P-x-y diagram for the n-pentane–propionaldehyde system at 40°C. The lines are the UNIFAC predictions, and the points are the experimental data of Eng and Sandler. (b) x-y diagram for the n-pentane–propionaldehyde system at 40°C. The solid line is the UNIFAC prediction, and the points are the experimental data of Eng and Sandler.

[5]R. Eng and S. I. Sandler, *J. Chem. Eng. Data*, **29**, 156 (1984).

Table 10.2-1 Vapor-Liquid Equilibrium Data for the n-Pentane (1)–Propionaldehyde (2) System at 40°C

x_1	y_1	P (bar)	x_1	y_1	P (bar)
0	0	0.7609	0.4463	0.5877	1.3354
0.0503	0.2121	0.9398	0.5031	0.6146	1.3494
0.1014	0.3452	1.0643	0.5610	0.6311	1.3568
0.1647	0.4288	1.1622	0.6812	0.6827	1.3636
0.2212	0.4685	1.2173	0.7597	0.7293	1.3567
0.3019	0.5281	1.2756	0.8333	0.7669	1.3353
0.3476	0.5539	1.2949	0.9180	0.8452	1.2814
0.4082	0.5686	1.3197	1.0	1.0	1.1541

n-pentane–propionaldehyde system, but having no experimental data, would be better to assume the UNIFAC model applies to this mixture than to assume that the system was ideal. Also, since propionaldehyde is strongly polar, the regular solution model could not be used for this mixture.

The dynamic still method of obtaining vapor-liquid equilibrium data has several disadvantages. First, it is a slow and tedious process. Second, the compositions of the vapor and liquid must be analyzed (usually by gas chromatography), which is less precise and direct than measuring temperature, pressure, or weight. Consequently, alternative methods of measuring partial vapor-liquid equilibrium data have been developed that do not require chemical analysis.

One method is to use a static cell, which consists of a small vessel that is evacuated and almost completely filled with a gravimetrically prepared liquid binary mixture. Such a device operated in the differential mode is shown in Fig. 10.2-9. In this equipment the pure solvent is placed in one cell and the gravimetrically prepared sample in the other. These vessels are then placed in a temperature bath, and the pressure dif-

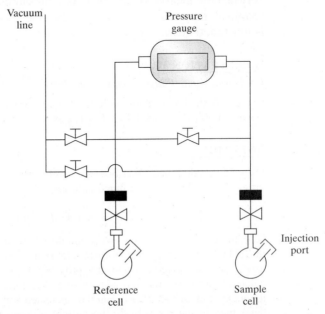

Figure 10.2-9 Schematic diagram of a differential static cell apparatus.

ference between the two vessels is measured after equilibrium has been reached. As weighing can be done very accurately, and since only a small amount of low-density vapor is formed, the liquid composition barely changes during the vaporization process and therefore is known very accurately. Also, by knowing the vapor pressure of the pure solvent and measuring the small pressure difference, we obtain an accurate pressure-liquid-composition point at the fixed temperature. By repeating this process with a number of prepared solutions, one obtains a set of P-T-x data.

These data can be studied in two ways. The first is to use the Gibbs-Duhem equation and numerical integration methods to calculate the vapor-phase mole fractions, as considered in Problem 10.2-6. A second method is to choose a liquid-phase activity coefficient model and determine the values of the parameters in the model that give the best fit of the experimental data. We have, from Eq. 10.2-2b, that at the jth experimental point

$$P_j = x_1^j \gamma_1^j P_1^{\text{vap}} + (1 - x_1^j)\gamma_2^j P_2^{\text{vap}} \tag{10.2-15}$$

The values of parameters in the activity coefficient model are chosen to minimize the sum-of-squares deviation between the measured and calculated pressures over all experimental points; that is, we want to find the parameters in the activity coefficient model that minimize the objective function

$$\min \sum_{\substack{\text{exp} \\ \text{pts} \\ j}} [P_j^{\text{exp}} - P_j^{\text{calc}}]^2 = \min \sum_{\substack{\text{exp} \\ \text{pts} \\ j}} [P_j^{\text{exp}} - x_1^j \gamma_1^j P_1^{\text{vap}} - (1 - x_1^j)\gamma_2^j P_2^{\text{vap}}]^2 \tag{10.2-16}$$

Thus, for example, if the van Laar equation is used to describe the liquid phase, then we want to determine the values of the parameters α and β that minimize the deviations between the measured and calculated pressures. Once these parameters have been determined, they can be used to calculate the vapor-phase compositions. This procedure is illustrated next.

ILLUSTRATION 10.2-5

Predicting Vapor-Phase Compositions from P-T-x Data

Using only the liquid-phase mole fraction and pressure data for the n-pentane–propionaldehyde system at 40°C given in Table 10.2-1, estimate the vapor compositions.

SOLUTION

Using the method just described, the van Laar equation, and a parameter estimation computer program with the objective function of Eq. 10.2-16, we find

$$\alpha = 1.4106 \quad \text{and} \quad \beta = 1.3438$$

With these parameter values, we obtain the calculated pressures and vapor mole fractions given in Table 10.2-2. It is clear from this table that the predictions are reasonably accurate. The azeotrope is predicted to occur at approximately the correct composition and pressure, the calculated vapor mole fractions usually agree to within ±0.015 of the measured composition, and the maximum difference between the calculated and measured pressures is only 0.0174 bar. Indeed, the calculated results for this system are so close to the experimental data as to be almost indistinguishable from them on x-y or P-x-y plots.   ∎

Table 10.2-2 Comparison of Measured Vapor-Phase Mole Fractions for the *n*-Pentane–Propionaldehyde System at 40°C with Values Predicted from *P-T-x* Data

x_1	y_1^{exp}	y_1^{calc}	P_1^{exp}	P_1^{calc}
0	0	0	0.7609	0.7609
0.0503	0.2121	0.2211	0.9398	0.9312
0.1014	0.3452	0.3424	1.0643	1.0555
0.1647	0.4288	0.4309	1.1622	1.1618
0.2212	0.4685	0.4810	1.2173	1.2257
0.3019	0.5281	0.5286	1.2756	1.2846
0.3476	0.5539	0.5484	1.2949	1.3067
0.4082	0.5686	0.5702	1.3197	1.3208
0.4463	0.5877	0.5824	1.3354	1.3379
0.5031	0.6146	0.5996	1.3494	1.3490
0.5610	0.6311	0.6173	1.3568	1.3566
0.6812	0.6827	0.6609	1.3636	1.3604
0.7597	0.7293	0.7005	1.3567	1.3500
0.8333	0.7669	0.7529	1.3353	1.3244
0.9180	0.8452	0.8455	1.2814	1.2640
1.0	1.0	1.0	1.1541	1.1541

These results suggest that, although not quite as good as *P-T-x-y* data, *P-T-x* data can be useful for estimating parameters in an activity coefficient model that can then be used to estimate the missing vapor compositions. An important disadvantage of *P-T-x* data, however, is that we cannot test its thermodynamic consistency since the activity coefficients are obtained from a model, not directly from experimental data.

A second method of obtaining partial vapor-liquid equilibrium information is by infinite-dilution ebulliometry. In this experiment a pure fluid of measured weight is boiled, and the vapor is condensed and returned to the boiling vessel. This is shown in Fig. 10.2-10. Then, after equilibrium is achieved, a very small measured weight of a second component is added, and the system is allowed to re-equilibrate. Then, depending on the equipment, one measures either the change in boiling pressure (between the pure fluid and the mixture) at fixed temperature or the change in boiling temperature at fixed pressure. Since a very small amount of the second component has been added, and the weights are known so that the mole fractions can be determined, one measures either

$$\left(\frac{\partial P}{\partial x_2}\right)_T \quad \text{or} \quad \left(\frac{\partial T}{\partial x_2}\right)_P$$

depending on the apparatus. Furthermore, if the amount of the added second component is small, these quantities have been determined in the limit of $x_2 \to 0$. (Alternatively, several weighed amounts of solute can be added, and then the rate of change of temperature or pressure with respect to mole fraction extrapolated to $x_2 \to 0$.)

To analyze the data from such an experiment, assuming an ideal vapor phase, we start from

$$P = x_1\gamma_1 P_1^{vap} + x_2\gamma_2 P_2^{vap}$$

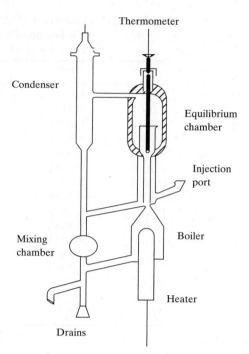

Figure 10.2-10 Schematic diagram of an ebulliometer.

For the constant-temperature experiment (noting that the pure-component vapor pressures depend only on temperature, which is being held fixed), we have

$$\left(\frac{\partial P}{\partial x_2}\right)_T = -\gamma_1 P_1^{\text{vap}} + x_1 \left(\frac{\partial \gamma_1}{\partial x_2}\right)_T P_1^{\text{vap}} + \gamma_2 P_2^{\text{vap}} + x_2 \left(\frac{\partial \gamma_2}{\partial x_2}\right)_T P_2^{\text{vap}}$$

Now in the limit of $x_2 \to 0$, we have $\gamma_1 \to 1$ and $(\partial \gamma_1 / \partial x_2) = 0$, as γ_1 is proportional to higher than a linear power of x_2 (i.e., see Eqs. 9.5-5, 9.5-7, etc.). Thus

$$\left(\frac{\partial P}{\partial x_2}\right)_{T,x_2 \to 0} = \gamma_2(x_2 \to 0) P_2^{\text{vap}} - P_1^{\text{vap}}$$

or

$$\gamma_2(x_2 \to 0) = \gamma_2^{\infty} = \frac{P_1^{\text{vap}} + \left(\dfrac{\partial P}{\partial x_2}\right)_{T,x_2 \to 0}}{P_2^{\text{vap}}} \tag{10.2-17}$$

In a similar fashion, for the constant-pressure ebulliometer, we have (Problem 10.2-13)

$$\gamma_2(x_2 \to 0) = \gamma_2^{\infty} = \frac{P_1^{\text{vap}} - \left(\dfrac{dP_1^{\text{vap}}}{dT}\right)\left(\dfrac{\partial T}{\partial x_2}\right)_{P,x_2 \to 0}}{P_2^{\text{vap}}} \tag{10.2-18}$$

It is, of course, possible to derive equations analogous to Eqs. 10.2-17 and 10.2-18 for a nonideal vapor phase.

Thus, from the ebulliometric experiment, one obtains the infinite-dilution activity coefficient directly. Now repeating the experiment by starting with pure component 2 and adding an infinitesimal amount of component 1, $\gamma_1(x_1 \to 0) = \gamma_1^\infty$ can be obtained. These two data points can then be used to determine the parameters in a two-constant activity coefficient model. For example, from the van Laar model of Eqs. 9.5-9, we have

$$\ln \gamma_1^\infty = \alpha \quad \text{and} \quad \ln \gamma_2^\infty = \beta \tag{10.2-19}$$

Thus once the infinite-dilution activity coefficients have been measured and the parameters in an activity coefficient model determined, the complete P-T-x-y behavior of the system can be estimated.

ILLUSTRATION 10.2-6
Predicting VLE from Infinite-Dilution Activity Coefficients Determined from Ebulliometry

A recent ebulliometric study of *n*-pentane–propionaldehyde at 40°C has found that $\gamma_1^\infty = 3.848$ and $\gamma_2^\infty = 3.979$. Use this information to prepare the P-x-y diagram for this system at 40°C.

SOLUTION

The van Laar activity coefficient model will be used. From Eqs. 10.2-19, we have

$$\alpha = \ln \gamma_1^\infty = \ln(3.848) = 1.3476 \quad \text{and} \quad \beta = \ln \gamma_2^\infty = \ln(3.979) = 1.3810$$

These values are in reasonable agreement with, but slightly different from, those found in the previous illustration. Using the values for the van Laar parameters, we obtain the y and P values in Table 10.2-3. Clearly, the agreement is excellent. ∎

The previous two illustrations demonstrate the utility of both P-T-x and ebulliometric γ^∞ data in determining values in activity coefficient models, and then using these

Table 10.2-3 Comparison of Measured Pressures and Vapor-Phase Mole Fractions for the *n*-Pentane–Propionaldehyde System at 40°C with Values Calculated Using γ^∞ Data

x_1	y_1^{exp}	y_1^{calc}	P_1^{exp}	P_1^{calc}
0	0	0	0.7609	0.7609
0.0503	0.2121	0.2131	0.9398	0.9214
0.1014	0.3452	0.3352	1.0643	1.0424
0.1647	0.4288	0.4267	1.1622	1.1493
0.2212	0.4685	0.4798	1.2173	1.2157
0.3019	0.5281	0.5307	1.2756	1.2784
0.3476	0.5539	0.5520	1.2949	1.3024
0.4082	0.5686	0.5752	1.3197	1.3255
0.4463	0.5877	0.5879	1.3354	1.3363
0.5031	0.6146	0.6056	1.3494	1.3483
0.5610	0.6311	0.6232	1.3568	1.3566
0.6812	0.6827	0.6652	1.3636	1.3616
0.7597	0.7293	0.7028	1.3567	1.3524
0.8333	0.7669	0.7528	1.3353	1.3283
0.9180	0.8452	0.8432	1.2814	1.2684
1.0	1.0	1.0	1.1541	1.1541

parameters to compute the complete P-T-x-y diagram. It should be remembered, however, that in the analysis of both static cell and ebulliometric measurements, an activity coefficient model that satisfies the Gibbs-Duhem equation has been used. Therefore, the calculated results must satisfy the thermodynamic consistency test. Consequently, there is no independent test of the quality of the results, as when complete P-T-x-y data have been measured. However, both static cell and ebulliometric measurements provide valuable data and measurements can be made quickly, which may be important for components that chemically react or decompose.

Table 10.2-4 Vapor-Liquid Equilibrium for the System Hexafluorobenzene (1)–Benzene (2) at 60°C

x_1	y_1	P (bar)	$\underline{G}^{\mathrm{ex}}$ (J/mol)
0.0000	0.0000	0.52160	0
0.0941	0.0970	0.52570	32
0.1849	0.1788	0.52568	40
0.2741	0.2567	0.52287	33
0.3648	0.3383	0.51818	16
0.4538	0.4237	0.50989	−4
0.5266	0.4982	0.50773	−21
0.6013	0.5783	0.50350	−35
0.6894	0.6760	0.49974	−44
0.7852	0.7824	0.49757	−45
0.8960	0.8996	0.49794	−30
1.0000	1.0000	0.50155	0

Source: Data of W. J. Gaw and F. L. Swinton, *Trans. Faraday Soc.,* **64,** 2023 (1968) with permission of The Royal Society of Chemistry.

The experimental data for the hexafluorobenzene-benzene system in Table 10.2-4 and Fig. 10.2-11 show a rarely encountered degree of complexity in low-pressure

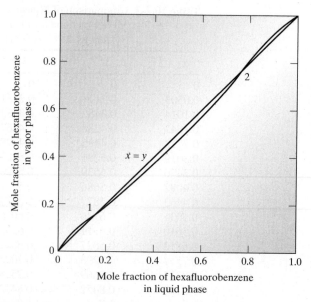

Figure 10.2-11 The x-y diagram for the hexafluorobenzene-benzene system at 60°C based on the data of Gaw and Swinton [*Trans. Faraday Soc.,* **64** 2023 (1968)] with permission of The Royal Society of Chemistry.

vapor-liquid equilibrium. This system exhibits both minimum and maximum boiling azeotropes. This occurs because the excess Gibbs energy for this system, though small, is first positive and then negative as the concentration of hexafluorobenzene is increased. Since the vapor pressures of hexafluorobenzene and benzene are almost identical, the solution nonidealities produce the double azeotrope.

Although the discussion and illustrations of this section have been concerned only with low-pressure vapor-liquid equilibria, phase equilibrium at somewhat higher pressures could have been considered also. The most important change in the analysis is that the gas phase can no longer be considered ideal or described by the Lewis-Randall rule; rather, an equation of state (the virial equation at low to moderate pressures, and more complicated equations at higher pressures) would have to be used. Also, the Poynting pressure correction of Eq. 5.4-18 may have to be used in the calculation of the pure liquid-phase fugacities. Both of these changes add some complexity to the calculations but improve their accuracy. For simplicity, these factors will not be considered here. We do, however, consider high-pressure vapor-liquid equilibria in the next section, a situation in which an equation of state must be used.

As the final example of the section, we consider the vapor-liquid equilibria of a polymer-solvent system.

ILLUSTRATION 10.2-7

Computing the Solvent Partial Pressure above a Polymer-Solvent Mixture

In the processing of polymers, and also for polymer devolatilization (the removal of the solvent from the polymer), it is important to be able to calculate the equilibrium partial pressure of a solvent above solvent-polymer mixtures of different compositions. Calculate the partial pressure of benzene in benzene + polyisobutylene (PIB) mixtures at 298.15 and 312.75 K. In this calculation you can assume that polyisobutylene has a negligible vapor pressure, and that the Flory-Huggins model describes the solution behavior of this polymer + solvent mixture. Do the calculations for values of the Flory-Huggins χ parameter equal to 0.5 to 1.0.

Data: The molar volume of benzene is 88.26 cm³/mol, its molecular weight is 78, and its vapor pressures are $P_B^{vap} = 0.1266$ bar at 298.15 K and 0.2392 bar at 312.75 K, respectively. The molecular weight of the PIB is 40,000, the monomeric unit in PIB has a molecular weight of 104, and the monomeric volume $\underline{V}_{PIB,m}$ is 131.9 cm³/mol monomer.

SOLUTION

The average number of monomer units, n, in the PIB polymer is computed as follows:

$$n = \frac{\text{Molecular weight of polymer}}{\text{Molecular weight of monomer}}$$

$$= \frac{40\,000}{104} = 384.6$$

The mole fraction x_B and the volume fraction ϕ_B of benzene in terms of its weight fraction W_B are

$$x_B = \frac{\dfrac{W_B}{78}}{\dfrac{W_B}{78} + \dfrac{W_{PIB}}{40\,000}} \quad \text{and} \quad \phi_B = \frac{\dfrac{W_B}{78} \times \underline{V}_B}{\dfrac{W_B}{78} \times \underline{V}_B + \dfrac{W_{PIB}}{40\,000} \times n \times \underline{V}_{PIB,m}}$$

$$= \frac{\dfrac{W_B}{78} \times 88.26}{\dfrac{W_B}{78} \times 88.26 + \dfrac{1 - W_B}{40\,000} \times 384.6 \times 131.9}$$

Also

$$m = \frac{\underline{V}_{PIB}}{\underline{V}_B} = \frac{\underline{V}_{PIB,m} \times n}{\underline{V}_B}$$

$$= \frac{131.9 \times 384.6}{88.26} = 574.8$$

Since the PIB is (assumed to be) involatile, we only have to equate the fugacity of benzene in the vapor and liquid phases. Further, since the total pressure will be low, we use

$$\bar{f}_B^L = \bar{f}_B^V \quad \text{or} \quad x_B \gamma_B P_B^{vap} = P_B = \text{partial pressure of benzene}$$

where the activity coefficient of benzene is calculated from the Flory-Huggins equation, Eq. 9.5-18,

$$\ln \gamma_B = \ln \frac{\phi_B}{x_B} + \left(1 - \frac{1}{m}\right)(1 - \phi_B) + \chi (1 - \phi_B)^2$$

Using this information, we obtain the following results:

	Partial pressure of benzene in PIB, bar					
	$T = 298.15$ K			$T = 312.75$ K		
wt % B	$\chi = 0.5$	$\chi = 1.0$	expt	$\chi = 0.5$	$\chi = 1.0$	expt
4.37						0.0715
5.00	0.0232	0.0367		0.0439	0.0693	
6.33						0.0971
9.45						0.1236
10.00	0.0428	0.0648		0.0809	0.1224	
15.00	0.0593	0.0861		0.1120	0.1626	
15.16						0.1681
18.42						0.1818
20.00	0.0730	0.1019		0.1378	0.1925	
23.43			0.1028			
25.37						0.2095
29.71						0.2182
29.98			0.1117			
30.00	0.0937	0.1217		0.1770	0.2299	
32.12						0.2207
33.51			0.1149			
34.57			0.1156			
37.30						0.2267
40.00	0.1075	0.1308		0.2031	0.2472	
47.62			0.1229			
50.00	0.1163	0.1338		0.2198	0.2572	
60.00	0.1217	0.1333		0.2299	0.2519	
70.00	0.1246	0.1314		0.2355	0.2482	
80.00	0.1260	0.1291		0.2381	0.2439	
90.00	0.1265	0.1273		0.2390	0.2405	
95.00	0.1266	0.1268		0.2392	0.2395	
100.00	0.1266	0.1266		0.2392	0.2392	

These results are plotted in Figures 1 and 2. These results show that the Flory-Huggins model with a constant value of $\chi = 1.0$ gives a reasonable representation of the experimental data of Eichinger and Flory [*Trans. Farad. Soc.*, **64**, 2053–2060 (1968)]. The results also show the significant effect of the value of the Flory χ parameter on the partial pressure predictions. ∎

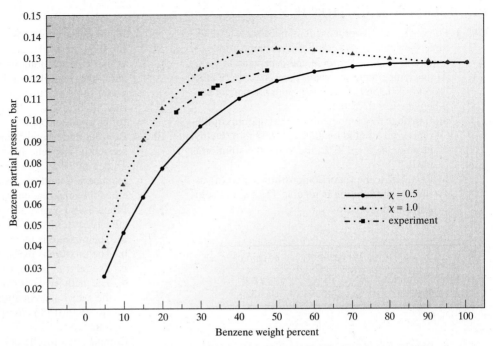

Figure 1 Partial pressure of benzene above benzene-polyisobutylene mixtures at 298.15 K. The experimental points are shown together with predictions of the Flory-Huggins model for $\chi = 0.5$ and 1.0.

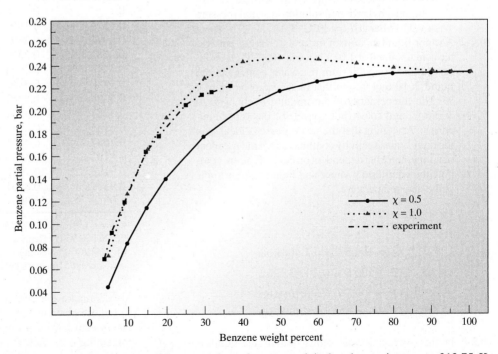

Figure 2 Partial pressure of benzene above benzene-polyisobutylene mixtures at 312.75 K. The experimental points are shown together with predictions of the Flory-Huggins model for $\chi = 0.5$ and 1.0.

PROBLEMS FOR SECTION 10.2

10.2-1 For a separations process it is necessary to determine the vapor-liquid equilibrium compositions for a mixture of ethyl bromide and n-heptane at 30°C. At this temperature the vapor pressure of pure ethyl bromide is 0.7569 bar, and the vapor pressure of pure n-heptane is 0.0773 bar.

   a. Calculate the composition of the vapor in equilibrium with a liquid containing 47.23 mol % ethyl bromide at $T = 30°C$, assuming the solution is ideal.

   b. Recalculate the vapor composition in part (a), assuming the solution is regular. The regular solution parameters are

	$\underline{V}^L$ (cc/mol)	δ (cal/cc)$^{1/2}$
Ethyl bromide	75	8.9
n-Heptane	148	7.4

   c. Recalculate the vapor composition of part (a) using the UNIFAC model.

   d. Recalculate the vapor composition of part (a) given that a vapor of composition 81.5 mol % ethyl bromide is in equilibrium with 28.43 mol % liquid ethyl bromide solution at a total pressure of 0.3197 bar at $T = 30°C$.

10.2-2 A vapor-liquid mixture of furfural ($C_5H_4O_2$) and water is maintained at 1.013 bar and 109.5°C. It is observed that at equilibrium the water content of the liquid is 10 mol % and that of the vapor is 81 mol %. The temperature of the mixture is changed to 100.6°C, and some (but not all) of the vapor condenses. Assuming that the vapor phase is ideal, and the liquid-phase activity coefficients are independent of temperature but dependent on concentration, compute the equilibrium vapor and liquid compositions at the new temperature.

Data:

$$P_{H_2O}^{vap}(T = 109.5°C) = 1.4088 \text{ bar}$$

$$P_{H_2O}^{vap}(T = 100.6°C) = 1.0352 \text{ bar}$$

$$P_{FURF}^{vap}(T = 109.5°C) = 0.1690 \text{ bar}$$

$$P_{FURF}^{vap}(T = 100.6°C) = 0.1193 \text{ bar}$$

10.2-3 In this section it was shown that if the equilibrium pressure versus mole fraction curve for a binary mixture has an interior extreme value at some liquid-phase mole fraction (i.e., if $(\partial P/\partial x)_T = 0$ for $0 < x < 1$), an azeotrope has been formed at that composition. Show that if, at constant pressure, the equilibrium temperature versus liquid-phase mole fraction has an interior extreme value (i.e., if $(\partial T/\partial x)_P = 0$ for $0 < x < 1$), the mixture forms an azeotrope.

10.2-4 Benzene and ethanol form azeotropic mixtures. Consequently, benzene is sometimes added to solvent grades of ethanol to prevent industrious chemical engineering students from purifying solvent-grade ethanol by distillation for use at an after-finals party. Prepare an x-y and a P-x diagram for the benzene-ethanol system at 45°C assuming, separately,

   a. The mixture is ideal.

   b. The mixture is regular.

   c. The mixture is described by the UNIFAC model.

   d. The activity coefficients for this system obey the van Laar equation and the datum point at $x_{EA} = 0.6155$ is used to obtain the van Laar parameters.

Compare the results obtained in parts (a)–(d) with the experimental data in the following table.

x_{EA}	y_{EA}	P (bar)
0	0	0.2939
0.0374	0.1965	0.3613
0.0972	0.2895	0.3953
0.2183	0.3370	0.4088
0.3141	0.3625	0.4124
0.4150	0.3842	0.4128
0.5199	0.4065	0.4100
0.5284	0.4101	0.4093
0.6155	0.4343	0.4028
0.7087	0.4751	0.3891
0.8102	0.5456	0.3615
0.9193	0.7078	0.3036
0.9591	0.8201	0.2711
1.00	1.00	0.2321

Source: I. Brown and F. Smith, *Aust. J. Chem.,* **7**, 264 (1954). Reprinted with permission of CSIRO Publishing. http://www.publish.csiro.au/nid/51/paper/CH9540264.htm

Also compare the computed van Laar coefficients with those given in Table 7.5-1.

10.2-5 The system toluene–acetic acid forms an azeotrope containing 62.7 mol % toluene and having a minimum boiling point of 105.4°C at 1.013 bar. The following vapor pressure data are available:

T (°C)	P^{vap} (bar)	
	Toluene	Acetic Acid
70	0.2699	0.1813
80	0.3863	0.2697
90	0.5395	0.3916
100	0.7429	0.5561
110	—	0.7744
110.7	1.0133	—
118.5	—	1.0133
120	—	1.0586

a. Calculate the van Laar constants for this system and plot $\ln \gamma_T$ and $\ln \gamma_A$ versus x_T. This is most easily done using the MATHCAD worksheet ACTCOEFF on the CD-ROM accompanying this book and discussed in Appendix B.III

b. Assuming that the activity coefficients (and van Laar constants) are independent of temperature over the limited range of temperatures involved, develop an x-y diagram for this system at $P = 1.013$ bar. Compare this with the x-y diagram that would be obtained if the system were ideal.

c. Assuming the activity coefficients are independent of temperature, develop an x-y diagram for this system at $P = 1.013$ bar using the UNIFAC model, and compare the results with those obtained in part (b).

10.2-6 The illustrations of this section were meant to demonstrate how one can determine activity coefficients from measurements of temperature, pressure, and the mole fractions in both phases of a vapor-liquid equilibrium system. An alternative procedure is, at constant temperature, to measure the total equilibrium pressure above liquid mixtures of known (or measured) composition. This replaces time-consuming measurements of vapor-phase compositions with a more detailed analysis of the experimental data and more complicated calculations.

a. Starting with the Gibbs–Duhem equation, show that at constant temperature,

$$RT \sum x_i \, d \ln(x_i \gamma_i) = \underline{V}^{ex} \, dP$$

and, for a binary mixture,

$$\left(\frac{x_1}{y_1} - \frac{x_2}{y_2} \right) dy_1 = \left(\frac{P\underline{V}^{ex}}{RT} - 1 \right) d \ln P$$

which can also be rewritten as

$$\frac{(y_1 - x_1)}{y_1(1 - y_1)} \frac{dy_1}{dx_1} = \frac{d \ln P}{dx_1}$$

since $P\underline{V}^{ex}/RT \ll 1$.

b. The equilibrium pressures above various mixtures of carbon tetrachloride and n-heptane at 50°C are given in the following table.

Mole Percent of CCl_4 in the Liquid	Pressure (bar)
0.0	0.1873
3.32	0.1956
9.83	0.2131
17.14	0.2320
30.24	0.2649
35.14	0.2765
43.24	0.2943
50.12	0.3097
57.00	0.3263
64.96	0.3425
73.23	0.3616
81.26	0.3765
89.92	0.3939
96.49	0.4055
100.0	0.4113

Source: Reprinted with permission of C. P. Smith and E. W. Engel, *J. Am. Chem. Soc.,* **51,** 2646 (1929). Copyright 1929 American Chemical Society.

Develop the x-y diagram for this system, and compute the liquid-phase activity coefficients of carbon tetrachloride and n-heptane.

c. Develop the x-y diagram for this system using the UNIFAC model.

10.2-7 For vapor-liquid equilibrium at low pressure (so the vapor phase is an ideal gas)

a. What is the bubble point pressure of an equimolar ideal liquid binary mixture?

b. What is the bubble point vapor composition of an equimolar ideal liquid binary mixture?

c. What is the bubble point pressure of an equimolar liquid binary mixture if the liquid mixture is nonideal and described by $\underline{G}^{ex} = Ax_1x_2$?

d. What is the bubble point vapor composition of an equimolar liquid binary mixture if the liquid mixture is nonideal and described by $\underline{G}^{ex} = Ax_1x_2$?

10.2-8 The following vapor-liquid equilibrium data for the n-pentane–acetone system at 1.013 bar were obtained by Lo, Beiker, and Karr [*J. Chem. Eng. Data,* **7**, 327 (1962)].

x_P	y_P	T (°C)	P_P^{vap} (bar)	P_A^{vap} (bar)
0.021	0.108	49.15	1.560	0.803
0.061	0.307	45.76	1.397	0.703
0.134	0.475	39.58	1.146	0.551
0.210	0.550	36.67	1.036	0.493
0.292	0.614	34.35	0.960	0.453
0.405	0.664	32.85	0.913	0.425
0.503	0.678	32.35	0.903	0.421
0.611	0.711	31.97	0.887	0.413
0.728	0.739	31.93	0.880	0.410
0.869	0.810	32.27	0.896	0.419
0.953	0.906	33.89	0.954	0.445

a. Are these data thermodynamically consistent?

b. Determine which of the activity coefficient models discussed in Chapter 9 best fits these data. This is most easily done using the MATHCAD worksheet ACTCOEFF on the CD-ROM accompanying this book and discussed in Appendix B.III

10.2-9 Use the UNIFAC model to predict the vapor-liquid behavior of the system in the previous problem, and compare the results with the experimental data.

10.2-10 Estimate, as best you can, the vapor-liquid equilibrium coexistence pressure and the composition of the vapor in equilibrium with a liquid containing 20 mol % ethanol, 40 mol % benzene, and 40 mol % ethyl acetate at 78°C.

10.2-11 The experimental data for the hexafluorobenzene-benzene mixture in Table 10.2-4 and Fig. 10.2-10 show a double azeotrope. Test the ability of common thermodynamic models, such as the equations of Wilson and van Laar, and the Redlich-Kister expansion to fit these data. Also, test whether the UNIFAC model predicts a double azeotrope for this system. (Note that hexafluorobenzene is also known as perfluorobenzene, and it is referred to as such in the Property program referred to in Chapter 6.)

10.2-12 Determine whether the data in Table 10.2-1 satisfy the Gibbs-Duhem integral consistency test.

10.2-13 Derive Eq. 10.2-26 for the constant-pressure ebulliometer.

10.2-14 Using the following data, estimate the total pressure and composition of the vapor in equilibrium with a 20 mol % ethanol (1) solution in water (2) at 78.15°C. *Data* (at 78.15°C):

Vapor pressure of ethanol (1) = 1.006 bar

Vapor pressure of water (2) = 0.439 bar

$$\lim_{x_1 \to 0} \gamma_1 = \gamma_1^\infty = 1.6931$$

$$\lim_{x_2 \to 0} \gamma_2 = \gamma_2^\infty = 1.9523$$

10.2-15 In vapor-liquid equilibrium the **relative volatility** α_{ij} is defined to be the ratio of the separation or K factor for species i to that for species j, that is,

$$\alpha_{ij} = \frac{K_i}{K_j} = \frac{y_i/x_i}{y_j/x_j}$$

In approximate distillation column calculations, the relative volatility is sometimes assumed to be a constant (independent of composition, temperature, and pressure). Test this assumption for the ethanol–ethyl acetate system using the following data:

$$RT \ln \gamma_i = 8.163 x_j^2 \text{ kJ/mol}$$

$$\ln P_{EOH}^{vap} = \frac{-4728.98}{T} + 13.4643 \quad P[=] \text{ bar}; \ T[=] \text{ K}$$

$$\ln P_{EAC}^{vap} = \frac{-3570.58}{T} + 10.4575$$

10.2-16 Based on vapor-liquid equilibrium data, some authors have claimed that the benzene (1)–cyclohexane (2) mixture is described by

$$\underline{G}^{ex} = A x_1 x_2$$

where A (J/mol) $= 3750 - 8T$ for T in degrees K.

a. Derive expressions for the activity coefficients of benzene and cyclohexane.

b. Determine the enthalpy and entropy changes on mixing when 1 mole of benzene and 2 moles of cyclohexane are mixed at $T = 300$ K and constant pressure.

c. Given the following vapor pressure data at $T = 320$ K, $P_B^{vap} = 0.3203$ bar and $P_C^{vap} = 0.3218$ bar, determine the bubble point pressure of the liquid in part (b) at $T = 320$ K, and the composition of the vapor in equilibrium with that liquid.

10.2-17 The relative volatility of component 2 to component 1 is defined to be

$$\alpha_{12} = \frac{y_2/x_2}{y_1/x_1}$$

a. Develop an expression for the relative volatility for a mixture described by the one-constant Margules equation. Discuss how the relative volatility depends upon temperature, pressure, and composition.

b. What would be the relative volatility at low pressures if the two components form an ideal mixture, and how would the relative volatility in this case depend on temperature, composition, and pressure?

10.2-18 a. Develop an expression for the vapor-phase mole fraction of species 1 in an ideal equimolar binary mixture at low pressures in terms of only the pure component vapor pressures.

   b. Repeat the derivation of part (a) for a nonideal mixture described by the one-constant Margules equation.

10.2-19 a. Develop an expression for how the bubble point pressure of a binary mixture changes with temperature at constant composition.

   b. How does the result for part (a) change if one of the components in the mixtures has such a low vapor pressure as to be considered involatile?

10.2-20 Estimate the vapor-liquid equilibrium of n-hexane and toluene as a function of composition at $T = 75°C$ using regular solution theory and the UNIFAC model.

10.2-21 After an esterification reaction, it is necessary to separate the product methyl acetate from the unreacted methanol to obtain both components at a high level of purity. It is known that at 20°C, this system forms an azeotrope with the following properties: $x_{MA} = y_{MA} = 0.754$, $x_{MeOH} = y_{MeOH} = 0.246$, and $P = 183.54$ mm Hg.

   A process engineer has suggested that the way to get two relatively pure components is to use two distillation columns operating at different conditions. The first column would be operated to get one of the relatively pure components and the azeotrope. This azeotrope would be fed into the second column, operating at a higher temperature, 50°C. By this argument, the higher temperature would shift the azeotrope, so that the products from the second column would be the other pure component and the new azeotrope (which would then be fed back to the first column). To determine whether this proposed process is possible, compute the azeotropic composition of a methyl acetate–methanol mixture at 50°C. For simplicity, assume that the parameters in the activity coefficient you use are independent of temperature.

   The pure component vapor pressures are as follows:

$$\log_{10} P_{MA}^{vap} \text{ (mm Hg)} = 7.065\,24 - \frac{1157.630}{T \text{ (K)} - 53.424}$$

$$\log_{10} P_{MeOH}^{vap} \text{ (mm Hg)} = 8.080\,97 - \frac{1582.271}{T \text{ (K)} - 33.424}$$

10.2-22 The n-octane (1)–tetraethyl methane (2) mixture has been studied at 50°C. From vapor pressure and calorimetric measurements, the following information is available:

$$\underline{G}^{ex} = 630 x_1 x_2 \text{ J/mol}$$

$$\underline{H}^{ex} = -335 x_1 x_2 \text{ J/mol}$$

$$\ln P_i^{vap} = A_i - \frac{B_i}{T + C_i}$$

where

i	A_i	B_i	C_i
1	9.3225	3120.29	−63.63
2	9.2508	3341.62	−57.57

with P in bar and T in K.

   a. Calculate the composition of the vapor that is in equilibrium with n-octane–tetraethyl methane mixtures at 380 K. In this calculation assume that the activity coefficients are independent of pressure.

   b. Repeat the calculation above at a fixed total pressure of 380 mm Hg.

   c. Repeat parts (a) and (b) using the predictive UNIFAC model.

10.2-23 Eichinger and Flory [*Trans. Farad. Soc.,* **64**, 2053–2060 (1968)] report the following data for the activity of benzene ($a_B = x_B \gamma_B$) in polyisobutylene (MW = 40 000) and 10°C as a function of the mass ratio of benzene to polyisobutylene, m_B/m_{PIB}:

m_C/m_{PIB}	$a_C = x_B \gamma_C$
0.8331	0.9811
0.5543	0.9595
0.291	0.8388

Using the data in Illustration 10.2-7, compare the predictions of the Flory-Huggins theory using $\chi = 1.0$ with the data above, and compute the equilibrium partial pressure of benzene over benzene-polyisobutylene mixtures.

10.2-24 Eichinger and Flory [*Trans. Farad. Soc.,* **64**, 2061–2065 (1968)] reported the following data for the activity of cyclohexane ($a_C = x_C \gamma_C$) in polyisobutylene (MW = 40 000) and 25°C as a function of the mass ratio of cyclohexane to polyisobutylene, m_C/m_{PIB}.

m_C/m_{PIB}	$a_C = x_C\gamma_C$	m_C/m_{PIB}	$a_C = x_C\gamma_C$
1.318	0.9598	0.307	0.6937
0.668	0.8758	0.232	0.6105
0.434	0.7836	0.198	0.5537
0.390	0.7623	0.147	0.4625

Determine the value of the Flory χ parameter in the Flory-Huggins model that gives the best fit of the data above, and compute the equilibrium partial pressure of cyclohexane over cyclohexane-polyisobutylene mixtures.

10.2-25 Eichinger and Flory [*Trans. Farad. Soc.*, **64**, 2066–2072 (1968)] reported the following data for the activity of *n*-pentane ($a_P = x_P\gamma_P$) in polyisobutylene (MW = 40 000) and 10°C as a function of the mass ratio of *n*-pentane to polyisobutylene, m_P/m_{PIB}.

m_P/m_{PIB}	$a_P = x_P\gamma_P$	m_P/m_{PIB}	$a_P = x_P\gamma_P$
1.405	0.9897	0.227	0.7684
0.476	0.9263	0.153	0.6434
0.363	0.8804	0.0786	0.4414
0.269	0.8093	0.0294	0.2120

Determine the value of the Flory χ parameter in the Flory-Huggins model that gives the best fit of the data above, and compute the equilibrium partial pressure of pentane over pentane-polyisobutylene mixtures.

10.2-26 The partial pressure of water above aqueous hydrochloric acid solutions can be represented by

$$\log_{10} P = A - \frac{B}{T}$$

where P is the pressure in bar and T is the temperature in K. The values of A and B are given in the table:

wt % HCl	A	B
10	6.123 57	2295
20	6.103 70	2334
30	6.126 10	2422
40	6.464 16	2647

Source: R. H. Perry, D. W. Green, and J. O. Maloney, eds., *The Chemical Engineers' Handbook*, 6th ed., McGraw-Hill, New York (1984), pp. 3–64.

The vapor pressure of pure water is given in Problem 7.12. Compute the activity coefficient of wa-

ter in each of the hydrochloric acid solutions in the table at 25°C. (*Hint:* hydrogen chloride can be assumed to completely ionize in aqueous solution.)

10.2-27 The following data for the partial pressure of water vapor over aqueous solutions of sodium carbonate at 30°C are given in *The Chemical Engineers' Handbook*, 5th ed. (R. H. Perry and C. H. Chilton, eds., McGraw-Hill, New York, 1973), pp. 3–68.

wt % Na_2CO_3	0	5	10	15	20	25	30
P_{H_2O} kPa	4.24	4.16	4.05	3.95	3.84	3.71	3.52

Compute the activity coefficient of water in each of these solutions. (*Hint:* Does sodium carbonate ionize in aqueous solution?)

10.2-28 In this section it was shown that the excess entropy and excess enthalpy can be determined from various temperature derivatives of the excess Gibbs energy. These and other excess thermodynamic functions can also be computed directly from derivatives of the activity coefficients. Show that in a binary mixture the following equations can be used for such calculations:

$$\underline{S}^{ex} = -\left(\frac{\partial \underline{G}^{ex}}{\partial T}\right)_{P,\underline{x}}$$
$$= -RT\left(x_1\frac{\partial \ln\gamma_1}{\partial T} + x_2\frac{\partial \ln\gamma_2}{\partial T}\right)$$
$$\quad - R(x_1\ln\gamma_1 + x_2\ln\gamma_2)$$

$$\underline{H}^{ex} = -T^2\frac{\partial(\underline{G}^{ex}/T)_{P,\underline{x}}}{\partial T}$$
$$= -RT^2\left(x_1\frac{\partial \ln\gamma_1}{\partial T} + x_2\frac{\partial \ln\gamma_2}{\partial T}\right)$$

$$\underline{V}^{ex} = \left(\frac{\partial \underline{G}^{ex}}{\partial P}\right)_{T,\underline{x}}$$
$$= RT\left(x_1\frac{\partial \ln\gamma_1}{\partial P} + x_2\frac{\partial \ln\gamma_2}{\partial P}\right)$$

$$\underline{U}^{ex} = -RT\left\{T\left(x_1\frac{\partial \ln\gamma_1}{\partial T} + x_2\frac{\partial \ln\gamma_2}{\partial T}\right)\right.$$
$$\left. + P\left(x_1\frac{\partial \ln\gamma_1}{\partial P} + x_2\frac{\partial \ln\gamma_2}{\partial P}\right)\right\}$$

$$C_P^{ex} = -2RT\left(x_1\frac{\partial \ln\gamma_1}{\partial T} + x_2\frac{\partial \ln\gamma_2}{\partial T}\right)$$
$$\quad - RT^2\left(x_1\frac{\partial^2 \ln\gamma_1}{\partial T^2}x_2\frac{\partial^2 \ln\gamma_2}{\partial T^2}\right)$$

Note that in the activity coefficient derivatives, all variables from the set T, P, $\underline{x}$, other than the one being varied, are fixed.

10.2-29 The following vapor-liquid equilibrium data have been reported[6] for the system 1,2-dichloroethane (1) + n-heptane (2) at 343.15 K.

P (mm Hg)	x_1	y_1
302.87	0.0000	0.0000
372.62	0.0911	0.2485
429.28	0.1979	0.4174
466.48	0.2867	0.5052
491.02	0.3674	0.5590
509.40	0.4467	0.6078
520.41	0.5044	0.6535
525.71	0.5733	0.6646
533.44	0.6578	0.7089
536.58	0.7644	0.7696
535.78	0.8132	0.7877
530.90	0.8603	0.8201
524.21	0.8930	0.8458
515.15	0.9332	0.8900
505.15	0.9572	0.9253
498.04	0.9812	0.9537
486.41	1.0000	1.0000

  a. Compute the activity coefficients for this system at each of the reported compositions.

  b. Calculate the excess Gibbs energy at each composition

  c. Obtain values of the van Laar parameters that best fit these data.

10.2-30 A binary liquid solution, having mole fraction x of component 1, is in equilibrium with a vapor that has mole fraction y of that component. Show that for this mixture the effect of a change in temperature on the equilibrium pressure at fixed liquid composition is approximately

$$\left(\frac{\partial \ln P}{\partial T}\right)_x = \frac{(\Delta_{vap}\underline{H})_{mix}}{RT^2}$$

where $(\Delta_{vap}\underline{H})_{mix}$ is the heat of vaporization of y moles of component 1 and $(1 - y)$ moles of component 2 from a large volume of solution.

10.2-31 Estimate the equilibrium pressure and isobutane vapor-phase mole fraction as a function of liquid composition for the isobutane-furfural system at 37.8°C. At this temperature the vapor pressure of furfural is 0.493 kPa and that of isobutane is 490.9 kPa. (*Hint:* See Illustration 11.2-2.)

10.2-32 Polyethylene of molecular weight 2800 is to be dissolved in compressed ethylene at 10°C. Assum-ing that as a result of the similarity of ethylene to polyethylene, there is no enthalpic interaction between these species, estimate the excess Gibbs energy and activity coefficients of ethylene and polyethylene as a function of composition.

10.2-33 A mixture of 80 mol % acetone and 20 mol % water is to be transported in a pipeline within a chemical plant. This mixture can be transported as either a liquid or a gas.

  a. A reciprocating pump would be used with a liquid mixture. Because of vapor lock, this pump will cease functioning if any vapor is present. Compute the minimum pressure that must be maintained at the pump inlet so that no vapor is formed when the liquid temperature is 100°C.

  b. A centrifugal pump would be used with a gas mixture. To prevent erosion of the pump blades, only vapor should be present. If the pump effluent is 100°C, compute the maximum pump effluent pressure so that no liquid is formed.

  Data:

Temperature (°C)	56.6	78.6	113.0
Vapor pressure of acetone (bar)	1.013	2.026	5.065

10.2-34 The figure that follows for the ethanol + water system is an unusual one in that it shows both vapor-liquid equilibrium and the enthalpy concentration diagrams on a single plot. This is done as follows. The lower collection of heavy lines give the enthalpy concentration data for the liquid at various temperatures and the upper collection of lines is the enthalpy-concentration data for the vapor, each at two pressures, 0.1013 and 1.013 bar. (There are also enthalpy-concentration lines for several other temperatures.) The middle collection of lines connect the equilibrium compositions of liquid and vapor. For example, at a pressure of 1.013 bar, a saturated vapor containing 71 wt % ethanol with an enthalpy of 1535 kJ/kg is in equilibrium with a liquid containing 29 wt % ethanol with an enthalpy of 315 kJ/kg at a temperature of 85°C. Note also that the azeotropes that form in the ethanol + water system are indicated at each pressure.

Consider the vapor-liquid equilibrium discussed above, in which 90 percent of the mixture by weight is liquid and 10 percent is vapor. If the pressure on this mixture is now reduced to 0.1013 bar by passing through a Joule-Thomson expansion valve, what are the compositions of the vapor and liquid phases in equilibrium, what is the

---

[6]Reprinted with permission of R. Eng and S. I. Sandler, *J. Chem. Eng. Data,* **29,** 156 (1984). Copyright 1984 American Chemical Society.

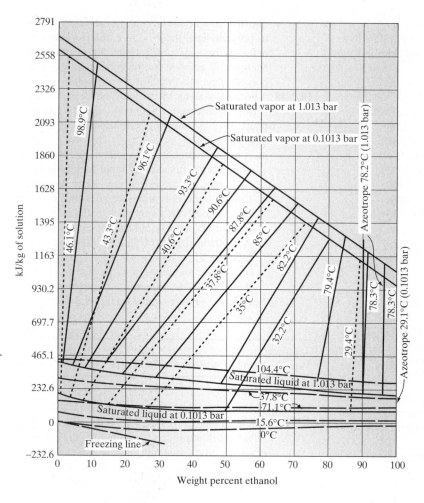

The enthalpy concentration diagram for ethanol + water system. The reference states are the pure liquids at 0°C. Based on data in *Chemical Process Principles, Part 1: Material and Energy Balances*, 2nd ed., by O. A. Hougen, K. M. Watson, and R. A. Ragatz, John Wiley & Sons, New York, 1954, p. 327. This figure appears as an Adobe PDF file on the CD-ROM accompanying this book, and may be enlarged and printed for easier reading and for use in solving problems.

equilibrium temperature, and what are the weight fractions of vapor and liquid?

10.2-35 A binary mixture of components 1 and 2 is in vapor-liquid equilibrium.

  a. At 90°C and 1.8505 bar pressure, a vapor of composition $y_1 = 0.3767$ coexists with a liquid of composition $x_1 = 0.4$. Use these data to determine the van Laar parameters for this mixture.

  b. Determine whether the mixture has an azeotrope at 90°C, and if so, determine its composition and whether it is a maximum-pressure or minimum-pressure azeotrope.

  c. Obtain the P-x-y and x-y diagrams for this system at 90°C.

  d. An equimolar mixture of species 1 and 2 at initially very low pressure is compressed at a constant temperature of 90°C. At what pressure does the first drop of liquid form, and what

is its composition? At what pressure does the last bubble of vapor disappear, and what was its composition?

Data: The vapor pressures of the components are given by

$$\log_{10} P_i^{\text{vap}} = A_i - \frac{B_i}{T}$$

for pressure in bar and T in K, where $A_1 = 4.125$, $B_1 = 1500$, $A_2 = 5.000$, and $B_{12} = 1750$.

10.2-36 The equilibrium total pressures above liquid mixtures (that is, P-T-x data) for the system ethylene bromide and 1-nitropropane at 75°C are given in the following table.

Develop an x-y diagram for this system, and compute the liquid-phase activity coefficients of ethylene bromide and 1-nitropropane. Use the UNI-

Mole Percent of Ethylene Bromide in Liquid	Pressure (bar)	Mole Percent of Ethylene Bromide in Liquid	Pressure (bar)
0.0	0.1533	50.85	0.1773
2.98	0.1556	65.62	0.1773
3.52	0.1568	76.48	0.1745
5.75	0.1597	88.01	0.1699
15.80	0.1659	94.31	0.1652
26.98	0.1719	100.00	0.1596
39.95	0.1760		

Source: Reprinted with permission from J.R. Lacher, W. B. Buck, and W. H. Parry, *J. Am. Chem. Soc.*, **63**, 2422. Copyright 1941 American Chemical Society.

FAC model to predict activity coefficients at other compositions. How do the two sets of activity coefficients compare? (*Hint*: See Problem 10.2-6.)

10.2-37 Use the UNIFAC model to predict the vapor-liquid equilibria for the benzene + 2,2,4-trimethylpentane system at 55°C, and compare the results with the experimental data in Illustration 10.2-4 and with VLE predictions using the Regular Solution model.

10.2-38 Use the UNIFAC model to predict the vapor-liquid equilibria for the acetone + water system at 25°C, and compare the results with the experimental data that can be found in the DECHEMA data series.

10.2-39 The following vapor-liquid equilibrium data have been reported[7] for the system water (1) + 1,4-dioxane (2) at 323.15 K.

P (mm Hg)	x_1	y_1
120.49	0.0000	0.0000
140.85	0.0560	0.1920
151.16	0.0970	0.2680
159.17	0.1700	0.3450
164.57	0.2160	0.3830
165.65	0.2980	0.4030
167.89	0.3660	0.4250
167.74	0.4400	0.4430
167.79	0.4460	0.4460
167.95	0.4840	0.4510
166.84	0.5390	0.4550
165.48	0.6290	0.4660
160.82	0.7490	0.4950
155.14	0.8110	0.5430
142.64	0.8900	0.6040
114.76	0.9670	0.7950
92.51	1.0000	1.0000

a. Compute the activity coefficients for this system at each of the reported compositions.
b. Are these data thermodynamically consistent?
c. Plot the excess Gibbs energy for this system as a function of composition.

Determining the activity coefficient parameters values in several of the following problems is most easily accomplished using the MATHCAD worksheet ACTCOEFF on the CD-ROM that accompanies this book.

10.2-40 Determine the parameter values in the two-constant Margules equation that best fit the data of Problem 10.2-39.

10.2-41 Determine the parameter values in the van Laar model that best fit the data of Problem 10.2-39.

10.2-42 Determine the parameter values in the Wilson model that best fit the data of Problem 10.2-39.

10.2-43 Determine the parameter values in the NRTL model that best fit the data of Problem 10.2-39.

10.2-44 Determine the parameter values in the UNIQUAC model that best fit the data of Problem 10.2-39.

10.2-45 Compare the predictions of the UNIFAC model with the data of Problem 10.2-39.

10.2-46 The following equilibrium pressure data for the system water (1) + 1,4-dioxane (2) at 353.15 K have been reported.[8]

P (mm Hg)	x_1	P (mm Hg)	x_1
382.8	0.000	575.5	0.600
476.0	0.100	569.5	0.700
526.5	0.200	550.0	0.800
556.5	0.300	501.5	0.900
571.0	0.400	355.1	1.000
576.5	0.500		

Fit the P versus x_1 data with a suitable polynomial. Then use the synthethic method; that is use the equation developed in Problem 10.2-6,

$$\frac{(y_1 - x_1)}{y_1(1 - y_1)} \frac{dy_1}{dx_1} = \frac{d \ln P}{dx_1}$$

to estimate the vapor compositions at each of the reported liquid-phase mole fractions.

10.2-47 Determine the parameter values in the two-constant Margules equation that best fit the P versus x_1 data of Problem 10.2-46, and then use this activity coefficient model to estimate the vapor-phase compositions.

[7] G. Kortum and V. Valent, *Ber. Bunsenges Phys. Chem.* **81**, 752 (1977).
[8] F. Hovorka, R. A. Schaefer, and D. Dreisbach, *J. Am. Chem. Soc.*, **58**, 2264 (1935).

10.2-48 Determine the parameter values in the van Laar model that best fit the the P versus x_1 data of Problem 10.2-46, and then use this activity coefficient model to estimate the vapor-phase compositions.

10.2-49 Determine the parameter values in the NRTL model that best fit the the P versus x_1 data of Problem 10.2-46, and then use this activity coefficient model to estimate the vapor-phase compositions.

10.2-50 Determine the parameter values in the Wilson equation that best fit the the P versus x_1 data of Problem 10.2-46, and then use this activity coefficient model to estimate the vapor-phase compositions.

10.2-51 Determine the parameter values in the UNIQUAC model that best fit the the P versus x_1 data of Problem 10.2-46, and then use this activity coefficient model to estimate the vapor-phase compositions.

10.2-52 Compare the predictions for total pressure from the UNIFAC model with the data of Problem 10.2-46.

10.2-53 **a.** Joe Udel argues that if in a low-pressure binary mixture,

$$\left(\frac{\partial P}{\partial x_1}\right)_{T,x_1=0} \quad \text{and} \quad \left(\frac{\partial P}{\partial x_1}\right)_{T,x_1=1}$$

have opposite signs, the system will have an azeotrope. Can you simply prove or disprove his contention? (The simpler the proof, the better.)

b. Will the system for which, at 65°C, $P_1^{\text{vap}} = 0.260$ bar and $P_2^{\text{vap}} = 0.899$ bar described by

$$\underline{G}^{\text{ex}} = 1.7RTx_1x_2$$

have an azeotrope?

10.2-54 The following excess Gibbs energy model describes the 1-propanol (1) + n-hexane (2) system

$$\underline{G}^{\text{ex}} = RTx_1x_2(A_1x_1 + A_2x_2)$$

In the temperature range near 65°C the parameter values in this equation are $A_1 = 1.867$ and $A_2 = 1.536$. The vapor pressure of 1-propanol at this temperature is 0.260 bar and that of n-hexane is 0.899 bar.

a. What is the excess enthalpy of mixing of this system as a function of temperature and composition?

b. What is the entropy of mixing of this system as a function of temperature and composition?

c. Obtain expressions for the activity coefficients of each compound.

d. Obtain values for the infinite-dilution activity coefficients for each compound in this mixture.

e. Does this system have an azeotrope?

10.2-55 Calculate the boiling point of aqueous sodium chloride solutions at ambient pressure over the range from 0 to 10 M NaCl. In this calculation, use the following expression for the vapor pressure of pure water:

$$\ln P^{\text{vap}}(T) = 13.149 - \frac{4903}{T}$$

for temperature in K and vapor pressure in bar.

10.3 HIGH-PRESSURE VAPOR-LIQUID EQUILIBRIA USING EQUATIONS OF STATE (ϕ-ϕ METHOD)

The discussion of the previous section was concerned with low-pressure vapor-liquid equilibria and involved the use of activity coefficient models. Here we are interested in high-pressure phase equilibrium in fluids in which both phases are describable by equations of state, that is, the ϕ-ϕ method. One example of the type of data we are interested in describing (or predicting) is shown in Fig. 10.3-1 for the ethane-propylene system. There we see the liquid (bubble point) and vapor (dew point) curves for this system at three different isotherms. At each temperature the coexisting vapor and liquid phases have the same pressure and thus are joined by horizontal tie lines, only one of which has been drawn. The intersections of these tie lines with the bubble and dew curves give the compositions of the coexisting equilibrium liquid and vapor phases, respectively.

Figure 10.3-1 shows some of the variety of phase behavior that occurs in hydrocarbon mixtures at high pressure. The lowest isotherm, 261 K, is below the critical temperatures of both ethane ($T_c = 305.4$ K) and propylene ($T_c = 365.0$ K). In this

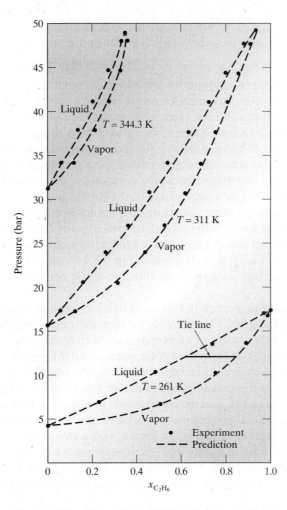

Figure 10.3-1 Constant-temperature vapor-liquid equilibrium data for the ethane-propylene system. [Reprinted with permission of R. A. McKay, H. H. Reamer, B. H. Sage, and W. N. Lacey, *Ind. Eng. Chem.*, **43**, 2112 (1951). Copyright 1951 American Chemical Society.]

case both vapor and liquid exist at all compositions, and the isotherm is qualitatively similar to the low-pressure isotherm for the *n*-hexane–triethylamine system of Fig. 10.1-1*c*. The next isotherm, at $T = 311$ K, is slightly above the critical temperature of ethane but below T_c for propylene. Thus, pure ethane cannot exist as a liquid at this temperature, nor can mixtures very rich in ethane. As a result, we see the bubble point and dew point curves join—not at pure ethane (as on the $T = 261$ K isotherm), but rather at an ethane mole fraction of 0.93. That is, at $T = 311$ K, both vapor and liquid can exist for ethane mole fractions in the range of 0 to 0.93, but at higher ethane mole fractions only a vapor is present, even at very high pressures. The point at which the bubble and dew curves intersect,

$$x_{C_2H_6} = 0.93$$
$$T = 311 \text{ K}$$
$$P = 49.8 \text{ bar}$$

is one critical point for the ethane-propylene mixture.

The highest temperature isotherm in the figure, 344.3 K, is well above T_c of ethane, so only liquids dilute in ethane are possible. Indeed, here we see that the bubble and dew point curves intersect at

$$x_{C_2H_6} = 0.35$$
$$T = 344.3 \text{ K}$$
$$P = 48.6 \text{ bar}$$

which is another critical point of the ethane-propylene mixture. As should be evident from this discussion, there is not a single critical point for a mixture, but rather a range of critical points at a different temperature and pressure for each composition.

In Fig. 10.3-2 we have plotted, for various fixed compositions, the bubble and dew point pressures of this mixture as a function of temperature. The leftmost curve in this figure is the vapor pressure of pure ethane as a function of temperature, terminating in the critical point of ethane (for a pure component, the coexisting vapor and liquid are necessarily of the same composition, so the bubble and dew pressures are identical and equal to the vapor pressure). Similarly, the rightmost curve is the vapor pressure of pure propylene, terminating at the propylene critical point. The intermediate curves (loops) are the bubble and dew point curves relating temperature and pressure for various fixed compositions. Finally, there is a line in Fig. 10.3-2 connecting the critical points of the mixtures of various compositions; this line is the **critical locus** of ethane-propylene mixtures.

High-pressure phase equilibria can be much more complicated than the cases shown in Figs. 10.3-1 and 10.3-2, especially for mixtures containing dissimilar components—for example, either water or carbon dioxide with hydrocarbons or oxygenated hydrocarbons. Examples of the pressure-temperature (P-T) projections of the various types of critical loci that have been observed for binary mixtures are shown in Fig. 10.3-3. When looking at these examples, remember that there are three independent variables (temperature, pressure, and the composition of one of the species), so that these figures are two-dimensional projections of surfaces in a three-dimensional figure. Thus, the mixture composition, which is not shown, varies along each of these critical lines.

The critical locus in Figure 10.3-3a is very much like that for the ethane-propylene system in Fig. 10.3-2; this is referred to as category I phase behavior. For such systems the critical line starts at the critical point of pure component 1 (C_1), and as the mixture

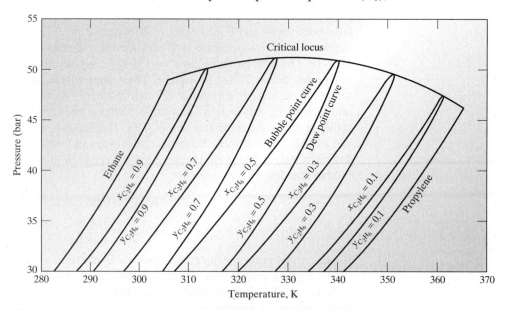

Figure 10.3-2 The bubble point, dew point, and critical locus for the ethane-propylene system.

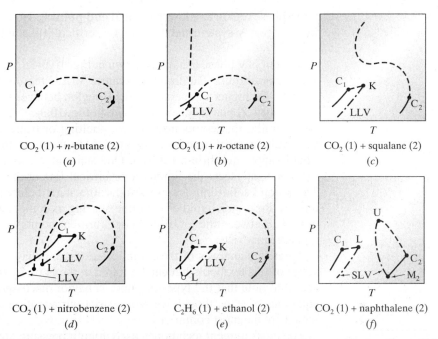

Figure 10.3-3 The six general categories of *P-T* projections of phase equilibrium lines of binary mixtures. In each case — is a pure component vapor pressure line; – – – is a binary mixture vapor-liquid or liquid-liquid critical line; and – – – is a mixture three-phase line (VLL in *a–e,* SLV in *f*). [Adapted from M. E. Paulaitis, V. J. Krukonis, R. T. Kurnick, and R. C. Reid, *Rev. Chem. Eng.*, **1**, 179 (1983). Used with permission.]

becomes richer in the second component, the critical line goes smoothly to the critical point C_2.

Category II behavior, an example of which is shown in Fig.10.3-3b, is slightly more complicated in that at low temperatures and high pressures there is a region of liquid-liquid equilibrium (LLE), whereas at low temperatures and low pressures there is a region where three phases, two liquids (of differing composition) and a vapor, are all in equilibrium (VLLE). Consequently, there are two sets of critical lines in a category II system: one for vapor-liquid behavior of the type we have already considered, and another that begins as a liquid-liquid-vapor line at low pressures and becomes a liquid-liquid critical line at high pressures, terminating at the highest temperature at which liquid-liquid equilibrium exists. This temperature is the upper **critical solution temperature** and will be considered in detail in Sec. 11.2. Figure 10.3-3b is an example of category II phase behavior. At very high pressures, usually beyond the range of interest to chemical engineers, the liquid-liquid equilibrium line intersects a region of solid-liquid-liquid equilibrium (SLLE). This is not shown in Fig. 10.3-3b.

Category III phase behavior, shown in Fig. 10.3-3c, is similar to category II behavior, except that the region of liquid-liquid-vapor equilibrium occurs at higher temperatures and, as the composition varies, intersects the vapor-liquid critical curve at a **critical end point** K. Thus, there is one vapor-liquid critical line originating at component 1 (C_1) and terminating at the critical end point, and a second starting at C_2 that merges smoothly into the liquid-liquid equilibrium line at high pressures.

Category V phase behavior, shown in Fig 10.3-3e, is similar to category III behavior, except that the critical line starting at component 2 intersects the liquid-liquid-vapor three-phase region at a **lower critical end point** L. Note that there is no region of LLE in a category V system and there are two critical end points (K and L) of the three-phase region.

Category IV phase behavior, shown in Fig. 10.3-3*d*, has two regions of liquid-liquid-vapor equilibrium. The low-temperature VLLE region exhibits category II behavior, whereas the higher-temperature VLLE region behaves like a category V system.

Category VI phase behavior, shown in Fig. 10.3-3*f*, occurs with components that are so dissimilar that component 2 has a melting or triple point (M_2) that is well above the critical temperature of component 1. In this case there are two regions of solid-liquid-vapor equilibrium (SLVE). One starts at the triple point of pure component 2 (M_2) and intersects the liquid-vapor critical line at the **upper critical end point** U. The second solid-liquid-vapor critical line starts below the melting point M_2 and intersects the vapor-liquid critical line starting at component 1 at the lower critical end point L. Between the lower and upper critical points only solid-vapor (or solid-fluid) equilibrium exists.

A detailed analysis of the complicated phase behavior of Fig. 10.3-3 is beyond the scope of this textbook, but can be found in the book by Rowlinson and Swinton.[9] It is useful to note that the types of phase behavior discussed here can be predicted using equations of state, though we will restrict our attention largely to category I systems, which are the most common.

Measurement of high-pressure vapor-liquid equilibria

The measurement techniques used at high pressure are similar in principle to those used at low pressure, but different in practice since leakproof metal tubing, fittings, and equilibrium cells (frequently with sapphire windows to enable one to see inside the cell) are used. Also, circulation of the vapor, liquid, or both to ensure that there is good contact between the phases and that equilibrium is obtained is usually done by pumps, rather than by heating to promote boiling, as is the case at low pressures. One example of a high-pressure dynamic VLE cell is shown in Fig. 10.3-4.

A static cell apparatus could also be used to measure high-pressure vapor-liquid equilibrium. Schematically, the equipment would look similar to that in Fig. 10.2-9, except that the glass flasks would be replaced with metal vessels.

We now turn from the qualitative description of high-pressure phase equilibria and its measurement to the quantitative description, that is, to the correlation or prediction of vapor-liquid equilibrium for hydrocarbon (and light gas) systems, of which the ethane-propylene system is merely one example. Our interest will be only in systems describable by a single equation of state for both the vapor and liquid phases, as the case in which the liquid is described by an activity coefficient model was considered in the previous section.

The starting point for any phase equilibrium calculation is, of course, the equality of fugacities of each species in each phase, that is,

Starting point for all phase equilibrium calculations

$$\boxed{\bar{f}_i^{\mathrm{L}}(T, P, \underline{x}) = \bar{f}_i^{\mathrm{V}}(T, P, \underline{y})} \tag{10.3-1}$$

Here, however, we will use an equation of state to calculate species fugacities in both phases. For example, when using the Peng-Robinson equation of state, Eq. 9.4-10 is used to compute the fugacity of a species in the vapor phase and Eq. 9.4-11 for the liquid phase (of course, with the composition and compressibility appropriate to each phase).

[9]J. S. Rowlinson and F. L. Swinton, *Liquids and Liquid Mixtures,* 3rd ed., Butterworths, London (1982), Chapter 6.

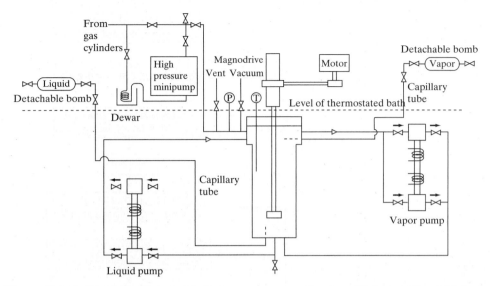

Figure 10.3-4 Schematic diagram of a high-pressure vapor-liqud equilibrium cell in which both the vapor and liquid are constantly circulated to ensure that the vapor and liquid phases are in equilibrium. The detachable bombs are used to collect samples for chemical analysis. Items P and T are ports for pressure and temperature measurement, respectively.

Phase equilibrium calculations with equations of state are iterative and sufficiently complicated to be best done on a digital computer. Consider, for example, the calculation of the bubble point pressure and vapor composition for a liquid of known composition at temperature T. One would need to make an initial guess for the bubble point pressure, P_B, and the vapor mole fractions (or, perhaps more easily, for the values of $K_i = y_i/x_i$), and then check to ensure that $\sum y_i = 1$ and that the equality of species fugacities (Eq. 10.3-1) are satisfied for each species with the fugacities calculated from the equation of state. If these restrictions are not satisfied, the pressure and K_i values must be adjusted and the calculation repeated. A flow diagram for one algorithm for solving this problem is given in Fig. 10.3-5.

The initial guesses for the bubble point pressure P_B and for the $K_i = y_i/x_i$ values for all species in the mixture do not affect the final solution to the problem, but may influence the number of iterations required to obtain the solution. One possible set of initial guesses is obtained by assuming ideal liquid and vapor mixtures so that

$$P_B = \sum x_i P_i^{\text{vap}}(T) \tag{10.3-2}$$

and

$$K_i = \frac{y_i}{x_i} = \frac{P_i^{\text{vap}}(T)}{P_B} \tag{10.3-3}$$

where the pure component vapor pressure can be estimated using the Antoine equation, Eq. 7.7-8, with parameters for the fluid of interest, or by using the equation of state as described in Sec. 7.5.

Although the algorithm described in Fig. 10.3-5 is specific to the computation of the bubble point pressure, slight changes make it applicable to other phase equilibrium calculations. For example, by specifying P and replacing $P_B = P_B \sum y_i'$ in the iteration sequence by $T_B = T_B / \sum y_i'$, an algorithm for the bubble point temperature calculation

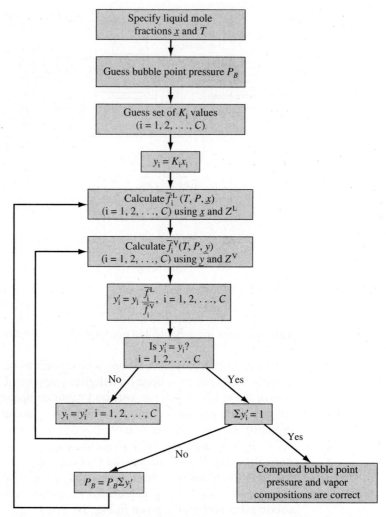

Figure 10.3-5 Flow diagram of an algorithm for the bubble point pressure calculation using an equation of state.

at fixed pressure is obtained. It is only slightly more difficult to change the calculational procedure so that the dew point pressure or temperature calculations can be made; this is left to you (Problem 10.3-5).

We consider only one additional type of phase equilibrium calculation here, the isothermal flash calculation discussed in Sec. 10.1. In this calculation one needs to satisfy the equality of species fugacities relation (Eq. 10.3-1) as in other phase equilibrium calculations and also the mass balances (based on 1 mole of feed of mole fractions $z_{i,F}$) discussed earlier,

$$x_i L + y_i V = z_{i,F} \quad i = 1, 2, \ldots, C \tag{10.1-17}$$

$$L + V = 1 \tag{10.1-18}$$

and the summation conditions (Eqs. 10.1-5 and 10.1-6)

$$\sum x_i = 1 \quad \text{and} \quad \sum y_i = 1 \tag{10.3-4}$$

In this calculation, T and P are known, but the liquid-phase mole fractions (x_i), the vapor-phase mole fractions (y_i), and the liquid-to-vapor split (L/V) are unknowns.

An algorithm for solving the flash problem is given in Fig. 10.3-6. It is based on making initial guesses for the equilibrium ratios $K_i = y_i/x_i$ (see Eq. 10.3-3) and for the fraction of liquid, L, and using the following equations obtained by simple rearrangement:

$$x_i = \frac{z_{i,F}}{L + K_i(1 - L)} \tag{10.3-5}$$

$$y_i = K_i x_i$$

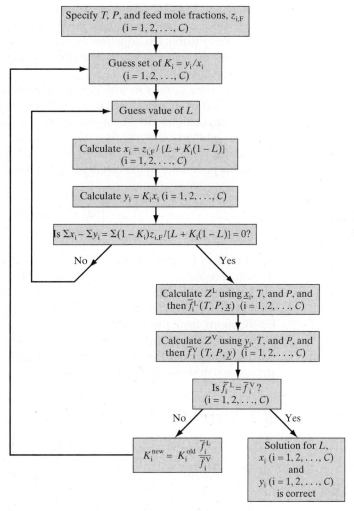

Figure 10.3-6 Flow diagram of an algorithm for the isothermal flash calculation using an equation of state.

Furthermore, from Eqs. 10.3-4, we have

$$\sum_{i=1} x_i = \sum \frac{z_{i,F}}{L + K_i(1 - L)} = 1 \qquad \textbf{(10.3-6a)}$$

and

$$\sum_{i=1} y_i = \sum_{i=1} \frac{K_i z_{i,F}}{L + K_i(1 - L)} = 1 \qquad \textbf{(10.3-6b)}$$

or equivalently,

$$\sum_{i=1} x_i - \sum_{i=1} y_i = \sum_{i=1} \frac{(1 - K_i)z_{i,F}}{L + K_i(1 - L)} = 0 \qquad \textbf{(10.3-6c)}$$

It is Eqs. 10.3-5 and 10.3-6 that are used in the algorithm of Fig. 10.3-5. Computer programs and MATHCAD worksheets for bubble point temperature, bubble point pressure, dew point temperature, dew point pressure, and isothermal flash calculations using the Peng-Robinson equation of state with generalized parameters (Eqs. 6.7-1 to 6.7-4), and the van der Waals one-fluid mixing rules (Eqs. 9.4-8 and 9.4-9) are discussed in Appendix B.

Now that the manner in which phase equilibrium calculations can be performed using equations of state has been discussed, we are in a position to consider the accuracy of such calculations. To begin, we again consider the ethane-propylene data in Fig. 10.3-1 using the isothermal flash algorithm, the Peng-Robinson equation of state, and the van der Waals one-fluid mixing rules (Eqs. 9.4-8 and 9.4-9), and setting the single adjustable parameter in the calculation, the **binary interaction parameter** k_{ij} in

Binary interaction parameter k_{ij}

$$\boxed{a_{ij} = \sqrt{a_{ii}a_{jj}}(1 - k_{ij})} \qquad \textbf{(10.3-7)}$$

equal to zero (so that, in fact, there are no adjustable parameters). Using the Peng-Robinson equation-of-state programs or MATHCAD worksheets discussed in Appendix B, the results indicated by the dashed lines in Fig. 10.3-1 are obtained. As can be seen in that figure, the predicted results are in excellent agreement with experiment. Furthermore, from the equation-of-state description, we get not only the compositions of the coexisting phases but also good estimates of the compressibilities or densities. Also, starting from Eqs. 6.4-27 and 6.4-28 and using the Peng-Robinson equation of state, one can show (Problem 10.3-4) that

$$\underline{H}(T, P, \underline{x}) - \underline{H}^{\text{IGM}}(T, P, \underline{x}) = RT(Z_{\text{mix}} - 1)$$

$$+ \frac{T\left(\dfrac{da_{\text{mix}}}{dT}\right) - a_{\text{mix}}}{2\sqrt{2}b_{\text{mix}}} \ln\left[\frac{Z_{\text{mix}} + \left(1 + \sqrt{2}\right)B_{\text{mix}}}{Z_{\text{mix}} + \left(1 - \sqrt{2}\right)B_{\text{mix}}}\right]$$

$$\textbf{(10.3-8a)}$$

and

$$\underline{S}(T, P, \underline{x}) - \underline{S}^{\text{IGM}}(T, P, \underline{x}) = R \ln(Z_{\text{mix}} - B_{\text{mix}})$$

$$+ \frac{\left(\dfrac{da_{\text{mix}}}{dT}\right)}{2\sqrt{2}b_{\text{mix}}} \ln \left[\frac{Z_{\text{mix}} + \left(1 + \sqrt{2}\right) B_{\text{mix}}}{Z_{\text{mix}} + \left(1 - \sqrt{2}\right) B_{\text{mix}}} \right]$$

(10.3-8b)

where the subscript mix denotes a mixture property; $B = bP/RT$; and $\underline{H}^{\text{IGM}}(T, P, \underline{x})$ and $\underline{S}^{\text{IGM}}(T, P, \underline{x})$ are the ideal gas mixture enthalpy and entropy, respectively, at the conditions of interest. Therefore, we can also compute the enthalpy and entropy of the vapor and liquid phases.

This example shows the real power of the equation-of-state description in that starting with relatively little information (T_c, P_c, and ω of the pure components), we can obtain the phase equilibrium, phase densities, and other thermodynamic properties.

In the ethane-propylene system calculation just described, we have used no adjustable parameters. A careful examination of the results shows that the predicted compositions of ethane in the liquid are systematically approximately 0.01 mole fraction too high. These predictions can be improved by using a regression procedure with the experimental data and the Peng-Robinson equation of state. Such detailed calculations have shown that setting $k_{ij} = 0.011$ in Eq. 10.3-7 improves the accuracy of the predictions, although this improvement would be barely visible on the scale of Fig. 10.3-1, and therefore is not shown. For other systems, containing species much more dissimilar in size and in types of molecular interactions, the importance of the binary interaction parameter k_{ij} is more apparent. This is evident in Fig. 10.3-7, which contains experimental vapor-liquid equilibrium data for the carbon dioxide–isopentane system at two temperatures, together with predictions (setting $k_{ij} = 0$) and correlations (regression of data to obtain $k_{ij} = 0.121$) using the Peng-Robinson equation of state. There we see that the predictions made with the binary interaction parameter equal to zero are not nearly as good as those with $k_{ij} = 0.121$. Therefore, in engineering applications, at least one experimental data point is needed, so that a value of k_{ij} can be obtained; better still is to have several data points and choose the binary interaction parameter to give an optimum fit of all the data points. The binary interaction parameters given in Table 9.4-1 were obtained by such a regression procedure.

Although the discussion so far has been concerned with binary mixtures, the calculational procedures described here are applicable to all multicomponent mixtures. However, for accurate predictions in this case one needs a value of the interaction parameter for each binary pair in the mixture.

As a further example of the complicated and unusual behavior that is possible in binary and multicomponent mixtures at high pressures, consider the vapor-liquid equilibrium for the ethane–n-heptane system shown in Fig. 10.3-8. The leftmost curve in this figure is the vapor-pressure line as a function of temperature for pure ethane, and the rightmost curve is the vapor-pressure curve for n-heptane. Between the two are the bubble point and dew point curves for a mixture at the constant composition of 58.71 mol % ethane. The bubble point curve is the locus of pairs of temperatures and pressures at which the first bubble of vapor will appear in a liquid of $x_{C_2H_6} = 0.5871$. (Note that the composition of this coexisting vapor cannot be found from the information in the figure. Why?) Similarly, the dew point curve is the locus of temperature-pressure pairs at which the first drop of liquid appears in a vapor of $y_{C_2H_6} = 0.5871$. (Again, the composition of the coexisting liquid cannot be found from this figure.) This fig-

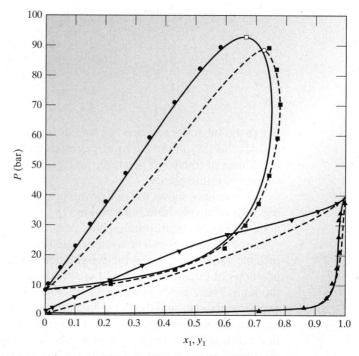

Figure 10.3-7 Vapor-liquid equilibrium of the carbon dioxide (1)–isopentane (2) system. The experimental data of G. J. Besserer and D. B. Robinson [*J. Chem. Eng. Data,* **20,** 93 (1976)] are shown at 277.59 K ($\blacktriangledown$ = liquid and $\blacktriangle$ = vapor) and 377.65 K ($\bullet$ = liquid and $\blacksquare$ = vapor). The dashed curves are the predictions using the Peng-Robinson equation of state and the van der Waals mixing rule with $k_{12} = 0$, and the solid lines are the correlation using the same equation of state with $k_{12} = 0.121$. The points $\circ$ and $\square$ are the estimated mixture critical points at 377.65 K using the same equation of state with $k_{12} = 0$ and 0.121, respectively.

ure shows how the bubble point temperature and dew point temperature change as a function of pressure for a mixture of fixed composition.

A number of interesting features are apparent from Fig. 10.3-8. First, the critical point of the mixture, which is the intersection of the bubble point and dew point curves (indicated by $\bullet$), is intermediate in temperature, but at a much higher pressure than the critical points of the pure components (also denoted by filled circles). Second, the critical point for the $x_{C_2H_6} = 0.5871$ mixture is neither the highest temperature nor the highest pressure along the phase boundary where the vapor and liquid coexist. The point of maximum pressure along the phase boundary (indicated by $\blacksquare$) is referred to as the **cricondenbar**, and the point of maximum temperature (denoted by $\square$) is called the **cricondentherm**.

Next note that if one has a vapor at $y_{C_2H_6} = 0.5871$, $P = 70$ bar, and $T = 475$ K (denoted by point *a*) and cools it at a constant pressure (following the line ----), first the dew point curve is intersected and a drop of liquid forms. Further cooling produces additional liquid, so that both the vapor and liquid will differ from the starting composition. Finally, at about $T = 408$ K, the last bit of vapor condenses, and a liquid with $x_{C_2H_6} = 0.5871$ is obtained. This is usual phase equilibrium behavior.

However, starting with a fluid of ethane mole fraction 0.5871, $P = 78$ bar, and temperature about 455 K (point *b*), which is above the critical point at this composition, and cooling at constant pressure, the bubble point curve is intersected. Consequently,

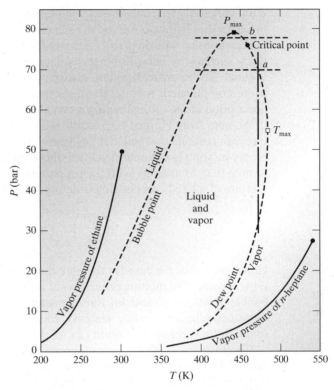

Figure 10.3-8 Vapor-liquid equilibrium for pure ethane, pure *n*-heptane (solid lines), and a mixture of fixed composition of 58.71 mol % ethane (dashed curve) as a function of temperature and pressure. [Data of W. B. Kay, *Ind. Eng. Chem.*, **30**, 459 (1938).] The symbols ● denote critical points, ■ is the cricondenbar, and □ is the cricondentherm.

the initial mixture was a liquid, and, at the bubble point curve, the first bubble of vapor is produced. Reducing the temperature still further produces more vapor, until a temperature of approximately 445 K is reached. On further cooling, some of this vapor condenses until, at approximately $T = 433$ K, all the vapor is condensed and the bubble point curve is crossed again into the region of all liquid. Vapor-liquid behavior such as this, in which, on traversing a path of either constant temperature or constant pressure, a phase first appears and then disappears so that the initial phase is again obtained, is referred to as **retrograde behavior**.

Another example of retrograde behavior occurs when one starts with the vapor at point *a* and reduces the pressure at constant temperature (following the line — · —). At 70 bar, the dew point curve is intersected, and a liquid appears. Further reductions in pressure first produce more liquid, but this liquid then begins to vaporize as the pressure decreases further. At about $P = 34.5$ bar, all the liquid vaporizes, as the dew point curve is crossed again. The behavior of an isotherm passing through two dew points is known as **retrograde behavior of the first kind**. Less common is an isotherm passing through two bubble points; this is referred to as **retrograde behavior of the second kind**. For such behavior to be observed, the mixture critical point must appear after both the cricondenbar and cricondentherm as one follows the phase boundary from the bubble point curve to the dew point curve.

(*Question*: The critical point may appear before both the cricondenbar and the cricondentherm, between them, or after both as one goes from the bubble point to the dew point curve. Sketch each of these phase boundaries, and discuss the type of retrograde behavior that will be observed in each case.)

Equations of state do predict retrograde behavior. However, since simple cubic equations, such as the Peng-Robinson equation used for illustration in this section, are not accurate in the critical region, retrograde predictions may not be of high accuracy. More complicated, multiterm equations of state are needed for reasonably accurate description of the critical region and retrograde behavior.

We note from Figures 10.3-2 and 10.3-3 that the shapes of the critical loci of mixtures are complicated and that, in general, the critical temperature and/or pressure of a binary mixture is not intermediate to those properties of the pure fluids. It is of interest to note that, in analogy with the properties of a pure fluid, a pseudocritical point of a mixture of a fixed composition is defined by the mechanical stability inflection point,

$$\left(\frac{\partial P}{\partial \underline{V}}\right)_{T,\underline{x}} = \left(\frac{\partial^2 P}{\partial \underline{V}^2}\right)_{T,\underline{x}} = 0$$

However, unlike the case for the pure fluid, this inflection point is not the real mixture critical point. The mixture critical point is the point of intersection of the dew point and bubble point curves, and this must be determined from phase equilibrium calculations, more complicated mixture stability conditions, or experiment, not simply from the criterion for mechanical stability as for a pure fluid.

We close this section by considering the application of equations of state to highly nonideal, polar mixtures. It is only recently, and by the use of combined equation-of-state–excess Gibbs energy models such as those described in Sec. 9.9, that it has been possible to use the equations of state for highly nonideal mixtures containing polar fluids. The procedures are similar to those used in other equation-of-state calculations as described earlier in this section, with the following changes. First, instead of using the generalized expression for the equation-of-state a parameter, Eq. 6.7-4, developed for hydrocarbons, Eqs. 7.5-1 and 7.5-2 should be used with the value of the parameter κ_1 fit to the vapor pressure for each pure compound. This is necessary to ensure that the pure component vapor pressures are correct. Second, one has to choose an activity coefficient model to be used for $\underline{G}^{ex}$ in the mixing rule of Eq. 9.9-9.

How one proceeds thereafter depends on the type of calculations in which one is interested. The following possibilities exist.

1. One can use the model in a completely correlative manner to fit vapor-liquid equilibrium data over a large range of temperatures and pressures using the parameters in the $\underline{G}^{ex}$ expression and k_{ij} of Eq. 9.9-10 as adjustable parameters. This is demonstrated in Fig. 10.3-9 for the water-acetone mixture using the NRTL expression (Eqs. 9.5-13 and 9.5-14) for $\underline{G}^{ex}$, setting $\alpha = 0.35$, and fitting τ_{12}, τ_{21}, and k_{12} to each isothermal data set.

2. Since it has been found that the parameters in the equation-of-state + $\underline{G}^{ex}$ model discussed in Sec. 9.9 are essentially independent of temperature, the model can be used to fit data at low temperatures, again by adjusting the $\underline{G}^{ex}$ and k_{ij} parameters, and then to make predictions at much higher temperatures and pressures, even temperatures that are 100 or 200°C higher than the fitted experimental data. This is demonstrated in Fig. 10.3-10, again for the water-acetone system. In this case, the NRTL model was used with $\alpha = 0.35$ and τ_{12}, τ_{21}, and k_{12} fit to the 298 K data, and the vapor-liquid equilibrium behavior at all other temperatures predicted.

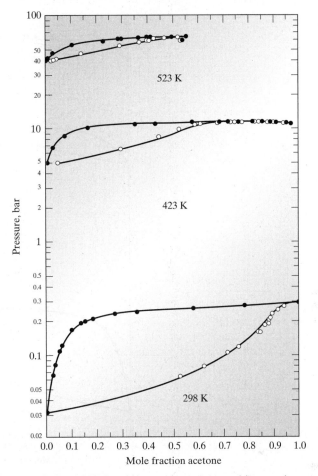

Figure 10.3-9 Vapor-liquid equilibria of the acetone + water binary mixture correlated using the combination of the Peng-Robinson equation of state, the Wong-Sandler mixing rule, and the NRTL activity coefficient model. The three parameters in this model have been fit to data for each isotherm.

3. An important characteristic of the mixing rule in Sec. 9.9 is that parameters obtained from the direct (γ-ϕ) correlation of vapor-liquid equilibrium data (as was done in Sec. 10.2) can be used in the mixing rule. Thus, all the activity coefficient model parameters reported, for example, in the vapor-liquid equilibrium collection of the DECHEMA Chemistry Data Series[10] can be used in the mixing rule without additional correlation of the vapor-liquid equilibrium data. Then only the binary parameter k_{ij} needs to be obtained, and this is done by adjusting its value so that the $\underline{G}^{ex}$ predicted by the equation-of-state + Gibbs energy model is as close as possible to the excess Gibbs energy at either a single data point (preferably near the midpoint in the composition range) or over the whole composition range along a single isotherm or isobar. Once this is done, the phase behavior at other temperatures and pressures can be predicted with reasonable accuracy,

---

[10]DECHEMA, Frankfurt, Germany (volumes appearing regularly from 1977 onward).

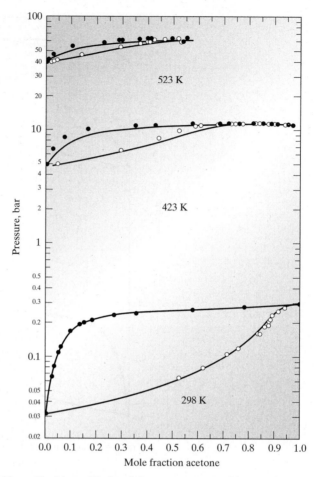

Figure 10.3-10 Vapor-liquid equilibria of the acetone-water binary mixture. The combination of the Peng-Robinson equation of state, the Wong-Sandler mixing rule, and the NRTL model fit to the experimental data at 298 K was used to obtain the model parameters. The solid lines at the higher temperatures are predictions using these parameters.

as demonstrated in Fig. 10.3-11 for the water-acetone mixture. Here the NRTL parameters reported in DECHEMA for 298 K were used to predict the behavior at all temperatures.

4. Finally, the mixing rule of Sec. 9.9 can be used to make phase equilibrium predictions even in the absence of experimental data by combining an equation of state with the UNIFAC group contribution model of Sec. 9.6. There are a number of ways in which this can be done. Perhaps the simplest is to use the UNIFAC model at some convenient temperature, for example, at 25°C, to predict the two infinite-dilution activity coefficients in a binary mixture and also the value of $\underline{G}^{ex}$ at $x_1 = 0.5$. Next, the two parameters in the UNIQUAC model of Eqs. 9.5-19 to 9.5-23 (which is algebraically equivalent to the UNIFAC model but simpler to use because molecules rather than functional groups are involved) are adjusted to give the same infinite-dilution coefficients. Finally, this UNIQUAC model is used in the mixing rule of Sec. 9.9 and the value of k_{ij} adjusted to reproduce the previously calculated value of $\underline{G}^{ex}(x_1 = 0.5)$. With the parameters determined in

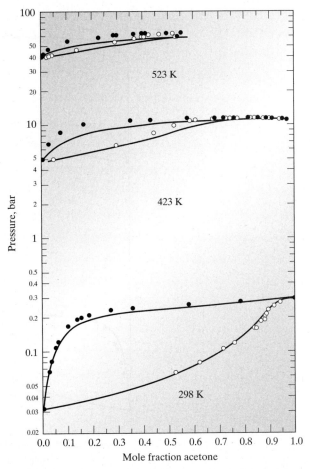

Figure 10.3-11 Vapor-liquid equilibria of the acetone + water binary mixture predicted using the combination of the Peng-Robinson equation of state, the Wong-Sandler mixing rule, and the NRTL model with parameters reported in the DECHEMA chemistry data series.

this way, predictions can be made at all other temperatures and pressures. This is shown in Fig. 10.3-12 for the acetone-water mixture.

The method just described is probably the simplest and most accurate extension of the UNIFAC prediction method to high temperatures and pressures. However, it is accurate only if the UNIFAC predictions (made by the direct activity coefficient or γ-ϕ method) for this system are accurate. If this is not the case, then the combination of the UNIFAC model with an equation of state as described here will also be inaccurate. That is, combining the UNIFAC model with an equation of state (the ϕ-ϕ method) does not improve its accuracy, but merely extends its range of applicability.

It should be pointed out that the results in Figs. 10.3-8 to 10.3-11 are examples of a successful application of the mixing rule of Sec. 9.9 to highly nonideal systems. For comparison, we show in Fig. 10.3-13 the results that would be obtained for the acetone-water system using the van der Waals one-fluid mixing rule, Eqs. 9.4-8 and 9.4-9, with the binary parameter k_{12} fit to the 298 K isotherm; the results in the figure at higher temperatures are predictions. Note that both the correlation at 298 K and predictions at higher temperatures with the van der Waals mixing rule are not very

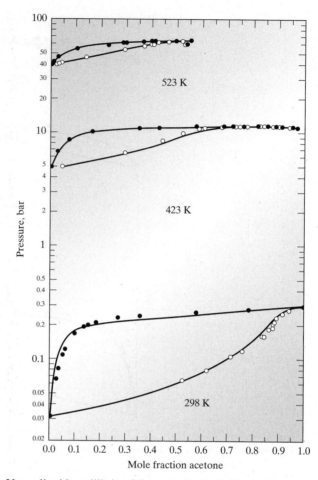

Figure 10.3-12 Vapor-liquid equilibria of the acetone-water binary mixture predicted using the combination of the Peng-Robinson equation of state, the Wong-Sandler mixing rule, and the UNIQUAC activity coefficient model with parameters obtained from the predictive UNIFAC model at 25°C.

good. In particular, the correlation at 298 K gives a false liquid-liquid phase separation or phase split (seen as a local maximum in pressure as a function of composition in the low acetone concentration range; liquid-liquid phase equilibria will be discussed in Sec. 11.2) and fails to properly represent the pressure versus composition behavior at all temperatures.

The dashed lines in the figure are the predictions at all temperatures for the acetone-water system that result from setting the binary parameter k_{12} equal to zero. Note that very nonideal behavior is predicted, which shows that setting $k_{12} = 0$ is not equivalent to assuming ideal solution behavior. In fact, such extreme nonideal behavior is predicted that vapor-liquid equilibrium calculations made with the program VLMU do not even converge for acetone mole fractions less than about 0.15.

PROBLEMS FOR SECTION 10.3

All problems in this section, except for Problems 10.3-4 and 10.3-11, can be solved using the Peng-Robinson equation-of-state programs for mixtures discussed in Appendix B and on the CD-ROM that accompanies this book.

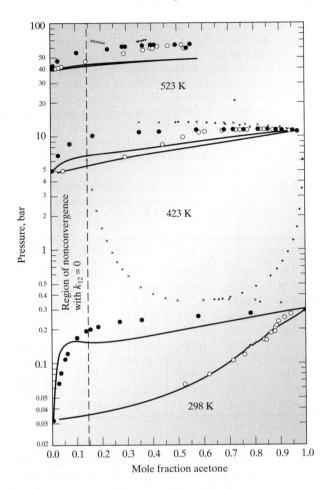

Figure 10.3-13 Vapor-liquid equilibria of the acetone-water binary mixture described by the Peng-Robinson equation of state and the van der Waals one-fluid mixing rules. The solid lines result from the value of the binary parameter k_{ij} being fit to the data at 298 K. The dotted lines are the highly nonideal (and unrealistic) behavior predicted by setting $k_{ij} = 0$.

10.3-1 The following mixture of hydrocarbons occurs in petroleum processing.

Component	Mole Percent
Ethane	5
Propane	57
n-Butane	38

Estimate the bubble point temperature and the composition of the coexisting vapor for this mixture at all pressures above 1 bar.

10.3-2 Estimate the dew point temperature and the composition of the coexisting liquid for the mixture in the previous problem at all pressures above 1 bar.

10.3-3 A liquid mixture of the composition given in Problem 10.3-1 is to be flashed at $P = 20$ bar and a collection of temperatures between the bubble point

temperature and the dew point temperature. Determine the compositions of the coexisting vapor and liquid, and the vapor-liquid equilibrium split for several temperatures in this range.

10.3-4 Derive Eqs. 10.3-8.

10.3-5 a. Develop an algorithm for the equation-of-state prediction of the dew point pressure.

   b. Develop an algorithm for the equation-of-state prediction of the dew point temperature.

10.3-6 a. For the adiabatic, steady-flow flash process from specified initial conditions of T_1 and P_1 to a specified final pressure P_2 shown here, develop the equilibrium and balance equations to compute the final temperature, the vapor-liquid split, and the compositions of the coexisting phases.

   b. Develop an algorithm for the calculation of part (a) using an equation of state.

   c. Use the Peng-Robinson equation of state for a multicomponent mixture to do the calculations for an adiabatic (Joule-Thomson) expansion of a liquid under pressure to produce a vapor-liquid mixture at ambient pressure. The results should include the outlet temperature, the mole fractions of each species in each phase, and the fractions of the outlet stream that are liquid and vapor.

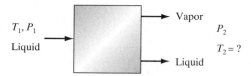

10.3-7 a. Make the best estimate you can of the composition of the vapor in equilibrium with a liquid containing 30.3 mol % ethane and 69.7 mol % ethylene at $-0.01°C$. Compare your results with the experimental data in the table.

b. Repeat the calculation in part (a) at other compositions for which the experimental data below are available.

Mole Percent of Ethane		
Liquid	Vapor	Pressure (bar)
7.8	6.2	39.73
22.8	19.7	37.07
30.3	25.5	35.60
59.0	53.1	32.13
89.0	85.4	25.45

10.3-8 Vapor-liquid equilibria in petroleum technology are usually expressed in terms of K factors $K_i = y_i/x_i$, where y_i and x_i are the mole fractions of species i in the vapor and liquid phases, respectively. Estimate the K values for methane and benzene in the benzene-methane system at 300 K and a total pressure of 30 bar.

10.3-9 In a petroleum refinery an equimolar stream containing propane and n-butane is fed to a flash separator operating at 40°C. Determine the pressure at which this separator should be operated so that an equal number of moles of liquid and vapor are produced.

10.3-10 A storage tank is known to contain the following mixture at 45°C and 15 bar:

Species	Overall Mole Fraction
Ethane	0.31
Propane	0.34
n-Butane	0.21
i-Butane	0.14

What is the composition of the coexisting vapor and liquid phases, and what fraction (by moles) of the contents of the tank is liquid?

10.3-11 Use the Peng-Robinson equation of state for a multicomponent mixture to do the calculations for an isentropic expansion of a liquid under pressure to produce a vapor-liquid mixture at ambient pressure. The output results should include the outlet temperature, the mole fractions of each species in each phase, and the fractions of the outlet stream that are liquid and vapor.

10.3-12 The following vapor-liquid equilibrium data are available for the system carbon dioxide (1) + isobutane (2) at 273.15 K.

P (bar)	x_1	y_1	P (bar)	x_1	y_1
1.57	0.000	0.000	14.793	0.317	0.877
2.736	0.022	0.422	16.718	0.378	0.899
3.546	0.037	0.541	17.63	0.403	0.913
4.256	0.038	0.623	18.34	0.416	0.914
5.167	0.053	0.690	21.379	0.513	0.922
6.282	0.073	0.744	23.811	0.598	0.925
7.498	0.098	0.783	26.445	0.697	0.947
8.511	0.149	0.800	29.485	0.817	0.963
9.93	0.166	0.830	32.423	0.917	0.981
11.855	0.224	0.863	34.855	1.000	1.000

a. Find the value of the binary interaction parameter in the Peng-Robinson equation of state with the van der Waals one-fluid mixing rules that best fits these data, and plot the correlated results and experimental data on the same graph.

b. Find the partition coefficients $K_i = y_i/x_i$ from both correlation and the experimental data for each species as a function of pressure, and plot all the partition coefficients as a function of pressure on a single graph.

10.3-13 To evaluate the potential use of carbon dioxide in tertiary oil recovery, it is necessary to estimate the vapor-liquid equilibrium between carbon dioxide and reservoir petroleum, which we will take to be n-hexane, at oil well conditions, typically 140 bar and 75°C. Make this estimate as best you can.

Chapter 11

Other Types of Phase Equilibria in Fluid Mixtures

In this chapter we continue the discussion of fluid phase equilibria by considering examples other than vapor-liquid equilibria. These other types of phase behavior include the solubility of a gas (a substance above its critical temperature) in a liquid, liquid-liquid and vapor-liquid-liquid equilibria, osmotic equilibria, and the distribution of a liquid solute between two liquids (the basis for liquid extraction). In each of these cases the starting point is the same: the equality of fugacities of each species in all the phases in which it appears,

$$\bar{f}_i^{\mathrm{I}}\left(T, P, \underline{x}^{\mathrm{I}}\right) = \bar{f}_i^{\mathrm{II}}\left(T, P, \underline{x}^{\mathrm{II}}\right) = \cdots$$

The difference in the various cases to be considered is how the fugacity of each species is computed, and this differs for vapor-liquid equilibrium, gas solubility, and liquid-liquid equilibrium.

INSTRUCTIONAL OBJECTIVES FOR CHAPTER 11

The goals of this chapter are for the student to:

- Be able to compute the solubility of a gas in a liquid (Sec. 11.1)
- Be able to compute the compositions when two partially miscible liquids are mixed (Sec. 11.2)
- Be able to compute the compositions when two partially miscible liquids and a vapor are in equilibrium (Sec. 11.3)
- Be able to compute the distribution coefficient of a solute between two liquid phases (Sec. 11.4)
- Be able to compute the osmotic pressure when a solute is dissolved in a solvent, or to use such data to determine the molecular weight of a solute (Sec. 11.5)

NOTATION INTRODUCED IN THIS CHAPTER

K Partition coefficient

$K_{\text{OW},i}$ Octanol-water partition coefficient of species i

K_C Concentration-based partition coefficient of species i

K_x Mole fraction–based partition coefficient of species i

T^{lc} Lower consolute or lower critical solution temperature (K)

T^{uc} Upper consolute or upper critical solution temperature (K)

$\underline{x}$ Set of liquid-phase mole fractions $x_1, x_2, \ldots$

Π Osmotic pressure (kPa)

11.1   THE SOLUBILITY OF A GAS IN A LIQUID

In the study of the solubility of a gas in a liquid one is interested in the equilibrium when the mixture temperature T is greater than the critical temperature of at least one of the components in the mixture, the gas. If the mixture can be described by an equation of state, no special difficulties are involved, and the calculations proceed as described in Sec. 10.3. Indeed, a number of cases encountered in Sec. 10.3 were of this type (e.g., ethane in the ethane-propylene mixture at 344.3 K). Consequently, it is not necessary to consider the equation-of-state description of gas solubility, as it is another type of equation-of-state vapor-liquid equilibrium calculation, and the methods described in Sec. 10.3 can be used.

However, the description of gas solubility using activity coefficient models does require some explanation, and this is what is discussed in this section. The activity coefficient description is of interest because it is applicable to mixtures that are not easily describable by an equation of state, and also because it may be possible to make simple gas solubility estimates using an activity coefficient model, whereas a computer program is required for equation-of-state calculations.

To study the solubility of a gas in a liquid using an activity coefficient model we start with the equilibrium relation

Starting point for all phase equilibrium calculations

$$\boxed{\bar{f}_i^{\text{L}}(T, P, \underline{x}) = \bar{f}_i^{\text{V}}(T, P, \underline{y}) \quad i = 1, 2, \ldots} \tag{11.1-1}$$

which, after using the Lewis-Randall rule for the gas phase and the definition of the activity coefficient, reduces to

$$x_i \gamma_i(T, P, \underline{x}) f_i^{\text{L}}(T, P) = y_i P \left(\frac{f}{P} \right)_i \tag{11.1-2}$$

The situation of interest here is when the mixture temperature T is greater than the critical temperature of one of the components, say component 1 (i.e., $T > T_{c,1}$), so that this species exists only as a gas in the pure component state. In this case the evaluation of the liquid-phase properties for this species, such as $f_1^{\text{L}}(T, P)$ and $\gamma_1(T, P, \underline{x})$, is not straightforward. (It is this complication that distinguishes gas solubility problems from those of vapor-liquid equilibrium, which were considered in Chapter 10.) We will refer to species that are in the liquid phase above their critical temperatures as the solutes. For those species below their critical temperatures, which we designate as the solvents, Eq. 11.1-2 is used just as in Sec. 10.2.

If the temperature of the mixture is only slightly greater than the critical temperature of the gaseous (solute) species, the (hypothetical) pure component liquid-phase fugac-

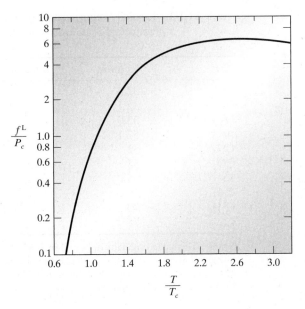

Figure 11.1-1 Extrapolated liquid-phase fugacity coefficients at 1.013 bar as a function of reduced temperature. [This figure originally appeared in J. M. Prausnitz and F. H. Shair, *AIChE J.*, **7**, 682 (1961). It appears here courtesy of the copyright owners, the American Institute of Chemical Engineers.]

ity $f_i^L(T, P)$ can be computed using the fugacity extrapolation scheme for nonsimple mixtures discussed in Sec. 9.7. In this case, the gas solubility problem is just like the vapor-liquid equilibrium problem of Sec. 10.2 and is treated in the same manner.

However, if the temperature of the mixture is well above $T_{c,1}$, the evaluation of $f_1^L(T, P)$ is more troublesome. A number of different procedures for estimating this hypothetical liquid fugacity have been proposed. Prausnitz and Shair[1] have suggested that the simple, approximate corresponding-states correlation of Fig. 11.1-1 be used to evaluate the liquid fugacity at 1.013 bar total pressure, that is, $f_1^L(T, P = 1.013 \text{ bar})$. To compute the hypothetical liquid-phase fugacity at any other pressure, a Poynting pressure correction is made to the value at 1.013 bar as follows:

$$f_1^L(T, P) = f_1^L(T, P = 1.013 \text{ bar}) \exp\left[\int_{1.013 \text{ bar}}^{P} \frac{V_1^L}{RT} \, dP\right]$$

$$\cong f_1^L(T, P = 1.013 \text{ bar}) \exp\left[\frac{V_1^L(P - 1.013 \text{ bar})}{RT}\right] \tag{11.1-3}$$

Equation 11.1-3 introduces another unknown quantity, the molar volume of the hypothetical liquid. These values have been tabulated by Prausnitz and Shair, and are given in Table 11.1-1.

The question of evaluating the liquid-phase activity coefficient of the solute species still remains. Although experimental data for γ_1 would be preferable, such data may not be available. Consequently, various liquid solution models and correlations are used. If the regular solution model is used, we have

$$\ln \gamma_1(x_1) = \frac{V_1^L \Phi_2^2 (\delta_1 - \delta_2)^2}{RT} \tag{11.1-4a}$$

---

[1] J. M. Prausnitz and F. H. Shair, *AIChE J.*, **7**, 682 (1961); also, J. M. Prausnitz, R. N. Lichtenthaler, and E. G. Azevedo, *Molecular Thermodynamics of Fluid Phase Equilibrium*, 2nd ed., Prentice Hall, Englewood Cliffs, N.J. (1986), p. 392ff.

Table 11.1-1 "Liquid" Volumes and Solubility Parameters for Gaseous Solutes at 25°C

Gas	$\underline{V}^L$ (cc/mol)	δ (cal/cc)$^{1/2}$
N_2	32.4	2.58
CO	32.1	3.13
O_2	33.0	4.0
Ar	57.1	5.33
CH_4	52	5.68
CO_2	55	6.0
Kr	65	6.4
C_2H_4	65	6.6
C_2H_6	70	6.6
Cl_2	74	8.7

Source: This table originally appeared in J. M. Prausnitz and F. H. Shair, *AIChE J.*, **7**, 682 (1961). It appears here courtesy of the copyright owners, the American Institute of Chemical Engineers.

for pure solvents, and

$$\ln \gamma_1(x_1) = \frac{\underline{V}_1^L(\delta_1 - \bar{\delta})^2}{RT} \tag{11.1-4b}$$

with

$$\bar{\delta} = \sum_j \Phi_j \delta_j \quad \text{and} \quad \Phi_j = \frac{x_j \underline{V}_j^L}{\sum_i x_i \underline{V}_i^L}$$

for mixed solvents. The Prausnitz and Shair estimates for the solubility parameters of the hypothetical liquids of several common gases at 25°C are also given in Table 11.1-1. (It is interesting to note that the values of this parameter for the hypothetical liquids at 25°C are quite different from those for the real liquids at 90 K given in Table 9.6-1.) Of course, any other solution model for which the necessary parameters are available can be used to evaluate γ_1. (However, as the UNIFAC model as presented in this book is applicable only to substances that are liquids at 25°C and 1.013 bar, it cannot be used.)

Using these estimates for the liquid-phase fugacity and the activity coefficient of the solute species, Eqs. 11.1-2 and 11.1-3 can be combined to give

Solute equilibrium relation

$$x_1 = \frac{y_1 P(f/P)_1}{\gamma_1(T, P, \underline{x}) f_1^L(T, P = 1.013 \text{ bar}) \exp[\underline{V}_1^L(P - 1.013 \text{ bar})/RT]} \tag{11.1-5a}$$

This equation is solved together with the equilibrium relations for the solvent species,

Solvent equilibrium relation

$$x_i = \frac{y_i P(f/P)_i}{\gamma_i(T, P, \underline{x}) P_i^{\text{vap}}(T)(f/P)_{\text{sat},i} \exp[\underline{V}_i^L(P - P_i^{\text{vap}})/RT]} \tag{11.1-6}$$

to compute the solute solubility in the liquid solvent and the solvent solubility in the gas.

For ideal solutions (i.e., solutions for which $\gamma_i = 1$), the solubility of the gas depends on its partial pressure (or gas-phase fugacity), and not on the liquid or liquid mixture into which it dissolves. This solubility is termed the ideal solubility of the gas and is given below.

Ideal solution equilibrium relation

$$x_1^{\mathrm{ID}} = \frac{y_1 P(f/P)_1}{f_1^{\mathrm{L}}(T, P = 1.013 \text{ bar}) \exp[\underline{V}_1^{\mathrm{L}}(P - 1.013 \text{ bar})/RT]} \qquad \textbf{(11.1-5b)}$$

Since the fugacity of the solute species, obtained either from the extrapolation of the vapor pressure or from Fig. 11.1-1, will be very large, the mole fraction x_1 of the gaseous species in the liquid is likely to be quite small. This observation may provide a useful simplification in the solution of Eqs. 11.1-5 and 11.1-6. Also, if one is merely interested in the solubility of the gas in the liquid for a given gas-phase partial pressure, only Eq. 11.1-5 need be solved.

When a gas is only sparingly soluble in a liquid or liquid mixture (i.e., as $x_1 \rightarrow 0$), it is observed that the liquid-phase mole fraction of the solute species is, at fixed temperature, linearly proportional to its gas-phase fugacity, that is,

$$x_1 H_1(T, P) = \bar{f}_1^{\mathrm{V}}(T, P, \underline{y}) = y_1 P \left(\frac{f}{P}\right)_1 \qquad \text{as } x_1 \rightarrow 0 \qquad \textbf{(11.1-7)}$$

where H_1 is the Henry's law constant[2] (see Sec. 9.7). Gas solubility measurements are frequently reported in terms of the Henry's law constant, which depends on both the gas and the solvent; values of H for many gas-liquid pairs appear in the chemical and chemical engineering literature.

To relate the Henry's law constant to other thermodynamic quantities, we recognize that since, at equilibrium,

$$\bar{f}_1^{\mathrm{L}}(T, P, \underline{x}) = \bar{f}_1^{\mathrm{V}}(T, P, \underline{y}) = x_1 H_1(T, P) \qquad \text{as } x_1 \rightarrow 0$$

we can take the following to be the formal definition of the Henry's law constant:

$$\lim_{x_1 \rightarrow 0} \frac{\bar{f}_1^{\mathrm{L}}(T, P, \underline{x})}{x_1} = H_1(T, P) \qquad \textbf{(11.1-8)}$$

Comparing Eqs. 11.1-2 and 11.1-8 yields

$$H_1(T, P) = \gamma_1(x_1 = 0) f_1^{\mathrm{L}}(T, P) \qquad \textbf{(11.1-9)}$$

where $\gamma_1(x_1 = 0)$ is the limiting value of the activity coefficient of the gas in the liquid at infinite dilution. Thus, Eqs. 11.1-3 and 11.1-4 and the correlation of Fig. 11.1-1 can be used to predict values of the Henry's law constant.

As the pressure (and hence the solute mole fraction x_1) increases, deviations from this simple limiting law are observed (see Fig. 9.7-3a). For appreciable concentrations of the gaseous species in the liquid phase, we write instead

---

[2]The Henry's law "constant," as defined in Sec. 9.7, is independent of concentration but is a function of temperature, pressure, and the solvent.

$$y_1 P \left(\frac{f}{P} \right)_1 = x_1 \gamma_1^* \left(\underline{x} \right) H_1 \left(T, P \right) \tag{11.1-10}$$

where

$$\gamma_1^*(x_1) = \frac{\gamma_1(x_1)}{\gamma_1(x_1 = 0)}$$

is the renormalized activity coefficient defined by Eq. 9.7-5. Clearly, $\gamma_1^* \to 1$ as $x_1 \to 0$, and γ_1^* departs from unity as the mole fraction of the solute increases. The regular solution theory prediction for γ_1^* (see Problem 9.10) is

$$\ln \gamma_1^*(x_1) = \ln \gamma_1(x_1) - \ln \gamma_1(x_1 = 0) = \frac{\underline{V}_1^{\mathrm{L}}(\delta_1 - \delta_2)^2(\Phi_2^2 - 1)}{RT} \tag{11.1-11}$$

ILLUSTRATION 11.1-1

Estimation of the Solubility of a Gas in a Liquid

Estimate the solubility and Henry's law constant for carbon dioxide in a liquid mixture of toluene and carbon disulfide as a function of the CS_2 mole fraction at 25°C and a partial pressure of CO_2 of 1.013 bar.

Data: See Tables 6.6-1, 9.6-1, and 11.1-1.

SOLUTION

Equation 11.1-5 provides the starting point for the solution of this problem. Since the partial pressure of carbon dioxide and the vapor pressures of toluene and carbon disulfide are so low, the total pressure must be low, and we can assume that

$$\left(\frac{f}{P} \right) = 1 \quad \text{and} \quad \exp\left[\frac{\underline{V}_{CO_2}^{\mathrm{L}}(P - 1.013 \text{ bar})}{RT} \right] = 1$$

Next, using the regular solution model for γ, we obtain

$$x_{CO_2} = \frac{y_{CO_2} P}{f_{CO_2}^{\mathrm{L}}(T, P = 1.013 \text{ bar}) \exp\left[\dfrac{\underline{V}_{CO_2}^{\mathrm{L}}(\delta_{CO_2} - \bar{\delta})^2}{RT} \right]}$$

with

$$\bar{\delta} = \sum_j \Phi_j \delta_j$$

The reduced temperature of CO_2 is $T_r = 298.15 \text{ K}/304.3 \text{ K} = 0.98$, so from the Shair-Prausnitz correlation $f^{\mathrm{L}}/P_c \approx 0.60$ and $f^{\mathrm{L}} \approx 0.60 \times 73.76 \text{ bar} = 44.26 \text{ bar}$. To calculate the activity coefficients we will assume that CO_2 is only slightly soluble in the solvents, so that its volume fraction is small; we will then verify this assumption. Thus, as a first guess, the contribution of CO_2 to $\bar{\delta}$ will be neglected.

To compute the solubility of CO_2 in pure carbon disulfide, we note that

$$\bar{\delta} \approx \delta_{CS_2} = 10 \text{ (cal/cc)}^{1/2} \quad \text{and} \quad (\delta_{CO_2} - \bar{\delta})^2 = 16 \text{ cal/cc} = 66.94 \text{ J/cc}$$

so that

$$x_{CO_2} = \frac{1.013 \text{ bar}}{44.26 \text{ bar} \times \exp\left\{\dfrac{55 \text{ cc/mol} \times 66.94 \text{ J/cc}}{8.314 \text{ J/(mol K)} \times 298.15 \text{ K}}\right\}}$$

$$= 5.18 \times 10^{-3}$$

(The experimental value is $x_{CO_2} = 3.28 \times 10^{-3}$.) Also,

$$H = P/x_{CO_2} = 1.013 \text{ bar}/(5.18 \times 10^{-3})$$

$$= 195.5 \text{ bar/mole fraction}$$

The solubility of CO_2 in pure toluene is computed as follows:

$$\bar{\delta} \approx \delta_T = 8.9 \text{ (cal/cc)}^{1/2} \quad \text{and} \quad (\delta_{CO_2} - \bar{\delta})^2 = 8.4 \text{ cal/cc} = 35.15 \text{ J/cc}$$

so that

$$x_{CO_2} = \frac{1.013 \text{ bar}}{44.26 \text{ bar} \times \exp\left\{\dfrac{55 \times 35.15}{8.314 \times 298.15}\right\}} = 1.05 \times 10^{-2}$$

and

$$H = 96.6 \text{ bar/mole fraction}$$

Finally, the solubility of CO_2 in a 50 mol % toluene, 50 mol % CS_2 mixture is found from

$$\underline{V}^L_{mix} = x_{CS_2}\underline{V}^L_{CS_2} + x_T\underline{V}^L_T$$

$$= 0.5 \times 61 \frac{cc}{mol} + 0.5 \times 107 \frac{cc}{mol}$$

$$= 84 \frac{cc}{mol}$$

$$\Phi_{CS_2} = \frac{0.5 \times 61}{84} = 0.363, \qquad \Phi_T = \frac{0.5 \times 107}{84} = 0.637$$

$$\bar{\delta} = 0.363 \times 10 \left(\frac{cal}{cc}\right)^{1/2} + 0.637 \times 8.9 \left(\frac{cal}{cc}\right)^{1/2}$$

$$= 9.30 \text{ (cal/cc)}^{1/2}$$

and

$$(\delta - \bar{\delta})^2 = 10.88 \text{ cal/cc} = 45.55 \text{ J/cc}$$

Thus

$$x_{CO_2} = \frac{1.013}{44.26 \times \exp\left\{\dfrac{55 \times 45.52}{8.314 \times 298.15}\right\}} = 8.33 \times 10^{-3}$$

and

$$H = 121.6 \text{ bar/mole fraction}$$

These results are plotted in Fig. 11.1-2. In all cases x_{CO_2} is small, as had initially been assumed, so that an iterative calculation is not necessary. ■

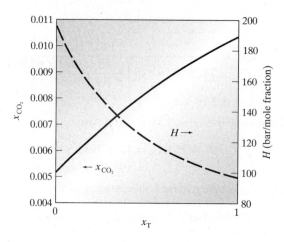

Figure 11.1-2 The solubility (x_{CO_2}) Henry's law constant (H) of carbon dioxide in carbon disulfide–toluene mixtures.

ILLUSTRATION 11.1-2

Prediction of the Solubility of a Gas in a Liquid Using an EOS

Predict the solubility of carbon dioxide in toluene at 25°C and 1.013 bar carbon dioxide partial pressure using the Peng-Robinson equation of state.

SOLUTION

The critical properties for both carbon dioxide and toluene are given in Table 6.6-1. The binary interaction parameter for the CO_2-toluene mixture is not given in Table 9.4-1. However, as the value for CO_2-benzene is 0.077 and that for CO_2–n-heptane is 0.10, we estimate that the CO_2-toluene interaction parameter will be 0.09. Using this value and the bubble point pressure calculation in either the programs or the MATHCAD worksheet for the Peng-Robinson equation of state for mixtures (described in Appendix B and on the CD-ROM accompanying this book), the following values were obtained:

x_{CO_2}	P_{tot} (bar)	y_{CO_2}	P_{CO_2} (bar) $= y_{CO_2} P_{tot}$
0.001	0.11	0.6579	0.072
0.002	0.20	0.7915	0.158
0.004	0.36	0.8834	0.318
0.006	0.51	0.9189	0.469
0.008	0.67	0.9378	0.628
0.010	0.83	0.9495	0.788
0.0125	1.03	0.9590	0.988
0.013	1.07	0.9605	1.028
0.015	1.23	0.9655	1.188

Therefore, using the Peng-Robinson equation of state, we estimate that at a partial pressure of 1.013 bar, carbon dioxide will be soluble in liquid toluene to the extent of 0.0128 mole fraction. This value differs from the value of 0.0077 computed in the last illustration using the Prausnitz-Shair correlation and regular solution theory. However, given the inaccuracy of both methods, this difference is not unreasonable.

COMMENT

Had we assumed that the CO_2-toluene binary interaction parameter was zero, the predicted CO_2 solubility in toluene at 1.013 bar CO_2 partial pressure would be 0.0221 mole fraction (Problem 11.1-6).

In order to compare the results with those of the previous illustration, we can also compute the solubility of carbon dioxide in carbon disulfide. There is no binary interaction parameter reported for the CO_2-CS_2 mixture, or for any similar mixtures. If we assume the binary interaction parameter is zero, we find that the CO_2 solubility in CS_2 at 1.013 bar CO_2 partial pressure is 0.0159 mole fraction, which is greater than the measured value by a factor of 5 (Problem 11.1-7). However, if we set $k_{CO_2-CS_2} = 0.2$, we obtain a CO_2 solubility of 3.4×10^{-3}, which is in excellent agreement with experiment. ∎

It is clear from Eq. 11.1-9 that the Henry's law constant will vary with pressure, since f_1^L and γ_1 are functions of pressure. The common method of accounting for this pressure variation is to define the Henry's law constant to be specific to a fixed pressure P_0 (frequently taken to be atmospheric pressure) and then include a Poynting correction for other pressures. Independent of whether we apply the correction to the fugacity of the solute species in solution $\bar{f}_1^L(T, P, x_1 \to 0)$ or separately to the pure component fugacity and the infinite-dilution activity coefficient (see Eq. 9.3-20), we obtain

$$\bar{f}_1^L(T, P, x_1 \to 0) = x_1 \gamma_1(T, P_0, x_1 \to 0) f_1^L(T, P_0) \exp\left[\int_{P_0}^{P} \frac{\bar{V}_1^L(x_1 = 0)}{RT} \, dP\right]$$

$$(11.1\text{-}12)$$

where $\bar{V}_1^L(x_1 = 0)$ is the partial molar volume of the gaseous species in the liquid at infinite dilution. Using this expression in Eq. 11.1-10 yields

$$y_1 P \left(\frac{f}{P}\right)_1 = x_1 \gamma_1^*(T, P_0) H_1(T, P_0) \exp\left[\int_{P_0}^{P} \frac{\bar{V}_1^L(x_1 = 0)}{RT} \, dP\right] \qquad (11.1\text{-}13)$$

Finally we note from Fig. 11.1-3 that the solubility in a liquid of some gases increases as the temperature increases, whereas for other gases it decreases. To explain this observation, we take the derivative of Eq. 11.1-2 with respect to temperature (at constant pressure and gas-phase composition) to get

$$0 = \left(\frac{\partial \ln x_1}{\partial T}\right)_P + \left(\frac{\partial \ln \gamma_1}{\partial T}\right)_P + \left(\frac{\partial \ln f_1^L}{\partial T}\right)_P \qquad (11.1\text{-}14)$$

where we have neglected the slight temperature dependence of the gas-phase fugacity coefficient. Now if we assume that the fugacity of the pure hypothetical liquid is obtained by extrapolating the vapor pressure of the real liquid, we have

$$\left(\frac{\partial \ln f_1^L}{\partial T}\right)_P = \left(\frac{\partial \ln P_1^{vap}}{\partial T}\right)_P = \frac{\Delta_{vap}\underline{H}_1}{RT^2} \qquad (11.1\text{-}15)$$

by the Clausius-Clapeyron equation, Eq. 7.7-5. Here again, we have neglected the temperature dependence of the fugacity coefficient. Next, from Eq. 9.3-21, we have

$$\left(\frac{\partial \ln \gamma_1}{\partial T}\right)_P = -\frac{\bar{H}_1^{ex}(T, P, \underline{x})}{RT^2} \qquad (11.1\text{-}16)$$

Combining Eqs. 11.1-14, 11.1-15, and 11.1-16 we get

$$\left(\frac{\partial \ln x_1}{\partial T}\right)_P = \frac{-\Delta_{vap}\underline{H}_1 + \overline{H}_1^{ex}}{RT^2}$$

$$= \frac{-(\underline{H}_1^V - \underline{H}_1^L) + (\overline{H}_1^L - \underline{H}_1^L)}{RT^2}$$

$$= \frac{-(\underline{H}_1^V - \overline{H}_1^L)}{RT^2}$$ **(11.1-17)**

$$\approx \frac{-\Delta_{vap}\underline{H}_1}{RT^2}$$

since $\overline{H}^{ex}$ is usually much smaller than $\Delta_{vap}\underline{H}_1$, the heat of vaporization of the pure solute. [Note that $(\underline{H}_1^V - \overline{H}_1^L)$ may be interpreted as the heat of vaporization of species 1 from the fluid *mixture*.]

For all fluids below their critical temperature, $\Delta_{vap}\underline{H}$ is positive; that is, energy is absorbed in going from the liquid to the gas. For $T > T_c$, but $P < P_c$, $\Delta_{vap}\underline{H}$ must be evaluated by extrapolation of the liquid-phase enthalpy into the vapor region. Here one finds that $\Delta_{vap}\underline{H}$ is positive in the vicinity of the critical temperature (but below the critical pressure), though its magnitude decreases as T increases. Finally, above some temperature T, where T is much greater than T_c, the extrapolated enthalpy change becomes negative (Problem 11.1-1). Therefore,

$$\left(\frac{\partial \ln x_1}{\partial T}\right)_P \begin{cases} > 0 & T \gg T_c \\ < 0 & \text{otherwise} \end{cases}$$

Thus the solubility of a gas increases with increasing temperature for gases very much above their critical temperature, and decreases with increasing temperature at temperatures near or only slightly above the critical temperature. This conclusion is in agreement with the experimental data of Fig. 11.1-3.

In the chemical literature the solubility of a gas at fixed partial pressure is frequently correlated as a function of temperature in the form

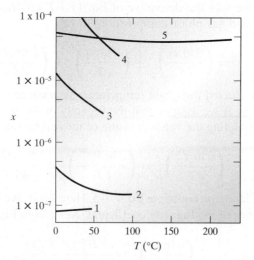

Figure 11.1-3 The solubility of several gases in liquids as a function of temperature. The solubility is expressed as mole fraction of the gas in the liquid at a gas partial pressure of 0.01 bar. *Curve 1:* Helium in water. *Curve 2:* Oxygen in water. *Curve 3:* Carbon dioxide in water. *Curve 4:* Bromine in water. *Curve 5:* Methane in *n*-heptane.

$$\ln x = A + \frac{B}{T} + C \ln T + DT + ET^2 \qquad \textbf{(11.1-18)}$$

where x is the gas mole fraction in the liquid, and T is the temperature in K. Table 11.1-2 contains the values for the constants of various gases in water at 1.013 bar partial pressure of the gas. Although we have used mole fractions throughout this section, other measures of gas solubility are also used. Some are listed in Table 11.1-3.

Air stripping

One method of removing a volatile contaminant from a liquid—for example, water—is by gas stripping, in which air or some other gas is bubbled through the liquid so that vapor-liquid equilibrium is achieved. If the contaminent is relatively volatile (as a result of a high value of its Henry's constant, vapor pressure, or activity coefficient), it will appear in the exiting air, and therefore its concentration in the remaining liquid is reduced. An example of this is given in the next illustration.

ILLUSTRATION 11.1-3

Air Stripping of Radon from Groundwater

Groundwater from some geological formations may contain radon, a gas that has been implicated in lung disease, so its concentration should be reduced. Air stripping is one method of doing this. Groundwater from a well is found to contain 10 parts per million by weight of radon, and it is desired to reduce its concentration to 0.1 parts per million. This is to be done by air stripping in the device shown here, which is open to the atmosphere, using previously humidifed air so that water does not evaporate in the air stripping process. Assuming that the air leaving the stripper is at 20°C and in equilibrium with the liquid, how many kilograms of air must be supplied per kilogram of water to reduce the radon content to the desired level? Radon has a molecular weight of 222, and its Henry's constant is $K = P_R/x_R = 5.2 \times 10^3$ bar/mole fraction, where P_R is the partial pressure of radon.

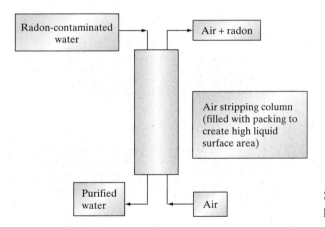

Schematic diagram of an air stripping apparatus.

SOLUTION

Since the air has been prehumidified, no water should evaporate. Also, since the concentration of radon in the water is so low, even if all of it was removed, there would be very little change (indeed, only one part in 100,000) in the original mass of liquid in the air stripper, so that the small change in total mass (or number of moles) can be neglected. Therefore, in the equations

Table 11.1-2 Mole Fraction Solubility of Gases in Water at 1.013 bar Partial Pressure as a Function of Temperature, $\ln x = A + B/T + C \ln T + DT + ET^2$; ($T$ in K)

Gas	T Range	A	B	C	D	E
Helium	273–348	−105.977	4 259.62	14.009 4	—	—
Neon	273–348	−139.967	6 104.94	18.915 7	—	—
Argon	273–348	−150.413	7 476.27	20.139 8	—	—
Krypton	273–353	−178.533	9 101.66	24.220 7	—	—
Xenon	273–348	−201.227	10 521.0	27.466 4	—	—
Radon	273–373	−251.751	13 002.6	35.004 7	—	—
Hydrogen	274–339	−180.054	6 993.54	26.312 1	−0.015 043 2	—
Deuterium	278–303	−181.251	7 309.62	26.178 0	−0.011 815 1	—
Nitrogen	273–348	−181.587 0	8 632.129	24.798 08	—	—
Oxygen	273–333	−1 072.489 02	27 609.261 7	191.886 028	−0.483 090 199	$2.244\,452\,61 \times 10^{-4}$
Ozone	277–293	−14.964 5	1 965.31	—	—	—
Carbon monoxide	278–323	−427.656 023	15 259.995 3	67.842 954 2	−0.070 459 535 6	—
Carbon dioxide	273–373	−4 957.824	105 288.4	933.170 0	−2.854 886	$1.480\,857 \times 10^{-3}$
Methane	275–328	−416.159 289	15 557.563 1	65.255 259 1	−0.061 697 572 9	—
Ethane	275–323	−11 268.400 7	221 617.099	2 158.421 79	−7.187 794 02	$4.050\,119\,24 \times 10^{-3}$
Ethylene	287–346	−176.910	9 110.81	24.043 6	—	—
Acetylene	274–343	−156.509	8 160.17	21.402 3	—	—
Propane	273–347	−316.460	15 921.2	44.324 3	—	—
Cyclopropane	298–361	326.902	−13 526.8	−50.901 0	—	—
n-Butane	276–349	−290.238 0	15 055.5	40.194 9	—	—
Isobutane	278–343	96.106 6	−2 472.33	−17.366 3	—	—
Neopentane	288–353	−437.182	21 801.4	61.889 4	—	—
CH_3F	273–353	−135.910	7 600.23	18.178 0	—	—
CH_3Cl	277–353	−172.503	9 768.67	23.424 1	—	—
CH_3Br	278–353	−163.745	9 641.71	22.039 7	—	—

CF_4	275–323	−342.437	16 250.6	48.344 1	—
CH_2FCl	283–352	−138.912	8 141.19	18.551 0	—
CHF_2Cl	297–352	−190.691	13 083.6	22.778 2	0.032 357
CHF_3	298–348	−19.103 7	3 214.07	—	—
C_2F_6	278–303	−644.222	29 933.6	93.026 9	—
C_2F_5Cl	298–348	−21.396 5	2 802.35	—	—
Vinyl chloride	273–358	−2 399.543	72 028.1	408.576	−0.594 578
C_2F_4	273–343	−184.958	10 843.3	23.145 8	0.020 961 1
C_3F_8	278–288	−10 180.5	438 919	1 526.05	—
C_3F_6	278–343	−66.582 0	4 491.54	6.909 81	—
$c\text{-}C_4F_8$	278–303	−759.717	35 705.7	110.033	—
COS	273–303	−221.211	12 025.1	30.365 9	—
CH_3NH_2	298–333	−9.191 7	2 607.10	—	—
$(CH_3)_2NH$	298–333	−14.098 0	4 026.14	—	—
$C_2H_5NH_2$	298–333	−12.623 1	3 628.65	—	—
NH_3	273–373	−81.746 6	1 096.82	16.560 3	0.060 246 9
N_2O	273–313	−158.620 8	8 882.80	21.253 1	—
NO	273–353	−328.097	12 541.9	50.761 6	−0.045 133 1
H_2S	273–333	−149.537	8 226.54	20.230 8	0.001 294 05
SO_2	283–306	−13.050 2	2 792.62	—	—
SF_6	275–323	−435.519	20 901.8	61.969 2	—
Cl_2	283–313	108.389	−2 428.63	−19.185 5	0.008 920 64
Cl_2O	273–293	−7.220 7	1 798.85	—	—
ClO_2	283–333	56.738 9	143.179	−10.745 4	—
Air	273–373	−388.760	14 097.6	61.201 8	−0.061 753 7

Experimental data were not corrected for nonideality of the gas phase and chemical reactions with the solvent. The quoted coefficients are valid in the temperature range given in the second column.

Source: S. Cabani and P. Gianni, in H. J. Hinz, ed., *Thermodynamic Data for Biochemistry and Biotechnology,* (1986), pp. 261–262 with kind permission from Springer Science+Business Media.

Table 11.1-3 Conversion Formulas of Various Expressions of Gas Solubility in Water to Mole Fraction (x) of the Dissolved Gas, Under a Gas Partial Pressure of 1.013 bar

Quantity	Symbol	Definition	Conversion Formula
Bunsen coefficient	α	Volume of gas, reduced to 273.15 K and 1.013 bar, absorbed by a unit volume of the absorbing solvent at the temperature of measurement under a gas partial pressure of 1.013 bar	$x = \left[1 + \dfrac{1.244 \times 10^{-3} \cdot \rho_w}{\alpha}\right]^{-1}$
Ostwald coefficient	L	Ratio of the volume of gas absorbed to the volume of absorbing liquid, both measured at the same temperature	$x = \left[1 + \dfrac{4.555 \times 10^{-6} T \cdot \rho_w}{L}\right]^{-1}$
Kuenen coefficient	S	Volume of gas (cm^3) at a partial pressure of 1.013 bar reduced to 273.15 K and 1.013 bar, dissolved by the quantity of solution containing 1 g of solvent	$x = \dfrac{S}{S + 1244.1}$
Henry's law constant	H	Limiting value of the ratio of the gas partial pressure to its mole fraction in solution as the latter tends to zero	$x = \dfrac{1.013}{H}$
Weight solubility	S_0	Grams of gas dissolved by 100 g of solvent, under a gas partial pressure of 1.013 bar	$x = \dfrac{S_0 \cdot 18.015}{100 m_S + S_0 \cdot 18.105}$

The ideal behavior of the solute in the gas phase is assumed. T = temperature (K); ρ_w = water density (g m^{-3}); m_s = solute molecular weight. The Henry's constant is expressed in bar. Based on a table in S. Cabani and P. Gianni, in H. J. Hinz, ed., *Thermodynamic Data for Biochemistry and Biotechnology*, (1986), p. 260 with kind permission from Springer Science+Business Media.

that follow, the total number of moles N is replaced by the number of moles of water N_W. The mass (mole) balance on the radon in the liquid in the air stripper is

$$\frac{dN_R}{dt} = \frac{d(N_W x_R)}{dt} = N_W \frac{dx_R}{dt} = -\frac{\dot{N}_{air} K_R x_R}{P_{atm}}$$

or

$$N_W \frac{dx_R}{dt} = -\frac{\dot{N}_{air} K_R x_R}{P_{atm}}$$

On integration from $t = 0$ we obtain

$$\ln\left[\frac{N_R(t)}{N_R(t=0)}\right] = -\frac{\dot{N}_{air} \cdot t \cdot K_R}{N_W \cdot P_{atm}} = \frac{N_{air} \cdot K_R}{N_W \cdot P_{atm}}$$

where $N_{air} = \dot{N}_{air} \cdot t$ is the number of moles of air that have passed through the radon-containing water in the stripping device. Note that this equation gives the amount of radon remaining in the water as a function of the number of moles of air that have been used. Consequently,

$$\ln(0.01) = -4.605 = \frac{N_{air} \cdot 0.52 \times 10^4}{N_W \cdot 1.103}$$

or

$$\frac{N_{air}}{N_W} = 8.97 \times 10^{-4}$$

so that 8.97×10^{-4} moles of air are needed for each mole of radon-contaminated water, or equivalently, 1.45×10^{-3} kg of air are needed for each kg of radon-contaminated water in the air stripper to reduce the radon content of the water to one-hundredth of its initial concentration. ■

The addition of a salt S to an aqueous solution of a gas (or uncharged organic compound) species i causes the solubility of the gas to change, usually to decrease. Empirically, this change in solubility is described by the simple relation

$$\log \frac{S_i\,(M_s)}{S_i\,(M_s = 0)} = \log \frac{S_i\,(M_s)}{S_{i,o}} = -K_{s,i} \cdot M_s \qquad \textbf{(11.1-19)}$$

where S and S_o are the solubilities of the gas in the aqueous salt solution and in pure water, respectively; M_s is the molal concentration of salt; and $K_{s,i}$ is referred to as the Setchenow coefficient, which depends on the gas and the salt. Some representative values are given below.

s/i	$K_{s,i}\ (\mathrm{M}^{-1})$			
	O_2	CO_2	Benzene	Naphthalene
NaCl	0.14	0.101	0.20	0.22
KCl	0.13	0.073	0.17	0.19
$NaC_6H_5SO_3$			−0.19	−0.15

Note that for some salts, such as sodium benzylsulfate in the table above, the Setchenow constant is negative, indicating that the solubility increases with salt concentration; this is referred to as **salting in**. A decrease in solubility with increasing salt concentration, **salting out** is the more common situation. Also, for mixed salts, the effects are generally assumed to be additive and Eq. 11.1-19 becomes

$$\log \frac{S_i}{S_{i,o}} = -\sum_{\text{salts,s}} K_{s,i} \cdot M_s \qquad \textbf{(11.1-20)}$$

ILLUSTRATION 11.1-4

The Solubility of Oxygen in an Aqueous Salt Solution, Blood, and Seawater

   a. Compare the solubility of oxygen in water containing 0.15 M NaCl to the solubility in pure water.
   b. Compare the solubility of oxygen in seawater and in pure water. Seawater can be approximated as containing 35 grams of NaCl per kg of water.

SOLUTION

   a. From Eq. 11.1-19 and data in the table, we have

$$\log \frac{S_{O_2}}{S_{O_2,o}} = -K_{\mathrm{NaCl},O_2} \cdot M_{\mathrm{NaCl}} = -0.14 \times 0.15 = -0.021$$

so that

$$\frac{S_{O_2}}{S_{O_2,o}} = 10^{-0.021} = 0.951$$

Therefore, the solubility of oxygen in an aqueous solution of 0.15 M NaCl is about 95 percent that in pure water. Since a 0.15 M NaCl solution has some of the same properties as blood (as we show elsewhere in this book, it has the same osmotic pressure and freezing point), we can expect that the solubility of oxygen in blood is also about 5 percent less than in pure water.

b. Since the molecular weight of NaCl is 58.44, the molality of seawater is 35/58.44 = 0.60 M. Thus,

$$\frac{S_{O_2}}{S_{O_2,o}} = 10^{-0.14 \times 0.6} = 0.824$$

Therefore, seawater contains appreciably less dissolved oxygen than pure water or blood.■

We next consider the dissolution of a mixture of gases in a liquid. We separate this situation into two different cases. First, if the concentrations of the dissolved gases in the liquid are relatively low, so that there are no nonideality departures from Henry's law, it is reasonable to assume that the solubility of each gas would be the same as if it were the only gas present at its gas-phase partial pressure. However, if the concentrations of the gases in the liquid are high enough that there are departures from the Henry's law limit, so that the activity coefficients γ_i^* need be included in the description, then the solubility of each species is affected by the presence of others through the values of the activity coefficients.

While using an activity coefficient model will provide a quantitative relationship between the mutual solubilities, we can get a qualitative understanding of how the presence of one dissolved species affects others by examining the interrelation between mixed second derivatives. In particular, the Maxwell equations in Chapter 8 and some of the pure fluid equations in Chapter 6 were derived by examining mixed second derivatives of thermodynamic functions. Another example of this is to start with the Gibbs energy and note that at constant temperature, pressure, and all other species mole numbers,

$$\left(\frac{\partial^2 G}{\partial N_i \partial N_j}\right)_{T,P} = \frac{\partial}{\partial N_i}\bigg|_{T,P,N_{k \neq i}} \left(\frac{\partial G}{\partial N_j}\right)_{T,P,N_{k \neq j}} = \left(\frac{\partial^2 G}{\partial N_j \partial N_i}\right)_{T,P}$$

$$= \frac{\partial}{\partial N_j}\bigg|_{T,P,N_{k \neq j}} \left(\frac{\partial G}{\partial N_i}\right)_{T,P,N_{k \neq i}}$$

or

$$\left(\frac{\partial \overline{G}_j}{\partial N_i}\right)_{T,P,N_{k \neq i}} = \left(\frac{\partial \overline{G}_i}{\partial N_j}\right)_{T,P,N_{k \neq j}} \tag{11.1-21}$$

Though this equation follows directly from the work of Gibbs, it was first used by Bjerrium (1923) and is usually referred to as the Bjerrium equation. As an example of its use, consider the solubility of two gases, such as carbon dioxide and oxygen in water. If adding additional carbon dioxide to the liquid (by increasing the CO_2 partial pressure and its partial molar Gibbs energy) increases the partial molar Gibbs energy of the oxygen in the liquid, then the inverse must also be true; that is, adding more oxygen will increase the partial molar Gibbs energy of the dissolved carbon dioxide. In fact, this is exactly what happens in water, in that if the oxygen partial pressure above water is increased at fixed partial pressure of carbon dioxide, additional oxygen is absorbed into the water, and since the CO_2 partial molar Gibbs energy in the liquid increases, some CO_2 desorbs for it to remain in equilibrium with the fixed partial pressure in the gas phase.

This thermodynamic coupling between oxygen and carbon dioxide exhibited in water will also occur between any two gases in any solvent. A physiological example of this is the solubility of oxygen and carbon dioxide in blood, where it is found that increasing the partial pressure of O_2 at fixed CO_2 partial pressure results in an increased oxygen concentration in blood and decreased carbon dioxide concentration. Also, the situation is reversed if the CO_2 partial pressure is increased at fixed O_2 partial pressure. This phenomenon was first experimentally observed in 1914 and is referred to as the Bohr effect.

Finally, we close this section by presenting the following approximate empirical model for the solubility of a gas or other compound in a solvent mixture:

$$\log(C_3)_{\text{mixedsolvent}} = \phi_1 \log(C_3)_{\text{solvent 1}} + \phi_2 \log(C_3)_{\text{solvent 2}} \qquad \textbf{(11.1-22)}$$

In this equation, known as the Yalkowsky-Rubino model, C is the molar concentration of the solute (component 3), and the ϕ terms are the solute-free volume fractions of the solvents (components 1 and 2). This equation should be used only for compounds of limited solubility.

PROBLEMS FOR SECTION 11.1

11.1-1 In this exercise we want to establish that the extrapolated value of $\Delta_{\text{vap}}\hat{H}$ is positive for a liquid at its critical temperature but below its critical pressure, and becomes negative as the temperature is increased significantly above the critical temperature. Use the data in the steam tables in Appendix A.III to prepare a plot of $\hat{H}^L$ and $\hat{H}^V$ for liquid water and steam at 0.1 MPa for all temperatures above 0°C, and extrapolate $\hat{H}^L$ to high temperatures from the low-pressure liquid-phase enthalpy data.
  a. Estimate the temperature at which $\Delta_{\text{vap}}\hat{H}$ will equal zero.
  b. Find the hypothetical $\Delta_{\text{vap}}\hat{H}$ of water at 1100°C and 1 bar.

11.1-2 Here we want to estimate the solubility of gaseous nitrogen in liquid carbon tetrachloride at 25°C and a partial pressure of nitrogen of 1 bar. The (hypothetical) liquid nitrogen fugacity at 25°C is 1000 bar.
  a. Calculate the mole fraction of nitrogen present in the liquid CCl_4 at equilibrium if the two species form an ideal solution.
  b. From regular solution theory it is estimated that

$$\ln \gamma_{N_2} = 0.526(1 - x_{N_2})^2$$

What is the equilibrium mole fraction of nitrogen in CCl_4 under these circumstances? What is the Henry's law constant for this system, if the excess volume for this system is zero?

11.1-3 The following data have been reported for the solubility of methane in various solvents at 25°C and 1.013 bar partial pressure of methane.[3]

Solvent	Mole Fraction of CH_4 in Saturated Liquid
Benzene	2.07×10^{-3}
Carbon tetrachloride	2.86×10^{-3}
Cyclohexane	2.83×10^{-3}
n-Hexane	$\begin{cases} 3.15 \times 10^{-3}* \\ 4.24 \times 10^{-3}\dagger \end{cases}$

*[A. S. McDaniel, *J. Phys. Chem.*, **15**, 587 (1911)]
†(D. Guerry, Jr., thesis, Vanderbilt Univ., 1944)

  a. Estimate the ideal solubility of methane at 25°C and a partial pressure of 1.013 bar.
  b. Calculate the activity coefficient of methane in each of the solvents in the table.
  c. Is the datum of either McDaniel or Guerry consistent with the regular solution model?

11.1-4 a. Estimate the vapor and liquid compositions of a nitrogen and benzene mixture in equilibrium at 75 bar and 100°C.

---

[3]The data for this problem were abstracted from J. H. Hildebrand and R. L. Scott, *The Solubility of Nonelectrolytes,* 3rd ed., Reinhold, New York (1950), Table 4, p. 243.

b. Estimate the vapor and liquid compositions at 100 bar and 100°C.

Data:

Vapor Pressure, bar	Temperature (°C)	
	Benzene	Nitrogen
1.013	80.1	−195.8
2.027	103.8	−189.2
5.067	142.5	−179.1
10.133	178.8	−169.8
30.399	249.5	−148.3
50.665	290.3	

11.1-5 One way to remove a dissolved gas from a liquid is by vaporizing a small amount of the liquid in such a way that the vapor formed is continually withdrawn from the system. This process is known as differential distillation or Rayleigh distillation, as discussed in Sec. 10.1.

a. If N is the total number of moles of liquid, x the instantaneous mole fraction of dissolved gas, and y the gas mole fraction of the vapor in equilibrium with that liquid, show that

$$\frac{dx}{dN} = \frac{y - x}{N}$$

b. If the mole fraction of the dissolved gas is low, then

$$y = Hx/P$$

where H is the Henry's law constant for the gas-solvent combination, and P is the total pressure above the liquid. Under these circumstances show that

$$\frac{x}{x_0} = \left(\frac{N}{N_0}\right)^{(H-P)/P}$$

or, equivalently

$$\frac{N}{N_0} = \left(\frac{x}{x_0}\right)^{P/(H-P)}$$

where N_0 and x_0 are the initial moles of solution and dissolved gas mole fraction, respectively.

c. The Henry's law constant for carbon dioxide in carbon disulfide was computed (in Illustration 11.1-1) to be 267 bar at 25°C. At a total pressure of 1 bar, how much liquid should be vaporized from a solution saturated in carbon dioxide to decrease the CO_2 concentration to 1% of its equilibrium value? To 0.01% of its equilibrium value?

11.1-6 Estimate the solubility of carbon dioxide in toluene at 25°C and 1.013 bar CO_2 partial pressure using the Peng-Robinson equation of state assuming $k_{CO_2-T} = 0.0$.

11.1-7 Estimate the solubility of carbon dioxide in carbon disulfide at 25°C and 1.013 bar CO_2 partial pressure using the Peng-Robinson equation of state, assuming

a. $k_{CO_2-CS_2} = 0.0$

b. $k_{CO_2-CS_2} = 0.2$

c. Compare the results of parts (a) and (b) with those obtained using the Prausnitz-Shair correlation.

11.1-8 Derive the conversion factors in Table 11.1-3.

11.1-9 From Eq. 11.1-18 derive equations for the Gibbs energy, enthalpy, entropy, and heat capacity for the transfer of gas from the ideal gas state at 1.013 bar pressure to the (hypothetical) ideal solution at unit mole fraction. Comment on the expected accuracy of the thermodynamic properties determined in this way.

11.1-10 The vapor pressure of pure bromine at 25°C is 28.4 kPa. The vapor pressure of bromine in dilute aqueous solution at that temperature obeys the equation $P = 147$ M, where M is the bromine molality, and the pressure is in kPa.

a. Calculate the Gibbs energy change for the dissolution process

$Br_2(\text{pure liquid, } 25°C)$

$\qquad \rightarrow Br_2(\text{aqueous solution of molality M, } 25°C)$

as a function of the bromine molality.

b. Compute the saturation molality of bromine in water at 25°C, and show that the Gibbs energy change for the process,

$Br_2(\text{pure liquid, } 25°C)$

$\qquad \rightarrow Br_2(\text{saturated aqueous solution, } 25°C)$

is zero.

11.1-11 Estimate the solubility of carbon dioxide in toluene at 25°C and 10 bar CO_2 partial pressure using

a. The Peng-Robinson equation of state with the van der Waals one-fluid mixing rules and $k_{CO_2-T} = 0$

b. The Peng-Robinson equation of state with the van der Waals one-fluid mixing rules and $k_{CO_2-T} = 0.09$

11.1-12 a. Water is carefully degassed, placed in an evacuated vessel that is closed, and heated to a temperature of 25°C. The vessel is approximately half full of liquid water at these conditions.

What will be the equilibrium pressure in the vessel?

b. The vessel of part (a) develops a small leak so that air enters the vessel (because of the pressure difference), but no water leaves. What will be the pressure in the vessel at equilibrium, the mole fraction of water in air, and the mole fractions of nitrogen and oxygen in the water?

Data: The Henry's law constants for oxygen and nitrogen in water at 25°C are

$$H_{O_2} = 4.35 \times 10^4 \frac{\text{bar}}{\text{mole fraction}}$$

and

$$H_{N_2} = 8.48 \times 10^4 \frac{\text{bar}}{\text{mole fraction}}$$

(*Hint:* The Henry's constants are so large that the mole fractions of oxygen and nitrogen in the water will be small.)

11.1-13 At $T = 60°C$ the vapor pressure of methyl acetate is 1.1260 bar, and the vapor pressure of methanol is 0.8465 bar. Their mixtures can be described by the one-constant Margules equation

$$\underline{G}^{ex} = Ax_1x_2 \quad \text{with} \quad A = 1.06RT$$

where R is the gas constant and T is temperature in K.

a. Plot the fugacity of methyl acetate and methanol in their mixtures as a function of composition at this temperature.

b. The Henry's law coefficient H_i is given by the equation

$$H_i = \lim_{x_i \to 0} \frac{P_i}{x_i}$$

Develop an expression for the Henry's law constant as a function of the A parameter in the Margules expression, the vapor pressure, and the composition. Compare the hypothetical pure component fugacity based on the Henry's law standard state with that for the usual pure component standard state.

11.1-14 The triple-point properties of a substance may be experimentally determined by measuring its temperature and pressure when the vapor, liquid, and solid phases all coexist at equilibrium. There will be an error in these measurements if air is trapped in the system; the entrapped air will be present in the vapor phase and dissolved in the liquid. Determine the error that would occur in the measurement of the triple-point temperature and pressure of water if, at equilibrium, the vapor contained 0.1333 kPa partial pressure of air.

Data: See Problem 7.12. The Henry's law expression for air in water at $T \sim 0°C$ is

$$P_{\text{air}} = 4.3 \times 10^4 x_{\text{air}}$$

where P_{air} is the partial pressure of air in the vapor phase, in bar, and x_{air} is the equilibrium mole fraction of air dissolved in water.

11.1-15 Derive the expression for the Henry's law coefficient of a species in a mixture that obeys the Peng-Robinson equation of state and the van der Waals one-fluid mixing rules.

11.1-16 Derive the expression for the Henry's law coefficient of a species in a mixture that obeys the Peng-Robinson equation of state and the Wong-Sandler mixing rules.

11.1-17 Estimate the solubility of oxygen in liquid bromine at 25°C and an oxygen partial pressure of 1 bar.

11.2 LIQUID-LIQUID EQUILIBRIUM

At low pressures all gases are mutually soluble in all proportions. The same is not always true for liquids, in that the equilibrium state of many binary liquid mixtures is two stable liquid phases of differing composition, rather than a single liquid phase, at least over certain ranges of temperature and composition. Our aim in this section is to obtain an understanding of why liquid-liquid phase separation occurs, and to develop thermodynamic equations that relate the properties of the two equilibrium phases. In this way we will be able to estimate the compositions of the coexisting phases when data are not available, or use available liquid-liquid phase equilibrium data to get information on the activity coefficients of each species in a mixture.

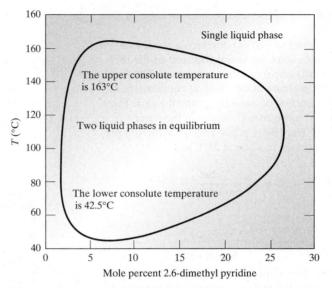

Figure 11.2-1 The liquid-liquid phase diagram of 2,6-dimethyl pyridine and water, with composition in mole percent.

Examples of liquid-liquid phase equilibrium behavior are given in Figs. 11.2-1 through 11.2-3.[4] These phase diagrams are to be interpreted as follows: If the temperature and overall composition of the mixture lie in the enclosed two-phase region, two liquid phases will form; the compositions of these phases are given by the intersection of the constant temperature (horizontal) line, or tie line, with the boundaries of the two-phase region. If, however, the temperature and composition are in the single-phase region, the equilibrium state is a single liquid phase. For example, if one attempts to prepare a mixture of 10 mol percent 2,6-dimethyl pyridine in water at 80°C, he will obtain, instead, two liquid phases, one containing 1.7 mol percent 2,6-dimethyl pyridine,

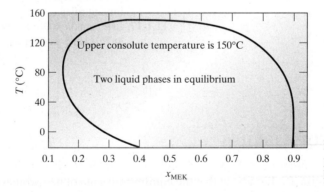

Figure 11.2-2 The liquid-liquid phase diagram for methyl ethyl ketone and water.

---

[4] The experimental data for the partial solubility of perfluoro-n-heptane in various solvents has been plotted as a function of both mole fraction and volume fraction in Fig. 11.2-3. It is of interest to notice that these solubility data are almost symmetric functions of the volume fraction and nonsymmetric functions of the mole fraction. Such behavior has also been found with other thermodynamic mixture properties; these observations suggest the use of volume fractions, rather than mole fractions or mass fractions, as the appropriate concentration variables for describing nonideal mixture behavior. Indeed, this is the reason that volume fractions have been used in both the regular solution model and the Wohl expansion of Eq. 9.5-8 for liquid mixtures.

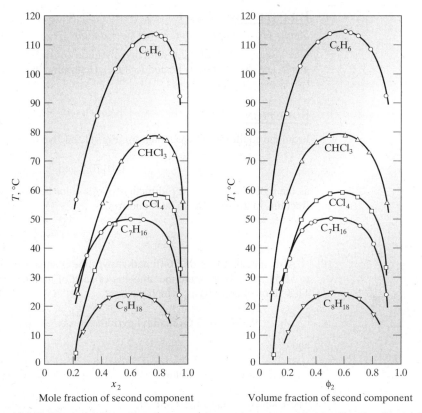

Figure 11.2-3 The solubility of perfluoro-*n*-heptane in various solvents. Species 1 is perfluoro-*n*-heptane; species 2 is the solvent. [Reprinted with permission from J. H. Hildebrand, B. B. Fisher, and H. A. Benesi, *J. Am. Chem. Soc.*, **72**, 4348 (1950). Copyright by the American Chemical Society.]

and the other containing 23.3 mol percent 2,6-dimethyl pyridine (see Fig. 11.2-1). The relative amounts of the two phases in such cases can be computed from the species mass balances

$$M_i = M_i^I + M_i^{II} \quad \text{(mass basis)}$$

$$N_i = N_i^I + N_i^{II} \quad \text{(molar basis)}$$

$$\text{(11.2-1a)}$$

Here M_i^J and N_i^J are the mass and number of moles of species i in phase J, respectively, and M_i and N_i are the total mass and number of moles of this species. Alternative forms of these equations are

$$M_i = \omega_i^I M^I + \omega_i^{II} M^{II} \quad \text{(mass basis)}$$

$$N_i = x_i^I N^I + x_i^{II} N^{II} \quad \text{(molar basis)}$$

$$\text{(11.2-1b)}$$

where ω_i^J and x_i^J are the mass fraction and mole fraction of species i in phase J, respectively, and M^J and N^J are the total weight and number of moles of that phase.

ILLUSTRATION 11.2-1

Simple Liquid-Liquid Equilibrium Calculation

One kilogram of liquid 2,6-dimethyl pyridine (C_7H_9N, MW = 107.16) is mixed with 1 kg of water, and the mixture is heated to 80°C. Determine the compositions and total amounts of the two coexisting liquid phases.

SOLUTION

From the equilibrium phase diagram, Fig. 11.2-1, we have

$$x_P^I = 0.0170 \quad \text{and} \quad x_P^{II} = 0.233$$

as the mole fractions of 2,6-dimethyl pyridine in the two equilibrium phases. The water mole fractions in the phases can be found from

$$x_{H_2O}^J = 1 - x_P^J$$

Also, note that 1 kg of 2,6-dimethyl pyridine is equal to 9.332 mol, and 1 kg of water is equal to 55.51 mol. To compute the amounts of each of the phases, we use the mass balances of Eq. 11.2-1b on a molar basis, which yield

$$9.332 \text{ mol 2,6-dimethyl pyridine} = x_P^I x^I + x_P^{II} x^{II} = 0.0170 N^I + 0.233 N^{II}$$

and

$$55.51 \text{ mol water} = (1 - x_P^I) N^I + (1 - x_P^{II}) N^{II} = 0.983 N^I + 0.767 N^{II}$$

These equations have the solution

$$N^I = 26.742 \text{ mol} \quad \text{and} \quad N^{II} = 38.10 \text{ mol}$$

Now 1 mol of a solution with $x_P = 0.0170$ weighs $0.0170 \times 107.66 + 0.987 \times 18.015 = 19.53$ g, and 1 mol of a solution with $x_P = 0.233$ weighs $0.233 \times 107.16 + 0.767 \times 18.015 = 38.786$ g. Therefore, there are 26.742 mol $\times$ 19.53 g/mol = 522 g = 0.522 kg of phase I and 38.10 mol $\times$ 38.786 g/mol = 1478 g = 1.478 kg of phase II. ∎

Generally, liquid-liquid phase equilibrium (or phase separation) occurs only over certain temperature ranges, bounded above by the **upper consolute** or **upper critical solution temperature**, and bounded below by the **lower consolute** or **lower critical solution temperature**. These critical solution temperatures are indicated on the liquid-liquid phase diagrams given here. All partially miscible mixtures should exhibit either one or both consolute temperatures; however, the lower consolute temperature may be obscured by the freezing of the mixture, and the upper consolute temperature will not be observed if it is above the bubble point temperature of the mixture, as vaporization will have instead occurred.

Measurement of liquid-liquid equilibrium

The measurement of liquid-liquid equilibrium at near ambient pressures can generally be done with relatively simple equipment, such as that shown in Fig. 11.2-4. The equilibrium cell shown consists of a small, glass, thermostated vessel with upper and lower ports for sampling the two liquid phases. Samples are taken using a syringe inserted through a rubber septum. Magnetic stirring is used until equilibrium is achieved; then the stirrer is turned off and the two liquid phases are allowed to disengage before samples are taken. The same general idea can be used at high pressures using a metal vessel.

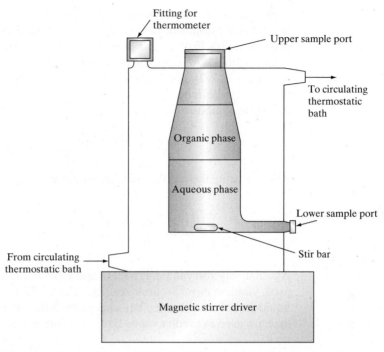

Figure 11.2-4 Schematic diagram of a simple liquid-liquid equilibrium cell.

The thermodynamic requirement for any type of phase equilibrium is that the composition of each species in each phase in which it appears be such that the equilibrium criterion

Starting point for all phase equilibrium calculations

$$\bar{f}_i^{\,I}(T, P, \underline{x}^I) = \bar{f}_i^{\,II}(T, P, \underline{x}^{II})$$

is satisfied. Introducing the activity coefficient definition into this equation yields

$$x_i^I \gamma_i^I(T, P, \underline{x}^I) f_i(T, P) = x_i^{II} \gamma_i^{II}(T, P, \underline{x}^{II}) f_i(T, P)$$

which reduces to[5]

General LLE relation using activity coefficients

$$x_i^I \gamma_i^I(T, P, \underline{x}^I) = x_i^{II} \gamma_i^{II}(T, P, \underline{x}^{II}) \quad i = 1, 2, \ldots, \mathcal{C}$$ **(11.2-2)**

Since the pure component liquid fugacities on both sides of the equation are the same, they cancel. (Why is this different from vapor-liquid equilibrium?) The compositions of the coexisting phases are the sets of mole fractions $x_1^I, x_2^I, \ldots, x_{\mathcal{C}}^I, x_1^{II}, x_2^{II}, \ldots, x_{\mathcal{C}}^{II}$ that simultaneously satisfy Eqs. 11.2-2 and

$$\sum_{i=1}^{\mathcal{C}} x_i^I = 1 \quad \text{and} \quad \sum_{i=1}^{\mathcal{C}} x_i^{II} = 1$$ **(11.2-3)**

---

[5]From this equation it is clear that liquid-liquid phase separation is a result of solution nonideality. If the solutions were ideal, that is, if $\gamma_i^I = \gamma_i^{II} = 1$, then $x_i^I = x_i^{II}$ for all species i and there would be only a single phase, not two liquid phases of different composition, present at equilibrium.

Equations 11.2-2 can be used with experimental phase equilibrium data to calculate the activity coefficient of a species in one phase from its known value in the second phase or, with Eqs. 11.2-3 and experimental activity coefficient data or appropriate solution models, to compute the compositions of both coexisting liquid phases (see Illustration 11.2-2). For example, using the one-constant Margules equation to represent the activity coefficients, we obtain from Eq. 11.2-2 the following relationship between the phase compositions:

$$x_i^{\mathrm{I}} \exp\left[\frac{A(1-x_i^{\mathrm{I}})^2}{RT}\right] = x_i^{\mathrm{II}} \exp\left[\frac{A(1-x_i^{\mathrm{II}})^2}{RT}\right] \qquad \textbf{(11.2-4a)}$$

whereas using the regular solution model leads to

$$x_i^{\mathrm{I}} \exp\left[\frac{\underline{V}_i^{\mathrm{L}}(\Phi_j^{\mathrm{I}})^2(\delta_1-\delta_2)^2}{RT}\right] = x_i^{\mathrm{II}} \exp\left[\frac{\underline{V}_i^{\mathrm{L}}(\Phi_j^{\mathrm{II}})^2(\delta_1-\delta_2)^2}{RT}\right] \qquad \textbf{(11.2-4b)}$$

Alternatively, the van Laar, NRTL, and UNIQUAC activity coefficient models could be used, yielding more accurate results. (The UNIFAC method can also be used to predict liquid-liquid equilibrium, but only with different main group interaction parameters than are used to predict vapor-liquid equilibrium.)

ILLUSTRATION 11.2-2

Prediction of LLE Using an Activity Coefficient Model

Use the van Laar equations to estimate the compositions of the coexisting liquid phases in an isobutane-furfural mixture at 37.8°C and a pressure of 5 bar. (You may assume that the van Laar constants for this system given in Table 9.5-1 are applicable at this pressure.)

SOLUTION

The compositions of the coexisting phases are the solution of the following set of algebraic equations:

$$x_1^{\mathrm{I}}\gamma_1^{\mathrm{I}} = x_1^{\mathrm{I}} \exp\left\{\frac{\alpha}{\left[1+\dfrac{\alpha x_1^{\mathrm{I}}}{\beta(1-x_1^{\mathrm{I}})}\right]^2}\right\} = x_1^{\mathrm{II}} \exp\left\{\frac{\alpha}{\left[1+\dfrac{\alpha x_1^{\mathrm{II}}}{\beta(1-x_1^{\mathrm{II}})}\right]^2}\right\} = x_1^{\mathrm{II}}\gamma_1^{\mathrm{II}} \qquad \textbf{(a)}$$

$$x_2^{\mathrm{I}}\gamma_2^{\mathrm{I}} = x_2^{\mathrm{I}} \exp\left\{\frac{\beta}{\left[1+\dfrac{\beta x_2^{\mathrm{I}}}{\alpha(1-x_2^{\mathrm{I}})}\right]^2}\right\} = x_2^{\mathrm{II}} \exp\left\{\frac{\beta}{\left[1+\dfrac{\beta x_2^{\mathrm{II}}}{\alpha(1-x_2^{\mathrm{II}})}\right]^2}\right\} = x_2^{\mathrm{II}}\gamma_2^{\mathrm{II}} \qquad \textbf{(b)}$$

$$x_1^{\mathrm{I}} + x_2^{\mathrm{I}} = 1 \qquad \textbf{(c)}$$

$$x_1^{\mathrm{II}} + x_2^{\mathrm{II}} = 1 \qquad \textbf{(d)}$$

Here isobutane will be taken as species 1 and furfural as species 2; thus, from Table 9.5-1, $\alpha = 2.62$ and $\beta = 3.02$. These equations may be solved directly using an equation-solving program such as MATHCAD. Alternatively, the following graphical procedure can be used in solving the equations:

i. A guess is made for the value of x_1^I.

ii. Equation a is then used to compute x_1^{II}, and Eq. c to compute x_2^I.

iii. x_2^{II} is then be computed using Eq. b.

iv. Finally, the check whether Eq. d is satisfied with the computed values of x_1^{II} and x_2^{II}. If it is, we have found the solution to the equations. If Eq. d is not satisfied, a new value of x_1^I is chosen, and the calculation is repeated.

The difficult part of this computation is the solution of Eqs. a and b for x_1^{II} and x_2^{II}. One way to solve these equations, if one does not wish to use an equation-solving program, is as follows. First, the values of the product $x_1\gamma_1$ are computed for various values of x_1 (see the table that follows). From a graph of $x_1\gamma_1$ versus x_1 (see Fig. 1), we can easily obtain the value of $x_1\gamma_1$ for any choice of x_1 and then locate other values of x_1 that have the same value of $x_1\gamma_1$. Thus, given the trial value of x_1^I, x_1^{II} is quickly found. In a similar fashion, computing and plotting $x_2\gamma_2$ as a function of x_2 (or $x_1 = 1 - x_2$) provides an easy method of finding x_2^{II} given $x_2^I = 1 - x_1^I$. We then need to check whether Eq. d is satisfied. If not, the calculation is repeated with another guess for x_1^I.

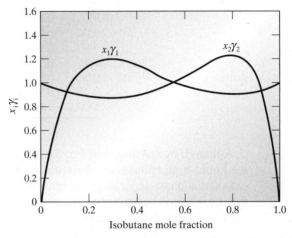

Figure 1 The product of species mole fraction and activity coefficients for the isobutane-furfural system at 37.8°C and 5 bar.

Using MATHCAD, we obtain the following compositions for the coexisting liquid phases:

$$x_1^I = 0.1128 \qquad x_1^{II} = 0.9284$$

and

$$x_2^I = 0.8872 \qquad x_2^{II} = 0.0716$$

At these compositions the activity coefficients are

$$\gamma_1^I = 8.375 \qquad \gamma_1^{II} = 1.018$$

$$\gamma_2^I = 1.030 \qquad \gamma_2^{II} = 12.77$$

and

$$x_1^I\gamma_1^I = 0.1128 \times 8.375 = 0.945 = x_1^{II}\gamma_1^{II} = 0.9284 \times 1.018$$

$$x_2^I\gamma_2^I = 0.8872 \times 1.030 = 0.914 = x_2^{II}\gamma_2^{II} = 0.0716 \times 12.77$$

There are no experimental liquid-liquid phase equilibrium data for the isobutane-furfural system with which we can compare our predictions. However, binary mixtures of furfural and, separately, 2,2-dimethyl pentane, 2-methyl pentane, and hexane, which are similar to the binary mixture here, all exhibit liquid-liquid phase separation,[6] with one phase containing between 89 and 90 mol % furfural (x_2^I) and the other between 6 and 7 mol % furfural (x_2^{II}).

$x_i \gamma_i$ versus x_1

x_1	$x_1 \gamma_1$	x_2	$x_2 \gamma_2$
0	0	1.0	1.0
0.05	0.5491	0.95	0.9555
0.10	0.8843	0.90	0.9231
0.15	1.0761	0.85	0.8965
0.20	1.1734	0.80	0.8805
0.30	1.2072	0.70	0.8739
0.40	1.1405	0.60	0.9000
0.50	1.0598	0.50	0.9594
0.60	0.9840	0.40	1.0506
0.70	0.9322	0.30	1.1607
0.80	0.9121	0.20	1.2343
0.85	0.9161	0.15	1.2071
0.90	0.9309	0.10	1.0731
0.95	0.9582	0.05	0.7325
1.00	1.0	0.0	0.0

∎

Although by starting with Eq. 11.2-2 one can proceed directly to the calculation of the liquid-liquid phase equilibrium state, this equation does not provide insight into the reason that phase separation and critical solution temperature behavior occur. To obtain this insight it is necessary to study the Gibbs energy versus composition diagram for various mixtures. For an ideal binary mixture, we have (Table 9.3-1)

$$\underline{G}^{IM} = x_1 \underline{G}_1 + x_2 \underline{G}_2 + RT(x_1 \ln x_1 + x_2 \ln x_2) \qquad (11.2\text{-}5)$$

Now since x_1 and x_2 are always less than unity, $\ln x_1$, $\ln x_2 \leq 0$, and the last term in Eq. 11.2-5 is negative. Therefore, the Gibbs energy of an ideal mixture is always less than the mole fraction–weighted sum of the pure component Gibbs energies, as illustrated by curve a of Fig. 11.2-5. For a real mixture, we have

$$\underline{G} = \underline{G}^{IM} + \underline{G}^{ex} \qquad (11.2\text{-}6)$$

where the excess Gibbs energy would be determined by experiment or approximated using a liquid solution model. For the purposes of illustration, assume that the one-constant Margules equation is satisfactory, so that

$$\underline{G}^{ex} = A x_1 x_2 \qquad (11.2\text{-}7)$$

with $A > 0$. The total Gibbs energy,

$$\underline{G} = x_1 \underline{G}_1 + x_2 \underline{G}_2 + RT(x_1 \ln x_1 + x_2 \ln x_2) + A x_1 x_2 \qquad (11.2\text{-}8)$$

for two values of A is plotted as curves b and c of Fig. 11.2-5.

---

[6]The reference for these data is H. Stephen and T. Stephen, eds., *Solubilities of Inorganic and Organic Compounds,* Vol. 1, *Binary Systems,* Macmillan, New York (1963).

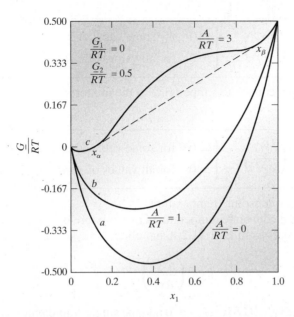

Figure 11.2-5 The molar Gibbs energy of ideal ($A = 0$) and nonideal ($A \neq 0$) binary mixtures if phase separation does not occur (solid line) and when phase separation occurs (dashed line).

The equilibrium criterion for a closed system at constant temperature and pressure is that the Gibbs energy of the system be a minimum.[7] For the mixture of curve c, with an overall composition between x_α and x_β, the lowest value of $\underline{G}$ is obtained when the mixture separates into two phases, one of composition x_α and the other of composition x_β. In this case the Gibbs energy of the mixture is a linear combination of the Gibbs energies of two coexisting equilibrium liquid phases (the lever rule, Eqs. 7.3-1) and is represented by the dashed line (representing different amounts of the two phases) rather than the solid line in Fig. 11.2-5. If, however, the total mixture mole fraction of species 1 is less than x_α or greater than x_β, only a single phase will exist. Of course, the phase equilibrium compositions x_α and x_β can be found directly from Eq. 11.2-2 in general, and from Eq. 11.2-4a for the case here, without this graphical construction.

The temperature range over which liquid-liquid phase separation occurs can be found by using the requirement for intrinsic fluid stability of Chapter 7 (see also Problem 8.25), that is,

$$d^2\underline{G} > 0 \qquad \text{at constant } N, \ T, \text{ and } P \qquad \textbf{(11.2-9)}$$

What, in fact, we will do is look at the second composition derivative of the Gibbs energy. If $(\partial^2\underline{G}/\partial x_1^2)_{T,P} > 0$ (which follows from Eq. 11.2-9), for a given temperature and composition, the single phase is stable; if, however, $(\partial^2\underline{G}/\partial x_1^2)_{T,P} < 0$ at the given values of T and x_1, the single phase is unstable and phase separation occurs. The compositions at which $(\partial^2\underline{G}/\partial x_1^2)_{T,P} = 0$, which are inflection points on the $\underline{G}$ versus x_1 curve, are the limits of stability of the single phase at the given temperature (though not the equilibrium compositions).[8] If a temperature T_{uc} exists for which

---

[7]Since the system is closed and nonreacting, the number of moles of each species is fixed. Therefore, in the analysis here we will consider the molar Gibbs energy of the mixture $\underline{G}$, rather than G.

[8]Remember from Sec. 7.3 that while the the condition $(\partial P/\partial V)_T = 0$ on the van der Waals loop of an equation of state gave the conditions of mechanical stability, it did not give the vapor-liquid equilibrium points (that is, the vapor pressure). That had to be determined from the equality of species fugacities in each phase. The situation is much the same here in that the limit of stability from Eq. 11.2-9 is not the equilibrium compositions found from the equality of species fugacities in the coexisting liquid phases.

Condition for upper consolute temperature

$$\left(\frac{\partial^2 \underline{G}}{\partial x_1^2}\right)_{T,P} \begin{cases} = 0 & \text{for some value of } x_1 \text{ at } T = T_{uc} \\ > 0 & \text{for all values of } x_1 \text{ at } T > T_{uc} \end{cases} \qquad \textbf{(11.2-10a)}$$

it is the upper consolute temperature of the mixture. Similarly, if there is a temperature T_{lc} for which

Condition for lower consolute temperature

$$\left(\frac{\partial^2 \underline{G}}{\partial x_1^2}\right)_{T,P} \begin{cases} = 0 & \text{for some value of } x_1 \text{ at } T = T_{lc} \\ > 0 & \text{for all values of } x_1 \text{ at } T < T_{lc} \end{cases} \qquad \textbf{(11.2-10b)}$$

it is the lower consolute temperature.

To find the consolute temperatures of a mixture obeying the one-constant Margules model, we start from Eqs. 11.2-8 and obtain[9]

$$\left(\frac{\partial^2 \underline{G}}{\partial x_1^2}\right)_{T,P} = \frac{RT}{x_1 x_2} - 2A \qquad \textbf{(11.2-11)}$$

Consequently, $(\partial^2 \underline{G}/\partial x_1^2)_{T,P} > 0$ and the single-liquid phase is the equilibrium state if

$$T > \frac{2A x_1 x_2}{R} \qquad \textbf{(11.2-12a)}$$

whereas $(\partial^2 \underline{G}/\partial x_1^2)_{T,P} < 0$ and phase separation occurs if

$$T < \frac{2A x_1 x_2}{R} \qquad \textbf{(11.2-12b)}$$

The limit of stability occurs at

$$T = \frac{2A x_1 (1 - x_1)}{R} \qquad \textbf{(11.2-13)}$$

and the highest temperature at which phase separation is possible at any composition for a Margules mixture occurs at $x_1 = x_2 = 0.5$, which gives the maximum value of the product $x_1(1 - x_1)$, so that

$$T_{uc} = \frac{A}{2R} \qquad \textbf{(11.2-14)}$$

This is the upper consolute temperature for a Margules mixture. Note that the Margules equation does not have a lower critical solution temperature (i.e., there is no solution of Eq. 11.2-11 for which Eq. 11.2-10b is satisfied). Thus the two partially miscible liquid phases of a Margules mixture cannot be made to combine by lowering the temperature.

Equations 11.2-11 through 11.2-14 are specific to the choice of the one-constant Margules equation to represent the excess Gibbs energy of the mixture. The use of more realistic models for $\underline{G}^{ex}$ will lead to other predictions for phase separation, such as the limit of stability at the upper consolute occurring away from $x_1 = 0.5$ (Problem 11.2-1).

For very nonideal mixtures, for example, aqueous solutions in which hydrogen bonding and other associative phenomena occur, the species activity coefficients may be

---

[9]In taking this derivative remember that $(\partial x_2/\partial x_1)_{T,P} = -1$.

very different in value from unity, asymmetric in composition, and temperature dependent. A detailed analysis of such systems is beyond the scope of this book, but it is useful to note here that such mixtures may have a lower consolute temperature. Also, in very nonideal mixtures the species activity coefficients can be large, so that the two species will be relatively insoluble, as will be shown in Illustration 11.2-3. (Since species activity coefficients are never infinite, it is evident from Eq. 11.2-2 that any liquid must be at least sparingly soluble in any other, so that no two liquids are truly immiscible. However, for some species the equilibrium solubilities are so low that it is convenient to refer to them as being immiscible.)

The Gibbs energy of a binary mixture is

$$\underline{G}(T, P, \underline{x}) = x_1\underline{G}_1(T, P) + x_2\underline{G}_2(T, P) + x_1RT\ln x_1\gamma_1 + x_2RT\ln x_2\gamma_2 \quad \textbf{(11.2-15)}$$

The change in Gibbs energy on forming this mixture from its pure components is

$$\begin{aligned}\Delta_{\mathrm{mix}}\underline{G}(T, P, \underline{x}) &= \underline{G}(T, P, \underline{x}) - x_1\underline{G}_1(T, P) - x_2\underline{G}_2(T, P) \\ &= x_1RT\ln x_1\gamma_1 + x_2RT\ln x_2\gamma_2\end{aligned} \quad \textbf{(11.2-16)}$$

Now consider the case of a very dilute solution of species 2 (an impurity or pollutant) in a solvent of species 1. In this case $x_1 \to 1$, so that $\gamma_1 \to 1$ and $\gamma_2 \to \gamma_2^\infty$. Therefore, the Gibbs energy change on forming a dilute mixture from the pure components is

$$\Delta_{\mathrm{mix}}\underline{G}(T, P, x_1 \to 1) = x_1RT\ln x_1 + x_2RT\ln x_2\gamma_2^\infty \quad \textbf{(11.2-17)}$$

Regardless of the value γ_2^∞, for sufficiently small mole fractions of an impurity or pollutant, the Gibbs energy change will be negative (since it consists of a sum of the logarithms of numbers less than 1). So a dilute solution has a lower Gibbs energy than its pure components at the same temperature and pressure, and therefore is the thermodynamically stable or preferred state. Thus a dilute mixture is preferred over the two pure species. There are several practical implications of this. One is that environmental pollutants, unless they are solids or present in very large amounts so as to saturate their surroundings, are much more likely to be found dissolved in water or present in the air than as a separate phase. Second, it is extremely unlikely that ultra high purity gaseous or liquid chemicals will be found in nature.

From Eq. 11.2-2 we have

$$x_i^{\mathrm{I}}\gamma_i^{\mathrm{I}}(T, P, \underline{x}^{\mathrm{I}}) = x_i^{\mathrm{II}}\gamma_i^{\mathrm{II}}(T, P, \underline{x}^{\mathrm{II}}) \quad \textbf{(11.2-2)}$$

Now consider the case in which compound 1 is essentially immiscible in the second compound, though the converse need not be true. (For example, *n*-octanol is essentially insoluble in water, but water is soluble in *n*-octanol to about 26 mol % at 25°C.) Taking phase I to be the one rich in component 1, we then have $x_1^{\mathrm{I}} \approx 1$ and $\gamma_1^{\mathrm{I}} \approx 1$, so that

$$1 = x_1^{\mathrm{II}}\gamma_1^{\mathrm{II}} \quad \text{or} \quad x_1^{\mathrm{II}} = \frac{1}{\gamma_1^{\mathrm{II}}} \quad \textbf{(11.2-18)}$$

This equation is useful for determining activity coefficients from liquid solubility data, as shown in the following illustration.

ILLUSTRATION 11.2-3

Determination of Activity Coefficients from Measured LLE Data for an Essentially Immiscible Mixture

From the data in Volume III of the *International Critical Tables* (McGraw-Hill, New York, 1929) we see that the equilibrium state in the carbon tetrachloride–water system at 25°C is two phases: one an aqueous phase containing 0.083 wt % carbon tetrachloride, and the other an organic phase containing 0.011 wt % water. Estimate the activity coefficient of CCl_4 in the aqueous phase and H_2O in the organic phase.

SOLUTION

Since the aqueous phase, which we designate as phase I, is 99.917 wt % water, it seems reasonable to assume that the activity coefficient of water in the aqueous phase is unity. Similarly, the activity coefficient of carbon tetrachloride in the organic phase will be taken to be unity. Therefore, from Eq. 11.2-2, we have

$$\gamma_{CCl_4}^{I} = \frac{x_{CCl_4}^{II}}{x_{CCl_4}^{I}} \cong \frac{1}{x_{CCl_4}^{I}}$$

$$\gamma_{H_2O}^{II} = \frac{x_{H_2O}^{I}}{x_{H_2O}^{II}} \cong \frac{1}{x_{H_2O}^{II}}$$

where we have used the superscripts I and II to denote the aqueous and organic phases, respectively.

The first calculation to be done is the conversion of the weight fraction data to mole fractions. Since the organic phase is almost pure carbon tetrachloride and the aqueous phase is essentially pure water, either the equations in this illustration or Eq. 11.2-18 can be used to compute the activity coefficients of carbon tetrachloride in water and water in carbon tetrachloride. The results are given in the table.

	x_{CCl_4}	x_{H_2O}	γ_{CCl_4}	γ_{H_2O}
Aqueous phase	0.9708×10^{-4}	0.9999	1.029×10^4	1 (assumed)
Organic phase	0.9991	0.9403×10^{-3}	1 (assumed)	1.063×10^3

COMMENT

The values of the activity coefficients here are very large, especially compared with those we found previously from vapor-liquid equilibrium data. Such values for the activity coefficient of the dilute species in a mixture with species of very different chemical nature, such as carbon tetrachloride and water, is quite common. Note that because of the concentrations involved, the activity coefficients found are essentially the values at infinite dilution. ∎

ILLUSTRATION 11.2-4

Approximate Values of Activity Coefficients from LLE Data for a Partially Miscible Mixture

When *n*-octanol and water are mixed, two liquid phases form. The water-rich phase is essentially pure water containing only 0.7×10^{-4} mole fraction *n*-octanol, while the octanol-rich phase contains approximately 0.26 mole fraction water. Estimate approximately the activity coefficient of *n*-octanol in water and water in *n*-octanol.

SOLUTION

The basic equation for the solution of this problem is Eq. 11.2-2:

$$x_i^{I}\gamma_i^{I}(\underline{x}^{I}) = x_i^{II}\gamma_i^{II}(\underline{x}^{II})$$

Since the water-rich phase (which we designate with superscript W) is essentially pure water, so that $x_W^W \cong 1$ and $\gamma_W^W \cong 1$, we have for water

$$1 = x_W^O \gamma_W^O$$

where the superscript O indicates the octanol-rich phase. Consequently, for water in the octanol phase,

$$\gamma_W^O = \frac{1}{x_W^O} = \frac{1}{0.26} = 3.846$$

For octanol we have

$$x_O^W \gamma_O^W = 0.7 \times 10^{-4} \gamma_O^W = x_O^O \gamma_O^O$$

Since the octanol-rich phase is not pure octanol, we do not know the value of γ_O^O. However, we see from Fig. 11.2-4 (for the system isobutane-furfural) that in the composition range of liquid-liquid equilibrium we do not introduce a serious error by assuming that the product $x_O^O \gamma_O^O$ is approximately equal to unity. Therefore, as an estimate, we have

$$0.7 \times 10^{-4} \gamma_O^W \approx 1 \quad \text{or} \quad \gamma_O^W \approx \frac{1}{0.7 \times 10^{-4}} = 1.43 \times 10^4 = 14\ 300$$

COMMENT

1. Since the concentration of n-octanol in water is so low, the value of γ_O^W above is essentially the value at infinite dilution.
2. While the value of the infinite-dilution activity coefficient of n-octanol in water above is large, in fact much larger activity coefficient values will be seen to occur in the next chapter. Consequently, although an activity coefficient represents a correction to ideal solution behavior, it can be a very large correction.
3. The value of the product $x_O^O \gamma_O^O$ cannot be greater than unity; otherwise a pure octanol phase would form. (Can you explain why this is so?) Therefore, the value of $x_O^O \gamma_O^O$ is likely to be somewhat less than unity, so our estimate above for the activity coefficient of octanol in water is too high, but probably only slightly so.   ∎

Note that we can rewrite Eq. 11.2-18 as

$$\ln x_i = -\ln \gamma_i = -\frac{\overline{G}_i^{ex}}{RT} \tag{11.2-19}$$

where we have omitted the superscript indicating the phase. Proceeding further and using Eq. 6.2-30, we have

$$\left(\frac{\partial \ln x_i}{\partial \left(\frac{1}{T} \right)} \right)_P = -\frac{1}{R} \left(\frac{\partial \left(\frac{\overline{G}_i^{ex}}{T} \right)}{\partial \left(\frac{1}{T} \right)} \right)_P = \frac{-\overline{H}_i^{ex}}{R} \tag{11.2-20}$$

Therefore, if the excess enthalpy is (approximately) independent of both temperature and composition, we have, as an indefinite integral,

$$\ln x_i(T) = A - \frac{\overline{H}_i^{ex}}{RT} \tag{11.2-21}$$

where A is the constant of integration. Alternatively, as a definite integral

$$\ln \frac{x_i(T_2)}{x_i(T_1)} = -\frac{\overline{H}_i^{ex}}{R}\left(\frac{1}{T_2} - \frac{1}{T_1}\right) \quad \text{or} \quad x_i(T_2) = x_i(T_1)\exp\left[-\frac{\overline{H}_i^{ex}}{R}\left(\frac{1}{T_2} - \frac{1}{T_1}\right)\right]$$

$$(11.2\text{-}22)$$

so that the solubility of the solute increases exponentially with temperature if the excess enthalpy of mixing is positive, and decreases exponentially with temperature if the excess enthalpy of mixing is negative. Remember, however, that Eqs. 11.2-18 to 11.2-22 are valid only for a solute that is of low solubility in a solvent.

Liquid-liquid equilibrium occurs when the species in a mixture are dissimilar. The most common situation is the one in which the species are of a different chemical nature. Such mixtures are best described by activity coefficient models, and that is the case considered in Illustration 11.2-3. However, liquid-liquid equilibrium may also occur when the two species are of similar chemical nature but differ greatly in size, as in the methane–n-heptane system, or when the species differ in both size and chemical nature, as in the carbon dioxide–n-octane system shown in Fig. 11.2-3b. Since both carbon dioxide and n-octane can be described by simple equations of state, liquid-liquid equilibrium in this system can be predicted or correlated using equations of state, though with some difficulty.

The calculation of liquid-liquid equilibrium using equations of state proceeds as in the equilibrium flash calculation described in Sec. 10.3 and illustrated in Fig. 10.3-6, with two changes. First, since both phases are liquids, the liquid root for the compressibility (which will differ in the two liquid phases since the compositions are different) must be found in each case and used in the fugacity calculation. Second, the initial guesses for $K_i = x_i^{II}/x_i^I$ are not made using the pure component vapor pressures as in vapor-liquid equilibrium, but are chosen arbitrarily. (For example, $K_1 = 10$ and $K_2 = 0.1$ when a prediction is to be made, or using the experimental data as the initial guess in a correlation to obtain the binary interaction parameter.)

ILLUSTRATION 11.2-5

Calculation of LLE Using an Equation of State

The experimental data for liquid-liquid equilibrium in the CO_2–n-decane system appear in the following table.

T (K)	P (bar)	$x_{CO_2}^I$	$x_{CO_2}^{II}$
235.65	10.58	0.577	0.974
236.15	10.75	0.582	0.973
238.15	11.52	0.602	0.970
240.15	12.38	0.627	0.965
242.15	13.19	0.659	0.960
244.15	14.14	0.695	0.954
246.15	15.10	0.734	0.942
248.15	16.11	0.783	0.916
248.74	16.38	0.850	0.850

Source: Reprinted with permission of A. A. Kulkarni, B. Y. Zarah, K. D. Luks, and J. P. Kohn, *J. Chem. Eng. Data,* **19,** 92 (1974). Copyright 1974 American Chemical Society.

Make predictions for the liquid-liquid equilibrium in this system using the Peng-Robinson equation of state with the binary interaction parameter equal to 0.114, as given in Table 9.4-1, as well as several other values of this parameter.

SOLUTION

Using one of the Peng-Robinson equation-of-state flash programs on the accompanying CD-ROM with the van der Waals one-fluid mixing rules, modified as just described for the liquid-liquid equilibrium calculation, gives the results shown in Fig. 11.2-6. There we see that no choice for the binary interaction parameter k_{12} will result in predictions that are in complete agreement with the experimental data. In particular, the value of the binary interaction parameter determined from higher-temperature vapor-liquid equilibrium data ($k_{12} = 0.114$) results in a much higher liquid-liquid critical solution temperature than is observed in the laboratory. Clearly, the Peng-Robinson equation-of-state prediction for liquid-liquid coexistence curve using the van der Waals one-fluid mixing rules is not of the correct shape for this system.

What should be stressed is not the poor accuracy of the equation-of-state predictions for the CO_2–n-decane system, but rather the fact that the same, simple equation of state can lead to good vapor-liquid equilibrium predictions over a wide range of temperatures and pressures, as well as a qualitative description of liquid-liquid equilibrium at lower temperatures.

It should be noted, however, that the equation-of-state predictions for this system could be greatly improved using the Wong-Sandler mixing rule rather than the van der Waals one-fluid mixing rules. Using the mixing rule of Sec. 9.9 with the UNIQUAC activity coefficient model and temperature-independent parameters that have been fit only to the data at 235.65 K, the very good predictions at all other temperatures shown in Fig. 11.2-6 are obtained. Note that if the UNIQUAC model were used directly (that is, not in the Wong-Sandler mixing rule), temperature-dependent parameters would be needed to obtain a fit of comparable quality. The success of this more complicated mixing rule with temperature-independent parameters results from the fact that there is a temperature dependence built into the equation of state. ∎

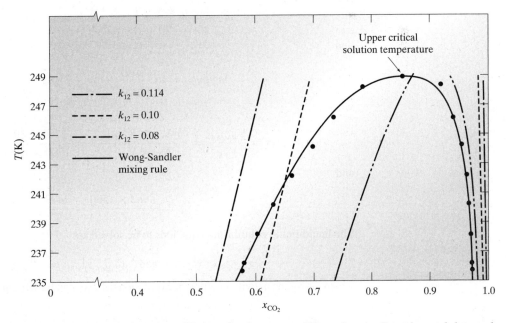

Figure 11.2-6 Liquid-liquid equilibrium for the system CO_2–n-decane. Experimental data and predictions using the Peng-Robinson equation of state and the simple van der Waals one-fluid mixing rule (dashed lines) and the Wong-Sandler mixing rule (solid line).

An interesting example of liquid–liquid equilibrium is the miscibility of molten polymers. This is especially important in the recycling of commingled plastics, as may occur when different types of used plastic containers are mixed before melting. If the plastics are compatible—that is, do not separate into two liquid phases when their mixture is melted—they can be recycled together. However, if there is a liquid-liquid phase separation in the molten state, the cooled and solidified plastic will have occlusions of the different phases, resulting in a product of little strength and unsuitable as a recycled plastic. Since polymer mixtures are usually described by the Flory-Huggins model, this equation can be used to determine whether polymers are compatible for recycling, as shown in the example below.

Polymer recycling

ILLUSTRATION 11.2-6

Determining the Compatibility of Polymers

Determine the liquid-liquid phase boundaries for the mixture of polymers polystyrene (PS) and polymethylmethacrylate (PMMA) over the temperature range from 25°C to 600°C. The polystyrene has a degree of polymerization (number of monomer units in the polymer), N_{PS}, of 1500, and the volume of a monomer unit, $\underline{V}_{PS,m}$, is 107.8 cm^3/mol. The polymethylmethacrylate has a degree of polymerization, N_{PMMA}, of 1700 and a monomer unit volume, $\underline{V}_{PMMA,m}$, of 89.7 cm^3/mol. The Flory parameter for the PS–PMMA mixture is given by

$$\chi = \frac{0.982 \times N_{PS}}{T}$$

where T is in K.

SOLUTION

The equations for the activity coefficients of PS and PMMA are, from Eqs. 9.5-18,

$$\ln \gamma_{PS} = \ln \frac{\phi_{PS}}{x_{PS}} + \left(1 - \frac{1}{m}\right) \phi_{PMMA} + \chi \phi_{PMMA}^2$$

and

$$\ln \gamma_{PMMA} = \ln \frac{\phi_{PMMA}}{x_{PMMA}} + (m - 1)\phi_{PS} + \chi \phi_{PS}^2$$

where

$$m = \frac{\underline{V}_{PMMA}}{\underline{V}_{PS}} = \frac{N_{PMMA} \times \underline{V}_{PMMA,m}}{N_{PS} \times \underline{V}_{PS,m}} = \frac{1700 \times 89.7}{1500 \times 107.8} = 0.943$$

and

$$\chi = \frac{0.982 \times 1500}{T} = \frac{1473}{T}$$

The liquid-liquid equilibrium equations to be solved are

$$x_{PS}^{I} \gamma_{PS}^{I} = x_{PS}^{II} \gamma_{PS}^{II} \quad \text{and} \quad x_{PMMA}^{I} \gamma_{PMMA}^{I} = x_{PMMA}^{II} \gamma_{PMMA}^{II}$$

which, using the activity coefficient expressions above, can be written as[10]

---

[10]Note that $\phi_i^J(x_i^J)$ is used to indicate that the volume fraction of species i in phase J is a function of its mole fraction, x_i^J.

$$\ln\left(\frac{\phi_{PS}^{I}(x_{PS}^{I})}{\phi_{PS}^{II}(x_{PS}^{II})}\right) + \left(1 - \frac{1}{m}\right)\left(\phi_{PMMA}^{I}(x_{PMMA}^{I}) - \phi_{PMMA}^{II}(x_{PMMA}^{II})\right)$$

$$+ \chi\left[\left(\phi_{PMMA}^{I}(x_{PMMA}^{I})\right)^{2} - \left(\phi_{PMMA}^{II}(x_{PMMA}^{II})\right)^{2}\right] = 0$$

and

$$\ln\left(\frac{\phi_{PMMA}^{I}(x_{PMMA}^{I})}{\phi_{PMMA}^{II}(x_{PMMA}^{II})}\right) - (m - 1)(\phi_{PS}^{I}(x_{PS}^{I}) - \phi_{PS}^{II}(x_{PS}^{II})) + \chi\left[\left(\phi_{PS}^{I}(x_{PS}^{I})\right)^{2} - \left(\phi_{PS}^{II}(x_{PS}^{II})\right)^{2}\right] = 0$$

These equations can be solved using MATHCAD or another equation-solving program. The results are given below.

T (°C)	x_{PS}^{I}	x_{PS}^{II}
25	7.71×10^{-3}	0.992
100	0.023	0.978
150	0.040	0.962
200	0.063	0.941
250	0.096	0.913
300	0.141	0.878
350	0.211	0.834
375	0.272	0.808
380	0.292	0.802
385	0.325	0.795
387	0.352	0.791
> 390	Complete miscibility	

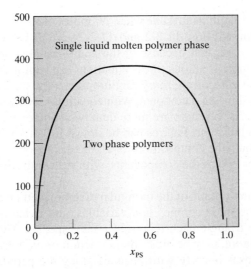

Polystyrene + PMMA phase diagram.

The decomposition temperature has been reported to be 364°C for polystyrene and lower than 327°C for PMMA. Consequently, any reprocessing of these polymers would have to be done at temperatures considerably below 327°C; at such temperatures there are only small regions of composition in which the polymers are compatible (that is, do not phase-separate) in the melt.

For example, at 250°C the polymers will be mutually soluble only for PS concentrations less than 0.096 mole fraction or greater than 0.913 mole fraction. Thus the two polymers can be commingled for recycling only in limited proportions.   ∎

So far we have considered liquid-liquid equilibrium only for binary mixtures. We next consider multicomponent mixtures. When two solvents are partially miscible (rather than immiscible), their mutual solubility will be affected by the addition of a third component. In this case the equilibrium conditions are

$$\bar{f}_i^{\mathrm{I}}(T, P, \underline{x}^{\mathrm{I}}) = \bar{f}_i^{\mathrm{II}}(T, P, \underline{x}^{\mathrm{II}}) \tag{11.2-23}$$

or, if an activity coefficient model is used,

$$x_i^{\mathrm{I}}\gamma_i(T, P, \underline{x}^{\mathrm{I}}) = x_i^{\mathrm{II}}\gamma_i(T, P, \underline{x}^{\mathrm{II}}) \tag{11.2-2}$$

for each of the species i noting that the addition of a third component affects the activity coefficients of all species in the mixture. (In writing these equations we have assumed the existence of only two liquid phases. It is left to the reader to generalize these equations for three coexisting liquid phases.) Thus, the addition of a third component may increase or decrease the equilibrium solubility of the two initially partially miscible solvents. An increase in solubility of this type is termed **salting in** and a decrease **salting out** (see Fig. 11.2-8).[11] In some cases the addition of a solute (usually an electrolyte) can so increase the mutual solubility of two partially miscible fluids that a completely miscible mixture is formed.

A typical liquid-liquid equilibrium problem is to determine the amounts and compositions of the two or more phases that are formed when known amounts of several chemicals are mixed. The equations to be solved are the equilibrium conditions of Eqs. 11.2-2 and the mass balances

$$N_i = N_i^{\mathrm{I}} + N_i^{\mathrm{II}} \tag{11.2-24}$$

Thus to compute the equilibrium state when N_1 moles of species 1, N_2 moles of species 2, and so on, are mixed, Eqs. 11.2-2 and 11.2-24 are to be solved. These equations can be difficult to solve, first because of the complicated dependence of the activity coefficients on the mole fractions (these equations are nonlinear) and second, because even in the simplest case of a ternary mixture, there are six coupled equations to be solved. This is best done on a computer with equation-solving software.

A difficulty that arises if there are more than two components is how to graphically represent the phase behavior. Figures can be drawn for a ternary system, either in triangular form (in which the compositions of all three components are represented) or in rectangular form for two of the components, with the composition of the third species obtained by difference.

Ternary systems and triangular diagrams

Figure 11.2-7 is an example of the triangular diagram method of the two-dimensional representation of the three composition variables, and is interpreted as follows. The three apexes of the triangle each represent a pure species. The composition (which, depending on the figure, may be either mass fraction or mole fraction) of each species in a mixture decreases linearly with distance along the perpendicular bisector from the apex for that species to the opposite side of the triangle. Thus each side of the

---

[11]Usually the terms *salting in* and *salting out* are used to describe the increase or decrease in solubility that results from the addition of a salt or electrolyte to a solute-solvent system. Their use here to describe the effects of the addition of a nonelectrolyte is a slight generalization of the definition of these terms.

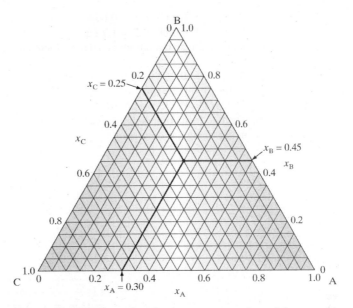

Figure 11.2-7 Triangular diagram representation of the compositions of a ternary mixture in two dimensions.

triangle is a binary mixture lacking the species at the opposite apex (i.e., the bottom of Fig. 11.2-7 represents an A+C binary mixture, the right side mixtures of A+B, and the left side B+C mixtures). For convenience, the fractional concentrations are usually indicated along one side of the triangle for each species.

The fractional concentration of each species at a specific point on the diagram is found by drawing a line through that point parallel to the side opposite the apex for that substance and noting the intersection of this line with the appropriate side of the triangle. This is illustrated in Fig. 11.2-7, where the filled point has the composition $x_A = 0.30$, $x_B = 0.45$, and $x_C = 0.25$, and the lines indicate how these compositions are to be read on this diagram.

A simple example of the use of a triangular diagram is given in the illustration that follows.

ILLUSTRATION 11.2-7

Mass Balance Calculation on a Triangular Diagram

One kilogram of a binary mixture containing 50 wt % of species A and 50 wt % of species B is mixed with two kilograms of a ternary mixture containing 15 wt % of A, 5 wt % of B, and 80 wt % of species C.

- **a.** What is the composition of the final mixture (assuming there is no liquid-liquid phase splitting)?
- **b.** Plot the compositions of the two initial mixtures and the final mixture on a triangular diagram.

SOLUTION

- **a.** The mass balance on each species is
  A: $0.5 \times 1 + 0.15 \times 2 = 0.8$ kg

B: $0.5 \times 1 + 0.05 \times 2 = 0.6$ kg
C: $0.0 \times 1 + 0.80 \times 2 = 1.7$ kg
Since, from an overall mass balance, there are 3 kg in the final mixture, the final composition is
A: $0.8/3 = 0.267$ weight fraction or 26.7 wt %
B: $0.6/3 = 0.200$ weight fraction or 20.0 wt %
C: $1.6/3 = 0.533$ weight fraction or 53.3 wt %

b. The two feed compositions and the final mixture composition are plotted on the accompanying triangular diagram.

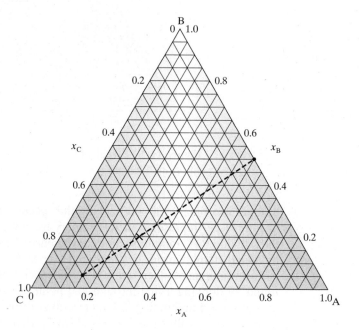

COMMENT

The final mixture composition is on a straight line connecting the two feed compositions. This is another example of the lever rule, and is merely a result of the mass balances being linear equations. Note also that the composition of the final mixture is found at two-thirds of the distance from the first feed to the second feed in accordance with their relative amounts. This graphical linear relation between the two feeds and the final mixture is the opposite case to that of a single feed that splits into two equilibrium streams, which is the case in liquid-liquid extraction. ∎

The liquid-liquid phase equilibrium data for the methyl isobutyl ketone (MIK) + acetone (A) + water (W) ternary mixture is shown Fig. 11.2-8. The inside of the dome-shaped region in this figure is a region of compositions in which liquid-liquid phase separation occurs, and tie lines are sometimes drawn within the phase separation region (as in this figure) to indicate the compositions of the coexisting phases. (As we will see shortly, phase diagrams can be more complicated than the one shown here.) From the intersection of the two-phase region with the base of the triangular diagram, we see that water and MIK are only slightly soluble in each other, while the binary mixtures of water + acetone and MIK + acetone are mutually soluble and so form only a single

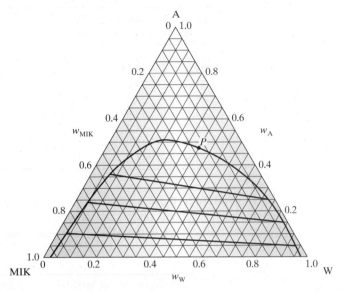

Figure 11.2-8 Liquid-liquid equilibrium compositions in weight fractions for the MIK + acetone + water system at 298.15 K.

liquid phase at all compositions. Note that there is a ternary composition at which the equilibrium tie line is of zero length, and the two equilibrium phases have the same composition. This composition is known as the **plait point** of the mixture and is indicated by the point labeled P in the figure.

Experimental liquid-liquid equilibrium data may be available as complete lists of the composition of each phase, so that triangular diagrams such as the one shown here can easily be prepared. However, it is also common for experimental data to be obtained in a less complete way that requires fewer chemical analyses. For example, if species A and B are mutually soluble, a mixture of a specified composition is prepared (gravimetrically, that is, by weighing each component before mixing), and then small weighed amounts of species C are added (and the solution mixed) until just enough of C has been added for the solution to become cloudy, indicating the formation of a second liquid phase. In this way the composition of the ternary mixture at one point of the liquid-liquid equilibrium (or binodal) curve is obtained without the need to do a chemical analysis. However, the composition of the coexisting equilibrium phase (that is, the other end of the tie line) is not known. (Note that the liquid-liquid equilibrium, or binodal, curve is also referred to as the cloud point curve, a name that is appropriate to the way it was determined. The cloud point curve in liquid-liquid equilibrium is analogous to the dew point or bubble point curve in vapor-liquid equilibrium in that it represents the saturation of a single phase and the formation of a second phase. The new phase is a second liquid at the cloud point, a vapor at the bubble point or a liquid at the dew point.)

The following data for the system methyl isobutyl ketone (MIK)–acetone–water[12] are an example of such data. These data form the boundary of the dome-shaped liquid-liquid coexistence region that was plotted in the triangular diagram of Fig. 11.2-8.

[12]D. F. Othmer, R. E. White, and E. Trueger, *Ind. Eng. Chem.*, vol. 33, 1240 (1941).

MIK (wt %)	Acetone (wt %)	Water (wt %)	MIK (wt %)	Acetone (wt %)	Water (wt %)
93.2	4.60	2.33	27.4	48.4	24.1
77.3	18.95	3.86	20.1	46.3	33.5
71.0	24.4	4.66	2.12	3.73	94.2
65.5	28.9	5.53	3.23	20.9	75.8
54.7	37.6	7.82	5.01	30.9	64.2
46.2	43.2	10.7	12.4	42.7	45.0
38.3	47.0	14.8	20.5	46.6	32.8
32.8	48.3	18.8	25.9	50.7	23.4

In addition to these data, usually several two-phase mixtures, with appreciable amounts of the second phase, are prepared and the solute composition measured. Such data for the acetone in the MIK + acetone + water system at 298.15 K are given in the following table.

Acetone (wt %) in MIK layer	Acetone (wt %) in water layer
10.66	5.58
18.0	11.83
25.5	15.35
30.5	20.6
35.3	23.8

These data are plotted in Fig. 11.2-9.

The data in Fig. 11.2-9 are used to add the tie lines in Fig. 11.2-8. The procedure is as follows.

1. A composition of the solute, acetone, is arbitrarily chosen for one of the phases—for example, 5 wt % in the water-rich phase. The point at which there is 5 wt % acetone on the water-rich portion of the binodal curve is identified in Fig. 11.2-8. This is one end of a tie line.
2. The acetone composition in the MIK-rich phase in equilibrium with 5 wt % acetone in the water-rich phase is found from Fig. 11.2-9 to be approximately 10 wt %.

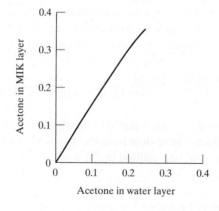

Figure 11.2-9 Distribution of acetone between the coexisting liquid phases in the MIK + acetone + water system at 298.15 K.

3. The point at which there is 10 wt % acetone on the MIK-rich portion of the binodal curve is identified in Fig. 11.2-8. This is the other end of the tie line.
4. The tie line connecting these two points is drawn in Figure 11.2-8.
5. This procedure is repeated for as many other tie lines as are desired.

Note that in this way the complete liquid-liquid equilibrium curve and the tie lines have been obtained with very few chemical analyses. In particular, the binodal curve was obtained gravimetrically, which is generally more accurate than chemical analysis, and the tie lines were obtained for only a few mixtures, and then only by analyzing for the solute (here acetone), and not for all three components in the mixture.

Since only two of the mole fractions or weight fractions are independent in a ternary system (since the three fractions must sum to unity), another way of presenting ternary liquid-liquid equilibrium data is as shown in Fig. 11.2-10 for the MIK + acetone + water system, in which only the MIK and acetone concentrations are presented, and the user must determine the water concentration by difference.

Triangular diagrams and other means of presenting ternary liquid-liquid equilibrium data can be used to design liquid-liquid extraction processes. Such applications are considered elsewhere in the chemical engineering curriculum. However, a brief introduction is given in the two illustrations that follow.

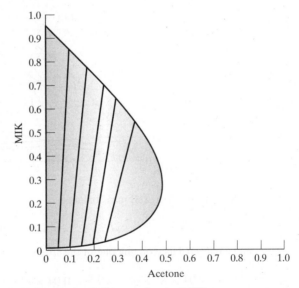

Figure 11.2-10 Liquid-liquid phase diagram for the MIK + acetone + water system at 298.15 K showing only the MIK and acetone weight fractions; the water weight fraction is obtained by difference.

Application to liquid-liquid extraction

ILLUSTRATION 11.2-8

Liquid-Liquid Extraction of an Organic Chemical from Aqueous Solution

It is desired to remove some of the acetone from a mixture that contains 60 wt % acetone and 40 wt % water by extraction with methyl isobutyl ketone (MIK). If 3 kg of MIK are contacted with 1 kg of the acetone-water mixture, what will be the amounts and compositions of the equilibrium phases?

SOLUTION

This problem is solved using the triangular diagram of Fig. 11.2-8. First the concentration of the combined streams is determined. It consists of 3 kg of MIK, 0.6×1 kg = 0.6 kg of acetone, and 0.4×1 kg of water. So the overall feed is 75 wt % MIK, 15 wt % acetone, and 10 wt % water. This point is located on the triangular diagram for this system, and is found to be in the two-liquid phase region. Next, a tie line is drawn through this feed point (indicated by the dashed line in the accompanying figure), and the compositions of the two coexisting phases are found at the two intersections of the tie line with the binodal curve. These compositions are given below.

	MIK (wt %)	Acetone (wt %)	Water (wt %)
MIK-rich	80.5	15.5	4.0
Water-rich	2.0	8.0	90.0

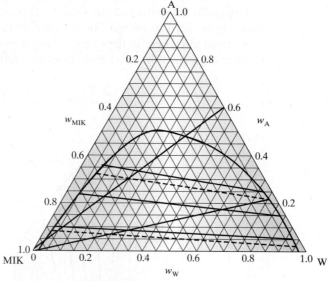

Next, from an overall mass balance,

$$L^{\mathrm{I}} + L^{\mathrm{II}} = 4 \text{ kg} \quad \text{so that} \quad L^{\mathrm{II}} = 4 - L^{\mathrm{I}}$$

and a mass balance on any one of the species, say water, we have

$$0.04 \times L^{\mathrm{I}} + 0.90 \times L^{\mathrm{II}} = 0.04 \times L^{\mathrm{I}} + 0.90 \times (4 - L^{\mathrm{I}}) = 0.4 \times 1 = 0.4 \text{ kg}$$

which has the solution $L^{\mathrm{I}} = 3.721$ and $L^{\mathrm{II}} = 0.279$ kg. So by liquid-liquid extraction, we have

	MIK (kg)	Acetone (kg)	Water (kg)
MIK-rich	2.979	0.574	0.148
Water-rich	0.005	0.024	0.27

COMMENT

By this single liquid-liquid extraction step we have been able to remove most of the acetone from the water. However, a great deal of methyl isobutyl ketone has been used. Consequently, this would not be a very useful way to recover acetone from aqueous solution. To reduce the amount of solvent used, an alternative would be to use a number of stages, each with a smaller amount of pure solvent. Such multistage extractions are frequently done in the chemistry laboratory. This is shown in the next illustration. ■

ILLUSTRATION 11.2-9

Staged Liquid-Liquid Extraction of an Organic Chemical from Aqueous Solution

The acetone-water mixture of the previous illustration is to be treated by a two-stage extraction with methyl isobutyl ketone (MIK). In the first stage 1 kg of MIK is contacted with 1 kg of the acetone-water mixture. The water-rich phase will go to a second stage, where it will be contacted with another 1 kg of pure MIK. What will be the amounts and compositions of the equilibrium phases at the exit of each stage?

SOLUTION

As in the previous illustration, this problem is solved using the triangular diagram of Fig. 11.2-8. First the concentration of the combined streams in stage 1 is determined. It consists of 1 kg MIK, 0.6×1 kg $= 0.6$ kg acetone, and 0.4×1 kg water. So the overall feed is 50 wt % MIK, 30 wt % acetone, and 20 wt % water. This point is located on the triangular diagram for this system, and found to be in the two-liquid phase region. Following the procedure used in the previous illustration, the compositions of the two phases are as follows.

	MIK (wt %)	Acetone (wt %)	Water (wt %)
MIK-rich	62	32	6
Water-rich	2	23	75

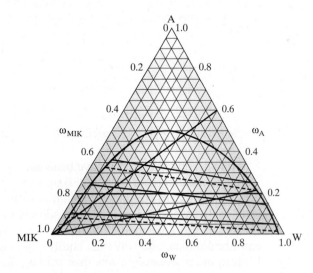

Next, from an overall mass balance,

$$L^{\mathrm{I}} + L^{\mathrm{II}} = 2 \text{ kg} \quad \text{so} \quad L^{\mathrm{II}} = 2 - L^{\mathrm{I}}$$

and a mass balance on water, we have

$$0.06 \times L^{\mathrm{I}} + 0.75 \times L^{\mathrm{II}} = 0.06 \times L^{\mathrm{I}} + 0.75 \times \left(2 - L^{\mathrm{I}}\right) = 0.4 \times 1 = 0.4 \text{ kg}$$

which has the solution $L^{\mathrm{I}} = 1.59$ and $L^{\mathrm{II}} = 0.41$ kg.

The feed to the second stage is the following: MIK = 1 + 0.02 × 0.41 = 1.008 kg, acetone = 0.23 × 0.41 = 0.094 kg, and water = 0.75 × 0.41 = 0.308 kg, for a total of 1.41 kg. Therefore, the composition of the combined feed to the second stage is 71.5 wt % MIK, 6.7 wt % acetone, and 21.8 wt % water, which is also in the two-phase region.

Using the tie line through this new feed point, we obtain the following:

	MIK (wt %)	Acetone (wt %)	Water (wt %)
MIK-rich	88.4	7.6	4
Water-rich	2	3	95

From the mass balances $L^{I} = 1.134$ and $L^{II} = 0.276$ kg., the water-rich stream leaving the second stage extraction unit contains 0.263 kg water, 0.0083 kg acetone, and 0.0055 kg MIK.

COMMENT

Note that by using two liquid extraction stages instead of a single stage, more acetone has been extracted (only 0.0083 kg in the exit water-rich stream, compared with 0.022 kg in the single-step process), and only 2 kg MIK have been used (of which 0.0055 kg is lost in the water stream), compared with 3 kg in the single-step process with a comparable MIK loss.

This example suggests that staging with smaller amounts of solvent (here MIK) will produce greater recovery than a single-stage process. However, this would involve greater costs since more equipment is needed. Clearly, a careful analysis, including costs, would be required to design the economically optimal process. In such a design, other extraction configurations would have to be considered, such as the countercurrent extraction process shown below and other, more complicated processes. Such designs are considered in a stagewise operations course elsewhere in the chemical engineering curriculum. The purpose of the illustrations here is merely to show the importance of thermodynamic equilibrium in the design of liquid-liquid extraction processes.

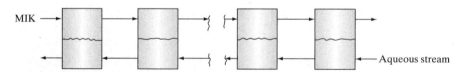

Schematic diagram of a staged liquid-liquid extraction system.

While graphical methods have been used in the illustrations above, that is not what would be done in more careful design, and especially in the design of a chemical process using modern computer simulation software. The procedure would be to use one of the activity coefficient models described in Chapter 9, frequently the NRTL, UNIQUAC, or van Laar model, with parameters adjusted to fit the available liquid-liquid equilibrium data, and solve the equilibrium equations, Eqs. 11.2-2, numerically. This leads to more accurate results than reading numbers from triangular diagrams, as was done in the illustrations here.

In this section we have considered liquid-liquid equilibrium in binary mixtures, and in ternary mixtures in which there was limited mutual solubility between only one pair of components (for example, the methyl isobutyl ketone + water binary mixture in the methyl isobutyl ketone + water + acetone system). In fact, liquid-liquid equilibria can be more complicated than this when two of three binary pairs in a ternary mixture, or all three of the binary pairs, have limited solubility. Such systems can be described by a generalization of the equations used in this section to allow for three (or more) liquid phases when multicomponent systems are considered:

$$x_i^I \gamma_i^I(T, P, \underline{x}^I) = x_i^{II} \gamma_i^{II}(T, P, \underline{x}^{II}) = x_i^{III} \gamma_i^{III}(T, P, \underline{x}^{III}) \quad i = 1, 2, \ldots, \mathcal{C}$$

(11.2-25)

Though we will not analyze such systems here, we conclude this section by showing examples of the types of liquid-liquid equilibria found to occur in ternary mixtures. Figure 11.2-11*a* is another example of a liquid-liquid phase diagram for a system

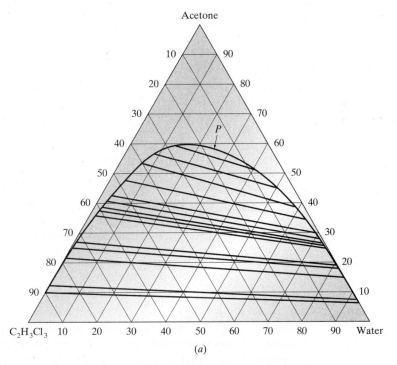

(*a*)

Figure 11.2-11 (*a*) Equilibrium diagram for the acetone–water–1,1,2-trichloroethane system. [Reprinted with permission from R. E. Treybal, L. D. Weber, and J. F. Daley, *Ind. Eng. Chem.*, **38,** 817 (1946). Copyright by American Chemical Society.] (*b*) Equilibrium diagram for the nitrobenzene-methanol-isooctane system at 15°C. [Reprinted with permission from A. W. Francis, *Liquid-Liquid Equilibriums,* John Wiley & Sons, New York (1963).] (*c*) Equilibrium diagram for the furfural–water–ethyl acetate system. [Reprinted with permission from A. W. Francis, *Liquid-Liquid Equilibriums,* John Wiley & Sons, New York (1963).] (*d*) Equilibrium diagram for the nitroethane–glycol–decyl alcohol system at 10°C. [Reprinted with permission from A. W. Francis, *J. Phys. Chem.,* **60,** 20 (1956). Copyright by the American Chemical Society.] (*e*) Equilibrium diagram for the nitromethane–glycol–lauryl alcohol system at 20°C, showing the presence of two and three coexisting solid and liquid phases. [Reprinted with permission from A. W. Francis, *J. Phys. Chem.,* **60,** 20 (1956). Copyright by the American Chemical Society.] (*f*) Equilibrium diagram for carbon dioxide with 10 pairs of other liquids, demonstrating the wide variety of liquid-liquid phase equilibria that occur at 0°C. Note that tie lines and plait points have been included in the diagrams. [Reprinted with permission from A. W. Francis, *J. Phys. Chem.,* **58,** 1099 (1954). Copyright by the American Chemical Society.]

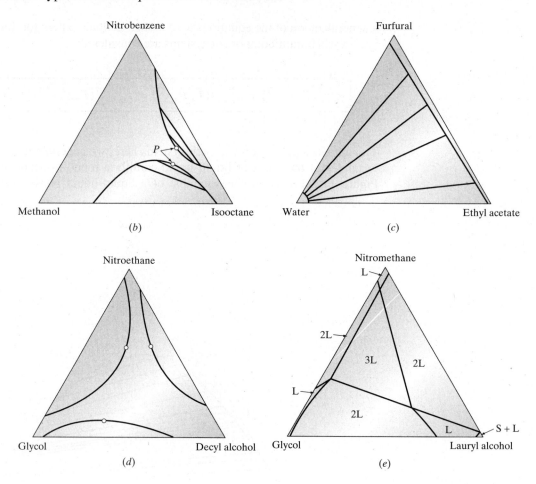

that has a single pair of components that are partially miscible. In the nitrobenzene–methanol–isooctane system of Fig. 11.2-11*b*, there are 2 two-phase regions, and in the furfural–water–ethyl acetate system of Fig. 11.2-11*c* the 2 two-phase regions merge into a band. The nitroethane–glycol–decyl alcohol system of Fig. 11.2-11*d* has three distinct two-phase regions, whereas Fig. 11.2-11*e*, for the nitromethane–glycol–lauryl alcohol system, shows the merging of these two-phase regions (denoted by 2*L*) into regions where three liquid phases coexist (denoted by 3*L*).

Figure 11.2-11*f*, for the liquid-liquid phase equilibrium behavior of liquid carbon dioxide with pairs of other liquids, has been included to illustrate the variety of types of ternary system phase diagrams the chemist and engineer may encounter. Complete discussions of these different types of phase diagrams are given in numerous places (including A. W. Francis, *Liquid-Liquid Equilibriums,* John Wiley & Sons, New York, 1963).

In a ternary mixture there is also the possibility of three or more liquid phases in equilibrium, which is allowed by a generalization of the two-phase equilibrium analysis of this section (see Problem 11.2-1). Indeed, note that some of the phase diagrams in Fig. 11.2-11 show regions of liquid-liquid-liquid equilibrium.

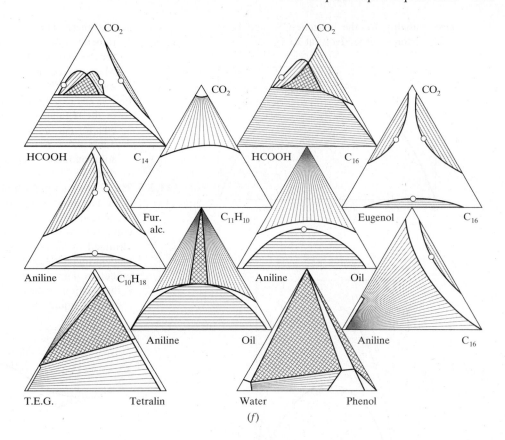

(f)

PROBLEMS FOR SECTION 11.2

11.2-1 a. Show that if two liquids form a regular solution, the critical temperature for phase separation is

$$RT_c = \frac{2x_1x_2\underline{V}_1^2\underline{V}_2^2}{(x_1\underline{V}_1 + x_2\underline{V}_2)^3}(\delta_1 - \delta_2)^2 = \frac{2\Phi_1\Phi_2\underline{V}_1\underline{V}_2}{(x_1\underline{V}_1 + x_2\underline{V}_2)}(\delta_1 - \delta_2)^2$$

b. Show that the composition at the upper consolute temperature is

$$x_1 = 1 - x_2 = \frac{(\underline{V}_1^2 + \underline{V}_2^2 - \underline{V}_1\underline{V}_2)^{1/2} - \underline{V}_1}{\underline{V}_2 - \underline{V}_1}$$

and develop an expression for the upper consolute temperature for the regular solution model.

11.2-2 Following is a portion of a phase diagram for two liquids that are only partially miscible. Note that the phase diagram contains both the coexistence curve (solid line) and a curve indicating the stability limit for each phase (dashed line) (i.e., the

dashed line represents the greatest concentration of the dilute species, or extent of supersaturation, that can occur in a metastable phase). This line is called the spinodal curve. For a binary mixture for which $\underline{G}^{\text{ex}} = Ax_1x_2$, develop the equations to be used to
a. Compute the liquid-liquid coexistence line.
b. Compute the spinodal curve.

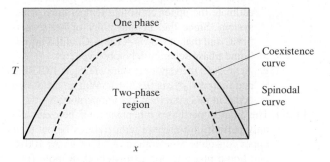

11.2-3 The two figures below have been obtained from measurements of the excess Gibbs energy and excess enthalpy for the benzene-CS_2 and benzene-CCl_4 systems, respectively, at 25°C.

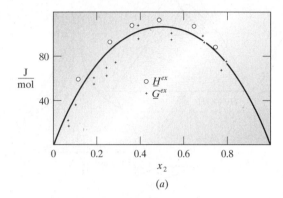

(a)

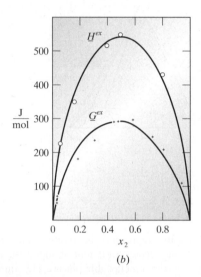

(b)

a. Comment on the applicability of the regular solution model to these two systems.

b. Assuming that the excess Gibbs energy for the C_6H_6–CS_2 system is temperature independent, estimate the upper consolute temperature of the system. Since the melting point of benzene is 5.5°C and that of CS_2 is −108.6°C, will a liquid-liquid phase separation be observed?

c. At 46.5°C, the vapor pressure of CS_2 is 1.013 bar and that of C_6H_6 is 0.320 bar. Will an azeotrope occur in this system at this temperature?

11.2-4 The liquids perfluoro-*n*-heptane and benzene are only partially miscible at temperatures below their upper consolute temperature of 113.4°C. At 100°C one liquid phase is approximately 0.48 mole fraction benzene and the other 0.94 mole fraction benzene (see Fig. 11.2-3). The liquid molar volume of perfluoro-*n*-heptane at 25°C is 0.226 m³/kmol.

a. Use these data to compute the interaction parameter A in a one-constant Margules equation for the excess Gibbs energy. Is the one-constant Margules equation consistent with the experimental data?

b. Use the experimental data to compute the value of the regular solution theory solubility parameter for perfluoro-*n*-heptane [the value given in Table 9.6-1 is $\delta = 6.0$ (cal/cc)$^{1/2}$]. Is regular solution theory consistent with the experimental data?

11.2-5 The Gibbs energy for highly nonideal solutions can behave as shown here. Prove that liquid-liquid phase separation will occur in this system, and that the compositions of the coexisting liquid phases are determined by the two intersections of the common tangent line with the $\underline{G}_{mix}$ curve. Points B and C are inflection points on the Gibbs energy curve. What is the relation of these points to the stability of the coexisting liquid phases?

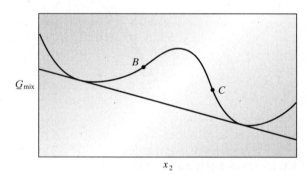

11.2-6 Show that the Wilson activity coefficient model of Eqs. 9.5-11 and 9.5-12 cannot predict the existence of two liquid phases for any values of its parameters.

11.2-7 Polymer-polymer and polymer-solvent systems are important in the chemical industry, and typically the Flory-Huggins model is used to describe the activity coefficients in such systems. Assuming the vapor phase is ideal, and that the vapor pressure of the polymer is negligible,

a. Develop the equations that should be solved for the bubble point pressure as a function of temperature for polymer-solvent mixtures using the Flory-Huggins model.

b. Develop the equations that should be solved for the molten polymer–molten polymer liquid-liquid immiscibility region as a function of temperature using the Flory-Huggins model.

What data would you need to do numerical calculations for the phase behavior with the equations you have developed above?

11.2-8 Explain why the activity (i.e., $x_i\gamma_i$) based on the pure component standard state of a species in a solution cannot be greater than unity.

11.2-9 Compute the range of temperatures and compositions over which the copolymer polystyrene-acrylonitrile (SAN) and polymethylmethacrylate (PMMA) are miscible if these polymers have the following properties. SAN has a molar volume, $\underline{V}_{SAN}$, of 1.6×10^5 cm^3/mol, and PMMA has a molar volume, $\underline{V}_{PMMA}$, of 1.5×10^5 cm^3/mol. The Flory parameter for the SAN-PMMA mixture is given by

$$\chi = \frac{1895}{T}$$

where T is in K.

11.2-10 The excess Gibbs energy for a mixture is given by

$$\underline{G}^{ex} = x_1 x_2 \left(a + bT + c \ln T\right)$$

For this system
a. Find the expressions for $\underline{H}^{ex}$, $\underline{S}^{ex}$, and C_P^{ex}.
b. Determine the range of values of the parameters a, b, and c that result in liquid-liquid equilibrium with only an upper consolute point.
c. Determine the range of values of the parameters a, b, and c that result in liquid-liquid equilibrium with only a lower consolute point.

11.2-11 a. Estimate the heat and work flows needed to reversibly and isothermally separate an equimolar mixture of two species into its pure components if the excess Gibbs energy for the mixture is given by

$$\underline{G}^{ex} = A x_1 x_2$$

where A is independent of temperature.
b. How does the temperature at which $W = 0$ compare with the upper consolute temperature of the mixture?
c. How would the answers to parts (a) and (b) change if A were a function of temperature?

11.2-12 The binary liquid mixture of nitromethane and n-nonane exhibit liquid-liquid equilibrium. At 70°C one phase has a nitromethane mole fraction of 0.131 and the other phase has an n-nonane mole fraction of 0.0247. At 90°C the mole fractions are 0.214 and 0.0469, respectively.
a. Estimate the excess Gibbs energy of mixing at each of the temperatures over the whole concentration range (that is, extrapolate into the liquid-liquid equilibrium region).

b. Estimate the excess enthalpy of mixing at 80°C over the whole concentration range.
c. Estimate the excess entropy of mixing at 80°C over the whole concentration range.

11.2-13 The diffusive flux is usually written as

$$j_1 = -cD\nabla x_1$$

where D is the diffusion coefficient and c is the concentration in mol/m^3; also, c_1 is the concentration of species 1 in mol/m^3. However, when written this way it is found that D depends on concentration. Also, this relation incorrectly predicts that if a concentration difference exists between two phases of different composition in equilibrium, there will be a diffusive flux. To correct for these problems, one can instead write an expression for the diffusive flux in terms of a gradient in chemical potential:

$$j_1 = -\frac{D_0 c_1}{RT}\nabla \mu_1$$

where D_0 is assumed to be independent of composition.
a. Show that D and D_0 are related via

$$D = D_0 \left(1 + \frac{\partial \ln \gamma_1}{\partial \ln x_1}\right)$$

b. Find the relation between D and D_0 if the system is described by the one-constant Margules equation.
c. Consider the diffusivity as described by the results in parts (a) and (b) in answering the following questions:
   i. If the diffusing component 1 is at infinite dilution, what is the relationship between D and D_0?
   ii. How does D vary with composition at the upper L-L critical point of a binary mixture?
   iii. Describe the effect on the value of D if the mixture exhibits negative deviations from Raoult's law.

11.2-14 The bubble point of a liquid mixture of an alcohol and water containing 2.0 mol % alcohol is 90°C at 1.013 bar. The vapor pressure of pure water is 0.7733 bar and that of the alcohol is 0.4 bar at this temperature.
a. Assuming that the activity coefficient of water is unity (since its concentration is 98 mol %), what is the composition of the vapor in equilibrium with the 2.0 mol % solution? What is the

activity coefficient of the alcohol in this solution?

b. At 90°C, the maximum amount of alcohol that can be dissolved in water is 2.0 mol %. When larger amounts are present, a second liquid phase forms that contains 65.0 mol % alcohol. What are the activity coefficients of the alcohol and the water in this second liquid phase?

11.2-15 The activity coefficients for a particular binary liquid mixture are given by

$$RT \ln \gamma_i = \alpha_i (1 - x_i)^2 + \beta_i (1 - x_i)^3 \quad \text{for } i = 1, 2$$

Determine whether a mixture containing 30 mol % of component 1 forms a single, stable liquid phase or two liquid phases at 245 K if the infinite-dilution activity coefficients of the two components at this temperature are

$$\ln \gamma_1^\infty = 2.0 \quad \text{and} \quad \ln \gamma_2^\infty = 3.0$$

11.2-16 The binary liquid mixture of nitromethane and 2,2,5-trimethylhexane exhibits liquid-liquid equilibrium. At 100°C one phase has a mole fraction of 0.361 of nitromethane, and the other phase has a mole fraction of 2,2,5-trimethylhexane of 0.0969. At 110°C the mole fractions are 0.580 and 0.198, respectively.

a. Estimate the excess Gibbs energy of mixing at each of the temperatures over the whole concentration range (that is, extrapolate into the liquid-liquid equilibrium region).

b. Estimate the excess enthalpy of mixing at 105°C over the whole concentration range.

c. Estimate the excess entropy of mixing at 105°C over the whole concentration range.

11.2-17 The greater the difference in the pure component vapor pressures in a binary mixture, the greater the solution nonideality must be in order for an azeotrope to form in vapor-liquid equilibrium. Assume that for a mixture liquid solution nonidealities can be represented by the simple one-constant Margules equation $G^{ex} = Ax_1x_2$.

a. Determine the values of the parameter A that will produce an azeotrope in terms of the ratio of vapor pressures and the mole fraction of species 1.

b. Compare the values of the parameter A required to form an azeotrope determined above with those necessary to produce a liquid-liquid

phase split as a function of the mole fraction of species 1.

The activity coefficient model parameters in Problems 11.2-18 to 11.2-25 are easily determined using the MATHCAD worksheet ACTCOEFF on the CD-ROM accompanying this book and described in Appendix B.III.

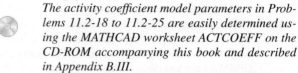

11.2-18 The following smoothed liquid-liquid equilibrium data have been reported[13] for the system nitromethane (1) + cyclohexane (2) as a function of temperature.

T (°C)	Mole percent, 1 in 2	Mole percent, 2 in 1
15	2.76	2.90
20	3.20	3.33
25	3.72	3.81
30	4.33	4.35
40	5.81	5.52
50	7.77	7.38
60	10.6	9.52

a. At each temperature find the value of the two parameters in the van Laar model that will fit these data.

b. Develop a correlation for the parameters found in part (a) as a function of temperature.

11.2-19 a. For the data in Problem 11.2-18 at each temperature find the value of the two parameters in the NRTL model (keeping $\alpha = 0.3$) that will fit these data.

b. Develop a correlation for the parameters found in part (a) as a function of temperature.

11.2-20 a. For the data in Problem 11.2-18 at each temperature find the value of the two parameters in the two-constant Margules equation that will fit these data.

b. Develop a correlation for the parameters found in part a as a function of temperature.

11.2-21 a. For the data in Problem 11.2-18 at each temperature find the value of the two parameters in the UNIQUAC model that will fit these data.

b. Develop a correlation for the parameters found in part a as a function of temperature.

11.2-22 The following smoothed liquid-liquid equilibrium data have been reported[14] for the system ethyl ester propanoic acid (1) + water (2) as a function of temperature.

---

[13] J.M. Sørensen and W. Arlt, *Liquid-Liquid Equilibrium Data Collection: I. Binary Systems, DECHEMA Chemistry Data Series,* Vol. V, part 1, 1979, Frankfurt/Main, p. 33.

[14] J.M. Sørensen and W. Arlt, *Liquid-Liquid Equilibrium Data Collection: I. Binary Systems, DECHEMA Chemistry Data Series,* Vol. V, part 1, 1979, Frankfurt/Main, p. 297.

T (°C)	Mole percent, 1 in 2	Mole percent, 2 in 1
20	0.347	8.17
25	0.351	8.67
30	0.358	9.30
40	0.379	10.9
50	0.409	12.7
60	0.452	14.8
70	0.519	17.0
80	0.659	19.2

a. At each temperature find the value of the two parameters in the van Laar model that will fit these data.

b. Develop a correlation for the parameters found in part a as a function of temperature.

11.2-23 a. For the data in Problem 11.2-22 at each temperature find the value of the two parameters in the NRTL model (keeping $\alpha = 0.3$) that will fit these data.

b. Develop a correlation for the parameters found in part a as a function of temperature.

11.2-24 a. For the data in Problem 11.2-22 at each temperature find the value of the two parameters in the two-constant Margules equation that will fit these data.

b. Develop a correlation for the parameters found in part (a) as a function of temperature.

11.2-25 a. For the data in Problem 11.2-22 at each temperature find the value of the two parameters in the UNIQUAC that will fit these data.

b. Develop a correlation for the parameters found in part (a) as a function of temperature.

11.2-26 It has been observed that an equimolar mixture of liquid oxygen and liquid propane has an upper critical solution temperature at 112 K. Assuming the one-constant Margules parameter for this system

is independent of temperature, compute the liquid-liquid equilibrium phase boundary for this system as a function of temperature.

11.2-27 a. If the Flory-Huggins χ parameter is equal to zero, will that model exhibit liquid-liquid equilibrium in a binary mixture? If so, what will be (i) the mole fraction and (ii) the volume fraction of the liquid-liquid critical point? (Note that if a liquid-liquid phase split occurs, it will be completely entropic in nature, that is, a result of only the size difference between the molecules.)

b. Assuming the χ parameter is equal to A/T, find the liquid-liquid critical temperature as a function of the parameters A and the volume ratio m. Also, find the mole fraction and volume fraction at the liquid-liquid critical point as a function of these parameters.

11.2-28 In Problem 10.2-54 it was mentioned that the following excess Gibbs energy model describes the 1-propanol (1) + n-hexane (2) system:

$$\underline{G}^{\mathrm{ex}} = RTx_1 x_2 \, (A_1 x_1 + A_2 x_2)$$

In the temperature range near 65°C the parameter values in this equation are $A_1 = 1.867$ and $A_2 = 1.536$. The vapor pressure of 1-propanol at this temperature is 0.260 bar and that of n-hexane is 0.899 bar. Does this system have an azeotrope or exhibit liquid-liquid phase splitting?

11.2-29 In determining the consolute temperature (critical temperature) for liquid-liquid equilibrium we used a shortcut method described by Eqs. 11.2-11 to 11.2-14. A more rigorous method to identify a critical point is to set both the second and third derivatives of the Gibbs energy with respect to the mole number of one of the species equal to zero. Prove that doing this gives the same result as Eq. 11.2-14.

11.3 VAPOR-LIQUID-LIQUID EQUILIBRIUM

Although the discussion of the previous section concerned only liquid-liquid equilibrium, the extension to vapor-liquid-liquid equilibrium is straightforward. As mentioned in Sec. 8.7 the condition for vapor–liquid–liquid equilibrium is

$$\bar{G}_{\mathrm{i}}^{\mathrm{I}} = \bar{G}_{\mathrm{i}}^{\mathrm{II}} = \bar{G}_{\mathrm{i}}^{\mathrm{V}} \tag{11.3-1}$$

or equivalently

$$\bar{f}_{\mathrm{i}}^{\mathrm{I}} = \bar{f}_{\mathrm{i}}^{\mathrm{II}} = \bar{f}_{\mathrm{i}}^{\mathrm{V}} \tag{11.3-2}$$

for each species distributed among the three phases (liquid I, liquid II, and vapor V). Consequently, one method of computing the three phases that are in equilibrium is to solve 2 two-phase problems. For example, one could first determine the compositions of the two liquids that are in equilibrium, and then use the methods of Secs. 10.2 or 10.3, as appropriate, to find the vapor that would be in equilibrium with either one of the liquids (since, by Eq. 11.3-2, a vapor in equilibrium with one of the coexisting equilibrium liquid phases will also be in equilibrium with the other). This is illustrated in the two examples that follow.

ILLUSTRATION 11.3-1

Vapor-Liquid-Liquid Equilibrium Calculation Using an Activity Coefficient Model

Since liquids are not very compressible, at low and moderate pressures liquid-liquid equilibrium compositions are almost independent of pressure. Therefore, assuming that the liquid-liquid equilibrium of the isobutane (1)–furfural (2) mixture at 37.8°C calculated in Illustration 11.2-2 is unaffected by pressure, compute the pressure at which the first bubble of vapor will form (i.e., compute the bubble point pressure of this system) and the composition of the vapor that forms. *Data:*

$$P_1^{\text{vap}}(T = 37.8°C) = 4.956 \text{ bar}$$

$$P_2^{\text{vap}}(T = 37.8°C) = 0.005 \text{ bar}$$

SOLUTION

Using the van Laar activity coefficient model as in Illustration 11.2-2 and the liquid-phase compositions found there, and assuming the vapor phase is ideal, we have the bubble point pressure of liquid phase I as

$$P = x_1^{\text{I}}\gamma_1^{\text{I}}P_1^{\text{vap}} + x_2^{\text{I}}\gamma_2^{\text{I}}P_2^{\text{vap}}$$

$$= 0.1128 \times 8.375 \times 4.956 + 0.8872 \times 1.030 \times 0.005 = 4.69 \text{ bar}$$

The bubble point pressure of liquid phase II is

$$P = x_1^{\text{II}}\gamma_1^{\text{II}}P_1^{\text{vap}} + x_2^{\text{II}}\gamma_2^{\text{II}}P_2^{\text{vap}}$$

$$= 0.9284 \times 1.018 \times 4.956 + 0.0716 \times 12.77 \times 0.005 = 4.69 \text{ bar}$$

which is the same as for liquid phase I (as it must be since $x_1^{\text{I}}\gamma_1^{\text{I}} = x_1^{\text{II}}\gamma_1^{\text{II}}$ and $x_2^{\text{I}}\gamma_2^{\text{I}} = x_2^{\text{II}}\gamma_2^{\text{II}}$). The composition of the vapor (computed using either liquid phase I or II) is obtained from $x_i\gamma_i P_i^{\text{vap}} = y_i P$, so that

$$y_1 = \frac{x_1\gamma_1 P_1^{\text{vap}}}{P} = \frac{0.1128 \times 8.375 \times 4.956}{4.69} = 0.999$$

and

$$y_2 = \frac{x_2\gamma_2 P_2^{\text{vap}}}{P} = \frac{0.8872 \times 1.030 \times 0.005}{4.69} = 0.001$$

Therefore, from the van Laar model, at $T = 37.8°C$ and $P = 4.69$ bar, the isobutane-furfural mixture has two liquid phases and a vapor phase all coexisting at equilibrium, with the following compositions:

Substance	Liquid I	Liquid II	Vapor
Isobutane	0.1128	0.9284	0.999
Furfural	0.8872	0.0716	0.001

COMMENT

The Gibbs phase rule for this nonreacting ($\mathcal{M} = 0$) system of two components ($\mathcal{C} = 2$) and three phases ($\mathcal{P} = 3$) establishes that the number of degrees of freedom is

$$\mathcal{F} = \mathcal{C} + 2 - \mathcal{P} - \mathcal{M}$$
$$= 2 + 2 - 3 - 0 = 1$$

Therefore, at each temperature, there is only a single pressure at which the three phases will coexist at equilibrium in this two-component system, and the compositions of these three phases are fixed. That is, as the feed changes within the boundaries of the three-phase region, the distribution of mass between the three phases will change, but not the composition of each of those phases

Also, at a pressure higher than the equilibrium pressure at each temperature only two liquid phases exist, while below the equilibrium pressure only a single liquid and a vapor exist. Therefore, in a binary mixture, at each temperature there is only a single pressure at which two liquids and a vapor are present, which may be difficult to determine experimentally. However, because of the extra degrees of freedom, states of liquid-liquid-vapor equilibrium are much easier to find in ternary and other multicomponent systems. ■

ILLUSTRATION 11.3-2

Vapor-Liquid-Liquid Equilibrium Calculation Using an EOS

Use the Peng-Robinson equation of state and the van der Waals one-fluid mixing rules, with $k_{12} = 0.114$, to compute the bubble point pressure and vapor composition in equilibrium with the two coexisting liquid phases in the CO_2–n-decane system of Illustration 11.2-5.

SOLUTION

Using the bubble point pressure option in the Visual Basic program of Appendix B.I-3, the DOS-based program VLMU, or the MATHCAD worksheet PRBUBP, and the computed composition for the liquid richer in n-decane, we obtain the following results for the three-phase coexistence region:

T (K)	P (bar)	$x^{I}_{CO_2}$	$x^{II}_{CO_2}$	$y_{C_{10}}$
235.65	10.84	0.539	0.997	1.2×10^{-6}
236.15	11.01	0.540	0.996	1.3×10^{-6}
238.15	11.82	0.552	0.996	1.6×10^{-6}
240.15	12.66	0.563	0.995	2.0×10^{-6}
242.15	13.56	0.575	0.995	2.4×10^{-6}
244.15	14.49	0.586	0.994	3.0×10^{-6}
246.15	15.46	0.597	0.993	3.6×10^{-6}
248.74	16.76	0.609	0.992	4.7×10^{-6}

As we can see from these results, there is very little n-decane in the vapor. This is because of the large volatility difference between carbon dioxide and n-decane. The three-phase experimental data for this system confirm this behavior. ∎

The predictions of three-phase equilibria considered so far were done as two separate two-phase calculations. Although applicable to the examples here, such a procedure cannot easily be followed in a three-phase flash calculation in which the temperature or pressure of a mixture of two or more components is changed so that three phases are formed. In this case the equilibrium relations and mass balance equations for all three phases must be solved simultaneously to find the compositions of the three coexisting phases. It is left to you (Problem 11.3-7) to develop the algorithm for such a calculation.

In Illustration 11.3-1, the bubble point pressure of a two-liquid-phase mixture of isobutane and furfural at 37.8°C was computed. A more complete phase diagram for this system, calculated with the van Laar model and the parameter values given in Illustration 11.2-2, is shown in Fig. 11.3-1. From the calculation in Illustration 11.3-1 we know that for isobutane mole fractions in the range $0.1128 < x_1 < 0.9284$ and above the bubble point pressure of 4.69 bar, two liquid phases will exist. This area is indicated as the LLE region in the figure. In the region of the figure indicated as L_2 only one liquid phase that is rich in furfural and dilute in isobutane exists; in the region L_1 only a liquid phase that is rich in isobutane is present. In the area denoted as the VLE region a liquid phase dilute in isobutane is in equilibrium with a vapor phase that is very rich in that component. (Indeed, at pressures above 1 bar, the vapor-phase mole fraction of isobutane is greater than 0.999 and appears coincident with the right axis on the scale of Fig. 11.3-1.) Neither the vapor-liquid tie lines nor the liquid-liquid tie lines are shown in this figure. Finally, there is a region of very low pressure and high isobutane mole fraction in which only a vapor exists; this region is indicated by V. This vapor-phase region starts at the pure furfural side of the diagram at pressures below 0.005 bar (the vapor pressure of furfural at 37.8°C) and continues to 4.956 bar (the

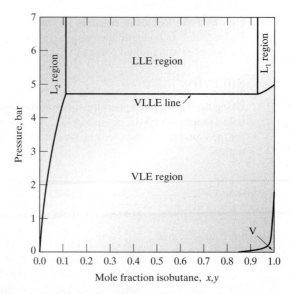

Figure 11.3-1 Predicted phase behavior of the isobutane-furfural mixture at 37.8°C. Note that regions of a single vapor (V), single liquid (L_1, L_2), vapor-liquid (VLE), liquid-liquid (LLE), and vapor-liquid-liquid (VLLE) are predicted to occur.

vapor pressure of isobutane at 37.8°C) for pure isobutane. However, for most of the composition range, the vapor region occurs at such low pressures as not to be visible on the scale of this figure.

One question that arises in doing vapor-liquid equilibrium calculations is, when should one be concerned about the possible existence of liquid-liquid equilibrium? Fortunately, we get some hints from the results of vapor-liquid equilibrium calculations. For example, the solid line in Fig. 11.3-2 shows the results of a calculation of the equilibrium pressure as a function of the isobutane liquid-phase mole fraction for this system at 37.8°C computed using the van Laar model and ignoring the possibility of two liquid phases. An unusual maximum and minimum in total pressure as a function of mole fraction is evident. Remembering from the discussion of Sec. 10.2 that a maximum or minimum in the pressure versus composition diagram can correspond to the occurrence of an azeotrope, one interpretation of the results in Fig. 11.3-2 is that the isobutane-furfural mixture has a double azeotrope, that is, both minimum-boiling and maximum-boiling azeotropes. There are some, but very few, mixtures that have double azeotropes, and the pure components in such mixtures have similar pure component vapor pressures, which is not the case here. (The benzene-hexafluorobenzene mixture of Fig. 10.2-10 was the first system observed to exhibit a double azeotrope.) Much more common when total pressure versus composition behavior as shown in Fig. 11.3-2 is found in calculations is the existence of two coexisting liquid phases. Consequently, when such behavior is seen, one should undertake a liquid-liquid phase equilibrium calculation. In the case here the correct total pressure versus liquid-phase composition from such a calculation in the liquid-liquid equilibrium region is given by the dashed line in the figure. Such an invariance of pressure with composition is what is found experimentally in liquid-liquid systems. (Note that since this figure plots pressure versus composition, not pressure versus volume as in Fig. 7.3-2, we are not doing a Maxwell construction as in Chapter 7, so that the areas above and below the dashed line need not be equal.)

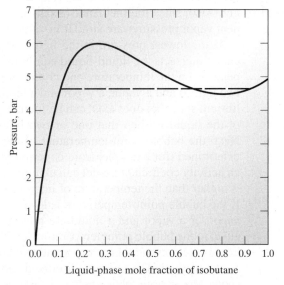

Figure 11.3-2 Predicted pressure of the isobutane-furfural system at 37.8°C ignoring the existence of liquid-liquid equilibrium (solid line) and allowing for LLE (dashed line).

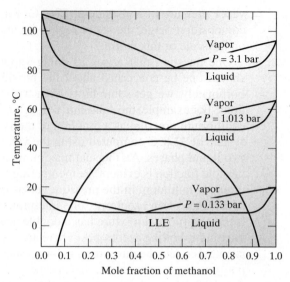

Figure 11.3-3 The vapor-liquid, liquid-liquid, and vapor-liquid-liquid behavior of the methanol–n-hexane mixture as a function of temperature.

Figure 11.3-3 shows the vapor-liquid and liquid-liquid equilibrium behavior computed for the system of methanol and n-hexane at various temperatures. Note that two liquid phases coexist in equilibrium to temperatures of about 43°C. Since liquids are relatively incompressible, the species liquid-phase fugacities are almost independent of pressure (see Illustrations 7.4-8 and 7.4-9), so that the liquid-liquid behavior is essentially independent of pressure, unless the pressure is very high, or low enough for the mixture to vaporize (this possibility will be considered shortly). The vapor-liquid equilibrium curves for this system at various pressures are also shown in the figure. Note that since the fugacity of a species in a vapor-phase mixture is directly proportional to pressure, the VLE curves are a function of pressure, even though the LLE curves are not. Also, since the methanol-hexane mixture is quite nonideal, and the pure component vapor pressures are similar in value, this system exhibits azeotropic behavior.

At the lowest pressure in the figure, $P = 0.133$ bar, the vapor-liquid equilibrium curve intersects the liquid-liquid equilibrium curve. Consequently, at this pressure, depending on the temperature and composition, we may have only a liquid, two liquids, two liquids and a vapor, a vapor and a liquid, or only a vapor in equilibrium. The equilibrium state that does exist can be found by first determining whether the composition of the liquid is such that one or two liquid phases exist at the temperature chosen. Next, the bubble point temperature of the one or either of the two liquids present is determined (for example, from experimental data or from known vapor pressures and an activity coefficient model calculation). If the liquid-phase bubble point temperature is higher than the temperature of interest, then only a liquid or two liquids are present. If the bubble point temperature is lower, then depending on the composition, either a vapor, or a vapor and a liquid are present. However, if the temperature of interest is equal to the bubble point temperature and the composition is in the range in which two liquids are present, then a vapor and two coexisting liquids will be in equilibrium.

The method of calculation discussed above was used to construct Fig. 11.3-4, which shows the various phase behavior regions for the methanol-hexane system at $P = 0.133$ bar. This diagram looks different from Fig. 11.3-1 for two reasons. First, we

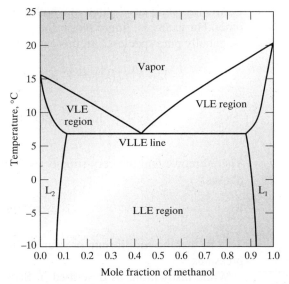

Figure 11.3-4 Phase behavior of the methanol–n-hexane system at $P = 0.133$ bar.

have plotted the phase behavior as a function of temperature at fixed pressure here, and as a function of pressure at fixed temperature in Fig. 11.3-1. Consequently, the vapor region is at the top of Fig. 11.3-4, corresponding to high temperature there, and at the bottom, low-pressure region of Fig. 11.3-1. (In fact, Fig. 11.3-4 would look more like Fig. 11.3-1 if it were turned upside down.) The second important difference is that the methanol-hexane system has an azeotrope, while the isobutane-furfural system does not. This is because the vapor pressures of isobutane and furfural are so different (indeed, they differ by three orders of magnitude) that even though that mixture is a very nonideal one, azeotropy does not occur (i.e., the vapor pressure difference of the species predominates over the effect of solution nonidealities for this mixture). The azeotropic behavior shown in Fig. 11.3-4 is interesting because the vapor-liquid equilibrium curve that exhibits azeotropy intersects the liquid-liquid phase boundary. The vapor composition in the vapor-liquid-liquid region is indicated by the common intersection of the two dew point curves with the liquid-liquid equilibrium line. Such a situation is referred to as **heterogeneous azeotropy**. Some distillation processes make use of heterogeneous azeotropy to purify mixtures.

Finally, we note that the equilibrium pressure above a two-phase liquid system, as indicated in Illustration 11.3-1, is computed from

$$P = x_1^{\mathrm{I}} \gamma_1^{\mathrm{I}}(\underline{x}^{\mathrm{I}}) P_1^{\mathrm{vap}}(T) + x_2^{\mathrm{I}} \gamma_2^{\mathrm{I}}(\underline{x}^{\mathrm{I}}) P_2^{\mathrm{vap}}(T) \tag{11.3-3a}$$

or

$$P = x_1^{\mathrm{II}} \gamma_1^{\mathrm{II}}(\underline{x}^{\mathrm{II}}) P_1^{\mathrm{vap}}(T) + x_2^{\mathrm{II}} \gamma_2^{\mathrm{II}}(\underline{x}^{\mathrm{II}}) P_2^{\mathrm{vap}}(T) \tag{11.3-3b}$$

which are equivalent from the requirement for liquid-liquid equilibrium:

$$x_1^{\mathrm{I}} \gamma_1^{\mathrm{I}}(\underline{x}^{\mathrm{I}}) = x_1^{\mathrm{II}} \gamma_1^{\mathrm{II}}(\underline{x}^{\mathrm{II}}) \quad \text{and} \quad x_2^{\mathrm{I}} \gamma_2^{\mathrm{I}}(\underline{x}^{\mathrm{I}}) = x_2^{\mathrm{II}} \gamma_2^{\mathrm{II}}(\underline{x}^{\mathrm{II}}) \tag{11.2-2}$$

An interesting case is that in which the two species are essentially insoluble in each other. For example, liquid phase I is essentially pure species 1, and liquid phase II is essentially pure species 2. In this case,

$$\lim_{x_1^I \to 1} x_1^I \gamma_1^I(\underline{x}^I) = 1 \cdot \gamma_1^I(x_1^I \to 1) = 1 = x_1^{II} \gamma_1^{II}(\underline{x}^{II}) \tag{11.3-4a}$$

$$x_2^I \gamma_2^I(\underline{x}^I) = \lim_{x_2^{II} \to 1} x_2^{II} \gamma_2^{II}(\underline{x}^{II}) = 1 \cdot \gamma_2^{II}(x_2^{II} \to 1) = 1 \tag{11.3-4b}$$

Therefore, we find the very simple result that in the limit of two essentially immisible phases,

$$P = P_1^{\text{vap}}(T) + P_2^{\text{vap}}(T) \tag{11.3-5}$$

That is, that the equilibrium pressure is just the sum of the two pure component vapor pressures.

Application to steam distillation

While distillation, as described in Secs. 10.1 and 10.2, is frequently used for the purification of chemicals, it may be difficult to use for a compound with a very high atmospheric pressure boiling point or a compound that decomposes at its boiling temperature. One way to avoid these problems is to distill at subatmospheric pressure (referred to as vacuum distillation), which can be analyzed using the methods briefly introduced in Secs. 10.1 and 10.2. An older method, especially for an organic chemical that is essentially insoluble in water, is to use steam distillation. Here steam is injected into a kettle or other device containing the mixture to be purified, the more volatile compounds are partially vaporized by the steam, and the steam–organic chemical vapor is then condensed. However, as the water and organic chemical are essentially insoluble as liquids, the condensed water is easily separated from the organic chemical. This process is shown schematically in Fig. 11.3-5.

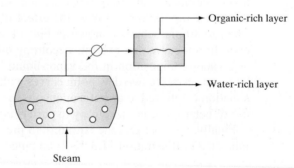

Figure 11.3-5 Schematic diagram of a steam distillation apparatus.

ILLUSTRATION 11.3-3

Steam Distillation of Turpentine

It is desired to recover and reuse turpentine that has been used as a paint remover. Turpentine can be considered insoluble in water, and the used turpentine contains small amounts of essentially involatile paint pigments and other impurities. Steam is injected into the bottom of a kettle that contains the used turpentine, and the temperature in the kettle is kept at 100°C by removing the vapor that forms. Assuming that equilibrium is achieved in the kettle, how many kilograms of turpentine are obtained for each kilogram of steam that is condensed?

The molecular weight of turpentine is 140, and its vapor pressure at 100°C is 0.177 bar.

SOLUTION

Since water and turpentine are essentially insoluble, the equilibrium pressure in the kettle is

$$P = P_W^{vap}(T) + P_T^{vap}(T) = 1.013 + 0.177 = 1.190 \text{ bar}$$

Therefore, the mole fraction of turpentine in the steam leaving the kettle is

$$y_T = \frac{0.177}{1.190} = 0.149$$

so that the vapor leaving the kettle contains 0.149 moles of turpentine for each 0.851 moles of water. Since the molecular weight of turpentine is 140 and that of water is 18.015, in the vapor 20.86 kg of turpentine are recovered for each 15.33 kg of steam used, or 1.36 kg of turpentine for each kilogram of steam.

COMMENT

Note that any inert and insoluble fluid could be used instead of steam for this sort of distillation process. However, there are several advantages to using steam, including its high heat of vaporization and its general availability in a chemical plant. ■

PROBLEMS FOR SECTION 11.3

11.3-1 The bubble point of a liquid mixture of *n*-butanol and water containing 4.0 mol % butanol is 92.7°C at 1.013 bar. At 92.7°C the vapor pressure of pure water is 0.784 bar and that of pure *n*-butanol is 0.427 bar.
  a. Assuming the activity coefficient of water in the 4.0 mol % butanol solution is about 1, what is the composition of the vapor in equilibrium with the 4.0 mol % butanol-water mixture, and what is the activity coefficient of *n*-butanol?
  b. At 92.7°C, the maximum amount of *n*-butanol that can be dissolved in water is 4.0 mol %. When larger amounts of *n*-butanol are present, a second liquid phase, containing 40.0 mol % *n*-butanol, appears. What are the activity coefficients for *n*-butanol and water in this second liquid phase?
  c. If the two coexisting liquids in part (b) are kept at 92.7°C, what will be the pressure when the first bubble of vapor is formed?
11.3-2 Estimate the pressure and vapor-phase composition in equilibrium with a liquid whose overall composition is 50 mol % furfural and 50 mol % isobutane at $T = 37.8°C$.

Data:

$$P_{iso}^{vap}(T = 37.8°C) = 4.909 \text{ bar}$$
$$P_{furf}^{vap}(T = 37.8°C) = 4.93 \times 10^{-3} \text{ bar}$$

Isobutane and furfural are only partially miscible in the liquid phase. At 37.8°C the two coexisting liquid phases contain 11.8 and 92.5 mol % isobutane, $\gamma_{iso}(x_{iso} = 0.925) = 1.019$, and $\gamma_{furf}(x_{furf} = 0.882) = 1.033$.

11.3-3 Ethyl alcohol and *n*-hexane are put into an evacuated, isothermal container. After equilibrium is established at 75°C, it is observed that two liquid phases and a vapor phase are in equilibrium. One of the liquid phases contains 9.02 mol % *n*-hexane. Activity coefficients for *n*-hexane–ethyl alcohol liquid mixtures can be represented by the equation

$$RT \ln \gamma_i = 8.163 x_j^2 \frac{\text{kJ}}{\text{mol}}$$

  a. Compute the equilibrium composition of the coexisting liquid phase.
  b. Compute the equilibrium pressure and vapor-

phase mole fractions.

Data: Vapor pressure equations (T in K, P in bar)

$$\ln P_{\mathrm{H}}^{\mathrm{vap}} = \frac{-3570.58}{T} + 10.4575$$

$$\ln P_{\mathrm{EOH}}^{\mathrm{vap}} = \frac{-4728.98}{T} + 13.4643$$

11.3-4 Glycerol and acetophenone are partially miscible at 140°C, one liquid phase containing 9.7 mol % acetophenone and the other containing 90.3 mol % acetophenone. The vapor pressure of pure glycerol at 140°C is 3.13×10^{-3} bar and that of acetophenone is 0.167 bar. Assuming that the activity coefficients for the glycerol-acetophenone system obey the one-constant Margules equation, compute the pressure versus liquid-phase mole fraction (P–x) diagram for this system.

11.3-5 For a separation process, it is necessary to ascertain whether water and methyl ethyl ketone form an azeotrope at 73.4°C. The engineer needing this information could not find any vapor-liquid equilibrium data for this system, but the following liquid-liquid equilibrium data are available:

	$T = 20°$ C	
	x_{MEK}	$x_{\mathrm{H_2O}}$
Liquid phase 1	0.0850	0.9150
Liquid phase 2	0.6363	0.3637

In addition, at 73.4°C

$$P_{\mathrm{H_2O}}^{\mathrm{vap}} = 0.3603 \text{ bar}$$

$$P_{\mathrm{MEK}}^{\mathrm{vap}} = 0.8337 \text{ bar}$$

Calculate, as best you can, the P-x-y diagram for MEK and water at 73.4°C and determine whether this system has an azeotrope, and if so, at what composition the azeotrope occurs.

11.3-6 Ethanol and n-hexane form two liquid phases at many temperatures. Assuming that the activity coefficients for n-hexane–ethyl alcohol mixtures can be represented by the equation

$$RT \ln \gamma_i = 8.163 x_j^2 \text{ kJ/mol}$$

and that the vapor pressures of the pure components are given by (P in bar and T in K)

$$\ln P_{\mathrm{H}}^{\mathrm{vap}} = \frac{-3570.58}{T} + 10.4575$$

and

$$\ln P_{\mathrm{EtOH}}^{\mathrm{vap}} = \frac{-4728.98}{T} + 13.4643$$

a. Compute the upper consolute temperature or upper critical solution temperature (UCST) for the ethanol–n-hexane mixture.

b. Develop a plot similar to Figure 11.3-3 for this system using pressures of 0.1013 bar, 1.013 bar, and 10.13 bar.

c. Develop a plot similar to Figure 11.3-4 for this system at 1.013 bar.

11.3-7 Develop an algorithm for solving vapor-liquid-liquid phase equilibrium calculations using an equation of state.

11.3-8 The infinite dilution activity coefficient of 1-propanol in n-heptane is 16.0 at 60°C, and that of n-heptane in 1-propanol is 6.34. At this temperature the vapor pressure of 1-propanol is 20.277 kPa, and the vapor pressure of n-heptane is 28.022 kPa. Prepare a P-x-y diagram for this system. Does the system exhibit azeotropy (either homogeneous or heterogeneous) or liquid-liquid phase splitting?

11.3-9 At 298 K it is found that component A has a vapor pressure of 0.8 bar, and component B has a vapor pressure of 1 bar. Also, by diluting component A with 0.01 mole fraction of B, the equilibrium pressure decreases to 0.7975 bar, while adding 0.01 mole fraction of A to B reduces the equilibrium pressure to 0.9945 bar. Determine the phase behavior of this mixture at 298 K over the whole composition range. (Is there an azeotrope, liquid-liquid equilibrium, vapor-liquid-liquid equilibrium?)

11.3-10 At 110°C, a saturated liquid solution of aniline in water contains 0.0162 mole fraction aniline, and a saturated liquid solution of water in aniline contains 0.587 mole fraction aniline. At this temperature the vapor pressure of pure aniline is 0.0923 bar, and that of pure water is 1.431 bar. Construct a P-x-y diagram for this mixture at 110°C.

11.3-11 The excess Gibbs energy of a certain liquid mixture containing components A and B is found to be given by the expression

$$\underline{G}^{\mathrm{ex}}(T, \underline{x}) = (4500 - 3T)x_A x_B \quad \text{J/mol}$$

where T is in K. The vapor pressures of the components at 320 K are

$$P_A^{\mathrm{vap}} = 1.3 \text{ bar} \quad \text{and} \quad P_B^{\mathrm{vap}} = 1.0 \text{ bar}$$

a. What is the expression for the excess enthalpy of this mixture?

b. What is the expression for the excess entropy of this mixture?

c. What is the expression for the excess constant-pressure heat capacity of this mixture?

One mole per second of A is isothermally mixed with two moles per second of B in a continuous mixing process at 1 bar and 275 K.

d. What is the rate of heat addition or removal for the mixing to be done isothermally?

e. What is the rate of entropy generation of this process?

f. It is thought possible that this mixture is sufficiently nonideal for a liquid-liquid phase split to occur at some temperatures. What is the upper consolute temperature for this mixture?

g. Compute the vapor-liquid equilibrium behavior of this mixture at 320 K. Does the system have an azeotrope at this temperature? If so, what is its composition?

h. A mixture of one mole of A and two moles of B undergoes an isothermal flash vaporization at 320 K and 1.5 bar. What are the compositions of the coexisting vapor and liquid phases, and what is the vapor-liquid split?

11.3-12 Using the liquid-liquid equilibrium data for the nitromethane + n-nonane system given in Problem 11.2-12, compute the vapor-liquid-liquid equilibrium for this system at 90°C.

11.3-13 At 333.15 K, Wu, Locke, and Sandler[15] measured the vapor pressure of pure pyrrolidine and cyclohexane to be 39.920 kPa and 51.886 kPa, respectively. After the isothermal addition of a small amount of pyrrolidine to pure cyclohexane, the total pressure was found to be 52.270 kPa when the pyrrolidine mole fraction was 0.0109. Similarly, after the isothermal addition of a small amount of cyclohexane to pure pyrrolidine, the total pressure was 40.280 kPa when the cyclohexane mole fraction was 0.0064.

a. Compute the infinite-dilution activity coefficients of pyrrolidine in cyclohexane, and of cyclohexane in pyrrolidine.

b. Compute the complete P versus x, and y versus x diagrams for this system at 333.15 K.

c. At 333.15 K, does this system have an azeotrope? Does it liquid-liquid phase split?

11.3-14 At 353.15 K the infinite-dilution activity coefficients of benzene in ethanol and of ethanol in benzene are both approximately 5.0 to within experimental accuracy. At this temperature the vapor pressure of benzene is 1.008 bar and that of ethanol is 1.086 bar. Compute the phase behavior

of this system at 353.15 K accounting for the fact that vapor-liquid, liquid-liquid, and vapor-liquid-liquid equilibrium and azeotropic behavior may be possible.

11.3-15 The change in vapor pressure of a pure liquid with temperature can be computed from the Clapeyron equation,

$$\frac{dP}{dT} = \frac{\Delta \underline{H}}{T \Delta \underline{V}}$$

where $\Delta \underline{H}$ and $\Delta \underline{V}$ are the enthalpy and volume changes on going from the liquid to the vapor. The vapor pressure and temperature must, of course, be related, since the Gibbs phase rule indicates that the system is univariant. Now consider two partially miscible liquids and the vapor in equilibrium with these two liquid phases. The system is also univariant (two components and three phases).

a. Derive a "generalized" Clapeyron equation relating the equilibrium pressure and temperature for this system.

b. Qualitatively discuss the variation of the equilibrium pressure with the overall two-phase liquid composition for this system at fixed temperature. How does the pressure variation with composition differ in the one- and two-liquid phase regions? Also, discuss the variation of the equilibrium temperature with overall liquid-phase composition at fixed pressure in the single-liquid phase and two-liquid phase regions.

11.3-16 An equimolar mixture of ethanol and ethyl acetate is maintained at 75°C and 10 bar. The pressure on the system is isothermally reduced to 1.12 bar. How many phases are present when equilibrium is achieved under these new conditions, and what are their compositions?

Data: For the ethanol–ethyl acetate system,

$$\ln \gamma_i = 0.896(1 - x_i)^2$$

$$P_{EAC}^{vap} = 0.9460 \text{ bar} \quad \text{and} \quad P_{EOH}^{vap} = 0.8905 \text{ bar}$$

11.3-17 Estimate the equilibrium pressure and isobutane vapor-phase mole fraction as a function of liquid composition for the isobutane-furfural system at 37.8°C. At this temperature the vapor pressure of furfural is 0.493 kPa and that of isobutane is 490.9 kPa. (*Hint:* See Illustration 11.3-1.)

11.3-18 At 20°C and 101.3 kPa dichloromethane (also referred to as the refrigerant R30) is found to be soluble in water to the extent of 2 wt %, while the

---

solubility limit for water in dichloromethane is 0.5 wt %. The vapor pressures of these compounds are

T (°C)	P_{R30}^{va} (bar)	$P_{H_2O}^{vap}$ (bar)
20	0.4685	0.0234
50	1.416	0.1235
80	3.471	0.4739
100	5.769	1.0138

a. Assuming that the van Laar parameters for this mixture are independent of temperature, obtain the phase diagram for this system at 50°C.

b. Assuming that the van Laar parameters for this mixture are independent of temperature, obtain the phase diagram for this system at 80°C.

c. Assuming that the van Laar parameters for this mixture are independent of temperature, obtain the phase diagram for this system at 100°C.

11.3-19 An equimolar mixture of liquid oxygen and liquid propane has a liquid-liquid equilibrium up-per critical solution temperature of 112 K. Assuming that this system can be described by a one-constant Margules expression with a temperature-independent A parameter, determine the liquid-liquid and liquid-liquid-vapor phase diagram for this system at a pressure of 0.1 bar.

11.3-20 Using the liquid-liquid equilibrium data for the nitromethane + 2,2,5-trimethylhexane system given in Problem 11.2-30, compute the vapor-liquid-liquid equilibrium for this system at 100°C.

11.3-21 The following data are available for the acetone + ethanol system at 49.3°C:

$$\gamma_{acetone}^{\infty} = 2.17 \quad \text{and} \quad \gamma_{ethanol}^{\infty} = 1.98$$

At this temperature the vapor pressure of ethanol is 0.2854 bar and the vapor pressure of acetone is 0.7710 bar. Compute the phase behavior of this system at 49.3°C. Does this mixture have a homogeneous or heterogeneous azeotrope?

11.4 THE PARTITIONING OF A SOLUTE AMONG TWO COEXISTING LIQUID PHASES; THE DISTRIBUTION COEFFICIENT

When a gas, liquid, or solid is added to two partially miscible or completely immiscible solvents, it will, depending on the amount of solute present, either partially or completely dissolve and be distributed unequally between the two liquid phases. Most chemists and chemical engineers first encounter this phenomenon in the organic chemistry laboratory, where diethyl ether, which is virtually immiscible with water, is used to extract organic chemical reaction products from aqueous solutions. The distribution of a solute among coexisting liquid phases is of industrial importance in purification procedures such as liquid extraction and partition chromatography, of pharmacological interest in the distribution of drugs between lipids and body fluids, and of environmental interest in determining how a pollutant is distributed between the air, water, and soil.

Experimental data on the partitioning of a solute between two liquid phases are usually reported in terms of a **distribution coefficient** K, defined to be the ratio of the solute concentration in the two phases:

Distribution coefficient

$$K = \frac{\text{Concentration of solute in phase I}}{\text{Concentration of solute in phase II}} \quad \textbf{(11.4-1)}$$

The purpose of this section is to study some aspects of the partitioning phenomenon and to relate the distribution coefficient to more fundamental thermodynamic quantities, so that we can (1) predict the distribution coefficient for a solute among two given solvents if experimental data are not available, or (2) use experimental distribution coefficient data to obtain information on liquid-phase activity coefficients.

However, in this section we are considering only the case in which the addition of the solute does not affect the miscibility of the solvents. The more general case of multicomponent liquid-liquid phase behavior in which the solute addition affects the mutual solubilities of the solvents was considered in Sec. 11.2. The situation considered here occurs when the mutual solubilities of the solvents are unchanged as a result of the solute addition, either because so little solute has been added or the solvents are so immiscible as to be unaffected by the solute.

The simplest type of solute distribution problem occurs when N_1 moles of a solute are completely dissolved and distributed between two immiscible solvents. The equilibrium distribution of the solute is determined from the single equilibrium relation

Starting point for all phase equilibrium calculations

$$\bar{f}_1^I(T, P, \underline{x}^I) = \bar{f}_1^{II}(T, P, \underline{x}^{II}) \tag{11.4-2}$$

and the constraint that the number of moles of solute be conserved

$$N_1 = N_1^I + N_1^{II} \tag{11.4-3}$$

where N_1^I is the number of moles of solute in phase I. Now, using the definition of the activity coefficient,

$$\bar{f}_1(T, P, \underline{x}) = x_1 \gamma_1(T, P, \underline{x}) f_1^L(T, P)$$

Eq. 11.4-2 can be rewritten as

$$x_1^I \gamma_1^I(T, P, \underline{x}^I) = x_1^{II} \gamma_1^{II}(T, P, \underline{x}^{II}) \tag{11.4-4}$$

since the pure component solute fugacity cancels. This last equation can be rearranged to

Liquid–liquid equilibrium relation

$$\frac{x_1^I}{x_1^{II}} = K_x = \frac{\gamma_1^{II}(T, P, \underline{x}^{II})}{\gamma_1^I(T, P, \underline{x}^I)} \tag{11.4-5}$$

which establishes that the distribution coefficient for solute mole fractions is equal to the reciprocal of the ratio of the solute activity coefficients in the two phases. Thus, given activity coefficient information for the solute in the two phases, one can compute the distribution of the solvent among the phases, or, given information about distribution of the solute, one can compute the ratio of the solute activity coefficients.

ILLUSTRATION 11.4-1

Calculation of Activity Coefficients from Distribution Coefficient Data

The following data are available for the concentration distribution coefficient K_c of bromine between carbon tetrachloride and water at 25°C.[16]

Concentration of Bromine in CCl₄ (kmol/m³)	$K_c = \left\{ \dfrac{\text{kmol Br}_2}{\text{m}^3 \text{ CCl}_4 \text{ solution}} \middle/ \dfrac{\text{kmol Br}_2}{\text{m}^3 \text{ H}_2\text{O solution}} \right\}$
0.04	26.8
0.1	27.2
0.5	29.0
1.0	30.4
1.5	31.4
2.0	33.4
2.5	35.0

[16]Reference for distribution coefficient data: *International Critical Tables,* Vol. 3, McGraw-Hill, New York, 1929.

Compute the ratio of the activity coefficient of bromine in water to that in carbon tetrachloride. Data for the pure species are:

Species	Molecular Weight	Density ρ, kg m^{-3}
CCl$_4$	153.84	$1.595 \times 10^{+3}$
H$_2$O	18.0	$1.0 \ \ \times 10^{+3}$
Br$_2$	159.83	$3.119 \times 10^{+3}$

SOLUTION

Since carbon tetrachloride and water are virtually immiscible (see Illustration 11.2-3), we will assume that the liquid bromine is distributed between water-free carbon tetrachloride and water free of carbon tetrachloride. Also, since solution volumetric data are not available, we will assume that there is no volume change on mixing bromine with either carbon tetrachloride or water.

In order to use Eq. 11.4-5 to evaluate the ratio of the activity coefficients, it is necessary to convert all the concentration distribution data to mole fractions. To do this, let C_B represent the concentration of bromine in kmol/m^3 of solution, which is known from the data, and C_S represent the concentration of solvent, again in kmol/m^3 of solution, which is also known. By the $\Delta_{mix}\underline{V} = 0$ assumption, we have

$$\underline{V}_{mix} = x_B\underline{V}_B + x_S\underline{V}_S = x_B\frac{m_B}{\rho_B} + (1 - x_B)\frac{m_S}{\rho_S}$$

where $\underline{V}_i$ is the molar volume of species i, and m and ρ are its molecular weight and density, respectively. The number of moles of bromine in one mole of solution is then computed from

$$\begin{array}{l}\text{Moles of bromine in one} \\ \text{mole of solution}\end{array} = C_B\underline{V}_{mix} = C_B\left[x_B\frac{m_B}{\rho_B} + (1 - x_B)\frac{m_S}{\rho_S}\right]$$

$$= \text{Mole fraction of bromine} = x_B$$

or

$$x_B = \frac{C_B \, m_S/\rho_S}{1 + C_B\left(\dfrac{m_S}{\rho_S} - \dfrac{m_B}{\rho_B}\right)} \tag{1}$$

Therefore, to compute the mole fraction of bromine in the CCl$_4$ phase, we use

$$x_{Br_2}^{CCl_4} = \frac{C_B\left(\dfrac{153.84 \text{ kg/mol}}{1595 \text{ kg/m}^3}\right)}{1 + C_B\left(\dfrac{153.84}{1595} - \dfrac{159.83}{3119}\right) \text{ m}^3/\text{kmol}} = \frac{0.096 \, 45C_B}{1 + 0.045 \, 21C_B} \tag{2}$$

where C_B is the concentration entries in the distribution coefficient table. Similarly, to compute the mole fraction of bromine in the aqueous phase, we again use Eq. 1, but now recognizing that if C_B is the concentration of bromine in the organic phase (i.e., the entries in the table), then C_B/K_c is the concentration of bromine in the aqueous phase. Therefore,

$$x_{Br_2}^{H_2O} = \frac{\dfrac{C_B}{K_c}\left(\dfrac{18 \text{ kg/mol}}{1000 \text{ kg/m}^3}\right)}{1 + \dfrac{C_B}{K_c}\left(\dfrac{18}{1000} - \dfrac{159.83}{3119}\right) \text{ m}^3/\text{kmol}} = \frac{0.018 \, C_B/K_c}{1 - 0.033 \, 24 \, C_B/K_c} \tag{3}$$

Using Eqs. 2 and 3, we obtain the following results:

Concentration Bromine in CCl$_4$	Mole Fraction of Br$_2$ in CCl$_4$ $x_{Br_2}^{CCl_4}$	Mole Fraction of Br$_2$ in H$_2$O $x_{Br_2}^{H_2O}$	$\dfrac{x_{Br_2}^{CCl_4}}{x_{Br_2}^{H_2O}} = \dfrac{\gamma_{Br_2}^{H_2O}}{\gamma_{Br_2}^{CCl_4}}$
0.04	3.85×10^{-3}	2.7×10^{-5}	142.6
0.10	9.60×10^{-3}	6.6×10^{-5}	145.4
0.5	4.70×10^{-2}	3.1×10^{-4}	151.6
1.0	9.23×10^{-2}	5.93×10^{-4}	155.6
1.5	0.135	8.62×10^{-4}	156.6
2.0	0.177	1.08×10^{-3}	163.9
2.5	0.217	1.29×10^{-3}	168.2

COMMENT

If, separately, we knew the activity coefficients for bromine in carbon tetrachloride, we could use the data in the table to evaluate the activity coefficient of bromine in water (Problem 11.4-3). Since the regular solution model can be used to represent the Br$_2$-CCl$_4$ mixture, we can then surmise that $\gamma_{Br_2}^{CCl_4}$ will be on the order of magnitude of unity. This suggests that $\gamma_{Br_2}^{H_2O}$ will be a large number, on the order of 100 or more. Such behavior for the activity coefficient of the minor component is not unusual in mixtures of species with such different molecular characteristics as strongly quadrupolar liquid bromine and strongly polar and hydrogen-bonded water. ∎

A slightly more complicated situation than the one just considered arises when there is some undissolved solute in equilibrium with two immiscible solvents. Here we will suppose that the undissolved solute is either a solid or a gas, and that neither solvent is present in the undissolved solute.[17]

In this case the equilibrium conditions are

$$f_1(T, P) = \bar{f}_1^{I}(T, P, \underline{x}^{I}) \quad \text{and} \quad f_1(T, P) = \bar{f}_1^{II}(T, P, \underline{x}^{II})$$

or

$$f_1(T, P) = \bar{f}_1^{I}(T, P, \underline{x}^{I}) = \bar{f}_1^{II}(T, P, \underline{x}^{II}) \tag{11.4-6}$$

since the undissolved solute must be in equilibrium with both liquid phases (which implies that both liquid phases are saturated with the solute). In this equation $f_1(T, P)$ is to be interpreted as the fugacity of the undissolved solute, which may be either a solid or a pure or mixed gas. Using the definition of the activity coefficient, Eq. 9.3-11, we obtain the equation

$$\frac{f_1(T, P)}{f_1^{L}(T, P)} = x_1^{I} \gamma_1(T, P, \underline{x}^{I}) = x_1^{II} \gamma_1(T, P, \underline{x}^{II}) \tag{11.4-7}$$

for the saturation mole fractions of the solute in each phase. If the undissolved solute is a solid, the fugacity ratio $f_1(T, P)/f_1^{L}(T, P)$ is equal to the ratio $f_1^{S}(T, P)/f_1^{L}(T, P)$,

---

[17]When the solute is a liquid, the problem is really one of multiple liquid-phase equilibrium. This is considered in Sec. 11.2 and Problem 11.4-1.

which can be evaluated using the prescription of Sec. 9.7 (Eq. 9.7-8a). If, on the other hand, the solute is a gas, $f_1(T, P)$ can be computed using an equation of state or the Lewis-Randall rule, and $f_1^L(T, P)$ is evaluated using the procedure discussed in Sec. 11.1. Note that Eqs. 11.4-5 and 11.4-7 require that the mole fraction distribution coefficient K_x again be equal to the reciprocal of solute activity coefficients in the two liquid phases; this is, in fact, a general result.

One interesting example of three-component liquid-liquid equilibrium is the partitioning of a small amount of a chemical in a mixture of water and *n*-octanol that has been allowed to equilibrate. As pointed out in Illustration 11.2-4, water is reasonably soluble in *n*-octanol, but *n*-octanol is virtually insoluble in water. Consequently, when these two chemicals are allowed to equilibrate at 25°C, two liquid phases form: a denser water-rich phase that is essentially pure water, and an octanol-rich phase that is approximately 74 mol % *n*-octanol and 26 mol % water. If a very small amount of a third chemical is added (so that the mutual solubility of water and *n*-octanol are unchanged), this chemical will then partition between the octanol-rich and water-rich phases. The octanol-water partition coefficient for chemical species i, $K_{OW,i}$ is defined as

$$K_{OW,i} = \frac{\text{Concentration of species i in the octanol-rich phase}}{\text{Concentration of species i in the water-rich phase}} = \frac{C_i^O}{C_i^W} \qquad \textbf{(11.4-8)}$$

A chemical that is hydrophilic (water-liking) will largely partition into the water-rich phase (resulting in a small value of K_{OW}), while a hydrophobic (water-disliking) compound will appear mainly in the octanol-rich phase, and K_{OW} will be large. Most organic chemicals produced by the chemical industry are very hydrophobic, so it is the logarithm of the octanol-water partition coefficient that is usually reported, as in Table 11.4-1. Note that the numeric values of the octanol-water partition coefficient for such chemicals can be very large, and the concentration of the added solute in the octanol-rich phase can be a factor of 10^5 or 10^6 or higher than in the water-rich phase.

The criterion for the equilibrium distribution of a small amount of solute between two coexisting octanol and water liquid phases is, from Eqs. 11.4-4 and 11.4-5,

$$\frac{x_i^O}{x_i^W} = \frac{\gamma_i^W(T, P, \underline{x}^W)}{\gamma_i^O(T, P, \underline{x}^O)} \qquad \textbf{(11.4-9)}$$

Now using C^O and C^W to represent the total molar concentrations of the octanol-rich and water-rich phases, respectively, we have

$$\frac{C^O x_i^O}{C^W x_i^W} = \frac{C_i^O}{C_i^W} = \frac{C^O \gamma_i^W(T, P, \underline{x}^W)}{C^W \gamma_i^O(T, P, \underline{x}^O)} = K_{OW,i} \qquad \textbf{(11.4-10)}$$

If only a very small amount of solute is added to the octanol-water mixture, so that the solute concentration is very low in both phases, then the activity coefficients that appear in this equation are essentially activity coefficients at infinite dilution, so that Eq. 11.4-10 can be written as

$$K_{OW,i} = \frac{C^O \gamma_i^{W,\infty}}{C^W \gamma_i^{O,\infty}} \qquad \textbf{(11.4-11)}$$

where the superscript ∞ indicates infinite dilution. In environmental engineering the pollutant species i is generally present at very low concentrations (which is fortunate), so that Eq. 11.4-11 is applicable.

The octanol-water partition coefficient of a chemical can be directly measured; however, the measurement is not an easy one since the concentration of a hydrophobic organic chemical in the water-rich phase will be very low (i.e., in the parts per million or parts per billion range, where accurate measurements are difficult). Further, the measurement can be subject to appreciable error because if any grease or extraneous organic matter is present in the equipment, a hydrophobic chemical will be absorbed from the water phase, resulting in an erroneously low concentration in that phase (and a high octanol-water partition coefficient).

A rough estimate of the octanol-water partition coefficient can be obtained by noting that since the density of n-octanol is 0.827 g/cc and its molecular weight is 130.22, and those of water are 1 g/cc and 18,

$$\frac{C^O}{C^W} \approx \frac{\dfrac{0.827 \text{ g/cc}}{130.22 \text{ g/mol}}}{\dfrac{1 \text{ g/cc}}{18 \text{ g/mol}}} = 0.114$$

where the "approximately equal" sign has been used since C^O should be the total molar concentration of the octanol-rich phase, though we have used the properties of pure octanol. Also, many organic chemicals that are environmental pollutants are very hydrophobic, so that $\gamma_i^{W,\infty}$ is very large (the chemical does not want to remain in the water-rich phase), while $\gamma_i^{O,\infty}$ is of order 5, since the octanol-rich phase is organic and hydrophobic. Therefore, to a reasonable approximation, we have

$$K_{OW,i} = 0.114 \frac{\gamma_i^{W,\infty}}{5} = 0.0228\gamma_i^{W,\infty}$$

or

$$\log_{10} K_{OW,i} = -1.642 + \log_{10} \gamma_i^{W,\infty} \tag{11.4-12}$$

From the correlation of data for a number of chemicals, a better approximation is

$$\log_{10} K_{OW,i} = -0.486 + 0.806 \log_{10} \gamma_i^{W,\infty} \tag{11.4-13}$$

Therefore, by knowing only the infinite-dilution activity coefficient of a very hydrophobic chemical in water, we can estimate its octanol-water partition coefficient from Eq. 11.4-13, and its saturation mole fraction or solubility as in Illustration 11.2-4. Alternatively, knowing any one among the infinite-dilution activity coefficient, octanol-water partition coefficient, and saturation solubility in water, the other two can be estimated.

ILLUSTRATION 11.4-2
Estimating the Value of the Octanol-Water Partition Coefficient for a Pollutant

In Illustration 12.1-3 it will be shown that the value of the infinite-dilution activity coefficient of benzo[*a*]pyrene in water is 3.76×10^8. Using the information, estimate the octanol-water partition coefficient of benzo[*a*]pyrene and compare the value obtained to the reported value of

$$\log_{10} K_{OW,BP} = 6.04 \quad \text{or} \quad K_{OW,BP} = 1.1 \times 10^6$$

Table 11.4-1 Octanol-Water Partition Coefficients and Other Properties of Some Organic Chemicals, Herbicides, and Pesticides

Substance	$\log_{10} K_{OW}$	Vapor Pressure (kPa) at 25°C	Henry's Constant (bar m^3/mol)	Melting Point (°C)	Solubility (mg/L or ppm)
Acrolein	−0.10	35.3	4.5×10^{-6}	−88	2.08×10^5
Aldicarb	1.13	1.3×10^{-5}	4.23×10^{-9}	99	6×10^3
Aniline	0.90	0.065	0.138	−6.3	3.607×10^4
1,3-Butadiene	1.99	281	20.6	−108.9	735
Chlorobenzene	2.80	1.38	3.82×10^{-3}	−45.6	484
Chlorodane	5.54	1.31×10^{-6}	4.92×10^{-5}	107	0.1
2-Chlorophenol	2.15	0.189	5.7×10^{-7}	9.3	2.8×10^4
2-Cresol	1.95	0.041	1.6×10^{-6}	30.9	30.8
2,4-D (herbicide)	2.81	8.0×10^{-8}	1.39×10^{-10}	140.5	690
1,2-Dichlorobenzene	3.38	0.196	1.2×10^{-3}	−17.0	156
1,4-Dichlorobenzene	3.52	0.235	1.5×10^{-3}	53.1	80
Dieldrin	4.32	5.00×10^{-7}	5.9×10^{-5}	~175	0.17
Ethylbenzene	3.15	1.27	8.19×10^{-3}	−95.0	152
Formaldehyde	0.35	517.6	3.31×10^{-7}	−92	~55%
Heptachlor	5.27	5.3×10^{-5}	1.50×10^{-3}	~95	0.18
Hexachlorobenzene	5.31	2.5×10^{-6}	1.3×10^{-3}	231	6.2×10^{-3}
Hexachloroethane	3.82	0.028	2.8×10^{-3}	~187	50
Lindane	3.61	7.43×10^{-6}	2.96×10^{-6}	112.5	7.3
Malathion	2.36	1.1×10^{-6}	2.8×10^{-8}	2.9	143
Methyl bromide	1.19	217.7	6.32×10^{-3}	−93.7	1.52×10^4
Methyl chloride	0.91	570	9.77×10^{-3}	−97	5235
Naphthalene	3.30	3.68×10^{-2}	4.3×10^{-4}	80.2	31.0
Nitrobenzene	1.85	0.020	2.47×10^{-5}	5.7	1.9×10^3
3-Nitrophenol	2.00	0.10	1.03×10^{-7}	97	1.35×10^4
Parathion	3.83	1.29×10^{-6}	5.73×10^{-7}	6	6.5
Phenol	1.46	0.0466	4.02×10^{-7}	41	8.7×10^4
Silvex	3.41	6.9×10^{-7}	1.33×10^{-8}	~180	140
Styrene	2.95	0.88	2.85×10^{-3}	−30.6	310
2,4,5-T	3.13	1.0×10^{-7}	8.80×10^{-9}	153	278
1,2,4-Trichlorobenzene	4.10	0.053	1.44×10^{-3}	17	48.8
2,4,5-Trichlorophenol	3.72	2.9×10^{-3}	5.90×10^{-6}	67	982
2,4,6-Trichlorophenol	3.69	1.12×10^{-3}	6.22×10^{-8}	69	900

SOLUTION

From Eq. 11.4-12, we have

$$K_{OW,BP} = 0.0228 \gamma_{BP}^{W,\infty}$$

$$= 0.0228 \times 3.76 \times 10^8 = 8.57 \times 10^6$$

or

$$\log_{10} K_{OW,BP} = 6.93$$

Alternatively, using Eq. 11.4-13,

$$\log_{10} K_{OW,BP} = -0.486 + 0.806 \log_{10}(3.76 \times 10^8) = 6.43$$

or

$$K_{\mathrm{OW,BP}} = 2.69 \times 10^6$$

This last result agrees within a factor of 2.5 to the reported benzo[*a*]pyrene octanol-water partition coefficient. ∎

ILLUSTRATION 11.4-3

Purification of an Antibiotic

Benzylpenicillin is an older antibiotic effective against pneumococcal and meningoccal infections, anthrax, and Lyme disease. As part of a purification process, 200 mg of benzylpenicillin is mixed with 25 mL of n-octanol and 25 mL of water. After equilibrium is established, there is a water-rich phase that contains essentially no n-octanol and an octanol-rich phase that contains 74 mol % n-octanol and 26 mol % water. Determine the concentrations of benzylpenicillin in each of these phases.

Data: The molecular weight of benzylpenicillin is 334.5, that of n-octanol is 130.23, the liquid density of n-octanol is 0.826 g/cc, and $K_{\mathrm{OW,B}} = 65.5$.

SOLUTION

First we need to determine the number of moles in each phase (on a benzylpenicillin-free basis). Since n-octanol is insoluble in water, it is all in the octanol-rich phase. The number of moles of n-octanol is

$$\frac{25 \text{ mL} \times 0.826 \text{ g/mL}}{130.23 \text{ g/mol}} = 0.1586 \text{ mol}$$

and the amount of water in the octanol-rich phase is

$$0.1586 \text{ mol octanol} \times \frac{0.26 \text{ mol water}}{0.74 \text{ mol octanol}} = 0.0557 \text{ mol water}$$

The volume of the n-octanol-rich phase, assuming no volume change of mixing, is

$$V_{\mathrm{O}} = \frac{0.1586 \text{ mol} \times 130.23 \text{ g/mol}}{0.826 \text{ g/mL}} + \frac{0.0557 \text{ mol} \times 18 \text{ g/mol}}{1 \text{ g/mL}} = 26.008 \text{ mL}$$

where the first term is from the amount of n-octanol and the second from the amount of water in the octanol-rich phase. The volume of the water-rich phase is

$$V_{\mathrm{W}} = 25 - 1.0028 = 23.997 \text{ mL}$$

The total number of moles of benzylpenicillin is $(0.2 \text{ g})/(334.4 \text{ g/mol}) = 5.981 \times 10^{-4}$ mol. Now to find the concentration of benzylpenicillin in each phase we use a mass balance:

$$5.981 \times 10^{-4} \text{ mol} = C_{\mathrm{B}}^{\mathrm{W}} \cdot V^{\mathrm{W}} + C_{\mathrm{B}}^{\mathrm{O}} \cdot V^{\mathrm{O}} = C_{\mathrm{B}}^{\mathrm{W}} \cdot V^{\mathrm{W}} + K_{\mathrm{OW,B}} \cdot C_{\mathrm{B}}^{\mathrm{W}} \cdot V^{\mathrm{O}}$$
$$= C_{\mathrm{B}}^{\mathrm{W}} \cdot (23.997 + 65.5 \times 26.008) = C_{\mathrm{B}}^{\mathrm{W}} \cdot 1727.52 \text{ mL}$$

Therefore,

$$C_{\mathrm{B}}^{\mathrm{W}} = 3.462 \times 10^{-7} \frac{\text{mol}}{\text{mL}} = 1.158 \times 10^{-4} \frac{\text{g}}{\text{mL}} = 0.1158 \frac{\text{mg}}{\text{mL}}$$

and

$$C_{\mathrm{B}}^{\mathrm{O}} = 3.462 \times 10^{-7} \times 65.5 = 2.268 \times 10^{-5} \frac{\text{mol}}{\text{mL}} = 7.585 \times 10^{-3} \frac{\text{g}}{\text{mL}} = 7.585 \frac{\text{mg}}{\text{mL}}$$

∎

An alternative method of considering the solubility of a gas in a solvent, or the distribution of a solute between two liquid phases, is by analyzing the Gibbs energy of transfer of a species from one phase to another, for example, from an ideal gas phase to a liquid, or from one liquid phase to another. This type of calculation is referred to a solvation free energy calculation. The basis is as follows. Consider first the transfer of N_S moles of a species that is a gas into N_1 moles of a pure liquid solvent. Before mixing, the total Gibbs energy of the separated gas and liquid is

$$G \text{ (before mixing)} = N_1 \underline{G}_1^L + N_S \underline{G}_S^V$$

and after mixing, assuming all the gas is dissolved,

$$G \text{ (after mixing)}$$
$$= N_1 \bar{G}_1^L + N_S \bar{G}_S^L = N_1 \left[\underline{G}_1^L + RT \ln (x_1 \gamma_1) \right] + N_S \left[\underline{G}_S^L + RT \ln (x_S \gamma_S) \right]$$

Therefore, the Gibbs energy change for this transfer process is

$$\triangle_{\text{tfr}} G = N_1 \left[\underline{G}_1^L + RT \ln (x_1 \gamma_1) \right] - N_1 \underline{G}_1^L + N_S \left[\underline{G}_S^L + RT \ln (x_S \gamma_S) \right] - N_S \underline{G}_S^V$$
$$= N_1 RT \ln (x_1 \gamma_1) + N_S RT \ln (x_S \gamma_S) + N_S \left(\underline{G}_S^L - \underline{G}_S^V \right)$$
$$= N_1 RT \ln \left(\frac{N_1 \gamma_1}{N_1 + N_S} \right) + N_S RT \ln \left(\frac{N_S \gamma_S}{N_1 + N_S} \right) + N_S \left(\underline{G}_S^L - \underline{G}_S^V \right)$$

$$\textbf{(11.4-14)}$$

Now assuming the solubility of the gas in the liquid solvent is low, so that $N_1 \gg N_S$, and $x_1 \gamma_1 \approx 1$, we obtain for the free energy of transfer from a gas to a liquid

$$\triangle_{\text{tfr}} G = N_S \left[\underline{G}_S^L - \underline{G}_S^V + RT \ln (x_S \gamma_S) \right]$$

or per mole of species transferred,

$$\frac{\triangle_{\text{tfr}} G}{N_S} = \triangle_{\text{tfr}} \underline{G} = \underline{G}_S^L - \underline{G}_S^V + RT \ln (x_S \gamma_S) = RT \left[\ln \left(\frac{f_S^L}{f_S^V} \right) + \ln (x_S \gamma_S) \right]$$

The ratio of the solute fugacities in the liquid and vapor phases can be computed as described in Chapter 9 (see Eq. 9.7-8b).

Next, consider the transfer of a very dilute solute from solvent 1 to solvent 2. The Gibbs energy of 1 mole of solvent 1 and δ moles of solute S is

$$G \text{ (in solvent 1)} = \bar{G}_1 + \delta \bar{G}_S = \underline{G}_1 + RT \ln (x_1 \gamma_1) + \delta \left[\underline{G}_S + RT \ln (x_S \gamma_S) \right]$$

and the Gibbs energy change of separating the solute from the solvent is

$$\triangle_1 G \text{ (in solvent 1)} = \underline{G}_1 + \delta \underline{G}_S - \left[\underline{G}_1 + RT \ln (x_1 \gamma_1) + \delta \underline{G}_S + \delta RT \ln (x_S \gamma_S) \right]$$
$$= -RT \ln (x_1 \gamma_1) - \delta RT \ln (x_S \gamma_S)_{\text{solvent 1}}$$

$$\textbf{(11.4-15)}$$

Similarly, the Gibbs energy change of adding δ moles of the solute to 1 mole of solvent 2 is

$$\triangle_2 G = RT \ln (x_2 \gamma_2) + \delta RT \ln (x_S \gamma_S)_{\text{solvent 2}}$$

Therefore, the total Gibbs energy change of transferring δ moles of a solute from 1 mole of solvent 1 to 1 mole of solvent 2 is

$$\triangle_{\text{tfr}} G = RT \ln \left(\frac{x_2 \gamma_2}{x_1 \gamma_1} \right) + \delta RT \ln \left[\frac{(x_S \gamma_S)_{\text{solvent 2}}}{(x_S \gamma_S)_{\text{solvent 1}}} \right]$$

$$= RT \ln \left(\frac{\gamma_2}{\gamma_1} \right) + \delta RT \ln \left[\frac{(\gamma_S)_{\text{solvent 2}}}{(\gamma_S)_{\text{solvent 1}}} \right] \tag{11.4-16}$$

Since in each of the liquids considered, there is 1 mole of solvent and δ moles of solute, the mole fractions of solvent and solute are equal in solvents 1 and 2, and have canceled in the equation above.

A case of special interest is if $\delta \ll 1$, so that the solute is present in the solvents at infinite dilution. In this case, the activity coefficients of both solvents, γ_1 and γ_2, are unity and

$$\triangle_{\text{tfr}} G = \delta RT \ln \left[\frac{\left(\gamma_S^{\infty} \right)_{\text{solvent 2}}}{\left(\gamma_S^{\infty} \right)_{\text{solvent 1}}} \right]$$

or

$$\frac{\triangle_{\text{tfr}} G}{\delta} = \triangle_{\text{tfr}} \underline{G} = RT \ln \left[\frac{\left(\gamma_S^{\infty} \right)_{\text{solvent 2}}}{\left(\gamma_S^{\infty} \right)_{\text{solvent 1}}} \right] \tag{11.4-17}$$

which is the Gibbs energy of transfer of 1 mole of solute from infinite dilution in solvent 1 to infinite dilution in solvent 2. Note that if

$$\left(\gamma_S^{\infty} \right)_{\text{solvent 2}} > \left(\gamma_S^{\infty} \right)_{\text{solvent 1}} \quad \text{then} \quad \triangle_{\text{tfr}} G > 0$$

and the transfer is unfavorable; that is, the Gibbs energy (or work) must somehow be supplied to accomplish the change. However, if

$$\left(\gamma_S^{\infty} \right)_{\text{solvent 2}} < \left(\gamma_S^{\infty} \right)_{\text{solvent 1}} \quad \text{then} \quad \triangle_{\text{tfr}} \underline{G} < 0$$

The transfer is favorable, and Gibbs energy would be released on the transfer.

ILLUSTRATION 11.4-4

The Gibbs Energy of Transfer of an Amino Acid

At 25°C a saturated aqueous solution of L-asparagine contains 0.186 M of the amino acid, while a saturated solution in ethanol is 2.3×10^{-5} M. What is the ratio of the activity coefficients of L-asparagine in these two solutions, and what is the Gibbs energy of transfer of a very small amount of L-asparagine from water to ethanol?

SOLUTION

We first need to estimate the activity coefficients, and especially the infinite-dilution activity coefficients for L-asparagine in water and ethanol. We can obtain this information from the saturated solution data. For solid L-asparagine in equilibrium with the two saturated liquid solutions, we have

$$\underline{G}_A^S = \overline{G}_A^W = \overline{G}_A^E$$

where the superscripts W and E indicate the water-rich and ethanol-rich phases, respectively, and the solubility concentrations are given above. Therefore,

$$\underline{G}_A^L + RT\ln\left(x_A^W\gamma_A^W\right) = \underline{G}_A^L + RT\ln\left(x_A^E\gamma_A^E\right)$$

and

$$x_A^W\gamma_A^W = x_A^E\gamma_A^E \quad \text{or} \quad \frac{x_A^W}{x_A^E} = \frac{\gamma_A^E}{\gamma_A^W}$$

To use this equation, we need to convert each of the molalities to mole fractions, and to do this we use the fact that 1 kg of water contains 55.51 moles, and there are 21.707 moles of ethanol in 1 kg, so that

$$x_A^W = \frac{0.186}{0.186 + 55.51} = 3.34 \times 10^{-3}$$

$$x_A^E = \frac{2.3 \times 10^{-5}}{2.3 \times 10^{-5} + 21.707} = 1.06 \times 10^{-6}$$

and

$$\frac{\gamma_A^E}{\gamma_A^W} = \frac{3.34 \times 10^{-3}}{1.06 \times 10^{-6}} = 3.15 \times 10^3$$

Note that the mole fractions of L-asparagine in each of the solvents is so low that we can assume the activity coefficient ratio we have obtained to be that at infinite dilution. Therefore,

$$\triangle_{\text{tfr}}\underline{G} = RT\ln\left(\frac{\gamma_A^E}{\gamma_A^W}\right) = 8.314 \times 298.15 \times \ln(3150) = 19\,970\ \frac{J}{\text{mol}}$$

and the transfer of an infinitesimal amount of L-asparagine from water to ethanol is unfavorable. ∎

The discussion in this section has been concerned with the distribution of a solute between two liquid phases whose equilibrium is unaffected by the added solute. This will occur if the amount of added solute is very small, or if the solvents are essentially immiscible at all conditions. However, if the amount of dissolved solute is so large as to affect the miscibility of the solvents, the solute addition can have a significant effect on the solvents, including the increase (salting in) or decrease (salting out) of the mutual solubility of the two solvents, as was discussed in Sec. 11.2. It is important to emphasize that such situations are described by the methods in Sec. 11.2 as a multicomponent liquid-liquid equilibrium problem, in contrast to the procedures in this section, which are based on the assumption that the partial or complete immiscibility of the solvents is unaffected by the addition of the partitioning solute.

PROBLEMS FOR SECTION 11.4

11.4-1 Write the equations that are to be used to compute the equilibrium compositions when three liquids are mixed and form

a. Two liquid phases

b. Two liquid phases in which species 1 is distributed among completely immiscible species 2

and 3
  c. Three partially miscible liquid phases
11.4-2 An important application of liquid-phase partitioning is partition chromatography. There a solvent containing several solutes is brought into contact with another solvent, usually one that is mutually immiscible with the first. When equilibrium is established, some solutes are concentrated in the first solvent and others in the second solvent. By repeating this process a number of times, it is possible to partially separate the solutes.

The concentration distribution coefficient K_c for gallic acid ($C_7H_6O_5$) between diethyl ether (phase I) and water (phase II) is 0.25, and that for p-hydroxybenzoic acid is 8.0. If 2 g each of gallic acid and p-hydroxybenzoic acid are present in one liter of water, how much of each component would be left in the aqueous phase if the aqueous phase were
  a. Allowed to achieve equilibrium with 5 liters of diethyl ether?
  b. Allowed to achieve equilibrium first with one batch of $2\frac{1}{2}$ liters of diethyl ether, and then with a second batch of the same amount?
  c. Allowed to achieve equilibrium with five successive 1-liter batches of pure diethyl ether?
11.4-3 Use regular solution theory to calculate the activity coefficients of bromine in carbon tetrachloride at the mole fractions found in Illustration 11.4-1, and use this information and that in the illustration to compute the activity coefficients of bromine in water.
11.4-4 Volume 3 of the *International Critical Tables* (McGraw-Hill, New York, 1928) gives the following data for the distribution coefficient of bromine between carbon disulfide and various aqueous solutions of potassium bromide at 25°C:

Using regular solution theory to estimate the activity coefficient of bromine in carbon disulfide, compute the activity coefficient of bromine in
  a. The $\frac{1}{16}$ molar KBr solution
  b. The $\frac{1}{4}$ molar KBr solution
  c. The $\frac{1}{2}$ molar KBr solution
  d. Try to infer a relationship between the salt concentration and the activity coefficient of bromine in aqueous solution.
11.4-5 a. For a liquid that is very slightly soluble in water, there is a consistency relationship between its Henry's law coefficient in water, its solubility in water, and its vapor pressure. Show for the Henry's law coefficient in the equation $x_i H_i = P_i$ that

$$H_i = \frac{P_i^{\text{vap}}}{x_i^{\text{sat}}}$$

and for the Henry's law coefficient in the equation $C_i H_i = P_i$ that

$$H_i = \frac{P_i^{\text{vap}}}{C_i^{\text{sat}}}$$

where C_i denotes the concentration of species i, and the superscript sat indicates the concentration at saturation.

  b. Test this relationship using the data in Table 11.4-1 for aniline, 1,2-dichlorobenzene, ethylbenzene, naphthalene, and styrene, all of which are above their melting point and only slightly soluble in water.
11.4-6 At 25°C a saturated aqueous solution contains 0.171 M of L-leucine, while a saturated solution in ethanol contains only 1.28×10^{-3} M of this amino acid. What is the ratio of the activity coefficients of L-leucine in these two solutions, and what is the

Bromine Concentration in Carbon Disulfide (mol Br$_2$/L of solution)	Distribution Coefficient $\left(\dfrac{\text{mol Br}_2/\text{L of aqueous phase}}{\text{mol Br}_2/\text{L of organic phase}}\right)$		
	$\frac{1}{16}$ Molar KBr	$\frac{1}{4}$ Molar KBr	$\frac{1}{2}$ Molar KBr
0.005	0.0696	0.1735	
0.02	0.0651	0.167	0.303
0.03	0.0621	0.163	0.298
0.04		0.159	0.293
0.05		0.155	0.288
0.06		0.151	0.283
0.095			0.271

Gibbs energy of transfer of L-leucine from water to ethanol?

11.4-7 Below is a table of the solubility of the family of amino acids $^+H_3NCHRCOO^-$ in water (S^W) and in ethanol (S^E) at 25°C in units of mol/dm^3. Compute the Gibbs energy of transfer of each of these amino acids from water to ethanol.

Amino Acid	R	S^W	S^E
Glycine	H	2.89	3.9×10^{-4}
α-Alanine	CH_3	1.66	7.6×10^{-4}
α-Amino-n-butyric acid	C_2H_5	1.80	2.6×10^{-3}
α-Amino-n-caproic acid	C_4H_9	0.0866	1.04×10^{-3}

11.5 OSMOTIC EQUILIBRIUM AND OSMOTIC PRESSURE

Next we consider the equilibrium state of liquid mixtures in two cells separated by a membrane that is permeable to some of the species present and impermeable to others. In particular, consider a solute-solvent system in which the solvent, but not the solute, can pass through the membrane. We will assume that cell I contains the pure solvent and cell II the solvent-solute mixture, and that the membrane separating the two cells is rigid, so that the two cells need not be at the same pressure. The equilibrium criterion for this system is

$$f^I_{solvent} = \bar{f}^{II}_{solvent} \tag{11.5-1}$$

or

$$f_{solvent}(T, P^I) = x^{II}_{solvent} \gamma^{II}_{solvent} f_{solvent}(T, P^{II}) \tag{11.5-2}$$

A similar equation is not written for the solute since it is mechanically constrained from passing through the membrane and thus need not be in thermodynamic equilibrium in the two cells.

Actually, the use of Eq. 11.5-1 as the equilibrium requirement for this case deserves some discussion. We have shown that this equation is the equilibrium criterion for a system at constant temperature and pressure. From the discussion of Sec. 8.7, it is clear that it is also valid for systems subject to certain other constraints. Here we are interested in the equilibrium criterion for a system divided into two parts, each of which is maintained at constant temperature and pressure, but not necessarily the same pressure (or, for that matter, the same temperature). Though it is not obvious that Eq. 11.5-1 applies to this case, it can be shown to be valid (Problem 11.5-1).

Since $P^I \neq P^{II}$, the fugacities of the pure solvent at the conditions in the two cells are related by the Poynting pressure correction (see Eq. 7.4-21),

$$f_{solvent}(T, P^{II}) = f_{solvent}(T, P^I) \exp\left[\frac{\underline{V}^L_{solvent}(P^{II} - P^I)}{RT}\right] \tag{11.5-3}$$

where we have assumed that the liquid is incompressible, so that $\int_{P^I}^{P^{II}} \underline{V}^L_{solvent} dP \simeq \underline{V}^L_{solvent}(P^{II} - P^I)$. Using Eq. 11.5-3 in Eq. 11.5-2 yields

$$1 = x^{II}_{solvent} \gamma^{II}_{solvent} \exp\left[\frac{\underline{V}^L_{solvent}(P^{II} - P^I)}{RT}\right]$$

or

Osmotic pressure equation

$$\boxed{\Pi = P^{II} - P^I = -\frac{RT}{\underline{V}^L_{solvent}} \ln(x^{II}_{solvent} \gamma^{II}_{solvent})} \tag{11.5-4}$$

where $\Pi = P^{II} - P^{I}$, called the **osmotic pressure**, is the pressure difference between the two cells needed to maintain thermodynamic equilibrium.

A very large pressure difference can be required to maintain osmotic equilibrium in a system with only a small concentration difference. Consider, for example, the very simple case of osmotic equilibrium at room temperature between an ideal aqueous solution containing 98 mol % water and pure water. Here

$$\Pi = P^{II} - P^{I} = -\frac{RT}{\underline{V}^{L}_{solvent}} \ln(x^{II}_{solvent})$$

$$= -\frac{8.314 \times 10^{-5} \dfrac{bar\ m^3}{mol\ K} \times 298.15\ K}{18 \times 10^{-6} \dfrac{m^3}{mol}} \ln 0.98 = 27.8\ bar$$

Thus if cell I is at atmospheric pressure, cell II would have to be maintained at 28.8 bar to prevent the migration of water from cell I to cell II. (A more accurate estimate of the osmotic pressure could be obtained using experimental data or appropriate liquid solution theories to evaluate the activity coefficient of water in aqueous solutions. See Problem 11.5-3.)

Since pressure measurements are relatively simple, Eq. 11.5-4 can be the basis for determining solvent activity coefficients in a solvent-solute system, provided a suitable leakproof membrane can be found. Osmotic pressure measurements are, however, more commonly used to determine the molecular weights of proteins and other macromolecules (for which impermeable membranes are easily found). In such cases an osmometer, such as the one shown in schematic form in Fig. 11.5-1, is used to measure the equilibrium pressure difference between the pure solvent and the solvent containing the macromolecules (which are too large to pass through the membrane); the pressure difference ΔP, which is the osmotic pressure Π, is equal to $\rho g h$, where ρ is the solution density and h is the difference in liquid heights. If the solute concentration is small, we have

$$\Pi = -\frac{RT}{\underline{V}_{solvent}} \ln(x_{solvent}) \approx \frac{RT}{\underline{V}_{solvent}}(1 - x_{solvent}) \qquad \textbf{(11.5-5)}$$

Furthermore, since $x_{solvent} + x_{solute} = 1$, this reduces to

$$\Pi = \frac{RT}{\underline{V}_{solvent}} x_{solute} \simeq \frac{RT\,C_{solute}/m_{solute}}{\underline{V}_{solvent}\,C_{solvent}/m_{solvent}} \qquad \textbf{(11.5-6)}$$

$$= RT\,C_{solute}/m_{solute}$$

Here the concentration C is in units of mass per volume, m is the molecular weight, and we have used the fact that since the solute mole fraction is low,

$$x_{solute} = \frac{Moles\ solute}{Moles\ solute + Moles\ solvent} \simeq \frac{Moles\ solute}{Moles\ solvent} = \frac{C_{solute}/m_{solute}}{C_{solvent}/m_{solvent}}$$

and

$$\frac{C_{solvent}}{m_{solvent}} = \frac{1}{\underline{V}_{solvent}}$$

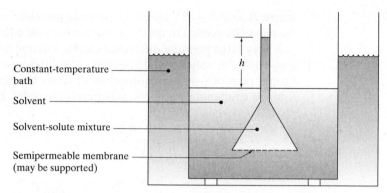

Figure 11.5-1 A schematic drawing of a simple osmometer.

Relatively small concentrations of macromolecules produce easily measurable osmotic-pressure differences. For example, suppose 1 g of a protein or polymer of molecular weight 60 000 is dissolved in 100 mL of water and placed in the osmometer of Fig. 11.5-1. At a temperature of 25°C, the osmotic-pressure difference would be

$$\Pi = \frac{0.01\ \dfrac{\text{g}}{\text{mL}} \times 298.15\ \text{K} \times 8.314 \times 10^{-5}\ \dfrac{\text{bar m}^3}{\text{mol K}} \times 10^6\ \dfrac{\text{mL}}{\text{m}^3}}{60\ 000\ \text{g/mol}}$$

$$= 4.13 \times 10^{-3}\ \text{bar} = 4.22\ \text{cm H}_2\text{O}$$

which is easily measurable.

To determine the molecular weight of a macromolecule, a known weight of the substance is added to a weighed amount of solvent so that the mass concentration of the solute C_{solute} is known. This mixture is then placed in the osmometer and Π measured. The molecular weight is then found from

$$m_{\text{solute}} = \frac{RT\,C_{\text{solute}}}{\Pi} \tag{11.5-7}$$

For high accuracy, this measurement is repeated several times at varying solute concentrations, and the limiting value of C_{solute}/Π as C_{solute} approaches zero is used in Eq. 11.5-7. This procedure allows for the fact that C_{solute} is imperfectly known (since additional solvent passes through the membrane until equilibrium is established) and that the simplifications in Eq. 11.5-6 become exact and solvent nonidealities vanish as $C_{\text{solute}} \to 0$.

The advantage of the osmotic-pressure difference method of determining the molecular weights of macromolecules over alternative methods, such as the freezing-point depression method, is evident from a comparison of the magnitudes of the effects to be measured. In Illustration 12.3-1 it will be shown that the addition of 0.01 g/mL of a 60 000 molecular-weight protein results in a freezing-point depression of water of only 0.00031 K, whereas here we find it results in an easily measureable osmotic pressure of 4.22 cm H_2O.

ILLUSTRATION 11.5-1

Determining the Molecular Weight of a Polymer

The polymer polyvinyl chloride (PVC) is soluble in the solvent cyclohexanone. At 25°C it is found that if a solution with 2 g of a specific batch of PVC per liter of solvent is placed in an

osmometer, the height h to which the pure cyclohexanone rises is 0.85 cm. Use this information to estimate the molecular weight of the PVC polymer.
Data: Cyclohexanone has a density of 0.98 g/cm^3.

SOLUTION

The osmotic pressure corresponding to 0.85 cm of cyclohexanone is

$$\Pi = \rho g h = 0.98 \; \frac{\text{g}}{\text{cm}^3} \times 0.85 \; \text{cm} \times 10^4 \; \frac{\text{cm}^2}{\text{m}^2} \times 9.81 \; \frac{\text{m}}{\text{s}^2} \times 10^{-3} \; \frac{\text{kg}}{\text{g}}$$

$$= 81.72 \; \frac{\text{kg}}{\text{m s}^2} = 81.72 \; \frac{\text{kg m}}{\text{s}^2} \frac{1}{\text{m}^2} = 81.72 \; \frac{\text{N}}{\text{m}^2} = 81.72 \; \text{Pa}$$

To determine the molecular weight of the polymer, we will use this value of the osmotic pressure and Eq. 11.5-7 and find

$$m_{\text{solute}} = \frac{8.314 \; \dfrac{\text{Pa m}^3}{\text{mol K}} \times 298.15 \; \text{K} \times 2 \; \dfrac{\text{g}}{\text{L}} \times 10^3 \; \dfrac{\text{L}}{\text{m}^3}}{81.72 \; \text{Pa}} = 60\,670 \; \frac{\text{g}}{\text{mol}} = 60.67 \; \frac{\text{kg}}{\text{mol}}$$

COMMENT

Osmometry is an important method of determining the molecular weight of macromolecules such as polymers and proteins. The advantages of this method for macromolecules are that membranes permeable to the solvent but not to the macromolecule are easily found, and the osmotic pressure (or height of the solvent) is large and easily measured. ■

ILLUSTRATION 11.5-2

Determining the Molecular Weight of a Protein

The following data have been reported for the osmotic pressure of aqueous serum albumin solutions at pH = 4.8 in an 0.05 M acetate buffer at 0°C.
Data:

Albumin Concentration C (g/cc)	Osmotic Pressure Π (cm H$_2$O)
0.78	2.39
1.25	4.01
2.79	8.53
3.38	10.80
4.18	13.01
8.98	27.84
12.45	37.69

SOLUTION

From the experimental data, we have

Albumin Concentration C (g/cc)	C/Π	Molecular Weight (g)
0.78	0.3264	75 600
1.25	0.3117	72 400
2.79	0.3271	75 700
3.38	0.3127	72 600
4.18	0.3213	74 400
8.98	0.3226	74 700
12.45	0.3303	76 600

The average of all these values is approximately 74 500. (Note that in the biological literature, this would be referred to as 74 500 daltons.) ■

While Eq. 11.5-4 is the correct expression for the concentration dependence of the osmotic pressure, most experimental data are reported in forms that look much like the virial equation of state for a gas. One such form is

$$\frac{\Pi}{RT} = \frac{C_S}{m_S} \left\{ 1 + B_2 \left(\frac{C_S}{m_S} \right) + B_3 \left(\frac{C_S}{m_S} \right)^2 + \cdots \right\} \qquad \text{(11.5-8)}$$

where C_S is the concentration of the solute in easily measurable units of mass per unit volume, m_S is the (generally unknown) molecular weight of the solute, and B_2, B_3, ... are referred to as the osmotic virial coefficients. Consequently, having data on the osmotic pressure as a function of solute concentration allows one to calculate the molecular weight and the osmotic virial coefficients.

ILLUSTRATION 11.5-3

Determining the Molecular Weight and Osmotic Virial Coefficients of a Protein

The following data have been reported for the osmotic pressure of α-chymotrypsin in water at pH = 4.0 in an 0.01 M potassium sulfate solution at 25°C.[18]
Data

C_S (g/L)	Π (Pa)	$\dfrac{\Pi}{C_S} \left(\dfrac{\text{Pa L}}{\text{g}} \right)$
0.902	74.8	82.92
1.798	142.3	79.14
3.574	276.2	77.29
5.331	398.0	74.65
7.027	513.3	73.05
8.835	625.3	70.77

Use these data to determine the molecular weight of α-chymotrypsin and the value of its osmotic second virial coefficient, B_2, at these conditions.

---

[18]Reprinted with permission of C. A. Haynes, T. Tamura, H. R. Korfer, H. W. Blanch, and J. M. Prausnitz, *J. Phys. Chem.*, **96**, 905 (1992). Copyright 1992 American Chemical Society.

SOLUTION

As shown in the figure, these data can be fit with the following equation:

$$\frac{\Pi}{C_S} = 83.961 - 2.168 \cdot C_S + 0.077 \cdot C_S^2$$

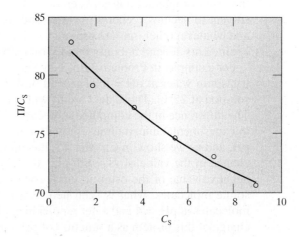

Therefore,

$$\lim_{C_S \to 0} \frac{\Pi}{C_S} = 83.961 \ \frac{\mathrm{Pa\,L}}{\mathrm{g}} = \frac{RT}{m_S}$$

so that the molecular weight is

$$m_S = \frac{8.314 \times 10^{-3} \frac{\mathrm{kPa\,m^3}}{\mathrm{mol\,K}} \times 298.15 \ \mathrm{K} \times 10^3 \frac{\mathrm{Pa}}{\mathrm{kPa}} \times 10^3 \frac{\mathrm{L}}{\mathrm{m^3}}}{83.961 \frac{\mathrm{Pa\,L}}{\mathrm{g}}} = 29\,528 \ \frac{\mathrm{g}}{\mathrm{mol}}$$

Next,

$$\frac{\Pi}{RT} = \frac{C_S}{m_S}$$

$$\frac{\Pi}{C_S} = \frac{RT}{m_S} \left\{ 1 + B_2 \frac{C_S}{m_S} + \cdots \right\}$$

so that

$$\lim_{C_S \to 0} \frac{d\left(\frac{\Pi}{C_S}\right)}{dC_S} = \frac{RT B_2}{m_S^2} = -2.168 \ \frac{\mathrm{Pa\,L^2}}{\mathrm{g^2}}$$

and

$$B_2 = -\frac{2.168 \ \frac{\mathrm{Pa\,L^2}}{\mathrm{g}} \times \left(2.9528 \times 10^4 \ \frac{\mathrm{g}}{\mathrm{mol}}\right)^2}{8.314 \times 10^3 \ \frac{\mathrm{Pa\,L}}{\mathrm{mol\,K}} \times 298.15 \ \mathrm{K}} = -762.4 \ \frac{\mathrm{L}}{\mathrm{mol}}$$

Note: When the protein (or other membrane-impermeable solute) has an appreciable net charge, and salts that can pass through the membrane are present, the analysis of osmotic equilibrium can be more complicated than discussed here as a result of the uneven partitioning of the ions. This phenomenon, referred to as Gibb-Donnan equilibrium, is discussed in Sec. 15.7.■

The virial coefficients in a gas are the result of the interactions between molecules that are separated by empty space. However, the osmotic virial coefficients here are the result of interactions between the solute molecules when they are separated by the solvent molecules (generally water) and, in the illustration above, potassium sulfate and whatever other ions are present to adjust the pH. So the values of the osmotic virial coefficients depend not only on the solute, but also on the solution conditions.

For example, in Problem 11.5-8 data are given for the osmotic pressure of α-chymotrypsin in water at pH = 8.0 (instead of 4.0 as above) in an 0.01 M potassium sulfate solution at 25°C. Those data are fit with $m_S = 22\,725$ g/mol and $B_2 = +257.8$ L/mol. The difference in the computed molecular weights at the two values of pH may be due to experimental uncertainties, though it is true that the charge on a protein varies with pH, and as we show in Chapter 15, the charge affects the osmotic pressure. The large change in the value of B_2 with pH, as well as its change in sign, is striking. Since a negative value of the osmotic virial coefficient indicates a net attraction between the solute molecules, what is seen here is a net attraction between the α-chymotrypsin molecules at pH = 4 and a net repulsion at pH = 8, which may be due to the change in charge of this protein as a function of pH. (see Chapter 15)

The discussion so far has been of the osmotic pressure resulting from a single solute in a solvent. If, instead, there are several solutes (proteins, polymers, salts, etc.), Eq. 11.5-4 still applies, though now the computation of the activity coefficient of the solvent may be more complicated because a multicomponent mixture is involved. However, if the solution is very dilute (so that the solvent activity coefficient is unity, or equivalently, the osmotic virial coefficients can be neglected), the analogue of Eq. 11.5-6 for a multicomponent mixture is

$$\Pi = \frac{RT}{\underline{V}_{\text{solvent}}} \sum_{\text{solutes i}} x_i = RT \sum_{\text{solutes i}} \frac{C_i}{m_i}$$

Osmotic pressure (times membrane area) can result in a significant force—indeed, one so large that it can cause the membrane to deform or even rupture. It is for this reason that any fluid used in medical applications (for example, kidney dialysis, blood replacement, organ flushing, or preserving fluids) must have approximately the same osmotic pressure as blood to ensure that the red blood cells do not burst. Empirically, it has been found that a solution that contains about 0.9 wt % sodium chloride is **isotonic** with, that is, has the same osmotic pressure as, blood.

ILLUSTRATION 11.5-4

Estimation of the Osmotic Pressure of Human Blood

Use the information in the paragraph above to estimate the osmotic pressure of blood.

SOLUTION

A 0.9 wt % solution of sodium chloride has 9 grams of NaCl per 1000 grams of solution that also contains 991 grams of water. Since the molecular weight of NaCl is 58.44, 9 grams of NaCl per 991 grams of water is equal to 9.0817 grams of NaCl per kg of water, or a molality of 0.1554

(0.1554 mol of NaCl in 55.51 mol of water). However, sodium chloride is a strong electrolyte and ionizes completely; therefore, the mole fraction of the solute water is

$$x_{\text{water}} = \frac{55.51}{2 \times 0.1554 + 55.51} = 0.994\,43$$

Assuming, for the moment, that the activity coefficient of water is unity, we have

$$\Pi = -\frac{8.314 \times 10^{-5} \,\dfrac{\text{bar m}^3}{\text{mol K}} \times 298.15 \text{ K}}{18 \times 10^{-6} \,\dfrac{\text{m}^3}{\text{mol}}} \ln(0.994\,43) = 7.694 \text{ bar}$$

However, in Appendix 9.3 we showed that the mean ionic activity coefficient in aqueous sodium chloride solutions is reasonably well correlated with

$$\ln \gamma_{\pm} = -\frac{\alpha \left| z_+ z_- \right| \sqrt{I}}{1 + \sqrt{I}} + 0.137 I$$

and provided the expression to compute the activity coefficient of water in such solutions. We obtain a more accurate estimate of the osmotic pressure of blood using this equation and find that $\gamma_{\text{water}} = 1.000\,13$, so that $x_{\text{water}} \gamma_{\text{water}} = 0.994\,57$. Using this value, we find $\Pi = 7.498$ bar.

COMMENT

We see from this illustration that the osmotic pressure of this sodium chloride solution, and therefore blood, is surprising high, above 7 bar, even though the solution is more than 99 percent water. Also, the activity coefficient of water is almost 1; however, in the case of the osmotic pressure, the activity coefficient correction should not be neglected. ∎

Finally, it should be pointed out that the Gibbs phase rule, as developed in Sec. 8.9, does not apply to osmotic equilibrium. This is because (1) the total pressure need not be the same in each phase (cell), so that Eq. 8.9-2 is not satisfied, and (2) the equality of partial molar Gibbs energies in each phase (cell) does not apply to all species, only to those that can pass through the membrane. A form of the Gibbs phase rule applicable to osmotic equilibrium can be easily developed (Problem 11.5-2).

PROBLEMS FOR SECTION 11.5

11.5-1 Show that the condition for equilibrium in a closed isothermal system, one part of which is maintained at P^{I} and the remainder at P^{II}, is that the function $G' = U - TS + P^{\text{I}}V^{\text{I}} + P^{\text{II}}V^{\text{II}}$ be a minimum (here V^{I} and V^{II} are the volumes of the portions of the system maintained at P^{I} and P^{II}, respectively). Then show that Eq. 11.5-1 is the condition for osmotic equilibrium.

11.5-2 Derive a form of the Gibbs phase rule that applies to osmotic equilibrium.

11.5-3 a. The osmotic coefficient of a solvent ϕ_S is defined by the relation

$$\bar{G}_S = \underline{G}_S + \phi_S RT \ln x_S$$

where $\phi_S \to 1$ as $x_S \to 1$. Develop an expression relating the activity coefficient γ_S to the osmotic coefficient ϕ_S.

b. Robinson and Wood[19] report the following interpolated values for the osmotic coefficient of seawater as a function of concentration at 25°C.

I	ϕ_S
0.5	0.9018
1.0	0.9155
2.0	0.9619
4.0	1.0938
6.0	1.2536

[19] R. A. Robinson and R. H. Wood, *J. Sol. Chem.*, **1**, 481 (1972).

Here I is the ionic strength, defined as

$$I = \frac{1}{2} \sum_i z_i^2 M_i$$

where z_i is the valence of the ith ionic species and M_i is its molality. For simplicity, seawater may be considered to be a mixture of only sodium chloride and water. Compute the water activity coefficient and the equilibrium osmotic pressure for each of the solutions in the table.

11.5-4 Bruno and coworkers [*Int. J. Thermophys.*, **7**, 1033 (1986)] describe an apparatus that contains a palladium-silver membrane that is permeable to hydrogen but not other gases. In their experiments after equilibration they measure the temperature T_1 in the apparatus, the pressure P_1 of pure hydrogen on one side of the semipermeable membrane, and the equilibrium composition of hydrogen x_{H_2} and other gases and pressure P_2 on the other side of the membrane. Next, by equating the hydrogen fugacities on both sides of the membrane,

$$f_{H_2}(T_1, P_1) = \bar{f}_{H_2}(T_1, P_2, x_{H_2})$$

and calculating the fugacity of the pure hydrogen using known virial coefficients from

$$\ln \frac{f_{H_2}}{P_1} = \frac{B P_1}{RT} + \frac{C - B^2}{2}\left(\frac{P_1}{RT}\right)^2$$

they have a direct measure of the fugacity of hydrogen in the mixture. Some of their results are given here:

Hydrogen + propane system at 3.45 MPa:

At 80°C

| $x_{H_2} =$ | 0.2801 | 0.4452 | 0.5935 | 0.7298 | 0.8215 |
| $\phi_{H_2} =$ | 1.283 | 1.106 | 1.058 | 1.028 | 1.033 |

At 130°C

| $x_{H_2} =$ | 0.2649 | 0.4715 | 0.5449 | 0.7827 | 0.8354 |
| $\phi_{H_2} =$ | 1.22 | 1.116 | 1.096 | 1.047 | 1.038 |

Hydrogen + methane system at 3.45 MPa:

At 80°C

| $x_{H_2} =$ | 0.2155 | 0.4594 | 0.5355 | 0.7901 | 0.8494 |
| $\phi_{H_2} =$ | 1.223 | 1.134 | 1.125 | 1.115 | 1.121 |

$$\phi_{H_2} = \bar{f}_{H_2}/x_{H_2} P$$

Compare these experimental results with the predicted fugacity coefficients for hydrogen in these mixtures calculated using the Peng-Robinson equation of state. Determine the sensitivity of the predictions to the value of the binary interaction parameter.

11.5-5 A rough rule of thumb in polymer solution theory is that a 4 molar aqueous polymer solution will have an osmotic pressure of approximately 100 bar. Is this rule of thumb in approximate agreement with Eq. 11.5-5?

11.5-6 Derive the form of the Gibbs phase rule that applies to a multicomponent system in osmotic equilibrium in which all but two of the components can pass through the membrane separating the two compartments.

11.5-7 Joe Udel lives on the second floor of a house that is adjacent to a well of pure water, but city water comes out of his indoor plumbing. He would rather have pure well water. So he has developed the following scheme. He will mount a pipe from the well up the side of his building into a tank in the third-floor attic. The bottom of the pipe will contain a membrane permeable to water, but not to the 1000-molecular-weight polymer he will add to the pipe (he is proposing to use a low-molecular-weight polymer so that the mixture viscosity will not be too high). He will add enough of the polymer so that water in the pipe will rise to the top of the tank, a height of 15 meters. The end of the pipe in the attic will have a second membrane of the same type at its base, so that water exiting the membrane will drip into a second tank open to the atmosphere. In this way the polymer will remain in the system, and Joe will have pure well water from the second tank. A diagram of the proposed process (with water dripping from tank 1 to tank 2) is shown.

a. How much polymer (in kg per kg water) is needed for this process?

b. Does the process violate the second law of thermodynamics? Will this process work indefinitely without any external power?

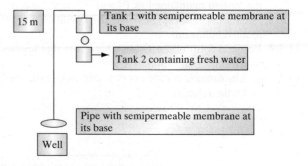

11.5-8 The following data have been reported for the osmotic pressure of α-chymotrypsin in water at pH = 8.0 in an 0.01 M potassium sulfate solution at 25°C.[20] Use these data to determine the molecular weight of α-chymotrypsin and the value of its osmotic second virial coefficient, B_2, at these conditions.

Data:

C_S (g/L)	Π (Pa)
0.901	58.7
1.841	121.2
4.070	271.3
5.558	376.9
7.211	476.7
9.298	615.0

---

[20]Reprinted with permission of C. A. Haynes, T. Tamura, H. R. Korfer, H. W. Blanch, and J. M. Prausnitz, *J. Phys. Chem.*, **96,** 905 (1992). Copyright 1992 American Chemical Society.

Chapter 12

Mixture Phase Equilibria Involving Solids

In this chapter we consider several other types of phase equilibria, mostly involving a fluid and a solid. This includes the solubility of a solid in a liquid, gas, and a supercritical fluid; the partitioning of a solid (or a liquid) between two partially soluble liquids; the freezing point of a solid from a liquid mixture; and the behavior of solid mixtures. Also considered is the environmental problem of how a chemical partitions between different parts of the environment. Although these areas of application appear to be quite different, they are connected by the same starting point as for all phase equilibrium calculations, which is the equality of fugacities of each species in each phase:

$$\bar{f}_i^{\mathrm{I}}(T, P, \underline{x}^{\mathrm{I}}) = \bar{f}_i^{\mathrm{II}}(T, P, \underline{x}^{\mathrm{II}})$$

Starting point for all phase equilibrium calculations

where the superscripts I and II represent the different phases.

INSTRUCTIONAL OBJECTIVES FOR CHAPTER 12

The goals of this chapter are for the student to:

- Be able to compute the solubility of a solute in a liquid, gas, or supercritical fluid (Sec. 12.1)
- Be able to compute the partitioning of a solid between two partially miscible liquids (Sec. 12.2)
- Be able to compute the freezing points of liquid mixtures (Sec. 12.3)
- Be able to compute the phase behavior of solid mixtures (Sec. 12.4)
- Be able to estimate the environmental distribution of a chemical (Sec. 12.5)

NOTATION INTRODUCED IN THIS CHAPTER

$\bar{f}_i^{\mathrm{S}}$ Fugacity of species i in a solid mixture (kPa)

$\bar{f}^{\mathrm{F}}$ Fugacity of species i in a fluid mixture (kPa)

$K_{\mathrm{OCW},i}$ Organic carbon-water partition coefficient of species i

$K_{\mathrm{AW},i}$ Air-water partition coefficient of species i

$K_{\mathrm{BW},i}$ Biota-water partition coefficient of species i

12.1 THE SOLUBILITY OF A SOLID IN A LIQUID, GAS, OR SUPERCRITICAL FLUID

We now want to consider the extent to which a solid is soluble in a liquid, a gas, or a supercritical fluid. (This last case is of interest for supercritical extraction, a new separation method.) To analyze these phenomena we again start with the equality of the species fugacities in each phase. However, since the fluid (either liquid, gas, or supercritical fluid) is not present in the solid, two simplifications arise. First, the equilibrium criterion applies only to the solid solute, which we denote by the subscript 1; and second, the solid phase fugacity of the solute is that of the pure solid.[1] Thus we have the single equilibrium relation

Equality of fugacities, the starting point for all phase equilibrium calculations

$$\boxed{f_1^S(T, P) = \bar{f}_1^F(T, P, \underline{x})} \tag{12.1-1}$$

where the superscripts S and F refer to the solid and fluid (liquid, gas, or supercritical fluid) phases, respectively.

We consider first the solubility of a solid in a liquid. Using Eq. 9.3-11 for the fugacity of the solute in a liquid, we obtain

$$f_1^S(T, P) = x_1 \gamma_1(T, P, \underline{x}) f_1^L(T, P) \tag{12.1-2}$$

where $f_1^S(T, P)$ and $f_1^L(T, P)$ refer to the fugacity of the pure species as a solid and as a liquid, respectively, at the temperature and pressure of the mixture, and x_1 is the saturation mole fraction of the solid solute in the solvent.

If the temperature of the mixture is equal to the normal melting temperature of the solid, T_m, then

$$f_1^S(T_m) = f_1^L(T_m) \tag{12.1-3}$$

by the pure-component phase equilibrium condition, so that at the melting point we have

$$x_1 = 1/\gamma_1(T_m, P, \underline{x}) \tag{12.1-4}$$

Thus, the solubility in a liquid of a solid at its melting point is equal to the reciprocal of its activity coefficient in the solute–solvent mixture.

If, as is usually the case, the solid is below its melting point, $f_1^L > f_1^S$, and Eq. 12.1-4 is not valid. To predict the solubility in this case, Eq. 12.1-2 must be used with some estimate for the ratio f_1^S/f_1^L. One way to make this estimate is to use the sublimation pressure for f_1^S (see Eq. 7.4-24), and then compute the fugacity for the "subcooled" liquid f_1^L by extrapolation of the liquid thermodynamic properties into the solid region. This can be done graphically as indicated in Fig. 9.7-2b if the temperature is not too far below the melting point. Alternatively, and most accurately, if the heat (enthalpy) of fusion at the melting point is available, the fugacity ratio can be computed from Eq. 9.7-8a:

$$f_1^L(T, P)$$

$$= f_1^S(T, P) \exp\left[\frac{1}{RT}\left[\triangle_{\text{fus}}\underline{H}(T)\left(1 - \frac{T}{T_m}\right) + \int_{T_m}^T \triangle C_P \, dT - T\int_{T_m}^T \frac{\triangle C_P}{T} dT\right]\right]$$

$$\tag{9.7-8a}$$

---

[1] In a true solid mixture use $\bar{f}_1^S(T, P, \underline{x}^S) = x_i^S \gamma_i^S(T, P, \underline{x}^S) f_1^S(T, P)$ for species 1 in the solid; see Sec. 12.4.

Using this result in Eq. 12.1-2 gives

**Equation for the
solubility of a solid in
a liquid**

$$\ln x_1 \gamma_1 = -\frac{\Delta_{\text{fus}}\underline{H}(T_m)}{RT}\left[1 - \frac{T}{T_m}\right] - \frac{1}{RT}\int_{T_m}^{T}\Delta C_{\text{P}}\,dT + \frac{1}{R}\int_{T_m}^{T}\frac{\Delta C_{\text{P}}}{T}\,dT$$

$$(12.1\text{-}5)$$

Equation 12.1-5 is the basic equation for predicting the saturation mole fraction of a solid in a liquid. This equation is almost exact.[2]

Two approximations can be made in Eq. 12.1-5 without introducing appreciable error. First, we assume that ΔC_{P} is independent of temperature, so that Eq. 12.1-5 becomes

$$\ln x_1 \gamma_1 = -\left\{\frac{\Delta_{\text{fus}}\underline{H}(T_m)}{RT}\left[1 - \frac{T}{T_m}\right] + \frac{\Delta C_{\text{P}}}{R}\left[1 - \frac{T_m}{T} + \ln\left(\frac{T_m}{T}\right)\right]\right\} \qquad (12.1\text{-}6)$$

Next, since the melting-point temperature at any pressure and the triple-point temperature (T_t) are only slightly different for most solids (see Fig. 9.7-2), we can rewrite Eq. 12.1-6, without significant error, as

$$\ln x_1 = -\ln \gamma_1 - \left\{\frac{\Delta_{\text{fus}}\underline{H}(T_t)}{RT}\left[1 - \frac{T}{T_t}\right] + \frac{\Delta C_{\text{P}}}{R}\left[1 - \frac{T_t}{T} + \ln\left(\frac{T_t}{T}\right)\right]\right\} \qquad (12.1\text{-}7)$$

Further, if heat capacity data for either the (real) solid or (hypothetical) liquid are not available, one can assume that they are approximately equal, so that $\Delta C_{\text{P}} = 0$ and

$$\ln x_1 = -\ln \gamma_1 - \left\{\frac{\Delta_{\text{fus}}\underline{H}(T_t)}{RT}\left[1 - \frac{T}{T_t}\right]\right\} \qquad (12.1\text{-}8)$$

If the liquid mixture is ideal, so that $\gamma_1 = 1$, we have the case of ideal solubility of a solid in a liquid, and the solubility can be computed from only thermodynamic data ($\Delta_{\text{fus}}\underline{H}$ and ΔC_{P}) for the solid species near the melting point. For nonideal solutions, γ_1 must be estimated from either experimental data or a liquid solution model, for example, UNIFAC. Alternatively, the regular solution theory estimate for this activity coefficient is

$$RT \ln \gamma_1 = \underline{V}_1^{\text{L}}(\delta_1 - \delta_2)^2 \Phi_2^2 \qquad (12.1\text{-}9)$$

where δ_1 and δ_2 are the solubility parameters of the solute as a liquid and the liquid solvent, respectively. The difficulty in using this relation is that we must be able to estimate both the solubility parameter and the liquid molar volume for a species whose pure-component state is a solid at the mixture temperature.

Usually one can neglect the thermal expansibility of both solids and liquids, so that $\underline{V}^{\text{L}}$ can be taken to be the molar volume of the liquid at the normal melting point, or it can be computed from

$$\underline{V}_1^{\text{L}} = \underline{V}_1^{\text{S}} + \Delta_{\text{fus}}\underline{V} \qquad (12.1\text{-}10)$$

---

[2]The approximations that have been made are to neglect the Poynting pressure correction terms to the solid and liquid free energies (or fugacities), and to assume $(f/p) = 1$.

where $\Delta_{fus}\underline{V}$ is the molar volume change on fusion at the triple point of the solid. To compute the solubility parameter δ_1, where

$$\delta = \left(\frac{\Delta_{vap}\underline{U}}{\underline{V}^L}\right)^{1/2} \tag{12.1-11}$$

it is necessary to estimate $\Delta_{vap}\underline{U}$, the molar internal energy change on vaporization of the subcooled liquid. The enthalpy change on vaporization of the subcooled liquid is

$$\Delta_{vap}\underline{H}(T) = \Delta_{sub}\underline{H}(T) - \Delta_{fus}\underline{H}(T) \tag{12.1-12}$$

where $\Delta_{sub}\underline{H}(T)$ is the heat of sublimation of the solid at the temperature T; this quantity can either be found in tables of thermodynamic properties or estimated from sublimation pressure data using the Clausius-Clapeyron equation (Eq. 7.7-5a), and $\Delta_{fus}\underline{H}(T)$ is presumed to be available from experimental data. Thus

$$\begin{aligned} \Delta_{vap}\underline{U} &= \Delta_{vap}\underline{H} - P\Delta_{vap}\underline{V} \\ &= \Delta_{sub}\underline{H}(T) - \Delta_{fus}\underline{H}(T_m) - \Delta C_P(T - T_m) - RT \end{aligned} \tag{12.1-13}$$

where we have assumed that $P\Delta_{vap}\underline{V} = P(\underline{V}^V - \underline{V}^L) \approx P\underline{V}^V = RT$, and that ΔC_P is independent of temperature.

Equations 12.1-8 through 12.1-13 can be used in Eq. 12.1-7 to compute the saturation solubility for any regular solution solute-solvent pair.

ILLUSTRATION 12.1-1

Solubility of a Solid in a Liquid

Estimate the solubility of solid naphthalene in liquid *n*-hexane at 20°C.

Data:[3]

Naphthalene ($C_{10}H_8$, molecular weight = 128.19)
Melting point: 80.2°C
Heat of fusion: 18.804 kJ/mol
Density of the solid: 1.0253 g/cc at 20°C
Density of the liquid: 0.9625 g/cc at 100°C
Sublimation pressure of the solid:

$$\log_{10} P^{sub} \text{ (bar)} = 8.722 - \frac{3783}{T} \quad (T \text{ in K})$$

The heat capacities of liquid and solid naphthalene may be assumed to be equal.

SOLUTION

The solubility parameter and liquid molar volume for *n*-hexane are given in Table 9.6-1 as $\delta_2 = 7.3$ and $\underline{V}_2^L = 132$ cc/mol, respectively. Since the liquid molar volume of naphthalene given in the data is for a temperature 80°C higher than the temperature of interest, and the volume change on melting of naphthalene is small, the molar volume of liquid naphthalene below its melting temperature will be taken to be that of the solid; that is,

$$\underline{V}_1^L = \frac{128.19 \text{ g/mol}}{1.0253 \text{ g/cc}} = 125 \text{ cc/mol}$$

[3]*Reference:* R. C. Weast, ed., *Handbook of Chemistry and Physics,* 68th ed., Chemical Rubber Publishing Co., Cleveland (1987), pp. C-357, D-214.

The heat of sublimation of naphthalene is not given. However, we can compute this quantity from the vapor pressure curve of the solid and the Clausius-Clapeyron equation (Eq. 7.7-5a) by taking P to be equal to the sublimation pressure P^{sub}, $\Delta \underline{H}$ to equal the heat of sublimation, and setting $\Delta \underline{V} = \underline{V}^V - \underline{V}^S = \Delta_{sub}\underline{V} \cong RT/P^{sub}$. Thus

$$\frac{\Delta_{sub}\underline{H}}{RT^2} = \frac{d \ln P^{sub}}{dT} = 2.303 \frac{d \log_{10} P^{sub}}{dT} = +2.303 \frac{(3783)}{T^2}$$

and

$$\Delta_{sub}\underline{H} = 2.303(3783)(8.314 \text{ J/mol}) = 72\,434 \text{ J/mol}$$

Next,

$$\Delta_{vap}\underline{U} = \Delta_{sub}\underline{H} - \Delta_{fus}\underline{H} - RT = 72\,434 - 18\,804 - 8.314 \times 293.15 = 51\,193 \text{ J/mol}$$

so that

$$\delta_1 = \left(\frac{51\,193 \text{ J/mol}}{125 \text{ cc/mol} \times 4.184 \text{ J/cal}} \right)^{1/2} = 9.9 \text{ (cal/cc)}^{1/2}$$

Now using Eq. 12.1-7 with $\Delta C_P = 0$, and the regular solution expression for the activity coefficient, we obtain

$$\ln x_1 = -\frac{\underline{V}_1^L (\delta_1 - \delta_2)^2 \Phi_2^2}{RT} - \frac{\Delta_{fus} H(T_m)}{RT} \left(1 - \frac{T}{T_m} \right)$$

As a first guess, assume that x_1 will be small, so that

$$\Phi_2 = \frac{x_2 \underline{V}_2^L}{x_1 \underline{V}_1^L + x_2 \underline{V}_2^L} \approx 1$$

In this case

$$\ln x_1 = \frac{-125 \frac{cc}{mol} \times (9.9 - 7.3)^2 \frac{cal}{cc} \times 4.184 \frac{J}{cal}}{8.314 \frac{J}{mol \, K} \times 293.15 \text{ K}}$$

$$- \frac{18\,804 \frac{J}{mol}}{8.314 \frac{J}{mol \, K} \times 293.15 \text{ K}} \left(1 - \frac{293.15}{353.35} \right)$$

$$= -1.451 - 1.314 = -2.765$$

$$x_1 = 0.063$$

With such a large value for x_1 we must go back and correct the value of Φ_2 for the presence of the solute and repeat the computation. Thus

$$\Phi_2 = \frac{0.937 \times 132}{0.937 \times 132 + 0.063 \times 125} = 0.94$$

and

$$\ln x_1 = 1.282 - 1.314 = 2.596$$
$$x_1 = 0.0746$$

The results of the next two iterations are $x_1 = 0.0768$ and $x_1 = 0.0772$, respectively. This last prediction is in reasonable agreement with the experimental result of $x_1 = 0.09$.[4]

COMMENT

Note that had we assumed ideal solution behavior, $\gamma_1 = 1$ and $\ln \gamma_1 = 0$, so that

$$\ln x_1 = -1.314$$
$$x_1 = 0.269$$

which is a factor of 3 too large. ∎

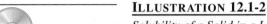

ILLUSTRATION 12.1-2

Solubility of a Solid in a Liquid Using UNIFAC

Repeat Illustration 12.1-1 using the UNIFAC group contribution model to estimate the naphthalene activity coefficient.

SOLUTION

Naphthalene has eight aromatic CH (ACH) and two aromatic C (AC) groups, and n-hexane has two CH_3 and four CH_2 groups. Since the output of the UNIFAC program is the list of activity coefficients, Eq. 12.1-12 is rewritten as

$$x_1 = \frac{\exp\left(-\left\{\dfrac{\Delta_{\mathrm{fus}}\underline{H}(T_t)}{RT}\left[1 - \dfrac{T}{T_t}\right] + \dfrac{\Delta C_{\mathrm{P}}}{R}\left[1 - \dfrac{T_t}{T} + \ln\left(\dfrac{T_t}{T}\right)\right]\right\}\right)}{\gamma_1} \tag{a}$$

or, using the same simplification as in the previous illustration,

$$x_1 = \frac{\exp\left(-\left\{\dfrac{\Delta_{\mathrm{fus}}\underline{H}(T_t)}{RT}\left[1 - \dfrac{T_t}{T}\right]\right\}\right)}{\gamma_1} = \frac{\exp(-1.314)}{\gamma_1} \tag{b}$$

Using $x_1 = 0.07$ as the first guess, we find $\gamma_1 = 3.2726$ and, from Eq. b, $x_1 = 0.0821$. Iterating several additional times yields $x_1 = 0.0856$, which is in very good agreement with the experimental value of 0.09 mentioned in the previous illustration. ∎

Although the discussion of this section has centered on the solubility of a solid in a pure liquid, the methods used can be easily extended to mixed solvents. In fact, to apply the equations developed in this section to mixed solvents, we need only recognize that the measured or computed value of γ_1, the activity coefficient for the dissolved solid, used in the calculations must be appropriate to the solid and mixed solvent combination being considered. In the regular solution model, for example, this means replacing Eq. 12.1-8 with

$$RT \ln \gamma_1 = \underline{V}_1^{\mathrm{L}} \left(\delta_1 - \bar{\delta}\right)^2 \tag{12.1-14}$$

Similarly, the UNIFAC model and program are applicable to both binary and multicomponent mixtures.

If the heat of fusion and the solubility of a solid in a liquid have been measured, this information can be used to compute the activity coefficient of the dissolved solid in the liquid. Sometimes the results can be quite surprising, as in the next illustration.

---

[4]G. Scatchard, *Chem. Rev.*, **8**, 329 (1931).

ILLUSTRATION 12.1-3

Calculation of Activity Coefficient from Data for the Solubility of a Solid

Haines and Sandler[5] reported the following data for benzo[*a*]pyrene and its solubility in water at 25°C:

Melting point: 178.1°C
Heat of fusion: 15.1 kJ/mol
Solubility in water: $x_{BP} = 3.37 \times 10^{-10}$

Estimate the activity coefficient of benzo[*a*]pyrene in water at 25°C.

SOLUTION

For this calculation we use Eq. 12.1-6 with the assumption that the contribution of the ΔC_P term is negligible (as we have no data for it). Rearranging Eq. 12.1-6, we have

$$
\gamma_{BP} = \frac{1}{x_{BP}} \exp\left\{ -\frac{\Delta_{fus}\underline{H}}{RT}\left[1 - \frac{T}{T_m} \right] \right\}
$$

$$
= \frac{1}{3.37 \times 10^{-10}} \exp\left\{ -\frac{15\,100\ \dfrac{J}{mol}}{8.314\ \dfrac{J}{mol\ K} \times (273.15 + 25)\ K}\left[1 - \frac{273.15 + 25}{273.15 + 178.1} \right] \right\}
$$

$$
= 3.76 \times 10^8
$$

COMMENT

Although one is accustomed to thinking of activity coefficients as being small corrections to ideal solution behavior, we see in this example of benzo[*a*]pyrene in water that the correction can be very large. This is especially important in environmental applications where one is interested in the distribution of a chemical between air, water, and soil, as will be seen in Sec. 12.5. Since the concentration of benzo[*a*]pyrene is so small, the value of the activity coefficient calculated above is effectively the activity coefficient at infinite dilution. ∎

To estimate the solubility of a solid in a gas, we again start from Eq. 12.1-1, which, using Eqs. 7.4-23 and 9.2-13, can be written as

$$
P_i^{sat}(T)\left(\frac{f}{P} \right)_{sat,T} \exp\left[\frac{\underline{V}_i^S (P - P_i^{sat})}{RT} \right] = y_i P\left(\frac{\bar{f}_i}{y_i P} \right) = y_i P \bar{\phi}_i \tag{12.1-15}
$$

where (in the Poynting factor) we have assumed the solid is incompressible. At low pressures, the Poynting factor and the fugacity coefficients in the solid and fluid phases can all be taken equal to unity. In this case we obtain the following expression for the ideal solubility of a solid in a gas:

$$
y_i^{ID} = \frac{P_i^{sat}(T)}{P} \tag{12.1-16}
$$

---

[5]R. I. S. Haines and S. I. Sandler, *J. Chem Eng. Data*, **40**, 833 (1995).

ILLUSTRATION 12.1-4

Solubility of a Solid in a Gas

Estimate the solubility of naphthalene in carbon dioxide at 1 bar and temperatures of 35.0 and 60.4°C using Eq. 12.1-16.

SOLUTION

Using the solid vapor pressure data in Illustration 12.1-1 and Eq. 12.1-16, we have

$T(°C)$	P_N^{sat} (bar)	y_N (at 1 bar pressure)
35.0	2.789×10^{-4}	2.8×10^{-4}
60.4	2.401×10^{-3}	2.4×10^{-3}

∎

ILLUSTRATION 12.1-5

Estimating the Solubility of a Solid in a Gas Using the Virial EOS

Estimate the solubility of naphthalene in carbon dioxide at 35°C and pressures ranging from 1 bar to 60 bar using the virial equation of state with the following values for the second virial coefficient

$$B_{CO_2\text{-}N} = -345 \text{ cc/mol}$$
$$B_{CO_2\text{-}CO_2} = -96.5 \text{ cc/mol}$$
$$B_{N\text{-}N} = -1850 \text{ cc/mol}$$

SOLUTION

We assume that CO_2 is insoluble in solid naphthalene, and therefore only equate the fugacities of naphthalene in the solid and vapor phases. The fugacity of solid naphthalene was calculated in Illustration 7.4-9. To calculate fugacity of naphthalene in the vapor phase, we use Eqs. 9.4-6 and 9.4-7. The procedure is to guess a vapor-phase mole fraction of naphthalene and then calculate its vapor-phase fugacity, which is compared with its solid-phase fugacity. If the two agree, the guessed composition is correct. If they do not, a new composition guess is made, and the calculation is repeated. The results are shown below.

P (bar)	$f_N^S = \bar{f}_N^V$ (bar)	y_N
2.789×10^{-4}	2.789×10^{-4}	1
1	2.803×10^{-4}	2.869×10^{-4}
10	2.933×10^{-4}	3.730×10^{-5}
20	3.083×10^{-4}	2.545×10^{-5}
30	3.241×10^{-4}	2.379×10^{-5}
40	3.408×10^{-4}	2.604×10^{-5}
50	3.582×10^{-4}	3.243×10^{-5}
60	3.766×10^{-4}	4.852×10^{-5}

Note that the solubility first decreases then increases with increasing total pressure at fixed temperature. ∎

At moderate and high pressures the solubility of a solid in a gas is computed from

Solubility of a solid in a gas using an EOS

$$y_i = \frac{P_i^{sat}(T)\left(\dfrac{f}{P}\right)_{sat,T}\exp\left[\dfrac{\underline{V}_i^S(P - P_i^{sat})}{RT}\right]}{P\bar{\phi}_i^V(T, P, \underline{y})} \qquad (12.1\text{-}17)$$

This equation usually must be solved by iteration since the solute mole fraction appears in the fugacity coefficient $\bar{\phi}_i^V(T, P, \underline{y})$. The vapor pressure of the solid is generally small, so that the term $(f/P)_{sat,T}$ is usually equal to unity. Common terminology is to define an ideal solubility y_i^{ID} and an **enhancement factor** E as

$$y_i^{ID} = \frac{P_i^{sat}}{P} \quad \text{and} \quad E(T, P, \underline{y}) = \frac{\left(\dfrac{f}{P}\right)_{sat,T}\exp\left[\dfrac{\underline{V}_i^S(P - P_i^{sat})}{RT}\right]}{\bar{\phi}_i^V(T, P, \underline{y})} \qquad (12.1\text{-}18a)$$

so that

$$y_i = \frac{P_i^{sat}(T)}{P}E(T, P, \underline{y}) = y_i^{ID} \times E(T, P, \underline{y}) \qquad (12.1\text{-}18b)$$

Note that the enhancement factor E has contributions from both the Poynting factor and the vapor-phase fugacity coefficient, both of which are important at high pressure, and that $E \to 1$ as $P \to P_i^{sat}$.

ILLUSTRATION 12.1-6

Solubility of a Solid in Supercritical Fluid (SCF) Using an EOS

McHugh and Paulaitis [*J. Chem. Eng. Data*, **25**, 326 (1980)] report the following data for the solubility of naphthalene in carbon dioxide at temperatures slightly above the CO_2 critical temperature and pressures considerably higher than its critical pressure.

$T = 35.0°C$		$T = 60.4°C$	
P (bar)	y_N	P (bar)	y_N
86.8	0.00750	108.4	0.00524
98.2	0.00975	133.8	0.01516
133.0	0.01410	152.5	0.02589
199.5	0.01709	164.2	0.04296
255.3	0.01922	192.6	0.05386
		206.0	0.06259

Assuming that the CO_2-naphthalene mixture obeys the Peng-Robinson equation of state with $k_{CO_2\text{-}N} = 0.103$, estimate the solubility of naphthalene in the CO_2 supercritical fluid (SCF). Also compute the predicted enhancement factors and the contribution of the Poynting factor to the enhancement factor.

SOLUTION

Using the data in Illustration 12.1-1 and Table 6.6-1, and one of the Peng-Robinson mixture programs discussed in Appendix B and on the CD-ROM to calculate the vapor-phase fugacities, the following results are obtained.

P (bar)	y_N	y_N^{ID} (Eq. 12.1-19)	E	Poynting Factor
$T = 35°C$:				
86.8	1.99×10^{-3}	3.21×10^{-6}	619	1.527
98.2	2.87×10^{-3}	2.84×10^{-6}	1 010	1.615
199.5	1.654×10^{-2}	1.40×10^{-6}	11 830	2.647
255.3	1.940×10^{-2}	1.09×10^{-6}	17 759	3.475
$T = 60.4°C$:				
108.4	2.80×10^{-3}	2.21×10^{-5}	126	1.630
133.8	1.05×10^{-2}	1.79×10^{-5}	584	1.828
152.5	1.96×10^{-2}	1.57×10^{-5}	1246	1.989
164.0	2.64×10^{-2}	1.46×10^{-5}	1806	2.096
192.6	5.25×10^{-2}	1.25×10^{-5}	4211	2.383
206.0	6.61×10^{-2}	1.17×10^{-5}	5671	2.531

The predicted and measured naphthalene mole fractions in supercritical carbon dioxide are plotted in Fig. 12.1-1.

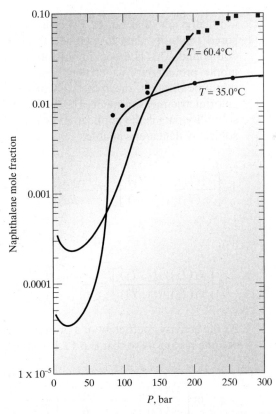

Figure 12.1-1 The solubility of naphthalene in supercritical carbon dioxide as a function of pressure. The points ● and ■ are the experimental data of McHugh and Paulaitis [*J. Chem. Eng. Data,* **25**, 326 (1980)] at $T = 35.0$ and $60.4°C$, respectively. The lines are the correlations of the data using the Peng-Robinson equation of state with $k_{CO_2\text{-}N} = 0.103$. Note the sharp increase in naphthalene solubility with pressure near the CO_2 critical pressure of 73.76 bar.

COMMENTS

1. Note that the predictions for these extreme conditions are good. Indeed, with only one adjustable parameter ($k_{CO_2-N} = 0.103$), reasonable predictions are obtained for the solubility of naphthalene in supercritical carbon dioxide for a range of temperatures in the near critical region, and to moderately high pressures.

2. The enhancement factors here are very large, in fact, among the larger nonideal corrections encountered in chemical engineering thermodynamics. (Enhancement factors for other mixtures at cryogenic conditions of 10^9 and larger have been reported.) Note that at $T = 35°C$ and $P = 255.3$ bar, the solubility of naphthalene is enhanced by a factor of more than 17 700 above its ideal value; however, its total solubility is still small at less than 2 mol %.

3. The solubility of naphthalene in supercritical carbon dioxide at $60.4°C$ increases from a mole fraction of 0.00240 at 1 bar (Illustration 12.1-4) to 0.098 at 291.3 bar. This illustrates the large increase in the solubility of a solute that may occur with increasing pressure, which is the basis of supercritical extraction to, for example, remove caffeine from coffee beans or fragrances and oils from plant material.

4. The Poynting corrections in this illustration are large, reaching values greater than 3. Consequently, the Poynting factor could not be ignored in this example. However, the main contribution to the enhancement factor arises from gas-phase nonidealities (the species fugacity coefficient, $\bar{\phi}_i$). ∎

Finally, note that Eq. 12.1-5 or its simplification, Eq. 12.1-8, can be used with measured solubility data to estimate the heat of fusion (or heat of crystallization) of a compound. This is especially useful for complex molecules and biomolecules (for example, proteins) for which the direct measurement of the heat of fusion in a calorimeter might be difficult, or, in the case of proteins, too little may be available for accurate calorimetric measurement. The basis for such a calculation of the heat of fusion from solubility data is the application of Eq. 12.1-8 at two different temperatures for which solubility data are available:

$$\ln\left[\frac{x_1(T_1)\gamma_1(\underline{x}, T_1)}{x_1(T_2)\gamma_1(\underline{x}, T_2)}\right] = -\frac{\triangle_{fus}\underline{H}_1}{RT_m}\left[\frac{T_m - T_1}{T_1} - \frac{T_m - T_2}{T_2}\right] = -\frac{\triangle_{fus}\underline{H}_1}{RT_m}\left[\frac{T_m}{T_1} - \frac{T_m}{T_2}\right]$$

or

$$\ln\left[\frac{x_1(T_1)\gamma_1(\underline{x}, T_1)}{x_1(T_2)\gamma_1(\underline{x}, T_2)}\right] = -\frac{\triangle_{fus}\underline{H}_1}{R}\left[\frac{1}{T_1} - \frac{1}{T_2}\right] = -\frac{\triangle_{fus}\underline{H}_1}{R}\left[\frac{T_2 - T_1}{T_1 T_2}\right] \quad \textbf{(12.1-19a)}$$

As the activity coefficient of a protein or other biomolecule may not be known, we will assume it cancels so that Eq. 12.1-19a can be simplified to

$$\ln\left[\frac{x_1(T_1)}{x_1(T_2)}\right] = -\frac{\triangle_{fus}\underline{H}_1}{R}\left[\frac{1}{T_1} - \frac{1}{T_2}\right] = -\frac{\triangle_{fus}\underline{H}_1}{R}\left[\frac{T_2 - T_1}{T_1 T_2}\right] \quad \textbf{(12.1-19b)}$$

in which case the heat of fusion that appears may be thought of as an apparent heat of fusion since the contribution of the activity coefficients and their temperature and composition dependence have been neglected.

ILLUSTRATION 12.1-7

Determination of the Heat of Fusion of Insulin

Insulin exists as a hexamer, both in solution and as a crystal. Bergeron et al.[6] report the solubility of insulin in aqueous solution to be about 0.122 mg/mL at 10°C and 0.182 mg/mL at 25°C, and the molecular weight of the insulin hexamer to be 34 800. Use these data to estimate the heat of fusion of the insulin hexamer.

SOLUTION

Since the molecular weight of the insulin hexamer is so high, its mole fraction at both temperatures is very small. For example, at the higher concentration of 0.182 mg/mL,

$$x = \frac{\frac{0.000182}{34800}}{\frac{0.000182}{34800} + \frac{1}{18}} = 9.414 \times 10^{-8}$$

Therefore, we can assume that the activity coefficients that appear in Eq. 12.1-19 at both temperatures are the values at infinite dilution and (assuming the temperature dependence is not very large) will cancel in the ratio term in the equation. Also, the solubilities S in the problem statement are so low that they are linearly related to the mole fraction by

$$x = \frac{S \times 18}{34\ 800}$$

so that the reported solubilities can be used in the ratio in Eq. 12.1-19, and we have

$$\ln\left[\frac{S_1}{S_2}\right] = -\frac{\triangle_{fus}\underline{H}}{R}\left[\frac{T_2 - T_1}{T_1 T_2}\right] \quad \text{or} \quad \triangle_{fus}\underline{H} = R\left[\frac{T_1 T_2}{T_1 - T_2}\right]\ln\left[\frac{S_1}{S_2}\right]$$

Consequently, for the crystallization of the insulin hexamer

$$\triangle_{fus}\underline{H} = 8.314\left[\frac{283.13 \times 298.15}{283.15 - 298.15}\right]\ln\left[\frac{0.122}{0.182}\right] = 18.7\ \frac{kJ}{mol}$$

Since the value of the heat of fusion (or crystallization) is positive, the solubility of insulin increases with increasing temperature. ∎

PROBLEMS FOR SECTION 12.1

12.1-1 Estimate the solubility of naphthalene in the following solvents at 20°C and compare with the experimental results.[7]

Solvent	Measured Solubility, x_N
Chlorobenzene	0.256
Benzene	0.241
Toluene	0.224
Carbon tetrachloride	0.205

12.1-2 Estimate the solubility of naphthalene in the mixed solvent *n*-hexane and carbon tetrachloride at 20°C as a function of the (initial) *n*-hexane concentration.

12.1-3 McHugh and Paulaitis (see Illustration 12.1-6) also measured the solubility of biphenyl in supercritical carbon dioxide. Some of their data appear here.

[6]L. Bergeron, L. F. Filobelo, O. Galkin, and P. G. Vekilov, *Biophys. J.*, **85**, 3935 (2003).

[7]Data for this problem were taken from G. Scatchard, *Chem. Rev.,* **8**, 329 (1931); and J. H. Hildebrand, J. M. Prausnitz, and R. L. Scott, *Regular and Related Solutions,* Prentice Hall, Englewood Cliffs, N.J. (1970), p. 152.

T (°C)	P (bar)	y_B
49.5	155.6	0.01782
49.5	204.5	0.02689
49.5	296.5	0.03605
49.5	379.4	0.03795
55.2	110.6	0.00447
55.2	132.6	0.01031
55.2	167.2	0.01829
55.2	252.5	0.03516
55.2	334.6	0.05615
55.2	412.8	0.07918
55.2	469.9	0.11054
55.2	482.7	0.12669
57.5	361.4	0.06365
57.5	430.3	0.09208

Choose a binary interaction parameter k_{CO_2-B} that gives a reasonable correlation for these data.

12.1-4 At moderate pressures, the virial equation of state can be used to estimate the solubility of a solid in a supercritical fluid.

a. Compute the solubility of naphthalene in carbon dioxide at 50°C and 60 bar using the data here:

At 50°C, the vapor pressure of naphthalene is 1.11×10^{-3} bar, its molar volume is 0.112 m³/kmol, and the virial coefficients are

$$B_{CO_2-CO_2} = -0.103 \text{ m}^3/\text{kmol}$$

$$B_{N-CO_2} = -0.405 \text{ m}^3/\text{kmol}$$

b. Repeat the computation in part (a) using the Peng-Robinson equation of state and compare the results.

12.1-6 The following data are available for the solubility of the crystalline amino acid leucine (molecular weight 131.17) in water at various temperatures:[8]

T (°C)	19.0	23.7	28.7	33.4	38.4
g/kg	21.20	21.54	22.19	22.81	23.81
T (°C)	43.5	48.4	53.4	58.8	
g/kg	24.88	26.03	27.63	28.84	

Determine the apparent heat of fusion of leucine over this temperature range.

12.1-7 The solubility of crystalline isoleucine in water is given in the table below (data from the same source as for the preceding problem). Its melting point has been reported to be 287°C, its heat of fusion has been estimated to be 5.825 kJ/mol, its molecular weight is 131.17, and there are no $\triangle_{fus}C_P$ data available.

T (°C)	19	30	40	50	58.4
g/kg	32.39	35.48	38.32	41.20	43.64

a. Determine the apparent heat of fusion of isoleucine over this temperature range.

b. Estimate the solubility of crystalline isoleucine in water over the temperature range from 19 to 58.4°C in g/kg, and compare your estimates with the experimental data.

12.1-8 An aqueous solution at 60°C contains leucine and isoleucine at concentrations of 28 g and 35 g per kg of water, respectively. This solution is to be cooled to crystallize the leucine, but not the isoleucine. To what temperature can the mixture be cooled before the isoleucine crystallizes, and what fraction of the leucine will be recovered in this crystallization process? Note that since the mole fractions of leucine and isoleucine are so low in this mixture, it can be assumed that each of these amino acids will crystallize independent of the other. Solubility data for leucine and isoleucine are given in the two preceding problems.

12.1-9 Nitrofurantoin is an antibiotic that is used to eliminate bacteria that cause urinary tract infections. The following data have been reported for its solubility in water[9]

T (°C)	24	30	37	45
$x_N \times 10^6$	6.01	8.57	13.16	18.99

Estimate the apparent heat of fusion of nitrofurantoin.

12.2 PARTITIONING OF A SOLID SOLUTE BETWEEN TWO LIQUID PHASES

In Sec. 11.2 we considered the distribution of a liquid or gaseous solute between two partially miscible or completely immiscible phases. Here we consider the case in which the solute is a solid at the temperature and pressure of the mixture. We will assume, as

[8]J. Givand, A. S. Teja, and R. W.Rousseau, *AIChE J.* **47**, 2705 (2001).
[9]L.-K. Chen, D. E. Cadwallader, and H. W. Jun, *J. Pharm. Sci.*, **65**, 868 (1976).

is generally the case, that some or all of the solid solute dissolves in the liquid solvents, but the solvents are not present in the undissolved solid, if any remains.

The starting point for the description of this phase equilibrium problem is again the equality of fugacities of each species in all phases in which that species appears,

Starting point for all phase equilibrium calculations

$$\boxed{\bar{f}_i^{I}(T, P, \underline{x}^I) = \bar{f}_i^{II}(T, P, \underline{x}^{II}) = \cdots}$$

(12.2-1)

and the constraint that the number of moles of each species is conserved:

$$N_{i,F} = N_i^{I} + N_i^{II} + \cdots$$

(12.2-2)

where N_i^{J} is the number of moles of species i in phase J, and $N_{i,F}$ is the number of moles of species i in the feed into the system.

Here it is useful to consider two cases: one case in which no undissolved solute remains, and a second in which there is undissolved solid in equilibrium with the two liquid phases. In the case in which no undissolved solid remains, the equilibrium condition for the solute and the two solvents is

$$\bar{f}_i^{I}(T, P, \underline{x}^I) = \bar{f}_i^{II}(T, P, \underline{x}^{II})$$

(12.2-3)

Now, using the definition of the activity coefficient,

$$\bar{f}_1(T, P, \underline{x}) = x_1 \gamma_1(T, P, \underline{x}) f_1^{L}(T, P)$$

Eq. 12.2-3 can be rewritten as

$$x_1^{I} \gamma_1^{I}(T, P, \underline{x}^I) = x_1^{II} \gamma_1^{II}(T, P, \underline{x}^{II})$$

(12.2-4)

since the pure-component fugacities cancel on both sides (for the solvents, which are real liquids, and for the solute, which is a hypothetical liquid). Equation 12.2-4 is then solved for the solvents and solutes simultaneously to obtain the compositions of both phases. As in Sec. 11.2, we can define the distribution coefficient or partition coefficient for the solute as

$$K_x = \frac{x_1^{I}}{x_1^{II}} = \frac{\gamma_1^{II}(T, P, \underline{x}^{II})}{\gamma_1^{I}(T, P, \underline{x}^I)}$$

(12.2-5)

so that again we see that the distribution coefficient for solute mole fractions is equal to the reciprocal of the ratio of the solute activity coefficients in the two phases. Thus, given activity coefficient information for the solute in the two phases, one can compute the distribution of the solvent among the phases, or, given information about distribution of the solute, one can compute the ratio of the solute activity coefficients.

The situation is slightly more complicated when there is some undissolved solute in equilibrium with two partially miscible solvents. In this case the equilibrium condition for the solvents (species 2, 3, . . .) is

$$\bar{f}_j^{I}(T, P, \underline{x}^I) = \bar{f}_j^{II}(T, P, \underline{x}^{II}) \quad \text{solvents } j = 2, 3, \ldots$$

(12.2-6a)

while for the solute species 1,

$$\bar{f}_1^{I}(T, P, \underline{x}^I) = \bar{f}_1^{II}(T, P, \underline{x}^{II}) = f_1^{S}(T, P)$$

(12.2-6b)

$$f_1^S(T, P) = \bar{f}_1^I(T, P, \underline{x}^I) = \bar{f}_1^{II}(T, P, \underline{x}^{II}) \qquad \textbf{(12.2-6c)}$$

since the undissolved solute must be in equilibrium with both liquid phases, which also implies that both liquid phases are saturated with the solute. Now using the activity coefficient formalism, Eq. 9.3-11, we obtain the following equations.

For the solid solute,

$$\frac{f_1^S(T, P)}{f_1^L(T, P)} = x_1^I \gamma_1(T, P, \underline{x}^I) = x_1^{II} \gamma_1(T, P, \underline{x}^{II}) \qquad \textbf{(12.2-7a)}$$

or

$$\ln \gamma_1^I x_1^I = \ln \gamma_1^{II} x_1^{II} = -\frac{\Delta_{\text{fus}}\underline{H}(T_m)}{RT}\left[1 - \frac{T}{T_m}\right] - \frac{1}{RT}\int_{T_m}^{T} \Delta C_P \, dT + \frac{1}{R}\int_{T_m}^{T}\frac{\Delta C_P}{T}\, dT$$

$$\textbf{(12.2-7b)}$$

and for the solvents,

$$x_j^I \gamma_j(T, P, \underline{x}^I) = x_j^{II} \gamma_j(T, P, \underline{x}^{II}) \quad j = 2, 3, \ldots \qquad \textbf{(12.2-8)}$$

where ΔC_P is usually set to zero.

These equations can be difficult to solve, first because of the complicated dependence of the activity coefficients on the mole fractions (these equations are nonlinear), and second because even in the simplest case of a single solute in a mixture of two partially miscible solvents, there are six coupled equations to be solved. This is best done on a computer with equation-solving software. Also, as mentioned in Sec. 11.2, the presence of a solute may result in the two solvents becoming more soluble (salting in) or less soluble (salting out) in each other.

Before we leave the subject of the partitioning of a substance between two liquid phases, it is useful to review the three situations that can arise.

Case 1: No undissolved solute. This is the usual liquid-liquid phase equilibrium calculation for all species, that is,

$$\bar{f}_i^I(T, P, \underline{x}^I) = \bar{f}_i^{II}(T, P, \underline{x}^{II}) \quad \text{or} \quad x_i^I \gamma_i(T, P, \underline{x}^I) = x_i^{II} \gamma_i(T, P, \underline{x}^{II})$$

In this case the calculation of the state of phase equilibrium is as in Sec. 11.2.

Case 2: The undissolved solute is a pure liquid. Here the liquid-liquid phase equilibrium calculation for the solvents is

$$\bar{f}_i^I(T, P, \underline{x}^I) = \bar{f}_i^{II}(T, P, \underline{x}^{II}) \quad \text{or} \quad x_i^I \gamma_i(T, P, \underline{x}^I) = x_i^{II} \gamma_i(T, P, \underline{x}^{II})$$

For the solute, however, the equation to be solved is

$$\bar{f}_S^I(T, P, \underline{x}^I) = \bar{f}_S^{II}(T, P, \underline{x}^{II}) = f_S(T, P) \quad \text{or} \quad x_S^I \gamma_S(T, P, \underline{x}^I) = x_S^{II} \gamma_S(T, P, \underline{x}^{II}) = 1$$

Case 3: The undissolved solute is a pure solid. Here the liquid-liquid phase equilibrium calculation for the solvents is

$$\bar{f}_i^I(T, P, \underline{x}^I) = \bar{f}_i^{II}(T, P, \underline{x}^{II}) \quad \text{or} \quad x_i^I \gamma_i(T, P, \underline{x}^I) = x_i^{II} \gamma_i(T, P, \underline{x}^{II})$$

However, for the solutes we use

$$\bar{f}_S^I\left(T, P, \underline{x}^I\right) = \bar{f}_S^{II}\left(T, P, \underline{x}^{II}\right) = f_S^S\left(T, P\right)$$

or

$$x_S^I \gamma_S\left(T, P, \underline{x}^I\right) = x_S^{II} \gamma_S\left(T, P, \underline{x}^{II}\right) = \frac{f_S^S\left(T, P\right)}{f_S^L\left(T, P\right)}$$

This is the calculation considered in this section in which the liquid-solid fugacity ratio is calculated using Eq. 12.3-2.

The remaining question is whether the concentration of the solute is sufficiently high to affect the mutual solubilities of the solvents. If not, the liquid-liquid equilibrium of the solvents is first calculated, and the solvent concentrations in each of the phases are then calculated with the solvent concentrations fixed. However, if it is likely that the solute concentration will affect the mutual solubility of the solvents, then all the equations must be solved simultaneously for all the species present.

PROBLEM FOR SECTION 12.2

12.2-1 Write the equations that are to be used to compute the equilibrium compositions when three liquids are mixed and form
a. Two liquid phases

b. Two liquid phases in which species 1 is distributed among completely immiscible species 2 and 3
c. Three partially miscible liquid phases

12.3   FREEZING-POINT DEPRESSION OF A SOLVENT DUE TO THE PRESENCE OF A SOLUTE; THE FREEZING POINT OF LIQUID MIXTURES

If a small amount of solute (gas, liquid, or solid) dissolves in a liquid solvent and the temperature of the mixture is lowered, a temperature T_f is reached at which the pure solvent begins to separate out as a solid. This temperature is lower than the freezing (or melting) point of the pure solvent, T_m. Here we are interested in estimating the depression of the solvent freezing point, $\Delta T = T_m - T_f$, due to the addition of the solute. This equilibrium state is, in a sense, similar to the one encountered in the solubility of a solid in a liquid in that there we were interested in the equilibrium between a solid and a dilute liquid mixture of that component, whereas here we are interested in the equilibrium between a solid and a liquid mixture that is concentrated in that component.

The starting point for the analysis of the freezing-point depression phenomenon is the observation that when a solid freezes out from a mixture, it generally is pure and it is the component that is concentrated in the solution, that is, the pure solvent. [Indeed, this is the basis for zone refining, in which melting and resolidification (or recrystallization) are used to purify metals.] The equilibrium condition when the first crystal of pure solvent forms is

Starting point for phase equilibrium calculation

$$\boxed{f_1^S(T_f, P) = \bar{f}_1^L(T_f, P, \underline{x}) = x_1 \gamma_1(T_f, P, \underline{x}) f_1^L(T_f, P)} \qquad \textbf{(12.3-1)}$$

where x_1 is the original liquid solution mole fraction of the component that precipitates out as a solid. In writing this equation we have assumed that the solid phase is pure solvent, and used the fact that the composition of the liquid phase is not significantly changed by the appearance of a very small amount of solid phase (in this regard the analysis here resembles that for dew point and bubble point phenomena considered in Chapter 10).

Combining Eq. 12.3-1 with Eq. 12.1-9 for $\ln[f_1^S(T_f, P)/f_1^L(T_f, P)]$ yields

General equation for the freezing point depression of a solvent

$$
\begin{aligned}
\ln x_1 \gamma_1 &= \ln \frac{f_1^S(T_f, P)}{f_1^L(T_f, P)} \\
&= -\frac{\Delta_{\text{fus}}\underline{H}(T_m)}{R}\left[\frac{T_m - T_f}{T_m T_f}\right] - \frac{1}{RT_f}\int_{T_m}^{T_f} \Delta C_P\, dT + \frac{1}{R}\int_{T_m}^{T_f} \frac{\Delta C_P}{T}\, dT \\
&= -\frac{\Delta_{\text{fus}}\underline{H}(T_m)}{R}\left[\frac{T_m - T_f}{T_m T_f}\right] - \frac{\Delta C_P}{R}\left[1 - \frac{T_m}{T_f} + \ln\left(\frac{T_m}{T_f}\right)\right]
\end{aligned}
$$

(12.3-2)

where $\Delta_{\text{fus}}\underline{H}(T_m)$ is the heat of fusion of the solid solvent at its normal melting-point temperature T_m. To obtain the last form of Eq. 12.3-2, we have assumed that ΔC_P is independent of temperature.

In many instances, T_m and T_f are sufficiently close that Eq. 12.3-2 can be simplified to

$$
\begin{aligned}
\ln x_1 \gamma_1 &\cong -\frac{\Delta_{\text{fus}}\underline{H}(T_m)}{R}\left(\frac{T_m - T_f}{T_m T_f}\right) \\
&\cong -\frac{\Delta_{\text{fus}}\underline{H}(T_m)\underline{H}}{RT_m^2}(T_m - T_f)
\end{aligned}
$$

(12.3-3)

so that

$$
\Delta T = T_m - T_f = -\frac{RT_m^2}{\Delta_{\text{fus}}\underline{H}(T_m)}\ln(x_1 \gamma_1)
$$

(12.3-4)

For the case of very dilute solutes we expect that $\gamma_1 \sim 1$ and $\ln x_1 = \ln(1 - x_2) \sim -x_2$, so that the freezing-point depression equation may be further simplified to

Simplified equation for a freezing-point depression of a solvent

$$
\Delta T = T_m - T_f \cong \frac{RT_m^2}{\Delta_{\text{fus}}\underline{H}(T_m)} x_2
$$

(12.3-5)

In the case of dilute mixed solutes the equation becomes

$$
\Delta T = T_m - T_f = \frac{RT_m^2}{\Delta_{\text{fus}}\underline{H}(T_m)}\sum_{i=2}^{\mathcal{C}} x_i
$$

(12.3-6)

where the sum, which extends from 2 to $\mathcal{C}$, is over all solute species.

ILLUSTRATION 12.3-1

Calculation of the Freezing-Point Depression of Water

Determine the freezing-point depression of water as a result of the addition of 0.01 g/cm³ of (a) methanol and (b) a protein whose molecular weight is 60 000.

SOLUTION

Since such a small amount of solute is involved, we will assume that the density of the solution is the same as that of pure water, 1 g/cm^3. Thus the solute mole fraction is

$$x_{solute} = \frac{\text{Moles of solute in 1 cm}^3 \text{ of solution}}{(\text{Moles of solute} + \text{moles of water}) \text{ in 1 cm}^3 \text{ of solution}}$$

$$= \frac{\dfrac{0.01}{m}}{\dfrac{0.99}{18} + \dfrac{0.01}{m}}$$

where m is the molecular weight of the solute. $\Delta_{fus}\underline{H}(T_m)$ for water is 6025 J/mol.

a. The molecular weight of methanol is 32. Thus, $x_{solute} = 0.005\,65$ and

$$\Delta T = \frac{8.314 \, \dfrac{J}{mol \, K} \times (273.15 \text{ K})^2 \times 5.65 \times 10^{-3}}{6025 \text{ J/mol}} = 0.58 \text{ K}$$

b. The molecular weight of the protein is 60 000, so $x_{solute} \sim 3 \times 10^{-6}$ and

$$\Delta T = \frac{8.314 \, \dfrac{J}{mol \, K} \times (273.15 \text{ K})^2 \times 3 \times 10^{-6}}{6025 \text{ J/mol}} = 3.1 \times 10^{-4} \text{ K}$$

This small depression of the freezing point is barely measurable. Compare this result with the osmotic pressure difference found in Illustration 11.5.1. ∎

One interesting application of Eq. 12.3-2 is the computation of the freezing point of a general binary liquid mixture, that is, a liquid mixture in which neither component is easily recognized to be either the solvent or the solute. This is considered in Illustration 12.3-2.

ILLUSTRATION 12.3-2

Calculation of Freezing-Point Depression and the Eutectic Point

Compute the freezing-point temperature versus composition curve for an ethyl benzene–toluene mixture. The physical properties for this system are given in the table.

	Toluene	Ethyl Benzene
δ (cal/cc)$^{1/2}$ at 25°C	8.9	8.8
$\underline{V}^L$ (cc/mol) at 25°C	107	123
Normal melting point (K)	178.16	178.2
Heat of fusion (J/mol)	6610.7	9070.9
C_P^S (J/mol K)	87.0	105.9
C_P^L (J/mol K)	135.6	157.4

SOLUTION

The solubility parameters for toluene and ethyl benzene are nearly equal, so that the liquid mixture can be considered to be ideal (γ_1, $\gamma_2 = 1$). Since we do not know a priori whether pure toluene or pure ethyl benzene will freeze out as the solid component from a given mixture, the calculational procedure we will follow is to first assume that toluene appears as the solid component and compute the freezing-point temperature T_f of all toluene–ethyl benzene mixtures.

The calculations will then be repeated assuming ethyl benzene appears as the solid phase. The freezing point of a given mixture, and the solid that appears, will then be determined by noting which of the two calculations leads to the higher freezing point, as it is that solid that will freeze out first.

If we assume that toluene freezes out, T_f is found from the solution of

$$\ln x_1 = -\frac{6610.7}{8.314}\left(\frac{178.16 - T_f}{178.16\, T_f}\right) - \frac{48.6}{8.314}\left(1 - \frac{178.16}{T_f} + \ln\frac{178.16}{T_f}\right)$$

where x is the mole fraction of toluene. Since ΔC_P is so large, none of the terms in this equation can be neglected. The results are given in the following table. Similarly, if we assume that the solid precipitate is ethyl benzene, the mixture freezing point is computed from

$$\ln x_2 = -\frac{9070.9}{8.314}\left(\frac{178.2 - T_f}{178.2\, T_f}\right) - \frac{51.5}{8.314}\left(1 - \frac{178.2}{T_f} + \ln\frac{178.2}{T_f}\right)$$

where x is now the mole fraction of ethyl benzene. The results of this calculation are also given in the following table. Both sets of freezing points are plotted in Fig. 12.3-1.

	T_f (K)	
x_T	Toluene as the Solid	Ethyl Benzene as the Solid
1.0	178.16	
0.9	174.0	122.2
0.8	169.4	137.0
0.7	164.3	146.4
0.6	158.6	153.4
0.5	151.9	159.1
0.4	144.0	163.9
0.3	134.2	168.1
0.2	121.0	171.8
0.1	99.8	175.1
0		178.2

■

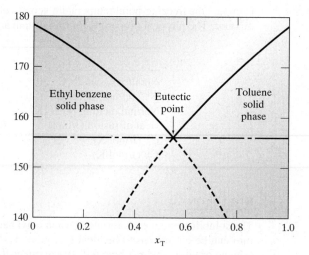

Figure 12.3-1 The liquid-solid phase diagram for ethyl benzene–toluene mixtures.

From the figure it is evident that below 55.8 mol % toluene, ethyl benzene precipitates as the solid phase, whereas at higher concentrations toluene is the solid precipitate. Furthermore, the minimum freezing point of the mixture is 155.9 K, which occurs at $x_T = 0.558$. This point is called the **eutectic point** of the mixture. At the eutectic temperature all the remaining liquid mixture solidifies into a mixed solid of the same composition (55.8 mol % toluene) as the liquid from which it was formed.

Although the phase diagram for the toluene–ethyl benzene mixture was constructed by determining the temperatures at which the first minute amount of solid appears in liquid mixtures of various compositions, it may now be used to analyze the complete solidification behavior of a liquid mixture at any composition. For example, suppose 1 mole of a 20 mol % toluene mixture is cooled. From the phase diagram we see that at 171.8 K a minute amount of pure ethyl benzene appears as the solid. As the temperature is lowered, more pure ethyl benzene precipitates out, and the remaining liquid becomes increasingly richer in toluene. The temperature and composition of this liquid, which is in equilibrium with the solid, follows the freezing-point curve in the figure; the relative amounts of the liquid and solid phases can be found by a species mass balance equation. Thus, at $T = 163.9$ K, the liquid contains 40 mol % toluene. Letting L represent the number of moles of liquid in equilibrium with the solid, and writing a mass (mole) balance on toluene using as a basis 1 mole of the initial mixture, we find

$$(0.2)(1) = 0.4L$$

or

$$L = 0.5 \text{ mol}$$

So when the temperature of the mixture is 163.9 K, half the original 20 mol % toluene mixture is present as pure, solid ethyl benzene, and the remainder appears as a liquid enriched in toluene.

As the temperature is lowered further, additional pure ethyl benzene precipitates out and the liquid becomes richer in toluene, until the eutectic composition and temperature is reached. At this point all the remaining liquid solidifies as two mixed solid phases.

It is interesting to compare Eq. 12.3-4 for the freezing point depression,

$$\Delta T = T_m - T_f = -\frac{RT_m{}^2}{\Delta_{\text{fus}}\underline{H}(T_m)} \ln(\gamma_1 x_1) \tag{12.3-4}$$

with Eq. 11.5-4 for the osmotic pressure of a solution:

$$\Pi = P^{\text{II}} - P^{\text{I}} = -\frac{RT}{\underline{V}^{\text{L}}_{\text{solvent}}} \ln(x^{\text{II}}_{\text{solvent}} \gamma^{\text{II}}_{\text{solvent}}) \tag{11.5-4}$$

Combining these two equations, we get

$$\Pi = \frac{\Delta_{\text{fus}}\underline{H}(T_m)}{\underline{V}_{\text{solvent}}} \frac{T_f}{T_m^2} (T_m - T_f)$$

which is strictly applicable if the osmotic pressure has been measured at the freezing point.

The interesting characteristic of this equation is that it does not involve any information about the solute. Therefore, two solutions involving the same solvent, but different solutes, that have the same osmotic pressure will have the same freezing-point depression, and vice versa.

ILLUSTRATION 12.3-3

Estimation of the Freezing-Point Depression of Blood

Use the information in Illustration 11.5-4 on the osmotic pressure of an aqueous sodium chloride solution and blood to estimate the freezing point of blood.

SOLUTION

To estimate the freezing point of blood, we will use the estimate of the osmotic pressure of blood at 25°C, 7.498 bar, computed in Illustration 11.5-4. While the freezing point of blood or the aqueous sodium chloride solution is not known, let us assume it is about the same as the freezing point of water, 273.15 K. The osmotic pressure of the sodium chloride solution at this temperature can be simply calculated from the value at 298.15 K by assuming the activity coefficient is independent of temperature, so that

$$\Pi\,(T = 273.15\text{ K}) = \Pi\,(T = 298.15\text{ K}) \times \frac{273.15\text{ K}}{298.15\text{ K}} = 6.869\text{ bar}$$

Then

$$6.869\text{ bar} \times 10^5\,\frac{\text{Pa}}{\text{bar}} \times \frac{\text{kg}}{\text{m s}^2\,\text{Pa}} = \frac{6025\,\frac{\text{J}}{\text{mol}} \times 1\,\frac{\text{kg}}{\text{s}^2\,\text{m}}}{18 \times 10^{-6}\,\frac{\text{m}^3}{\text{mol}}}\frac{T_f}{(273.15\text{ K})^2}(273.15 - T_f)$$

which has the solution

$$T_f = 272.481\text{ K} = -0.67°\text{C}$$

which is in good agreement with the experimental value of −0.52°C. Note also that the small amount of solute (0.9 wt % NaCl) results in a very large osmotic pressure, but a relatively small freezing-point depression. ∎

PROBLEMS FOR SECTION 12.3

12.3-1 The addition of a nonvolatile solute, such as a protein or a salt, to a pure solvent will raise its normal boiling point as well as lower its freezing point. Develop an expression relating the boiling-point elevation of a solvent to the solute concentration, heat of vaporization, and normal boiling temperature.

12.3-2 Chemical additives are used in automotive cooling systems to lower the freezing point of the water used as the engine coolant. Estimate the number of grams of methanol, ethanol, and glycerol that, when separately added to 1 kg of pure water, will lower its freezing point to −12°C.

12.3-3 Ethylene glycol, which forms nonideal mixtures with water, is used as an antifreeze to lower the freezing point of water in automotive engines in the winter. Using the UNIFAC model to estimate activity coefficients, develop a figure similar to Fig. 12.3-1 for the water–ethylene glycol system.

    Data: The heat of fusion for water is 6003 J/mol, and that of ethylene glycol is 10 998 J/mol.

12.3-4 Figure 12.3-1 shows the freezing-point curves for an ideal mixture with components with approximately equal melting points. Draw a freezing-point diagram for a mixture in which components 1 and 2 have freezing points of 185 K and 200 K, respectively, and for which the excess Gibbs energy of mixing is described by the one-constant Margules equation with $A = 3250$ J/mol. Assume that the heats of fusion of both components are 10 000 J/mol. (*Hint:* Does this mixture exhibit liquid-liquid equilibrium?)

12.3-5 The following table gives the boiling point of liquid mixtures containing x grams of tartaric acid ($C_4H_6O_6$, molecular weight = 150) per 100 g of water at 1.013 bar, as well as the vapor pressure of pure water at the boiling-point temperatures. The vapor pressures of liquid water and ice are given in Problem 7.12.

Boiling point (°C)	105	110	115
x (g/100 g H_2O)	87.0	177.0	272.0
Vapor pressure of pure water (bar)	1.1848	1.4050	1.6580

    a. What is the activity coefficient of water in each of these solutions?

b. Assuming that the activity coefficients (be they unity or not) are independent of temperature, calculate the freezing-point depression for each of the mixtures in the table.

12.3-6 McHugh and Paulaitis [*J. Chem. Eng. Data*, **25**, 326 (1980)], while measuring the solubilities of solid naphthalene in supercritical carbon dioxide at elevated pressures, frequently observed that the solid would melt at temperatures well below its normal melting temperature due to the high solubility of CO_2 in liquid naphthalene. Estimate the pressure at which naphthalene will melt in the presence of supercritical CO_2 at 65°C assuming that an ideal liquid mixture is formed.

Data: The Henry's law constant for CO_2 in liquid naphthalene at 65°C is 50 MPa and is independent of pressure. The normal melting point of naphthalene is 80.2°C, its heat of fusion is 18.804 kJ/mol, its critical temperature is 304.2 K, and its critical pressure is 7.376 MPa.

12.3-7 In Problem 11.2-26 it was observed that an equimolar mixture of liquid oxygen and liquid propane has an upper critical solution temperature of 112 K. You were asked to compute the liquid-liquid equilibrium phase boundary for this system as a function of temperature assuming the one-constant Margules parameter for this system is independent of temperature. However, on further study it was found that the melting point of pure oxygen is 64.4 K and that propane melts at 85.5 K. Therefore, recompute the phase diagram for the oxygen + propane system as a function of temperature, including solid-liquid and liquid-liquid equilibria. The heat of fusion of oxygen is 4448 J/mol and that of propane is 3515 J/mol.

12.3-8 Calculate the freezing point of water in a sodium chloride solution as a function of the sodium chloride molality over the range from 0 to 10 molal NaCl.

12.3-9 It is necessary to determine the molecular weight of a soluble, but essentially involatile component. The methods that can be used include dissolution in a solvent and then measurement of (1) the freezing-point depression of a solvent, (2) the boiling-point elevation of a solvent, or (3) the osmotic pressure of the solvent. Comment on these alternatives for

a. A solute of molecular weight of about 100.
b. A solute of molecular weight of about 1000.
c. A solute of molecular weight of about 1,000,000.

12.4 PHASE BEHAVIOR OF SOLID MIXTURES

Solids are different from gases and liquids in that the atoms are immobile, and the solid may exist as a well-ordered crystalline phase. Each atomic and molecular species has a unique size (which may or may not fit into the crystalline structure of other species), electronic charge, and interaction energy with other species. Because of the distortion of the crystal that would be needed to accommodate a solute, especially a molecular solute (unless the two species are very similar), the two will not be miscible in the crystalline phase. That is, one species will not appear in the crystal phase of the other, except in small concentrations as impurities. This is generally the case for organic chemicals, salts, and some metals. It is for this reason that so far in this book we have considered solids to be pure (or at least composed of macroscopically sized regions that are pure), so that their fugacity in a mixture can be taken to be equal to the pure component fugacity. However, there are some groups of atomic species that do form mixtures, and these will be considered here.

Solid solutions can form in metals if the atoms of which they are composed are similar; also, compounds can form. In such cases, expressions for the excess Gibbs energy of solid mixtures should contain a strain or mechanical energy term (which results from distorting the crystal structure to accommodate an atom of different size), a valence or coulombic term to account for the difference in charge between the solute atom and the atoms of the host crystal, the noncoulombic interactions of the type we considered in discussing molecular fluids in Sec. 9.5, and perhaps a chemical reaction term to account for compound formation. Alloys, amalgams, and intermetallic compounds can occur in solids; these more complicated situations will not be considered here.

The simplest model of a solid solution is the regular solution in which the excess entropy ($\underline{S}^{ex}$) and excess volume ($\underline{V}^{ex}$) of mixing are both zero. Though we will not try to give an explanation here, an excess entropy of zero implies that the solution is random in the sense of the distribution of the species on the crystal lattice (see Appendix 9.1). The excess enthalpy of mixing is still unspecified. The simplest model for the excess enthalpy of mixing is based on the assumptions that the lattice of the host species is not affected by the presence of the other species (that is, a species in its crystalline lattice can be replaced with the other species without affecting the lattice spacing), that only nearest-neighbor energetic interactions need be considered, and, as mentioned above, that the arrangement of atoms on the lattice sites is completely random. In this case the internal energy change on forming a binary solid mixture from the pure solids is

$$\Delta_{mix}\underline{U} = \underline{U}^{ex} = \frac{x_1 x_2 Z\bar{\omega}}{2} = x_1 x_2 \Omega \quad \text{where} \quad \Omega = \frac{Z\bar{\omega}}{2} = \frac{Z}{2}(2\varepsilon_{12} - \varepsilon_{11} - \varepsilon_{22})$$

$$(12.4\text{-}1)$$

Here Z is the coordination number of a lattice site, ε_{ij} is the interaction between a single species i atom with a single species j atom expressed on a molar basis, and $\bar{\omega} = 2\varepsilon_{12} - (\varepsilon_{11} + \varepsilon_{22})$ is the exchange energy, that is, the energy of replacing a 1-1 and a 2-2 interaction with two 1-2 interactions. (In chemical terms the exchange energy can be thought of as the energy change on forming two 1-2 molecules in an exchange reaction from one 1-1 molecule and one 2-2 molecule.) Therefore, for this model we have

$$\Delta_{mix}\underline{G} = \Delta_{mix}\underline{U} + P\Delta_{mix}\underline{V} - T\Delta_{mix}\underline{S} = x_1 x_2 \Omega + RT\sum_{i=1}^{2} x_i \ln x_i$$

and

$$(12.4\text{-}2)$$

$$\underline{G}^{ex} = x_1 x_2 \Omega$$

In solid-state thermodynamics, this is referred to as the regular solution model, and is related to the regular solution model for liquids of Sec. 9.6 in that both assume that $\underline{S}^{ex}$ and $\underline{V}^{ex}$ are each zero.

The simplest extension of this model is to assume that the arrangements of the atoms on the lattice sites are not random, but rather are proportional to the Boltzmann factors of the interaction energies. Linearizing the exponential terms that result, one finds

$$\Delta_{mix}\underline{U} = x_1 x_2 \Omega \left(1 - x_1 x_2 \frac{2\Omega}{ZRT}\right) \qquad \Delta_{mix}\underline{S} = -x_1^2 x_2^2 \frac{\Omega^2}{ZRT^2} - R\sum_{i=1}^{2} x_i \ln x_i$$

$$\Delta_{mix}\underline{G} = \Delta_{mix}\underline{U} + P\Delta_{mix}\underline{V} - T\Delta_{mix}\underline{S} = x_1 x_2 \Omega \left(1 - x_1 x_2 \frac{\Omega}{ZRT}\right) + RT\sum_{i=1}^{2} x_i \ln x_i$$

and

$$\underline{S}^{ex} = -x_1^2 x_2^2 \frac{\Omega^2}{ZRT^2}$$

$$(12.4\text{-}3)$$

so that

$$\underline{G}^{\text{ex}} = x_1 x_2 \Omega \left(1 - x_1 x_2 \frac{\Omega}{ZRT} \right)$$

This is referred to as the quasi-chemical model. More detailed solid solution models can be found in textbooks on metallurgical thermodynamics, for example, *Chemical Thermodynamics of Materials,* by C. H. P. Lupis (Elsevier Science Publishers, Amsterdam, 1983).

With this background, we can now consider several examples of phase equilibria involving solid solutions. The first example is a solid mixture consisting of atoms on a lattice with holes or lattice vacancies. Isolated lattice vacancies are considered to be point defects in the crystal, and we are interested in computing the equilibrium number of lattice vacancies. We can visualize the formation of lattice vacancies as starting with a well-ordered lattice of N atoms and mixing it with n holes (lattice vacancies). After the mixing process there will be N atoms and n holes distributed over $N + n$ lattice sites. When such vacancies are formed, there is an enthalpy change $\Delta_{\text{vac}}\underline{H}$ (on a per-mole-of-holes basis) resulting from the fact that the interactions between atoms in the crystal around the vacancy have been broken to accommodate the vacancy. Since the interactions between atoms in a crystal are attractive, energy must be supplied to add a vacancy, so $\Delta_{\text{vac}}\underline{H}$ is positive. There are also two contributions to the entropy change on forming a vacancy. One is the configurational contribution of the mixing process, which we indicate as $\Delta_{\text{conf}}\underline{S}$, which on a molar basis for n holes and N atoms is equal to

$$\Delta_{\text{conf}}\underline{S} = -R \left[\frac{N}{N+n} \ln \frac{N}{N+n} + \frac{n}{N+n} \ln \frac{n}{N+n} \right]$$

$$= -R[x_{\text{A}} \ln x_{\text{A}} + x_{\text{H}} \ln x_{\text{H}}] \tag{12.4-4}$$

where x_{A} is the fraction of lattice sites occupied by atoms, and x_{H} is the fraction of vacant lattice sites. In writing this it is assumed that the vacancies are randomly distributed on the lattice. The second contribution arises because the entropy of a crystal depends on the vibrational frequencies of the atoms on the lattice (this can be derived from statistical mechanics), and these vibrations are changed for the atoms in the immediate vicinity of a vacancy due to the change in intermolecular interactions and, therefore, the force field. We will refer to this entropy change (again on a per-mole-of-holes basis) as $\Delta_{\text{vac}}\underline{S}$. From mechanics (to relate the vibrational frequency change to the change in force field) and statistical mechanics (to relate the change in vibrational frequencies to the entropy change), both of which are beyond the scope of this book, it can be shown that $\Delta_{\text{vac}}\underline{S}$ is positive.

Therefore, the Gibbs energy of a crystal with N atoms and n lattice vacancies is

$$G(N, n) = N\underline{G}_{\text{A}} + n(\Delta_{\text{vac}}\underline{H} - T\Delta_{\text{vac}}\underline{S}) + RT \left[N \ln \frac{N}{N+n} + n \ln \frac{n}{N+n} \right]$$

$$\tag{12.4-5}$$

The first term on the right-hand side of this equation is the Gibbs free energy of the N atoms of species A on a perfect lattice; the second term, which results from the introduction of vacancies, is generally positive; and the third, mixing term is always negative. Regardless of the magnitude of the second term, there will always be some small value of the fraction of holes in the lattice that will result in the sum of the second and third terms being negative, so that the Gibbs free energy of the imperfect lattice will be less than that of the perfect lattice. Consequently, a crystal with lattice vacancies will be thermodynamically more stable and is the preferred state (in the sense of being more likely to occur naturally) than a perfect crystal. This is not good news for those in the electronics or other solid-state industries that require perfect crystals.

Since holes do not have a fugacity, we cannot do the usual phase equilibrium calculation of equating the fugacities of a pure fluid of holes to the fugacity of holes in the crystal. Instead, to determine the equilibrium number of holes in a lattice, we need to determine the minimum value of the Gibbs energy of the imperfect lattice with respect to the number of lattice vacancies, n. This is done by setting the derivative of the total Gibbs energy with respect to the number of vacancies equal to zero,

$$\left(\frac{\partial G}{\partial n}\right)_{N,T,P} = 0$$

$$= (\Delta_{\text{vac}}\underline{H} - T\,\Delta_{\text{vac}}\underline{S}) + RT\left[\ln\frac{n}{N+n} + \frac{n}{n} - \frac{n}{N+n} - \frac{N}{N+n}\right]$$

$$= (\Delta_{\text{vac}}\underline{H} - T\,\Delta_{\text{vac}}\underline{S}) + RT\,\ln\frac{n}{N+n}$$

$$(12.4\text{-}6)$$

so that at equilibrium

$$\frac{n}{N+n} = \text{Fraction of lattice sites vacant}$$

$$= \exp\left(-\frac{[\Delta_{\text{vac}}\underline{H} - T\,\Delta_{\text{vac}}\underline{S}]}{RT}\right) = \exp\left(-\frac{\Delta_{\text{vac}}\underline{G}}{RT}\right) \quad (12.4\text{-}7)$$

where $\Delta_{\text{vac}}\underline{G} = \Delta_{\text{vac}}\underline{H} - T\,\Delta_{\text{vac}}\underline{S}$ is the Gibbs energy change on adding 6×10^{23} holes to a very large perfect lattice (i.e., $N \gg 6 \times 10^{23}$, so that there are no adjacent lattice vacancies or holes).

ILLUSTRATION 12.4-1

Estimating the Number of Lattice Vacancies

It has been estimated that the Gibbs energy change on forming vacancies in a crystal of copper is approximately 126 kJ/mol and is independent of temperature. Estimate the fraction of the lattice sites that are vacant at 500, 1000, and 1500 K.

SOLUTION

From Eq. 12.4-4 we have

$$\text{Fraction of lattice sites that are vacant} = \exp\left(-\frac{\Delta_{\text{vac}}\underline{G}}{RT}\right)$$

$$= \exp\left(-\frac{126\,000\,\dfrac{\text{J}}{\text{mol}}}{8.314\left(\dfrac{\text{J}}{\text{mol K}}\right)T\,\text{K}}\right)$$

Using this equation, we find that the fraction of vacant lattice sites is 6.9×10^{-14} at 500 K, 2.6×10^{-7} at 1000 K, and 4.1×10^{-5} at 1500 K. ∎

Another type of crystal defect is an impurity in which a foreign atom has replaced the host atom on a lattice site. The analysis of the number of impurity defects is similar to that used above for vacancy defects except that in this case impurity atoms instead of holes appear in a lattice that contains N atoms of species A. Since the impurity atoms have a free energy, the analysis of this process is as follows. We start by considering a perfect crystal lattice containing N atoms of species A, and another region of M atoms of the impurity. We want to know the number n of these impurity atoms in the lattice of species A when equilibrium has been achieved. At equilibrium we then have a lattice with N atoms of species A and n impurity atoms in contact with a region containing the remaining $M - n$ impurity atoms. The Gibbs energy for this system, assuming the impurity atoms in the species A crystal are randomly distributed, is

$$
\begin{aligned}
G(N, n, M) = {} & G(\text{impure crystal consisting of } N \text{ atoms of species A} \\
& \text{and } n \text{ impurity atoms}) \\
& + G(\text{pure region consisting of } M - n \text{ impurity atoms}) \\
= {} & N\underline{G}_A + n\underline{G}_{\text{imp}} + n(\Delta_{\text{imp}}\underline{H} - T\Delta_{\text{imp}}\underline{S}) \\
& + RT \left[N \ln \frac{N}{N+n} + n \ln \frac{n}{N+n} \right] + (M - n)\underline{G}_{\text{imp}} \\
= {} & N\underline{G}_A + M\underline{G}_{\text{imp}} + n(\Delta_{\text{imp}}\underline{H} - T\Delta_{\text{imp}}\underline{S}) \\
& + RT \left[N \ln \frac{N}{N+n} + n \ln \frac{n}{N+n} \right]
\end{aligned}
$$
(12.4-8)

where $\underline{G}_A$ and $\underline{G}_{\text{imp}}$ are the molar Gibbs energies of the pure host and impurity atoms, and $\Delta_{\text{imp}}\underline{H}$ and $\Delta_{\text{imp}}\underline{S}$ are the enthalpy and vibrational entropy changes (on a molar basis) that result from the addition of 1 mole of impurities into a very large crystal. Generally these two terms are smaller than the analogous terms above for the introduction of vacancies. The equilibrium concentration of impurities is found from

$$
\begin{aligned}
\left(\frac{\partial G}{\partial n} \right)_N &= 0 \\
&= (\Delta_{\text{imp}}\underline{H} - T\Delta_{\text{imp}}\underline{S}) + RT \left[\ln \frac{n}{N+n} + \frac{n}{n} - \frac{n}{N+n} - \frac{N}{N+n} \right] \\
&= (\Delta_{\text{imp}}\underline{H} - T\Delta_{\text{imp}}\underline{S}) + RT \ln \frac{n}{N+n}
\end{aligned}
$$
(12.4-9)

so that the equilibrium fractional concentration of impurities is

$$
\begin{aligned}
\frac{n}{N+n} &= \exp \left(-\frac{[\Delta_{\text{imp}}\underline{H} - T\Delta_{\text{imp}}\underline{S}]}{RT} \right) \\
&= \exp \left(-\frac{\Delta_{\text{imp}}\underline{G}}{RT} \right)
\end{aligned}
$$
(12.4-10)

Following the discussion in the lattice vacancy case, we conclude that the impure crystal is also the thermodynamically preferred state over that of a perfect crystal. Note here that the term $\Delta_{\text{imp}}\underline{G}$ is the molar Gibbs energy change of placing 1 mole of impurities into a very large crystal.

Another type of crystal defect is a dislocation in which some lattice sites are shifted with respect to the perfect lattice configuration. In such a dislocation the enthalpy change is very large, there is no entropy-of-mixing term, and the remaining entropy change due to changes in vibrational frequencies is small. In this case the enthalpy term dominates, the free energy change is positive, and dislocation defects therefore are thermodynamically unstable and not preferred states. Consequently, dislocations are much easier to eliminate from a crystal—for example, by annealing—than are vacancies or impurities.

If, in contrast to the situation so far considered, both substances are at least partially miscible in each other so that solid solutions form, a variety of solid-liquid phase behaviors are possible. The simplest case is that of two species that form ideal solutions in both the liquid and the solid. The equilibrium relation then is

$$\bar{f}_i^S(T, P, \underline{x}^S) = \bar{f}_i^L(T, P, \underline{x}^L)$$

Using the ideal solution assumption, we have

$$x_i^S f_i^S(T, P) = x_i^L f_i^L(T, P)$$

Now from Eq. 9.7-8a,

$$f_i^L(T, P) = f_i^S(T, P) \exp\left[\frac{1}{RT}\left\{\Delta_{\text{fus}}\underline{H}_i(T_{m,i})\left[1 - \frac{T}{T_{m,i}}\right]\right.\right.$$
$$\left.\left. + \int_{T_{m,i}}^T \Delta C_{\text{P},i}\, dT - T\int_{T_{m,i}}^T \frac{\Delta C_{\text{P},i}}{T}\, dT\right\}\right]$$

so that

$$x_i^S = x_i^L \exp\left[\frac{1}{RT}\left\{\Delta_{\text{fus}}\underline{H}_i(T_{m,i})\left[1 - \frac{T}{T_{m,i}}\right] + \int_{T_{m,i}}^T \Delta C_{\text{P},i}\, dT - T\int_{T_{m,i}}^T \frac{\Delta C_{\text{P},i}}{T}\, dT\right\}\right]$$

$$(12.4\text{-}11)$$

In many cases the ΔC_{P} terms can be neglected, so the equation reduces to

$$x_i^S = x_i^L \exp\left[\frac{\Delta_{\text{fus}}\underline{H}_i(T_{m,i})}{RT}\left\{1 - \frac{T}{T_{m,i}}\right\}\right] \qquad (12.4\text{-}12)$$

which relates the composition in the liquid to that in the solid.

Since in a binary mixture we have

$$x_1^L + x_2^L = 1 \quad \text{and} \quad x_1^S + x_2^S = 1$$

we have that for the liquidus line

$$x_1^L \exp\left[\frac{\Delta_{fus}\underline{H}_1(T_{m,1})}{RT}\left\{1-\frac{T}{T_{m,1}}\right\}\right] + x_2^L \exp\left[\frac{\Delta_{fus}(\underline{H}_2 T_{m,2})}{RT}\left\{1-\frac{T}{T_{m,2}}\right\}\right] = 1$$

$$= x_1^L \exp\left[\frac{\Delta_{fus}\underline{H}_1(T_{m,1})}{RT}\left\{1-\frac{T}{T_{m,1}}\right\}\right]$$

$$+ (1-x_1^L)\exp\left[\frac{\Delta_{fus}\underline{H}_2(T_{m,2})}{RT}\left\{1-\frac{T}{T_{m,2}}\right\}\right]$$

so that

$$x_1^L = \frac{1 - \exp\left[\frac{\Delta_{fus}\underline{H}_2(T_{m,2})}{RT}\left\{1-\frac{T}{T_{m,2}}\right\}\right]}{\exp\left[\frac{\Delta_{fus}\underline{H}_1(T_{m,1})}{RT}\left\{1-\frac{T}{T_{m,1}}\right\}\right] - \exp\left[\frac{\Delta_{fus}\underline{H}_2(T_{m,2})}{RT}\left\{1-\frac{T}{T_{m,2}}\right\}\right]}$$

(12.4-13)

Similarly, for the solidus line we have

$$x_1^S = \frac{1 - \exp\left[-\frac{\Delta_{fus}\underline{H}_2(T_{m,2})}{RT}\left\{1-\frac{T}{T_{m,2}}\right\}\right]}{\exp\left[-\frac{\Delta_{fus}\underline{H}_1(T_{m,1})}{RT}\left\{1-\frac{T}{T_{m,1}}\right\}\right] - \exp\left[-\frac{\Delta_{fus}\underline{H}_2(T_{m,2})}{RT}\left\{1-\frac{T}{T_{m,2}}\right\}\right]}$$

(12.4-14)

For substances that form nonideal solutions in both the liquid and solid phases, the analogous results (Problem 12.4-2) are, for the liquidus line,

$$x_1^L = \frac{1 - \frac{\gamma_2^L(x^L)}{\gamma_2^S(x^S)}\exp\left[\frac{\Delta_{fus}\underline{H}_2(T_{m,2})}{RT}\left\{1-\frac{T}{T_{m,2}}\right\}\right]}{\frac{\gamma_1^L(x^L)}{\gamma_1^S(x^S)}\exp\left[\frac{\Delta_{fus}\underline{H}_1(T_{m,1})}{RT}\left\{1-\frac{T}{T_{m,1}}\right\}\right] - \frac{\gamma_2^L(x^L)}{\gamma_2^S(x^S)}\exp\left[\frac{\Delta_{fus}\underline{H}_2(T_{m,2})}{RT}\left\{1-\frac{T}{T_{m,2}}\right\}\right]}$$

(12.4-15)

and for the solidus line,

$$x_1^S = \frac{1 - \frac{\gamma_2^S(x^S)}{\gamma_2^L(x^L)}\exp\left[-\frac{\Delta_{fus}\underline{H}_2(T_{m,2})}{RT}\left\{1-\frac{T}{T_{m,2}}\right\}\right]}{\frac{\gamma_1^S(x^S)}{\gamma_1^L(x^L)}\exp\left[-\frac{\Delta_{fus}\underline{H}_1(T_{m,1})}{RT}\left\{1-\frac{T}{T_{m,1}}\right\}\right] - \frac{\gamma_2^S(x^S)}{\gamma_2^L(x^L)}\exp\left[-\frac{\Delta_{fus}\underline{H}_2(T_{m,2})}{RT}\left\{1-\frac{T}{T_{m,2}}\right\}\right]}$$

(12.4-16)

Note that in this equation there are activity coefficients for each species in each phase. In general, the activity coefficient models and the values of the activity coefficients will be different in each phase. In particular, the activity coefficient models of Secs. 9.5 and 9.6 can be used for the liquid phase, and those of this section for the solid phase. Alternatively, the regular solution model of this section can be used for the solid and liquid phases, but with different values of the exchange energy or Ω parameter (i.e., with Ω^L in the liquid phase and Ω^S in the solid phase).

Figure 12.4-1 shows the results predicted for three different mixtures. In the first case the solid and liquid mixtures are ideal, and in the second case the liquid mixture

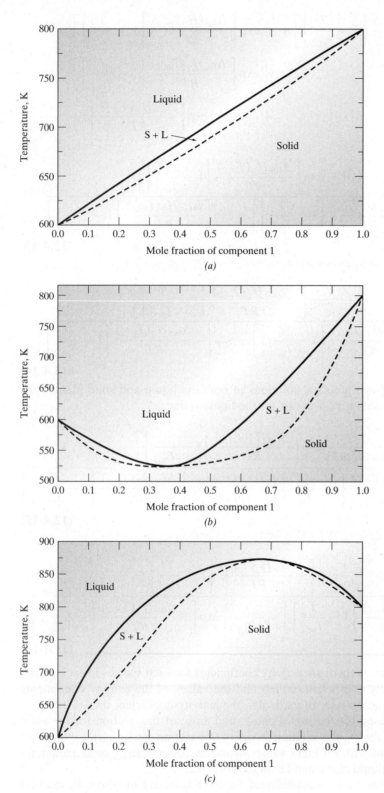

Figure 12.4-1 Solid-liquid phase behavior of a model system. The pure-component properties in this figure are $T_{m,1} = 800$ K, $\Delta_{\text{fus}}\underline{H}_1 = 6200$ J/mol, $T_{m,2} = 600$ K, $\Delta_{\text{fus}}\underline{H}_2 = 4900$ J/mol, and $C_P^L = C_P^S$ for both species. (a) $\Omega^L = \Omega^S = 0$ (ideal mixture in both the solid and liquid phases). (b) $\Omega^S = 5000$ J/mol and $\Omega^L = 0$ (ideal liquid phase, nonideal solid phase). (c) $\Omega^L = 5000$ J/mol and $\Omega^S = 0$ (ideal solid phase, nonideal liquid phase). Figures b and c have congruent points in which the solid and liquid phases in equilibrium have the same composition.

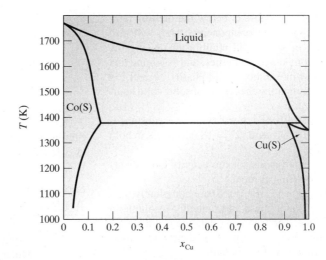

Figure 12.4-2 Solid-liquid phase diagram for the cobalt-copper system. The melting point of copper is 1356.6 K, and the melting point of cobalt is 1768 K.

follows the regular solution model and the solid is an ideal mixture. In the third case both the liquid and solid mixtures form regular solutions, but with different exchange energies. We see that a variety of phase behaviors are predicted to occur. In Fig. 12.4-1*a* ideal solid-liquid behavior is predicted, similar to the vapor-liquid equilibrium phase behavior seen in Secs. 10.1 and 10.2. In Fig. 12.4-1*b* and *c* nonideal solid-liquid phase behavior is shown. Indeed, in these plots we see azeotropic behavior in which the liquid and solid in equilibrium are of the same composition. In materials science, such mixtures are said to solidify **congruently** in that the solid has the same composition as the liquid from which it was formed.

Figure 12.4-2 is an example of the liquid-solid phase diagram for the copper-cobalt system. There we see that copper and cobalt are partially miscible in the solid phase, and that there is a region of temperature and composition in which solid cobalt is in equilibrium with molten copper-cobalt solutions. Above its melting point, cobalt is completely miscible with copper.

The phase diagrams of solid mixtures can be considerably more complicated than in the figures shown here. The reasons for this are as follows:

1. The species in the mixture may be only partially miscible in each other.
2. Several solids of different crystal structure can form, and the solid-phase transitions may occur in the range of the liquid-solid-phase transition.
3. The species can react to form new compounds, which affects the overall phase behavior.

PROBLEMS FOR SECTION 12.4

12.4-1 Prove that Eqs. 12.4-13 and 12.4-14 are correct.

12.4-2 Derive the equations for the liquidus and solidus lines for two partially miscible solids for two substances that
  a. Form a regular solution in the liquid phase, while the solid phase is ideal

  b. Form an ideal solution in the liquid phase and a regular solution in the solid phase
  c. Form regular solutions in both the solid and liquid phases, but with different regular solution parameters in the two phases

12.4-3 Figure 12.4-1*b* is incomplete. Since the congruent point in this mixture is below the melting temperature of both pure components, there will also be a region of solid-solid equilibrium. Using the same pure-component properties, and assuming that $\Omega^S = 5000$ J/mol in both solid phases, correct Fig 12.4-1b by computing the region of solid-solid equilibrium.

12.4-4 Draw the phase boundaries for the two substances of Fig 12.4-1b, but with $\Omega^S = 10\ 000$ J/mol in both solid phases.

12.4-5 Derive the expressions for the activity coefficients in the quasi-chemical model.

12.4-6 Antimony (Sb) and silicon (Si) are completely miscible in the liquid phase and mutually insoluble in the solid state. The following data are available for the solidification temperature of their liquid mixtures as a function of temperature.

Mole Fraction Silicon	Solidification Temperature (°C)
1	1410
0.9	1385
0.8	1350
0.7	1316
0.6	1290
0.5	1278
0.4	1261
0.3	1242
0.2	1215
0.1	1090
0	630.5

The following additional data are available for silicon.

$\Delta_{fus}\underline{H}(T = 1683$ K$) = 50.626$ kJ/mol

$C_P^S = 23.932 + 2.469 \times 10^{-3}T$

$\qquad - 4.142 \times 10^{-5}T^2$ J/(mol K), T in K

$C_P^L = 25.606$ J/(mol K)

Use these data to compute the activity coefficient of silicon in the antimony-silicon mixtures as a function of the liquid composition of silicon.

12.4-7 Assuming oxygen and nitrogen form ideal liquid mixtures and that their solid phases are immiscible, compute the complete solid-liquid-vapor phase dia-

gram for this mixture for the temperature range of 46 to 100 K.
Data:
Solid oxygen[10]
    Molar volume $= 0.024\ 62$ m^3/kmol
    Melting point $= 54.35$ K

T (K)	46	48	50
Sublimation pressure (Pa)	5.252	13.02	29.86
$\Delta_{sub}\underline{H}$ (kJ/mol)	8.350	8.316	8.281

T (K)	52	54	54.35
Sublimation pressure (Pa)	64.25	130.1	146.4
$\Delta_{sub}\underline{H}$ (kJ/mol)	8.248	8.213	8.207

Liquid oxygen
    Molar volume $= 0.027\ 35$ m^3/kmol

T (°C)[11]	−219.1	−213.4	−210.6	−204.1
Vapor pressure (kPa)	0.1333	0.6665	1.333	5.332

T (°C)	−198.8	−188.8	−183.1
Vapor pressure (kPa)	13.33	53.32	101.3

Solid nitrogen[12]
    Molar volume $= 0.031\ 86$ m^3/kmol
    Melting point $= 63.2$ K
    Sublimation pressure: $\log_{10} P = -\dfrac{381.6}{T}$
    $- 0.006\ 237\ 2T + 7.535\ 88$ (P in kPa, T in K)
    $\Delta_{fus}\underline{H}$ (melting point) $= 720.9$ J/mol

Liquid nitrogen
    Molar volume $= 0.02414/(1 - 0.0039T)$ m^3/kmol
    Vapor pressure: $\log_{10} P = -\dfrac{339.8}{T}$
    $- 0.005\ 628\ 6T + 6.835\ 40$ (P in kPa, T in K)

12.4-8 One intriguing problem in atmospheric physics is the relatively long persistence of contrails emitted from high-altitude jet aircraft when the sky is clear. An explanation of this phenomenon is that as the water vapor emitted from the jet engines cools, some of it will condense to form water droplets and, on further cooling, will form ice crystals. The claim is that the ice crystals, although they are

---

[10]J. C. Mullins, W. T. Ziegler, and B. S. Kirk, *Adv. Cryog. Eng.,* **8**, 126 (1963).

[11]R. H. Perry, D. W. Green, and J. O. Maloney, eds., *Chemical Engineer's Handbook,* 6th ed., McGraw-Hill, New York (1984), pp. 3–48.

[12]*Kirk-Othmer Encyclopedia of Chemical Technology,* Vol. 15, 3rd ed., John Wiley & Sons, New York (1981), p. 933.

formed when the air is saturated with respect to water, will persist as long as the air is saturated with respect to ice. Therefore, the ice crystal contrail can persist in equilibrium even though all the water droplets have either crystallized or evaporated, and the partial pressure of water in the air is less than the liquid water vapor pressure (i.e., the relative humidity of the air is less than 100 percent).

a. Establish, by the principles of thermodynamics, the validity or fallacy of the explanation given here.

b. Estimate the relative humidity above which ice crystals will be stable at $-25°C$ and 0.5 bar pressure.

The vapor pressure of liquid water and the sublimation pressure of ice are given in Problem 7.12.

12.4-9 The temperature of a liquid mixture is reduced so that solids form. However, unlike the illustrations in Section 12.3, on solidification, a solid mixture (rather than pure solids) is formed. Also, the liquid phase is not ideal. Assuming that the nonideality of the liquid and solid mixtures can be described by the same one-constant Margules excess Gibbs energy expression, derive the equations for the compositions of the coexisting liquid and solid phases as a function of the freezing point of the mixture and the pure-component properties.

12.4-10 The temperature of a liquid mixture is reduced so that solids form. However, as in the previous problem, on solidification a nonideal solid mixture (rather than pure solids) is formed, and the liquid phase is not ideal. The following data are available.

	Species 1	Species 2	Mixture
T_f (K)	150	200	
$\Delta_{fus}\underline{H}$ (J/mol)	6300	8200	
$\Delta_{fus}C_P$ [J/(mol·K)]	0	0	
$\underline{G}^{ex}$ (J/mol, liquid)			$1.2RTx_1x_2$
$\underline{G}^{ex}$ (J/mol, solid)			$1.2RTx_1x_2$

Determine and plot the equilibrium solid compositions and freezing point as a function of the liquid composition.

12.4-11 It has been argued that since solids form in a well-defined crystalline structure with preferred intermolecular atom-atom contacts—unlike liquids, the molecules of which are free to move and rotate—that while the excess Gibbs energies of liquids and solids may be described by the same expression, the constants should be different in both phases. Using this idea, recompute the solid-liquid phase diagram of the previous problem if

$$\underline{G}^{ex}\left(\frac{J}{mol}, \text{solid}\right) = 2.0RTx_1x_2$$

12.4-12 Benzene and cyclohexane are very similar compounds, but their mixture is sufficiently nonideal that an azeotrope is formed in vapor-liquid equilibrium. At $40°C$, the vapor pressure of cyclohexane is 0.246 bar, and that of benzene is 0.244 bar. At this temperature, the azeotrope occurs at 0.51 mole fraction of cyclohexane and a total pressure of 0.2747 bar.

a. Obtain the complete dew point and bubble point curves for this mixture at $40°C$.

b. What is the liquid-liquid upper critical solution temperature for this mixture?

c. The melting point of benzene is $5.53°C$ and its heat of fusion is 9953 J/mol. The melting point of cyclohexane is $6.6°C$ and its heat of fusion is 2630 J/mol. The solid and liquid heat capacities of each compound can be assumed to be equal. Determine the eutectic point of this mixture on freezing, assuming that the solution nonideality is independent of temperature.

12.5 THE PHASE BEHAVIOR MODELING OF CHEMICALS IN THE ENVIRONMENT

When some chemicals are released into the environment they rapidly degrade due to microbial action, hydrolysis, photochemical reactions, and other processes. However, other chemicals degrade very slowly in the environment. For example, the insecticide DDT has an environmental half-life of more than 20 years. Such long-lived chemicals generally reach phase equilibrium among the different parts of the local environment in which they are in contact. The different parts of the environment are generally referred to as compartments and include air, water, soil, sediment, suspended sediment, aquatic biota (fish), and terrestrial biota (animals, plants). The purpose of this section is to show that phase equilibrium calculations can be used to make approximate but reasonable estimates of the distribution of long-lived organic chemicals between these environmental compartments.

The distribution of a chemical between different environmental compartments is usually described in terms of concentration-based partition coefficients that are ratios of the concentrations of the chemical in two compartments. For example, the air-water partition coefficient of species i, $K_{AW,i}$, is

$$K_{AW,i} = \frac{C_i^A}{C_i^W} \qquad (12.5\text{-}1)$$

where the superscripts A and W indicate air and water, respectively. Similarly the biota-water partition coefficient, $K_{BW,i}$, is

$$K_{BW,i} = \frac{C_i^B}{C_i^W} \qquad (12.5\text{-}2)$$

By convention, partition coefficients are usually given relative to the concentration in water as above, since the water phase is usually the easiest to reliably sample (it is well mixed compared with, say, the soil, and the concentrations are higher than in the air). However, if we wanted a numerical value for the air-biota partition coefficient of the species, $K_{AB,i}$, we could easily compute it from $K_{AW,i}$ and $K_{BW,i}$ as follows:

$$K_{AB,i} = \frac{C_i^A}{C_i^B} = \frac{C_i^A}{C_i^W}\frac{C_i^W}{C_i^B} = \frac{K_{AW,i}}{K_{BW,i}} \qquad (12.5\text{-}3)$$

Similar relations are valid for other partition coefficients to be introduced shortly.

From the equality of fugacities, we have, for a component distributed between air and water,

$$\bar{f}_i^W = \bar{f}_i^A \quad \text{or} \quad x_i\gamma_i P_i^{\text{vap}}(T) = y_i P = y_i 1.013 \text{ bar} \qquad (12.5\text{-}4)$$

Here we have recognized that at normal environmental conditions the total pressure is 1.013 bar, which is so low that all fugacity coefficients can be neglected. Also, the activity coefficient that appears, since pollutants are usually present only at very low concentrations, is the activity coefficient at infinite dilution. Therefore,

$$\frac{y_i}{x_i} = \frac{\gamma_i^\infty P_i^{\text{vap}}(T)}{1.013 \text{ bar}} = \frac{H_i(T)}{1.013 \text{ bar}} \qquad (12.5\text{-}5)$$

since, from Eq. 9.7-9 (with $f_i^L = P_i^{\text{vap}}$), the product $\gamma_i^\infty P_i^{\text{vap}}$ is equal to the Henry's law constant.

By convention, pollutant concentrations are frequently given in units of g/m^3, and partition coefficients are given as a ratio of such concentrations. The concentration of a chemical in air is given above as a mole fraction; its concentration in g/m^3 is

$$C_i^A \left[\frac{g_i}{m^3} \right]$$

$$= y_i \left(\frac{mol_i}{mol_{air}} \right) \times \frac{1}{V} \left(\frac{mol_{air}}{m^3} \right) \times MW_i \left(\frac{g_i}{mol_i} \right)$$

$$= y_i \frac{P}{RT} MW_i = y_i \frac{1.013 \text{ bar}}{8.314 \times 10^{-5} \dfrac{\text{bar m}^3}{\text{mol K}} T} MW_i \frac{g_i}{mol_i} = \frac{1.218 \times 10^4}{T} y_i MW_i$$

$$\text{(12.5-6)}$$

where, since the pressure is low, the ideal gas law has been used. Similarly, given the mole fraction in the aqueous phase, we have

$$C_i^W \left[\frac{g}{m^3} \right] = x_i \left(\frac{mol_i}{mol_{water}} \right) \times \frac{1 \text{ mol}_{water}}{18 \text{ g}_{water}} \times \frac{10^6 \text{ g}_{water}}{1 \text{ m}^3} \times MW_i \frac{g}{mol} \qquad \text{(12.5-7)}$$

$$= 5.556 \times 10^4 x_i MW_i$$

Therefore,

$$K_{AW,i} \left[\frac{\frac{g}{m^3}}{\frac{g}{m^3}} \right] = \frac{C_i^A}{C_i^W} = \frac{1.218 \times 10^4 \times MW_i \times y_i}{T \times 5.556 \times 10^4 \times MW_i \times x_i} = \frac{0.2192}{T} \frac{y_i}{x_i} \qquad \text{(12.5-8)}$$

$$= 0.2164 \frac{\gamma_i^\infty P_i^{vap}}{T} = 0.2164 \frac{H_i}{T}$$

for vapor pressure in bar, T in K, and H_i in bar/mole fraction.

ILLUSTRATION 12.5-1

Calculation of the Air-Water Partition Coefficient

Given that the extrapolated vapor pressure of liquid benzo[*a*]pyrene is 2.13×10^{-5} Pa at 25°C and the value of its infinite-dilution activity coefficient in water, $\gamma_{BP}^\infty = 3.76 \times 10^8$, found in Illustration 12.1-3, determine its air-water partition coefficient.

SOLUTION

$$K_{AW,BP} \left[\frac{\frac{g_{BP}}{m_A^3}}{\frac{g_{BP}}{m_W^3}} \right] = 0.2164 \frac{\gamma_{BP}^\infty P_{BP}^{vap}}{T} = 0.2164 \times \frac{3.76 \times 10^8 \times 2.13 \times 10^{-5} \text{ Pa}}{298.15} \times \frac{1 \text{ bar}}{10^5 \text{ Pa}}$$

$$= 5.844 \times 10^{-5}$$

The value of the biota-water partition coefficient, $K_{BW,i}$, is more difficult to compute since we do not know how to chemically characterize a fish, a plant, or other biota. For animals or fish the assumption generally made is that the hydrophobic chemicals of interest partition mainly into the organic lipid (fat) portion of the biota, rather than the aqueous fluid, inorganic bone, or fibrous tissue of the body. Further, it is assumed that the lipids are chemically similar to octanol, so that the lipid-water partition coefficient is equal to the octanol-water partition coefficient. Also, the densities of water and biota are approximately the same (which is why people barely float). Therefore, $K_{\mathrm{BW,i}}$, the biota-water partition coefficient, is taken to be

$$K_{\mathrm{BW,i}} \left[\frac{\dfrac{\mathrm{g_i}}{\mathrm{m_B^3}}}{\dfrac{\mathrm{g_i}}{\mathrm{m_W^3}}} \right] = w_{\mathrm{B}} K_{\mathrm{OW,i}} \qquad (12.5\text{-}9)$$

where w_{B} is the weight fraction of lipids in the biota. For fish, w_{B} is typically 0.05, or 5 percent.

ILLUSTRATION 12.5-2

Estimation of the Distribution of PCBs in the Environment

Polychlorinated biphenyls (PCBs) were manufactured as a liquid transformer oil. Due to leakage, PCBs are now found throughout the environment, and this long-lived chemical represents a health hazard. PCBs have been found to be present in the St. Lawrence River at concentrations of about 0.3 ppb (parts per billion by weight). Estimate the likely concentration of PCBs in fish in this river. The average octanol-water partition coefficient for PCBs is

$$\log_{10} K_{\mathrm{OW,PCB}} = 5.5$$

SOLUTION

From Eq. 12.5-9, we have

$$K_{\mathrm{BW,PCB}} = w_{\mathrm{B}} K_{\mathrm{OW,PCB}} = 0.05 \times 10^{5.5} = 1.58 \times 10^4$$

Therefore,

$$\begin{aligned}
C_{\mathrm{B,PCB}} &= K_{\mathrm{BW,PCB}} \times C_{\mathrm{W,PCB}} \\
&= 1.58 \times 10^4 \times 0.3 \text{ ppb} \\
&= 4740 \text{ ppb} = 4.74 \text{ ppm}
\end{aligned}$$

Actual field testing has found the PCB concentration in eels in the St. Lawrence River to be 7.9 ppm.

COMMENT

The PCB concentration of eels in this river is a factor of 26 300 higher than the PCB concentration in the river water. The very simple phase equilibrium model for the distribution of the chemical used here is able to predict this very large **bioconcentration** effect to within a factor of 2. ∎

It is also assumed that hydrophobic organic chemicals partition only into the organic matter in soils, sediment, and suspended sediment. Further, the organic matter in soils and sediment is not as well represented by octanol as is the lipids in biota. Empirically it has been found that the organic carbon (in soils and sediment)–water partition coefficient, $K_{\mathrm{OCW,i}}$, defined as

$$K_{\mathrm{OCW,i}} \left[\frac{\dfrac{\mathrm{g_i}}{10^6 \text{ g}}}{\dfrac{\mathrm{g_i}}{\mathrm{m_{water}^3}}} \right] = \frac{\text{grams i per } 10^6 \text{ grams organic matter}}{\text{grams i per m}^3 \text{ of water}}$$

is approximately 40 percent of the octanol-water partition coefficient; that is,

$$K_{OCW,i} = 0.4 K_{OW,i}$$

Therefore, estimates for the soil-water, $K_{SW,i}$, and the sediment-water, $K_{DW,i}$, partition coefficients are

$$K_{SW,i} \left[\frac{\dfrac{g_i}{10^6 \, g_{soil}}}{\dfrac{g_i}{m^3_{water}}} \right] = 0.4 w_S K_{OW,i} \qquad \textbf{(12.5-10)}$$

and

$$K_{DW,i} \left[\frac{\dfrac{g_i}{10^6 \, g_{sediment}}}{\dfrac{g_i}{m^3_{water}}} \right] = 0.4 w_D K_{OW,i} \qquad \textbf{(12.5-11)}$$

where w_S and w_D are the weight fractions of organic matter in the soil and sediment, respectively. While soils and sediments are very heterogeneous and can vary greatly in properties, for the purposes here we will use $w_S = 0.02$ and $w_D = 0.05$ as average values.

ILLUSTRATION 12.5-3

Computing the Concentration of a Pollutant in the Different Environmental Compartments

The concentration of benzo[*a*]pyrene in water in southern Ontario has been reported to be 2.82×10^4 ng/m^3. Compute the likely concentration of this chemical in ng/m^3 in the air, soil, sediment, and fish.

Data: For benzo[*a*]pyrene $\log_{10} K_{OW,BP} = 6.04$, and from Illustration 12.5-1 we have $K_{AW,BP} = 5.884 \times 10^{-5}$. Also, the density of soil is approximately 1500 kg/m^3 and that of sediment is about 1420 kg/m^3.

SOLUTION

Starting from

$$C^A_{BP} = 5.884 \times 10^{-5} C^W_{BP} = 5.884 \times 10^{-5} \times 2.82 \times 10^4 = 1.66 \, \frac{ng}{m^3}$$

and given that $K_{OW,BP} = 10^{6.04} = 1.096 \times 10^6$, we have

$$K_{SW,BP} = 0.4 \times 0.02 \times 1.096 \times 10^6 = 8768 \, \frac{\dfrac{g_{BP}}{10^6 \, g_{soil}}}{\dfrac{g_{BP}}{m^3_{water}}}$$

so that

$$C_{BP}^{S} = 8768 \times C_{BP}^{W} = 8768 \frac{\dfrac{g_{BP}}{10^6 \, g_{soil}}}{\dfrac{g_{BP}}{m_{water}^3}} \times 2.82 \times 10^4 \frac{ng_{BP}}{m_{water}^3} = 2.47 \times 10^8 \frac{ng_{BP}}{10^6 \, g_{soil}}$$

$$= 2.47 \times 10^8 \frac{ng_{BP}}{10^6 \, g_{soil}} \times \frac{1 \, mg}{10^6 \, ng} \times \frac{10^3 \, g}{1 \, kg} = 0.247 \frac{mg_{BP}}{kg_{soil}}$$

$$= 0.247 \text{ ppm (by weight)}$$

and

$$C_{BP}^{S} = 0.247 \frac{mg_{BP}}{kg_{soil}} \times 1500 \frac{kg_{soil}}{m_{soil}^3} \times 10^6 \frac{ng}{mg} = 3.71 \times 10^8 \frac{ng_{BP}}{m_{soil}^3}$$

Also,

$$K_{DW,BP} = 0.4 \times 0.05 \times 1.096 \times 10^6 = 21\,920 \frac{\dfrac{g_{BP}}{10^6 \, g_{sediment}}}{\dfrac{g_{BP}}{m_{water}^3}}$$

and

$$C_{BP}^{D} = 21\,920 \times C_{BP}^{W} = 21\,920 \frac{\dfrac{g_{BP}}{10^6 \, g_{sediment}}}{\dfrac{g_{BP}}{m_{water}^3}} \times 2.82 \times 10^4 \frac{ng_{BP}}{m_{water}^3}$$

$$= 6.18 \times 10^8 \frac{ng_{BP}}{10^6 \, g_{sediment}}$$

$$= 6.18 \times 10^8 \frac{ng_{BP}}{10^6 \, g_{sediment}} \times \frac{1 \, mg}{10^6 \, ng} \times \frac{10^3 \, g}{1 \, kg}$$

$$= 0.618 \frac{mg_{BP}}{kg_{sediment}} = 0.618 \text{ ppm (by weight)}$$

so that

$$C_{BP}^{D} = 0.618 \frac{mg_{BP}}{kg_{sediment}} \times 1420 \frac{kg_{sediment}}{m_{sediment}^3} \times 10^6 \frac{ng}{mg} = 8.78 \times 10^8 \frac{ng_{BP}}{m_{sediment}^3}$$

Finally, in the biota we have

$$C_{BP}^{B} = 0.05 \times K_{OW,BP} \times C_{BP}^{W} = 0.05 \times 1.096 \times 10^6 \times 2.82 \times 10^4 = 1.55 \times 10^9 \frac{ng}{m^3}$$

$$= 1.55 \frac{mg_{BP}}{kg_{biota}} = 1.55 \text{ ppm (by weight)}$$

Since the density of biota is approximately 1 g/cc or 1000 kg/m^3, this concentration is also 1.55×10^9 ng/m^3. (Note that in these calculations we have used the fact that 1 m^3 of water is equal to 10^3 kg.)

The values computed above together with reported values of Mackay and Paterson[13] are listed below.

---

[13] D. Mackay and S. Paterson, *Environ. Sci. Technol.*, **25**, 427–436 (1991).

Environmental Compartment	Concentration (ng/m^3)	
	Calculated	Reported
Water		2.82×10^4
Air	1.66	1.3 to 7.1
Soil	3.71×10^8	1.1×10^8
Sediment	8.78×10^8	0.8×10^8 to 3×10^8
Fish	1.55×10^9	1.4×10^8

COMMENT

The agreement between the results of the calculation here and reported data, while not perfect, is certainly reasonable given the simplicity of the thermodynamic model used and the complexity of environmental processes. In particular, the calculated results show that the large, orders-of-magnitude difference in the concentration of a long-lived pollutant in the various environmental compartments can be explained using simple phase equilibrium concepts. ∎

PROBLEMS FOR SECTION 12.5

12.5-1 Estimate the concentration of PCBs in the soil and sediment of the St. Lawrence River, and in the air above that river.
Data: Vapor pressure of PCBs at atmospheric temperature is approximately 2×10^{-7} bar, PCB solubility in water is about 0.1 mg/L, molecular weight is 250, density of soil is 1500 kg/m^3, and density of sediment is 1300 kg/m^3.

12.5-2 To assess bioaccumulation, four fish tanks are prepared such that the water in each is saturated with one of the four insecticides listed below, and each fish tank contains only a single fish. Compute the expected concentration of the insecticide in each fish after a long period of time.

Insecticide	Solubility in Water (μg/L)	$\log_{10} K_{OW}$
Aldrin	27	5.52
Dieldrin	140	4.32
Lindane	7 000	3.61
Diazinon	40 000	3.31

12.5-3 a. A closed terrarium is 10 m^3 in total volume, of which 4 m^3 is water, 3 m^3 is soil, 200 cm^3 are fish and other biota, and the remainder is air. The terrarium is accidentally contaminated with 10 mg of benzene. What is the concentration of benzene in each of the four compartments, and what is the fraction of the total amount of benzene present in each of the compartments?

b. Repeat the preceding calculation for the case where 10 mg of the pesticide DDT is the contaminant.
Data: For benzene, $\log_{10} K_{OW} = 2.13$, water solubility is 0.0405 mol %, and vapor pressure is 0.127 bar. For DDT, $\log_{10} K_{OW} = 6.20$ and $K_{AW} = 9.5 \times 10^{-4}$. Also, the average density of soil is 1.5 g/cc.

12.5-4 The chemical 1-chloro-2-nitrobenzene is an intermediate in dye manufacture. It is very resistant to hydrolysis and biodegradation, and so can be assumed to be persistent in the environment. At 20°C its water solubility is 440 mg/L, its vapor pressure is 4×10^{-5} bar, and K_{OW} is 224. If 100 kg of this chemical is released into a model environment that consists of 6×10^9 m^3 of air, 7×10^6 m^3 of water, 4.5×10^4 m^3 of soil, and 2.1×10^4 m^3 of sediment, determine its concentration in each compartment. The soil can be assumed to contain 2 wt % organic matter and to have a density of 1500 kg/m^3, while the sediment contains 5 wt % organic matter and has a density of 1300 kg/m^3.

12.6 PROCESS DESIGN AND PRODUCT DESIGN

The traditional role of thermodynamics in chemical engineering has been in the area of process design. In such cases the starting feed streams are known, and the role of the chemical engineer is to design a process that can produce the desired final product in a manner that is economical, is safe for workers and those in the vicinity of the plant, and has minimal environmental impact and waste effluents. Thus, vapor-liquid equilibrium, considered in Chapter 10, is used to design distillation, absorption, and stripping columns; liquid-liquid equilibrium information is used to design extraction systems, as mentioned in Chapter 11; and multicomponent mass and energy balances (Chapters 8 and 14) are used in the design of chemical reactors and heat exchangers, for example. Indeed, thermodynamics plays a very central role in all of process design.

However, in recent years chemical engineers have been involved not only in the design of chemical plants, that is, process design, but also in developing new or replacement products. This is referred to as product design. Many of these efforts involve the identification of a new compound or the formulation of a mixture with properties tailored for a specific application—anything from a new refrigerant (discussed below), to solvent mixtures used in the home or in medicine, to consumer products. Product design is increasingly being considered in the chemical engineering curriculum, and a detailed study of this subject is beyond the scope of this introductory thermodynamics book. However, as an introduction to product design, we briefly consider some simple examples here. There are several recent books containing information on product engineering.[14–16]

ILLUSTRATION 12.6-1

A Drop-in Replacement for the Antifreeze Ethylene Glycol

Water (with additives to protect against corrosion) circulates between the engine block and the radiator to cool automobile engines. Since water expands on freezing, if the water were to freeze in the engine block of a parked car in a cold-weather climate, severe damage would occur. To prevent this, a chemical antifreeze agent is added to the cooling water to lower the freezing temperature of the mixture; the analysis of freezing-point depression was discussed in Sec. 12.3, from which it is evident that the addition of any solute will lower the freezing point of water. Ethylene glycol is most commonly used. However, some automobile radiators leak, dripping the water + ethylene glycol mixture under parked cars. Because ethylene glycol has a pleasing, sweet taste but is toxic, it has been implicated in the deaths of pet cats.

Your job as a "product engineer" is to find a drop-in replacement for ethylene glycol as an antifreeze agent to ensure that the engine cooling fluid will not freeze at or above $-25°C$. By the term *drop-in replacement*, the following is meant. There are more than 100 million cars in current use in the United States, and we are not interested in making major alterations to the automobile engines currently in use; rather, we want something that we can use as a replacement antifreeze that does not require any change to existing automobiles.

SOLUTION

The solution to product design problems usually involves first thinking about various properties needed to narrow down the classes of chemicals or mixtures that might be appropriate, then using predictive methods to identify specific candidate chemicals, and finally obtaining data (hopefully from the literature or the use of predictive methods, and perhaps by measurement) to further discriminate between candidate compounds to choose the optimum that meets the constraints. Of course, any choice would have to be verified by testing before being commercialized.

---

[14]E. L. Cussler and G. D. Moggridge, *Chemical Product Design*, Cambridge University Press, New York (2001).

[15]J. Wei, *Molecular Structure-Property: Product Engineering*, Oxford University Press, New York (2006)

[16]W. D. Seider, J. D. Seader, and D. R. Levin, *Product and Process Design Principles*, 2nd ed. John Wiley & Sons, New York (1999).

For a replacement antifreeze mixture, some of the following constraints could be important:

1. Should be noncorrosive
2. Should have similar thermodynamic properties to ethylene glycol + water mixtures (so that the size of the radiator will not have to be changed)
3. Should have similar viscosity and thermal conductivity to ethylene glycol + water mixtures (so the water pump, plumbing, and radiator do not have to be changed)
4. Should be economical
5. Should not be very toxic or produce irritations or allergic reactions

The first decision is whether to look for a completely different fluid from the water–ethylene glycol mixture, or to keep water as the heat transfer fluid and look only for a replacement for ethylene glycol. If we were designing a new automobile, or at least a new automobile engine, there would be many possible heat transfer fluids to choose from. For example, silicon oils or mineral (petroleum-based) oils are frequently used as heat transfer fluids. However, we are looking only for a drop-in replacement, so that constraints 2 and 3 need to be satisfied. Therefore, we will restrict our search to a replacement for ethylene glycol in an aqueous mixture.

We next consider the classes of possible additives. We would not want to use a strong acid or base, such as sulfuric acid or sodium hydroxide, as they would be too corrosive. Indeed, most inorganic, soluble salts (for example, sodium chloride) are corrosive and further, could deposit on heat transfer surfaces if boiling occurred. Consequently, we will restrict our attention to organic liquids. Clearly, only organic liquids with a reasonably high water solubility should be considered. This excludes, for example, the alkanes and other hydrocarbons (benzene derivatives, cyclohydrocarbons, alkenes, alkynes), silicon oils, and other relatively nonpolar organic compounds of known low solubility in water. Also, only organic liquids with vapor pressures equal to or less than water should be considered, so as not to result in a significant increase in pressure in the automobile radiator. Water-soluble polymers—for example, polyethylene glycol, already used in foods and beauty and medicinal products—result in mixtures with significantly greater viscosity and lower heat transfer coefficients than pure water; therefore, polymers also are not candidate additives.

While these considerations have greatly narrowed the list of possible replacements, a large number of candidate antifreeze agents still exist. As an illustration of the methods that could be used, we will consider only two compounds here: propylene glycol, which is very similar to ethylene glycol but is much less toxic, and *n*-pentanol, which is toxic to animals and humans. However, since it has a bitter rather than a sweet taste, it is unlikely that lethal amounts would be ingested.

To proceed, we need to consider whether the two chemicals we have chosen have sufficient solubility to be used, and how much of each would be needed to obtain a desired freezing-point depression. The two pieces of data we need are the solubility of the two compounds we have chosen, and the activity coefficient of water in mixtures with these compounds as a function of composition at $-25°C$. The most reliable information is obtained from experimental data, and these should be obtained (from the literature or the laboratory) before a final decision is made. However, for a preliminary study of possible compounds, it is more common to use approximate predictive methods, such as UNIFAC, which we will do here.

Using the UNIFAC predictive method, the infinite-dilution activity coefficients in each of the three solutions at $-25°C$ are as follows:

Water + ethylene glycol mixture: $\gamma_W^\infty = 1.104$, $\gamma_{EG}^\infty = 2.204$

Water + *n*-pentanol mixture: $\gamma_W^\infty = 3.783$, $\gamma_{EG}^\infty = 223.9$

Water + propylene glycol mixture: $\gamma_W^\infty = 1.250$, $\gamma_{EG}^\infty = 6.133$

The starting point for the analysis of the freezing point depression is:

$$\ln \gamma_1 x_1 = -\frac{\Delta_{fus}\underline{H}(T_m)}{R}\left[\frac{T_m - T_f}{T_m T_f}\right] - \frac{\Delta C_P}{R}\left[1 - \frac{T_m}{T_f} + \ln\left(\frac{T_m}{T_f}\right)\right] \quad \textbf{(12.3-2)}$$

where the properties on the right-hand side of the equation are those of pure water. Using the heat of fusion of water given in Illustration 12.3-1 (neglecting the heat capacity term), we obtain the following condition to obtain a freezing point of $-25°C$:

$$\ln x_W \gamma_W = -0.267 \quad \text{or} \quad x_W \gamma_W = 0.765$$

Therefore, if we want to achieve the same freezing-point depression with any component X as we obtain with ethylene glycol, which we designate as EG, it is necessary that

$$\ln\left[x_W^{EG}\gamma_W^{EG}\left(x_W^{EG}\right)\right] = \ln\left[x_W^X\gamma_W^X\left(x_W^X\right)\right] = -0.267$$

Also, we want to ensure that the compound is sufficiently soluble in water, which we can check by solving the liquid-liquid equilibrium equations

$$x_W^I\gamma_W^I\left(x_W^I\right) = x_W^{II}\gamma_W^{II}\left(x_W^{II}\right) \quad \text{and} \quad x_X^I\gamma_X^I\left(x_X^I\right) = x_X^{II}\gamma_X^{II}\left(x_X^{II}\right) \qquad \textbf{(12.6-1)}$$

While we could do UNIFAC calculations at all compositions for each of the candidate compounds, we will adopt a further simplification here. We will use the simpler van Laar equations with the parameters obtained to match the infinite-dilution activity coefficients from the UNIFAC model. From such a calculation, we find that the mole fractions and weight fractions of the additives needed to attain a freezing point of $-25°C$ are as follows:

Compound	Mole fraction	Weight fraction
Ethylene glycol	0.277	0.569
Propylene glycol	0.338	0.665
n-Pentanol	0.729	0.929

To test for the possibility of liquid-liquid phase separation, which is needed to determine the solubility of each antifreeze candidate in water, we can solve Eq. 12.6-1 or simply examine whether the activity of water is a monotonic (i.e., steadily increasing) function of composition. If it is, then from an analysis like that in Fig. 11.2-4, a liquid-liquid phase split will not occur, as solubility is assured. Using the UNIFAC-based prediction discussed above, we find that all the candidate antifreezes are predicted to be completely soluble in water.

Based on these results, it appears that propylene glycol is a better choice than *n*-propanol as the antifreeze (freezing-point lowering agent) as it is less toxic, and much less is needed. Somewhat more propylene glycol than ethylene glycol will be needed, on a weight basis; however, propylene glycol has the advantage of not being as toxic as ethylene glycol.

Of course, all the calculations here are approximate. In particular, we have used UNIFAC, which is not meant to be applicable to liquid-liquid equilibrium, and further, $-25°C$ is below the temperature at which the UNIFAC parameters were obtained, so the results we have obtained are not expected to be accurate. Therefore, all the conclusions should be checked against experimental data. As examples of the uncertainty of such predictions, from experimental data it is found that the amount of ethylene glycol required to result in a mixture freezing point of $-25°C$ is closer to 45 wt % than the ~57 wt % estimated here, and that *n*-pentanol is actually only soluble in water to about 2.5 wt %, rather than the complete miscibility found here. This should serve as a warning concerning the use of any completely predictive method, even the UNIFAC model, which is currently the best.

Though our results are approximate, these simple calculations suggest that propylene glycol should be studied further as an antifreeze replacement through laboratory measurements. In fact, it is propylene glycol that is being used. ∎

As another example of product engineering, we return to the consideration of power and refrigeration cycles discussed in Chapter 5. There we did not consider the choice of the working fluid in these cycles. Here we consider the thermodynamic basis for the choice of working fluid in refrigeration cycles.

Choice of a refrigerant: the CFC problem

Among the considerations in choosing a working fluid or refrigerant when refrigeration was first developed were the following:

1. The fluid should have moderate vapor pressures between about 0°C and 50°C, the approximate operating range in the evaporator and condenser, respectively, of a home refrigerator, so that very high-pressure piping would not be needed and the probability of leaks would be lessened.
2. The fluid should have a reasonably high heat of vaporization, to minimize the amount of operating fluid circulating and therefore the size of the refrigeration system, and especially the size of the compressor.
3. The refrigerant should be noncorrosive and nontoxic. Other considerations, such as cost, availability, chemical stability, and safety, were also considered.

Below is a list of the properties of some candidate refrigerant fluids.

Compound	MW	T_b (°C)	$\triangle_{vap}\underline{H}$ (kJ/mol)	P^{vap} (0°C)	(bar) (50°C)	Flammable?	Toxic?	Reactive?
Ammonia (NH_3)	17	−33.4	24	4.3	20	Yes	Yes	Yes
CO_2	44	−78	25	34.9	$> T_c$	No	No	No
SO_2	64	−10	25	1.6	8.5	No	Yes	No
CF_2Cl_2 (Freon12)	121	−28.0	22	3.1	12.2	No	No	No

From these data, it is clear that carbon dioxide is not an appropriate working fluid for household refrigerators or automobile air conditioners because of the high pressures needed at the temperatures at which such devices operate, and that it could not be liquefied in the condenser of a home refrigerator since its critical temperature is exceeded. However, carbon dioxide was used for a while on some naval combat vessels, where toxicity and flammability considerations were of utmost importance, and the operating temperature range was different. Very early refrigerators used ammonia or sulfur dioxide. However, people died as a result of toxic leaks of both of these fluids in home refrigeration systems, or from fires as a result of ammonia leaks. Therefore, a safer refrigerant was needed.

Based on clever insight, chemical synthesis, and trials of new compounds, chlorinated fluorinated hydrocarbons (CFCs) were chosen as the class of working fluids of choice. As can be seen from the data in the table above, CF_2Cl_2, also known as Freon12, has thermodynamic properties very similar to those of ammonia. This refrigerant was developed in 1928, and Freon12 and other CFCs became the mostly widely used fluids for residential and commercial refrigeration. Other uses of CFCs were as propellants in spray cans, cleaning solvents for electronic circuit boards, and blowing agents in the manufacture of foams and expanded plastics. These latter uses resulted in the release of large amounts of CFCs into the atmosphere.

In 1974, in a classic Nobel Prize–winning paper by Rowland and Molina [*Nature*, **249**, 810 (1974)], it was shown that CFCs in the atmosphere could be responsible for destroying the ozone (O_3) layer that protects our planet from the strong ultraviolet (UV) radiation from the sun that is known to cause cancer in humans. This unsuspected problem arises from the lack of chemical reactivity of the CFCs under normal conditions, which results in their remaining in the atmosphere for a century or more, allowing time for their diffusion into the upper atmosphere. Once in the stratosphere, UV radiation from the sun breaks down the CFCs, releasing chlorine atoms that react in a continuing cycle (shown below) with ozone and oxygen radicals (produced by the effect of UV radiation on oxygen atoms).

$$Cl + O_3 \longrightarrow ClO + O_2$$

$$ClO + O \longrightarrow O_2 + Cl$$

So the net reaction is the chlorine-catalyzed process

$$O_3 + O \longrightarrow 2O_2$$

which destroys the ozone layer and regenerates the chlorine radical so that the process is repeated. It has been estimated that these reactive chlorine atoms remain in the stratosphere for between 60 and 400 years, during which time the 80 grams of a CFC in a typical household refrigerator could destroy as much as 6000 kilograms of ozone.

In 1987 the Montreal Protocol was agreed upon to drastically reduce the use of CFCs, and in 1990 this treaty was strengthened to call for the complete elimination of CFCs by the year 2000. This presented two different technological problems: finding a drop-in replacement for existing refrigeration equipment, and finding a long-term replacement for the CFCs. As for the first of these problems, there is a huge installed capacity of refrigerators, air conditioners, and related equipment with relatively long lifetimes so that it is not economically feasible to immediately replace them with new equipment. Therefore, there is the problem of developing a drop-in replacement refrigerant that can be used in existing equipment—an alternate working fluid that would be environmentally safe and have thermodynamic properties similar to the CFCs in use. In fact, the fluids with the most similar thermodynamic properties are HCFCs, hydrochlorofluorocarbons, which, as a result of replacing a halogen with a hydrogen atom, have greater chemical reactivity and consequently less stability, so their environmental lifetime is shorter and their impact on the ozone layer is less. (Note that they are not simply drop-in replacements in the sense that one only has to drain the CFC from a refrigeration system and replace it with an appropriate HCFC, as fittings, seals, and lubricants must also be replaced.) The HCFCs, in Montreal Protocol language, are transitional refrigerants and must be phased out by 2015. Among the other refrigerants in use are the HFCs, hydrofluorocarbons, which do not harm the ozone layer. However, the HCFCs and some of the HFCs are very strong contributors to global warming and, on a per-mass basis, are predicted to have an effect 100 to about 12 000 times greater than that of carbon dioxide, depending on the specific compound (though their concentration in the atmosphere is much lower than that of CO_2, so that their overall impact will be less). A separate problem is choosing working fluids for new refrigeration equipment designed for those fluids.

There are various thermodynamic considerations involved in the choice of a refrigerant. We have already discussed some of them (for example, the properties in the table above). However, other thermodynamic considerations arise, especially in the design of drop-in replacements. As there is no chemical species with exactly the same thermodynamic properties (vapor pressure, heat of vaporization, etc.) as the refrigerant it is to replace, a mixture or blend of compounds could be used instead. And, indeed, many of the new refrigerants that have been brought to market are blends of CFC, HCFC, and HFC compounds. However, refrigeration involves vapor-liquid equilibrium (vaporization in the evaporator and condensation in the condenser). With a pure fluid at a fixed operating pressure, these phase changes occur at a single temperature. A potential problem with a blend is that at fixed pressure vaporization occurs over a range of temperatures between the bubble temperature and dew point temperature (as is generally true for mixtures). This would lead to some uncertainty in how the refrigeration system will behave as a result of changes in load, and whether unanticipated concentration differences (and therefore temperature variations) will occur in different parts of the system. One way to avoid this is to use a blend that is an azeotrope at the operating pressures of the system. Indeed, some of the refrigerant blends used are azeotropic (or near azeotropic) mixtures.

The important message from this example is that it is the thermodynamic properties and phase behavior that determine which fluids could serve as refrigerants. Once the range of possible fluids has been determined, questions of cost, safety, ozone depletion, and global warming must also be taken into account.

Choice of a solvent

Choosing a solvent for a process—for example, as a degreasing agent, a liquid-liquid extraction agent, a paint, or another coating solvent—involves some of the same considerations as the choice of refrigerant, including toxicity, reactivity, moderate to low volatility (so there is not too much evaporative loss), environmental impact, and, perhaps most important, compatibility in the sense of high solubility for some components and low solubility for others. To be specific, in the choice of a solvent for degreasing electronic circuit boards during their manufacture, a desirable property is high solubility of grease, largely a moderate molecular weight hydrocarbon, while having no effect on the plastic in the circuit board. To ensure high solubility, one would want a solvent in which the solute (grease) had a low to moderate activity coefficient. Therefore, predictive methods, such as regular solution theory (frequently used in the paint industry) or UNIFAC (see Chapter 9), could be used to estimate values of infinite-dilution activity coefficients; in this way one could quickly consider many possible solvents and develop a list of solvent candidates, which could then be tested. Then, among the candidate solvents that have the appropriate thermodynamic properties for a specific application, the final choice would be made on the basis of other issues, such as cost, safety, environmental impact, and ease of recovery and reuse.

PROBLEM FOR SECTION 12.6

12.6-1 The more environmentally friendly refrigerants are made of hydrogen-containing compounds that have a short environmental lifetime. However, these compounds are also combustible. Therefore, one strategy is to design less flammable refrigerant mixtures by forming a ternary mixture of a flammable hydrocarbon refrigerant, such as FC152a, with two other, nonflammable components, one that is more volatile and another that is less volatile than the flammable component. One such refrigerant blend is a mixture containing FC152a, FC114, and R22. What composition of this liquid mixture will have an equilibrium pressure of 1.034 MPa at 50°C, but result in a vapor-phase FC152a mole fraction of only 0.043 so that the vapor will not be combustible? For this calculation, you can assume that the liquid mixture is ideal. Also, what are the vapor-phase mole fractions of the other components? The vapor pressures at 50°C are as follows:

$$P_{FC152a}^{vap} = 1.18 \text{ MPa}$$
$$P_{FC114}^{vap} = 0.43 \text{ MPa}$$
$$P_{R22}^{vap} = 2.03 \text{ MPa}$$

12.7 CONCLUDING REMARKS ON PHASE EQUILIBRIA

In this chapter and the previous two, many different types of phase equilibrium were considered. It is our hope that by first presenting the thermodynamic basis of phase equilibrium in Chapters 7 and 8, followed by the models for activity coefficients and mixture equations of state in Chapter 9, and then considering many types of phase equilibria in Chapters 10, 11, and 12, the reader will appreciate the essential unity of this subject and its wide applicability in chemical engineering.

It should be evident from the examples in Chapters 10, 11, and 12 that the evaluation of species fugacities or partial molar Gibbs energies (or chemical potentials) is central to any phase equilibrium calculation. Two different fugacity descriptions have been used, equations of state and activity coefficient models. Both have adjustable parameters. If the values of these adjustable parameters are known or can be estimated, the phase equilibrium state may be predicted. Equally important, however, is the observation that measured phase equilibria can be used to obtain these parameters. For example, in Sec. 10.2 we demonstrated how activity coefficients could be computed directly from P-T-x-y data and how activity coefficient models could be fit to such data. Similarly, in Sec. 10.3 we pointed out how fitting equation-of-state predictions to experimental high-pressure phase equilibrium data could be used to obtain a best-fit value of the binary interaction parameter.

In a similar fashion, solubility measurements (of a gas in a liquid, a liquid in a liquid, or a solid in a liquid) can be used to determine the activity coefficient of a solute in a solvent at saturation. Also, measurements of the solubility of a solid solute in two liquid phases can be used to relate the activity coefficient of the solute in one liquid to a known activity coefficient in another liquid, and freezing-point depression or boiling-point elevation measurements are frequently used to determine the activity of the solvent in a solute-solvent mixture. We have also showed that osmotic-pressure measurements can be used to determine solvent activity coefficients, or to determine the molecular weight of a large polymer or protein.

The body of thermodynamic information determined in the ways just described provide a base of knowledge for making the estimates and predictions needed for engineering design; for testing equations of state, their mixing rules, and liquid solution theories; and for extending our knowledge of the way molecules in mixtures interact. Perhaps more surprisingly, thermodynamics also allows us to make estimates of how some long-lived pollutants distribute in the environment.

Chapter 13

Chemical Equilibrium

Our interest in this chapter is with changes of state involving chemical reactions. (Biochemical reactions will be considered in Chapter 15.) Of concern here is the prediction of the equilibrium state in both simple and complicated chemical reaction systems. The mass and energy balance equations for such systems will be discussed in Chapter 14. Here we consider first chemical equilibrium in a single-phase, single-reaction system to indicate the underlying chemistry and physics of reaction equilibrium. This simple case is then generalized, in steps, to the study of equilibrium in multiphase, multireaction systems. For simplicity, the discussion of this chapter is largely of reaction equilibrium in ideal mixtures. Chemical equilibrium computations for nonideal systems are tedious and may require equation-solving programs and considerable iteration; several examples are given in this chapter.

INSTRUCTIONAL OBJECTIVES FOR CHAPTER 13

The goals of this chapter are for the student to:

- Be able to compute the equilibrium state of a single-phase chemical reaction (Secs. 13.1 and 13.3)
- Be able to compute the equilibrium state of a multiphase chemical reaction (Secs. 13.2 and 13.4)

IMPORTANT NOTATION INTRODUCED IN THIS CHAPTER

$\triangle_{\text{rxn}} G$ Gibbs energy change on reaction (J)

$\triangle_{\text{rxn}} G°$ Standard-state Gibbs energy change on reaction (J)

$\triangle_{\text{rxn}} H$ Enthalpy change on reaction, also called heat of reaction (J)

$\triangle_{\text{rxn}} H°$ Standard-state enthalpy change on reaction, also called standard heat of reaction (J)

K_a Chemical equilibrium constant

K_c Concentration chemical equilibrium ratio (units depend on reaction stoichiometry)

K_p Partial-pressure chemical equilibrium ratio (units depend on reaction stoichiometry)

K_x Liquid-phase mole fraction chemical equilibrium ratio
K_y Vapor-phase mole fraction chemical equilibrium ratio
$\triangle_{\mathrm{rxn}} V$ Volume change on reaction (m^3)

13.1   CHEMICAL EQUILIBRIUM IN A SINGLE-PHASE SYSTEM

In Chapter 7 the criteria for equilibrium were found to be

$$S = \text{maximum for a closed system at constant } U \text{ and } V$$
$$A = \text{minimum for a closed system at constant } T \text{ and } V \qquad \textbf{(13.1-1)}$$
$$G = \text{minimum for a closed system at constant } T \text{ and } P$$

For chemical reaction equilibrium in a single-phase, single-reaction system, these criteria lead to (see Sec. 8.8)

Starting point for chemical equilibrium calculations

$$\boxed{\sum v_i \bar{G}_i = 0} \qquad \textbf{(13.1-2)}$$

which, *together with the set of mass balance and state variable constraints on the system,* can be used to identify the equilibrium state of a chemically reacting system. The constraints on the system are important since, as will be seen shortly, a system in a given initial state, but subject to different constraints (e.g., constant T and P, or constant T and V), may have different equilibrium states (see Illustration 13.1-4). In each case the equilibrium state will satisfy both Eq. 13.1-2 and the constraints, and it will correspond to an extreme value of the thermodynamic function appropriate to the imposed constraints.

As an introduction to the application of Eqs. 13.1-1 and 13.1-2 to chemical equilibrium problems, consider the prediction of the equilibrium state for the low-pressure, gas-phase reaction

$$CO_2 + H_2 = CO + H_2O$$

occurring in a closed system at constant pressure and a temperature of 1000°C. The total Gibbs energy for this gaseous system is

$$G = \sum N_i \bar{G}_i(T, P, \underline{y})$$

$$= \sum N_i \underline{G}_i(T, P) + \sum N_i \{\bar{G}_i(T, P, \underline{y}) - \underline{G}_i(T, P)\} \qquad \textbf{(13.1-3a)}$$

$$= \sum N_i \underline{G}_i(T, P) + RT \sum N_i \ln\{\bar{f}_i(T, P, \underline{y})/f_i(T, P)\}$$

$$= \sum N_i \underline{G}_i(T, P) + RT \sum_i N_i \ln \left\{ \frac{y_i P \left(\dfrac{\bar{f}_i(T, P, \underline{y})}{y_i P} \right)}{P \left(\dfrac{f_i(T, P)}{P} \right)} \right\} \qquad \textbf{(13.1-3b)}$$

where $\bar{G}_i(T, P, \underline{y})$ and $\underline{G}_i(T, P)$ are the partial molar and pure component Gibbs energies for gaseous species i evaluated at the reaction conditions, and $(\bar{f}_i/y_i P)$ and (f_i/P) are the fugacity coefficients for species i in the mixture (Eq. 9.2-13) and as a pure component (Eq. 7.4-8), respectively. Since the pressure is low in this case, we assume that the gas phase is ideal, so that all fugacity coefficients are set equal to unity and Eq. 13.1-3 reduces to

$$G = \sum N_i \underline{G}_i(T, P) + RT \sum_i N_i \ln y_i \qquad (13.1\text{-}4)$$

As the reaction proceeds, the mole numbers and mole fractions of all species, and the total Gibbs energy of the reacting mixture, change. The number of moles of each reacting species in a closed system is not an independent variable (i.e., it cannot take on any value), but is related to the mole numbers of the other species and the initial mole numbers through the reaction stoichiometry. This is most easily taken into account using the molar extent of reaction variable of Chapters 2 and 8

$$N_i = N_{i,0} + \nu_i X$$

or

$$X = \frac{N_i - N_{i,0}}{\nu_i}$$

In this case, the initial and final mole numbers of the species are related as follows:

$$\begin{aligned} X &= (N_{CO} - N_{CO,0}) = (N_{H_2O} - N_{H_2O,0}) \\ &= -(N_{CO_2} - N_{CO_2,0}) = -(N_{H_2} - N_{H_2,0}) \end{aligned} \qquad (13.1\text{-}5)$$

where $N_{i,0}$ is the initial number of moles of species i. The number of moles of each species and the gas-phase mole fractions at any extent of reaction X are given in Table 13.1-1 for the case in which $N_{CO_2,0} = N_{H_2,0} = 1$ mole and $N_{CO,0} = N_{H_2O,0} = 0$. Balance tables such as this one are used throughout this chapter in the solution of chemical equilibrium problems.

The first term on the right side of Eq. 13.1-4 is the sum of the Gibbs energies of the pure components at the temperature and pressure of the mixture. The second term is the Gibbs energy change on forming a mixture from the pure components; it arises here because the reaction takes place not between the pure components, but between components in a mixture. The solid curve in Fig. 13.1-1 is a plot of the Gibbs energy of this reacting mixture as a function of extent of reaction for the initial mole numbers given in the table. That is, the solid line is a plot of

$$\begin{aligned} G &= \sum_i N_i \underline{G}_i + RT \sum_i N_i \ln y_i \\ &= (1 - X)(\underline{G}_{CO_2} + \underline{G}_{H_2}) + X(\underline{G}_{CO} + \underline{G}_{H_2O}) \\ &\quad + 2RT[(1 - X)\ln\{(1 - X)/2\} + X\ln\{X/2\}] \end{aligned} \qquad (13.1\text{-}6)$$

Table 13.1-1 Mass Balance for the Reaction $CO_2 + H_2 = CO + H_2O$

Species	Initial Number of Moles	Final Number of Moles	Mole Fractions
CO_2	1	$1 - X$	$(1 - X)/2$
H_2	1	$1 - X$	$(1 - X)/2$
CO	0	X	$X/2$
H_2O	0	X	$X/2$
Total	2	2	

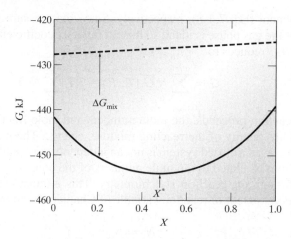

Figure 13.1-1 The Gibbs energy for the system $CO_2 + H_2 = CO + H_2O$ at 1000 K and 1.013 bar ($\sim$0.1 MPa) relative to each atomic species in its standard state at 298.15 K (25°C).

The dashed line in the figure is the Gibbs energy change as a function of the extent of reaction *neglecting* the Gibbs energy of mixing (that is, the logarithmic terms in X). The important feature of Fig. 13.1-1 is that because of the Gibbs energy of mixing term, there is an intermediate value of the reaction variable X for which the total Gibbs energy of the mixture is a minimum; this value of X has been denoted X^*. Since the condition for equilibrium at constant temperature and pressure is that G be a minimum, X^* is the equilibrium extent of reaction for this system.

This equilibrium state can be mathematically (rather than graphically) identified using the criterion that at equilibrium in a closed system at constant T and P, the Gibbs energy G is a minimum, or $dG = 0$, for all possible mole number variations *consistent* with the reaction stoichiometry. This implies that

$$\left(\frac{\partial G}{\partial X}\right)_{T,P} = 0 \tag{13.1-7}$$

since X is the single variable describing the mole number variations consistent with the stoichiometry and the initial amount of each species. Using this result in Eq. 13.1-6 yields

$$(\underline{G}_{CO} + \underline{G}_{H_2O} - \underline{G}_{CO_2} - \underline{G}_{H_2}) + 2RT[-\ln\{(1 - X^*)/2\} + \ln(X^*/2)] = 0$$

or

$$\frac{-(\underline{G}_{CO} + \underline{G}_{H_2O} - \underline{G}_{CO_2} - \underline{G}_{H_2})}{RT} = \ln\left[\frac{X^{*2}}{(1 - X^*)^2}\right] \tag{13.1-8}$$

To find the equilibrium mole fractions, Eq. 13.1-8 is first solved for the equilibrium extent of reaction X^*, and this is used with the initial mole numbers and stoichiometric information (i.e., Table 13.1-1) to find the mole fractions.

It is also of interest to note that using the stoichiometric relations between X and the species mole fractions, we obtain

$$\frac{-(\underline{G}_{CO} + \underline{G}_{H_2O} - \underline{G}_{CO_2} - \underline{G}_{H_2})}{RT} = \ln\left[\frac{y_{CO}y_{H_2O}}{y_{CO_2}y_{H_2}}\right] \tag{13.1-9a}$$

or, equivalently (and more generally),

$$\frac{-\sum_i \nu_i \underline{G}_i}{RT} = \sum_i \ln y_i^{\nu_i} = \ln\left[\prod_i y_i^{\nu_i}\right] \tag{13.1-9b}$$

(The symbol Π_i, used here denotes a product of numbers, that is, $\Pi_i y_i^{\nu_i} = y_A^{\nu_A} y_B^{\nu_B} y_C^{\nu_C} \ldots$) Thus, the equilibrium product of species mole fractions, each taken to the power of its stoichiometric coefficient, is related to the sum over the species of the stoichiometric coefficients times the *pure component* Gibbs energies at the temperature and pressure of the reacting mixture.

Whereas Eq. 13.1-8 is specific to the reaction and initial mole numbers of the example being considered, Eq. 13.1-9b (of which Eq. 13.1-9a is a special case) is generally valid for single-phase reactions in ideal mixtures; furthermore, Eq. 13.1-9b can be obtained directly from the general equilibrium relation, Eq. 13.1-2:

$$\sum_i \nu_i \overline{G}_i(T, P, \underline{x}) = 0$$

We show this here by first considering the more general case of chemical reactions in either nonideal vapor or liquid systems, for which

$$\overline{G}_i(T, P, \underline{x}) = \underline{G}_i(T, P) + RT \ln \frac{\bar{f}_i(T, P, \underline{x})}{f_i(T, P)} \tag{13.1-10}$$

Thus

$$\sum_i \nu_i \overline{G}_i(T, P, \underline{x}) = \sum_i \nu_i \underline{G}_i(T, P) + RT \sum_i \nu_i \ln\left(\frac{\bar{f}_i(T, P, \underline{x})}{f_i(T, P)}\right) = 0$$

or

$$\frac{-\sum_i \nu_i \underline{G}_i(T, P)}{RT} = \sum_i \nu_i \ln \frac{\bar{f}_i(T, P, \underline{x})}{f_i(T, P)}$$

$$= \ln\left[\prod_i \left(\frac{\bar{f}_i(T, P, \underline{x})}{f_i(T, P)}\right)^{\nu_i}\right] \tag{13.1-11}$$

For a gas-phase mixture,

$$\bar{f}_i(T, P, \underline{y}) = y_i P\left(\frac{\bar{f}_i}{y_i P}\right) \quad \text{and} \quad f_i(T, P) = P\left(\frac{f}{P}\right)_i$$

where the fugacity coefficients can be evaluated from an equation of state (or the principle of corresponding states). Thus Eq. 13.1-11 reduces to

Reactions in gas phase

$$\frac{-\sum \nu_i \underline{G}_i}{RT} = \ln\left[\prod_i \left(\frac{y_i(\bar{f}_i/y_i P)}{(f/P)_i}\right)^{\nu_i}\right] \tag{13.1-12a}$$

which, for an ideal gas (all fugacity coefficients equal to unity), becomes

Reactions in ideal gas phase

$$\frac{-\sum v_i \underline{G}_i}{RT} = \ln\left[\prod_i y_i^{v_i}\right] \qquad \text{(13.1-12b)}$$

The last equation is identical to Eq. 13.1-9b. (Note that Eq. 13.1-12a also applies to liquid mixtures describable by an equation of state by x replacing y.)

For a liquid mixture described by an activity coefficient model,

$$\bar{f}_i(T, P, \underline{x}) = x_i \gamma_i f_i(T, P)$$

so that

Reactions in liquid phase

$$\frac{-\sum v_i \underline{G}_i}{RT} = \ln\left[\prod_i (x_i \gamma_i)^{v_i}\right] \qquad \text{(13.1-13a)}$$

which, for an ideal liquid mixture (all activity coefficients equal to unity), reduces to

Reactions in ideal liquid phase

$$\frac{-\sum v_i \underline{G}_i}{RT} = \ln\left(\prod_i x_i^{v_i}\right) \qquad \text{(13.1-13b)}$$

For all future chemical equilibrium calculations in single-phase, single-reaction systems, we will start from Eqs. 13.1-12 or 13 as appropriate, rather than starting at Eq. 13.1-2 and repeating the derivation each time.

Figures like Fig. 13.1-1 provide some insight into the direction of progress of a chemical reaction. To be specific, the equilibrium and stability analysis of Chapter 7 establishes that a system at constant temperature and pressure evolves toward a state of minimum Gibbs energy. Here this is the state for which Eqs. 13.1-2, 13.1-12, or 13.1-13 are satisfied. Therefore, if, at any instant, the mole fractions of the reacting species in an ideal mixture are such that

$$\prod_i x_i^{v_i} < \exp\left[-\frac{\sum_i v_i \underline{G}_i(T, P)}{RT}\right]$$

(which, for the example in Fig. 13.1-1, occurs when the molar extent of reaction is to the left of X^*), the reaction proceeds as written. That is, reactants (species with negative stoichiometric coefficients) are consumed to form the reaction products (species with positive stoichiometric coefficients) until the equilibrium composition is reached. Conversely, if the species mole fractions are such that

$$\prod_i x_i^{v_i} > \exp\left[-\frac{\sum_i v_i \underline{G}_i(T, P)}{RT}\right]$$

the reaction proceeds in a manner such that what heretofore had been designated as the reaction products would be consumed to form reactants, until equilibrium was achieved, that is, until

$$\prod_i x_i^{v_i} = \exp\left[-\frac{\sum_i v_i \underline{G}_i(T, P)}{RT}\right]$$

Thus, our choice of reactants and products in a chemical reaction is arbitrary, in that the reaction can proceed in either direction. Indeed, replacing each stoichiometric coefficient (ν_i) by its negative ($-\nu_i$) in the previous equations, which is equivalent to interchanging the choice of reactants and products, leads to the same equilibrium state—again a state of minimum Gibbs energy. This freedom to arbitrarily choose which species are the reactants and which are the products is also evident from the equilibrium relation

$$\sum_i \nu_i \bar{G}_i(T, P, \underline{x}) = 0$$

since multiplying this equation by the constant -1 leaves the equation unchanged. In fact, multiplying this equation by any constant, either positive or negative, affects neither the equation nor the equilibrium state computed from it. Thus, whether we choose to write a reaction as

$$\alpha A + \beta B + \cdots \rightarrow \rho R + \cdots$$

$$2\alpha A + 2\beta B + \cdots \rightarrow 2\rho R + \cdots$$

or

$$\rho R + \cdots \rightarrow \alpha A + \beta B + \cdots$$

has no effect on what is ultimately predicted to be the equilibrium state of the system. Since chemical reactions can proceed either in the direction written or in the opposite direction, they are said to be **reversible**. (Note that here the word *reversible* is being used in a different sense than in Chapter 4.)

A chemical reaction goes to **completion** if it proceeds until one of the reactants is completely consumed. In principle, no homogeneous (i.e., single-phase) reaction goes to completion because the balance between the Gibbs energy of mixing and the $\Delta_{rxn}G = \sum_i \nu_i \underline{G}_i$ terms (see Fig. 13.1-1) forces the equilibrium value of X (that is, X^*) to lie between 0 and complete reaction. However, there are many instances when $\Delta_{rxn}G$ is so large in magnitude compared with the Gibbs energy of mixing term, which is of the order of magnitude of RT (see Eq. 13.1-4), that the reaction either goes essentially to completion ($-\Delta_{rxn}G/RT \gg 1$) or does not measurably occur ($\Delta_{rxn}G/RT \gg 1$). The room temperature oxidation of hydrogen to form water vapor,

$$H_2 + \tfrac{1}{2}O_2 = H_2O$$

for which

$$\frac{\Delta_{rxn}G}{RT} = \frac{-228\,570 \text{ J/mol}}{8.314 \text{ J/(mol K)} \times 298.15 \text{ K}} = -92.21$$

is an example of a reaction that goes essentially to completion. To see this, note that starting with stoichiometric amounts of hydrogen and oxygen, we have

Species	Initial Number of Moles	Final Number of Moles	Mole Fraction
H_2	1	$1 - X$	$(1 - X)/(1.5 - 0.5X)$
O_2	0.5	$0.5 - 0.5X$	$0.5(1 - X)/(1.5 - 0.5X)$
H_2O	0	X	$X/(1.5 - 0.5X)$
Total	1.5	$1.5 - 0.5X$	

so that

$$\exp\left(-\frac{\sum \nu_i \underline{G}_i}{RT}\right) = \prod x_i^{\nu_i}$$

$$\exp(+92.21) = 1.11 \times 10^{40} = \frac{x_{H_2O}}{x_{H_2} x_{O_2}^{0.5}} = \frac{X(1.5 - 0.5X)^{0.5}}{0.5^{0.5}(1.0 - X)^{1.5}}$$

The solution to this equation is $X \sim 1$ (actually $1.0 - 3 \times 10^{-27}$), so that, for all practical purposes, the reaction goes to completion. It is left to you to show that the room temperature oxidation of nitrogen

$$\tfrac{1}{2}N_2 + \tfrac{1}{2}O_2 = NO$$

for which

$$\frac{\Delta_{rxn}G}{RT} = \frac{86\,550 \text{ J/mol}}{8.314 \text{ J/(mol K)} \times 298.15 \text{ K}} = 34.92$$

does not, for practical purposes, occur at all.

The identification of the equilibrium extent of reaction in the carbon dioxide–hydrogen reaction discussed earlier was straightforward for two reasons. First, the mixture was ideal, so that there were no fugacity corrections or activity coefficients to consider. Second, it was presumed that the pure component Gibbs energies were available for the reacting species at the same temperature, pressure, and state of aggregation as the reacting mixture. Most chemical equilibrium calculations are, unfortunately, more complicated because few mixtures (other than low-pressure gas mixtures) are ideal, and pure component thermodynamic properties are usually tabulated only at a single temperature and pressure (25°C and 1 bar, as in Appendix A.IV). Thus, for most chemical equilibrium computations one needs fugacity or activity coefficient data (or adequate mixture models) and one must be able to estimate Gibbs energies for any pure component state.

One possible starting point for the analysis of general single-phase chemical equilibrium problems is the observation that the partial molar Gibbs energy of a molecular species can be written as

$$\begin{aligned}
\overline{G}_i(T, P, \underline{x}) = {} & \overline{G}_i^\circ(T = 25°C, \ P = 1 \text{ bar}, \ x_i^\circ) \\
& + [\overline{G}_i(T, P, \underline{x}) - \overline{G}_i^\circ(T = 25°C, \ P = 1 \text{ bar}, x_i^\circ)]
\end{aligned} \tag{13.1-14}$$

where $\overline{G}_i^\circ$ denotes the Gibbs energy of species i at 25°C, 1 bar, and composition x_i° (usually taken to be the pure component state of $x_i^\circ = 1$, the infinite-dilution state of $x_i^\circ = 0$, or the ideal 1 molal solution, depending on the species). However, starting from Eq. 13.1-14 requires estimates of the changes in species partial molar Gibbs energy with temperature, pressure, and composition—a difficult task. (The formulas of Chapter 9 account mainly for the composition and pressure variations of the species fugacity and partial molar Gibbs energy.) A more practical procedure is to choose the **standard state** of each species to be the species at composition x_i°, the *temperature of interest T,* and a pressure of 1 bar. In this case we have

Definition of standard state

$$\bar{G}_{\mathrm{i}}(T, P, \underline{x}) = \bar{G}_{\mathrm{i}}^{\circ}(T, P = 1 \text{ bar}, x_{\mathrm{i}}^{\circ})$$
$$+ [\bar{G}_{\mathrm{i}}(T, P, \underline{x}) - \bar{G}_{\mathrm{i}}^{\circ}(T, P = 1 \text{ bar}, x_{\mathrm{i}}^{\circ})]$$
$$= \bar{G}_{\mathrm{i}}^{\circ}(T, P = 1 \text{ bar}, x_{\mathrm{i}}^{\circ}) + RT \ln \left\{ \frac{\bar{f}_{\mathrm{i}}(T, P, \underline{x})}{\bar{f}_{\mathrm{i}}^{\circ}(T, P = 1 \text{ bar}, x_{\mathrm{i}}^{\circ})} \right\} \quad \textbf{(13.1-15)}$$
$$= \bar{G}_{\mathrm{i}}^{\circ}(T, P = 1 \text{ bar}, x_{\mathrm{i}}^{\circ}) + RT \ln a_{\mathrm{i}}$$

where we have introduced the **activity** of species i, denoted by a_{i}, defined to be the ratio of the species fugacity in the mixture to the fugacity in its standard state,

Definition of activity

$$a_{\mathrm{i}} = \frac{\bar{f}_{\mathrm{i}}(T, P, \underline{x})}{\bar{f}_{\mathrm{i}}^{\circ}(T, P = 1 \text{ bar}, x_{\mathrm{i}}^{\circ})} = \exp \left(\frac{\bar{G}_{\mathrm{i}}(T, P, \underline{x}) - \bar{G}_{\mathrm{i}}^{\circ}(T, P = 1 \text{ bar}, x_{\mathrm{i}}^{\circ})}{RT} \right)$$

$$\textbf{(13.1-16)}$$

With this formulation of the Gibbs energy function, the activity or fugacity for each species is a function of pressure and composition only, and its value can be computed using equations of state, or liquid solution data or models, all of which are discussed in Chapter 9. Note that once the standard state is chosen, the standard state Gibbs energy $\bar{G}_{\mathrm{i}}^{\circ}$ is a function of temperature only. The calculation of its value for any temperature will be considered shortly.

For convenience we have listed in Table 13.1-2 species activities for several common choices of the standard state.

With the notation introduced here, the equilibrium relation, Eq. 13.1-2, can now be written as

$$0 = \sum_{i=1}^{\mathcal{C}} \nu_{\mathrm{i}} \bar{G}_{\mathrm{i}} = \sum_{i=1}^{\mathcal{C}} \nu_{\mathrm{i}} \bar{G}_{\mathrm{i}}^{\circ}(T, P = 1 \text{ bar}, x_{\mathrm{i}}^{\circ}) + RT \sum_{i=1}^{\mathcal{C}} \nu_{\mathrm{i}} \ln a_{\mathrm{i}}$$

$$= \Delta_{\mathrm{rxn}} G^{\circ} + RT \sum_{i=1}^{\mathcal{C}} \nu_{\mathrm{i}} \ln a_{\mathrm{i}}$$

or

$$-\frac{\Delta_{\mathrm{rxn}} G^{\circ}}{RT} = \ln \prod_{i=1}^{\mathcal{C}} a_{\mathrm{i}}^{\nu_{\mathrm{i}}} \quad \textbf{(13.1-17)}$$

where we have used $\Delta_{\mathrm{rxn}} G^{\circ}$ to denote the quantity $\sum \nu_{\mathrm{i}} \bar{G}_{\mathrm{i}}^{\circ}(T, P = 1 \text{ bar}, x_{\mathrm{i}}^{\circ})$, the Gibbs energy change on reaction with each species (both reactants and products) in its standard state or state of unit activity. Finally, the **equilibrium constant** K_a is defined by the relation

Definition of the chemical equilibrium constant

$$\boxed{K_a(T) = \exp \left(-\frac{\Delta_{\mathrm{rxn}} G^{\circ}}{RT} \right)} \quad \textbf{(13.1-18)}$$

Table 13.1-2 Species Activity Based on Different Choices of Standard-State Activity[a]

State (All at Temperature T and Pressure P)	Standard State (All at Temperature T and 1 bar Pressure)	General	Simplification for Low and Moderate Pressures
Pure gas	Pure gas $\bar{G}_i^\circ = G_i^V(T, P = 1\text{ bar})$	$a_i = \dfrac{f_i^V(T,P)}{f_i^V(T,P=1\text{ bar})} = \dfrac{P(f/P)}{1\text{ bar}}$	$a_i = \dfrac{P}{1\text{ bar}}$
Species in a gaseous mixture	Pure gas[b] $\bar{G}_i^\circ = G_i^V(T, P = 1\text{ bar})$	$a_i = \dfrac{\bar{f}_i^V(T,P,\underline{y})}{f_i^V(T,P=1\text{ bar})} = \dfrac{y_i P(\bar{f}_i/y_i P)}{1\text{ bar}} = \dfrac{P_i(\bar{f}_i/y_i P)}{1\text{ bar}} = \dfrac{P_i\bar{\phi}_i(T,P,\underline{y})}{1\text{ bar}}$	$a_i = \dfrac{y_i P}{1\text{ bar}} = \dfrac{P_i}{1\text{ bar}}$
Pure liquid[c]	Pure liquid $\bar{G}_i^\circ = G_i^L(T, P = 1\text{ bar})$	$a_i = \dfrac{f_i^L(T,P)}{f_i^L(T,P=1\text{ bar})}$ $= \dfrac{P^{vap}(T)(f/P)_{sat}\exp\{(\underline{V}_i^L/RT)(P-P^{vap})\}}{P^{vap}(T)(f/P)_{sat}} = \exp\left\{\dfrac{\underline{V}^L}{RT}(P-P^{vap})\right\}$	$a_i = 1$
	Pure liquid[b] $\bar{G}_i^\circ = G_i^L(T, P = 1\text{ bar})$	$a_i = \dfrac{x_i\gamma_i P^{vap}(f/P)_{sat}\exp\{(\underline{V}_i/RT)(P-P^{vap})\}}{P^{vap}(f/P)_{sat}} = x_i\gamma_i\exp\left\{\dfrac{\underline{V}_i}{RT}(P-P^{vap})\right\}$	$a_i = x_i\gamma_i$ $(\gamma_i \to 1$ as $x_i \to 1)$
Species in a liquid mixture[c]	Species in a 1 molal ideal solution $\bar{G}_i^\circ = \bar{G}_i^\square(T, P = 1\text{ bar}, M_i = 1)$ (see Eq. 7.8-15)	$a_i = \dfrac{M_i\mathcal{H}_i(T,P=1\text{ bar})\gamma_i^\square\exp\left\{\dfrac{\bar{V}_i^\infty}{RT}(P-1\text{ bar})\right\}}{(M_i=1)\mathcal{H}_i(T,P=1\text{ bar})} = \dfrac{M_i\gamma_i^\square}{(M_i=1)}\exp\left\{\dfrac{\bar{V}_i^\infty}{RT}(P-1\text{ bar})\right\}$	$a_i = M_i\gamma_i^\square/(M_i=1)$ $(\gamma_i^\square \to 1$ as $M_i \to 0)$
	Species as a pure liquid with infinite-dilution properties $\bar{G}_i^\circ = \bar{G}_i^*(T, P = 1\text{ bar}, x_i = 1)$	$a_i = \dfrac{x_i H_i(T,P=1\text{ bar})\gamma_i^*\exp\left\{\dfrac{\bar{V}_i^\infty}{RT}(P-1\text{ bar})\right\}}{H_i(T,P=1\text{ bar})} = x_i\gamma_i^*\exp\left\{\dfrac{\bar{V}_i^\infty}{RT}(P-1\text{ bar})\right\}$	$a_i = x_i\gamma_i^*$ $(\gamma_i^* \to 1$ as $x_i \to 0)$
Pure solid	Pure solid $\bar{G}_i^\circ = G_i^S(T, P = 1\text{ bar})$	$a_i = \dfrac{f_i^S(T,P)}{f_i^S(T,P=1\text{ bar})} = \dfrac{P^{sat}(T)(f/P)_{sat}\exp\{(\underline{V}^S/RT)(P-P^{sat})\}}{P^{sat}(T)(f/P)_{sat}} = \exp\left\{\dfrac{\underline{V}^S}{RT}(P-P^{sat})\right\}$	$a_i = 1$
Species in a solid mixture	Pure solid $\bar{G}_i^\circ = G_i^S(T, P = 1\text{ bar})$	$a_i = \exp\left\{\dfrac{\underline{V}^S}{RT}(P-1\text{ bar})\right\}$ as above	$a_i = 1$
Dissolved electrolyte in solution	Dissolved electrolyte, each ion at unit molality in an ideal solution $\bar{G}_i^\circ = \nu_+\bar{G}_{A^+}(T, P = 1\text{ bar}, M_{A^+} = 1) + \nu_-\bar{G}_{B^-}(T, P = 1\text{ bar}, M_{B^-} = 1) = \bar{G}_{AB,D}^\square(T, P = 1\text{ bar})$		$a_i = \dfrac{M_{A^+}^{\nu+}M_{B^-}^{\nu-}(\gamma_\pm)^\nu}{(M_\pm=1)^\nu} = \dfrac{(M_\pm\gamma_\pm)^\nu}{(M_\pm=1)^\nu}$

[a] In this table we have neglected the Poynting correction to the pure component standard states.

[b] If a gaseous or liquid mixture is describable by an equation-of-state we have

$$a_i = \dfrac{\bar{f}_i(T,P,\underline{x})}{f_i(T,P=1\text{ bar})} = \dfrac{x_i P\bar{\phi}_i(T,P,\underline{x})}{(1\text{ bar})\phi_i(T,P=1\text{ bar})}$$

where the fugacity coefficients $\bar{\phi}_i(T,P,\underline{x})$ and $\phi_i(T,P=1\text{ bar})$ are computed from the equation of state.

[c] If a liquid or liquid mixture is describable by equation of state, the same equations as for gases or gaseous mixtures are used.

Thus, Eq. 13.1-17 can be rewritten as

Relation between
the equilibrium
constant and the
species activities at
equilibrium

$$K_a(T) = \prod_{i=1}^{\mathcal{C}} a_i^{\nu_i} = \frac{a_R^\rho \cdots}{a_A^\alpha a_B^\beta \cdots}$$

(13.1-19)

This equation is equivalent to Eq. 13.1-2 and can be used in the prediction of the chemical equilibrium state, provided that we can calculate a value for the equilibrium constant K_a and the species activities a_i.

If the standard state of each component is chosen to be $T = 25°C$, $P = 1$ bar, and the state of aggregation given in Appendix A.IV, then, following Eq. 8.5-2,

Calculation of the
standard state Gibbs
energy change on
reaction

$$\Delta_{rxn}G°(T = 25°C) = \sum_i \nu_i \, \Delta_f\underline{G}_i°(T = 25°C)$$

where $\Delta_f\underline{G}°$, the standard state Gibbs energy of formation, discussed in Sec. 8.5, is given in Appendix A.IV.

ILLUSTRATION 13.1-1
Gas-Phase Chemical Equilibrium Calculation

Calculate the equilibrium extent of decomposition of nitrogen tetroxide as a result of the chemical reaction $N_2O_4(g) = 2NO_2(g)$ at 25°C and 1 bar.

SOLUTION

The equilibrium relation is

$$K_a = \exp\left\{-\frac{\Delta_{rxn}G°(T = 25°C, P = 1 \text{ bar})}{RT}\right\} = \frac{a_{NO_2}^2}{a_{N_2O_4}} = \frac{\left(\dfrac{y_{NO_2}P}{1 \text{ bar}}\right)^2}{\left(\dfrac{y_{N_2O_4}P}{1 \text{ bar}}\right)} = \frac{y_{NO_2}^2}{y_{N_2O_4}}$$

Here we have assumed the gas phase is ideal. Furthermore, since the reaction and standard state pressures are both 1 bar, the pressure cancels out of the equation. Using the entries in Appendix A.IV, we have

$$\Delta_{rxn}G°(T = 25°C, \ P = 1 \text{ bar}) = 2\Delta_f\underline{G}_{NO_2}° - \Delta_f\underline{G}_{N_2O_4}°$$
$$= (2 \times 51.31 - 97.89) \text{ kJ/mol} = 4730 \text{ J/mol}$$

so that

$$K_a = \exp\left\{\frac{-4730 \text{ J/mol}}{8.314 \text{ J/(mol K)} \times 298.15 \text{ K}}\right\} = 0.1484$$

Next, we write the mole fractions of both NO_2 and N_2O_4 in terms of a single extent of reaction variable. This is most easily done using the mass balance table:

	Initial Number of Moles of Each Species	Final Number of Moles of Each Species	Equilibrium Mole Fraction
N_2O_4	1	$1 - X$	$y_{N_2O_4} = (1 - X)/(1 + X)$
NO_2	0	$2X$	$y_{NO_2} = (2X)/(1 + X)$
Total		$1 + X$	

Therefore,

$$K_a = 0.1484 = \frac{(2X/(1+X))^2}{(1-X)/(1+X)} = \frac{4X^2}{(1+X)(1-X)} = \frac{4X^2}{(1-X^2)}$$

or

$$X = \sqrt{\frac{K_a}{4+K_a}} = 0.1926$$

so that the nitrogen tetroxide is 19.3 percent decomposed at the conditions given, and

$$y_{N_2O_4} = 0.677$$
$$y_{NO_2} = 0.323$$

∎

Equilibrium compositions in a chemically reacting system are affected by changes in the state variables (i.e., temperature and pressure), the presence of diluents, or variations in the initial state of the system. These effects are considered in this discussion and the illustrations in the remainder of this section.

Effect of inert diluent

If an inert diluent is added to a reacting mixture, it may change the equilibrium state of the system, not as a result of a change in the value of the equilibrium constant (which depends only on the standard states and temperature), but rather as a result of the change in the concentration, and hence the activity, of each reacting species. This effect is illustrated in the following example.

ILLUSTRATION 13.1-2

Ideal Gas-Phase Chemical Equilibrium Calculation

Pure nitrogen tetroxide at a low temperature is diluted with nitrogen and heated to 25°C and 1 bar. If the initial mole fraction of N_2O_4 in the N_2O_4–nitrogen mixture before dissociation begins is 0.20, what is the extent of the decomposition and the mole fractions of NO_2 and N_2O_4 present at equilibrium?

SOLUTION

As in the preceding illustration,

$$K_a = 0.1484 = \frac{y_{NO_2}^2}{y_{N_2O_4}}$$

Here, however, we have

	Initial Number of Moles	Final Number of Moles	Equilibrium Mole Fraction
N_2O_4	0.2	$0.2 - X$	$(0.2 - X)/(1 + X)$
NO_2	0	$2X$	$(2X)/(1 + X)$
N_2	0.8	0.8	$0.8/(1 + X)$
Total		$1 + X$	

Therefore,

$$0.1484 = \frac{4X^2}{(0.2 - X)(1 + X)}$$

so that

$$X = 0.0715 \qquad y_{N_2O_4} = 0.1199 \qquad y_{NO_2} = 0.1334 \quad \text{and} \quad y_{N_2} = 0.7466$$

COMMENT

The fractional decomposition of N_2O_4, which is equal to

$$\frac{\text{Number of moles of } N_2O_4 \text{ reacted}}{\text{Initial Number of moles of } N_2O_4} = \frac{X}{0.2} = 0.3574$$

is higher here than in the case of undiluted nitrogen tetroxide (preceding illustration). At a fixed extent of reaction, the presence of the inert diluent nitrogen decreases the mole fractions of NO_2 and N_2O_4 equally. However, since the mole fraction of NO_2 appears in the equilibrium relation to the second power, the equilibrium must shift to the right (more dissociation of nitrogen tetroxide). (*Questions for the reader*: How would the presence of an inert diluent affect an association reaction, e.g., $2A \rightarrow B$? How would the presence of an inert diluent affect a gas-phase reaction in which $\sum \nu_i = 0$?) ∎

To compute the value of the equilibrium constant K_a at any temperature T, given the Gibbs energies of formation at $25°C$, we start with the observation that

$$\frac{\partial}{\partial T}\left(\frac{\bar{G}_i}{T}\right)_P = \frac{1}{T}\left(\frac{\partial \bar{G}_i}{\partial T}\right)_P - \frac{\bar{G}_i}{T^2} = -\frac{\bar{S}_i}{T} - \frac{\bar{H}_i}{T^2} + \frac{\bar{S}_i}{T} = -\frac{\bar{H}_i}{T^2} \qquad \textbf{(13.1-20a)}$$

and use the fact that $\ln K_a = -\sum \nu_i \Delta_f \underline{G}_i^\circ / RT$ to obtain

Variation of the equilibrium constant with temperature

$$\left(\frac{d \ln K_a}{dT}\right)_P = -\frac{1}{R}\frac{d}{dT}\left[\frac{\sum_i \nu_i \Delta_f \underline{G}_i^\circ}{T}\right] = \frac{1}{RT^2}\sum_i \nu_i \Delta_f \underline{H}_i^\circ = \frac{\Delta_{rxn} H^\circ(T)}{RT^2}$$

$$\textbf{(13.1-20b)}$$

(Note that these are total derivatives since the equilibrium constant K_a and the standard state Gibbs energy change are functions only of temperature.) Here $\Delta_{rxn} H^\circ = \sum \nu_i \Delta_f \underline{H}_i^\circ$ is the heat of reaction in the standard state, that is, the heat of reaction if the reaction took place with each species in its standard state at the reaction temperature. Equation 13.1-20b is known as the **van't Hoff equation**. If a reaction is **exothermic**, that is, if energy is released on reaction so that $\Delta_{rxn} H^\circ$ is negative, the equilibrium constant and the equilibrium conversion from reactants to products decrease with increasing temperature. Conversely, if energy is absorbed as the reaction proceeds, so that $\Delta_{rxn} H^\circ$ is positive, the reaction is said to be **endothermic**, and both the equilibrium constant and the equilibrium extent of reaction increase with increasing temperature. These facts are easily remembered by noting that reactions that release energy are favored at lower temperatures, and reactions that absorb energy are favored at higher temperatures.

The standard state heat of reaction $\Delta_{rxn} H^\circ$ at $25°C$ and 1 bar can be computed, as was pointed out in Chapter 8, from

$$\Delta_{rxn} H^\circ(T = 25°C) = \sum_i \nu_i \Delta_f \underline{H}_i^\circ(T = 25°C)$$

and the standard state heat of formation data in Appendix A.IV. At temperatures other than 25°C, we start from

$$\underline{H}_i(T) = \underline{H}_i(T = 25°C) + \int_{T=25°C}^{T} C_{P,i}(T') \, dT'$$

(where T' is a dummy variable in integration) and obtain

$$\Delta_{rxn}H°(T) = \sum \nu_i \underline{H}_i°(T) = \Delta_{rxn}H°(T = 25°C) + \int_{\substack{T=25°C \\ (298.15 \text{ K})}}^{T} \Delta_{rxn}C_P°(T') \, dT'$$

$$(13.1\text{-}21)$$

with $\Delta_{rxn}C_P° = \sum \nu_i C_{P,i}°$ where $C_{P,i}°$ is the heat capacity of species i in its standard state. Note that in this integration $\Delta C_P°$ may be a function of temperature.

Equation 13.1-20b can be integrated between any two temperatures T_1 and T_2 to give

$$\ln \frac{K_a(T_2)}{K_a(T_1)} = \int_{T_1}^{T_2} \frac{\Delta H_{rxn}°(T)}{RT^2} \, dT \qquad (13.1\text{-}22a)$$

so that if the equilibrium constant K_a is known at one temperature, usually 25°C, its value at any other temperature can be computed if the standard state heat of reaction is known as a function of temperature. If $\Delta_{rxn}H°$ is temperature independent, or if T_1 and T_2 are so close that $\Delta_{rxn}H°$ may be assumed to be constant over the temperature range, we obtain

Simplified equation for the variation of the equilibrium constant with temperature

$$\boxed{\ln \frac{K_a(T_2)}{K_a(T_1)} = -\frac{\Delta_{rxn}H°}{R} \left(\frac{1}{T_2} - \frac{1}{T_1} \right)} \qquad (13.1\text{-}22b)$$

Equation 13.1-22b suggests that the logarithm of the equilibrium constant should be a linear function of the reciprocal of the absolute temperature if the heat of reaction is independent of temperature and, presumably, an almost linear function of $1/T$ even if $\Delta_{rxn}H°$ is a function of temperature. (Compare this behavior with that of the vapor pressure of a pure substance in Sec. 7.7, especially Eq. 7.7-6.) Consequently, it is common practice to plot the logarithm of the equilibrium constant versus the reciprocal of temperature. Figure 13.1-2 gives the equilibrium constants for a number of reactions as a function of temperature plotted in this way. (Can you identify those reactions that are endothermic and those that are exothermic?)

For the general case in which $\Delta_{rxn}H°$ is a function of temperature, we start from the observation that the constant-pressure heat capacity is usually given in the form[1]

$$C_{P,i} = a_i + b_i T + c_i T^2 + d_i T^3 + e_i T^{-2}$$

(see Appendix A.II) and obtain, from Eq. 13.1-21 with $T_1 = 298.15$ K,

$$\begin{aligned}
\Delta_{rxn}H°(T) = \Delta_{rxn}H°(T_1) &+ \Delta a(T - T_1) + \frac{\Delta b}{2}(T^2 - T_1^2) \\
&+ \frac{\Delta c}{3}(T^3 - T_1^3) + \frac{\Delta d}{4}(T^4 - T_1^4) - \Delta e(T^{-1} - T_1^{-1})
\end{aligned} \qquad (13.1\text{-}23a)$$

---

[1] The last term, $e_i T^{-2}$, is usually present in expressions for the heat capacity of solids and is included here for generality.

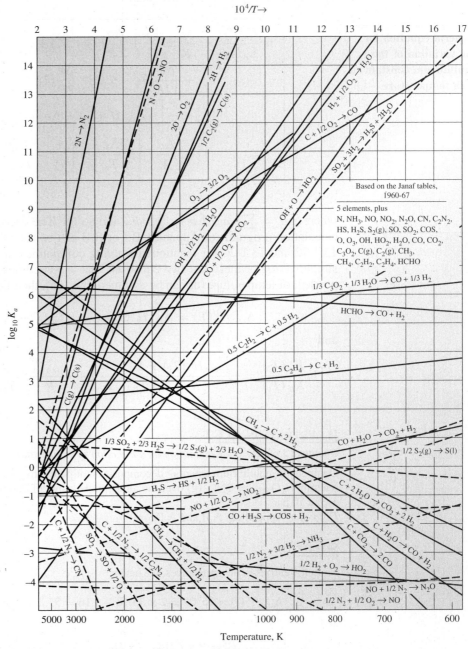

Figure 13.1-2 Chemical equilibrium constants as a function of temperature. MODELL & REID, THERMODYNAMICS AND ITS APPLICATIONS IN CHEMICAL ENGINEERING., 1st Edition, © 1974. Reprinted by permission of Pearson Education, Inc., Upper Saddle River, NJ. This figure appears as an Adobe PDF file on the CD-ROM accompanying this book, and may be enlarged and printed for easier reading and for use in solving problems. Also, the values of the chemical equilibrium constants can be calculated using Visual Basic and DOS-Basic Programs on the CD-ROM accompanying this book. These programs are discussed in Appendix B.I and B.II.]

Further, from Eq. 13.1-22a, we obtain

General equation for the variation of the equilibrium constant with temperature

$$
\begin{aligned}
\ln \frac{K_a(T_2)}{K_a(T_1)} = {} & \frac{\Delta a}{R} \ln \frac{T_2}{T_1} + \frac{\Delta b}{2R}(T_2 - T_1) + \frac{\Delta c}{6R}(T_2^2 - T_1^2) \\
& + \frac{\Delta d}{12R}(T_2^3 - T_1^3) + \frac{\Delta e}{2R}(T_2^{-2} - T_1^{-2}) \\
& + \frac{1}{R}\left[-\Delta_{\text{rxn}}H^\circ(T_1) + \Delta a T_1 + \frac{\Delta b}{2}T_1^2 + \frac{\Delta c}{3}T_1^3 \right. \\
& \left. + \frac{\Delta d}{4}T_1^4 - \frac{\Delta e}{T_1} \right] \times \left[\frac{1}{T_2} - \frac{1}{T_1} \right]
\end{aligned}
$$

(13.1-23b)

where $\Delta a = \sum v_i a_i$, $\Delta b = \sum v_i b_i$, and so on. The chemical equilibrium constant calculation programs of Appendix B.I or B.II can be used for calculations using Eqs. 13.1-22 and 13.1-23 for reactions involving compounds in their database. There is a simplification of these last two equations that is sometimes useful. If $\Delta_{\text{rxn}} C_P$ is independent of temperature (or can be assumed to be so), then Eq. 9.1-23a reduces to

$$
\Delta_{\text{rxn}}H^\circ(T) = \Delta_{\text{rxn}}H^\circ(T_1) + \Delta_{\text{rxn}}C_P(T - T_1) \tag{13.1-23c}
$$

and

$$
\ln \frac{K_a(T_2)}{K_a(T_1)} = -\frac{\Delta_{\text{rxn}}H^\circ(T_1)}{R}\left(\frac{1}{T_2} - \frac{1}{T_1} \right) + \frac{\Delta_{\text{rxn}}C_P}{R}\ln\frac{T_2}{T_1} + \frac{\Delta_{\text{rxn}}C_P}{R}\left(\frac{T_1}{T_2} - 1 \right)
$$

(13.1-23d)

ILLUSTRATION 13.1-3
Chemical Equilibrium Calculation as a Function of Temperature

Compute the equilibrium extent of decomposition of pure nitrogen tetroxide due to the chemical reaction $N_2O_4(g) = 2NO_2(g)$ over the temperature range of 200 to 400 K, at pressures of 0.1, 1, and 10 bar.
Data: See Appendices A.II and A.IV.

SOLUTION

From the entries in the appendices, we have

$$
\Delta_{\text{rxn}}H^\circ(T = 25^\circ C) = 2 \times 33.18 - 9.16 = 57.2 \text{ kJ/mol} = 57\,200 \text{ J/mol}
$$

and

$$
\begin{aligned}
\Delta_{\text{rxn}}C_P = \Delta_{\text{rxn}}C_P^* &= 2 \times C_P^*|_{NO_2} - C_P^*|_{N_2O_4} \\
&= 12.80 - 7.239 \times 10^{-2}T + 4.301 \times 10^{-5}T^2 + 1.573 \times 10^{-8}T^3 \text{ J/(mol K)}
\end{aligned}
$$

Thus

$$
\Delta a = 12.80 \qquad \Delta b = -7.239 \times 10^{-2} \qquad \Delta c = 4.301 \times 10^{-5}
$$

and

$$
\Delta d = 1.573 \times 10^{-8}
$$

Using these values in Eqs. 13.1-23a and b (with $T_1 = 298.15$ K), we obtain

$$\Delta_{rxn}H^\circ(T) = 56\,189 + 12.80T - 3.62 \times 10^{-2}T^2 + 1.434 \times 10^{-5}T^3 + 3.933 \times 10^{-9}T^4$$

and

$$
\begin{aligned}
\ln\left(\frac{K_a(T)}{K_a(T=25^\circ C)}\right) = \ln\left(\frac{K_a(T)}{0.154}\right) &= \int_{298.15\,K}^{T} \frac{\Delta_{rxn}H^\circ(T)}{RT^2}\,dT \\
&= -6758.4\left(\frac{1}{T} - \frac{1}{298.15\,K}\right) + 1.54\ln\frac{T}{298.15} \\
&\quad - 0.435 \times 10^{-2}(T - 298.15) + 0.862 \times 10^{-6}(T^2 - 298.15^2) \\
&\quad + 0.1577 \times 10^{-9}(T^3 - 298.15^3)
\end{aligned}
$$

The numerical values for the standard state heat of reaction $\Delta_{rxn}H^\circ(T)$ and the equilibrium constant K_a calculated from these equations are plotted in Fig. 1.

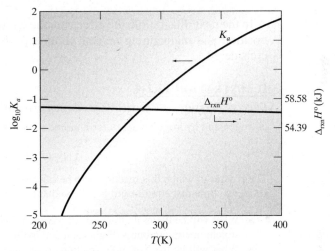

Figure 1 K_a versus pressure for the reaction $\frac{1}{2}N_2 + \frac{3}{2}H_2 \rightarrow NH_3$.

Now assuming that the gas phase is ideal, we have

$$K_a = \frac{a_{NO_2}^2}{a_{N_2O_4}} = \frac{(y_{NO_2}P/P = 1\,bar)^2}{(y_{N_2O_4}P/P = 1\,bar)} = \frac{y_{NO_2}^2}{y_{N_2O_4}}\left(\frac{P}{1\,bar}\right)$$

where, as in Illustration 13.1-1, $y_{NO_2} = 2X/(1 + X)$ and $y_{N_2O_4} = (1 - X)/(1 + X)$, so that

$$K_a = \frac{4X^2}{(1 - X^2)}\left(\frac{P}{1\,bar}\right)$$

and

$$X = \sqrt{\frac{K_a/P}{4 + K_a/P}} \quad P \text{ in bar}$$

The extent of reaction X and the mole fraction of nitrogen dioxide as a function of temperature and pressure are plotted in Fig. 2. (Note that the equilibrium constant is independent of pressure, but the equilibrium activity ratio increases linearly with pressure. Therefore, the extent of reaction for this reaction decreases as the pressure increases.)

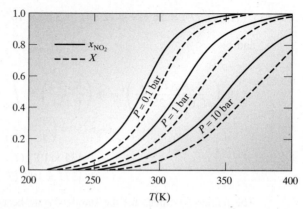

Figure 2 Equilibrium mole fraction x_{NO_2} and molar extent of reaction X for the reaction $N_2O_4 = 2NO_2$ as a function of temperature and pressure.

■

In the next illustration, the effects on the equilibrium composition of both feed composition and maintaining reactor volume (rather than reactor pressure) constant are considered.

ILLUSTRATION 13.1-4

Effect of Pressure on Chemical Equilibrium in an Ideal Gas Mixture

Nitrogen and hydrogen react to form ammonia in the presence of a catalyst,

$$\tfrac{1}{2}N_2 + \tfrac{3}{2}H_2 \rightarrow NH_3$$

The reactor in which this reaction is to be run is maintained at 450 K and has a sufficiently long residence time that equilibrium is achieved at the reactor exit.

- **a.** What will be the mole fractions of nitrogen, hydrogen, and ammonia exiting the reactor if stoichiometric amounts of nitrogen and hydrogen enter the reactor, which is kept at 4 bar?
- **b.** What will be the exit mole fractions if the reactor operates at 4 bar and the feed consists of equal amounts of nitrogen, hydrogen, and an inert diluent?
- **c.** The reaction is to be run in an isothermal, constant-volume reaction vessel with a feed consisting of stoichiometric amounts of nitrogen and hydrogen. The initial pressure of the reactant mixture (before any reaction has occurred) is 4 bar. What is the pressure in the reactor and the species mole fractions when equilibrium is achieved?

SOLUTION

- **a.** The starting point for this problem is the evaluation of the equilibrium constant for the ammonia production reaction at 450 K. From Appendices A.II and A.IV, we have

$$\Delta_f \underline{H}^\circ_{NH_3}(T = 25^\circ C, \ P = 1 \text{ bar}) = -46\ 100 \text{ J/(mol NH}_3) = \Delta_{rxn} H^\circ$$

$$\Delta_f \underline{G}^\circ_{NH_3}(T = 25^\circ C, \ P = 1 \text{ bar}) = -16\ 500 \text{ J/(mol NH}_3) = \Delta_{rxn} G^\circ$$

and

$$\Delta_{rxn} C_P^* = -30.523 + 2.928 \times 10^{-2} T - 1.40 \times 10^{-7} T^2 - 3.9465 \times 10^{-9} T^3 \text{ J/(mol K)}$$

Thus,

$$\Delta_{rxn}H^\circ(T) = -46\,100 + \int_{298.15\,K}^{T} \Delta_{rxn}C_P^* \, dT$$

and

$$\ln \frac{K_a(T = 450\,K)}{K_a(T = 298.15\,K)} = \int_{298.15\,K}^{450\,K} \frac{\Delta_{rxn}H^\circ(T)}{RT^2} \, dT = -6.459$$

Also,

$$\ln K_a(T = 298.15\,K) = -\frac{\Delta_{rxn}G^\circ}{RT} = \frac{16\,500}{8.314 \times 298.15} = 6.656$$

so that

$$\ln K_a(T = 450\,K) = 6.656 - 6.459 = 0.197$$

and

$$K_a(T = 450\,K) = 1.218 = \frac{a_{NH_3}}{a_{N_2}^{1/2} a_{H_2}^{3/2}}$$

At the low pressure here we will assume the gas phase is ideal, so that

$$a_i = \frac{y_i P_{rxn}}{P = 1\,bar}$$

where P_{rxn} is the reaction pressure. The mole fraction of each species is related to the inlet mole numbers and the molar extent of reaction as indicated in the following table.

	Initial Mole Number	Final Mole Number	Mole Fraction
NH_3	0	X	$X/(2 - X)$
N_2	$\frac{1}{2}$	$\frac{1}{2}(1 - X)$	$\frac{1}{2}(1 - X)/(2 - X)$
H_2	$\frac{3}{2}$	$\frac{3}{2}(1 - X)$	$\frac{3}{2}(1 - X)/(2 - X)$
Total		$2 - X$	

Therefore,

$$K_a = 1.218 = \frac{X(2 - X)}{\left[\frac{1}{2}(1 - X)\right]^{1/2} \left[\frac{3}{2}(1 - X)\right]^{3/2} \left(\dfrac{P_{rxn}}{1\,bar}\right)}$$

$$= \frac{X(2 - X)}{\left(\dfrac{1}{2}\right)^{1/2} \left(\dfrac{3}{2}\right)^{3/2} (1 - X)^2 \left(\dfrac{P_{rxn}}{1\,bar}\right)}$$

and at $P_{rxn} = 4$ bar

$$\left(\frac{1}{2}\right)^{1/2} \left(\frac{3}{2}\right)^{3/2} \left(\frac{P_{rxn}}{1\,bar}\right) K_a = 6.329 = \frac{X(2 - X)}{(1 - X)^2}$$

This equation has the solution

$$X = 0.6306$$

so that

$$y_{NH_3} = 0.4605 \qquad y_{N_2} = 0.1349 \qquad \text{and} \qquad y_{H_2} = 0.4046$$

b. Here we have

	Initial Mole Number	Final Mole Number	Mole Fraction
NH_3	0	X	$X/(3-X)$
N_2	1	$1 - \frac{1}{2}X$	$(1 - \frac{1}{2}X)/(3 - X)$
H_2	1	$1 - \frac{3}{2}X$	$(1 - \frac{3}{2}X)/(3 - X)$
Diluent	1	1	$1/(3 - X)$
Total		$3 - X$	

and

$$K_a = \frac{X(3 - X)}{\left(1 - \dfrac{1}{2}X\right)^{1/2} \left(1 - \dfrac{3}{2}X\right)^{3/2} \left(\dfrac{P_{rxn}}{1\,\text{bar}}\right)}$$

with $K_a = 1.218$ and $P_{rxn} = 4$ bar. The solution is $X = 0.4072$, and

$$y_{NH_3} = 0.157 \qquad y_{H_2} = 0.1501$$
$$y_{N_2} = 0.3072 \qquad y_{Dil} = 0.3857$$

c. As in part (a), we have

$$\left(\frac{1}{2}\right)^{1/2} \left(\frac{3}{2}\right)^{3/2} K_a \left(\frac{P_{rxn}}{1\,\text{bar}}\right) = \frac{X(2 - X)}{(1 - X)^2}$$

since the equilibrium criterion $\sum v_i \bar{G}_i = 0$ holds for reactions at constant T and V, just as it does for reactions at constant T and P (see Problem 8.4). Here, however, P_{rxn} is not fixed at 4 bar but depends on the number of moles of gas through the ideal gas law. Since the volume and temperature are fixed,

$$P_{rxn} = \frac{N_{eq}}{N_0} P_0 = \frac{2 - X}{2} P_0 = 2(2 - X) \,\text{bar}$$

where P_0 is the initial-state pressure, 4 bar. Thus,

$$2\left(\frac{1}{2}\right)^{1/2} \left(\frac{3}{2}\right)^{3/2} K_a = 3.1645 = \frac{X}{(1 - X)^2}$$

since $K_a = 1.218$. This equation has the solution $X = 0.5741$, so that

$$y_{NH_3} = 0.4206$$
$$P = 2.8519 \,\text{bar} \quad \text{and} \quad y_{N_2} = 0.1494$$
$$y_{H_2} = 0.4481$$

This solution should be compared with that obtained in part (a) for a reactor maintained at constant pressure. Can you explain why these answers differ? ■

The algebra involved in solving for the molar extent of reaction in general chemical equilibrium calculations can be tedious, especially if several reactions occur simultaneously, because of the coupled, nonlinear equations that arise. It is frequently possible, however, to make judicious simplifications based on the magnitude of the equilibrium constant. This is demonstrated in the next illustration.

ILLUSTRATION 13.1-5

Chemical Equilibrium at High Temperatures

At high temperatures, hydrogen sulfide dissociates into molecular hydrogen and sulfur:

$$2H_2S = 2H_2 + S_2$$

At $700°C$ all species are gases and the equilibrium constant for this reaction, K_a, is equal to 2.17×10^{-5}, based on standard states of the pure gases at the reaction temperature and a pressure of 1 bar.

    a. Estimate the extent of dissociation of pure hydrogen sulfide at $700°C$ and $P = 1$ bar.
    b. Show that the extent of dissociation is proportional to $P^{-1/3}$, and that if N moles of an inert diluent are added to 1 mole of hydrogen sulfide,

$$\frac{X(\text{with diluent})}{X(\text{without diluent})} = (1 + N)^{1/3}$$

SOLUTION

The starting point for the solution of this problem is the construction of a species balance table relating the mole fractions of each species to the molar extent of reaction. For brevity, this table is presented in a form that is applicable to parts (a) and (b) of this illustration.

Species	Initial Number of Moles	Final Number of Moles	Mole Fraction $N = 0$	Mole Fraction $N \neq 0$
H_2S	1	$1 - 2X$	$(1 - 2X)/(1 + X)$	$(1 - 2X)/(1 + X + N)$
H_2	0	$2X$	$2X/(1 + X)$	$2X/(1 + X + N)$
S_2	0	X	$X/(1 + X)$	$X/(1 + X + N)$
Diluent	N	N		
Total		$1 + X + N$		

The chemical equilibrium relation is

$$K_a = 2.17 \times 10^{-5} = \frac{a_{S_2} a_{H_2}^2}{a_{H_2S}^2} = \frac{y_{S_2} y_{H_2}^2 (P/1 \text{ bar})}{y_{H_2S}^2} = \frac{4X^3 (P/1 \text{ bar})}{(1 - 2X)^2 (1 + X + N)} \quad (1)$$

since $a_i = y_i P/(1 \text{ bar})$ for low-pressure gas mixtures. Because K_a is so small, we expect that $X \ll 1$. Therefore, it is reasonable to assume that

$$1 - 2X \approx 1$$

and

$$1 + X + N \approx 1 + N$$

so that

$$\frac{4X^3 P}{(1 - 2X)^2 (1 + X + N)} \approx \frac{4X^3 P}{(1 + N)} = 2.17 \times 10^{-5}$$

or

$$X = \left(\frac{2.17 \times 10^{-5}(1 + N)}{4P} \right)^{1/3}$$

Thus, it is clear that the equilibrium extent of reaction X is proportional to $P^{-1/3}$, and that

$$\frac{X(N \neq 0)}{X(N = 0)} = (1 + N)^{1/3}$$

At $P = 1$ bar and $N = 0$, we have

$$X = \left(\frac{2.17 \times 10^{-5}}{4}\right)^{1/3} = 0.0176$$

so that the fraction of H_2S dissociated, $2X/1$, is equal to 0.035. (Had we not made the simplification, the solution, by trial and error, would be $X = 0.0173$.)

COMMENT

At high temperatures K_a can become large and we can expect X to be close to 0.5. In this case we would assume that

$$1 + X + N \approx 1.5 + N$$

so

$$X^3 \approx (0.5)^3 = 0.125$$

However, $(1 - 2X)$, which appears in the denominator of Eq. 1, cannot be set equal to zero since the equilibrium relation is not satisfied in this case. Instead, at high temperatures (large values of K_a) we obtain an estimate of X by solving the equation

$$(1 - 2X)^2 \approx \frac{4(0.125)P}{(1.5 + N)K_a}$$

or

$$X \approx 0.5 - \sqrt{\frac{0.125P}{(1.5 + N)K_a}}$$

∎

A number of other equilibrium ratios that are more easily measured than K_a frequently appear in the scientific literature—for example, the concentration equilibrium ratio K_c, defined to be

$$K_c = \prod_i C_i^{\nu_i} \tag{13.1-24a}$$

where C_i is the concentration of species i (in $kmol/m^3$ or similar units); the mole fraction equilibrium ratios

$$K_x = \prod_i x_i^{\nu_i} \quad \text{and} \quad K_y = \prod_i y_i^{\nu_i} \tag{13.1-24b}$$

and the partial-pressure equilibrium ratio

$$K_p = \prod_i P_i^{\nu_i} \tag{13.1-24c}$$

In Table 13.1-3 each of these quantities is related to the equilibrium constant K_a and the quantities

Nonideal gas and nonideal solution contributions to chemical equilibrium

$$K_v = \prod_i \left(\frac{\bar{f}_i}{y_i P} \right)^{v_i} \quad \text{and} \quad K_\gamma = \prod_i \gamma_i^{v_i}$$ (13.1-24d)

which account for gas-phase and solution nonidealities, respectively. [Note that if the Lewis-Randall rule, $\bar{f}_i(T, P, \underline{y}) = y_i f_i(T, P)$, is used, then

$$K_v = \prod_i \left(\frac{f_i}{P} \right)^{v_i}$$ (13.1-24e)

This approximate expression, together with Fig. 7.4-1 for f/P is useful for making rapid estimates of the importance of gas-phase nonidealities.]

There are important differences between the true equilibrium constant K_a and the equilibrium ratios defined here. First, K_a depends only on temperature and the standard states of the reactants and products; the equilibrium ratios, such as K_c, K_x, and K_y, however, depend on mixture nonidealities (through K_v and K_γ) and on the total pressure or the total molar concentration. Consequently, while the thermodynamic equilibrium constant K_a can be used to study the same reaction with different diluents or solvents, the ratios K_c, K_x, and K_y have meaning only in the situation in which they were obtained. Finally, K_a is nondimensional, whereas K_p has units of (pressure)$^{\sum v_i}$ and K_c has units of (concentration)$^{\sum v_i}$.

ILLUSTRATION 13.1-6

Comparing Chemical Equilibrium Ratios

Compare the numerical values of K_a, K_p, and K_y for the ammonia production reaction of Illustration 13.1-4 at $P = 4$ bar and $T = 450$ K.

SOLUTION

From Illustration 13.1-4, $K_a = 1.218$ for the pure gas, $T = 450$ K, $P = 1$ bar standard state. From Table 13.1-3,

$$K_a = 1.218 = (P = 1 \text{ bar})^{-\sum v_i} K_p = \left(\frac{P}{P = 1 \text{ bar}} \right)^{\sum v_i} K_y$$

where we have set $K_v = 1$. For the reaction being considered,

$$\sum v_i = 1 - \frac{1}{2} - \frac{3}{2} = -1$$

so that

$$K_p = \frac{P_{NH_3}}{P_{N_2}^{1/2} P_{H_2}^{3/2}} = K_a(1 \text{ bar})^{-1} = 1.218 \text{ bar}^{-1}$$

and

$$K_y = K_a \left(\frac{P}{1 \text{ bar}} \right)^{-\sum v_i} = 1.218 \left(\frac{4}{1} \right)^1 = 4.872 \quad \text{(unitless)}$$ ∎

The prediction of the reaction equilibrium state for nonideal gas-phase reactions is more complicated than for ideal gas-phase reactions. This is demonstrated in the

Table 13.1-3 Chemical Equilibrium Ratios

Gaseous Mixture at Moderate or High Density
Standard state: Pure gases at $P = 1$ bar

$$K_a = \prod_i a_i^{\nu_i} = \prod_i \left[\frac{y_i P\left(\dfrac{\bar{f}_i}{y_i P} \right)}{1 \text{ bar}} \right]^{\nu_i} = \prod_i \left[\frac{P_i \left(\dfrac{\bar{f}_i}{y_i P} \right)}{1 \text{ bar}} \right]^{\nu_i} = \prod_i \left[\frac{P_i \bar{\phi}_i(T, P, \underline{y})}{1 \text{ bar}} \right]^{\nu_i}$$

$$= (1 \text{ bar})^{-\sum \nu_i} K_p K_\nu = \left(\frac{P}{1 \text{ bar}} \right)^{\sum \nu_i} K_y K_\nu$$

Gaseous Mixture at Low Density
Standard state: State of unit activity

$$\left(\frac{f}{P} \right) \cong 1 \qquad K_\nu \cong 1$$

and

$$K_a = (1 \text{ bar})^{-\sum \nu_i} K_p = \left(\frac{P}{1 \text{ bar}} \right)^{\sum \nu_i} K_y$$

Liquid Mixture
Standard state: State of unit activity*

$$K_a = \prod_i a_i^{\nu_i} = \prod_i (x_i \gamma_i)^{\nu_i} = K_x K_\gamma$$

Using $x_i = C_i / C$, where C_i is the molar concentration of species i, and C is the total molar concentration of the mixture, we have

$$K_a = \prod_i (x_i \gamma_i)^{\nu_i} = \prod_i \left(\frac{C_i}{C} \gamma_i \right)^{\nu_i} = C^{-\sum \nu_i} K_c K_\gamma$$

For an ideal mixture $\gamma_i = 1$, and

$$K_a = C^{-\sum \nu_i} K_c$$

Standard state: Ideal 1-molal solution

$$K_a = \prod_i a_i^{\nu_i} = \prod_i \left(\frac{M_i \gamma_i^{\square}}{M = 1} \right)^{\nu_i} = (M = 1)^{-\sum \nu_i} \prod_i \left(M_i \gamma_i^{\square} \right)^{\nu_i}$$

For an ideal mixture $\gamma_i^{\square} = 1$, and

$$K_a = (M = 1)^{-\sum \nu_i} \prod_i (M_i)^{\nu_i}$$

---

*The expressions here have been written assuming that the standard state of each component is the pure component state or the ideal 1-molal state. Analogous expressions can be written using the other of the Henry's law standard states for each component or, more generally, for the case in which the standard state for some species in the reaction is the pure component state and for others it is the infinite-dilution or ideal 1 molal state. One has to be careful to be consistent in that the standard state for the Gibbs energy of formation used for each component in the computation of the chemical equilibrium constant must also be considered to be the state of unit activity of each component.

following two illustrations, which show that there is an effect of pressure on reaction equilibrium due to gas-phase nonideality (as a result of the fugacity coefficient ratio K_ν), in addition to the primary pressure effect that was shown in Illustration 13.1-3 when there is a mole number change.

ILLUSTRATION 13.1-7

Effect of Pressure on an Ideal Gas-Phase Chemical Equilibrium

Compute the equilibrium mole fraction of each of the species in the gas-phase reaction

$$CO_2 + H_2 = CO + H_2O$$

at 1000 K and (a) 1 bar total pressure and (b) 500 atm (506.7 bar) total pressure. The equilibrium constant for this reaction K_a experimentally has been found to be equal to 0.693 at 1000 K (standard state is the pure gases at $T = 1000$ K, $P = 1$ bar), and initially there are equal amounts of carbon dioxide and hydrogen present.

SOLUTION

For this reaction, we have

$$K_a = 0.693 = \frac{a_{CO}a_{H_2O}}{a_{CO_2}a_{H_2}} = \frac{y_{CO}y_{H_2O}}{y_{CO_2}y_{H_2}} \times \frac{\left(\dfrac{\bar{f}_{CO}}{y_{CO}P}\right)\left(\dfrac{\bar{f}_{H_2O}}{y_{H_2O}P}\right)}{\left(\dfrac{\bar{f}_{CO_2}}{y_{CO_2}P}\right)\left(\dfrac{\bar{f}_{H_2}}{y_{H_2}P}\right)} \times \left(\frac{P}{1\text{ bar}}\right)^{\sum \nu_i}$$

$$= K_y K_\nu \left(\frac{P}{1\text{ bar}}\right)^{\sum \nu_i}$$

Since $\sum \nu_i = 0$, there is no primary effect of pressure on the extent of reaction; there is, however, a secondary effect through the pressure dependence of K_ν. Eliminating the species mole fractions in terms of the extent of reaction, we obtain

	Initial Mole Numbers	Final Mole Numbers	Mole Fractions
CO_2	1	$1 - X$	$(1 - X)/2$
H_2	1	$1 - X$	$(1 - X)/2$
CO	0	X	$X/2$
H_2O	0	X	$X/2$
Total		2	

and

$$K_a = 0.693 = \frac{X^2}{(1 - X)^2} K_\nu$$

a. At 1 bar $K_\nu = 1$, so that to predict the equilibrium state we need to solve

$$0.693 = X^2/(1 - X)^2$$

The solution to this equation is $X = 0.4543$, so that

$$y_{CO} = y_{H_2O} = 0.227 \quad \text{and} \quad y_{CO_2} = y_{H_2} = 0.273$$

b. At 506.7 bar, K_ν cannot be assumed to be equal to unity since each of the fugacity coefficients will have values different from unity; instead, the value of K_ν will be computed by using the Peng-Robinson equation of state to calculate each of the fugacity coefficients. However, since, in the fugacity coefficient calculation, the mole fractions are needed (first to compute the a and b parameters in the mixture, and then for evaluation of the fugacity coefficients after the compressibility factor has been found), the computation of the equilibrium state will be iterative. That is, a value of K_ν will be assumed (unity for the first iteration), and the equilibrium compositions will be computed. The mole fractions that result will then be used in the Peng-Robinson equation to compute the species fugacity coefficients and a new value of K_ν, which is then used to compute new equilibrium mole fractions. This procedure is repeated until the calculation has converged. The final result is

$$\left(\frac{\bar{f}}{yP}\right)_{H_2O} = 0.9606 \qquad \left(\frac{\bar{f}}{yP}\right)_{CO} = 1.1805$$

$$\left(\frac{\bar{f}}{yP}\right)_{CO_2} = 1.1283 \qquad \left(\frac{\bar{f}}{yP}\right)_{H_2} = 1.0992$$

and

$$K_\nu = 0.9144$$

Thus $X = 0.4654$, so that

$$y_{CO} = y_{H_2} = 0.2327 \quad \text{and} \quad y_{CO_2} = y_{H_2O} = 0.2673$$

which differs slightly from the low-pressure result. Note that $\prod_i P^{\nu_i} = P^{\sum_i \nu_i}$ is equal to unity for the reaction being considered since $\sum \nu_i = 0$. Therefore, for this reaction, largely because the pressures are not very high, the effect of this gas-phase nonideality is not very large. The only effect of pressure on this reaction equilibrium is as a result of gas-phase nonidealities. Such an effect may be much smaller than the direct effect of pressure when $\sum \nu_i \neq 0$ (and $\prod_i P^{\nu_i} \neq 1$), as found in Illustration 13.1-4. ∎

The following illustration is an example of a reaction where the effect of gas-phase nonideality is of greater importance.

ILLUSTRATION 13.1-8
Calculation of High-Pressure Chemical Equilibrium

The nitrogen fixation reaction to form ammonia, considered in Illustration 13.1-4, is run at higher temperatures in commercial reactors to take advantage of the faster reaction rates. However, since the reaction is exothermic, at a fixed pressure the equilibrium conversion (extent of reaction) decreases with increasing temperature. To overcome this, commercial reactors are operated at high pressures. The operating range of commercial reactors is pressures of about 350 bar and temperatures from 350°C to 600°C.

In trying to find the economically optimal operating conditions for design, it is frequently necessary to consider a wide range of conditions. Therefore, determine the equilibrium extent of reaction over the temperature range of 500 to 800 K and the pressure range from 1 bar to 1000 bar.

SOLUTION

The equilibrium constant for this reaction over the temperature range can be computed using the programs in Appendix B.I or B.II, or manually using the data in Appendix A. The result is

T (K)	K_a
500	0.258
600	0.02946
700	0.005754
800	0.001595

The equilibrium relationship is

$$K_a = \frac{a_{NH_3}}{a_{N_2}^{1/2} a_{H_2}^{3/2}} = \frac{y_{NH_3} \left(\dfrac{P_{rxn}}{1\ bar}\right) \bar{\phi}_{NH_3}}{\left[y_{N_2} \left(\dfrac{P_{rxn}}{1\ bar}\right) \bar{\phi}_{N_2}\right]^{1/2} \left[y_{H_2} \left(\dfrac{P_{rxn}}{1\ bar}\right) \bar{\phi}_{H_2}\right]^{3/2}}$$

or

$$K_a = \frac{y_{NH_3}}{y_{N_2}^{1/2} y_{H_2}^{3/2} \left(\dfrac{P_{rxn}}{1\ bar}\right)} \frac{\bar{\phi}_{NH_3}}{\bar{\phi}_{N_2}^{1/2} \bar{\phi}_{H_2}^{3/2}} = \frac{y_{NH_3}}{y_{N_2}^{1/2} y_{H_2}^{3/2} \left(\dfrac{P_{rxn}}{1\ bar}\right)} K_\nu$$

where $\bar{\phi}_i$ is the fugacity coefficient of species i in the mixture, and K_ν is the product of the species fugacity coefficients to the power of their stoichiometric coefficients.

Using the same feed as in Illustration 13.1-4, the equation to be solved is

$$K_a = \frac{X(2-X)}{\left[\dfrac{1}{2}(1-X)\right]^{1/2} \left[\dfrac{3}{2}(1-X)\right]^{3/2} \dfrac{P_{rxn}}{1\ bar}} K_\nu$$

The difficulty in solving this equation is that the fugacity coefficients, and therefore K_ν (computed using, for example, the Peng-Robinson equation of state), depend upon the mole fractions that must be obtained from the equilibrium relation. Consequently, the calculation is a complicated, iterative one. The results obtained using a specially prepared MATHCAD worksheet are shown in Figures 1 and 2.

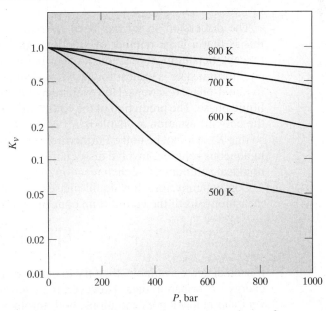

Figure 1 K_ν versus pressure for the reaction $\frac{1}{2}N_2 + \frac{3}{2}H_2 \rightarrow NH_3$.

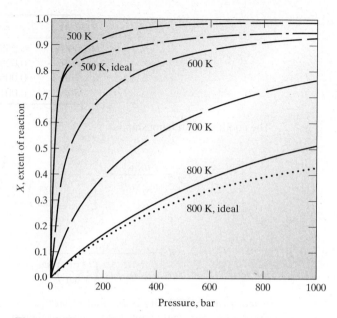

Figure 2 Extent of reaction versus pressure at several temperatures.

COMMENTS

There are several things to be noted from these figures. The first is that at any temperature, the equilibrium extent of reaction increases rapidly with increasing pressure. Second, at any pressure, the equilibrium conversion decreases with increasing temperature. Next, the fugacity ratio K_ν can be quite small in value at high pressures, especially at low temperatures. Consequently, the gas-phase nonidealities have an important effect on the equilibrium conversion. This is evident in Figure 2, which shows the equilibrium extent of reaction versus temperature and pressure, and where we have also plotted the ideal gas results ($K_\nu = 1$) at 500 and 800 K.■

The discussion so far has been restricted to gas-phase reactions. One reason for this is that a large number of reactions, including many high-temperature reactions (except metallurgical reactions), occur in the gas phase. Also, the identification of the equilibrium state is easiest for gases, as nonidealities are generally less important than in liquid-phase reactions. However, many reactions of interest to engineers occur in the liquid phase. The prediction of the equilibrium state in such cases can be complicated if the only information available is K_a or $\Delta_{rxn}G^\circ$, since liquid solutions are rarely ideal, so that K_γ will not be unity. Furthermore, some liquid-phase reactions, especially those in aqueous solution, involve dissociation of the reactants into ions (which form highly nonideal solutions), and then reaction of the ions (see Illustration 13.1-10).

Consequently, for many liquid-phase reactions one is more likely to use experimental measurements of the equilibrium concentration ratio

$$K_c = \prod_i C_i^{\nu_i} = C^{\sum \nu_i} K_a / K_\gamma \qquad (13.1\text{-}25)$$

if such data are available, than try to calculate the equilibrium state from K_a or Gibbs energy of formation data. However, the equilibrium ratio K_c is a function of the solvent and reactant concentrations, both through the $C^{\sum \nu_i}$ term and the activity coefficient ratio K_γ. Therefore, at a given temperature and pressure, several values of K_c may be given, each corresponding to different solvents or molar concentrations of reactants and diluents. (Frequently, as in Illustration 13.1-10, we will be satisfied with relating the value of K_c in one solution to its value in another, rather than trying to predict its value a priori.)

Given a value of K_c for the reaction conditions of interest, equilibrium calculations for liquid-phase reactions become straightforward, as indicated next.

ILLUSTRATION 13.1-9

Chemical Equilibrium in a Liquid Mixture

The ester ethyl acetate is produced by the reaction

$$CH_3COOH + C_2H_5OH = CH_3COOC_2H_5 + H_2O$$

In aqueous solution at 100°C, the equilibrium ratio K_c for this reaction is 2.92 (which is unitless since $\sum \nu_i = 0$). We will assume that the value of K_c is independent of concentration. Compute the equilibrium concentrations of each species in an aqueous solution that initially contains 250 kg of acetic acid and 500 kg of ethyl alcohol in each 1 m^3 of solution. The density of the solution may be assumed to be constant and equal to 1040 kg/m^3.

SOLUTION

The initial concentration of each species is

$$C_A = \frac{250 \text{ kg/m}^3}{60 \text{ g/mol}} = 4.17 \text{ kmol/m}^3$$

$$C_E = \frac{500 \text{ kg/m}^3}{46 \text{ g/mol}} = 10.9 \text{ kmol/m}^3$$

$$C_W = \frac{(1040 - 250 - 500) \text{ kg/m}^3}{18 \text{ g/mol}} = 16.1 \text{ kmol/m}^3$$

and the concentration of each species at an extent of reaction $\hat{X}$, in units of kmol/m^3, is

$$C_A = 4.17 - \hat{X}$$
$$C_E = 10.9 - \hat{X}$$
$$C_W = 16.1 + \hat{X}$$
$$C_{EA} = \hat{X}$$

Therefore, at equilibrium, we have

$$K_c = 2.92 = \frac{(16.1 + \hat{X})\hat{X}}{(10.9 - \hat{X})(4.17 - \hat{X})}$$

This equation has the solution $\hat{X}^* = 2.39$ kmol/m^3, so that

$$C_A = 1.78 \text{ kmol/m}^3$$
$$C_E = 8.51 \text{ kmol/m}^3$$
$$C_W = 18.49 \text{ kmol/m}^3$$
$$C_{EA} = 2.39 \text{ kmol/m}^3$$

Consequently, 57.3 percent of the acid has reacted at equilibrium.   ■

The partial ionization of weak acids, bases, and salts in solution, most commonly in aqueous solution, can be considered to be a chemical equilibrium process. Examples include the ionization of acetic acid,

$$CH_3COOH = CH_3COO^- + H^+$$

the ionization of water,

$$H_2O = H^+ + OH^-$$

and the dissociation of organic acids and alcohols. In the discussion here the ionization reaction will be designated as

$$A_{\nu_+}B_{\nu_-} = \nu_+ A^{z+} + \nu_- B^{z-} \tag{13.1-26}$$

where ν_+ and ν_- are the stoichiometric coefficients of the cation and anion, respectively, and z_+ and z_- are their valences.

The chemical equilibrium relation for this reaction is

$$K_a = \exp\left[-\frac{1}{RT}\{\nu_+ \Delta_f \underline{G}^\circ_{A^{z+}} \text{ (ideal 1-molal solution)}\right.$$
$$+ \nu_- \Delta_f \underline{G}^\circ_{B^{z-}} \text{ (ideal 1-molal solution)}$$
$$\left. - \Delta_f \underline{G}^\circ_{A_{\nu_+}B_{\nu_-}} \text{ (ideal 1-molal solution)}\}\right]$$

$$= \frac{a^{\nu_+}_{A^{z+}} a^{\nu_-}_{B^{z+}}}{a_{A_{\nu_+}B_{\nu_-}}} = \frac{\left(\dfrac{M_{A^{z+}} \gamma^\square_{A^{z+}}}{1 \text{ molal}}\right)^{\nu_+} \left(\dfrac{M_{B^{z-}} \gamma^\square_{B^{z-}}}{1 \text{ molal}}\right)^{\nu_-}}{\left(\dfrac{M_{A_{\nu_+}B_{\nu_-}} \gamma^\square_{A_{\nu_+}B_{\nu_-}}}{1 \text{ molal}}\right)}$$

Chemical equilibrium constant for an ionic dissociation reaction

$$K_a = \frac{\left(\dfrac{M_{A^{z+}}}{1 \text{ molal}}\right)^{\nu_+} \left(\dfrac{M_{B^{z-}}}{1 \text{ molal}}\right)^{\nu_-} (\gamma_\pm)^{\nu_+ + \nu_-}}{\left(\dfrac{M_{A_{\nu_+}B_{\nu_-}} \gamma^\square_{A_{\nu_+}B_{\nu_-}}}{1 \text{ molal}}\right)} \tag{13.1-27}$$

Here the standard state for the ionic species is a 1-molal ideal solution; the enthalpies and Gibbs energies of formation for some ions in this standard state at 25°C are given in Table 13.1-4. In Eq. 13.1-27 the standard state for the undissociated molecule has also been chosen to be the ideal 1-molal solution (see Eq. 9.7-20), although the pure component state could have been used as well (with appropriate changes in $\Delta_f \underline{G}^\circ_{A_{\nu_+}B_{\nu_-}}$ and $a_{A_{\nu_+}B_{\nu_-}}$). Finally, we have used the mean molal activity coefficient, $\gamma_\pm$, of Eq. 9.10-11. Also remember that for the 1-molal standard state, $\gamma^\square \to 1$ as the solution becomes very dilute in the component.

From the electrical conductance measurements it is possible to determine the total ionic concentration in a solution of a weak electrolyte and thus calculate the degree of ionization. This information is usually summarized in terms of the equilibrium concentration ratio

$$K_c = \frac{(C_{A^{z+}})^{\nu_+} (C_{B^{z-}})^{\nu_-}}{C_{A_{\nu_+}B_{\nu_-}}} \tag{13.1-28}$$

Historically, the equilibrium ratio K_c for ionization reactions has been called the **ionization constant** or the **dissociation constant**; clearly, its value will depend on both the total electrolyte concentration and the solvent.

Table 13.1-4 The Heats and Gibbs Energies of Formation for Ions in an Ideal 1-Molal Solution at 25°C

Ion	$\Delta_f\underline{H}^\circ$ (kJ/mol)	$\Delta_f\underline{G}^\circ$ (kJ/mol)	Ion	$\Delta_f\underline{H}^\circ$ (kJ/mol)	$\Delta_f\underline{G}^\circ$ (kJ/mol)
Ag^+	105.90	77.11	K^+	−251.21	−282.28
Al^{+++}	5470.33		Li^+	−278.40	−293.80
Ba^{++}	−538.36	−560.66	Mg^{++}	−461.96	−456.01
Ca^{++}	−542.96	−553.04	Mn^{++}	−218.82	−223.43
Cl^-	−167.46	−131.17	Na^+	−239.66	−261.88
Cs^+	−247.69	−282.04	NH_4^+	−132.80	−79.50
Cu^+	51.88	50.21	Ni^{++}	−64.02	−46.44
Cu^{++}	64.39	64.98	OH^-	−229.94	−157.30
F^-	−329.11	−276.48	Pb^{++}	1.63	−24.31
Fe^{++}	−87.86	−84.94	SO_3^{--}	−624.25	−497.06
Fe^{+++}	−47.70	−10.54	SO_4^{--}	−907.51	−741.99
Hg^{++}		164.77	Sr^{++}	−545.51	−557.31
HSO_3^-	−627.98	−527.31	Tl^+	5.77	−32.45
HSO_4^-	−885.75	−752.87	Tl^{+++}	115.90	209.2
I^-	−55.94	−51.67	Zn^{++}	−152.42	−147.21

Source: Based on data in G. N. Lewis, M. Randall, K. S. Pitzer, and L. Brewer, *Thermodynamics,* 2nd ed., McGraw-Hill, New York (1961).

Comparing Eqs. 13.1-27 and 13.1-28, and neglecting the difference between concentration and molality, yields

$$K_c = K_a \left(\frac{\gamma^{\square}_{A_{\nu_+}B_{\nu_-}}}{(\gamma_\pm)^{\nu_++\nu_-}} \right) (1 \text{ molal})^{\nu_++\nu_--1} \qquad \textbf{(13.1-29)}$$

Generally, we will be interested in very dilute solutions, so that $\gamma^{\square}_{A_{\nu_+}B_{\nu_-}}$ can be taken to unity. Also, we define an equilibrium constant K_c° by

$$K_c^\circ = K_a (1 \text{ molal})^{\nu_++\nu_--1} \qquad \textbf{(13.1-30)}$$

and obtain

$$\ln K_c = \ln K_c^\circ - (\nu_+ + \nu_-) \ln \gamma_\pm \qquad \textbf{(13.1-31)}$$

(Note that K_c° would be the ionization constant if the ions formed an ideal solution.) Depending on the total ionic strength of the solution, one of Eqs. 9.10-15, 9.10-17, 9.10-18, or 9.10-19 will be used to compute $\ln \gamma_\pm$. At very low ionic strengths, the Debye-Hückel relation applies so that

Simplified equation for the variation of the chemical equilibrium constant with ionic strength

$$\boxed{\ln K_c = \ln K_c^\circ + (\nu_+ + \nu_-)|z_+z_-|\alpha\sqrt{I}} \qquad \textbf{(13.1-32)}$$

where α is given in Table 9.10-1 and the ionic strength I is

$$I = \frac{1}{2} \sum_{\text{ions}} z_i^2 M_i$$

The important feature of Eq. 13.1-31 is that it can be used to predict the ionization constant of a molecule when data on the Gibbs energy of formation of its ion fragments are available (so that K_a and K_c° can be computed), or, when such data are not available, it can at least be used to interrelate the ionization constants for the same molecule at different ionic strengths, as in Illustration 13.1-10.

ILLUSTRATION 13.1-10

Calculation of the Chemical Equilibrium Constant for an Ionic Dissociation Reaction From Concentration Data

MacInnes and Shedlovsky [*J. Am. Chem. Soc.*, **54**, 1429 (1932)] report the following data for the ionization of acetic acid in water at 25°C:

Total Amount of Acetic Acid Added, C_T, kmol/m³	CH_3COO^- Concentration, kmol/m³
0.028×10^{-3}	0.1511×10^{-4}
0.1532×10^{-3}	0.4405×10^{-4}
1.0283×10^{-3}	1.273×10^{-4}
2.4140×10^{-3}	2.001×10^{-4}
5.9115×10^{-3}	3.139×10^{-4}
20.000×10^{-3}	5.975×10^{-4}

Establish that, at low acetic acid concentrations, these data satisfy Eq. 13.1-32, and compute K_c° and the standard Gibbs energy change for this reaction.

SOLUTION

The ionization reaction is

$$CH_3COOH = CH_3COO^- + H^+$$

so that $v_+ = v_- = 1$, $z_+ = 1$, $z_- = -1$ and

$$\ln K_c = \ln K_c^\circ + 2.356\sqrt{I}$$

Also, $C_{H^+} = C_{CH_3COO^-}$, $C_{CH_3COOH} = C_T - C_{CH_3COO^-}$, and

$$I = \frac{1}{2}\sum z_i^2 C_i = \frac{1}{2}\{C_{H^+} + C_{CH_3COO^-}\} = C_{CH_3COO^-}$$

The dissociation constant K_c is related to ionic strength as follows

$$K_c = \frac{C_{CH_3COO^-} C_{H^+}}{C_{CH_3COOH}} = \frac{(C_{CH_3COO^-})^2}{C_T - C_{CH_3COO^-}} = \frac{I^2}{C_T - I}$$

The dissociation constant and ionic strength are tabulated and plotted here:

Total Amount of Acetic Acid Added C_T, kmol/m³	$I = C_{CH_3COO^-}$ kmol/m³	K_c	$\ln K_c$
0.028×10^{-3}	0.1511×10^{-4}	1.768×10^{-5}	-10.943
0.1532×10^{-3}	0.4405×10^{-4}	1.778×10^{-5}	-10.938
1.0283×10^{-3}	1.273×10^{-4}	1.799×10^{-5}	-10.926
2.4140×10^{-3}	2.001×10^{-4}	1.809×10^{-5}	-10.920
5.9115×10^{-3}	3.193×10^{-4}	1.823×10^{-5}	-10.912
20.000×10^{-3}	5.975×10^{-4}	1.840×10^{-5}	-10.903

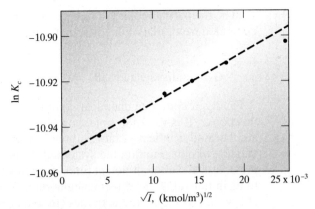

Figure 13.1-3 $\ln K_c$ versus $\sqrt{I}$ for the dissociation of acetic acid.

From Fig. 13.1-3 we see that Eq. 13.1-32 does fit the low-ionic-strength data well, and that $\ln K_c^\circ = -10.9515$. Thus, $K_c^\circ = 1.753 \times 10^{-5}$ kmol/m^3 and $K_a = 1.753 \times 10^{-5}$. Finally, since $\Delta_{rxn}G^\circ = -RT \ln K_a$, we have that at 25°C

$$\Delta_{rxn}G^\circ = \bar{G}_{CH_3COO^-}(\text{ideal 1 molal solution}) + \bar{G}_{H^+}(\text{ideal 1 molal solution})$$
$$- \bar{G}_{CH_3COOH}(\text{ideal 1 molal solution}) = 27.15 \text{ kJ/mol}$$

∎

Before leaving the subject of ionization equilibrium in electrolyte solutions, it is useful to note that the total Gibbs energy of a partially ionized electrolyte in a solvent is

$$G = N_S \bar{G}_S + N_{AB} \bar{G}_{AB} + N_{AB,D} \bar{G}_{AB,D} \qquad (9.10\text{-}8)$$

where $\bar{G}_{AB,D}$ is given by Eq. 9.10-14. To find the equilibrium degree of dissociation (or ionization) at fixed temperature, pressure, and number of moles of solvent, for which the total Gibbs energy is a minimum, we start from

$$dG = \bar{G}_S \, dN_S + \bar{G}_{AB} \, dN_{AB} + \bar{G}_{AB,D} \, dN_{AB,D}$$

(since $N_S \, d\bar{G}_S + N_{AB} \, d\bar{G}_{AB} + N_{AB,D} \, d\bar{G}_{AB,D} = 0$ by the Gibbs-Duhem equation) and set

$$\left(\frac{\partial G}{\partial N_{AB}}\right)_{T,P,N_S} = 0 = \bar{G}_{AB}\left(\frac{\partial N_{AB}}{\partial N_{AB}}\right)_{T,P,N_S} + \bar{G}_{AB,D}\left(\frac{\partial N_{AB,D}}{\partial N_{AB}}\right)_{T,P,N_S}$$

However, $N_{AB,D} = N_{AB}^\circ - N_{AB}$, where N_{AB}° is the initial fixed amount of the electrolyte AB added to the solution. Therefore,

$$\left(\frac{\partial N_{AB}}{\partial N_{AB}}\right)_{T,P,N_S} = 1 \quad \text{and} \quad \left(\frac{\partial N_{AB,D}}{\partial N_{AB}}\right)_{T,P,N_S} = \left(\frac{\partial \left(N_{AB}^\circ - N_{AB}\right)}{\partial N_{AB}}\right)_{T,P,N_S} = -1$$

and we obtain

$$0 = \bar{G}_{AB} - \bar{G}_{AB,D} \quad \text{or} \quad \bar{G}_{AB} = \bar{G}_{AB,D}$$

That is, the partial molar Gibbs energy of the undissociated salt is equal to, and can be computed from, that of the dissociated ions in solution:

$$\bar{G}_{AB}(\text{undissociated salt}) = \bar{G}^{\square}_{AB,D} + \nu RT \ln[(M_{\pm}\gamma_{\pm})/(M = 1)] \qquad \textbf{(13.1-33)}$$

Finally, a complete chemical equilibrium stability analysis is beyond the scope of this book.[2] It is useful to note, however, that generally states of chemical equilibrium are thermodynamically stable in that if a small change is made in the value of a state variable or constraint on the system, the equilibrium will shift, but it will return to its previous state, after sufficient time, if the altered state variable or constraint is restored to its initial value. By direct calculation involving either the equilibrium constant (for changes in temperature), activities (for changes in pressure and species concentration), or the system constraints, we can determine the shift in the equilibrium state in response to any external change. For example, Illustration 13.1-3 shows that both the equilibrium constant K_a and the molar extent of reaction X increase with increasing temperature for an endothermic ($\Delta_{rxn}H > 0$) reaction, whereas Illustration 13.1-4 shows that K_a decreases with increasing temperature for an exothermic ($\Delta_{rxn}H < 0$) reaction. Illustration 13.1-3 can also be interpreted as demonstrating that an increase in pressure decreases the extent of reaction for a reaction in which $\Delta_{rxn}V$ is positive, and would increase X if $\Delta_{rxn}V$ were negative (note that in an ideal gas phase reaction $\Delta_{rxn}V$ is proportional to $\Sigma\nu_i$). Adding an inert diluent to a reaction at constant temperature and pressure reduces the concentration of the reactant species, and its effect can be found accordingly. (In a gas-phase reaction the partial pressure of each species is reduced, so that the direction of the shift in equilibrium is the same as that which accompanies a reduction in total pressure.) Adding a diluent at constant temperature and volume, however, has no effect on the equilibrium, except through mixture nonidealities.

These observations and others on the direction of the shift in equilibrium in response to a given change are usually summarized by a statement referred to as the Principle of LeChatelier[3] and Braun[4] (but also known as the Principle of Moderation or the Principle of Spite[5]):

A system in chemical equilibrium responds to an imposed change in any of the factors governing the equilibrium (for example, temperature, pressure, or the concentration of one of the species) in such a way that, had this same response occurred without the imposed variation, the factor would have changed in the opposite direction.

Although this simple statement is a good rule of thumb for inferring the effects of changes in an equilibrium system, it is not universally valid, and exceptions to it do occur (Problem 13.17). A more general, universally valid statement of this principle is best given within the context of a complete thermodynamic stability analysis for a multicomponent reacting system. For many situations, however, it may be more useful, and even more expeditious, to ascertain the equilibrium shift by direct computation of

[2]See Chapter 9 of *Thermodynamics and Its Applications,* 2nd ed., by M. Modell and R. C. Reid, Prentice Hall, Englewood Cliffs, N.J., 1983; and *Chemical Thermodynamics,* by I. Prigogine and R. Defay (transl. by D. H. Everett), Longmans Green, New York, 1954.

[3]H. LeChatelier, "Recherches sur les equilibres chimique," *Annales des mines,* **13,** 200 (1888).

[4]F. Braun, *Z. Phys. Chem.,* **1,** 259 (1887).

[5]This term was used by J. Kestin in *A Course in Thermodynamics,* vol. 2, John Wiley & Sons, New York (1968), p. 304.

the new equilibrium state, rather than merely surmising the direction of that shift from a detailed stability analysis.

13.2   HETEROGENEOUS CHEMICAL REACTIONS

The discussion of the previous section was concerned with chemical reactions that occur in a single phase. Here our interest is with reactions that occur among species in different phases but that do not involve combined chemical and phase equilibrium. Examples of such reactions are

$$CaF_2(s, \text{ fluorspar}) + H_2SO_4(l, \text{ pure}) \rightarrow CaSO_4(s) + 2HF(g) \quad \textbf{(13.2-1)}$$
$$CH_4 \rightarrow C(s) + 2H_2 \quad \textbf{(13.2-2)}$$

and

$$CaCO_3(s) \rightarrow CaO(s) + CO_2 \quad \textbf{(13.2-3)}$$

where s, l, and g indicate the solid, liquid, and gas phases, respectively. In the first of these reactions, the solubility of hydrogen fluoride in sulfuric acid is negligible, and in all these reactions the gaseous and solid species can be considered to be mutually insoluble. Consequently, the determination of the equilibrium state for each of these systems involves considerations of chemical, but not phase, equilibrium.

The most important characteristic that differentiates these heterogeneous reactions from the homogeneous reactions considered in the previous section is that in the calculation of the equilibrium state for heterogeneous reactions, the activity of each species is affected by either dilution or mixture nonideality only by other components that appear **in the same phase**. Thus, to analyze the production of hydrogen fluoride gas by the reaction of Eq. 13.2-1, we have

$$K_a = \frac{a_{CaSO_4} a_{HF}^2}{a_{CaF_2} a_{H_2SO_4}}$$

which, at low pressures and using the 1 bar pure component gaseous standard state for HF, the pure component liquid standard state for H_2SO_4, and the pure component solid standard states for CaF_2 and $CaSO_4$, reduces to

$$K_a = \left(\frac{y_{HF}P}{1 \text{ bar}}\right)^2 = \left(\frac{P_{HF}}{1 \text{ bar}}\right)^2$$

since the activities of the pure liquid (H_2SO_4) and the unmixed solids ($CaSO_4$ and CaF_2) are each unity (see Table 13.1-2), and

$$a_{HF} = \frac{y_{HF}P}{(P = 1 \text{ bar})}$$

Thus the equilibrium pressure of hydrogen fluoride is easily calculated—clearly much more easily than if all reaction species were present in the same phase.

ILLUSTRATION 13.2-1

Calculation of Chemical Equilibrium for a Reaction Involving a Gas and a Solid

Carbon black is to be produced from methane in a reactor maintained at a pressure of 1 bar and a temperature of 700°C (the reaction of Eq. 13.2-2). Compute the equilibrium gas-phase conversion and the fraction of pure methane charged that is reacted. The equilibrium constant K_a for this reaction at $T = 700°C$ is 7.403 based on the pure component standard states (gaseous for CH_4 and H_2, and solid for C) at the temperature of the reaction and 1 bar.

SOLUTION

Basis of the calculation: 1 mole of methane

Species	Initial Amount	Final Amount	Present in Gas Phase	Gas-Phase Mole Fraction
CH_4	1	$1 - X$	$1 - X$	$(1 - X)/(1 + X)$
C	0	X		
H_2	0	$2X$	$2X$	$(2X)/(1 + X)$
Total			$1 + X$	

Thus

$$K_a = 7.403 = \frac{a_C a_{H_2}^2}{a_{CH_4}} = \frac{y_{H_2}^2}{y_{CH_4}} = \frac{(2X)^2}{(1 + X)(1 - X)}$$

since the activity of solid carbon is unity and therefore does not appear in the equilibrium calculation. Also, the reaction and standard-state pressures are both 1 bar. Therefore,

$$X = \sqrt{\frac{K_a}{4 + K_a}} = 0.806 \quad \text{and} \quad \begin{matrix} y_{CH_4} = 0.108 \\ y_{H_2} = 0.892 \end{matrix}$$ ■

An important observation from the equilibrium calculation just performed, or, for that matter, of any chemical equilibrium calculation in which both reactants *and* products appear in the same gaseous or liquid phase, is that the reaction will not go to completion (i.e., completely consume one of the reacting species) unless K_a is infinite. This is because of the contribution to the total Gibbs energy from the Gibbs energy of mixing reactant and product species, as was the case in homogeneous liquid- or gas-phase chemical reactions (see Sec. 13.1 and Fig. 13.1-1).

The decomposition of calcium carbonate (Eq. 13.2-3), or any other reaction in which the reaction products and reactants do not mix in the gas or liquid phase, represents a fundamentally different situation from that just considered, and such a reaction *may* go to completion. To see why this occurs, consider the reaction of Eq. 13.2-3 in a constant temperature and constant pressure reaction vessel, and let $N_{CaCO_3,0}$ and $N_{CO_2,0}$ represent the number of moles of calcium carbonate and carbon dioxide, respectively, before the decomposition has started. Also, since none of the species in the reaction mixes with the others, we use pure component molar than partial molar Gibbs energies in the analysis. Then,

$$\begin{pmatrix} \text{Initial Gibbs} \\ \text{energy of} \\ \text{the system} \end{pmatrix} = G_0 = N_{CaCO_3,0} \underline{G}_{CaCO_3,0}(T, P) + N_{CO_2,0} \underline{G}_{CO_2,0}(T, P)$$

and

$$\begin{pmatrix} \text{Gibbs energy of} \\ \text{the system at molar} \\ \text{extent of reaction } X \end{pmatrix} = (N_{CaCO_3,0} - X)\underline{G}_{CaCO_3}(T, P) + X\underline{G}_{CaO}(T, P)$$

$$+ (N_{CO_2,0} + X)\underline{G}_{CO_2}(T, P) \qquad \textbf{(13.2-4)}$$

$$= G_0 + X \sum v_i \underline{G}_i$$

$$= G_0 + X \left\{ \Delta_{rxn}G^\circ + RT \ln\left(\frac{P_{CO_2}}{P = 1 \text{ bar}}\right) \right\}$$

where $\Delta_{rxn}G^\circ$ is the standard-state Gibbs energy change of reaction, that is, the Gibbs energy change if the reaction took place between the pure components at the temperature T and a pressure of 1 bar. Notice that the actual (not standard state) Gibbs energy is a function of the molar extent of reaction both explicitly and, since the system is closed and the CO_2 partial pressure depends on X, implicitly.

If the initial partial pressure of carbon dioxide is so low that

$$\Delta_{rxn}G^\circ + RT \ln\left(\frac{P_{CO_2}}{P = 1 \text{ bar}}\right) < 0$$

the minimum value of the total Gibbs energy of the system occurs when enough calcium carbonate has decomposed to raise the partial pressure of carbon dioxide so that

$$\Delta_{rxn}G^\circ + RT \ln\left(\frac{P_{CO_2}}{P = 1 \text{ bar}}\right) = 0$$

or, if there is insufficient calcium carbonate present to do this, when all the calcium carbonate has been consumed and the reaction has gone to completion (i.e., $X = N_{CaCO_3}$). That is, if

$$-\frac{\Delta_{rxn}G^\circ}{RT} > \ln\left(\frac{P_{CO_2}}{P = 1 \text{ bar}}\right)$$

or, equivalently, if

$$K_a = \exp\left(-\frac{\Delta_{rxn}G^\circ}{RT}\right) > a_{CO_2}$$

where $a_{CO_2} = P_{CO_2}/1$ bar, the reaction will proceed as written until either the equilibrium partial pressure of CO_2 is achieved or all the calcium carbonate is consumed, whichever occurs first.

On the other hand, if initially P_{CO_2} is so large that

$$\Delta_{rxn}G^\circ + RT \ln\left(\frac{P_{CO_2}}{P = 1 \text{ bar}}\right) > 0$$

or

$$K_a = \exp\left\{-\frac{\Delta_{rxn}G^\circ}{RT}\right\} < a_{CO_2}$$

the minimum Gibbs energy occurs when $X = 0$ (i.e., when there is no decomposition of calcium carbonate). In systems containing both $CaCO_3$ and CaO, when the partial pressure of carbon dioxide is such that

$$K_a = a_{CO_2} = \left\{ \frac{P_{CO_2}}{P = 1 \text{ bar}} \right\}$$

equilibrium is achieved; that is, calcium carbonate will neither decompose to, nor be formed from, calcium oxide.

These various possibilities are shown in Fig. 13.2-1, where we plot, as a function of the imposed partial pressure of carbon dioxide, the quantity

$$\Delta G = \left(\begin{array}{c} \text{Gibbs energy of} \\ \text{system at molar extent} \\ \text{of reaction } X \end{array} \right) - \left(\begin{array}{c} \text{Initial Gibbs} \\ \text{energy of system} \end{array} \right)$$

$$= \left[G_0 + X \left\{ \Delta_{rxn} G^\circ + RT \ln \left(\frac{P_{CO_2}}{P = 1 \text{ bar}} \right) \right\} \right] - G_0$$

$$= X \left\{ \Delta_{rxn} G^\circ + RT \ln \left(\frac{P_{CO_2}}{P = 1 \text{ bar}} \right) \right\}$$

at $T = 1200$ K. At this temperature $\Delta_{rxn} G^\circ = -8706$ J/mol (see Illustration 13.2-2). In Fig. 13.2-1 we see that for carbon dioxide partial pressures of less than 2.393 bar, the Gibbs energy change on reaction and the system Gibbs energy decrease as the molar extent of reaction increases. Therefore, the reaction will proceed until all the calcium carbonate decomposes to calcium oxide. Conversely, for carbon dioxide partial pressures above 2.393 bar, the Gibbs energy change on reaction and the system Gibbs energy increase as the reaction proceeds. Therefore, the state of minimum Gibbs energy of the system, which is the equilibrium state, is when $X = 0$ and no dissociation of the calcium carbonate occurs. However, if the partial pressure of carbon dioxide is maintained exactly at 2.393 bar and the system temperature at 1200 K, there is no Gibbs energy change on reaction, and any extent of reaction is allowed. Thus, independent of whether the reaction cell, because of previous history, contained calcium carbonate in an undecomposed or partially or fully decomposed state, that state would remain, as there is no driving force for change. (You should compare Figs. 13.1-1 and 13.2-1 and understand the differences between the two.)

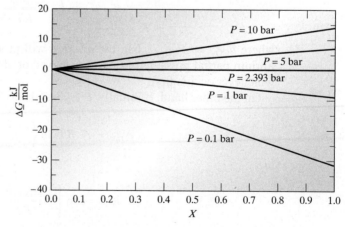

Figure 13.2-1 Gibbs energy change for the heterogeneous reaction $CaCO_3 \rightarrow CaO + CO_2$ as a function of extent of reaction at $T = 1200$ K.

The equilibrium partial pressure of a gaseous species that results from the dissociation of a solid is called the **decomposition pressure** of the solid. In the foregoing discussion, this is 2.393 bar for calcium carbonate at 1200 K. As is evident from Illustration 13.2-2, the decomposition pressure of a solid is a strong function of temperature.

ILLUSTRATION 13.2-2

Calculation of the Decomposition Pressure of a Solid

Compute the decomposition pressure of calcium carbonate over the temperature range of 298.15 K to 1400 K.

Data:

$$C_{P,CaCO_3} = 82.34 + 0.049\,75T - 1\,287\,000/T^2 \quad \dfrac{J}{mol\ K}$$
$$C_{P,CaO} = 41.84 + 0.020\,25T - 451\,870/T^2$$

SOLUTION

From the preceding discussion, we have

$$K_a = a_{CO_2} = P_{CO_2}/(P = 1\ bar)$$

so that here the calculation of the decomposition pressure, P_{CO_2}, is equivalent to the calculation of the equilibrium constant as a function of temperature. From Appendix A.IV, we have

$$\Delta G^\circ(T = 25^\circ C) = \sum \nu_i \Delta_f \underline{G}_i^\circ = 130.40\ kJ/mol$$

and

$$K_a(T = 25^\circ C) = \exp\left\{-\frac{130\,401\ J/mol}{8.314\ J/mol\ K \times 298.15\ K}\right\} = 1.424 \times 10^{-23}$$

To calculate the equilibrium constant at other temperatures, Eq. 13.1-23b and the heat capacity data given in this problem and in Appendix A.II are used. The results for both the equilibrium constant and the decomposition pressure are:

T (K)	K_a	P_{CO_2} (bar)	$\Delta_{rxn}G^\circ$ (kJ)
298.15	1.424×10^{-23}	1.424×10^{-23}	130.401
400	1.259×10^{-15}	1.259×10^{-15}	114.095
600	6.538×10^{-8}	6.538×10^{-8}	82.524
800	4.313×10^{-4}	4.313×10^{-4}	51.539
1000	0.0788	0.0788	21.122
1200	2.393	2.393	−8.706
1400	25.959	25.959	−37.905

Note that the decomposition pressure changes by 25 orders of magnitude over the temperature range. Also, as seen in the figure, the plot of $\ln K_a$ versus reciprocal temperature is almost a straight line.

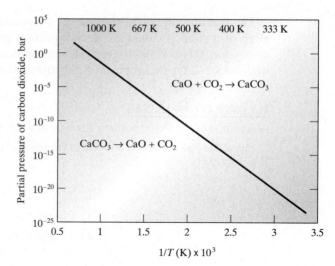

Calcium carbonate–calcium oxide–carbon dioxide equilibrium.

■

Another example of heterogeneous chemical equilibrium is the dissolution and dissociation of a weak salt in solution, most commonly aqueous solution. This process can be represented as

$$A_{\nu_+}B_{\nu_-}(s) = \nu_+ A^{z+}(aq) + \nu_- B^{z-}(aq)$$

where the notation (s) and (aq) indicate solid and aqueous solution, respectively, and z_+ and z_- are the valences of the cation and anion. The equilibrium relation for this ionization process is

$$
\begin{aligned}
K_a &= \exp\left[-\frac{1}{RT}\{\nu_+ \Delta_f \underline{G}^{\circ}_{A^{z+}}(\text{ideal, 1 molal}) + \nu_- \Delta_f \underline{G}^{\circ}_{B^{z-}}(\text{ideal, 1 molal})\right. \\
&\left. \qquad - \Delta_f \underline{G}^{\circ}_{A_{\nu_+}B_{\nu_-}}(\text{solid})\}\right] \\
&= \frac{(a_{A^{z+}})^{\nu_+}(a_{B^{z-}})^{\nu_-}}{(a_{A_{\nu_+}B_{\nu_-}})} = (a_{A^{z+}})^{\nu_+}(a_{B^{z-}})^{\nu_-} \\
&= \frac{(M_{A^{z+}}\gamma^{\square}_{A^{z+}})(M_{B^{z-}}\gamma^{\square}_{B^{z-}})^{\nu_-}}{(M = 1\ \text{molal})^{\nu_+ + \nu_-}} = \frac{(M_{A^{z+}})^{\nu_+}(M_{B^{z-}})^{\nu_-}\gamma_{\pm}^{(\nu_+ + \nu_-)}}{(M = 1\ \text{molal})^{\nu_+ + \nu_-}}
\end{aligned}
$$

(13.2-5)

since $a_{A_{\nu_+}B_{\nu_-}}$, the activity of the pure, undissociated solid is unity, and we have taken the standard state of the ions to be an ideal 1-molal solution. Also, the mean ionic activity coefficient, discussed in Sec. 9.10, has been used.

Common notation is to define the **solubility product** K_s by

Definition of the solubility product

$$\boxed{K_s = (C_{A^{z+}})^{\nu_+}(C_{B^{z-}})^{\nu_-}}$$

so that K_s and K_a are related as follows:[6]

[6]In writing this expression we have neglected the difference between concentration in $kmol/m^3$ = moles/liter and molality.

$$K_a = K_s \gamma_\pm^{(\nu_+ + \nu_-)} / (M = 1 \text{ molal})^{(\nu_+ + \nu_-)}$$

or

$$K_s = K_a (M = 1 \text{ molal})^{(\nu_+ + \nu_-)} / \gamma_\pm^{(\nu_+ + \nu_-)}$$

If we now define K_s° to be the solubility in an ideal solution, we have

$$K_s^\circ = K_a (M = 1 \text{ molal})^{(\nu_+ + \nu_-)}$$

and

$$K_s = K_s^\circ / \gamma_\pm^{(\nu_+ + \nu_-)}$$

or

$$\ln K_s = \ln K_s^\circ - (\nu_+ + \nu_-) \ln \gamma_\pm \qquad (13.2\text{-}6)$$

where one of Eqs. 9.10-15, 9.10-17, 9.10-18, or 9.10-19 is used for $\ln \gamma_\pm$, depending on the ionic strength. (You should compare the relation in Eq. 13.2-6 with Eq. 13.1-31.) At low ionic strengths, the Debye-Hückel limiting law applies, so that

Simplified expression for the variation of the solubility product with ionic strength

$$\boxed{\ln K_s = \ln K_s^\circ + (\nu_+ + \nu_-)|z_+ z_-| \alpha \sqrt{I}} \qquad (13.2\text{-}7)$$

where the ionic strength I is given by

$$I = \frac{1}{2} \sum_{\text{ions}} z_i^2 M_i$$

The important feature of Eq. 13.2-7 is that it can be used to predict the solubility product of salts when data on the Gibbs energy of formation of the ions are available (so that K_a and K_s° can be computed) or, when such data are not available, it can be used to interrelate the solubility products for the same salt at different ionic strengths. These two types of calculations are demonstrated in Illustration 13.2-3.

ILLUSTRATION 13.2-3

Calculation of the Solubility Product from Solubility Data

The following data give the solubility of silver chloride in aqueous solutions of potassium nitrate at 25°C:

Concentration of KNO_3 (kmol/m³)	Concentration of AgCl at Saturation (kmol/m³)
0.0	1.273×10^{-5}
0.000509	1.311×10^{-5}
0.009931	1.427×10^{-5}
0.016431	1.469×10^{-5}
0.040144	1.552×10^{-5}

Source: S. Popoff and E. W. Neumann, *J. Phys. Chem.*, **34**, 1853 (1930); E. W. Neumann, *J. Am. Chem. Soc.*, **54**, 2195 (1932).

a. Compute the solubility product for silver chloride in each of these solutions.

b. Make a prediction of the solubility product of silver chloride in the absence of any potassium nitrate without using the data above.

c. Show that the silver chloride solubility data satisfy Eq. 13.2-7, at least at low potassium nitrate concentrations, and find the numerical value of K_s°.

SOLUTION

a. The solubility product for silver chloride is

$$K_s = C_{Ag^+} C_{Cl^-} = (C_{AgCl})^2$$

since here the molar concentration of silver ions and chloride ions are each equal to the molar concentration of dissolved silver chloride; that is, $C_{Ag^-} = C_{Cl^-} = C_{AgCl}$. The values of K_s and $\ln K_s$ are given in the table that follows.

b. To independently predict the solubility product for silver chloride, we first compute the thermodynamic equilibrium constant K_a from Eq. 13.2-5, Appendix A.IV, and Table 13.1-4. Thus

$$K_a = \exp\left[-\frac{\{77\,110 + (-131\,170) - (-109\,800)\}}{8.314 \times 298.15}\right] = 1.715 \times 10^{-10}$$

and $K_s^\circ = 1.715 \times 10^{-10}\ (\text{kmol/m}^3)^2$. Now using Eq. 13.2-7, we have

$$\ln K_s = 2 \ln C_{Ag^+} = \ln(1.715 \times 10^{-10}) + 2 \times 1.178 \sqrt{C_{Ag^+}}$$

since, for this case $I = \frac{1}{2}(C_{Ag^+} + C_{Cl^-}) = C_{Ag^+}$, and $K_s = C_{Ag^+} C_{Cl^-} = (C_{Ag^+})^2$. This equation has the solution that, at saturation, $C_{Ag^+} = C_{Cl^-} = 1.315 \times 10^{-5}\ \text{kmol/m}^3$, and $K_s = 1.73 \times 10^{-10}\ (\text{kmol/m}^3)^2$. Thus our prediction leads to a silver-ion concentration and a chloride ion concentration, at saturation, that are each only 3.3 percent too large, and a solubility product that is 6.7 percent too large.

c. For the situation here

$$I = \frac{1}{2}\sum z_i^2 C_i = \frac{1}{2}\{C_{Ag^+} + C_{Cl^-} + C_{K^+} + C_{NO_3^-}\}$$
$$= \frac{1}{2}\{2C_{AgCl} + 2C_{KNO_3}\} = C_{AgCl} + C_{KNO_3}$$

and

$$\sqrt{I} = \sqrt{C_{AgCl} + C_{KNO_3}}$$

C_{KNO_3}, kmol/m^3	C_{AgCl}, kmol/m^3	K_s	$\ln K_s$	$\sqrt{I}$, (kmol/m^3)$^{1/2}$
0	1.273×10^{-5}	1.621×10^{-10}	-22.543	3.568×10^{-3}
0.000 509	1.311×10^{-5}	1.719×10^{-10}	-22.484	2.285×10^{-2}
0.009 931	1.427×10^{-5}	2.036×10^{-10}	-22.315	9.973×10^{-2}
0.016 431	1.469×10^{-5}	2.158×10^{-10}	-22.257	1.282×10^{-1}
0.040 144	1.552×10^{-5}	2.409×10^{-10}	-22.147	2.004×10^{-1}

The quantity $\ln K_s$ is plotted versus $\sqrt{I}$ in Fig. 13.2-2, together with a dashed line, which has a slope of 2α, that was drawn to pass through the $C_{KNO_3} = 0$ datum point. Clearly, Eq. 13.2-7 is valid up to an ionic strength of about 0.0225 kmol/m^3; at higher ionic strengths there are deviations from the Debye-Hückel limiting law (Eq. 9.10-15) and, therefore, from Eq. 13.2-7.

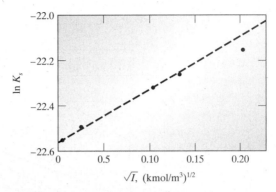

Figure 13.2-2 The solubility product of silver chloride as a function of the square root of the ionic strength in aqueous potassium nitrate solutions.

Using Eq. 13.2-7 and the $C_{KNO_3} = 0.0$ datum point, we find that $\ln K_s^\circ = -22.5514$, so that the ideal solution solubility product is 1.607×10^{-10} (kmol/m^3)^2. ■

An important chemical reaction for metallurgical processing is the oxidation of a metal which, to be perfectly general, we will represent as

$$\text{Metal} + O_2 = \text{Oxide} \quad (\text{rxn 1}) \qquad (13.2\text{-}8)$$

We have written this reaction in generic form since the stoichiometry for oxidation reactions can be quite different depending on the valence of the metal. For example, the oxide could be PbO, Al_2O_3, Fe_2O_3, Fe_3O_4, P_2O_5, SnO_2, Na_2O, etc. The equilibrium relation for this reaction is

$$K_a(\text{rxn 1}, T) = \exp\left(-\frac{\Delta G^\circ(\text{rxn 1})}{RT}\right) = \frac{a_{\text{oxide}}}{a_{\text{metal}} a_{O_2}} = \frac{1}{1 \cdot \dfrac{P_{O_2}}{1 \text{ bar}}} = \frac{1 \text{ bar}}{P_{O_2}} \quad (13.2\text{-}9)$$

where in writing this last form of the equation we have recognized that since the metal and the oxide are solids and do not mix, they will be pure, and thus in their standard states of unit activity. Alternatively, the equation above can be written as

$$\Delta G^\circ(\text{rxn 1}) = RT \ln P_{O_2} = -RT \ln K_a(\text{rxn 1}, T) \qquad (13.2\text{-}10)$$

where here the partial pressure of oxygen is taken to be in units of bar. Thus here, as with other heterogeneous reactions, if the partial pressure of oxygen exceeds a value equal to

$$P_{O_2}(\text{bar}) = \exp\left(\frac{\Delta G^\circ(\text{rxn 1})}{RT}\right) \qquad (13.2\text{-}11)$$

the reaction will proceed until either the partial pressure of oxygen is reduced to this value (if the supply of oxygen is limiting) or until all the metal is oxidized (if the amount of metal is limiting). Alternatively, if the partial pressure of oxygen is below this value, oxidation will not occur; rather, some of the oxide will be reduced to the metal, releasing oxygen. If the partial pressure of oxygen exactly equals the value given by Eq. 13.2-11, no reaction will occur.

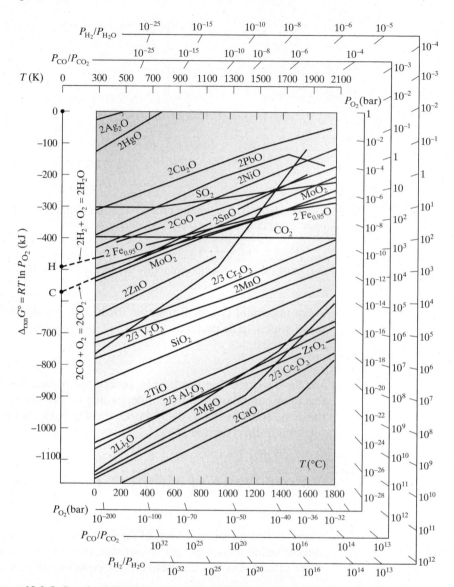

Figure 13.2-3 Standard Gibbs energies of formation of selected oxides, or Ellingham diagram. (Based on figure in *Chemical Thermodynamics of Materials* by C. H. P. Lupis, 1983. Reprinted by permission of Prentice-Hall, Inc. Upper Saddle River, NJ. This figure appears as an Adobe PDF file on the CD-ROM accompanying this book, and may be enlarged and printed for easier reading and for use in solving problems.)

While the calculation of the standard-state Gibbs energy change as a function of temperature needed to use Eq. 13.2-11 could be done using programs such as those in Appendix B and on the CD-ROM accompanying this book, to permit simple hand calculations, plots such as Fig. 13.2-3 have been used in the metallurgical industry (though the standard-state Gibbs energy changes on reaction can more accurately be computed from heat capacity data and the Gibbs energies of formation at 25°C). These plots are used as follows. First the intersection of the line corresponding to the metal oxide that is being formed with the vertical line corresponding to the temperature of interest is located. The standard-state Gibbs energy change is then read from the left-hand axis. Next, drawing a straight line through the point $\Delta_{rxn}G° = 0$ on the left-hand axis to the intersection found in the first step and extending this line to the line along the right edge or bottom of the plot marked P_{O_2} (bar) gives the equilibrium oxygen partial pressure for the oxidation reaction at the temperature chosen.

The reason this last graphical construction gives the equilibrium partial pressure is as follows. The extreme left side of Fig. 13.2-3 corresponds to $T = 0$ K, for which $RT \ln P_{O_2} = 0$. The slope of the line drawn in the construction above is

$$\text{Slope} = \frac{[RT \ln P_{O_2}]_T - [RT \ln P_{O_2}]_{T=0}}{T - 0} = R \ln P_{O_2}(T) \qquad \textbf{(13.2-12)}$$

so that the slope of the line drawn above is simply related to the equilibrium partial pressure. It is the slope of the line, reported in units of the partial pressure of oxygen, that is given in the figure.

ILLUSTRATION 13.2-4

Calculation of the Equilibrium Partial Pressures for the Oxidation of a Metal

Estimate the equilibrium partial pressures of oxygen for the oxidation of chromium and aluminum at 1000°C.

SOLUTION

From Fig. 13.2-3 and the construction described above, we find that the equilibrium partial pressure of oxygen for the oxidation of aluminum is approximately 10^{-35} bar and that for chromium is almost 10^{-22} bar. ∎

It is interesting to interpret and generalize the results of the preceding illustration. For aluminum we see that if the partial pressure of oxygen exceeds 10^{-35} bar at 1000 K, the metal will oxidize. Alternatively, the oxygen partial pressure must be reduced below 10^{-35} bar for the oxide to be reduced to the metal. Since this partial pressure is low, Al_2O_3 is a very stable oxide. For the oxide of chromium the corresponding oxygen partial pressure is 10^{-22} bar. Therefore, the oxide of aluminum is more stable than that of chromium, and conversely, metallic chromium is more stable to oxidation than aluminum; that is, a higher oxygen partial pressure is needed for oxidation. The oxide is more stable and its metal less stable to oxidation as one goes vertically down the components in Fig. 13.2-3.

Most of the standard-state Gibbs energy changes as a function of temperature in Fig. 13.2-3 are nearly linear. Since

$$\left(\frac{\partial G}{\partial T}\right)_P = -S \qquad \textbf{(13.2-13)}$$

the fact that the Gibbs energy change lines are nearly straight indicates that the standard-state entropy change on reaction is almost independent of temperature (which also means that the standard-state enthalpy change on reaction is almost independent of temperature; that is, the contribution of the heat capacity terms is quite small. How can you ascertain this?). Also, the slopes of most of the lines are very similar; that is because the dominant contribution to the entropy change arises from oxygen being transformed from a gas to a solid oxide, and this is almost independent of the particular oxide formed. However, for the oxidation of carbon, the change is from oxygen gas to carbon dioxide gas, which results in a much smaller entropy change. As a consequence, the standard-state Gibbs energy change for the oxidation of carbon to carbon dioxide is almost independent of temperature. There are some oxides in Fig. 13.2-3 for which the standard-state Gibbs energy change shows an abrupt change of slope. These are a result of a phase change (generally either melting or boiling, but perhaps also a solid phase change) in either the metal or its oxide.

Metals may also be oxidized by oxygen-containing compounds such as water or even carbon dioxide according to the reactions

$$\text{Metal} + 2H_2O = \text{Oxide} + 2H_2 \quad (\text{rxn 2}) \tag{13.2-14a}$$

and

$$\text{Metal} + 2CO_2 = \text{Oxide} + 2CO \quad (\text{rxn 3}) \tag{13.2-15a}$$

The equilibrium relations for these reactions are

$$K_a(\text{rxn 2}, T) = \exp\left(-\frac{\Delta_{\text{rxn}}G^\circ(\text{rxn 2})}{RT}\right) = \frac{a_{\text{Oxide}}a_{H_2}^2}{a_{\text{Metal}}a_{H_2O}^2} = \frac{1\left(\dfrac{P_{H_2}}{1\text{ bar}}\right)^2}{1\left(\dfrac{P_{H_2O}}{1\text{ bar}}\right)^2} = \left(\frac{P_{H_2}}{P_{H_2O}}\right)^2 \tag{13.2-14b}$$

and

$$K_a(\text{rxn 3}, T) = \exp\left(-\frac{\Delta_{\text{rxn}}G^\circ(\text{rxn 3})}{RT}\right) = \frac{a_{\text{Oxide}}a_{CO}^2}{a_{\text{Metal}}a_{CO_2}^2} = \frac{1\left(\dfrac{P_{CO}}{1\text{ bar}}\right)^2}{1\left(\dfrac{P_{CO_2}}{1\text{ bar}}\right)^2} = \left(\frac{P_{CO}}{P_{CO_2}}\right)^2 \tag{13.2-15b}$$

Therefore

$$\Delta_{\text{rxn}}G^\circ(\text{rxn 2}) = 2RT \ln\frac{P_{H_2}}{P_{H_2O}} = -RT \ln K_a(\text{rxn 2}, T) \tag{13.2-14c}$$

and

$$\Delta_{\text{rxn}}G^\circ(\text{rxn 3}) = 2RT \ln\frac{P_{CO}}{P_{CO_2}} = -RT \ln K_a(\text{rxn 3}, T) \tag{13.2-15c}$$

Because of the similarity of these equations to Eqs. 13.2-10 and 13.2-11, it is possible to construct Gibbs energy–partial pressure diagrams similar to Fig. 13.2-3. In particular, by subtracting the standard-state Gibbs energy change for the reaction

$$2H_2 + O_2 = 2H_2O \quad (\text{rxn 4}) \tag{13.2-16}$$

from rxn 1 (Eq. 13.2-8) we obtain the standard-state Gibbs energy change for rxn 2, and a new diagram could be prepared that would give the equilibrium partial pressure ratio P_{H_2}/P_{H_2O} as a function of temperature for this reaction. In fact, it is common practice to also include this information in Fig. 13.2-3. Note, however, that to use this diagram to compute the equilibrium value of the partial pressure ratio P_{H_2}/P_{H_2O}, one must start the construction of the straight line from the point marked H on the left-hand axis, rather than the point 0 used for rxn 1 (the difference corresponding to the Gibbs energy change of rxn 4). Also, a different scale is used to obtain the ratio P_{H_2}/P_{H_2O}.

Similarly, by subtracting the standard-state Gibbs energy change for the reaction

$$2CO + O_2 = 2CO_2 \quad \text{(rxn 5)} \tag{13.2-17}$$

from rxn 1 (Eq. 13.2-8) we obtain the standard-state Gibbs energy for rxn 3. A new diagram could be prepared that would give the equilibrium partial pressure ratio P_{CO}/P_{CO_2} as a function of temperature; however, again this has been combined with Fig. 13.2-3. To use this diagram to compute the equilibrium value of the partial pressure ratio P_{CO}/P_{CO_2}, one must start the construction of the straight line from the point marked C on the left-hand axis, and use the third scale to obtain the equilibrium partial pressure ratio P_{CO}/P_{CO_2}.

Of course, an alternative to using these figures is to use a program described in Appendix B on the CD-ROM that accompanies this book to calculate the equilibrium constant and then compute the equilibrium state in the usual manner.

ILLUSTRATION 13.2-5

Determining Whether an Oxidation Reaction Will Occur

Lead oxide at 1400 K is placed in an atmosphere that initially contains 20 mol % carbon dioxide, with the remainder being nitrogen. It is possible that the carbon dioxide will dissociate into carbon monoxide and oxygen, and that the lead oxide will be reduced to elemental lead. Determine whether the lead oxide will be reduced in this environment.

SOLUTION

Since the reaction between the oxygen, carbon monoxide, and carbon dioxide is fast at this temperature, it can be assumed to be in equilibrium, so the first step is to calculate the equilibrium composition. Using the program CHEMEQ, we find that at 1400 K for the reaction

$$CO_2 = CO + \tfrac{1}{2}O_2$$

the equilibrium constant is $K_a = 1.046 \times 10^{-6}$. Next the mass balance table below is used.

Species	Initial	Final	Mole Fraction
CO_2	0.2	$0.2 - X$	$\dfrac{0.2 - X}{1 + 0.5X}$
N_2	0.8	0.8	
CO	0	X	$\dfrac{X}{1 + 0.5X}$
O_2	0	$0.5X$	$\dfrac{0.5X}{1 + 0.5X}$
Total		$1 + 0.5X$	

Consequently,

$$K_a = 1.046 \times 10^{-6} = \frac{a_{CO}a_{O_2}^{0.5}}{a_{CO_2}} = \frac{\dfrac{X}{1+0.5X}\dfrac{P}{1\,\text{bar}}\left(\dfrac{0.5X}{1+0.5X}\dfrac{P}{1\,\text{bar}}\right)^{0.5}}{\dfrac{0.2-X}{1+0.5X}\dfrac{P}{1\,\text{bar}}}$$

$$= \frac{X(0.5X)^{0.5}}{(0.2-X)(1+0.5X)^{0.5}}\left(\frac{P}{1\,\text{bar}}\right)^{0.5}$$

At $P = 1$ bar, this equation has the solution $X = 9.78 \times 10^{-5}$, which gives $P_{CO_2} = 0.199\ 90$ bar, $P_{CO} = 9.8 \times 10^{-5}$ bar, and $P_{O_2} = 4.9 \times 10^{-5}$ bar. Therefore,

$$\frac{P_{CO}}{P_{CO_2}} = 4.89 \times 10^{-5}$$

From Fig. 13.2-3 we see that at 1400 K the equilibrium ratio for the reaction

$$Pb + CO_2 = PbO + CO$$

is P_{CO}/P_{CO_2} equal to about 2×10^{-3}. As a result of rxn 5, the partial pressure of carbon monoxide is below the equilibrium value for this reaction, so that lead oxide would not be reduced to elemental lead in the presence of the reacting carbon monoxide–carbon dioxide mixture considered here.   ■

13.3   CHEMICAL EQUILIBRIUM WHEN SEVERAL REACTIONS OCCUR IN A SINGLE PHASE

The state of chemical equilibrium for $\mathcal{M}$ independent reactions occurring in a single phase is the state that satisfies the constraints on the system, the set of stoichiometric relations

Mass balance equations for simultaneous reactions

$$\boxed{N_i = N_{i,0} + \sum_{j=1}^{\mathcal{M}} \nu_{ij} X_j \quad i = 1, 2, \ldots, \mathcal{C}}$$

and the $\mathcal{M}$ equilibrium relations (cf. Sec. 8.8)

$$\sum_{i=1}^{\mathcal{C}} \nu_{ij}\overline{G}_i = 0 \quad j = 1, 2, \ldots, \mathcal{M} \tag{13.3-1}$$

Using the notation introduced in this chapter, the last equation can be rewritten as

$$K_{a,j} = \prod_{i=1}^{\mathcal{C}} a_i^{\nu_{ij}} \quad j = 1, 2, \ldots, \mathcal{M} \tag{13.3-2}$$

where $K_{a,j}$ is the equilibrium constant for the jth reaction. Since a collection of equations (which are nonlinear, and usually coupled) are to be solved here, rather than merely one equation, as when only a single reaction occurs, equilibrium calculations for multiple-reaction systems can be complicated.

Before proceeding to a sample calculation of a multireaction equilibrium state, it is useful to consider the simplifications that can be made in the analysis. First, the number of reactions that must be considered (and hence the number of simultaneous equations that must be solved) can frequently be reduced by taking into account the accuracy desired in the calculations and eliminating from consideration reactions that occur to such a small extent as to produce products with concentrations that are below our level of interest. Such reactions are usually identified by very small equilibrium constants, that is, equilibrium constants that are several orders of magnitude smaller than those of the other reactions being considered, or by the fact that their reactants include the products of reactions that occur only to a small extent. In fact, such reasoning is used here in writing only 5 reactions among carbon, hydrogen, and oxygen, rather than the 20 reactions among these substances that appear in Fig. 13.1-2 or, worse, all the reactions that occur in organic chemistry involving these three atomic species.

We can further reduce the number of reactions by identifying, and then only including in the equilibrium analysis, the independent reactions among the species. To see that the independent reactions and no others need be studied, consider the following set of reactions that occur between steam and coal (which we take to be pure carbon) at temperatures below 2000 K:

$$C + 2H_2O \rightarrow CO_2 + 2H_2 \quad \text{(rxn 1)}$$
$$C + H_2O \rightarrow CO + H_2 \quad \text{(rxn 2)}$$
$$C + CO_2 \rightarrow 2CO \quad \text{(rxn 3)}$$
$$C + 2H_2 \rightarrow CH_4 \quad \text{(rxn 4)}$$
$$CO + H_2O \rightarrow CO_2 + H_2 \quad \text{(rxn 5)}$$

There are only three independent chemical reactions among these five reactions [see Problem 8.9(c)]. This is easily demonstrated using Denbigh's method (cf. Sec. 8.3); we start by writing the reactions

$$2H + O \rightarrow H_2O \quad \text{(a)}$$
$$C + 2O \rightarrow CO_2 \quad \text{(b)}$$
$$C + O \rightarrow CO \quad \text{(c)}$$
$$C + 4H \rightarrow CH_4 \quad \text{(d)}$$
$$2H \rightarrow H_2 \quad \text{(e)}$$

Since neither atomic hydrogen nor atomic oxygen are present below several thousand degrees, these two species are to be eliminated from the equations. First using reaction (e) to eliminate atomic hydrogen (i.e., $H = \frac{1}{2}H_2$) and then reactions (a) and (e) to eliminate atomic oxygen (i.e., $O = H_2O - H_2$) yields

$$C + 2H_2O \rightarrow CO_2 + 2H_2$$
$$C + H_2O \rightarrow CO + H_2 \quad \text{(13.3-3)}$$
$$C + 2H_2 \rightarrow CH_4$$

These three reactions form a set of independent reactions among the six chemical species being considered.

Now the claim is that only these three reactions (or, equivalently, any other set of three independent reactions) need be considered in computing the equilibrium state for this system. This is easily verified by an examination of the five equilibrium relations

$$K_{a,1} = \exp\left[-\frac{(\Delta_f\underline{G}^{\circ}_{CO_2} + 2\Delta_f\underline{G}^{\circ}_{H_2} - \Delta_f\underline{G}^{\circ}_C - 2\Delta_f\underline{G}^{\circ}_{H_2O})}{RT}\right] = \frac{a_{CO_2}a^2_{H_2}}{a_C a^2_{H_2O}} \quad \textbf{(13.3-4a)}$$

$$K_{a,2} = \exp\left[-\frac{(\Delta_f\underline{G}^{\circ}_{CO} + \Delta_f\underline{G}^{\circ}_{H_2} - \Delta_f\underline{G}^{\circ}_C - \Delta_f\underline{G}^{\circ}_{H_2O})}{RT}\right] = \frac{a_{CO}a_{H_2}}{a_C a_{H_2O}} \quad \textbf{(13.3-4b)}$$

$$K_{a,3} = \exp\left[-\frac{(2\Delta_f\underline{G}^{\circ}_{CO} - \Delta_f\underline{G}^{\circ}_C - \Delta_f\underline{G}^{\circ}_{CO_2})}{RT}\right] = \frac{a^2_{CO}}{a_C a_{CO_2}} \quad \textbf{(13.3-4c)}$$

$$K_{a,4} = \exp\left[-\frac{(\Delta_f\underline{G}^{\circ}_{CH_4} - \Delta_f\underline{G}^{\circ}_C - 2\Delta_f\underline{G}^{\circ}_{H_2})}{RT}\right] = \frac{a_{CH_4}}{a_C a^2_{H_2}} \quad \textbf{(13.3-4d)}$$

and

$$K_{a,5} = \exp\left[-\frac{(\Delta_f\underline{G}^{\circ}_{CO_2} + \Delta_f\underline{G}^{\circ}_{H_2} - \Delta_f\underline{G}^{\circ}_{CO} - \Delta_f\underline{G}^{\circ}_{H_2O})}{RT}\right] = \frac{a_{CO_2}a_{H_2}}{a_{CO}a_{H_2O}} \quad \textbf{(13.3-4e)}$$

It is clear that if Eqs. 13.3-4a, b, and d are satisfied, Eqs. 13.3-4c and e will also be satisfied, because these two equations are merely ratios of the other three; that is,

$$K_{a,3} = \frac{(K_{a,2})^2}{(K_{a,1})} = \frac{\left(\dfrac{a_{CO}a_{H_2}}{a_C a_{H_2O}}\right)^2}{\left(\dfrac{a_{CO_2}a^2_{H_2}}{a_C a^2_{H_2O}}\right)} = \frac{a^2_{CO}}{a_C a_{CO_2}} \quad \textbf{(13.3-5a)}$$

and

$$K_{a,5} = \frac{K_{a,1}}{K_{a,2}} = \frac{\left(\dfrac{a_{CO_2}a^2_{H_2}}{a_C a^2_{H_2O}}\right)}{\left(\dfrac{a_{CO}a_{H_2}}{a_C a_{H_2O}}\right)} = \frac{a_{CO_2}a_{H_2}}{a_{CO}a_{H_2O}} \quad \textbf{(13.3-5b)}$$

Thus we are led to the conclusion that it is not necessary to consider all five reactions when computing the equilibrium state of this reaction system, but merely the three independent reactions of Eqs. 13.3-3. In fact, this result could have been anticipated by observing that reactions 3 and 5 are linear combinations of the other reactions, that is

$$\text{rxn } 3 = 2(\text{rxn } 2) - \text{rxn } 1$$

and

$$\text{rxn } 5 = \text{rxn } 1 - \text{rxn } 2$$

In all the chemical equilibrium calculations that follow, we limit our attention to the independent chemical reactions among the reacting species.

ILLUSTRATION 13.3-1

Chemical Equilibrium When Several Reactions Occur

Compute the equilibrium mole fractions of H_2O, CO, CO_2, H_2, and CH_4 in the steam-carbon system at a total pressure of 1 bar and over the temperature range of 600 to 1600 K.

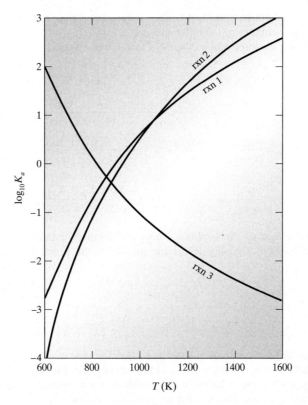

Figure 13.3-1 The equilibrium constants for the three independent reactions between coal and steam.

Data:

The equilibrium constants (based on the pure component standard states at $P = 1$ bar and the reaction temperature) for the set of independent reactions

$$C + 2H_2O = CO_2 + 2H_2 \qquad \textbf{(rxn 1)}$$
$$C + H_2O = CO + H_2 \qquad \textbf{(rxn 2)}$$
$$C + 2H_2 = CH_4 \qquad \textbf{(rxn 3)}$$

are given in Fig. 13.3-1 or can be calculated using the chemical equilibrium constant calculation programs in Appendix B.I or B.II.

SOLUTION

X_1, X_2, and X_3 will be used as the molar extents of reaction for the three reactions being considered. Basing all calculations on 1 mole of steam, we can construct the following mole balance table for the *gaseous* species (as long as there is sufficient solid carbon present to ensure equi-

	Number of Moles in the Gas Phase		
	Initial	Final	Equilibrium Mole Fraction
H_2O	1	$1 - 2X_1 - X_2$	$(1 - 2X_1 - X_2)/\Sigma$
CO_2	0	X_1	X_1/Σ
CO	0	X_2	X_2/Σ
H_2	0	$2X_1 + X_2 - 2X_3$	$(2X_1 + X_2 - 2X_3)/\Sigma$
CH_4	0	X_3	X_3/Σ
Total	1	$\Sigma = 1 + X_1 + X_2 - X_3$	

librium, it need not be considered, since it does not appear in the gas phase and therefore does not affect the activities of the other species).

To solve for the three molar extents of reaction at 1 bar pressure, we use the three equilibrium equations

$$K_{a,1} = \frac{a_{CO_2} a_{H_2}^2}{a_C a_{H_2O}^2} = \frac{(y_{CO_2})(y_{H_2})^2}{(y_{H_2O})^2} = \frac{X_1(2X_1 + X_2 - 2X_3)^2}{(1 - 2X_1 - X_2)^2(1 + X_1 + X_2 - X_3)} \tag{a}$$

$$K_{a,2} = \frac{a_{CO} a_{H_2}}{a_C a_{H_2O}} = \frac{y_{CO} y_{H_2}}{y_{H_2O}} = \frac{X_2(2X_1 + X_2 - 2X_3)}{(1 - 2X_1 - X_2)(1 + X_1 + X_2 - X_3)} \tag{b}$$

and

$$K_{a,3} = \frac{a_{CH_4}}{a_C a_{H_2}^2} = \frac{y_{CH_4}}{y_{H_2}^2} = \frac{X_3(1 + X_1 + X_2 - X_3)}{(2X_1 + X_2 - 2X_3)^2} \tag{c}$$

Since these equations are nonlinear, there will be more than one set of solutions for the molar extents of reaction. The only acceptable solution to this problem is one for which

$$2X_1 + X_2 \leq 1$$

and

$$0 \leq 2X_3 \leq 2X_1 + X_2$$

The first of these restrictions ensures that we do not use more steam than was supplied, and the second that no more hydrogen is used than has been produced.

Clearly, finding the solution to this problem is a nontrivial computational task. The results obtained using an equation-solving program are given in Fig. 13.3-2. ■

The dissolution and ionization of a mixture of electrolytes provides another example of equilibrium in a multireaction system. To be specific, suppose two electrolytes $A_{\nu_A} B_{\nu_B}$ and $G_{\nu_G} H_{\nu_H}$ ionize in solution as follows:

$$A_{\nu_A} B_{\nu_B} = \nu_A A^{z_A} + \nu_B B^{z_B}$$
$$G_{\nu_G} H_{\nu_H} = \nu_G G^{z_G} + \nu_H H^{z_H}$$

The equilibrium relations for these ionization processes (see Eqs. 13.1-31 and 13.2-6) are

$$\ln K_{AB} = \ln(C_A^{\nu_A} C_B^{\nu_B}) = \ln K_{AB}^{\circ} - (\nu_A + \nu_B) \ln \gamma_{\pm} \tag{13.3-6}$$

and

$$\ln K_{GH} = \ln(C_G^{\nu_G} C_H^{\nu_H}) = \ln K_{GH}^{\circ} - (\nu_G + \nu_H) \ln \gamma_{\pm} \tag{13.3-7}$$

where K_{AB} and K_{GH} are the equilibrium ratios of Eqs. 13.1-31 and 13.2-6. The difficulty in solving these equations to find the equilibrium state is that even if there is no common ion among the electrolytes, the equations are coupled by the fact that the activity coefficient $\gamma_{\pm}$ is a function of the total ionic strength I,

$$I = \frac{1}{2} \sum_{ions} z_i^2 C_i = \frac{1}{2}[z_A^2 C_A + z_B^2 C_B + z_G^2 C_G + z_H^2 C_H] \tag{13.3-8}$$

and thus depends on the concentration of all the ions present.

If there is an ion that is common to both electrolytes, the solubility and extent of ionization of each can be much more strongly affected by the presence of the other than

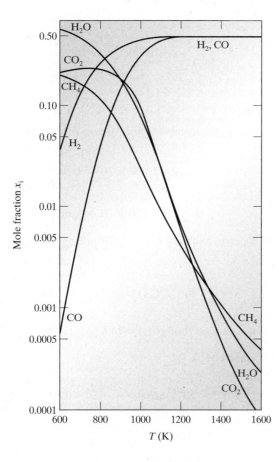

Figure 13.3-2 The equilibrium mole fractions for the coal-steam reactions considered in Illustration 13.3-1.

would be the case with only an ionic-strength coupling. This phenomenon, known as the **common ion effect**, is demonstrated in Illustration 13.3-2.

ILLUSTRATION 13.3-2

Common Ion Effect in Chemical Reaction of Electrolytes

The solubility of silver chloride in water at 25°C is 1.273×10^{-5} kmol/m^3, and that of thallium chloride is 0.144 kmol/m^3. Estimate the simultaneous solubility of AgCl and TlCl in water.

SOLUTION

The first step in the calculation is to compute the constants K°_{AgCl} and K°_{TlCl} using the solubility data for each of the pure salts. Recognizing that $\nu_{\text{Ag}} = \nu_{\text{Tl}} = \nu_{\text{Cl}} = 1$ and that $z_{\text{Ag}} = z_{\text{Tl}} = +1$ and $z_{\text{Cl}} = -1$, we write for the pure salts

$$\ln K^\circ_{\text{AgCl}} = \ln(C_{\text{Ag}} C_{\text{Cl}}) + 2 \ln \gamma_\pm = 2 \ln C_{\text{AgCl}} + 2 \ln \gamma_\pm \tag{a}$$

and

$$\ln K^\circ_{\text{TlCl}} = \ln(C_{\text{Tl}} C_{\text{Cl}}) + 2 \ln \gamma_\pm = 2 \ln C_{\text{TlCl}} + 2 \ln \gamma_\pm \tag{b}$$

Also, for the pure salts, I(silver chloride) $= 1.273 \times 10^{-5}$ kmol/m^3 and I(thallium chloride) $=$ 0.144 kmol/m^3. Because of the high ionic strengths, Eq. 9.10-18 will be used to compute $\gamma_\pm$;

that is,

$$\ln \gamma_\pm = \frac{-1.178|z_+z_-|\sqrt{I}}{1+\sqrt{I}} + 0.1|z_+z_-|I$$

$$= \frac{-1.178\sqrt{I}}{1+\sqrt{I}} + 0.1I$$

(c)

Using Eq. c and the experimental solubility data in Eqs. a and b yields

$$K^\circ_{\text{AgCl}} = 1.607 \times 10^{-10} \ (\text{kmol/m}^3)^2$$

and

$$K^\circ_{\text{TlCl}} = 1.116 \times 10^{-2} \ (\text{kmol/m}^3)^2$$

To find the simultaneous solubility of silver chloride and thallium chloride, we must solve the equations

$$\ln(C_{\text{Ag}}C_{\text{Cl}}) = \ln K^\circ_{\text{AgCl}} - 2\ln\gamma_\pm$$

and

$$\ln(C_{\text{Tl}}C_{\text{Cl}}) = \ln K^\circ_{\text{TlCl}} - 2\ln\gamma_\pm$$

or, using Eq. c, the equations

$$1.607 \times 10^{-10} \exp\left[\frac{2.356\sqrt{I}}{1+\sqrt{I}} - 0.2I\right] = C_{\text{Ag}}C_{\text{Cl}}$$

$$1.116 \times 10^{-2} \exp\left[\frac{2.356\sqrt{I}}{1+\sqrt{I}} - 0.2I\right] = C_{\text{Tl}}C_{\text{Cl}}$$

where

$$I = \tfrac{1}{2}(C_{\text{Ag}} + C_{\text{Tl}} + C_{\text{Cl}}) = C_{\text{AgCl}} + C_{\text{TlCl}}$$

since $C_{\text{Ag}} + C_{\text{Tl}} = C_{\text{Cl}}$. These nonlinear equations can be solved by trial and error or using an equation-solving program.

Because of the low solubility of silver chloride, as a first guess, we will assume that the solubility of thallium chloride is unaffected by the presence of silver chloride, and that the ionization of silver chloride has little effect on either the ionic strength I or the total chloride ion concentration. In this case we have

$$C_{\text{Tl}} = C_{\text{Cl}} = 0.144 \ \text{kmol/m}^3$$

$$I = 0.144 \ \text{kmol/m}^3 \quad \text{and} \quad \sqrt{I} = 0.3975 \ (\text{kmol/m}^3)^{1/2}$$

The silver ion concentration (and the solubility of silver chloride) can then be computed from

$$1.607 \times 10^{-10} \exp\left[\frac{2.356 \times 0.3975}{1.3975} - 0.2 \times 0.144\right] = C_{\text{Ag}} \times 0.144$$

so that

$$C_{\text{Ag}} = C_{\text{AgCl}} = 2.119 \times 10^{-9} \ \text{kmol/m}^3$$

Therefore, as we assumed, AgCl is so slightly soluble in the aqueous TlCl solution that neither the ionic strength nor the solubility of thallium chloride is greatly affected by its presence. On the other hand, owing to the common ion effect, the solubility of silver chloride is reduced by almost four orders of magnitude from its value when only AgCl is present.

COMMENT

If, instead of thallium chloride, some other salt that does not have a common ion with AgCl (for example, $NaNO_3$) had been introduced into the aqueous solution, the solubility of silver chloride would have increased because of the ionic-strength dependence of the ionic activity coefficient. Thus, if $NaNO_3$ were present in the same concentration as, but instead of, thallium chloride, we would have

$$C_{Ag}C_{Cl} = 1.607 \times 10^{-10} \exp\left\{\frac{2.356 \times 0.3975}{1.3975} - 0.2 \times 0.144\right\} = C_{Ag}^2 = C_{Cl}^2$$

or

$$C_{Ag} = C_{Cl} = C_{AgCl} = 1.75 \times 10^{-5} \text{ kmol/m}^3$$

which is an increase of 37 percent in the solubility of silver chloride over the value when only AgCl is present. ∎

There is a point about multiple chemical reactions that at first can be a bit confusing, and so is useful to mention it here. Consider, as an example, the two reactions

Comment on multiple reactions

$$A + B \rightleftharpoons C + D \quad \text{for which } K_{a,1} = \frac{a_C \cdot a_D}{a_A \cdot a_B} \tag{13.3-9a}$$

and

$$C + E \rightleftharpoons F \quad \text{for which } K_{a,2} = \frac{a_F}{a_C \cdot a_E} \tag{13.3-9b}$$

Now if our interest is in the equilibrium concentrations of all the species present, the equilibrium relations for the two reactions must be solved simultaneously. Note that the overall reaction, obtained from the sum of the two reactions above, is

$$A + B + E \rightleftharpoons D + F \quad \text{for which } K_{a,3} = \frac{a_D \cdot a_F}{a_A \cdot a_B \cdot a_E} \tag{13.3-9c}$$

However, our interest may be in the equilibrium compositions of only the reactants A, B, and E, and the products D and F, not in the intermediate product C. In fact, almost all reactions occur by a number of underlying intermediate steps (referred to as the reaction mechanism) that produce intermediate products that are immediately consumed in another reaction in the sequence. Usually, the person doing the measurements or the engineer designing a process may not even know such intermediate reactions are occurring. For the case here, the individual may be unaware that the reaction of Eq. 13.3-9c is actually the result of the reactions of Eqs. 13.3-9a and b. Now note that

$$RT \ln K_{a,1} = \Delta_f \underline{G}_C^\circ + \Delta_f \underline{G}_D^\circ - \Delta_f \underline{G}_A^\circ - \Delta_f \underline{G}_B^\circ$$
$$RT \ln K_{a,2} = \Delta_f \underline{G}_F^\circ - \Delta_f \underline{G}_C^\circ - \Delta_f \underline{G}_E^\circ$$
$$RT \ln K_{a,3} = \Delta_f \underline{G}_D^\circ + \Delta_f \underline{G}_F^\circ - \Delta_f \underline{G}_A^\circ - \Delta_f \underline{G}_B^\circ - \Delta_f \underline{G}_E^\circ$$
$$= \Delta_f \underline{G}_D^\circ + \Delta_f \underline{G}_F^\circ - \Delta_f \underline{G}_A^\circ - \Delta_f \underline{G}_B^\circ - \Delta_f \underline{G}_E^\circ + \Delta_f \underline{G}_C^\circ - \Delta_f \underline{G}_C^\circ$$
$$= \Delta_f \underline{G}_D^\circ + \Delta_f \underline{G}_C^\circ - \Delta_f \underline{G}_A^\circ - \Delta_f \underline{G}_B^\circ + \Delta_f \underline{G}_F^\circ - \Delta_f \underline{G}_C^\circ - \Delta_f \underline{G}_E^\circ$$
$$= RT \ln K_{a,1} + RT \ln K_{a,2}$$

or simply

$$K_{a,3} = K_{a,1} \cdot K_{a,2} \tag{13.3-10}$$

which is the important result.

What Eq. 13.3-10 tells us is that we will obtain the same equilibrium ratio of the activities of the products D and F to the activities of the reactants A, B, and E if we solve Eqs. 13.3-9a and b, or only Eq. 13.3-9c with the equilibrium constant obtained from Eq. 13.3-10. Though it is easier to solve, in the last case we will not get any information about the intermediate component C, so there could be some uncertainty in the mass balance. However, the most common situation in which this analysis would be used is that in which the intermediate species (here C) is of very low concentration, for example, in intermediate that is almost completely consumed by the second reaction after production by the first reaction.

In some cases, Eq. 13.3-10 can be used to advantage. For example, consider the case in which a reaction has such a large and negative standard-state Gibbs energy change (and therefore such a large equilibrium constant) that the reaction goes almost to completion, and that it is not possible to accurately measure the remaining very small concentrations of the reactants needed to obtain an accurate value for the equilibrium constant. In such a circumstance it may be possible to determine the equilibrium constant indirectly by measuring the equilibrium constants for two (or more) reactions that include additional species but do not go to completion—provided that the sum of the reactions is equal to the initial reaction, so that by Eq. 13.3-10 the product of the equilibrium constants (and the sum of the standard-state Gibbs energies of the reactions) will equal that of the initial reaction.

One example of using intermediate reactions is for the physiologically important adenosine triphosphate (ATP) to adenosine diphosphate reaction

$$\text{ATP} + \text{H}_2\text{O} \overset{K}{\rightleftharpoons} \text{ADP} + \text{P}$$

which has such a large equilibrium constant that it cannot be determined by experiment with any accuracy. However, it is possible to accurately measure the chemical equilibrium of the following reactions:

$$\text{glutamate} + \text{ammonia} + \text{ATP} \overset{K_1}{\rightleftharpoons} \text{glutamine} + \text{ADP} + \text{P}$$

and

$$\text{glutamine} + \text{H}_2\text{O} \overset{K_2}{\rightleftharpoons} \text{glutamate} + \text{ammonia}$$

for which

$$K_1 = \frac{a_{\text{glutamine}} \cdot a_{\text{ADP}} \cdot a_{\text{P}}}{a_{\text{glutamate}} \cdot a_{\text{NH}_3} \cdot a_{\text{ATP}}} \quad \text{and} \quad K_2 = \frac{a_{\text{glutamate}} \cdot a_{\text{NH}_3}}{a_{\text{glutamine}} \cdot a_{\text{H}_2\text{O}}}$$

Since the reaction of interest is the sum of the two measurable reactions, it follows that

$$K = K_1 \cdot K_2$$

and for the ATP hydrolysis reaction,

$$\triangle_{\text{rxn}} G^\circ = -RT \ln K = -RT \ln (K_1 \cdot K_2)$$

In this way Rosing and Slater [*Biochim. Biophys. Acta* **267**, 275 (1972)] were able to measure the standard-state Gibbs free energy change for this reaction to be -34 kJ/mol at one set of conditions.

ILLUSTRATION 13.3-3

Free Energy Change of the ATP $\rightleftharpoons$ ADP *Reaction*

Rosing and Slater [*Biochim. Biophys. Acta* **267**, 275 (1972)] reported the following free energy change extrapolated to zero ionic strength for the reaction

$$\text{ATP} + \text{H}_2\text{O} \overset{K}{\rightleftharpoons} \text{ADP} + \text{P}$$

pH	$\triangle_{rxn}G^\circ(25^\circ C)$ (kJ/mol)	$\triangle_{rxn}G^\circ(37^\circ C)$ (kJ/mol)
6.0	-31.77	-32.16
6.5	-32.27	-32.69
7.0	-33.51	-34.00
7.5	-35.77	-36.40
8.0	-38.67	-39.49

Determine the standard-state heat of reaction as a function of pH.

SOLUTION

Since

$$\ln K(T_2) - \ln K(T_1) = -\frac{\triangle_{rxn}\underline{H}^\circ}{R}\left(\frac{1}{T_2} - \frac{1}{T_1}\right) = \left(-\frac{\triangle_{rxn}\underline{G}^\circ(T_2)}{RT_2}\right) - \left(-\frac{\triangle_{rxn}\underline{G}^\circ(T_1)}{RT_1}\right)$$

and

$$\triangle_{rxn}\underline{H}^\circ = \frac{\dfrac{\triangle_{rxn}\underline{G}^\circ(T_2)}{T_2} - \dfrac{\triangle_{rxn}\underline{G}^\circ(T_1)}{T_1}}{\left(\dfrac{1}{T_2} - \dfrac{1}{T_1}\right)}$$

we obtain

pH	$\triangle_{rxn}\underline{H}^\circ$ (kJ/mol)
6.0	-22.08
6.5	-21.83
7.0	-46.17
7.5	-52.04
8.0	-18.30

Note that there is considerable scatter in the results, which is an indication of the difficulty in obtaining accurate thermodynamic data for some biochemical reactions. It is also interesting to notice that the results for the equilibrium constants, Gibbs energies of reaction, and heats of reaction are all affected by pH. This is typical of reactions involving compounds that can ionize, as discussed earlier and will be considered in greater detail in Chapter 15. ∎

13.4 COMBINED CHEMICAL AND PHASE EQUILIBRIUM

In Sec. 8.8 we established that the conditions for equilibrium when chemical reaction and phase equilibrium occur simultaneously are that (1) each species must be in phase equilibrium among all the phases, that is,

$$\bar{G}_i^I = \bar{G}_i^{II} = \cdots = \bar{G}_i^P = \bar{G}_i \quad i = 1, 2, \ldots, \mathcal{C} \tag{13.4-1}$$

(2) each chemical reaction must be in chemical equilibrium in each phase

$$\sum_{i=1}^{\mathcal{C}} \nu_{ij} \bar{G}_i = 0 \quad \begin{array}{l} \text{for all independent} \\ \text{reactions } j = 1, 2, \ldots, \mathcal{M} \end{array} \tag{13.4-2}$$

and (3) the stoichiometric and state-variable constraints on the system must be satisfied. In fact, the equality of the partial molar Gibbs energy of each species in all phases at equilibrium (Eq. 13.4-1) ensures that if the chemical equilibrium criterion is satisfied in any one phase, it will be satisfied in all phases. Thus, in computations it is necessary only to seek a solution for which

$$\bar{f}_i^I = \bar{f}_i^{II} = \cdots = \bar{f}_i^P = \bar{f}_i \quad i = 1, 2, \ldots, \mathcal{C} \tag{13.4-3}$$

in all phases (which follows from Eq. 13.4-1), and for which

$$K_{a,j} = \prod_{i=1}^{\mathcal{C}} a_i^{\nu_{ij}} \quad j = 1, 2, \ldots, \mathcal{M} \tag{13.4-4}$$

is satisfied in any one phase[7] (which follows from Eq. 13.4-2).

The prediction of a state of combined chemical and phase equilibrium using these equations can be complicated because of the large number of nonlinear equations that must be solved simultaneously. Frequently, the equations involved can be simplified or reduced in number by recognizing that some species are only slightly soluble in certain phases, and that some reactions may go virtually to completion or do not measurably proceed at all in some phases. However, even with such simplifications, the calculations are likely to be tedious, as indicated in Illustration 13.4-1.

ILLUSTRATION 13.4-1

Combined Chemical and Vapor-Liquid Equilibrium

One mole of nitrogen, 3 moles of hydrogen, and 5 moles of water are placed in a closed container maintained at 25°C and 13.33 kPa and, using the appropriate catalyst and stirring, allowed to attain phase and chemical equilibrium. Assuming that ammonia is formed by chemical reaction and that the liquid and vapor phases are ideal, compute the amount and composition of each phase at equilibrium (neglecting the aqueous-phase reaction of ammonia to form NH_4OH, and its subsequent ionization). The following data are available:

Vapor pressure of water at 25°C = 3.167 kPa
Henry's law constants:

---

[7]In certain instances it may be convenient to choose different states of aggregation as the standard states for the various species in a reaction, usually because of the availability of $\Delta_f G^\circ$ data. In such cases the activity of each species is evaluated in that phase that corresponds most closely to the state of aggregation of the standard state. (See the discussion of Eqs. 13.4-5 and 13.4-9.)

$$N_2 \text{ in } H_2O: \quad H_{N_2} = 13.224 \times 10^7 \text{ kPa/mole fraction}$$
$$H_2 \text{ in } H_2O: \quad H_{H_2} = 7.158 \times 10^7 \text{ kPa/mole fraction}$$
$$NH_3 \text{ in } H_2O: \quad H_{NH_3} = 97.58 \text{ kPa/mole fraction}$$

(The Henry's law constant for ammonia was computed from experimental data using the equation $P_{NH_3} = H_{NH_3} x_{NH_3}$ and assuming all of the ammonia absorbed to be present as NH_3.)

SOLUTION

The only reaction that can occur between the species at 25°C is the formation of ammonia,

$$\tfrac{1}{2}N_2 + \tfrac{3}{2}H_2 = NH_3$$

for which

$$\Delta_{rxn}G = \sum \nu_i \Delta_f \underline{G}_i^\circ$$
$$= \Delta_f \underline{G}_{NH_3}^\circ - \tfrac{1}{2}\Delta_f \underline{G}_{N_2}^\circ - \tfrac{3}{2}\Delta_f \underline{G}_{H_2}^\circ = -16\,450 \text{ J/mol}$$

and

$$K_a = \exp\left[-\frac{\Delta_{rxn}G}{RT}\right] = 762.2$$

The chemical equilibrium equation is

$$K_a = \frac{a_{NH_3}}{a_{N_2}^{1/2} a_{H_2}^{3/2}} = \frac{y_{NH_3}}{y_{N_2}^{1/2} y_{H_2}^{3/2} \left(\dfrac{P}{1 \text{ bar}}\right)} = 762.2 \qquad (1)$$

The mass balance constraints are most easily taken into account using the mole numbers of each species in each phase as the independent variables, rather than the mole fractions. Also, to reduce the number of variables we need to solve for, the mass balances will be used to eliminate the number of moles of each species in the liquid phase in terms of the vapor-phase mole numbers and the molar extent of reaction. Thus we have

$$N_{H_2O} = 5 \text{ mol} = N_{H_2O}^L + N_{H_2O}^V \quad \text{or} \quad N_{H_2O}^L = 5 \text{ mol} - N_{H_2O}^V \qquad (2)$$
$$N_{NH_3} = X = N_{NH_3}^L + N_{NH_3}^V \quad \text{or} \quad N_{NH_3}^L = X - N_{NH_3}^V \qquad (3)$$
$$N_{N_2} = 1 - \tfrac{1}{2}X = N_{N_2}^L + N_{N_2}^V \quad \text{or} \quad N_{N_2}^L = 1 - \tfrac{1}{2}X - N_{N_2}^V \qquad (4)$$

and

$$N_{H_2} = 3\left(1 - \frac{X}{2}\right) = N_{H_2}^L + N_{H_2}^V \quad \text{or} \quad N_{H_2}^L = 3\left(1 - \frac{X}{2}\right) - N_{H_2}^V \qquad (5)$$

The total number of moles in the liquid phase is then

$$N^L = N_{N_2}^L + N_{H_2}^L + N_{NH_3}^L + N_{H_2O}^L$$
$$= \left(1 - \tfrac{1}{2}X - N_{N_2}^V\right) + \left(3 - \tfrac{3}{2}X - N_{H_2}^V\right) + \left(X - N_{NH_3}^V\right) + \left(5 - N_{H_2O}^V\right) \qquad (6)$$
$$= 9 - X - N_{N_2}^V - N_{H_2}^V - N_{NH_3}^V - N_{H_2O}^V$$

Also, the phase equilibrium relation $\bar{f}_i^L = \bar{f}_i^V$ must be satisfied for each species in each phase. For N_2, H_2, and NH_3, this leads to the following equation:

$$P_i = y_i P = H_i x_i \quad \text{or} \quad y_i = H_i x_i / P$$

Thus

$$y_{N_2} = \frac{9.224 \times 10^7}{P \text{ (kPa)}} x_{N_2}$$

or in terms of mole numbers,

$$
\frac{N_{N_2}^V}{N_{N_2}^V + N_{H_2}^V + N_{NH_3}^V + N_{H_2O}^V} = \frac{9.224 \times 10^7}{P} \frac{N_{N_2}^L}{N_{N_2}^L + N_{H_2}^L + N_{NH_3}^L + N_{H_2O}^L}
$$
$$
= \frac{9.224 \times 10^7}{P} \frac{1 - \frac{1}{2}X - N_{N_2}^V}{9 - X - N_{N_2}^V - N_{H_2}^V - N_{NH_3}^V - N_{H_2O}^V}
\tag{7}
$$

Similarly, for hydrogen and ammonia, we have

$$
\frac{N_{H_2}^V}{N_{N_2}^V + N_{H_2}^V + N_{NH_3}^V + N_{H_2O}^V} = \frac{7.158 \times 10^7}{P} \frac{3 - \frac{3}{2}X - N_{H_2}^V}{9 - X - N_{N_2}^V - N_{H_2}^V - N_{NH_3}^V - N_{H_2O}^V}
\tag{8}
$$

and

$$
\frac{N_{NH_3}^V}{N_{N_2}^V + N_{H_2}^V + N_{NH_3}^V + N_{H_2O}^V} = \frac{97.58}{P} \frac{X - N_{NH_3}^V}{9 - X - N_{N_2}^V - N_{H_2}^V - N_{NH_3}^V - N_{H_2O}^V}
\tag{9}
$$

For water, the equilibrium relation is

$$
y_{H_2O}P = x_{H_2O}P_{H_2O}^{\text{vap}} \quad \text{or} \quad y_{H_2O} = \frac{x_{H_2O}P_{H_2O}^{\text{vap}}}{P} = \frac{3.167x_{H_2O}}{P}
\tag{10}
$$

so that

$$
\frac{N_{H_2O}^V}{N_{N_2}^V + N_{H_2}^V + N_{NH_3}^V + N_{H_2O}^V} = \frac{3.167}{P} \frac{5 - N_{H_2O}^V}{9 - X - N_{N_2}^V - N_{H_2}^V - N_{NH_3}^V - N_{H_2O}^V}
\tag{11}
$$

Finally, in terms of mole numbers the equilibrium relation, Eq. 1, is

$$
K_a = \frac{\dfrac{N_{NH_3}^V}{N^V}}{\left(\dfrac{N_{N_2}^V}{N^V}\right)^{0.5} \left(\dfrac{N_{H_2}^V}{N^V}\right)^{1.5} \left(\dfrac{P}{1 \text{ bar}}\right)}
$$
$$
= \frac{N_{NH_3}^V N^V}{(N_{N_2}^V)^{0.5}(N_{H_2}^V)^{1.5}\left(\dfrac{P}{1 \text{ bar}}\right)}
\tag{12}
$$
$$
= \frac{N_{NH_3}^V (N_{N_2}^V + N_{H_2}^V + N_{NH_3}^V + N_{H_2O}^V)}{(N_{N_2}^V)^{0.5}(N_{H_2}^V)^{1.5}\left(\dfrac{P}{1 \text{ bar}}\right)} = 762.2
$$

There are five unknowns in these equations: the extent of reaction X and the number of moles of each species in the vapor phase. There are also five equations to be solved: the four phase equilibrium relations (Eqs. 7, 8, 9, and 11) and one vapor-phase chemical equilibrium relation (Eq. 12). It is possible, with some difficulty, to solve these equations for the five unknowns.

If the computations are to be done by hand, one can simplify these equations by approximation. Since the Henry's law constants for both hydrogen and nitrogen in water are so high, a reasonable approximation is that the amount of these gases in the liquid phase are small enough to be neglected. With this assumption, we have

$$N_{N_2}^V = 1 - \tfrac{1}{2}X \quad \text{and} \quad N_{H_2}^V = 3 - \tfrac{3}{2}X = 3N_{N_2}^V$$

Also,

$$N^L = N_{N_2}^L + N_{H_2}^L + N_{NH_3}^L + N_{H_2O}^L$$
$$= 0 + 0 + (X - N_{NH_3}^V) + (5 - N_{H_2O}^V) = 5 + X - N_{NH_3}^V - N_{H_2O}^V$$

and the three equations to be solved (since we no longer require that nitrogen and hydrogen be in phase equilibrium) are

$$\frac{N_{NH_3}^V}{N_{N_2}^V + N_{H_2}^V + N_{NH_3}^V + N_{H_2O}^V} = \frac{97.58}{P} \frac{X - N_{NH_3}^V}{5 + X - N_{NH_3}^V - N_{H_2O}^V} \tag{13}$$

$$\frac{N_{H_2O}^V}{N_{N_2}^V + N_{H_2}^V + N_{NH_3}^V + N_{H_2O}^V} = \frac{3.167}{P} \frac{5 - N_{H_2O}^V}{5 + X - N_{NH_3}^V - N_{H_2O}^V} \tag{14}$$

and

$$K_a \left(\frac{P}{1 \text{ bar}} \right) = \frac{N_{NH_3}^V (N_{N_2}^V + N_{H_2}^V + N_{NH_3}^V + N_{H_2O}^V)}{(N_{N_2}^V)^{1/2} (N_{H_2}^V)^{3/2}}$$
$$= \frac{N_{NH_3}^V (4 - 2X + N_{NH_3}^V + N_{H_2O}^V)}{\left(1 - \dfrac{1}{2}X\right)^{1/2} \left(3 - \dfrac{3}{2}X\right)^{3/2}} \tag{15}$$

for the three unknowns, X, $N_{NH_3}^V$, and $N_{H_2O}^V$.

The solution, using the complete equations (without the simplification mentioned above) and MATHCAD, is

Species	Liquid Phase		Vapor Phase	
	Moles	Mole Fraction	Moles	Mole Fraction
N_2	2.52×10^{-8}	5.08×10^{-9}	0.0771	0.0351
H_2	9.73×10^{-8}	1.96×10^{-8}	0.2312	0.1054
NH_3	0.4355	0.0878	1.4103	0.6428
H_2O	4.5245	0.9122	0.4755	0.2167

As can be seen, the quantities of nitrogen and hydrogen in the liquid phase are indeed very small. Therefore, a virtually identical solution is obtained with the simplified model of Eqs. 13, 14, and 15. ∎

If the pressure is changed on a system in which phase and chemical equilibrium occur simultaneously, the equilibrium state of the system will shift due to changes in both the phase behavior and in the molar extent of reaction. This is illustrated in the next example.

ILLUSTRATION 13.4-2

Combined Chemical and Vapor-Liquid Equilibrium (Continued)

Repeat the calculations of the previous illustration for a range of pressures.

SOLUTION

The formulation of the problem in the previous illustration permits solutions for various pressures, and the same MATHCAD program was used to obtain solutions for most of the pressure range. The results of such calculations are shown in the accompanying figure in terms of the molar extent of reaction X and the molar fraction of the total mixture that is liquid, denoted by f. There are several things to notice in the solution to this problem. First, for pressures above about 33 kPa there is no equilibrium vapor phase (i.e., $f = 1$). That is, the bubble point pressure of this equilibrium reacting mixture at 25°C is 33 kPa, so for higher pressures only a liquid phase is present. Also, and only coincidentally, the reaction is essentially at completion ($X = 2$) at this pressure. At other temperatures and certainly for other reactions, the disappearance of the vapor phase and the point of essentially complete reaction would not coincide.

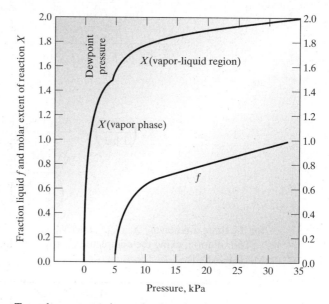

Two-phase ammonia production reaction.

Second, at pressures below about 4.8 kPa there is no equilibrium liquid phase (i.e., $f = 0$); 4.8 kPa is the dew point pressure of this reacting mixture at 25°C. Consequently, to solve for the equilibrium state of this system at pressures lower than 4.8 kPa, only vapor-phase chemical equilibrium needs to be considered without any phase equilibrium constraints. In this case, the mass balance equations are

$$N_{H_2O} = 5 \qquad N_{NH_3} = X \qquad N_{N_2} = 1 - \tfrac{1}{2}X$$

$$N_{H_2} = 3\left(1 - \frac{X}{2}\right) = 3N_{N_2} \quad \text{and} \quad N = 9 - X$$

where N is the total number of moles present. The species mole fractions that appear in the chemical equilibrium relation are

$$y_{NH_3} = \frac{X}{9 - X} \qquad y_{N_2} = \frac{1 - \tfrac{1}{2}X}{9 - X} \quad \text{and} \quad y_{H_2} = \frac{3\left(1 - \tfrac{1}{2}X\right)}{9 - X}$$

The single equilibrium relation to be solved for the molar extent of reaction X is, then,

$$K_a = \frac{a_{NH_3}}{a_{N_2}^{1/2} a_{H_2}^{3/2}} = \frac{y_{NH_3}}{y_{N_2}^{1/2} y_{H_2}^{3/2} \left(\dfrac{P}{1\ bar}\right)} = \frac{X(9-X)}{\left(1 - \dfrac{1}{2}X\right)^{1/2} \left(3 - \dfrac{3}{2}X\right)^{3/2} \left(\dfrac{P}{1\ bar}\right)}$$

$$= \frac{X(9-X)}{3^{3/2}\left(1 - \dfrac{1}{2}X\right)\left(\dfrac{P}{1\ bar}\right)} = 762.2$$

The solution to this equation in terms of the molar extent of reaction X is also plotted in the figure. There we see that the molar extent of reaction decreases rapidly with decreasing pressure. Notice, however, that while the molar extent of reaction curve is continuous at the dew point, its derivative with respect to pressure is not. That is, while the molar extents of reaction approaching the dew point pressure from lower pressures (only vapor present) and from higher pressures (vapor-liquid mixture present) converge to the same value at the dew point pressure, the two molar extent-of-reaction curves have different slopes at the dew point pressure. Mathematically, this is because there are different constraining equations in these two regions. At low pressures, when only vapor is present, one needs to solve only the chemical equilibrium equation to obtain the equilibrium molar extent of reaction. However, at higher pressures, the equations of chemical and phase equilibria must be simultaneously solved to find the equilibrium extent of reaction. One should not expect the solutions for the molar extents of reaction to have the same dependence on pressure in these two regions.

Physically what is happening is that in the high-pressure region water is condensing and ammonia is being absorbed into this water. Therefore, at any molar extent of reaction X this results in higher mole fractions of nitrogen and hydrogen in the vapor when a liquid is present (since there are fewer moles of water and ammonia in the vapor) and a lower mole fraction of ammonia as it is dissolving into the condensed water. By the LeChatelier-Braun principle, we would expect that at a given pressure, there would be a greater conversion of nitrogen and hydrogen to ammonia when a liquid phase is also present than if only a vapor phase exists. Consequently, if we extrapolated the molar extent-of-reaction curve when only a vapor is present to higher pressures, it should be below the values when both the vapor and liquid are present. This is indeed what we see, and it explains why the two molar extent-of-reaction curves should have different slopes. ∎

For relatively simple combined chemical and phase equilibrium problems, such as the ones considered here, hand calculations or calculations using MATHCAD or other equation-solving programs are possible. More complicated problems, especially those involving multiple reactions (even in a single phase) and both multiple reactions and multiple phases, require a large database of thermophysical properties and specially prepared computer programs. Several programs are available[8] to both generate all the thermodynamic data for the reactant and product species and identify the state of thermodynamic equilibrium. The computation algorithm used in these programs is different from the procedures discussed here in that the chemical equilibrium constant concept is not used. Instead, a general expression is written for the Gibbs energy in terms of all the species and phases that may be present, and this function is minimized by a direct search subject to the state-variable and mass balance constraints on the system. Computationally, this method is more efficient than setting up and solving a large collection of nonlinear, coupled equilibrium equations. The two procedures are theoretically equivalent and must lead to the same result. (For the reaction considered in Fig. 13.1-1, the direct search would involve an iterative calculation of X^* by searching

---

[8]For a general discussion, see *Chemical Reaction Equilibrium Analysis: Theory and Algorithms* by W. R. Smith and R. W. Missen, John Wiley & Sons, New York, 1982. Also see *Fortran IV Computer Program for Calculation of Thermodynamic and Transport Properties of Complex Chemical Systems* by R.A. Svehla and B. J. McBride, National Aeronautics and Space Administration Technical Note D-7056, January 1973; *Rand's Chemical Composition Program*, Rand Corporation, Santa Monica, Calif.; and others.

for the value of X for which G was a minimum, whereas in the equilibrium constant approach X^* is found as the solution to a nonlinear algebraic equation that results from analytically identifying the state of minimum Gibbs energy.)

We conclude this section by noting that, in some cases, states involving both chemical and phase equilibrium can be considered to be chemical equilibrium problems only, but with different states of aggregation for the standard states of the various species present. Consider, for example, the dissolution of gaseous ammonia in water, and its subsequent reaction to form ammonium hydroxide. This process can be considered either to occur in two steps, the first involving phase equilibrium,

$$NH_3(g) = NH_3(aq)$$

and the second single-phase chemical equilibrium

$$NH_3(aq) + H_2O(l) = NH_4OH(aq)$$

or to occur as a single-step multiphase chemical equilibrium process:

$$NH_3(g) + H_2O(l) = NH_4OH(aq)$$

We now establish that the same equilibrium state will be found independent of which of the two descriptions is used.

In the second case we take the standard state of ammonia to be the pure gas, of water to be the pure liquid, and of ammonium hydroxide to be the ideal 1-molal solution, and obtain

$$
K_{a,\mathrm{II}} = \exp\left\{-\frac{1}{RT}[\Delta_f\underline{G}^\circ_{NH_4OH}\ (\text{ideal, 1 molal}) - \Delta_f\underline{G}^\circ_{H_2O}(\text{liquid})\right.
$$

$$
\left. - \Delta_f\underline{G}^\circ_{NH_3}\ (\text{gas, 1 bar})]\right\}
$$

$$
= \frac{a_{NH_4OH}}{a_{NH_3}a_{H_2O}} = \frac{\left(\dfrac{M_{NH_4OH}\gamma^\square_{NH_4OH}}{M = 1\ \text{molal}}\right)}{\left(\dfrac{P_{NH_3}}{1\ \text{bar}}\right)(x_{H_2O}\gamma_{H_2O})}
\tag{13.4-5}
$$

In the two-step description, the standard state of ammonia would be taken to be the ideal 1-molal solution, so that

$$
K_{a,\mathrm{I}} = \exp\left\{-\frac{1}{RT}[\Delta_f\underline{G}^\circ_{NH_4OH}\ (\text{ideal, 1 molal}) - \Delta_f\underline{G}^\circ_{H_2O}(\text{liquid})\right.
$$

$$
\left. - \Delta_f\underline{G}^\circ_{NH_3}\ (\text{ideal, 1 molal})]\right\}
$$

$$
= \frac{a_{NH_4OH}}{a_{NH_3}a_{H_2O}} = \frac{\left(\dfrac{M_{NH_4OH}\gamma^\square_{NH_4OH}}{M = 1\ \text{molal}}\right)}{\left(\dfrac{M_{NH_3}\gamma^\square_{NH_3}}{M = 1\ \text{molal}}\right)(x_{H_2O}\gamma_{H_2O})}
\tag{13.4-6}
$$

Clearly $K_{a,\text{I}} \neq K_{a,\text{II}}$. However, Eq. 13.4-6 must be solved together with the phase equilibrium requirement

$$\bar{f}^{\text{V}}_{\text{NH}_3} = \bar{f}^{\text{L}}_{\text{NH}_3}$$

or equivalently,

$$\bar{G}^{\text{V}}_{\text{NH}_3} = \bar{G}^{\text{L}}_{\text{NH}_3}$$

We will use the latter of these two relations. Next, we note that

$$\bar{G}^{\text{V}}_{\text{NH}_3}(T, P, y_{\text{NH}_3}) = \bar{G}^{\text{V}}_{\text{NH}_3}(T, P_{\text{NH}_3}) = \underline{G}^{\text{V}}_{\text{NH}_3}(T, P = 1 \text{ bar}) + RT \ln \left[\frac{P_{\text{NH}_3}}{1 \text{ bar}} \right]$$

$$(13.4\text{-}7)$$

$$\bar{G}^{\text{L}}_{\text{NH}_3}(T, M_{\text{NH}_3}) = \bar{G}^{\text{L}}_{\text{NH}_3}(T, \text{ideal 1 molal}) + RT \ln \left[\frac{M_{\text{NH}_3} \gamma^{\square}_{\text{NH}_3}}{M_{\text{NH}_3} = 1 \text{ molal}} \right]$$

$$(13.4\text{-}8)$$

and that

$$\Delta_{\text{f}} \underline{G}^{\circ}_{\text{NH}_3}(T, \text{ideal 1 molal}) + \underline{G}^{\text{V}}_{\text{NH}_3}(T, P = 1 \text{ bar}) - \bar{G}^{\text{L}}_{\text{NH}_3}(T, \text{ideal 1 molal})$$
$$= \bar{G}^{\text{L}}_{\text{NH}_3}(T, \text{ideal 1 molal}) - \tfrac{1}{2}\underline{G}^{\text{V}}_{\text{N}_2}(T, P = 1 \text{ bar}) - \tfrac{3}{2}\underline{G}^{\text{V}}_{\text{H}_2}(T, P = 1 \text{ bar})$$
$$+ \underline{G}^{\text{V}}_{\text{NH}_3}(T, P = 1 \text{ bar}) - \bar{G}^{\text{L}}_{\text{NH}_3}(T, \text{ideal 1 molal})$$
$$= \Delta_{\text{f}} \underline{G}^{\circ}_{\text{NH}_3}(\text{gas}, T, P = 1 \text{ bar})$$

$$(13.4\text{-}9)$$

Using Eqs. 13.4-7, 13.4-8, and 13.4-9 in the equilibrium relation of Eq. 13.4-6 yields Eq. 13.4-5. Thus, precisely the same equilibrium relation is found between the ammonia partial pressure in the gas phase and the ammonium hydroxide concentration in the liquid phase, independent of the manner in which we presume the absorption-reaction process to take place. Consequently, it is a matter of convenience whether we consider multiphase reactions such as the one here to be chemical equilibrium problems or problems of combined chemical and phase equilibrium.[9]

PROBLEMS

(*Note:* The *Chemical Engineer's Handbook,* McGraw-Hill, New York, contains a comprehensive list of standard-state Gibbs energies and enthalpies of formation.)

13.1 Isopropyl alcohol is to be dehydrogenated in the gas phase to form propionaldehyde according to the reaction

$$(\text{CH}_3)_2\text{CHOH(g)} = \text{CH}_3\text{CH}_2\text{CHO(g)} + \text{H}_2(\text{g})$$

For this reaction,

$$\Delta_{\text{rxn}} G^{\circ}(T = 298.15 \text{ K}) = 17.74 \text{ kJ/(mol } i\text{-propanol reacted)}$$
$$\Delta_{\text{rxn}} H^{\circ}(T = 298.15 \text{ K}) = 55.48 \text{ kJ/(mol } i\text{-propanol reacted)}$$

and

$$\Delta_{\text{rxn}} C_{\text{P}} = \sum_{i} \nu_i C_{\text{P},i} = 16.736 \text{ J/(mol } i\text{-propanol reacted K)}$$

---

[9]In general, considering the process to be a combined chemical and phase equilibrium problem will result in more complicated calculations, but will yield somewhat more information—here the concentration of ammonia in the liquid phase.

Compute the equilibrium fraction of isopropyl alcohol that would be dehydrogenated at 500 K and 1.013 bar.

13.2 The dissociation pressure of calcium oxalate in the reaction

$$CaC_2O_4(s) = CaCO_3(s) + CO$$

at various temperatures is

T (°C)	375	388	403
Dissociation pressure (kPa)	1.09	4.00	17.86

T (°C)	410	416	418
Dissociation pressure (kPa)	33.33	78.25	91.18

Source: J. H. Perry, ed., *Chemical Engineers' Handbook,* 4th ed., McGraw-Hill, New York, 1963, pp. 3–69.

Compute the standard-state Gibbs energy change, enthalpy change, and entropy change for this reaction for the temperature range in the table.

13.3 Carbon dioxide can react with graphite to form carbon monoxide,

$$C(graphite) + CO_2(g) = 2CO(g)$$

and the carbon monoxide formed can further react to form carbon and oxygen:

$$2CO(g) = 2C(s) + O_2(g)$$

Determine the equilibrium composition when pure carbon dioxide is passed over a hot carbon bed maintained at 1 bar and (a) 2000 K or (b) 1000 K.

13.4 The extent of reaction generally depends on pressure as well as temperature. For the reaction (or phase transition)

$$C(graphite) = C(diamond)$$

the standard-state Gibbs energy change at 25°C is 2866 J/mol. The density of graphite is 2.25 g/cc and that of diamond is approximately 3.51 g/cc; both solids may be considered to be incompressible. To help chemical engineering students pay their tuition, we are thinking of setting up equipment to run this reaction in the senior laboratory. Estimate what pressure, at room temperature ($T \sim 25$°C), must be used to convert old pencil "leads" (graphite) into diamonds.

13.5 The production of NO by the direct oxidation of nitrogen,

$$\tfrac{1}{2}N_2 + \tfrac{1}{2}O_2 = NO$$

occurs naturally in internal combustion engines. This reaction is also used to commercially produce nitric

oxide in electric arcs in the Berkeland-Eyde process. If air is used as the feed, compute the equilibrium conversion of oxygen at 1 atm (1.013 bar) total pressure over the temperature range of 1500 to 3000°C. Air contains 21 mol % oxygen and 79 mol % nitrogen.

13.6 Crystalline sodium sulfate, in the presence of water vapor, may form a decahydrate,

$$Na_2SO_4(s) + 10H_2O(g) = Na_2SO_4 \cdot 10H_2O(s)$$

a. Estimate the minimum partial pressure of water at which the decahydrate will form at 25°C.

b. Make a rough estimate of the minimum water partial pressure for decahydrate formation at 15°C.

13.7 Carbon disulfide is produced from the high-temperature reaction of carbon and sulfur:

$$C(s) + S_2(g) = CS_2(g)$$

This reaction is carried out in a retort at low pressure, and in the absence of oxygen and other species that may react with either the carbon or the sulfur. Compute the equilibrium percentage conversion of sulfur at 750°C and 1000°C.

13.8 The data in the following table give the solubility of silver chloride in various aqueous solutions at 25°C. Show that these data can be plotted on the same $\ln K_s$ versus $\sqrt{I}$ curve as used in Illustration 13.2-3.

Electrolyte Added	Concentration of Added Salt (mol/m^3)	Concentration of Silver Chloride (mol/m^3)
Ba(NO$_3$)$_2$	0.2111	1.309×10^{-2}
	0.7064	1.339
	4.402	1.450
	5.600	1.467
La(NO$_3$)$_3$	0.1438	1.317×10^{-2}
	0.5780	1.367
	1.660	1.432
	2.807	1.477

Source: E. W. Newman, *J. Am. Chem. Soc.,* **54**, 2195 (1932).

13.9 The following data are available for the solubility of barium sulfate in water:

Temperature (°C)	5	10	15
Solubility (mol m^{-3})	0.0156	0.0167	0.0183

Temperature (°C)	20	25
Solubility (mol m^{-3})	0.0198	0.0216

The mean activity coefficient $\gamma_\pm$ for ions of a salt at low ionic strength is given by

$$\ln \gamma_\pm = -\alpha |z_+ z_-| \sqrt{I}$$

where values for the parameter α for water are given in Table 9.10-1.

a. Compute K_s°, the ideal solution solubility product, for barium sulfate at each of the temperatures in the table.

b. At each of the temperatures in the table calculate the Gibbs energy change for the reaction

$$BaSO_4(s) = Ba^{2+}(aq, M = 1, ideal)$$
$$+ SO_4^{2-}(aq, M = 1, ideal)$$

Here (s) denotes the pure solid state, and (aq, $M = 1$, ideal) indicates the ion in a hypothetical ideal aqueous solution at 1-molal concentration.

c. Compute the entropy and enthalpy changes for the reaction in part (b) at 5°C, 15°C, and 25°C.

13.10 Hydrogen gas can be produced by the following reactions between propane and steam in the presence of a nickel catalyst:

$$C_3H_8 + 3H_2O = 3CO + 7H_2$$
$$C_3H_8 + 6H_2O = 3CO_2 + 10H_2$$

a. Compute the standard heat of reaction and the standard-state Gibbs energy change on reaction for each of the reactions at 1000 and 1100 K.

b. What is the equilibrium composition of the product gas at 1000 K and 1 bar if the inlet to the catalytic reactor is pure propane and steam in a 1-to-10 ratio?

c. Repeat calculation (b) at 1100 K.

13.11 An important step in the manufacture of sulfuric acid is the gas-phase oxidation reaction

$$SO_2 + \tfrac{1}{2}O_2 = SO_3$$

Compute the equilibrium conversion of sulfur dioxide to sulfur trioxide over the temperature range of 0 to 1400°C for a reactant mixture consisting of initially pure sulfur dioxide and a stoichiometric amount of air at a total pressure of 1.013 bar. (Air contains 21 mol % oxygen and 79 mol % nitrogen.)

13.12 Ethylene dichloride is produced by the direct chlorination of ethylene using small amounts of ethylene dibromide as a catalyst:

$$C_2H_4 + Cl_2 = C_2H_4Cl_2$$

If stoichiometric amounts of ethylene and chlorine are used, and the reaction is carried out at 50°C and 1

bar, what is the equilibrium conversion of ethylene? (*Note:* The normal boiling point of ethylene dichloride is 83.47°C.)

13.13 Polar molecules interact more strongly at large distances than do nonpolar molecules, and generally form nonideal solutions. One model for solution nonidealities in a binary mixture consisting of a nonpolar species, which we denote by A, and a polar substance, designated by the symbol B, is based on the supposition that the polar substance partially dimerizes,

$$2B \rightleftharpoons B_2$$

Thus, although the mixture is considered to be a binary mixture with mole fractions

$$x_A = \frac{N_A^\circ}{N_A^\circ + N_B^\circ} \quad \text{and} \quad x_B = \frac{N_B^\circ}{N_A^\circ + N_B^\circ}$$

where N_A° and N_B° are the initial number of moles of A and B, respectively, it is, according to the supposition, really a ternary mixture with mole fractions

$$x_A^\ddagger = \frac{N_A^\circ}{N_A^\circ + N_B + N_{B_2}} \qquad x_B^\ddagger = \frac{N_B}{N_A^\circ + N_B + N_{B_2}}$$

and

$$x_{B_2}^\ddagger = \frac{N_{B_2}}{N_A^\circ + N_B + N_{B_2}}$$

where, by conservation of B molecules,

$$N_B^\circ = N_B + 2N_{B_2}$$

It is further assumed that the ternary mixture is ideal, and that the apparent nonidealities in mixture properties result from considering the A-B solution to be a binary mixture with mole fractions x_i, rather than a ternary mixture with true fractions $x_i^\ddagger$. Show that the apparent activity coefficients for the binary mixture that result from this model are

$$\gamma_A = 2k/\delta$$

$$\gamma_B = \left(\frac{2}{x_B}\right) \frac{(-x_A + (x_A^2 + 2kx_Ax_B + kx_B^2)^{1/2})}{\delta}$$

and

$$\delta = (2k - 1)x_A + kx_B + (x_A^2 + 2kx_Ax_B + kx_B^2)^{1/2}$$

where $k = 4K_a + 1$, and K_a is the equilibrium constant for the dimerization reaction.

13.14 Acetaldehyde is produced from ethanol by the following gas-phase reactions:

$$C_2H_5OH + \tfrac{1}{2}O_2 = CH_3CHO + H_2O \qquad \textbf{(a)}$$
$$C_2H_5OH = CH_3CHO + H_2 \qquad \textbf{(b)}$$

The reactions are carried out at 540°C and 1 bar pressure using a silver gauze catalyst and air as an oxidant. If 50% excess air [sufficient air that 50 percent more oxygen is present than is needed for all the ethanol to react by reaction (**a**)] is used, calculate the equilibrium composition of the reactor effluent.

13.15 When propane is heated to high temperatures, it pyrolyzes or decomposes. Assume that the only reactions that occur are

$$C_3H_8 = C_3H_6 + H_2$$
$$C_3H_8 = C_2H_4 + CH_4$$

and that these reactions take place in the gas phase.
a. Calculate the composition of the equilibrium mixture of propane and its pyrolysis products at a pressure of 1 bar and over a temperature range of 1000 to 2000 K.
b. Calculate the composition of the equilibrium mixture of propane and its pyrolysis products over a temperature range of 1000 to 2000 K if pure propane at 25°C and 1 bar is loaded into a constant-volume (bomb) reactor and heated.

13.16 As part of the process of glycogen breakdown and utilization in muscles, glucose 1-phosphate is converted to glucose 6-phosphate (that is, the phosphate group on a glucose molecule moves from the carbon 1 atom to the carbon 6 atom) as a result of the action of the enzyme phosphoglumutase. An in vitro analysis shows that at 25°C and a pH of 7.0 adding the enzyme to a solution containing 0.020 M glucose 1-phosphate reduces its concentration to 0.001 M. Determine the apparent equilibrium constant for this reaction, and the actual Gibbs energy change at the reaction conditions.

13.17 The simple statement of the LeChatelier-Braun principle given in Sec. 13.1 leads one to expect that if the concentration of a reactant were increased, the reaction would proceed so as to consume the added reactant. This, however, is not always true. Consider the gas-phase reaction

$$N_2 + 3H_2 = 2NH_3$$

Show that if the mole fraction of nitrogen is less than 0.5, the addition of a small amount of nitrogen to the system at constant temperature and pressure results in the reaction of nitrogen and hydrogen to form ammonia, whereas if the mole fraction of nitrogen is greater than 0.5, the addition of a small amount of nitrogen leads to the dissociation of some ammonia to form more nitrogen and hydrogen. Why does this occur?

13.18 By catalytic dehydrogenation, 1-butene can be produced from *n*-butane,

$$C_4H_{10} = C_4H_8 + H_2$$

However, 1-butene may also be dehydrogenated to form 1,3-butadiene,

$$C_4H_8 = C_4H_6 + H_2$$

Compute the equilibrium conversion of *n*-butane to 1-butene and 1,3-butadiene at 1 bar and
a. 900 K
b. 1000 K

13.19 Silver, when exposed to air, tarnishes. The following reactions have been proposed for this tarnishing:

$$2Ag + \tfrac{1}{2}O_2 \rightarrow Ag_2O \qquad 2Ag + H_2O \rightarrow Ag_2O + H_2$$

$$2Ag + H_2S \rightarrow Ag_2S + H_2 \qquad 2Ag + SO_2 \rightarrow Ag_2S + O_2$$

The following data are available:

Species	$\triangle_f\underline{G}$, kJ/mol
Ag_2O	−9.33
Ag_2S	−31.80
$SO_2(g)$	−299.91
$H_2O(g)$	−228.59

Air can be assumed to contain 0.5 ppm H_2, 0.03 ppm (80 mg/m³) SO_2, and 0.1 H_2S mg/m³ and water at a partial pressure of 2.0 kPa.

Which of these reactions are likely to occur at normal conditions and result in the tarnishing of silver?

13.20 The Soviet *Venera VII* probe, which reached Venus on December 15, 1970, found the following conditions on the planet surface:

$$T \approx 747 \pm 20 \text{ K}$$
$$P \approx 90 \pm 15 \text{ (earth) atmospheres}$$

Interferometric measurements indicate that there is only a 10 to 20 K temperature variation across the planet, and that the coldest region lies at the equator. Radar astronomy measurements suggest that the dielectric constant of the Venus surface is typical of mineral silicates. Furthermore, from spectroscopic observations it has been concluded that the approximate atmospheric composition of Venus is

	Species	Mole Percent
	CO_2	99.9–
	H_2O	0.01
	CO	0.01
	HCl	1×10^{-1}
	HF	1×10^{-6}
	O_2	trace
Sulfur-bearing constituents	$\{ H_2S, COS$ $\ SO_2, SO_3$	0

Since the temperature of Venus is so high, and the planet is quite old, it may be assumed that all chemical reactions occurring between the atmosphere and the surface minerals are in chemical equilibrium. The following reactions are thought to occur:

$$CaCO_3 + SiO_2 = CaO{\cdot}SiO_2 + CO_2$$
$$3FeO{\cdot}SiO_2 + CO_2 = 3SiO_2 + Fe_3O_4 + CO$$
$$3FeO{\cdot}SiO_2 + \tfrac{1}{2}O_2 = Fe_3O_4 + 3SiO_2$$

Determine whether the chemical equilibrium assumption is consistent with the reported data.
Data: For $FeO{\cdot}SiO_2$, $\Delta_f\underline{G}^\circ = -1.060$ MJ/mol and $\Delta_f\underline{H}^\circ = -1.144$ MJ/mol; other Gibbs energy of formation data are given in Appendix A.IV and the *Chemical Engineers' Handbook.*[10] The heat capacity of the solid species is

$$C_P = a + bT + e/T^2 \text{ J/(mol K)}; \quad T[=]\text{K}$$

Species	a	$b \times 10^2$	$e \times 10^{-4}$
$CaO{\cdot}SiO_2$	116.94	0.8602	-3.120
$CaCO_3$	82.34	4.975	-1.287
Fe_3O_4	172.26	7.874	-4.098
$FeO{\cdot}SiO_2$	98.28	4.269	-1.215
SiO_2	45.48	3.645	-1.009

Since the temperature is so high, you may assume that the gas phase is ideal.

13.21 Styrene can be hydrogenated to ethyl benzene at moderate conditions in both the liquid phase and the gas phase. Calculate the equilibrium compositions in the vapor and liquid phases of hydrogen, styrene, and ethyl benzene at each of the following conditions:
  a. 3-bar pressure and 25°C, with a starting mole ratio of hydrogen to styrene of 2 to 1
  b. 3-bar pressure and 150°C, with a starting mole ratio of hydrogen to styrene of 2 to 1
  Data:
  Reaction stoichiometry: $C_6H_5HC{=}CH_2 + H_2 \longrightarrow C_6H_5CH_2CH_3$
  Physical properties:

Species	δ $(cal/cc)^{1/2}$	$\underline{V}^L$ (cc/mol)	$\Delta_f\underline{G}^\circ$ (gas, 1 bar and 25°C)
Styrene C_8H_8	9.3	116	213.9 kJ/mol
Ethyl benzene C_8H_{10}	10.1	123	130.9 kJ/mol
Hydrogen	3.25	3.1	0.0

  Heat capacity: See Appendix A.II.
  Vapor pressure:

	1.333 kPa	13.33	53.32	101.3	202.6	506.5
C_8H_8	30.8°C	82.0	122.5	145.2		
C_8H_{10}	25.9	74.1	113.8	136.2	163.5	207.5

13.22 If 1 mole of a gas in a constant-volume system is heated, and both the heat flow and the gas temperature are measured as a function of time, the constant-volume heat capacity can be computed from

$$C_V = \left(\frac{\partial U}{\partial T}\right)_V = \frac{\dot{Q}}{\left(\dfrac{\partial T}{\partial t}\right)_V}$$

This equation can also be used to calculate the effective heat capacity $C_{V,eff}$ of a gas that is undergoing a chemical reaction, such as nitrogen tetroxide dissociating to form nitrogen dioxide,

$$N_2O_4 \rightleftharpoons 2NO_2$$

In such cases $C_{V,eff}$ can be much larger than the heat capacity of the nonreacting gas.
  a. Develop an expression for $C_{V,eff}$ for dissociating nitrogen tetroxide, and comment on the depen-

[10]R. H. Perry, D. W. Green, and J. O. Maloney, eds., *Chemical Engineers' Handbook,* 6th ed., McGraw-Hill, New York (1984), pp. 3-129–3-135.

dence of $C_{V,eff}$ on the internal energy change on chemical reaction.

b. Compute the molar effective heat capacity $C_{V,eff}$ as a function of temperature for nitrogen tetroxide over the temperature range of 300 to 600 K, if pure N_2O_4 is loaded into a constant-volume reactor at 300 K at a pressure of 1.013 bar.

13.23 Consider a gaseous mixture containing the three isomers *n*-pentane (1), *iso*-pentane (2), and *neo*-pentane (3). These species may interconvert by reaction.

a. Determine the number of independent chemical reactions and the number of degrees of freedom for this system.

b. Calculate the equilibrium concentration of the three isomers at 400 K and 1 bar.

c. What is the effect of increasing the pressure of the system on the equilibrium concentration of the isomers?

Data: The Gibbs energies of formation of these compounds at 400 K are 40.2 kJ/mol, 34.4 kJ/mol, and 37.6 kJ/mol, respectively.

13.24 Methane gas hydrates are formed from liquid water by the following reaction:

$$CH_4\,(g) + 5.75H_2O\,(l) \rightarrow CH_4 \cdot 5.75H_2O\,(s)$$

a. Calculate the Gibbs energy of formation of the hydrate at 278 K and 283 K using the information that the methane partial pressure in equilibrium with the hydrate at 278 K is 4.2 MPa and at 283 K is 6.8 MPa.

b. One common way to prevent hydrates from forming is by adding an inhibitor to the system, usually methanol or a salt (e.g., NaCl). If 10 wt % methanol is added to water, what will be the equilibrium partial pressure for methane for hydrate formation at 278 K and 283 K (assume no methanol is present in the vapor)?

c. What phases and components are present at equilibrium at 273.15 K? How many degrees of freedom are there for this system?

13.25 When pure hydrogen iodide gas enters an evacuated cylinder, the following reactions may occur:

$$HI(g) = \tfrac{1}{2}H_2(g) + \tfrac{1}{2}I_2(g)$$
$$I_2(g) = I_2(s)$$

(Note that since Gibbs energies of formation data are available for iodine in both the gaseous and solid phases, it is more convenient to think of the solid-vapor iodine phase equilibrium as a chemical equilibrium.) If the reaction mixture is gradually compressed at 25°C, a pressure is reached at which the first bit of solid iodine appears. What is the pres-

sure at which this occurs, and what is the vapor composition at this pressure?

13.26 The calcination of sodium bicarbonate takes place according to the reaction

$$2NaHCO_3(s) = Na_2CO_3(s) + CO_2(g) + H_2O(g)$$

When this reaction was run in the laboratory by placing sodium bicarbonate in an initially evacuated cylinder, it was observed that the equilibrium total pressure was 0.826 kPa at 30°C and 166.97 kPa at 110°C. The heat of reaction for the calcination can be assumed to be independent of temperature.

a. What is the heat of reaction for this reaction?

b. Develop an equation for the equilibrium constant at any temperature.

c. At what temperature will the partial pressure of carbon dioxide in the reaction vessel be exactly 1 bar?

13.27 Carbon is deposited on a catalytic reactor bed as a result of the cracking of hydrocarbons. Periodically, hydrogen gas is passed through the reactor in an effort to remove the carbon and to preserve the reduced state of the catalyst. It has been found, by experiment, that the effluent gas contains about 10 mol % methane and 90 mol % hydrogen when the temperature is 1000 K and the pressure is 1 bar. Is this conversion thermodynamically limited? Can a higher concentration of methane be produced by reducing the rate of hydrogen flow through the reactor, thereby removing more carbon for a given amount of hydrogen?

13.28 A gas mixture containing equimolar quantities of carbon dioxide and hydrogen is to be "reformed" by passing it over a catalyst. The pressure in the reformer will be determined by the possibility of solid carbon deposition. Although a large number of reactions are possible, only the following are believed to occur:

$$C + H_2O = CO + H_2$$
$$C + 2H_2O = CO_2 + 2H_2$$
$$CO_2 + C = 2CO$$
$$CO + H_2O = CO_2 + H_2$$

a. At temperatures between 600 and 1000 K, over what range of pressure will carbon deposit if each of the reactions is assumed to achieve equilibrium?

b. For this feed, what pressure should be maintained for exactly 30 percent of the carbon present in the feed to precipitate as solid carbon at each temperature between 600 and 1000 K?

13.29 A process is being developed to produce high-purity titanium. As part of the proposed process, titanium will be kept in a quartz (silicon dioxide) crucible at 1273 K. A chemical engineer working on the process is concerned that the titanium could reduce the silicon dioxide, producing titanium dioxide and elemental silicon, which would lower the purity of the titanium. Is this concern justified?

Data: At 1273 K

$$Ti(s) + O_2(g, 1\ bar) = TiO_2(s) \qquad \Delta_{rxn}G^\circ = -674\ kJ/mol$$
$$Si(s) + O_2(g, 1\ bar) = SiO_2(s) \qquad \Delta_{rxn}G^\circ = -644\ kJ/mol$$

13.30 The liquid in a two-phase, binary mixture of benzene and cyclohexane has a composition of 20 mol % of benzene and 80 mol % of cyclohexane at $T = 80°C$.
  a. Find the pressure of the system and the composition of the vapor phase.
  b. The vapor is removed, heated isobarically to 550 K, and passed through a reactor, where the following dehydrogenation reaction occurs:

$$C_6H_{12}\,(g) \rightarrow C_6H_6\,(g) + 3H_2\,(g)$$

  If the system is in chemical equilibrium when leaving the reactor at 550 K, what is the composition of the species leaving the reactor?
  c. The reactor stream is very quickly cooled (quenched) to 80°C so that the overall mixture composition remains the same as in part (b). Find the dew point pressure of this stream at 80°C, and the composition of the first drop of liquid that is formed.
  Data:

$$\underline{G}^{ex}\,(J/mol) = 100 x_{bnz} x_{cC6}$$

$$\log_{10} P_{bnz}^{vap}\,(bar) = 4.018\,16 - \frac{1203.828}{T\,(K) - 53.229}$$

  and

$$\log_{10} P_{cC6}^{vap}\,(bar) = 3.980\,22 - \frac{1202.299}{T\,(K) - 49.623}$$

13.31 At 452.2 K, a total pressure of 95.9 kPa, and with an appropriate catalyst, the equilibrium extent of dissociation of pure isopropanol to acetone and hydrogen is found to be 56.4 percent [H. J. Kolb and R. L. Burwell, Jr., *J. Am. Chem. Soc.,* **67,** 1084 (1945)]. Use this information to calculate the standard-state Gibbs energy change for this reaction at this temperature.

13.32 Liquid benzene can be catalytically hydrogenated, with cyclohexene, cyclohexane, and 1,3-cyclohexadiene being among the products. De-

termine the product distribution as a function of the hydrogen partial pressure of 1 bar at 298.15 K. Assume that the vapor phase contains only hydrogen, and that its partial pressure is maintained at 1 bar as the reaction proceeds.

13.33 Consider a closed vessel in which selenium is in equilibrium with its vapor. Selenium, in the vapor phase, polymerizes to Se_i species where i = 1, 2, 3, 5, 7, 8.
  a. At any temperature above the melting point of selenium, compute the degrees of freedom in the system. Note that the vapor-phase reactions can be considered to be

$$iSe \rightleftharpoons Se_i \quad \text{and} \quad Se + Se_i \rightleftharpoons Se_{i+1}$$

  b. Develop an interrelationship between the partial molar Gibbs energies of each of the selenium species in the vapor phase.

13.34 Formaldehyde is an important industrial solvent and also a raw material in chemical manufacture. At elevated temperatures, it dissociates into ammonia and carbon monoxide in the following gas-phase reaction:

$$HCONH_2 \rightleftharpoons NH_3 + CO$$

  a. Compute the equilibrium constant for this reaction over the temperature range of 400 to 500 K.
  b. One mole of formaldehyde is placed in an evacuated 25-liter vessel and heated. Compute the equilibrium pressure and mole fractions of all species over the temperature range of 400 to 500 K.
  c. Repeat the calculation of part (b) if, instead of being evacuated, the 25-liter vessel initially contains air (21 mol % O_2 and 79 mol % N_2) at ambient conditions. Assume there is no reaction of oxygen with any of the components in the formaldehyde dissociation reaction at these temperatures.

13.35 Repeat the calculations of Illustrations 13.4-1 and 13.4-2 for 50°C. The Henry's law constant for ammonia in water at this temperature is $H_{HN_3} = 384.5$ kPa/mole fraction, and you can assume that the Henry's constants for nitrogen and hydrogen are unchanged from the values given in Illustration 13.4-1.

13.36 The industrial solvent carbon disulfide can be made from the reaction between methane and hydrogen sulfide at elevated temperatures using the following reaction gas-phase reaction:

$$2H_2S + CH_4 \rightleftharpoons CS_2 + 4H_2$$

  a. Compute the equilibrium constant for this reaction from 500 to 800°C.
  b. Starting with equal number of moles of H_2S and CH_4, compute the equilibrium compositions of

all components in the reaction at 1 bar and over the temperature range of part (a).

c. Repeat the calculation of part (b) for a pressure of 10 bar. Ideal gas mixture behavior may be assumed at this pressure at these high temperatures.

13.37 At an appropriate temperature, a larger alkane will "crack" to form a smaller alkane and an olefin. One example is the cracking of propane to form methane and ethylene:

$$C_3H_8(g) \rightarrow CH_4(g) + C_2H_4(g)$$

a. Calculate the equilibrium compositions that would result at 298.15 K and a total pressure of 1 bar by starting with pure propane.

b. Assuming that the standard-state heat of reaction for this reaction is independent of temperature, repeat the calculation above at 650 K.

c. Repeat the calculation of part (b) for a total pressure of 10 bar, assuming that an ideal gas mixture is formed at this high temperature.

13.38 Consider the reaction

$$A(l) + B(s) \rightarrow C(l) + D(g)$$

where l, s, and g indicate species in the liquid, solid, and gas phases, respectively, and each species appears only in the phase indicated. For this reaction, $\Delta_{rxn}\underline{G}° = -2.4$ kJ/mol A at 25°C for all pure components in the states of aggregation indicated above.

a. Determine the equilibrium amounts and compositions for substances A and C in this system if the liquid mixture formed is ideal, there is an excess of B present, and the partial pressure of component D is maintained at 0.5 bar.

b. Repeat the calculation in part (a) if the two components present in the liquid, A and C, form a nonideal mixture described by

$$\underline{G}^{ex}_{AC} = 0.3RT\, x_A x_C$$

c. What will the equilibrium amounts and concentrations of all substances be if the reaction is carried out in an initially evacuated 4-liter constant-volume bomb, one mole of A and twice the stoichiometric amount of B are initially present, and the liquid mixture formed is ideal? (You can assume that the volume of the liquid and solid are negligible, and that the temperature is constant.)

13.39 It is possible that hydrogen and other gases dissociate when adsorbed on a solid surface, and in catalysis it is important to know whether such a dissociation occurs. If hydrogen did not dissociate, that is, the adsorption process was

$$H_2 \rightarrow H_2(ad)$$

the amount of hydrogen adsorbed would be given by

$$\text{Amount of } H_2 \text{ adsorbed} = K_1 a_{H_2}$$

where a_{H_2} is the activity of molecular hydrogen in the gas phase. However, if hydrogen dissociates, the following two-step process occurs at the surface:

$$H_2 \rightarrow 2H \quad \text{and} \quad H \rightarrow H(ad)$$

and

$$\text{Amount of } H_2 \text{ adsorbed}$$
$$= \tfrac{1}{2} \times (\text{Amount of H adsorbed}) = \tfrac{1}{2} \times K_3 a_H$$

Using the notation that K_2 is the activity-based equilibrium constant for the dissociation reaction, develop expressions for the amount of molecular hydrogen adsorbed as a function of the equilibrium constants and the hydrogen partial pressure for the two cases (adsorption without dissociation and adsorption with dissociation). How would you discern which process was occurring if you had experimental data on the total hydrogen adsorption as a function of its partial pressure?

13.40 The proteolytic enzyme α-chymotrypsin is known to dimerize. The following data are available for this reaction at 25°C and pH = 7.8.

$$\Delta_{rxn}\underline{G}° = -19.6 \frac{kJ}{mol} \quad \text{and} \quad \Delta_{rxn}\underline{S}° = -430.7 \frac{J}{mol\,K}$$

based on ideal 1 M standard states.

a. Determine the values of the equilibrium constant for this reaction over the physiologically important temperature range of 0 to 40°C.

b. If the initial concentration of α-chymotrypsin is 0.001 M, determine the fraction that is dimerized over the temperature range of part (a).

13.41 Gases can be (very) slightly soluble in molten metals. For example, the solubility of nitrogen in liquid iron is well correlated by the empirical expression

$$\text{wt \% nitrogen} = 0.045\sqrt{P_{N_2}(bar)}$$

One question that arises is whether the gas dissociates on the dissolution, that is, whether the dissolution process is

$$N_2(gas) \rightarrow N_2(in\ metal)$$

or whether the process is

$$N_2(gas) \rightarrow N(in\ metal)$$

Use the equilibrium correlation above to decide which process is occurring. (*Hint:* Since the concentration of either molecular or atomic nitrogen in molten iron is so low, you can assume that both would obey Henry's law with activity coefficients equal to unity.)

13.42 a. One mole of calcium carbonate is placed in an evacuated 10-liter cylinder and heated to 1150 K. Compute the extent of dissociation of calcium carbonate to calcium oxide and the pressure in the cylinder as a function of temperature. The volume of one mole of calcium carbonate is 35 cc, and you can assume the solids volume does not change in the course of the reaction.

 b. Repeat the calculation if a 100-liter cylinder is used.

13.43 The dissociation of hydrogen selenide gas to produce pure selenium may be approximated as occurring by the reaction

$$2H_2Se(g) \rightleftharpoons 2H_2(g) + Se_2(s)$$

for which $\triangle_{rxn}G = 89.38 - 0.0879T$ in kJ and T in K. Over the temperature range of 1000 to 1250 K, compute the equilibrium number of moles of Se_2 that will be formed for each mole of hydrogen selenide that enters the reactor.

13.44 At high temperatures—for example, in a combustion process—nitrogen and oxygen in air can react to form nitrous oxide,

$$N_2 + \tfrac{1}{2}O_2 \rightleftharpoons N_2O$$

nitric oxide,

$$N_2 + O_2 \rightleftharpoons 2NO$$

and/or nitrogen dioxide,

$$\tfrac{1}{2}N_2 + O_2 \rightleftharpoons NO_2$$

 Starting with air (79 mol % nitrogen and 21 mol % oxygen), compute the equilibrium concentrations of all the oxides of nitrogen at atmospheric pressure over the temperature range from 1000 to 2000 K. [The oxides of nitrogen are referred to collectively as NO_x compounds, and are smog-forming air pollutants.]

13.45 a. Use the data in Table 13.1-4 and the information that $\Delta_f\underline{G}^\circ_{AgCl} = -108.7$ kJ/mol to predict the solubility product K°_{AgCl} of silver chloride in water, and compare your predictions with the data given in Illustration 13.3-2.

 b. Use the data in Table 13.1-4 and the information that $\Delta_f\underline{G}^\circ_{TlCl} = -186.02$ kJ/mol to predict the solubility product K°_{TlCl} of thallium chloride in

water, and compare your predictions with the data given in Illustration 13.3-2.

13.46 While ethanol can be made by biological fermentation, for large-scale production the following non-biological reaction starting with ethylene and water can be used instead:

$$C_2H_4 + H_2O \rightleftharpoons C_2H_5OH$$

If stoichiometric amounts of ethylene and water are used, compute the equilibrium constant and the equilibrium extent of reaction at

 a. 1 bar and 25°C, assuming the liquid solution is ideal

 b. Repeat the calculation assuming a nonideal liquid solution if formed.

13.47 The reaction

$$SO_2 + \tfrac{1}{2}O_2 \rightleftharpoons SO_3$$

is used as a step in the process to convert waste sulfur dioxide to sulfuric acid. Starting with stoichiometric amounts of sulfur dioxide and oxygen, determine the equilibrium conversion to sulfur trioxide at 1000 K and

 a. 1.013 bar

 b. 101.3 bar

13.48 The chemical reaction for the dissociation of nitrogen tetroxide is

$$N_2O_4(g) \rightarrow 2NO_2(g)$$

The reported standard-state Gibbs energy change for this reaction over a limited temperature range is

$$\Delta_{rxn}G^\circ(T) = 57.33 - 0.17677T \text{ kJ/mol of } N_2O_4 \text{ reacted}$$

for the pure component, ideal gas at 1 bar standard state and T in kelvins.

 a. What is the standard-state heat of reaction for the dissociation of nitrogen tetroxide?

 b. Determine the equilibrium composition of this mixture at 50°C and 0.1, 1, and 10 bar.

 c. Repeat the calculation at 200°C.

13.49 Very polar molecules may associate in the gas phase. One example is acetic acid, which, because of its structure, can form dimers but not higher polymers. For the reaction

$$2CH_3COOH \rightarrow (CH_3COOH)_2$$

the following information is available:

$$\Delta_{rxn}H^\circ = -58.62 \text{ kJ/mol of dimer}$$

and

$$\Delta_{rxn}S^\circ = -138.2 \text{ J/mol of dimer}$$

These values are for the ideal gas, 1 bar standard state, and can be assumed to be independent of temperature.

a. Compute the degree of dimerization of acetic acid at 25°C and at 0.1, 1, and 10 bar.

b. Compute the degree of dimerization of acetic acid at 100°C and at 0.1, 1, and 10 bar.

c. If one assumes that only the monomer is present, acetic acid does not satisfy the ideal gas law. However, it is thought that at the temperatures and pressures considered above, the acetic acid monomer and dimer form an equilibrium ideal gas mixture, and that the apparent nonideality that is found when only monomer is assumed to be present is actually the result of the mole number change due to dimerization. Develop an equation of state for acetic acid that takes into account dimerization but has temperature, pressure, volume, and the number of moles of acetic acid if no dimerization occurred as the independent variables.

13.50 At low temperatures mixtures of water and methane can form a hydrate, that is, a solid containing trapped methane. Hydrates have both positive and negative features. For example, they are potentially a very large source of underground trapped methane in the Arctic and Antarctic regions that could be used to meet future energy needs. However, in cold areas such as the North Slope of Alaska and in the North Sea, hydrates can form in pipelines, blocking the flow of natural gas. The approximate stoichiometry of hydrate formation is

$$CH_4(g) + 5.75H_2O(s) \rightarrow CH_4 \cdot 5.75H_2O(s)$$

The equilibrium partial pressure of methane for hydrate formation at 267 K is approximately 2.0 MPa, and at 255 K it is 1.5 MPa. Using the fact that the standard states for this reaction are pure methane as a gas at 1 bar, and water and the hydrate as a pure solid:

a. Find $\Delta_{rxn}G°$ for hydrate formation per mole of methane at 267 K and 255 K.

b. Assuming that $\Delta_{rxn}H°$ and $\Delta_{rxn}S°$ are independent of temperature, calculate values for these quantities.

c. Use the information in the problem statement to calculate the equilibrium methane partial pressure for hydrate formation at 273 K.

13.51 Assume two species can associate in the vapor phase according to the reactions

$$iA_1 = A_i \quad i = 1,2, \text{ etc.}$$
$$jB_1 = B_j \quad j = 1,2, \text{ etc.}$$
$$\left.\begin{array}{c}\text{but not all}\\ \text{integers need}\\ \text{be included}\end{array}\right)$$

$$iA_1 + jB_1 = A_iB_j$$

Using the notation that $\bar{G}_A$ and $\bar{G}_B$ are the partial molar Gibbs energies of the total species A and B (in all its forms, unassociated and associated), prove that

$$\bar{G}_A = \bar{G}_{A_1} \quad \text{and} \quad \bar{G}_B = \bar{G}_{B_1}$$

where $\bar{G}_{A_1}$ and $\bar{G}_{B_1}$ are the partial molar Gibbs energies of the monomeric species, respectively. Also prove that

$$\bar{f}_A = \bar{f}_{A_1} \quad \text{and} \quad \bar{f}_B = \bar{f}_{B_1}$$

13.52 The description of components that associate or hydrogen-bond is difficult. An alternative model to the one considered in the previous problem is the continuous association model, in which

$$A_1 + A_1 = A_2$$
$$A_1 + A_2 = A_3$$
$$A_1 + A_n = A_{n+1}$$
$$\left.\begin{array}{c}\text{here all integers}\\ \text{to } \infty \text{ are included}\end{array}\right)$$
etc.

Assume that this associating fluid is described by the van der Waals equation, and in the equation-of-state representation the parameters of the j-mer are gotten from

$$a_j = j^2 a_1$$

and

$$b_j = jb_1$$

Further, the value of the equilibrium constant for the association

$$K_{j+1} = \frac{a_{j+1}}{a_j a_1}$$

will be assumed independent of the degree of association, that is,

$$K_2 = K_3 = \cdots = K_n = K$$

Using the notation

N_0 = number of moles if no association occurs

N_T = number of moles when association occurs

and assuming the binary interaction parameters are all zero since the species are so similar,

a. Obtain expressions for a and b for the mixture in terms of a_1, b_1, N_0, and N_T only.

b. Show that the ratio

$$\frac{P\bar{\phi}_j\bar{\phi}_1}{\bar{\phi}_{j+1}}$$

is independent of index j and obtain an explicit expression for this ratio. (Note that $\bar{\phi}_j$ is the fugacity coefficient of species j.)

c. Obtain an expression for the equation of state of this associating fluid that contains only P, V, T, N_0, a_1, b_1, and the equilibrium constant K. (Does your equation reduce to the van der Waals equation for a nonassociating one-component system in the limit of $K \to 0$?)

13.53 The behavior of hydrogen fluoride is unusual! For example, here are the critical properties of various hydrogen halides:

	MW	T_c (K)	P_c (bar)	Z_c	ω
HF	20	461.0	64.88	0.12	0.372
HCl	36.46	324.6	83.07	0.249	0.12
HBr	80.91	363.2	85.50	0.283	0.063
HI	127.9	424.0	83.07	0.309	0.05

We see that HF has a very high critical temperature and acentric factor for its molecular weight; it also has the lowest reported critical compressibility of any species. Experimental data for the vapor pressure and the apparent molecular weight of saturated HF vapor are given here:

T (K)	P^{vap} (bar)	(MW)$_{\text{apparent}}$
227.3	0.0519	92.8
243.9	0.1265	85.0
256.4	0.2328	79.4
277.8	0.5780	69.8
303.0	1.4353	58.4
322.6	2.6178	50.3

Apparent molecular weight at a fixed pressure of 0.993 bar:

T (K)	(MW)$_{\text{apparent}}$	
227.3	117.6	
250.0	110.7	Calculated
270.3	95.7	
285.7	74.6	
294.1	59.8	
303.0	43.0	Measured
312.5	28.8	
322.6	21.8	

In each case the apparent molecular weight has been found by measuring the mass density of the vapor and comparing that with an ideal gas of molecular weight 20. One possible explanation for this behavior is that hydrogen fluoride associates according to the following set of reactions:

$$2HF = (HF)_2$$
$$6HF = (HF)_6$$
$$8HF = (HF)_8$$

Describe how you would use these data to develop a model for HF so that you could determine the vapor-liquid equilibrium of HF and a component that did not associate.

Chapter **14**

The Balance Equations for Chemical Reactors and Electrochemistry

Our interest in this chapter is with the mass and energy balances for chemical reactors, and in electrochemical cells. We consider first the mass and energy balances for tank and tubular reactors, and then for a general "black-box" chemical reactor, since these balance equations are an important application of the thermodynamic equations for reacting mixtures and the starting point for practical reactor design and analysis. Finally, we consider equilibrium and the energy balance for electrochemical systems such as batteries and fuel cells, and the use of electrochemical cells for thermodynamic measurements.

INSTRUCTIONAL OBJECTIVES FOR CHAPTER 14

The goals of this chapter are for the student to:

- Be able to use the energy balance for a tank-type chemical reactor (Sec. 14.1)
- Be able to use the energy balance for a tube-type chemical reactor (Sec. 14.2)
- Be able to use the energy balance for a general black-box chemical reactor and compute the adiabatic reactor temperature (Sec. 14.3)
- Be able to compute the voltage produced by an electrochemical cell (Sec. 14.6)
- Be able to use electrochemical cell voltages to compute pH (Sec. 14.6)

NOTATION INTRODUCED IN THIS CHAPTER

E Electrochemical cell potential difference (V)
E° Zero current cell potential (V)
F Faraday constant (C/mol)
T_{ad} Adiabatic reaction temperature (K)
τ Reactor residence time $= \dfrac{V}{q}$ (s)

$\hat{X}$ Molar extent of reaction per unit volume $= \frac{X}{V}$ (mol/m^3)

Ψ Availability

14.1 THE BALANCE EQUATIONS FOR A TANK-TYPE CHEMICAL REACTOR

The design of a chemical reactor, that is, the choice of its size, shape, and conditions of operation, is largely determined by the kinetics of the chemical reactions[1] (i.e., the rates at which reactions occur and whether a catalyst is needed) and the rate at which heat is produced or absorbed during the reaction. Thermodynamics can be useful in reactor design. In particular, the multicomponent mass and energy balances of thermodynamics can provide useful information on the energy requirements and temperature programming in reactor operations. Also, through the use of computational techniques discussed in the previous chapter, thermodynamics can provide information on the maximum (equilibrium) conversions that can be obtained with any reactor at the given operating conditions. Although it is not our intention to consider reactor design in great detail here, in this section and the next we will establish the relationship between the thermodynamic balance equations and those of reaction engineering; these are the equations that will be used in a course in chemical reaction engineering, which frequently follows thermodynamics in the chemical engineering curriculum.

In this section we are concerned with the analysis of tank-type reactors used for liquid-phase reactions. A schematic diagram of a tank reactor is given in Fig. 14.1-1. To develop a quantitative description of such reactors, we will assume that its contents are well mixed, as is the case with many industrial reactors, so that the species concentrations and the temperature are uniform throughout the reactor. *We do not assume that the reactor exit stream is in chemical equilibrium, since industrial reactors generally do not operate in such a manner.* Using the entries of Table 8.4-1, the species mass and total energy balances for a reactor with one inlet and one outlet stream are, respectively,

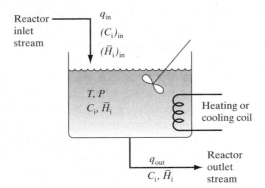

Figure 14.1-1 A schematic diagram of a simple stirred-tank reactor.

---

[1] The relationship between the rate of a chemical reaction and the species concentrations cannot be predicted and must be determined from experiment. Detailed discussions of the analysis of experimental reaction rate data to get the constitutive equation relating the reaction rate to species concentrations are given in *Kinetics and Mechanism,* 3rd ed., by J. W. Moore and R. G. Pearson, John Wiley & Sons, New York (1981), Chaps. 2 and 3; *Chemical Reaction Engineering,* 2nd ed., by O. Levenspiel, John Wiley & Sons, New York (1972), Chap. 3; and *Introduction to Chemical Engineering Analysis,* by T. W. F. Russell and M. M. Denn, John Wiley & Sons, New York (1972), Chap. 5.

Mass and energy balances for a stirred-tank reactor

$$\frac{dN_i}{dt} = (\dot{N}_i)_{in} - (\dot{N}_i)_{out} + \sum_{j=1}^{\mathcal{M}} \nu_{ij}\dot{X}_j \qquad \textbf{(14.1-1)}$$

and

$$\frac{dU}{dt} = \sum_{i=1}^{\mathcal{C}}(\dot{N}_i\bar{H}_i)_{in} - \sum_{i=1}^{\mathcal{C}}(\dot{N}_i\bar{H}_i)_{out} + \dot{Q} \qquad \textbf{(14.1-2)}$$

(The generalization of these equations to multiple feed streams is simple, and is left to you.) In writing the energy balance equation, the kinetic and potential energy, shaft work, and $P(dV/dt)$ terms have been neglected, because these terms are usually of little importance compared with heats of reaction and temperature change terms. Also, since the contents of the reactor are of uniform temperature and composition (by the "well-mixed" assumption), the species concentrations and temperature of the exit stream are the same as those of the reactor contents.

With several small changes in notation, the mass and energy balances of Eqs. 14.1-1 and 14.1-2 can be made to look more like those commonly used in reactor analysis. By letting q_{in} and q_{out} represent the volumetric flow rates into and out of the reactor, C_i be the molar concentration of species i, V be the fluid volume in the reactor (so that $C_iV = N_i$), and $r_j = \dot{X}_j/V$ be the specific reaction rate (reaction rate per unit volume) for the jth reaction, Eqs. 14.1-1 and 14.1-2 can be rewritten as

$$\frac{d}{dt}(C_iV) = (C_i)_{in}q_{in} - C_iq_{out} + V\sum_{j=1}^{\mathcal{M}}\nu_{ij}r_j \qquad \textbf{(14.1-3)}$$

and

$$\frac{d}{dt}\left(V\sum_i^{\mathcal{C}}C_i\bar{U}_i\right) = \left(\sum_i^{\mathcal{C}}C_i\bar{H}_i\right)_{in}q_{in} - \left(\sum_i^{\mathcal{C}}C_i\bar{H}_i\right)q_{out} + \dot{Q} \qquad \textbf{(14.1-4)}$$

Here $(C_i)_{in}$ and $(\bar{H}_i)_{in}$ are the concentration and partial molar enthalpy, respectively, of species i in the inlet stream, and C_i is the concentration of species i and $\bar{H}_i$ its partial molar enthalpy in *both* the reactor and in the outlet stream (again, this is the "well-mixed" assumption).

It is possible that the volume of fluid in a reactor is a function of time, in which case a total mass balance for the reactor would have to be written to obtain this time dependence. However, few reactors are operated in this manner, so we will neglect this complication, as well as volume changes that may occur on mixing and reaction, and assume that $q_{in} = q_{out} = q$. Also, since the tank-type reactor is used almost exclusively for liquid-phase reactions, we may safely assume that $\bar{U}_i = \bar{H}_i$ (since $P\bar{V}_i \ll RT$ for liquids). Thus Eqs. 14.1-3 and 14.1-4 can be rewritten as

Design equations for a stirred-tank reactor

$$V\frac{dC_i}{dt} = q\{(C_i)_{in} - C_i\} + V\sum_{j=1}^{\mathcal{M}}\nu_{ij}r_j \qquad \textbf{(14.1-5)}$$

$$V\frac{d}{dt}\left(\sum_i C_i\bar{H}_i\right) = q\left(\sum_i(C_i\bar{H}_i)_{in} - \sum_i C_i\bar{H}_i\right) + \dot{Q} \qquad \textbf{(14.1-6)}$$

which are the equations generally used in the design of tank reactors.

An important special case of these equations is their application to the steady-state operation of a continuous-flow reactor. At steady-state the contents of the reactor do not change with time, so that $dC_i/dt = 0$ and $dU/dt = dH/dt = 0$, and the design equations reduce to

Steady-state mass balance for a stirred-tank reactor

$$C_i = (C_i)_{in} + \frac{V}{q} \sum_{j=1}^{\mathcal{M}} \nu_{ij} r_j \qquad (14.1\text{-}7)$$

and

$$\dot{Q} = q \left(\sum_i C_i \overline{H}_i - \sum_i (C_i \overline{H}_i)_{in} \right) \qquad (14.1\text{-}8)$$

The first of these equations relates the exit composition of the reactor to its volume, the inlet composition and flow rate, and the reaction kinetics (i.e., constitutive relations for the reaction rates), whereas the second equation is used to determine the heat load for steady isothermal operation of the reactor.

Actually, the form of Eq. 14.1-8 is a little deceptive because one expects the heat of reaction to play an important role in determining the reactor heat load, yet it does not seem to appear explicitly in the equation. As was indicated in Chapter 8, the heat of reaction is imbedded in this equation. To see this, we use Eq. 14.1-7 to eliminate either C_i or $(C_i)_{in}$ from Eq. 14.1-8 and obtain

$$\dot{Q} = q \sum_i (C_i)_{in} \{ \overline{H}_i - (\overline{H}_i)_{in} \} + V \sum_j r_j \Delta_{rxn,j} H \qquad (14.1\text{-}9a)$$

if C_i is eliminated, or if $(C_i)_{in}$ is eliminated,

$$\dot{Q} = q \sum_i C_i \{ \overline{H}_i - (\overline{H}_i)_{in} \} + V \sum_j r_j (\Delta_{rxn,j} H)_{in} \qquad (14.1\text{-}9b)$$

Here $\Delta_{rxn,j} H = \sum_i \nu_{ij} \overline{H}_i$, and $(\Delta_{rxn,j} H)_{in} = \sum_i \nu_{ij} (\overline{H}_i)_{in}$ are both heats of reaction, the first being the heat of reaction at the reactor outlet conditions (temperature and composition), and the second at the reactor inlet conditions.

One simplification that is usually made in writing the energy balance for a reacting system is to neglect solution nonidealities with respect to the much larger energy changes that accompany the chemical reaction and temperature changes in the system [i.e., to neglect the difference between $\overline{H}_i(T, \underline{x})$ and $\underline{H}_i(T)$]. With this assumption we have

$$\Delta_{rxn,j} H(T, \underline{x}) = \sum_i \nu_{ij} \overline{H}_i(T, \underline{x}) = \sum_i \nu_{ij} \underline{H}_i(T)$$
$$= \Delta_{rxn,j} H(T)$$

so that Eq. 14.1-9a becomes

Two forms of simplified energy balance for stirred-tank reactor

$$\dot{Q} = q \sum_{i=1}^{\mathcal{C}} (C_i)_{in} \{ \underline{H}_i(T) - \underline{H}_i(T_{in}) \} + V \sum_{j=1}^{\mathcal{M}} \Delta_{rxn,j} H(T) r_j \qquad (14.1\text{-}10a)$$

The two terms in this equation have simple physical interpretations. The first term represents the rate of flow of energy into the reactor required to heat the inlet fluid from T_{in} to T without reaction, and the second term is the energy requirement for all chemical reactions to occur isothermally at the reactor exit temperature. Similarly, neglecting solution nonidealities in Eq. 14.1-9b yields

$$\dot{Q} = V \sum_{j=1}^{\mathcal{M}} \Delta_{rxn,j} H(T_{in}) r_j + q \sum_i C_i \{\underline{H}_i(T) - H_i(T_{in})\}$$

(14.1-10b)

Here the first term is the energy required for isothermal chemical reaction at the reactor feed temperature, and the second term is the energy required to heat the reaction mixture, without further reaction, from the inlet temperature to the reactor operating temperature.

The form of Eq. 14.1-10a suggests that the first process in the reactor is the heating of the feed to the reactor operating temperature, and then isothermal chemical reactions occur, whereas Eq. 14.1-10b indicates that the chemical reactions occur, followed by fluid heating. Of course, chemical reaction and fluid heating occur simultaneously. These two alternative, and seemingly contradictory, descriptions of the process illustrate again that the enthalpy change between two given states is independent of path, and therefore any convenient path may be used for its calculation. In this regard the equations here are similar to those used in Sec. 6.4, where the change in thermodynamic properties accompanying a change in state of a real fluid were computed along any convenient path.

ILLUSTRATION 14.1-1

Design of a Steady-State Stirred-Tank Reactor

The ester ethyl acetate is produced by the reversible reaction

$$CH_3COOH + C_2H_5OH \underset{k'}{\overset{k}{\rightleftharpoons}} CH_3COOC_2H_5 + H_2O$$

in the presence of a catalyst such as sulfuric or hydrochloric acid. The rate of ethyl acetate production has been found, from the analysis of chemical kinetics data, to be given by the following equation:

$$\frac{dC_{EA}}{dt} = k C_A C_E - k' C_{EA} C_W$$

where the subscripts EA, A, E, and W denote ethyl acetate, acetic acid, ethanol, and water, respectively. The values of the reaction rate constants at $100°C$ and the catalyst concentration of interest are

$$k = 4.76 \times 10^{-4} \text{ m}^3/(\text{kmol min})$$

and

$$k' = 1.63 \times 10^{-4} \text{ m}^3/(\text{kmol min})$$

The feed stream is an aqueous solution containing 250 kg of acetic acid and 500 kg of ethyl alcohol per m^3 of solution (including catalyst). The density of the solution is 1040 kg/m^3 and constant. The reaction will be carried out at $100°C$, the feed is at $100°C$, and the reactor will be operated at a sufficiently high pressure that a negligible amount of reactants or products vaporizes.

If a continuous-flow stirred-tank reactor is used for the reaction, determine

a. The size of the reactor needed to produce 1250 kg/hr of the ester, if 37.2 percent of the acid reacts.
b. The heat load on the reactor for this extent of reaction.

SOLUTION

a. Using the equations in Illustration 13.1-9, and the fact that 37.2 percent of the acid (or $0.372 \times 4.17 = 1.55$ kmol/m^3) reacts, we have for the reactor contents and outlet concentrations

$$C_A = 4.17 - 1.55 = 2.62 \text{ kmol/m}^3$$
$$C_E = 10.9 - 1.55 = 9.35 \text{ kmol/m}^3$$
$$C_{EA} = 0 + 1.55 = 1.55 \text{ kmol/m}^3$$
$$C_W = 16.1 + 1.55 = 17.65 \text{ kmol/m}^3$$

Next, using Eq. 14.1-7 yields

$$C_{EA,in} = \frac{V}{q}r = \frac{V}{q}\{kC_A C_E - k'C_{EA}C_W\}$$

$$r = \{4.76 \times 10^{-4} \times 2.62 \times 9.35 - 1.63 \times 10^{-4} \times 1.55 \times 17.65\}$$
$$= 7.20 \times 10^{-3} \frac{\text{kmol}}{\text{m}^3 \text{ min}} = 7.20 \frac{\text{mol}}{\text{m}^3 \text{ min}}$$

and

$$1.55 \frac{\text{kmol}}{\text{m}^3} = \frac{V}{q} 7.20 \times 10^{-3} \frac{\text{kmol}}{\text{m}^3 \text{ min}}$$

so that

$$\tau = \frac{V}{q} = 215.3 \text{ min} = 3.588 \text{ hr}$$

is required to obtain the desired extent of reaction. (Note that τ has units of time and can be interpreted to be an average time that an element of fluid is in the reactor; this is referred to as the average residence time for fluid in the reactor.) The volumetric flow rate into and out of the reactor must be such that

$$q \times C_{EA} \times \text{mol wt of ester} = 1250 \text{ kg/hr}$$

or

$$q = \frac{1250 \text{ kg/hr}}{1.55 \frac{\text{kmol}}{\text{m}^3} \times 88 \frac{\text{g}}{\text{mol}}} = 9.164 \text{ m}^3/\text{hr}$$

Therefore, the reactor volume should be

$$V = \tau q = 3.588 \text{ hr} \times 9.164 \text{ m}^3/\text{hr} = 32.88 \text{ m}^3$$

to obtain the desired production rate.

b. To determine the heat load on the reactor, we can use either Eqs. 14.1-9 or 14.1-10. Assuming that liquid-phase nonidealities are unimportant, and that the reactor effluent is at the reactor temperature (100°C), these equations reduce to

$$Q = V \Delta_{rxn} H \cdot r$$

since the inlet stream and reactor temperatures are equal. From part (a), $V = 32.88$ m^3 and $r = 7.20$ mol/(m^3 min), and from Appendix A.IV we have

$$\Delta_f \underline{H}^\circ_A (T = 25^\circ C) = -486.1 \text{ kJ/mol}$$
$$\Delta_f \underline{H}^\circ_E (T = 25^\circ C) = -277.7 \text{ kJ/mol}$$
$$\Delta_f \underline{H}^\circ_{EA} (T = 25^\circ C) = -463.3 \text{ kJ/mol}$$
$$\Delta_f \underline{H}^\circ_W (T = 25^\circ C) = -285.8 \text{ kJ/mol}$$

Therefore,

$$\Delta_{rxn} H^\circ (T = 25^\circ C) = \sum \nu_i \Delta_f \underline{H}^\circ_i = (-463.3) + (-285.8) - (-486.1) - (-277.7)$$
$$= 14.7 \text{ kJ/mol ester produced}$$

and

$$\Delta_{rxn} H^\circ (T = 100^\circ C) = \Delta_{rxn} H^\circ (T = 25^\circ C) + \int_{T=25^\circ C}^{T=100^\circ C} \Delta_{rxn} C_P \, dT$$

where $\Delta_{rxn} C_P = -13.39$ J/(mol K). Therefore,

$$\Delta_{rxn} H^\circ (T = 100^\circ C) = 14\,700 \text{ J/mol} - 13.39 \text{ J/(mol K)} \times 75 \text{ K}$$
$$= 14\,700 - 1004 \cong 13\,700 \text{ J/mol}$$

Neglecting solution nonidealities, we have $\Delta_{rxn} H (T = 100^\circ C) = \Delta H^\circ_{rxn} (T = 100^\circ C)$, so that

$$\dot{Q} = 32.88 \text{ m}^3 \times 7.20 \text{ mol/(m}^3 \text{ min)} \times 13.7 \text{ kJ/mol} = 3243 \text{ kJ/min}$$

(*Note:* The fact that $\dot{Q}$ is positive indicates that the reaction is endothermic, so that heat must be supplied to the reactor to maintain isothermal operation.)   ∎

Batch reactor

A different type of tank reactor is the batch reactor. In the batch reactor the total charge is introduced initially, and the reaction then proceeds without mass flows into or out of the reactor until some later time, when the contents of the reactor are discharged. The fluid leaving the reactor may or may not be in chemical equilibrium, and the reactor may or may not be operated in an isothermal and/or an isobaric manner.

The mass (mole) and energy balances for the batch reactor are (see Table 8.4-1)

$$\frac{dN_i}{dt} = \sum_{j=1}^{\mathcal{M}} \nu_{ij} \dot{X}_j \tag{14.1-11}$$

and

$$\frac{dU}{dt} = \frac{d}{dt} \left(\sum_{i=1}^{\mathcal{C}} N_i \bar{U}_i \right) = \dot{Q} \tag{14.1-12}$$

Note that here, as before, the kinetic energy and shaft work terms have been neglected. The term $P(dV/dt)$ has also been neglected since it is small for liquid-phase reactions and identically zero for gaseous reactions where the reactant mixture fills the total volume of the reactor. Making the same substitutions as were used in going from Eqs. 14.1-5 and 14.1-6 to Eqs. 14.1-7 and 14.1-8 here yields

$$\frac{dC_i}{dt} = \sum_{j=1}^{\mathcal{M}} \nu_{ij} r_j \tag{14.1-13}$$

and

$$V\frac{d}{dt}\sum_{i=1}^{\mathcal{C}}(C_i\bar{U}_i) = \dot{Q} \tag{14.1-14}$$

where V is the fluid volume in the reactor. The term on the left side of Eq. 14.1-14 can be written as

$$V\frac{d(\sum C_i\bar{U}_i)}{dt} = V\sum_{i=1}^{\mathcal{C}}\bar{U}_i\frac{dC_i}{dt} + V\sum_{i=1}^{\mathcal{C}}C_i\frac{d\bar{U}_i}{dt}$$

$$= V\sum_{i=1}^{\mathcal{C}}\bar{U}_i\left(\sum_{j=1}^{\mathcal{M}}\nu_{ij}r_j\right) + V\sum_{i=1}^{\mathcal{C}}C_i\bar{C}_{V,i}\frac{dT}{dt}$$

$$= V\sum_{j=1}^{\mathcal{M}}\Delta_{\mathrm{rxn},j}U(T,P,\underline{x})r_j + \mathbf{C}_V\frac{dT}{dt}$$

where $\Delta_{\mathrm{rxn},j}U(T,P,\underline{x}) = \sum_i \nu_{ij}\bar{U}_i(T,P,\underline{x})$ is the internal energy change for the jth reaction at the reaction conditions (i.e., the temperature, pressure, and composition of reactant mixture), and $\mathbf{C}_V = V\sum C_i\bar{C}_{V,i}$ is the total heat capacity of the fluid in the reactor. Thus, Eq. 14.1-14 becomes

$$\mathbf{C}_V\frac{dT}{dt} = -V\sum_{j=1}^{\mathcal{M}}\Delta_{\mathrm{rxn},j}U(T)r_j + \dot{Q} \tag{14.1-15}$$

For a liquid mixture, and again neglecting solution nonidealities, $\underline{U} \approx \underline{H}$ and $\mathbf{C}_V \approx \mathbf{C}_P$, so that this equation can be rewritten as

$$\mathbf{C}_P\frac{dT}{dt} = -V\sum_{j=1}^{\mathcal{M}}\Delta_{\mathrm{rxn},j}H(T)r_j + \dot{Q} \tag{14.1-16}$$

Equations 14.1-13 and 14.1-16 are used by chemical engineers for the design of batch reactors.

ILLUSTRATION 14.1-2
Design of a Batch Reactor

Ethyl acetate is to be produced in a batch reactor operating at 100°C and at 37.2 percent conversion of acetic acid. If the reactor charge of the preceding example is used and 20 minutes are needed to discharge, clean, and charge the reactor, what size reactor is required to produce, on average, 1250 kg of the ester per hour? What heat program should be followed to ensure the reactor remains at 100°C?
Data: See the preceding example.

SOLUTION

The starting point here is Eq. 14.1-13, which, for the present case, can be written

$$\frac{dC_{EA}}{dt} = kC_A C_E - k'C_{EA}C_W \tag{a}$$

where, at any extent of reaction $\hat{X} = X/V$, we have

	Initial Concentration	Concentration* at the Extent of Reaction $\hat{X}$
Acetic acid	4.17 kmol/m^3	$4.17 - \hat{X}$ kmol/m^3
Ethanol	10.9	$10.9 - \hat{X}$
Ethyl acetate	0	$\hat{X}$
Water	16.1	$16.1 + \hat{X}$

*Note that $\hat{X} = X/V$ is the molar extent of reaction per unit volume and has units of concentration.

Thus, Eq. a becomes

$$r \equiv \frac{d\hat{X}}{dt} = k(4.17 - \hat{X})(10.9 - \hat{X}) - k'(16.1 + \hat{X})\hat{X}$$
$$= 4.76 \times 10^{-4}(4.17 - \hat{X})(10.9 - \hat{X}) - 1.63 \times 10^{-4}(16.1 + \hat{X})\hat{X} \tag{b}$$
$$= 2.163 \times 10^{-2}(1 - 0.4528\hat{X} - 0.01447\hat{X}^2) \; \text{kmol/(m}^3 \, \text{min)}$$

which can be rearranged to

$$2.163 \times 10^{-2} \, dt = \frac{d\hat{X}}{1 - 0.4528\hat{X} + 0.014\,47\hat{X}^2}$$

Integrating this equation between $t = 0$ and the time t yields[2]

$$2.163 \times 10^{-2} \int_0^t dt = 2.163 \times 10^{-2} \frac{\text{kmol}}{\text{m}^3 \, \text{min}} \times t = \int_0^{\hat{X}} \frac{d\hat{X}}{1 - 0.4528\hat{X} + 0.014\,47\hat{X}^2}$$

or

$$\ln\left\{\frac{0.028\,94\hat{X} - 0.8364}{0.028\,94\hat{X} - 0.0692}\right\} - \ln\left(\frac{0.8364}{0.0692}\right) = 0.8297 \times 10^{-2}t \tag{c}$$

and for $\hat{X} = 1.55$ (37.2 percent conversion of the acid) we have $0.8297 \times 10^{-2}t = 0.9896$ min or $t = 119.3$ min.

Thus, the total cycle time for the batch reactor is $119.3 + 20 = 139.3$ min $= 2.322$ hr. Therefore, 2.322 hr $\times$ 1250 kg/hr $= 2902.5$ kg of ester that must be produced in each reactor cycle. Since, in this example, the conversion and feed in the batch reactor are the same as those in the previous illustration, the effluent concentration of the ester is again

$$C_{EA} = 1.55 \; \text{kmol m}^3 = 136.4 \; \text{kg/m}^3$$

---

[2]Note that
$$\int \frac{dX}{a+bX+cX^2} = \frac{1}{\sqrt{-q}} \ln\left[\frac{2cX+b-\sqrt{-q}}{2cX+b+\sqrt{-q}}\right] \quad \text{where } q = 4ac - b^2$$

Therefore, the reactor volume is

$$V = \frac{2902.5 \text{ kg}}{136.4 \text{ kg/m}^3} = 21.28 \text{ m}^3$$

To compute the reactor heat program, that is, the heating rate as a function of time to keep the reactor at constant temperature, Eq. 14.1-16 is used, noting that for the isothermal case $dT/dt = 0$, so that

$$\dot{Q} = V(\Delta_{\text{rxn}}H) \cdot r \qquad \qquad \textbf{(d)}$$

where r is given by Eq. b and $\hat{X}$ is given as a function of time by Eq. c.

The instantaneous values of $\hat{X}$ and $\dot{Q}$ as a function of time are plotted in the following figure.

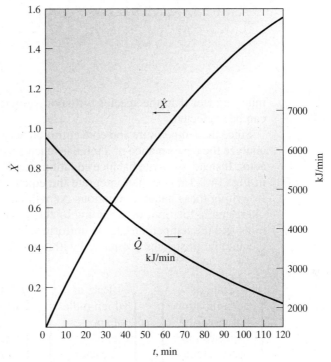

The heat flow rate $\dot{Q}$ and extent of reaction $\hat{X}$ for ethyl acetate production in a batch reactor.

■

14.2 THE BALANCE EQUATIONS FOR A TUBULAR REACTOR

Tubular reactors are commonly used in the chemical process industry. This type of reactor, as the name implies, is a tube that may be packed with catalytic or other material and through which the reactant mixture flows (see Fig. 14.2-1a). Such reactors are frequently used for gas-phase reactions and for reactions in which good heat transfer or contact with a heterogeneous catalyst is needed. Our interest here will be in the "plug-flow" tubular reactor, that is, a reactor in which the fluid flow is sufficiently turbulent that there are no radial gradients of concentration, temperature, or fluid velocity; however, as a result of chemical reactions, concentrations and temperatures change along the reactor from the inlet to the outlet. Also, we will assume that the fluid velocity is

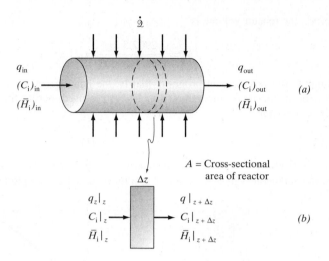

Figure 14.2-1 The tubular reactor.

much greater than the species diffusion velocity, so that diffusion along the tube axis can be neglected.

Since the temperature and concentration vary along the length of the tube, we cannot analyze the performance of a tubular reactor using balance equations for the whole reactor. Instead, we write balance equations for the differential reactor element Δz shown in Fig. 14.2-1b, and then integrate the equations so obtained over the reactor length. In writing these balance equations Δz will be taken to be so small (in fact, we will be interested in the limit of Δz going to zero) that we can assume the species concentrations and the temperature to be uniform within the element. The balance equation for the amount of species i contained within the reactor element in Fig. 14.2-1b is

$$
\begin{pmatrix} \text{Rate of} \\ \text{accumulation} \\ \text{of moles of} \\ \text{species i in} \\ \text{reactor element} \\ \text{of length } \Delta z \end{pmatrix} = \begin{pmatrix} \text{Rate at} \\ \text{which moles} \\ \text{of species i} \\ \text{enter the} \\ \text{reactor} \\ \text{element} \end{pmatrix} - \begin{pmatrix} \text{Rate at} \\ \text{which moles} \\ \text{of species i} \\ \text{leave the} \\ \text{reactor} \\ \text{element} \end{pmatrix} + \begin{pmatrix} \text{Rate at} \\ \text{which species i is} \\ \text{produced in} \\ \text{the reactor} \\ \text{element by chemical} \\ \text{reaction} \end{pmatrix}
$$

$$
A \Delta z \frac{\partial C_i}{\partial t} = (q C_i)_z - (q C_i)_{z+\Delta z} + A(\Delta z) \sum_{j=1}^{\mathcal{M}} \nu_{ij} r_j
$$

where A is the cross-sectional area of the tubular reactor and q is the volumetric flow rate in the reactor. Now dividing by $A \Delta z$ and taking the limit as $\Delta z \to 0$ gives

$$
\frac{\partial C_i}{\partial t} = -\frac{1}{A} \frac{\partial (q C_i)}{\partial z} + \sum_{j=1}^{\mathcal{M}} \nu_{ij} r_j \tag{14.2-1}
$$

Similarly, the energy balance for the differential element of reactor volume is

$$\begin{pmatrix}\text{Rate of accumulation}\\\text{of energy in reactor}\\\text{element of length }\Delta z\end{pmatrix}=\begin{pmatrix}\text{Rate at which}\\\text{energy enters}\\\text{reactor element}\\\text{due to flow}\end{pmatrix}-\begin{pmatrix}\text{Rate at which}\\\text{energy leaves}\\\text{reactor element}\\\text{due to flow}\end{pmatrix}$$

$$+\begin{pmatrix}\text{Rate of}\\\text{energy input}\\\text{by heat flow}\end{pmatrix}$$

$$A\,\Delta z\frac{\partial}{\partial t}\left(\sum C_i\bar{U}_i\right)=q\left(\sum_i C_i\bar{H}_i\right)\bigg|_z-q\left(\sum_i C_i\bar{H}_i\right)\bigg|_{z+\Delta z}+\dot{Q}\Delta z$$

where $\dot{Q}$ is the heat flow rate per unit length of the reactor. Again dividing by $A\,\Delta z$ and taking the limit as $\Delta z\to 0$ gives

$$\frac{\partial}{\partial t}\left(\sum_i C_i\bar{U}_i\right)=-\frac{1}{A}\frac{\partial}{\partial z}\left(q\sum_i C_i\bar{H}_i\right)+\frac{1}{A}\dot{Q}\qquad(14.2\text{-}2)$$

Equations 14.2-1 and 14.2-2 are the design equations for a general tubular reactor.

For liquid-phase reactions, $\bar{U}_i=\bar{H}_i$, and assuming that the volume change on reaction is small, q is independent of position. Also, recognizing that q/A is equal to the mass average velocity v, and defining the derivative with respect to the fluid motion D/Dt by

$$\frac{D}{Dt}=\frac{\partial}{\partial t}+v\frac{\partial}{\partial z}$$

(see Sec. 2.4) we have, from Eqs. 14.2-1 and 14.2-2, that

Mass and energy balances for a tubular reactor

and

$$\frac{DC_i}{Dt}=\sum_{j=1}^{\mathcal{M}}v_{ij}r_j\qquad(14.2\text{-}3)$$

$$\frac{D(\sum C_i\bar{U}_i)}{Dt}=\frac{1}{A}\dot{Q}\qquad(14.2\text{-}4)$$

Equations 14.2-3 and 14.2-4 bear a striking resemblance to the mass and energy balances for a batch reactor, Eqs. 14.1-13 and 14. There is, in fact, good physical reason why these equations should look very much alike. Our model of a plug-flow reactor, which neglects diffusion and does not allow for velocity gradients, assumes that each element of fluid travels through the reactor with no interaction with the fluid elements before or after it. Therefore, if we could follow a small fluid element in a tubular reactor, we would find that it had precisely the same behavior in time as is found in a batch reactor. This similarity in the physical situation is mirrored in the similarity of the descriptive equations.

For gas-phase reactions this tubular reactor–batch reactor analogy may not be valid since the volumetric flow rate of the gas, and therefore the gas velocity, can vary along the reactor length as a result of mole number changes accompanying chemical reaction, the thermal expansibility of the gas if the reactor temperature varies, and the expansion of the gas accompanying the hydrodynamic pressure drop in the reactor. Consequently, for gas-phase reactions there may not be a simple relationship between distance traversed down the reactor and residence time. Thus, it is not surprising that in this case Eqs. 14.2-1 and 14.2-2 cannot be rewritten in the form of Eqs. 14.2-3 and 14.2-4.

ILLUSTRATION 14.2-1

Design of a Tubular Reactor

Ethyl acetate is to be produced at a rate of 1250 kg/hr in a tubular reactor operating at 100°C. If the feed is the same as that used in the previous examples, how large should the reactor be to produce the desired amount of ester by achieving the same conversion as the batch reactor? What are the heat transfer specifications for this reactor?

SOLUTION

The steady-state mass and energy balances for the tubular reactor are

$$\frac{q}{A}\frac{dC_i}{dz} = r_i \tag{a}$$

and

$$\frac{q}{A}\frac{d}{dz}\sum_i C_i \overline{H}_i = \frac{q}{A}\left[\sum_i \frac{dC_i}{dz}\overline{H}_i + \sum_i C_i \frac{d\overline{H}_i}{dz}\right] = \frac{\dot{Q}}{A} \tag{b}$$

Since we are neglecting solution nonidealities, and the temperature is constant, $d\overline{H}_i/dz = 0$, so that Eq. b can be rewritten as

$$\frac{q}{A}\sum_i \overline{H}_i \frac{dC_i}{dz} = \frac{\dot{Q}}{A} \tag{c}$$

Setting $\tau = Az/q$, we obtain

$$\frac{dC_{EA}}{d\tau} = r_{EA} = kC_A C_E - k'C_{EA}C_W \tag{d}$$

and using Eq. a in Eq. c yields

$$\sum \frac{dC_i}{d\tau}\overline{H}_i = \sum r_i \overline{H}_i = \frac{d\hat{X}}{d\tau}\sum \nu_i \overline{H}_i = \Delta_{rxn}H\frac{d\hat{X}}{d\tau} = \frac{\dot{Q}}{A} \tag{e}$$

If we equate the time variable t with the distance variable τ, Eqs. d and e become identical to Eqs. a and c of the preceding batch reactor illustration. That is, the composition in a batch reactor t minutes after start-up will be the same as that at a distance z feet down a tubular reactor, where z is the distance traversed by the fluid in the time t. That is, $z = qt/A$ or $t = Az/q$.

Thus, to convert 37.2 percent of the acid to ester, the reaction time τ in the tubular reactor must be such that $\tau = AL/q = V/q = 119.3$ min, where L is the total reactor length and $V = AL$ is its volume. Therefore, $V = (119.3/139.3)V_{batch\ reactor} = 18.22$ m³, since we do not have to allow time for discharging, cleaning, and charging of the tubular reactor, as we did with the batch reactor. Also

$$\frac{\dot{Q}}{A} = \Delta_{rxn}H\frac{d\hat{X}}{d\tau} \quad \text{or} \quad \dot{Q} = A\Delta_{rxn}H\frac{d\hat{X}}{d\tau}$$

Since $\hat{X}$ is known as a function of t from the previous illustration, we can use the same figure to obtain $\hat{X}$, and hence $d\hat{X}/d\tau$, as a function of τ (or distance down the reactor). Thus, the heat flux at each point along the tubular reactor needed to maintain isothermal conditions can be computed from the heat program as a function of time in a batch reactor. ■

14.3 OVERALL REACTOR BALANCE EQUATIONS AND THE ADIABATIC REACTION TEMPERATURE

The mass and energy balance equations developed in Secs. 14.1 and 14.2 are the basic equations used in reactor design and analysis. In many cases, however, our needs are much more modest than in engineering design. In particular, we may not be interested in such details as the type of reactor used and the concentration and temperature profiles or time history in the reactor, but merely in the species mass and total energy balances for the reactor. In such situations one can use the general "black-box" equations of Table 8.4-1:

$$\frac{dN_i}{dt} = (\dot{N}_i)_{in} - (\dot{N}_i)_{out} + \sum_{j=1}^{\mathcal{M}} \nu_{ij} \dot{X}_j$$

$$= (\dot{N}_i)_{in} - (\dot{N}_i)_{out} + V \sum_{j=1}^{\mathcal{M}} \nu_{ij} r_j \tag{14.3-1}$$

and

$$\frac{d(\sum_i N_i \overline{U}_i)}{dt} = \sum_{i=1}^{\mathcal{C}} (\dot{N}_i \overline{H}_i)_{in} - \sum_{i=1}^{\mathcal{C}} (\dot{N}_i \overline{H}_i)_{out} + \dot{Q} + \dot{W} \tag{14.3-2}$$

The difficulty that arises is that to evaluate each of the terms in these equations we need chemical reaction rate data, reactor heat programs, and so forth, information we may not have. To avoid this difficulty, instead of using these differential equations directly, we can use the balance equations obtained by integrating these equations over the time interval t_1 to t_2, chosen so that the reactor is in the same state at t_2 as it was in t_1. If the reactor is a flow reactor, this corresponds to any period of steady-state operation, whereas if it is a batch reactor, the time interval is such that the initially empty reactor is charged, the reaction run, and the reactor contents discharged over the time interval. In either of these cases the results of the integrations are

Overall steady-state mass and energy balances for a reactor

and

$$\boxed{\begin{array}{c} (N_i)_{out} = (N_i)_{in} + \sum_{j=1}^{\mathcal{M}} \nu_{ij} X_j \qquad\qquad (14.3\text{-}3) \\[2em] Q + W = \sum_{i=1}^{\mathcal{C}} (N_i \overline{H}_i)_{out} - \sum_{i=1}^{\mathcal{C}} (N_i \overline{H}_i)_{in} \qquad (14.3\text{-}4) \end{array}}$$

where Q and W are the total heat and work flows into the reactor over the time interval t_1 to t_2, and X is the total molar extent of reaction in this time period. Comparing Eqs. 14.3-1 and 14.3-2 with Eqs 14.3-3 and 14.3-4, we find

$$Q = \int_{t_1}^{t_2} \dot{Q} \, dt \qquad W = \int_{t_1}^{t_2} \dot{W} \, dt$$

$$X_j = \int_{t_1}^{t_2} \dot{X}_j \, dt = \int_{t_1}^{t_2} \int_V r_j \, dV \, dt$$

$$(N_i)_{\text{in}} = \int_{t_1}^{t_2} (\dot{N}_i)_{\text{in}} \, dt \quad \text{and} \quad (N_i)_{\text{out}} = \int_{t_1}^{t_2} (\dot{N}_i)_{\text{out}} \, dt$$

The important feature of Eqs. 14.3-3 and 14.3-4 is that they contain only the total mass, heat, and work flows into the system, and the total molar extents of reaction, rather than the flow rates and rates of change of these quantities. Therefore, although Q, W, and X_j can be evaluated from integrals here, Eqs. 14.3-3 and 14.3-4 can also be used to interrelate Q, W, and X_j even when the detailed information needed to do these integrations is not available. This is demonstrated in Illustration 14.3-1.

ILLUSTRATION 14.3-1

Comparison of the Calculated Heat Loads with the Overall and Specific Energy Balances

Ethyl acetate is to be produced from the acetic acid and ethyl alcohol feed used in the illustrations of Secs. 14.1 and 14.2. If the feed and the product streams are both at 100°C, and 37.2 percent of the acid is converted to the ester, determine the total heat input into any reactor to produce 2500 kg of the ester. Show that this heat input is equivalent to that obtained in the illustrations in the previous sections.

SOLUTION

From Eqs. 14.3-3 and 14.3-4 we can immediately write

$$(N_i)_{\text{out}} = (N_i)_{\text{in}} + \nu_i X$$

and

$$Q = \sum (N_i \underline{H}_i)_{\text{out}} - \sum (N_i \underline{H}_i)_{\text{in}}$$

$$= \sum_i (N_i)_{\text{in}} [\underline{H}_i(T_{\text{out}}) - \underline{H}_i(T_{\text{in}})] + X \sum \nu_i \underline{H}_i(T_{\text{out}})$$

$$= X \Delta_{\text{rxn}} H(T_{\text{out}}) = 13.746 X \text{ kJ}$$

In writing this last equation we have neglected solution nonidealities; that is, we have set $\overline{H}_i = \underline{H}_i$, and noted from Illustration 14.1-1 that $\Delta_{\text{rxn}} H(T = 100°\text{C}) = 13.746$ kJ/mol.

The total molar extent of reaction can be computed by writing the species mass balance, Eq. 14.3-3, for ethyl acetate to obtain

$$X = \frac{(N_i)_{\text{out}} - (N_i)_{\text{in}}}{\nu_i} = \frac{2500 \text{ kg} \times 1000 \text{ g/kg} \times (1 \text{ mol})/(88 \text{ g}) - 0}{+1}$$

$$= 2.841 \times 10^4 \text{ mol}$$

Thus $Q = 2.841 \times 10^4 \text{ mol} \times 13.7 \text{ kJ/mol} = 3.892 \times 10^5$ kJ.

From Illustration 14.1-1 we have that for the continuous-flow stirred-tank reactor $\dot{Q} = 3254$ kJ/min and that 120.0 min are required to produce 2500 kg of ethyl acetate. Therefore, the total heat load is

$$Q = 3.243 \times 10^3 \text{kJ/min} \times 1.20 \times 10^2 \text{ min} = 3.892 \times 10^5 \text{ kJ}$$

To obtain the total input to either the batch or tubular reactor, we must integrate the area under the $\dot{Q}$ versus t curve given in Illustration 14.1-2. The result is

$$Q = 3.892 \times 10^5 \text{ kJ}$$

Thus, the energy required to produce a given amount of product from specified amounts of reactants all at the same temperature is independent of the type of reactor used to accomplish the transformation. This is still another demonstration that the change in thermodynamic state properties of a system, here ΔH, between any two states is independent of the path between the states.

COMMENT

This example illustrates both the advantages and disadvantages of the black-box style of thermodynamic analysis. If we are interested in merely computing the total heat requirement for the reaction, the black-box or overall analysis is clearly the most expeditious and does not require any kinetic data. However, black-box thermodynamics gives us no information about the reactor size or the details of the heat program (that is, the heat flow as a function of time in the batch reactor or as a function of distance in the tubular reactor). The decision of whether to use the black-box or more detailed thermodynamic analysis will largely depend on the amount of kinetic information available and the degree of detail desired in the final solution. ∎

In order to put Eq. 14.3-4 into a form analogous to Eq. 14.1-9b, we first multiply Eq. 14.3-3 by $(\overline{H}_i)_{in}$, sum over all species i, and then use the result in Eq. 14.3-4 to obtain

$$Q + W = \sum_{i=1}^{C}(N_i)_{out}\left[(\overline{H}_i)_{out} - (\overline{H}_i)_{in}\right] + \sum_{i=1}^{C}\sum_{j=1}^{M}(\overline{H}_i)_{in}\nu_{ij}X_j$$

$$= \sum_{i=1}^{C}(N_i)_{out}\left[(\overline{H}_i)_{out} - (\overline{H}_i)_{in}\right] + \sum_{j=1}^{M}(\Delta_{rxn,j}\overline{H})_{in}X_j \qquad \textbf{(14.3-5)}$$

Similarly, multiplying Eq. 14.3-3 by $(\overline{H}_i)_{out}$, and following the same procedure as the one that led to Eq. 14.1-15, yields the analogue of Eq. 14.1-9a,

$$Q + W = \sum_{i=1}^{C}(N_i)_{in}\left[(\overline{H}_i)_{out} - (\overline{H}_i)_{in}\right] + \sum_{j=1}^{M}(\Delta_{rxn,j}\overline{H})_{out}X_j \qquad \textbf{(14.3-6)}$$

For simplicity we will again neglect the effects of solution nonidealities with respect to the heats of reaction. With this assumption we have

$$\overline{H}_i(T, P, \underline{x}) = \underline{H}_i(T, P)$$

and

$$(\overline{H}_i)_{out} - (\overline{H}_i)_{in} = (\underline{H}_i)_{out} - (\underline{H}_i)_{in} = \int_{T_{in}}^{T_{out}} C_{P,i}\, dT$$

so that

$$(\Delta_{rxn,j}\overline{H})_{out} = \Delta_{rxn,j}H(T_{out})$$

and

$$(\Delta_{rxn,j}\overline{H})_{in} = \Delta_{rxn,j}H(T_{in})$$

Using these results in Eqs. 14.3-5 and 14.3-6 yields

$$Q + W = \sum_{j=1}^{\mathcal{M}} X_j \Delta_{\mathrm{rxn},j} H(T_{\mathrm{in}}) + \sum_{i=1}^{\mathcal{C}} (N_i)_{\mathrm{out}} \int_{T_{\mathrm{in}}}^{T_{\mathrm{out}}} C_{\mathrm{P},i}\, dT \tag{14.3-7}$$

and

$$Q + W = \sum_{i=1}^{\mathcal{C}} (N_i)_{\mathrm{in}} \int_{T_{\mathrm{in}}}^{T_{\mathrm{out}}} C_{\mathrm{P},i}\, dT + \sum_{j=1}^{\mathcal{M}} X_j \Delta_{\mathrm{rxn},j} H(T_{\mathrm{out}}) \tag{14.3-8}$$

Equations 14.3-7 and 14.3-8 have the same dual interpretation concerning the occurrence of fluid heating and chemical reaction as do Eqs. 14.1-10, and the comments made previously about those equations are equally valid here.

Equations 14.3-7 and 14.3-8 can be used to compute the steady-state outlet temperature of an *adiabatic* reactor whose effluent is in chemical equilibrium; this temperature is called the **adiabatic reaction temperature**[3] and will be designated here by T_{ad}. By definition, the adiabatic reaction temperature must satisfy (1) the equilibrium relations

$$K_{a,j}(T_{ad}) = \prod_{i=1}^{\mathcal{C}} a_i^{\nu_{ij}} \quad j = 1, 2, \ldots, \mathcal{M} \tag{14.3-9}$$

(2) one of the energy balances

Equations to calculate the adiabatic flame temperature

$$0 = \sum_{i=1}^{\mathcal{C}} (N_i)_{\mathrm{in}} \int_{T_{\mathrm{in}}}^{T_{ad}} C_{\mathrm{P},i}\, dT + \sum_{j=1}^{\mathcal{M}} [\Delta_{\mathrm{rxn},j} H(T_{ad})] X_j \tag{14.3-10a}$$

or

$$0 = \sum_{j=1}^{\mathcal{M}} [\Delta_{\mathrm{rxn},j} H(T_{\mathrm{in}})] X_j + \sum_{i=1}^{\mathcal{C}} (N_i)_{\mathrm{out}} \int_{T_{\mathrm{in}}}^{T_{ad}} C_{\mathrm{P},i}\, dT \tag{14.3-10b}$$

and (3) the mass balance and state variable constraints on the system. Since the equilibrium constants $K_{a,j}$ are nonlinear functions of temperature through the van't Hoff relation

$$\ln \frac{K_{a,j}(T_{ad})}{K_{a,j}(T_0)} = \int_{T_0}^{T_{ad}} \frac{\Delta_{\mathrm{rxn},j} H^\circ(T)}{RT^2}\, dT \tag{14.3-11}$$

the computation of T_{ad} can be tedious. The graphical procedure illustrated next can be used in solving Eqs. 14.3-9 through 14.3-11 for the adiabatic reaction temperature when only a single chemical reaction is involved. More commonly, and especially for multiple reactions, T_{ad} is found by computer calculation using an equation-solving program such as MATHCAD.

---

[3]The computation of the temperature of an adiabatic combustion flame is the most common adiabatic reaction temperature calculation (see Problems 14.8 and 14.9).

ILLUSTRATION 14.3-2

Adiabatic Flame Temperature Calculation

An equimolar gaseous mixture of benzene and ethylene at 300 K is fed into a reactor, where ethyl benzene is formed. If the reactor is operated adiabatically at a pressure of 1 bar, and the reactor effluent is in chemical equilibrium, find the reactor effluent temperature and species concentrations.

Data: See Appendices A.II and A.IV.

SOLUTION

Since the reactor effluent is in equilibrium, we can write

$$K_a(T_{ad}) = \frac{a_{EB}}{a_E a_B} = \frac{y_{EB}\left(\dfrac{P}{1\text{ bar}}\right)}{y_E\left(\dfrac{P}{1\text{ bar}}\right) y_B\left(\dfrac{P}{1\text{ bar}}\right)} = \frac{y_{EB}}{y_E y_B}$$

where we have assumed ideal gas mixture behavior and used the fact that the pressure in the reactor is 1 bar. Using the following table, the three mole fractions in this equation can be replaced by the single molar extent-of-reaction variable X.

	Inlet	Outlet	Outlet Mole Fraction
C_6H_6	1	$1 - X$	$(1 - X)/(2 - X)$
C_2H_4	1	$1 - X$	$(1 - X)/(2 - X)$
$C_6H_5C_2H_5$	0	X	$X/(2 - X)$
		$\overline{2 - X}$	

Thus, we have

$$K_a(T_{ad}) = \frac{X(2 - X)}{(1 - X)^2}$$

so that

$$X = 1 - \sqrt{\frac{1}{1 + K(T_{ad})}}$$

With the data given in Appendix A, and Eq. 13.1-23b, it is possible to compute the value of K_a at any temperature. Once K_a is known, X can be computed. The equilibrium values of X calculated in this manner for various values of T_{ad} are plotted in the following figure.

The adiabatic reaction temperature also must satisfy the energy balance equation; using Eq. 14.3-10a, we have

$$0 = \int_{T_{in}}^{T_{ad}} (C_{P,B} + C_{P,E})\, dT + \Delta_{rxn}H(T_{ad})X$$

or

$$X = \frac{-\displaystyle\int_{T_{in}}^{T_{ad}} (C_{P,B} + C_{P,E})\, dT}{\Delta_{rxn}H(T_{ad})}$$

where the heat capacity data for benzene and ethylene are given in Appendix A.II. Using this equation, the molar extent of reaction needed to satisfy the energy balance for fixed values of T_{in}

and T_{ad} can be computed. Since the inlet temperature is fixed at 300 K, we will use this equation to compute values of the molar extent of reaction X for a number of different choices of T_{ad}; the results of this calculation are plotted in the figure. For illustration, energy balance X-T_{ad} curves for several other values of the reactor inlet temperature are given as dashed lines.

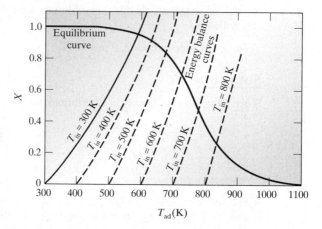

Graph for computing the adiabatic reaction temperature for the formation of ethyl benzene from ethylene and benzene.

The adiabatic reaction temperature and the equilibrium extent of reaction X that satisfy both the equilibrium and the energy balance relations for a given value of T_{in} are found at the intersection of the equilibrium and energy balance curves in the figure. For the reactor feed at 300 K this occurs at

$$T_{ad} \simeq 588 \text{ K}$$
$$X \simeq 0.955$$

Thus

$$y_B = y_E = 0.045/1.045 = 0.043$$
$$y_{EB} = 0.955/1.045 = 0.914$$

The results for other reactor feed temperatures T_{in} can be computed in a similar fashion from the data in the figure. ∎

In Illustration 14.3-2 the reaction was exothermic ($\Delta_{rxn}H < 0$). This implies that (1) energy is released on reaction, which must appear as an increase in temperature of the reactor effluent since the reactor is adiabatic, and (2) the equilibrium conversion of reactants to products decreases with increasing temperature. Consequently, the equilibrium extent of reaction achieved in an adiabatic reactor is less than that which would be obtained had the reactor been operating isothermally at the reactor inlet conditions. For an endothermic reaction ($\Delta_{rxn}H > 0$), the equilibrium conversion decreases with decreasing temperature and, since energy is absorbed on reaction, the reactor effluent is at a lower temperature than the feed stream. Thus, the equilibrium and energy balance curves are mirror images of those in Illustration 14.3-2 (see Problem 14.13), and again the equilibrium extent of reaction is less than would be obtained with an isothermal reactor operating at the feed temperature.

Another interesting application of the multicomponent balance equations for reacting systems is in the estimation of the maximum work that can be obtained from an isothermal chemical reaction that occurs in a steady-state fuel cell or other work-producing device. The starting point here is the steady-state isothermal energy and entropy balances of Table 8.4-1:

$$0 = \sum_{i=1}^{\mathcal{C}}(\dot{N}_i \overline{H}_i)_{\text{in}} - \sum_{i=1}^{\mathcal{C}}(\dot{N}_i \overline{H}_i)_{\text{out}} + \dot{Q} + \dot{W}_s \qquad \textbf{(14.3-12)}$$

and

$$0 = \sum_{i=1}^{\mathcal{C}}(\dot{N}_i \overline{S}_i)_{\text{in}} - \sum_{i=1}^{\mathcal{C}}(\dot{N}_i \overline{S}_i)_{\text{out}} + \frac{\dot{Q}}{T} + \dot{S}_{\text{gen}} \qquad \textbf{(14.3-13)}$$

Solving the second of these equations for the heat flow $\dot{Q}$,

$$\dot{Q} = T\sum_{i=1}^{\mathcal{C}}(\dot{N}_i \overline{S}_i)_{\text{out}} - T\sum_{i=1}^{\mathcal{C}}(\dot{N}_i \overline{S}_i)_{\text{in}} - T\dot{S}_{\text{gen}} \qquad \textbf{(14.3-14)}$$

and then eliminating the heat flow from Eq. 14.3-12 yields

$$\dot{W}_s = \sum_{i=1}^{\mathcal{C}}[\dot{N}_i(\overline{H}_i - T\overline{S}_i)]_{\text{out}} - \sum_{i=1}^{\mathcal{C}}[\dot{N}_i(\overline{H}_i - T\overline{S}_i)]_{\text{in}} + T\dot{S}_{\text{gen}}$$

$$= \sum_{i=1}^{\mathcal{C}}[\dot{N}_i \overline{G}_i]_{\text{out}} - \sum_{i=1}^{\mathcal{C}}[\dot{N}_i \overline{G}_i]_{\text{in}} + T\dot{S}_{\text{gen}} \qquad \textbf{(14.3-15)}$$

Since a fuel cell is a work-producing device, $\dot{W}_s$ will be negative. The maximum work (i.e., the largest negative value of $\dot{W}_s$ for given inlet conditions) is obtained when the terms $\sum_{i=1}^{\mathcal{C}}[\dot{N}_i \overline{G}_i]_{\text{out}}$ and $T\dot{S}_{\text{gen}}$ are as small as possible. Since $\dot{S}_{\text{gen}} \geq 0$, with the equality holding only for reversible processes, one condition for obtaining the maximum work is that the chemical reaction and mechanical energy production process be carried out reversibly, so that $\dot{S}_{\text{gen}}$ is equal to zero. The first term on the right side of Eq. 14.3-15, $\sum_{i=1}^{\mathcal{C}}[\dot{N}_i \overline{G}_i]_{\text{out}}$, is the flow of Gibbs energy accompanying the mass flow out of the system. The minimum value of this term for a given mass flow at fixed temperature occurs when the exit stream is in chemical equilibrium (so that the Gibbs energy per unit mass is a minimum). Thus, another condition for obtaining maximum work from an isothermal flow reactor is that the exit stream be in chemical equilibrium. When both these conditions are met,

$$\dot{W}_{\text{max}} = \sum_{i=1}^{\mathcal{C}}(\dot{N}_i \overline{G}_i)_{\text{out}} - \sum_{i=1}^{\mathcal{C}}(\dot{N}_i \overline{G}_i)_{\text{in}} \qquad \textbf{(14.3-16a)}$$

and

Maximum work from a fuel cell at constant T and P

$$\boxed{W_{\text{max}} = \sum_{i=1}^{\mathcal{C}}(N_i \overline{G}_i)_{\text{out}} - \sum_{i=1}^{\mathcal{C}}(N_i \overline{G}_i)_{\text{in}}} \qquad \textbf{(14.3-16b)}$$

It is also possible to compute the maximum work attainable from an isothermal, constant-volume batch reactor. It is left to you to show that

$$W_{\text{max}} = \sum_{i=1}^{\mathcal{C}}(N_i \overline{A}_i)_{\text{final}} - \sum_{i=1}^{\mathcal{C}}(N_i \overline{A}_i)_{\text{initial}} \qquad \textbf{(14.3-17)}$$

ILLUSTRATION 14.3-3

Calculation of Energy Produced in a Fuel Cell

The November 1972 issue of *Fortune* magazine discussed the possibility of the future energy economy of the United States being based on hydrogen. In particular, the use of hydrogen fuel in automobiles was considered. Assuming that pure hydrogen gas and twice the stoichiometric amount of dry air, each at 400 K and 1 bar, are fed into a catalytic fuel cell, and that the reaction products leave at the same temperature and pressure, compute the maximum amount of work that can be obtained from each mole of hydrogen.

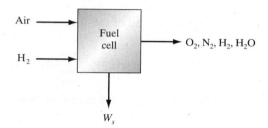

Data: Air may be considered to be 21 mol % oxygen and 79 mol % nitrogen; the heats and Gibbs energies of formation are given in Appendix A.IV, and the heat capacities in Appendix A.II.

SOLUTION

From Eq. 14.3-16 we have

$$W_{max} = \sum_{i=1}^{c} \{(N_i \bar{G}_i)_{out} - (N_i \bar{G}_i)_{in}\}$$

where the outlet compositions are related by the fact that chemical equilibrium exists. Here, since both the reactants and products are gases (presumably ideal gases at the reaction conditions), we have

$$\bar{G}_i(T, P, \underline{y}) = \underline{G}_i(T, P) + RT \ln y_i$$

Since hydrogen enters the fuel cell as a separate, pure stream, we have

Species	$(y_i)_{in}$	$(N_i)_{in}$ (mol)	$(N_i)_{out}$ (mol)	$(y_i)_{out}$
H_2	1	1	$1 - X$	$(1-X)/\Sigma$
O_2	0.21	1	$1 - \frac{1}{2}X$	$(1 - \frac{1}{2}2X)/\Sigma$
N_2	0.79	$\frac{0.79}{0.21} \times 1 = 3.762$	3.762	$3.762/\Sigma$
H_2O	0		X	X/Σ
Total			$\Sigma = 5.762 - \frac{1}{2}X$	

Now, from the equilibrium relation, we have

$$K_a = \frac{a_{H_2O}}{a_{H_2} a_{O_2}^{1/2}}$$

$$= \frac{y_{H_2O}}{y_{H_2} y_{O_2}^{1/2} (P/1 \text{ bar})^{1/2}} = \frac{X(5.762 - \frac{1}{2}X)^{1/2}}{(1 - X)(1 - \frac{1}{2}X)^{1/2}}$$

Using the data in the problem statement and Eq. 13.1-16, we find that $K_a = 1.754 \times 10^{+29}$; a huge number. Therefore, $X \sim 1$. Also, using Eq. 13.1-23b and the heat capacity data, the values of the pure-component Gibbs energies at 400 K and 1 bar, relative to the reference states in Appendix A.IV, can be computed; the results are

Species	$\underline{G}_i(T = 400 \text{ K}, P = 1 \text{ bar})$	$(N_i)_{out}$	$(y_i)_{out}$
H_2	−453.1 J/mol	0	0
O_2	−466.9	0.5	0.095
N_2	−457.5	3.762	0.715
H_2O	−224 600	1	0.190

Therefore,

$$
\begin{aligned}
W_{max}(T = 400 \text{ K}) &= \sum_i (N_i \bar{G}_i)_{out} - \sum_i (N_i \bar{G}_i)_{in} \\
&= \sum_i [N_i(\underline{G}_i + RT \ln y_i)]_{out} - \sum_i [N_i(\underline{G}_i + RT \ln y_i)]_{in} \\
&= \{0 + 0.5[-466.9 + 8.314 \times 400 \ln(0.095)] \\
&\quad + 3.762[-457.5 + 8.314 \times 400 \ln(0.715)] \\
&\quad + 1[-224,600 + 8.314 \times 400 \ln(0.190)]\} \\
&\quad - \{1[-453.1 + 8.314 \times 400 \ln(1)] \\
&\quad + 1[-466.9 + 8.314 \times 400 \ln(0.21)] \\
&\quad + 3.762[-457.5 + 8.314 \times 400 \ln(0.79)] + 0\} \\
&= (0 - 4147.5 - 5918.2 - 230\,122.9) - (-453.1 - 5657.0 - 4670.1) \text{ J/mol} \\
&= -229\,408.4 \text{ J/mol} = -229.4 \text{ kJ/mol}
\end{aligned}
$$

COMMENTS

1. In this illustration the Gibbs energy of mixing terms make a relatively small contribution to W_{max}. This is only because for the $H_2 + \frac{1}{2}O_2 = H_2O$ reaction, the dominant term in the calculation is $\Delta_{rxn}G$. For many reactions $\Delta_{rxn}G$ is not so large, and the Gibbs energy of mixing terms can be important.

2. Although the simple black-box thermodynamic analysis here permits us to calculate the maximum work obtainable from the fuel cell, it does not provide any indication as to how to design the fuel cell (i.e., as an internal combustion engine, electrolytic cell, etc.) to obtain this work. The design of efficient fuel cells is currently an important unsolved engineering problem, much like the design of heat engines at the time of Sadi Carnot. ∎

14.4 THERMODYNAMICS OF CHEMICAL EXPLOSIONS

In Sec. 5.3 we considered the thermodynamics of explosions that did not involve chemical reaction. Here we extend this discussion to explosions with chemical reaction. Since the energy released on an exothermic reaction may be very large, chemical explosions are generally more devastating than purely mechanical explosions. In a chemical explosion there are extremely rapid temperature and pressure rises as some or all of the reactants are oxidized, followed by an expanding shock wave resulting in a decrease in the temperature and pressure of the combustion products. It is the rapid overpressurization at the expanding shock-wave front that causes most of the damage in an explosion. In order for a chemical explosion to occur, the heat of reaction must be reasonably large and negative ($\Delta_{rxn}H < 0$), that is, the reaction must be exothermic, and the equilibrium constant for the reaction should be large ($K_{eq} \gg 0$, or $\Delta_{rxn}G < 0$). In this case, the reaction will proceed to an appreciable extent and a significant amount of energy released.

The equations describing a chemical explosion are the same as those for a mechanical explosion (see Sec. 5.3); that is, to compute the maximum energy released in an explosion, we assume the process within the region bounded by the shock wave is reversible and adiabatic so that

$$U^f - U^i = U(T^f, P^f, \underline{N}^f) - U(T^i, P^i, \underline{N}^i) = W \qquad \textbf{(14.4-1)}$$

$$S^f - S^i = S(T^f, P^f, \underline{N}^f) - S(T^i, P^i, \underline{N}^i) = S_{gen} = 0 \qquad \textbf{(14.4-2)}$$

Here the notation $\underline{N}$ has been used to indicate all the mole numbers $\underline{N} = (N_1, N_2, \ldots, N_C)$, and we have written these equations in terms of total thermodynamic properties rather than molar or specific properties since, as a result of the chemical reaction, there is a change in the number of moles of each species. Also, as before, we assume that the entropy generation term is zero, thereby obtaining an upper bound on the energy released in an explosion.

Multiplying Eq. 14.4-2 by the initial temperature T^i and subtracting it from Eq. 14.4-1 gives

$$U(T^f, P^f, \underline{N}^f) - T^i S(T^f, P^f, \underline{N}^f) - [U(T^i, P^i, \underline{N}^i) - T^i S(T^i, P^i, \underline{N}^i)]$$
$$= U(T^f, P^f, \underline{N}^f) - T^i S(T^f, P^f, \underline{N}^f) - A(T^i, P^i, \underline{N}^i) = W$$

$$\textbf{(14.4-3)}$$

where $A(T^i, P^i, \underline{N}^i) = U(T^i, P^i, \underline{N}^i) - T^i S(T^i, P^i, \underline{N}^i)$ is the Helmholtz energy of the initial state. [Why is $U(T^f, P^f, \underline{N}^f) - T^i S(T^f, P^f, \underline{N}^f)$ not the Helmholtz energy of the final state?]

As in Sec. 5.3 the final pressure can be taken to be 1.013 bar when the shock wave has dissipated, and all the damage has been done. What are unknown are the final temperature and the composition of the reacting mixture. The composition is especially difficult to estimate because the rates of chemical reaction can be slow compared with the rates of change of temperature and pressure in an explosion. Consequently, it is not certain that chemical equilibrium will be achieved in a chemical explosion. Further, because of the rapid drop in temperature in the expanding shock wave, and as the rate of reaction decreases with decreasing temperature, even if chemical equilibrium were achieved, it is not clear to which temperature this equilibrium would correspond. That is, chemical equilibrium could be achieved at some high or intermediate temperature, and then the reaction may be quenched by the rapid temperature drop at a composition corresponding to that unknown temperature.

Another question that arises is the amount of oxidant available for the reaction. In a vapor cloud explosion one can generally assume that there is sufficient oxygen present for the complete oxidation of all carbon to carbon dioxide and all hydrogen to water. In this case the final composition is known. The situation for explosions involving liquids and solids in which there is a limited amount of oxidizing agent available is more complicated and will be considered shortly. The final temperature of the explosive mixture, after the shock wave expands so that the internal pressure is 1.013 bar, will generally be higher than ambient. Since the quantity $U^f - T^i S^f$ increases with increasing temperature, to obtain an upper bound on the energy released in a chemical explosion, we will assume that the final temperature and pressure are ambient, as are the initial temperature and pressure. In this case we obtain from Eq. 14.4-3

$$U(25°C, 1.013 \text{ bar}, \underline{N}^f) - (298.15 \text{ K}) \times S(25°C, 1.013 \text{ bar}, \underline{N}^f)$$
$$- U(25°C, 1.013 \text{ bar}, \underline{N}^i) - (298.15 \text{ K}) \times S(25°C, 1.013 \text{ bar}, \underline{N}^i)$$
$$= A(25°C, 1.013 \text{ bar}, \underline{N}^f) - A(25°C, 1.013 \text{ bar}, \underline{N}^i) = W$$
$$\textbf{(14.4-4)}$$

Since Gibbs energy data are more readily available than Helmholtz energies, we note that

$$G = A + PV \approx A \quad \text{for solids and liquids}$$

and

$$G = A + PV = A + N^V RT \quad \text{for vapors}$$

where in the last equation we have used the ideal gas law (as the pressure is low) and the notation N^V to indicate the total number of moles in the gas phase. With this substitution we have

$$W = G(25°C, 1.013 \text{ bar}, \underline{N}^f) - G(25°C, 1.013 \text{ bar}, \underline{N}^i) - [N^{V,f} - N^{V,i}]RT \quad \textbf{(14.4-5)}$$

This is the equation that will be used to compute the energy released in an explosion. In using this equation, we will make one further simplification. The total Gibbs energies in this equation should be computed from

$$G(T, P, \underline{N}) = \sum N_i \bar{G}_i(T, P, \underline{x}) \qquad \textbf{(14.4-6)}$$

For simplicity, we assume here that vapors and liquids form ideal solutions, since the contribution of the solution nonideality to the energy is small compared with the chemical reaction term (you should verify this), so that

$$\bar{G}_i = \Delta_f \underline{G}_i^\circ + RT \ln x_i \qquad \textbf{(14.4-7)}$$

and that solid phases are pure so that

$$\bar{G}_i = \Delta_f \underline{G}_i^\circ \qquad \textbf{(14.4-8)}$$

In a vapor-phase explosion it is generally assumed that sufficient oxygen is present for the substance to be completely oxidized. This makes estimating the explosive energy release simpler, as the reaction stoichiometry is known. Also, because of the large amount of nitrogen present before and after the explosion (air contains about 79 percent nitrogen), the entropy change on mixing is small and frequently can be neglected, as is shown in the following illustration.

ILLUSTRATION 14.4-1

Estimating the Energy Released in a Vapor-Phase Explosion

Estimate the energy released in a vapor-phase explosion of 1 kg of ethylene with a stoichiometric amount of air.

SOLUTION

The reaction stoichiometry is

$$C_2H_4 + 3O_2 + 3 \times \frac{0.79}{0.21}N_2 = 2CO_2 + 2H_2O + 3 \times \frac{0.79}{0.21}N_2$$

where we have recognized that with 1 mole of oxygen in air there is 0.79/0.21 moles of nitrogen. To proceed further we note that $N^{V,i} = 1 + 3 + 3 \times (0.79/0.21) = 15.29$ moles of gas present initially, and $N^{V,f} = 2 + 2 + 3 \times (0.79/0.21) = 15.29$ moles of gas present finally, so that there is no contribution to the total energy release as a result of a change in the number of moles present in the gas phase. Next,

$$\begin{aligned}
G(\underline{N}^i) &= \Delta_f\underline{G}^\circ_{C_2H_4} + 3 \times \Delta_f\underline{G}^\circ_{O_2} + 11.29 \times \Delta_f\underline{G}^\circ_{N_2} \\
&\quad + RT\left[1 \times \ln\left(\frac{1}{1+3+11.29}\right) + 3 \times \ln\left(\frac{3}{1+3+11.29}\right)\right. \\
&\quad \left. + 11.29 \times \ln\left(\frac{11.29}{1+3+11.29}\right)\right] \\
&= 68.5 + 3 \times 0 + 11.29 \times 0 + 8.314 \times 10^{-3} \times 298.15 \times (-2.7 - 4.9 - 3.4) \\
&= 68.5 - 27.3 = 41.2 \text{ kJ/mol}
\end{aligned}$$

and

$$\begin{aligned}
G(\underline{N}^f) &= 2 \times \Delta_f\underline{G}^\circ_{CO_2} + 2 \times \Delta_f\underline{G}^\circ_{H_2O} + 11.29 \times \Delta_f\underline{G}^\circ_{N_2} \\
&\quad + RT\left[2 \times \ln\left(\frac{2}{2+2+11.29}\right) + 2 \times \ln\left(\frac{2}{2+2+11.29}\right)\right. \\
&\quad \left. + 11.29 \times \ln\left(\frac{11.29}{2+2+11.29}\right)\right] \\
&= 2 \times (-394.4) + 2 \times (-228.6) + 11.29 \times 0 \\
&\quad + 8.314 \times 10^{-3} \times 298.15 \times (-4.1 - 4.1 - 3.4) \\
&= -488.8 - 457.2 - 28.8 = -1274.8 \text{ kJ/mol}
\end{aligned}$$

Therefore,

$$W = -1274.8 - 41.2 = -1316.0 \frac{\text{kJ}}{\text{mol}}$$

and the total energy released by the explosion of 1 kg of ethylene is

$$W = -1316.0 \frac{\text{kJ}}{\text{mol}} \times \frac{1 \text{ mol}}{28.054 \text{ g}} \times \frac{1000 \text{ g}}{1 \text{ kg}} = -46\,910 \frac{\text{kJ}}{\text{kg}}$$

Note that the energy released on the explosion of ethylene, 46 910 kJ/kg, is more than 10 times that for TNT, which is 4570 kJ/kg. It is of interest to note that in this example of the total energy release of -1316 kJ/mol, the amount -1314.5 [$= 2 \times (-394.4) + 2 \times (-228.6) - 68.5$]

kJ/mol results from the standard-state Gibbs energy change on reaction, which is the dominant contribution, and only $-28.8 - (-27.3) = -1.5$ kJ/mol is from the entropy-of-mixing term. ■

As indicated earlier, one of the difficulties in chemical explosions is ascertaining the final composition of the exploding mixture. In explosions of a solid or a liquid, because of mass transfer limitations usually only the oxygen present in the original compound is involved in the reaction. In this case there may be an infinite number of partial oxidation reaction stoichiometries possible.

ILLUSTRATION 14.4-2

Examining Different Reaction Stoichiometries

List some of the possible reactions for TNT, $C_7H_5N_3O_6$.

SOLUTION

Among the many possible reactions are

$$C_7H_5N_3O_6 \rightarrow 1.0C + 2.5H_2 + 1.5N_2 + 6.0CO$$
$$\rightarrow 5.25C + 1.75CO_2 + 1.5N_2 + 2.5H_2O$$
$$\rightarrow 0.96C + 6.0CO + 1.48N_2 + 0.04HCN + 2.48H_2$$
$$\rightarrow 3.7C + 0.4CO + 2.2CO_2 + 1.5N_2 + 1.1H_2O + 0.7CH_4$$

However, there is some experimental evidence that the reaction that occurs is actually

$$C_7H_5N_3O_6 \rightarrow 3.65C + 1.98CO + 1.25CO_2 + 1.32N_2 + 0.16NH_3 + 0.02HCN$$
$$+ 0.46H_2 + 1.60H_2O + 0.10CH_4 + 0.004C_2H_6$$

which has an explosive energy release of about 4570 kJ/kg. ■

Generally what is done in dealing with reactions involving liquids or solids to make a conservative (which in this context means worst-case) estimate of the explosive energy release of a chemical is to examine the different reaction stoichiometries, and then choose the reaction with the largest negative Gibbs energy change (or equivalently, the largest equilibrium constant). However, this assumes that the reaction is a rapid one.

ILLUSTRATION 14.4-3

Estimating the Energy Released on the Explosion of a Liquid

Nitromethane (a liquid) undergoes the following reaction on explosion:

$$CH_3NO_2 \rightarrow 0.2CO_2 + 0.8CO + 0.8H_2O + 0.7H_2 + 0.5N_2$$

Estimate the energy released on the explosion of nitromethane.

Data:
$$\Delta_f \underline{G}^\circ_{NM} = -13.0 \text{ kJ/mol}$$
$$\Delta_f \underline{H}^\circ_{NM} = -111.7 \text{ kJ/mol}$$

SOLUTION

There are a number of terms that contribute to the total energy released. First we note that $N^{V,f} = 0.2 + 0.8 + 0.8 + 0.7 + 0.5 = 3$ moles of gas present finally, and $N^{V,i} = 0$ moles of gas initially. Next,

$$G(\underline{N}^i) = \Delta_f G^\circ_{NM} = -13.0 \text{ kJ/mol}$$

$$G(\underline{N}^f) = 0.2\Delta_f \underline{G}^\circ_{CO_2} + 0.8\Delta_f \underline{G}^\circ_{CO} + 0.8\Delta_f \underline{G}^\circ_{H_2O} + 0.7\Delta_f \underline{G}^\circ_{H_2} + 0.5\Delta_f \underline{G}^\circ_{N_2}$$

$$+ RT \left[0.2 \ln \frac{0.2}{3} + 0.8 \ln \frac{0.8}{3} + 0.8 \ln \frac{0.8}{3} + 0.7 \ln \frac{0.7}{3} + 0.5 \ln \frac{0.5}{3} \right]$$

$$= -371.5 \frac{\text{kJ}}{\text{mol}} + 8.314 \frac{\text{J}}{\text{mol K}} \times 298.15 \text{ K} \times (-4.571) \times \frac{1 \text{ kJ}}{1000 \text{ J}}$$

$$= -371.5 - 11.3 = -382.8 \frac{\text{kJ}}{\text{mol}}$$

Therefore,

$$W = -382.8 - (-13.0) - 3 \times \frac{8.314 \times 298.15}{1000} = -377.2 \frac{\text{kJ}}{\text{mol}}$$

COMMENTS

It is again interesting to assess the contributions of the various terms to the total energy release on explosion. By far the largest contribution is the difference in Gibbs energies of the components, that is

$$\Delta G = 0.2\Delta_f \underline{G}^\circ_{CO_2} + 0.8\Delta_f \underline{G}^\circ_{CO} + 0.8\Delta_f \underline{G}^\circ_{H_2O} + 0.7\Delta_f \underline{G}^\circ_{H_2} + 0.5\Delta_f \underline{G}^\circ_{N_2} - \Delta_f \underline{G}^\circ_{NM}$$

$$= -371.5 - (-13.0) = -358.5 \text{ kJ/mol}$$

The Gibbs energy change due to the entropy of mixing is -11.1 kJ/mol, and the energy released due to the vapor-phase mole number change is -7.4 kJ/mol. Therefore, the majority of the explosive energy release comes from the Gibbs energies of formation of the species involved. A reasonable approximation in many cases, accurate to ± 10 percent, is to neglect the gas-formation and entropy-of-mixing terms and merely use $W = \Delta_{rxn} G$.

A second condition for an explosion is that the heat of reaction be negative (i.e., the reaction should be exothermic). For the reaction here, we have

$$\Delta_{rxn} H = 0.2\Delta_f \underline{H}^\circ_{CO_2} + 0.8\Delta_f \underline{H}^\circ_{CO} + 0.8\Delta_f \underline{H}^\circ_{H_2O} + 0.7\Delta_f \underline{H}^\circ_{H_2} + 0.5\Delta_f \underline{H}^\circ_{N_2} - \Delta_f \underline{H}^\circ_{NM}$$

$$= 0.2 \times (-393.5) + 0.8 \times (-110.5) + 0.8 \times (-241.8)$$

$$+ 0.7 \times (0) + 0.5 \times (0) - (-111.7)$$

$$= -78.7 - 88.4 - 193.4 + 111.7 = -248.8 \frac{\text{kJ}}{\text{mol}}$$

which is large and negative. Therefore, nitromethane is a likely candidate for a chemical explosion. ∎

ILLUSTRATION 14.4-4

Estimating the Energy Released on the Explosion of a Solid

One possible reaction for 2,4,6-trinitrotoluene (TNT) is as follows:

$$C_7H_5O_6N_3 \rightarrow C + 6CO + 2.5H_2 + 1.5N_2$$

Estimate the energy released on an explosion of TNT.

Data: $\Delta_f G^\circ_{TNT} = 273$ kJ/mol

SOLUTION

Since TNT is a solid, $N^{V,i} = 0$; also $N^{V,f} = 10$. Using the standard-state Gibbs energy of formation tables in Appendix A.IV and Eqs. 14.4-7 and 14.4-8, we have at 25°C

$$\underline{G}_{TNT} = 273 \ \frac{kJ}{mol}, \qquad \underline{G}_C = 0$$

$$\bar{G}_{CO} = -110.5 + \frac{8.314 \ \frac{J}{mol \ K} \times 298.15 \ K}{1000 \ \frac{J}{kJ}} \ln \frac{6}{10}$$

$$= -110.5 - 1.3 = -111.8 \ \frac{kJ}{mol}$$

$$\bar{G}_{H_2} = 0 + \frac{8.314 \ \frac{J}{mol \ K} \times 298.15 \ K}{1000 \ \frac{J}{kJ}} \ln \frac{2.5}{10} = -3.4 \ \frac{kJ}{mol}$$

and

$$\bar{G}_{N_2} = 0 + \frac{8.314 \ \frac{J}{mol \ K} \times 298.15 \ K}{1000 \ \frac{J}{kJ}} \ln \frac{1.5}{10} = -4.7 \ \frac{kJ}{mol}$$

Therefore,

$$W = [1 \times 0 + 6 \times (-111.8) + 2.5 \times (-3.4) + 1.5 \times (-4.7)]$$
$$- [273] - (10 - 0) \times \frac{8.314 \times 298.15}{1000}$$

$$= -670.8 - 8.5 - 7.1 - 273 - 2.48 = -961.9 \ \frac{kJ}{mol \ TNT}$$

$$= -961.9 \ \frac{kJ}{mol \ TNT} \times \frac{1 \ mol \ TNT}{237 \ g} = -4.059 \ \frac{kJ}{g \ TNT} = -4059 \ \frac{kJ}{kg \ TNT}$$

The experimentally measured energy release on a TNT explosion is about 4570 kJ/kg. The reason for the discrepancy is that, as discussed in Illustration 14.4-2, the actual reaction stoichiometry is different from that assumed here.

It is of interest to note that in this example of the total energy release of -961.9 kJ, $-6 \times 110.5 - 273 = -936$ kJ arises from the Gibbs energy change on reaction, which is the dominant contribution; -2.48 kJ is a result of the formation of 10 moles of gas per mole of TNT; and $RT \times [6 \times \ln(0.6) + 2.5 \times \ln(0.25) + 1.5 \times \ln(0.15)] = -23.24$ kJ results from the entropy-of-mixing term. ∎

14.5 AVAILABILITY AND AVAILABLE WORK IN CHEMICALLY REACTING SYSTEMS

Computing the work that can be obtained in a chemical reacting system is, in general, somewhat more complicated than in the nonreacting systems that we considered in, for example, Chapters 4 and 5. Of particular interest here is computing the maximum work that can be obtained in a chemically reacting system.

We start by considering the steady-state mass, energy, and entropy balances for a chemically reacting system that can exchange heat and work with its environment:

$$0 = N_{i,\text{in}} + N_{i,\text{out}} + \nu_i X$$

$$0 = \sum_i N_{i,\text{in}} \bar{H}_{i,\text{in}} + \sum_i N_{i,\text{out}} \bar{H}_{i,\text{out}} + Q + W = \sum_i N_{i,\text{in}} \bar{H}_{i,\text{in}} + X \sum_i \nu_i \bar{H}_{i,\text{out}} + Q + W$$

and

$$0 = \sum_i N_{i,\text{in}} \bar{S}_{i,\text{in}} + \sum_i N_{i,\text{out}} \bar{S}_{i,\text{out}} + \frac{Q}{T} + S_{\text{gen}} = \sum_i N_{i,\text{in}} \bar{S}_{i,\text{in}} + X \sum_i \nu_i \bar{S}_{i,\text{out}} + \frac{Q}{T} + S_{\text{gen}}$$

Since ultimately we are interested in the maximum amount of work that can be obtained from a chemically reacting system, one consideration is that all heat transferred between the system and its surroundings should occur at ambient temperature. The reason for this is that if this energy flow were at a higher temperature, useful work that could have been extracted by making use of the temperature difference between the system and the environment using a Carnot cycle or other heat engine has been lost. (The same argument applies if the heat flow is below the ambient temperature.) Denoting the ambient temperature by T_o, the entropy balance becomes

$$0 = \sum_i N_{i,\text{in}} \bar{S}_{i,\text{in}} + \sum_i N_{i,\text{out}} \bar{S}_{i,\text{out}} + \frac{Q}{T_o} + S_{\text{gen}}$$

Now combining the energy and entropy balances, we have

$$0 = \sum_i N_{i,\text{in}} \left(\bar{H}_{i,\text{in}} - T_o \bar{S}_{i,\text{in}} \right) + \sum_i N_{i,\text{out}} \left(\bar{H}_{i,\text{out}} - T_o \bar{S}_{i,\text{out}} \right) + W - T_o S_{\text{gen}}$$

At first glance, the terms in parentheses might appear to be Gibbs energies; however, this is not the case unless the temperature of the flow stream is equal to the ambient temperature T_o. That is, $G(T, P) = H(T, P) - TS(T, P)$ and not $H(T, P) - T_o S(T, P)$.

To proceed, it is convenient to define a new function, the availability Ψ, defined on a molar basis as

$$\underline{\Psi}(T, P) = \underline{H}(T, P) - T_o \underline{S}(T, P)$$

and the partial molar availability by

$$\bar{\Psi}_i(T, P) = \left(\frac{\partial N \underline{\Psi}}{\partial N_i} \right)_{T, P, N_{j \neq i}} = \bar{H}_i(T, P) - T_o \bar{S}_i(T, P)$$

Since

$$\bar{H}_i(T, P, \underline{x}) = \underline{H}_i(T, P) + \bar{H}_i^{\text{ex}}(T, P, \underline{x})$$

and

$$\bar{S}_i(T, P, \underline{x}) = \underline{S}_i(T, P) + \bar{S}_i^{\text{ex}}(T, P, \underline{x}) - R \ln x_i$$

it follows that

$$\bar{\Psi}_i(T, P, \underline{x}) = \underline{H}_i(T, P) + \bar{H}_i^{\text{ex}}(T, P, \underline{x}) - T_o \left[\underline{S}_i(T, P) + \bar{S}_i^{\text{ex}}(T, P, \underline{x}) - R \ln x_i \right]$$

Note that it is only if the stream temperature and pressure are the same as the ambient temperature and pressure, that the Gibbs energy is equal to the availability.

Consequently, the combined energy and entropy balance is

$$0 = \sum_i N_{i,in} \overline{\Psi}_{i,in} - \sum_i N_{i,out} \overline{\Psi}_{i,out} + W - T_o S_{gen}$$

This equation is exact. If the mixtures (gas or liquid) are ideal, the excess (or non-ideality) terms are zero. Also, if the excess terms are small compared with the energy changes on chemical reaction (which will almost always be the case), these terms can usually be neglected. (Also, there will be some cancellation between the inlet and outlet nonideality terms, further diminishing the extent of their contribution.) Another approximation is that the contributions of the mixing terms $(RT_o \sum_i N_i \ln x_i)$ are also usually small compared with the energy changes on chemical reaction, and again there will be some cancellation of these terms between between the inlet and outlet streams. Therefore, for most engineering calculations it is sufficient to use

$$- W + T_o S_{gen} = \sum_i N_{i,in} \underline{\Psi}_{i,in} - T_o \sum_i N_{i,out} \underline{\Psi}_{i,out}$$

$$= \sum_i N_{i,in} \underline{\Psi}_i (T_{in}, P_{in}) - T_o \sum_i N_{i,out} \underline{\Psi}_i (T_{out}, P_{out})$$

though one should check whether a significantly different result is obtained using the more complete equation.

To obtain the maximum work output that can be obtained from a process, the following conditions should be met:

a. As already discussed, all heat flows to the environment should occur at the ambient temperature T_o.
b. All mass flows out of the system also should be at the ambient temperature T_o. Exhausting a stream at a higher temperature loses the potential to do work by, for example, operating a heat engine between the stream temperature and the ambient temperature.
c. All mass flows out of the system should occur at the ambient pressure P_o. Exhausting a stream at a higher pressure loses the potential to do work by, for example, flowing the stream through a work-producing turbine.
d. All processes should be reversible so that $S_{gen} = 0$.
e. The exiting streams should be in chemical equilibrium, as that is the state of minimum Gibbs energy; and since, from items (b) and (c) above, the exit streams should be at ambient conditions, the availability is equal to the Gibbs energy in this state.

Therefore, the maximum work that can be obtained is

$$-W^{max} = \sum_i N_{i,in} \underline{\Psi}_i (T_{in}, P_{in}, \underline{x}_{in}) - T_o \sum_i N_{i,out}^{eq} \underline{\Psi}_i (T_o, P_o, \underline{x}^{eq})$$

where the superscript eq has been used to indicate that the mole numbers and mole fractions of the outlet stream should be such that the stream is in chemical equilibrium at ambient conditions.

One measure of the efficiency of a work-producing process is the actual work obtained (regardless of the actual final temperature and pressure of the exit stream, which could be, for example, the exhaust of an automobile) compared with the maximum work that could have been obtained if the process was reversible, and the exhaust streams had been at ambient conditions, that is

$$\eta = \frac{-W^{\text{act}}}{-W^{\text{max}}} = \frac{-W^{\text{max}}}{\sum_i N_{i,\text{in}} \underline{\Psi}_i(T_{\text{in}}, P_{\text{in}}, \underline{x}_{\text{in}}) - T_o \sum_i N_{i,\text{out}}^{\text{eq}} \underline{\Psi}_i(T_o, P_o, \underline{x}^{\text{eq}})}$$

Note that since

$$\left(\frac{\partial \underline{H}}{\partial P}\right)_T = C_P \quad \text{and} \quad \left(\frac{\partial \underline{S}}{\partial P}\right)_T = \frac{C_P}{T}$$

it follows that

$$\left(\frac{\partial \underline{\Psi}}{\partial T}\right)_P = C_P \left(1 - \frac{T_o}{T}\right)$$

and

$$\underline{\Psi}(T_2, P) - \underline{\Psi}(T_1, P) = \int_{T_1}^{T_2} C_P \left(1 - \frac{T_o}{T}\right) dT$$

Also,

$$\left(\frac{\partial \underline{H}}{\partial P}\right)_T = \underline{V} - T\left(\frac{\partial \underline{V}}{\partial P}\right)_T \quad \text{and} \quad \left(\frac{\partial \underline{S}}{\partial P}\right)_T = -\left(\frac{\partial \underline{V}}{\partial P}\right)_T$$

so that

$$\left(\frac{\partial \underline{\Psi}}{\partial P}\right)_T = \underline{V} + (T_o - T)\left(\frac{\partial \underline{V}}{\partial P}\right)_T$$

and

$$\underline{\Psi}(T, P_2) - \underline{\Psi}(T, P_1) = \int_{P_1}^{P_2} \left[\underline{V} + (T_o - T)\left(\frac{\partial \underline{V}}{\partial P}\right)_T\right] dP$$

ILLUSTRATION 14.5-1

Computation of the Maximum Work That Can Be Obtained from a Chemical Reaction

In part of a sulfuric acid plant, a stream of pure SO_2 at 2 bar and 600 K flowing at a rate of 100 mol/s is mixed with a pure oxygen stream at 293 K, 1 bar, and a flow rate of 50 mol/s and reacted to form SO_3. The stream exiting that part of the plant is in chemical equilibrium at 1 bar and 293 K. Joe Udel is interested in improving the energy efficiency of this part of the process and suggests that instead a new type of work-producing reactor (i.e., a fuel cell) be installed that would accept the two pure streams, and work would be extracted. What is the maximum rate at which work could be obtained from this process if such a reactor could be developed?

SOLUTION

Using the program CHEMEQ (or, by hand calculations, using the data in Appendix A.IV) the equilibrium constant for the reaction

$$SO_2 + \tfrac{1}{2}O_2 \rightleftharpoons SO_3$$

is so large at 298.15 K that the reaction will go to completion, and only SO_3 will be present in the reactor exit stream. The steady-state energy and entropy balances for the reactor are

$$0 = 100\underline{H}_{SO_2}\ (2\ \text{bar},\ 600\ \text{K}) + 50\underline{H}_{O_2}\ (2\ \text{bar},\ 298\ \text{K}) - 100\underline{H}_{SO_3}\ (1\ \text{bar},\ 298\ \text{K}) + \dot{Q} + \dot{W}$$

and

$$0 = 100\underline{S}_{SO_2}\ (2\ \text{bar},\ 600\ \text{K}) + 50\underline{S}_{O_2}\ (1\ \text{bar},\ 298\ \text{K}) - 100\underline{S}_{SO_3}\ (1\ \text{bar},\ 298\ \text{K}) + \frac{\dot{Q}}{T} + \dot{S}_{\text{gen}}$$

For the maximum work that can be obtained, all heat transfer should occur at ambient temperature, $T_o = 298$ K, and there should be no irreversibilities (i.e., $\dot{S}_{\text{gen}} = 0$). Using this in the entropy balance, and then eliminating the heat flow between the energy and entropy balances gives

$$-\dot{W}^{\text{max}} = 100\left[\underline{H}_{SO_2}\ (2\ \text{bar},\ 600\ \text{K}) - T_o\underline{S}_{SO_2}\ (2\ \text{bar},\ 600\ \text{K})\right]$$
$$+ 50\left[\underline{H}_{O_2}\ (1\ \text{bar},\ 298\ \text{K}) - T_o\underline{S}_{O_2}\ (1\ \text{bar},\ 298\ \text{K})\right]$$
$$- 100\left[\underline{H}_{SO_3}\ (1\ \text{bar},\ 298\ \text{K}) - T_o\underline{S}_{SO_3}\ (1\ \text{bar},\ 298\ \text{K})\right]$$

or in terms of the availability function,

$$-\dot{W}^{\text{max}} = 100\underline{\Psi}_{SO_2}\ (2\ \text{bar},\ 600\ \text{K}) + 50\underline{\Psi}_{O_2}\ (1\ \text{bar},\ 298\ \text{K}) - 100\underline{\Psi}_{SO_3}\ (1\ \text{bar},\ 298\ \text{K})$$

For oxygen,

$$\underline{\Psi}_{O_2}\ (1\ \text{bar},\ 298\ \text{K}) = \underline{G}_{O_2}\ (1\ \text{bar},\ 298\ \text{K}) = \triangle_{f}\underline{G}_{O_2}\ (1\ \text{bar},\ 298\ \text{K}) = 0$$

and for SO_3,

$$\underline{\Psi}_{SO_3}\ (1\ \text{bar},\ 298\ \text{K}) = \underline{G}_{SO_3}\ (1\ \text{bar},\ 298\ \text{K}) = \triangle_{f}\underline{G}_{SO_3}\ (1\ \text{bar},\ 298\ \text{K}) = -371.1\ \frac{\text{kJ}}{\text{mol}}$$

since they are at ambient conditions. For sulfur dioxide, we have from Appendix A.IV

$$\underline{G}_{SO_2}\ (1\ \text{bar},\ 298\ \text{K}) = \triangle_{f}\underline{G}_{SO_2}\ (1\ \text{bar},\ 298\ \text{K}) = -300.2\ \frac{\text{kJ}}{\text{mol}}$$

and

$$\underline{H}_{SO_2}\ (1\ \text{bar},\ 298\ \text{K}) = \triangle_{f}\underline{H}_{SO_2}\ (1\ \text{bar},\ 298\ \text{K}) = -296.8\ \frac{\text{kJ}}{\text{mol}}$$

so that

$$\underline{S}_{SO_2}(1\ \text{bar},\ 298\ \text{K}) = \frac{\underline{H}_{SO_2}(1\ \text{bar},\ 298\ \text{K}) - \underline{G}_{SO_2}(1\ \text{bar},\ 298\ \text{K})}{T = 298\ \text{K}}$$
$$= \frac{-296.8 - (-300.2)}{298} = 11.41\ \frac{\text{J}}{\text{mol K}}$$

Now assuming ideal gas behavior,

$$\underline{H}_{SO_2}\ (2\ \text{bar},\ 600\ \text{K}) = \underline{H}_{SO_2}\ (1\ \text{bar},\ 298\ \text{K}) + \int_{298}^{600} C_{P}^{*}\ dT = -283.54\ \frac{\text{kJ}}{\text{mol}}$$

and

$$\underline{S}_{SO_2}(2 \text{ bar, } 600 \text{ K}) = \underline{S}_{SO_2}(1 \text{ bar, } 298 \text{ K}) + \int_{298}^{600} \frac{C_P^*}{T} \, dT - R \ln \frac{2 \text{ bar}}{1 \text{ bar}} = 36.67 \frac{J}{\text{mol K}}$$

so that

$$\underline{\Psi}_{SO_2}(2 \text{ bar, } 600 \text{ K}) = -283.54 - 298 \times 36.67 \times 10^{-3} = -294.46 \frac{\text{kJ}}{\text{mol}}$$

Finally,

$$-\dot{W}^{\max} = 100 \left(-2.9446 \times 10^5\right) + 50 \times 0 - 100 \left(-3.717 \times 10^5\right) = 0.7724 \times 10^7 = 7724 \frac{\text{kJ}}{\text{s}}$$

which is 7725 kW.

Finally, it is instructive to return to the energy balance to compute the amount of low-level (298 K) heat that would be released while obtaining the maximum work.

$$0 = 100 \times (-283.54) + 50 \times 0 - 100(-395.7) + \dot{Q} - 7724 \frac{\text{kJ}}{\text{s}}$$

so that $\dot{Q} = -3492$ kJ/s. Without the fuel cell, no work is produced and the heat released on this reaction would be $\dot{Q} = -(3491 + 7724) = -11215$ kJ/s. ∎

14.6 INTRODUCTION TO ELECTROCHEMICAL PROCESSES

Electrical work can be obtained from carefully controlled chemical reactions, but not if the reaction is allowed to proceed spontaneously. For example, in the standard state of pure gases of hydrogen and chlorine at 1 bar and 298 K, the reaction to form hydrogen chloride in aqueous solution

$$\tfrac{1}{2}H_2(g) + \tfrac{1}{2}Cl_2(g) = HCl(aq)$$

has a Gibbs energy change of

$$\Delta_{rxn}G° = -13.12 - [0 + 0] = -13.12 \frac{\text{kJ}}{\text{mol HCl}}$$

Therefore, there is a possibility of the reaction occurring spontaneously; however, this will produce no useful work. An alternative is to run the reaction in an electrolytic cell in which hydrogen and chlorine gases are metered into separate electrodes and the electromotive force (EMF) or voltage produced is almost balanced by the application of an external voltage. In this way the reaction will occur at a very slow rate and electrical work can be obtained from the cell. In this process chemical energy is directly converted to electrical energy.

Electrochemical processes occur in batteries, fuel cells, electrolysis, electrolytic plating, and corrosion (generally an undesirable process). Electrochemical processes can be used to produce electricity, to recover metals from solution, and for the measurement of the thermodynamic properties of electrolyte solutions. The device used to study electrochemical reactions is an **electrochemical cell**, which consists of two electrodes (metallic conductors) in electrolytes that are usually liquids containing salts, but may be solids, as in solid-state batteries. The two electrodes may be in the same electrolyte, as shown in Fig. 14.6-1a, or each electrode may be in a separate compartment with its

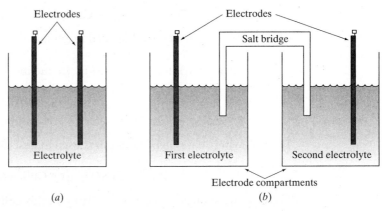

Figure 14.6-1 Two types of electrochemical cells. (*a*) A cell with two electrodes and a shared electrolyte. One example of such a cell contains a copper electrode, a zinc electrode, and a zinc sulfate and copper sulfate electrolyte solution. The overall cell reaction is $Cu^{2+}(aq) + Zn(s) \rightarrow Cu(s) + Zn^{2+}(aq)$. (*b*) A cell with two separate compartments connected by a salt bridge. If the same electrodes as in the previous case were used, one compartment would contain a copper electrode and a $CuSO_4$ solution, the other would have a zinc electrode and a $ZnSO_4$ solution as the electrolyte, and the two compartments would be connected by a bridge containing, for example, a sodium chloride solution.

own electrolyte, as in Fig. 14.6-1*b*. In this case the two compartments are connected by a salt bridge, an electrolyte that completes the electrical circuit. A third alternative, not shown, is for the two compartments to be in direct contact through a porous membrane.

When the two electrodes are connected through a potentiometer or electrical resistance, and electricity is produced by the chemical reaction that occurs spontaneously, the electrochemical cell is referred to as a galvanic cell; it is considered to be a fuel cell if the reagents are continually supplied to the cell. Batteries are galvanic cells. The term **electrolytic cell** is used to indicate an electrochemical cell operated in the reverse manner to that just described, in that an external voltage is used to cause a nonspontaneous reaction to occur, as in the electrolysis of water. An automobile battery and other storage batteries can be considered galvanic cells when they are supplying electricity, and electrolytic cells when they are being recharged. There are several ways that thermodynamics is used to analyze electrochemical cells. One is to compute the work, or equivalently the voltage, that can be produced by a galvanic cell. Alternatively, measured cell voltages can be used to determine the equilibrium constant of the reaction taking place within the cell. A third use of electrochemical cells is to measure the thermodynamic activity or activity coefficients of the ions in electrolyte solutions.

The main processes occurring in electrochemical cells are simultaneous oxidation and reduction reactions,[4] or redox reactions. At one electrode, the **anode**, a reduced species is oxidized here meaning to release electrons, while at the other electrode, the **cathode**, an oxidized species absorbs electrons and is reduced. It is common to think of an electrochemical cell as consisting of two half-cells (one containing the anode and the second containing the cathode) and to describe the processes in terms of half-cell reactions. For example, one common cell consists of a copper cathode in a copper sulfate solution, and a zinc anode in a zinc sulfate solution. The overall reaction is

---

[4]Here the term oxidation is being used as in general chemical terminology to indicate a release of electrons. In this sense it is not necessary for oxygen to be involved in an oxidation reaction, as unusual as this may seem.

$$Cu^{2+}(aq) + Zn(s) \rightarrow Cu(s) + Zn^{2+}(aq) \quad \text{(redox reaction)}$$

which is the sum of the two half-cell reactions

$$Cu^{2+}(aq) + 2e^- \rightarrow Cu(s) \quad \text{(reduction, absorption of electrons)}$$

and

$$Zn(s) \rightarrow Zn^{2+}(aq) + 2e^- \quad \text{(oxidation, release of electrons)}$$

The following abbreviated notation is used to describe the complete cell

$$Zn(s) \mid ZnSO_4(aq) \parallel CuSO_4(aq) \mid Cu(s)$$

The electrochemical processes occurring in this cell are the oxidation of zinc and the production of zinc sulfate and electrons at the anode, the absorption of electrons and the reduction and deposition of copper at the cathode, the flow of electrons through an external electrical circuit (resulting in electrical work), and a balancing flow of sulfate ions through the salt bridge.

The metallic electrode described above is the simplest of the electrode types. Another type of electrode is the insoluble-salt electrode, in which a metal is covered with one of its insoluble salts; silver chloride deposited on silver is one such example. The oxidation reaction in this case is

$$AgCl(s) + e^- \rightarrow Ag(s) + Cl^-(aq)$$

and the half-cell description is

$$Ag \mid AgCl \mid Cl^-$$

A third type of electrode is the gas electrode, in which a gas is in equilibrium with a solution of its ions in a half-cell that contains an inert metal conductor. The hydrogen electrode, in which one bubbles hydrogen through a solution and across a platinum electrode, is perhaps the best known of this type. The half-cell chemical reaction is

$$H_2(g) \rightarrow 2H^+(aq) + 2e^-$$

and the description used is

$$Pt \mid H_2 \mid H^+(aq)$$

Since hydrogen must be continually supplied from outside the cell, an electrochemical device using a hydrogen gas electrode would be considered a fuel cell.

From Sec. 4.3 we have that for any process occurring at constant temperature and pressure, the manner in which electrochemical cells are operated, the maximum work that can be obtained is equal to the change in Gibbs energy of the process, that is,

$$W^{max} = \Delta G \tag{14.6-1}$$

This maximum work is obtained if the process is sufficiently slow that there are no irreversibilities, for example, no resistive heating as a result of the current flow. This implies that the rate of reaction is very slow, and that the electrical potential produced is just balanced by an external potential so that the current flow is infinitesimal. This electrical potential produced by the cell (or of the balancing external potential) will be referred to as the zero-current cell potential and designated by E. The work done by the electrical cell W_{elec} in moving n moles of electrons across a potential difference of E is

$$W_{elec} = -nFE \qquad (14.6\text{-}2)$$

where $F = 96\,485$ C/mol is the Faraday constant, and the negative sign indicates that work is done by the cell on the surroundings if the cell potential is positive. [A coulomb (abbreviated C) is a unit of electrical charge; moving one coulomb of charge through a potential difference of one volt requires one joule of energy. Also, for later reference we note that at 25°C, the quantity RT/F is equal to 25.7 mV.] Consequently, we have

$$\Delta G = W^{max} = W_{elec} = -nFE$$

or simply

$$\Delta G = -nFE \qquad (14.6\text{-}3)$$

We write a generic electrochemical reaction as

$$\nu_{1+} M_{1+}^{z_1^+}(aq) + \nu_2 M_2(s) + \nu_1 M_1(s) + \nu_{2+} M_{2+}^{z_2^+}(aq) = 0$$

or, more generally (using our standard notation for chemical reactions),

$$\sum_i \nu_i M_i = 0$$

For example, the reaction

$$Cu^{2+}(aq) + Zn(s) \rightarrow Cu(s) + Zn^{2+}(aq)$$

will be written as $Zn^{2+}(aq) + Cu(s) - Zn(s) - Cu^{2+}(aq) = 0$, so that $\nu_{Zn^{2+}} = 1$, $\nu_{Cu} = 1$, $\nu_{Zn} = -1$, and $\nu_{Cu^{2+}} = -1$. Also, the Gibbs energy of any species can be written as

$$\bar{G}_i(T, P, \underline{x}) = \bar{G}_i^\circ(T, P^\circ, x_i^\circ) + RT \ln \frac{\bar{f}_i(T, P, \underline{x})}{\bar{f}_i^\circ(T, P^\circ, x_i^\circ)} \qquad (14.6\text{-}4)$$

$$= \bar{G}_i^\circ(T, P^\circ, x_i^\circ) + RT \ln a_i(T, P, \underline{x})$$

where P° and x_i° are the standard-state pressure and composition (in units appropriate to the standard state chosen), $\bar{f}_i^\circ(T, P^\circ, x_i^\circ)$ and $\underline{G}_i(T, P^\circ, x_i^\circ)$ are the standard-state fugacity and Gibbs energy of species i, and a_i is its activity. Combining these last two equations, we have

$$\Delta_{rxn}G = \Delta_{rxn}G^\circ + RT \ln \prod_i a_i^{\nu_i} = -nFE = -nFE^\circ + RT \ln \prod_i a_i^{\nu_i} \qquad (14.6\text{-}5)$$

where E° would be the zero-current cell potential if the ions were in their standard states, while E is the actual (measurable) zero-current cell potential with the ions at the concentration of the cell. The last relation is known as the **Nernst equation**.

Remembering from Eq. 13.1-18 that

$$-\frac{\Delta_{rxn}G^\circ}{RT} = \ln K_a \qquad (13.1\text{-}8)$$

Table 14.6-1 Standard Half-Cell Potentials at 25°C

Oxidizing Agent or Oxidant		Reducing Agent or Reductant	$E°$ (V)
Au^+	$+e^-$	Au	+1.69
Cl_2	$+2e^-$	$2Cl^-$	+1.36
Br_2	$+2e^-$	$2Br^-$	+1.09
Ag^+	$+e^-$	Ag	+0.80
Hg_2^{2+}	$+2e^-$	2Hg	+0.79
Fe^{3+}	$+e^-$	Fe^{2+}	+0.77
Cu^{2+}	$+2e^-$	Cu	+0.34
AgCl	$+e^-$	$Ag + Cl^-$	+0.22
$2H^+$	$+2e^-$	H_2	0.0 (by definition)
Fe^{3+}	$+3e^-$	Fe	−0.04
Pb^{2+}	$+2e^-$	Pb	−0.13
Zn^{2+}	$+2e^-$	Zn	−0.76
Al^{3+}	$+3e^-$	Al	−1.66
Mg^{2+}	$+2e^-$	Mg	−2.36
Na^+	$+e^-$	Na	−2.71
Li^+	$+e^-$	Li	−3.05

where K_a is the chemical equilibrium constant, we have

$$\ln K_a = \frac{nFE°}{RT} \tag{14.6-6}$$

This equation allows one to compute the chemical equilibrium constant from measured standard-state electrochemical cell potentials (usually referred to as standard cell potentials). Some standard half-cell potentials are given in Table 14.6-1. The standard potential of an electrochemical cell is obtained by combining the two relevant half-cell potentials.

One point to note when using this table is that

$$\Delta_{rxn}G° = -nFE° \quad \text{or} \quad E° = -\frac{\Delta_{rxn}G°}{nF}$$

where n is the number of electrons transferred. Consequently, for example, for the reaction $Cl_2 + 2e^- \rightarrow 2Cl^-$, $E°$ is 1.36 V, and $\Delta_{rxn}G° = -2 \times 1.36F = -2.72F$. However, if instead we wrote the reaction as $\frac{1}{2}Cl_2 + e^- \rightarrow Cl^-$, then $\Delta_{rxn}G°$ would be one-half its previous value, or $-1.36F$. However, now n equals unity, so that again

$$E° = -\frac{\Delta_{rxn}G°}{nF} = -\frac{(-1.36F)}{1 \cdot F} = 1.36 \text{ V}$$

Consequently, if we multiply any of the half-cell reactions in Table 14.6-1 by an integer or fractional constant, the standard-state Gibbs energy change will change by that same factor, but the standard-state half-cell potential will be unchanged.

ILLUSTRATION 14.6-1

Computation of the Standard Cell Potential

Compute the standard cell potential and the equilibrium constant for the reaction

$$Cu^{2+}(aq) + Zn(s) \rightarrow Cu(s) + Zn^{2+}(aq)$$

SOLUTION

The reaction above is the sum of the two half-cell reactions

$$Cu^{2+}(aq) + 2e^- \rightarrow Cu(s) \quad \text{half-cell standard potential} = +0.34 \text{ V}$$
$$Zn(s) \rightarrow Zn^{2+}(aq) + 2e^- \quad \text{half-cell standard potential} = -(-0.76 \text{ V}) = +0.76 \text{ V}$$

where the negative of the reported half-cell standard potential has been used for the second reaction, since its direction is opposite to that given in the table. Therefore, the standard cell potential for the overall reaction is

$$E^\circ = +0.34 + 0.76 = +1.10 \text{ V}$$

The equilibrium constant for this reaction is then

$$\ln K_a = \frac{nFE^\circ}{RT} = \frac{2 \times 9.6485 \times 10^4 \dfrac{C}{mol} \times 1.10 \text{ V}}{8.314 \dfrac{J}{mol \text{ K}} \times 298.15 \text{ K} \times 1 \dfrac{C \text{ V}}{mol}} = 85.6$$

or

$$K_a = 1.5 \times 10^{37}$$

or equivalently,

$$\ln K_a = \frac{nFE^\circ}{RT} = \frac{nE^\circ}{\left(\dfrac{RT}{F}\right)} = \frac{2 \times 1.10 \text{ V} \times \dfrac{1000 \text{ mV}}{V}}{25.7 \text{ mV}} = 85.6$$

or

$$K_a = 1.5 \times 10^{37}$$

∎

ILLUSTRATION 14.6-2

Calculation of the Equilibrium Constant from Standard Half-Cell Potentials

Determine the equilibrium constant for the dissolution and dissociation of silver chloride in water, and the silver chloride solubility in water.

SOLUTION

The reaction is

$$AgCl \rightarrow Ag^+ + Cl^-$$

which in terms of half-cell reactions we write as

$$AgCl + e^- \rightarrow Ag + Cl^- \quad \text{half-cell standard potential} = +0.22 \text{ V}$$
$$Ag \rightarrow Ag^+ + e^- \quad \text{half-cell standard potential} = -(+0.80 \text{ V})$$

Thus $E^\circ = +0.22 - (+0.80) = -0.58$ V, and

$$\ln K_a = \frac{nFE^\circ}{RT} = \frac{9.6485 \times 10^4 \dfrac{C}{mol} \times (-0.58) \text{V}}{8.314 \dfrac{J}{mol \text{ K}} \times 298.15 \text{ K} \times 1 \dfrac{C \text{ V}}{mol}} = -22.568$$

or

$$K_a = 1.58 \times 10^{-10}$$

which compares well with the value of 1.607×10^{-10} computed in Illustration 13.3-2. The equilibrium relation is then

$$K_a = 1.58 \times 10^{-10} = \frac{a_{Ag^+} a_{Cl^-}}{a_{AgCl}} = M_{Ag^+} M_{Cl^-} = \left(M_{Ag^+}\right)^2$$

so that

$$M_{Ag^+} = M_{AgCl} = 1.257 \times 10^{-5} \text{ M} = 1.257 \times 10^{-5} \frac{\text{mol}}{\text{kg water}}$$

In writing these last equations we have recognized that the activity of the pure silver chloride solid is unity, assumed that the ion concentrations will be so low that the activity coefficients would be unity, and used the fact that by stoichiometry the concentrations of the silver and chloride ions must be equal. ∎

Following upon the illustration above, quite generally we can write

$$
\begin{aligned}
\Delta G &= \Delta_{\text{rxn}} G^\circ + RT \ln \prod_i a_i^{\nu_i} = -nFE \\
&= \Delta_{\text{rxn}} G^\circ + RT \ln a_{M_1(s)}^{\nu_1} a_{M_2^+(aq)}^{\nu_2+} a_{M_1^+(aq)}^{\nu_1+} a_{M_2(s)}^{\nu_2} \\
&= -RT \ln K_a + RT \ln a_{M_1(s)}^{\nu_1} a_{M_2^+(aq)}^{\nu_2+} a_{M_1^+(aq)}^{\nu_1+} a_{M_2(s)}^{\nu_2} \\
&= \Delta_{\text{rxn}} G^\circ + RT \ln a_{M_2^+(aq)}^{\nu_2+} a_{M_1^+(aq)}^{\nu_1+} \\
&= -RT \ln K_a + RT \ln a_{M_2^+(aq)}^{\nu_2+} a_{M_1^+(aq)}^{\nu_1+} \\
&= -nFE^\circ + RT \ln a_{M_2^+(aq)}^{\nu_2+} a_{M_1^+(aq)}^{\nu_1+}
\end{aligned}
\tag{14.6-7}
$$

In writing these equations we have recognized that the activities of the pure metal electrodes are unity. Note that if the electrochemical cell is in chemical equilibrium, that is, if

$$K_a = a_{M_2^+(aq)}^{\nu_2+} a_{M_1^+(aq)}^{\nu_1+} \tag{14.6-8}$$

then E is equal to zero (there is no voltage produced by the cell even though E° is nonzero), and $\Delta G = 0$, as must be the case for a process at equilibrium.

We have shown earlier that it is possible to produce work if there is a temperature difference between two subsystems (for example, by connecting them through a Carnot or other cycle). Similarly, if there is a pressure difference between two subsystems, this can be used to drive a fluid through a turbine, again producing work. An electrochemical cell is one way of producing work if there is a concentration difference between two subsystems. To see how this could be done, suppose we had two beakers containing a salt, say copper sulfate, at different concentrations. We could put a metallic copper electrode connected to a potentiometer or other electrical circuit in each beaker and then connect the electrolytes in the beakers by a salt bridge. In the beaker containing the dilute copper sulfate solution, which we designate as beaker 1, the reaction would be

$$Cu \rightarrow Cu^{2+} + 2e^-$$

while in the beaker containing the concentrated solution, which we will refer to as beaker 2, the reaction would be

$$Cu^{2+} + 2e^- \rightarrow Cu$$

In this case the two half-cell potentials cancel, but as a result of the concentration difference, we have

$$\Delta G = nFE = RT \ln \frac{\left[a_{Cu^{2+}} a_{SO_4^{2-}} \right]_1}{\left[a_{Cu^{2+}} a_{SO_4^{2-}} \right]_2} = RT \ln \frac{\left[M_{Cu^{2+}} M_{SO_4^{2-}} \gamma_{Cu^{2+}} \gamma_{SO_4^{2-}} \right]_1}{\left[M_{Cu^{2+}} M_{SO_4^{2-}} \gamma_{Cu^{2+}} \gamma_{SO_4^{2-}} \right]_2}$$

$$= RT \ln \frac{\left[M_{Cu^{2+}} M_{SO_4^{2-}} \gamma_{\pm}^2 \right]_1}{\left[M_{Cu^{2+}} M_{SO_4^{2-}} \gamma_{\pm}^2 \right]_2}$$

$$(14.6\text{-}9)$$

To proceed, we use that by stoichiometry and from the fact that copper sulfate is fully dissociated into ions,

$$M_{Cu^{2+}} = M_{SO_4^{2-}} = M_{CuSO_4}$$

Also, for the purposes of illustration, we will use the simple Debye-Hückel limiting law of Eq. 9.10-15 for the mean ionic activity coefficient, which here becomes

$$\ln \gamma_{\pm} = -\alpha |z_+ z_-| \sqrt{I} = -\alpha 4 \sqrt{4 M_{CuSO_4}} = -\alpha 8 \sqrt{M_{CuSO_4}}$$

since

$$I = \frac{1}{2}(M_+ z_+^2 + M_- z_-^2) = \frac{1}{2} M_{CuSO_4} \left((2)^2 + (-2)^2 \right) = 4 M_{CuSO_2}$$

Putting this all together gives

$$\Delta G = -nFE = -2FE = RT \ln \frac{\left[M_{Cu^{2+}} M_{SO_4^{2-}} \gamma_{\pm}^2 \right]_1}{\left[M_{Cu^{2+}} M_{SO_4^{2-}} \gamma_{\pm}^2 \right]_2} = RT \ln \frac{\left[M_{CuSO_4}^2 \gamma_{\pm}^2 \right]_1}{\left[M_{CuSO_4}^2 \gamma_{\pm}^2 \right]_2}$$

$$= 2RT \ln \frac{\left[M_{CuSO_4} \right]_1}{\left[M_{CuSO_4} \right]_2} + 2RT \ln \frac{\left[\gamma_{\pm} \right]_1}{\left[\gamma_{\pm} \right]_2}$$

$$= 2RT \ln \frac{\left[M_{CuSO_4} \right]_1}{\left[M_{CuSO_4} \right]_2} - RT\alpha 8 \left(\sqrt{[I]_1} - \sqrt{[I]_2} \right)$$

$$= 2RT \ln \frac{\left[M_{CuSO_4} \right]_1}{\left[M_{CuSO_4} \right]_2} - RT\alpha 16 \left(\sqrt{\left[M_{CuSO_4} \right]_1} - \sqrt{\left[M_{CuSO_4} \right]_2} \right)$$

$$(14.6\text{-}10)$$

ILLUSTRATION 14.6-3

Computing the Cell Voltage That Is Produced as a Result of the Concentration Difference

One beaker contains copper sulfate at a concentration of 0.0001 M and another contains a 0.01-M solution of the same salt. Compute the maximum voltage that could be obtained at 25°C with an electrochemical cell that used these two solutions as electrolytes.

SOLUTION

The starting point is the preceding equation written as follows:

$$
\begin{aligned}
-E &= \frac{RT}{F} \ln \frac{[M_{CuSO_4}]_1}{[M_{CuSO_4}]_2} - \frac{RT}{F} \cdot \alpha \cdot 8 \cdot \left(\sqrt{[M_{CuSO_4}]_1} - \sqrt{[M_{CuSO_4}]_2} \right) \\
&= 25.7 \text{ mV} \left[\ln \frac{0.0001 \text{ M}}{0.01 \text{ M}} - 1.178 \cdot 8 \cdot \left(\sqrt{0.0001} - \sqrt{0.01} \right) \right] \\
&= 25.7 \text{ mV} \left[\ln 0.01 - 1.178 \cdot 8 \cdot (0.01 - 0.1) \right] \\
&= 25.7 \text{ mV}[-4.605 + 0.848] = -96.5 \text{ mV}
\end{aligned}
$$

so that $E = 96.5$ mV. ∎

When a half-cell reaction involves hydrogen ions, the cell potential will depend upon the hydrogen ion concentration of the solution, or the pH, where the **pH scale** is defined as follows:

$$
\text{pH} = -\log a(\text{H}^+) = -\log a(\text{H}_3\text{O}^+) \approx -\log \left[\frac{M_{\text{H}^+}}{1 \text{ molal}} \right] \tag{14.6-11}
$$

In writing this equation we have recognized that in solution the hydrogen ion is actually present as a hydronium ion, and the last expression is valid only if the hydrogen-ion concentration is so low that its activity coefficient is unity. It is interesting to compute how the cell potential varies with changes in pH. To do this we consider an electrochemical cell that contains a hydrogen electrode, and leave the other half-cell reaction unspecified. The overall cell reaction is written as

$$
\text{H}_2(\text{g}) + 2\text{M}^+ = 2\text{H}^+ + 2\text{M}(\text{s})
$$

for which the cell potential will be

$$
\begin{aligned}
E &= E^\circ + \frac{RT}{2F} \ln \left[\frac{a_{\text{H}^+}^2 a_{\text{M}}^2}{a_{\text{H}_2} a_{\text{M}^+}^2} \right] = E^\circ + \frac{RT}{2F} \ln \left[\frac{a_{\text{H}^+}^2}{a_{\text{H}_2} a_{\text{M}^+}^2} \right] \\
&= E^\circ + \frac{RT}{F} \ln [a_{\text{H}^+}] - \frac{RT}{2F} \ln[a_{\text{H}_2} a_{\text{M}^+}^2] \\
&= E^\circ - 2.303 \frac{RT}{F} \text{pH} - \frac{RT}{2F} \ln[a_{\text{H}_2} a_{\text{M}^+}^2]
\end{aligned} \tag{14.6-12}
$$

At 25°C we have

$$
E = E^\circ - 59.2 \cdot \text{pH} - 12.86 \cdot \ln[a_{\text{H}_2} a_{\text{M}^+}^2] \quad \text{mV} \tag{14.6-13}
$$

We see from this equation that the actual potential produced by an electrochemical cell involving a hydrogen (or hydronium) ion depends linearly on the pH of the solution. The total cell potential also depends on the activity of the metal ion in the other half-cell, which usually would be approximately constant, and the activity of molecular hydrogen, which can be controlled by its partial pressure. Consequently, the primary variation of the cell potential is with pH. This result suggests that an electrochemical cell can be used to measure the pH of a solution. This is actually done in the laboratory, but by using specially chosen liquid electrodes rather than a hydrogen gas electrode, which is not convenient to use.

By the appropriate choice of an electrochemical cell, it is possible to measure the mean ionic activity coefficient of an electrolyte. As an example, consider a cell consisting of a hydrogen electrode and a silver–silver chloride electrode, both in the same solution with hydrochloric acid as the electrolyte. It is the mean ionic activity coefficient of hydrochloric acid that can be measured. The overall cell reaction is

$$\tfrac{1}{2}H_2(g) + AgCl(s) = HCl(aq) + Ag(s)$$

The cell potential for this reaction, since $n = 1$, is

$$E = E^\circ - \frac{RT}{2F} \ln\left[\frac{a_{HCl(aq)} a_{Ag(s)}}{a_{H_2(g)}^{0.5} a_{AgCl(s)}}\right] = E^\circ - \frac{RT}{F} \ln\left[\frac{a_{HCl(aq)}}{a_{H_2(g)}^{0.5}}\right] \qquad (14.6\text{-}14)$$

since the solids silver and silver chloride are at unity activity. Further, the activity of hydrogen is regulated by its partial pressure, so that we can consider it to be independently fixed and known. Therefore,

$$E = E^\circ + \frac{RT}{2F} \ln\left[a_{H_2(aq)}\right] - \frac{RT}{F} \ln\left[a_{HCl(aq)}\right] \qquad (14.6\text{-}15)$$

But

$$a_{H_2(g)} = \frac{P_{H_2}}{1\ \text{bar}}$$

and

$$a_{HCl} = (a_{H^+})(a_{Cl^-}) = \frac{M_{H^+} M_{Cl^-} \gamma_{H^+} \gamma_{Cl^-}}{(M = 1\ \text{molal})^2}$$

$$= \left(\frac{M_\pm}{M = 1\ \text{molal}}\right)^2 \gamma_\pm^2 = \left(\frac{M_{HCl}}{M = 1\ \text{molal}}\right)^2 \gamma_\pm^2$$

since the hydrochloric acid, being a strong electrolyte, is fully ionized. Consequently,

$$E = E^\circ + \frac{RT}{2F} \ln\left[\frac{P_{H_2}}{1\ \text{bar}}\right] - \frac{2RT}{F} \ln\left[\frac{M_{HCl}}{M = 1\ \text{molal}}\right] - \frac{2RT}{F} \ln \gamma_\pm \qquad (14.6\text{-}16)$$

or

$$\ln \gamma_\pm = \frac{F}{2RT}\left[E^\circ - E + \frac{RT}{2F} \ln\left[\frac{P_{H_2}}{1\ \text{bar}}\right] - \frac{2RT}{F} \ln\left[\frac{M_{HCl}}{M = 1\ \text{molal}}\right]\right]$$

$$= \frac{F}{2RT}(E^\circ - E) + \frac{1}{4} \ln\left[\frac{P_{H_2}}{1\ \text{bar}}\right] - \ln\left[\frac{M_{HCl}}{M = 1\ \text{molal}}\right] \qquad (14.6\text{-}17)$$

So that by fixing the value of the partial pressure of hydrogen and measuring the cell potential E, the value of the mean ionic activity coefficient of hydrogen chloride can be determined.

PROBLEMS

(*Note:* The *Chemical Engineer's Handbook,* McGraw-Hill, New York, contains a comprehensive list of standard-state Gibbs energies and enthalpies of formation.)

14.1 The flame temperature attained in a torch or a burner can be computed using the adiabatic reaction temperature analysis of Sec. 14.3 if it is assumed that the radiant heat loss from the flame is negligible. Compute the flame temperature and exit composition in a hydrogen torch if

a. A stoichiometric amount of pure oxygen is used as the oxidant.

b. Oxygen is the oxidant, but a 100 percent excess is used.

c. Twice the stoichiometric amount of air is used as the oxidant.

In each case the hydrogen and oxidant entering the torch are at room temperature (298.15 K), and the torch pressure is 1.013 bar.

14.2 Compute the flame temperature of an oxyacetylene torch using pure acetylene and 50 percent more pure oxygen than is needed to convert all the acetylene to carbon dioxide and water. Both the oxygen and acetylene are initially at room temperature and atmospheric pressure. The following reactions may occur:

$$HCCH + \tfrac{3}{2}O_2 = 2CO + H_2O$$
$$HCCH + \tfrac{5}{2}O_2 = 2CO_2 + H_2O$$
$$CO + \tfrac{1}{2}O_2 = CO_2$$
$$CO + H_2O = CO_2 + H_2$$
$$HCCH = 2C(s) + H_2$$

14.3 One mole of ethylene and one mole of benzene are fed to a constant-volume batch reactor and heated to 600 K. On the addition of a catalyst, an equilibrium mixture of ethylbenzene, benzene, and ethylene is formed:

$$C_6H_6(g) + C_2H_4(g) \rightleftharpoons C_6H_5C_2H_5(g)$$

The pressure in the reactor, before addition of the catalyst (i.e., before any reaction has occurred), is 1.013 bar. Calculate the equilibrium conversion and the heat that must be removed to maintain the reactor temperature constant at 600 K.

14.4 How would you determine the entropy change and the enthalpy change of an electrochemical cell reaction?

14.5 One of the purposes of the kidneys is to transfer useful chemicals from the urine to the blood, and toxins from the blood to the urine. In the transport of glucose from the urine to the blood, the kidneys are transporting glucose against a concentration gradient (that is, the direction of transport is from a low concentration to a high concentration). This can occur only because the transport is coupled with a chemical reaction, a process called active transport. If the concentration of glucose initially in the urine is 5×10^{-5} mol/kg and after leaving the kidney is 5×10^{-6} mol/kg, and the concentration of glucose in the blood (which has a much larger volume) is approximately constant at 5×10^{-3} mol/kg, compute the minimum work

that must be done (or Gibbs energy supplied) per mole of glucose transported across the kidney.

14.6 There are two beakers, each one liter in total volume. One of the beakers contains copper sulfate at a concentration of 0.0001 M and the other contains a 0.01-M solution of the same salt. Compute the maximum total work that can be obtained at 25°C with an electrochemical cell that used these two solutions as electrolytes.

14.7 Estimate the maximum amount of work that can be obtained from the combustion of gasoline, which we will take to be represented by n-octane (C_8H_{18}), in an automobile engine. For this calculation, assume that n-octane vapor and a stoichiometric amount of air (21 vol % oxygen, 79 vol % nitrogen) initially at 1 bar and 25°C react to completion, and that the exit gas is at 1 bar and 150°C. (You may want to compare this result with that of Problem 8.38.)

14.8 Methane is to be burned in air. Determine the adiabatic flame temperature as a function of the methane-to-air ratio at a pressure of 1 bar.

14.9 Estimate the maximum amount of work that can be obtained from the controlled combustion of methane in air at 1 bar as a function of the methane-to-air ratio. Assume that the process is as follows. First the methane is burned adiabatically so that the adiabatic flame temperature is obtained. Next a Carnot cycle is used to extract heat from the combustion products until they are cooled to 25°C. Note that the work cannot all be extracted from the combustion gases at the adiabatic flame temperature, but rather is extracted over a range of temperatures starting at the adiabatic flame temperature and ending at 25°C.

14.10 Compare your answer above with the maximum amount of work that could be obtained from methane if a nonthermal energy conversion route, such as a fuel cell, was used.

14.11 **a.** What is the solubility product for silver sulfate, Ag_2SO_4, in water?

b. An electrolytic silver-producing cell consists of a copper cathode, a silver anode, and a solution that initially contains Ag_2SO_4 at its solubility limit and 0.5 M $CuSO_4$. Compute the minimum electrical potential that must initially be applied to electrolytically deposit silver in this cell.

14.12 Redo Problem 14.3 with the change that the reactor is to operate at a constant pressure of 1.0 bar (instead of at constant volume) and that the initial pressure is 1.0 bar.

14.13 Equal amounts of pure nitrogen and pure oxygen, each at 3000 K and 1 bar, are continuously fed into a chemical reactor, and the reactor effluent, consisting of the reaction product nitric oxide and unreacted nitrogen and oxygen, is continually withdrawn.

Assuming that the reactor is adiabatic and that the reactor effluent is in chemical equilibrium, determine the temperature and composition of the effluent.

14.14 An unsecured tank contains 20 kg of *n*-butane at its vapor pressure at 25°C. The cylinder falls over,

breaking off the valve and releasing the entire contents of the cylinder, and the resulting vapor cloud comes in contact with an ignition source and explodes. Estimate the energy released.

14.15 Derive Eq. 14.3-17.

Chapter 15

Some Biochemical Applications of Thermodynamics

Though most of the applications of thermodynamics in this textbook have dealt with chemicals and petrochemicals, there have been a few examples dealing with biochemical processes. In this chapter we focus on the use of the principles of thermodynamics, that is, mass, energy and entropy balances, and the concept of the equilibrium state, to some applications involving biochemical reactions, biochemical processing, and processes occurring in living cells. These cases are somewhat more complicated to deal with than those of traditional chemical processing because of the large number of chemical species that are involved, and because many of the species (cells, proteins, enzymes, etc.) may be incompletely specified. Also, aqueous solutions of electrolytes are generally involved, and pH and ionic strength have significant effects on the thermodynamics of such systems, including the solubility and biological activity of proteins, cells, and other biomaterials, and on the extent of biological reactions. Therefore, we start with a discussion of the acidity of solutions and pH, which you may have encountered in courses in general chemistry and physical chemistry. We then move on to a number of applications of thermodynamics to biological, physiological, and biochemical processes. There are several very good references on the subject of biothermodynamics.[1,2]

INSTRUCTIONAL OBJECTIVES FOR CHAPTER 15

The goals of this chapter are for the student to:

- Understand the concepts of pH and acidity (Sec. 15.1)
- Be able to calculate the pH of strong and weak acids and bases (Sec. 15.1)
- Be able to compute the extent of ionization of chemicals and biochemicals as a function of pH (Secs. 15.1 and 15.2)

---

[1] E. J. Cohn and J. T. Edsall, *Proteins, Amino Acids and Peptides,* Reinhold, New York (1943).

[2] J. T. Edsall and H. Gutfreund, *Biothermodynamics: The Study of Biochemical Processes at Equilibrium,* John Wiley & Sons, New York (1983).

- Be able to calculate the solubility of amino acids, pharmaceuticals, and other ionizable compounds as a function of pH (Sec. 15.3)
- Be able to compute the equilibrium state in biochemical reactions such as ligand binding and the denaturation (unfolding) of proteins (Secs. 15.4 and 15.5)
- Be able to calculate the effect of pH on biochemical reactions (Sec. 15.5)
- Understand how proteins can be concentrated in an ultracentrifuge (Sec. 15.6)
- Be able to compute the equilibrium state, osmotic pressure, and membrane potentials of proteins and other charged species (Gibbs-Donnan equilibrium) (Sec. 15.7)
- Be able to develop a thermodynamic description of fermenters and other biochemical reactors (Sec. 15.9)

NOTATION INTRODUCED IN THIS CHAPTER

$\mathcal{H}$	Stoichiometric coefficient of hydrogen in a biochemical compound
$\mathcal{O}$	Stoichiometric coefficient of oxygen in a biochemical compound
$\mathcal{N}$	Stoichiometric coefficient of nitrogen in a biochemical compound
$\mathcal{C}$	Stoichiometric coefficient of carbon in a biochemical compound
K	An apparent chemical equilibrium constant based on concentration ratios
$\mathcal{K}$	A product of equilibrium constants
pH	$= -\log (a_{H+})$
pOH	$= -\log (a_{OH-})$
pK	$= -\log (K)$
M	Molality concentration unit, as in a 1-M solution
M_i	Concentration of species i in molality units of moles per kilogram of solvent (Note that if the solvent is water, the solution is dilute, and the temperature is near 25°C, molality is equal to molarity, which is moles per liter of solution.)
S_o	Saturation solubility of an electrically neutral substance (M)
S_T	Total solubility of the neutral and ionized forms of a substance (M)
$Y_{B/S}$	C-moles of biomass produced per C-mole of substrate consumed
$Y_{P/S}$	C-moles of product per C-mole of substrate consumed
$Y_{N/S}$	Moles of nitrogen source consumed per C-mole of substrate consumed
$Y_{O_2/S}$	Moles of O_2 consumed per C-mole of substrate consumed
$Y_{W/S}$	Moles of water consumed per C-mole of substrate consumed
$Y_{C/S}$	Moles of CO_2 produced per C-mole of substrate consumed
$Y_{Q/S}$	Heat flow per C-mole of substrate consumed
Z	Charge on a protein
θ	Extent of coverage in ligand binding
ω	Rotational speed in an ultracentrifuge (1/s)
ξ	Generalized degree of reduction

15.1 ACIDITY OF SOLUTIONS

Water, though very polar, is only very slightly ionized. The ionization of water can be considered a chemical reaction that occurs as follows:

$$H_2O \rightleftharpoons H^+ + OH^-$$

or

$$2H_2O \rightleftharpoons H_3O^+ + OH^- \tag{15.1-1}$$

For simplicity, we will use the first of these expressions; the thermodynamic equilibrium constant for that reaction is written as

$$K_{a,W} = \frac{a_{H^+} \cdot a_{OH^-}}{a_W} \tag{15.1-2a}$$

where the symbol a_i is, as usual, the activity of species i. The pure-component standard state is used for water. As many cases of interest in biochemical processing involve solutions that are mostly water and relatively dilute in the other species, the activity of water is usually taken to be unity (i.e., $a_W = 1$).

The ideal one-molal (1 M) standard state is used for the ions.[3] For solutions that are very dilute in the ions (i.e., water without an added electrolyte), we can neglect the solution nonidealities and replace the activities with concentrations in terms of molalities,

$$K_{a,W} = \left(\frac{M_{H^+}}{M=1}\right)\left(\frac{M_{OH^-}}{M=1}\right) = \frac{M_{H^+} \cdot M_{OH^-}}{(M=1)^2} \tag{15.1-2b}$$

though we can correct for this assumption using the methods in Sec. 9.10. For mathematical simplicity we will do so in some, but not all, of the illustrations in this chapter. Values of $K_{a,W}$ are given in Table 15.1-1 as a function of temperature.

Since the values of the equilibrium constants are so small, it is common to use the following notation:

$$pK_W = -\log(K_{a,W}) \tag{15.1-3}$$

Table 15.1-1 Values of $K_{a,W}$ and pK_W as a Function of Temperature

T (°C)	$K_{a,W}$	pK_W
0	1.15×10^{-15}	14.94
10	2.88×10^{-15}	14.54
20	6.76×10^{-15}	14.17
25	1.00×10^{-14}	14.00
30	1.48×10^{-14}	13.83
40	2.88×10^{-14}	13.54
50	5.50×10^{-14}	13.26
60	9.55×10^{-14}	13.02

[3]In this chapter (as in Chapter 9) molality, which is number of moles of solute per kilogram of solvent and indicated by the symbol M, will be used. Molarity, defined as the number of moles of solute per liter of solution, is another commonly used concentration unit, but can be somewhat more difficult to deal with since the volume of a solution varies with composition and temperature. However, if the solvent is water and the solution is dilute in solute (so that one liter of solution contains one kilogram of water), as is generally the case in this chapter, molality and molarity are equal. Therefore, in some of the calculations that follow, especially the titration calculations in this section, we may ignore the distinction between molality (moles of solute per kilogram of water) and molarity (moles of solute per liter of solution.)

where a base 10 logarithm is used, and the minus sign is included in the definition so that the value of pK_W is positive. Values of pK_W are also given in the table.

Since in water without any added acids (e.g., HCl or H_2SO_4) or added bases (e.g., NaOH, NH_4OH, or sodium acetate) the number of hydrogen ions equals the number of hydroxyl ions (i.e., $M_{H+} = M_{OH-}$), we have

$$10^{-14} = \frac{M_{H+} \cdot M_{OH-}}{(M = 1)^2} = \frac{(M_{H+})^2}{(M = 1)^2}$$

so that

$$M_{H+} = M_{OH-} = 10^{-7} \text{ M}$$

Another definition frequently used is that of the pH,

pH

$$\text{pH} = -\log(a_{H+}) \qquad \textbf{(15.1-4)}$$

Again, simplifying by neglecting activity coefficient departure from unity for a solution of pure water, which has an ionic strength of 10^{-7} M at 25°C, we have

$$\text{pH} = -\log\left(\frac{10^{-7} \text{ M}}{1 \text{ M}}\right) = 7.0$$

Similarly,

pOH

$$\text{pOH} = -\log(a_{OH-}) = -\log\left(\frac{10^{-7} \text{ M}}{1 \text{ M}}\right) = 7.0 \qquad \textbf{(15.1-5)}$$

Note that in the presence of added acids, bases, or salts, the ionic strength I will be much larger. In these cases it is generally necessary to correct for solution nonideality so that the pH of a solution is related to the hydrogen ion activity, and is not simply its molality.

A strong acid or base is one that completely ionizes when added to water. Examples include

$$\text{NaOH} \rightarrow \text{Na}^+ + \text{OH}^-$$

$$\text{HCl} \rightarrow \text{H}^+ + \text{Cl}^-$$

and

$$\text{H}_2\text{SO}_4 \rightarrow 2\text{H}^+ + \text{SO}_4^{2-}$$

Consequently, when a strong acid or base is added to water, the water ionization equilibrium shifts, and the pH of the solution changes due to the added H^+ ions (for an acid) or hydroxyl ions OH^- (for a base).

ILLUSTRATION 15.1-1

The pH of a Solution of a Strong Acid and a Strong Base

What is the pH for each of the following solutions at 25°C?

a. 0.05 M solution of HCl
b. 0.05 M solution of NaOH

SOLUTION

a. Hydrochloric acid ionizes completely, producing a solution of 0.05 M in H^+ and 0.05 M in Cl^-. The ionization of water also produces some hydrogen ions, but, by Le Chatelier's principle for the common ion effect discussed in Chapter 13, the concentration of hydrogen ions produced from water will be less than 10^{-7}. Therefore, we will neglect this source of hydrogen ions, so $M_{H^+} = 0.05$ M, and as an approximation $a_{H^+} = 0.05$ M, so that pH = 1.30.

To be more accurate, we should use

$$pH = -\log\left(\frac{M_{H^+}\gamma_\pm}{M = 1}\right)$$

Now using the Debye-Hückel limiting law with $I = 0.05$, we find

$$\gamma_\pm = 0.768 \quad \text{and} \quad pH = 1.42$$

So we see that even in a 0.05-M solution, the correction for electrolyte solution nonideality is significant.

b. The ionization of the strong base sodium hydroxide produces a solution of 0.05 M OH^- ions, and here we will neglect the number of OH^- ions produced by the ionization of water. Therefore, first neglecting the solution nonideality (setting $\gamma_\pm = 1$), we have $M_{OH^-} = 0.05$ M so that

$$10^{-14} = M_{OH^-} \cdot M_{H^+}$$

or

$$M_{H^+} = \frac{10^{-14}}{M_{OH^-}}$$

and

$$M_{H^+} = \frac{10^{-14}}{M_{OH^-}} = \frac{10^{-14}}{0.05 \text{ M}} = 2 \times 10^{-13} \text{ M}$$

Therefore,

$$pH = 12.70$$

Including solution nonideality using the Debye-Hückel limiting law gives

$$M_{H^+} = \frac{10^{-14}}{0.05 \times 0.768} = 2.604 \times 10^{-13} \text{ M} \quad \text{and} \quad pH = 12.58$$

∎

A more accurate treatment of the ionization of a strong acid involves the two reactions

$$HA \rightarrow H^+ + A^-$$

$$H_2O \rightleftharpoons H^+ + OH^-$$

where the first reaction goes to completion, and the extent of the second is determined by the value of the equilibrium constant

$$K_{a,W} = \frac{M_{H^+} \cdot M_{OH^-}\gamma_\pm^2}{(M = 1)^2}$$

Here the total concentration of hydrogen ions M_{H^+} is the result of both the ionization of the acid $[M_{H^+}]_{HA}$ and the ionization of water $[M_{H^+}]_W$. Also, since hydroxyl ions present are only the result of the dissociation of water,

$$[M_{H^+}]_W = [M_{OH^-}]_W$$

Therefore,

$$
\begin{aligned}
K_{a,W}\,(M=1)^2 &= \left\{ [M_{H^+}]_{HA} + [M_{H^+}]_W \right\} \cdot [M_{H^+}]_W\, \gamma_\pm^2 \\
&= \left\{ M_{HA} + [M_{H^+}]_W \right\} \cdot [M_{H^+}]_W\, \gamma_\pm^2
\end{aligned}
\tag{15.1-6}
$$

since $[M_{H^+}]_{HA}$ is equal to the known initial concentration of the strong, completely ionized acid M_{HA}. The solution to this equation is

$$[M_{H^+}]_W = \frac{-M_{HA} + \sqrt{(M_{HA})^2 + \frac{4 K_{a,W}}{\gamma_\pm^2}\,(M=1)^2}}{2} \tag{15.1-7a}$$

and

$$M_{H^+} = M_{HA} + [M_{H^+}]_{HA} = \frac{1}{2} \left\{ M_{HA} + \sqrt{(M_{HA})^2 + \frac{4 K_{a,W}}{\gamma_\pm^2}\,(M=1)^2} \right\} \tag{15.1-7b}$$

As $K_{a,W}$ is so small in value $(\sim 10^{-14})$, the solution to this equation, except very close to the neutral point, reduces to $M_{H^+} = M_{HA}$, which is what we use in the illustration that follows.

ILLUSTRATION 15.1-2

Titration of a Strong Acid with a Strong Base

A common analytical procedure is titration, in which a solution of a strong base is quantitatively added to an acidic solution (or a strong acid added to a basic solution) until neutrality (pH = 7) is obtained, as indicated by a pH meter or chemical indicator. As an example of titration, a 0.20-M solution of NaOH is added to 10 mL of a 0.20-M HCl solution at 25°C. Compute the pH of the solution as a function of the amount of NaOH added, neglecting the solution nonideality of the ions.

SOLUTION

The solution originally contains 0.20 M × 0.010 L = 0.002 mol or 2 mmol of HCl, or more correctly, 2.00 mmol of H^+ and 2.00 mmol of Cl^-. If 1 mL of the NaOH solution is added, then 0.20 × 0.001 = 0.0002 mmol of NaOH, that is, 0.0002 mmol of Na^+ and 0.0002 mmol of OH^-, will have been added, and the volume of the mixture will now be 11 mL. The 0.0002 mmol of OH^- will combine with 0.0002 mmol of H^+ to form water (since the water ionization constant is so small in value), so after the NaOH addition, 0.0018 mol of H^+ will remain in 11 mL of water. Therefore,[4]

$$pH = -\log\left(\frac{0.0018\ \text{mol}}{0.011\ \text{L}} \right) = 0.786$$

This is to be compared with the pH of the initial solution, pH = $-\log(0.20) = 0.70$.

---

[4]Here we are neglecting the small difference between moles per liter of solution and moles per kilogram of water; that is, we are assuming that the addition of a small amount of electrolyte to water results in a solution containing 1 kg of water per liter.

This calculation of the effect of adding sodium hydroxide can now be continued until 10 mL and 0.002 mol of NaOH (or 0.002 mol of OH⁻) have been added, at which point the only hydrogen ions present would be the result of the water ionization equilibrium, and the pH of the solution would be 7.

Continuing further, suppose 15 mL of the NaOH solution is added. At this point in the titration the volume of the mixture is 25 mL and 0.003 mol of NaOH (and OH⁻) have been added to the mixture. However, of the 0.003 mol of OH⁻, 0.002 mol have reacted with the H⁺ from HCl, so that only 0.001 mol of OH⁻ remain in solution. Therefore,

$$M_{OH^-} = \frac{0.001 \text{ mol}}{0.025 \text{ mL}} = 0.04 \text{ M}$$

But from the water ionization equilibrium (neglecting solution nonideality)

$$K_{a,W} = \frac{M_{H^+} \cdot M_{OH^-}}{(1 \text{ M})^2} = 10^{-14}$$

so that

$$M_{H^+} = \frac{10^{-14}}{0.04} \text{ M} = 0.25 \times 10^{-12} \text{ M}$$

and

$$pH = -\log\left(0.25 \times 10^{-12}\right) = 12.60$$

The complete titration curve is shown in the accompanying figure as a function of the amount X of sodium hydroxide solution added. Note that there is a very sharp pH change around the neutral (pH = 7) point. Consequently, the indicator used to determine the pH change does not have to be a very sensitive one.

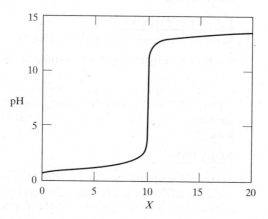

Titration curve for a strong acid (HCl) with a strong base (NaOH).

COMMENT

In this calculation, we have neglected the hydrogen ions resulting from the ionization of water, except at the neutrality point, where we have stated that pH = 7, and above the neutrality point, where the hydrogen ion concentration was computed from the calculated hydroxyl ion concentration. In fact, the calculations could be done more carefully by (1) considering the water ionization reaction, and (2) taking into account the ion solution nonideality, which we leave to the reader. ∎

Using the more complete Eq. 15.1-7b in this illustration yields results that are insignificantly different from those obtained with the simpler equations, except at the neutral point. However, Eq. 15.1-7b correctly reduces to pH $= -\log\left(\sqrt{K_{a,w}}\right)$ ($= 7$ at $25°C$) at the neutral point, while the simpler equations lead to $M_{H^+} = 0$ and a singularity in the pH at these conditions.

In a similar manner for a strong base,

$$BOH \longrightarrow B^+ + OH^-$$

it is easily shown (Problem 15.2) that

$$M_{OH^-} = \frac{1}{2}\left\{ M_{BOH} + \sqrt{(M_{BOH})^2 + \frac{4K_{a,W}}{\gamma_\pm^2}(M=1)^2} \right\} \qquad \textbf{(15.1-8a)}$$

and, except at the neutral point, $M_{OH^-} \approx M_{BOH}$. Consequently,

$$M_{H^+} = \frac{K_{a,W}(M=1)^2}{M_{OH^-}\gamma_\pm^2} = \frac{K_{a,W}(M=1)^2}{\frac{1}{2}\left[M_{BOH} + \sqrt{(M_{BOH})^2 + 4\frac{K_{a,W}}{\gamma_\pm^2}(M=1)^2}\right]\gamma_\pm^2}$$

$$\approx \frac{K_{a,W}(M=1)^2}{M_{BOH}\gamma_\pm^2} \qquad \textbf{(15.1-8b)}$$

The ionization of a weak electrolyte is somewhat more difficult to describe since it is only partially dissociated. Further, if on ionization the electrolyte produces either hydrogen ions or hydroxide ions, there is a common ion effect that partially suppresses the ionization of water. For example, acetic acid, a weak acid, dissociates as follows:

$$CH_3COOH \rightleftharpoons CH_3COO^- + H^+ \quad \text{(partially ionized acid)}$$

with the following thermodynamic equilibrium constant (generally referred to as the dissociation constant)

$$K_{a,HA} = \frac{a_{CH_3COO^-} \cdot a_{H^+}}{a_{CH_3COOH}} = \frac{M_{CH_3COO^-} \cdot M_{H^+}\gamma_\pm^2}{M_{CH_3COOH}\cdot\gamma_{CH_3COOH}\cdot(M=1)}$$

where we have used the ideal 1-molal standard states. These equations can be written as

$$K_{a,HA} = \frac{M_{CH_3COO^-} \cdot M_{H^+}}{M_{CH_3COOH}} \cdot \frac{\gamma_\pm^2}{\gamma_{CH_3COOH}\cdot(M=1)} = K_{HA}\cdot\frac{\gamma_\pm^2}{\gamma_{CH_3COOH}\cdot(M=1)}$$

where

$$K_{HA} = \frac{M_{CH_3COO^-}\cdot M_{H^+}}{M_{CH_3COOH}}$$

Apparent equilibrium constant

Note the absence of the subscript a on K in the equation above, indicating that this is not a true equilibrium constant. Instead, K_{HA} is referred to as an **apparent equilibrium constant** and is a measured quantity that depends on the solution conditions (ionic strength, pH, etc.), unlike the thermodynamic equilibrium constant, which depends only on the standard state and temperature. This apparent equilibrium constant is one of the concentration chemical equilibrium ratios defined in Table 13.1-3. As it is the concentrations (molalities) that have been measured to determine K_{HA}, no correction for solution nonidealities is made when its value is used to calculate the equilibrium state. However, the calculated results will only be approximate if the apparent equilibrium constant is used at concentrations or other solution conditions that are very different from those at which its value was determined.

As the apparent equilibrium constant is more easily measured in the laboratory than the thermodynamic equilibrium constant, most reported data on biochemical reactions are apparent equilibrium constants. Since the thermodynamic equilibrium constant here is based on ideal 1-molal standard states, so that the activity coefficients are unity at infinite dilution, we have

$$\lim_{\text{inf dilution}} \frac{K_{HA}}{(M = 1)} \to K_{a,HA}$$

[Also note that as a result of the (M=1) term in the denominator, $K_{a,HA}$ is unitless, while for the ionization reaction here K_{HA} has units of M^{-1}.] At infinite dilution, where all the activity coefficients are unity (based on using ideal 1-molal standard states), the apparent equilibrium constant will have the same numerical value as the true thermodynamic equilibrium constant. Throughout much of this chapter, the values of the equilibrium and dissociation constants given are from experimental data for the apparent or concentration-based equilibrium constants K rather than the thermodynamic equilibrium constants K_a. Thus solution nonidealities will frequently (but not always) be ignored.

In general, we can consider the following two ionization reactions to occur in a solution of a weak acid (neglecting the activity coefficient of the ions):

Weak acid

$$HA \rightleftharpoons H^+ + A^- \quad \text{with} \quad K_{a,HA} = \frac{a_{H^+} \cdot a_{A^-}}{a_{HA}}$$

and

$$H_2O \rightleftharpoons H^+ + OH^- \quad \text{with} \quad K_{a,W} = a_{H^+} \cdot a_{OH^-}$$

However, instead of using the thermodynamic equilibrium constant we will use the apparent equilibrium constant (we leave the analysis of this more general case to the reader as Problem 15.3) and therefore write

$$K_{HA} = \frac{M_{H^+} \cdot M_{A^-}}{M_{HA}} \quad \text{and} \quad K_W = M_{H^+} \cdot M_{OH^-} \tag{15.1-9}$$

We consider several special cases. The first case is that in which the dissociation of the acid produces a sufficiently large number of ions that the number of hydrogen ions resulting from the dissociation of water can be neglected, but that the extent of dissociation is still small enough that the molality of the weak acid is essentially unchanged from that added to the solution, $M_{HA,o}$. In this case the only equilibrium relation to be considered is

$$K_{HA} = \frac{M_{H^+} \cdot M_{A^-}}{M_{HA,o}}$$

But by stoichiometry $M_{H^+} = M_{A^-}$, so that

$$K_{HA} = \frac{M_{H^+}^2}{M_{HA,o}}$$

$$\log K_{HA} = 2 \cdot \log (M_{H^+}) - \log (M_{HA,o}) \tag{15.1-10}$$

Now, by analogy with the definition of pH, we define[5]

$$pK = -\log K$$

so that

$$-pK_{HA} = -2 \cdot pH - \log \left(M_{HA,o} \right)$$

or

$$2pH = pK_{HA} - \log \left(M_{HA,o} \right)$$

which gives the final result

$$pH = \tfrac{1}{2} \cdot pK_{HA} - \tfrac{1}{2} \cdot \log \left(M_{HA,o} \right) \tag{15.1-11}$$

The next special case is that in which the dissociation of the acid occurs to an extent such that the amount of undissociated acid is reduced below its initial value, but the dissociation of water can be neglected due to the presence of the hydrogen ions resulting from the dissociation of the acid. In this case, we have

$$M_{HA} = M_{HA,o} - M_{H^+} \quad \text{and} \quad M_{A^-} = M_{H^+}$$

so that

$$K_{HA} = \frac{M_{H^+} \cdot M_{A^-}}{M_{HA}} = \frac{M_{H^+}^2}{\left(M_{HA,o} - M_{H^+} \right)}$$

and

$$M_{H^+}^2 + K_{HA} \cdot M_{H^+} - K_{HA} \cdot M_{HA,o} = 0$$

which has the solution

$$M_{H^+} = \frac{-K_{HA} + \sqrt{K_{HA}^2 + 4K_{HA} \cdot M_{HA,o}}}{2} = \frac{K_{HA}}{2} \left(\sqrt{1 + \frac{4 \cdot M_{HA,o}}{K_{HA}}} - 1 \right) \tag{15.1-12a}$$

and

$$pH = pK_{HA} - \log \left(\sqrt{1 + \frac{4 \cdot M_{HA,o}}{K_{HA}}} - 1 \right) + \log (2) \tag{15.1-12b}$$

ILLUSTRATION 15.1-3

The pH of Solution of a Weak Acid

The value of pK_{HA} for acetic acid in water at 298 K is 4.76. Determine the pH of solutions of 0.01 M, 0.1 M, and 1 M acetic acid in water at that temperature.

---

[5] Also we can define $pK_a = -\log K_a$ based on the true thermodynamic equilibrium constant.

SOLUTION

The equilibrium constant is sufficiently small that the equation

$$pH = \tfrac{1}{2} \cdot pK_{HA} - \tfrac{1}{2} \cdot \log\left(M_{HA,o}\right)$$

can be used. Therefore,

$$pH = \tfrac{1}{2}(4.76) - \tfrac{1}{2}\log\left(M_{HA,o}\right) = 2.38 - \tfrac{1}{2}\log\left(M_{HA,o}\right)$$

$$pH = \begin{cases} 2.38 + 1 \;\; = 3.38 & \text{at } M_{HA,o} = 0.01 \text{ M} \\ 2.38 + 0.5 = 2.88 & \text{at } M_{HA,o} = 0.1 \text{ M} \\ 2.38 + 0 \;\; = 2.38 & \text{at } M_{HA,o} = 1 \text{ M} \end{cases}$$

Suppose instead we use the more accurate

$$pH = pK_{HA} - \log\left(\sqrt{1 + \frac{4 \cdot M_{HA,o}}{K_{HA}}} - 1\right) + \log(2)$$

then

$$pH = \begin{cases} = 3.389 & \text{at } M_{HA,o} = 0.01 \text{ M} \\ = 2.883 & \text{at } M_{HA,o} = 0.1 \text{ M} \\ = 2.381 & \text{at } M_{HA,o} = 1 \text{ M} \end{cases}$$

We see that the difference between using Eq. 15.1-12b and the simpler Eq. 15.1-10 is quite small. ∎

In the most general case the total molality of hydrogen ions includes that due to the dissociation of the weak acid and of the water,

$$M_{H^+} = (M_{H^+})_{acid} + (M_{H^+})_{water} = M_{A^-} + M_{OH^-}$$

since $(M_{H^+})_{acid} = M_{A^-}$ and $(M_{H^+})_{water} = M_{OH^-}$. Now from the dissociation of water,

$$M_{H^+} = \frac{K_W}{M_{OH^-}}$$

and by stoichiometry,

$$M_{HA} = M_{HA,o} - M_{A^-}$$

so that

$$K_{HA} = \frac{M_{H^+} \cdot M_{A^-}}{M_{HA}} = \frac{M_{H^+} \cdot M_{A^-}}{\left(M_{HA,o} - M_{A^-}\right)}$$

$$M_{A^-} = \frac{K_{HA} \cdot M_{HA,o}}{K_{HA} + M_{H^+}}$$

Therefore,

$$M_{H^+} = \frac{K_{HA} \cdot M_{HA,o}}{K_{HA} + M_{H^+}} + \frac{K_W}{M_{H^+}}$$

which results in the following cubic equation for the hydrogen ion molality:

$$M_{\mathrm{H+}}^3 + K_{\mathrm{HA}} M_{\mathrm{H+}}^2 - M_{\mathrm{H+}} \left(K_{\mathrm{HA}} M_{\mathrm{HA,o}} + K_{\mathrm{W}} \right) = K_{\mathrm{HA}} K_{\mathrm{W}} \qquad \text{(15.1-13)}$$

In fact, this equation is rarely used since usually if there is sufficient dissociation of the acid for the change in its initial molality to be taken into account, the contribution to the hydrogen ion molality from the dissociation of water can be neglected (that is, K_{W} can be taken to be equal to zero), and Eqs. 15.1-12 can be used.

ILLUSTRATION 15.1-4

The pH *of Solutions of Weak Acids*

Compute the pH of 0.01-M solutions of the following compounds at 25°C.

    a. Amino acid glycine ($pK = 2.34$)
    b. Amino-*n*-capriotic acid ($pK = 4.43$)
    c. Acetic acid ($pK_a = 4.76$)

[These weak acids ionize as $\mathrm{R-COOH} \rightleftharpoons \mathrm{R-COO^-} + \mathrm{H^+}$, where $\mathrm{R} = \mathrm{H_2NCH_2}$ for glycine, $\mathrm{R} = \mathrm{H_2NCH_3(CH_2)_4}$ for amino-*n*-capriotic acid, and $\mathrm{R} = \mathrm{CH_3}$ for acetic acid. Glycine is the only protein-forming amino acid without a center of chirality, and amino-*n*-capriotic acid is the amino acid that is used to treat hematological problems.]

SOLUTION

Solving Eq. 15.1-12 all three cases (ignoring the difference between pK and pK_a), we obtain

    a. Glycine $M_{\mathrm{H+}} = 4.85 \times 10^{-3}$ M and pH $= 2.314$
    b. Amino-*n*-capriotic acid $M_{\mathrm{H+}} = 5.91 \times 10^{-4}$ M and pH $= 3.228$
    c. Acetic acid $M_{\mathrm{H+}} = 4.08 \times 10^{-4}$ M and pH $= 3.389$

However, while the equilibrium constants for glycine and amino-*n*-capriotic acid are apparent ones (that is, K and pK in terms of concentrations), that for acetic acid is a true equilibrium constant, K_a and pK_a. It is of interest for this case case to determine the effect of solution nonidealities, as the more correct starting point is

$$K_{a,\mathrm{HA}} = \frac{a_{\mathrm{H+}} \cdot a_{\mathrm{A-}}}{a_{\mathrm{HA}}} = \frac{M_{\mathrm{H+}} \gamma_\pm \cdot M_{\mathrm{A-}} \gamma_\pm}{M_{\mathrm{HA}} \cdot (M = 1)} = \frac{(M_{\mathrm{H+}} \gamma_\pm)^2}{M_{\mathrm{HA}} \cdot (M = 1)}$$

Effect of electrolyte nonideality

where ideal 1-molal standard states have been used for all species, and the activity coefficient for the undissociated acid has been set to unity because of the high dilution of the charged species. For this analysis we will use the extended Debye-Hückel expression

$$\ln \gamma_\pm = -\frac{a |z_+ z_-| \sqrt{I}}{1 + \sqrt{I}} = -\frac{1.178 \sqrt{I}}{1 + \sqrt{I}}$$

with $I = \frac{1}{2} \sum_{\mathrm{ions\ i}} z_i^2 M_i = \frac{1}{2} (M_{\mathrm{H+}} + M_{\mathrm{A-}}) = M_{\mathrm{H+}}$. So the equation to be solved is

$$10^{-4.76} = \frac{(M_{\mathrm{H+}})^2 \cdot \exp\left(-\dfrac{2 \times 1.178 \times \sqrt{M_{\mathrm{H+}}}}{1 + \sqrt{M_{\mathrm{H+}}}} \right)}{(M_{\mathrm{HA,o}} - M_{\mathrm{H+}}) \cdot (M = 1)}$$

which has the solution

$$M_{\mathrm{H+}} = 4.29 \times 10^{-4} \text{ M} \quad \text{and} \quad \text{pH} = 3.368$$

This more exact value is only slightly different from pH $= 3.389$ found by neglecting the solution nonideality of the ions (that is, assuming $\gamma_\pm = 1$ or neglecting the difference between K and K_a). As a result of the small difference between the approximate and exact results, and also because of the uncertainties in the measurement and characterization of biological processes, in the biochemical literature it is common to use apparent (or measured) equilibrium constants and to neglect the effects of solution nonidealities. ∎

ILLUSTRATION 15.1-5

Estimating the pK_A of an Amino Acid from Measurements of Its pH in Solution

The pH of a 0.15-M solution of the amino acid serine is found to be 1.56. Estimate its pK_A. (Serine is one of the amino acids in proteins.)

SOLUTION

Since the pH is known, $M_{H^+} = 10^{-1.56} = 2.754 \times 10^{-2}$ M, and is equal to the molality of the ionized serine (since at such a low pH the number of hydrogen ions resulting from the ionization of water can be neglected). Therefore,

$$K_A = \frac{(2.754 \times 10^{-2} \text{ M})^2}{(0.15 - 2.754 \times 10^{-2}) \text{ M}} = 6.195 \times 10^{-3} \text{ M}$$

and

$$pK_A = -\log (6.195 \times 10^{-3}) = 2.21$$

Note that using the more approximate Eq. 15.1-10 (since the amino acid concentration in the denominator has not been corrected for the amount ionized), we obtain the less accurate

$$pK_A = 2\,pH + \log (M_{HA,o}) = 2 \times 1.56 + \log (0.15) = 2.296$$

∎

ILLUSTRATION 15.1-6

The Effect of an Added Salt on the Dissociation and pH of a Solution of a Weak Acid

The $pK_{a,HA}$ of acetic acid is 4.76. Assuming the activity coefficient of undissociated acetic acid is unity, compute the extent of dissociation of acetic acid and the pH in the following solutions of acetic acid and sodium chloride:

M_A (M)	M_{NaCl} (M)
1×10^{-4}	0
0.1	0
1.0	0
5.0	0
1×10^{-4}	1
0.1	1
1.0	1
5.0	1
1×10^{-4}	5
0.1	5
1.0	5
5.0	5

Consider two cases:

a. The activity coefficients of the ions are assumed to be unity.

b. The following extended Debye-Hückel expression is used:

$$\ln \gamma_{\pm} = -\frac{1.178\sqrt{I}}{1 + \sqrt{I}}$$

where I is the ionic strength due to all the ions present.

SOLUTION

The equilibrium relation is

$$K_{a,\text{HA}} = 10^{-pK_{\text{HA}}} = 10^{-4.76} = \frac{a_{\text{CH}_3\text{COO}^-} \cdot a_{\text{H}^+}}{a_{\text{CH}_3\text{COOH}}}$$

Using the ideal 1-molal standard states,

$$10^{-4.76} = \frac{M_{\text{CH}_3\text{COO}^-} \cdot M_{\text{H}^+} \cdot \gamma_{\pm}^2}{M_{\text{CH}_3\text{COOH}} \cdot (1\ \text{M})}$$

Letting α be the fractional dissociation,

$$M_{\text{CH}_3\text{COOH}} = (1 - \alpha)\, M_{\text{CH}_3\text{COOH,o}} \quad \text{and} \quad M_{\text{CH}_3\text{COO}^-} = M_{\text{H}^+} = \alpha M_{\text{CH}_3\text{COOH,o}}$$

the equilibrium relation is

$$10^{-4.76} = \frac{\alpha^2 \cdot M_{\text{CH}_3\text{COOH,o}}}{(1 - \alpha)(1\ \text{M})} \gamma_{\pm}^2$$

a. Therefore, if the activity coefficients of the ions are assumed to be unity, the equation to be solved for α is

$$10^{-4.76} = \frac{\alpha^2 \cdot M_{\text{CH}_3\text{COOH,o}}}{(1 - \alpha)(1\ \text{M})}$$

and the value of α so obtained is used in

$$\text{pH} = -\log\left(\frac{\alpha \cdot M_{\text{CH}_3\text{COOH,o}}}{1\ \text{M}}\right)$$

The results appear in the following table. Note that in this case the sodium chloride concentration does not affect the calculated results since the ion activity coefficients have been neglected.

b. Including the activity coefficients of the ions, the equilibrium equation is

$$10^{-4.76} = \frac{\alpha^2 \cdot M_{\text{CH}_3\text{COOH,o}}}{(1 - \alpha)(1\ \text{M})} \cdot \exp\left(-\frac{2 \cdot 1.178 \cdot \sqrt{I}}{1 + \sqrt{I}}\right)$$

where I is the ionic strength due to all the ions, that is,

$$I = \frac{1}{2} \sum_{\text{ions } i} z_i^2 M_i = \frac{1}{2}\left(M_{\text{H}^+} + M_{\text{CH}_3\text{COO}^-} + M_{\text{Na}^+} + M_{\text{Cl}^-}\right)$$

$$= \frac{1}{2}\left(2 \cdot \alpha \cdot M_{\text{CH}_3\text{COOH,o}} + 2 \cdot M_{\text{NaCl}}\right) = \alpha \cdot M_{\text{CH}_3\text{COOH,o}} + M_{\text{NaCl}}$$

Therefore, the equation to be solved for α is

$$10^{-4.76} = \frac{\alpha^2 \cdot M_{\mathrm{CH_3COOH,o}}}{(1-\alpha)\,(1\ \mathrm{M})} \cdot \exp\left(-\frac{2 \times 1.178 \times \sqrt{\alpha \cdot M_{\mathrm{CH_3COOH,o}} + M_{\mathrm{NaCl}}}}{1 + \sqrt{\alpha \cdot M_{\mathrm{CH_3COOH,o}} + M_{\mathrm{NaCl}}}}\right)$$

and the value of α is then used in

$$\mathrm{pH} = -\log\left[\frac{\alpha \cdot M_{\mathrm{CH_3COOH,o}}}{1\ \mathrm{M}} \cdot \exp\left(-\frac{1.178 \times \sqrt{\alpha \cdot M_{\mathrm{CH_3COOH,o}} + M_{\mathrm{NaCl}}}}{1 + \sqrt{\alpha \cdot M_{\mathrm{CH_3COOH,o}} + M_{\mathrm{NaCl}}}}\right)\right]$$

The results of this calculation also appear in the table.

Perhaps the most surprising result from the calculation is that while the extent of dissociation of acetic acid is strongly affected by the amount of sodium chloride added, there is relatively little effect on the pH of the solution, as the effect on the shift in the equilibrium due to the ion activity coefficient is largely compensated for by the appearance of the same activity coefficient in the definition of the pH.

The Extent of Dissociation and pH of Solutions of Acetic Acid and Sodium Chloride

$M_{\mathrm{CH_3COOH,o}}$	M_{NaCl}	α $(\gamma_\pm = 1)$	pH	α $(\gamma_\pm \neq 1)$	pH
0.0001	0	0.339	4.470	0.339	4.473
0.1	0	0.013	2.883	0.014	2.880
1	0	4.160×10^{-3}	2.381	4.368×10^{-3}	2.391
5	0	1.863×10^{-3}	2.031	2.046×10^{-3}	2.037
0.0001	1	0.339	4.470	0.520	4.540
0.1	1	0.013	2.883	0.023	2.885
1	1	4.160×10^{-3}	2.381	7.493×10^{-3}	2.382
5	1	1.863×10^{-3}	2.031	3.362×10^{-3}	2.031
0.0001	5	0.339	4.470	0.597	4.577
0.1	5	0.013	2.883	0.029	2.886
1	5	4.160×10^{-3}	2.381	9.366×10^{-3}	2.382
5	5	1.863×10^{-3}	2.031	4.201×10^{-3}	2.031

∎

Based on the results of this illustration, if we are interested only in computing the pH of a solution containing a weak acid (or base), it may be possible to neglect the ion activity coefficients. However, if our interest is in the extent of dissociation of the weak acid (or base), the activity coefficients of the ions should be included. What happens in the calculation here (and in other examples later in this chapter) is that there is some cancellation between the effect of the ion nonideality on the calculation of the equilibrium and on the calculation of the pH. This is especially true for a 1:1 acid (that is, an acid that on ionization produces a cation of charge $+1$ and an anion of charge -1).

The calculation of the pH of a mixture of a weak acid and a strong base is considered next, and is slightly more complicated. (The development of the equations for a weak base and a strong acid is left for the reader as Problem 15.5.) For simplicity of presentation, we will neglect the ionization of water, except at the neutral point (pH = 7) or when there is an excess of base, and also neglect solution nonidealities. Of course, electrolyte solution nonideality can be included following a procedure such as that in the illustration above. The dissociation reactions of a weak acid and a strong base are

Weak acid + strong base

$$\text{HA} \overset{K_{\text{HA}}}{\rightleftharpoons} \text{H}^+ + \text{A}^- \quad \text{and} \quad \text{BOH} \rightarrow \text{B}^+ + \text{OH}^- \tag{15.1-14}$$

The apparent equilibrium constant for the dissociation of the weak acid is

$$K_{\text{HA}} = \frac{M_{\text{H}^+} \cdot M_{\text{A}^-}}{M_{\text{HA}}}$$

Now using α to represent the fraction of the acid that has dissociated, we have from the mass balances

$$M_{\text{HA}} = M_{\text{HA,o}}(1-\alpha) \qquad M_{\text{A}^-} = \alpha M_{\text{HA,o}} \quad \text{and} \quad M_{\text{H}^+} = \alpha M_{\text{HA,o}} - M_{\text{BOH}}$$

The last relation arises from the fact that as long as the amount of base added does not exceed the amount of acid initially present, each hydroxyl ion formed from the dissociation of the base will react with a hydrogen ion produced from the dissociation of the weak acid to form a neutral water molecule.

Consequently, the equilibrium relation is

$$K_{\text{HA}} = \frac{\left(M_{\text{HA,o}}\alpha - M_{\text{BOH}}\right) \cdot \alpha M_{\text{HA,o}}}{M_{\text{HA,o}}(1-\alpha)}$$

which has the solution

$$\alpha = \frac{-(K_{\text{HA}} - M_{\text{BOH}}) + \sqrt{(K_{\text{HA}} - M_{\text{BOH}})^2 + 4K_{\text{HA}} \cdot M_{\text{HA,o}}}}{2M_{\text{HA,o}}} \tag{15.1-15}$$

and

$$M_{\text{H}^+} = \alpha M_{\text{HA,o}} - M_{\text{BOH}} = \frac{-(K_{\text{HA}} + M_{\text{BOH}}) + \sqrt{(K_{\text{HA}} - M_{\text{BOH}})^2 + 4K_{\text{HA}} \cdot M_{\text{HA,o}}}}{2}$$

$$\tag{15.1-16}$$

ILLUSTRATION 15.1-7

The pH of a Solution of a Weak Acid and a Strong Base

What is the pH of a solution that contains 0.1 M acetic acid ($\text{p}K_{\text{HA}} = 4.76$) and 0.07 M sodium hydroxide?

SOLUTION

Using the equation above, we have

$$M_{\text{H}^+} = \frac{-(10^{-4.76} + 0.07) + \sqrt{(10^{-4.76} + 0.07)^2 + 4 \cdot 10^{-4.76} \cdot 0.1}}{2} = 7.445 \times 10^{-6} \text{ M}$$

and pH = 5.128. ∎

ILLUSTRATION 15.1-8

The Titration of a Weak Acid and a Strong Base

A 10-mL solution of a 0.2-M acetic acid solution is titrated with a 0.2-M sodium hydroxide solution. What is the pH of the solution as a function of the amount of sodium hydroxide solution added?

SOLUTION

When 10 mL of the 0.2-M sodium hydroxide solution has been added, the solution will be neutral with pH = 7.0. For lesser amounts of sodium hydroxide added, Eq. 15.1-16 is used with the following concentrations for the addition of X mL of the sodium hydroxide solution to the 10 mL of acid solution:

$$M_{\text{HA},o} = \frac{10 \cdot 0.2}{10 + X} \, \text{M} \quad \text{and} \quad M_{\text{NaOH}} = \frac{0.2 \cdot X}{10 + X} \, \text{M} \quad \text{for } X < 10 \text{ mL}$$

(since the addition of the sodium hydroxide solution dilutes the initial acid concentration.)

For additions of the sodium hydroxide solution beyond the point of neutrality, the molality of hydroxide ions is

$$M_{\text{OH}^-} = \frac{0.2 \cdot X - 2}{10 + X} \, \text{M} \quad \text{for } X > 10 \text{ mL}$$

since the first 0.2 M × 10 mL = 2 mmol of hydroxyl ions neutralize the hydrogen ions resulting from the dissociation of the weak acid. Therefore,

$$M_{\text{H}^+} = \frac{10^{-14}}{M_{\text{OH}^-}} = \frac{(10 + X) \cdot 10^{-14}}{0.2 \cdot X - 2} \, \text{M} \quad X > 10 \text{ mL}$$

and

$$\text{pH} = -\log \left[\frac{(10 - X) \cdot 10^{-14}}{2 \cdot X - 2} \right]$$

The results are shown in the following figure.

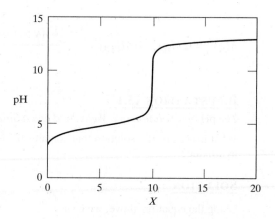

It is of interest to compare the titration curves for the weak acid with a strong base with the titration curve for a strong acid with a strong base. In Illustration 15.1-2 and here we have used equal amounts of acidic solutions of equal concentrations, and titrated both with the same sodium hydroxide solution. However, we see that the initial parts of the titration curve look somewhat different while the parts beyond neutrality are identical. In particular, the titration curve for the strong acid starts at a much lower pH than the titration curve for the weak acid.

Also, the initial shape of the titration curve has more curvature for the weak acid than is the case here for the strong acid. ∎

A buffer is a compound that stabilizes the pH of a solution in spite of the addition of an acid or a base. To be specific, sodium acetate, a salt that dissociates completely, can be used to buffer an acidic solution such as acetic acid. For example, in an aqueous solution of sodium acetate (which we will represent as NaAc), which has a pH of 7, and acetic acid (HAc), the following reactions occur:

$$NaAc \rightarrow Na^+ + Ac^-$$

$$HAc \overset{K_{HA}}{\rightleftharpoons} H^+ + Ac^-$$

and

$$H_2O \overset{K_W}{\rightleftharpoons} H^+ + OH^- \tag{15.1-17}$$

In addition, we have the mass balances

$$M_{Na^+} = (M_{Ac^-})_{salt} \equiv b$$

$$\text{Total acetate} = M_{HAc} + (M_{Ac^-})_{\text{from acid}} + (M_{Ac^-})_{\text{from salt}}$$
$$= M_{HAc} + M_{Ac^-} \equiv a + b$$

Note that with these definitions, a is the initial (before dissociation) molal concentration of acetic acid $M_{HAc,o}$, and b is the molal concentration of sodium acetate. Finally, from the condition that the solution must be electrically neutral, we have

$$M_{H^+} + M_{Na^+} = M_{OH^-} + M_{Ac^-} \quad \text{or} \quad M_{Na^+} = b = M_{OH^-} + M_{Ac^-} - M_{H^+}$$

so that

$$M_{Ac^-} = b + M_{H^+} - M_{OH^-}$$

and

$$a = M_{HAc} + M_{Ac^-} - b = M_{HAc} + M_{H^+} - M_{OH^-}$$

or

$$M_{HAc} = a - M_{H^+} + M_{OH^-}$$

These expressions are used in the equilibrium relations as follows:

$$K_{HA} = \frac{M_{H^+} \cdot M_{Ac^-}}{M_{HAc}} = \frac{M_{H^+} \cdot (b + M_{H^+} - M_{OH^-})}{(a + M_{OH^-} - M_{H^+})}$$

Also,

$$K_W = M_{H^+} \cdot M_{OH^-} \quad \text{or} \quad M_{OH^-} = \frac{K_W}{M_{H^+}}$$

and

$$K_{HA} = \frac{M_{H^+} \cdot \left(b + M_{H^+} - \frac{K_W}{M_{H^+}}\right)}{\left(a + \frac{K_W}{M_{H^+}} - M_{H^+}\right)} = \frac{M_{H^+} \cdot \left(bM_{H^+} + M_{H^+}^2 - K_W\right)}{\left(aM_{H^+} + K_W - M_{H^+}^2\right)}$$

Therefore,

$$pK_{HA} = pH - \log\left[\frac{b + 10^{-pH} - 10^{(pH-pK_W)}}{a + 10^{(pH-pK_W)} - 10^{-pH}}\right]$$

$$= pH - \log\left[\frac{M_{NaAc} + 10^{-pH} - 10^{(pH-pK_W)}}{M_{HAc,o} + 10^{(pH-pK_W)} - 10^{-pH}}\right]$$

or

$$pH = pK_{HA} + \log\left[\frac{M_{NaAc} + 10^{-pH} - 10^{(pH-pK_W)}}{M_{HAc,o} + 10^{(pH-pK_W)} - 10^{-pH}}\right] \tag{15.1-18}$$

Since pK_W is of the order of 14, for values of pH in the approximate range of $4 \leq pH \leq 10$, this equation reduces to

Henderson-Hasselbalch equation

$$pK_{HA} = pH - \log\left(\frac{b}{a}\right)$$

or

$$pH = pK_{HA} + \log\left(\frac{b}{a}\right) = pK_{HA} + \log\left(\frac{M_{NaAc}}{M_{HAc,o}}\right) \tag{15.1-19}$$

which is referred to as the **Henderson-Hasselbalch equation**.

The result of Eq. 15.1-18 for the pH of a solution in the presence of a buffer is analytically and numerically very different from Eq. 15.1-12b, obtained earlier when no buffering salt is present:

$$pH = pK_{HA} - \log\left\{\sqrt{1 + \frac{4 \cdot M_{HA,o}}{K_{HA}}} - 1\right\} + \log 2 \tag{15.1-12b}$$

The calculated results for the pH of acetic acid solutions with various amounts of sodium acetate are given in the table below. There we see that the presence of sodium acetate keeps the pH of acetic acid solutions closer to neutrality (pH = 7), though greater amounts of the salt are needed as the acid concentration is increased.

$M_{HAc,o}$	M_{NaAc}	pH	
1.0	0.0	2.375	(no buffer)
1.0	0.5	4.449	
1.0	1.0	4.750	
1.0	2.0	5.051	
1.0	5.0	5.449	
5.0	0.0	2.026	(no buffer)
5.0	0.5	3.750	
5.0	1.0	4.051	
5.0	2.0	4.352	
5.0	5.0	4.750	

(At these concentrations, we should include the effect of electrolyte solution nonideality. That would make the analysis is more difficult, and consequently this is left to the reader as an exercise.)

15.2 IONIZATION OF BIOCHEMICALS

There are chemicals, especially biochemicals, that have more than a single ionization state. Indeed, proteins, composed of many amino acids, have numerous ionization sites and therefore multiple ionization states with different net charges. As a simple introduction to species with multiple ionization sites, consider phthalic acid, a so-called dibasic acid that ionizes as follows:

and

with $pK_1 = 2.95$ and $pK_2 = 5.41$. To be general, we will write the reactions for an acidic species with two ionization sites as

Ionization of a dibasic acid

$$H_2A \underset{}{\overset{K_1}{\rightleftharpoons}} H^+ + HA^- \quad \text{and} \quad HA^- \underset{}{\overset{K_2}{\rightleftharpoons}} H^+ + A^{2-} \tag{15.2-1}$$

The reaction equilibrium relations using the apparent equilibrium constants are

$$K_1 = \frac{M_{H^+} \cdot M_{HA^-}}{M_{H_2A}} \quad \text{and} \quad K_2 = \frac{M_{H^+} \cdot M_{A^{2-}}}{M_{HA^-}} \tag{15.2-2a}$$

Combining these two equations,

$$K_1 K_2 = \frac{(M_{H^+})^2 \cdot M_{A^{2-}}}{M_{H_2A}} \tag{15.2-2b}$$

Our interest is in determining the fraction of the initial amount of the dibasic acid in each of the states (H_2A, HA^-, and A^{2-}) at each pH of the solution, which we assume is being adjusted by adding a strong acid or base. That is, we will take the pH to be known, and we want to compute the fraction of acid in each ionization state. The total concentration of acidic species (that is, of all forms of the dibasic acid) M_{acid} is

$$M_{acid} = M_{H_2A} + M_{HA^-} + M_{A^{2-}} = M_{H_2A} + \frac{M_{H_2A} \cdot K_1}{M_{H^+}} + \frac{M_{H_2A} \cdot K_1 K_2}{(M_{H^+})^2}$$

$$= M_{H_2A} \left[1 + \frac{K_1}{M_{H^+}} + \frac{K_1 \cdot K_2}{(M_{H^+})^2} \right] \tag{15.2-3}$$

Therefore, the fraction of undissociated dibasic acid is

$$f_{H_2A} = \frac{M_{H_2A}}{M_{acid}} = \frac{1}{\left[1 + \dfrac{K_1}{M_{H^+}} + \dfrac{K_1 \cdot K_2}{(M_{H^+})^2} \right]} = \frac{(M_{H^+})^2}{\left[(M_{H^+})^2 + K_1 M_{H^+} + K_1 \cdot K_2 \right]}$$

$$= \frac{10^{-2pH}}{10^{-2pH} + 10^{-(pK_1 + pH)} + 10^{-(pK_1 + pK_2)}} \tag{15.2-4a}$$

Similarly,

$$f_{HA^-} = \frac{M_{HA^-}}{M_{acid}} = \frac{K_1 \cdot M_{H_2A}}{M_{H^+}} \cdot \frac{1}{M_{H_2A} \cdot \left[1 + \dfrac{K_1}{M_{H^+}} + \dfrac{K_1 \cdot K_2}{(M_{H^+})^2} \right]}$$

$$= \frac{K_1 \cdot M_{H^+}}{(M_{H^+})^2 + K_1 \cdot M_{H^+} + K_1 \cdot K_2}$$

$$= \frac{10^{-(pK_1 + pH)}}{10^{-2pH} + 10^{-(pK_1 + pH)} + 10^{-(pK_1 + pK_2)}} \tag{15.2-4b}$$

and

$$f_{A^{2-}} = \frac{K_1 \cdot K_2}{(M_{H^+})^2 + K_1 \cdot M_{H^+} + K_1 \cdot K_2}$$

$$= \frac{10^{-(pK_1 + pK_2)}}{10^{-2pH} + 10^{-(pK_1 + pH)} + 10^{-(pK_1 + pK_2)}} \tag{15.2-4c}$$

ILLUSTRATION 15.2-1

The Ionization of a Dibasic Acid

Compute the fraction of phthalic acid in each of its ionization states as a function of pH. Also, compute the average charge on the phthalic acid species as a function of pH.

SOLUTION

Using the pK_1 and pK_2 values given earlier and the equations above, we obtain the results in Fig. 1 for the fraction of each species present as a function of pH. Though each molecule of phthalic acid can have a charge of only 0, -1, or -2, the average charge on the phthalic acid components (which need not be an integer) is computed as the sum of the fractions of molecules in each of these charge states times the integer charge in that state, that is,

$$\text{Charge} = 0 \cdot f_{H_2A} - 1 \cdot f_{HA^-} - 2 \cdot f_{A^{2-}}$$

The results are shown in Fig. 2. There we see that at low pH, where there is no ionization of phthalic acid, it has no net charge. However, as the pH is increased, there is an increasingly negative charge on the acid until a pH of 7 is reached, at which point the acid is fully dissociated with a charge of -2.

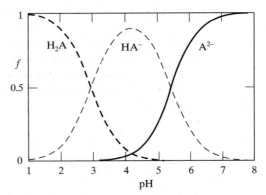

Figure 1 The fractions of undissociated f_{H_2A} (dashed line), singly ionized f_{HA^-} (dotted line), and completely dissociated $f_{A^{2-}}$ (solid line) phthalic acid as a function of pH.

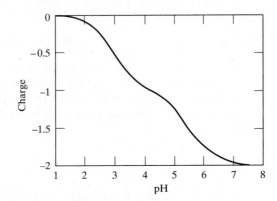

Figure 2 The net charge on phthalic acid as a function of pH.

■

Amino acids consist of acidic and basic groups on the same molecule, and at neutral pH can be considered to be neutral or to have a negatively charged anionic part and a positively charged cationic part. This is shown in Fig. 15.2-1. In the figure different

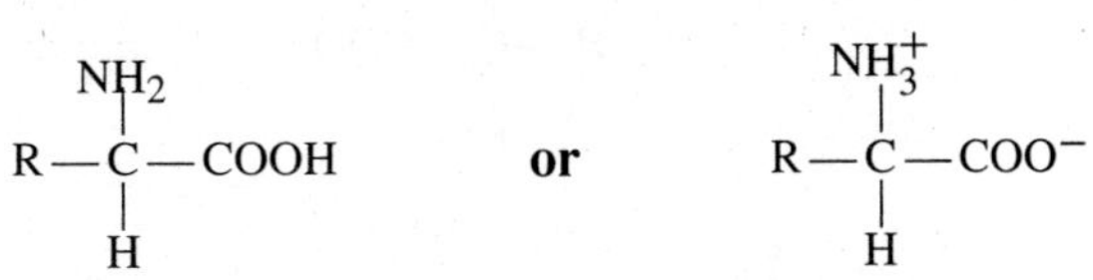

NH$_2$ NH$_3^+$

R—C—COOH   **or**   R—C—COO$^-$

H   H

Figure 15.2-1 Amino acid as a zwitterion.

amino acids have different R groups as side chains. For example, if R = H the amino acid is glycine, if $R = CH_3$ the amino acid is alanine, and in lysine the side group is $R = (CH_2)NH_2$.

Compounds such as amino acids with both anionic and cationic parts are referred to as **zwitterions** (coming from the German word *zwitter* meaning "hybrid"), and their charge changes with pH as a result of the reactions shown in Fig. 15.2-2 for glycine. The sequence shown starts at low pH (high hydrogen ion concentration), where the amino acid has a positive charge. As the pH is raised (the hydrogen ion concentration is lowered) the first reaction occurs, releasing a hydrogen ion from the amine group and resulting in no net charge on the amino acid. As the pH increases further, the second reaction occurs releasing a second hydrogen ion, now from the carboxylic acid, and resulting in the negatively charged form of the amino acid.

The coupled set of equations that result from the coupled ionization reactions, using an apparent equilibrium constant, with the species denoted as A, B, and C as indicated in Fig. 15.2-2, are

$$K_1 = \frac{M_B M_{H^+}}{M_A} \tag{15.2-5a}$$

and

$$K_2 = \frac{M_C M_{H^+}}{M_B} \tag{15.2-6a}$$

Using $M_{B,o}$ to represent the initial molar concentration of the uncharged amino acid in neutral solution (pH = 7), we have

$$M_{B,o} = M_A + M_B + M_C$$

A simplification that we can use in solving these equations is that since the equilibrium constants are generally so different in value for amino acids that one reaction will have gone to completion before the other reaction has occurred to an appreciable extent. For example, for glycine we have $pK_1 = 2.34$ or $K_1 = 4.57 \times 10^{-3}$ and $pK_2 = 9.6$ or $K_2 = 2.51 \times 10^{-10}$, so that the positively charged (A) form of the acid is present

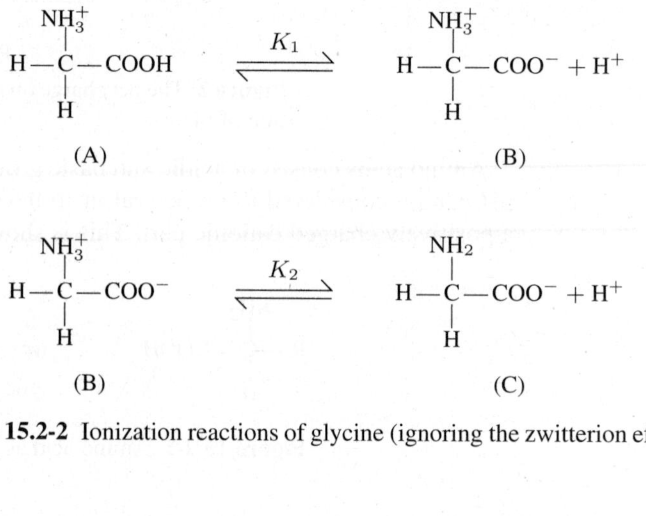

Figure 15.2-2 Ionization reactions of glycine (ignoring the zwitterion effect).

to an appreciable extent only below pH = 7, and the negatively charged (C) form is present to an appreciable extent only above pH = 7. $M_B = M_{B,o} - M_A$, so that

$$K_1 = \frac{\left(M_{B,o} - M_A\right) M_{H^+}}{M_A} \tag{15.2-5b}$$

which has the solution

$$\frac{M_A}{M_{B,o}} = \frac{M_{H^+}/K_1}{M_{H^+}/K_1 + 1} = \frac{10^{(pK_1 - pH)}}{1 + 10^{(pK_1 - pH)}} \tag{15.2-5c}$$

The equilibrium relation for the second reaction is

$$K_2 = \frac{M_C M_{H^+}}{M_B} \tag{15.2-6a}$$

Now using $M_B = M_{B,o} - M_C$, we obtain

$$\frac{M_C}{M_{B,o}} = \frac{1}{1 + M_{H^+}/K_2} = \frac{1}{1 + 10^{(pK_2 - pH)}} \tag{15.2-6b}$$

The average charge f on an amino acid, then, is

$$f(pH) = (+1) \times \frac{M_A}{M_{B,o}} + (-1) \times \frac{M_C}{M_{B,o}} = \frac{10^{(pK_1 - pH)}}{1 + 10^{(pK_1 - pH)}} - \frac{1}{1 + 10^{(pK_2 - pH)}} \tag{15.2-7a}$$

Also, it is easy to show (and left to the reader) that for the case here of an amino acid with a single carboxylic acid group and a single amine group, the **isoelectric point, pI,** that is, the pH at which the the amino acid has zero net charge, is

$$pH = \frac{pK_1 + pK_2}{2} \tag{15.2-7b}$$

(Here the isoelectric point has been taken to be the pH at which the substance has no net charge. The experimental definition is the pH at which the substance has zero mobility in an electrophoresis cell. In fact, these two values of pH differ so slightly, they are frequently used interchangeably, as is done here.)

ILLUSTRATION 15.2-2

The Charge on an Amino Acid as a Function of pH

The ionization equilibrium constants for glycine are $pK_1 = 2.34$ and $pK_2 = 9.6$. Determine the fraction of glycine molecules in the positively charged, negatively charged, and neutral states as a function of pH. Also, determine the average net charge on glycine as a function of pH, and the isoelectric point of glycine.

SOLUTION

Using Eqs. 15.2-5c and 15.2-6b, and the pK values given above, we can easily calculate the fraction of glycine in each of the ionization states, and by difference, the fraction that is uncharged. The results are shown in Figs. 1 and 2.

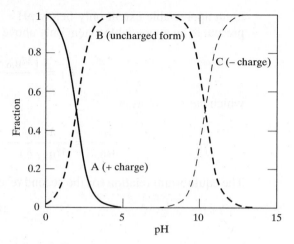

Figure 1 Fractions of neutral and charged forms of glycine as a function of pH.

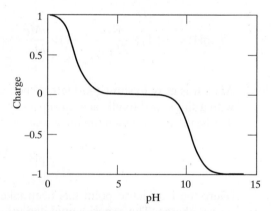

Figure 2 Average net charge on glycine as a function of pH.

Note that, as assumed, the positively charged (A) species have essentially disappeared at pH = 5, and the negatively charged (C) species begin to appear only above about pH = 8. So our assumption that the first reaction goes to completion before the second reaction occurs is justified. Because the charge on glycine is near zero over the pH range between about 3 and 9.5, it is very difficult to determine the isoelectric point from Fig. 2. However, from Eq. 15.2-7b, the isoelectric point is calculated to be pH = 5.97. ∎

A more detailed description of the ionization states of the amino acid glycine (H_2NCH_2COOH) is shown below, and is a bit more complicated than discussed above since the neutral form should be considered to be a zwitterion that exists in two different states.

$$^+\text{H}_3\text{NCH}_2\text{COOH} \qquad (\text{GH}_2{}^+)$$

$$K_1 \qquad\qquad K_2$$

$$(\text{GH}) \qquad \text{H}_2\text{NCH}_2\text{COOH} \qquad\qquad\qquad ^+\text{H}_3\text{NCH}_2\text{COO}^- \qquad (\text{GH}')$$

$$K_3 \qquad\qquad K_4$$

$$\text{H}_2\text{NCH}_2\text{COO}^- \qquad (\text{G}^-)$$

$$(15.2\text{-}8)$$

In this diagram the letters in parentheses indicate the notation that will be used to represent each of these species. The equilibrium relations for the glycine reactions, neglecting solution nonidealities, are

$$K_1 = \frac{M_{\text{GH}} \cdot M_{\text{H}^+}}{M_{\text{GH}_2^+}} \qquad K_2 = \frac{M_{\text{GH}'} \cdot M_{\text{H}^+}}{M_{\text{GH}_2^+}} \qquad K_3 = \frac{M_{\text{G}^-} \cdot M_{\text{H}^+}}{M_{\text{GH}}}$$

and

$$K_4 = \frac{M_{\text{G}^-} \cdot M_{\text{H}^+}}{M_{\text{GH}'}}$$

Consequently,

$$K_1 K_3 = \frac{M_{\text{GH}} \cdot M_{\text{H}^+}}{M_{\text{GH}_2^+}} \cdot \frac{M_{\text{G}^-} \cdot M_{\text{H}^+}}{M_{\text{GH}}} = \frac{M_{\text{G}^-} \cdot (M_{\text{H}^+})^2}{M_{\text{GH}_2^+}} \qquad (15.2\text{-}9\text{a})$$

and

$$K_2 K_4 = \frac{M_{\text{GH}'} \cdot M_{\text{H}^+}}{M_{\text{GH}_2^+}} \cdot \frac{M_{\text{G}^-} \cdot M_{\text{H}^+}}{M_{\text{GH}'}} = \frac{M_{\text{G}^-} \cdot (M_{\text{H}^+})^2}{M_{\text{GH}_2^+}} \qquad (15.2\text{-}9\text{b})$$

so that $K_1 K_3 = K_2 K_4$.

In laboratory measurements we cannot distinguish between the GH and GH' states, as they contain the same number of protons, and by hydrogen transfer, the molecule will jump between these two states. Therefore, we define new equilibrium constants that can be measured as follows:

$$\mathcal{K}_1 = \frac{(M_{\text{GH}} + M_{\text{GH}'}) \cdot M_{\text{H}^+}}{M_{\text{GH}_2^+}} = \frac{M_{\text{GH}} \cdot M_{\text{H}^+}}{M_{\text{GH}_2^+}} + \frac{M_{\text{GH}'} \cdot M_{\text{H}^+}}{M_{\text{GH}_2^+}} = K_1 + K_2$$

$$(15.2\text{-}10\text{a})$$

and

$$\mathcal{K}_2 = \frac{M_{\text{G}^-} \cdot M_{\text{H}^+}}{(M_{\text{GH}} + M_{\text{GH}'})}$$

so that

$$\frac{1}{\mathcal{K}_2} = \frac{(M_{\text{GH}} + M_{\text{GH}'})}{M_{\text{G}^-} \cdot M_{\text{H}^+}} = \frac{M_{\text{GH}}}{M_{\text{G}^-} \cdot M_{\text{H}^+}} + \frac{M_{\text{GH}'}}{M_{\text{G}^-} \cdot M_{\text{H}^+}} = \frac{1}{K_3} + \frac{1}{K_4} = \frac{K_3 + K_4}{K_3 K_4}$$

or

$$\mathcal{K}_2 = \frac{K_3 K_4}{K_3 + K_4} \qquad\qquad \textbf{(15.2-10b)}$$

Thus the glycine reaction system can be represented as

$$GH_2^+ \overset{\mathcal{K}_1}{\rightleftharpoons} GH + H^+ \quad \text{and} \quad GH \overset{\mathcal{K}_2}{\rightleftharpoons} G^- + H^+$$

with GH now representing the sum of the two neutral forms of glycine, GH and GH′. It is this description that is generally used in describing amino acids, as was done in Illustration 15.2-2.

Proteins are long-chain molecules formed by polymerization reactions between amino acids. For example, the first step in such a reaction to form a peptide is shown below.

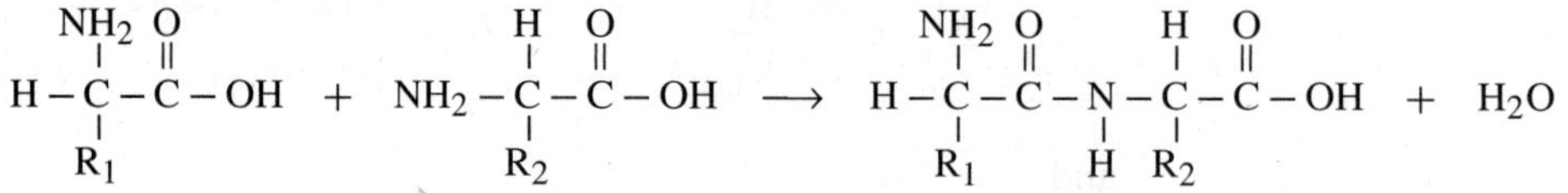

$$\underset{R_1}{\overset{NH_2}{\underset{|}{H-\overset{|}{\underset{|}{C}}}}} \overset{O}{\underset{}{\overset{\|}{C}}} - OH \;+\; \underset{R_2}{\overset{H}{\underset{|}{NH_2-\overset{|}{\underset{|}{C}}}}} \overset{O}{\underset{}{\overset{\|}{C}}} - OH \;\longrightarrow\; \underset{R_1}{\overset{NH_2}{\underset{|}{H-\overset{|}{\underset{|}{C}}}}} \overset{O}{\underset{}{\overset{\|}{C}}} - \underset{H}{\overset{H}{\underset{|}{N}}} - \underset{R_2}{\overset{}{\underset{|}{C}}} \overset{O}{\underset{}{\overset{\|}{C}}} - OH \;+\; H_2O$$

The carbon-nitrogen bond is referred to as the peptide bond. A polypeptide chain is the result of a number of such reactions, and the characteristic of all peptides, like the amino acids from which they have been formed, is that one end of the molecule is an amino group and the other end is a carboxylic acid. Different peptides are characterized by their lengths and their side chains (R groups). In fact, only twenty types of side chains, with different sizes, shapes, charges, chemical reactivity, and hydrogen-bonding capacity, have so far been found in the proteins on all species on earth, from bacteria to man.

Since proteins are large polypeptides molecules and are made up of a number of acidic (usually carboxylic acid) and basic groups (usually simple amino groups, but also histidine, lysine, arginine, and guanidine), including on their side chains, proteins have many ionization states. Therefore, the charge on a protein varies greatly with pH; however, like the amino acids from which they are formed, they will have a positive charge at low pH and a negative charge at high pH. As an example, the charge on the protein hen egg white lysozyme as a function of pH is shown in Fig. 15.2-3. [Lysozyme is a small protein that protects mammals (including us) from bacterial infection by attacking the protective cell walls of bacteria.] It has 32 ionizable groups, some anionic and others cationic, and as a result, its charge changes from +19 in very acidic solutions to −13 in very basic solutions. Notice that at a pH of 11.2, there is zero net charge on a lysozyme molecule, so this is the isoelectric point of this protein.

A rule of thumb is that an amino acid or protein with multiple ionization sites will always include a primary carboxylic acid (−COOH) group that ionizes at low pH (usually 2 to 3), a primary amino (−NH$_2$) group that ionizes in the pH range between 8 and 10, and other ionizable groups on side chains.

As a result of their highly polar, ionizable groups, amino acids are very soluble in polar solvent such as water, and only slightly soluble in nonpolar organic liquids. Also, by releasing hydrogen ions at high pH and accepting hydrogen ions at low pH, zwitterions act as good buffers by keeping a solution closer to neutrality upon the addition of an acid or a base.

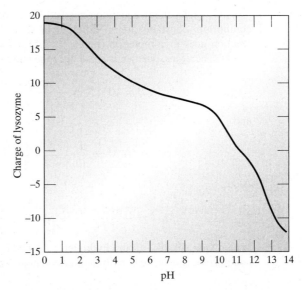

Figure 15.2-3 Average net charge on hen egg white lysozyme as a function of pH.

ILLUSTRATION 15.2-3

The pH of a Solution Containing an Amino Acid as a Buffer

Twenty-five milliliters of a 0.1-M aqueous solution of glycine is to be titrated with a 0.1-M sodium hydroxide solution. Determine the pH of the solution, the fraction of glycine in the GH_2^+, GH, and G^- ionization states, and the charge on glycine as a function of the amount of sodium hydroxide added.

SOLUTION

Initially there are 25 mL of a 0.1-M solution, so the titration starts with 2.5 mmol of glycine. Let α be the number of mmoles of GH and β the number of mmoles of G^- present at any time. Also, as the 0.1-M NaOH solution is added, the OH^- ions resulting from the sodium hydroxide ionization react with the H^+ ions from the ionization of glycine to produce water. Therefore, the amount of each species present after the addition of X mL of NaOH is

Species	Initial (mmol)	Final (mmol)
GH_2^+	2.5	$2.5 - \alpha$
GH	0	$\alpha - \beta$
G^-	0	β
H^+	0	$\alpha + \beta - 0.1X$

Also, the volume of the solution is $25 + X$, where X is the milliliters of the 0.1-M NaOH solution added. Therefore, the equilibrium relations are

$$\mathcal{K}_1 = \frac{M_{GH} \cdot M_{H^+}}{M_{GH_2^+}} = \frac{(\alpha - \beta) \cdot (\alpha + \beta - 0.1X)}{(2.5 - \alpha) \cdot (25 + X)}$$

and

$$\mathcal{K}_2 = \frac{M_{G^-} \cdot M_{H^+}}{M_{GH}} = \frac{\beta \cdot (\alpha + \beta - 0.1X)}{(\alpha - \beta) \cdot (25 + X)}$$

In principle, these equations need to be solved simultaneously. As already discussed, the equilibrium constants are so different in value ($pK_1 = 2.34$ and $pK_2 = 9.6$) that the first reaction goes to completion (i.e., $\alpha = 2.5$) before the second reaction occurs to a significant extent. Therefore, one needs solve only the simpler equations

$$\mathcal{K}_1 = 10^{-2.34} = \frac{\alpha \cdot (\alpha - 0.1X)}{(2.5 - \alpha) \cdot (25 + X)} \quad \text{for } X \le 25 \text{ mL}$$

and

$$\mathcal{K}_2 = 10^{-9.6} = \frac{\beta \cdot (2.5 + \beta - 0.1X)}{(2.5 - \beta) \cdot (25 + X)} \quad \text{for } 25 < X \le 50 \text{ mL}$$

Also,

$$M_{H^+} = \frac{\alpha + \beta - 0.1X}{25 + X} \text{ M} \quad \text{and} \quad pH = -\log\left(\frac{\alpha + \beta - 0.1X}{25 + X}\right)$$

When $X > 50$ mL, all the hydrogen ions resulting from the ionization of glycine have been neutralized, and the only hydrogen ions remaining are from the ionization of water, so that

$$M_{OH^-} = \frac{0.1X}{25 + X} \quad \text{and} \quad M_{H^+} = \frac{10^{-14}}{M_{OH^-}} = \frac{10^{-14}(25 + X)}{0.1X}$$

and

$$pH = -\log M_{H^+} = -\log\left(\frac{10^{-14}(25 + X)}{0.1X}\right) = 14 + \log\left(\frac{0.1X}{25 + X}\right) \quad \text{for } X > 50 \text{ mL}$$

Finally, the fractions f of glycine molecules in the GH_2^+, GH, and G^- states can be computed from

$$f_{GH_2^+} = \frac{2.5 - \alpha}{2.5} \qquad f_{GH} = \frac{\alpha - \beta}{2.5} \quad \text{and} \quad f_{GH} = \frac{\beta}{2.5}$$

The results are shown in Figs. 1 and 2. For comparison, note that in water without any added sodium hydroxide, 81 percent of the glycine is in the GH_2^+ state, 19 percent is in the GH or GH′ state, and the pH of the solution is 1.72.

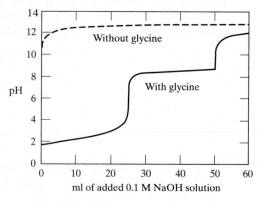

Figure 1 The pH of a glycine solution as a function of added sodium hydroxide (solid line). The dashed line is the pH of water upon NaOH addition without glycine, showing the effectiveness of the zwitterion glycine as a buffer for pH values near 2 and near 8.

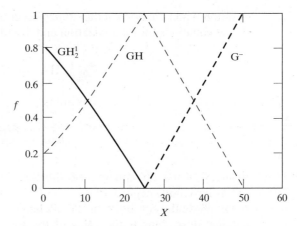

Figure 2 Fractions f of glycine in the GH_2^+, GH, and G^- states as a function of the milliliters of 0.1-M sodium hydroxide solution added.

If the glycine were not present, so that the starting solution was only 25 mL of pure water, the molality of hydrogen ions (neglecting solution nonideality) would be given by

$$M_{H^+} = \frac{K_W}{M_{OH^-}} = \frac{10^{-14}}{M_{NaOH}} \quad \text{so that} \quad pH = 14 + \log M_{NaOH}$$

since the strong base, sodium hydroxide, is fully ionized. The molality of sodium hydroxide when X mL of a 0.1-M NaOH solution is added is

$$M_{NaOH} = \frac{0.1X}{25 + X}$$

and

$$pH = 14 + \log\left(\frac{0.1X}{25 + X}\right)$$

The pH of the solution without glycine is shown in Fig. 1. Note that without glycine the solution is initially at pH 7, and that the pH initially rises rapidly upon the addition of sodium hydroxide and then more gradually. However, with glycine present the initial pH is low and, while gradually increasing, remains low until 25 mL of the NaOH solution has been added, as the hydrogen ion released by carboxylic acid portion of the glycine reacts with the hydroxide ion of NaOH to form water. Further addition of NaOH results in a sharp increase in pH to slightly above pH = 8. Then continued additions of the strong base lead to the release of hydrogen ions from the amino groups that react with the sodium hydroxide, resulting in only a slight increase in pH until 50 mL of NaOH solution has been added. At that point, there are no additional hydrogen ions that can be released, and the pH of the solution increases rapidly. From Fig. 1, we see that glycine, being a zwitterion, is an effective buffer in the pH range near 2 and also in the pH range slightly above 8. ■

15.3 SOLUBILITIES OF WEAK ACIDS, WEAK BASES, AND PHARMACEUTICALS AS A FUNCTION OF pH

We consider next some acids, bases, proteins, pharmaceuticals, or other compounds that ionize, and in which the undissociated species is only partially soluble in aqueous solution while the portion that ionizes is completely soluble. We start with an analysis of the equations for the ionization and dissolution of an acid. Thus we consider

$$HA \overset{K_{HA}}{\rightleftharpoons} H^+ + A^-$$

for which the equilibrium constant is

$$K_{HA} = \frac{M_{H^+} \cdot M_{A^-}}{M_{HA}}$$

where, as usual, the apparent equilibrium constant has been used and solution nonidealities have been neglected.

Let S_o be the solubility of the uncharged acidic species, which could be a pharmaceutical drug. This is the value of the solubility at such a low pH that ionization is suppressed. As long as there is sufficient undissolved acid for the solution to remain saturated, we have

$$S_o = M_{HA} \tag{15.3-1}$$

We will use S_T to represent total solubility of the acid, that is, the concentration of both undissociated and ionized acid:

$$S_T = M_{HA} + M_{A^-} = S_o + M_{A^-} \tag{15.3-2}$$

or

$$M_{A^-} = S_T - S_o$$

Now

$$K_{HA} = \frac{M_{H^+} \cdot M_{A^-}}{M_{HA}} = \frac{M_{H^+} \cdot (S_T - S_o)}{S_o}$$

or

$$pK_{HA} = pH - \log\left[\frac{(S_T - S_o)}{S_o}\right]$$

Then the total acid solubility at any pH, S_T (pH), is

$$S_T (pH) = S_o \cdot \left(1 + 10^{-[pK_{HA} - pH]}\right)$$

or

$$S_T (pH) = S_o + \frac{S_o \cdot K_{HA}}{M_{H^+}} \tag{15.3-3}$$

Clearly, if

$$M_{H^+} \gg K_{HA} \quad \text{then} \quad S_T = S_o$$

while if

$$K_{HA} \gg M_{H^+}$$

there is a significant increase in solubility. In particular, at pH = 7 (pure water), we have

$$S_T\,(pH = 7) = S_o \cdot \left(1 + 10^{-[pK_{HA}-7]}\right) \tag{15.3-4a}$$

so that the total solubility of a compound at any pH is related to its solubility in pure water (which is the solubility that is usually reported) by

Solubility of a weak acid

$$S_T\,(pH) = S_T\,(pH = 7) \cdot \frac{\left(1 + 10^{-[pK_{HA}-pH]}\right)}{\left(1 + 10^{-[pK_{HA}-7]}\right)} \tag{15.3-4b}$$

ILLUSTRATION 15.3-1

The Solubility of a Weak Acid as a Function of pH

Hexanoic acid ($C_6H_{12}O_2$), in biology usually referred to as caproic acid, has a solubility in water of 9.67 g/kg at pH = 7 and 25°C, and its pK_{HA} is 4.85. Estimate the solubility of hexanoic acid in the following body fluids:

Blood	pH = 7.4
Saliva	pH = 6.4
Stomach contents	pH = 1.0 to 3.0
Urine	pH = 5.8

SOLUTION

Using Eq. 15.3-4b, the solubility of hexanoic acid in blood is found to be 24.2 g/kg, in saliva it is 2.48 g/kg, in stomach contents the solubility is 0.068 g/kg, and in urine the solubility of hexanoic acid is 0.68 g/kg. The observation from this calculation is that there is very different solubility of an organic chemical (or pharmaceutical) in different bodily fluids. This can be especially important in the formulation of a drug, where one wants to ensure its solubility and bioavailability.

The complete solubility versus pH curve is shown in the figure below. Note that the solubility of hexanoic acid begins to increase when the value of the pH is close to the pK_{HA} value. This is a general result for acids, since, as we can see from Eqs. 15.3-4, there will be a significant change in the slope of the solubility curve of a partially soluble acid at a pH that is close to its pK_{HA} value.

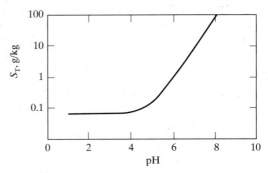

Solubility of hexanoic acid (g/kg) in aqueous solutions as a function of pH.

Next we consider the solubility of a partially soluble weak base. The extent of ionization of the weak base

$$\text{BOH} \overset{K_{\text{BOH}}}{\rightleftharpoons} \text{B}^+ + \text{OH}^-$$

can be computed from the equilibrium relation

$$K_{\text{BOH}} = \frac{M_{\text{B}^+} \cdot M_{\text{OH}^-}}{M_{\text{BOH}}}$$

Now, using the notation

$$S_{\text{o}} = M_{\text{BOH}} \qquad S_{\text{T}} = M_{\text{BOH}} + M_{\text{B}^+} = S_{\text{o}} + M_{\text{B}^+} \quad \text{and} \quad M_{\text{OH}^-} = \frac{K_{\text{W}}}{M_{\text{H}^+}} \qquad \textbf{(15.3-5)}$$

we obtain

$$K_{\text{BOH}} = \frac{(S_{\text{T}} - S_{\text{o}}) \cdot K_{\text{W}}/M_{\text{H}^+}}{S_{\text{o}}}$$

and

$$S_{\text{T}} = S_{\text{o}} \cdot \left[1 + \frac{K_{\text{BOH}} \cdot M_{\text{H}^+}}{K_{\text{W}}} \right] = S_{\text{o}} \cdot \left(1 + 10^{-[\text{p}K_{\text{BOH}} + \text{pH} - \text{p}K_{\text{W}}]} \right) \qquad \textbf{(15.3-6)}$$

or, using pH = 7 (pure water) as the reference,

Solubility of a weak base

$$S_{\text{T}}(\text{pH}) = S_{\text{T}}(\text{pH} = 7) \cdot \frac{\left(1 + 10^{-[\text{p}K_{\text{BOH}} + \text{pH} - \text{p}K_{\text{W}}]} \right)}{\left(1 + 10^{-[\text{p}K_{\text{BOH}} + 7 - \text{p}K_{\text{W}}]} \right)} \qquad \textbf{(15.3-7)}$$

In the biological literature, equilibrium constants are sometimes written for the reaction

$$\text{BOH} + \text{H}^+ \overset{K_{\text{BOH}}^*}{\rightleftharpoons} \text{B}^+ + \text{H}_2\text{O} \qquad \textbf{(15.3-8)}$$

which can be considered the combination of

$$\text{BOH} \overset{K_{\text{BOH}}}{\rightleftharpoons} \text{B}^+ + \text{OH}^-$$

and

$$\text{H}^+ + \text{OH}^- \overset{1/K_{\text{W}}}{\rightleftharpoons} \text{H}_2\text{O}$$

Summing these last two reactions gives the reaction of Eq. 15.3-8. Also,

$$K_{\text{BOH}} = \frac{M_{\text{B}^+} \cdot M_{\text{OH}^-}}{M_{\text{BOH}}} \quad \text{and} \quad \frac{1}{K_{\text{W}}} = \frac{M_{\text{H}_2\text{O}}}{M_{\text{H}^+} \cdot M_{\text{OH}^-}}$$

so that

$$K^*_{\text{BOH}} = \frac{K_{\text{BOH}}}{K_{\text{W}}} \quad \text{or} \quad pK^*_{\text{BOH}} = pK_{\text{BOH}} - pK_{\text{W}}$$

Therefore,

$$S_{\text{T}}(pH = 7) = S_{\text{o}} \cdot \left(1 + 10^{-\left[pK^*_{\text{BOH}} + 7\right]}\right) \qquad \textbf{(15.3-9a)}$$

so that an alternative expression is

$$S_{\text{T}}(pH) = S_{\text{T}}(pH = 7) \cdot \frac{\left(1 + 10^{-\left[pK^*_{\text{BOH}} + pH\right]}\right)}{\left(1 + 10^{-\left[pK^*_{\text{BOH}} + 7\right]}\right)} \qquad \textbf{(15.3-9b)}$$

ILLUSTRATION 15.3-2

The Solubility of a Weak Base as a Function of pH

The pyrimidine base thymine ($C_5H_6N_2O_2$, also known as 5-methyluracil) has a solubility in water of 4.0 g/kg (pH = 7) at 25°C and its pK_{BOH} is 9.9 (there is a further dissociation at higher pH that we will neglect here). Estimate the solubility of thymine in aqueous solutions over the pH range of 1 to 13.

SOLUTION

The results in the following figure calculated with Eq. 15.3-7 show a significant decrease of thymine solubility with increasing pH until about pH = 4, beyond which the solubility remains approximately constant at 4.0 g/kg.

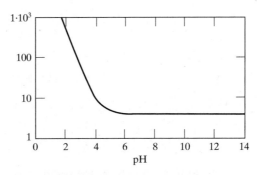

Thymine solubility (g/kg) in aqueous solutions as a function of pH

In this figure there is a marked change in the slope of the solubility curve of a base when the $pOH = (pK_{\text{W}} - pH) = (14 - pH)$ is equal to the pK_{BOH} value. (This is seen here in the change in slope at about pH = 4, corresponding to pOH = 10, which is close to the pK_{BOH} value of 9.9.)

<u>NOTE</u>

As an aside, it is of interest to note that deoxyribonucleic acid, or DNA, which has been called the molecule of heredity, consists of a repeated backbone with a collection of side chain groups, the sequence of which is different in different species (and in individuals within the same species) and contains the genetic information. The side chains on the DNA backbone consist of only four types of bases, the two purines adenine (A) and guanine (G), and the two pyrimidines thymine (T) and cytosine (C). What is commonly referred to as the genetic code is a long sequence of the four letters A, G, T, and C in the order in which these bases appear on the DNA chain. ∎

Some slightly soluble amino acids and other weak acids and bases have more than a single ionization state. For example, at low pH the simple amino acid glycine is positively charged; it is uncharged around pH = 7 and has a negative charge at higher pH, as shown in Fig. 2 of Illustration 15.2-2. Using the A (positively charged), B (neutral), and C (negatively charged) species identifications of that illustration, we have

$$K_1 = \frac{M_B \cdot M_{H^+}}{M_A}$$

and

$$K_2 = \frac{M_C \cdot M_{H^+}}{M_B}$$

Letting S_o represent the equilibrium saturation solubility of the uncharged amino acid, that is, using $S_o = M_B$, we have

$$K_1 = \frac{S_o \cdot M_{H^+}}{M_A} \quad \text{and} \quad K_2 = \frac{M_C \cdot M_{H^+}}{S_o}$$

Therefore, the total solubility

$$S_T = S_o + M_A + M_C$$

$$= S_o + S_o \frac{M_{H^+}}{K_1} + S_o \frac{K_2}{M_{H^+}} = S_o \left(1 + \frac{M_{H^+}}{K_1} + \frac{K_2}{M_{H^+}}\right)$$

is

$$S_T = S_o \cdot \left[1 + 10^{(pK_1 - pH)} + 10^{-(pH - pK_2)}\right] \qquad \textbf{(15.3-10a)}$$

and

$$\frac{S_T}{S_o} = 1 + 10^{(pK_1 - pH)} + 10^{-(pH - pK_2)} \qquad \textbf{(15.3-10b)}$$

Note that while S_o is the solubility of the uncharged species, it is usually the solubility of the amino acid in water pH = 7 that is reported, and that may be somewhat different from the solubility of the unchanged amino acid, unless its isoelectric point is pH = 7. At other pH values the total solubility will be the result of the solubility of the uncharged and charged species. The total solubility at pH = 7 is

$$S_T (pH = 7) = S_o \left(1 + 10^{(pK_1 - 7)} + 10^{-(7 - pK_2)}\right) \qquad \textbf{(15.3-11a)}$$

Now eliminating the unknown solubility of the uncharged amino acid gives the final result for the total solubility at any pH value in terms of the measured solubility in neutral water at pH = 7:

$$S_T(pH) = S_T(pH = 7) \cdot \frac{\left(1 + 10^{(pK_1 - pH)} + 10^{-(pH - pK_2)}\right)}{\left(1 + 10^{(pK_1 - 7)} + 10^{-(7 - pK_2)}\right)} \qquad \textbf{(15.3-11b)}$$

The discussion that led to this equation is easily extended to amino acids, proteins, and other biomolecules with more than two ionizable sites (Problems 15.7 and 15.11).

ILLUSTRATION 15.3-3
The Solubility of a Weak Amino Acid as a Function of pH

Tyrosine ($C_9H_{11}NO_3$), another amino acid in proteins, has two dominant ionizable groups with pK_{HA} values of 2.24 and 9.04 at 25°C (and a third, less easily ionizable group with a pK value of 10.10, which we will neglect here). Its water solubility at this temperature is 0.46 g/kg of water.

a. Estimate the total solubility of tyrosine in aqueous solutions over the pH range of 0.4 to 11.
b. Estimate the solubility of the uncharged tyrosine.

SOLUTION

a. Using Eq. 15.3-11b, we obtain the results shown in the following figure. Note that the solubility of tyrosine varies little (only between 0.46 and 0.56 g/kg of water) over the pH range of 2.9 of 8.4, and then increases rapidly as the pH either decreases below 2.9 or increases above 8.9. The pH range of limited solubility is also where tyrosine has approximately zero average net charge (Problem 15.20).

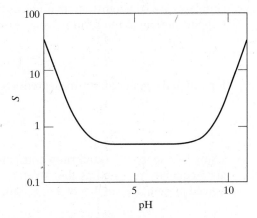

The solubility S of tyrosine (g/kg) in aqueous solutions as a function of pH.

b. Using Eq. 15.3-11a, we have

$$S_T(pH = 7) = 0.46 \frac{g}{kg} = S_o \times 1.009 \quad \text{and} \quad S_o = 0.456 \frac{g}{kg}$$

Understanding the solubility of weak acids and bases as a function of pH is important in the formulation of pharmaceuticals for general use. From Illustration 15.3-1 we see that if we had to deliver a drug with low solubility in water that had a pK value similar to hexanoic acid, it would have to be delivered in a slightly basic solution to have appreciable solubility or concentration in the aqueous solution. For example, at pH = 7.4 (the pH of blood), its solubility would be 2.5 times greater than in water, while its solubility would be considerably less in the fluids in the stomach, and may indeed precipitate out of solution there. However, for a drug with a pK similar to that of thymine (9.9), the solubility decreases with increasing pH at low pH, and is essentially unchanged from its neutral water solubility by increasing the pH to that of blood.

Understanding the solubility of proteins and other biochemicals is also important in developing methods for their purification. For example, an important method for the purification of antibiotics is by liquid-liquid extraction. The procedure is as follows. The pH of the solution is first adjusted to the isoelectric point at which the desired biochemical compound has no net charge and limited solubility, while the other biochemicals in the mixture are charged and are more soluble. This solution is then contacted with an organic solvent, which results in the extraction of the desired biochemical compound into the organic phase. Then by contacting this organic solution with an aqueous phase at a pH at which desired biochemical has a charge, it is extracted back into an aqueous phase. This method is used to separate components with different isoelectric points. Also, the differing solubilities of species as a function of pH can be used to recover one biochemical from a mixture by selective precipitation (or crystallization).

15.4 BINDING OF A LIGAND TO A SUBSTRATE

Another type of chemical reaction that occurs, especially among biological molecules, is the reversible binding of a smaller molecule (referred to as a ligand) to a binding site on a protein or other large molecule (referred to as a substrate). We consider first the simplest case of a substrate that has only a single binding site. The reaction is ligand (L) plus protein (P) to form a ligated protein (PL):

$$L + P \overset{K}{\rightleftharpoons} PL \qquad \text{(15.4-1)}$$

for which the apparent (or measured) concentration-based equilibrium constant is

$$K = \frac{M_{PL}}{M_L \cdot M_P} \qquad \text{(15.4-2)}$$

Using this apparent equilibrium constant, we have $M_{PL} = K \cdot M_L \cdot M_P$. Further, data are frequently reported as the coverage θ, which is the ratio of concentration of the ligated protein M_{PL} to the total concentration of protein in solution $M_P + M_{PL}$. That is,

Fractional ligand coverage

$$\theta = \frac{M_{PL}}{M_P + M_{PL}} = \frac{K \cdot M_L \cdot M_P}{M_P + K \cdot M_L \cdot M_P} = \frac{K \cdot M_L}{1 + K \cdot M_L} \qquad \text{(15.4-3)}$$

With increasing ligand concentration, the coverage initially increases linearly (though the linear range may be small if the value of K is large), and at high ligand concentration θ saturates (for this case of a single binding site) to unity. This form of the coverage versus concentration relation was originally derived for molecules adsorbing on a solid surface, and is referred to as the **Langmuir isotherm** (isotherm designating a process occurring at constant temperature), after Irving Langmuir, who developed the equation to describe the adsorption of gases on metal surfaces.

Consider next a protein molecule with two different binding sites. We will designate proteins with a single bound ligand as PL and LP, depending on which of the two sites binds the ligand, and use LPL to represent the protein with the two ligands. The binding reactions are

$$L + P \underset{}{\overset{K_1}{\rightleftharpoons}} LP \qquad \text{Reaction 1}$$

$$L + P \underset{}{\overset{K_2}{\rightleftharpoons}} PL \qquad \text{Reaction 2}$$

$$L + LP \underset{}{\overset{K_3}{\rightleftharpoons}} LPL \qquad \text{Reaction 3}$$

and

$$L + PL \underset{}{\overset{K_4}{\rightleftharpoons}} LPL \qquad \text{Reaction 4}$$

with

$$K_1 = \frac{M_{LP}}{M_L \cdot M_P} \qquad K_2 = \frac{M_{PL}}{M_L \cdot M_P} \qquad K_3 = \frac{M_{LPL}}{M_L \cdot M_{LP}} \quad \text{and} \quad K_4 = \frac{M_{LPL}}{M_L \cdot M_{PL}}$$

The coverage θ, defined as the moles of bound ligand per mole of protein in any form, is

$$\theta = \frac{\text{moles of bound ligand}}{\text{moles of protein}} = \frac{M_{LP} + M_{PL} + 2 \cdot M_{LPL}}{M_P + M_{LP} + M_{PL} + M_{LPL}} \qquad \textbf{(15.4-4)}$$

where the factor of 2 in the numerator arises because two ligands are attached in each LPL combination resulting from reactions 3 and 4. In this case the saturation coverage θ will be 2 when each protein binds two ligands. Now using the equilibrium constant relations, we obtain

$$\theta = \frac{K_1 \cdot M_L \cdot M_P + K_2 \cdot M_L \cdot M_P + K_3 \cdot M_L \cdot M_{LP} + K_4 \cdot M_L \cdot M_{PL}}{M_P + K_1 \cdot M_L \cdot M_P + K_2 \cdot M_L \cdot M_P + \frac{1}{2}(K_3 \cdot M_L \cdot M_{LP} + K_4 \cdot M_L \cdot M_{PL})}$$

$$= \frac{K_1 \cdot M_L \cdot M_P + K_2 \cdot M_L \cdot M_P + K_3 \cdot K_1 \cdot M_L^2 \cdot M_P + K_4 \cdot K_1 \cdot M_L^2 \cdot M_P}{M_P + K_1 \cdot M_L \cdot M_P + K_2 \cdot M_L \cdot M_P + \frac{1}{2}(K_3 \cdot K_1 \cdot M_L^2 \cdot M_P + K_4 \cdot K_1 \cdot M_L^2 \cdot M_P)}$$

$$= \frac{K_1 \cdot M_L + K_2 \cdot M_L + K_3 \cdot K_1 \cdot M_L^2 + K_4 \cdot K_1 \cdot M_L^2}{1 + K_1 \cdot M_L + K_2 \cdot M_L + \frac{1}{2}(K_3 \cdot K_1 \cdot M_L^2 + K_4 \cdot K_1 \cdot M_L^2)} \qquad \textbf{(15.4-5)}$$

However, also note that from the equilibrium relations,

$$M_{LPL} = K_3 \cdot M_L \cdot M_{LP} = K_3 \cdot K_1 \cdot M_L^2 \cdot M_P = K_4 \cdot M_L \cdot M_{PL} = K_4 \cdot K_2 \cdot M_L^2 \cdot M_P \quad \textbf{(15.4-6)}$$

so that $K_3 \cdot K_1 = K_4 \cdot K_2$, and therefore

$$\theta = \frac{(K_1 + K_2) \cdot M_L + 2 \cdot K_1 \cdot K_3 \cdot M_L^2}{1 + (K_1 + K_2) \cdot M_L + K_1 \cdot K_3 \cdot M_L^2}$$

Of special interest is the case of a protein with four indistinguishable binding sites (i.e., a case where experimentally we cannot distinguish between the LP and PL states as above), but because of steric interference at binding sites or attractions between bound ligands on the formation of successive ligand bonds, the equilibrium constants for each successive ligand binding are different. The reason this is interesting is because this is exactly the case of oxygen binding to hemoglobin in our blood. The oxygen + hemoglobin (Hb) binding reactions are

Hemoglobin + oxygen binding

$$Hb + O_2 \overset{K_1}{\rightleftharpoons} HbO_2$$

$$HbO_2 + O_2 \overset{K_2}{\rightleftharpoons} Hb(O_2)_2$$

$$Hb(O_2)_2 + O_2 \overset{K_3}{\rightleftharpoons} Hb(O_2)_3$$

$$Hb(O_2)_3 + O_2 \overset{K_4}{\rightleftharpoons} Hb(O_2)_4$$

and the reported apparent equilibrium constants[6] are

$$K_1 = \frac{M_{HbO_2}}{M_{Hb} \cdot M_{O_2}} = 4.0 \times 10^4 \text{ M}^{-1}$$

$$K_2 = \frac{M_{Hb(O_2)_2}}{M_{HbO_2} \cdot M_{O_2}} = 2.73 \times 10^4 \text{ M}^{-1}$$

$$K_3 = \frac{M_{Hb(O_2)_3}}{M_{Hb(O_2)_2} \cdot M_{O_2}} = 5.94 \times 10^4 \text{ M}^{-1}$$

and

$$K_4 = \frac{M_{Hb(O_2)_4}}{M_{Hb(O_2)_3} \cdot M_{O_2}} = 2.02 \times 10^6 \text{ M}^{-1}$$

Therefore

$$\theta = \frac{\text{Number of bound oxygen molecules}}{\text{Number of hemoglobin molecules}}$$

$$= \frac{M_{HbO_2} + 2 \cdot M_{Hb(O_2)_2} + 3 \cdot M_{Hb(O_2)_3} + 4 \cdot M_{Hb(O_2)_4}}{M_{Hb} + M_{HbO_2} + M_{Hb(O_2)_2} + M_{Hb(O_2)_3} + M_{Hb(O_2)_4}}$$

which, after some algebra, becomes

$$\theta = \frac{K_1 \cdot M_{O_2} + 2 \cdot K_1 \cdot K_2 \cdot M_{O_2}^2 + 3 \cdot K_1 \cdot K_2 \cdot K_3 \cdot M_{O_2}^3 + 4 \cdot K_1 \cdot K_2 \cdot K_3 \cdot K_4 \cdot M_{O_2}^4}{1 + K_1 \cdot M_{O_2} + K_1 \cdot K_2 \cdot M_{O_2}^2 + K_1 \cdot K_2 \cdot K_3 \cdot M_{O_2}^3 + K_1 \cdot K_2 \cdot K_3 \cdot K_4 \cdot M_{O_2}^4}$$

$$(15.4\text{-}7)$$

Clearly, at saturation $\theta = 4$.

The binding isotherm as a function of oxygen concentration is shown in Fig. 15.4-1. Note that as a result of the sigmoidal shape of the curve, there is a large change in the number of oxygen molecules bound to hemoglobin in the range between 0.5×10^{-5} M and 1.5×10^{-5} M oxygen. This is especially important in human and animal physiology since, as a result of the reverse reactions, a small decrease in oxygen concentration (or

---

[6]G. K. Ackers, *Biophys. J.* **32**, 331 (1980).

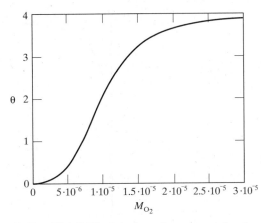

Figure 15.4-1 The number of oxygen molecules (θ) bound to a molecule of hemoglobin as a function of oxygen concentration.

partial pressure) will lead to a significant release of oxygen from hemoglobin in blood. As the arterial blood passes through the body, it experiences a decrease in oxygen partial pressure due to the metabolic use of oxygen and also the small pressure drop due to flow in the capillaries, resulting in the release of a significant amount of oxygen. It is in this way that oxygen is delivered to cells throughout the body.

The efficiency of this four-site adsorption/desorption process can be seen by noting that for a physiologically important change from 5×10^{-6} M oxygen to 5×10^{-5} M, 3.51 molecules of oxygen are released for each molecule of hemoglobin. In contrast, if there were only a single binding site with $K = 4.0 \times 10^4$ M (the value of the first equilibrium constant of the actual four sites), the shape of the isotherm would be of the form given by Eq. 15.4-3, with a release of only 0.50 oxygen molecules per hemoglobin molecule for the same change in oxygen concentration. Further, if we considered the case of single-site adsorption and used the largest value of the four equilibrium constants, $K = 2.02 \times 10^6$ M, the isotherm would also have the shape given by Eq. 15.4-3, and the oxygen release would be only 0.080 molecules of oxygen per molecule of hemoglobin since the isotherm is near saturation over this range of oxygen concentrations, as shown in Figure 15.4-2. Thus, it is a result of the cooperative four-site oxygen binding to hemoglobin that large animals (such as humans) are possible.

This sigmoidal shape of the ligand-binding isotherm, which occurs in hemoglobin-oxygen binding, is a result of what is referred to as cooperative binding. That is, as a result of the increasing values of the apparent equilibrium constants (i.e., $K_3 > K_1$ and K_2, and $K_4 > K_1$, K_2, and K_3), the binding of oxygen molecules beyond the first becomes more favorable. The mechanistic reason that this occurs is that the hemoglobin molecule, which is actually a tetramer, undergoes structural rearrangements as each oxygen binds, resulting in the binding of the third and fourth oxygens being more favorable than the first.

NOTE:

Hemoglobin consists of four chains, each with a heme or iron group that can bind an oxygen molecule. Myoglobin is a simpler molecule consisting of a single chain that is of the same general structure as each of the chains in hemoglobin. The oxygen-

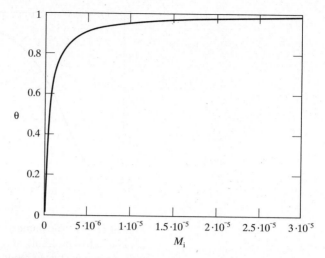

Figure 15.4-2 The number of oxygen molecules (θ) bound to hemoglobin if it had a single binding site with a high value of the binding constant.

carrying and -releasing capacity of myoglobin, which has only a single oxygen binding site, is much like that shown in Figure 15.4-2, and is much less than hemoglobin. Consequently, the association of myoglobin chains into hemoglobin was important to the development of life as we know it.

ILLUSTRATION 15.4-1

The Binding of Warfarin to Human Albumin

Warfarin, also known as coumadin, is used to prevent blood clots in humans and also, at high body doses (referred to as a high body burden in the biomedical literature), as a rat poison by promoting internal bleeding. Warfarin binds with human plasma albumin in the reaction

$$W + A \overset{K}{\rightleftharpoons} WA$$

The following measured or apparent (not standard-state) data at 25°C are available for this binding reaction, $\Delta_{bnd}G = -30.8$ kJ mol^{-1}, $\Delta_{bnd}H = -13.1$ kJ mol^{-1}, and $\Delta_{bnd}C_P \sim 0$, where these apparent changes are for the following equilibrium constant:

$$\Delta_{bnd}G = RT \ln K \quad \text{where} \quad K = \frac{M_{WA}}{M_W \cdot M_A}$$

where M is the concentration in molality.

a. What is the entropy change for this reaction at 25°C?
b. Starting with concentrations of warfarin and human plasma albumin of 0.0001 M, what is the fraction of unbound (unreacted) albumin over the temperature range of 0 to 50°C?

SOLUTION

a. Since $\Delta G = \Delta H - T\Delta S$, at 25°C

$$\Delta S = \frac{\Delta H - \Delta G}{T} = \frac{-13\,100 - (-30\,800) \text{ kJ mol}^{-1}}{298.15 \text{ K}} = 59.37 \frac{\text{J}}{\text{mol K}}$$

b. Since $\Delta_{bnd}C_P = 0$, $\Delta_{bnd}H$ is independent of temperature, and the equilibrium constant as a function of temperature is computed from

$$\ln\frac{K(T)}{K(T = 298.15 \text{ K})} = \int_{298.15 \text{ K}}^{T} \frac{\Delta_{bnd}H}{RT^2}dT = \frac{\Delta_{bnd}H}{R} \cdot \left(\frac{1}{T} - \frac{1}{298.15}\right)$$

The results are given in the table that follows. Also, letting α be the fraction of the initial warfarin (or albumin) that has reacted, we have

$$M_{WA} = \alpha \cdot M_{A,o} = 0.0001\alpha \quad \text{and} \quad M_W = M_A = 0.0001(1 - \alpha)$$

so that

$$K = \frac{0.0001\alpha}{(0.1)^2 \cdot (1 - \alpha)^2} = \frac{\alpha}{0.0001 \cdot (1 - \alpha)^2}$$

the solutions of which are also given in the table.

T (°C)	$K(T)$	α	$1 - \alpha$
0	4.039×10^5	0.855	0.145
10	3.295×10^5	0.840	0.160
20	2.725×10^5	0.826	0.174
30	2.282×10^5	0.811	0.189
40	1.933×10^5	0.797	0.203
50	1.654×10^5	0.783	0.217

Consequently, we see that warfarin is strongly bound to human plasma albumin at all temperatures. ∎

15.5 SOME OTHER EXAMPLES OF BIOCHEMICAL REACTIONS

In this section we consider several other examples of physiologically important biochemical reactions.

ILLUSTRATION 15.5-1

The Trimerization of a Pancreatic Hormone

Glucagon is a 29-amino peptide hormone of the pancreas. At pH = 10.6 it is known that glucagon can associate or trimerize, and the following standard-state (ideal 1-molal solution at 25°C) information is available:

$$G \overset{K}{\rightleftharpoons} \tfrac{1}{3}G_3$$

$$\Delta_{rxn}G° = -30.6 \frac{\text{kJ}}{\text{mol}} \qquad \Delta_{rxn}H° = -131 \frac{\text{kJ}}{\text{mol}} \quad \text{and} \quad \Delta_{rxn}C_P° = -1.8 \frac{\text{kJ}}{\text{mol K}}$$

If the initial concentration of glucagon is 0.01 M (which is sufficiently low for the solution to be considered ideal), what is the fraction of glucagon trimerized over the temperature range from 0 to 50°C?

SOLUTION

The equilibrium relation is

$$K_a = e^{-(\Delta_{rxn}G^\circ/RT)} = \frac{(a_{G_3})^{1/3}}{a_G} \cong \frac{(M_{G_3})^{1/3} \cdot (1\ M)^{2/3}}{M_G}$$

and the mass balances are

$$M_G = M_{G,o} - \alpha = 0.01 - \alpha \quad \text{and} \quad M_{G_3} = \frac{\alpha}{3}$$

so that

$$K_a = \frac{\left(\dfrac{\alpha}{3}\right)^{1/3} \cdot (1\ M)^{2/3}}{0.01 - \alpha}$$

To calculate the equilibrium constant at any temperature, we use

$$\ln \frac{K_a(T)}{K_a(T = 298.15\ \text{K})} = \int_{298.15\ \text{K}}^{T} \frac{\Delta_{rxn}H^\circ(T)}{RT^2}\,dT$$

with

$$\Delta_{rxn}H^\circ(T) = \Delta_{rxn}H^\circ(T = 298.15\ \text{K}) + \int_{298.15\ \text{K}}^{T} \triangle_{rxn}C_P^\circ\,dT$$

$$= \Delta_{rxn}H^\circ(T = 298.15\ \text{K}) + \triangle_{rxn}C_P^\circ \cdot (T - 298.15\ \text{K})$$

so that

$$\ln \frac{K_a(T)}{K_a(T = 298.15\ \text{K})} = \int_{298.15\ \text{K}}^{T} \frac{\Delta_{rxn}H^\circ(T = 298.15\ \text{K})}{RT^2}\,dT$$

$$+ \int_{298.15\ \text{K}}^{T} \frac{\Delta_{rxn}C_P^\circ \cdot (T - 298.15\ \text{K})}{RT^2}\,dT$$

$$= -\Delta_{rxn}H^\circ(T = 298.15\ \text{K}) \cdot \left(\frac{1}{T} - \frac{1}{298.15\ \text{K}}\right)$$

$$+ \frac{\Delta_{rxn}C_P^\circ}{R}\left[\ln \frac{T}{298.15\ \text{K}} + (298.15\ \text{K}) \cdot \left(\frac{1}{T} - \frac{1}{298.15\ \text{K}}\right)\right]$$

We obtain the following values:

T (°C)	$K_a(T)$	α (M)	Percent Trimerized
0	1.237×10^7	0.01000	100.0
25	2.297×10^5	0.009999	99.99
50	1.944×10^3	0.009923	99.23

Therefore, over the temperature range of 0 to 50°C glucagon is almost completely trimerized.■

Polymers are formed by the polymerization of monomers, and proteins can aggregate (or associate) as in the illustration above. It is sometimes of use to know the relationship between the partial molar Gibbs energies of the polymer and the monomer or protein from which it is formed. To be completely general, suppose species A (whether it be a protein or polymer) polymerizes according to the reaction

$$\beta A \rightleftharpoons A_\beta \tag{15.5-1}$$

The total Gibbs energy of the system is

$$G = N_A \bar{G}_A + N_{A_\beta} \bar{G}_{A_\beta}$$

If the initial number of moles of species A is N, after chemical equilibrium is established we have

$$N_A = N - \alpha \quad \text{and} \quad N_{A_\beta} = \frac{\alpha}{\beta}$$

where α is the number of moles of species A that has reacted. Therefore,

$$G = (N - \alpha)\, \bar{G}_A + \frac{\alpha}{\beta} \bar{G}_{A_\beta}$$

At equilibrium at constant temperature and pressure $\left(\frac{\partial G}{\partial \alpha}\right)_{T,P} = 0$, and using the Gibbs-Duhem equation (Eq. 8.2-9a) gives

$$-\bar{G}_A + \frac{1}{\beta}\bar{G}_{A_\beta} = 0 \quad \text{or} \quad \bar{G}_{A_\beta} = \beta\bar{G}_A \qquad \textbf{(15.5-2)}$$

That is, the partial molar Gibbs energy of a polymer containing β monomeric units is just β times the partial molar Gibbs energy of the monomer. Note that this is true whether or not the solution is ideal.

We next consider the pH dependence of a biochemical reaction. To be specific we examine the oxidation reaction of ethanol by dehydrogenase enzyme and the oxidized form of nicotinamide adenine dinucleotide (NAD^+) to produce acetaldehyde and the reduced form NADH:

pH dependence of a biochemical reaction

$$NAD^+ + CH_3CH_2OH \overset{K_a}{\rightleftharpoons} CH_3CHO + NADH + H^+$$

The equilibrium relation for this reaction is

$$\Delta_{\text{rxn}} G^\circ = -RT \ln K_a = -RT \ln \frac{a_{CH_3CHO} \cdot a_{NADH} \cdot a_{H^+}}{a_{NAD^+} \cdot a_{CH_3CH_2OH}}$$

based on ideal, 1-molal standard states. This equilibrium relation can be rewritten as

$$-RT \ln K_a = -RT \ln \frac{a_{CH_3CHO} \cdot a_{NADH}}{a_{NAD^+} \cdot a_{CH_3CH_2OH}} - RT \ln a_{H^+}$$

$$= -RT \ln \frac{a_{CH_3CHO} \cdot a_{NADH}}{a_{NAD^+} \cdot a_{CH_3CH_2OH}} - RT \ln(10) \log(a_{H^+})$$

$$= -RT \ln \frac{a_{CH_3CHO} \cdot a_{NADH}}{a_{NAD^+} \cdot a_{CH_3CH_2OH}} + 2.303 RT \cdot \text{pH} \qquad \textbf{(15.5-3)}$$

At reaction conditions, the pH will normally be controlled externally—for example, by a buffer solution—rather than being determined by the ionization of water or by hydrogen ion production from the reaction. Consequently, we can write this equation as

$$\ln K_a + 2.303 \cdot \text{pH} = \ln \frac{a_{CH_3CHO} \cdot a_{NADH}}{a_{NAD^+} \cdot a_{CH_3CH_2OH}}$$

So increasing the pH of the solution results in an increased equilibrium dehydrogenation of ethanol to acetaldehyde. (*Note:* The dehydrogenation of ethanol is an enzymatic reaction. In this reaction NAD$^+$ is what is referred to as a cofactor, that is, a substance that is needed for the reaction and that will be regenerated by another reaction, but is not a catalyst for the reaction; the enzymes involved are the catalysts. Vitamins and minerals are common cofactors.)

ILLUSTRATION 15.5-2

The pH *Dependence of a Biochemical Reaction*

At 25°C and pH = 7, $\Delta_{rxn}G^\circ$ for the reaction above based on ideal 1-molal standard states is +23.8 kJ/mol. Assuming an initial concentration of both NAD$^+$ and ethanol of 0.1 M, compute the extent of reaction as a function of the externally controlled pH.

SOLUTION

The mass balance, using α as the fraction of ethanol oxidized, is

$$M_{NAD^+} = M_{CH_3CH_2OH} = 0.1(1 - \alpha) \quad \text{and} \quad M_{NADH} = M_{CH_3CHO} = 0.1\alpha$$

Now neglecting solution nonidealities (or equivalently, assuming they cancel in the numerator and denominator of the equilibrium relation), we have

$$\ln K_a + 2.303 \cdot \text{pH} = \ln \frac{a_{CH_3CHO} \cdot a_{NADH}}{a_{NAD^+} \cdot a_{CH_3CH_2OH}} = \ln \frac{0.1^2 \alpha^2}{0.1^2 (1 - \alpha)^2} = \ln \frac{\alpha^2}{(1 - \alpha)^2}$$

and

$$\frac{\alpha^2}{(1 - \alpha)^2} = K_a \cdot e^{2.303 \cdot \text{pH}} \quad \text{or} \quad \frac{\alpha}{1 - \alpha} = \sqrt{K_a} \cdot e^{(2.303 \cdot \text{pH}/2)}$$

Alternatively, this can be written as

$$\frac{\alpha}{1 - \alpha} = e^{\frac{2.303}{2}(\text{pH} - \text{p}K_a)}$$

and

$$\alpha = \frac{8.224 \times 10^{-3} \times e^{(2.303 \cdot \text{pH}/2)}}{1 + 8.224 \times 10^{-3} \times e^{(2.303 \cdot \text{pH}/2)}}$$

The figure below shows how the extent of this reaction, and the fraction of ethanol oxidized, change dramatically with the pH of the solution, with very little reaction at low pH (acid solution), a rapid increase with pH between pH = 2 and pH = 6, and the reaction being essentially complete above pH = 7.

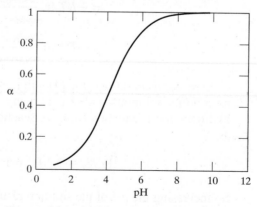

Extent of the reaction α as a function of pH

Note that $\alpha = 0.5$ when pH $= pK_a$, also $\alpha < 0.5$ when pH $< pK_a$, and the extent of reaction $\alpha > 0.5$ when pH $> pK_a$. Thus in the case of reactions, as already mentioned in the discussion on solubility, large changes occur at pH values near the pK_a value. ∎

Protein unfolding (denaturation)

The unfolding of proteins can be considered to be a type of biochemical reaction. Proteins in the native state exist in a specific folded state that is necessary for their biological catalytic activity. With changes in temperature, pressure, or chemical environment (e.g., changes in pH or ionic strength, or the addition of denaturants such as urea), the protein will unfold or denature and lose its biological activity. This unfolding and denaturation, depending on the protein, may be reversible or irreversible. Though many intermediate steps are involved, protein unfolding is frequently treated as a two-state process (folded or unfolded). Also, since the transition occurs over a range of conditions (for example, over a temperature range rather than at a single temperature), the process is considered to be a unimolecular chemical reaction, not a first-order phase transition (such as the boiling of a pure liquid that occurs at a single temperature). In the biological literature the temperature at which half of the initial amount of protein has unfolded is referred to as the "melting point" of the protein, though that is a misnomer since the protein does not melt (which would be a first-order phase transition). Also, this melting point can be a function of state conditions such as pH, pressure, ionic strength, etc.

ILLUSTRATION 15.5-3

Unfolding of a Protein as a Function of Temperature

G-actin is a globular dimer of actin, a cytoskeleton protein. In solutions at pH $= 7.5$ the following data are available for the unfolding of G-actin at 25°C:

$$\Delta_{\text{unf}}\underline{G} = 27\ \frac{\text{kJ}}{\text{mol}} \qquad \Delta_{\text{unf}}\underline{H} = 197\ \frac{\text{kJ}}{\text{mol}} \quad \text{and} \quad \Delta_{\text{unf}}C_{\text{P}} = 27\ \frac{\text{J}}{\text{mol K}}$$

Determine the temperature at which half of the G-actin is unfolded (that is, find its "melting temperature").

SOLUTION

The apparent equilibrium constant for this unimolecular chemical reaction is

$$K(T) = \frac{M_{\text{unfolded}}(T)}{M_{\text{folded}}(T)} = \frac{\alpha(T) \cdot M_{\text{o}} \cdot \gamma_{\text{unfolded}}(T)}{(M_{\text{o}} - \alpha(T) \cdot M_{\text{o}}) \cdot \gamma_{\text{folded}}(T)} = \frac{\alpha(T)}{(1 - \alpha(T))}$$

where $\alpha(T)$ is the fraction of the initial G-actin that has been denatured at the temperature T, and since there is only a conformational change between the folded and unfolded protein, we have assumed that their activity coefficients are equal, and therefore cancel from the equation. Also, we are interested in the melting temperature, that is, the temperature T_{m} at which $\alpha = 1/2$, so that $K = 1$ and $\ln K = 0$. Therefore, the equation to be solved is

$$\ln K(T_{\text{m}}) = 0$$

However,

$$\ln K(T_{\text{m}}) = -\frac{\Delta_{\text{unf}}G(25°C)}{R \cdot 298.15} - \frac{\Delta_{\text{unf}}H(25°C)}{R}\left(\frac{1}{T_{\text{m}}} - \frac{1}{298.15}\right)$$
$$+ \frac{\Delta_{\text{unf}}C_{\text{P}}}{R}\left[\ln\left(\frac{T_{\text{m}}}{298.15}\right) + \frac{298.15}{T_{\text{m}}} - 1\right]$$

or

$$0 = -\frac{27\,000}{8.314 \cdot 298.15} - \frac{197\,000}{8.314}P\left(\frac{1}{T_\mathrm{m}} - \frac{1}{298.15}\right)$$
$$+\frac{54\,000}{R}\left[\ln\left(\frac{T_\mathrm{m}}{298.15}\right) + \frac{298.15}{T_\mathrm{m}} - 1\right]$$

which has the solution of $T_\mathrm{m} = 330.0 \mathrm{~K} = 56.8°C$, which is close to the measured value of 57.1°C.

It is also of interest to determine the fraction of G-actin that is unfolded at any temperature. To do this we solve the equation

$$K(T) = \frac{\alpha}{1 - \alpha} \quad \text{or} \quad \alpha = \frac{K(T)}{1 + K(T)}$$

where

$$K(T) = \exp\left[-\frac{27\,000}{8.314 \cdot 298.15} - \frac{197\,000}{8.314}\left(\frac{1}{T} - \frac{1}{298.15}\right)\right.$$
$$\left. + \frac{54\,000}{R}\left[\ln\left(\frac{T}{298.15}\right) + \frac{298.15}{T} - 1\right]\right]$$

The results are shown in the following figure. Note that the unfolding of G-actin occurs over a temperature interval of more than 20°C.

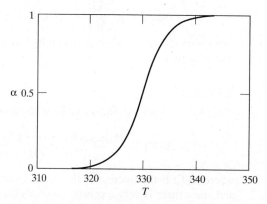

Fraction α of the protein G-actin unfolded ("melted") as a function of temperature.

Proteins may also unfold or denature as a result of increasing external pressure, presumably because a pressure increase results in an increase in solution density that produces a structural rearrangement of the protein. Consequently, the melting temperature of a protein will also be a function of pressure.

To determine how the melting temperature of a protein changes with pressure, we start from the observation that at the melting temperature at any pressure

$$\triangle_\mathrm{unf}\underline{G}(T_\mathrm{m}) = -RT \ln K_a(T_\mathrm{m}) = 0 = \triangle_\mathrm{unf}\underline{H}(T_\mathrm{m}) - T_\mathrm{m} \cdot \triangle_\mathrm{unf}\underline{S}(T_\mathrm{m}) \qquad \textbf{(15.5-4a)}$$

since $K_a(T_m) = 1$. Therefore,

$$\triangle_{unf}\underline{S}(T_m) = \frac{\triangle_{unf}\underline{H}(T_m)}{T_m} \tag{15.5-4b}$$

Since, along the melting curve

$$\underline{G}_{nat}(T, P) = \underline{G}_{unf}(T, P)$$

where the subscripts nat and unf refer to the native (folded) and unfolded proteins, respectively. Therefore, for a small change in pressure following the melting curve, we have

$$d\underline{G}_{nat}(T, P) = d\underline{G}_{unf}(T, P)$$

$$\underline{V}_{nat}\,dP - \underline{S}_{nat}\,dT = \underline{V}_{unf}\,dP - \underline{S}_{unf}\,dT$$

or

$$\left(\frac{dP}{dT}\right)_{\underline{G}_{nat}=\underline{G}_{unf}} = \frac{\underline{S}_{unf} - \underline{S}_{nat}}{\underline{V}_{unf} - \underline{V}_{nat}} = \frac{\triangle_{unf}\underline{S}}{\triangle_{unf}\underline{V}} = \frac{\triangle_{unf}\underline{H}}{T_m \cdot \triangle_{unf}\underline{V}} \tag{15.5-5}$$

(Compare this equation with the Clapeyron equation, Eq. 7.7-4.) Consequently, along the melting curve

$$T_m = \frac{\triangle_{unf}\underline{H}}{\triangle_{unf}\underline{V}} \cdot \left(\frac{dT}{dP}\right)_{\underline{G}_{nat}=\underline{G}_{unf}} \tag{15.5-6}$$

ILLUSTRATION 15.5-4

Unfolding of a Protein as a Function of Pressure

It has been observed[7] that at 1.013 bar the cell membrane compound dipalmitoylphosphatidylcholine (DPPC) melts at 313 K, and that the melting temperature increases by 0.04 K for each increase of 1 bar in pressure. Also, when this compound unfolds, there is a volume change of 26 cm^3/mol. Use these data to estimate

a. The enthalpy change on the denaturation of DPPC
b. The entropy change on this pressure-induced denaturation of DPPC

SOLUTION

a. To obtain the enthalpy change, we use

$$\triangle_{unf}\underline{H} = T_m \cdot \triangle_{unf}\underline{V}\left(\frac{dP}{dT}\right)_{\underline{G}_{nat}=\underline{G}_{unf}} = 313\text{ K} \times 26\,\frac{\text{cm}^3}{\text{mol}} \times \frac{1\text{ bar}}{0.04\text{ K}} \times 10^{-6}\,\frac{\text{m}^3}{\text{cm}^3} \times \frac{10^5\text{ J}}{\text{bar m}^3}$$

$$= 20\,345\,\frac{\text{J}}{\text{mol}} = 20.345\,\frac{\text{kJ}}{\text{mol}}$$

---

[7]D. B. Mountcastle, R. L. Biltonen, and M. J. Halsey, *Proc. Nat'l. Acad. Sci. USA* **75**, 4906 (1978).

b. To compute the entropy change, we use

$$\triangle_{unf}\underline{S}\,(T_m) = \frac{\triangle_{unf}\underline{H}\,(T_m)}{T_m} = \frac{20\,345\,\frac{J}{mol}}{313\,K} = 65\,\frac{J}{mol\,K}$$

■

Proteins can also be unfolded or denatured as a result of chemical interaction or changes in solution conditions. Common chemical denaturants include urea and guanidine hydrochloride; we do not consider this type of denaturation here.

15.6   PROTEIN CONCENTRATION IN AN ULTRACENTRIFUGE

While we have generally neglected the effect of kinetic and potential energies in this textbook, there are some cases in which such contributions can be important. One example is the case of an ultracentrifuge, a very-high-speed centrifuge that can be used to concentrate or separate proteins and other biological molecules based on their molecular weights.

Figure 15.6-1 is a very simple schematic diagram of a short centrifuge tube in an ultracentrifuge. The steady-state mass, energy, and entropy balance equations will be written for the region of the centrifuge tube between distances r and $r + \Delta r$ from the axis of rotation of the ultracentrifuge tube. The steady-state mass balance for each species i is

$$0 = \dot{N}_i\big|_r + \dot{N}_i\big|_{r+\triangle r} \quad \text{or} \quad \dot{N}_i\big|_{r+\triangle r} = -\dot{N}_i\big|_r \qquad (15.6\text{-}1)$$

The energy balance for this same volume element is

$$0 = \sum_i \dot{N}_i \left(\overline{H}_i - \frac{m_i\omega^2 r^2}{2} \right)_r + \sum_i \dot{N}_i \left(\overline{H}_i - \frac{m_i\omega^2 r^2}{2} \right)_{r+\triangle r} + \dot{Q} \qquad (15.6\text{-}2)$$

where ω is the angular velocity, m_i is the molecular weight of species i (and $\dot{N}_i m_i$ is its mass flow rate), and $-m_i\omega^2 r^2/2$ is the potential energy per unit mass of species i at a distance r from the center of a centrifuge with a rotational velocity ω.

The entropy balance on this same volume element is

$$0 = \sum_i \dot{N}_i \bar{S}_i\big|_r + \sum_i \dot{N}_i \bar{S}_i\big|_{r+\triangle r} + \frac{\dot{Q}}{T} + \dot{S}_{gen} \qquad (15.6\text{-}3)$$

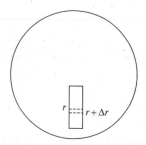

Figure 15.6-1 Schematic diagram of a tube containing a protein solution in an ultracentrifuge.

Now recognizing that during operation the temperature is constant, and combining the mass, energy, and entropy balances, we have

$$0 = \sum_i \dot{N}_i \left[\left(\bar{H}_i - T\bar{S}_i - m_i \frac{\omega^2 r^2}{2} \right)_r - \left(\bar{H}_i - T\bar{S}_i - m_i \frac{\omega^2 r^2}{2} \right)_{r+\triangle r} \right] - T\dot{S}_{\text{gen}}$$

Therefore, at equilibrium

$$0 = \sum_i \dot{N}_i \left[\left(\bar{G}_i - m_i \frac{\omega^2 r^2}{2} \right)_r - \left(\bar{G}_i - m_i \frac{\omega^2 r^2}{2} \right)_{r+\triangle r} \right] \qquad \textbf{(15.6-4)}$$

Since this equation must be satisfied for all values of $\dot{N}_i$, it then follows that the equilibrium condition for each species is

$$\left(\bar{G}_i - m_i \frac{\omega^2 r^2}{2} \right)_r = \left(\bar{G}_i - m_i \frac{\omega^2 r^2}{2} \right)_{r+\triangle r} \qquad \textbf{(15.6-5a)}$$

where r and $r + \Delta r$ are any two points along the centrifuge tube. Indeed, to emphasize this we will write the equilibrium relation as

$$\left(\bar{G}_i - m_i \frac{\omega^2 r^2}{2} \right)_1 = \left(\bar{G}_i - m_i \frac{\omega^2 r^2}{2} \right)_2 \qquad \textbf{(15.6-5b)}$$

Therefore, the basic equation describing the concentration of each species in an ultracentrifuge at any two points is

$$\bar{G}_{i,1}(T, P_1, \underline{x}_1) - \bar{G}_{i,2}(T, P_2, \underline{x}_2) = m_i \frac{\omega^2}{2} \left(r_1^2 - r_2^2 \right) \qquad \textbf{(15.6-6)}$$

but by definition

$$\bar{G}_{i,1}(T, P_1, \underline{x}_1) - \bar{G}_{i,2}(T, P_2, \underline{x}_2) = RT \ln \left[\frac{\bar{f}_{i,1}(T, P_1, \underline{x}_1)}{\bar{f}_{i,2}(T, P_2, \underline{x}_2)} \right] \qquad \textbf{(15.6-7)}$$

Now using the Poynting factor to correct for the effect of pressure on fugacity, assuming that the partial molar volume is not dependent on concentration or pressure,

$$\frac{\bar{f}_i(T, P_1, \underline{x}_1)}{\bar{f}_i(T, P_2, \underline{x}_2)} = \frac{\bar{f}_i(T, P_1, \underline{x}_1)}{\bar{f}_i(T, P_1, \underline{x}_2)} \exp \left[\frac{(P_1 - P_2)\bar{V}_i}{RT} \right] = \frac{x_{i,1}\gamma_{i,1}}{x_{i,2}\gamma_{i,2}} \exp \left[\frac{(P_1 - P_2)\bar{V}_i}{RT} \right]$$
$$\textbf{(15.6-8)}$$

we obtain

$$RT \ln \left(\frac{x_{i,1}\gamma_{i,1}}{x_{i,2}\gamma_{i,2}} \right) + (P_1 - P_2)\bar{V}_i = m_i \frac{\omega^2}{2}(r_1^2 - r_2^2) \qquad \textbf{(15.6-9)}$$

As a result of the centrifugal force, the pressure varies at each point in the centrifuge as follows:

$$(P_1 - P_2)\underline{V} = m \frac{\omega^2}{2} \left(r_1^2 - r_2^2 \right)$$

or

$$(P_1 - P_2) = \frac{m}{\underline{V}} \frac{\omega^2}{2} \left(r_1^2 - r_2^2\right)$$

(Note that we can derive this equation simply by taking the pure-component limit of Eq. 15.6-9. That is, for a pure component $x_i = 1$ and $\gamma_i = 1$, so the logarithmic term vanishes, and $m/\underline{V}$ becomes the density of the pure component, which is the solvent.) For the case of water as the solvent, $\rho = 1$ g/cc, we have

$$RT \ln \left(\frac{x_{i,1}\gamma_{i,1}}{x_{i,2}\gamma_{i,2}}\right) = m_i \frac{\omega^2}{2}(r_1^2 - r_2^2) - \rho_i \overline{V}_i \frac{\omega^2}{2}(r_1^2 - r_2^2)$$

$$= m_i \left(1 - \frac{\rho \overline{V}_i}{m_i}\right) \frac{\omega^2}{2}(r_1^2 - r_2^2)$$

$$= m_i \left(1 - \frac{\rho}{\rho_i}\right) \frac{\omega^2}{2}(r_1^2 - r_2^2) \qquad (15.6\text{-}10)$$

where $\rho_i = m_i/\overline{V}_i$ is the effective mass density of the protein in solution. As a simplification, we will neglect the departures from ideality (either because the system is ideal, or particularly for systems with biochemical compounds, there are so many other uncertainties due to their complexity, that the effects of nonidealities are frequently neglected.) Consequently, the final equation is

Ultracentrifuge equation

$$RT \ln \left(\frac{x_{i,1}}{x_{i,2}}\right) = m_i \left(1 - \frac{\rho}{\rho_i}\right) \frac{\omega^2}{2}\left(r_1^2 - r_2^2\right) \qquad (15.6\text{-}11)$$

ILLUSTRATION 15.6-1
Use of an Ultracentrifuge to Concentrate a Protein

The protein lysozyme has a molecular weight (m_i) of about 14 000 and a partial molar volume ($\overline{V}_i$) of $0.75m_i$, for an effective density (ρ_i) of 1.333 g/cc (which is a typical value for proteins). An aqueous solution of lysozyme is placed in an ultracentrifuge tube, the top of which is 2 cm and the bottom of which is 6 cm from the axis of rotation of the ultracentrifuge. The ultracentrifuge is at room temperature and operating at 10 000 revolutions per minute (rpm). What will be the ratio of the lysozyme concentration at the bottom of the tube to that at the top of the tube at equilibrium?

SOLUTION

Since the lysozyme is in aqueous solution, the solvent density $\rho \approx 1$ g/cc. Also,

$$\omega = 2\pi \times \text{(revolutions per second)} = \frac{2\pi \times 10\,000}{60} = 1047 \text{ s}^{-1}$$

Therefore,

$$\ln\left(\frac{x_{i,1}}{x_{i,2}}\right) = \frac{m_i\left(1 - \frac{\rho}{\rho_i}\right)\frac{\omega^2}{2}(r_1^2 - r_2^2)}{RT}$$

$$= \frac{14\,000\,\frac{g}{mol} \times \left(1 - \frac{1}{1.333}\right)\frac{g}{cc} \times 1047^2\,s^{-2} \times (2^2 - 6^2)cm^2}{2 \times 8.314\,\frac{J}{mol\,K} \times 298.15\,K}$$

$$= \frac{14\,000\,\frac{g}{mol} \times 0.2498 \times 1.0962 \times 10^6\,s^{-2} \times (2^2 - 6^2)cm^2 \times 10^{-4}\,\frac{m^2}{cm^2}}{2 \times 8.314\,\frac{J}{mol\,K} \times 298.15\,K \times 1\,\frac{kg\,m^2}{s^2\,J} \times 1000\,\frac{g}{kg}}$$

$$= -2.475$$

$$\ln\left(\frac{x_{i,1}}{x_{i,2}}\right) = -2.475 \quad \text{or} \quad x_{i,2} = x_{i,1}e^{2.475} = 11.87x_{i,1}$$

Thus, at 10 000 rpm, the lysozyme equilibrium concentration at the bottom of the centrifuge tube should be a factor of 11.87 higher than at the top of the centrifuge tube. ∎

15.7 GIBBS-DONNAN EQUILIBRIUM AND MEMBRANE POTENTIALS

In Sec. 11.5 we considered the osmotic pressure that arose between two cells separated by a membrane that was permeable to some components (there the solvent) but not permeable to the solute. Here we revisit the phenomenon of osmotic pressure for a somewhat more complicated case in which the solvent (usually water) contains a strong electrolyte, such as sodium hydroxide

$$NaOH \longrightarrow Na^+ + OH^-$$

the ions of which can pass through the membrane, and a protein that ionizes to a net charge of Z^-

$$P(H)_Z \Longleftrightarrow P^{Z-} + ZH^+$$

that cannot pass through the membrane, though the hydrogen ions produced can.

There are also hydrogen and hydroxyl ions present that result from the ionization of water; however, those concentrations are very low and can be neglected. Consequently, the hydrogen ions formed by the ionization of the protein can react with the hydroxyl ions to form water. By charge neutrality the molality of the free hydroxyl ions is related to the molality of the sodium ions and the molality of the ionized protein as follows:

$$M_{OH} = M_{Na} - ZM_P$$

For notational simplicity, we will not use the superscripts + and − to indicate the charges of the ions. The situation we will consider is that in which the water, sodium ions, and hydroxide ions can pass through the membrane, but the protein cannot.

We will designate the protein-free side of the membrane with the superscript I and the protein-containing side by II. Each of the compartments must itself be electrically neutral (because considerable energy is required to create a charge imbalance). Therefore, in compartment I, in which there is no protein, we have (using molality for the unit of concentration)

$$M_{Na}^{I} = M_{OH}^{I} \tag{15.7-1}$$

and in compartment II

$$M_{Na}^{II} = Z M_{P}^{II} + M_{OH}^{II} \tag{15.7-2}$$

Because both the sodium and hydroxyl ions are free to pass through the membrane, their partial molar Gibbs energies are the same in both compartments:

$$\bar{G}_{Na}^{I} = \bar{G}_{Na}^{II} \quad \text{and} \quad \bar{G}_{OH}^{I} = \bar{G}_{OH}^{II} \tag{15.7-3a}$$

However, here as in Sec. 9.10, the partial molar Gibbs energies of each species cannot be separately determined, so in fact we use

$$\bar{G}_{Na}^{I} + \bar{G}_{OH}^{I} = \bar{G}_{Na}^{II} + \bar{G}_{OH}^{II} \tag{15.7-3b}$$

which leads to

$$M_{Na}^{I} M_{OH}^{I} \left(\gamma_{\pm}^{I} \right)^{2} = M_{Na}^{II} M_{OH}^{II} \left(\gamma_{\pm}^{II} \right)^{2} \tag{15.7-3c}$$

For simplicity, we will neglect the activity coefficients (or assume they cancel from both sides of the equation) so that

$$M_{Na}^{I} M_{OH}^{I} = \left(M_{Na}^{I} \right)^{2} = M_{Na}^{II} M_{OH}^{II} \tag{15.7-4}$$

since the sodium ion and hydroxyl ion concentrations are equal in compartment I.

Combining Eq. 15.7-2 and 15.7-4, we obtain

$$M_{Na}^{II} = M_{Na}^{I} \cdot \frac{M_{OH}^{I}}{M_{OH}^{II}} = \frac{\left(M_{Na}^{I} \right)^{2}}{M_{Na}^{II} - Z \cdot M_{P}^{II}}$$

or

$$\left(M_{Na}^{II} \right)^{2} - Z \cdot M_{P}^{II} \cdot M_{Na}^{II} - \left(M_{Na}^{I} \right)^{2} = 0 \tag{15.7-5}$$

which has the solution

Gibbs-Donnan equilibrium

$$M_{Na}^{II} = \frac{Z \cdot M_{P}^{II}}{2} + \left[\sqrt{\left(\frac{Z \cdot M_{P}^{II}}{2} \right)^{2} + \left(M_{Na}^{I} \right)^{2}} \right] \tag{15.7-6a}$$

and

$$M_{OH}^{II} = \sqrt{\left(\frac{Z \cdot M_{P}^{II}}{2} \right)^{2} + \left(M_{Na}^{I} \right)^{2}} - \frac{Z \cdot M_{P}^{II}}{2} \tag{15.7-6b}$$

Therefore, if the sodium hydroxide molality in compartment I and the protein molality in compartment II are known, the sodium and hydroxyl ion molalities in the protein-containing compartment II can be computed. We see from these equations that if the protein is not ionized (i.e., $Z = 0$), the sodium ion concentrations and the hydroxide ion concentrations will be the same in compartments I and II. Also, if the sodium hydroxide concentration is much greater than the protein concentration (and especially if $M_{Na}^I = M_{NaOH}^I \gg Z \cdot M_P^{II}$), the sodium ion concentrations and the hydroxyl ion concentrations will be the same in compartments I and II. Under any other circumstance, the sodium and hydroxyl ion concentrations will be different in the two compartments. This is the Gibbs-Donnan equilibrium effect. (It is left to the reader to develop the equations for the case in which a positively charged protein is produced on ionization. See Problem 15.25.)

In this analysis it has been assumed that the activity coefficients are unity in going from Eq. 15.7-3c to Eq. 15.7-4. Including the activity coefficients would make the analysis and final results more complicated. However, the correction is usually small since the activity coefficients in both compartments may be close in value, and so may cancel in Eq. 15.7-3c.

ILLUSTRATION 15.7-1

Gibbs-Donnan Equilibrium of a Lysozyme Solution

One compartment in an osmotic cell contains lysozyme, which has a molecular weight of 14 000 at a concentration of 50 mg/g water.

  a. The second compartment contains a sodium hydroxide solution that is maintained at 1.585×10^{-3} M, which keeps the pH at 11.2, the isoelectric point of lysozyme. Find the concentration of sodium and hydroxyl ions in the lysozyme-containing compartment.
  b. The second compartment contains a sodium hydroxide solution that is maintained at 0.1 M, which keeps the pH at 13.0. Find the concentration of sodium and hydroxyl ions in the lysozyme-containing compartment.

SOLUTION

  a. At a pH = 11.2, which is the isoelectric point of lysozyme, its net charge is zero, so that from the equations above, the sodium and hydroxide ion concentrations in compartment II will be the same as in compartment I, that is,

$$M_{Na}^{II} = 1.585 \times 10^{-3} \text{ M} \quad \text{and} \quad M_{OH}^{II} = 1.585 \times 10^{-3} \text{ M}$$

  b. At pH = 13, from Fig. 15.2-3, the average charge on lysozyme is −9, so that in the equation above, $Z = 9$. Also, at 50 mg/g water, the protein molality is

$$M_P = \frac{50 \, \frac{\text{mg}}{\text{g water}} \times \frac{1000 \text{ g}}{\text{kg}} \times \frac{1 \text{ g}}{1000 \text{ mg}}}{14000 \, \frac{\text{g lysozyme}}{\text{mol}}} = 3.571 \times 10^{-3} \, \frac{\text{mol}}{\text{kg}} = 3.571 \times 10^{-3} \text{ M}$$

The solutions to Eqs. 15.7-6a and b (with $M_{Na}^I = M_{OH}^I = 0.1$ M) are

$$M_{Na}^{II} = 0.117 \, 352 \text{ M} \quad \text{and} \quad M_{OH}^{II} = 0.085 \, 213 \text{ M}$$

COMMENT

Note that the sodium ion concentrations in the two compartments are quite different in this last case, as are the hydroxyl ion concentrations. This is the Gibbs-Donnan equilibrium effect. ■

The fact that each ion partitions unevenly between the two compartments results in a contribution to the osmotic pressure from the ions in addition to that which results from the protein. Also, the fact that neither of the two phases is pure water (or another pure solvent) requires a modification to the derivation of the osmotic pressure of Sec. 11.5. The starting point now is

$$\bar{f}^I_{solvent} = \bar{f}^{II}_{solvent} \tag{15.7-7}$$

which leads to

$$x^I_{solvent} \gamma^I_{solvent} f_{solvent}(T, P^I) = x^{II}_{solvent} \gamma^{II}_{solvent} f_{solvent}(T, P^{II}) \tag{15.7-8}$$

Next, neglecting the activity coefficients of the solvent (which are usually close to 1 in value unless the solutions are concentrated in the solutes) and using the Poynting pressure correction, we obtain

$$x^I_{solvent} = x^{II}_{solvent} \exp\left[\frac{\underline{V}_{solvent}(P^{II} - P^I)}{RT}\right] \tag{15.7-9a}$$

or

Osmotic pressure in Gibbs-Donnan equilibrium

$$P^{II} - P^I = \Pi = \frac{RT}{\underline{V}_{solvent}} \ln\left(\frac{x^I_{solvent}}{x^{II}_{solvent}}\right) \tag{15.7-9b}$$

ILLUSTRATION 15.7-2
Osmotic Pressure of a Lysozyme Solution

   a. One compartment of an osmotic cell contains lysozyme at a concentration of 50 mg/g of water, and the second cell contains an aqueous solution of NaOH maintained at a concentration of 1.585×10^{-3} M, which has a pH of 11.2. Compute the osmotic pressure difference between the two cells.
   b. One compartment of an osmotic cell contains lysozyme at a concentration of 50 mg/g of water, and the second cell contains an aqueous solution of NaOH maintained at a concentration of 0.1 M, which has a pH of 13. Compute the osmotic pressure difference between the two cells.

SOLUTION

   a. Since at pH = 11.2, lysozyme is at its isoelectric point and therefore uncharged. Consequently, the concentration of sodium ions and hydroxyl ions in each phase is the same, 1.585×10^{-3} moles per kg of water. Also, from the previous illustration, the lysozyme concentration is 3.517×10^{-3} moles per kg of water, so that the water mole fractions in each cell are

$$x^I_{water} = \frac{55.51}{2 \times 1.585 \times 10^{-3} + 55.51} = 0.999\,943$$

and

$$x^{II}_{water} = \frac{55.51}{2 \times 1.585 \times 10^{-3} + 55.51 + 3.571 \times 10^{-3}} = 0.999\,879$$

Therefore, the osmotic pressure is

$$\Pi = \frac{RT}{\underline{V}_{\text{solvent}}} \ln\left(\frac{x_{\text{solvent}}^{\text{I}}}{x_{\text{solvent}}^{\text{II}}}\right) = \frac{8.314 \times 10^{-5} \dfrac{\text{bar m}^3}{\text{mol K}} \times 298.15 \text{ K}}{18 \times 10^{-6} \dfrac{\text{m}^3}{\text{mol}}} \ln\left(\frac{0.999\,943}{0.999\,879}\right) = 0.881 \text{ bar}$$

b. Using the results from Illustration 15.7-1, the water mole fraction in compartment I is

$$x_{\text{water}}^{\text{I}} = \frac{55.51}{2 \times 0.1 + 55.51} = 0.996\,410$$

the water mole fraction in compartment II is

$$x_{\text{water}}^{\text{II}} = \frac{55.51}{0.085\,213 + 0.117\,352 + 55.51 + 3.571 \times 10^{-3}} = 0.996\,300$$

and the osmotic pressure difference is

$$\Pi = \frac{RT}{\underline{V}_{\text{solvent}}} \ln\left(\frac{x_{\text{solvent}}^{\text{I}}}{x_{\text{solvent}}^{\text{II}}}\right) = \frac{8.314 \times 10^{-5} \dfrac{\text{bar m}^3}{\text{mol K}} \times 298.15 \text{ K}}{18 \times 10^{-6} \dfrac{\text{m}^3}{\text{mol}}} \ln\left(\frac{0.996\,410}{0.996\,300}\right) = 1.520 \text{ bar}$$

COMMENT

If the sodium ions partitioned equally in this last case, the concentrations in compartment II would be

$$M_{\text{lys}}^{\text{II}} = 3.571 \times 10^{-3} \text{ M} \qquad M_{\text{Na}}^{\text{II}} = 0.1 \text{ M}$$

and

$$M_{\text{OH}}^{\text{II}} = 0.1 - 9 \times 3.571 \times 10^{-3} = 0.067\,861 \text{ M}$$

$$x_{\text{water}}^{\text{II}} = \frac{55.51}{0.1 + 0.067\,861 + 55.51 + 3.571 \times 10^{-3}} = 0.996\,921$$

and the osmotic pressure difference would be

$$\Pi = \frac{8.314 \times 10^{-5} \dfrac{\text{bar m}^3}{\text{mol K}} \times 298.15 \text{ K}}{18 \times 10^{-6} \dfrac{\text{m}^3}{\text{mol}}} \ln\left(\frac{0.996\,410}{0.996\,921}\right) = -7.060 \text{ bar}$$

We see from this the importance of including the Gibbs-Donnan equilibrium effect when determining the partitioning of ions across a membrane when there is an ionic species (here the protein) that cannot pass through the membrane. In this case we see that by neglecting this effect, not only is the magnitude of the osmotic pressure changed, but so is its sign. ■

If there is a difference in ion concentrations across a membrane, an electrical potential difference will arise. To compute the magnitude of that potential difference, we start with the Nernst equation, Eq. 14.6-5, which can be applied to any one of the ions that can pass through the membrane:

$$\triangle G = RT \ln\frac{a_i^{\text{II}}}{a_i^{\text{I}}} = W = Z_i F \left(E^{\text{II}} - E^{\text{I}}\right) = Z_i F \triangle E \qquad \textbf{(14.6-5)}$$

where a_i is the activity of ion species i, Z_i is its charge, and F is the Faraday constant. If, as we have frequently done, we neglect the activity coefficients of the ions, this equation reduces to

Electrostatic potential in Gibbs-Donnan equilibrium

$$\left(E^{II} - E^{I}\right) = \frac{RT}{Z_i F} \ln \frac{M_i^{II}}{M_i^{I}}$$ (15.7-10)

ILLUSTRATION 15.7-3
Electrostatic Potential of Lysozyme Solutions in an Osmometer

a. One compartment of an osmotic cell contains lysozyme at a concentration of 50 mg/g of water, and the second cell contains an aqueous solution of NaOH maintained at a concentration of 1.585×10^{-3} M, which has a pH of 11.2. Compute the electrostatic potential difference between the two cells.

b. One compartment of an osmotic cell contains lysozyme at a concentration of 50 mg/g of water, and the second cell contains an aqueous solution of NaOH maintained at a concentration of 0.1 M, which has a pH of 13. Compute the electrostatic potential difference between the two cells.

SOLUTION

a. At pH $= 11.2$, lysozyme is at its isoelectric point and therefore uncharged. Consequently, the concentration of sodium ions and hydroxyl ions in each phase is the same, and there is no electrostatic potential difference between the two compartments.

b. From Illustration 15.7-1, we have for the sodium ions that

$$M_{Na}^{I} = 0.1 \text{ M and } M_{Na}^{I} = 0.117\,352 \text{ M}$$

Therefore,

$$E^{II} - E^{I} = \frac{25.7 \text{ mV}}{1} \ln \frac{0.117\,352}{0.1} = 4.11 \text{ mV}$$

As a check on this calculation, we can also calculate the potential difference resulting from the difference in hydroxyl ion concentrations:

$$E^{II} - E^{I} = \frac{25.7 \text{ mV}}{-1} \ln \frac{0.085\,213}{0.1} = 4.11 \text{ mV}$$

which agrees with the calculation based on the sodium ion concentration, as must be the case based on Eq. 15.7-3c or 15.7-4. ∎

Gibbs-Donnan equilibrium determines the concentration difference across simple membranes made of polymers, porous ceramic media, and other ultrafiltration devices. However, the difference of ion concentrations across the membranes of living cells and nerves is more complicated because of the existence of "ion pumps" as a result of carrier-mediated or facilitated diffusion, so that the concentrations of some ions are not in thermodynamic equilibrium. For example, there is a much higher sodium concentration outside cells than there is inside, while the reverse is true for potassium ions. This occurs because there is a carrier (probably a lipoprotein) that binds with a sodium ion inside the cell, transports the ion across membrane, and then releases it into the fluid outside the cell. The carrier is then transformed and binds with a potassium ion, which is then transported into the cell. This mechanism is discussed in courses in biology and physiology. The resulting concentration difference leads to diffusion of the sodium ion across the membrane, so that the steady-state (not equilibrium) sodium ion concentration gradient is a result of rapid active transport by the ion pump out of the cell, and much slower diffusion into the cell.

This pump does not operate equally in both directions, and two to three sodium ions are transported out of the cell for each potassium ion that enters the cell. Also, cell membranes are almost 100 times more permeable to potassium ions than to sodium ions, so that it is reasonable to assume the potassium ion concentration difference across a cell is an equilibrium state. On average, for a resting nerve fiber, the sodium ion concentrations are 10 millimolar (mM) within cells and 142 mM outside the cell, while the potassium ion concentrations are 5 mM outside the cell (in the extracellular fluid, ECF) and 140 mM inside the cell. (Note that the actual situation is further complicated by the fact that the ion channels open and close depending on the physiological situation. For example, during a nerve impulse the permeability of the membrane to sodium ions can increase by a factor of a thousand, almost eliminating the sodium ion concentration difference.)

ILLUSTRATION 15.7-4

Estimate of the Cell Membrane Potential

Using the potassium ion concentrations mentioned above, estimate the cell membrane potential in a nerve fiber.

SOLUTION

The potassium ion concentrations are approximately 140 mM within the cell and 5 mM in the plasma outside the cell. Consequently, at body temperature of 37°C the electrostatic potential across the cell membrane, referred to as the membrane potential, is

$$\triangle E = \frac{RT}{F} \ln \frac{M_K^{cell}}{M_K^{ECF}} = \frac{8.314 \dfrac{J}{mol\ K} \cdot 310.15\ K}{96\ 485 \dfrac{C}{mol} \cdot 1\dfrac{J}{C\ V}} \ln \frac{140}{5} = 0.0891\ V = 89.1\ mV$$

(At a laboratory temperature of 25°C, the electrostatic potential difference is 85.6 mV.) ∎

Chloride ions (and many other anions) diffuse easily through cell membranes, even though there is not a chloride ion pump. Consequently, the chloride ion concentration difference inside and outside of a cell is determined by the electrostatic potential.

ILLUSTRATION 15.7-5

Chloride Ion Concentration within a Nerve Cell

The chloride ion concentration in extracellular fluid is approximately 100 mM. Using the cell membrane potential as a result of potassium ions from the illustration above, estimate the chloride ion concentration within a resting nerve cell.

SOLUTION

Using the Nernst equation, we have

$$\triangle E = -\frac{RT}{F} \ln \frac{M_{Cl}^{cell}}{M_{Cl}^{ECF}} = -0.0891\ V = \frac{8.314 \dfrac{J}{mol\ K} \cdot 310.15\ K}{96\ 485 \dfrac{C}{mol} \cdot 1\dfrac{J}{C\ V}} \ln \frac{M_{Cl}^{cell}}{100}$$

which has the solution $M_{Cl}^{cell} = 5.0$ mM, which agrees with reported values of 4 to 5 mM. ∎

This illustration shows how the unequal distribution of a dominant ion, here K^+, across a cell membrane can establish an electrostatic potential difference that then results in the unequal distribution of other ions, many of which are vital to processes in the human body.

15.8 COUPLED CHEMICAL REACTIONS: THE ATP-ADP ENERGY STORAGE AND DELIVERY MECHANISM

The most important energy storage method in cells is as the chemical adenosine triphosphate (ATP). In the dissociation of ATP to adenosine diphosphate (ADP) and a phosphate ion, schematically shown as

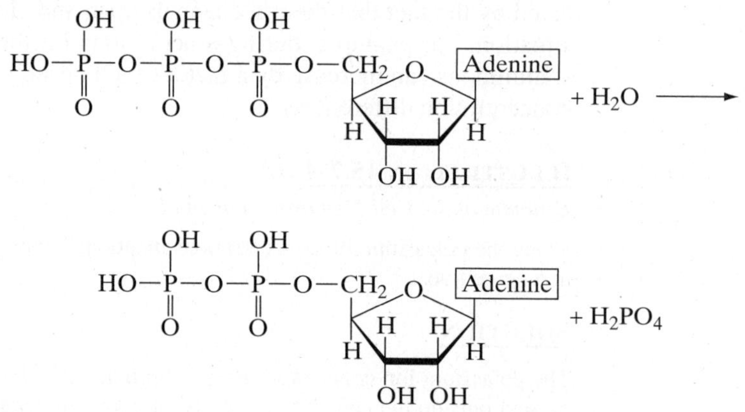

which is usually written simply as

$$ATP \Longleftrightarrow ADP + \text{phosphate} \tag{15.8-1}$$

a large amount of Gibbs energy is released that, for example, muscles convert to mechanical work by contraction, and by coupled reactions is used elsewhere in the body to provide the Gibbs energy for the synthesis of physiologically important molecules. In fact, ATP, which is present in the cytoplasm and nucleoplasm, provides the energy for most processes in cells. For example, as mentioned earlier, the sodium ion concentration within a cell is usually about 10 mM, while the concentration in the external plasma outside the cell membrane is much higher, about 140 mM. This concentration difference is maintained by a set of biological reactions and transport processes, the sodium ion pump, that uses Gibbs energy from ATP hydrolysis to expel sodium ions from the cell. Also, ATP hydrolysis provides energy for the functioning of nerves and muscle. Consequently, ATP has been called the energy currency of biological systems; it can be used to store energy in the body, supply energy when necessary, and be regenerated by a series of chemical reactions coupled to the partial oxidation of carbohydrates.

For this and other biological reactions the standard state chosen depends on temperature, the ionic strength, pH of the solution, and the presence of other ions (especially magnesium ions for this reaction). At very low ionic strength, pH = 7, and in the absence of magnesium ions, the values[8] of the standard-state Gibbs energy of reaction,

---

[8] As reported in *Biothermodynamics: The Study of Biochemical Processes at Equilibrium*, by J. T. Edsall and H. Gutfreund, John Wiley & Sons, New York (1983).

based on ideal 1-molal aqueous solutions, is 33.5 kJ/mol at 25°C and 34.0 kJ/mol at 37°C. However, physiological conditions are quite different, and the actual free energy change, neglecting solution nonidealities, is computed from

$$\triangle_{rxn}\underline{G} = \triangle_{rxn}\underline{G}^\circ + RT\ \ln\left[\frac{a_{ADP}\cdot a_P}{a_{ATP}}\right] = \triangle_{rxn}\underline{G}^\circ + RT\ \ln\left[\frac{\left(\frac{M_{ADP}}{M=1}\right)\cdot\left(\frac{M_P}{M=1}\right)}{\left(\frac{M_{ATP}}{M=1}\right)}\right]$$

(15.8-2)

While the concentrations of ATP, ADP, and the phosphate ion are neither constant nor known exactly, it is possible to make an estimate of the Gibbs energy released on the ATP hydrolysis reaction in muscles at body temperature (37°C) as in the illustration below.

ILLUSTRATION 15.8-1
Energy Release by ATP Hydrolysis in the Body

Assuming the following concentrations in human muscle,

$$M_{ATP} = 1 \times 10^{-3}\ \text{M} \qquad M_{ADP} = 1 \times 10^{-5} \quad \text{and} \quad M_P = 1 \times 10^{-3}\ \text{M}$$

estimate the Gibbs energy released on the hydrolysis of 1 mol of ATP at body temperature (37°C).

SOLUTION

Using these values of the concentration, we have

$$\triangle_{rxn}\underline{G} = -34.0 + 8.314 \times 10^{-3} \times 310.15 \times \ln\left[\frac{1 \times 10^{-5} \times 1 \times 10^{-3}}{1 \times 10^{-3}}\right] = -63.7\ \frac{\text{kJ}}{\text{mol}}$$

which is almost twice that of the standard-state Gibbs energy release. (It has been suggested[9] that the Gibbs energy change for this reaction in vivo is about −50 kJ/mol, though the concentrations of the species were not reported, and undoubtedly are different from the estimates used here.) Nonetheless, we can conclude that 50 to 60 kJ of Gibbs energy becomes physiologically available per mole of ATP hydrolyzed in muscles or other parts of the body. ∎

It should be pointed out that the mechanism by which the reaction occurs is much more complicated than Eq. 15.8-1 implies. Many intermediate compounds, and also enzymes acting as biological catalysts, are involved. The precise mechanism of this reaction is discussed in basic biology courses. However, as the sum of all the reactions is as shown in Eq. 15.8-1, from the point of view of thermodynamics, that is the only reaction we need to consider if our interest is solely in the net energy release (Sec. 13.1).

Given that there is such a large Gibbs energy release on the hydrolysis reaction of ATP to form ADP, we can ask the question of how ATP can be formed by the reverse reaction so that, for example, a relaxed muscle can regenerate ATP in preparation for the next contraction. We consider this next, but first have to introduce the concept of coupled chemical reactions.

---

[9]A. L. Lehninger, *Bioenergetics*, 2nd ed., Benjamin, London (1971).

From the second law of thermodynamics, any spontaneous process must occur with a net production of entropy. As shown in Chapter 5, a result of the second law is that the Gibbs energy of a system must decrease for a spontaneous process to occur at constant temperature and pressure. At first glance it might appear that this would mean that every process that occurs spontaneously must lower the Gibbs energy of the system. However, if there is a sequence of coupled processes, and they must be coupled, then as long as the total Gibbs energy of the system is lowered, the collection of coupled processes can occur even if some of the intermediate steps by themselves would increase the Gibbs energy of the system.

Before considering how this may occur in chemical reactions, we consider an analogy with the simple mechanical system shown in Fig. 15.8-1a. There we see two weights; weight A is at a low level, and the somewhat heavier weight B is at a higher level. Raising weight A to the higher level would increase its potential energy (and had we included mechanical potential and kinetic energy in our thermodynamic description, would increase its Gibbs energy as well); we know that this process will not occur spontaneously. Moving weight B to the lower level will decrease its potential energy and could occur easily, for example, if the weight were given a slight sideways push off the ledge. Now suppose that we couple the two weights, for example, by connecting them with the rope-and-pulley system shown in Fig. 15.8-1b. Since weight B is somewhat heavier than weight A, in this arrangement it is possible to raise weight A to the higher level, lower weight B, and still decrease the total potential (or Gibbs) energy of the system. Thus by coupling the two processes, it is possible to make one process occur that has a positive Gibbs energy change by driving it with a coupled process that has a larger negative free energy change.

A similar coupling of processes occurs frequently in nature. In fact, many chemical reactions consist of a number of molecular-level steps, the so-called elementary steps, one or more of which have a positive Gibbs energy change and would not occur spontaneously. However, the total reaction process, which consists of coupled elementary steps in which the products of one elementary step are the reactants for the next step, results in a net decrease in the Gibbs energy of the system.

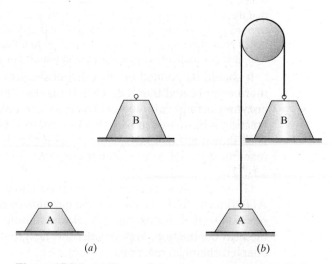

(a)   (b)

Figure 15.8-1 (a) Two weights that are not coupled. (b) Two weights that are coupled.

As a simple example of coupled reactions, consider an electrochemical cell consisting of lead and magnesium electrodes in a lead sulfate electrolyte. The half-cell reactions are

$$Pb^{2+} + 2e^- \rightarrow Pb \qquad E = -0.13 \text{ V} \qquad \Delta G^\circ = 0.13 \text{ F} = 12.5 \text{ kJ}$$

and

$$Mg \rightarrow Mg^{2+} + 2e^- \qquad E = +2.36 \text{ V} \qquad \Delta G^\circ = -2.36 \text{ F} = -227.6 \text{ kJ} \quad \textbf{(15.8-3)}$$

In this case the standard-state Gibbs energy change for the first half-cell reaction is positive, so that the reaction is not favorable, while the standard-state Gibbs energy change for the second half-cell reaction is strongly negative. For the overall process

$$Pb^{2+} + Mg \rightarrow Pb + Mg^{2+} \qquad \Delta G^\circ = -215.1 \text{ kJ} \qquad \textbf{(15.8-4)}$$

The Gibbs energy change for this overall process is very negative, so that the reaction can (and will) occur, even though it is the result of one process with a positive Gibbs energy change coupled with a second process with a larger negative Gibbs energy change. It is only by looking at the details of the process, here the two half-cell reactions, that we see that a process with a positive Gibbs energy change is being driven by another process with a larger negative Gibbs energy change. This is not evident by looking only at the overall process of Eq. 15.8-4.

Many naturally occurring processes are of the type discussed above, in that the overall process results in a negative Gibbs energy change, even though one or more of the elementary steps that are part of the overall process result in a positive Gibbs energy change. Some of the most important examples of this occur in biological processes. Of interest here is the phosphorylation of adenosine diphosphate (ADP) to regenerate adenosine triphosphate (ATP). The reaction is

$$ADP + \text{phosphate} \rightarrow ATP \qquad \Delta G^\circ = 33.5 \text{ kJ} \qquad \textbf{(15.8-5)}$$

Since the standard-state Gibbs energy change is positive, one would expect that this reaction would occur to only a very limited extent in biological systems. (Of course, the reason it would occur to any extent is that the reacting components are not in their standard states.) However, through a complicated series of enzymatic reactions, the phosphorylation of ATP is coupled to a reaction in which glucose is oxidized with an extremely large negative Gibbs energy change,

$$C_6H_{12}O_6 + 6O_2 \rightarrow 6CO_2 + 6H_2O \qquad \Delta G^\circ = -2870.2 \text{ kJ} \qquad \textbf{(15.8-6)}$$

Indeed, the Gibbs energy change is so large that in many biological systems the oxidation of one molecule of glucose can result in the production of 38 molecules of ATP, that is,

$$C_6H_{12}O_6 + 6O_2 + 38ADP + 38 \text{ phosphate} \rightarrow 6CO_2 + 6H_2O + 38ATP \quad \textbf{(15.8-7)}$$

for which $\Delta G^\circ = -1756.8$ kJ. Written in this manner, we would conclude that this reaction is thermodynamically possible and will occur to an appreciable extent. One measure of the efficiency of this process is to note that $38 \times 33.5 = 1273$ kJ of Gibbs energy (in the standard state) is stored as ATP for each mole of glucose oxidized, which releases 2870.2 kJ, so that only about 44 percent of the Gibbs energy released on oxidation is stored, the rest going to heat. The actual "in vivo" efficiency will be somewhat different, as the reactants and products are not in their standard states. The important observation here is that it is from coupled reactions resulting from the oxidation of glucose that ATP is regenerated in cells.

As we have seen by looking one level deeper, the reaction of Eq. 15.8-7 is composed of two reactions, one of which is thermodynamically favored and the second of which is not. If we examined the mechanism still further by considering all the coupled reactions, as biologists do, we would find that there is a very large collection of enzymatically catalyzed intermediate reactions that occur, and that some of them have negative Gibbs energy changes and others have positive Gibbs energy changes. Thermodynamics provides us with the information that the overall reaction is possible regardless of the internal mechanisms that might be necessary to drive this reaction. Nature has developed a coupled reaction mechanism in biological systems that is more clever than any developed by humans to make this reaction occur. However, even nature is constrained by the fact that the overall Gibbs energy change for any process at constant temperature and pressure must be negative! It is only within this limitation that interesting biological processes, such as active transport (the transport of species against a concentration gradient), occur.

The discussion of ATP synthesis above is for the case in which sufficient oxygen is present for the reaction to occur as described. That is, the glucose oxidation reaction occurs aerobically. It is also possible that as a result of extreme exercise or circulatory impairment there is not sufficient oxygen for the oxidation of glucose to occur. Under anaerobic conditions glucose can be used to form ATP from ADP, but by a different biochemical pathway (referred to as anaerobic glycolysis). In this process only about 2 percent of the energy in a glucose molecule is stored in ATP. However, even this small amount of ATP generated by anaerobic glycolysis can be important to the survival of a cell (and of organisms) during a temporary disruption of its oxygen supply.

While this book is not the appropriate place for a detailed discussion of the ADP-ATP cycle, perhaps a bit more mechanistic detail is useful. The aerobic metabolism of glucose to produce ATP from ADP can be considered to consist of three parts: the fermentation of glucose to form pyruvate (and a small amount of ATP from ADP), the conversion of the pyruvate to carbon dioxide in the citric acid cycle in which NAD^+ and FAD (the oxidized form of flavin adenine dinucleotide) are converted to NADH and $FADH_2$, and their oxidation in the respiratory chain, resulting in the formation of a larger quantity of ATP from ADP. If oxygen is not available, so the reaction is anaerobic, only the first step, formation of pyruvate, occurs with a small amount of ATP production.

In this discussion the concept of the reactions being coupled has been stressed, and this is an important point. If there were two reactions that were not coupled—for example, by the product of one reaction not being the reactant of the second—then the reactions would be independent, and the Gibbs energy of the system would be minimized when the Gibbs energy of each reaction is separately a minimum. That is, each reaction would achieve an equilibrium state independent of the other (except for the possibility of a weak coupling as a result of solution nonidealities that depend on the compositions of all species present). However, if the two reactions are coupled as in the examples here, the equilibrium state of minimum Gibbs energy is a function of both reactions and is in general different from the state in which each reaction separately goes to equilibrium.

15.9   THERMODYNAMIC ANALYSIS OF FERMENTERS AND OTHER BIOREACTORS

A bioreactor or fermenter is a chemical reactor in which microbes (e.g., bacteria or yeast) act on an organic material (referred to as a substrate) to produce additional microbes and other desired or undesired products. A schematic diagram of a bioreactor is given in Fig. 15.9-1. Mass balances for a biochemical reactor or fermenter are slightly more complicated than those for a usual chemical reactor because one of the products is cells or biomass (in the case of a wastewater or sewage treatment plant, which is another form of bioreactor, is referred to as sludge, which is a complicated chemically-undefined mixture of biochemical species). Likewise, other products of a bioreactor may also not be well-defined compounds of known molecular weight. The usual procedure, then, is that an elemental analysis is done on such products, and reported as the number of atoms of hydrogen, oxygen, and nitrogen (perhaps also sulfur, phosphorus, and other trace atoms that we will neglect here) for each carbon atom. For example, Roels[10] has suggested that in the analysis of bioreactors, one can use an "average biomass" of $CH_{1.8}O_{0.5}N_{0.2}$ to represent cells and bacteria. An elemental analysis can also be used for the substrate, especially if it contains a mixture of compounds, as, for example, in the mixed organic wastes entering a sewage treatment facility from an industrial or farming process. In these cases we do not have well-defined chemical species, but cells, other products, or substrates identified only by their atom ratios (usually referred to carbon). Consequently, mass balances for bioreactors are usually done for each atomic element, rather than for molecules, as was the case elsewhere in this book.

As a simple example, consider the fermentation of carbon monoxide in aqueous solution using the bacterium *Methylotrophicum* to produce acetic acid, butyric acid, and additional cells, which has the reported stoichiometry

$$CO + xO_2 + yN_2 + zH_2O \rightarrow 0.52CO_2 + 0.21CH_3COOH$$
$$+ 0.006CH_3CH_2CH_2COOH + 0.036 \text{ C-mole cells}$$

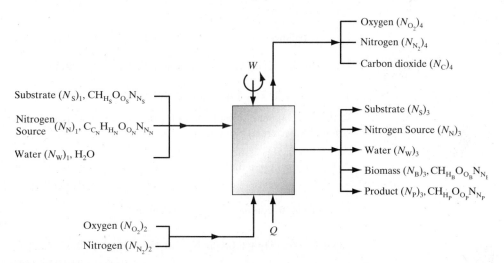

Figure 15.9-1 A schematic diagram of a fermenter (bioreactor) showing mass, heat, and work flows. The heat flow is generally to keep the reactor at constant temperature, and the work flow, which is usually small, is for stirring the reactor contents.

---

[10]J. A. Roels, *Energetics and Kinetics in Biotechnology*, Elsevier Biomedical Press, Amsterdam (1983).

where we have used the abbreviated notation C-mole cells to indicate the number of moles of elemental carbon present in cells. In the biochemical literature it is common notation to write stoichiometric equations in terms of each species on a C-mole basis, even though familiar compounds then appear in a less familiar way. For example, the reaction above would be written as

$$CO + xO_2 + yN_2 + zH_2O \rightarrow 0.52CO_2 + 0.42CH_2O$$
$$+ 0.024CH_2O_{0.5} + 0.036 \text{ C-mole cells}$$

where in this second equation the stoichiometric coefficients have changed since acetic acid contains two carbon atoms (so its initial stoichiometric coefficient is multiplied by 2 in this abbreviated notation), and butyric acid has four carbon atoms.

ILLUSTRATION 15.9-1

Mass Balance on a Bioreactor

Assuming that cells produced in this biochemical reaction can be represented by the Roels average biomass (that is, 1 C-mole cells = $CH_{1.8}O_{0.5}N_{0.2}$), how many moles of oxygen, nitrogen, and water are consumed for each mole of carbon monoxide consumed?

SOLUTION

First, we test the consistency of the reported stoichiometry by checking that the number of moles of carbon atoms balance on each side of the equation.

$$\text{C-balance:}\quad 1 = 0.52 + 0.42 + 0.024 + 0.036 = 1$$

So the carbon-atom balance is correct. The other atom balances are as follows.

$$\text{O-balance:}\quad 1 + 2x + z = 0.52 \cdot 2 + 0.42 + 0.024 \cdot 0.5 + 0.036 \cdot 0.5 = 1.49$$

$$\text{N-balance:}\quad 2 \cdot y = 0.036 \cdot 2 = 0.0072 \quad \text{or} \quad y = 0.0036$$

$$\text{H-balance:}\quad 2 \cdot z = 0.42 \cdot 2 + 0.024 \cdot 2 + 0.036 \cdot 1.8 = 0.9528 \quad \text{or} \quad z = 0.4764$$

Using this last result in the oxygen-atom balance, we obtain

$$2 \cdot x = 1.49 - 1 - z = 0.49 - 0.4764 = 0.0136 \quad \text{or} \quad x = 0.0068$$

so that the complete stoichiometry for this biochemical reaction (based on the assumed composition of the average biomass) is

$$CO + 0.0068 \cdot O_2 + 0.0036 \cdot N_2 + 0.4764 \cdot H_2O \rightarrow 0.52 \cdot CO_2 + 0.42 \cdot CH_2O$$
$$+ 0.024 \cdot CH_2O_{0.5} + 0.036CH_{1.8}O_{0.5}N_{0.2} \quad \textbf{(a)}$$

or, returning to common chemical notation,

$$CO + 0.0068 \cdot O_2 + 0.0036 \cdot N_2 + 0.4764 \cdot H_2O \rightarrow 0.52 \cdot CO_2 + 0.21 \cdot CH_3COOH$$
$$+ 0.006 \cdot CH_3CH_2CH_2COOH + 0.036CH_{1.8}O_{0.5}N_{0.2}$$

∎

In the analysis of biochemical reactors, it is common to report results in terms of yield factors that relate the production or consumption of one species relative to that

Table 15.9-1. Yield Factors Based on Substrate Consumed

$$Y_{B/S} = \frac{\text{Number of C-moles of biomass produced}}{\text{Number of C-moles of substrate consumed}}$$

$$Y_{i/S} = \frac{\text{Number of C-moles of product i produced}}{\text{Number of C-moles of substrate consumed}}$$

$$Y_{N/S} = \frac{\text{Number of C-moles of nitrogen source consumed}}{\text{Number of C-moles of substrate consumed}}$$

$$Y_{O_2/S} = \frac{\text{Number of moles of } O_2 \text{ consumed}}{\text{Number of C-moles of substrate consumed}}$$

$$Y_{W/S} = \frac{\text{Number of moles of water consumed}}{\text{Number of C-moles of substrate consumed}}$$

$$Y_{C/S} = \frac{\text{Number of moles of } CO_2 \text{ produced}}{\text{Number of C-moles of substrate consumed}}$$

of another. For example, the yield factor for biomass produced per mole of substrate consumed $Y_{B/S}$ is defined to be

$$Y_{B/S} = \frac{\text{Number of C-moles of biomass produced}}{\text{Number of C-moles of substrate consumed}}$$

The substrate is the carbonaceous material consumed, carbon monoxide in the example above. Some common yield factors are listed in Table 15.9.1.

ILLUSTRATION 15.9-2
Yield Factors for a Biochemical Reaction

What are the yield factors for the reaction in the previous illustration?

SOLUTION

From Eq. (a) of the previous illustration, we have $Y_{B/S} = 0.036$, $Y_{CH_2O/S} = 0.42$, $Y_{CH_2O_{0.5}/S} = 0.024$, $Y_{N/S} = 0.0036$, $Y_{O_2/S} = 0.0068$, $Y_{C/S} = 0.52$, and $Y_{W/S} = 0.4764$. ∎

Note that for generality, the reaction so far considered in this section could be written as

$$CO + Y_{O_2/S} \cdot O_2 + Y_{N_2/S} \cdot N_2 + Y_{W/S} \cdot H_2O \rightarrow Y_{C/S} \cdot CO_2 + Y_{CH_2O/S} \cdot CH_2O$$
$$+ Y_{CH_2O_{0.5}/S} \cdot CH_2O_{0.5} + Y_{B/S} \text{ C-mole cells}$$

Consider next the fermentation of some nonnitrogenous, but otherwise unspecified, substrate (for example, a sugar such as glucose) in aqueous solution with dissolved ammonia as the nitrogen source to produce "average" biomass, carbon dioxide, and water. On a carbon mole basis, this reaction is

$$\text{Substrate} + Y_{O_2/S} \cdot O_2 + Y_{N/S} \cdot NH_3 + Y_{W/S} \cdot H_2O \rightarrow Y_{B/S} \cdot CH_{1.8}O_{0.5}N_{0.2} + Y_{C/S} \cdot CO_2$$

The carbon balance is

$$1 = Y_{B/S} + Y_{C/S} \quad \text{or} \quad Y_{B/S} = 1 - Y_{C/S}$$

and the nitrogen balance is

$$Y_{N_2/S} = 0.2 \cdot Y_{B/S} \quad \text{or} \quad Y_{B/S} = 5 \cdot Y_{N_2/S}$$

Since we do not know the oxygen or hydrogen contents of the substrate, we cannot write any other complete balance equations for this system or solve for the other yield factors. All that we can say with certainty is (1) the number of C-moles of biomass produced per C-mole of substrate consumed will be less than unity by the number of moles of carbon dioxide produced per mole of substrate consumed, and (2) five C-moles of biomass will be produced for each mole of ammonia substrate consumed.

We now want to consider the completely general steady-state mass balance equations for the bioreactor of Fig. 15.9-1. The balances will be written for each species of atoms C, H, N, and O, and we will use the following notation:

[The equations to follow equally well be written for one cycle of a batch reactor in which case $(\dot{N}_I)_i$ is replaced with $(N_I)_i$]

Flows:
$(\dot{N}_I)_i$ = molar flow rate of species I in flow stream i

Subscripts:
S = substrate
N = nitrogen source substance
B = biomass produced
P = other product
C = carbon dioxide
W = water
N_2 = nitrogen
O_2 = oxygen

Atomic stoichiometric notation:
$\mathcal{C}_S$ = moles of carbon per C-mole of substrate = 1
$\mathcal{C}_N$ = moles of carbon per C-mole of nitrogen source (e.g., 0 in ammonia)
$\mathcal{C}_B$ = moles of carbon per C-mole of biomass = 1
$\mathcal{C}_P$ = moles of carbon per C-mole of product = 1
$\mathcal{C}_C$ = moles of carbon per mole of carbon dioxide = 1
$\mathcal{N}_S$ = moles of nitrogen atoms per C-mole of substrate
$\mathcal{N}_N$ = moles of nitrogen atoms per C-mole of nitrogen source (e.g., 1 in ammonia)
$\mathcal{N}_B$ = moles of nitrogen atoms per C-mole of biomass (e.g., 0.2 in average biomass)
$\mathcal{N}_P$ = moles of nitrogen atoms per C-mole of product
$\mathcal{N}_{N_2}$ = moles of nitrogen atoms per mole of molecular nitrogen = 2
$\mathcal{H}_S$ = moles of hydrogen atoms per C-mole of substrate
$\mathcal{H}_N$ = moles of hydrogen atoms per C-mole of nitrogen source (e.g., 3 in ammonia)
$\mathcal{H}_B$ = moles of hydrogen atoms per C-mole of biomass (e.g., 1.8 in average biomass)
$\mathcal{H}_P$ = moles of hydrogen atoms per C-mole of product
$\mathcal{H}_W$ = moles of hydrogen atoms per mole of water = 2
$\mathcal{O}_S$ = moles of oxygen atoms per C-mole of substrate
$\mathcal{O}_N$ = moles of oxygen atoms per C-mole of nitrogen source (e.g., 0 in ammonia)
$\mathcal{O}_B$ = moles of oxygen atoms per C-mole of biomass (e.g., 1.8 in average biomass)
$\mathcal{O}_P$ = moles of oxygen atoms per C-mole of product
$\mathcal{O}_W$ = moles of oxygen atoms per mole of water = 1
$\mathcal{O}_{O_2}$ = moles of oxygen atoms per mole of molecular oxygen = 2

With this notation, we have for the steady-state carbon balance

$$0 = (\dot{N}_S)_1 + (\dot{N}_N)_1 \cdot \mathcal{C}_N - (\dot{N}_S)_3 - (\dot{N}_N)_3 \cdot \mathcal{C}_N - (\dot{N}_B)_3 - (\dot{N}_P)_3 - (\dot{N}_C)_4 \cdot \mathcal{C}_C$$

or

$$0 = \left[(\dot{N}_S)_1 - (\dot{N}_S)_3\right] + \left[(\dot{N}_N)_1 - (\dot{N}_N)_3\right] \cdot \mathcal{C}_N - (\dot{N}_B)_3 - (\dot{N}_P)_3 - (\dot{N}_C)_4 \cdot \mathcal{C}_C$$

and

$$0 = 1 + \frac{(\dot{N}_N)_1 - (\dot{N}_N)_3}{(\dot{N}_S)_1 - (\dot{N}_S)_3} \cdot \mathcal{C}_N - \frac{(\dot{N}_B)_3}{(\dot{N}_S)_1 - (\dot{N}_S)_3}$$
$$- \frac{(\dot{N}_P)_3}{(\dot{N}_S)_1 - (\dot{N}_S)_3} - \frac{(\dot{N}_C)_4}{(\dot{N}_S)_1 - (\dot{N}_S)_3} \cdot \mathcal{C}_C$$

Finally, in terms of the yield factors, we obtain the simpler equation

Carbon balance

$$\boxed{1 + Y_{N/S} \cdot \mathcal{C}_S = Y_{B/S} + Y_{P/S} + Y_{C/S}} \qquad (15.9\text{-}1)$$

since in the notation used here to describe the flows into and out of the reactor,

$$Y_{N/S} = \frac{(\dot{N}_N)_1 - (\dot{N}_N)_3}{(\dot{N}_S)_1 - (\dot{N}_S)_3} \qquad Y_{B/S} = \frac{(\dot{N}_B)_3}{(\dot{N}_S)_1 - (\dot{N}_S)_3} \qquad Y_{P/S} = \frac{(\dot{N}_P)_3}{(\dot{N}_S)_1 - (\dot{N}_S)_3}$$

$$Y_{C/S} = \frac{(\dot{N}_C)_4}{(\dot{N}_S)_1 - (\dot{N}_S)_3} \qquad Y_{W/S} = \frac{(\dot{N}_W)_1 - (\dot{N}_W)_3 - (\dot{N}_W)_4}{(\dot{N}_S)_1 - (\dot{N}_S)_3}$$

and

$$Y_{O_2/S} = \frac{(\dot{N}_{O_2})_2 - (\dot{N}_{O_2})_4}{(\dot{N}_S)_1 - (\dot{N}_S)_3}$$

Note that defined in this way, the yield factors should be positive, with the exception of $Y_{W/S}$, which may be positive if water is consumed in the reaction or negative if water is produced in the reaction.

The hydrogen balance is as follows:

$$0 = (\dot{N}_S)_1 \cdot \mathcal{H}_S + (\dot{N}_N)_1 \cdot \mathcal{H}_N + (\dot{N}_W)_1 \cdot \mathcal{H}_W - (\dot{N}_S)_3 \cdot \mathcal{H}_S - (\dot{N}_N)_3 \cdot \mathcal{H}_N$$
$$- (\dot{N}_B)_3 \cdot \mathcal{H}_B - (\dot{N}_P)_3 \cdot \mathcal{H}_P - (\dot{N}_W)_3 \cdot \mathcal{H}_W - (\dot{N}_W)_4 \cdot \mathcal{H}_W$$

or

$$0 = \mathcal{H}_S + \frac{(\dot{N}_N)_1 - (\dot{N}_N)_3}{(\dot{N}_S)_1 - (\dot{N}_S)_3} \cdot \mathcal{H}_N + \frac{(\dot{N}_W)_1 - (\dot{N}_W)_3 - (\dot{N}_W)_4}{(\dot{N}_S)_1 - (\dot{N}_S)_3} \cdot \mathcal{H}_W$$
$$- \frac{(\dot{N}_B)_3}{(\dot{N}_S)_1 - (\dot{N}_S)_3} \cdot \mathcal{H}_B - \frac{(\dot{N}_P)_3}{(\dot{N}_S)_1 - (\dot{N}_S)_3} \cdot \mathcal{H}_P$$

and

$$\mathcal{H}_S + Y_{N/S} \cdot \mathcal{H}_N + Y_{W/S} \cdot \mathcal{H}_W = Y_{B/S} \cdot \mathcal{H}_B + Y_{P/S} \cdot \mathcal{H}_P$$

or

Hydrogen balance

$$\boxed{\mathcal{H}_S + Y_{N/S} \cdot \mathcal{H}_N + Y_{W/S} \cdot 2 = Y_{B/S} \cdot \mathcal{H}_B + Y_{P/S} \cdot \mathcal{H}_P}$$ (15.9-2)

The nitrogen balance is

$$0 = (\dot{N}_S)_1 \cdot \mathcal{N}_S + (\dot{N}_N)_1 \cdot \mathcal{N}_N + (\dot{N}_{N_2})_2 \cdot \mathcal{N}_{N_2} - (\dot{N}_S)_3 \cdot \mathcal{N}_S$$
$$- (\dot{N}_N)_3 \cdot \mathcal{N}_N - (\dot{N}_B)_3 \cdot \mathcal{N}_B - (\dot{N}_P)_3 \cdot \mathcal{N}_P - (\dot{N}_{N_2})_4 \cdot \mathcal{N}_{N_2}$$

which reduces to

Nitrogen balance

$$\boxed{\mathcal{N}_S + Y_{N/S} \cdot \mathcal{N}_N = Y_{B/S} \cdot \mathcal{N}_B + Y_{P/S} \cdot \mathcal{N}_P}$$ (15.9-3)

In the last form of this equation we have assumed that the number of moles of N_2 entering the reactor (for example, in the air) equals that leaving, since molecular nitrogen is not usually produced in a biochemical reaction.

Finally, the oxygen balance is

$$0 = (\dot{N}_S)_1 \cdot \mathcal{O}_S + (\dot{N}_N)_1 \cdot \mathcal{O}_N + (\dot{N}_W)_1 \cdot \mathcal{O}_W + (\dot{N}_{O_2})_2 \cdot \mathcal{O}_{O_2}$$
$$- (\dot{N}_S)_3 \cdot \mathcal{O}_S - (\dot{N}_N)_3 \cdot \mathcal{O}_N - (\dot{N}_B)_3 \cdot \mathcal{O}_B - (\dot{N}_P)_3 \cdot \mathcal{O}_P$$
$$- (\dot{N}_W)_3 \cdot \mathcal{O}_W - (\dot{N}_{O_2})_4 \cdot \mathcal{O}_{O_2} - (\dot{N}_W)_4 \cdot \mathcal{O}_W$$

which reduces to

Oxygen balance

$$\boxed{\mathcal{O}_S + Y_{N/S} \cdot \mathcal{O}_N + 2 \cdot Y_{O_2/S} + Y_{W/S} = 2 \cdot Y_{C/S} + Y_{B/S} \cdot \mathcal{O}_B + Y_{P/S} \cdot \mathcal{O}_P}$$

(15.9-4)

The advantage of using the yield factor notation can be seen by comparing the rather complicated initial forms of each of these balance equations for the atomic species in terms of the flows or flow rates to the much simpler final equations in terms of the yield factors. Note that while the steady-flow fermenter was considered here, precisely the same equations, Eqs. 15.9-1, 2, 3 and 4, would be obtained from the analysis of one cycle of a batch fermenter.

The following two illustrations are based on data in the book by Roels.

ILLUSTRATION 15.9-3

Mass Balance Using the Yield Factors

The yeast *Saccharomyces cerevisiae* is used in beer making and other fermentation processes to produce ethanol from the common plant sugar glucose in an anaerobic (no added oxygen) process in which dissolved ammonia is used as the nitrogen source. The production of biomass also occurs to the extent of 0.14 C-moles per C-mole of glucose. The biomass has an elemental composition of $CH_{1.8}O_{0.5}N_{0.2}$. What is the fractional conversion of glucose to ethanol (on a C-mole basis), the amount of ammonia used, and the amount of carbon dioxide produced per C-mole of glucose consumed?

SOLUTION

The chemical formula for glucose is $C_6H_{12}O_6$, which on a C-mole basis is CH_2O. The formula for ethanol is C_2H_5OH or $CH_3O_{0.5}$. Therefore, the set of algebraic equations to be solved is

$$\text{C-balance:} \quad 1 = Y_{B/S} + Y_{P/S} + Y_{C/S}$$

$$\text{H-balance:} \quad 2 + 3 \cdot Y_{N/S} + 2 \cdot Y_{W/S} = 1.8 \cdot Y_{B/S} + 3 \cdot Y_{P/S}$$

$$\text{N-balance:} \quad Y_{N/S} = 0.2 \cdot Y_{B/S}$$

and

$$\text{O-balance:} \quad 1 + Y_{W/S} = 2 \cdot Y_{C/S} + 0.5 \cdot Y_{B/S} + 0.5 \cdot Y_{P/S}$$

The solution to this set of algebraic equations is

$Y_{P/S} = 0.569$; that is, 0.569 C-moles of the product ethanol are produced for each C-mole of glucose consumed, or 1.707 ($= 0.569 \times 6/2$) moles of ethanol are produced per mole of glucose consumed.

$Y_{C/S} = 0.291$; that is, 0.291 C-moles of carbon dioxide are produced for each C-mole of glucose consumed, or 1.748 ($= 0.291 \times 6$) moles of carbon dioxide are produced per mole of glucose consumed.

$Y_{W/S} = -0.063$; that is, 0.063 moles of water are produced for each C-mole of glucose consumed, or 0.378 ($= 0.063 \times 6$) moles of water are produced per mole of glucose consumed.

$Y_{N/S} = 0.028$; that is, 0.028 moles of ammonia are consumed for each C-mole of glucose consumed, or 0.168 ($= 0.028 \times 6$) moles of ammonia are consumed per mole of glucose consumed. ∎

ILLUSTRATION 15.9-4

Production of Saccharomyces cerevisiae

The yeast *Saccharomyces cerevisiae* of the previous illustration is grown in a continuous, aerobic (oxygen-consuming) fermentation using glucose as the substrate and dissolved ammonia as the nitrogen source. It has been observed that the consumption of 4.26 C-moles of glucose results in 1 C-mole of the yeast and 1.92 C-moles of ethanol. Determine how much carbon dioxide is evolved and how much oxygen, ammonia, and water are consumed per C-mole of yeast produced. The elemental composition of the yeast is $CH_{1.8}O_{0.56}N_{0.17}$.

SOLUTION

From the problem statement, we have

$$Y_{B/S} = \frac{1}{4.26} = 0.235 \quad \text{and} \quad Y_{P/S} = \frac{1.92}{4.26} = 0.451$$

To calculate the other information needed, we use the four balance equations.

$$\text{C-balance:} \quad 1 = Y_{B/S} + Y_{P/S} + Y_{C/S}$$

$$\text{H-balance:} \quad 2 + 3 \cdot Y_{N/S} + 2 \cdot Y_{W/S} = 1.8 \cdot Y_{B/S} + 3 \cdot Y_{P/S}$$

$$\text{N-balance:} \quad Y_{N/S} = 0.17 \cdot Y_{B/S}$$

and

$$\text{O-balance: } 1 + 2 \cdot Y_{O_2/S} + Y_{W/S} = 2 \cdot Y_{C/S} + 0.56 \cdot Y_{B/S} + 0.5 \cdot Y_{P/S}$$

The solution to this set of equations is

$Y_{C/S} = 0.3145$; that is, 0.3145 C-moles of carbon dioxide are produced for each C-mole of glucose consumed, or 1.887 ($= 0.3145 \times 6$) moles of carbon dioxide are produced per mole of glucose consumed.

$Y_{N/S} = 0.0399$; that is, 0.0399 moles of ammonia are consumed for each C-mole of glucose consumed, or 1.748 ($= 0.291 \times 6$) moles of ammonia are consumed per mole of glucose consumed.

$Y_{W/S} = -0.1725$; that is, 0.1725 moles of water are produced for each C-mole of glucose consumed, or 1.035 ($= 0.1725 \times 6$) moles of water are produced per mole of glucose consumed.

$Y_{O_2/S} = 0.07923$; that is, 0.00792 moles of molecular oxygen are consumed for each C-mole of glucose consumed, or 0.475 ($= 0.0792 \times 6$) moles of O_2 are consumed per mole of glucose consumed.

Available experimental data indicate that about 0.35 moles of O_2 are consumed for each mole of yeast produced. Since 4.26 C-moles of glucose are required to produce 1 C-mole of yeast, it follows that $0.0792 \times 4.26 = 0.3375$ moles of molecular oxygen are required to produce 1 C-mole of yeast, which is close to the experimental value. ∎

The energy balance for a steady-state bioreactor is

$$0 = (\dot{N}\underline{H})_1 + (\dot{N}\underline{H})_2 + (\dot{N}\underline{H})_3 + (\dot{N}\underline{H})_4 + \dot{Q} + \dot{W} \qquad \text{(15.9-5)}$$

where we have recognized that at steady state the reactor volume is not changing. Proceeding further, we have

$$
\begin{aligned}
0 = & (\dot{N}_S\overline{H}_S)_1 + (\dot{N}_N\overline{H}_N)_1 + (\dot{N}_W\overline{H}_W)_1 + (\dot{N}_{O_2}\overline{H}_{O_2})_2 + (\dot{N}_{N_2}\overline{H}_{N_2})_2 \\
& - (\dot{N}_S\overline{H}_S)_3 - (\dot{N}_N\overline{H}_N)_3 - (\dot{N}_W\overline{H}_W)_3 - (\dot{N}_B\overline{H}_B)_3 - (\dot{N}_P\overline{H}_P)_3 \\
& - (\dot{N}_{O_2}\overline{H}_{O_2})_4 - (\dot{N}_{N_2}\overline{H}_{N_2})_4 - (\dot{N}_{CO_2}\overline{H}_{CO_2})_4 + \dot{Q} + \dot{W}
\end{aligned}
$$

Because of the number of flow streams and components involved, here we are departing from the usual notation in this textbook by writing the energy balance so that each flow rate of each species $\dot{N}$ is positive, and the sign in front of the term indicates whether it is a flow into the reactor ($+$) or a flow out of the reactor ($-$).

Generally, in chemical reactions, the effect of solution nonidealities on the enthalpy is small compared with the very much larger heats of reaction. Therefore, as an approximation, the partial molar enthalpies can be replaced by the pure-component enthalpies. Also, for simplicity, we will assume that the temperatures of the streams entering and leaving the fermenter are the same. (This may not be exactly true, but this effect will also be small compared with the heat of reaction, and could easily be included, if needed.) Finally, unless the solution is very viscous, the energy input from stirring is small compared with the heat of reaction. Therefore, we will also neglect the work term $\dot{W}$.

With these assumptions, the energy balance is

$$
\begin{aligned}
0 = & \left[(\dot{N}_S)_1 - (\dot{N}_S)_3\right] \underline{H}_S + \left[(\dot{N}_N)_1 - (\dot{N}_N)_3\right] \underline{H}_N + \left[(\dot{N}_W)_1 - (\dot{N}_W)_3\right] \underline{H}_W \\
& + \left[(\dot{N}_{O_2})_2 - (\dot{N}_{O_2})_4\right] \underline{H}_{O_2} + \left[(\dot{N}_{N_2})_2 - (\dot{N}_{N_2})_4\right] \underline{H}_{N_2} \\
& - (\dot{N}_B)_3 \underline{H}_B - (\dot{N}_P)_3 \underline{H}_P - (\dot{N}_{CO_2})_4 \underline{H}_{CO_2} + \dot{Q}
\end{aligned}
$$

Note that in this equation, the enthalpies of oxygen, nitrogen, carbon dioxide, and water are on a per-mole basis, while that of the substrate, the product, and the biomass are on a per–C-mole basis. The enthalpy of the nitrogen source is on a per-mole basis if the source does not contain carbon (for example, ammonia), and on a per–C-mole basis for a carbon-containing nitrogen source (for example, glutamic acid $C_5H_9NO_4$, which on a C-mole basis is $CH_{1.8}N_{0.2}O_{0.8}$).

Next we use the yield factors defined earlier and introduce one additional yield factor

$$Y_{Q/S} = \frac{\dot{Q}}{(\dot{N}_S)_1 - (\dot{N}_S)_3} \qquad (15.9\text{-}6)$$

to obtain the energy balance written on a per C-mole of substrate consumed basis and in terms of yield factors,

Bioreactor energy balance I

$$H_S + Y_{N/S}H_N + Y_{O_2/S}\underline{H}_{O_2} + Y_{W/S}H_W + Y_{N_2/S}\underline{H}_{N_2} + Y_{Q/S}$$
$$= Y_{B/S}H_B + Y_{P/S}H_P + Y_{CO_2/S}\underline{H}_{CO_2} \quad (15.9\text{-}7)$$

The first difficulty in using the energy balance is that enthalpy of formation data may not be available, especially for the biochemical species that are not completely defined. Also, the substrate may be a mixture of substances, and an incompletely characterized mixture of products may be produced in a fermenter. The information that is more likely to be available are the elemental analysis and the heat of combustion (it is relatively easy to put any substance in a calorimeter and measure the heat released on combustion). As a simple example, suppose we wanted to know the enthalpy of benzene at 25°C. We could put one mole of benzene in an isothermal calorimeter with 7.5 moles of oxygen and provide an ignition source to cause the reaction

$$C_6H_6 + 7.5O_2 \longrightarrow 6CO_2 + 3H_2O$$

The energy balance is

$$0 = \underline{H}_{C_6H_6} + 7.5\underline{H}_{O_2} - 6\underline{H}_{CO_2} - 3\underline{H}_{H_2O} + Q$$

where Q is the heat flow as a result of the combustion process, which we know from experience will be a negative number since heat is released on combustion, that is, $Q = -\triangle_c H$. Here a negative sign is used because while there is an energy release on combustion, the heat of combustion is reported as a positive number and applies to complete combustion—that is, for all carbon being converted to carbon dioxide, all hydrogen to water, and all nitrogen to N_2 (and other elements, if present, to their oxides)[11]:

$$0 = \underline{H}_{C_6H_6} + 7.5\underline{H}_{O_2} - 6\underline{H}_{CO_2} - 3\underline{H}_{H_2O} - \triangle_c\underline{H}_{C_6H_6}$$

It is important to note that when using heats of combustion, the reference states (i.e., the states of zero enthalpy) are nitrogen as molecular nitrogen, hydrogen as water, carbon as carbon dioxide, oxygen as molecular oxygen, and other elements that are present in their oxidized states (e.g., sulfur as SO_2). This is different from the reference

---

[11] In fact, two heats of combustion may be reported for a compound, one in which all the water formed is present as liquid, and a smaller value in which water is a vapor. The two values will differ by the product of heat of vaporization of water times the number of moles of water produced. We will generally use the liquid water reference state here.

states used for heats of formation, in which each element is in its simplest pure, stable state. Therefore, in the example being considered, if all species enter and leave at 25°C,

$$\underline{H}_{O_2} = 0 \quad \underline{H}_{CO_2} = 0 \quad \text{and} \quad \underline{H}_{H_2O} = 0 \quad \text{so that} \quad \underline{H}_{C_6H_6} = \triangle_c \underline{H}_{C_6H_6}$$

Now using the heat of combustion in the general energy balance for a bioreactor, we have

$$\triangle_c H_S + Y_{N/S} \triangle_c H_N + Y_{Q/S} = Y_{B/S} \triangle_c H_B + Y_{P/S} \triangle_c H_P \qquad \textbf{(15.9-8a)}$$

or

Bioreactor energy balance II

$$Y_{Q/S} = Y_{B/S} \triangle_c H_B + Y_{P/S} \triangle_c H_P - \triangle_c H_S - Y_{N/S} \triangle_c H_N \qquad \textbf{(15.9-8b)}$$

The next difficulty in using the energy balance is that thermodynamic property data, such as the heat of combustion, may not be known for some identified molecular species compounds, and is not likely to been measured for incompletely defined species such as the "average biomass" $CH_{1.8}O_{0.5}N_{0.2}$. To estimate the properties in these cases, we will use the generalized degree of reduction,[12] ξ, defined as follows:

Generalized degree of reduction

$$\xi_i = \frac{4 \times \mathcal{C}_i + \mathcal{H}_i - 2 \times \mathcal{O}_i}{\mathcal{C}_i} \qquad \text{for carbon-containing compounds}$$

and

$$\xi_i = \mathcal{H}_i - 2 \times \mathcal{O}_i \quad \text{for compounds without carbon but with H or O atoms} \quad \textbf{(15.9-9)}$$

where the numerical coefficients in Eq. 15.9-9 are the valences of the atoms. (Note that there are alternative definitions of the generalized degree of reduction. In the one used here, nitrogen does not appear, so that $\xi = 0$ for N_2 and $\xi = 3$ for NH_3. Also, $\xi = 0$ for O_2.) To a reasonable approximation, based on the data in Table 15.9-2 for the organic compounds,

$$\Delta_c G = 112 \cdot \xi \, \frac{kJ}{\text{C-mole}} \quad \text{and} \quad \Delta_c H = 110.9 \cdot \xi \, \frac{kJ}{\text{C-mole}} \qquad \textbf{(15.9-10)}$$

Equations 15.9-10 are referred to as the energy regularity principles in the biochemical literature.

ILLUSTRATION 15.9-5

Properties of Average Biomass

The "average biomass" used to represent many yeasts and bacteria has an elemental composition, on a C-mole basis, of $CH_{1.8}O_{0.5}N_{0.2}$. Compute its generalized degree of reduction, and estimate its Gibbs energy and heat of combustion.

SOLUTION

The generalized degree of reduction for the average biomass is

$$\xi = \frac{4 \times \mathcal{C} + \mathcal{H} - 2 \times \mathcal{O}}{\mathcal{C}} = \frac{4 \times 1 + 1.8 - 2 \times 0.5}{1} = 4.8$$

---

[12]V. K. Eroshin and I. J. Minkevich, *Biotech. Bioeng.*, **24**, 2263 (1982).

Table 15.9-2 The Degree of Reduction ξ, Gibbs Energy, and Heat of Combustion of Biochemicals and Hydrocarbons on a Carbon Mole Basis (kJ/C-mole)[a,b]

Compound	ξ	$\triangle_c G$ (kJ/C-mole)	$\triangle_c H$ (kJ/C-mole)	$\triangle_c G/\xi$	$\triangle_c H/\xi$
Acetic acid	4.00	447.0	438.0	111.8	109.5
Propionic acid	4.67	511.0	509.7	109.5	109.2
Butyric acid	5.00	543.3	548.5	108.7	109.7
Valeric acid	5.20	562.6	568.2	108.3	109.3
Palmitic acid	5.75	612.5	624.3	106.5	108.6
Lactic acid	4.00	459.0	456.3	114.8	114.1
Oxalic acid	1.00	163.5	123.0	163.5	123.0
Succinic acid	3.50	399.8	373.3	114.2	106.6
Fumaric acid	3.00	362.0	334.3	120.7	111.4
Malic acid	3.00	361.0	332.3	120.3	110.8
Citric acid	3.00	357.8	327.2	119.3	109.1
Glucose	4.00	478.7	467.8	119.7	117.0
Ethanol	6.00	659.5	684.5	109.9	114.1
i-Propanol	6.00	648.7	663.0	108.1	110.5
n-Butanol	6.00	648.0	670.0	108.0	111.7
Ethylene glycol	5.00	585.0	590.5	117.0	118.1
Glycerol	4.67	547.7	554.3	117.4	118.8
Glucitol	4.33	514.0	508.2	118.6	117.3
Acetone	5.33	578.0	597.7	108.4	112.1
Acetaldehyde	5.00	561.5	584.0	112.3	116.8
Alanine	5.00	547.3	569.0	109.5	113.8
Arginine	5.67	631.0	624.0	111.4	110.1
Asparagine	4.50	499.8	484.0	111.1	107.6
Glutamic acid	4.20	463.0	450.0	110.2	107.1
Aspartic acid	3.75	421.5	402.0	112.4	107.2
Glutamine	4.80	525.6	514.0	109.5	107.1
Glycine	4.50	505.5	487.0	112.3	108.2
Leucine	5.50	594.2	598.0	108.0	108.7
Isoleucine	5.50	594.0	598.0	108.0	108.7
Phenylalanine	4.78	516.3	517.0	108.0	108.2
Serine	4.33	500.7	485.0	115.6	111.9
Threonine	4.75	532.5	526.0	112.1	110.7
Trytophane	4.73	513.6	512.0	108.6	108.3
Tyrosine	4.56	498.1	493.0	109.3	108.2
Valine	5.40	584.0	584.0	108.2	108.2
Guanine	4.60	522.4	500.0	113.6	108.7
n-Hexane (l)	6.33	670.5	693.9	105.9	109.6
n-Heptane (l)	6.29	665.7	688.1	105.9	109.5
n-Octane (l)	6.25	662.2	683.9	106.0	109.4
n-Decane (l)	6.20	657.1	677.7	106.0	109.3
n-Dodecane (l)	6.17	653.7	673.5	106.0	109.2
n-Hexadecane (l)	6.13	649.5	667.9	106.0	109.0
n-Eicosane (l)	6.10	646.8	663.9	106.0	108.8
Toluene (l)	5.14	547.7	563.0	106.6	109.7
Cyclohexane (l)	6.00	636.1	652.0	106.0	108.7
Ethyl acetate (l)	5.00	552.0	563.6	110.0	112.7
Average				112.0	110.9

Table 15.9-2 (*Continued*) The Degree of Reduction ξ, Gibbs Energy, and Heat of Combustion of Biochemicals and Hydrocarbons on a Carbon Mole Basis (kJ/C-mole)[a,b]

Compound	ξ	$\triangle_c G$ (kJ/C-mole)	$\triangle_c H$ (kJ/C-mole)	$\triangle_c G / \xi$	$\triangle_c H / \xi$
Other compounds					
Ammonia(l, 0.01 M)	3.00	391.9	348.1	130.6	116.0
Hydrogen	2.00	238.0	286.0	119.0	143.0
Carbon monoxide	2.00	257.0	283.0	128.5	141.5
Nitric acid	−5.00	7.3	−30.0	−1.5	6.0
Hydrazine	4.00	602.4	622.0	150.6	155.6
Hydrogen sulfide	2.00	323.0	247.0	161.5	123.5
Sulfuric acid	−6.00	−507.4	−602.0	84.6	100.3

[a]The heat and Gibbs energy of combustion are based on liquid water and gaseous carbon dioxide and nitrogen as combustion products. Also, the Gibbs energy is based on liquid phase reactants and products at unit molality.

[b]Based on a table in Roels.

Therefore, from the energy regularity approximation

$$\Delta_c G = 112 \cdot \xi = 112 \cdot 4.8 = 537.6 \ \frac{\text{kJ}}{\text{C-mole}}$$

and

$$\Delta_c H = 110.9 \cdot \xi = 110.9 \cdot 4.8 = 532.3 \ \frac{\text{kJ}}{\text{C-mole}}$$

∎

While measured enthalpies or heats of combustion should be used whenever possible, the energy regularity estimate for the heat of combustion can be used in the energy balance, Eq. 15.9-8, for those species for which such data are not available. The most approximate form for the energy balance obtained when using the energy regularity estimate for all species is

Approximate bioreactor energy balance

$$Y_{Q/S} = 110.9 \left(Y_{B/S} \xi_B + Y_{P/S} \xi_P - \xi_S - Y_{N/S} \xi_N \right) \text{kJ} \qquad \textbf{(15.9-11)}$$

for each C-mole of substrate consumed.

ILLUSTRATION 15.9-6

Energy Balance on an Isothermal Fermenter

In Illustration 15.9-4 we considered the production of ethanol from glucose using the yeast *Saccharomyces cerevisiae*. In that illustration 0.451 C-moles of ethanol and 0.235 C-moles of biomass of elemental composition $CH_{1.8}O_{0.56}N_{0.17}$ were produced per C-mole of glucose, using 0.0399 moles of ammonia. Assuming the inlet and outlet streams are maintained at 25°C, and that the work input is negligible, what is the heat load on the reactor?

SOLUTION

From Table 15.9-2, the heat of combustion of ammonia is 383.0 kJ/mol, that of glucose is 467.8 kJ/C-mole, and ethanol is 684.5 kJ/mol. The heat of combustion of the biomass is unknown, so the energy regularity method will be used:

$$\xi = 4 \times 1 + 1 \times 1.8 - 2 \times 0.56 = 4.68$$

Therefore,

$$\Delta_c H = 110.9 \cdot \xi = 110.9 \cdot 4.68 = 519.0 \ \frac{\text{kJ}}{\text{C-mole}}$$

So that, for each C-mole of glucose consumed,

$$Y_{Q/S} = 0.235 \times 519.0 + 0.451 \times 684.5 - 467.8 - 0.0399 \times 348.1 = -51.0 \ \text{kJ}$$

of heat must be removed.

NOTE

It is useful to compare this answer with the more approximate one that is obtained using the generalized degree of reduction for all components:

$$Y_{Q/S} = 110.9 \, (0.235 \times 4.68 + 0.451 \times 6.00 - 4.00 - 0.0399 \times 3) = -34.8 \ \text{kJ}$$

per C-mole of glucose consumed. The difference between the two results largely arises from the error in the generalized degree of reduction approximation for the heat of combustion of glucose ($4 \times 110.9 = 443.6$ kJ/C-mole, compared with the experimental value of 467.8). ∎

There is one linear combination of the carbon, nitrogen, and oxygen balances that is especially useful. If we take the sum of four times the carbon balance plus the hydrogen balance and subtract twice the oxygen balance (note that the multipliers are the valences, and are the same numbers that appear in the definition of the generalized degree of reduction), we obtain

Alternative form of oxygen balance

$$(4 + \mathcal{H}_S - 2\mathcal{O}_S) + (4 + \mathcal{H}_N - 2\mathcal{O}_N) \, Y_{N/S} - 4Y_{O_2/S}$$
$$= (4 + \mathcal{H}_B - 2\mathcal{O}_B) \, Y_{B/S} + (4 + \mathcal{H}_P - 2\mathcal{O}_P) \, Y_{P/S}$$

or

$$\xi_S + \xi_N Y_{N/S} - 4Y_{O_2/S} = \xi_B Y_{B/S} + \xi_P Y_{P/S}$$

and

$$Y_{O_2/S} = \frac{\xi_S + \xi_N Y_{N/S} - \xi_B Y_{B/S} - \xi_P Y_{P/S}}{4} \qquad \textbf{(15.9-12)}$$

This last equation has several uses, as we will see shortly.

Now we move on to the entropy balance. Following the same type of analysis used for the energy balance on a bioreactor, we obtain the following equation for the entropy balance:

$$0 = \left[(\dot{N}_S)_1 - (\dot{N}_S)_3 \right] \bar{S}_S + \left[(\dot{N}_N)_1 - (\dot{N}_N)_3 \right] \bar{S}_N + \left[(\dot{N}_W)_1 - (\dot{N}_W)_3 \right] \bar{S}_W$$
$$+ \left[(\dot{N}_{O_2})_2 - (\dot{N}_{O_2})_4 \right] \bar{S}_{O_2} + \left[(\dot{N}_{N_2})_2 - (\dot{N}_{N_2})_4 \right] \bar{S}_{N_2} - (\dot{N}_B)_3 \bar{S}_B$$
$$- (\dot{N}_P)_3 \bar{S}_P - (\dot{N}_{CO_2})_4 \bar{S}_{CO_2} + \frac{\dot{Q}}{T} + \dot{S}_{gen} \qquad \textbf{(15.9-13)}$$

For an isothermal system, we can eliminate the heat flow between the energy and entropy balances, and use the partial molar Gibbs energy $\bar{G}_i = \bar{H}_i - T\bar{S}_i$ to obtain

$$0 = \left[(\dot{N}_S)_1 - (\dot{N}_S)_3\right]\bar{G}_S + \left[(\dot{N}_N)_1 - (\dot{N}_N)_3\right]\bar{G}_N + \left[(\dot{N}_W)_1 - (\dot{N}_W)_3\right]\bar{G}_W$$
$$+ \left[(\dot{N}_{O_2})_2 - (\dot{N}_{O_2})_4\right]\bar{G}_{O_2} + \left[(\dot{N}_{N_2})_2 - (\dot{N}_{N_2})_4\right]\bar{G}_{N_2} - (\dot{N}_B)_3\bar{G}_B$$
$$- (\dot{N}_P)_3\bar{G}_P - (\dot{N}_{CO_2})_4\bar{G}_{CO_2} - T\dot{S}_{gen} \quad \textbf{(15.9-14)}$$

The use of this equation is more complicated than using the energy balance because the partial molar Gibbs energy contains both the activity coefficient and the mole fraction of each component (which in the case of biomass is really a complex mixture); that is,

$$\bar{G}_i = \underline{G}_i + RT \ln(x_i\gamma_i) \quad \textbf{(15.9-15)}$$

Evaluation of this logarithmic term is very difficult since the mole fractions are not likely to be known (we may know only the elemental compositions of the biomass and perhaps other species, but not their molecular formulae), and little or no information is available on solution nonidealities in the mixtures containing yeast, bacteria, or other incompletely characterized substances. Consequently, we will make what might appear to be a rather severe assumption, that the logarithmic term can be neglected—that is, that we can replace the partial molar Gibbs energies with the pure-component Gibbs energies. Why is such an assumption not unreasonable here? There are four reasons:

1. There are an equal number of inlet and outlet streams, and each one has a term of this form, so there will be some degree of cancellation of this term among the four streams.
2. The contribution of any one species to the Gibbs energy of the stream is $N_i(\underline{G}_i + RT \ln x_i\gamma_i)$. When $\dot{N}_i$ is large, the mole fraction of that species will be near 1, so that the logarithmic term will be small. When the mole fraction of the species is very small (so that the logarithmic term is larger), the flow term $\dot{N}_i$ multiplying that term is small. Therefore, in both cases the contribution will not be large.
3. The value of RT that is the multiplier for the logarithmic term has a value of 2.5 kJ/mole, which, while not insignificant, is much smaller in value than the pure-component Gibbs energy terms.
4. The logarithmic terms being neglected here are of great importance in determining the equilibrium state, as was shown in Chapter 13. However, in manufacturing processes chemical reactors, fermenters, and other bioreactors are not generally operated to achieve equilibrium as the residence time required, and therefore the reactor size, would be so large as to be uneconomical. For the reasons stated in points 1, 2, and 3, these logarithmic terms are of less importance away from the equilibrium state.

(Note that the comments made here about the contributions of the solution nonidealities and the mixing terms are similar to those made in Sec. 14.5 in discussing calculations involving the availability function.)

Therefore, it is reasonable to simplify the equation to

$$0 = \left[(\dot{N}_S)_1 - (\dot{N}_S)_3\right]G_S + \left[(\dot{N}_N)_1 - (\dot{N}_N)_3\right]G_N + \left[(\dot{N}_W)_1 - (\dot{N}_W)_3\right]G_W$$
$$+ \left[(\dot{N}_{O_2})_2 - (\dot{N}_{O_2})_4\right]\underline{G}_{O_2} + \left[(\dot{N}_{N_2})_2 - (\dot{N}_{N_2})_4\right]\underline{G}_{N_2} - (\dot{N}_B)_3 G_B$$
$$- (\dot{N}_P)_3 G_P - (\dot{N}_{CO_2})_4\underline{G}_{CO_2} - T\dot{S}_{gen} \quad \textbf{(15.9-16)}$$

(where G is a pure-component Gibbs energy on a per–C-mole basis). Now rewriting this equation in terms of the yield factors gives

$$0 = G_S + Y_{N/S}G_N + Y_{O_2/S}\underline{G}_{O_2} + Y_{W/S}G_W + Y_{N_2/S}\underline{G}_{N_2}$$
$$- Y_{B/S}G_B - Y_{P/S}G_P - Y_{CO_2/S}\underline{G}_{CO_2} - T\underline{S}_{gen} \quad \textbf{(15.9-17)}$$

and $\underline{S}_{gen}$ is the entropy generated per C-mole of substrate consumed. Here, as with the energy balance, we can use Gibbs energies of combustion (see Table 15.9-2) in place of the Gibbs energies and simplify the equation to

Second-law balance

$$\triangle_c G_S + Y_{N/S}\triangle_c G_N = Y_{B/S}\triangle_c G_B + Y_{P/S}\triangle_c G_P + T\underline{S}_{gen} \quad \textbf{(15.9-18)}$$

Our interest in this equation will be in determining the constraint that thermodynamics (and, in particular, the second law or entropy balance) places on the maximum conversion possible of substrate to product and additional biomass. Based on discussions elsewhere in this textbook, the maximum production of product will occur when the process is operated such that $S_{gen} = 0$. However, for real processes, $S_{gen} \geq 0$. Therefore, the form of the equation that will be used is

Second-law constraint

$$\triangle_c G_S + Y_{N/S}\triangle_c G_N \geq Y_{B/S}\triangle_c G_B + Y_{P/S}\triangle_c G_P \quad \textbf{(15.9-19)}$$

Finally, for the purposes of making an approximate calculation of the maximum possible conversion, we can use the energy regularity correlation and obtain the following constraint based on a C-mole of substrate consumed:

Approximate second-law constraint

$$\xi_S + \xi_N Y_{N/S} \geq \xi_B Y_{B/S} + \xi_P Y_{P/S} \quad \textbf{(15.9-20)}$$

The interesting and important result of Eq. 15.9-19 (and its simplification, Eq. 15.9-20) is that it provides a thermodynamic limit on the amount of product and biomass that can be produced from a given substrate and nitrogen source based on the Gibbs energy (or availability) of the reactants.

One should recognize that the constraint of Eq. 15.9-19 has a simple interpretation, which is that the Gibbs energy of the substrate and nitrogen source consumed must exceed (or at least equal) that of the biomass and products produced.

ILLUSTRATION 15.9-7

Second-Law Limitation on a Fermenter

In Illustration 15.9-4, 0.451 C-moles of ethanol and 0.235 C-moles of biomass were produced per C-mole of glucose consumed. How does this compare with the second-law limit on the production of ethanol and biomass?

SOLUTION

Using the data in Table 15.9-2, we have per C-mole of glucose consumed

$$478.7 + 0.0399 \times 391.9 \geq 0.235 \times 4.68 \times 112 + 0.451 \times 659.5 \text{ kJ}$$

or

$$494.3 > 420.6 \text{ kJ}$$

So we conclude that the fermentation satisfies the second law of thermodynamics. Further, we can say that $\frac{420.6}{494.3} \times 100 = 85.1$ percent of the Gibbs energy in the glucose and dissolved ammonia appears in the products of ethanol and biomass. Presumably the rest is a result of inefficiencies in the conversion process, and in the respiration and metabolism processes to keep the yeast alive.

Since the second-law constraint has been satisfied, the data reported for the production of ethanol from glucose using the yeast *Saccharomyces cerevisiae* are thermodynamically consistent.

NOTE

It is of interest to examine the simplified version of the second-law constraint in terms of the generalized degrees of reduction. In this case we have

$$4.00 + 3 \times 0.0399 = 4.1197 > 4.68 \times 0.235 + 6.00 \times 0.451 = 3.8058$$

so again the second-law constraint is satisfied, though this approximate result gives a somewhat higher thermodynamic efficiency of $\frac{3.8058}{4.1197} \times 100 = 92.4$ percent. ∎

ILLUSTRATION 15.9-8
The Thermodynamic Analysis of Fermentation Data

In Illustration 15.9-3 the yeast *Saccharomyces cerevisiae* was produced anaerobically (without oxygen) from the fermentation of glucose using ammonia as the nitrogen source. In that illustration it was found using the atom balance that

$$Y_{P/S} = 0.569 \qquad Y_{B/S} = 0.14 \qquad Y_{W/S} = -0.063 \quad \text{and} \quad Y_{N/S} = 0.028$$

Determine the heat load required to keep the fermenter at a constant temperature of 25°C, with the reactants entering at that temperature; whether the data satisfy the second-law consistency requirement; and the fraction of the Gibbs energy of the reactants that appears in the products. All calculations should be on a per C-mole of glucose consumed basis.

SOLUTION

The heat load on the fermenter is found using the energy balance, Eq. 15.9-8,

$$Y_{Q/S} = 0.14 \times 532.3 + 0.569 \times 684.5 - 467.8 - 0.028 \times 348.1 = -13.5 \text{ kJ}$$

So 13.5 kJ per C-mole of glucose consumed, or $6 \times 14.5 = 81$ kJ per mole of glucose consumed, must be removed from the fermenter to keep the temperature constant. Note that if the glucose was merely burned, 467.8 kJ per C-mole of glucose would have to be removed to keep the temperature at 25°C.

To see if the second-law constraint is satisfied, Eq. 15.9-19 is used:

$$478.7 + 0.028 \times 391.9 \overset{?}{>} 0.14 \times 537.6 + 0.569 \times 659.5$$

or

$$489.6 > 450.2 \text{ kJ}$$

Therefore, the second-law constraint is satisfied. Finally, $\frac{450.5}{489.6} \times 100 = 92.0$ percent of the Gibbs energy of the feed appears in the two products, ethanol and the yeast. ∎

A complete thermodynamic analysis of a fermenter involves the following:

a. Elemental mass balances for each of the atoms present (usually carbon, hydrogen, nitrogen, and oxygen, but perhaps also including sulfur, phosphorus, and other trace atoms that may be present in proteins and other biomass);

b. The energy balance (Eq. 15.9-8 or 15.9-11); and
c. The second-law constraint (Eq. 15.9-19 or 15.9-20).

All of these can be used together to decide on the effect of using different substrates to carry out a fermentation to produce a specified product and biomass. What we expect is that different amounts of product and biomass will be produced depending on the substrate chosen, and that the thermodynamic analysis in this section can provide some guidance. This is demonstrated in the illustration below.

ILLUSTRATION 15.9-9
Determining the Maximum Amount of Product That Can Be Obtained from a Substrate

A biologist is trying to genetically engineer bacteria to produce valuable organic chemicals from the inexpensive organic sugar xylose found in a variety of plants and berries. One such organism appears capable of producing 2,3-butanediol. The following data are available.

Chemical	Formula	$\Delta_c \underline{H}$, J/mol
2,3-butanediol	$C_4H_{10}O_2$	2461.0
Xylose	$C_5H_{10}O_5$	2345.2

a. Test the generalized degree of reduction approximation for each of these compounds using the reported heat of combustion data.
b. Estimate the maximum amount of 2,3-butanediol that can be produced per mole of xylose consumed assuming no additional biomass production.
c. Assuming that no biomass is produced (and therefore no nitrogen source is needed), determine the amount of carbon dioxide produced and oxygen and water consumed as a function of the amount of 2,3-butanediol produced.
d. Determine the amount of heat produced as a function of the amount of 2,3-butanediol produced.

SOLUTION

a. The definition of the generalized degree of reduction is

$$\xi = \frac{4 \cdot \mathcal{C} + \mathcal{H} - 2 \cdot \mathcal{O}}{\mathcal{C}}$$

so that

$$\xi_X = \frac{4 \times 5 + 10 - 2 \times 5}{5} = 4 \quad \text{and} \quad \xi_B = \frac{4 \times 4 + 10 - 2 \times 2}{4} = 5.5$$

Using the correlation of Eq. 15.9-10, we have for 2,3-butanediol

$$\Delta_c \underline{H}_B = 110.9 \times 5.5 \, \frac{\text{kJ}}{\text{C-mole}} \times 4 \, \frac{\text{C-mole}}{\text{mole}} = 2439.8 \, \frac{\text{kJ}}{\text{mole}}$$

compared with the measured value of 2461.0 kJ/mole, and for xylose

$$\Delta_c \underline{H}_X = 110.9 \times 4 \, \frac{\text{kJ}}{\text{C-mole}} \times 5 \, \frac{\text{C-mole}}{\text{mole}} = 2218 \, \frac{\text{kJ}}{\text{mole}}$$

compared with the measured value of 2345.2 kJ/mole. So that while the correlation is not exact, the estimates are reasonable.

b. Since there is no nitrogen source, no products other than 2,3-butanediol, and no biomass produced, the second-law constraint, Eq. 15.9-20, reduces to

$$\xi_S \geq \xi_P Y_{P/S} \quad \text{so that here} \quad Y_{P/S} \leq \frac{\xi_S}{\xi_P} = \frac{4}{5.5} = 0.727$$

So there is a limit of 0.727 C-moles of 2,3-butanediol produced per C-mole of xylose consumed. Since xylose has five carbons and a molecular weight of 150, one C-mole of xylose is 30 g. Similarly 2,3-butanediol has four carbon atoms and a molecular weight of 90, so one C-mole is 22.5 g. Therefore, 0.727 C-moles of 2,3-butanediol per C-mole of xylose equals $0.727 \times 22.5/30 = 0.545$ g 2,3-butanediol per g of xylose, or 545 g/kg of xylose.

c. The overall fermentation reaction can be written as

$$CH_2O + Y_{O_2/S}O_2 + Y_{W/S}H_2O \rightarrow Y_{P/S}CH_{2.5}O_{0.5} + Y_{C/S}CO_2$$

The carbon mass balance is

$$1 = Y_{P/S} + Y_{C/S} \quad \text{or} \quad Y_{C/S} = 1 - Y_{P/S}$$

The hydrogen mass balance is

$$2 + 2Y_{W/S} = 2.5Y_{P/S} \quad \text{or} \quad Y_{W/S} = 1.25Y_{P/S} - 1$$

The oxygen mass balance can be developed two different but equivalent ways. The first is

$$1 + Y_{W/S} + 2Y_{O/S} = 0.5Y_{P/S} + 2Y_{C/S}$$

which, substituting in the carbon and hydrogen balances, reduces to

$$Y_{O/S} = 1 - 1.375Y_{P/S}$$

The second method of obtaining the oxygen balance comes from Eq. 15.9-12, which here reduces to

$$Y_{O/S} = \frac{\xi_S - \xi_P Y_{P/S}}{4} = \frac{4 - 5.5Y_{P/S}}{4} = 1 - 1.375Y_{P/S}$$

The results are shown in the figure below. Note that this figure and the one that follows end at $Y_{P/S} = 0.727$, which is the second-law (or availability) limit for the maximum production of 2,3-butanediol.

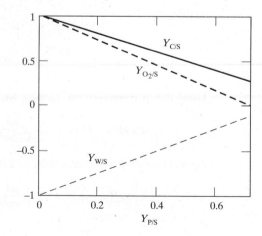

d. The energy balance, from Eq. 15.9-11, is

$$Y_{Q/S} = 110.9 \left(Y_{B/S}\xi_B + Y_{P/S}\xi_P - \xi_S - Y_{N/S}\xi_N \right) = 110.9 \left(5.5 Y_{P/S} - 4 \right) \text{kJ}$$

The result is shown in the figure below.

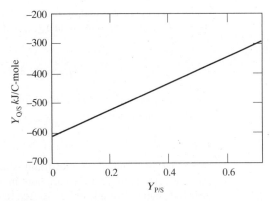

Heat flow as a function of 2,3-butanediol formed.

ILLUSTRATION 15.9-10

Choosing a Substrate for a Fermentation

An unspecified substrate that does not contain nitrogen, characterized only by its generalized degree of reduction ξ_S, together with dissolved ammonia as the nitrogen source is to be fermented to produce "average biomass" and no other products. As a function of the generalized degree of reduction of the substrate, determine the following:

a. The maximum product yield per C-mole of substrate consumed if the amount of ammonia is limited
b. The maximum product yield per C-mole of substrate consumed if the amount of oxygen is limited
c. The second-law limitation on the amount of product that can be formed
d. The heat that must be removed to keep the fermenter at a constant temperature per C-mole of substrate consumed

SOLUTION

This fermentation is written as

$$\text{Substrate} + Y_{N/S}\text{NH}_3 + Y_{O_2/S}\text{O}_2 + Y_{W/S}\text{H}_2\text{O} \rightarrow Y_{B/S}\text{CH}_{1.8}\text{O}_{0.5}\text{N}_{0.2} + Y_{C/S}\text{CO}_2$$

a. The nitrogen balance is

$$Y_{N/S} = Y_{B/S} \cdot 0.2 \quad \text{or} \quad Y_{B/S} = 5 \cdot Y_{N/S}$$

so that if the nitrogen source, ammonia, is limited, there is a mass balance limitation on the amount of biomass (cells) that can be produced. This limit is shown as solid lines in the figure that follows.

b. For the oxygen balance, we use Eq. 15.9-12, which here becomes

$$Y_{O_2/S} = \frac{\xi_S + \xi_N Y_{N/S} - \xi_B Y_{B/S}}{4} = \frac{\xi_S + 3Y_{N/S} - 4.8Y_{B/S}}{4} = \frac{\xi_S + \frac{3}{5}Y_{B/S} - 4.8Y_{B/S}}{4}$$
$$= \frac{\xi_S - 4.2Y_{B/S}}{4}$$

Values of $Y_{B/S}$ as a function of ξ_S and $Y_{O_2/S}$ are also plotted in the figure as dashed-dotted lines. Of special interest is the $Y_{O_2/S} = 0$ line (corresponding to anaerobic fermentation), which is

$$Y_{B/S} = 0.2381 \cdot \xi_S \quad (Y_{O_2/S} = 0)$$

We will come back to this result shortly.
   The carbon balance is

$$1 = Y_{B/S} + Y_{C/S} \quad \text{or} \quad Y_{B/S} = 1 - Y_{C/S}$$

Note that the result of this constraint is independent of the properties of the substrate, and merely requires that for the case here in which there is no other product, the carbon in the substrate can be completely converted to biomass if there is no carbon dioxide evolved. If there is CO_2 evolution, only a fraction of the substrate will appear as biomass.
   (We cannot do a hydrogen balance because we do not know the hydrogen content of the substrate; only its degree of reduction is being specified.)

c. The last restriction is the second-law constraint, which for the case here of no product, and with the Gibbs energy of the dissolved ammonia known but that of the substrate and biomass unknown, has the form

$$112 \cdot \xi_S + Y_{N/S} \cdot 391.9 \geq 112 \cdot \xi_B \cdot Y_{B/S}$$

Now using the nitrogen balance, the second-law maximum conversion of substrate to biomass is

$$112 \cdot \xi_B \cdot Y_{B/S} = 112 \cdot \xi_S + 0.2 \cdot Y_{B/S} \cdot 391.9$$

or

$$(112 \cdot 4.8 - 0.2 \cdot 391.9) \, Y_{B/S} = 112 \cdot \xi_S \quad \text{or} \quad Y_{B/S} = 0.2439 \cdot \xi_S$$

This constraint is also plotted in the following figure. Note that it is almost identical to the $Y_{O_2/S} = 0$ line, and differs from it only to the extent that the dissolved ammonia is not exactly described by the energy regularity principle.
   The interpretation of this figure is as follows. For any substrate, characterized only by its degree of reduction ξ_S, the maximum conversion to biomass, $Y_{B/S}$, is governed by whichever is the limiting constraint:
1. The availability of a nitrogen source;
2. The availability of oxygen; or
3. The second law constraint.
In particular, assuming complete availability of a nitrogen source, the maximum conversion of a substrate of a given value of ξ_S to biomass for this system, $Y_{B/S}^{max}$, can be determined from the second-law constraint line. However, since no fermentation or other biological process operates without irreversibilities (i.e., the generation of entropy), the actual biomass production will be less than the theoretical maximum, that is, $Y_{B/S}^{max} \geq Y_{B/S}$.
d. Finally, the energy balance on the fermenter (using the nitrogen balance) is

$$Y_{Q/S} = 110.9 Y_{B/S}\xi_B - 110.9\xi_S - Y_{N/S}348.1$$
$$= 532.3 Y_{B/S} - 110.9\xi_S - 0.2 Y_{B/S}348.1$$
$$= 462.7 Y_{B/S} - 110.9\xi_S \text{kJ}$$

The relationship between $Y_{Q/S}$, $Y_{B/S}$, and ξ_S is also plotted in the figure as the light dotted line.

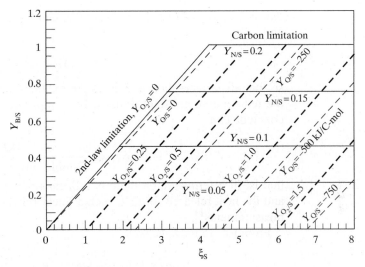

Figure 1 Second-law limitation, oxygen $Y_{O_2/S}$ and nitrogen source $Y_{N/S}$ consumption, and heat release $Y_{Q/S}$ as a function of degree of reduction of the substrate ξ_S and biomass production $Y_{B/S}$.

◼

Before leaving this section, it is useful to consider two further implications of the second-law constraint on bioreactors. To do this, we will consider the approximate forms of the energy balance and the second-law constraint based on the energy regularity assumption. While the simple conclusions obtained will not be exact, the errors should not be great. For simplicity, we first repeat the oxygen and energy balances and the second-law constraint here.

Oxygen balance

$$Y_{O_2/S} = \frac{\xi_S + \xi_N Y_{N/S} - \xi_B Y_{B/S} - \xi_P Y_{P/S}}{4} \tag{15.9-12}$$

Energy balance

$$Y_{Q/S} = 110.9 \left(Y_{B/S}\xi_B + Y_{P/S}\xi_P - \xi_S - Y_{N/S}\xi_N \right) \text{ kJ/C-mole} \tag{15.9-11}$$

and the second-law constraint written as

$$\xi_S + \xi_N Y_{N/S} \geq \xi_B Y_{B/S} + \xi_P Y_{P/S} \tag{15.9-20}$$

$$\xi_S + \xi_N Y_{N/S} - \xi_B Y_{B/S} - \xi_P Y_{P/S} \geq 0 \tag{15.9-21}$$

The important thing to notice is that the same term that appears in the second-law constraint and must be greater than or at best equal to zero also appears in the oxygen balance and the energy balance. By comparing these equations, we conclude that as a result of the second law of thermodynamics (and the energy regularity assumption),

$$Y_{O_2/S} \geq 0$$

that is, oxygen can only be consumed, and not produced by a fermentation process, and

$$Y_{Q/S} \leq 0$$

so that all fermentation processes are exothermic; that is, cooling is needed to keep the temperature constant. Finally, by comparing Eqs. 15.9-11 and 15.9-12, we also conclude that

$$Y_{Q/S} = -443.6 Y_{O_2/S} \ \frac{\text{kJ}}{\text{C-mole substrate consumed}} \qquad \textbf{(15.9-22)}$$

that is, in a fermentation there is a linear relation between the oxygen consumption and the heat released. In fact, such a linear relationship has been found in laboratory measurements[13].

From Eq. 15.9-18 we have

$$\underline{S}_{\text{gen}} = \frac{\left(\triangle_c G_S + Y_{N/S} \triangle_c G_N - Y_{B/S} \triangle_c G_B - Y_{P/S} \triangle_c G_P \right)}{T} \qquad \textbf{(15.9-23a)}$$

or more approximately, using the energy regularity principle,

$$\underline{S}_{\text{gen}} = \frac{\left(\xi_S + \xi_N Y_{N/S} - \xi_B Y_{B/S} - \xi_P Y_{P/S} \right)}{T} = 4 \frac{Y_{O_2/S}}{T} = -\frac{Y_{Q/S}}{110.9 T} \qquad \textbf{(15.9-23b)}$$

These equations can be used to compute the entropy generation of a biochemical process.

ILLUSTRATION 15.9-11

Entropy Generation in a Fermentation

The following data have been reported by von Stockar and Birou[14] for the fermentation of glucose by *K. fragilis* at 25°C to produce ethanol as the product and biomass of elemental composition $CH_{1.75}O_{0.52}N_{0.15}$:

$$Y_{B/S} = 0.572 \qquad Y_{CO_2/S} = 0.398 \quad \text{and} \quad Y_{Q/S} = -180.9 \ \frac{\text{kJ}}{\text{C-mole}}$$

Compute the following per C-mole of substrate consumed:

    a. The amount of product formed ($Y_{P/S}$)
    b. The amount of ammonia consumed ($Y_{N/S}$)
    c. The amount of oxygen consumed ($Y_{O_2/S}$)
    d. The amount of heat released ($Y_{Q/S}$)
    e. The amount of entropy generated (S_{gen})

---

[13]C. L. Cooney, D. I. C. Wang, and R. I. Mateles, *Biotech. Bioeng.*, **11**, 269 (1968).
[14]U. von Stockar and B. Birou, *Biotech. Bioeng.*, **34**, 86 (1989).

SOLUTION

To begin, we compute the generalized degree of reduction of the biomass produced:

$$\xi_B = 4 \times \mathcal{C} + \mathcal{H} - 2 \times \mathcal{O} = 4 + 1.75 - 2 \times 0.52 = 4.71$$

Therefore, for the biomass,

$$\triangle_c G_B = 112 \times \xi_B = 527.5 \frac{kJ}{C\text{-mole}} \quad \text{and} \quad \triangle_c H_B = 110.9 \times \xi_B = 522.3 \frac{kJ}{C\text{-mole}}$$

The heats and Gibbs energies of all other components are available in Table 15.9-2.

a. By the carbon balance,

$$1 = Y_{B/S} + Y_{P/S} + Y_{CO_2/S}$$

so that

$$Y_{P/S} = 1 - Y_{B/S} - Y_{CO_2/S} = 1 - 0.572 - 0.398 = 0.030$$

b. By the nitrogen balance,

$$Y_{N/S} = 0.15 \times Y_{N/S} = 0.15 \times 0.572 = 0.086$$

c. From Eq. 15.9-12, we have

$$Y_{O_2/S} = \frac{\xi_S + \xi_N \cdot Y_{N/S} - \xi_B \cdot Y_{B/S} - \xi_P \cdot Y_{P/S}}{4}$$

$$= \frac{4 + 0.086 \times 3 - 4.71 \times 0.572 - 0.030 \times 6}{4} = 0.346$$

d. From the energy balance,

$$Y_{Q/S} = Y_{B/S} \cdot \triangle_c H_B + Y_{P/S} \cdot \triangle_c H_P - \triangle_c H_S - Y_{N/S} \cdot \triangle_c H_N$$

$$= 0.572 \cdot 522.3 + 0.030 \cdot 684.5 - 467.8 - 0.086 \cdot 348.1 = -178.5 \frac{kJ}{C\text{-mole}}$$

e. From the entropy balance, we have

$$\underline{S}_{gen} = \frac{1}{T}(\triangle_c G_S + Y_{N/S} \cdot \triangle_c G_N - Y_{B/S} \cdot \triangle_c G_B - Y_{P/S} \cdot \triangle_c G_P)$$

$$= \frac{(478.7 + 0.086 \cdot 391.9 - 0.572 \cdot 527.5 - 0.030 \cdot 684.5)}{298.15} = 0.640 \frac{kJ}{C\text{-mole K}}$$

Finally, it is of interest to compare the heat release calculated in part (d) above, $Y_{Q/S} = -178.5$ kJ/ C-mole, with the more approximate result from Eq. 15.9-22:

$$Y_{Q/S} = -443.6 \times Y_{O/S} = -443.6 \times 0.346 = -153.5 \frac{kJ}{C\text{-mole}}$$

Though the approximate result is not very accurate, it is not unreasonable given the simplicity of the calculation. ■

The results of this illustration, and previous ones in this section, show that the methods of thermodynamics are just as applicable to biochemical processes as they are to more classical problems in chemical engineering.

As a final comment to this chapter and to end this book, it is interesting to note that Eq. 15.9-22 established a linear relationship between the heat released in fermentation and the oxygen consumed. The following similar linear correlation exists in the mammalian animal world between the average metabolic rate (directly proportional to the energy released as heat, muscular work, and in basic metabolic processes) and the rate of oxygen consumed:

$$\text{Metabolic rate}\left(\frac{\text{kJ}}{\text{day}}\right) = 21\ 800 \times \text{Oxygen consumption}\left(\frac{\text{m}^3}{\text{day}}\right)$$

This correlation is shown in Figure 15.9-2 for a number of warm-blooded (39°C) mammals, including a shrew (average body weight of 0.0048 kg), a cat (2.5 kg), dog (11.7 kg), a human (70 kg), a horse (650 kg), and an elephant (3833 kg). It is amusing to note that this correlation spans animals whose body weight varies by almost a factor of 1 000 000! [There are similar but different correlations for cold-blooded (20°C) animals, which have lower heat losses and slower metabolic rates.] Using this correlation, since the average human male consumes about 15 liters of oxygen per hour (0.36 m^3 per day), we estimate that the average human metabolic energy release is about 7850 kJ/day or 1875 kcal/day, which is close to the recommended daily food allowance for weight control.

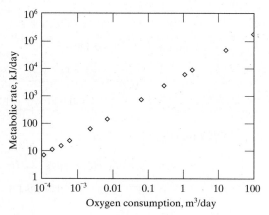

Figure 15.9-2 Correlation between average metabolic rate and oxygen consumption for warm-blooded animals.

PROBLEMS

15.1 Using the values for the equilibrium constant for the ionization of water in Table 15.1-1, estimate the standard-state heat of ionization of water as a function of temperature. Also, determine the pH of water at each of the temperatures in this table.

15.2 Derive Eqs. 15.1-8a and b.

15.3 Derive the equation that replaces Eq. 15.1-11 if the thermodynamic equilibrium constant (rather than the apparent equilibrium constant) is used and electrolyte solution nonideality is included.

15.4 Derive the equations that replace Eqs. 15.1-12a and b if the thermodynamic equilibrium constant (rather than the apparent equilibrium constant) is used and electrolyte solution nonideality is included.

15.5 Derive the equation that replaces Eq. 15.1-16 for the case of a weak base and a strong acid.

15.6 Derive the equation that replaces Eq. 15.1-16 if the thermodynamic equilibrium constant (rather than the apparent equilibrium constant) is used and electrolyte solution nonideality is included.

15.7 Derive the equation that replaces Eq. 15.3-3 for a protein that has six ionizable sites.

15.8 Redo Illustration 15.1-2 including the ionization of water and including the effect of solution nonidealities.

15.9 The following values are known for the amino acid serine C$_3$H$_7$NO$_2$S:

$$pK_1 = 2.21 \quad \text{and} \quad pK_2 = 9.15$$

Determine the charge on this amino acid as a function of pH. How does your result compare with the reported isoelectric point (pI) for cysteine of 5.68?

15.10 The compound levodopa $C_9H_{11}NO_4$ (usually referred to a L-dopa) is used in the treatment of Parkinson's disease. The chemical structure (not showing its carbon atoms and attached hydrogens) is

L-dopa

The following values are known for its equilibrium constants for the release of hydrogen ions: $pK_1 = 2.20$ from the carboxylic acid group, $pK_2 = 8.75$ from the NH_3^+ group, $pK_3 = 9.81$ from the first OH group, and $pK_4 = 13.40$ for a hydrogen ion release from the remaining OH group. Determine the charge on this amino acid as a function of pH.

15.11 Creatine $C_4H_9N_3O_2$ has been used by body builders and athletes to increase muscle mass and strength. Its chemical structure (not showing its carbon atoms and attached hydrogens) is

Its solubility in water is 16 g/kg of water, and the following values are known for its dissociation constants:

$$pK_1 = 2.63 \quad \text{and} \quad pK_2 = 14.30$$

a. Determine the charge on this amino acid as a function of pH and find its isoelectric point.
b. Determine the solubility of creatine in water as a function of pH.

15.12 Oxalic acid $C_2H_2O_4$ is a dibasic acid with the following values for its ionization constants:

$$pK_1 = 1.2 \quad \text{and} \quad pK_2 = 4.2$$

Determine the charge on oxalic acid as a function of pH.

15.13 Oxaloacetic acid $C_4H_4O_5$ has the following values for its ionization constants:

$$pK_1 = 2.55 \qquad pK_2 = 4.37 \quad \text{and} \quad pK_3 = 13.03$$

Determine the charge on oxaloacetic acid as a function of pH.

15.14 L-cystine $C_6H_{12}N_2O_4S_2$ has four ionization states with the following values for its ionization constants:

$$pK_1 = 1.0 \qquad pK_2 = 2.1 \qquad pK_3 = 8.02$$
$$\text{and} \quad pK_3 = 8.71$$

Determine the charge on L-cystine as a function of pH.

15.15 An ultracentrifuge is found to have a lysozyme concentration of 1 mg/g of water 2 cm from the axis of rotation. Determine the equilibrium concentration profile over the range from 2 cm to 4 cm from the rotation axis if the temperature is 25°C and
a. The rotational speed is 10 000 rpm;
b. The rotational speed is 15 000 rpm.

15.16 A centrifuge operating at 5000 rpm and 25°C is to being used to partially separate antipneumococcus serum globulin (molecular weight 195 000 and partial molar volume of 0.745 cc/g) from hemoglobin (molecular weight 68 000 and partial molar volume of 0.749 cc/g) in aqueous solution. If the concentrations of hemoglobin and antipneumococcus serum globulin are both 0.1 mg/g of water 2 cm from the centrifuge axis of rotation, what are their concentrations at 4 cm from the axis of rotation?

15.17 Derive the equations that replace Eqs. 15.3-3 and 15.3-4b if electrolyte solution nonideality is included.

15.18 Repeat Illustration 15.1-6 for the dissociation of acetic acid in the presence of sodium chloride using the simple Debye-Hückel limiting law (instead of the extended version used in the illustration) and compare the results.

15.19 a. Prove that the isoelectric point for an amino acid with two ionization sites is

$$pH = \frac{pK_1 + pK_2}{2}$$

b. Develop an equation for the calculation of the isoelectric point of an amino acid or protein with three ionization sites.

15.20 Determine the charge on tyrosine as a function of pH. (See Illustration 15.3-3 for the necessary data.)

15.21 From data in the literature it has been estimated that when the bacteria *K. fragilis* is used in the aerobic (that is, with added oxygen) fermentation of glucose $C_6H_{12}O_6$, using dissolved ammonia as a nitrogen source produces a biomass of atomic composition $CH_{1.75}O_{0.52}N_{0.15}$ and ethanol C_2H_5OH. The following yield factors have been reported: $Y_{B/S} = 0.569$ and $Y_{C/S} = 0.407$. Determine the other yield factors for this fermentation. What fraction of the carbon atoms present in glucose are converted to ethanol?

15.22 When a yeast of average biomass composition is grown on an aqueous solution of methanol with dissolved nitrogen, it is found that 400 g of dry yeast are produced for each kilogram of methanol consumed. No other products are formed. Compute the oxygen consumed and carbon dioxide produced per kilogram of methanol consumed.

15.23 When the bacterium *Zymomonas mobilis* is used to produce ethanol from glucose in an anaerobic process with ammonia as the nitrogen source, it is found that bacteria biomass is produced to the extent of 0.06 C-moles per C-mole of glucose. The biomass has an elemental composition of $CH_{1.8}O_{0.5}N_{0.2}$. What are the fractional conversion of glucose to ethanol (on a C-mole basis), the amount of ammonia used, and the amount of carbon dioxide produced per C-mole of glucose consumed?

15.24 The following data have been reported by von Stockar and Birou for biomass production and heat release for the fermentation of glucose by *K. fragilis*, which also produces ethanol as the product.

$Y_{B/S}$	$Y_{C/S}$	$Y_{Q/S}$ (kJ/C-mole)
0.569	0.407	−186.7
0.567	0.403	−199.8
0.584	0.375	−192.0
0.552	0.373	−166.7
0.492	0.339	−129.0
0.456	0.329	−108.9
0.345	0.320	−83.6
0.294	0.307	−59.4

Assuming the biomass has the elemental composition $CH_{1.75}O_{0.52}N_{0.15}$ and dissolved ammonia is used as the nitrogen source, compute the following for each data point:
a. The ethanol yield factor, $Y_{P/S}$.
b. The amount of oxygen consumed, $Y_{O_2/S}$.
c. The amount of heat released, $Y_{Q/S}$. Compare the predicted values with those reported in the table above.
d. The amount of entropy generated.

15.25 Repeat Illustration 15.9-10 if nitric acid $C_5H_9NO_4$ is used as the nitrogen source instead of ammonia.

15.26 Derive the analogue of Eqs. 15.7-6 for Gibbs-Donnan equilibrium involving a negatively charged protein.

15.27 Starting with an equimolar mixture of lactate and NAD^+ at 25°C, calculate and plot the extent of the following reaction as a function of pH:

$$R-O_2C-CHOH-CH_3 + NAD^+$$
$$\text{lactate}$$
$$\rightarrow R-O_2C-CO-CH_3 + NADH + H^+$$
$$\text{pyruvate}$$

where R is a side group. For this reaction at solution conditions, the apparent Gibbs energy change is $\triangle_{rxn}G = 25.9$ kJ. Also determine the pH at which half the lactate will have reacted at equilibrium.

15.28 Starting with an equimolar mixture of methanol and NAD^+ at 25°C, calculate and plot the extent of the following reaction as a function of pH:

$$CH_3OH + NAD^+ + H_2O \rightarrow HCOO^- + NADH + 3H^+$$
$$\text{methanol} \qquad\qquad\qquad \text{formate}$$

For this reaction at solution conditions, the apparent Gibbs energy change is $\triangle_{rxn}G = -15.1$ kJ. Also, assume that water is present in great excess, so that its concentration does not change in the course of the reaction.

15.29 Starting with an equimolar mixture of acetaldehyde and NAD^+ at 25°C, calculate and plot the extent of the following reaction as a function of pH. Also determine the pH at which half the acetaldehyde will have reacted at equilibrium.

$$CH_3CHO + NAD^+ + H_2O \rightarrow CH_3CO_2^- + NADH + 2H^+$$
$$\text{acetaldehyde} \qquad\qquad\qquad \text{acetate}$$

For this reaction at solution conditions, the apparent Gibbs energy change is $\triangle_{rxn}G = 50.2$ kJ. Also, assume that water is present in great excess, so that its concentration does not change in the course of the reaction.

15.30 Glucose and fructose both have the composition $C_6H_{12}O_6$, but differ slightly in structure, as shown below, and more so in sweetness. Repeat Illustration 15.9-9 using these sugars instead of xylose.

Glucose   Fructose

a. Estimate the maximum amount of 2,3-butanediol that can be produced per kg of glucose and fructose consumed, assuming no additional biomass production.

b. Estimate the amount of water and oxygen consumed, and carbon dioxide produced per C-mole of 2,3-butanediol produced assuming no additional biomass production.

c. Estimate the amount of heat released to keep the fermenter at 25°C per C-mole of 2,3-butanediol produced assuming no additional biomass production.

15.31 The biologist of Illustration 15.9-9 is examining other genetically engineered bacteria to produce 2,3-butanediol from lignins, which are derived from abundant and renewable resources such as trees, plants, and agricultural crops. While lignin polymerizes, the repeating unit is believed to have the following general formula: $C_{10}H_{12}O_3$.

a. Estimate the maximum amount of 2,3-butanediol that can be produced per kg of lignin consumed, assuming no additional biomass production.

b. Estimate the amount of water and oxygen consumed, and carbon dioxide produced per C-mole of 2,3-butanediol produced, assuming no additional biomass production.

c. Estimate the amount of heat released to keep the fermenter at 25°C per C-mole of 2,3-butanediol produced, assuming no additional biomass production.

15.32 Succinic acid, which has the structure HOOC—CH_2—CH_2—COOH and on a C-mole basis is represented by $CH_{3/2}O$, is useful as a starting material in the manufacture of solvents and polymers. It is typically made by the catalytic oxidation of butane. However, by that route only about 40 percent of the carbon initially present in the butane is converted to succinic acid. It has been suggested instead to use glucose as a starting material and a biochemical pathway with the following proposed reaction stoichiometry:

$$CH_2O + \frac{1}{7}CO_2 \rightarrow \frac{8}{7}CH_{3/2} + \frac{1}{7}H_2O$$

a. Is this reaction stoichiometry possible?

b. Determine the maximum kg of succinic acid that can be produced per kg of glucose.

c. Estimate the heat released per C-mole of glucose consumed.

d. What fraction of the Gibbs energy of the glucose appears in the succinic acid?

15.33 A considerable amount of methane is produced in Norway from its North Sea oil and gas wells. Some

of this methane is converted to methanol by partial oxidation, and then biochemically converted to biomass that is used as animal feed. The reported stoichiometry for the biochemical reaction is

$$CH_3OH + 0.731 \cdot O_2 + 0.146 \cdot NH_3$$
$$\rightarrow 0.269 \cdot CO_2 + 0.731 \text{C-mol biomass}$$

a. What is the atomic composition of the biomass produced?

b. How much heat is released per mole of methanol consumed?

c. What fraction of the Gibbs energy of the reactants is present in the biomass?

15.34 Another possibility is to convert the methane directly to biomass that is used as animal feed. The stoichiometry for this biochemical reaction is

$$CH_4 + Y_{O_2/S} \cdot O_2 + Y_{N/S} \cdot NH_3 + Y_{W/S} \cdot H_2O$$
$$\rightarrow Y_{B/S} \cdot CH_{1.8}O_{0.5}N_{0.2} + Y_{C/S} \cdot CO_2$$

It has been found that the consumption of 2.5 m³ of methane (measured at 273 K and 1.013 bar) results in the production of 1 kg of biomass.

a. Determine all the yield factors for this reaction.

b. How much heat is released per m³ of methane consumed?

c. What fraction of the Gibbs energy of the methane is present in the biomass?

15.35 The compound 1,3-propanediol, which has the structure HO—CH_2—CH_2—CH_2—OH, is a monomer used in the production of polyester fibers and in the manufacture of polyurethanes. 1,3 Propanediol can be produced by chemical processes and also by the fermentation of glycerol in a two-step process using certain bacterial strains. However, neither the chemical nor biological method is well suited for industrial production, because the chemical processes are energy intensive and the biological processes are based on glycerol, an expensive starting material. The DuPont Company has developed a process using a microorganism containing the dehydratase enzyme that can convert other carbon sources, such as carbohydrates or sugars, to 1,3-propanediol.

a. What is the maximum number of kg of 1,3-propanediol that can be biologically produced from each kg of glycerol, assuming no other energy source is used? What is the stoichiometry of the reaction at the maximum product yield?

b. What is the maximum number of kg of 1,3-propanediol that can be biologically produced from each kg of glucose, assuming no other energy source is used? What is the stoichiometry of the reaction at the maximum product yield?

15.36 Various different substrates are being considered for the biochemical production of 1,3-propanediol (see previous problem) using dissolved ammonia as the nitrogen source. Determine the maximum amount of 1,3-propanediol that can be produced as a function of

the generalized degree of reduction of the substrate if there are no other by-products. Also, when less than the maximum amount of 1,3-propanediol is produced, determine the oxygen and heat flow yield factors as a function of the generalized degree of reduction of the substrate and the 1,3-propanediol yield factor.

Appendix A.I

Conversion Factors to SI Units

To Convert from:	To:	Multiply by:
atmosphere (standard)	Pa	$1.013\,25 \times 10^5$
bar	Pa	1×10^5
British thermal unit*	J	1.054×10^3
BTU/lb	$J\,kg^{-1}$	2.324×10^3
BTU/lb °F	$J\,kg^{-1}\,K^{-1}$	4.184×10^3
calorie*	J	4.184
cal/g	$J\,kg^{-1}$	4.184×10^3
cal/g °C	$J\,kg^{-1}\,K^{-1}$	4.184×10^3
cm of mercury (0°C)	Pa	1.33×10^3
cm^3	m^3	1×10^{-6}
dyne	N	1×10^{-5}
dyne cm	N m	1×10^{-7}
dyne/cm^2	Pa	1×10^{-1}
erg	J	1×10^{-7}
foot	m	3.048×10^{-1}
gallon (U.S., liquid)	m^3	3.785×10^{-3}
g/cm^3	$kg\,m^{-3}$	1×10^3
g/liter	$kg\,m^{-3}$	1
horsepower	W	7.457×10^2
inch	m	2.540×10^{-2}
inch water (60°F)	Pa	2.488×10^2
kilogram-force	N	9.807
kilowatt hour	J	3.600×10^6
liter	m^3	1×10^{-3}
millibar	Pa	1×10^2
millimeter of mercury (0°C)	Pa	1.333×10^2
ounce (avoirdupois)	kg	2.835×10^{-2}
pound (avoirdupois)	kg	4.536×10^{-1}
lb/ft^3	$kg\,m^{-3}$	1.602×10^1
lb/gal (U.S.)	$kg\,m^{-3}$	1.198×10^2
lb-force/ft^2	Pa	4.788×10^1
lb-force/in^2 (psi)	Pa	6.895×10^3
ton (refrigeration)	W	3.517×10^3
ton (2000 lb)	kg	9.072×10^2
torr (mm Hg, 0°C)	Pa	1.333×10^2
yard	m	9.144×10^{-1}
degree Celsius	K	$T(\mathrm{K}) = T(^{\circ}\mathrm{C}) + 273.15$
degree Fahrenheit	K	$T(\mathrm{K}) = [T(^{\circ}\mathrm{F}) + 459.67]/1.8$
degree Rankine	K	$T(\mathrm{K}) = T(^{\circ}\mathrm{R})/1.8$

*Thermochemical unit that is used, for example, in the tables in the *Chemical Engineers Handbook*. For International Table units use the constant 4.1868 instead of 4.184, and for mean calorie use 4.190 02.

Appendix A.II

The Molar Heat Capacities of Gases in the Ideal Gas (Zero-Pressure) State*

		a	$b \times 10^2$	$c \times 10^5$	$d \times 10^9$	Temperature Range (K)
Paraffinic Hydrocarbons						
Methane	CH_4	19.875	5.021	1.268	−11.004	273–1500
Ethane	C_2H_6	6.895	17.255	−6.402	7.280	273–1500
Propane	C_3H_8	−4.042	30.456	−15.711	31.716	273–1500
n-Butane	C_4H_{10}	3.954	37.126	−18.326	34.979	273–1500
i-Butane	C_4H_{10}	−7.908	41.573	−22.992	49.875	273–1500
n-Pentane	C_5H_{12}	6.770	45.398	−22.448	42.259	273–1500
n-Hexane	C_6H_{14}	6.933	55.188	−28.636	57.657	273–1500
Monoolefinic Hydrocarbons						
Ethylene	C_2H_4	3.950	15.628	−8.339	17.657	273–1500
Propylene	C_3H_6	3.151	23.812	−12.176	24.603	273–1500
1-Butene	C_4H_8	−1.004	36.193	−21.381	50.502	273–1500
i-Butene	C_4H_8	6.904	32.226	−16.657	33.557	273–1500
cis-2-Butene	C_4H_8	−7.439	33.799	−17.046	33.013	273–1500
trans-2-Butene	C_4H_8	9.791	30.209	−14.239	25.398	273–1500
Cycloparaffinic Hydrocarbons						
Cyclopentane	C_5H_{10}	−54.213	54.757	−31.159	68.661	273–1500
Methylcyclopentane	C_6H_{12}	−50.686	64.352	−37.301	83.808	273–1500
Cyclohexane	C_6H_{12}	−66.674	68.845	−38.506	80.628	273–1500
Methylcyclohexane	C_7H_{14}	−63.054	79.381	−45.979	100.795	273–1500
Aromatic Hydrocarbons						
Benzene	C_6H_6	−36.193	48.444	−31.548	77.573	273–1500
Toluene	C_7H_8	−34.364	55.887	−34.435	80.335	273–1500
Ethylbenzene	C_8H_{10}	−35.138	66.674	−41.854	100.209	273–1500
Styrene	C_8H_8	−24.971	60.059	−38.285	92.176	273–1500
Cumene	C_9H_{12}	−39.548	78.184	−49.661	120.502	273–1500
Oxygenated Hydrocarbons						
Formaldehyde	CH_2O	22.791	4.075	0.713	−8.695	273–1500
Acetaldehyde	C_2H_4O	17.531	13.239	−2.155	−15.900	273–1000
Methanol	CH_4O	19.038	9.146	−1.218	−8.034	273–1000
Ethanol	C_2H_6O	19.875	20.946	−10.372	20.042	273–1500
Ethylene oxide	C_2H_4O	−4.686	20.607	−9.996	13.176	273–1000
Ketene	C_2H_2O	17.197	12.410	−7.502	17.657	273–1500

*Constants are for the equation $C_P^* = a + bT + cT^2 + dT^3$, where T is in kelvins and C_P^* in J (mol K)$^{-1}$.

Appendix II (*Continued*)

		a	$b \times 10^2$	$c \times 10^5$	$d \times 10^9$	Temperature Range (K)
Miscellaneous Hydrocarbons						
Cyclopropane	C_3H_6	−27.117	34.335	−23.335	65.314	273–1000
Isopentane	C_5H_{12}	−9.511	52.025	−29.695	66.360	273–1500
Neopentane	C_5H_{12}	16.172	55.670	−33.548	78.787	273–1500
o-Xylene	C_8H_{10}	−15.854	59.795	−34.954	78.661	273–1500
m-Xylene	C_8H_{10}	−27.335	62.364	−36.950	83.891	273–1500
p-Xylene	C_8H_{10}	−22.318	59.498	−33.406	71.255	273–1500
C_3 Oxygenated Hydrocarbons						
Carbon suboxide	C_3O_2	34.322	12.858	−8.707	21.682	273–1500
Acetone	C_3H_6O	6.799	27.870	−15.636	34.757	273–1500
i-Propyl alcohol	C_3H_8O	3.321	35.573	−20.987	48.368	273–1500
n-Propyl alcohol	C_3H_8O	−5.469	38.640	−24.268	59.163	273–1500
Allyl alcohol	C_3H_6O	2.177	29.799	−17.820	41.623	273–1500
Chloroethenes						
Chloroethene	C_2H_3Cl	10.046	17.866	−11.511	28.439	273–1500
1,1-Dichloroethene	$C_2H_2Cl_2$	24.682	18.339	−13.314	35.632	273–1500
cis-1,2-Dichloroethene	$C_2H_2Cl_2$	18.142	19.628	−14.213	37.699	273–1500
trans-1,2-Dichloroethene	$C_2H_2Cl_2$	23.686	17.971	−12.644	33.017	273–1500
Trichloroethene	C_2HCl_3	38.494	18.900	−15.063	42.259	273–1500
Tetrachloroethene	C_2Cl_4	63.222	15.895	−13.301	38.029	273–1500
Nitrogen Compounds						
Ammonia	NH_3	27.551	2.563	0.990	−6.687	273–1500
Hydrazine	N_2H_4	16.276	14.870	−9.640	25.063	273–1500
Methylamine	CH_5N	12.534	15.105	−6.881	12.345	273–1500
Dimethylamine	C_2H_7N	−1.151	27.679	−14.572	29.921	273–1500
Trimethylamine	C_3H_9N	−8.778	40.246	−23.217	52.017	273–1500
Halogens and Halogen Acids						
Fluorine	F_2	25.586	2.454	−1.752	4.099	273–2000
Chlorine	Cl_2	28.541	2.389	−2.137	6.473	273–1500
Bromine	Br_2	33.686	1.030	−0.890	2.680	273–1500
Iodine	I_2	35.582	0.550	−0.447	1.308	273–1800
Hydrogen fluoride	HF	30.130	−0.493	0.659	−1.573	273–2000
Hydrogen chloride	HCl	30.310	−0.762	1.326	−4.335	273–1500
Hydrogen bromide	HBr	29.996	−0.671	1.387	−4.858	273–1500
Hydrogen iodide	HI	28.042	0.190	0.509	−2.014	273–1900
Chloromethanes						
Methyl chloride	CH_3Cl	12.762	10.862	−5.205	9.623	273–1500
Methylene chloride	CH_2Cl_2	17.573	14.305	−9.833	25.389	273–1500
Chloroform	$CHCl_3$	31.841	14.481	−11.163	30.728	273–1500
Carbon tetrachloride	CCl_4	51.213	14.226	−12.531	36.937	273–1500
Phosgene	$COCl_2$	43.305	6.916	−3.518		273–1000
Thiophosgene	$CSCl_2$	45.188	7.778	−4.372		273–1000

Appendix II (*Continued*)

		a	$b \times 10^2$	$c \times 10^5$	$d \times 10^9$	Temperature Range (K)
Cyanogens						
Cyanogen	$(CN)_2$	41.088	6.217	−2.749		273–1000
Hydrogen cyanide	HCN	26.527	3.504	−1.093		273–1500
Cyanogen chloride	CNCl	33.347	4.496	−2.203		273–1000
Cyanogen bromide	CNBr	36.904	3.801	−1.827		273–1000
Cyanogen iodide	CNI	40.544	3.018	−1.366		273–1000
Acetonitrile	CH_3CN	21.297	11.562	−3.812		273–1200
Acrylic nitrile	CH_2CHCl	19.038	17.171	−7.087		273–1000
Oxides of Nitrogen						
Nitric oxide	NO	27.034	0.987	−0.322	0.365	273–3800
Nitric oxide	NO	29.322	−0.094	0.974	−4.184	273–1500
Nitrous oxide	N_2O	24.092	5.859	−3.560	10.569	273–1500
Nitrogen dioxide	NO_2	22.929	5.711	−3.519	7.866	273–1500
Nitrogen tetroxide	N_2O_4	33.054	18.661	−11.339		273–600
Acetylenes and Diolefins						
Acetylene	C_2H_2	21.799	9.208	−6.523	18.197	273–1500
Methylacetylene	C_3H_4	17.615	17.042	−9.172	19.720	273–1500
Dimethylacetylene	C_4H_6	14.812	24.427	−11.548	20.812	273–1500
Propadiene	C_3H_4	10.167	19.636	−11.636	27.130	273–1500
1,3-Butadiene	C_4H_6	−5.398	34.937	−23.356	59.582	273–1500
Isoprene	C_5H_8	−1.841	43.590	−28.293	70.837	273–1500
Combustion Gases (Low Range)						
Nitrogen	N_2	28.883	−0.157	0.808	−2.871	273–1800
Oxygen	O_2	25.460	1.519	−0.715	1.311	273–1800
Air		28.088	0.197	0.480	−1.965	273–1800
Hydrogen	H_2	29.088	−0.192	0.400	−0.870	273–1800
Carbon monoxide	CO	28.142	0.167	0.537	−2.221	273–1800
Carbon dioxide	CO_2	22.243	5.977	−3.499	7.464	273–1800
Water vapor	H_2O	32.218	0.192	1.055	−3.593	273–1800
Ammonia	NH_3	24.619	3.75	−0.138		300–1500
Combustion Gases (High Range)†						
Nitrogen	N_2	27.318	0.623	−0.095		273–3800
Oxygen	O_2	28.167	0.630	−0.075		273–3800
Air		27.435	0.618	−0.090		273–3800
Hydrogen	H_2	26.879	0.435	−0.033		273–3800
Carbon monoxide	CO	27.113	0.655	−0.100		273–3800
Water vapor	H_2O	29.163	1.449	−0.202		273–3800
Sulfur Compounds						
Sulfur	S_2	27.193	2.217	−1.627	3.983	273–1800
Sulfur dioxide	SO_2	25.762	5.791	−3.809	8.607	273–1800
Sulfur trioxide	SO_3	16.393	14.573	−11.193	32.402	273–1300
Hydrogen sulfide	H_2S	29.582	1.309	0.571	−3.292	273–1800
Carbon disulfide	CS_2	30.921	6.230	−4.586	11.548	273–1800
Carbonyl sulfide	COS	26.034	6.427	−4.427	10.711	273–1800

†The equation for CO_2 in the temperature range of 273–3800 K is $C_P^* = 75.464 − 1.872 \times 10^{-4}T − 661.42/\sqrt{T}$.

Based on data in O. Hougen, K. Watson, and R. A. Ragatz, *Chemical Process Principles,* Part 1, John Wiley & Sons, New York (1954). Used with permission.

Appendix A.III

The Thermodynamic Properties of Water and Steam[1]

THERMODYNAMIC PROPERTIES OF STEAM

Saturated Steam: Temperature Table

Temp (°C)	Press. (kPa)	Specific Volume		Internal Energy			Enthalpy			Entropy		
		Sat. Liquid	Sat. Vapor	Sat. Liquid	Evap.	Sat. Vapor	Sat. Liquid	Evap.	Sat. Vapor	Sat. Liquid	Evap.	Sat. Vapor
T	P	$\hat{V}^{\text{L}}$	$\hat{V}^{\text{V}}$	$\hat{U}^{\text{L}}$	$\Delta\hat{U}$	$\hat{U}^{\text{V}}$	$\hat{H}^{\text{L}}$	$\Delta\hat{H}$	$\hat{H}^{\text{V}}$	$\hat{S}^{\text{L}}$	$\Delta\hat{S}$	$\hat{S}^{\text{V}}$
0.01	0.6113	0.001 000	206.14	0.00	2375.3	2375.3	0.01	2501.3	2501.4	0.0000	9.1562	9.1562
5	0.8721	0.001 000	147.12	20.97	2361.3	2382.3	20.98	2489.6	2510.6	0.0761	8.9496	9.0257
10	1.2276	0.001 000	106.38	42.00	2347.2	2389.2	42.01	2477.7	2519.8	0.1510	8.7498	8.9008
15	1.7051	0.001 001	77.93	62.99	2333.1	2396.1	62.99	2465.9	2528.9	0.2245	8.5569	8.7814
20	2.339	0.001 002	57.79	83.95	2319.0	2402.9	83.96	2454.1	2538.1	0.2966	8.3706	8.6672
25	3.169	0.001 003	43.36	104.88	2304.9	2409.8	104.89	2442.3	2547.2	0.3674	8.1905	8.5580
30	4.246	0.001 004	32.89	125.78	2290.8	2416.6	125.79	2430.5	2556.3	0.4369	8.0164	8.4533
35	5.628	0.001 006	25.22	146.67	2276.7	2423.4	146.68	2418.6	2565.3	0.5053	7.8478	8.3531
40	7.384	0.001 008	19.52	167.56	2262.6	2430.1	167.57	2406.7	2574.3	0.5725	7.6845	8.2570
45	9.593	0.001 010	15.26	188.44	2248.4	2436.8	188.45	2394.8	2583.2	0.6387	7.5261	8.1648
50	12.349	0.001 012	12.03	209.32	2234.2	2443.5	209.33	2382.7	2592.1	0.7038	7.3725	8.0763
55	15.758	0.001 015	9.568	230.21	2219.9	2450.1	230.23	2370.7	2600.9	0.7679	7.2234	7.9913
60	19.940	0.001 017	7.671	251.11	2205.5	2456.6	251.13	2358.5	2609.6	0.8312	7.0784	7.9096
65	25.03	0.001 020	6.197	272.02	2191.1	2463.1	272.06	2346.2	2618.3	0.8935	6.9375	7.8310
70	31.19	0.001 023	5.042	292.95	2176.6	2469.6	292.98	2333.8	2626.8	0.9549	6.8004	7.7553
75	38.58	0.001 026	4.131	313.90	2162.0	2475.9	313.93	2321.4	2635.3	1.0155	6.6669	7.6824
80	47.39	0.001 029	3.407	334.86	2147.4	2482.2	334.91	2308.8	2643.7	1.0753	6.5369	7.6122
85	57.83	0.001 033	2.828	355.84	2132.6	2488.4	355.90	2296.0	2651.9	1.1343	6.4102	7.5445
90	70.14	0.001 036	2.361	376.85	2117.7	2494.5	376.92	2283.2	2660.1	1.1925	6.2866	7.4791
95	84.55	0.001 040	1.982	397.88	2102.7	2500.6	397.96	2270.2	2668.1	1.2500	6.1659	7.4159
	MPa											
100	0.101 35	0.001 044	1.6729	418.94	2087.6	2506.5	419.04	2257.0	2676.1	1.3069	6.0480	7.3549
105	0.120 82	0.001 048	1.4194	440.02	2072.3	2512.4	440.15	2243.7	2683.8	1.3630	5.9328	7.2958
110	0.143 27	0.001 052	1.2102	461.14	2057.0	2518.1	461.30	2230.2	2691.5	1.4185	5.8202	7.2387
115	0.169 06	0.001 056	1.0366	482.30	2041.4	2523.7	482.48	2216.5	2699.0	1.4734	5.7100	7.1833
120	0.198 53	0.001 060	0.8919	503.50	2025.8	2529.3	503.71	2202.6	2706.3	1.5276	5.6020	7.1296
125	0.2321	0.001 065	0.7706	524.74	2009.9	2534.6	524.99	2188.5	2713.5	1.5813	5.4962	7.0775
130	0.2701	0.001 070	0.6685	546.02	1993.9	2539.9	546.31	2174.2	2720.5	1.6344	5.3925	7.0269

$\hat{V}$ [=] m^3/kg; $\hat{U}, \hat{H}$ [=] J/g = kJ/kg; $\hat{S}$ [=] kJ/kg K

[1]From G. J. Van Wylen and R. E. Sontag, *Fundamentals of Classical Thermodynamics, S. I. Version.* 2nd ed., John Wiley & Sons, New York (1978). Used with permission.

Saturated Steam: Temperature Table (*Continued*)

Temp (°C)	Press. (MPa)	Specific Volume		Internal Energy			Enthalpy			Entropy		
		Sat. Liquid	Sat. Vapor	Sat. Liquid	Evap.	Sat. Vapor	Sat. Liquid	Evap.	Sat. Vapor	Sat. Liquid	Evap.	Sat. Vapor
T	P	$\hat{V}^L$	$\hat{V}^V$	$\hat{U}^L$	$\Delta\hat{U}$	$\hat{U}^V$	$\hat{H}^L$	$\Delta\hat{H}$	$\hat{H}^V$	$\hat{S}^L$	$\Delta\hat{S}$	$\hat{S}^V$
135	0.3130	0.001 075	0.5822	567.35	1977.7	2545.0	567.69	2159.6	2727.3	1.6870	5.2907	6.9777
140	0.3613	0.001 080	0.5089	588.74	1961.3	2550.0	589.13	2144.7	2733.9	1.7391	5.1908	6.9299
145	0.4154	0.001 085	0.4463	610.18	1944.7	2554.9	610.63	2129.6	2740.3	1.7907	5.0926	6.8833
150	0.4758	0.001 091	0.3928	631.68	1927.9	2559.5	632.20	2114.3	2746.5	1.8418	4.9960	6.8379
155	0.5431	0.001 096	0.3468	653.24	1910.8	2564.1	653.84	2098.6	2752.4	1.8925	4.9010	6.7935
160	0.6178	0.001 102	0.3071	674.87	1893.5	2568.4	675.55	2082.6	2758.1	1.9427	4.8075	6.7502
165	0.7005	0.001 108	0.2727	696.56	1876.0	2572.5	697.34	2066.2	2763.5	1.9925	4.7153	6.7078
170	0.7917	0.001 114	0.2428	718.33	1858.1	2576.5	719.21	2049.5	2768.7	2.0419	4.6244	6.6663
175	0.8920	0.001 121	0.2168	740.17	1840.0	2580.2	741.17	2032.4	2773.6	2.0909	4.5347	6.6256
180	1.0021	0.001 127	0.194 05	762.09	1821.6	2583.7	763.22	2015.0	2778.2	2.1396	4.4461	6.5857
185	1.1227	0.001 134	0.174 09	784.10	1802.9	2587.0	785.37	1997.1	2782.4	2.1879	4.3586	6.5465
190	1.2544	0.001 141	0.156 54	806.19	1783.8	2590.0	807.62	1978.8	2786.4	2.2359	4.2720	6.5079
195	1.3978	0.001 149	0.141 05	828.37	1764.4	2592.8	829.98	1960.0	2790.0	2.2835	4.1863	6.4698
200	1.5538	0.001 157	0.127 36	850.65	1744.7	2595.3	852.45	1940.7	2793.2	2.3309	4.1014	6.4323
205	1.7230	0.001 164	0.115 21	873.04	1724.5	2597.5	875.04	1921.0	2796.0	2.3780	4.0172	6.3952
210	1.9062	0.001 173	0.104 41	895.53	1703.9	2599.5	897.76	1900.7	2798.5	2.4248	3.9337	6.3585
215	2.104	0.001 181	0.094 79	918.14	1682.9	2601.1	920.62	1879.9	2800.5	2.4714	3.8507	6.3221
220	2.318	0.001 190	0.086 19	940.87	1661.5	2602.4	943.62	1858.5	2802.1	2.5178	3.7683	6.2861
225	2.548	0.001 199	0.078 49	963.73	1639.6	2603.3	966.78	1836.5	2803.3	2.5639	3.6863	6.2503
230	2.795	0.001 209	0.071 58	986.74	1617.2	2603.9	990.12	1813.8	2804.0	2.6099	3.6047	6.2146
235	3.060	0.001 219	0.065 37	1009.89	1594.2	2604.1	1013.62	1790.5	2804.2	2.6558	3.5233	6.1791
240	3.344	0.001 229	0.059 76	1033.21	1570.8	2604.0	1037.32	1766.5	2803.8	2.7015	3.4422	6.1437
245	3.648	0.001 240	0.054 71	1056.71	1546.7	2603.4	1061.23	1741.7	2803.0	2.7472	3.3612	6.1083
250	3.973	0.001 251	0.050 13	1080.39	1522.0	2602.4	1085.36	1716.2	2801.5	2.7927	3.2802	6.0730
255	4.319	0.001 263	0.045 98	1104.28	1496.7	2600.9	1109.73	1689.8	2799.5	2.8383	3.1992	6.0375
260	4.688	0.001 276	0.042 21	1128.39	1470.6	2599.0	1134.37	1662.5	2796.9	2.8838	3.1181	6.0019
265	5.081	0.001 289	0.038 77	1152.74	1443.9	2596.6	1159.28	1634.4	2793.6	2.9294	3.0368	5.9662
270	5.499	0.001 302	0.035 64	1177.36	1416.3	2593.7	1184.51	1605.2	2789.7	2.9751	2.9551	5.9301
275	5.942	0.001 317	0.032 79	1202.25	1387.9	2590.2	1210.07	1574.9	2785.0	3.0208	2.8730	5.8938
280	6.412	0.001 332	0.030 17	1227.46	1358.7	2586.1	1235.99	1543.6	2779.6	3.0668	2.7903	5.8571
285	6.909	0.001 348	0.027 77	1253.00	1328.4	2581.4	1262.31	1511.0	2773.3	3.1130	2.7070	5.8199
290	7.436	0.001 366	0.025 57	1278.92	1297.1	2576.0	1289.07	1477.1	2766.2	3.1594	2.6227	5.7821
295	7.993	0.001 384	0.023 54	1305.2	1264.7	2569.9	1316.3	1441.8	2758.1	3.2062	2.5375	5.7437
300	8.581	0.001 404	0.021 67	1332.0	1231.0	2563.0	1344.0	1404.9	2749.0	3.2534	2.4511	5.7045
305	9.202	0.001 425	0.019 948	1359.3	1195.9	2555.2	1372.4	1366.4	2738.7	3.3010	2.3633	5.6643
310	9.856	0.001 447	0.018 350	1387.1	1159.4	2546.4	1401.3	1326.0	2727.3	3.3493	2.2737	5.6230
315	10.547	0.001 472	0.016 867	1415.5	1121.1	2536.6	1431.0	1283.5	2714.5	3.3982	2.1821	5.5804
320	11.274	0.001 499	0.015 488	1444.6	1080.9	2525.5	1461.5	1238.6	2700.1	3.4480	2.0882	5.5362
330	12.845	0.001 561	0.012 996	1505.3	993.7	2498.9	1525.3	1140.6	2665.9	3.5507	1.8909	5.4417
340	14.586	0.001 638	0.010 797	1570.3	894.3	2464.6	1594.2	1027.9	2622.0	3.6594	1.6763	5.3357
350	16.513	0.001 740	0.008 813	1641.9	776.6	2418.4	1670.6	893.4	2563.9	3.7777	1.4335	5.2112
360	18.651	0.001 893	0.006 945	1725.2	626.3	2351.5	1760.5	720.5	2481.0	3.9147	1.1379	5.0526
370	21.03	0.002 213	0.004 925	1844.0	384.5	2228.5	1890.5	441.6	2332.1	4.1106	.6865	4.7971
374.14	22.09	0.003 155	0.003 155	2029.6	0	2029.6	2099.3	0	2099.3	4.4298	0	4.4298

$\hat{V}$ [=] m³/kg; $\hat{U}, \hat{H}$ [=] J/g = kJ/kg; $\hat{S}$ [=] kJ/kg K

Saturated Steam: Pressure Table

Press. (kPa)	Temp (°C)	Specific Volume		Internal Energy			Enthalpy			Entropy		
		Sat. Liquid	Sat. Vapor	Sat. Liquid	Evap.	Sat. Vapor	Sat. Liquid	Evap.	Sat. Vapor	Sat. Liquid	Evap.	Sat. Vapor
P	T	$\hat{V}^L$	$\hat{V}^V$	$\hat{U}^L$	$\Delta \hat{U}$	$\hat{U}^V$	$\hat{H}^L$	$\Delta \hat{H}$	$\hat{H}^V$	$\hat{S}^L$	$\Delta \hat{S}$	$\hat{S}^V$
0.6113	0.01	0.001 000	206.14	0.00	2375.3	2375.3	0.01	2501.3	2501.4	0.0000	9.1562	9.1562
1.0	6.98	0.001 000	129.21	29.30	2355.7	2385.0	29.30	2484.9	2514.2	0.1059	8.8697	8.9756
1.5	13.03	0.001 001	87.98	54.71	2338.6	2393.3	54.71	2470.6	2525.3	0.1957	8.6322	8.8279
2.0	17.50	0.001 001	67.00	73.48	2326.0	2399.5	73.48	2460.0	2533.5	0.2607	8.4629	8.7237
2.5	21.08	0.001 002	54.25	88.48	2315.9	2404.4	88.49	2451.6	2540.0	0.3120	8.3311	8.6432
3.0	24.08	0.001 003	45.67	101.04	2307.5	2408.5	101.05	2444.5	2545.5	0.3545	8.2231	8.5776
4.0	28.96	0.001 004	34.80	121.45	2293.7	2415.2	121.46	2432.9	2554.4	0.4226	8.0520	8.4746
5.0	32.88	0.001 005	28.19	137.81	2282.7	2420.5	137.82	2423.7	2561.5	0.4764	7.9187	8.3951
7.5	40.29	0.001 008	19.24	168.78	2261.7	2430.5	168.79	2406.0	2574.8	0.5764	7.6750	8.2515
10	45.81	0.001 010	14.67	191.82	2246.1	2437.9	191.83	2392.8	2584.7	0.6493	7.5009	8.1502
15	53.97	0.001 014	10.02	225.92	2222.8	2448.7	225.94	2373.1	2599.1	0.7549	7.2536	8.0085
20	60.06	0.001 017	7.649	251.38	2205.4	2456.7	251.40	2358.3	2609.7	0.8320	7.0766	7.9085
25	64.97	0.001 020	6.204	271.90	2191.2	2463.1	271.93	2346.3	2618.2	0.8931	6.9383	7.8314
30	69.10	0.001 022	5.229	289.20	2179.2	2468.4	289.23	2336.1	2625.3	0.9439	6.8247	7.7686
40	75.87	0.001 027	3.993	317.53	2159.5	2477.0	317.58	2319.2	2636.8	1.0259	6.6441	7.6700
50	81.33	0.001 030	3.240	340.44	2143.4	2483.9	340.49	2305.4	2645.9	1.0910	6.5029	7.5939
75	91.78	0.001 037	2.217	384.31	2112.4	2496.7	384.39	2278.6	2663.0	1.2130	6.2434	7.4564
MPa												
0.100	99.63	0.001 043	1.6940	417.36	2088.7	2506.1	417.46	2258.0	2675.5	1.3026	6.0568	7.3594
0.125	105.99	0.001 048	1.3749	444.19	2069.3	2513.5	444.32	2241.0	2685.4	1.3740	5.9104	7.2844
0.150	111.37	0.001 053	1.1593	466.94	2052.7	2519.7	467.11	2226.5	2693.6	1.4336	5.7897	7.2233
0.175	116.06	0.001 057	1.0036	486.80	2038.1	2524.9	486.99	2213.6	2700.6	1.4849	5.6868	7.1717
0.200	120.23	0.001 061	0.8857	504.49	2025.0	2529.5	504.70	2201.9	2706.7	1.5301	5.5970	7.1271
0.225	124.00	0.001 064	0.7933	520.47	2013.1	2533.6	520.72	2191.3	2712.1	1.5706	5.5173	7.0878
0.250	127.44	0.001 067	0.7187	535.10	2002.1	2537.2	535.37	2181.5	2716.9	1.6072	5.4455	7.0527
0.275	130.60	0.001 070	0.6573	548.59	1991.9	2540.5	548.89	2172.4	2721.3	1.6408	5.3801	7.0209
0.300	133.55	0.001 073	0.6058	561.15	1982.4	2543.6	561.47	2163.8	2725.3	1.6718	5.3201	6.9919
0.325	136.30	0.001 076	0.5620	572.90	1973.5	2546.4	573.25	2155.8	2729.0	1.7006	5.2646	6.9652
0.350	138.88	0.001 079	0.5243	583.95	1965.0	2548.9	584.33	2148.1	2732.4	1.7275	5.2130	6.9405
0.375	141.32	0.001 081	0.4914	594.40	1956.9	2551.3	594.81	2140.8	2735.6	1.7528	5.1647	6.9175
0.40	143.63	0.001 084	0.4625	604.31	1949.3	2553.6	604.74	2133.8	2738.6	1.7766	5.1193	6.8959
0.45	147.93	0.001 088	0.4140	622.77	1934.9	2557.6	623.25	2120.7	2743.9	1.8207	5.0359	6.8565
0.50	151.86	0.001 093	0.3749	639.68	1921.6	2561.2	640.23	2108.5	2748.7	1.8607	4.9606	6.8213
0.55	155.48	0.001 097	0.3427	655.32	1909.2	2564.5	655.93	2097.0	2753.0	1.8973	4.8920	6.7893
0.60	158.85	0.001 101	0.3157	669.90	1897.5	2567.4	670.56	2086.3	2756.8	1.9312	4.8288	6.7600
0.65	162.01	0.001 104	0.2927	683.56	1886.5	2570.1	684.28	2076.0	2760.3	1.9627	4.7703	6.7331
0.70	164.97	0.001 108	0.2729	696.44	1876.1	2572.5	697.22	2066.3	2763.5	1.9922	4.7158	6.7080
0.75	167.78	0.001 112	0.2556	708.64	1866.1	2574.7	709.47	2057.0	2766.4	2.0200	4.6647	6.6847
0.80	170.43	0.001 115	0.2404	720.22	1856.6	2576.8	721.11	2048.0	2769.1	2.0462	4.6166	6.6628
0.85	172.96	0.001 118	0.2270	731.27	1847.4	2578.7	732.22	2039.4	2771.6	2.0710	4.5711	6.6421
0.90	175.38	0.001 121	0.2150	741.83	1838.6	2580.5	742.83	2031.1	2773.9	2.0946	4.5280	6.6226
0.95	177.69	0.001 124	0.2042	751.95	1830.2	2582.1	753.02	2023.1	2776.1	2.1172	4.4869	6.6041

$\hat{V}$ [=] m³/kg; $\hat{U}, \hat{H}$ [=] J/g = kJ/kg; $\hat{S}$ [=] kJ/kg K

Saturated Steam: Pressure Table (*Continued*)

Press. (MPa)	Temp (°C)	Specific Volume		Internal Energy			Enthalpy			Entropy		
		Sat. Liquid	Sat. Vapor	Sat. Liquid	Evap.	Sat. Vapor	Sat. Liquid	Evap.	Sat. Vapor	Sat. Liquid	Evap.	Sat. Vapor
P	T	$\hat{V}^{L}$	$\hat{V}^{V}$	$\hat{U}^{L}$	$\Delta\hat{U}$	$\hat{U}^{V}$	$\hat{H}^{L}$	$\Delta\hat{H}$	$\hat{H}^{V}$	$\hat{S}^{L}$	$\Delta\hat{S}$	$\hat{S}^{V}$
1.00	179.91	0.001 127	0.194 44	761.68	1822.0	2583.6	762.81	2015.3	2778.1	2.1387	4.4478	6.5865
1.10	184.09	0.001 133	0.177 53	780.09	1806.3	2586.4	781.34	2000.4	2781.7	2.1792	4.3744	6.5536
1.20	187.99	0.001 139	0.163 33	797.29	1791.5	2588.8	798.65	1986.2	2784.8	2.2166	4.3067	6.5233
1.30	191.64	0.001 144	0.151 25	813.44	1777.5	2591.0	814.93	1972.7	2787.6	2.2515	4.2438	6.4953
1.40	195.07	0.001 149	0.140 84	828.70	1764.1	2592.8	830.30	1959.7	2790.6	2.2842	4.1850	6.4693
1.50	198.32	0.001 154	0.131 77	843.16	1751.3	2594.5	844.89	1947.3	2792.2	2.3150	4.1298	6.4448
1.75	205.76	0.001 166	0.113 49	876.46	1721.4	2597.8	878.50	1917.9	2796.4	2.3851	4.0044	6.3896
2.00	212.42	0.001 177	0.099 63	906.44	1693.8	2600.3	908.79	1890.7	2799.5	2.4474	3.8935	6.3409
2.25	218.45	0.001 187	0.088 75	933.83	1668.2	2602.0	936.49	1865.2	2801.7	2.5035	3.7937	6.2972
2.5	223.99	0.001 197	0.079 98	959.11	1644.0	2603.1	962.11	1841.0	2803.1	2.5547	3.7028	6.2575
3.00	233.90	0.001 217	0.066 68	1004.78	1599.3	2604.1	1008.42	1795.7	2804.2	2.6457	3.5412	6.1869
3.5	242.60	0.001 235	0.057 07	1045.43	1558.3	2603.7	1049.75	1753.7	2803.4	2.7253	3.4000	6.1253
4	250.40	0.001 252	0.049 78	1082.31	1520.0	2602.3	1087.31	1714.1	2801.4	2.7964	3.2737	6.0701
5	263.99	0.001 286	0.039 44	1147.81	1449.3	2597.1	1154.23	1640.1	2794.3	2.9202	3.0532	5.9734
6	275.64	0.001 319	0.032 44	1205.44	1384.3	2589.7	1213.35	1571.0	2784.3	3.0267	2.8625	5.8892
7	285.88	0.001 351	0.027 37	1257.55	1323.0	2580.5	1267.00	1505.1	2772.1	3.1211	2.6922	5.8133
8	295.06	0.001 384	0.023 52	1305.57	1264.2	2569.8	1316.64	1441.3	2758.0	3.2068	2.5364	5.7432
9	303.40	0.001 418	0.020 48	1350.51	1207.3	2557.8	1363.26	1378.9	2742.1	3.2858	2.3915	5.6772
10	311.06	0.001 452	0.018 026	1393.04	1151.4	2544.4	1407.56	1317.1	2724.7	3.3596	2.2544	5.6141
11	318.15	0.001 489	0.015 987	1433.7	1096.0	2529.8	1450.1	1255.5	2705.6	3.4295	2.1233	5.5527
12	324.75	0.001 527	0.014 263	1473.0	1040.7	2513.7	1491.3	1193.6	2684.9	3.4962	1.9962	5.4924
13	330.93	0.001 567	0.012 780	1511.1	985.0	2496.1	1531.5	1130.7	2662.2	3.5606	1.8718	5.4323
14	336.75	0.001 611	0.011 485	1548.6	928.2	2476.8	1571.1	1066.5	2637.6	3.6232	1.7485	5.3717
15	342.24	0.001 658	0.010 337	1585.6	869.8	2455.5	1610.5	1000.0	2610.5	3.6848	1.6249	5.3098
16	347.44	0.001 711	0.009 306	1622.7	809.0	2431.7	1650.1	930.6	2580.6	3.7461	1.4994	5.2455
17	352.37	0.001 770	0.008 364	1660.2	744.8	2405.0	1690.3	856.9	2547.2	3.8079	1.3698	5.1777
18	357.06	0.001 840	0.007 489	1698.9	675.4	2374.3	1732.0	777.1	2509.1	3.8715	1.2329	5.1044
19	361.54	0.001 924	0.006 657	1739.9	598.1	2338.1	1776.5	688.0	2464.5	3.9388	1.0839	5.0228
20	365.81	0.002 036	0.005 834	1785.6	507.5	2293.0	1826.3	583.4	2409.7	4.0139	0.9130	4.9269
21	369.89	0.002 207	0.004 952	1842.1	388.5	2230.6	1888.4	446.2	2334.6	4.1075	0.6938	4.8013
22	373.80	0.002 742	0.003 568	1961.9	125.2	2087.1	2022.2	143.4	2165.6	4.3110	0.2216	4.5327
22.09	374.14	0.003 155	0.003 155	2029.6	0.0	2029.6	2099.3	0.0	2099.3	4.4298	0.0	4.4298

$\hat{V}\ [=]\ \text{m}^3/\text{kg};$ $\hat{U}, \hat{H}\ [=]\ \text{J/g} = \text{kJ/kg};$ $\hat{S}\ [=]\ \text{kJ/kg K}$

Superheated Vapor‡

T (°C)	$\hat{V}$	$\hat{U}$	$\hat{H}$	$\hat{S}$	$\hat{V}$	$\hat{U}$	$\hat{H}$	$\hat{S}$	$\hat{V}$	$\hat{U}$	$\hat{H}$	$\hat{S}$
	P = 0.010 MPa (45.81)				P = 0.050 MPa (81.33)				P = 0.10 MPa (99.63)			
Sat.	14.674	2437.9	2584.7	8.1502	3.240	2483.9	2645.9	7.5939	1.6940	2506.1	2675.5	7.3594
50	14.869	2443.9	2592.6	8.1749	—	—	—	—	—	—	—	—
100	17.196	2515.5	2687.5	8.4479	3.418	2511.6	2682.5	7.6947	1.6958	2506.7	2676.2	7.3614
150	19.512	2587.9	2783.0	8.6882	3.889	2585.6	2780.1	7.9401	1.9364	2582.8	2776.4	7.6134
200	21.825	2661.3	2879.5	8.9038	4.356	2659.9	2877.7	8.1580	2.172	2658.1	2875.3	7.8343
250	24.136	2736.0	2977.3	9.1002	4.820	2735.0	2976.0	8.3556	2.406	2733.7	2974.3	8.0333
300	26.445	2812.1	3076.5	9.2813	5.284	2811.3	3075.5	8.5373	2.639	2810.4	3074.3	8.2158
400	31.063	2968.9	3279.6	9.6077	6.209	2968.5	3278.9	8.8642	3.103	2967.9	3278.2	8.5435
500	35.679	3132.3	3489.1	9.8978	7.134	3132.0	3488.7	9.1546	3.565	3131.6	3488.1	8.8342
600	40.295	3302.5	3705.4	10.1608	8.057	3302.2	3705.1	9.4178	4.028	3301.9	3704.7	9.0976
700	44.911	3479.6	3928.7	10.4028	8.981	3479.4	3928.5	9.6599	4.490	3479.2	3928.2	9.3398
800	49.526	3663.8	4159.0	10.6281	9.904	3663.6	4158.9	9.8852	4.952	3663.5	4158.6	9.5652
900	54.141	3855.0	4396.4	10.8396	10.828	3854.9	4396.3	10.0967	5.414	3854.8	4396.1	9.7767
1000	58.757	4053.0	4640.6	11.0393	11.751	4052.9	4640.5	10.2964	5.875	4052.8	4640.3	9.9764
1100	63.372	4257.5	4891.2	11.2287	12.674	4257.4	4891.1	10.4859	6.337	4257.3	4891.0	10.1659
1200	67.987	4467.9	5147.8	11.4091	13.597	4467.8	5147.7	10.6662	6.799	4467.7	5147.6	10.3463
1300	72.602	4683.7	5409.7	11.5811	14.521	4683.6	5409.6	10.8382	7.260	4683.5	5409.5	10.5183
	P = 0.20 MPa (120.23)				P = 0.30 MPa (133.55)				P = 0.40 MPa (143.63)			
Sat.	0.8857	2529.5	2706.7	7.1272	0.6058	2543.6	2725.3	6.9919	0.4625	2553.6	2738.6	6.8959
150	0.9596	2576.9	2768.8	7.2795	0.6339	2570.8	2761.0	7.0778	0.4708	2564.5	2752.8	6.9299
200	1.0803	2654.4	2870.5	7.5066	0.7163	2650.7	2865.6	7.3115	0.5342	2646.8	2860.5	7.1706
250	1.1988	2731.2	2971.0	7.7086	0.7964	2728.7	2967.6	7.5166	0.5951	2726.1	2964.2	7.3789
300	1.3162	2808.6	3071.8	7.8926	0.8753	2806.7	3069.3	7.7022	0.6548	2804.8	3066.8	7.5662
400	1.5493	2966.7	3276.6	8.2218	1.0315	2965.6	3275.0	8.0330	0.7726	2964.4	3273.4	7.8985
500	1.7814	3130.8	3487.1	8.5133	1.1867	3130.0	3486.0	8.3251	0.8893	3129.2	3484.9	8.1913
600	2.013	3301.4	3704.0	8.7770	1.3414	3300.8	3703.2	8.5892	1.0055	3300.2	3702.4	8.4558
700	2.244	3478.8	3927.6	9.0194	1.4957	3478.1	3927.1	8.8319	1.1215	3477.9	3926.5	8.6987
800	2.475	3663.1	4158.2	9.2449	1.6499	3662.9	4157.8	9.0576	1.2372	3662.4	4157.3	8.9244
900	2.706	3854.5	4395.8	9.4566	1.8041	3854.2	4395.4	9.2692	1.3529	3853.9	4395.1	9.1362
1000	2.937	4052.5	4640.0	9.6563	1.9581	4052.3	4639.7	9.4690	1.4685	4052.0	4639.4	9.3360
1100	3.168	4257.0	4890.7	9.8458	2.1121	4256.8	4890.4	9.6585	1.5840	4256.5	4890.2	9.5256
1200	3.399	4467.5	5147.3	10.0262	2.2661	4467.2	5147.1	9.8389	1.6996	4467.0	5146.8	9.7060
1300	3.630	4683.2	5409.3	10.1982	2.4201	4683.0	5409.0	10.0110	1.8151	4682.8	5408.8	9.8780
	P = 0.50 MPa (151.86)				P = 0.60 MPa (158.85)				P = 0.80 MPa (170.43)			
Sat.	0.3749	2561.2	2748.7	6.8213	0.3157	2567.4	2756.8	6.7600	0.2404	2576.8	2769.1	6.6628
200	0.4249	2642.9	2855.4	7.0592	0.3520	2638.9	2850.1	6.9665	0.2608	2630.6	2839.3	6.8158
250	0.4744	2723.5	2960.7	7.2709	0.3938	2720.9	2957.2	7.1816	0.2931	2715.5	2950.0	7.0384
300	0.5226	2802.9	3064.2	7.4599	0.4344	2801.0	3061.6	7.3724	0.3241	2797.2	3056.5	7.2328
350	0.5701	2882.6	3167.7	7.6329	0.4742	2881.2	3165.7	7.5464	0.3544	2878.2	3161.7	7.4089
400	0.6173	2963.2	3271.9	7.7938	0.5137	2962.1	3270.3	7.7079	0.3843	2959.7	3267.1	7.5716
500	0.7109	3128.4	3483.9	8.0873	0.5920	3127.6	3482.8	8.0021	0.4433	3126.0	3480.6	7.8673
600	0.8041	3299.6	3701.7	8.3522	0.6697	3299.1	3700.9	8.2674	0.5018	3297.9	3699.4	8.1333
700	0.8969	3477.5	3925.9	8.5952	0.7472	3477.0	3925.3	8.5107	0.5601	3476.2	3924.2	8.3770
800	0.9896	3662.1	4156.9	8.8211	0.8245	3661.8	4156.5	8.7367	0.6181	3661.1	4155.6	8.6033
900	1.0822	3853.6	4394.7	9.0329	0.9017	3853.4	4394.4	8.9486	0.6761	3852.8	4393.7	8.8153
1000	1.1747	4051.8	4639.1	9.2328	0.9788	4051.5	4638.8	9.1485	0.7340	4051.0	4638.2	9.0153
1100	1.2672	4256.3	4889.9	9.4224	1.0559	4256.1	4889.6	9.3381	0.7919	4255.6	4889.1	9.2050
1200	1.3596	4466.8	5146.6	9.6029	1.1330	4466.5	5146.3	9.5185	0.8497	4466.1	5145.9	9.3855
1300	1.4521	4682.5	5408.6	9.7749	1.2101	4682.3	5408.3	9.6906	0.9076	4681.8	5407.9	9.5575

‡*Note:* Number in parentheses is temperature of saturated steam at the specified pressure.

$\hat{V}$ [=] m³/kg; $\hat{U}, \hat{H}$ [=] J/g = kJ/kg; $\hat{S}$ [=] kJ/kg K

Superheated Vapor (*Continued*)

T (°C)	$\hat{V}$	$\hat{U}$	$\hat{H}$	$\hat{S}$	$\hat{V}$	$\hat{U}$	$\hat{H}$	$\hat{S}$	$\hat{V}$	$\hat{U}$	$\hat{H}$	$\hat{S}$
	P = 1.00 MPa (179.91)				P = 1.20 MPa (187.99)				P = 1.40 MPa (195.07)			
Sat.	0.194 44	2583.6	2778.1	6.5865	0.163 33	2588.8	2784.8	6.5233	0.140 84	2592.8	2790.0	6.4693
200	0.2060	2621.9	2827.9	6.6940	0.169 30	2612.8	2815.9	6.5898	0.143 02	2603.1	2803.3	6.4975
250	0.2327	2709.9	2942.6	6.9247	0.192 34	2704.2	2935.0	6.8294	0.163 50	2698.3	2927.2	6.7467
300	0.2579	2793.2	3051.2	7.1229	0.2138	2789.2	3045.8	7.0317	0.182 28	2785.2	3040.4	6.9534
350	0.2825	2875.2	3157.7	7.3011	0.2345	2872.2	3153.6	7.2121	0.2003	2869.2	3149.5	7.1360
400	0.3066	2957.3	3263.9	7.4651	0.2548	2954.9	3260.7	7.3774	0.2178	2952.5	3257.5	7.3026
500	0.3541	3124.4	3478.5	7.7622	0.2946	3122.8	3476.3	7.6759	0.2521	3121.1	3474.1	7.6027
600	0.4011	3296.8	3697.9	8.0290	0.3339	3295.6	3696.3	7.9435	0.2860	3294.4	3694.8	7.8710
700	0.4478	3475.3	3923.1	8.2731	0.3729	3474.4	3922.0	8.1881	0.3195	3473.6	3920.8	8.1160
800	0.4943	3660.4	4154.7	8.4996	0.4118	3659.7	4153.8	8.4148	0.3528	3659.0	4153.0	8.3431
900	0.5407	3852.2	4392.9	8.7118	0.4505	3851.6	4392.2	8.6272	0.3861	3851.1	4391.5	8.5556
1000	0.5871	4050.5	4637.6	8.9119	0.4892	4050.0	4637.0	8.8274	0.4192	4049.5	4636.4	8.7559
1100	0.6335	4255.1	4888.6	9.1017	0.5278	4254.6	4888.0	9.0172	0.4524	4254.1	4887.5	8.9457
1200	0.6798	4465.6	5145.4	9.2822	0.5665	4465.1	5144.9	9.1977	0.4855	4464.7	5144.4	9.1262
1300	0.7261	4681.3	5407.4	9.4543	0.6051	4680.9	5407.0	9.3698	0.5186	4680.4	5406.5	9.2984
	P = 1.60 MPa (201.41)				P = 1.80 MPa (207.15)				P = 2.00 MPa (212.42)			
Sat.	0.123 80	2596.0	2794.0	6.4218	0.110 42	2598.4	2797.1	6.3794	0.099 63	2600.3	2799.5	6.3409
225	0.132 87	2644.7	2857.3	6.5518	0.116 73	2636.6	2846.7	6.4808	0.103 77	2628.3	2835.8	6.4147
250	0.141 84	2692.3	2919.2	6.6732	0.124 97	2686.0	2911.0	6.6066	0.111 44	2679.6	2902.5	6.5453
300	0.158 62	2781.1	3034.8	6.8844	0.140 21	2776.9	3029.2	6.8226	0.125 47	2772.6	3023.5	6.7664
350	0.174 56	2866.1	3145.4	7.0694	0.154 57	2863.0	3141.2	7.0100	0.138 57	2859.8	3137.0	6.9563
400	0.190 05	2950.1	3254.2	7.2374	0.168 47	2947.7	3250.9	7.1794	0.151 20	2945.2	3247.6	7.1271
500	0.2203	3119.5	3472.0	7.5390	0.195 50	3117.9	3469.8	7.4825	0.175 68	3116.2	3467.6	7.4317
600	0.2500	3293.3	3693.2	7.8080	0.2220	3292.1	3691.7	7.7523	0.199 60	3290.9	3690.1	7.7024
700	0.2794	3472.7	3919.7	8.0535	0.2482	3471.8	3918.5	7.9983	0.2232	3470.9	3917.4	7.9487
800	0.3086	3658.3	4152.1	8.2808	0.2742	3657.6	4151.2	8.2258	0.2467	3657.0	4150.3	8.1765
900	0.3377	3850.5	4390.8	8.4935	0.3001	3849.9	4390.1	8.4386	0.2700	3849.3	4389.4	8.3895
1000	0.3668	4049.0	4635.8	8.6938	0.3260	4048.5	4635.2	8.6391	0.2933	4048.0	4634.6	8.5901
1100	0.3958	4253.7	4887.0	8.8837	0.3518	4253.2	4886.4	8.8290	0.3166	4252.7	4885.9	8.7800
1200	0.4248	4464.2	5143.9	9.0643	0.3776	4463.7	5143.4	9.0096	0.3398	4463.3	5142.9	8.9607
1300	0.4538	4679.9	5406.0	9.2364	0.4034	4679.5	5405.6	9.1818	0.3631	4679.0	5405.1	9.1329
	P = 2.50 MPa (223.99)				P = 3.00 MPa (233.90)				P = 3.50 MPa (242.60)			
Sat.	0.079 98	2603.1	2803.1	6.2575	0.066 68	2604.1	2804.2	6.1869	0.057 07	2603.7	2803.4	6.1253
225	0.080 27	2605.6	2806.3	6.2639	—	—	—	—	—	—	—	—
250	0.087 00	2662.6	2880.1	6.4085	0.070 58	2644.0	2855.8	6.2872	0.058 72	2623.7	2829.2	6.1749
300	0.098 90	2761.6	3008.8	6.6438	0.081 14	2750.1	2993.5	6.5390	0.068 42	2738.0	2977.5	6.4461
350	0.109 76	2851.9	3126.3	6.8403	0.090 53	2843.7	3115.3	6.7428	0.076 78	2835.3	3104.0	6.6579
400	0.120 10	2939.1	3239.3	7.0148	0.099 36	2932.8	3230.9	6.9212	0.084 53	2926.4	3222.3	6.8405
450	0.130 14	3025.5	3350.8	7.1746	0.107 87	3020.4	3344.0	7.0834	0.091 96	3015.3	3337.2	7.0052
500	0.139 98	3112.1	3462.1	7.3234	0.116 19	3108.0	3456.5	7.2338	0.099 18	3103.0	3450.9	7.1572
600	0.159 30	3288.0	3686.3	7.5960	0.132 43	3285.0	3682.3	7.5085	0.113 24	3282.1	3678.4	7.4339
700	0.178 32	3468.7	3914.5	7.8435	0.148 38	3466.5	3911.7	7.7571	0.126 99	3464.3	3908.8	7.6837
800	0.197 16	3655.3	4148.2	8.0720	0.164 14	3653.5	4145.9	7.9862	0.140 56	3651.8	4143.7	7.9134
900	0.215 90	3847.9	4387.6	8.2853	0.179 80	3846.5	4385.9	8.1999	0.154 02	3845.0	4384.1	8.1276
1000	0.2346	4046.7	4633.1	8.4861	0.195 41	4045.4	4631.6	8.4009	0.167 43	4044.1	4630.1	8.3288
1100	0.2532	4251.5	4884.6	8.6762	0.210 98	4250.3	4883.3	8.5912	0.180 80	4249.2	4881.9	8.5192
1200	0.2718	4462.1	5141.7	8.8569	0.226 52	4460.9	5140.5	8.7720	0.194 15	4459.8	5139.3	8.7000
1300	0.2905	4677.8	5404.0	9.0291	0.242 06	4676.6	5402.8	8.9442	0.207 49	4675.5	5401.7	8.8723

$\hat{V}$ [=] m³/kg; $\hat{U}, \hat{H}$ [=] J/g = kJ/kg; $\hat{S}$ [=] kJ/kg K

Superheated Vapor (*Continued*)

T (°C)	\hat{V}	\hat{U}	\hat{H}	\hat{S}	\hat{V}	\hat{U}	\hat{H}	\hat{S}	\hat{V}	\hat{U}	\hat{H}	\hat{S}
	\multicolumn P = 4.0 MPa (250.40)				P = 4.5 MPa (257.49)				P = 5.0 MPa (263.99)			
Sat.	0.049 78	2602.3	2801.4	6.0701	0.044 06	2600.1	2798.3	6.0198	0.039 44	2597.1	2794.3	5.9734
275	0.054 57	2667.9	2886.2	6.2285	0.047 30	2650.3	2863.2	6.1401	0.041 41	2631.3	2838.3	6.0544
300	0.058 84	2725.3	2960.7	6.3615	0.051 35	2712.0	2943.1	6.2828	0.045 32	2698.0	2924.5	6.2084
350	0.066 45	2826.7	3092.5	6.5821	0.058 40	2817.8	3080.6	6.5131	0.051 94	2808.7	3068.4	6.4493
400	0.073 41	2919.9	3213.6	6.7690	0.064 75	2913.3	3204.7	6.7047	0.057 81	2906.6	3195.7	6.6459
450	0.080 02	3010.2	3330.3	6.9363	0.070 74	3005.0	3323.3	6.8746	0.063 30	2999.7	3316.2	6.8186
500	0.086 43	3099.5	3445.3	7.0901	0.076 51	3095.3	3439.6	7.0301	0.068 57	3091.0	3433.8	6.9759
600	0.098 85	3279.1	3674.4	7.3688	0.087 65	3276.0	3670.5	7.3110	0.078 69	3273.0	3666.5	7.2589
700	0.110 95	3462.1	3905.9	7.6198	0.098 47	3459.9	3903.0	7.5631	0.088 49	3457.6	3900.1	7.5122
800	0.122 87	3650.0	4141.5	7.8502	0.109 11	3648.3	4139.3	7.7942	0.098 11	3646.6	4137.1	7.7440
900	0.134 69	3843.6	4382.3	8.0647	0.119 65	3842.2	4380.6	8.0091	0.107 62	3840.7	4378.8	7.9593
1000	0.146 45	4042.9	4628.7	8.2662	0.130 13	4041.6	4627.2	8.2108	0.117 07	4040.4	4625.7	8.1612
1100	0.158 17	4248.0	4880.6	8.4567	0.140 56	4246.8	4879.3	8.4015	0.126 48	4245.6	4878.0	8.3520
1200	0.169 87	4458.6	5138.1	8.6376	0.150 98	4457.5	5136.9	8.5825	0.135 87	4456.3	5135.7	8.5331
1300	0.181 56	4674.3	5400.5	8.8100	0.161 39	4673.1	5399.4	8.7549	0.145 26	4672.0	5398.2	8.7055
	P = 6.0 MPa (275.64)				P = 7.0 MPa (285.88)				P = 8.0 MPa (295.06)			
Sat.	0.032 44	2589.7	2784.3	5.8892	0.027 37	2580.5	2772.1	5.8133	0.023 52	2569.8	2758.0	5.7432
300	0.036 16	2667.2	2884.2	6.0674	0.029 47	2632.2	2838.4	5.9305	0.024 26	2590.9	2785.0	5.7906
350	0.042 23	2789.6	3043.0	6.3335	0.035 24	2769.4	3016.0	6.2283	0.029 95	2747.7	2987.3	6.1301
400	0.047 39	2892.9	3177.2	6.5408	0.039 93	2878.6	3158.1	6.4478	0.034 32	2863.8	3138.3	6.3634
450	0.052 14	2988.9	3301.8	6.7193	0.044 16	2978.0	3287.1	6.6327	0.038 17	2966.7	3272.0	6.5551
500	0.056 65	3082.2	3422.2	6.8803	0.048 14	3073.4	3410.3	6.7975	0.041 75	3064.3	3398.3	6.7240
550	0.061 01	3174.6	3540.6	7.0288	0.051 95	3167.2	3530.9	6.9486	0.045 16	3159.8	3521.0	6.8778
600	0.065 25	3266.9	3658.4	7.1677	0.055 65	3260.7	3650.3	7.0894	0.048 45	3254.4	3642.0	7.0206
700	0.073 52	3453.1	3894.2	7.4234	0.062 83	3448.5	3888.3	7.3476	0.054 81	3443.9	3882.4	7.2812
800	0.081 60	3643.1	4132.7	7.6566	0.069 81	3639.5	4128.2	7.5822	0.060 97	3636.0	4123.8	7.5173
900	0.089 58	3837.8	4375.3	7.8727	0.076 69	3835.0	4371.8	7.7991	0.067 02	3832.1	4368.3	7.7351
1000	0.097 49	4037.8	4622.7	8.0751	0.083 50	4035.3	4619.8	8.0020	0.073 01	4032.8	4616.9	7.9384
1100	0.105 36	4243.3	4875.4	8.2661	0.090 27	4240.9	4872.8	8.1933	0.078 96	4238.6	4870.3	8.1300
1200	0.113 21	4454.0	5133.3	8.4474	0.097 03	4451.7	5130.9	8.3747	0.084 89	4449.5	5128.5	8.3115
1300	0.121 06	4669.6	5396.0	8.6199	0.103 77	4667.3	5393.7	8.5473	0.090 80	4665.0	5391.5	8.4842
	P = 9.0 MPa (303.40)				P = 10.0 MPa (311.06)				P = 12.5 MPa (327.89)			
Sat.	0.020 48	2557.8	2742.1	5.6772	0.018 026	2544.4	2724.7	5.6141	0.013 495	2505.1	2673.8	5.4624
325	0.023 27	2646.6	2856.0	5.8712	0.019 861	2610.4	2809.1	5.7568	—	—	—	—
350	0.025 80	2724.4	2956.6	6.0361	0.022 42	2699.2	2923.4	5.9443	0.016 126	2624.6	2826.2	5.7118
400	0.029 93	2848.4	3117.8	6.2854	0.026 41	2832.4	3096.5	6.2120	0.020 00	2789.3	3039.3	6.0417
450	0.033 50	2955.2	3256.6	6.4844	0.029 75	2943.4	3240.9	6.4190	0.022 99	2912.5	3199.8	6.2719
500	0.036 77	3055.2	3386.1	6.6576	0.032 79	3045.8	3373.7	6.5966	0.025 60	3021.7	3341.8	6.4618
550	0.039 87	3152.2	3511.0	6.8142	0.035 64	3144.6	3500.9	6.7561	0.028 01	3125.0	3475.2	6.6290
600	0.042 85	3248.1	3633.7	6.9589	0.038 37	3241.7	3625.3	6.9029	0.030 29	3225.4	3604.0	6.7810
650	0.045 74	3343.6	3755.3	7.0943	0.041 0*l*	3338.2	3748.2	7.0398	0.032 48	3324.4	3730.4	6.9218
700	0.048 57	3439.3	3876.5	7.2221	0.043 58	3434.7	3870.5	7.1687	0.034 60	3422.9	3855.3	7.0536
800	0.054 09	3632.5	4119.3	7.4596	0.048 59	3628.9	4114.8	7.4077	0.038 69	3620.0	4103.6	7.2965
900	0.059 50	3829.2	4364.8	7.6783	0.053 49	3826.3	4361.2	7.6272	0.042 67	3819.1	4352.5	7.5182
1000	0.064 85	4030.3	4614.0	7.8821	0.058 32	4027.8	4611.0	7.8315	0.046 58	4021.6	4603.8	7.7237
1100	0.070 16	4236.3	4867.7	8.0740	0.063 12	4234.0	4865.1	8.0237	0.050 45	4228.2	4858.8	7.9165
1200	0.075 44	4447.2	5126.2	8.2556	0.067 89	4444.9	5123.8	8.2055	0.054 30	4439.3	5118.0	8.0987
1300	0.080 72	4662.7	5389.2	8.4284	0.072 65	4460.5	5387.0	8.3783	0.058 13	4654.8	5381.4	8.2717

$\hat{V}$ [=] m³/kg; $\hat{U}, \hat{H}$ [=] J/g = kJ/kg; $\hat{S}$ [=] kJ/kg K

Superheated Vapor (*Continued*)

T (°C)	$\hat{V}$	$\hat{U}$	$\hat{H}$	$\hat{S}$	$\hat{V}$	$\hat{U}$	$\hat{H}$	$\hat{S}$	$\hat{V}$	$\hat{U}$	$\hat{H}$	$\hat{S}$
	P = 15.0 MPa (342.24)				P = 17.5 MPa (354.75)				P = 20.0 MPa (365.81)			
Sat.	0.010 337	2455.5	2610.5	5.3098	0.007 920	2390.2	2528.8	5.1419	0.005 834	2293.0	2409.7	4.9269
350	0.011 470	2520.4	2692.4	5.4421	—	—	—	—	—	—	—	—
400	0.015 649	2740.7	2975.5	5.8811	0.012 447	2685.0	2902.9	5.7213	0.009 942	2619.3	2818.1	5.5540
450	0.018 445	2879.5	3156.2	6.1404	0.015 174	2844.2	3109.7	6.0184	0.012 695	2806.2	3060.1	5.9017
500	0.020 80	2996.6	3308.6	6.3443	0.017 358	2970.3	3274.1	6.2383	0.014 768	2942.9	3238.2	6.1401
550	0.022 93	3104.7	3448.6	6.5199	0.019 288	3083.9	3421.4	6.4230	0.016 555	3062.4	3393.5	6.3348
600	0.024 91	3208.6	3582.3	6.6776	0.021 06	3191.5	3560.1	6.5866	0.018 178	3174.0	3537.6	6.5048
650	0.026 80	3310.3	3712.3	6.8224	0.022 74	3296.0	3693.9	6.7357	0.019 693	3281.4	3675.3	6.6582
700	0.028 61	3410.9	3840.1	6.9572	0.024 34	3398.7	3824.6	6.8736	0.021 13	3386.4	3809.0	6.7993
800	0.032 10	3610.9	4092.4	7.2040	0.027 38	3601.8	4081.1	7.1244	0.023 85	3592.7	4069.7	7.0544
900	0.035 46	3811.9	4343.8	7.4279	0.030 31	3804.7	4335.1	7.3507	0.026 45	3797.5	4326.4	7.2830
1000	0.038 75	4015.4	4596.6	7.6348	0.033 16	4009.3	4589.5	7.5589	0.028 97	4003.1	4582.5	7.4925
1100	0.042 00	4222.6	4852.6	7.8283	0.035 97	4216.9	4846.4	7.7531	0.031 45	4211.3	4840.2	7.6874
1200	0.045 23	4433.8	5112.3	8.0108	0.038 76	4428.3	5106.6	7.9360	0.033 91	4422.8	5101.0	7.8707
1300	0.048 45	4649.1	5376.0	8.1840	0.041 54	4643.5	5370.5	8.1093	0.036 36	4638.0	5365.1	8.0442
	P = 25.0 MPa				P = 30.0 MPa				P = 35.0 MPa			
375	0.001 973	1798.7	1848.0	4.0320	0.001 789	1737.8	1791.5	3.9305	0.001 700	1702.9	1762.4	3.8722
400	0.006 004	2430.1	2580.2	5.1418	0.002 790	2067.4	2151.1	4.4728	0.002 100	1914.1	1987.6	4.2126
425	0.007 881	2609.2	2806.3	5.4723	0.005 303	2455.1	2614.2	5.1504	0.003 428	2253.4	2373.4	4.7747
450	0.009 162	2720.7	2949.7	5.6744	0.006 735	2619.3	2821.4	5.4424	0.004 961	2498.7	2672.4	5.1962
500	0.011 123	2884.3	3162.4	5.9592	0.008 678	2820.7	3081.1	5.7905	0.006 927	2751.9	2994.4	5.6282
550	0.012 724	3017.5	3335.6	6.1765	0.010 168	2970.3	3275.4	6.0342	0.008 345	2921.0	3213.0	5.9026
600	0.014 137	3137.9	3491.4	6.3602	0.011 446	3100.5	3443.9	6.2331	0.009 527	3062.0	3395.5	6.1179
650	0.015 433	3251.6	3637.4	6.5229	0.012 596	3221.0	3598.9	6.4058	0.010 575	3189.8	3559.9	6.3010
700	0.016 646	3361.3	3777.5	6.6707	0.013 661	3335.8	3745.6	6.5606	0.011 533	3309.8	3713.5	6.4631
800	0.018 912	3574.3	4047.1	6.9345	0.015 623	3555.5	4024.2	6.8332	0.013 278	3536.7	4001.5	6.7450
900	0.021 045	3783.0	4309.1	7.1680	0.017 448	3768.5	4291.9	7.0718	0.014 883	3754.0	4274.9	6.9886
1000	0.023 10	3990.9	4568.5	7.3802	0.019 196	3978.8	4554.7	7.2867	0.016 410	3966.7	4541.1	7.2064
1100	0.025 12	4200.2	4828.2	7.5765	0.020 903	4189.2	4816.3	7.4845	0.017 895	4178.3	4804.6	7.4057
1200	0.027 11	4412.0	5089.9	7.7605	0.022 589	4401.3	5079.0	7.6692	0.019 360	4390.7	5068.3	7.5910
1300	0.029 10	4626.9	5354.4	7.9342	0.024 266	4616.0	5344.0	7.8432	0.020 815	4605.1	5333.6	7.7653
	P = 40.0 MPa				P = 50.0 MPa				P = 60.0 MPa			
375	0.001 641	1677.1	1742.8	3.8290	0.001 559	1638.6	1716.6	3.7639	0.001 503	1609.4	1699.5	3.7141
400	0.001 908	1854.6	1930.9	4.1135	0.001 731	1788.1	1874.6	4.0031	0.001 634	1745.4	1843.4	3.9318
425	0.002 532	2096.9	2198.1	4.5029	0.002 007	1959.7	2060.0	4.2734	0.001 817	1892.7	2001.7	4.1626
450	0.003 693	2365.1	2512.8	4.9459	0.002 486	2159.6	2284.0	4.5884	0.002 085	2053.9	2179.0	4.4121
500	0.005 622	2678.4	2903.3	5.4700	0.003 892	2525.5	2720.1	5.1726	0.002 956	2390.6	2567.9	4.9321
550	0.006 984	2869.7	3149.1	5.7785	0.005 118	2763.6	3019.5	5.5485	0.003 956	2658.8	2896.2	5.3441
600	0.008 094	3022.6	3346.4	6.0114	0.006 112	2942.0	3247.6	5.8178	0.004 834	2861.1	3151.2	5.6452
650	0.009 063	3158.0	3520.6	6.2054	0.006 966	3093.5	3441.8	6.0342	0.005 595	3028.8	3364.5	5.8829
700	0.009 941	3283.6	3681.2	6.3750	0.007 727	3230.5	3616.8	6.2189	0.006 272	3177.2	3553.5	6.0824
800	0.011 523	3517.8	3978.7	6.6662	0.009 076	3479.8	3933.6	6.5290	0.007 459	3441.5	3889.1	6.4109
900	0.012 962	3739.4	4257.9	6.9150	0.010 283	3710.3	4224.4	6.7882	0.008 508	3681.0	4191.5	6.6805
1000	0.014 324	3954.6	4527.6	7.1356	0.011 411	3930.5	4501.1	7.0146	0.009 480	3906.4	4475.2	6.9127
1100	0.015 642	4167.4	4793.1	7.3364	0.012 496	4145.7	4770.5	7.2184	0.010 409	4124.1	4748.6	7.1195
1200	0.016 940	4380.1	5057.7	7.5224	0.013 561	4359.1	5037.2	7.4058	0.011 317	4338.2	5017.2	7.3083
1300	0.018 229	4594.3	5323.5	7.6969	0.014 616	4572.8	5303.6	7.5808	0.012 215	4551.4	5284.3	7.4837

$\hat{V}$ [=] m³/kg; $\hat{U}, \hat{H}$ [=] J/g = kJ/kg; $\hat{S}$ [=] kJ/kg K

Compressed Liquid

T (°C)	$\hat{V}$	$\hat{U}$	$\hat{H}$	$\hat{S}$	$\hat{V}$	$\hat{U}$	$\hat{H}$	$\hat{S}$	$\hat{V}$	$\hat{U}$	$\hat{H}$	$\hat{S}$
	P = 5 MPa (263.99)				P = 10 MPa (311.06)				P = 15 MPa (342.24)			
Sat.	0.001 285 9	1147.8	1154.2	2.9202	0.001 452 4	1393.0	1407.6	3.3596	0.001 658 1	1585.6	1610.5	3.6848
0	0.000 997 7	0.04	5.04	0.0001	0.000 995 2	0.09	10.04	0.0002	0.000 992 8	0.15	15.05	0.0004
20	0.000 999 5	83.65	88.65	0.2956	0.000 997 2	83.36	93.33	0.2945	0.000 995 0	83.06	97.99	0.2934
40	0.001 005 6	166.95	171.97	0.5705	0.001 003 4	166.35	176.38	0.5686	0.001 001 3	165.76	180.78	0.5666
60	0.001 014 9	250.23	255.30	0.8285	0.001 012 7	249.36	259.49	0.8258	0.001 010 5	248.51	263.67	0.8232
80	0.001 026 8	333.72	338.85	1.0720	0.001 024 5	332.59	342.83	1.0688	0.001 022 2	331.48	346.81	1.0656
100	0.001 041 0	417.52	422.72	1.3030	0.001 038 5	416.12	426.50	1.2992	0.001 036 1	414.74	430.28	1.2955
120	0.001 057 6	501.80	507.09	1.5233	0.001 054 9	500.08	510.64	1.5189	0.001 052 2	498.40	514.19	1.5145
140	0.001 076 8	586.76	592.15	1.7343	0.001 073 7	584.68	595.42	1.7292	0.001 070 7	582.66	598.72	1.7242
160	0.001 098 8	672.62	678.12	1.9375	0.001 095 3	670.13	681.08	1.9317	0.001 091 8	667.71	684.09	1.9260
180	0.001 124 0	759.63	765.25	2.1341	0.001 119 9	756.65	767.84	2.1275	0.001 115 9	753.76	770.50	2.1210
200	0.001 153 0	848.1	853.9	2.3255	0.001 148 0	844.5	856.0	2.3178	0.001 143 3	841.0	858.2	2.3104
220	0.001 186 6	938.4	944.4	2.5128	0.001 180 5	934.1	945.9	2.5039	0.001 174 8	929.9	947.5	2.4953
240	0.001 226 4	1031.4	1037.5	2.6979	0.001 218 7	1026.0	1038.1	2.6872	0.001 211 4	1020.8	1039.0	2.6771
260	0.001 274 9	1127.9	1134.3	2.8830	0.001 264 5	1121.1	1133.7	2.8699	0.001 255 0	1114.6	1133.4	2.8576
280					0.001 321 6	1220.9	1234.1	3.0548	0.001 308 4	1212.5	1232.1	3.0393
300					0.001 397 2	1328.4	1342.3	3.2469	0.001 377 0	1316.6	1337.3	3.2260
320									0.001 472 4	1431.1	1453.2	3.4247
340									0.001 631 1	1567.5	1591.9	3.6546
	P = 20 MPa (365.81)				P = 30 MPa				P = 50 MPa			
Sat.	0.002 036	1785.6	1826.3	4.0139								
0	0.000 990 4	0.19	20.01	0.0004	0.000 985 6	0.25	29.82	0.0001	0.000 976 6	0.20	49.03	0.0014
20	0.000 992 8	82.77	102.62	0.2923	0.000 988 6	82.17	111.84	0.2899	0.000 980 4	81.00	130.02	0.2848
40	0.000 999 2	165.17	185.16	0.5646	0.000 995 1	164.04	193.89	0.5607	0.000 987 2	161.86	211.21	0.5527
60	0.001 008 4	247.68	267.85	0.8206	0.001 004 2	246.06	276.19	0.8154	0.000 996 2	242.98	292.79	0.8502
80	0.001 019 9	330.40	350.80	1.0624	0.001 015 6	328.30	358.77	1.0561	0.001 007 3	324.34	374.70	1.0440
100	0.001 033 7	413.39	434.06	1.2917	0.001 029 0	410.78	441.66	1.2844	0.001 020 1	405.88	456.89	1.2703
120	0.001 049 6	496.76	517.76	1.5102	0.001 044 5	493.59	524.93	1.5018	0.001 034 8	487.65	539.39	1.4857
140	0.001 067 8	580.69	602.04	1.7193	0.001 062 1	576.88	608.75	1.7098	0.001 051 5	569.77	622.35	1.6915
160	0.001 088 5	665.35	687.12	1.9204	0.001 082 1	660.82	693.28	1.9096	0.001 070 3	652.41	705.92	1.8891
180	0.001 112 0	750.95	773.20	2.1147	0.001 104 7	745.59	778.73	2.1024	0.001 091 2	735.69	790.25	2.0794
200	0.001 138 8	837.7	860.5	2.3031	0.001 130 2	831.4	865.3	2.2893	0.001 114 6	819.7	875.5	2.2634
220	0.001 169 3	925.9	949.3	2.4870	0.001 159 0	918.3	953.1	2.4711	0.001 140 8	904.7	961.7	2.4419
240	0.001 204 6	1016.0	1040.0	2.6674	0.001 192 0	1006.9	1042.6	2.6490	0.001 107 2	990.7	1049.2	2.6158
260	0.001 246 2	1108.6	1133.5	2.8459	0.001 230 3	1097.4	1134.3	2.8243	0.001 203 4	1078.1	1138.2	2.7860
280	0.001 296 5	1204.7	1230.6	3.0248	0.001 275 5	1190.7	1229.0	2.9986	0.001 241 5	1167.2	1229.3	2.9537
300	0.001 359 6	1306.1	1333.3	3.2071	0.001 330 4	1287.9	1327.8	3.1741	0.001 286 0	1258.7	1323.0	3.1200
320	0.001 443 7	1415.7	1444.6	3.3979	0.001 399 7	1390.7	1432.7	3.3539	0.001 338 8	1353.3	1420.2	3.2868
340	0.001 568 4	1539.7	1571.0	3.6075	0.001 492 0	1501.7	1546.5	3.5426	0.001 403 2	1452.0	1522.1	3.4557
360	0.001 822 6	1702.8	1739.3	3.8772	0.001 626 5	1626.6	1675.4	3.7494	0.001 483 8	1556.0	1630.2	3.6291
380					0.001 869 1	1781.4	1837.5	4.0012	0.001 588 4	1667.2	1746.6	3.8101

$\hat{V}$ [=] m³/kg; $\hat{U}, \hat{H}$ [=] J/g = kJ/kg; $\hat{S}$ [=] kJ/kg K

Saturated Solid–Vapor

Temp (°C)	Press. (kPa)	Specific Volume		Internal Energy			Enthalpy			Entropy		
		Sat. Solid	Sat. Vapor	Sat. Solid	Evap.	Sat. Vapor	Sat. Liquid	Evap.	Sat. Vapor	Sat. Liquid	Evap.	Sat. Vapor
T	P	$\hat{V}^S \times 10^3$	$\hat{V}^V$	$\hat{U}^S$	$\Delta\hat{U}$	$\hat{U}^V$	$\hat{H}^S$	$\Delta\hat{H}$	$\hat{H}^V$	$\hat{S}^S$	$\Delta\hat{S}$	$\hat{S}^V$
0.01	0.6113	1.0908	206.1	−333.40	2708.7	2375.3	−333.40	2834.8	2501.4	−1.221	10.378	9.156
0	0.6108	1.0908	206.3	−333.43	2708.8	2375.3	−333.43	2834.8	2501.3	−1.221	10.378	9.157
−2	0.5176	1.0904	241.7	−337.62	2710.2	2372.6	−337.62	2835.3	2497.7	−1.237	10.456	9.219
−4	0.4375	1.0901	283.8	−341.78	2711.6	2369.8	−341.78	2835.7	2494.0	−1.253	10.536	9.283
−6	0.3689	1.0898	334.2	−345.91	2712.9	2367.0	−345.91	2836.2	2490.3	−1.268	10.616	9.348
−8	0.3102	1.0894	394.4	−350.02	2714.2	2364.2	−350.02	2836.6	2486.6	−1.284	10.698	9.414
−10	0.2602	1.0891	466.7	−354.09	2715.5	2361.4	−354.09	2837.0	2482.9	−1.299	10.781	9.481
−12	0.2176	1.0888	553.7	−358.14	2716.8	2358.7	−358.14	2837.3	2479.2	−1.315	10.865	9.550
−14	0.1815	1.0884	658.8	−362.15	2718.0	2355.9	−362.15	2837.6	2475.5	−1.331	10.950	9.619
−16	0.1510	1.0881	786.0	−366.14	2719.2	2353.1	−366.14	2837.9	2471.8	−1.346	11.036	9.690
−18	0.1252	1.0878	940.5	−370.10	2720.4	2350.3	−370.10	2838.2	2468.1	−1.362	11.123	9.762
−20	0.1035	1.0874	1128.6	−374.03	2721.6	2347.5	−374.03	2838.4	2464.3	−1.377	11.212	9.835
−22	0.0853	1.0871	1358.4	−377.93	2722.7	2344.7	−377.93	2838.6	2460.6	−1.393	11.302	9.909
−24	0.0701	1.0868	1640.1	−381.80	2723.7	2342.0	−381.80	2838.7	2456.9	−1.408	11.394	9.985
−26	0.0574	1.0864	1986.4	−385.64	2724.8	2339.2	−385.64	2838.9	2453.2	−1.424	11.486	10.062
−28	0.0469	1.0861	2413.7	−389.45	2725.8	2336.4	−389.45	2839.0	2449.5	−1.439	11.580	10.141
−30	0.0381	1.0858	2943	−393.23	2726.8	2333.6	−393.23	2839.0	2445.8	−1.455	11.676	10.221
−32	0.0309	1.0854	3600	−396.98	2727.8	2330.8	−396.98	2839.1	2442.1	−1.471	11.773	10.303
−34	0.0250	1.0851	4419	−400.71	2728.7	2328.0	−400.71	2839.1	2438.4	−1.486	11.872	10.386
−36	0.0201	1.0848	5444	−404.40	2729.6	2325.2	−404.40	2839.1	2434.7	−1.501	11.972	10.470
−38	0.0161	1.0844	6731	−408.06	2730.5	2322.4	−408.06	2839.0	2430.9	−1.517	12.073	10.556
−40	0.0129	1.0841	8354	−411.70	2731.3	2319.6	−411.70	2838.9	2427.2	−1.532	12.176	10.644

$\hat{V}$ [=] m^3/kg; $\hat{U}, \hat{H}$ [=] J/g = kJ/kg; $\hat{S}$ [=] kJ/kg K

Appendix A.IV

Enthalpies and Gibbs Energies of Formation

Standard Enthalpies and Gibbs Energies of Formation at 298.15 K for One Mole of Each Substance from Its Elements

Chemical Species		State (See Note)	$\Delta_f \underline{H}^\circ$	$\Delta_f \underline{G}^\circ$
			kJ/mol of the Substance Formed	
Paraffinic Hydrocarbons				
Methane	CH_4	(g)	−74.5	−50.5
Ethane	C_2H_6	(g)	−83.8	−31.9
Propane	C_3H_8	(g)	−104.7	−24.3
n-Butane	C_4H_{10}	(g)	−125.8	−16.6
n-Pentane	C_5H_{12}	(g)	−146.8	−8.7
n-Pentane	C_5H_{12}	(l)	−173.1	−9.2
n-Hexane	C_6H_{14}	(g)	−166.9	0.2
n-Heptane	C_7H_{16}	(g)	−187.8	8.3
n-Octane	C_8H_{18}	(g)	−208.8	16.3
n-Octane	C_8H_{18}	(l)	−255.1	—
Unsaturated Hydrocarbons				
Acetylene	C_2H_2	(g)	227.5	210.0
Ethylene	C_2H_4	(g)	52.5	68.5
Propylene	C_3H_6	(g)	19.7	62.2
1-Butene	C_4H_8	(g)	1.2	70.3
1-Pentene	C_5H_{10}	(g)	−21.3	78.4
1-Hexene	C_6H_{12}	(g)	−42.0	86.8
1-Heptene	C_7H_{14}	(g)	−62.8	—
1,3-Butadiene	C_4H_6	(g)	109.2	149.8
Aromatic Hydrocarbons				
Benzene	C_6H_6	(g)	82.9	129.7
Benzene	C_6H_6	(l)	49.1	124.5
Ethylbenzene	C_8H_{10}	(g)	29.9	130.9
Naphthalene	$C_{10}H_8$	(s)	78.5	—
Styrene	C_8H_8	(g)	147.4	213.9
Toluene	C_7H_8	(g)	50.2	122.1
Toluene	C_7H_8	(l)	12.2	113.6
Cyclic Hydrocarbons				
Cyclohexane	C_6H_{12}	(g)	−123.1	31.9
Cyclohexane	C_6H_{12}	(l)	−156.2	26.9
Cyclopropane	C_3H_6	(g)	53.3	104.5
Methylcyclohexane	C_7H_{14}	(g)	−154.8	27.5
Methylcyclohexane	C_7H_{14}	(l)	−190.2	20.6
Cyclohexene	C_6H_{10}	(l)		106.9

Chemical Species		State (See Note)	$\Delta_f \underline{H}°$ kJ/mol of the	$\Delta_f \underline{G}°$ Substance Formed
Oxygenated Hydrocarbons				
Acetaldehyde	C_2H_4O	(g)	−166.2	−128.9
Acetic Acid	CH_3COOH	(l)	−484.5	−389.9
Acetic Acid	CH_3COOH	(aq)	−486.1	−396.5
1,2-Ethanediol	$C_2H_6O_2$	(l)	−454.8	−323.1
Ethanol	C_2H_6O	(g)	−235.1	−168.5
Ethanol	C_2H_6O	(l)	−277.7	−174.8
Ethyl acetate	$CH_3COOC_2H_5$	(l)	−463.3	−318.4
Ethylene oxide	C_2H_4O	(g)	−52.6	−13.0
Formaldehyde	CH_2O	(g)	−108.6	−102.5
Formic acid	$HCOOH$	(l)	−424.7	−361.4
Methanol	CH_4O	(g)	−200.7	−162.0
Methanol	CH_4O	(l)	238.7	−166.3
Phenol	C_6H_5OH	(s)	−165.0	−50.9
Inorganic Compounds				
Aluminum oxide	Al_2O_3	(s, α)	−1675.7	−1582.3
Aluminum chloride	$AlCl_3$	(s)	−704.2	−628.8
Ammonia	NH_3	(g)	−46.1	−16.5
Ammonia	NH_3	(aq)	—	−26.5
Ammonium nitrate	NH_4NO_3	(s)	−365.6	−183.9
Ammonium chloride	NH_4Cl	(s)	−314.4	−202.9
Barium oxide	BaO	(s)	−553.5	−525.1
Barium chloride	$BaCl_2$	(s)	−858.6	−810.4
Bromine	Br_2	(l)	0	0
Bromine	Br_2	(g)	30.9	3.1
Calcium carbide	CaC_2	(s)	−59.8	−64.9
Calcium carbonate	$CaCO_3$	(s)	−1206.9	−1128.8
Calcium chloride	$CaCl_2$	(s)	−795.8	−748.1
Calcium chloride	$CaCl_2$	(aq)	—	−8101.9
Calcium chloride	$CaCl_2 \cdot 6H_2O$	(s)	−2607.9	—
Calcium hydroxide	$Ca(OH)_2$	(s)	−986.1	−898.5
Calcium hydroxide	$Ca(OH)_2$	(aq)	—	−868.1
Calcium oxide	CaO	(s)	−635.1	−604.0
Carbon dioxide	CO_2	(g)	−393.5	−394.4
Carbon dioxide	CO_2	(aq)	−413.8	−386.0
Carbon disulfide	CS_2	(l)	89.7	65.3
Carbon monoxide	CO	(g)	−110.5	−137.2
Carbon tetrachloride	CCl_4	(l)	−135.4	−65.2
Carbonic acid	H_2CO_3	(aq)	−699.7	−623.1
Hydrochloric acid	HCl	(g)	−92.3	−95.3
Hydrochloric acid	HCl	(aq)	−167.2	−131.2
Hydrogen bromide	HBr	(g)	−36.4	−53.5
Hydrogen cyanide	HCN	(g)	135.1	124.7
Hydrogen cyanide	HCN	(l)	108.9	125.0
Hydrogen flouride	HF	(g)	−271.1	−273.2
Hydrogen iodide	HI	(g)	26.5	1.7
Hydrogen peroxide	H_2O_2	(l)	−187.8	−120.4
Hydrogen sulfide	H_2S	(g)	−20.6	−33.6
Hydrogen sulfide	H_2S	(aq)	−39.7	−27.8
Iodine	I_2	(g)	62.3	19.8
Iodine	I_2	(s)	0	0
Iron (II)	FeO	(s)	−272.0	—
Iron (III) oxide (hematite)	Fe_2O_3	(s)	−824.2	−742.2

Chemical Species		State (See Note)	$\Delta_f\underline{H}^\circ$	$\Delta_f\underline{G}^\circ$
			kJ/mol of the Substance Formed	
Iron (II) sulfide	FeS	(s, α)	−100.0	−100.4
Iron sulfide (pyrite)	FeS$_2$	(s)	−178.2	−166.9
Lead oxide	PbO	(s, yellow)	−217.3	−187.9
Lead oxide	PbO	(s, red)	−219.0	−188.9
Lead dioxide	PbO$_2$	(s)	−277.4	−217.3
Lithium chloride	LiCl	(s)	−408.6	—
Lithium chloride	LiCl·H$_2$O	(s)	−712.6	—
Lithium chloride	LiCl·2H$_2$O	(s)	−1012.7	—
Lithium chloride	LiCl·3H$_2$O	(s)	−1311.3	—
Magnesium oxide	MgO	(s)	−601.7	−569.4
Magnesium carbonate	MgCO$_3$	(s)	−1095.8	−1012.1
Magnesium chloride	MgCl$_2$	(s)	−641.3	−591.8
Mercury (I) chloride	Hg$_2$Cl$_2$	(s)	−265.2	−210.8
Mercury (II) chloride	HgCl$_2$	(s)	−224.3	−178.6
Nitric acid	HNO$_3$	(l)	−174.1	−80.7
Nitric acid	HNO$_3$	(aq)	—	−111.3
Nitric oxide	NO	(g)	90.3	86.6
Nitrogen dioxide	NO$_2$	(g)	33.2	51.3
Nitrous oxide	N$_2$O	(g)	82.1	104.2
Nitrogen tetroxide	N$_2$O$_4$	(g)	9.2	97.9
Potassium chloride	KCl	(s)	−436.8	−409.1
Silicon dioxide	SiO$_2$	(s, α)	−910.9	−856.6
Silver bromide	AgBr	(s)	−100.4	−96.9
Silver chloride	AgCl	(s)	−127.1	−109.8
Silver nitrate	AgNO$_3$	(s)	−124.4	−33.4
Sodium bicarbonate	NaHCO$_3$	(s)	−945.6	−847.9
Sodium carbonate	Na$_2$CO$_3$	(s)	−1130.7	−1044.4
Sodium carbonate	Na$_2$CO$_3$·10H$_2$O	(s)	−4081.3	—
Sodium chloride	NaCl	(s)	−411.2	−384.1
Sodium chloride	NaCl	(aq)	—	−393.1
Sodium hydroxide	NaOH	(s)	−425.6	−379.5
Sodium hydroxide	NaOH	(aq)	—	−419.2
Sodium sulfate	Na$_2$SO$_4$	(s)	−1382.8	−1265.2
Sodium sulfate·decahydrate	Na$_2$SO$_4$·10H$_2$O	(s)	−4322.5	−3642.3
Sulfur	S$_2$	(g)	129.8	81.0
Sulfur	S$_2$	(l)	1.1	0.3
Sulfur	S$_2$	(s)	0	0
Sulfur dioxide	SO$_2$	(g)	−296.8	−300.2
Sulfur trioxide	SO$_3$	(g)	−395.7	−371.1
Sulfur trioxide	SO$_3$	(l)	−441.0	—
Sulfuric acid	H$_2$SO$_4$	(l)	−814.0	−690.0
Sulfuric acid	H$_2$SO$_4$	(aq)	—	−744.5
Water	H$_2$O	(g)	−241.8	−228.6
Water	H$_2$O	(l)	−285.8	−237.1
Zinc oxide	ZnO	(s)	−348.3	−318.3

Taken from *TRC Thermodynamic Tables—Hydrocarbons,* Thermodynamics Research Center, Texas A&M University System, College Station, Tex.; "The NBS Tables of Chemical Thermodynamic Properties," *I. Physical and Chemical Reference Data,* Vol. 11, Suppl. 2, 1982; and a collection of other sources. Because of the variety of sources used here and in the PROPERTY program, there may be very small differences between the data here and in that program.

Note: All standard states are at 25°C. The standard state for a gas, designated by (g), is the pure ideal gas at 1 bar. For liquids (l) and solids (s), the standard state is the substance in that state of aggregation at 1 bar and 25°C. For solutes in aqueous solution, denoted by (aq), the standard state is the hypothetical ideal 1-molal solution of the solute in water at 1 bar and 25°C.

Appendix A.V

Heats of Combustion

Compound	Formula	State	Heat of Combustion in kJ/mol	
			H_2O (l) CO_2 (g)	H_2O (g) CO_2 (g)
Hydrogen	H_2	(g)	285.840	241.826
Carbon	C	(s, graphite)	393.513	—
Carbon monoxide	CO	(g)	282.989	—
Methane	CH_4	(g)	890.35	802.32
Ethane	C_2H_6	(g)	1559.88	1427.84
Propane	C_3H_8	(g)	2220.05	2044.00
Propane	C_3H_8	(l)	2204.06	2028.00
n-Butane	C_4H_{10}	(g)	2878.52	2658.45
n-Butane	C_4H_{10}	(l)	2857.02	2636.95
2-Methylpropane (isobutane)	C_4H_{10}	(g)	2871.65	2651.58
2-Methylpropane (isobutane)	C_4H_{10}	(l)	2851.92	2631.85
n-Pentane	C_5H_{12}	(g)	3536.15	3272.06
n-Pentane	C_5H_{12}	(l)	3509.54	3245.45
n-Hexane	C_6H_{14}	(g)	4194.75	3886.64
n-Hexane	C_6H_{14}	(l)	4163.12	3855.01
n-Heptane	C_7H_{16}	(g)	4853.48	4501.36
n-Heptane	C_7H_{16}	(l)	4816.91	4464.79
n-Octane	C_8H_{18}	(g)	5512.21	5116.07
n-Octane	C_8H_{18}	(l)	5470.71	5074.56
n-Nonane	C_9H_{20}	(g)	6170.98	5730.82
n-Nonane	C_9H_{20}	(l)	6124.54	5684.38
n-Decane	$C_{10}H_{22}$	(g)	6829.71	6345.58
n-Decane	$C_{10}H_{22}$	(l)	6778.33	6294.20
n-Dodecane	$C_{12}H_{26}$	(g)	8147.21	7575.05
n-Dodecane	$C_{12}H_{26}$	(l)	8085.96	7513.79
n-Hexadecane	$C_{16}H_{34}$	(g)	10782.17	10033.94
n-Hexadecane	$C_{16}H_{34}$	(l)	10701.17	9952.94
n-Eicosane	$C_{20}H_{42}$	(g)	13417.13	12492.84
n-Eicosane	$C_{20}H_{42}$	(l)	13316.37	12392.09
Benzene	C_6H_6	(g)	3301.51	3169.46
Benzene	C_6H_6	(l)	3267.62	3135.57
Methylbenzene (toluene)	C_7H_8	(g)	3947.94	3771.88
Methylbenzene (toluene)	C_7H_8	(l)	3909.95	3733.89

Appendix A.V (*Continued*)

Compound	Formula	State	Heat of Combustion in kJ/mol	
			H_2O (l) CO_2 (g)	H_2O (g) CO_2 (g)
Ethylbenzene	C_8H_{10}	(g)	4607.13	4387.05
Ethylbenzene	C_8H_{10}	(l)	4564.87	4344.79
1,2-Dimethylbenzene (*o*-xylene)	C_8H_{10}	(g)	4596.29	4376.21
1,2-Dimethylbenzene (*o*-xylene)	C_8H_{10}	(l)	4552.86	4332.78
1,3-Dimethylbenzene (*m*-xylene)	C_8H_{10}	(g)	4594.53	4374.46
1,3-Dimethylbenzene (*m*-xylene)	C_8H_{10}	(l)	4551.86	4331.78
1,4-Dimethylbenzene (*p*-xylene)	C_8H_{10}	(g)	4595.25	4375.17
1,4-Dimethylbenzene (*p*-xylene)	C_8H_{10}	(l)	4552.86	4332.78
Cyclopentane	C_5H_{10}	(g)	3319.54	3099.47
Cyclopentane	C_5H_{10}	(l)	3290.88	3070.80
Methylcyclopentane	C_6H_{12}	(g)	3969.44	3705.35
Methylcyclopentane	C_6H_{12}	(l)	3937.73	3673.64
Cyclohexane	C_6H_{12}	(g)	3953.00	3688.91
Cyclohexane	C_6H_{12}	(l)	3919.91	3655.81
Methylcyclohexane	C_7H_{14}	(g)	4600.68	4292.57
Methylcyclohexane	C_7H_{14}	(l)	4565.29	4257.18
Ethene (ethylene)	C_2H_4	(g)	1410.99	1322.96
Propene (propylene)	C_3H_6	(g)	2058.47	1926.43
1-Butene	C_4H_8	(g)	2718.58	2542.53
Ethyne (acetylene)	C_2H_2	(g)	1299.61	1255.60
Propyne (methylacetylene)	C_3H_4	(g)	1937.65	1849.62
1-Butyne (ethylacetylene)	C_4H_6	(g)	2597.68	2465.64
2-Butyne (dimethylacetylene)	C_4H_6	(g)	2579.57	2447.53

Appendix B

Brief Descriptions of Computer Aids
for Use with This Book

Programs and worksheets to enhance the use of this book are on the accompanying CD-ROM and also on the Web site *www.wiley.com/college/sandler*, with instructions on how to use them. It is my intention to add to and update these programs on this Web site.

The most easily understood computational aids are the MATHCAD worksheets because in this computer-algebra package equations are written, in general, in the exactly same form as in this book. Consequently, the MATHCAD worksheets are easily changed to accommodate a different equation of state or activity coefficient model. The suite of MATHCAD worksheets allow calculations of the thermodynamic properies and phase behavior for both pure fluids and mixtures, including various flash calculations, activity coefficient fitting and phase behavior calculations, and chemical equilibrium constant and adiabatic flame temperature computations. However, the MATHCAD package is required to use these worksheets. Included on the CD-ROM is a 120-day evaluation version of MATHCAD, should the software not otherwise be available to the reader. Academic group purchases of this software for colleges and universities are available from MATHSOFT, Inc.

VISUAL BASIC programs are provided in two forms. One is as easy-to-use stand-alone Windows-based executable programs. The second form is as VISUAL BASIC code that can be modified if the user has access to a VISUAL BASIC compiler. The first VISUAL BASIC program, called PROPERTY, retrieves pure component data from a database of over 600 compounds, and can also be used to calculate vapor pressures and heat capacities at a given temperature or over a range of temperatures. This program is also called by the others in this suite to provide the pure component data needed in those calculations. The pure fluid Peng-Robinson equation of state program can be used to calculate thermodynamic properies and phase behavior at any temperature and pressure. The Peng-Robinson mixture program is for the calculation of mixture vapor-liquid equilibrium. The UNIFAC program uses the modified UNIFAC model for vapor-liquid equilibrium calculations with the most recent parameters available at the time of publication of this book. The final program in this set is used to calculate chemical equilibrium constants at a given temperature or over a temperature range.

The DOS BASIC programs were provided with the third edition, and are also supplied in stand-alone executable and BASIC code forms. They are a similar suite of VISUAL BASIC programs, except that there is no PROPERTY database. The stand-alone versions run in a DOS window, and have less functionality and are not as user-friendly as the VISUAL BASIC programs. Also, the DOS BASIC UNIFAC program uses the original version of UNIFAC (as a result of DOS BASIC memory limitations). The BASIC code programs can be modified if a BASIC or QUICKBASIC compiler is available.

We also supply three MATLAB programs: a pure fluid Peng-Robinson program, a mixture Peng-Robinson program and a modified UNIFAC program. As there is no property database for the MATLAB programs, all component data must be entered manually. To run these either MATLAB 7.0 (or higher) must be installed on the computer or the file MCRInstaller.exe must first be run (as described in Appendix B.IV). The MCRInstaller is on the CD-ROM and is also available from MATHWORKS, Inc.

Appendix C

Answers to Selected Problems

This appendix contains answers to selected problems so that the user can check his or her work. However, only the final answers are provided, not the complete solution. Therefore, these answers must be used with some discretion. Unlike a simple algebra problem, for which there will be an exact solution, here the user should remember that if the thermodynamic property chart is read somewhat differently than was done here, a slightly different answer will be obtained. Likewise, if you have used one equation of state and I have used another in the solution, or if you have used one activity coefficient model and I have used another, or even if you have fit the parameters in an activity coefficient or equation-of-state model somewhat differently than I have, the final answers will be somewhat different. Also, since there are slight differences in the databases used, there may be some differences in the chemical equilibria calculated using the DOS- and Windows-based programs. Therefore, except for the simplest problems (for example, those involving ideal gases), you may not get exactly the same answers as I did. However, the differences should not be large.

Finally, because the world is not an ideal gas, equations of state or activity coefficients frequently must be used to describe a real system of interest. This results in considerable mathematical complexity. Consequently, many problems are best solved using the computer software I have provided that is discussed in the previous appendix, or with MATHCAD, MATLAB, MATHEMATICA, POLYMATH, or equation-solving software and programs you develop on your own.

Chapter 1

1.2 a. Water is inappropriate as a thermometric fluid between $0°C$ and $10°C$, since the volume is not a unique function of temperature in this range. **b.** The thermal expansion of mercury is not linear with temperature, which introduces an error. For example, at $10°C$ the error is $-0.068°C$.

Chapter 2

2.1 a. 0.1 g/m^3. **b.** 32.35 minutes.
2.4 a. 0.5 moles of NO_2. **b.** 1.32 hours.

Chapter 3

3.1 a. $h = 1.57$ m. **b.** $v_f = 134.9$ m/s, which is much too high because of neglected air and road resistance.
3.5 With steam the final temperature is about 383 K or $656°C$, while with an ideal gas it is $600°C$.
3.14 Initial amount of steam in the tank is 194.9 kg. At end of the process there is 215.3 kg of liquid and 67.5 kg of steam. Fraction of contents by weight that is liquid is 0.761.
3.22 At 5 minutes $T = 152.5$ K, $P = 0.69$ bar, and the rate of temperature change is -1.15 K/s.
3.25 Final pressure in both tanks is 133.3 bar; the final

temperature in tank 1 is 223.4 K; and in tank 2 is 328.0 K.
3.32 a. Using the pressure-enthalpy diagram, $\hat{H} = 153$ kJ/kg, leading to $T = {\sim}90$ K with 55 percent of the nitrogen being liquid. **b.** $T = 135$ K, and no liquid is formed. Why?

Chapter 4

4.1 a. $T = 8.31°C$. **b.** $\triangle S = S_{gen} = 64.6$ J/K.
4.4 a. $T = 308°C$. **b.** $\underline{S}_{gen} = 1.62$ kJ/kg $\cdot$ K.
4.9 a. 627 kW. **b.** $\underline{S}_{gen} = 843$ kJ/kg $\cdot$ h. **c.** 415 kW.
4.12 a. $\triangle N = -1768$ mol. **b.** $\triangle N = -1152$ mol.
4.16 a. $W = 9334$ J/mol and $T = 549.4$ K. **b.** $W = 9191$ J/mol and $T = 520.9$ K.
4.22 $P_1^f = P_2^f = 133.3$ bar, $T_1^f = 226$ K and $T_2^f = 285$ K.
4.23 a. $P_1^f = P_2^f = 5$ bar, $T_1^f = 275.6°C$ and $T_2^f = 497.7°C$. **b.** $P_1^f = P_2^f = 5$ bar, $T_1^f = T_2^f = 366°C$.

Chapter 5

5.1 a. If the turbine does not drive the compressor, the work is 30 kJ/kg, and C.O.P. $= 4.30$. **b.** The compressor work is 34.8 kJ/kg and C.O.P. $= 3.18$.
5.8 a. Work $= 533$ kJ/kg methane through compressors. **b.** Work is 1713 kJ for each kg of LNG, and

the fraction of vapor is 0.689. **c.** Work is 1168 kJ/kg of LNG produced.

5.9 0.535 kg of TNT.

5.12 a. 610 kJ/kg. **b.** 630 kJ/kg. **c.** 685 kJ/kg.

Chapter 6

6.1 a. 4.463°C/MPa. **b.** 17.02°C/MPa. 0.262.

6.12 3186 J/mol.

6.17 From P-R EOS, $\underline{V} = 1.667 \times 10^{-3}$ m^3/mol, $\underline{H} = 21,033$ J/mol, $\underline{S} = 7.061$ J/mol K; from corresponding states $\underline{V} = 0.967 \times 10^{-3}$ m^3/mol, $\underline{H} = 7253$ J/mol, $\underline{S} = 27.5$ J/mol K.

6.21 a. $Q = -14088$ J/mol, $W = 11\ 540$ J/mol. **b.** $Q = -12\ 136$ J/mol, $W = 12\ 319$ J/mol.

6.31 a. $T = 300°$C. **b.** $T = 301.9°$C. **c.** $T = 299.9°$C. **d.** $T = 267°$C.

Chapter 7

7.1 a. $\hat{G} = -307.8$ J/g in both phases. **b.** At 225°C $\hat{G} = -314.1$ J/g. **c.** At 210°C $\hat{G} = -273.8$ J/g.

7.13 a. $\triangle_{\text{vap}}\underline{H} \sim 42.7$ kJ/mol. **b.** $\triangle_{\text{vap}}\underline{H} \sim 313.6$ kJ/mol.

7.15 a. $f = 15.3$ bar. **b.** $f = 69.25$ kPa at 100 bar and 129.6 kPa at 1000 bar.

7.19 a. 31.8%. **b.** 28.3%.

7.24 $P_{\text{TP}} = 0.483$ bar, $T_{\text{TP}} = 278.7$ K.

7.35 a. Using the Peng-Robinson equation of state, by trial and error, $T = 225.35$ K. **b.** 47.4 wt % liquid and 52.6 wt % vapor (which corresponds to only 3.1 vol % liquid).

Chapter 8

8.9 a. 2. **b.** 3. **c.** 3.

8.10 a. 1. **b.** 4. **c(i)** 2, **c(ii)** 1.

8.13 a. 5. **b.** 3. **c.** 2.

8.15 a. -59.6 kJ. **b.** -6.59 kJ. **c.** -357 kJ. **e.** $\overline{H}_1 - \underline{H}_1 = -46.0$ kJ/mol and $\overline{H}_2 - \underline{H}_2 = -2.3$ kJ/mol.

8.29 170.5 kg ethylene glycol per minute.

8.35 59.9 kJ/mol of nitrogen entering the reactor.

Chapter 9

9.12 a. $f_{\text{O}_2} = 820$ bar from corresponding states and 735.1 bar from Peng-Robinson program. **b.** $f_{\text{N}_2} = 1088$ bar from corresponding states and 1043 bar from Peng-Robinson program. **c.** Lewis-Randall rule with pure component corresponding states $\overline{f}_{\text{N}_2} = 761.6$ and $\overline{f}_{\text{N}_2} = 246.0$ bar. **d.** $\overline{f}_{\text{N}_2} = 732.3$ and $\overline{f}_{\text{N}_2} = 224.9$ bar.

9.23 a. $\gamma_{\text{CH}_4} = 1.080$ and $\gamma_{\text{N}_2} = 1.073$. **b.** 88.8 J/mol. **c.** 5.984 J/mol K.

Chapter 10

Section 10.1

10.1-1 a. $T = 293.7$ K, $y_{\text{ET}} = 0.4167$, $y_{\text{P}} = 0.3730$, $y_{\text{NB}} = 0.1610$, and $y_{\text{MP}} = 0.2502$. **b.** $T = 314.2$ K, $y_{\text{ET}} = 0.0039$, $y_{\text{P}} = 0.0337$, $y_{\text{NB}} = 0.5215$, and $y_{\text{MP}} = 0.4409$. **c.** $L = 0.8667$, $x_{\text{ET}} = 0.0225$, $x_{\text{P}} = 0.0858$, $x_{\text{NB}} = 0.4258$, and $x_{\text{MP}} = 0.4659$. $V = 0.1333$, $y_{\text{ET}} = 0.2289$, $y_{\text{P}} = 0.1921$, $y_{\text{NB}} = 0.2326$, and $y_{\text{MP}} = 0.3464$.

10.1-3 a. $P_{\text{dew}} = 0.768$ bar, $y_5 = 0.071$, $y_6 = 0.338$, $y_7 = 0.592$. **b.** $P_{\text{bub}} = 1.258$ bar, $y_5 = 0.541$, $y_6 = 0.36$, $y_7 = 0.093$.

10.1-6 21.8%.

Section 10.2

10.2-2 Using the van Laar expression, $x_{\text{W}} = 0.075$, $x_{\text{FURF}} = 0.925$, $y_{\text{W}} = 0.867$ and $y_{\text{FURF}} = 0.129$.

10.2-5 a. $\alpha = 1.075$, $\beta = 1.029$.

10.2-14 $P = 0.6482$ bar, $y_1 = 0.4483$ and $y_2 = 0.5517$.

10.2-33 a. 3.658 bar. **b.** 3.601 bar.

Section 10.3

10.3-1 At 1 bar, $T = 231.31$ K, $y_{\text{C}_2} = 0.3444$, $y_{\text{C}_3} = 0.5907$, and $y_{\text{C}_4} = 0.0649$. At 20 bar, $T = 341.99$ K, $y_{\text{C}_2} = 0.1261$, $y_{\text{C}_3} = 0.6642$, and $y_{\text{C}_4} = 0.2097$. At 40 bar, $T = 384.56$ K, $y_{\text{C}_2} = 0.0699$, $y_{\text{C}_3} = 0.6135$, and $y_{\text{C}_4} = 0.3168$.

10.3-3 At 20 bar, at $T = 350$ K, $x_{\text{C}_2} = 0.0258$, $x_{\text{C}_3} = 0.4928$, and $x_{\text{C}_4} = 0.4814$. $y_{\text{C}_2} = 0.0690$, $y_{\text{C}_3} = 0.6304$, and $y_{\text{C}_4} = 0.3007$; also $V/L = 0.5612/0.4388$.

10.3-8 $K_{\text{M}} = 13.95$ and $K_{\text{B}} = 0.009\ 08$.

Chapter 11

Section 11.1

11.1-3 a. $x_N = 0.0001$. **b.** $x_N = 0.000\ 059\ 1$, $H_N = 1692$ bar.

11.1-6 $x_{\text{CO}_2} = 0.022\ 15$.

11.1-12 a. From the Steam Tables $P = 3.169$ kPa. **b.** $P = 101.3$ kPa, $y_{\text{W}} = 0.0313$.

Section 11.2

11.2-4 a. One-constant Margules is not consistent with the data; using the liquid compositions of both species gives two different A values. **b.** $\delta_{\text{P}} = 12.0$ or 6.4.

Section 11.3

11.3-1 a. $\gamma_{\text{B}} = 15.24$, $\gamma_{\text{B}} = 0.257$ and $y_{\text{W}} = 0.743$. **b.**

$\gamma_B = 1.524$ and $\gamma_W = 1.60$. **c.** 1.013 bar.

11.3-3 a. $x_H^I = 0.0902$ and $x_H^{II} = 0.9098$. **b.** $P = 1.9657$ bar, $y_H = 0.5795$, and $y_{EtOH} = 0.4205$.

Section 11.4

11.4-3 Sample answer: At $x_{Br} = 0.047$, $\gamma_{Br}^{CCl_4} = 1.990$ and $\gamma_{Br}^{H_2O} = 298.8$.

Section 11.5

11.5-5 a. Freezing pt. depression (K) = 0.1868 (MW = 100), 0.1871×10^{-1} (MW = 1000), and 0.1871×10^{-4} (MW = 1×10^6). **b.** Boiling pt. elevation (K) = 0.050 86 (MW = 100), 5.094×10^{-3} (MW = 1000), and 5.095×10^{-6} (MW = 1×10^6). **c.** Osmotic pressure (mm Hg) = 1843 (MW = 100), 184.6 (MW = 1000), and 0.1846 (MW = 1×10^6).

Chapter 12

Section 12.1

12.1-1 Using the regular solution model, x_N is predicted to be 0.2655 in chlorobenzene, 0.256 in benzene, 0.239 in toluene, and 0.221 in CCl_4.

12.1-3 The 49.5°C isotherm is fit reasonably well with $k_{CO_2-B} = 0.08$.

12.1-4 a. $y_N = 0.006\ 92$. **b.** $y_N = 0.002\ 25$.

12.1-7 a. $\gamma_L^\infty = 71.7$. **b.** $S(58.4°C) = 42.9$ g/kg.

Section 12.3

12.3-2 221.6 g of methanol, 318.6 g of ethanol, or 636.9 g of glycerol, assuming the solutions are ideal. These correspond to 16.5 wt % (vs. 18.1 wt % exp) methanol, 21 wt % (24.4 wt % exp) ethanol, and 35.5 wt % (38.9 wt % exp) glycerol.

12.3-5 a. $\gamma_W = 0.9445$ (87 g), 0.8744 (177 g), and 0.8105 (272 g). **b.** $\triangle T$ (K) = −15.95 (87 g),

−33.76 (177 g), and −49.99 (272 g) including the $\triangle C_P$ term.

Section 12.4

12.4-4 Sample result: At $z = 0.5$, $T_S = 393.54$ K, $T_L = 372.61$ K, $T_{LLE} = 500$ K, $x_1^I = 0.256$, $x_1^I = 0.744$.

12.4-7 Sample result: At 85 K, $x_N = 0.158$, $y_N = 0.512$, $P_N^{vap} = 3.284$ bar, and $P_O^{vap} = 0.5867$ bar.

Section 12.5

12.5-3 a. $C_{B,H_2O} = 1.05 \times 10^{-3}$ g/m³, $C_{B,soil} = 1.070 \times$

10^{-3} g/m³, $C_{B,air} = 0.239 \times 10^{-3}$ g/m³, $C_{B,fish} = 0.707 \times 10^{-3}$ g/m³.

Section 12.6

12.6-1 b. $x_{152} = 0.0377$, $x_{114} = 0.6025$, and $x_{22} = 0.3598$.

Chapter 13

(Note that all the calculations were done with the DOS-based CHEMEQ program. Slightly different answers will be obtained if the Windows-based program is used to calculate the chemical equilibrium constants.)

13.1 94.4%.

13.4 14970 bar.

13.6 a. 0.0253 bar. **b.** 0.0122 bar.

13.14 $y_{C_2H_5OH} = y_{H_2} \sim 0$, $y_{CH_3CHO} = y_{H_2O} \sim 0.197$, and $y_{O_2} \sim 0.049$.

13.18 a. $K_{a,1} = 0.9731$, $K_{a,2} = 0.1191$, $y_{C_4H_{10}} = 0.147$, $y_{C_4H_{10}} = 0.308$, and $y_{H_2} = 0.446$.

13.32 $\triangle_{rxn}G° = 30.2$ kJ/mol.

13.50 a. $\triangle_{rxn}G°(267$ K$) = 6.65$ kJ, $\triangle_{rxn}G°(255$ K$) = 5.47$ kJ. **b.** $\triangle_{rxn}H° = 13.57$ kJ, $\triangle_{rxn}S° = -75.74$ kJ/mol. **c.** 288.8 kPa.

Chapter 14

14.1 a. $T_{ad} = 3553$ K, $X = 0.659$, $y_{H_2} = 0.291$, $y_{O_2} = 0.146$, $y_{H_2O} = 0.563$. **b.** $T_{ad} = 3343$ K, $X = 0.835$, $y_{H_2} = 0.104$, $y_{O_2} = 0.368$, $y_{H_2O} = 0.528$. **c.** $T_{ad} = 1646$ K, $X = 1.0$, $y_{H_2} = 0$, $y_{O_2} = 0.095$, $y_{H_2O} = 0.190$, $y_{N_2} = 0.715$.

14.3 Using the CHEMEQ program, $K_a = 345$, $X = 0.927$, $P = 0.5434$ bar, and $Q = -96.7$ kJ.

14.5 11.87 kJ/mol at a body temperature of 310.1 K.

Chapter 15

15.9 Sample result: At pH = 5.6 the charge is calculated to be $+1.255 \times 10^{-4}$.

15.11 Sample result: At pH = 2 the solubility is 84.35 g/kg of water.

15.15 Sample results: At 4 cm the concentration is 0.649 mg/g. **b.** 6.763 mg/g.

15.18 Sample result: At 5 M acetic acid and 5 M NaCl, $\alpha = 0.00186$ and pH = 2.031 in an ideal solution ($\gamma_\pm = 1$), and $\alpha = 0.027$ and pH = 2.036 using simple Debye-Hückel model.

15.23 $Y_{P/S} = 0.024$, $Y_{N/S} = 0.08535$, $Y_{W/S} = -0.594$, $Y_{O/S} = 0.3579$, and only 2.4% of carbon in glucose is converted to ethanol.

Index

SUBSCRIPTS

Symbol	Designates
A, B, C, ...,	species
AB, D	dissociated electrolyte AB
ad	adiabatic process
conf	configurational
c	critical property
EOS	equation of state
eq	equilibrium state
i	ith species, i $= 1, \ldots, \mathcal{C}$
imp	impurities
in	inlet conditions
j	generally denotes jth reaction, $j = 1, \ldots, \mathcal{M}$
k	kth flow stream, $k = 1, \ldots, K$
m	mixture property
mix	mixing or mixture
R	reference property
rxn	reaction
sat	property along a two-phase coexistence line
x, y, z	coordinate direction
vac	vacancy

SUPERSCRIPTS

Symbol	Designates
I, II	phase
calc	calculated property
ex	excess property on mixing
exp	measured property
fus	property change on melting or fusion
i, f	initial and final states, respectively
ID	ideal property
IG	ideal gas property
IGM	ideal gas mixture property
IM	ideal mixture property
max	maximum
res	residual
rev	reversible process
sat	property along vapor–liquid coexistence line
sub	property change on sublimation
V, L, S	vapor, liquid, and solid phase, respectively
vap	property change on vaporization
z_+, z_-	charge on an ionic species

The Gas Constant

$R = 8.314$	J/mol K
$= 8.314$	N m/mol K
$= 8.314 \times 10^{-3}$	kPa m^3/mol K
$= 8.314 \times 10^{-5}$	bar m^3/mol K
$= 8.314 \times 10^{-2}$	bar m^3/kmol K
$= 8.314 \times 10^{-6}$	MPa m^3/mol K

Notation

Standard, generally accepted notation has been used throughout this text. This list contains the important symbols, their definition, and, when appropriate, the page of first occurrence (where a more detailed definition is given). Symbols used only once, or within only a single section are not listed. In a few cases it has been necessary to use the same symbol twice. These occurrences are rare and widely separated, so it is hoped no confusion will result.

SPECIAL NOTATION

Symbol	Designates
$\hat{\ }$ (caret as in $\hat{H}$)	property per unit mass (enthalpy per unit mass)
$\bar{\ }$ (overbar as in $\bar{H}_i$)	partial molar property (partial molar enthalpy)
$\_$ (underscore as in $\underline{H}$)	property per mole (enthalpy per mole)
$\pm$ (as in $M_\pm$)	mean ionic property (mean ionic molality)
$*$ (as in G_i^*)	property (Gibbs energy) in hypothetical pure component state extrapolated from infinite dilution behavior
$\square$ (as in $\bar{G}_i^\square$)	ideal unit molal property (Gibbs energy) extrapolated from infinite dilution behavior
$\circ$	standard state

GENERAL NOTATION

Symbol	Designates
A	Helmholtz energy (110)
a_i	activity of species i (711)
$a, b, c, \ldots$	constants in heat capacity equation, equation of state, etc.
$B(T), C(T), \ldots$	virial coefficients (210)
$\mathcal{C}$	number of components (337)
$^\circ C$	degrees Celsius (13)
C_i	concentration of species i (638)
C_V, C_P	constant-volume and constant-pressure heat capacities (60)
C_V^*, C_P^*	ideal gas heat capacity (45)
∂, d, D	partial, total, and substantial derivative symbols
D	diffusion coefficient (22)
$\mathcal{F}$	degrees of freedom (314)
$^\circ F$	degrees Fahrenheit (13)
f	pure component fugacity (291)
$\bar{f}_i$	fugacity of a species in a mixture (405)
F_{fr}	frictional forces (84)
g	acceleration of gravity (11)
G	Gibbs energy (110)
$\Delta_{fus}G, \Delta_{rxn}G, \Delta_{mix}G$	Gibbs free energy changes on fusion (435), reaction (363), and mixing (403)
$\Delta_f G_i^\circ$	molar Gibbs free energy of formation of species i (363)
H	enthalpy (50)
$\mathcal{H}, \mathcal{O}, \mathcal{N}, \mathcal{C}$	stoichiometric coefficient in a biochemical compound (888)
$H_i, \mathcal{H}_i$	Henry's law constants (456, 458)
$\Delta_{fus}H, \Delta_{mix}H, \Delta_{rxn}H$	enthalpy changes on fusion (318), mixing (338), and reaction (364)

Symbol	Designates
$\Delta_{sub}H$, $\Delta_{vap}H$	enthalpy changes on sublimation (321) and vaporization (319)
$\Delta_c H_i^\circ$	standard heat of combustion of species i (365)
$\Delta_f H_i^\circ$	molar heat of formation of species i (363)
$\Delta_s H$	molar integral heat solution (393)
I	ionic strength (468) in Chapters 9–15
K	number of flow streams (28)
K	degrees Kelvin (11)
K, K_c, K_x	distribution coefficients (505, 587, 588)
K_a	chemical equilibrium constant (711)
K_c, K_p, K_x, K_y	chemical equilibrium ratios (724) in Chapter 13
K_c°, K_s°	ideal solution ionization (733) and solubility (743) products
k_{fr}	coefficient of sliding friction (85)
K_i	K-factor, y_i/x_i (505)
K_{OW}	octanol-water position coefficient (640)
K_s	solubility product (742)
K_γ	product of activity coefficients (725)
K_ν	product of fugacity coefficients (725)
k_{ij}	equation of state binary interation parameter (423)
L	moles of liquid phase (504)
M	mass (30)
$\dot{M}$	mass flow rate (30)
mw	molecular weight (30)
$\mathcal{M}$	number of independent chemical reactions (353)
ΔM_k	amount of mass that entered from kth flow stream (31)
M_1	mass of system in state 1 (31)
N	number of moles (30)
$N_{i,0}$	initial number of moles of species i (353)
P	pressure (11)
$\mathcal{P}$	number of phases (314)
P^{vap}, P^{sub}, P^{sat}	vapor (318), sublimation (318), and saturation pressures (318)
P_{atm}	atmospheric pressure (11)
P_c	critical pressure (239)
pH	measure of solution acidity (825)
pK_a	$-\log(K_a)$ (831)
P_t	triple point pressure (288)
P_i	partial pressure of species i (480)
P_r	reduced pressure (239)
P_{rxn}	reaction pressure (721)
Q, $\dot{Q}$	heat flow (22) and heat flow rate (47)
q	volumetric flow rate (780)
R	gas constant (13)
r	drop radius (325) or specific reaction rate (353)
S_o	solubility (852)
S_T	total solubility (852)
S	entropy (100)
S_{gen}, $\dot{S}_{gen}$	entropy generated (101) and entropy generation rate (101)
S_{UV}, S_{UN}, etc.	second partial derivatives of entropy (277)
T	temperature (13)
T_c	critical temperature (239)
T_f	mixture freezing temperature (674)
T_{lc}, T_{uc}	lower and upper consolute temperatures (602)
T_m	melting temperature (674)
T_r	reduced temperature (239)